Biophysics

Biophysics

Physical Processes Underlying the Living State

Wayne F. Reed
Tulane University
New Orleans, Louisiana, USA

WILEY

Published by John Wiley & Sons, Inc., Hoboken, New Jersey.
Published simultaneously in Canada.

Library of Congress Cataloging-in-Publication Data Applied for:

Hardback: 9781119696407
ePDF: 9781119696377
epub: 9781119696421

Cover Design: Wiley
Cover Image: Created by Linnette F. Reed

Set in 9.5/12.5pt STIX Two Text by Straive, Pondicherry, India

Dedicated to my family, past, present, and future

Contents

Preface

Biophysics is not a canonical or settled field, unlike traditional areas of physics, such as electromagnetic theory, thermodynamics, and classical and quantum mechanics, for which a unified body of knowledge exists and is treated in approximately the same form for each subject in many different texts. All would agree that biophysics involves both physics and biology, but then perspectives, fields, and texts quickly diverge after that.

It is appealing to attempt to establish the "living state" of matter as a subspace of the vast phase diagram for multicomponent systems, which can be unified by fundamental principles: living state physics. While the living state obeys all known physical laws, there is no universal agreement about whether the currently known laws are adequate for the description of the emergence and properties of the living state, or whether there are other operative principles yet to be discovered.

Beyond the search for fundamentals, biophysics has come to encompass an enormous space of research activity at all length scales, including exoplanets, local ecosystems, macroscopic biomechanics, neuroimaging of the brain, approaches to organ function, electrophysiology, information and systems integration, cellular and membrane level functions, and structure and dynamics of biomacromolecules. Some also consider biophysics to be the transfer to life sciences of existing physical theories, such as thermodynamics, statistical mechanics, fluid mechanics, and polymer physics, and methods, such as nuclear magnetic resonance, superconducting quantum interference devices (SQUIDs), scattering techniques, and a host of optical spectroscopies, to solve biologically and medically related problems. Yet another dimension deals with the interaction of ionizing radiation with living tissue, that is, radiation biophysics. With rapid advances in neuroscience, attempts at establishing the biophysics of consciousness are underway, intense, and controversial. While quantum phenomena underlie all atomic and molecular processes, the area of quantum biology is relatively untrammeled but is quickly pushing into areas of vision, photosynthesis, enzymatic action, theories of mutation, and others.

The preface to many biophysics texts often mentions "this book grew out of lecture notes prepared while I was teaching a course in biophysics at the University." Many biophysics instructors cannot find the constellation of topics they want to present in any single text and so weave together their own narrative. It is fascinating to peruse the wide range of available biophysics texts and see how they range from dispassionate presentations of "established fact," to highly intuitive, integrated approaches for student discovery of principles, to mathematically dense forays into very complex, specific biological systems or functions. There is virtually no unity among biophysics texts, unlike, say, texts on classical mechanics or electromagnetism.

This work tries to capture the broad outlines of the well-understood processes and principles in the living state, while taking a more detailed look into the area of macromolecular science. No single text can adequately embrace the vast scope of knowledge in all the dimensions of the living state and macromolecular science. Virtually every topic in this book roots back to scientific developments that took centuries to establish, and for which there are entire books and even dedicated journals. Most areas are still under active investigation. As a colleague once said, "if you woke up early and studied all day long, by the end of the day the amount of new knowledge produced in the world on that day will have dwarfed what you learned."

Regarding macromolecular science, while it is also highly interdisciplinary, the situation is more focused than for biophysics. The historical foundations of polymer research are well documented, and a certain logical flow in terms of chemical structure, reactions, and resulting properties can be established. There are well-regarded foundational texts, such as "Principles of Polymer Chemistry" by Paul Flory and "Physical Chemistry of Macromolecules" by Charles Tanford which lay out large parts of the field and continue to be relevant. Further work by researchers such as John Kirkwood, Peter Debye, Bruno Zimm, Walter Stockmayer, Hiromi Yamakawa, and Pierre-Gilles de Gennes, among many others, built the foundation for much of modern polymer physics. Hence, many of the introductory polymer science texts will follow a similar pattern while emphasizing different aspects – chemistry, physics, engineering, applications, liquid state, solid state, rheology, materials properties – according to the author's interests.

The organization of this book is as follows:

Part I. Scientific Overview, Biological and Biochemical Surveys

Chapter 1 starts with the evolution of scientific thought and considers theories that have been successful and others that have fallen on the trash heap of discarded ideas. A fair amount of historical background is provided to give an idea of the millennia of thought and myriad of investigators behind fields, such as atomic science, evolution, and biophysics itself. There is a brief survey of the ongoing attempts to define fundamentals in biophysics, including the notion of the living state as a far from equilibrium dissipative system, in which the phenomena of self-organization, auto-catalysis, cooperativity, and information storage and processing figure largely.

Chapters 2 and 3 provide very brief overviews of biology and biochemistry, respectively, and present some of the underlying basics behind biophysics, including a condensed summary of metabolism. These chapters are intended for the physical science majors who may have had little or no exposure to these life science fields.

Part II. Physical Processes underlying the Living State

Since the living state involves chemically driven thermodynamic structures and involves the creation of order and information, principles of thermodynamics are introduced first in Chapter 4. Elementary situations are worked through to get a feel for heat flow, entropy, free energy, and chemical potential. These lead naturally to entropy-based forces, chemical kinetics, and basic nonequilibrium thermodynamics. While probability distributions and averaging recur throughout the book, no attempt is made to use statistical thermodynamics beyond the introduction of some of the most important ideas, such as the partition function and statistical interpretation of entropy. A very lean background on information theory is given, used subsequently for a simple analysis of the genetic code.

Because the living state occurs largely in aqueous phase and involves electrically charged biological molecules, attention next turns to electrostatics in solution in Chapter 5, which begins by assessing the different types of electrostatic interactions in the liquid state, including how electric potentials and ions arrange themselves around charged objects. Introductory notions of ionic equilibria across membranes are also introduced. This is followed in Chapter 6 by fluid mechanics and transport properties in solution, manifestations of which abound from the intracellular level to the largest creatures. Throughout Part II, wherever possible, the physical principles are illustrated with examples from the living state.

Part III. Macromolecular Science

The many functions of the living state are enabled by macromolecules, so considerable attention is devoted to this field. Chapter 7 provides a brief overview of important organic functional groups, along with an overview of general polymer types and properties, the polymer manufacturing industry, and details on polymerization kinetics, and macromolecular characterization methods.

There is a fascinating elision of macromolecular science and molecular biology occurring throughout the twentieth century and continuing into the current one. Accordingly, the common ground between biological and synthetic macromolecules is laid out in Chapter 8, which treats the physics of macromolecules, with an emphasis on conformations, polymer chain statistics, interactions, molar mass distributions, transport properties, and electrically charged macromolecules. To accompany the focus on macromolecules, Chapter 9 is devoted to light scattering and other scattering and optical methods for their characterization. The portion on light scattering reflects one of the author's abiding interests, and is elaborated in detail, and subsequently used to explore a variety of processes in some of the succeeding chapters.

While the complexity and capabilities of biomacromolecules vastly surpass those of any simple, human-created polymers, the latter have still taught us many things about the mechanisms by which they are produced. A profound difference exists between the behavior of synthetic polymers acting on their own or in combination with other components in a blend or nanostructure, and the complex processes which organize, fold, stretch, edit, chemically modify, activate, and deactivate biomacromolecules performing astoundingly specialized functions.

Part IV. Examples of Specific Living State Phenomena

While Part III explored the common ground between bio- and synthetic macromolecules, biomacromolecules got a head start on Earth approximately 4 billion years ahead of the latter, and have evolved a complexity of structure and function far beyond synthetic polymers. Accordingly, Chapters 10–13 give a sampling of some of the unique properties, structures, and processes enabled by the major classes of biomacromolecules introduced in Chapter 3.

A feature of the living state that distinguishes it from nonliving forms of matter is that it is held far from equilibrium using a continuous supply of energy harvested from the environment. This permits complex, controlled dynamics, such as maintaining nonequilibrium ion gradients across membranes to produce nerve impulses, folding DNA into tiny spaces under complex 3D mechanical stresses that allows it to activate and deactivate different genetic sequences to produce proteins through feedback mechanisms, and keeping rhodopsin in an isomeric state to absorb photons and initiate the biochemical events that produce vision. Biological processes are exquisitely controlled, perhaps even optimized, and few processes proceed spontaneously. The living state is imbued with energetic cycles that allow processes to run "downhill" energetically, but then use complex mechanisms to bring the process energetically back "uphill." Chapters 10–13 try to capture and analyze a small subset of these processes.

The many crucial living state functions of proteins, first delineated in Chapter 3, lead to Chapter 10 and the complex issue of how the amino acid sequence delivered by DNA leads to specific protein folding. This is followed by consideration of other processes, including protein aggregation, enzyme kinetics, cooperativity, phase transitions, and internal protein motions that underlie their behavior.

In Chapter 11, macromolecular properties of RNA and DNA are outlined, such as persistence length and bending energies. The genetic code is introduced and discussed in terms of information theory and protein synthesis, and the organization of DNA and associated proteins into chromatin is discussed qualitatively.

While carbohydrates might sound less glamorous than the proteins and nucleic acids, their macromolecules take on many vital functions besides their well-known energy storage. As a class of macromolecules, they can take on a wide variety of conformations, and those that are polyelectrolytes share similar physics with other charged macromolecules. An analysis of their polymer behavior in Chapter 12 is based on characterization principles from Chapter 7. Glycoproteins wed saccharides and proteins and are involved in a wide variety of signaling, immune, and control mechanisms.

Lipids, as well as the phospholipids they produce, are not in themselves macromolecules, but the principles of entropically driven self-organization and supramolecular assembly lead to the cell membrane. Chapter 13 expands on the physical properties of membranes. The membrane is a cell's integument, its protective layer, separating it from the rest of the organism or its environment. A vast array of membrane-bound channels and receptors modulate cell functions and control interactions between the cell and its environment, including other cells. The main focus of Chapter 13 is the generation and propagation of electrical action potentials in the context of neural cell membranes.

To get a feel for how the biomacromolecules and their supporters come together to enable the living state, some integrated processes are introduced in Chapter 14. Because of the vastness of the biophysics field, I have included brief cameos of deep and fascinating areas that may whet a student's appetite for further discovery. These include basic principles of motility, and the interaction of light with the living state, including elements of vision and image formation, several dimensions of theoretical biology, including a closer look at models of far-from-equilibrium systems, such as Turing patterns and population dynamics.

The book concludes in Chapter 15 with a glimpse of several rapidly evolving fields, including prebiotic origins of life, quantum biology, neuroscience, and exobiology. A brief survey of exoplanetary discovery is given as a crucial step in the search for extraterrestrial life.

By the end of the course, students will have at least peered through many portals onto this vast field of human endeavor. Because of the breadth and depth of the field, students can be required to make a deeper research dive into a specific area of their interest, developing a report and class presentation for the end of the semester.

W.F. Reed
January 2023

Student Audience and Particulars of This Book

This book is aimed at students at any level and field who are comfortable with basic calculus and physics. A background in basic biology and chemistry is helpful but not a prerequisite. This course has been given in various ways at Tulane University since 1986. It is normally double-listed as a graduate/undergraduate course, where the graduate students are given more challenging problems and tests to solve than the undergraduates. Students are typically graduate and undergraduate students in physics, engineering physics, chemical and biomolecular engineering, chemistry, and cell and molecular biology.

When working across disciplines, the issue of physical units can be vexing. Although the preference in this text is to use the Systême International (SI), also referred to as the Meter–Kilogram–Second–Ampere system (MKSA), the physical calculations involved in aqueous solutions, the most important milieu of the living state, have traditionally been developed in the Centimeter–Gram–Second system (CGS) and the corresponding CGS Gaussian units. To strike a compromise, MKSA units will be used where traditional, e.g. in electrostatic force and potential calculations, and thermodynamic calculations (most people are used to Newtons, Joules, Volts, and Coulombs, whereas Dynes, Ergs, Statvolts, and Electrostatic Units are less familiar for most), and CGS units will be used in the traditional areas of fluid mechanics and physical chemistry of solutions ... essentially, "when in Rome, do as the Romans."

As regards notation, no attempt has been made to use the same symbols consistently throughout the book. Students are accustomed to the arbitrary nature of using symbols to represent different physical quantities and so are comfortable with adapting convenient symbols to specific situations. This avoids the use of many subscripts, boldfaces, and fonts, such as block and cursive. However, there are some quantities that are so pervasive that their symbols are always kept the same, unless otherwise noted: T = temperature, k_B = Boltzmann's constant, N_A = Avogadro's number, e = the elementary charge, and $\hbar$ = Planck's constant (divided by 2π).

Usage of the Book

Like many texts, this one is too lengthy to cover completely in one semester. It is suggested that the instructor identify the topics of most interest and lay down a path, incorporating as many sections as desired, while omitting portions, even whole chapters, in order to optimize its use.

Because of the vast expanse of research underway in biophysical and macromolecular areas, at the end of each chapter, there is a list of "Further Reading," a wide selection of reviews, articles, and books, which can get the interested reader pointed in promising directions.

Part II and Chapter 9 in Part III can be used as a stand-alone course in some of the areas that traditionally receive less attention in physics curricula; fluid mechanics, scattering of electromagnetic radiation, electrostatics in solution, polymer physics, information theory, and certain less treated aspects of thermodynamics.

Acknowledgments

I thank many people for their contributions to this book. Colleagues and others who critically reviewed content and style include Dan Purrington, André Striegel, David Mullin, Larry Byers, Nick Sparks, Kathryn Stone, and Curt Jarand. Marissa Calko provided many of the figures and illustrations, while Linnette Reed created several original paintings. Sara Wrobel and Priscilla Mandujano obtained permissions for many figures.

I especially acknowledge the following students, who both took the course using a draft of this book, and who each critically reviewed several chapters: Stephen Graf, Elizabeth H. Gregory, Ben Vasquez, Brittany Kay Simone, Jackson Smith, Allen H.H. Li, Lanie McLeod, Julia S. Siqueira, and Gwen Leifer.

Support from the Murchison-Mallory Endowed Chair in physics at Tulane University is also gratefully acknowledged.

About the Companion Website

This book is accompanied by a companion website which includes a number of resources created by author for students and instructors that you will find helpful.

www.wiley.com/go/Reed/BiophysicsandMacromolecules

The instructor website includes the following resources.

- Answers to Selected Problems

Part I

Scientific Overview, Biological and Biochemical Surveys

Biophysics: Physical Processes Underlying the Living State, First Edition. Wayne F. Reed.

Companion website: www.wiley.com/go/Reed/BiophysicsandMacromolecules

1

Background Notions, Histories, and Fundamental Issues in Physics and Biophysics

There is nothing quite as frightening as someone who knows they are right.

Michael Faraday

Interactions are the basis of life. For a thermodynamically "ideal" system, there is no interaction among constituents, beyond point-like elastic collisions. It is only through interaction of constituents, which is the basis of "nonideal" behavior, that phase behavior and cooperativity arise, and hence life. Nonideality manifests itself at basic levels; water freezes and boils, molten metals cool into crystals, and at advanced levels, where complex mixtures of chemicals stimulated with energy can self-organize into living structures.

1.1 The Evolution of Scientific Thought

The luminiferous aether has definite mechanical properties.

John Tyndall

Nature is reluctant to reveal her secrets. The history of science is a running drama of misconceptions seized upon as correct and final, often for centuries, before a new clarification results, itself subject to the relentless experimental tests upon which continued, provisional acceptance of scientific knowledge is based. Some of the great misconceptions in science include the geocentric theory, heat conceived as a substance called "phlogiston," later replaced with the erroneous idea it was a fluid termed "caloric," spontaneous generation of life, the crystalline spheres of the heavens, the luminiferous aether, absolute determinism, and many more.... One person's theory is another's laughingstock.

On the other hand, an imperfect theory is often perfectly adequate for some purposes and is erroneous only when pushed beyond the limits of approximation in which its validity holds. Newtonian mechanics is one of the best-known such cases, where it is valid for computing terrestrial ballistics and the greater part of planetary motion but fails dramatically on the subnanometer scale, near the speed of light, and in strong gravitational fields.

In this sense, scientific theories are maps onto physical reality, useful constructs with predictive, calculational, and manipulative power within their domain of application. While the claim is frequently made that all observable phenomena in the universe are linked to fundamental properties of matter, the existence of the living state is currently not predictable from these properties. In analogy to maps, actual map makers do not refer to distant, or even all-encompassing maps in their endeavors; making a local topographic map does not require a map of the known universe, nor could the latter predict the former. Hence, those studying chemical and biochemical reactions have no need for the current standard model of particle physics in their endeavors, nor are they stymied by our lack of understanding of cosmic dark energy and matter, even though the phenomena they study may ultimately emerge from these fundamentals.

Living state physics emerges from the fundamental properties of matter, but its manifestations seem much further removed from these fundamentals than, say, the quantum mechanical description of a molecular reaction. A current three-way split in thought (and there are more than these three) is (1) life may be viewed as perched upon a pyramid of vanishingly small thermodynamic probabilities and hence is likely quite rare throughout the universe, and (2) the opposing notion that life is actually *inevitable* if the right abundance of matter and energy is present under the right thermodynamic

Biophysics: Physical Processes Underlying the Living State, First Edition. Wayne F. Reed.

Companion website: www.wiley.com/go/Reed/BiophysicsandMacromolecules

conditions; in this sense, there are "living state portions" of the greater multicomponent phase diagram, in which life will arise and subsequently evolve. This view derives from the ubiquity of self-organization of matter, such as inorganic crystals and spontaneous micellization of surfactants in solution, and from the fact that even on a tiny planet, such as Earth, millions of vastly different life forms exist. It is probable that there are principles of self-organization yet to be discovered. Time may tell. Meanwhile, phase diagrams that predict the existence of life-facilitating atmospheres and liquid water on exoplanets (those planets orbiting a star other than the Sun) are being constructed by astronomers and astrobiologists, using both the latest detection instruments, modeling, and theory (Chapter 15.4).

Then, there is perspective (3) that the information content of physical laws is much smaller than the information content of even the simplest living entity so that the living state cannot be predicted from the physical laws. Hubert Yockey (United States, 1916–2016), who helped pioneer the information-theoretic basis of the genetic code and associated proteins, takes the argument of (3) even a step further and claims that the genetic code must be taken as axiomatic, a given, and that the existence of its coding is "a-thermodynamic," that is, it does not depend on temperature and pressure, and is unrelated to physical laws, and that it is not "merely a physico-chemical phenomenon," and that no set of physical experiments will ever lead to a paradigm necessary to produce a genetic code. Part of his argument is that, while physics elucidates the principles and workings of the physical world, nobody knows how the matter, energy, and principles arose, and they are taken as a given, and theories must explain the given behavior of matter and energy. Similarly, humans have discovered the way the genetic code works, but there is currently no explanation of how it arose.

The improbable pyramid of the living state on Earth. *At the base there is no life, but eventually random prebiotic evolution gives way to genetically-based life, and evolution, in the face of huge improbabilities, leads to higher organisms, including humans.* (Painting by Linnette F. Reed).

Phase diagram of the living state. *Life will inevitably arise under the right conditions of temperature, pressure, abundance of elements, and energy input.* (Painting by Linnette F. Reed).

The heyday of Classical Science. Learned societies in the nineteenth century. (Painting by Linnette F. Reed).

In the meantime, we will have to make do with the knowledge that exists and which new knowledge we can create. There is controversy in every area of science, and biophysics is no exception. It is riddled with competing views, from its large-picture perspectives down to the minutiae of its many subfields, such as specific enzyme mechanisms, and membrane dynamics. Concepts, models, and methods will continue to be judged by what inductive, analytical, explanatory, and predictive power they bring with them. Meanwhile, almost every new idea and theory is subjected to constant scrutiny and often becomes the object of intense ridicule and schisms in thought, which frequently produce schools with competing ideas.

To add to the tension, scientific ideas almost always derive from a ferment of hot debate involving many individuals, often over several continents and decades or more. Frequently, one individual crystallizes the prevailing debate into a new enlightenment and gets their name associated with the discovery, but the fact remains that the latter has arisen from a human community of thought and experimentation.

Another feature of the fraught progress of science is that many, if not most scientific statements can be challenged for being too sweeping, used out of context, employing notation too specific or arbitrary, contravening a convention, being too vague, for inappropriate use of a specific word(s), and so on. Many a sentence in this book can likewise be called into question, and I apologize in advance for all of the inevitable flaws.

All the misconceptions, mistakes, and flaws, are a constant reminder of the fact that many of our scientific notions are fragile, often broken, and replaced by larger, more fruitful perspectives and concepts. These misconceptions and mistakes highlight the fact that we have not found "perfect theories," and that the scientist best gird himself with a circumspect humility, and be ready for wholly unexpected surprises.

1.1.1 Error in the Laboratory

When you can measure what you're talking about, and express it in numbers, you know something about it.

Lord Kelvin

Scientific inquiry ultimately rests on experimental evidence. If theories and models in a certain field are to be fully accepted, they must be able to explain existing experimental evidence and make predictions about the results other experiments would yield. While wide-ranging conceptual errors have been made throughout the history of science, and are ultimately debunked on the basis of experimental evidence, the experiments themselves are prone to many sources of error. So much so that students in science are instructed from their earliest laboratory experiments to attempt to account for all sources of random and systematic error they can think of, and formal structures for statistical analysis of data and estimating error bars have been developed. Sources of error can include faulty or miscalibrated instrumentation, inherent noise in the instrument, instrumentation utilized outside of its range of sensitivity, validity or response time, bias in data analysis due to inappropriate reduction schemes or theories, etc. Erroneous data can lead to serious problems of interpretation and hinder development of applicable theories and models.

Random error limits the precision of measurements and is often referred to as *noise* in the data. Figure 1.1 shows raw light scattering intensity data for two experiments from solutions containing aggregating proteins at two different concentrations. There is no analytical function that can account for every point in each of the two data sets, although a phenomenological single exponential of the form $A(1 - e^{-\alpha t}) + 1$ fits the general shapes very well, as seen by the solid lines in the figure. The noise in the data can arise from thermal noise that is ubiquitous in every detector, fluctuations in the incident laser intensity, and heterogeneities and fluctuations in the solution itself. The isolated points above the main signal band are due to large particles, such as "dust," drifting through the detected scattering volume. The noise phenomena occur on the time scale of seconds, which was the sampling rate of the data in Figure 1.1. There can also be longer term error in the form of signal drift, which is also often thermal in origin, e.g. heating up or cooling down of instrument components, and changes in ambient temperature. The amplitude of the signal divided by the amplitude of the noise is a measure of the *signal-to-noise ratio*, or S/N, and we attempt to maximize it through instrument design and operation, and experimental conditions and procedures. The signal amplitude runs through the middle of the noise band, as seen in Figure 1.1.

When the data below are analyzed to yield absolute values of protein aggregate masses, systematic errors can also intervene, such as miscalibration of the instrument, "stray" light not from protein scattering that adds to the signal, errors in knowledge of the protein concentration, and in its optical contrast in the solution (differential index of refraction), and extrapolation errors. These systematic errors limit the experimental *accuracy*, while the random errors limit the *precision*.

1.1.2 Error Versus Approximation

"Error" has several meanings. We saw how unsuccessful theories are largely discarded and sometimes invoked as cautionary tales. But we have also seen how error, both systematic and random, is an inevitable part of experimental work using

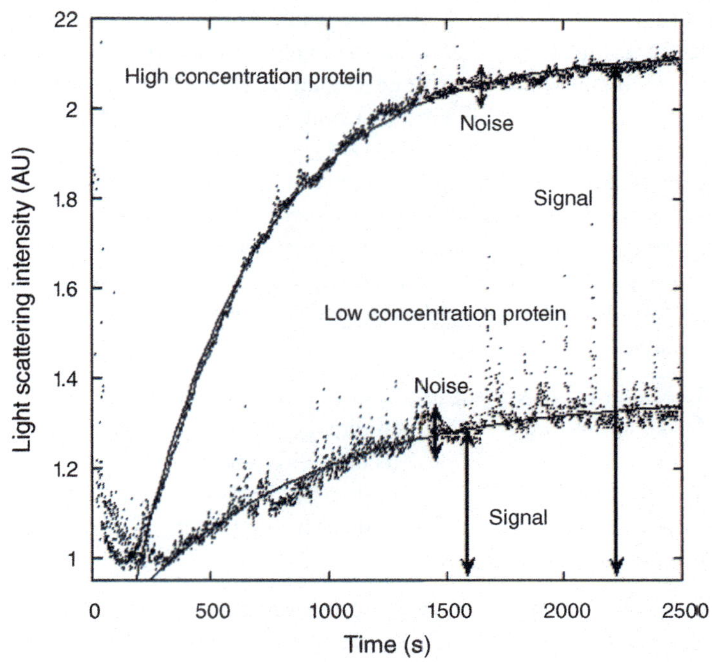

Figure 1.1 Raw light scattering intensity signals for aggregating proteins at high and low concentrations. The noise bands are indicated and isolated points above the bands are due to large scattering impurities drifting through the field of detection. Phenomenological first-order fits are shown by the solid lines. (W.F. Reed unpublished results).

well-accepted instrumental practices within the framework of accepted theory. These types of error cannot be eliminated, but they can be minimized, managed, and accounted for.

Error is implicit as soon as we use the word "approximation" in science. We normally weigh and assess the amount of error that is likely when we choose to simplify a situation with an approximation. Approximations can be quite useful, for example, simplifying an equation such that an analytical solution can be found, rather than a numerical one. While considerable numerical/computational power is abundant and inexpensive nowadays, finding an analytical result usually lends to a more fundamental understanding of a phenomenon, e.g. oscillatory, damped, or asymptotic behavior, electrostatic shielding, and so forth. But it is critical that one remain constantly aware of the assumptions and limits inherent in an approximation, so that the legitimacy and error level due to its use can be assessed.

1.2 Historical Sketch of Atomic Theory and Evolutionary and Genetic Thought

So, how far have we come in our understanding of the natural world, and how did we get here? Before getting into the thick of biophysics and its foundations, we should step back for a few moments and look at some of the history.

1.2.1 Historical Sketch of Atomic Theory

Historically, there was an ongoing argument about whether matter was continuous and infinitely divisible, or was composed of tiny, indivisible particles. Leucippus (Greece, ~500 BC–?) and his student Democritus (Greece, ~460 to ~370 BC) in the fifth century BC proposed an atomistic theory with early glimmerings of molecules built from these. They considered that there were a vast number of types of atoms, that they were indivisible, indestructible, had always existed, and that all things in the universe were composed of atoms and the void. Anaxagoras (Greece, ca. 510–428 BC) conceived of a cosmological evolution of elemental entities under some type of organizational principle, termed "nous" or "mind."

Even earlier, there arose early atomistic thought in India. Aruni (India eighth century BC), a famous Vedic philosopher in the Hindu tradition, held that "particles too small to see aggregate together into the substances and objects of our experience."[1] Later, over several centuries, Vaisheshika and Nyaya schools in India formalized ideas about how atoms combine into dyads, triads, and higher structures.

Throughout the Middle Ages, there were schools of atomistic thought in India, Central Asia, and the Middle East, albeit in many different and often conflicting forms. But in medieval Europe atomism was essentially abandoned, largely due to the continuing influence of Aristotle (Greece, 384–322 BC), who considered matter to be infinitely divisible and composed

1 From McEvilley, T. (2012). *The Shape of Ancient Thought: Comparative Studies in Greek and Indian Philosophies*. New York: Constable & Robinson.

of the elements earth, air, fire, and water. Atomistic thought in Europe was slowly rekindled by the likes of Galileo Galilei (Italy, 1564–1642), Francis Bacon (England, 1561–1626), and Thomas Hobbes (England, 1588–1679). Added momentum for atomism came from René Descartes (France, 1596–1650) and Robert Boyle (Ireland/England, United Kingdom, 1627–1691). Robert Boscovich (Croatia/Italy, 1711–1787) produced a mathematical atomic theory based on the laws of Isaac Newton (England, 1643–1727). Daniel Bernoulli (Holland/Switzerland, 1700–1782) in 1738 posited that gases consisted of atoms moving in many directions and that their impact on surfaces caused pressure.

Despite these ancient and early modern insights, atomic science in its modern form did not emerge until the late eighteenth century. Joseph Louis Proust (France, 1754–1826) discovered the law of constant composition in 1794. While greeted skeptically by most scientists[2] at first, John Dalton (England, 1766–1844) developed an atomic theory, which may be considered the springboard to the modern view, and offered five important propositions around 1803:

Dalton's propositions:

1) Matter is made up of indivisible, indestructible atoms
2) All atoms of a given element are identical
3) Atoms of different elements have different weights and chemical properties
4) Atoms of different elements combine in ratios of whole numbers to form compounds
5) Atoms cannot be created, and according to (1) they also cannot be destroyed. Atoms are recovered, unchanged, when a compound decomposes.

One can see that ideas one and five coincide with the ancient ideas of Leucippus and Democritus. Dalton's theory was also greeted with much skepticism throughout the nineteenth century. Amedeo Avogadro (Italy, 1776–1856) strongly bolstered molecular theory by positing in 1811 that equal volumes of gas, under the same pressure and temperature, of different gaseous substances contained the same number of atoms or molecules, even though the mass in the volumes differed. Avogadro's proposition was largely ignored for decades.

Despite persistent skepticism, developments of atomic theory in the nineteenth century were rapid. James Clerk Maxwell (Scotland, United Kingdom, 1831–1879) assumed the existence of atoms and molecules in developing the kinetic theory of gases. Ludwig Boltzmann (Austria, 1844–1906) used the notion of atoms throughout his early development of statistical mechanics.

As the atomic theory was becoming universally accepted in the latter part of the nineteenth century, J.J. Thomson (England, 1856–1940) made a startling discovery. In 1897, using a cathode ray tube with the velocity selector he invented, he discovered that he could dislodge a negative particle – the electron – from any gaseous or metallic material he used. Just when the scientific community was becoming fully adjusted to the atomic theory, Thomson discovered that the atom is not indivisible, after all! This broke with the ancient Greek theory of Leucippus and Democritus and Dalton's first proposition. Naturally, the discovery met with much opposition. But the experiment was widely reproduced and the same electron with the same charge-to-mass ratio, e/m, was found in many different materials. After so much verification the electron rose to the status of the first *subatomic particle* to be discovered. In 1909 Robert Millikan (United States, 1868–1953) in his famous experiment measuring the fall of charged oil droplets found the value of the elementary charge $e = 1.602 \times 10^{-19}$ C, which hence also gave the mass of the electron, $m_{electron} = 9.11 \times 10^{-31}$ kg.

Discovery accelerated rapidly in the late nineteenth and early twentieth centuries. The mass spectrometer allowed elements and their isotopes to be classified by their atomic masses. Henri Becquerel (France, 1852–1908) in 1896 discovered natural radioactivity and was amazed that the radiation was given off from uranium salts without having to put energy into them. Using a mass spectrometer, he discovered that the radiation was deflected, and so could not be the recently discovered X-rays, which are undeflected by magnetic fields.

The radioactive elements thorium, polonium, and radium were discovered by Marie Skłodowska Curie (Poland/France, 1867–1934) and Pierre Curie (France, 1859–1906). Ernest Rutherford (New Zealand/England, 1871–1937) discovered that the positive portion of the atomic charge was contained in a tiny nucleus, 10^{-5} times smaller in diameter than the atom itself. This immediately created the field of nuclear physics, in which a few years later Rutherford both created the first human-made nuclear reaction and discovered another subatomic particle: the proton.

The discovery of the tiny nucleus spurred Niels Bohr (Denmark, 1885–1962), who worked in Rutherford's lab, to develop the first semi-quantum mechanical model of the hydrogen atom, making use of both Planck's constant, h, determined by Max Planck (Germany, 1858–1947) in 1900, and Albert Einstein's (Germany/USA, 1879–1955) 1905 notion of the quantized light particle/wave, later named the *photon*. In 1923, Louis de Broglie (France, 1892–1987) proposed a wave nature of the electron, in addition to its status as a particle. He based this on the success of quantized momentum in the Bohr

2 Until the twentieth century, scientists considered themselves "natural philosophers" and those working in physical science based on mathematics considered themselves "geometers."

atom with its discrete electron energy levels. Furthermore, de Broglie reasoned that in nature one seldom finds physical phenomena with integers attached, the most prominent example being *standing waves*. He conceptualized the quantized energy levels belonging to 3D standing electron waves in the atom. The de Broglie hypothesis was serendipitously confirmed in 1927 by Clinton Davisson (United States, 1881–1958) and Lester Germer (United States, 1896–1971) at Western Electric Company (later Bell Labs). While studying the surface of nickel for practical applications, they unwittingly annealed the nickel into a crystalline lattice and found that electrons were diffracted from the nickel in a manner similar to X-rays, i.e. the electrons were behaving as waves. Using the wave properties of the electron, the first working electron microscope was built in 1931 by Max Knoll (Germany, 1897–1969) and Ernst Ruska (Germany, 1906–1988).

Building on over a century of discoveries, quantum mechanics was essentially born in 1925, with important contributors at that time being Werner Heisenberg (Germany, 1901–1976), Max Born (Poland/Germany, 1882–1970), Pascal Jordan (Germany, 1902–1980), and Erwin Schrödinger (Austria, 1887–1961), among others. Heisenberg is best known for the uncertainty principle, which is a radical break from deterministic classical science, since it states that the position and momentum of a particle at any instant cannot both be exactly known, and hence its future trajectory cannot be computed with certainty. Classical mechanics requires exact knowledge of a particle's position and momentum in order to exactly predict its trajectory under known forces. The ramifications of the uncertainty principle are still being explored today.

Schrödinger is best known for the nonrelativistic quantum mechanical wave equation which bears his name. The Schrödinger wave equation is the "$F = ma$" of quantum mechanics, the starting point for computing a system's properties. The Solvay Conference in Belgium in 1927 brought together the pioneers of quantum mechanics and important conventions, formalisms, and interpretations were agreed upon. At around this time, Paul Dirac (England/United States, 1902–1984) developed a relativistic formulation of the Schrödinger equation, which led to the startling discovery of antimatter. Since then, quantum mechanics has spawned an amazing number of new theories and devices. It led to the invention of lasers and solid-state devices, such as diodes, transistors, and integrated circuits, which revolutionized electronics, and led to computers. Quantum mechanics was immediately incorporated into chemistry by scientists such as Friedrich Hund (Germany, 1896–1997), John C. Slater (United States, 1900–1976), Douglas R. Hartree (England, 1897–1958), Vladimir Aleksandrovich Fock (Russia, 1898–1974), Wolfgang Pauli (Austria/Switzerland, 1900–1958), and Linus Pauling (United States, 1901–1994). Pierre Hohenberg (France/United States, 1934–2017) and Walter Kohn (Austria/United States, 1923–2016) developed a more efficient density functional approach to computation of electronic structure, which has developed rapidly in the past 20 years with advances by John Perdew (United States, 1943–), and others.

1.2.2 Evolutionary Thought

> *Life is not a miracle. It is a natural phenomenon, and can be expected to appear whenever there is a planet whose conditions duplicate those of the Earth.*
>
> Harold Urey

There is an "either/or" character to the historical debate on evolution that parallels the dichotomy of the atomist viewpoint versus the notion of infinite divisibility of matter. The diametric viewpoints were the change and transformation of species versus a timeless, unchanging nature of each species. The former won the long debate and there is no evidence of the latter. At the same time, while evolutionary theory can explain outcomes, it is not well suited for predicting what specific life forms might arise.

The theory of evolution and cell and molecular biology are bedrocks of modern biological science. Concepts of biological evolution existed at least as far back as Anaximander of Miletus (Greece, ~610–546 BC), in sixth century BC, pre-Socratic Greece, who conceived of animals deriving from ancestors living in water in an earlier geological epoch. Empedocles (Greece, ~490–430 BC) held the idea of what resembled a random evolutionary process, where living elements and parts were mixed and joined in embryonic development, hence producing the variety of living forms. Importantly, major philosophers, such as Plato (Greece, ~429–347 BC) and his student, Aristotle, were decidedly anti-evolutionary, maintaining that all living and inanimate things were permanently fixed and of a divine design and purpose. Considering the influence of Aristotle's views in Europe, such as his anti-atomistic thought, this might help explain the late re-appearance of evolutionary thought there. Meanwhile, Chinese philosophers, such as Zhuang Zhou (China, ~369–286 BC) considered that everything was constantly in a state of transformation, including plants and animals, as well as the earth and cosmos themselves. The notion of flux and continuous change also pervades Hindu thought.

While Charles Darwin (England, 1809–1882) is credited with having established the foundation of modern evolutionary theory, there were actually several formulations espoused by earlier scientists and philosophers that contained some of the key elements of Darwin's theory. In the late eighteenth century, George-Louis Leclerc (Compte de Buffon, France,

1707–1788) proposed that what were considered separate species might be descendants of common ancestors, and that a small number of original species might have led to the much greater current variety of species. Denis Diderot (France, 1713–1784) went further and posited that species were constantly changing through essentially trial and error changes, out of which the most fit survived. Interestingly, Charles Darwin's own grandfather, Erasmus Darwin (England, 1731–1802), conjectured that the diversity of modern species might have originated from tiny, primitive organisms.

Jean-Baptiste Lamarck (France, 1744–1829) worked out a highly developed theory of evolution starting in 1800. He proposed that evolution was driven by chemical and physical principles of organization of matter toward ever more complex organisms, and that the environment helped shape the changes and adaptations, thus leading to large-scale differentiation among species. While his theory was coherent and detailed, he made what is considered a significant error, which was that characteristics acquired during an individual's life could be transmitted to succeeding generations. This view was discredited in the nineteenth century and carries the unfortunate name of *Lamarkism* or *neo-Lamarckism*. This is unfortunate, because this idea of *inheritance of acquired characteristics* formed, at most, only a minor part of his evolutionary theory and was not his idea in the first place. Darwin also thought that acquired characteristics could be passed on to offspring. There is a resurgence of the notion of certain hereditable phenotypes in the relatively new field of *epigenetics* (see Section 1.2.2.1).

All the ferment in ongoing evolutionary thought came to its most powerful formulation by Darwin. As a young man of 22, he embarked upon a legendary five-year circumnavigation of the globe, which concentrated on South America, and returned to England via Tahiti and Australia. The trip was made on HMS Beagle from 1831 to 1836, and its primary mission was to collect mapping data for the then imperfectly charted portions of South America. The captain, Robert FitzRoy (England, 1805–1865), an English naval officer and an early pioneer of meteorology and earth science, wanted to enrich the mission by having a "gentleman"[3] geologist aboard. Darwin, a son of the rich, dynastic Darwin–Wedgwood family was given the opportunity. Darwin made copious collections of fossils, skeletons, plants, and other specimens and made lengthy notes on his observations, both biological and geological. Specimens and notes were periodically dispatched back to England during the five years. Greeted as a famous geologist and naturalist upon his return, his scientific career exploded and he was introduced into the most prominent scientific circles, in which he met Charles Lyell (Scotland, United Kingdom, 1797–1875). Lyell advanced the theory of the Earth's surface as a gradually changing landscape of oceans, mountains, valleys, and other formations, worked upon over vast stretches of time by natural forces. The ideas of inorganic evolution and vast geological epochs helped stimulate Darwin's thoughts on the evolution of species. Interestingly, Darwin's chief work for 15 years after his return was concentrated in geology, even as he developed his evolutionary ideas alongside these efforts.

In what seems an odd conjuncture, Darwin was spurred in 1838 to the idea of natural selection by the book of Thomas Malthus (England, 1766–1834), *An essay on the principle of population*. Malthus explained that populations should increase exponentially with succeeding generations and hence eventually overrun a habitat. But famine, illness, and want curtail such exponential explosions. Darwin realized that there was continuous selective pressure on populations and that vast numbers of individuals perished prematurely, leaving those best adapted to their environment to survive and procreate. Because environments change there will be a continuous selection of the best adaptations, and hence species evolve over time into other forms. In his view, there was a tree of life, stemming from a single, successful original life form, which then evolved into the many branches of the tree of life. In his theory, a species' variations are spontaneous, gradual, and accidental and survive or perish through natural selection.

Often treated as a footnote to Darwin's work, Alfred Russel Wallace (Wales/England, 1823–1913) independently discovered the notion of natural selection in evolution during a lengthy naturalist expedition to Indonesia. According to Wallace, the idea came to him as he lay sick with a fever and contemplating, no less, Malthus's *An essay on the principle of population*. He and Darwin co-authored the first public summary of the theory of natural selection at the Linnean Society of London on July 1, 1858. Darwin's book "*On the Origin of Species*" followed soon after, in 1859, and had an immediate impact on the scientific community and the popular imagination. Wallace was 14 years Charles Darwin's junior, and Darwin had arrived at his most important notions of evolution at least a decade before Wallace's independent discovery. Darwin was gratified that Wallace had independently come to similar conclusions, based on lengthy fieldwork. The two exchanged their ideas, mutually influencing each other, until Darwin's death. Wallace was a polymath who worked in many fields, including engineering and biological sciences, anthropology, political science, economics, and civics. He published the equivalent of 22 books and 747 essays and other contributions in several of these fields. He had a financially precarious life and railed publicly against nineteenth-century European capitalism and colonialism and was an early voice for nature conservancy.

---

3 "Gentleman," i.e. someone who could pay their own way on the voyage, as a supernumerary. Darwin's father, Robert Darwin, and his maternal uncle, Josiah Wedgwood II funded Darwin's expenses for the expedition. When Darwin returned to England as a prominent naturalist his father funded his work as a "gentleman scientist."

Ancient notions of evolution. *"For before this I was born once a boy, and a maiden, and a plant, and a bird, and a darting fish in the sea," Empedocles, ca. 400 BC.* (Painting by Linnette F. Reed).

While Darwinian theory is still held as a bedrock of modern biology, progress in the field has been so stunning that many discoveries, unforeseen by Darwin, are still aggregated under this aegis, sometimes also termed *Darwinism*.[4] Darwin was unaware of Gregor Mendel's (Czechia, 1822–1884) groundbreaking work in genetics (Mendel is often considered the "father of genetics"). This latter involved the statistical study of the transmission of hereditable traits, mostly using pea plants, and in crossing different strains he was led to the identification of dominant, and recessive genes. The results were published in 1866 and originally ignored or misunderstood for 35 years, after which the results were rediscovered by several scientists. When finally resurrected, this work became foundational for the field of genetics. In contrast, Darwin had proposed what he termed *pangenesis* to describe transmission of traits. Ironically, this admits to inheritance of acquired characteristics as one of several mechanisms of variations, the notion which was subsequently discarded by Darwinists.

The essential feature of Darwinian theory is that all species arise through natural selection of small, inherited variations that enhance an individual's ability to survive and reproduce. The thinking that developed and became virtually dogmatic for a long period was that random mutations of DNA alone dictated heredity and hence evolution. Developments in epigenetics and horizontal gene transfer, however, provide additional mechanisms for heredity. It is worth briefly considering these here.

1.2.2.1 Epigenetics

While there are historic precedents for notions of *epigenetics* and differences in meaning among different researchers, a consensus definition was formulated by a group of researchers at a meeting in December 2008 at the Cold Spring Harbor

4 A term coined by Thomas Huxley (England, 1825–1895), a prominent biologist, anatomist, and anthropologist, who was a strong supporter of Darwin's theory.

Laboratory: "An epigenetic trait is a stably heritable phenotype resulting from changes in a chromosome without alterations in the DNA sequence."[5] This has some deep implications for understanding the development of an organism, including disease conditions, and hence development of this field may positively impact medicine.

Phenotype is defined by the Oxford English Dictionary as "the set of characteristics of a living thing, resulting from its combination of genes and the effect of its environment." *Genotype* is defined as "the combination of genes that a particular living thing carries, some of which may not be noticed from its appearance."

It is remarkable that a heritable phenotype can exist which involves no alteration of an organism's covalent DNA sequence, because heredity has generally been viewed as the passing along of information carried in the DNA code. Thus, epigenetic processes involve how genes are expressed, and this can be controlled in many ways which do not involve any changes in the DNA sequence. Histones, discussed in Chapter 10.6 in relation to DNA superfolding, are proteins intimately involved in regulation of DNA so that changes in histones and their interactions with DNA can cause differences in gene expression. Other processes, such as methylation of DNA, which inserts methyl groups into the groove of the DNA double helix and inhibits transcription, differential action of "silencer" portions of DNA, and many others can cause a wide variety of epigenetic traits.

A stably heritable epigenetic pathway may be triggered by changes in a cell's external environment. The environments and events that trigger the initiation of an epigenetic pathway, which culminate in a modification of a specific *chromatin* location, have been termed the *epigenator*. Triggering events may involve toxins, radiation, hormones, presence of specific nutrients, and many others. An *initiator* translates the epigenator signal into action at a specific *chromatin site*. The initiator may be a protein that can bind to DNA, a special type of noncoding RNA, or some other agent. *Maintainers* then stabilize and perpetuate the modifications to the chromatin site, to the extent that the modifications can be transmitted to offspring. Chromatin is the complex of DNA and proteins that chromosomes are composed of (Chapter 11.4).

Epigenetics is relatively new and has many controversial aspects, among them a possible rethinking of the discarded "inheritance of acquired traits," which has been a particular laughingstock of Darwinists since the nineteenth century. Epigenetic researchers have proposed analogs to all things genetic; hence, the *epigenome*, which is the overall epigenetic constitution of an organism, *epigenomics*, which is the analysis of epigenome characteristics, and the *epigenetic code*.

Epigenetics is fascinating because it suggests almost a type of interactive software atop the hard-wired DNA sequence of an individual, in that it allows the individual to use the genetic DNA hardware in a way that is responsive to environmental factors, producing and inhibiting protein production as conditions warrant. In this sense, the hard-wired DNA sequence is an instrument, whose notes and stanzas are wrung from it, at least partially, by epigenetics. Hence, epigenetic mutations can happen quickly, during an individual's lifetime, whereas change in DNA in offspring is through mutation and hereditary transmission of the altered DNA, a much slower process. The fact that the means in which the genetic hardware is used by an individual, its epigenome, can be heritable still seems astonishing. Some have suggested that what is known as "instinct" in animals, manifestation of unlearned behavior (e.g. a Border Collie's propensity to herd animals), may be partially the result of epigenetic transmission, as well as hardwired behaviors encoded in DNA sequences. It is also thought that epigenetic inheritance lasts fewer generations than hard DNA changes, which makes it a very supple mechanism, responsive to rapidly changing environments.

While there is still much uncertainty about the modes of transmission of epigenetic information, the information likely must be present in germ cells and survive meiosis, which produces the fertilized egg, or zygote. One possibility is that certain parts of the complex chromatin structure involved in DNA dynamics may survive meiosis and be passed on during fertilization. DNA is always associated with proteins and other molecules, including during meiosis, and is never completely free or 'bare'.

1.2.2.2 Horizontal Gene Transfer

This astonishing mechanism leads to the exchange of genes among organisms of different types and species, and even across the three domains of life: eukaryota, bacteria, and archaea (Chapter 14.1.1). It is astonishing because it breaks with the older notion that genes were transmitted "vertically," from parent to offspring. The fact that genes can be transmitted among species, without the need for inheritance, adds a different dimension to evolution. It complicates the tracing of evolution of at least *prokaryote*[6] and single-cell *eukaryote* species and suggests that the "tree of life" for these may be more of a "web of life," with many interconnections among species, and not just branching through selection of inherited genes. It also provides a "shortcut" to the lengthy process of genetic evolution through mutations and generational transfer, by

5 Berger, S. L., Kouzarides, T., Shiekhattar, R., Shilatifard, A. (2009). An operational definition of epigenetics. *Genes and Development*, *23*(7): 781–3.
6 A prokaryote has no cell nucleus, whereas eukaryotes have a distinct cell nucleus containing its genome.

allowing whole genetic functions to "jump" from one organism to a different type and become incorporated in the latter's genome. In this sense, the ever-deepening and expanding field of *genetic engineering* is also a form of horizontal gene transfer, in that humans can take DNA sequences from one organism and incorporate them into the DNA of other organisms, in what is termed *genome editing*. The advent of CRISPR/Cas9[7] has enormously increased the possibilities for inheritable genome editing. Use of this methodology allows engineered changes made to the human germ cells inheritable. This vast new power has naturally stirred many ethical and philosophical concerns.

The number of examples of horizontal gene transfer in nature is large, and several mechanisms have been identified. Viruses, for example, can transfer genetic material among bacteria and horizontal gene transfer is thought to be the origin of antibiotic resistance in bacteria. Other transfers include fungi to plants, bacteria and viruses to plants, and humans to protozoans.

1.2.2.3 Debate Continues Concerning Purpose or Direction in Nature, or Lack Thereof

The notion that there is purpose in nature is termed *teleology*,[8] and Aristotle is often considered its founder. Some gasp at this word as heretical, others find it an inevitable or fundamental part of nature. This is not a settled debate, and we won't wade into it here. A *teleologist* might say that an eye evolved *in order* to take advantage of the light in the environment and allow perception of objects, their qualities, dimensions, and distances in 3D. A *reductionist* insists that phenomena can be decomposed into, and built up from, simpler parts and principles. In this debate, a reductionist might maintain that the eye arose through random mutation and selection and led to perception of objects, but the eye did not arise "on purpose," or "for a purpose," or "in order" to do anything. The reductionist might similarly maintain that the totality of a mental process, such as adding numbers or sensing fear, is explained entirely by the corresponding neural electrochemical events in the brain. The teleologist might argue that the neural electrochemical events are necessary but not sufficient to fully characterize the mental process. A prominent reductionist argument against purpose is that it requires *design*, and then design can lead to the notion of a designer, which tends to correspond to the God notion, which then tends to transcend the current bounds of the scientific method. The teleologist might refuse to be called a teleologist and defend against this argument by maintaining there are as yet unknown principles immanent in nature, especially those involving self-organization of matter, that lead to what seem purposeful biological features in an organism, but don't require any notion of design, a designer, or a prior contemplation of a feature or a blueprint; e.g. that there is possibly a biased random walk in evolution leading toward optimal use of energetic resources in the environment. The teleologist might further insist that complex biological features cannot arise through random mutation and selection, simply because a few billion years and the limited material available on the outer crust of our small planet are insufficient to test and select enough mutations to arrive at the sophisticated functions found in organisms. At which point the reductionist might claim that, yes indeed, it is highly improbable that life arose randomly, but it beat the odds and *did* arise randomly. We are now back to the earlier mentioned divide of life being *improbable* versus life being *inevitable* when certain conditions of elemental abundance and energy availability are met.

1.2.3 Genetics and Molecular Biology

The aim of this book is to interweave discoveries and methods from different fields, rather than to divide natural phenomena and practical applications into rigid subfields. Nevertheless, the areas of biochemistry, genetics, and molecular biology overlap greatly and are underpinned by evolutionary theory, and yet they parcel themselves out as distinct fields with different goals and methods. While Darwin is considered the "father of modern evolutionary theory," Mendel is usually considered the "father of genetics." But neither of them were involved in the underlying chemistry of evolution and heredity.

The search for the molecular basis of living state processes began in the eighteenth century and encompasses many very different research questions using very different approaches. Just a few examples: Antoine Lavoisier (France, 1743–1794) worked on the chemistry of respiration and fermentation, the discovery of the first enzyme was by Anselme Payen (France, 1795–1871); in 1833, the first abiotic synthesis of a living state molecule, urea, by Friedrich Wöhler (Germany, 1800–1882); in 1828, the discovery of the citric acid cycle by Hans Krebs (Germany/England, 1900–1981); and the naming of the distinct field "biochemistry" by modern contributors Carl Neuberg (Germany/United States, 1877–1956) and Franz Hofmeister (Czechia, 1850–1922). Arguably, two of the strongest areas in biochemistry are metabolism and enzymology, but biochemistry overlaps with just about every other life science area, including biomacromolecules, cell biology, genetics, immunology, and structural biology.

---

7 CRISPR = clustered regularly interspaced short palindromic repeats. Palindrome = a word which is the same when spelled backward. Cas9 is a bacterial defense protein which can cut DNA.

8 Teleology is the study of the purpose or design of natural occurrences, not to be confused with "theology."

In 1871, Friedrich Miescher (Switzerland, 1844–1895) published his discovery of phosphate-rich nucleic acids isolated from white blood cells. This established the foundation of the chemical origin of genes, even though Miescher and others did not recognize its genetic role at the time, and the theory that proteins were the origin of heredity generally prevailed. However, the discovery of nucleic acids added an entirely new class of biochemical to the ones already known at that time: lipids, polysaccharides, and proteins.

Chromosomes were discovered as subcellular structures by Karl Wilhelm von Nägeli (Switzerland, 1817–1891) in 1842, without realizing their role. Edouard van Beneden (Belgium, 1846–1910) found that the chromosomes of parents' organisms don't fuse together, but remain as distinct entities after fertilization. This connects directly to Mendel's genetic laws, but it wasn't realized at the time. In 1903, Theodor Boveri (Germany, 1862–1915) and Walter Sutton (United States, 1877–1916) independently demonstrated that the chromosome was the basic hereditary unit. Hugo DeVries (Holland, 1848–1935) crystallized many previous notions about mutations in his book *Species and Varieties: Their Origin by Mutation*, published in 1905. DeVries's theory of mutation held that variations could arise in a single mutational step, related to changes in genes, and manifested according to Mendel's laws, and not gradually over many generations, as Darwin had thought. This discovery fell midway in the period of the *eclipse of Darwinism*, which lasted for a few decades in the late nineteenth and early twentieth centuries, as certain erroneous Darwinian ideas of gradual variations and inheritance of acquired characteristics were shed. This continued until various discoveries illuminated relationships among genes, proteins, and evolution. The *modern synthesis* reconciled Darwin's evolutionary theory with Mendel's genetic laws. Thomas Hunt Morgan (United States, 1866–1945) discovered that genes reside on specific locations in chromosomes, and that mutations involved physical changes in genes.

Because chromosomes contain both nucleic acids and proteins it was widely held until 1941 that proteins carried genetic information. A series of discoveries by several researchers led to the definitive conclusion in 1944 that a nucleic acid, deoxyribose nucleic acid (DNA, see Chapter 11), is the genetic material, and that genes code for proteins. Using the X-ray diffraction patterns of DNA captured by Rosalind Franklin (England, 1920–1958), James Watson (United States, 1928–) and Francis Crick (England, 1916–2004) elucidated the double helical form of the DNA molecule in 1953, and it was discovered that there were complementary pairings of just four base pairs involved in the protein coding.

The biochemistry and biophysics of DNA, and associated genetics, went through explosive developments after the discovery of the DNA double helix. A well-known nuclear physicist, George Gamow (Soviet Union/United States, 1906–1968), postulated that there were three pairings of bases to produce $4^3 = 64$ *codons*, the sequence of three nucleotides that specify an amino acid. He also postulated that since there are only 20 amino acids (AA) used in living organisms there was *degeneracy* in the code, such that more than one codon could specify the same amino acid. Crick and colleagues found the codon for phenyl alanine, and Severo Ochoa (Spain/United States, 1905–1993) found the codes for the amino acids lysine and cytosine. Finally, after contributions by many researchers, Har Gobind Khorana (India/United States, 1922–2011) completed the table of the *genetic code* (Chapter 11.2) and was also the first to synthesize *oligonucleotides*, a polymer with a small number of *nucleotides*, which are the basic structural units of RNA and DNA.

There were many researchers involved in the development of protein and nucleic acid sequencing. Frederic Sanger (England, 1918–2013) was seminal in both fields. He coupled chromatography in one dimension with electrophoresis in a second dimension and separated and identified hydrolyzed fragments of the insulin molecule, and then puzzled them back together to obtain the entire sequence. He succeeded in sequencing the amino acids in bovine insulins in 1951–1952 and proved that proteins have a defined amino acid sequence, and hence a defined chemical composition. The group of Robert Holley (United States, 1922–1993) at Cornell University sequenced tRNA from yeast in 1964. Sanger's group had sequenced tRNA from *Escherichia coli* bacteria by 1967. Many researchers were involved in the next step, the sequencing of DNA. Ray Wu (China/United States, 1928–2008) and colleagues developed methods based on nucleotide labeling and DNA polymerization over 1970–1973. The first full *genome*, the entire genetic content of an organism, was sequenced in 1977. This latter was for a *bacteriophage*, a virus that attacks bacteria. Since then, sequencing has become commonplace and rapid. The *human genome project* was an ambitious international project whose goal was to sequence all the genes in the human genome. It was started in 1990 and completed in 2003. It was the largest international biological cooperative project in history.

Genetic engineering has undergone exponential growth since the late twentieth century and into the twenty-first. Two main methods are used. One is the *recombinant DNA method*, which, as mentioned, is a humanmade horizontal gene transfer, where desired sequences of DNA are taken from one organism and allowed to recombine with DNA from another organism, which then expresses the desired protein. The second method, even more powerful, is direct *de novo* synthesis of any desired nucleotide sequence. Genetic engineering is used to create novel organisms and characteristics to produce new medicines, fermentation products, and plant species resistant to herbicides and insecticides. Cloning, the creation of identical copies of an organism by taking and expressing its genetic material, has unleashed both tremendous possibilities and huge ethical challenges and controversies. It has furnished the fuel for many science fiction novels and movies.

1.3 Historical Developments in Biophysics

Because biophysics is interpreted in many different ways it is worth starting by revisiting the opening remarks in the Preface. Biophysics is not a canonical, unitary field, but all would agree that biophysics involves both physics and biology, but then perspectives diverge after that. It is appealing to attempt to establish the "living state" of matter as a phase unified by fundamental principles: living state physics.

Because of its interdisciplinarity and these various interpretations it is hard to discern a continuous thread of development in biophysics.

From the point of view that biophysics is the transfer of devices from physics to biology, Antonie van Leeuwenhoek (Holland, 1632–1723) invented the first practical, high-power, single lens microscopes, a milestone in optics, which is traditionally a part of physics. His microscopes revealed the existence of microbes, a discovery which shook the scientific world and led Leeuwenhoek to be considered the "father of microbiology."

1.3.1 Electrophysiology

Luigi Galvani (Italy, 1737–1798) discovered that the legs of a dead frog could be induced to move by shocking them with an electrically charged object. This gave rise to the concept that animal motion was coupled to electrical phenomena, and this might be considered the beginning of *electrophysiology*. John Walsh (England, 1726–1795) in 1773 published his extensive studies of the properties of electrical fish, including conduction of their electricity through water, metals, and people with joined hands. Carlo Matteucci (Italy, 1811–1868) made continuous contributions to bioelectricity and even constructed a battery based on frog legs in series, and demonstrated that the currents summed up in the same way as the inorganic piles (batteries) of Alessandro Volta (Italy, 1745–1827).

Electrophysiology grew rapidly throughout the late nineteenth and the entire twentieth centuries. Augustus Waller (France/England, 1856–1922) created the first *electrocardiogram* of the human heart in 1887. Willem Einthoven (Holland, 1860–1921) produced the first high-quality electrocardiogram, ECG or EKG, in the 1890s, and spent the rest of his life improving and standardizing the method, including interpretation of the signals and relation to normal and disease conditions. The oscilloscope was a foundational instrument in this field, and the first one using a cathode ray tube was invented by Karl Ferdinand Braun (Germany, 1850–1918).

While researchers were devising means to detect electrical processes in the heart, others were fascinated by electrical activity in the brain and this led to the field of *electroencephalography*. The use of electrodes to probe bioelectricity was increasing in refinement and applications in the nineteenth century. Using electrodes placed directly on the brain surfaces of dogs and rabbits Adolf Beck (Poland, 1863–1942) discovered rhythmic electrical oscillations in the brains, or *brain waves* in 1890–1891. The first animal *electroencephalogram* (EEG), a recording of brain electrical activity was published in 1912 by Vladimir Vladimirovich Pravdich-Neminsky (Russia, 1879–1952). Through an invention of Hans Berger (Germany, 1873–1941), the first human EEG was recorded in 1924. Encephalography as a research field and clinical tool rapidly grew out of this. EEGs are used clinically to detect epilepsy, brain tumors, strokes, and other brain injuries. In research, they are important in neuroscience and cognitive science.

Electrical phenomena associated with nerve conduction were studied both theoretically and experimentally throughout the nineteenth century. Julius Bernstein (Germany, 1839–1917) invented the differential rheotome and was able to measure nerve impulses with it. He developed a remarkable hypothesis in which the impulses are caused by membranes of nerve cells having a process involving differential permeability to K^+ and other ions, thus changing electrical potential across the membrane, which could travel as a pulse along axons. This fruitful idea powered intense experimental activity right up to WWII. After WWII, in the late 1940s to early 1950s, Alan Hodgkin (England, 1914–1998) and Andrew Huxley (England, 1917–2012) made their breakthrough series of experiments, and corresponding mathematical interpretation, that put nerve impulse propagation on a modern electrochemical and quantitative basis (Chapter 13).

1.3.2 Hermann von Helmholtz

Hermann von Helmholtz (Germany, 1821–1894) was something of a latter-day Renaissance Man. A physicist, physician, physical psychologist, inventor, musician, and philosopher, he contributed to the understanding of visual and auditory perception, the physics of the eye, formulation of the principle of conservation of energy, and made landmark discoveries in electrodynamics, fluid dynamics, and chemical thermodynamics. While the conservation of energy is usually thought of as a bedrock principle of physics, his insight into energy conservation actually came from the study of muscle movement, and his philosophical intent to rid physiology of *vitalism* (Section 1.7). He unified the notion of energy having manifestations in mechanical, thermal, light, and magnetic energies.[9] His biophysical inventions included the *opthalmoscope*, which allows the inside structure of the human eye to be viewed, a galvanometer so sensitive he made the first measurements of the speed of nerve conduction (ca. 30 m/s, Chapter 13), and the *Helmholtz resonator*, which isolates selected frequencies from complex sounds, such as music, and can be used for mechanical Fourier analysis of sound. Among Helmholtz's many doctoral students: Max Planck; Albert Michelson (Germany/United States, 1852–1938) who accurately measured the speed of light using interferometric principles and could find no difference in the speed due to the Earth's motion, which both preceded and corroborated Einstein's special theory of relativity; and Heinrich Hertz (Germany, 1857–1894), known for the first demonstration of electromagnetic waves and the discovery of the photoelectric effect.

1.3.3 Osmosis and Diffusion

Two other entry points to biophysics were osmotic pressure (Chapter 4.8) and diffusion (Chapter 6.8). Jean-Antoine Nollet (France, 1700–1770) discovered osmotic flow through natural membranes in 1748. The semipermeable membranes used for succeeding osmotic studies were also from natural membranes. Nollet also deduced that electricity flows at very high speed, using the curious experiment of having hundreds of monks connecting their hands in a circle thousands of feet in circumference via pieces of iron, then discharging a battery of Leyden jars,[10] and observing that all the monks reacted painfully to the shock at the same time. This also further proved that electricity could flow both through biological organisms and inorganic metals in the same circuit. The flurry of activity involving osmotic flow led quickly to notions of diffusion. Adolf Fick

Nollet shocks his monks. (Painting by Linnette F. Reed).

9 The term "energy" came later, Helmholtz referred to these linked phenomena with the word Kraft (German for "force").

10 Leyden jars accumulate large amounts of electrical charge between metals on the inside and outside of a jar, a capacitor in essence. The charge can be released when the inside and outside oppositely charged metals are joined by a conducting path.

(Germany, 1829–1901), who had been formally trained in mathematics, physics, and medicine, published the basic laws of diffusion in 1856 in a text on medical physics (*Die medizinische Physik*), which is perhaps the first in biophysics.

Work on osmosis and diffusion gave the impetus for the field of *physical chemistry*. The latter is immediately relevant to the living state, because its purview includes reaction kinetics, thermodynamics and electrochemistry in solutions and on surfaces, and intermolecular forces. It was recognized as a field by two of its founders, Jacobus van't Hoff (Holland, 1852–1911) and Wilhelm Ostwald (Latvia/Germany, 1853–1932), with the founding in 1887 of *Zeitschrift für physikalische Chemie*.

1.3.4 X-rays

Radioactivity and X-rays were both discovered at the end of the nineteenth century. Wilhelm Röntgen (Germany 1845–1923) discovered X-rays in 1895. He serendipitously discovered the rays while using a Crooke's tube[11] and realized it was a new form of radiation, to which he gave the name X-rays (X-Strahlen), the "X" being a place holder for the unknown rays until they could be identified. The name stuck, although within days it was suggested that they be called *Röntgenstrahlen*, the name still used in the German-speaking world. Others had observed the glows coming from Crooke's tubes, but Röntgen was the first to systematically study their properties, in particular their penetrating power through various materials. He made the first X-ray photograph, or simply "X-ray," of his own hand, and then of his wife's, Figure 1.2. When she saw it she declared "Ich habe meinen Tod gesehen" (I have seen my death).

Within a year, by 1896, X-rays were already being used clinically in Europe, the United States, and Japan, in what was probably the nineteenth century's most rapid major technology transfer from laboratory discovery to practical use. X-ray technology developed throughout the twentieth century and culminated in the invention in 1972 of *computed axial tomography scans* (CAT scans, or simply CT scan), which involve taking X-ray images at many different angles and combining them computationally into a 3D image. Many other types of *tomography*[12] exist, using gamma rays, visible light, infrared, electrical resistivity and capacitance, neutrons, and even muons, as well as antimatter.

There were arguments over the nature of X-rays for over a decade after the 1895 discovery, until it became clear that they exhibited polarization and other properties of electromagnetic waves and were hence a part of the electromagnetic wave spectrum. The polarization of X-rays was demonstrated in 1906 by Charles Barkla (England/Scotland, 1877–1944).

Max von Laue (Germany, 1879–1960) and his assistants performed the first experiment in X-ray diffraction in 1912,[13] obtaining an X-ray diffraction pattern from copper sulfate crystals. The father and son team of William Henry Bragg (England, 1862–1942) and William Lawrence Bragg (Australia/England, 1890–1971) was inspired by Laue's results, confirmed them, and then focused much of their career on the growing field of X-ray diffraction and crystallography. The methods came quickly into use in many countries. The structures of large numbers of crystals were determined, and increasingly sophisticated methods of data interpretation developed, including the use of Fourier transforms in the late 1920s. The work had immediate impact in physics, chemistry, and materials science. The detailed structure of different alloys could quickly be determined, greatly advancing metallurgy. Working with Lawrence Bragg, Max Perutz (Austria/England, 1914–2002) turned efforts toward the difficult task of determining the structure of biological macromolecules. His laboratory attracted Watson and Crick and their famed DNA experiments. In 1953, he invented a method of substituting a heavy metal (mercury) into a crystal of the protein hemoglobin and used X-ray diffraction for the world's first determination of the 3D structure of a protein. Since then, X-ray diffraction of proteins has become a mainstay of both molecular biology and pharmaceutical science, and the structures of thousands of proteins are now determined each year.

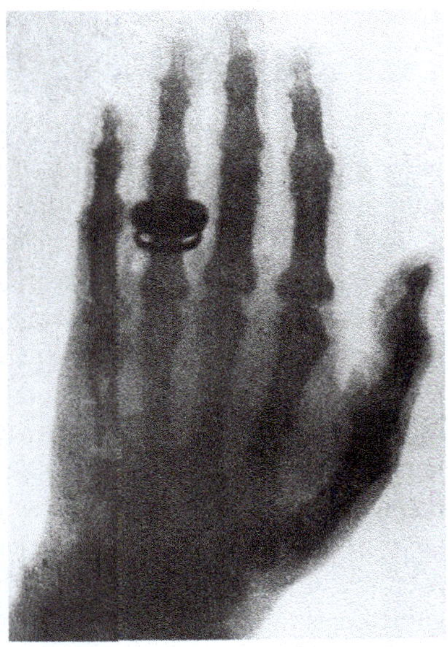

Figure 1.2 The famous X-ray of Röntgen's wife's hand, made by Röntgen in 1895. (Courtesy of Science History Images / Alamy Stock Photo).

In addition to X-ray crystallography, *X-ray spectroscopy* became a major tool shortly after Röntgen's discovery and is a major component of modern chemical analysis. Using the principle of X-ray diffraction with crystal lattices of known

11 A tube with low-pressure gas inside across which a high-voltage discharge is made. High-energy electrons (KeV) striking the end of the tube, typically a metal target, cause X-ray emission through breaking radiation (Bremsstrahlung) and inner electron ionization of the target atoms. This stands in contrast to a cathode ray tube which uses a heated filament and thermionic emission of low energy electrons (eV) which strike a phosphor that emits light.

12 Greek τομοζ "tomos," a "slice" or "section," and γραφω, "grapho," to write or describe.

13 Max von Laue and the discovery of x-ray diffraction in 1912. *Annals of Physics (Berlin)* 2012, *524* (5): A83–5. doi:10.1002/andp.201200724.

atomic spacing a spectrometer can be made that determines the wavelength of X-rays produced in a process. X-ray absorption spectroscopy measures the wavelength of X-rays associated with energy transitions from electron ground states in elements to higher excited states, whereas X-ray emission spectroscopy measures the wavelength of X-rays corresponding to energy released when an electron falls from a highly excited state to a lower one. It quickly became a method for identifying elements.

In 1913, Henry Moseley (England, 1887–1915) discovered that X-ray spectra from one element to the next massive one followed a very regular pattern, and from this deduced the notion of *atomic number*, which was later understood to correspond to the number of protons in an element's nucleus. This led to a modification of the periodic table according to atomic number rather than atomic mass, as had first been systematized by Dmitri Mendeleev (Russia, 1834–1907). Moseley, seeing gaps in his periodic table, also predicted four missing elements, which were subsequently found. These had atomic numbers 43 (technetium), 61 (promethium), 72 (hafnium), and 75 (rhenium). Moseley left his research at Oxford to volunteer as a telecommunications officer. He was killed in action in WWI in 1915, in the ghastly and devastating battle of Gallipoli, Turkey.[14]

1.3.5 Imaging Methods

Besides X-ray CT imaging, some other major imaging methods in modern medicine trace back to developments in physics. Nuclear magnetic resonance (NMR) was first demonstrated by Isidor Rabi (United States, 1898–1988) and colleagues in 1937. They measured the magnetic moment of the proton and of several lithium compounds. The method quickly expanded to become one of the greatest workhorses of modern chemistry and materials science. It has advanced by the availability of ever higher magnetic fields, and use of radio frequency waves to interrogate the spin states of nuclei. Because it uses no high-energy radiation, such as X-rays, it is biologically benign and it has been adapted to 3D imaging of whole organs in patients in *nuclear magnetic resonance imaging*, or MRI, where "nuclear" has been dropped, presumably so as not to frighten patients.

Positron emission tomography (PET) is used to pinpoint tumors or infections. It relies on the annihilation of electrons and their antiparticle, the positron, which produces two gamma rays which can then be traced back to their location of annihilation in tissues (Chapter 9.17). As mentioned above, antimatter was discovered from Dirac's formulation of the relativistic quantum mechanical wave equation.

1.3.6 Radiation Biophysics

The effects of radiation on living organisms, cells, biological tissues, and biological macromolecules is the traditional purview of *radiation biophysics*, sometimes also termed *radiobiology*. Research on biological effects of radiation began soon after the discovery of X-rays, gamma rays (electromagnetic waves with wavelengths on the order of 10^{-11} m and less), and alpha (ionized helium nuclei) and beta particle (electrons and positrons) radiation.

Several early radioactivity researchers and their assistants lost their hair, sustained skin injuries, or suffered worse fates. Marie Curie and her daughter, Irène Joliot-Curie (France, 1897–1956), herself a Nobel-prize-winning chemist, both died of leukemia, likely due to their long exposure to radioactive substances. Pierre Curie found that exposing tadpoles to radiation from radium caused developmental abnormalities. Thomas Edison (United States, 1847–1931) became a major force in improving and commercializing medical X-ray technology, via the General Electric Co. He observed that some assistants fell gravely ill from working with X-rays.

Hermann Heineke (Germany, 1873–1922) made the first published studies on the effect of radiation on animals and internal organs, starting in 1902. Heinrich Albers-Schönberg (Germany, 1865–1921) reported the sexual sterilization of guinea pigs due to prolonged exposure to radiation in 1903. Efim Simenovic London (Lithuania/Russia, 1869–1939) subjected guinea pigs to prolonged exposure of radium and reported in 1903 that this led to loss of appetite, lethargy, weakness of response to stimuli, and death in four to five days. Friedrich Dessauer (German/Switzerland, 1881–1963) made one of the first quantitative radiation studies, establishing statistical relationships between doses and responses in 1925. Radiation biophysics blossomed after WWII. Methods for producing cloned cell lines of mammalian tissue, and for creating mutations, were developed in the 1950s. With these new tools, it was soon realized that chromosomes could be severed by X-rays.

---

14 The Gallipoli campaign lasted eight months, and pitted allies Britain, France, and Russia against the Ottoman Empire, which was allied with Germany. One of the main goals was for the Western allies to control the Suez Canal and secure routes through to the Black Sea. There were over 500,000 casualties in the conflict!

1.4 Subfields in Biophysics

Biophysics can be divided into subfields along a number of different axes. Most subfields connect to other fields and subfields in biological and medical sciences, such as neuroscience, electrophysiology, enzymology, cell and molecular biology, and genetics. Below, some of the most often used designations for biophysical subfields are outlined.

1.4.1 Molecular Biophysics

Molecular biophysics involves the physical structure and properties of nucleic acids, proteins, lipids, carbohydrates, hormones, and many other molecules. Molecular biophysics seeks the physical origins of their biofunctionality. The large molecules are examples of *polymers*, also termed *macromolecules*. Macromolecules are central to genetics, and to biochemistry and its explanations of processes such as metabolic cycles. Macromolecules, whether biological or human-made, share many fundamental properties and many of the same characterization methods. We shall see, however, that the functionality and capabilities of biological macromolecules are mind-bogglingly more sophisticated than any that humans have created to-date.

Macromolecules are sometimes viewed as a branch of *soft matter physics*, which is itself a subfield of *condensed matter physics*. Many features of the living state fall into focus areas of soft matter physics. Among other phenomena soft matter physics involves mesoscopic structures – larger than the component atoms and molecules, but much smaller than the entire system – and their sensitivity to mechanical, thermal, optical, electromechanical, and other stresses. Other soft matter examples besides macromolecules include liquids, foams, colloids, gels, and many biological materials, such as cartilage and membranes.

The largest single focus in this book is on macromolecules, both biological and synthetic, the properties common to both types, and the experimental and theoretical approaches to understanding them. Chapters 7 and 8 are concerned exclusively with macromolecules, while Chapter 9 introduces scattering and cognate methods used to characterize them. Selected properties and phenomena involving the main classes of biomacromolecules are treated in Chapters 10–13.

1.4.2 Cellular Level and Integrative Biophysics

Macromolecules, small molecules, ions, and water come together to create functional entities. These include specific subcellular structures, such as mitochondria, the Golgi complex, and lysosomes, as well as whole-cell organisms, such as prokaryotic and eukaryotic microbes. Many specific processes are enabled at the cellular level. Some examples include cell and membrane structures which support nerve impulses and propagation, as well as active and passive transport of molecules and structures. Highly specialized cell types also lead to muscular contraction, photosynthesis, vision, tactility, olfaction, gustation, and acoustic perception. Also of major importance is how cells interact and integrate with each other, including regulatory and feedback mechanisms. We will touch on some of these topics in several of the chapters in Part IV.

1.4.3 Macroscopic Biophysics

This generally deals with large-scale aspects of higher organisms. Biomechanics involves locomotion, stresses and strains, muscular–skeletal systems, and mechanics and rheological properties of biofluids, bone, wood, and other biomaterials. It also deals with systems such as fluid transport in plants, blood circulation in animals, and gas/liquid exchange processes.

1.4.4 Theoretical Biophysics

This is a vast area and often is meant to include theoretical biology, in general, where many topics are dealt with mathematically; population dynamics, symbiotic and ecological models, entire biomes, and the interaction between living organisms and geology, the oceans, and the atmosphere. Some of its organizing principles come from thermodynamic concepts such as autocatalysis, stability, irreversibility, nonequilibrium processes, self-assembly, dynamic order, synergetics, and phase transitions. Other fundamentals come from fluid mechanics: viscosity, diffusion, electrophoresis, laminar and turbulent flow, and heat and mass transport.

Other areas that make use of principles more akin to engineering and computer science include network and information theory, and negative and positive feedback control theory.

Modeling and computational efforts are often subsumed under theoretical biophysics. Many simulations have been developed to test theories on the prebiotic origins of life (Chapter 15.1). Importantly, there is no universal consensus on how life arose. Among many other applications, computer simulations are also used to predict mutations, to understand how proteins fold, and to explore embryonic and development pathways. Simulations are sometimes termed *in silico experiments*.

The beginnings of *quantum biophysics* or *quantum biology* were hinted at by the founders of quantum physics, including Schrödinger, Delbrück, and Bohr. Emerging areas in the field include energy transfer and charge separation in photosynthesis, tunneling in enzyme catalysis, and roles in sensing processes, such as magnetoreception and olfaction, and possible quantum mechanical roles in cognition and the origin of life. There is a short treatment of some of these in Chapter 15.

1.4.5 Instrumental and Medical Biophysics

Section 1.3.6 introduced some of the central concerns of radiation and medical biophysics. Instrumental biophysics involves the large-scale adaptation of the experimental tools of physics to biophysical and medical research and practice. Instruments frequently invented in physics laboratories to probe the underlying structure of matter have become the bedrock of many disciplines.

1.5 Interdisciplinarity

Until just the last few centuries scientific inquiry was an open field, and thinkers typically expounded their ideas in many different areas, such as philosophy, history, political science, and the physical and life sciences. Many of the most famous Greek philosophers were multidisciplinary, and during the European Renaissance, from the fourteenth to seventeenth century, there was a tremendous rediscovery of classical art, architecture, literature, philosophy, and science. The term "Renaissance man" denotes someone talented in many different fields, e.g. Leonardo da Vinci (Italy, 1452–1519). In Renaissance Italy, the term was *uomo universale*, or universal man, someone educated and productive in many areas. In that era, knowledge was so sparse that a vigorous thinker could wade into several fields, find its boundaries and make contributions. As more knowledge accrued it became more and more difficult to be in the lead of any given area of inquiry, and so specialization inevitably set in. The modern Ph.D., or doctor of philosophy (*Philosophiae Doctor* in Latin), with its emphasis on original research in a particular field didn't arise until the early nineteenth century in Germany, spearheaded by Wilhelm von Humboldt (Germany, 1767–1835). It spread to the United States by the mid-nineteenth century, as well as to other countries.

While there is much current emphasis on "interdisciplinarity" in modern research and teaching, scientific fields became disciplinary for a very good reason; the set of phenomena and the methods used to understand and structure them have traditionally been very different. Disciplinary journals are the norm rather than the exception. The traditional and current departmental structure of universities in North America and elsewhere strongly embodies this disciplinarity, and there are frequently very high barriers to communication and cooperation between such departments.

Nonetheless, there are some areas of inquiry which require so many different perspectives and approaches that they have necessarily become interdisciplinary endeavors, gathering from physics, chemistry, life sciences, medicine, the engineering disciplines, and physiological psychology. Examples include programs and departments in biophysics, materials science and engineering, neuroscience, polymer science and engineering, and planetary science.

1.5.1 A Simplified Net of Scientific Knowledge

Caminante no hay camino, se hace el camino al andar.[15]

Antonio Machado

Figure 1.3 is a simplified attempt to capture the evolution of scientific thought. In the early period – ancient Greece, China, and India – the thread of knowledge is quite thin. Knowledge was limited enough for philosophers to contribute in many

15 A loose translation "Traveler, there is no trail, you make the trail as you go forth."

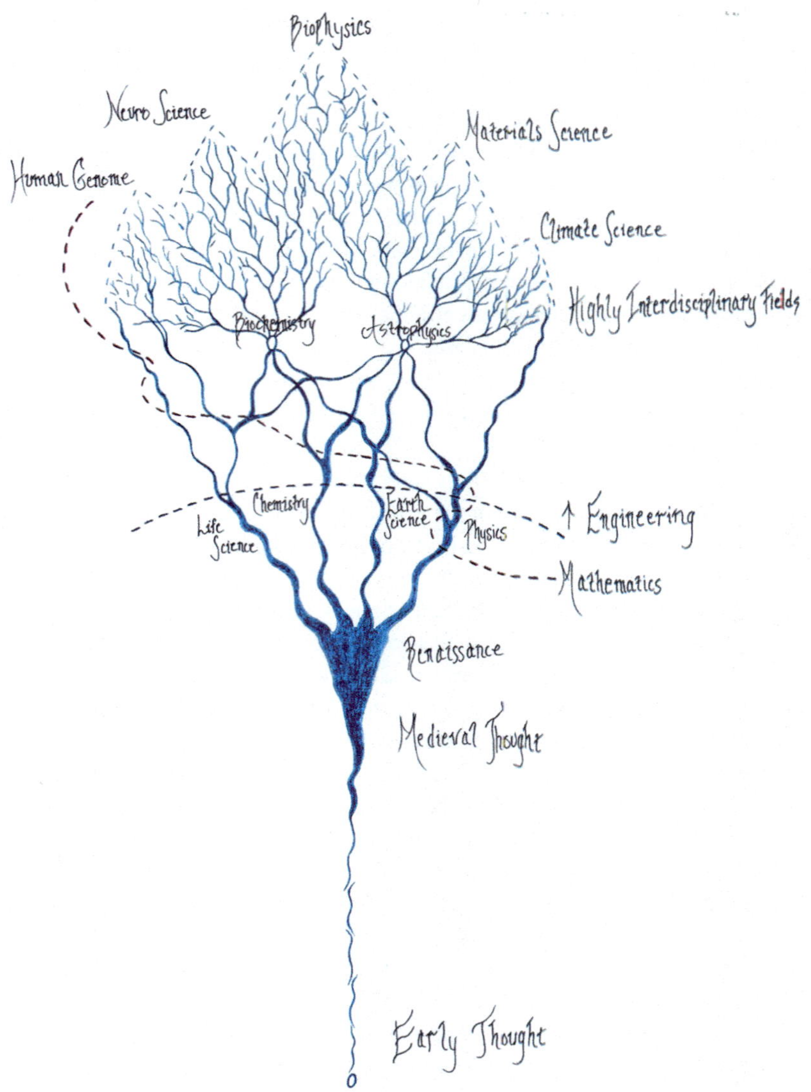

Figure 1.3 Artist's depiction of a web of science. Expansion of science from early thought to highly branched and specialized fields, gathering into complex interdisciplinary fields. (Drawing by Cynthia Satzinger).

very different fields over a lifetime. The rise of the Roman Empire, which imported much learning from Greece, and even used enslaved Greek philosophers as tutors for the ruling class, had different priorities. It was natural to emphasize applied science, in the form of great civil engineering projects, of roads, dams, and aqueducts, of improving means of terrestrial and ocean-going transport and communication, of weaponry and the machines of war; in short, engineering. Meanwhile, it was critical to develop civil and political mechanisms to administer a vast empire with up to 70 million people from radically different customs, languages, and cultures.

It is no surprise that the thread of knowledge stays thin and even diminishes in the West through the European middle ages, as ancient knowledge was often lost. The torch was passed during this time to the burgeoning Islamic world, which made fundamental advances in mathematics, astronomy, chemistry, geophysics, and medicine during the *Islamic Golden Age*, from the 8th to 14th centuries. The net of knowledge in Figure 1.3 thickens during this period. The Islamic civilization also had classical Arabic translations of the ancient works from Greece, India, Persia, and China. In Spain, the Toledo School of Translators (*Escuela de Traductores de Toledo*) was a meeting place of Islamic, Jewish, and Christian scholars, who translated the ancient texts into Classical Arabic, Latin, and early forms of Castilian (the root of modern Spanish). Scholars from medieval northern Europe frequently traveled to Toledo to gain access to ancient knowledge.

There is no doubt that the advent of the printing press in 1439 by Johannes Gutenberg (Germany, ~1400–1468) was one of the key inventions of the modern world and enhanced the vigor of the European Renaissance. With the advent of the Renaissance not only were the best legacies of the ancient world rediscovered and expanded upon, but a new wave of critical thinking and experimentation arose starting in the sixteenth century. Nicolaus Copernicus (Poland/Prussia, 1473–1543) resurrected the heliocentric theory of Aristarchus of Samos (Greece, ~310 to ~230 BC) and underpinned it with his astronomical observations and mathematical calculations. Once again, Aristotle's influence, which maintained the false geocentric theory, predominated in Europe. Copernicus's book *De revolutionibus orbium coelestium* (*On the Revolutions of the Celestial Spheres*) was published in 1543, shortly before his death. It is thought that Copernicus feared both scientific and theological ostracism from this radical theory. In fact, the book unleashed a torrent of criticism from both areas.

Famously, the consummate Renaissance man, Galileo, strongly supported the Copernican theory and fell into conflict with the Catholic Inquisition, which accused and convicted him of heresy, forced him to publicly recant his support, and condemned him to house arrest for the rest of his life.[16] His major contributions to physics, astronomy, mathematics, and engineering are too numerous to begin listing, but a major feature of his work was his mathematical description of phenomena, and his belief that mathematics could describe natural phenomena. This was a major step in separating natural science from theology and speculative philosophy.

Francis Bacon (England, 1561–1626) was influential in advocating for observation, hypothesis, inductive thinking, and objective testing of nature, rather than relying on ancient authority. This type of skeptical empiricism still underlies most of modern science.

It is notable that all these Renaissance men worked in several very different fields during their lifetime, and so the thread of knowledge in Figure 1.3 is now becoming a trunk, but still does not branch. Moving out of the Renaissance there are two distinct tendencies; Figure 1.3 shows mathematics beginning to permeate the sciences around the seventeenth century, and remains intertwined thereafter, with the strongest vine encircling physics, but growing into chemistry and earth sciences, and much later into biological sciences. Second, the separate disciplines of life and physical sciences become defined in their own right, and the physical sciences become subdivided into physics and chemistry, while earth sciences draw from all the other fields. Branching and specialization have begun.

Isaac Newton is the towering figure that powered centuries of discovery and applications in physics and mathematics. His 1687 book *Philosophia Naturalis Principia Mathematica (Mathematical Principles of Natural Philosophy)* provided the first quantitative theory of gravitation and founded classical mechanics. He developed calculus in order to elaborate his mechanical theories. In his 1704 book, *Opticks*, he gave the first detailed theory of light as tiny particles, which foreshadowed Einstein's photon theory presented in 1905, and he also invented the first known reflecting telescope.

Increasingly, in the seventeenth and eighteenth centuries, many scientists were simultaneously physicists and mathematicians. Johannes Kepler (Germany, 1571–1630) derived mathematical laws for the motion of planets, based on the observations of Tycho Brahe (Denmark/Sweden/Czechia, 1546–1601), which supported the heliocentric theory of Copernicus. Pierre de Fermat (France, 1601–1665) foreshadowed calculus and established an important extremalization principle for how light travels. Blaise Pascal (France, 1623–1662) was a pioneer in probability and number theory, contributed to the theory of hydraulics and gases, insisted on the reality of the vacuum, invented the hydraulic press and several mechanical calculators, and was also an eminent philosopher of mathematics and science, and an eloquent prose writer. Leonhard Euler (Switzerland/Russia/Germany, 1707–1783) contributed to almost every branch of physics and is often considered the most prolific mathematician in history. Carl Friedrich Gauss (Germany, 1777–1855) made wide-ranging contributions in mathematics, including non-Euclidean geometry and probability theory, as well as in astronomy and electromagnetic theory. Daniel Bernoulli (Holland/Switzerland, 1700–1782) made enduring contributions in pure math, statistics, and fluid mechanics, including his equation which predicts the principle of aerodynamic lift, eventually leading to the airplane.

Throughout the nineteenth century, the various disciplines begin to subdivide, e.g. physical chemistry, organic chemistry, biochemistry, electrophysiology, physical psychology, fluid mechanics, acoustics, thermodynamics, and optics. At the same time, some of their mutual interests make them link up to each other, forming a web of science: organic chemistry, biochemistry, and physiology find shared areas of inquiry. Electrochemistry and electrophysiology join in the quest to understand nerve conduction.

In the twentieth century and into the twenty-first century, there is ever-increasing specialization, sometimes so deep that practitioners in one specialty have little or no knowledge of even proximate fields. At the same time, philosophical musings have been eradicated from the goals and expression of mainstream scientific research and literature. But, starting roughly in the twentieth century, research areas were becoming too complex to be approached by a single specialty. Materials science, polymer science, neuroscience, genetic engineering, and other areas increasingly required knowledge and methods from previously disjoint specialties. Figure 1.3 shows several areas and how they reconvene the disconnected specialties and bring them to bear.

In Figure 1.3, there is a wave propagating outwards with the development of science: practical applications which shape entire civilizations. This is engineering, and virtually every field and subfield has an engineering component

16 Galileo is said to have uttered "Eppur si muove," upon his sentencing to house arrest by the Inquisition, meaning "yet it moves," referring to the Earth moving in space and not being the stationary focal point of the solar system (or of the entire universe, as was often thought!).

associated with it. Civil engineering, one of the oldest applied sciences, gathers together mechanics, earth science, materials science, and others to create structures appropriate for their designated functions and environments. Mechanical engineering arguably begins with bows and arrows, works through the innovations of the ancient Greek and other civilizations, and makes a huge leap after the establishment of Newtonian mechanics and its consequences. Chemical engineering combines chemistry, fluid mechanics, thermodynamics, metallurgy, and a host of uniquely industrial process, operational, and control concepts to enable production of modern products. Biomolecular engineering emerges from the latter, where the engineering now focuses on the creation of biologically derived molecules and products. Electrical engineering fleshes out the innumerable inventions deriving from classical electromagnetic theory and has harnessed quantum physics to produce the diodes, transistors, and integrated circuits underlying modern computers and a vast number of other devices. Biomedical engineering combines modern concepts from electrical, chemical, and mechanical engineering, fuses them with neuroscience, cardiology, and other fields of medicine to develop complex, effective new devices. The list goes on.

1.5.2 Engineering Vis-à-vis Science

The American Engineers' Council for Professional Development (ECPD, the predecessor of ABET, the Accreditation Board for Engineering and Technology) has defined "engineering" as:

> The creative application of scientific principles to design or develop structures, machines, apparatus, or manufacturing processes, or works utilizing them singly or in combination; or to construct or operate the same with full cognizance of their design; or to forecast their behavior under specific operating conditions; all as respects an intended function, economics of operation or safety to life and property.

Most people have this notion of engineering: reducing scientific concepts, theories, and mathematical representations to practice and applications. The notion that engineering traditionally stands apart from the sciences can again be seen in the departmental and college structure of North American and other universities. Frequently, a college of engineering is separate from the sciences and is considered a "professional school," producing graduates equipped for immediate employment in the traditional sectors of chemical, civil, electrical, and mechanical engineering, as well as more recent fields of biomolecular, environmental, computer, and software engineering. In the sciences, in contrast, graduates serious about practicing as researchers in their profession are invariably oriented toward multiyear Ph.D. programs, which are largely research apprenticeships.

A certain blurring between science and engineering has also occurred with the recognition that "engineering science" is a discernible aspect of engineering, different from practice. This can be seen in the nature of research in engineering schools, and publication of work from many academic engineers in traditionally scientific rather than engineering journals.

1.6 Disciplinary Physics

A common rejoinder to the question "What is physics?" is "It's what physicists do." Through the last centuries, physics was implicitly "not biological." In fact, so separate did life seem to earlier scientists that the question of ultimate explanation in terms of physical principles raged for centuries, as briefly discussed above.

Physics describes the structure, properties, interactions, and transformations of mass–energy in the arena of space-time. This includes

- Unified theory of fundamental forces: gravitational, electromagnetic, strong, and weak.
- Structure and transformation of matter at the level of elementary particles (hadrons/quarks, leptons, photons, etc.): the Standard Model.
- Mechanics, whether classical or quantum mechanical, of the motion of two or more objects under the influence of other objects and fields.
- Average behavior and distributions of ensembles of particles: statistical mechanics/thermodynamics.
- Applications of fundamental theories to particular systems, including nuclei, atoms, molecules, macromolecules, hard and soft solids, liquids, gases, plasmas, ballistics, celestial mechanics, optics, and more.
- Astrophysics, the application of physical principles to the interpretation of astronomical data on stars, galaxies, and other celestial objects.
- Cosmology, the origin, structure, and evolution of the universe (or multiverse).

1.6.1 Fundamental Forces

Physics to-date has recognized four fundamental forces: gravitational, electromagnetic, strong, and weak. Gravitational and electromagnetic forces share an inverse square distance law between interacting particles and proportionality to the product of masses and charges, respectively. The forces are mediated by gravitons and photons, respectively. The gravitational force is exclusively attractive and cannot be "canceled" out. As such, its reach is cosmic and it is responsible for the formation of planets, solar systems, galaxies, and galactic clusters. The electromagnetic force is responsible for the binding of electrons to nuclei to form atoms and molecules. It is hence the most important of the fundamental forces in chemical and biological processes. The electromagnetic force is enormously stronger than gravitation; the electrostatic attraction between electron and proton is on the order of 2.3×10^{39} times greater than that of their mutual gravitational attraction, and hence the latter force has no discernible role in atomic and molecular structure. On the other hand, because positive and negative charges can neutralize or "cancel" electromagnetic effects, and there is a general, overall electroneutrality of objects, the electromagnetic force does not normally compete with gravity on celestial and cosmic scales. In fact, barring specific instances such as atmospheric accumulation of charge leading to lightning discharge, possible net charge in the ionosphere, and other effects, the Earth is essentially electroneutral everywhere, although there is debate over whether and how much net charge the Earth may have. Compared to the number of elementary charges in and on the earth, any net charge would be an exceedingly tiny fraction of this number. You can easily compute how little net charge it would take on the Earth and, say, Mars to rip each other out of their gravitational orbits about the sun.

The electrostatic attraction between the atomic nucleus and atomic electrons is responsible for the atom's existence. Size scales for atoms and small molecules are on the order of 0.1 nm, and the energy levels for electrons are on the order of 1–10 eV. The Pauli exclusion principle for fermions, electrons in the atomic case, forbids electrons from sharing the same set of quantum numbers. This results in electrons pairing with opposite spins, yielding the atomic energy level structure of the periodic table, $1s^2 2s^2 2p^6$.... Without the exclusion principle electrons would merely fall into the 1s energy ground state for all elements and chemistry as we know it would not exist.

The strong force, or *strong interaction*, mediated by *gluons*, whose range is on the order of 10^{-15} m, binds *quarks* together into the *baryons* called neutrons and protons, also termed *nucleons*, as well as a menagerie of more massive, unstable baryons. All free baryons, including the neutron, will decay into protons. A "residual" of the short-range strong force leads to nuclear structure, with binding energies on the order of 8 MeV/nucleon,[17] mediated by mesons. The stabilization of the neutron in nuclei, leads, together with protons, to the existence of the chemical elements. The short-range nature of the strong force puts a limit on the size of stable nuclei. Bismuth, atomic number 83, is the most massive stable element.[18] All of the isotopes of the next element, Polonium, number 84, are unstable, with relatively short half-lives.

The weak force has an even shorter range than the strong force, on the order of 10^{-18} m. It is responsible for the transformation of many elementary particles into each other. For example, the change of an up quark to a down quark is responsible for beta-decay of the neutron into a proton, with release of an electron and electron anti-neutrino. The force is mediated by the Z and W bosons. This leads to the current picture of the Standard Model.

1.6.2 The Standard Model

There are two major classes of matter; the quarks and the leptons. Each of these classes has particles falling into three generations. All particles have a corresponding *antiparticle*. The photon, and several other electrically neutral particles, such as the Z and Higgs bosons, and neutral mesons are their own antiparticles.

The leptons include the *electron*, the *muon*, and the *tau-lepton*, in increasing order of mass. The electron is the only stable one of these three, and the other two will spontaneously decay into electrons and corresponding *neutrinos*, an electrically neutral particle with extremely low mass. There is a neutrino type for each of these – electron, muon, and tau neutrinos – although the neutrino is found to oscillate between these forms as it propagates through space. Lepton interactions are mediated by photons.

The *hadrons* are composed of quarks. There are three generations of quarks, each with two quark members; the *up and down quarks*, the *strange and charm quarks*, and the *bottom and top quarks*, in increasing order of mass. Baryons are composed of three quarks and, as mentioned, the proton, two up and one down quark (uud), is the only stable one. The higher baryons such as the *sigma particles* and *hyperons*, to name just two, all decay down to the proton, principally via the weak force. The majority of detectable matter in the universe is baryonic. *Mesons* are the other family of hadrons. They are composed of quark/anti-quark pairs and are hence unstable.

---

17 MeV = megaelectron volt, that is, 10^6 eV.

18 Its isotope ^{209}Bi has a half-life over 10^{19} years, i.e. much longer than the age of the known universe.

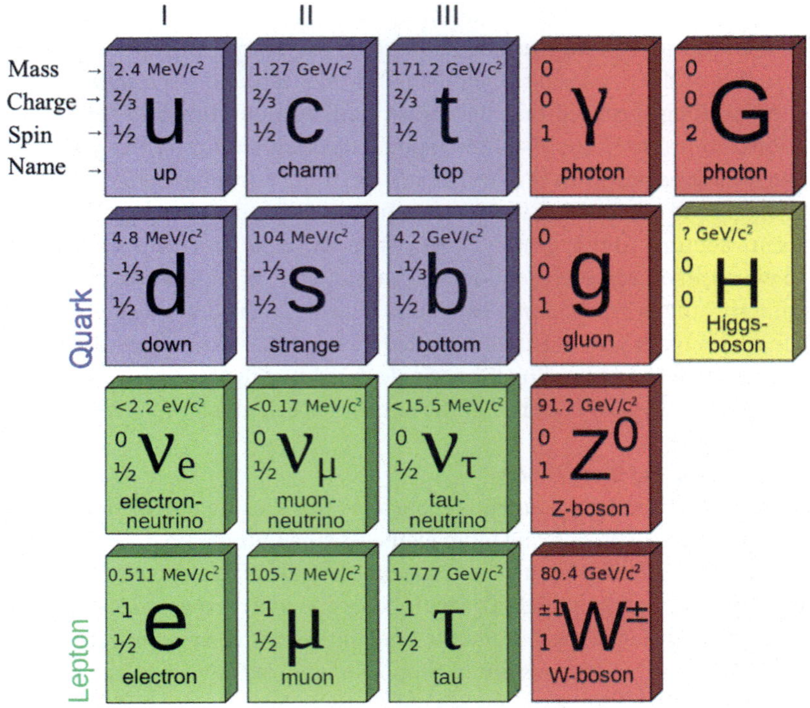

Figure 1.4 The Standard Model. (Wikimedia Commons (2013) / Licensed under CC BY-SA 3.0).

All the mediating (force-carrying) particles are *bosons*. Figure 1.4 summarizes the basic particles of the standard model. The *Higgs boson* responsible for the specific masses of particles in the standard model has recently been detected at CERN's Large Hadron Collider. The mass of the Higgs Boson is around 125 GeV/c^2 .

A very simplistic view on elementary particle stability is to consider Einstein's iconic equation $E_0 = m_0c^2$, which assigns a rest mass energy, E_0, to any given particle of rest mass m_0. Given nature's propensity to seek the lowest energy level, this drives particles in a given family to decay toward the less massive particles in their family. The proton is the least massive baryon and the only stable form. Even the neutron, which is only slightly more massive than the proton, will decay into a proton (and electron and electron antineutrino) when removed from a stable nucleus, with a half-life of about ten minutes. Similarly, the electron is the only stable lepton. The muon is a couple hundred times more massive than an electron, and decays to an electron (and electron anti-neutrino and muon neutrino) in its own reference frame with a half-life of a couple microseconds.

The elementary particles are locked into the structure of atoms and molecules, and hence into the fabric of the living state. In this sense, the properties of the living state could be "reduced" through a long chain of time and length scales to the properties and interactions of elementary particles, but this would likely be unrevealing of the living state properties themselves. Once the elementary particle properties are locked into atoms and molecules, then chemistry, and hence biochemistry and the living state, no longer need any recourse to their detailed properties. This is analogous to the construction of a house; it is unnecessary to know the microscopic details of wood fiber in beams or of the atomic metallic lattice structure in nails and screws in order to produce a macroscopic house whose shape, properties, and functions are unrelated to these microscopic and atomic details.

The reverse challenge is even more extreme and virtually impossible: Could the existence of the living state and its properties be induced starting from the properties of elementary particles? From the information point of view, this is impossible, because the amount of information contained in the standard model is far less than the information contained in even the simplest living organism. As discussed in Chapter 4.19, complete information can only flow from a system with higher information to lower information, and not vice versa. *Crick's Central Dogma*, presented in Chapter 4.19.4 is an example of this; complete information can only flow from DNA and RNA (with a 64-letter alphabet from the genetic code) to proteins (with only 20 amino acids that can be specified), but not vice versa.

1.6.3 Cosmology

The sky is the daily bread of the eyes.

Ralph Waldo Emerson

Before discussing conjectures on the prebiotic origins of life (Chapter 15.1) one might ask about the origins of the known universe itself. While much progress in understanding has been made in cosmology there are many daunting mysteries and there is no definitive cosmology that successfully deals with many of the open issues. When one considers that it is less than 400 years since Galileo was found heretical in his support of the solar-centered Copernican system, which ran afoul of then-prevailing religious dogma in much of Europe, the results of modern astrophysics have catapulted human understanding of the cosmos far beyond anything conceivable just a short time ago.

It was long surmised that stars were similar to our Sun, and such was suggested to some by the work of Copernicus. Giordano Bruno (Italy, 1548–1600) and René Descartes were likely the first to assert that the Sun is just one of countless stars. By the eighteenth century, there was strong astronomical evidence to support this, and the discovery of nebulae, star clusters, and other celestial objects eventually led to the idea of galaxies, which became firmly established in the late nineteenth century. At that time, a "steady state universe" was conceived, in which stars and planets could be born and die, but in which an overall average state was maintained. This notion was so ingrained that Einstein famously added a "cosmological term" to his gravitational equations in order to ensure a steady universe that neither expanded nor contracted. When the Astronomer Edwin Hubble (United States, 1889–1953) discovered, via the red-shift of light from distant galaxies, that the universe was expanding, Einstein declared that the cosmological term might have been his greatest blunder. In light of recent astrophysical evidence of the outward acceleration of the known universe, it might not have been a blunder.

1.6.3.1 The Big Bang Theory

In the early twentieth century, Hubble and others, using the best telescopes and spectrometers of the day, found that spectral lines from hydrogen and other elements emanating from galaxies were *red-shifted*.[19] This was found to be generally the case and that the farther away the galaxy, the greater the redshift. Einstein's relativistic Doppler shift equation[20] could be used to compute the speed of galactic recession with respect to earthbound observers, and it was found, within the limits of celestial distance measurements available, that the speed of recession was proportional to the distance of the galaxy away from earth. The constant of proportionality is *Hubble's constant*, and its reciprocal is the age of the known universe from the time of the presumed Big Bang; approximately 13.8 billion years.

The term "Big Bang" was coined sardonically by astronomer Fred Hoyle (England, 1915–2001), who maintained the Steady State universe to be a more accurate cosmology. The evidence for the Big Bang comes from several sources; the Hubble redshift just mentioned, the ubiquitous *2.7 K cosmic microwave background*, the abundance of light nuclei, and the formation and structure of galaxies. It is also linked to several unsolved quandaries; because several leading theories predict equal abundances of baryonic matter and antimatter, there is no compelling reason that the visible universe is now dominated by matter. Then there is *dark matter*, thought to be responsible for celestial motion for which the quantity of visible matter is insufficient. Visible baryonic matter is thought to be only a small amount of the actual matter, the rest being dark. It is not known what form(s) the dark matter may exist in, though there are several hypotheses, e.g. MACHOS (massively compact halo objects, such as nonluminous brown dwarf stars), WIMPS (weakly interacting massive particles, which carry large mass but don't interact with other matter except gravitationally), possible huge amount of neutrino mass, even though the rest mass energy of the neutrino is thought to be less than 1 eV, *axions*, and several other conjectures. Finally, *dark energy* is conjectured to be behind the observed acceleration of the expanding universe. The nature of dark energy is also currently unknown. The latest estimates are that 68% of the universe is dark energy, 27% dark matter, and only 5% "normal matter." If these estimates are even approximately true, it suggests that we know pitifully little about our universe. On the other hand, the existence of dark energy and matter may result from flaws in our current best theories, especially gravitational theory.

Further uncertainties include the notion that the ongoing expansion of space leads to a cosmic horizon[21], in which we would perceive ever less and less of the visible universe around us, leading to a sort of "island universe." Another dimension to this is the possibility of the "multi-verse," that many, perhaps an uncountable number of universes exist and we are the inhabitants of just one of them, i.e. the big bang may not be so "big" after all. Because there is no current means of experimentally testing the multiverse idea, it is often treated as metaphysical, even though its proponents can elaborate thought-provoking ideas, theories, and equations.

With these caveats, a thumbnail sketch of the Big Bang follows. Extrapolation of the measured expansion of the known universe implies an initial gravitational singularity, a point at which density and temperature are infinite, around 13.8 billion years ago. Taken less literally, it may have been extreme but finite density and temperature over a small region. In either case, nobody knows where this mass/energy may have come from, but the idea is that it exploded and then rapidly expanded,

---

19 Red-shifted light is measured to have a longer wavelength than is measured in the rest-frame of the light source.

20 The Doppler shift derived in special relativity does not take account of the expansion of space-time itself. While reasonably accurate at explaining redshifts from galaxies at short cosmological distances, it begins to diverge at large distance from the *cosmological redshift*, which uses general relativity and the expansion of spacetime.

21 Cosmic horizon is defined as 'the maximum distance from which light from particles could have traveled to the observer in the age of the universe. It represents the boundary between the observable and the unobservable regions of the universe, so its distance at the present epoch defines the size of the observable universe.

reducing density and temperature as the matter/energy expanded outwards, *creating* space in the process. All four forces – gravitational, electromagnetic, electroweak, and strong nuclear – were unified initially, but at around 10^{-43} s into the expansion, the temperature reduced enough for gravity to split off from the other forces. It is postulated that at around 10^{-37} s, there was *cosmic inflation*, in which space, the universe, expanded rapidly at greater than light speeds.[22] The theory then posits that quantum fluctuations in density, due to the Heisenberg uncertainty principle, were quickly magnified to macroscopic dimensions due to this inflation, setting the stage for local inhomogeneity in the macro-universe, and the subsequent formation of galaxies, and the stars within them. During this period the strong nuclear force split off from the electromagnetic and weak nuclear forces. The universe's volume may have expanded by a factor of 10^{78}, or more, during this inflationary period.

Inflation stopped after about 10^{-32} s, and the universe was still exceedingly hot and could sustain *quark-gluon plasmas*. These latter occur when the temperature reaches the *Hagedorn temperature* of about 2×10^{12} K. Baryons, especially protons and neutrons have so much internal energy at this temperature that they essentially "boil" and "evaporate" into a plasma of their constituent quarks and mediating particles, the gluons. There was a huge density of photons which scattered strongly from all the free, charged particles. Atoms did not yet exist. Only as the universe continued to expand and the temperature dropped sufficiently was it possible for a proton to combine with a neutron to form a deuteron, the deuterium nucleus, and for two neutrons and a proton to form a triton, the tritium nucleus, and for two protons and two neutrons to form the alpha particle, the helium nucleus. The measured abundance of these and other light isotopes add support to the Big Bang theory. They are thought to be the only nuclei formed outside of nucleosynthesis inside stars. Nonetheless, the majority of the baryonic matter continued to be in the form of protons.

As expansion and cooling continued electrons could be captured by the protons to form hydrogen atoms, and the light nuclei formed helium and an isotope of lithium, about 400,000 years after the Big Bang. These neutral atoms are dramatically less efficient at scattering light than charged particles[23] so that, with the formation of atoms, the universe became "transparent." The photons from the previous epoch, no longer scattered and absorbed so intensely as before the formation of atoms, expanded with the rest of the universe as a remnant of the Big Bang. Due to the *cosmological Doppler redshift* this remnant's wavelength lengthened extremely and became the *cosmic microwave background radiation*, discovered serendipitously in 1964 at Bell Telephone Laboratories by Robert Wilson (United States, 1936–) and Arno Penzias (Germany/United States, 1933–). This added major strength to the Big Bang theory. In 1989, the United States National Aeronautics and Space Administration (NASA) launched the Cosmic Background Explorer (COBE), which confirmed that the background radiation conformed within experimental error to a perfect black body at $T = 2.7$ K. Additionally, slight anisotropies in the radiation were found throughout the entire cosmic field mapped, which strengthened the idea that the rapid expansion of space due to inflation had magnified primordial quantum fluctuations.

Once hydrogen atoms were formed, "the rest is history." The universe continued to expand and the gas density inhomogeneities became nucleation sites for proto-galaxies, where portions of the pervasive hydrogen gas condensed locally under gravitation. Within the protogalaxies, further anisotropies caused the gravitational condensation of hydrogen gas into proto-stars. In a protostar, the gravitational collapse led to heating of the gas, reaching the minimum temperature for nuclear fusion reactions to begin at above 10^7 K, at which point, the collapsing hydrogen mass lit up with thermal radiation and became a star. There is a fascinating ancestry to stars. Those formed early began to generate heavy elements in their exceedingly hot cores. When these stars exploded they sent heavy elements throughout their galaxy, causing "second generation stars," such as our Sun, to have small amounts of heavier nuclei, in addition to the preponderant hydrogen and helium. These heavy elements ejected by the massive stars may also have led to the formation of "iron" planets, such as the Earth, as well as gas giants, such as Jupiter, and ice giants, such as Neptune (see Chapter 15.5).

While cosmology is still a speculative and unsettled subject, the results of astrophysics are very well established. Hence, such celestial issues as the life cycle of stars and their various fates as red giants, white dwarves, neutron stars, and black holes are well described theoretically and backed by massive and increasing observational data. *Gravitational lensing*, detection of extrasolar planets, origins of *quasars*,[24] and more, are active and substantial areas of research.

22 Special Relativity shows why matter cannot exceed the speed of light in "flat" spacetime. But if space itself is growing – think of a spatial grid expanding equally in all directions – its rate of growth is unrelated to the relative motion of matter through the grid, i.e. particles at rest in the grid will move away from each other due to the grid's expansion, which is not matter and not constrained by the speed of light.

23 Think of the charged particles as "mirrors," analogous to free electrons in a metal, and the neutral atoms as a gas of dielectric particles. Light reflects intensely in every which direction as it bounces off the charged particles, but scatters weakly from the gas, and mostly just passes through the gas undeflected.

24 Quasars=quasi stellar sources of radiation, are supermassive black holes presumed to be *galactic nuclei*. Gas pulled into a quasar accelerates and emits enormous amounts of electromagnetic radiation.

1.7 Are Currently Known Physical Laws Adequate for Understanding Living State Phenomena?

There are currently no known biological phenomena that disobey physical laws and principles. However, this is not to say that there are not necessarily other physical principles, as yet undiscovered. For example, an automobile is built entirely on physical principles, but knowledge of all those principles does not allow one to deduce the existence of an automobile. This is not to introduce an "intelligent design" substrate into the discussion, and this book will remain uninvolved concerning current debates and controversies in this area. But the automobile analogy at least leaves the door open to the likelihood that we have not yet discovered all the principles involved in the self-organization of matter into the living state. After all, the history of science is replete with incidences when people thought that all the major principles were known, only to be entirely controverted by new insights and principles, e.g. the famous fall from ultimate glory of nineteenth-century physical science, with its great achievements in electromagnetism, mechanics, and thermodynamics, was a particularly poignant era of hubris, which was humbled by special and general relativity, quantum mechanics, and quantum statistical physics. Nonlinearity in systems, often previously avoided as intractable, has been shown to lead both to chaotic behavior, as well as to the formation of far-from-equilibrium structures.

Concerning the issue of whether physical laws, whether known or not yet known, are an adequate basis for understanding the living state, many ventured that biology has its own set of principles. This was a long-held view, often termed *vitalism*, and there were many attempts to find the "spark of life," or the "élan vital." It was often assumed that this distinct "life force" is foreign or even contradictory to physics. Niels Bohr in his reflections on the nature of life extended his notion of wave/particle complementarity to life, conjecturing that the principles of life stand in complementary relationship to those of physics. Bohr incorporated a "yin and yang" symbol into his coat of arms to show the central meaning that complementarity had for him. Vitalism reigned for millennia until certain facts began to reveal that biomolecules and biological processes obey physical laws.

Luigi Galvani appears to be the first to discover bioelectricity while producing kicking motion of dead frog legs with electrical currents. He initially thought of it as "animal electricity" and that living tissue was needed to transmit electricity. Volta, a rival, soon demonstrated that metals and other inorganic substances could conduct electricity and so the distinction between "inanimate" electricity and "bioelectricity" disappeared.

It was widely believed that the molecules of life were distinct from those found in nonliving entities. As mentioned, in 1828 Friedrich Wöhler found that urea, a byproduct of living organisms, could be synthesized without the use of any biological precursor molecules, using silver cyanate and ammonium chloride.

Spontaneous generation was first formulated by Aristotle and held that life could arise spontaneously where a suitable nutrient source existed, e.g. maggots could appear spontaneously around rotted food. Louis Pasteur (France, 1822–1895) disproved this idea in his famous sterilization experiments. Yet Pasteur himself held that fermentation was a "vital function" that could only occur via living organisms. Eduard Buchner (Germany, 1860–1917) found that sugar could be fermented from yeast extract free of any living cells. Numerous abiological synthetic routes to end products of fermentation have since been found. Yet, two more examples of absolute concepts are overturned by careful experimentation and analysis.

In 1967, Francis Crick, the codiscoverer of the structure of DNA, stated "And so to those of you who may be vitalists I would make this prophecy: what everyone believed yesterday, and you believe today, only cranks will believe tomorrow."[25]

The prevailing "mainstream" scientific view is that biological and biochemical phenomena can be traced to underlying physical laws. Another controversy, of course, is whether we currently understand physical laws well enough to describe biology. One school of thought holds that quantum mechanics and relativity are final theories which ultimately span all possible physical phenomena, and hence it is only a matter of finding the right ways of applying them in order to explain any physically perceivable phenomenon. Another school of thought exists that says that throughout human history new theories have always arisen that have shrunken or falsified existing theories and that to claim that we have finally reached conclusive, all-encompassing theories is to merely continue the vanity of bygone eras when similar pronouncements of universality were made concerning the state of erstwhile, often debunked knowledge.

The view that biology is ultimately explainable by physics does not imply that biology "reduces" to physics, only that the hierarchy of complex phenomena are based on physical principles. Still, are biological, and hence physical laws an adequate basis for understanding the phenomenon of consciousness? Whereas vitalism is almost extinct as a school of thought there is still at least philosophical controversy about the reducibility of consciousness to physical mechanisms,

25 Francis Crick In 'The Prospect Before Us', of Molecules and Men (1969), 99.

and to whether we do or do not have free will. Some of the unsettled issues of biophysics of consciousness are introduced in Chapter 15.3.

Neurological activity, electrochemical processes, is exhibited in all activities of the brain, including emotions, thought, etc. Hence, it is accepted that "conscious phenomena," such as emotions and thinking, depend on neurophysiology for their existence, and that consciousness as we know it ceases when the corresponding neurophysiological processes cease.

What is still controversial is whether or not the neurophysiological processes which are necessary for consciousness are sufficient to explain what consciousness actually is. For example, are the physiological processes merely a projection onto the subspace of spacetime/matter for a process occurring in a greater space which would have corresponding projections into other spaces besides spacetime that would yield different characterizations of consciousness? The projection of any 3D shape onto a single axis is a line, but no one would declare that the line *is* the 3D object, or to be a complete description of it. At this point, such conjectures are metaphysical because there is currently no knowledge of what the other dimensions of consciousness might be nor how they could be measured. "Holistic" approaches might maintain that the other dimensions of consciousness can in fact be experienced by proper meditative and other self-regulating approaches.

Finally, no matter how triumphant anyone proclaims evolutionary or any other reigning theory to be, we still run into the hard fact that nobody knows the ultimate origin of the universe, of matter/energy, of the laws that imbue it, or of what, if anything, lies beyond our own particular physically detectable universe.

This is also where "modern science" as it is currently practiced diverges from its earlier forms of "natural philosophy," in which scientists and mathematicians (e.g. Newton, Descartes, Pascal) expounded on their philosophic and even religious views in addition to their concrete scientific advances. Modern science seeks to discover underlying principles in nature, to formulate them mathematically or symbolically, and to use them for greater unified understanding of phenomena and for technological manipulation. But science is not required to answer where the principles come from, or if there is any "meaning" to them beyond what they represent for understanding and manipulation. A practicing scientist can successfully apply the principles of quantum mechanics and statistical mechanics, without ever questioning whence they come or if there is any "intent" in these principles.

Extremalization principles, such as Fermat's principle, and Lagrangian and Hamiltonian mechanics, provide tremendously successful scientific workhorses, as does minimization of Gibbs free energy in chemistry. It has long been debated whether these principles imply an "intent" of nature or are just "the way things are" with no additional explanations needed. Moreau de Maupertuis (France/Switzerland, 1698–1759), who formulated a least action principle actually interpreted it theologically, as the "wisdom of God" in following the "thriftiest" of paths in nature's processes.

The applications of least action principles are vast indeed, encompassing huge swaths of optics, mechanics, chemistry, biophysics, and more. And the sweep of these principles is stunning. Biochemistry makes excellent predictions based on them. Couple the exclusion principle for fermions to the least energy principle and one arrives at the notion of electronic structure in the periodic table, the seat of all chemistry. Apply least energy and exclusion to electrons in a metal and you get enormously high electron speeds, even at absolute zero. Take away exclusion from this and you get things like superconductivity and superfluidity. Add the directionality of superposed quantum mechanical electron states and you get the self-organizational principles involved in crystallization of substances. One wonders what as yet undiscovered self-organizational principles may exist that insert into the least energy and other principles that may impel biological self-organization and evolution.

1.8 Unifying Characteristics of the Living State

There was a time when it was thought that life could be characterized by a few simple qualitative attributes, such as reproduction, irritability, consumption and excretion, etc. Biophysics tends to define these and other living state characteristics based on physical principles. Despite the vast variety and extraordinary complexity of living organisms, there are nonetheless unifying properties common to all life forms. Some of these are:

- Proteins are made of 20 main amino acids, all of which have L-chirality, except glycine, which is achiral (Chapter 3.1).
- All DNA is made of four nucleotides (Chapter 3.2). Virtually unlimited combinations are possible.
- The nucleotide adenosine triphoshpate (ATP) is a universal source of free energy. This may be derived from sources such as sunlight, inorganic chemical reactions, or use of organic materials produced by other organisms.
- All organisms reproduce copies or new versions of themselves.
- The genetic code is carried by DNA in all organisms (some viruses use RNA), and, with few variations, mediates the accumulation and transmission of biological information; all organisms produce proteins using the genetic code.

- All biomolecules and processes are the result of evolutionary adaptation of organisms to their environment, primarily through natural selection.
- Biochemical processes and structures are defined by very precise compositions, conformations and interactions. Life shows extreme sensitivity to small chemical differences and minor changes can have drastic biological effects, e.g. methanol is toxic, ethanol far less so.
- Life is stereoselective and can process only a specific chirality type for given families of biomolecules.
- Organisms are compartmentalized by membranes or other structures that mediate matter and energy exchange with the environment.
- The biological milieu is aqueous; most biopolymers are charged for solubility, and are *polyelectrolytes*.
- Biological molecules are built of light atoms: C, H, O, N, P, S ions Na^+, K^+, Ca^{2+}, Mg^{2+}, and a small amount of heavier atoms, such as Fe, Co, and Mn.
- Life represents far from equilibrium, open systems. Equilibrium = Death.
- Life entails highly organized structures that are maintained far from equilibrium due to a flow of energy from a source (e.g. sunlight or food), through the organism, which maintains its organization and excretes waste heat and chemical byproducts.
- The highly organized state of life is due to accumulated, high information content in genetic material.
- Living state processes involve low energies, typically in the 0.01–5 eV range.
- Living organisms are self-regulating and feedback controlled.
- Reproduction involves conservation and transmission of accumulated information.
- Large numbers of molecules are involved ($>10^9$ even in the smallest bacterium). Large numbers act as a buffer against random thermal fluctuations.
- Life involves great material heterogeneity; there is a very wide variety of molecules involved in life, including many types of small molecules and an amazing sweep of biopolymers.
- Nonergodicity: In statistical mechanics the averages pertaining to a large ensemble of systems will yield the same values as the long-term average of a single particle's trajectory through phase space (Ergodic theorem). As an example of non-ergodicity; the response of a collection of neurons in a neural structure will not yield the same result as a single neuron stimulated many times.

The so-called *central dogma of molecular biology* should be mentioned.[26] This principle states that information flows in a single direction from DNA to proteins, mediated by RNA, but the reverse does not occur; information does not pass back from proteins to DNA.[27]

Another important observation is that developmental strategies are conserved in evolution, e.g. human hand, bat wing, bird wing. Dinosaur bone structural motifs are similar to modern bones.

1.9 Summary

Scientific discovery over the ages has been a growing activity, launched in antiquity and burgeoning in modern times. Its path has been strewn with countless wrong turns and misconceptions, but it corrects itself through experimental verification and new, more comprehensive theories which subsume or replace older ones. Everything in science can be questioned, although some things more than others, e.g. the certainty of the electron mass, charge, and spin versus the uncertainty of precisely how a huge number of interacting components, effects, and principles affect the Earth's climate. Disciplinary specialization was needed over the past centuries because the growing depth of knowledge in all fields required concentration on a huge amount of information in any given field to make progress. Biophysics represents one of the areas which requires a multidisciplinary approach, where the disparate strands of biology, physics, chemistry, physiology, and many others, need be woven together to approach the mind-boggling complexities of the living state. The living state is highly organized, maybe even optimized in many instances, and it is not known yet whether this order arises from purely random events, out of nonlinear chaos, or from informational and self-organizational principles of matter we haven't yet discovered ... and perhaps may never discover.

---

26 The use of the word "dogma" in this is unfortunate, since it implies something that is incontrovertible and provided by a higher authority. While the principle is well established, it is risky in any field of science to declare something "incontrovertible," since the essence of science is falsifiability; all laws and principles in science are always open to upset and modification, should there be evidence to falsify or reduce the sweep of a law.

27 *Retroviruses* use the enzyme *reverse transcriptase*, which turns its RNA into DNA, which becomes inserted and active in the host's genome via the enzyme *integrase*.

For this author, science is not a belief system. I don't "believe" in science, physics, evolution, the Big Bang. Science is just the best map currently available to make sense of the bewildering phenomena we experience and are a part of. But it does not contain the central element necessary for an ideological or religious belief system, which posit unquestionable, untestable dogma and concepts as their basis. If what was considered to be a "law of science" is found to actually not be universal, scientists do not typically murder the discoverers, commit suicide, or reject the entire scientific method. Rather, they pour heavy skepticism on the new findings and work exhaustively to test their experimental veracity, and then just move with excitement to the next phase of discovery.

References and Further Reading

Adams, F.C. (2019). The degree of fine-tuning in our universe - and others. *Physics Reports* 807: 1–111.

Adli, M. (2018). The CRISPR tool kit for genome editing and beyond. *Nature Communications* 9 (1): 1–13.

Bertman, S. (2010). *The Genesis of Science: The Story of Greek Imagination*. Prometheus Books.

Bonnet-Bidaud, J.-M. (2017). The diffuse light of the universe: on the microwave background before and after its discovery: open questions. *Physics* 1–22. arXiv.org, e-Print Archive.

Brax, P. (2018). What makes the Universe accelerate? A review on what dark energy could be and how to test it. *Reports on Progress in Physics* 81 (1): 016902/1–016902/52.

Chadwick, B.P. (ed.). (2015). *Epigenetics: Current Research and Emerging Trends*. Caister Academic Press.

Chen, Z. and Natarajan, R. (2022). Epigenetic modifications in metabolic memory: what are the memories, and can we erase them. *American Journal of Physiology* 323 (2): C570–C582.

Cooper, L.N. (1968). *An Introduction to the Meaning and Structure of Physics*. Harper Row.

Cossart, P. (2018). *New Microbiology: From Microbiomes to CRISPR*. Washington, USA: Am Soc. Microbiology.

Cromer, A.H. (1993). *Uncommon Sense: The Heretical Nature of Science*. Oxford University Press.

Crump, T. (2002). *A Brief History of Science*. Running Press.

Dalafi, H.R. and Hassan, M.H.A. (1994). *Renaissance of Sciences in Islamic Countries*. Singapore, New Jersey, London: World Scientific.

Darwin, C. (1859). *The Origin of Species*. available from various publishers; e.g. Prometheus Books, Signet, Random House, and others.

DeWitt, R. (2018). *Worldviews: An Introduction to the History and Philosophy of Science*, 3e. Wiley Blackwell.

Dvorkin, C., Mishra-Sharma, S., Nord, B. et al. (2022). Machine learning and cosmology. *High Energy Physics–Phenomenology* 1–32. arXiv.org, e-Print Archive.

Eded, A. and Cormack, L.B. (2012). *A History of Science in Society: From Philosophy to Utility*. University of Toronto Press.

Fernández-Jambrina, L. and Lazkoz, R. (2022). New futures for cosmological models. *Philosophical Transactions. Series A, Mathematical, Physical, and Engineering Sciences* 380 (2222): 20210333.

Feynman, R. (1967). *The Character of Physical Law*. MIT Press.

Gasperini, M. (2022). From pre-to post-big bang: an (almost) self-dual cosmological history. *Philosophical Transactions of the Royal Society, A: Mathematical, Physical & Engineering Sciences* 380 (2230): 20210179.

Goldberg, D. (2017). *The Standard Model in a Nutshell*. Princeton University Press.

Huang, S., Litt, M., and Blakely, A.C. (eds.). (2016). *Epigenetic Gene Expression and Regulation*. London: Elsevier.

Kafedzic-Briga, A. and Dzaferovic-Masic, E. (2021). How far have we searched for a dark matter so far. *Journal of Physics: Conference Series* 1814: 012007.

Kuhn, T.S. (1962). *The Structure of Scientific Revolutions*. University of Chicago Press.

Kumar, A. (2014). *Sciences of the Ancient Hindu*. CreateSpace Independent Publishing Platform.

Lebedev, O. (2021). The Higgs portal to cosmology. *Progress in Particle and Nuclear Physics* 120: 103881.

Lindberg, D.C. (1992). *Beginnings of Western Science: The European Scientific Tradition in Philosophical, Religious, and Institutional Context, 600B.C To A.D. 1450*. University of Chicago Press.

Natividad, C. (2021). *Epigenetics: Beyond Genetics*. Arcler Education, Inc.

O'Raifeartaigh, C., O'Keeffe, M., Nahm, W., and Mitton, S. (2017). One hundred years of the cosmological constant: from 'superfluous stunt' to dark energy. *Physics* 1–62. arXiv.org, e-Print Archive.

Popper, K.R. (1959). *The Logic of Scientific Discovery*. Routledge, Taylor & Francis Group, Hutchinson & Co. from original "Logik der Forshcung," Julius Springer, Vienna, Austria, 1935.

Raby, S. (2021). *Introduction to the Standard Model and Beyond; Quantum Field Theory, Symmetries and Phenomenology*. Cambridge University Press.

Ringrose, L. (ed.). (2017). *Epigenetics and Systems Biology*. London: Elsevier.

Sami, M. and Gannouji, R. (2021). Spontaneous symmetry breaking in the late Universe and glimpses of the early Universe phase transitions a la baryogenesis. *International Journal of Modern Physics D: Gravitation, Astrophysics and Cosmology* 30 (13): 2130005.

Sarton, G. (1952). *A History of Science: Ancient Science through the Golden Age of Greece*. Harvard University Press.

Schwerdtfeger, P., Smits, O.R., and Pyykko, P. (2020). The periodic table and the physics that drives it. *Nature Reviews Chemistry* 4 (7): 359–380.

Staggs, S., Dunkley, J., and Page, L. (2018). Recent discoveries from the cosmic microwave background: a review of recent progress. *Reports on Progress in Physics* 81 (4): 044901/1–044901/35.

Tacconi, L.J., Genzel, R., and Sternberg, A. (2020). The Evolution of the Star-Forming Interstellar Medium Across Cosmic Time. *Annual Review of Astronomy and Astrophysics* 58: 157–203.

Van Proeyen, A. (2020). Supergravity. *AAPPS Bulletin* 30 (2): 66–75.

Xue, Y., Wang, J., He, Y. et al. (2022). Multi-scale gene regulation mechanism: spatiotemporal transmission of genetic information. *Current Opinion in Structural Biology* 77: 102487.

Other Biophysics Texts

Ashrafuzzaman, M. (2024). *Introduction to Modern Biophysics*. CRC Press.

Bialek, W. (2012). *Biophysics, Searching for Principles*. Princeton.

Clacomb, J. and Tran, J. (2011). *Introductory Biophysics, Perspectives on the Living State*. Jones and Bartlett.

Cotterill, R. (2002). *Biophysics, an Introduction*. Wiley.

Glaser, R. (2012). *Biophysics: An Introduction*. Berlin: Humboldt Universität.

Jackson, M.B. (2006). *Molecular and Cellular Biophysics*. Cambridge: Cambridge University Press.

Jue, T. (ed.). (2009). Fundamental Concepts in Biophysics. In: *Handbook of Modern Biophysics*. Humana Press. series.

Jue, T. and Davis, U.C. (ed.). (2009). *Fundamental Concepts in Biophysics*, vol. 1. Humana Press. Handbook of Modern Biophysics.

Nadeau, J.L. (2017). *Introduction to Experimental Biophysics*. CRC Press.

Nelson, P. (2004). *Biological Physics; Energy, Information, Life*. U. Penn. Freeman.

Noelting, B. (2006). *Methods in Modern Biophysics*. Springer.

Nossal, R., NIH and Lecar, H., Berkeley (1991). *Molecular & Cell Biophysics*. Addison Wesley.

Parke, W.C. (2020). *Biophysics. A Student's Guide to the Physics of the Life Sciences and Medicine*. Springer.

Scherer, P.O.J. and Fischer, S.F. (2017). *Theoretical Molecular Biophysics*, 2e. Springer.

Schiessel, H. (2014). Biophysics for Beginners, a journey through the cell nucleus.

Srivastava, P.K. (2005). *Elementary Biophysics*. Alpha Science International Ltd.

Sybesma, C. (1977). *Biophysics, an Introduction*, 2e. 1989.

Tuszynski, J.A. and Kurzynski, M. (2024). *Introduction to Molecular Biophysics*. CRC Press.

Volkenstein, M.V. (1977). *Molecular Biophysics*. Academic Press (from MIR, USSR).

Waigh, T.A. (2014). *The Physics of Living Processes*. Wiley.

Yeargers, E.K. (1992). *Basic Biophysics for Biology*. CRC press.

Zuckerman, D.M. (2010). *Statistical Physics of Biomolecules; an Introduction*. CRC Press, Taylor and Francis Group.

Problems: Review of Basic Physics

These elementary problems are diagnostic in nature. They are a reminder of the type of physics problems you will have solved in earlier courses.

1 A singly ionized CH_4 molecule in a vacuum is accelerated across a potential of 5×10^4 V.
   a) Compute the time it takes to traverse a 50 cm straight line path after exiting the acceleration stage.
   b) The ion enters a uniform magnetic field B, at right angles in which it undergoes a circular orbit of radius $R = 5$cm. Compute the magnitude of the magnetic field.
   c) If the ion is allowed to continue to circulate in the above orbit, find the frequency (Hz) and angular frequency (radians/s) of the orbit.
   d) If the ion enters the B field at 45° to a B field of 3 T, find the radius of its orbit.
   e) Describe the trajectory of the ion in part (e).

2 The energy levels in a hydrogen atom are given by

$$E_n = -\frac{13.6(\text{eV})}{n^2}$$

a) Find the wavelength of light emitted when an electron makes a transition from $n = 3$ to $n = 1$.
b) Find the fraction of H atoms in the $n = 2$ level at $T = 300$ K.
c) Consider the emission spectrum of a gas consisting of H atoms and H ions, and an H-atom gas at room temperature through which a light containing continuous ultraviolet and visible wavelengths shines. Does the emission spectrum of the former have more, less, or the same number of lines as the absorption spectrum of the latter? Explain.

3 a) A proton is in a 2-T magnetic field. Find the frequency of electromagnetic radiation that will drive it between the spin-up and spin-down states in this field. (This is NMR.)
b) Consider an electron in the same field. Find the frequency of electromagnetic radiation that will drive it between the spin-up and spin-down states in this field. (This is electron spin resonance, ESR.)

4 The spacing between atoms in a certain metal crystal is $d = 0.12$ nm. If X-rays incident at 45° to the normal to the face of the crystal reflect and create a first-order maximum ($m = 1$), what is the energy of the X-rays?

5 Say the Earth and Mars have equal and opposite net charge, what net charge at their distance of closest approach would be enough to overcome the gravitational attraction between the Earth and the Sun?

6 Particles of α-, β-, and γ-radiation all have energy of 700 keV and enter a uniform magnetic field of 3 T. Find the speed of each particle and make a diagram to show the trajectory of each when in the B field.

7 ${}_6C^{14}$ has a half-life of 5730 years. In radioactive dating, the standard used is that there are 14 disintegrations per minute per gram of "fresh" carbon (i.e. carbon on the earth's surface participating in the carbon cycle, and not buried in the earth).
a) Compute the fraction of ${}_6C^{14}$ isotopes in fresh carbon according to this disintegration standard
b) If a 300-g archeological (buried) carbon specimen is unearthed and found to yield 1200 disintegrations per second, estimate the age of the specimen.

8 Air rushes past an airfoil of 30 m^2, going over the airfoil at 150 m/s and below the airfoil at 80 m/s. Find the maximum load this airfoil can lift up.

9 An electron and positron moving directly toward each other have 800 keV kinetic energy in the lab frame just before they annihilate. Two gamma rays are produced in the annihilation. Compute the energy of each.

2

Overview of Biological Cell Structure

Over these past billions of years, evolution has led to certain functions and motifs that are conserved across most, and sometimes, all life forms. Equipped with an arsenal of genetic tools, evolutionary scientists have identified possible branch points in evolution leading to the concept of the *most recent common ancestor* or *last common ancestor*, which can be defined at almost any level, the last common ancestor of a certain gene, two species, plant and animal kingdom, humans, etc. The identity and period in time of the last universal common ancestor (LUCA) for all life forms is unknown and is a subject of research and debate. It may well have been a single-cell organism similar to a bacterium. The last common ancestor between humans and chimpanzees is also unknown and is speculated to have lived some time between 4 million and 14 million years ago. Figure 2.1 is a schematic showing some of the evolutionary paths and branch points among plants, fungi, and animals.

To keep a vastly complex subject manageable, we want to assemble a very brief overview of the two major cell groups, prokaryote and eukaryote. Bacteria and archaea are prokaryotes, and all other organisms are eukaryotes. The place of viruses in the living state is unique from the latter, sharing very few characteristics except genetic reproduction and use of proteins to carry out its functions. All living entities are composed of one or more cells. Viruses are not cells. Archaea are briefly presented in Chapter 14.1.1.

Some have compared the inside of cells to miniature cities, in which organelles with specialized functions reside in different locations, producing and trading "molecular goods" produced with other organelles and structures, with the cytoskeleton acting as dynamic highways and paths, for distribution and delivery of these molecular goods. Meanwhile, the cell membrane, like the enclosing walls of a medieval city, but dynamically shifting the borders, strictly controls the ingress and egress of molecular agents and interactions with other "cities" (cells) and the "no man's land" expanses of extracellular matrix and the external world.

2.1 The Prokaryotic Cell

These comprise single-cell microorganisms without nuclei, and without compartments and rarely any internal membranes around organelles. There is no nuclear membrane to cross during reproductive and other genetic events. The basic motif of a bilayer membrane,[1] common to all cells, is found, as well as the ribosomes as sites of protein synthesis, adenosine triphosphate, ATP, as the energy carrier, the basic composition of the cytoplasm, and, of course, the use of DNA as the mechanism of information storage and expression. Such cells often also contain a mechanism for motion, such as flagella, and special proteins to operate them. Figure 2.2 provides a sketch of a prokaryotic cell, with some of the major structures and organelles. The bacterial flagellar motor is termed the *basal body*. Prokaryotic and eukaryotic flagella are completely different from each other.

1 Some Archaea have single-layer membranes.

Biophysics: Physical Processes Underlying the Living State, First Edition. Wayne F. Reed.

Companion website: www.wiley.com/go/Reed/BiophysicsandMacromolecules

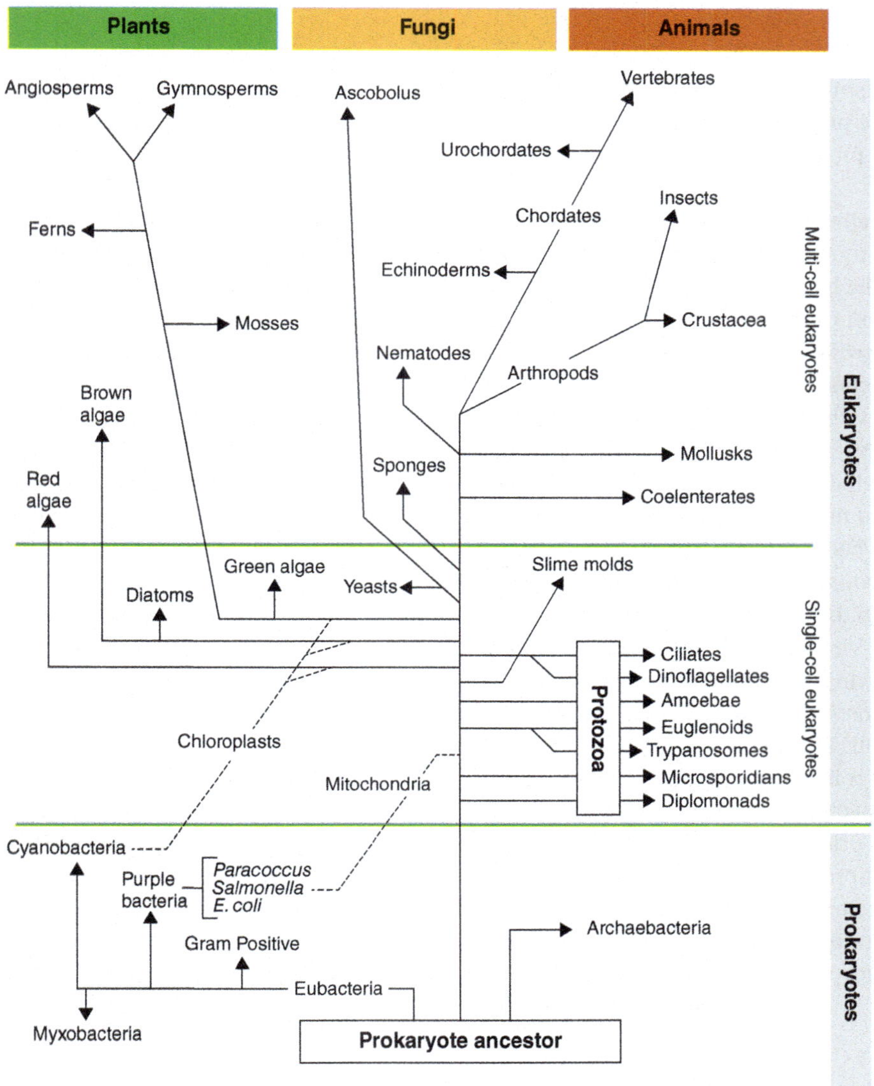

Figure 2.1 Schematic of evolution and branch points among plants, fungi, and animals. (Courtesy of Marissa Calko).

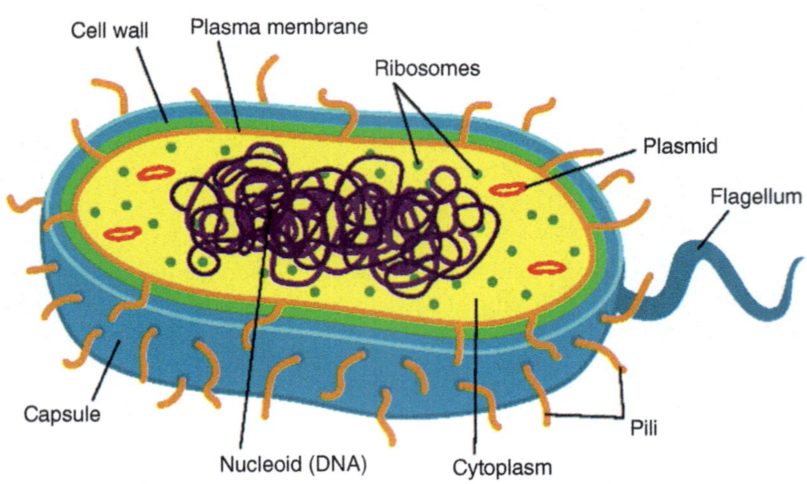

Figure 2.2 Sketch of a prokaryotic cell. (Courtesy of Marissa Calko).

2.2 The Eukaryotic Cell

Eukaryotes start with unicellular organisms, often termed *protists*, and work up into more specialized and complex multicellular organisms. Eukaryotes include protists, fungi, plants, animals, and many types of algae. Eukaryotic cells have more subcellular structures than the prokaryotic cells. Figure 2.3 is a schematic of a eukaryotic cell and major components.

The *cytosol* comprises around 70% of cell volume and is mostly water but contains a significant concentration of proteins and small molecules. Many of the proteins are involved in the formation of the *cytoskeleton*, a dynamic skein of filaments, fibers, and microtubules that maintain the balance of internal and external forces on a cell, act as sensors and transmitters of external forces, signal the nucleus and other organelles, and provide channels for material flow within the cell. The protein families of *tubulin*, *actin*, and *myosin* are chief components of the cytosol. Together with the various organelles, usually excluding the nucleus, the cytosol is often termed the *cytoplasm*.

Mitochondria are complex organelles with their own genome, membranes, and compartmentalization, and have several functions. Pre-eminently, they are localized sources of ATP production, the cell's "power plant," but they also have other functions in cell signaling, regulation, cell death, and differentiation. Some cells contain a single mitochondrion, whereas others can have thousands. One theory of the origin of mitochondria is that they were bacteria that survived incorporation into a different species of eukaryote and became integral to the organism.

The *nucleus* of eukaryotic cells is surrounded by a double-layer *nuclear membrane* and contains DNA complexed with histones (proteins) and other proteins to form *chromatin*, which packs together to form *chromosomes*, the cell's genetic material (Chapter 11.3). The nucleus is viewed as the cell's control center, since it controls gene expression to produce proteins, which in turn carry out most living state functions, including feedback-controlled metabolic cycles.

Rough endoplasmic reticulum and smooth endoplasmic reticulum are structures that are attached to the outer nuclear membrane and form networks in the cell. The latter is involved in lipid and carbohydrate metabolism, while the former normally contains ribosomes and hence is involved in events occurring during protein synthesis.

Camillo Golgi (Italy, 1843–1926) discovered the organelle now termed the *Golgi apparatus* or *Golgi complex*. It stores, sorts, and chemically modifies proteins prior to their distribution, either inside the cell or to the exterior of the cell via *exocytosis*. It receives newly synthesized proteins from the endoplasmic reticulum and packages them in vesicles which bind to the Golgi complex of flattened membrane disks. The vesicle bound proteins pass through the Golgi complex which has discrete compartments with different enzymes that can make post-translational modifications to the freshly minted proteins. These include sulfation and phosphorylation of amino acids, and adding and modifying carbohydrates. A major

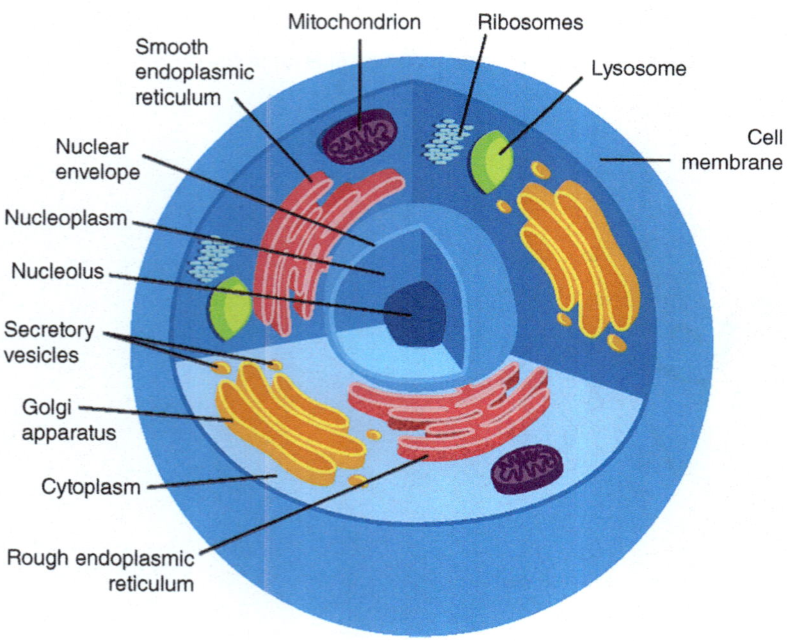

Figure 2.3 Sketch of an animal eukaryotic cell. (Courtesy of Marissa Calko).

function is the formation of *proteoglycans*, which have protein backbones to which polysaccharides, *glycosaminoglycans*, are covalently bound. The proteoglycans have a wide range of functions (Chapters 12.2 and 12.3). There are various models for how the Golgi complex functions. Its intake and delivery functions have been compared to a post office, which packages and sorts incoming items (proteins) then prepares and stamps them for delivery to selected locations. Of course, the heart of the Golgi function, the chemical modifications, are not part of this simple analogy. The modifications might be analogous to a host of sites, including food processing plants, restaurants, hospitals and surgical units, and gyms.

Lysosomes contain a wide array of digestive and other enzymes at low pH (pH 4.5–7, as compared to ~ 7.2 in the cytosol), and this dangerous environment is walled off from the rest of the cell by the lysosomal membrane. One of their chief functions is as the cell's waste disposal system, since they can break down the major macromolecules: polynucleic acids, polypeptides, and polysaccharides, as well as small molecules such as lipids. There are around 40 classes of hydrolytic enzymes, including proteases, lipases, phosphatases, phospholipases, nucleases, and glycosidases. The molecules produced by these different hydrolyses, such as amino acids, nucleotides, and sugars, can be excreted through the lysosomal membrane into the cytosol, where they can either be re-used in synthesizing new products or excreted from the cell. In addition to the lysosomal enzymes within the lysosome the membrane contains its own unique set of proteins. Particularly important is the H^+-pump (proton pump) which drives H^+ into the lysosome to keep it acidified.

The lysosome has other functions. It helps defend the cell from pathogens as it surveys molecules and agents brought into the cell. It can hydrolyze macromolecules the pathogen needs to reproduce in the host cell. Diminished lysosomal activity can result in greater susceptibility to viral infections. Lysosomes can fuse with other organelles to digest larger structures, such as bacteria. Lysosomes are also part of the process called *apoptosis* in which a cell commits suicide because it is damaged, stressed, or otherwise weakened beyond repair. It is a normal process, and it is estimated that an average human adult loses around 60 billion cells per day via apoptosis.

2.3 Plant Cell

Some of the similarities and major contrasts between plant and animal cells can be seen by comparing Figures 2.3 and 2.4. First of all, plant cells are eukaryotic, like animal cells. The overall organization of the nucleus and its contents, the ribosomes, endoplasmic reticulum, mitochondria, Golgi apparatus, cytoplasm, peroxisomes, and other features closely parallel other eukaryotes. The biggest major difference is the existence of chloroplasts for carrying out photosynthesis, allowing plants to produce biological molecules from light, water, carbon dioxide, and other inorganic substances, which makes plants *autotrophs*. Photosynthesis is also carried out by prokaryotic and eukaryotic algae and prokaryotic cyanobacteria. Another major difference is the plant's *cell wall*, which envelopes its lipid membrane. The cell wall is composed mainly of the polysaccharide *cellulose*, which, by shear mass, is the most abundant biomacromolecule on the planet. Animals don't have cell walls. The plant cell has a large *vacuole* for storing different types of molecules and helping to maintain the cell's internal pressure and shape, the latter of which is typically more rectangular. In nonplant eukaryotes, vacuoles are typically much smaller and more numerous and move about the cytoplasm.

2.4 Where Do Viruses Fit In?

Viruses are the most abundant life form on this planet. But, are they really living? There is an ongoing debate on whether they meet enough criteria to be considered living. After all, they have no metabolic functions, cannot reproduce themselves by dividing, do not significantly respond to the environment, and require the genetic machinery of a host cell to duplicate their genetic material. All they have is their own little genome, either RNA or DNA, and a *capsid*, a protein shell that houses the virus's genetic material.[2] It seems all they do is invade a host cell, impugn their genome into that of the host and force the host to churn out copies of the genome and express the associated viral proteins. These latter then self-assemble into copies of the virus. While hosts have several strategies for defending themselves against viruses, once a virus has successfully gained entrance to a host cell and commandeered its genetic apparatus, that cell will usually die. Some viruses, M13, a bacterial virus, for example, infect cells without killing them.

The protein self-assembly process, together with the packaging of the genetic material in the capsid may give some of the most fundamental clues about biological self-assembly of macromolecules. This self-assembly leads to remarkably regular

---

2 Some viruses, such as HIV (human immunodeficiency virus), also have a bilayer membrane.

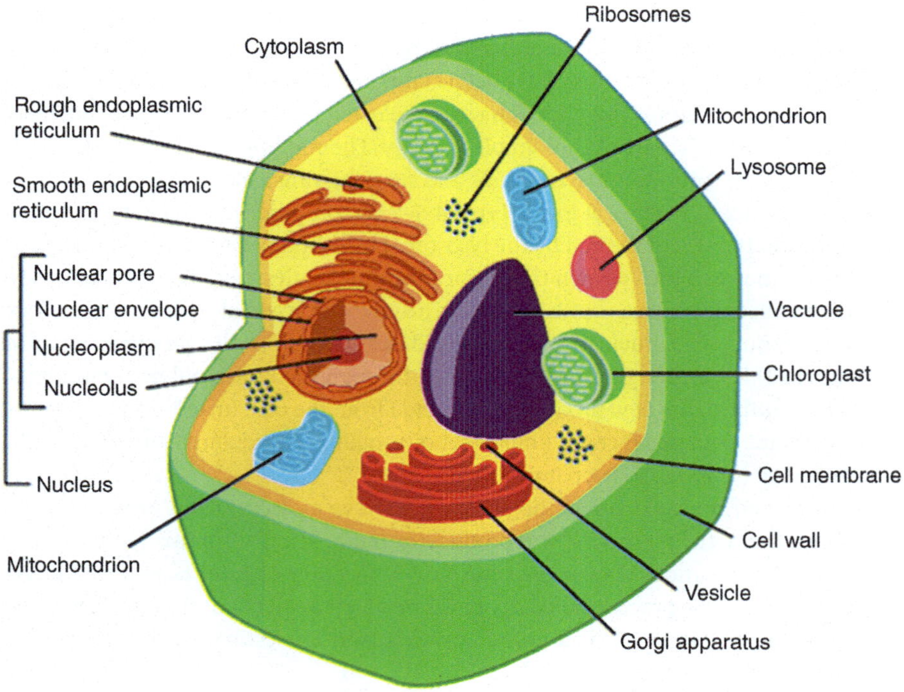

Figure 2.4 Schematic of a plant cell. (Courtesy of Marissa Calko).

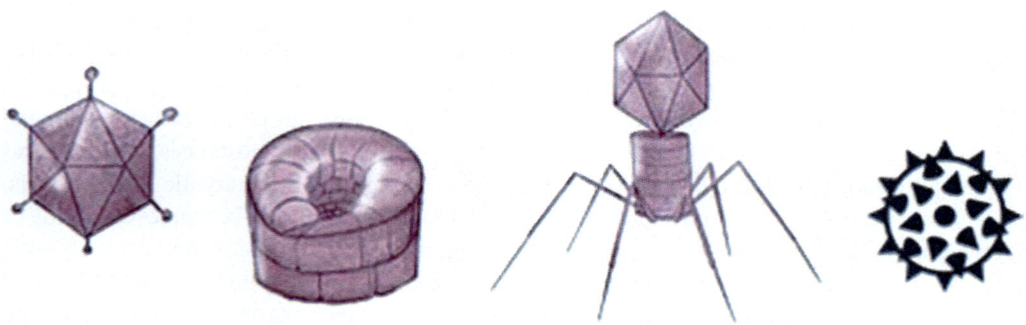

Figure 2.5 Some viral shapes. Left to right: icosahedral, helical, complex, and enveloped. (Wikimedia Commons (2020) / Wikimedia Commons / Licensed under CC BY-SA 4.0).

geometrical forms, with a predominance of icosahedral, spherical, and helical forms, as if forming a bridge between the world of inorganic crystals and "edge of life" self-organization. Some of these shapes are pictured in Figure 2.5. An overview of soft matter physics aspects of viruses is given in Chapter 10.10.

Besides their ruinous effect on host cells, there is increasing evidence that horizontal gene transfer between viruses and both prokaryotes and eukaryotes occurs. Some of this leads to permanent incorporation of portions of viral genomes into the host genome. Viruses are stripped down, hyperefficient, nanoscopic genetic monsters.

It is thought that millions of types of viruses exist and the complete nucleotide sequence of the genomes for several thousand have been determined. They attack virtually every life form, including bacteria, fungi, archaea, plants, and animals. Their origin is still hotly debated. One theory is that they descend from once viable, independent organisms, which were parasites of other organisms, relying on hosts for much of their metabolic and other biofunctions, until they lost all living state functionality except for infection and raw reproduction via a host. Another theory speculates that viruses arose from fragments of genetic material that escaped from a host, so-called "jumping genes", possibly during horizontal gene transfer events. Yet, a third theory suggests that viruses are extremely ancient and co-evolved with nucleic acids and proteins, at the same time these latter were developing into more sophisticated life forms. In this theory, viruses represent a sort of "stunted" evolution that involved early self-assembling biopolymers that did not evolve into more complex organisms with autonomous metabolic and reproductive functions.

2.5 Overview of Cell Functions

The biological cell is the basic unit of life. By this definition, viruses are not alive since they don't consist of one or more cells. Microorganisms are usually unicellular, while cells of the same type can group together to form macroscopic entities, such as slime molds. Higher organisms have a large number of different cell types with different functions. Humans have on the order of 4×10^{13} cells, of which about 0.5% are in the brain. A very brief review of some cellular functions is given next, and some of the associated macromolecular phenomena are explored in subsequent chapters. The focus is on humans, although many of the functions are shared by a very wide range of organisms.

2.5.1 Energy Production

Adenosine triphosphate, ATP, is the universal energy source driving biological functions. Hydrolysis of the γ-phosphate converts ATP to *adenosine diphosphate*, ADP, releasing about 30 kJ/mol, depending on concentrations in its environment. Most ATP comes from the recycling (rephosphorylation) of ADP, although there are also pathways for *de novo* ATP synthesis. There are a variety of metabolic cycles for ATP to be recycled, including photosynthesis in plants, and aerobic and anaerobic mechanisms in animals.

2.5.2 Gene Expression

The critical central processes in cells are the expression of gene-specified macromolecules – proteins and several types of RNA – and the control of reproductive events. In short, prompted by a host of signals and feedback mechanisms, an organism's genes, and its coded DNA sequences, must synthesize the right macromolecules at the right time, at the right place, and in the right amount. Evolution has developed stunningly complex, yet efficient means for carrying out this very tall order. It involves the highly coordinated action of enzyme-mediated processes and macromolecular chemical and physical processes of these agents.

2.5.3 Motion

We usually think of biomotion as a cell or organism moving through space (translational motion), and this occurs on many length scales. In Chapter 4, we examine the effect of length scale and speed on laminar and turbulent flow via the Reynolds number, from the scale of microns to tens of meters.

However, there are also myriad forms of *intracellular motion*, and many agents that carry it out for different purposes. Since there is always random thermal motion of small molecules and macromolecules, we are interested here only in *organized motion*. *Cytoplasmic streaming* in eukaryotic cells is a prominent example, wherein the flow of cytosol and entrained components within the cell is driven by the cytoskeleton, which hence implicates myosin, actin, and tubulin, the main components of the cytoskeleton. Such flow allows for the rapid and directional delivery of materials within large cells, where diffusion would be too slow.

Kinesins and *dyneins* are families of proteins that can bear cargo, such as vesicles and membrane components, and "walk" along cytoskeletal microtubules for delivery. The polymerization of actin and microtubules create structures that can push and move cell components about within the cell. These are just a few examples of intracellular motion. A cartoon of kinesin action is shown in Figure 2.6, where a burdened protein bears its cargo atop a microtubule.

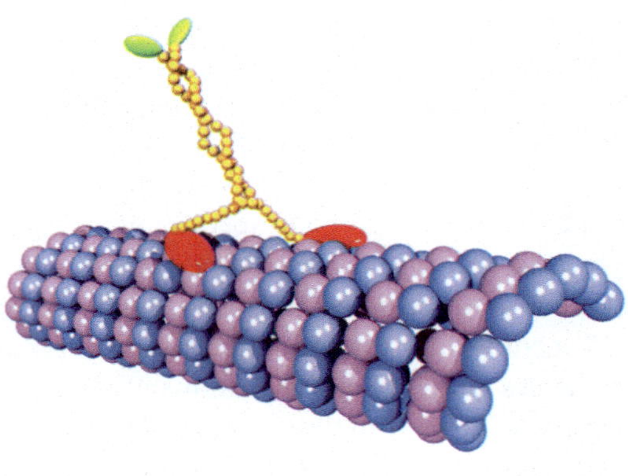

Figure 2.6 A kinesin bears its load as it marches along a microtubule. (By Tomoki Fukushima / Wikimedia Commons).

In terms of translational motion, due to the force produced by actin and microtubule polymerization, some amoebas can push their cytoplasm into protruding *pseudopodia* ("false feet"), thereby changing their shape and allowing a "crawl" through space in *amoeboid movement*. The smallest and simplest motor device is the *flagellum*, a single lash that many bacteria and *protozoa* (single-cell eukaryotes, such as *euglena*) have. Made into a tubular assembly formed from globular *flagellin* proteins, the flagellum is driven by a rotational motor fueled by a proton pump. This rotational motion contrasts to the motion of a *cilium*, which is structurally identical to a flagellum, but microscopically very different, doesn't use a proton gradient for power, and it uses a reciprocating motion with a power stroke and relaxed return stroke, conceptually analogous to a canoe paddle cycle, where the paddle's surface normal is opposite the direction of motion of the power stroke, and the paddle is "feathered," with the surface normal perpendicular to the direction of motion on the return stroke.

Moving up the length scale, the actin and myosin protein families really come into their own as the chief players in muscle contraction, see Section 2.6.4.

2.5.4 Cell Reproduction

Reproduction is common to all organisms. Single-cell organisms, such as *Escherichia coli*, can duplicate themselves in under 20 min, depending on strain, temperature, and nutrient conditions, whereas humans can only reproduce typically after about 13 years. In complex organisms with many different types of cells, there can be widely varying time scales involved in their cellular turnover. Consider the case of humans where blood neutrophils turn over every 1–5 days, red blood cells every 4 months, fat cells every 8 years, osteocytes in bones every 50 years, other skeletal cells 10% per year, and eye lens cells and female oocytes (eggs) never reproduce during the entire lifetime of the individual.

The way the long turnover rates were determined is quite fascinating and serendipitous: In the nuclear bomb testing era, from the mid-1940s to mid-1960s, the concentration of ^{14}C in the atmosphere increased considerably, and then began to fall exponentially, after testing in open air and in water was banned, due to absorption of the isotope by large quantities of organic matter on the earth. With a half-life due to absorption of 10 years for atmospheric ^{14}C concentration, it was possible to analyze ^{14}C content in different human cell types for people born in the late 1970s onward. By 2020 or so, the ^{14}C level in the atmosphere has subsided to prenuclear bomb testing levels, so this method can no longer be used.

Several distinctly different strategies for reproduction have evolved. The simplest form that is used by prokaryotes and archaea is asexual and involves *fission* of the cell into two identical daughter cells. In the parent cell, the DNA is tightly coiled prior to beginning fission. The first step in fission is the uncoiling of the DNA and its duplication into two identical strands which are transported to opposite ends of the cell, after which a cell wall begins to grow in the middle of the cell, leading to the eventual separation of the two halves of the cell into independent, identical daughter cells.[3] The uncoiled DNA in each cell coils back up. The DNA in each cell is identical, but in subsequent divisions, mutations can occur independently in each cell or their further offspring.

Besides mutations, there are several other fascinating mechanisms for modifying the DNA in prokaryotes: Horizontal gene transfer, which some consider "sexual," allows for combining genes of different individuals and even different organisms (Chapter 1.2.2.2). *Conjugation*, one mechanism for horizontal gene transfer, involves a donor cell transmitting all or a portion of its DNA into another cell of its own or other species, which is incorporated into the recipient cell's genome and transmitted in subsequent binary fission events. The transfer normally takes place via an intercellular bridge that transiently grows between the two cells and allows the DNA transfer to take place. *Transformation* is the incorporation into a prokaryote cell of DNA in the environment from other organisms of the same species. In this, there is no direct cell–cell contact, as happens in conjugation. Rather, the receiving cell must be in a receptive or *competent* state where it can produce all the proteins needed to absorb the DNA in the environment. This state usually occurs due to stress on the cell, such as starvation, damaged DNA, or overcrowding in the environment. Transformation can also occur between dissimilar species including eukaryotes and has become an important tool in genetic engineering. *Transduction* involves the uptake of DNA in a cell from a viral agent that enters the cell. Transduction has likewise become a powerful tool in genetic engineering and gene therapy.

For eukaryotes, the reproductive process is more complex and requires more steps, in part, due to the location of the chromosomes in the cell nucleus. In many eukaryotes, reproduction is also asexual, such as the budding of a hydra to produce new individuals with the same genetics. This asexual reproductive process is *mitosis*, which is similar to prokaryotic fission and, although involving many more steps and phases, has the same result of two identical copies of the parent cell.

---

3 Some bacteria, such as *Caulobacter crescentus*, produce two different cell types upon division.

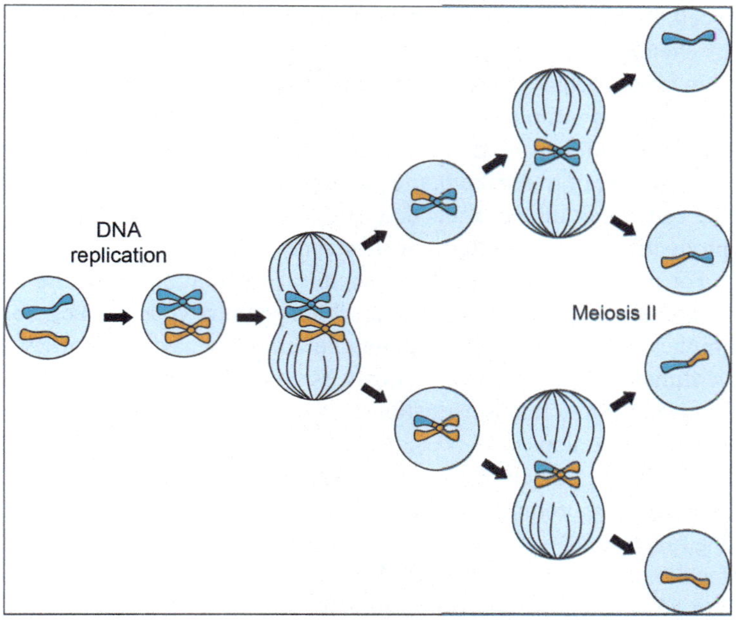

Figure 2.7 A simplified scheme for meiosis. (Courtesy of Marissa Calko).

In the following, we concentrate on humans, although the processes are shared with many other species. Most human cells are produced by mitosis as they turn over during an individual's lifetime. The important exception is the initiation of a new entity by sexual reproduction, in which fertilization joins together and combines the genes of the *gametes*, which are the sperm and egg cells. These latter are produced in the process of *meiosis*.

Meiosis produces the gametes from the *germ cell*. A cell with its full complement of 23 *chromosome pairs* is termed *diploid*. Each chromosome pair has one paternal chromosome and one maternal chromosome, which are termed *homologs* of each other. The essence of meiosis is to turn a diploid germ cell into four gametes, each of which has only half the number of chromosomes as the parent gamete and is termed a *haploid* cell. The process of meiosis is very complex and involves many phases. We omit the details and sketch the essentials in the following.

Figure 2.7 gives a highly simplified schematic of a few of the main phases. The first step in meiosis is the dissociation of the 23 chromosome pairs into the maternal and paternal homologs. Each of these homolog chromosomes is then duplicated and each identical pair binds physically at a point, often shown as an "X," as schematized in Figure 2.7. The bound pairs are termed *sister chromatids*. The sister chromatids of the maternal homolog then associate with the sister chromatids of the paternal homolog. This is where the main genetic action takes place. In a process of *recombination*, the associated pairs of sister chromatids physically exchange genes with each other, thus creating many unique genetic combinations. Next, these genetically recombined sister chromatids line up in a random order, furthering genetic mixing, and are drawn apart by a spindle which forms from the cytoskeleton, and cell division occurs. There are now two diploid cells each with 23 pairs of sexually mixed chromosomes. The final step is for the sister chromatids to separate from each other in each diploid cell, and each of these diploid cells divides, yielding four haploid gametes. There are many processes involved in these steps, but the four haploid gametes from the original germ cell is the net effect. In fertilization, the haploid sperm cell and haploid egg cell, each with 23 chromosomes join with each other to form the 23 chromosome pairs and a full diploid cell, termed the *zygote*.

2.5.5 Chemical Sensing and Signaling

If you could place your olfactory epithelium into the soil, it would feel like the performance of a jazz group, with the players listening, interacting, responding to one another in real time.

– Merlin Sheldrake *in Entangled Lives*

Cells sense their own internal environment, their neighboring cells, interstitial spaces, and the external environment, and react in a variety of ways. Within the cell, concentrations of ATP, specific proteins, nucleic acids, sugars, and other substances are sensed and acted upon in response to metabolic and gene expression processes. Functions, such as repair of a cell, its immunity against invasion, and maintenance of its optimal operating state, or *homeostasis* are controlled largely by chemical sensing.

In terms of sensing the external environment, chemical recognition is one of the most profound, ancient, and widespread mechanisms. Simple organisms, such as bacteria, sense chemical gradients and move toward food, e.g. glucose, in *positive chemotaxis* and away from toxins in *negative chemotaxis*. As usual, such processes are controlled by proteins. There are a vast number of transmembrane proteins that can bind to specific chemicals in the environment and transmit a signal into the cell which directs a reaction, such as motion, polymerization of actin filaments to affect cytoskeleton functions, or production of other proteins or substances.

For more complex organisms, chemical recognition forms the basis of taste, or *gustation*, and smell, or *olfaction*. The former plays a vital role in sensing which potential food sources are viable and which may be harmful or fatal. Olfaction is implicated in a vast array of plant and animal behaviors. Truffles, a fungus which lives fairly deep underground (10s of centimeters), have no way to release its spores into the air as other fungi do. It has evolved a very complex mixture of aromatic molecules that is irresistible to certain animals, such as hogs, who then dig them up, eat them, and spread the spores through excretion.

Pheromones are chemicals produced and secreted by an organism to produce a response in other individuals of its species. They are produced by simple organisms, such as bacteria, as well as by many animals, and have a wide range of signaling functions. Dogs secrete pheromones in their urine to mark territory. Ants secrete pheromones that can mark a trail to food for other ants or signal an abandoned trail when the food is exhausted. Pheromones in exceedingly low concentration can attract mates; some butterflies can sense sex pheromones of a potential mate kilometers away. Other animals emit pheromones as a danger alert to others of their kind. Synthetic pheromones are finding increasing use in agricultural pest control, where release of these can confuse male insects trying to find and follow true pheromone trails from "calling" females.

2.5.6 Excitable Cells and Receptors

While there are vast numbers of proteins and cell responses to chemical stimulation, there are more complex mechanisms that involve one or more specialized cells for reception of acoustic waves, light, mechanical pressure, and electrostatic environment.

The sense of hearing, or *audition*, extends to a wide variety of waves and pulses that can travel through gas, liquid, and solids. The human frequency range of sensing is taken as 20–20,000 Hz. Frequencies below 20 Hz are *infrasonic* and are produced and sensed by whales and are sensed by snakes through their jaws. Sounds above 20,000 Hz are *ultrasonic* and can be heard by dogs, and sensed by other species, and are used by bats, whales and other species for *echolocation*. The reception of sound is usually connected to a series of mechanical and cellular events leading to its sensing and processing.

The mechanism of audition in humans, found also in other species, is quite complex: Sound waves from the environment are guided through the ear canal and make the eardrum, or *tympanic membrane*, vibrate at the sound's composition of frequencies. The vibrating eardrum is in touch with a series of tiny bones, the *ossicles*, which transduce the frequency of the vibrating eardrum mechanically to the *cochlea* in the inner ear, a snail-shaped, fluid-filled organ about one-half centimeter in size. The fluid in the cochlea imparts the vibrations to specialized hair cells which have placements and variations among themselves that make them respond to different frequencies. These vibrating hairs release neurotransmitters at the auditory nerve synapses, which create nerve impulses that travel to the brain stem for reception and further processing in different parts of the brain.

Mechanisms in simpler species can involve simply acoustically sensitive hairs on the body that vibrate directly from sounds in the environment. Other species, such as some flying insects, have a tympanal membrane covering a hollow, resonant cavity in their legs. Some are sensitive to ultrasound which allows them to avoid bats.

Light reception and sensing are widespread in the living state. In addition to the primary function of photosynthesis, plants use the reception of light to signal other processes. Some plants, *heliotropes*, rotate to follow the sun. The length of sunlight and how it changes can signal plants and trees to progress toward fall and lose their leaves. While humans cannot sense ultraviolet light, many other species can.

Many animals have evolved light receptors, culminating in a wide variety of eyes in insects, fish, birds, and almost all vertebrates. These have visual pigments of related structures that capture light, and through a biochemical cascade, transduce it into nerve impulses, resulting in images interpreted by the brain. Vision will be briefly covered in Chapter 14.2. There are a surprising number of complex biochemical events between reception of light in a visual receptor cell, a rod or cone in the human eye, and the transmission of the reception through bipolar cells into neuronal ganglion cells which carry the signal to the optic nerve which in turn sends it on to the brain for processing.

The sense of touch, or tactility, involves different types of mechanoreceptors and thermoreceptors to sense changes in pressure and temperature on an organism's surface. In humans and many other species, these specialized receptors reside within the skin structure and are integrally connected to a neuron which carries the signal to the local nerve network.

2.5.7 Self-defense

Organisms are loaded with mechanisms to counteract threats. These begin with threats inside the cell itself, where sophisticated mechanisms exist to edit gene expression products, and either repair or destroy defective proteins and polynucleic acids.

In terms of external threats all species, from bacteria on up have developed elaborate defenses. Bacteria are under constant attack from *bacteriophages* (or "phages" for short), which are viruses, and they have developed defenses to block the virus from commandeering its genetic apparatus at each step of the attack. The virus must first attach to receptors in the bacterial outer membrane.[4] Bacteria can modify receptor structure, deploy a variety of different receptors, or hide the receptors with physical barriers. Once a phage anchors, defeating the first line of defense, several different bacterial proteins are able to marshal at the attachment site and prevent injection of the viral DNA. If the viral DNA nonetheless enters, the bacteria can hydrolyze it into deactivated fragments using restriction endonucleases (enzymes that can cleave DNA chains). When these mechanisms fail, the bacteria can also resort to shutting itself down (e.g. depleting its ATP), so the phage cannot propagate to other healthy members of its bacterial community. It can also interfere with the gene expression of the invading DNA, thus limiting the amount of phage reproduction. The attack mechanisms of phages and defenses of bacteria are constantly evolving, creating a sort of biochemical "arms race."

Seeing the diversity and complexity of simple bacteria against phage attack, it is no wonder that higher organisms have immensely evolved immune systems with numerous specialized cells and strategies for warding off attacks by viruses, bacteria, and other malignant organisms. *Antibodies* are specialized proteins that can be "custom produced" when an organism is exposed to a new invader – viral, bacterial, fungal, or other – for the first time. These antibodies can then destroy the particular type of invader. This is termed *adaptive immunity*. An organism has a large repertoire of antibodies (estimated to be in the range of 10^9–10^{12}) which, after being produced for the first time, can be re-activated when new attacks are launched by previously encountered foes. This is the principle of *vaccination*;[5] a "classical" vaccine is a formulation of de-activated or weakened viruses, or macromolecular parts of a virus, which is introduced into an organism so that the antibody can be developed, at which point the organism gains immunity to the virus.

In its rawest manifestation, vaccination can even be achieved by introducing viral pus removed from the sores of infected victims into individuals, usually via an incision. While Louis Pasteur discovered the microbial basis of vaccination and developed purified vaccines, direct viral inoculation of this type had been practiced long before, approximately 1000 years ago in China and India. It is well known that, historically, far more soldiers succumb to disease than to enemy fire during war. Throughout history, there are countless accounts of mass epidemics among troops, and also attempts at direct vaccination, or "variolation," of troops.

George Washington (United States, 1732–1799) was infected in Barbados with smallpox at age 19 but survived after a month-long struggle. Later, American troops in the Revolutionary War were falling ill *en masse* to smallpox. John Adams wrote in June 1776 "The smallpox is ten times more terrible than Britons, Canadians, and Indians together." Having made attempts at quarantines and testing with only weak results, Washington in February 1777 made the bold decision to inoculate 40,000 American troops, which was actually prohibited by the American Continental Congress at the time. The results, however, were stunning. The infection rate in the American Continental Army dropped from 17% before the vaccination program to 1%. The Continental Congress promptly legalized vaccination after this.

A radically new form of vaccination was introduced and given to hundreds of millions of people worldwide during the struggle against *severe acute respiratory syndrome coronavirus 2* (SARS-CoV-2) which produces the coronavirus disease termed *COVID-19*, which first entered humans around 2019 and reached pandemic proportions in 2020–2021. Whereas traditional vaccines used de-activated and modified forms of a virus, the new method that became available in late 2020

4 Gram-negative bacteria have an outer wall, whereas gram-positive bacteria do not.

5 The word "vaccination" comes from the Latinate word "vacca" for cow, due to the fact that the first vaccines were derived from cowpox viruses.

involves the use of messenger RNA (mRNA) encapsulated in lipid nanoparticles, containing a formulation of lipids and polymers, which is injected directly into the person. The mRNA codes for the spike proteins on the surface of SARS-CoV-2, which the virus needs to anchor itself to the receptor on its victim's cell surfaces. The body uses the mRNA to produce spike proteins, harmless in and of themselves, and thus trains the immune system to make antibodies against these, so that when the actual virus with spike proteins invades the immune system can combat it. Long before this emergency, global introduction of the vaccine, research and development were ongoing for many years to develop the mRNA encapsulation and delivery methods with the intent of using them for a wide spectrum of diseases beyond viral infections.

2.6 Specialized Cell Types and Structures

There are many types of specialized cells. Vertebrates have well over 200 different types of specialized cells, many of which have several varieties in each individual organism. The intent here is not to be encyclopedic, rather to look at a few important classes.

Stem cells are a transcendent class of cells found throughout the animal and plant living state. There are *pluripotent stem cells* that can develop into all the cell types of a given organism. *Embryonic stem cells* are pluripotent and drive much of embryonic development. Stem cells can also continue to reproduce themselves indefinitely. *Multipotent stem cells* can develop only into a limited subset of an organism's cell types.

2.6.1 Nerve Cells

Nerve cells, or *neurons*, are exotic, complex cells, unlike any other type. They are excitable cells found in sensory, motor, and deep brain systems and are capable of generating electrical pulses (Chapter 13). There is a wide variety of neuron types in a given species. The history of neurons is fascinating because it is the forerunner of the vast interdisciplinary field of neuroscience. In the nineteenth century, a great debate raged about the nature of the nervous system. The so-called *reticularists* considered the nervous system to be a large network of tissue of fused nerve cells and the so-called *neuronists* considered the nervous system was composed of individual, discrete cells. Golgi discovered a staining method that illuminated the highly dendritic nature of nerve cells. He himself was a reticularist. Santiago Ramón y Cajal (Spain, 1852–1934) provided the definitive graphical evidence, in the late 1880s, that nerve cells are discrete and individual and not part of a larger, fused tissue. He is often considered the "father of neuroscience," and his, and the work of many others, led to the *neuron doctrine*, stating that the neuron is a discrete cell and the fundamental cellular unit of the nervous system, which consists of a cell body, axon, and dendrites, and that information flows through the neuron from the dendrites through the cell body and to the axon.

There is a certain antiparallel of the reticular versus neuron controversy to the one concerning polymers, which spanned the late nineteenth and early twentieth centuries. One group considered polymers as small molecules linked by as yet unknown (noncovalent) colloidal forces, similar to the neuronists in their notion of individual cells, and the other group considered them as giant covalent molecules, similar to the reticularists. The comparison is antiparallel because the covalent group won the polymer debate, while the neuronists won the nervous system debate.

Figure 2.8 shows a schematic of a typical neuron. The cell body, also called the *soma*, contains all the eukaryotic organelles needed for its function, including the nucleus. Neurons do not divide and hence do not directly reproduce. In vertebrate embryonic development, they are produced from pluripotent embryonic stem cells. Later, neurons are produced only from multipotent neural stem cells, which develop into *radial glial cells*, which in turn develop into *glial cells* or neurons. Interestingly, *Planaria* can completely regenerate their (tiny) brain from any severed portion of their body, even portions containing no brain tissue. The process of *neurogenesis* continues throughout the adult life of humans. If it weakens or stops, cognitive decline is inevitable.

Glial cells themselves form a large class. Glial cells are an integral part of the nervous system but do not have the electrical signaling capabilities of neurons. Most of them can divide, unlike neurons. They exist mostly to nurture neurons by surrounding them for structural stability, keeping neurons autonomous and isolated from each other (*satellite cells*), producing the myelin sheath around axons (*oligodendrocytes* and *Schwann cells*) supplying nutrients and oxygen, connecting them to their blood supply (*astrocytes*), participating in synaptic transmission, while fighting off pathogens and helping to remove dead neurons.

Emanating from the cell body in Figure 2.8 are many dendrites. These branch, often extensively, as shown in the schematic. The dendrites can send their filaments out for hundreds of microns and are used principally to receive signals from other cells, via *synapses*. The *axon* protrudes from the cell body and while only a micron or so thick in mammals, it can

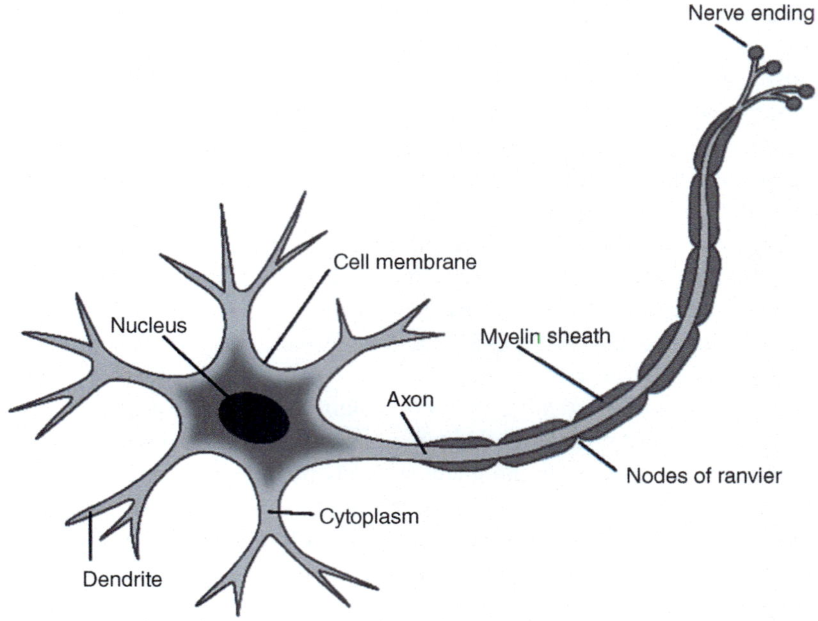

Figure 2.8 Schematic of a neuron. (Courtesy of Marissa Calko).

reach lengths over 1 m in humans, all part of a single neuron. The squid giant axon is up to 1 mm in diameter and up to 1 m in length and provided the single most important subject for building the modern understanding of electrical nerve impulse formation and conduction. The axon is responsible for sending signals from the neuron to other neurons, almost always via a synaptic connection with one of another neuron's dendrites, and occasionally directly to another axon. Dendrites can also form synaptic connections among themselves. The propagating electrical nerve impulse is termed *the action potential* (Chapter 13).

The huge diameter of the squid giant axon seems to be one of the culminations of evolution's attempt at high-speed nerve impulse transmission. In the case of squid, it allows it to very rapidly jet-propel itself away from a menace. But a more efficient route to nerve impulse speed was evolved via *myelin*, widespread among vertebrates. In Figure 2.8, the myelin sheath coats the axon and is seen to have constriction points, making it have a segmented appearance. In 1878, Louis-Antoine Ranvier (France, 1835–1922) discovered these constrictions, now called the *nodes of Ranvier*.

Remarkably, the myelin sheath is an insulator against the action potential, so that, instead of sending the pulse down the entire axon, as in the squid axon, the pulse jumps from one node of Ranvier to the next, imparting high-speed pulse transmission in an axon 1/1000 the diameter of the squid axon, and with 10^{-6} of the cross-sectional area. This allows huge numbers of neurons to be grouped together in vastly smaller volumes than would be possible without myelin, allowing for dense, complex brain development.

The myelin sheath is composed mostly of lipids and has a white appearance, which is the "white matter" associated with large nerves in both the central and peripheral nervous systems. Nerves in the former include the optic nerve and the *corpus callosum*, the large nerve bundle that connects the two halves of the brain in *placental mammals*. Large nerves in the peripheral nervous system include the auditory nerve and the sciatic nerve, the longest nerve in the human body, running from the lower vertebrae to the foot.

The transmission of an action potential along an axon, whether myelinated or not, is electrochemical in nature. It involves axon membrane depolarization due to a transient flux of mainly K^+ and Na^+ through transmembrane channels, which open and close in response to the voltage across the membrane.

How does the action potential get communicated to the neighboring dendrite or axon, once it reaches the *axon terminal*, which has many mitochondria, indicating high energy demand? There are both *electrical synapses* and *chemical synapses*. The electrical synapses are relatively simple and involve a direct transfer of ionic current from one cell to the next via a *gap junction* at the synapse only about 3 nm in length, which is spanned by a protein bridge that allows ion transfer across the gap. These electrical synapses are bi-directional, whereas chemical synapses generally have a one-way transmission path. Electrical synapses are found in invertebrates and also in the vertebrate brain, especially during embryonic development.

Chemical synapses are the major transmission mechanism in mammalian brains. The synaptic cleft is 20–50 nm wide, an order of magnitude greater than gap junctions in electrical synapses and filled with extracellular fibrous protein. One-way transmission from one cell to another is mediated by a *neurotransmitter*, of which over 200 have so far been identified in humans. When the nerve pulse arrives at the synapse through the axon of the *presynaptic cell*, a barrage of neurotransmitters leaves the presynaptic cell and crosses the synapse to be received by the *postsynaptic cell*, a dendrite, or occasionally an axon. Highly specialized protein *neurotransmitter receptors*, or *neuroreceptors*, in the postsynaptic cell membrane selectively receive the neurotransmitter and re-initiate the action potential, sending it on to its next destination. This mind-boggling process of chemical transmission, with so many different neurotransmitters, neuroreceptors, and associated structures and proteins lends itself to highly evolved and complex sensory, motor, memory, and behavioral characteristics.

One classification of the bewildering variety of neurotransmitters is into three groups: (i) amino acids, (ii) amines, and (iii) peptides. The first two types are small molecules, whereas the peptides are much larger and follow a different pathway for synthesis and transmission.

The most important *excitatory neurotransmitter* is the amino acid *glutamate*, where it is used in over 90% of human brain synapses. The most prevalent *inhibitory neurotransmitter* is γ-aminobutyric acid. *Serotonin* is derived from the amino acid tryptophan and appears to modulate such phenomena as mood, memory, learning, and cognition. *Acetylcholine* is an important neurotransmitter implicated in voluntary movement, sleep/waking functions, and time memory in the *circadian rhythm*, or *circadian clock*, which is roughly a 24-h cycle. Most organisms, including fungi and many bacteria, have their own circadian rhythms, often temperature compensated. *Dopamine* is a hormonal neurotransmitter synthesized from the amino acid tyrosine. It is implicated in motor control and reward-motivated behavior. Loss of dopamine secretion in the brain can lead to *Parkinson's disease*, characterized by tremors and loss of motor control. Certain psychiatric conditions are linked to dopamine activity, and many antipsychotic drugs function by blocking dopamine receptors.

Figure 2.9 shows a schematic of neurotransmitter action for amino acid and amine neurotransmitters. *Synaptic vesicles* are assembled near the membrane of the presynaptic cell and protein transporters load them with neurotransmitter. The vesicles themselves are bilayer phospholipid membranes (Chapter 13), rich in proteins, and structurally akin to cell membranes, but on a much smaller scale, with a diameter ~40 nm. The vesicles with their cargo dock to release sites on the end of the neuron membrane where they are prepared for release by proteins. When the action potential arrives at the axon terminal voltage-gated calcium ion channels in its membrane open and Ca^{2+} flows into the cytoplasm of the axon terminal from the extracellular fluid, which cause vesicles to fuse together and to the axon terminal membrane. Vesicles then release their neurotransmitter cargo through the neuron membrane and into the synaptic cleft, which the neurotransmitters cross and are received by the neuroreceptors of the postsynaptic cell. Meanwhile, after the release, vesicles in the presynaptic cell are recycled and prepared for the next load of neurotransmitter. Neurotoxic proteins such as *botulinum toxin* (or *botox*) and bacterial infections such as tetanus can impede the vesicle fusion and release process, thereby leading to muscle spasms, paralysis, and sometimes death.

The specific binding of a neurotransmitter to a neuroreceptor is analogous to a flexible key entering a lock, which leads to a change in protein conformation and function. Whereas the majority of neurotransmitters are small molecules the neuroreceptors are proteins, and hence inherently more complex. There are a large number of these, but fall into essentially two classes: (i) ion channels that are opened by neurotransmitters and (ii) *G-protein-coupled receptors*. Ion channel neuroreceptor proteins receive and bind with their specific type of neurotransmitter(s) and create a channel through the membrane through which ions can flow, re-creating and sending the action pulse in the postsynaptic cell. The receptor can also inhibit further propagation of the action potential. The G-protein coupled receptors, GPCR, do not themselves form channels, but actuate other cell regulation responses in response to their binding to a neurotransmitter. GPCR form a very large class of membrane-bound receptors that can recognize and signal cell functions for many types of molecules besides neurotransmitters. These other molecules include hormones, pheromones, and many more, and thus make GPCR involved in functions such as vision, taste, smell, regulation of the immune system, and maintenance of homeostasis.

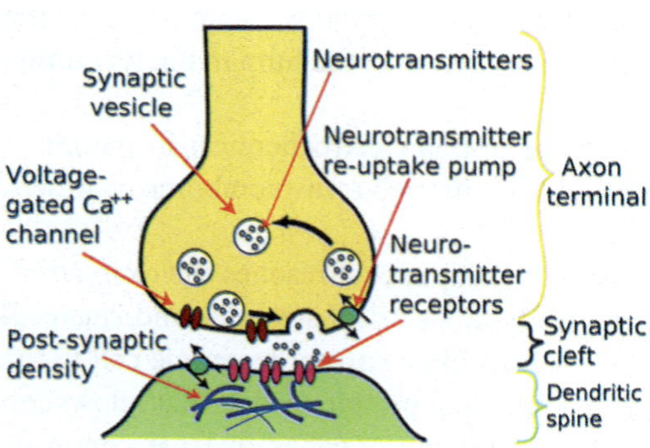

Figure 2.9 Schematic of neurotransmitter function, including the cycle of vesicle action involved. (Nrets (2006) / Wikimedia Commons / Licensed under CC BY-SA 3.0).

Unlike neurotransmitters which cluster near the presynaptic cell membrane near the synapse, neuroreceptors can be found in any part of the neuronal membrane, including dendrites, axons, and the cell body. Neuroreceptors in the presynaptic cell can sense levels of a particular neurotransmitter and hence provide feedback regulation of the neurotransmitter's release.

Besides connecting to each other, neurons also make synaptic junctions with other types of cells. Neurons can connect to and activate glands.

Synaptic junctions form between neurons and smooth and cardiac muscles in the *autonomic nervous system*, which controls involuntary functions, such as heart rate, digestion, and the response of the eye's pupil to light. *Neuromuscular junctions* connect motor neurons in the spinal cord to skeletal muscle, where acetylcholine provides fast neurotransmission.

A brain in its most commonly used sense involves massive number of neurons interconnected via synapses, forming a *neural network*. Figure 2.10 shows many dendrites each of which can connect to another neuron via a synapse. The axon has branched ends, each of which can also connect to other neurons via synapses. There are typically 80×10^9 neurons in the human brain, each of which can connect to up to 100,000 other neurons. This amounts to the order of a staggering 10^{15} synapses in the brain. From a simplified viewpoint the brain's neural network provides *weighting* between neuronal signals in the network, via *synaptic strength*. Signals are combined and processed and can lead to *activation functions*, determining whether a given neuron fires or not. These can involve non-linear responses. The connection between natural neural networks and artificial neural networks, used in artificial intelligence and elsewhere, is briefly considered in Chapter 15.4.1.

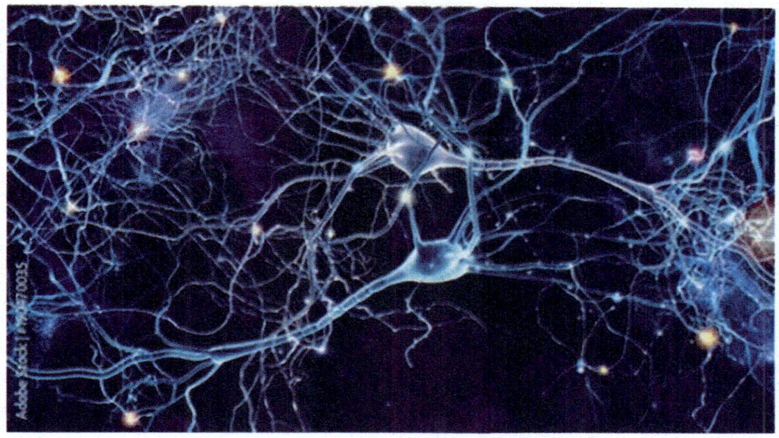

Figure 2.10 A neural network.

2.6.2 Bone

Bone is a stunning example of how the living state has leveraged inorganic matter into crystalline structures riddled with proteins, protein-producing cells, and regulatory cells. It gives enormous mechanical strength and mobility to an organism, protects organs and other soft tissue, and is a storehouse of calcium and phosphate.

The bones in the human body are continuously undergoing *bone remodeling*, which repairs and replaces tiny fragments of bone and helps to maintain ion concentrations in the rest of the body. The process in which bone is dissolved, releasing calcium and phosphate into the bloodstream is *bone resorption*. Specialized cells, *osteoclasts*, are produced and activated under hormone control by the *parathyroid gland*, which monitors calcium levels. The reverse process of building new bone is *ossification*, carried out by groups of cooperating *osteoblast* cells, which synthesize the fibrous *collagen* and noncollagenous proteins. Osteoblasts don't divide, rather are produced by progenitor cells. In adults, bone resorption and ossification are kept in balance. In infant and childhood growth, ossification proceeds at a greater rate than resorption. In some maladies, such as *osteoporosis*, resorption dominates over ossification.

Osteocytes are closely related to osteoblasts and are distributed throughout the bone interior, which allow them to act as mechanical sensors and participate in bone resorption and ossification. They reside in small gaps in the bone matrix and connect to each other to form a communication and regulatory system that has been described as somewhat analogous to nerve networks. The hard bone matrix is made principally of *osteons*, approximately millimeter-sized cylinders of bone tissue, with a central canal for blood flow.

Keeping the focus on human bone, 60% by mass is composed of inorganic phosphate and calcium in the form of *hydroxyapatite crystals*, $Ca_{10}(PO_4)_6(OH)_2$, conjugated to small amounts of magnesium carbonate, and sodium and potassium ions. The protein matrix is dominated by the fibrous protein collagen in long bundles. Proteoglycans are an important component of the noncollagenous protein and are involved in 3D construction of collagenous fibers, signaling, and as receptors of immune agents.

Figure 2.11 shows some of the major features of a long bone. The *articular cartilage* is an extracellular matrix which covers the end parts of the bone that connect to other bones, and provide lubrication and wear resistance. Breakdown of cartilage leads to arthritis. The interior of the end portions of the bone are filled with *spongy* (or *cancellous* or *trabecular*) *bone*, made up of a porous matrix of collagenous, vascular, and other material, whose interstitial space is filled largely with *red bone marrow*. This marrow contains *hematopoietic stem cells*, which are the progenitors of red blood cells (*erythrocytes*), white blood cells (*leukocytes*), and platelets (*thrombocytes*). Red marrow is also involved in the breakdown of blood cells. The *medullary cavity*, or *marrow cavity* runs the length of the bone's main body and contains some red marrow but also *yellow bone marrow*. This latter contains *mesenchymal stem cells* (or *stromal cells*), which produce fat cells (*adipocytes*), bone cells (osteoblasts), and cartilage cells (*chondrocytes*). Red bone marrow can become yellow. In fact in newborn

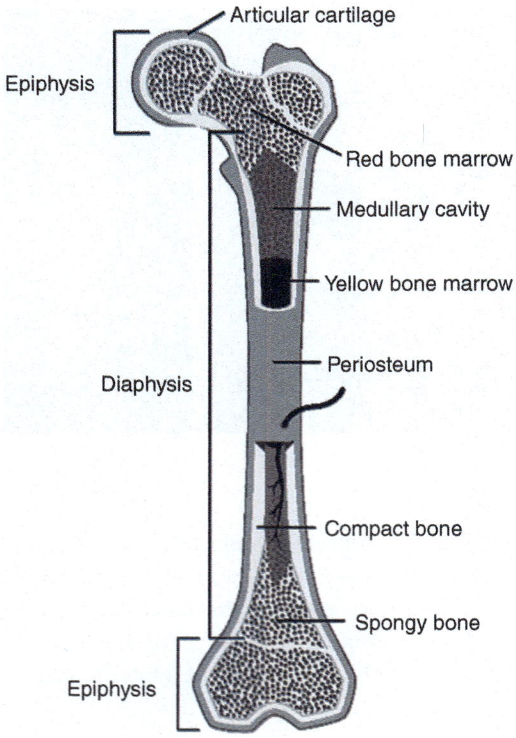

Figure 2.11 Schematic of a long bone. (Courtesy of Marissa Calko).

humans, all the marrow is in the red form and converts to a higher percentage of yellow as an individual ages. The medullar canal is surrounded by *compact bone*, which is much denser than trabecular bone.

2.6.3 Blood Cells

Red blood cells, *erythrocytes*, make up the largest blood cell fraction and typically comprise 40–45% of total blood volume in humans. Via hemoglobin, they bind and distribute oxygen throughout the body and accumulate and discharge waste CO_2. *Platelets* and red blood cells are unique in that they have no nucleus. They are biconcave, disk-like structures. *Blood type* in humans refers to the presence or absence of antigens in plasma and red blood cells. There are over 300 antigens involved and include proteins, carbohydrates, and combinations of these. Blood compatibility for transfusion reduces to the presence of so-called antigen combinations A, B, and Rh, leading to a total of eight blood types; positive and negative forms of A, B, AB, and O, where positive indicates the presence of Rh factors and negative their absence.

White blood cells are part of the immune system and compose only about 1% of blood volume. They are also present in the *lymphatic system*, which is an open circulatory system that delivers white blood cells for defense and also feeds filtered blood plasma back into the closed-loop blood circulatory system.

While all white blood cells have a nucleus there are several significantly different classes of these cells, each with different targets and mechanisms of defense. In classifying their immune properties there exists *innate immunity* which includes wide varieties of mechanisms and proteins that are inherent in an organism's genome, available on short notice, and *adaptive* or *acquired immunity*. In this latter, specific pathogens are recognized and specific antibodies, *immunoglobulin* proteins, are developed, on the order of ten days to two weeks. These proteins combat the pathogen, and an *immunological memory* of the pathogen is formed. This allows the immune system to ramp up antibody production if the same pathogen is encountered later in life. Acquired immunity is the basis of vaccination.

Neutrophils are the most abundant of the white blood cells and respond rapidly to sites of bacterial or fungal entry into an organism, guided by chemotaxis provided by *cytokines*, which are glycoproteins, such as the *interleukins*. Neutrophils are *phagocytes*,[6] engulfing the invading species and killing it with antibiotics, or trapping it for other processing. As "first responders" to attack, neutrophils are highly mobile and also have a short lifetime, typically tens of hours.

Eosinophils can defend against certain multicellular organisms, larger than bacteria and fungi. Since they cannot engulf these organisms they have a variety of biochemical means of attack, such as packets of degranulating protein they carry and excrete against attackers. These proteins include peroxidases that can directly destroy tissue and cationic proteins that act as neurotoxins and can also attack viruses. Eosinophils are also involved with pathogens that create asthma and other respiratory ailments. They last for days before dying and being replaced.

Basophils' main tactics come from their ability to release *histamine*, a small molecule, $C_5H_9N_3$, and *heparin*, an anionic polysaccharide. These are stored in granules within the basophil and are released by signals. Histamine can dilate blood vessels to allow neutrophils and other immune agents to rapidly travel to affected sites, and heparin is a blood anticoagulant that can keep areas of an infection clear enough from coagulation to permit neutrophils and other white blood cells access. Basophils last a few days before dying and replacement.

Lymphocytes are in charge of acquired immunity. They are further classified as *T cells* (or *thymus cells*), *B cells* (or *bone marrow-derived cells*), and *natural killer cells*. T cells mature in the thymus gland, a part of the lymphatic system, after being produced in the red marrow hemopoietic process. T cells and B cells react differently in building acquired immunity. T cells themselves break down into different types. *Killer T cells* can recognize virus-infected cells and directly kill them through their *cytotoxicity*. These cells can also summon cytokines to marshal other parts of the immune system to mount an attack. *Helper T cells* can't directly destroy pathogens but work by signaling other parts of the immune system including B cells.

6 In Greek "Phago" "eating" and kutos (cyte) "vessel." "Cyte" is used widely in naming type of cells.

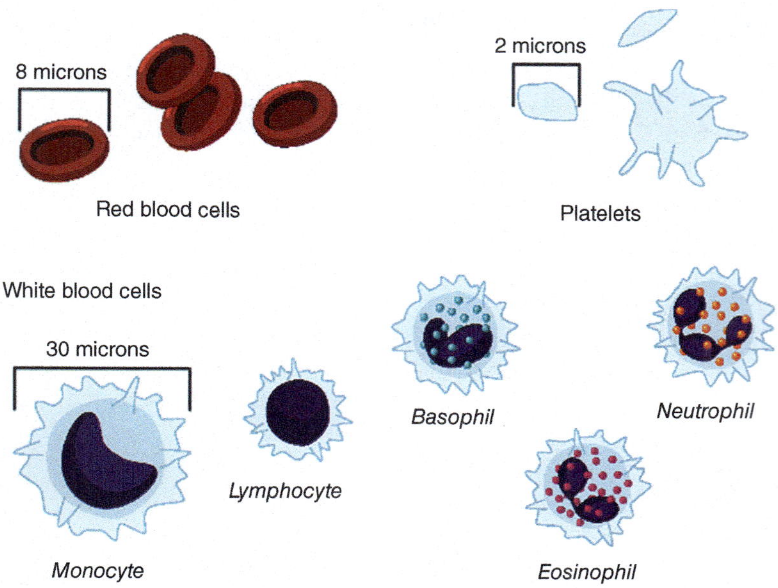

Figure 2.12 Blood cell types produced in the hemopoietic process. Note the differences in scale: 30 μm white cell diameter, red blood cell 7.5 μm, platelets 2.5 μm. (Courtesy of Marissa Calko).

Regulatory T cells distinguish between invading cells and those of the host. Breakdown of this function can lead to autoimmune diseases such as rheumatoid arthritis, lupus, celiac disease, type 1 diabetes, and many others.

B cells have the distinction of being the only cells to produce antibodies. This is in large part due to *B cell receptors* on their cell membrane which allows them to bind to specific antigens with the release of specific antibodies. *Memory B cells* allow B cells to produce antibodies to a known antigen in future attacks. The memory B cells may live for years, whereas the other types of lymphocytes last for weeks.

Finally, *monocytes*, round out the white blood cell lineup. Like neutrophils, they are phagocytes, and, with diameters reaching 30 μm, are the largest of the white blood cells, where the others have diameters of about 7–15 μm.

Platelets are the third type of blood cell and are only found in mammals. They are small, 2–3 μm in diameter, and clump together in the amazingly complex blood clotting process. There are nearly 200 proteins involved in this process and over 300 interactions.

Figure 2.12 summarizes the different blood cell types, along with length scales to illustrate the large difference in sizes among them.

2.6.4 Muscle Cells

There are many forms of muscle cells among higher animals. The main types are skeletal muscle, cardiac muscle, and smooth muscle. One of the common motifs is their hierarchical nature, which can be seen in Figure 2.13. Looking at an exposed muscle its fibrous nature is evident, and the individual fibers can be separated out, which in turn are composed of arrays of basic muscle elements, *myofibrils*, composed of actin and myosin and held in place by the giant protein *titin*,[7] the largest known protein ($M \sim 3 \times 10^6$ g/mol, consisting of over 200 folded subunit proteins), which also imparts passive stiffness. Finally, the myofibrils consist of myofilaments, which are intertwined thin actin filaments and thick myosin filaments. The actin filaments form a sort of track to which the myosin can attach and move along during muscle contraction.

Titin also has random coil segments between subunits. This leads to an entropic random coil type restoring force, since titin helps with the passive stiffness of muscle and the return of fibers to resting state after being extended. At high levels of stretching, there are also enthalpic contributions to titin's restoring force. The basic theory of elasticity of random coils is presented in Chapter 8.1.

7 Derived from Greek "titan."

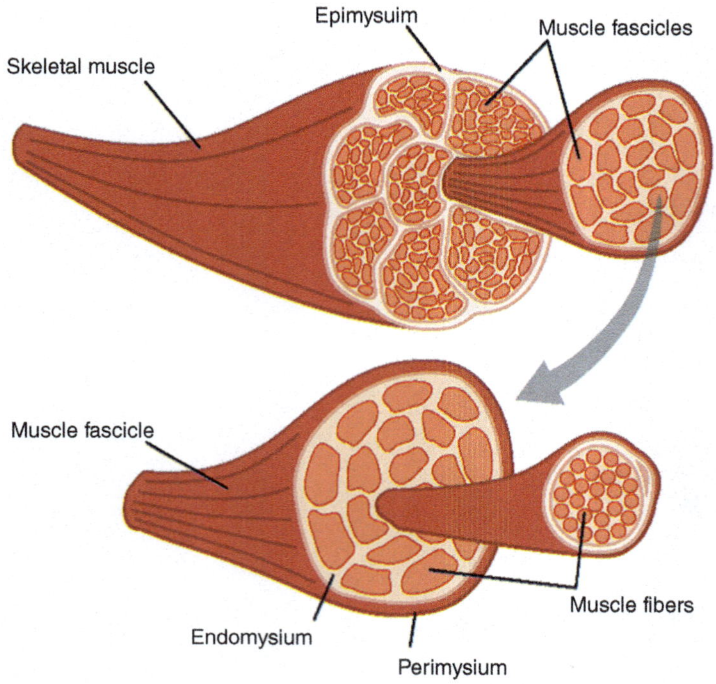

Figure 2.13 Skeletal muscle cells. (Courtesy of Marissa Calko).

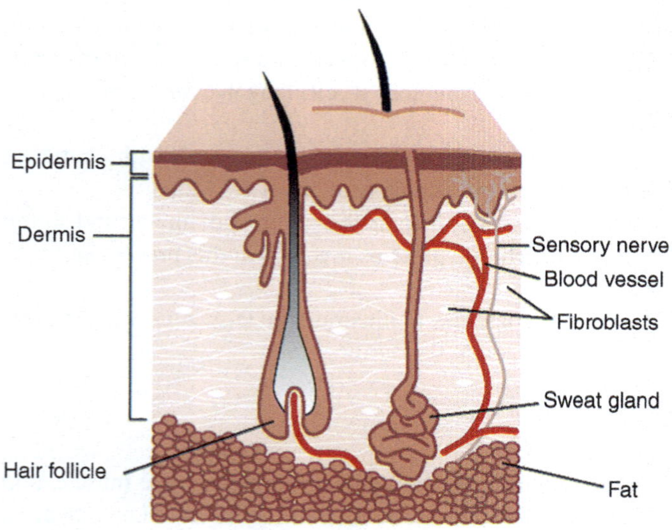

Figure 2.14 Dermal tissue. (Courtesy of Marissa Calko).

2.6.5 Skin

Skin is the outer barrier of an animal which protects its organs, bones, and other internal structures, regulates the loss of water, provides thermal insulation, acts as a first barrier against pathogens, and imbues the organism with the sensation of touch of the outside world. It is chief organ of the *integumentary system*,[8] which includes ancillary tissues such as hooves, feathers, and nails. There are large differences in types of skins among species, for example, between amphibians and mammals, but the essential functions are similar, and there are shared structural motifs, especially that of *layers*.

Figure 2.14 is an example of dermal tissue.

2.7 Molecular Biology Methods

We will cover some of the physical methods useful in characterizing both synthetic and biomacromolecules in Chapters 6, 7, and 9. But, because the macromolecules of life contain intricate codes, and associated complex 3D structures, absent so far in human-made polymers, methods have been developed that are uniquely suited to the former and generally have no application for the latter. We will not go into these here in detail, but it is worth mentioning a few of them that have rapidly expanded our knowledge of proteins and nucleic acids, and hence of cell biology.

2.7.1 DNA Polymerase Chain Reaction

Polymerase chain reaction (PCR) is an amplification method to rapidly produce millions or billions of replicas of a given DNA molecule, thus producing enough material for many different types of biochemical tests. A thermally stable DNA polymerase is used in thermal cycles of the target DNA to produce ever-increasing copies of the target DNA.

2.7.2 Gel Electrophoresis

In *gel electrophoresis*, electrically charged biomacromolecules move under the influence of an external electric field through a gel which is a sort of obstacle course in which the smallest molecules pass most quickly and the largest more slowly, thus separating the molecules out into bands perpendicular to the electric field. This action is the reverse of size exclusion chromatography (SEC),[9] covered in Chapter 7.13, where entropic diffusion in the tiny pores of the GPC gel slows down small molecules and lets larger ones through more quickly. Agarose gel is normally used to separate nucleic acids and polyacrylamide for proteins and small DNA. Electrophoretic mobility in a solution is discussed in Chapter 6.7.

---

8 *Integument*, from the Latin *integer, "to cover in."*

9 SEC is often referred to also as gel permeation chromatography (GPC).

Two-dimensional gel electrophoresis is particularly useful for separating proteins, whose net charge is a function of the pH of its aqueous environment. For a given protein its charge will be zero at a very specific pH, called the *isoelectric point*, IP (Chapter 3.1.2). Hence, applying an electric field across a gel with a pH gradient will make a given protein stop motion and accumulate where the pH = IP. In the second electrophoresis, where the electric field is perpendicular to the first electric field and pH gradient, the gel contains the anionic surfactant sodium dodecyl sulfate (SDS), which coats the proteins, and ensures they all move in the same direction. Separation in this dimension is approximately controlled by the protein molar mass.

2.7.3 Ion Exchange Chromatography

Ion exchange chromatography (IEC) is also a method for separating molecules, especially proteins, according to their charge, and bears a conceptual similarity to the pH-gradient method mentioned above. In this approach, a column is packed with a resin with immobilized charged groups, such as diethylamino ethyl moieties. Oppositely charged sample polymer molecules pass through and electrostatically bind to the immobilized resin charges. After this, the ionic strength of the column medium is increased gradually with a salt solution, whose small ions of the same sign as the sample molecules compete for the column binding sites, thus releasing the charged polymers, the least charged eluting first. A *fraction collector*, often a moving conveyor belt with vials, gathers aliquots of the eluting molecules in synchronization with the increasing ionic strength.

2.7.4 Affinity Chromatography

This is an extraordinarily powerful method because of its specificity. Instead of relying on electrostatic interactions of charged biomacromolecules, the *affinity chromatography column* is packed with a material that contains an *affinity reagent*, which only the sample molecule, usually a protein, recognizes and binds to. As an example, the resin may contain immobilized antibodies that recognize and capture a specific protein while all the others pass through with no binding at all. Later, the protein is removed from the column by a reagent that will interfere or compete with the binding of the protein to the affinity reagent. This allows protein purification in a single step. Often the column is not even used, rather the affinity-specific resin is mixed with the unseparated protein solution and the mixture centrifuged. The nonspecific proteins are free in the supernatant liquid, which is removed, and the resin in the centrifuge pellet has the target protein bound, after which is removed and treated with the unbinding reagent and the mixture is again centrifuged. In this step, the released protein of interest is in the supernatant, where it is recovered in its purified form.

2.7.5 Radioactive Tracer Techniques

Methods such as light scattering, UV and visible absorption, refractometry, and others require a significant number of molecules to obtain measurable signals. They have limited applicability inside living organisms. *Tracer techniques* are used where very high detection sensitivity, or specific spatial and location information are required. Radioactive tracers have been the primary type used because their radioactive decay products, generally electrons, positrons, or γ-rays are very energetic and easy to detect, including single events. They can also be localized in space for imaging purposes. Because radioactive decay must take place over short periods of time, half-lives on the order of days or tens of days are normally used. To have readily available short-lifetime isotopes, it is necessary to artificially produce them in nuclear facilities. Many isotopes are used. ^{123}I is produced by proton bombardment of ^{124}Xe, and decays by electron capture followed by γ-ray emission, and has a half-life of 13.2 h. Technetium in the form of ^{99m}Te ("m" in the superscript stands for "metastable") decays by γ-emission and is the most widely used medical isotope. It is the β-decay daughter of ^{99}Mo, which latter has a useful storage life of two weeks, while the ^{99m}Tc has a half-life of 6 h so that its radioactivity in the body only lasts a few days. Among the most widely used isotopes in biochemistry and molecular biology are ^{32}P, ^{31}P, ^{14}C, ^{35}S, and ^{125}I. Some nonradioactive tracers also now exist in the form of enzymes that can produce substrate products in sufficient quantity for detection.

2.7.6 Nucleic Acid Hybridization

This has spawned some of the most important methods in molecular biology. The basic notion is the ability of one strand of DNA to form a double helix with another strand with complementary nucleotide sequence. One of the first methods, *Southern blotting*, enzymatically degraded target DNA into fragments, spread these out into bands by gel electrophoresis, and then labeled fragments of known sequence were added. Wherever double-strand formation occurred due to a labeled fragment's complementarity to a target DNA fragment, the band for that DNA fragment would be detected by the label's presence. More sophisticated means of producing the fragments have been developed. The most widely used method, which is also the basis of *automated DNA sequencing*, uses a *primer* of known sequence that will bind in complementary fashion to a desired location on the target DNA, and then this primer is extended following the sequence of the target DNA, and different *chain termination agents* (2′,3′-dideoxy nucleotides) in different batches yield a final collection of fragments to be separated by gel electrophoresis and identified with a detection means, whose sequences are then pieced together with highly developed computer algorithms.

2.7.7 Chromosome Conformation Capture

These methods are used to analyze spatial organization of chromatin. They cross-link the genome to proteins with formaldehyde which introduces bonds that crosslink and "freeze" interactions between genomic loci and proteins. The genome is then cut with restriction endonuclease. The size of the fragments determines the resolution. Fragments are then ligated with T4 DNA ligase. Interacting loci are subsequently quantified by amplifying ligated junctions with PCR (polymerase chain reaction). There are various levels of conformation capture, 3C, 4C, 5C, and Hi–C. In Hi–C all possible pairwise interactions are determined. These powerful methods have resulted in elucidation of chromosome structural features, characterizing chromatin loops, and helped with understanding of transcriptional regulation.

2.8 Summary

This chapter is chiefly for physical scientists and engineers who have not taken a biology course. It is meant only as a very brief overview of prokaryotic and eukaryotic cells, and the main functions carried out by them, such as replication, nerve transmission, and defense. Important classes of specialized cells, such as blood, bone, and nerve cells are surveyed. Some important methods in molecular biology are introduced, which are quite different in approach from the optical and physical methods of Chapter 9.

References

Alberts, B., Johnson, A., Lewis, J. et al. (2008). *Molecular Biology of the Cell*, 5e. Taylor and Francis Group.

Bouwman, B.A.M., Crosetto, N., and Bienko, M. (2022). The era of 3D and spatial genomics. *Trends in Genetics* 38 (10): 1062–1075.

Davis, R.R., Campbell, M., Izumiya, Y., Tepper, C.G. (2020). Capture Hi-C: characterization of chromatin contacts. *Translational Epigenetics Series* 18 (Epigenetics Methods): 419–444.

Holley, D. (2017). *General Biology I: Molecules, Cells, and Genes*. Dog Ear Publishing LLC.

Krup, G. (2012). *Cell and Molecular Biology: Concepts and Experiments*, 7e. Wiley.

Lodish, H., Berk, A., Kaiser, C.A. et al. (2016). *Molecular Cell Biology*. W.H. Freeman & Co., Ltd.

Raven, P., Johnson, G., Mason, K. et al. (2020). *Biology*, 12e. McGraw Hill.

"Metabolic Mapping," Casland, J.S. and Graczyk, G.M. (2001). *Current protocols in Neuroscience* Chapter 1, Unit1.6.

(2021). *Trends in Biochemical Sciences* 46 (11): 902–917.

Weaver, R.F. (2012). *Molecular Biology*, 5e. McGraw Hill.

Further Reading

Coleman, W.B., Tsongalis, G.J. (2006). Polymerase chain reaction: (PCR). In: *Molecular Diagnostics*, 2e (ed. W.B. Coleman and G.J. Tsongalis), 47–55.

Iovine, J.C., Claypool, S.M., Alder, N.N. Mitochondrial compartmentalization: emerging themes in structure and function.

Pingoud, A., Hahn, M., and Wilhelm, J. (2009). Polymerase chain reaction (PCR) in chemical biology. In: *Wiley Encyclopedia of Chemical Biology*, 3 (ed. T.P. Begley), 772–783.

Vijesh, K. and Lenhoff, A.M. (2020). Mechanistic modeling of preparative column chromatography for biotherapeutics. *Annual Review of Chemical and Biomolecular Engineering* 11: 235–255.

3

Biochemistry Survey

Life is a relationship among molecules and not a property of any molecule.

Linus Pauling

The major classes of biomolecules include *amino acids*, *nucleic acids*, *carbohydrates*, and *lipids*. The first three of these can exist in polymeric form, whereas lipids can form membranes. DNA is the biopolymer responsible for accumulation and transmission of the genetic code. Proteins are biopolymers built from amino acids that carry out all the catalytic functions that control metabolism, in addition to having other major functions in structure, regulation, transcription, immune responses, etc. Carbohydrates and their polymers provide energy storage mechanisms and structural building blocks and also have essential roles in glycoproteins, biopolymers that combine carbohydrates and proteins. Lipids in the form of diglycerides are a major source of energy storage and the essential building blocks of membranes. Other, relatively small molecules have vital functions, such as porphyrins (bases for cytochrome and hemoglobin), chlorophyll, adenosine triphosphate, diphosphate, and monophosphate (ATP, ADP, and AMP, respectively), nicotinamide, hormones, and vitamins, and others.

The aim of the current chapter is to introduce the basics of these molecules. A more in-depth look at selected properties and functions of these are given in Chapters 10–13.

3.1 Amino Acids and Proteins

While all biomolecules are important, proteins hold a special place. They work in more contexts and with more functions than any other group. Even the information-storing molecules, RNA and DNA, exist chiefly in order to produce proteins.

Let's take a brief inventory of protein functions. All known enzymes are proteins. Enzymes are catalysts that can increase reaction rates by over a trillion-fold. As a result, virtually all biochemical reactions are mediated by enzymes. While catalyzing reactions, proteins also exert feedback control on metabolic functions and rates. An enzyme typically has the suffix *-ase*. Examples include *polymerases*, which catalyze biopolymer synthesis; *depolymerases*, which break down biopolymers; *lipases*, which hydrolyze triglycerides; *kinases*, which transfer phosphate groups to substrates; *isomerases*, which catalyze isomeric transformations within single molecules; *ligases*, which join two molecular substrates. There are many other classes of enzymes.

Proteins are central to transport processes, for example, the transport of oxygen by the protein *hemoglobin* from the lungs to tissues and carbon dioxide from tissues to lungs. Platelets in blood, composed largely of protein components, are central to blood clotting and protection of vascular tissue. Proteins also transport many other small molecules. They form selective channels in membranes that allow the cell to exploit concentration and electrostatic differences to drive energy-producing, nutrient-transfer, and nerve impulse formation and propagation processes.

While the chief function of blood is the transport of oxygen via hemoglobin in red blood cells, it is also the chief carrier of antibodies, which are proteins that are soluble in blood plasma, or bound to the membranes of the B cells that produce them, and recognize and help to destroy antigens of bacteria and fungi. The principle of molecular recognition that allows antibodies to identify invading microorganisms is also used in self-identification, and for many other processes. One of the most amazing features of the immune system is its ability to distinguish self from nonself. Failures in this self-recognition can result in a wide variety of autoimmune diseases.

Biophysics: Physical Processes Underlying the Living State, First Edition. Wayne F. Reed.

Companion website: www.wiley.com/go/Reed/BiophysicsandMacromolecules

Proteins are fundamental to motion in the animal kingdom. At the most primitive level, protein-derived flagella and cilia provide motility for microorganisms. In eukaryotes, the proteins actin and myosin cooperate to provide muscle function, which in turn drives locomotion across species, on the land, in the sea, and in the air.

Proteins enable the generation and propagation of nerve impulses. They are also an important component of myelin, the insulating sheath around nerve cell axons, the "wires" of the nervous system.

A major function of proteins is to provide structural materials. Collagen and elastin form the basis for many structures, such as hair, skin, tendon, cartilage, and bone. Collagen derives its stability and strength from its triple helical macromolecular structure. These can both form into higher order helices and fibers, as well as provide sites for mineral binding, especially in bone.

In addition to the above functions, proteins are also integrally involved in their own production. Some proteins bind to elements of DNA and hence participate in gene regulation and expression. Protein *chaperones* are vitally important for ensuring the proper folding of newly synthesized proteins into their proper functional form.

Proteins are built from 20 amino acids. Virtually endless combinations of these amino acids are possible. The *proteome* refers to the set of proteins that are produced in a given sub-system of an organism, e.g. blood, the entire organism itself, a species, such as humans, or even the entire list of all proteins on the planet. The human proteome includes at least 20,000 proteins.

It is mindboggling that such diverse and complex functions can be carried out by stringing together a base set of 20 amino acids in over hundreds of thousands of sequences, and that new proteins are constantly evolving into new sequences with new capabilities. While only these same 20 amino acids make up proteins, there are over 400 other types of naturally occurring amino acids.

All the amino acids in a protein have the peptide motif in common, which involves an amino group on one end, a carboxyl on the other end, and 1 of 20 distinct pendant functional groups. A covalent *peptide bond* is formed when two amino acids react, with the elimination of a water molecule. While the peptide linkage has a partial double bond and hence does not have free rotational flexibility – it is planar and rigid – the bonds on either side of it do have rotational flexibility. Hence, protein molecules can "wrap up" into many different configurations leading to different net morphologies, such as globules, fibers, and rods.

At physiological pH the amino group is positively charged and the carboxyl group negatively charged. Only the L-form of each amino acid exists in proteins in living state systems, with the exception of glycine, which is achiral. The origin of this specific chirality[1] has not been explained, similar to the situation in cosmology, where no explanation of the preponderance of matter over antimatter has been definitively found.

3.1.1 The Peptide Bond and the Amino Acids

The peptide bond is formed between two amino acids with the elimination of a water molecule, as seen in Figure 3.1. By convention, the ordering of the sequence of amino acids in a protein starts at the terminal amino group and progresses toward the terminal carboxyl group.

Figure 3.2 shows the 20 amino acids (AA) grouped according to type. They are distinguished from one another by their side groups, which vary widely in chemical composition, conferring unique and multifold properties to the individual amino acids. The genetic code, discussed in Chapter 11.2, consists of the sequence of nucleic acids, taken three at a time in series, which specify the order of the amino acids in a protein.

Amino acids which are polar but uncharged have a variety of chemically distinct side groups conferring unique properties on each. Glycine, for example, has only an H-group and so is the least bulky of the amino acids. Cysteine has sulfur in a thiol group, which gives it the unique ability to form covalent disulfide bridges in proteins helping to stabilize their 3D structures. Glutamine, the most abundant excitatory neurotransmitter in the human nervous system, was discussed in Chapter 2.6.1.

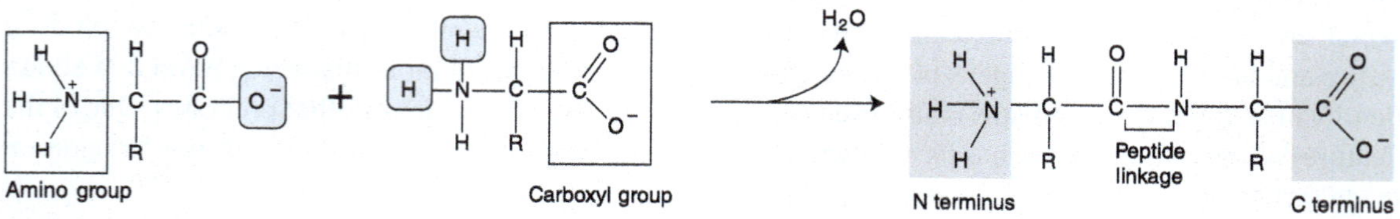

Figure 3.1 Formation and components of the peptide bond.

1 See Chapter 7.3.1 for a molecular explanation of chirality.

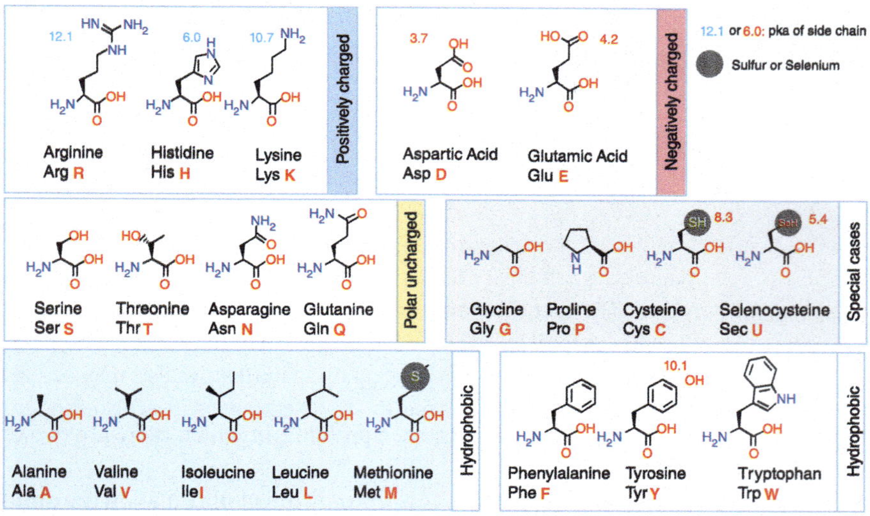

Figure 3.2 The 20 amino acids, their formula, classification, and where applicable the pK_a of the side chain. (Wikipedia, Thomas.ryckmans68).

The nonpolar, hydrophobic amino acids have highly differentiated side groups among them. They are usually found in the interior of folded proteins, although they are often also found on the surface of proteins bound in membranes, including membrane channel formers. Alanine has one CH_3 group and is the least hydrophobic of the series of aliphatic amino acids which ascend in hydrophobicity from alanine to valine, with an isopropyl side group with three carbons, to leucine with an isobutyl group with four carbons, and isoleucine, with a branched hydrocarbon chain with four carbons.

The ring structure on tryptophan's group makes it highly hydrophobic and its conjugated structure also gives it an ultraviolet absorption band with a peak around 280 nm, with a fluorescence emission band between 300 nm and 350 nm, which is very sensitive to the local polarity, an effect known as *solvatochromism*. It is often used as an intrinsic UV marker and fluorescent probe in protein biophysical studies, and its solvatochromism helps to determine a protein's state of folding and unfolding. Similarly, tyrosine, with its conjugated ring structure also absorbs around 280 nm and is moderately hydrophobic.

The pK_a of the carboxyl and amino moieties in a lone amino acid are around 2.3 and 9.5, respectively.[2] Figure 3.2 provides the pK_a for isolated amino acids with ionizable acid and base side groups, in addition to their structures and functional classes.

There are two polar amino acids with carboxylic acid groups, aspartic acid, and glutamic acid. There are three polar basic amino acids: lysine, arginine, and histidine. Histidine has an aromatic ring and a pK_a of 6.0, giving it a partially positive character at physiological pH and allowing it to act as a "proton shuttle" in many catalytic reactions. In soluble proteins, the charged polar amino acids are usually found on the protein surface.

3.1.2 Protein Isoelectric Point

Most proteins contain some pH-dependent negative and positive side groups, for which there will be a pH at which the net charge is zero. This is termed the *isoelectric point* (IP) of the protein. Each isolated amino acid has a pK_a for its –COOH and $-NH_3^+$ end groups, as well as a third pK_a for those amino acids with an ionizable R-group. All the pK_a values for the amino acids and ionizable side groups are known. It is tempting to think that one could predict both the IP of a given protein, as well as its net charge at any given pH by simply using the known pK_a values for the isolated amino acids in order to determine the protein's IP. Research has shown, however, that the IP of proteins can only be approximated by this approach, because there are many interactions that affect ionizability, and hence the pK_a of an amino acid when in a protein. In fact, even simple homopolymeric polyelectrolytes with pH-dependent ionizable side groups do not retain the pK_a's of the isolated charged monomers and depend on solution ionic strength and temperature. This is because any ionizable unit in the polymer chain experiences a net summation of repulsive and attractive potentials due to all the other charges in the chain, and this potential depends strongly on electrostatic screening due to ionic strength, and other solution conditions (Chapter 4.6).

---

2 pK_a is defined in Chapter 4.15.1

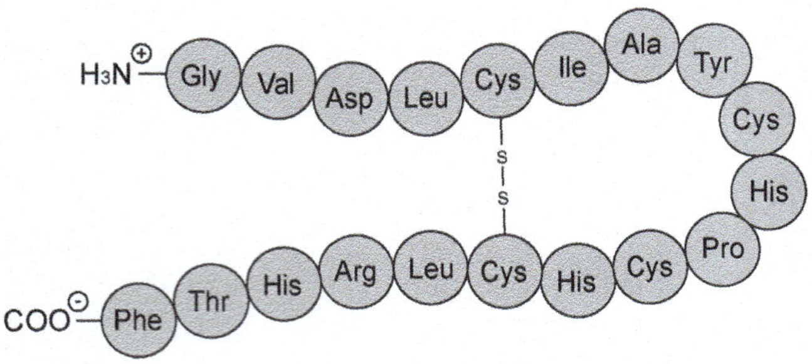

Figure 3.3 A generic AA sequence in a very small protein.

3.1.3 Levels of Protein Organization

Proteins have three principal levels of organization. A fourth level also exists for some proteins, in which there are two or more proteins held together by covalent or noncovalent associations.

Primary structure refers to the sequence of amino acids specifying the protein. It is the only direct information transmitted by the DNA to the protein. Figure 3.3 shows a typical amino acid sequence in a protein, specifying its primary structure. An example of a stabilizing covalent link between sulfur atoms on a pair of cysteine amino acids is also shown.

Protein secondary structure refers to regularities in the 3D substructures formed in proteins. The main secondary structural motifs are the α-helix and the β-sheet. These structures are highly cooperative and are stabilized by hydrogen bonds. The secondary structure is determined by the local H-bonds and is not significantly affected by other types of H-bonds, nor by electrostatic effects, such as van der Waals forces.

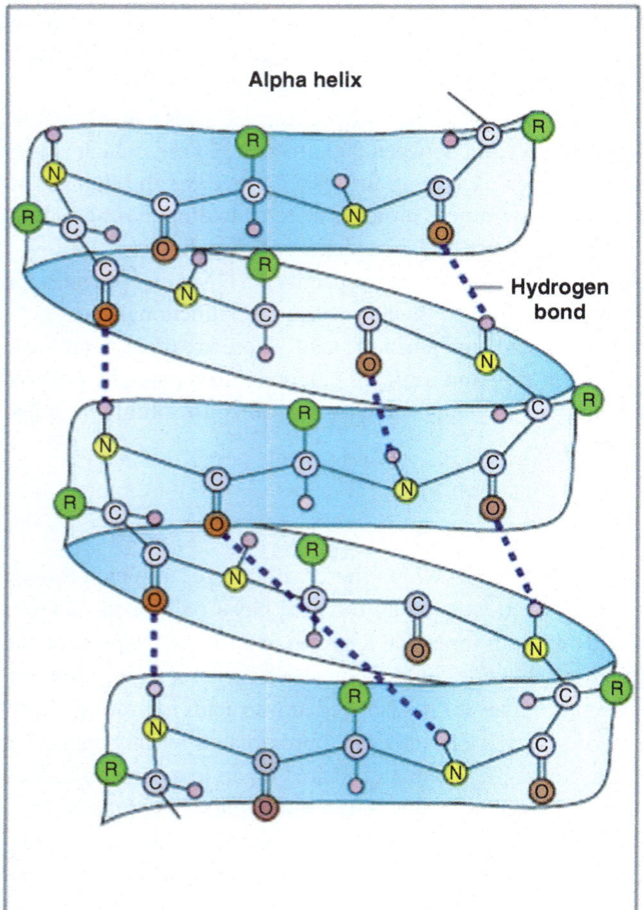

Figure 3.4 The alfa helix. (Jashari, A. (2012) / Wikimedia Commons / Licensed under CC BY-SA 3.0).

The α-helix is generally right-handed, although many proteins have short stretches of left-handed α-helix. A schematic is shown in Figure 3.4. Its backbone is the polypeptide chain, whose functional groups are arranged outwards. All the main chain CO and NH groups are H-bonded and are staggered by four units. The helix has a 0.15-nm vertical rise and 100° of turn between successive AA, which gives 3.6 AA per turn, or a helical pitch of 0.54 nm per turn. Two or more helices can wind together to form fibrous bundles, e.g. keratin in hair, myosin in muscles, collagen in bone.

β-Pleated sheets are formed by pairing of polypeptide chains in their fully extended all-trans configuration, and are stabilized by H-bonds on adjacent chains of the type C=O....H–N. There is a 0.35-nm spacing between α-carbons of adjacent AA. In the *parallel β-pleated sheet*, the two polypeptide strands are arranged in the same direction, with the amino-terminal groups next to each other. This produces a configuration in which the R groups alternately both point above and then below the chain to the right on the next AA pair. It also has the effect of making the stabilizing H-bond point diagonally to the parallel strands, in zigzag pattern. This is illustrated in Figure 3.5.

The two polypeptides in the *antiparallel β-pleated sheet* run in opposite directions, one with its carboxyl terminus next to the amino group of the other. This has the effect of making the H-bonds between opposite chains perpendicular to the two chains. Adjacent R groups alternately point above and below the sheet.

β-Pleated sheets have a number of structural motifs. The most important one is the formation of a hairpin turn where a small number of AA in antiparallel formation

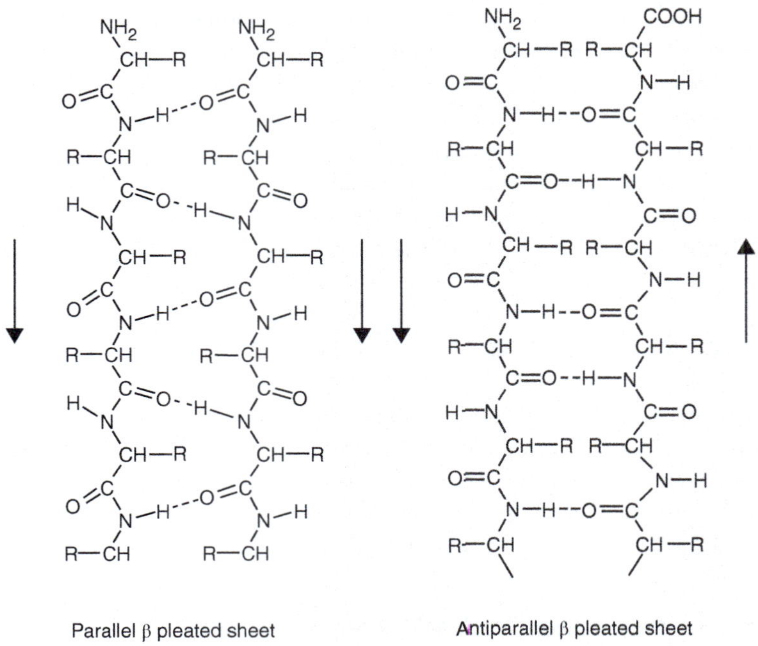

Figure 3.5 The parallel and antiparallel sheets.

allow a sharp turn, resulting in a change in direction of the protein strand. β-Pleated sheets can also occur in conjunction with α helices and are widely involved in tertiary structure folding.

Tertiary structure is the overall 3D structure of the protein. Such structures have been deciphered by X-ray scattering, solution phase NMR, and cryo-electronmicroscopy. A representation of 3D structure is shown in Figure 3.6. Understanding how the protein arrives at its tertiary structure based solely on its primary AA sequence continues to be a challenging issue and constitutes the basis of the protein folding problem, discussed in Chapter 10.1.

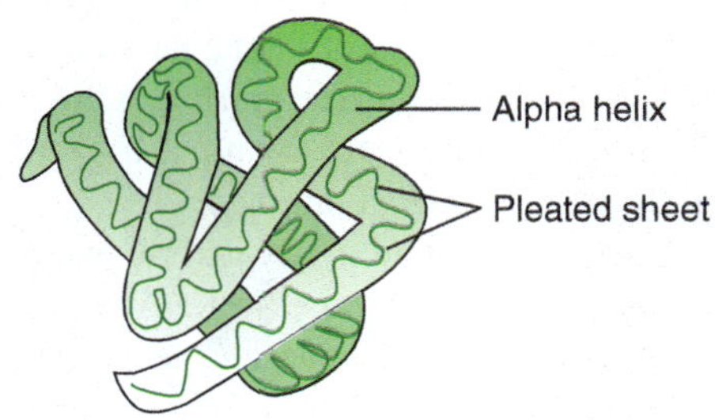

Figure 3.6 Low resolution schematic of a fully folded protein.

Quaternary structure refers to noncovalent associations among proteins or portions of proteins. Figure 3.7 illustrates an example of this. Such associations are widespread and also encompass *protein complexes*. These structures are found extensively and exhibit flexible modularity. Depending on their function and environment they have various degrees of stability. An essential notion of the modularity is that proteins that are a member of a complex can add their specialized functionality to the overall function of the complex, which produces a process or product that is synergistic, i.e. greater than the sum of the individual protein capabilities. Proteins can also participate in other complexes with other functionalities.

Often times the formation of quaternary structure is what turns on an enzyme's catalytic activity. *Allosterism* is often related to quaternary structure and involves the binding of an *effector molecule* at a site other than the active site. This binding leads to either a cooperative or anticooperative effect. Hemoglobin is the most classic example of this, where, as an allosteric unit comprised of four heme subgroups, it can bind up to four O_2 molecules. Each O_2 molecule binds more strongly than the first. This type of cooperative binding will be covered in more detail in Chapter 10.4.

Figure 3.7 An illustration of two proteins interacting noncovalently to form a quaternary structure. (Sasaki, W. (2021) / Wikimedia Commons / Licensed under CC BY 4.0).

3.1.4 Multidomain Proteins

Biological evolution is conservative and finds ways to recycle and re-adapt successful motifs. *Multidomain proteins* are a striking example of this principle. The majority of proteins in higher organism proteomes are of the multidomain type. A *protein domain* consists of an amino acid sequence that is autonomous in the sense that by itself it can fold into a well-defined structure, have active functions, and is subject to further evolution. Domains contain up to 250 amino acids, hence up about 25,000 g/mol molar mass. Autonomous domains can link together to form multidomain proteins which can then have various functionalities, some of them synergistic. Some of the functionalities gathered in a multidomain protein include substrate binding, subcellular targeting, and catalytic activity.

Evolution uses domains as building blocks to create new functions via multidomain proteins. Similarly, genetic engineering makes use of domains to create new proteins. Multidomain proteins are to be distinguished from protein complexes, mentioned above, in that these latter are modular, noncovalent associations of individual proteins.

3.2 Nucleic Acids

Nucleic acids have several functions. They play the central role in storage and transmission of genetic information in the form of the macromolecules: *deoxyribonucleic acid (DNA)* and *ribonucleic acid (RNA)*. *Adenosine triphosphate (ATP)* is the universal energy currency in living cells and releases energy when hydrolyzed to *adenosine diphosphate (ADP)*, or further to *adenosine monophosphate (AMP)*. *Nicotinamide* and *flavin dinucleotide* have fundamental roles in oxidation reactions and corresponding energy and respiratory processes.

3.2.1 Nucleobases, Nucleosides, and Nucleotides

The different molecules and structures involving nucleic acids are built off the *nucleobases*. The four nucleobases involved in DNA are *guanine* and *adenine*, the purines, and *cytosine* and *thymine*, the pyrimidines, shown in Figure 3.8. They are nitrogen-containing, aromatic heterocycles, which provide stacking and H-bond properties, allowing formation of single helices in RNA, and base pairing in the double-stranded DNA helix. Some DNA viruses have single-stranded genomes. In RNA, *uracil* substitutes thymine. The only difference between thymine and uracil is that the CH_3 group on the thymine is replaced simply by H in uracil.

Nucleobases bond to five carbon pentose sugar rings to form the *nucleosides*. There are two forms of pentose occurring in the nucleosides, ribose (in RNA) and 2′-deoxyribose (in DNA), shown in Figure 3.9.

An example of a nucleoside, *deoxyadenosine*, is shown in Figure 3.10.

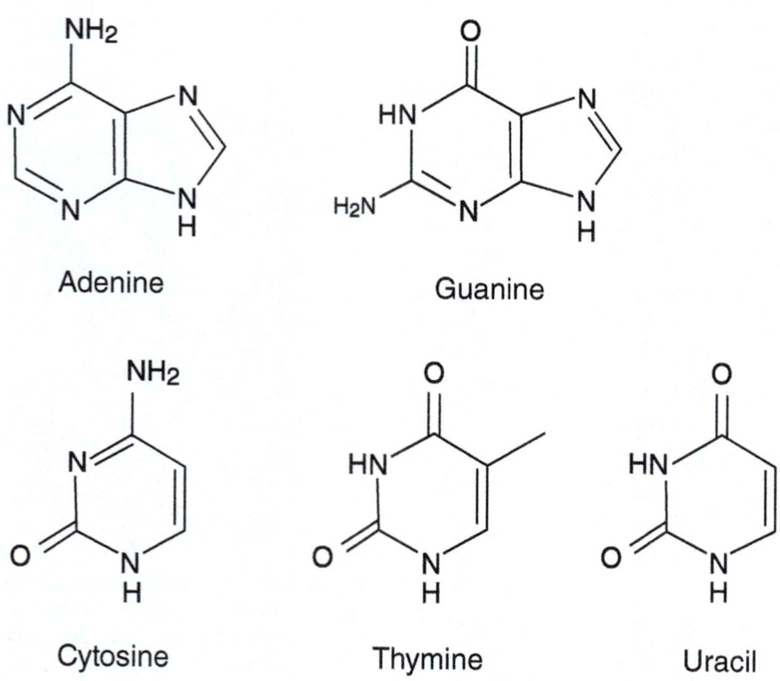

Figure 3.8 The four nucleobases. The purines are in the top row, and the pyrimidines on the bottom. Uracil is found in RNA, rather than thymine.

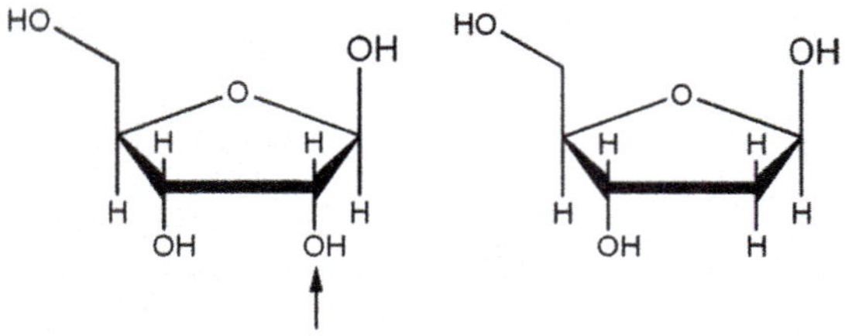

Figure 3.9 Ribose and deoxyribose, which, when added to nucleobases form nucleosides.

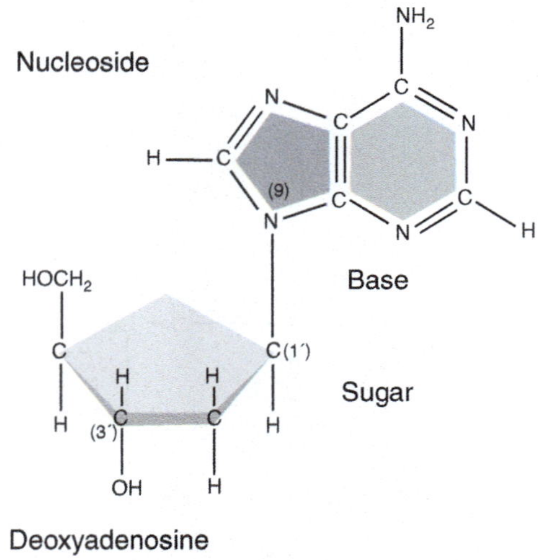

Figure 3.10 Deoxyadenosine, an example of a nucleoside.

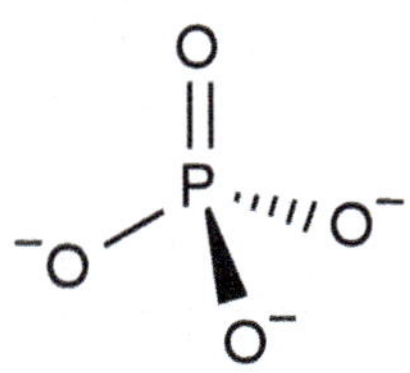

Figure 3.11 The inorganic phosphate molecule.

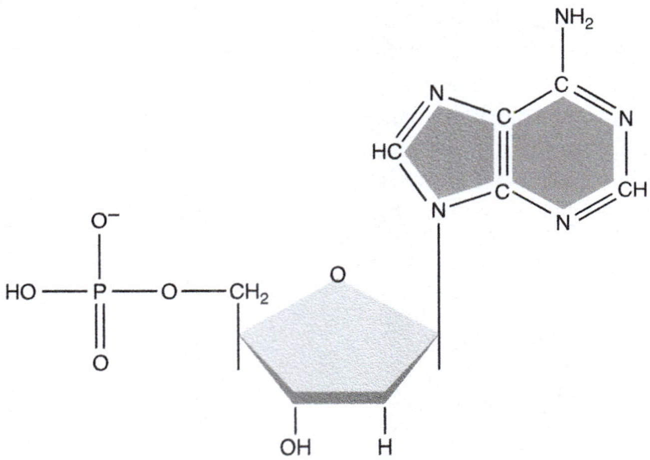

Figure 3.12 Deoxyadenosine monophosphate (dAMP), a nucleotide, composed of a nucleobase, a pentose (deoxyribose), and a phosphate group.

Inorganic phosphate, Figure 3.11, plays an essential role in DNA and RNA structure and activities. So much so that one of the primary questions in the search for the prebiotic origins of life is to determine where the phosphorous may have come from, given that it is relatively rare in the solar system, but more abundant on Earth.

When an inorganic phosphate couples to a nucleoside a *nucleotide* is formed, an example of which is shown in Figure 3.12. The phosphate group in the nucleotide allows two nucleotides to be covalently bonded via a phosphodiester linkage.

3.2.2 Nucleotides for Energy Currency

As mentioned, ATP is the universal energy currency. As the name implies, it is the adenine nucleoside, adenosine, coupled to a triphosphate group. A driving force for this is the release of electrostatic energy that occurs when the charged phosphate groups are separated from each other. Figure 3.13 shows ATP, with the three phosphate groups circled.

The change in Gibbs free energy in a system at constant pressure and volume, ΔG, such as applies in most living state processes, is given by:

$$\Delta G = \Delta U - T\Delta S \tag{3.1}$$

where ΔG is the change in the system's internal energy, T its temperature, and ΔS its change in entropy. (see Chapter 4.2 for an introduction and explanation of entropy and Gibbs free energy)

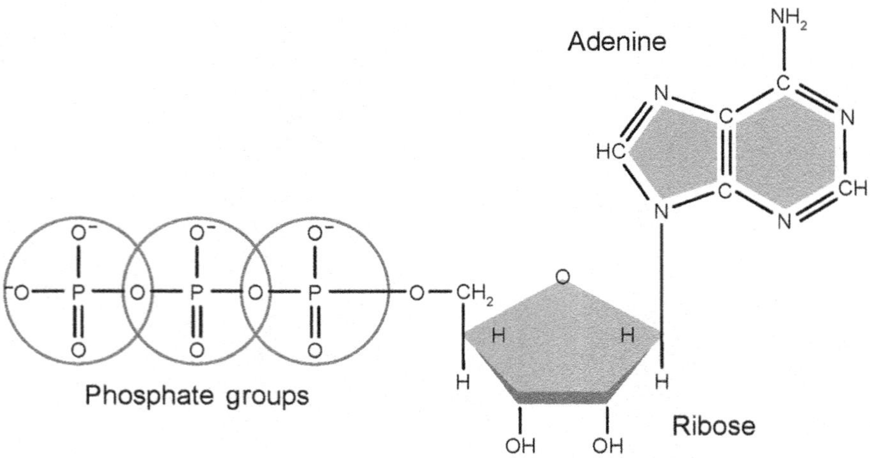

Figure 3.13 ATP.

A reaction is exothermic and occurs spontaneously when $\Delta G < 0$ and is endothermic when $\Delta G > 0$. Figure 4.26 shows energy as a function of reaction sequence. The energy difference between products and reactants is ΔG^0, which determines K_{eq} in Equation 4.164. The *free energy of activation* $\Delta G^{\ddagger}$, is an energy barrier between the reactants and products that determines the reaction rate. Activation barriers are generally due to the fact that molecules must make collisions with each other and pass through transitional and sometimes intermediate states on their way to forming products. The activation barrier corresponds to the highest energy transitional state that must be passed through in order to go downhill toward product formation. Catalysis lowers $\Delta G^{\ddagger}$ and hence increases the reaction rate, according to Arrhenius equation 4.194.

Consider the equilibrium concentrations between reactants A and B, and products C and D.

$$A + B \overset{K_{eq}}{\leftrightarrow} C + D \qquad 3.2$$

with an equilibrium constant K_{eq}

$$K_{eq} = \frac{[C][D]}{[A][B]} \qquad 3.3$$

As will be shown in Chapter 4, from the Boltzmann point of view the initial and final states are related via the energy difference between them

$$K_{eq} = \frac{[C][D]}{[A][B]} = \frac{\exp\left(-G_f/RT\right)}{\exp\left(-G_i/RT\right)} = \exp\left(\frac{-\Delta G}{RT}\right) \qquad 3.4$$

Hydrolysis of ATP to ADP and inorganic phosphate P_i is highly exothermic, and this energy can couple to other reactions to strongly shift their equilibria toward products.

$$ATP + H_2O \overset{K_A}{\rightarrow} ADP + P_i, \quad \Delta G = -7.3\ \text{kcal/mol} \qquad 3.5$$

At $T = 300$ K, with $R = 8.31$ J/K-mol, and 1 cal = 4.184 J, RT = 2,490 J/mol = 0.59 kcal/mol:

$$K_A = \frac{[ADP][P_i]}{[ATP]} = \exp\left(\frac{-\Delta G_A}{RT}\right) = 2.36\times 10^5 \qquad 3.6$$

(H_2O is omitted in the denominator because it is the solvent with an approximately constant value).

Now consider a generic reaction that is endothermic, requiring 4 kcal/mol of energy input:

$$A \overset{K_B}{\rightarrow} B, \quad \Delta G_B = 4\ \text{kcal/mol} \qquad 3.7$$

The equilibrium ratio of product/reactant is

$$K_B = \exp\left(\frac{-\Delta G_B}{RT}\right) = 1.14\text{e}-3 \qquad 3.8$$

In the presence of ATP the complete reaction is

$$ATP + \text{H}_2\text{O} + A \xrightarrow{K_{net}} ADP + P_i + B \qquad 3.9$$

Hence

$$K_{net} = K_A K_B = \frac{[ADP][P_i][B]}{[ATP][A]} = 270 \qquad 3.10$$

So that

$$\frac{[B]}{[A]} = 270\frac{[ATP]}{[ADP][P_i]} \qquad 3.11$$

$[B]/[A]$ depends on the balance of ATP and ADP at any given time in a cell, which depends on its metabolic state (ATP is always being actively created and destroyed and so does not "rest" in the equilibrium value A).

A typical value is

$$\frac{[ATP]}{[ADP][P_i]} \approx 500 \qquad 3.12$$

So that

$$\frac{[B]}{[A]_{w/ATP}} \approx 1.35 \times 10^5 \sim 10^8 \frac{[B]}{[A]_{w_0/ATP}} \qquad 3.13$$

In this case, the presence of ATP in the excess shown yields a factor of 10^8 shift in equilibrium for $[B]/[A]$.

ADP can be hydrolyzed to AMP (the monophosphate) to release energy. An index for the "energy charge state" of a cell is a composite of the concentrations of ATP, ADP, and AMP. AMP has no energy value and ADP has half the energy value of ATP, so the energy charge, EC, is defined as

$$\text{Energy charge} = \frac{[ATP] + 1/2\,[ADP]}{[ATP] + [ADP] + [AMP]} \qquad 3.14$$

When ATP predominates EC approaches 1 and the metabolism for synthesizing ATP is inhibited. When ATP is depleted EC approaches zero and ATP synthesis is stimulated. As is true, in general, for metabolic pathways, each process is feedback controlled by the abundance of the substances produced in a pathway or cycle.

3.2.3 Nucleotides for Reducing Power: ATP, FAD, and NAD

Electron transfer reactions, or *oxidation–reduction (redox) reactions* abound in metabolic pathways and other processes. A simplified notion of the "energy trail" of a photon absorbed in photosynthesis is that it produces a high-energy electron in a photoreceptor, which then cascades through a series of electron transfers, transferring energy each time to the buildup of biologically useful molecules, before arriving at a ground state where no more energy can be transferred. The electron transfer can be accompanied by the making or breaking of a covalent bond, or can just change the oxidation state of a molecule, setting it up for other types of reactions.

Nicotinamide adenine dinucleotide, NAD, is one of the major *redox agents* and can exist in its oxidized form, denoted NAD^+ and in its reduced form NADH. It consists of two nucleotides, the adenine nucleotide and the nicotinamide nucleotide. They are joined together via a phosphate linkage, seen in Figure 3.14.

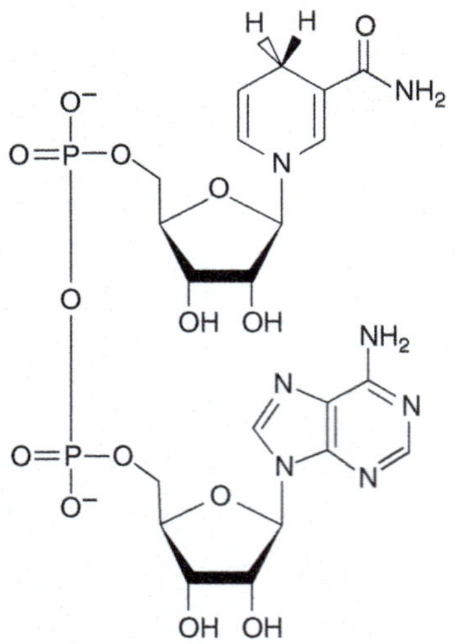

Figure 3.14 Nicotinamide adenine dinucleotide, NAD. (By Shoyrudude555 at English Wikipedia – Transferred from en.wikipedia to Commons, Public Domain, https://commons.wikimedia.org/w/index.php?curid=2814855).

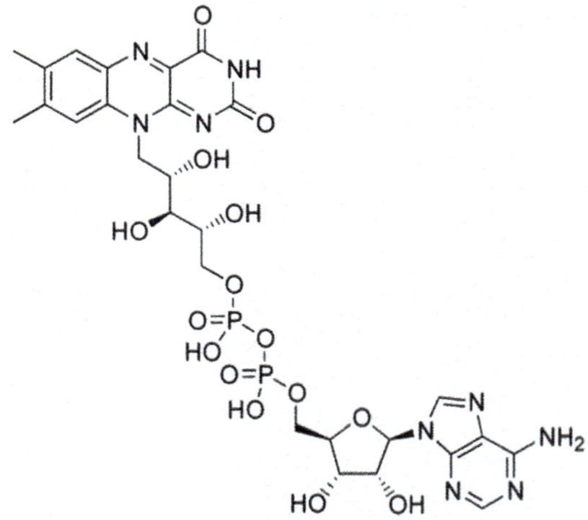

Figure 3.15 Flavin adenine dinucleotide, FAD. (Edgar181/Public domain).

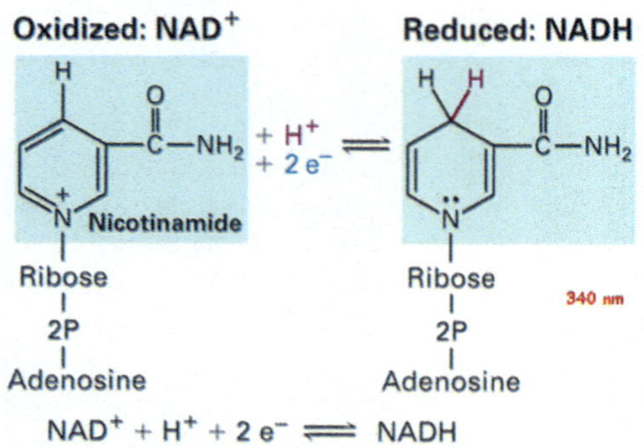

Figure 3.16 Oxidized and reduced forms of NAD.

Flavin adenine dinucleotide, FAD, shown in Figure 3.15, also makes use of the adenine nucleoside, but is coupled via a phosphate linkage to a flavin group rather than multiple phosphate groups (ATP, ADP). Its oxidized form is FAD^+ which can accept two electrons and two protons, which converts the low energy state aromatic ring into an electron donor non-aromatic ring in the $FADH_2$ form. Similarly, NAD, is the adenine nucleotide coupled to nicotinamide. The oxidized form of this is NAD^+ and can accept two electrons into a lone pair on the nitrogen in the heterocycle, and one proton. The high-energy electrons in $FADH_2$ and NADH are subsequently available for reduction reactions, notably in the respiratory chain, as well as many other reactions. The oxidized and reduced forms of FAD are shown in Figure 3.16.

3.2.4 Polynucleic Acids, DNA, and RNA

The focus on nucleotides here has been in energy production and redox reactions. The polymer of the nucleotides, DNA and RNA are the basis for genetics and the storage of biological information. In these polymers, nucleotides are taken three at a time, to form 64 *codons*. Since there are 20 amino acids in the living state, each of these can be specified by more than one codon. *The genetic code* is the alphabet linking each codon to an amino acid. DNA, RNA, and the genetic code are considered further in Chapter 11.

3.3 Carbohydrates

Carbohydrates mostly contain C, O, and H (roughly as multiples of $C_1H_2O_1$). They are a readily available energy source and can store energy by converting into both macromolecular carbohydrates, or *polysaccharides*, and fat. We have already seen that they are an integral part of the nucleic acids involved in DNA and RNA, as well as the energy and reactive molecules ATP, FAD, and NAD. Besides energy storage, polysaccharides have many structural roles, including cellulose in plants, and chitin in the shells of mollusks and cell walls of some fungi. Cellulose constitutes the majority of biomass on Earth.

Carbohydrates combine with many other elements and groups beyond C, O, and H. *Amino sugars* substitute an amino group for a hydroxyl group in simple sugars. A sugar which has a carboxyl group substituting a hydroxyl group is termed a *uronic sugar acid*. The class of polysaccharides termed *glycosaminoglycans* (GAGs), contain repeating disaccharide units consisting of one amino sugar and one uronic sugar. In some GAGs, the uronic sugar is replaced by a sulfate group, e.g. heparan sulfate and keratan sulfate. GAGs have important structural roles in connective tissues, contributing greatly to

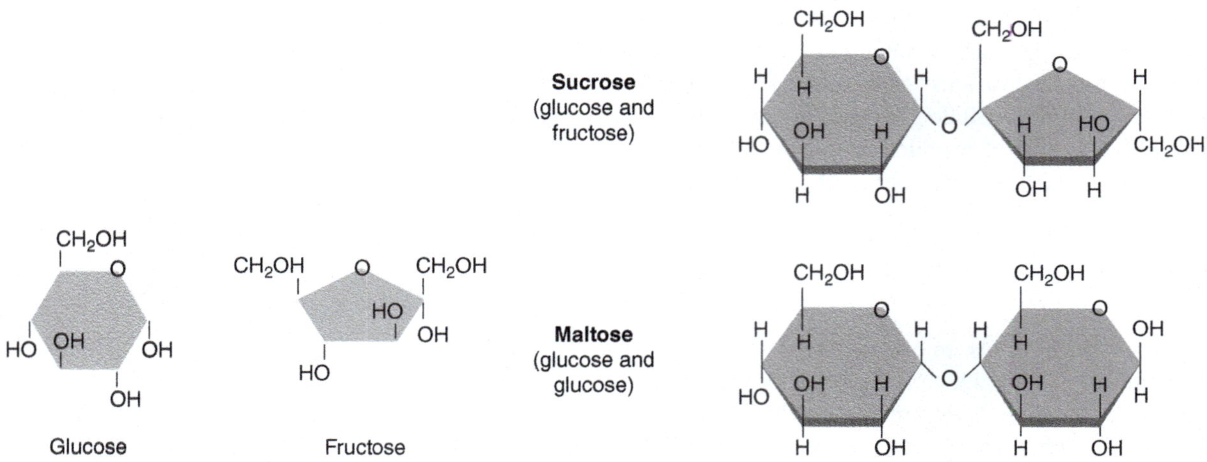

Figure 3.17 Mono- and disaccharides.

elasticity and compressibility, detailed in Chapter 12.2. *Glycoproteins* are proteins to which carbohydrates are covalently attached, either during protein translation, or as a modification after translation. Glycoproteins have many roles in the immune system, in membrane function, in signaling, and as hormones. *Proteoglycans* are a subclass of glycoprotein, where the attached oligo- or polysaccharides are GAGs. They control many events in the extracellular matrix, including embryonic development.

3.3.1 Mono- and Disaccharides

The simplest carbohydrates are *monosaccharides* and *disaccharides*. Examples of these are shown in Figure 3.17. Sugars are generally chiral. Glucose and fructose, for example, are dextrorotatory.

3.3.2 Oxidation of Carbohydrates

There are several metabolic pathways which oxidize carbohydrates to produce ATP, $FADH_2$, and NADH. These processes involve dozens of enzymatically catalyzed steps. When using stored polysaccharides, glucose must first be extracted from them using other specialized enzymes in order to enter the oxidative pathway. The process of bringing glucose to final products is termed *cellular respiration*, which can occur *aerobically* (using O_2) and *anaerobically* (without O_2). Some organisms carry out this process by *fermentation* rather than cellular respiration. Complete aerobic oxidation of a glucose molecule can be represented as

$$C_6H_{12}O_6 + 6O_2 \rightarrow\rightarrow\rightarrow 6CO_2 + 6H_2O, \quad \Delta G = -686 \text{ kcal/mol} \qquad 3.15$$

Complete aerobic oxidation of a single glucose yields a net production of approximately 30 ATP molecules. The number is approximate, since, although there is a theoretical yield of 38 ATP per glucose molecule, transport, mitochondrial leakage, heat and other processes consume several ATP, lowering the net yield. Hydrolysis of a single ATP to ADP releases 7.3 kcal/mol; 30 ATP molecules can release about 219 kcal/mol, so that carbohydrate conversion efficiency in aerobic respiration at neutral pH is

$$\text{Carbohydrate conversion efficiency} = \frac{219\,\text{kcal/mol}}{686\,\text{kcal/mol}} = 32\% \qquad 3.16$$

Anaerobic respiration occurs when no O_2 is available to be the final electron acceptor and is less efficient that aerobic respiration. Final electron acceptors include nitrate and sulfate.

3.3.3 Polysaccharides

While simple carbohydrates are essential in energy production and structure of nucleic acids, polysaccharides include a vast array of intermediate to high molar mass molecules.

Plant polysaccharides include cellulose, in interlocking strands, amylose as long single chains of glucose, amylopectin, branched chains of glucose, and many other types. Some of the plant polysaccharide polyelectrolytes include alginates, carrageenans, pectins, and various "gums," such as xanthan.

Animal polysaccharides include glycogen, highly branched glucose polymers stored in granules 10–40 nm in the liver and muscles, and polyelectrolytes, such as the GAGs. GAGs in extracellular matrix impart compressibility, control diffusion of solutes, produce streaming potentials, and participate in embryonic development.

3.4 Lipids

There are many types of lipids. In contrast to proteins, nucleic acids, and the majority of carbohydrates, many lipids are relatively insoluble in water, which is the basis for membrane formation. Membranes form the crucial interface between the cell and its surrounding environment, whether the external world or other cells. Membranes regulate a host of functions including the maintenance of electrical and chemical potentials, transport of nutrients and other substances, molecular recognition, and the production of nerve signals.

In addition to their membrane functions, lipids also store energy in triglycerides, or "fat," which also serves as thermal insulation for organisms. Lipids have other functions, such as signaling, and are also the constituents of vitamins A, D, E, and K, cholesterol, and steroid hormones.

3.4.1 Classification of Lipids

Lipids are soluble in organic solvents, such as chloroform. There are a wide variety of lipids, but we briefly discuss only some of the major types. *Triacylglycerols*, also termed *triglycerides*, are an important example. Figure 3.18 shows the chemical motif, which is that of an ester derived from glycerol, also shown in Figure 3.18. The functional groups R_1, R_2, and R_3 are *fatty acids*. Triglycerides can be decomposed by *lipases* back into glycerol and the constituent fatty acids. Triglycerides store energy and the lipases, often hormone activated, can free up the fatty acids for metabolic purposes.

Fatty acids consist of a carboxylic acid end group and a long aliphatic chain, which ranges in even number of carbon atoms from 4 to 28. They can be *saturated*, where H is bonded to all available sites on C, and *unsaturated*, where there are double bonds in the C–C structure that can accept H substitutions. Figure 3.19 shows two typical fatty acids: *stearic acid* which is saturated and oleic acid which is *unsaturated*. The *cis* double bond in oleic acid produces the bend, or kink, seen in the chain.

Palmitic acid is the most abundant fatty acid found in plants and animals and is a major component of plant oils and animal fat deposits.

Phospholipids are the major structural elements of eukaryotic and bacterial membranes. The generic phospholipid structural motif is shown in Figure 3.20, where R_1 and R_2 are two long-chain fatty acids, and R_3 has a phosphate link to a hydrophilic headgroup. Hence, phospholipids have surfactant character. Because they have two long chains instead of one, they cannot form surfactant micelles (e.g. such as SDS micelles) because of geometrical packing constraints. Rather the

Triglycerides

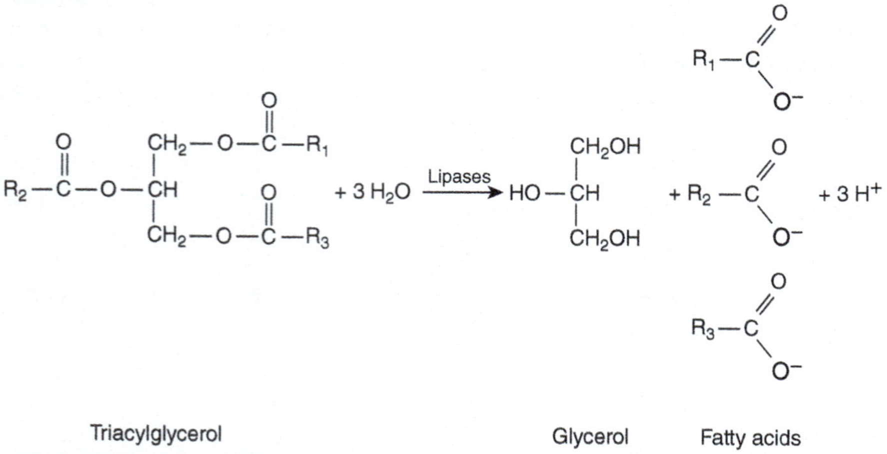

Figure 3.18 Structure of a triglyceride and its decomposition by a lipase.

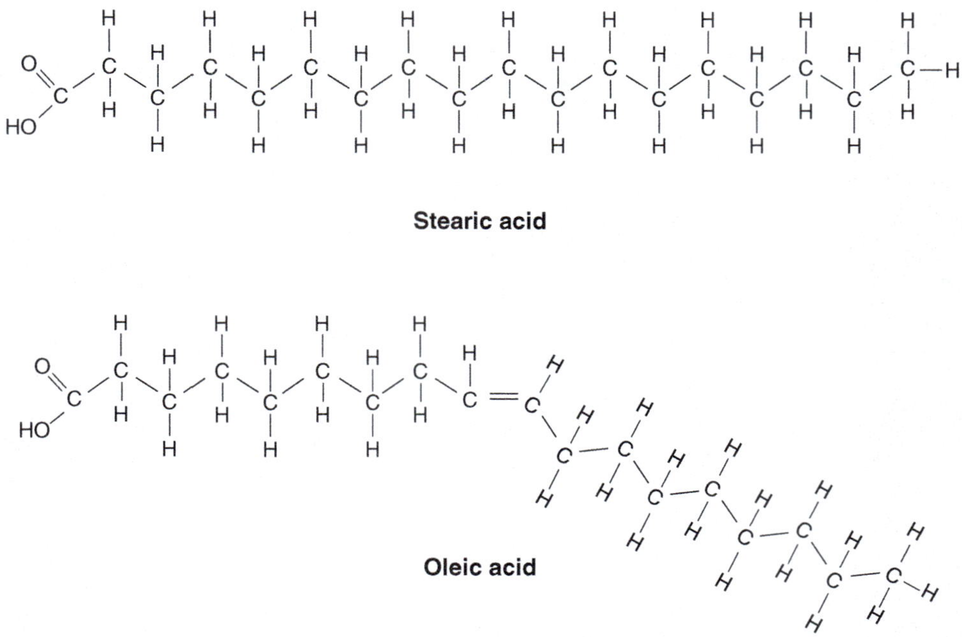

Figure 3.19 A saturated fatty acid, stearic acid, and an unsaturated fatty acid, oleic acid.

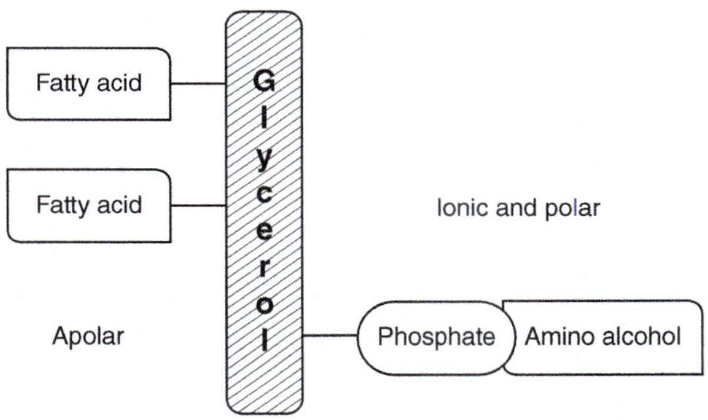

Figure 3.20 Generic structural motif of a phospholipid.

hydrophobic effect, that frees water from structuring around the long hydrophobic diglyceride chains, leads to a bilayer structure, where a back-to-back packing of the hydrophobic portions create the bilayer, and the hydrophilic groups face the water on the inside and outside of the membrane.

Some phospholipids are shown in Figure 3.21. Several important amino alcohols involved in the phosphate moiety of the phospholipid are shown in Figure 3.22.

The lipid *cholesterol* has an especially important function in the cell membranes of animals, of which it composes around 30%. Its structure is shown in Figure 3.23. The –OH group on the end of the four-ring structure is hydrophilic and so can reside near the phosphate groups of phospholipids on membrane surfaces, while the hydrophobic rings embed into the aliphatic hydrocarbon layer of the membrane. The ring structure is rigid which allows the aliphatic chains to slide along it, which gives fluidity to the membrane. This increases the elasticity of cells and allows them to change shape, which is required when stresses propagate through an organism due to its own active motion. In contrast, plants lack significant cholesterol in their cell membrane, which is hence rigid.

Hormones are a class of biomolecule with active functions (from Greek participle, Ορμον, "setting into motion"). They influence the rate of synthesis of enzymes and other proteins, affect rate of enzymatic catalysis, alter permeability of cell membranes, and have wide-ranging signaling functions. They are produced in glands that make up the *endocrine signaling system*. Steroid hormones are derived from cholesterol. Several of the cholesterol-derived steroids are shown in Figure 3.24. *Estradiol* is an estrogen hormone and the chief female sex hormone, while testosterone is that of the male.

There are also hormones which are polypeptides, such as insulin, and small molecule hormones derived directly from amino acids, such as epinephrine and thyroxine.

3.4.2 Membrane Formation from Lipids

Because phospholipids are double-tail amphiphiles, they form a bilayer instead of a micelle. Because two tails are involved the energy involved in membrane formation should roughly be on the order of twice that of micellization of a single chain amphiphile. Since the Boltzmann weighting (Chapter 4.5) exponentiates this factor, the bilayer is even more wont to form

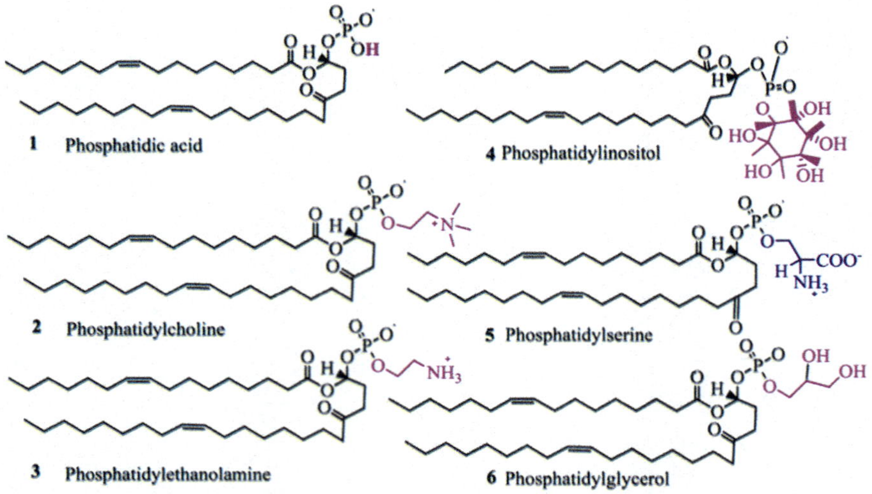

Figure 3.21 Various phospholipids.

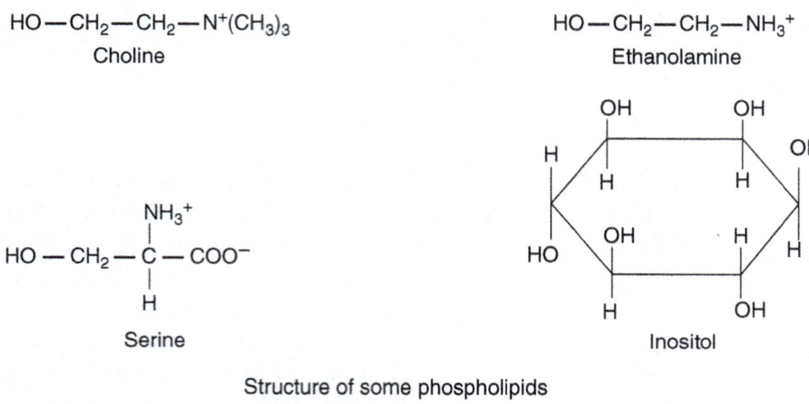

Figure 3.22 Some important alcohol moieties of phosphoglycerides.

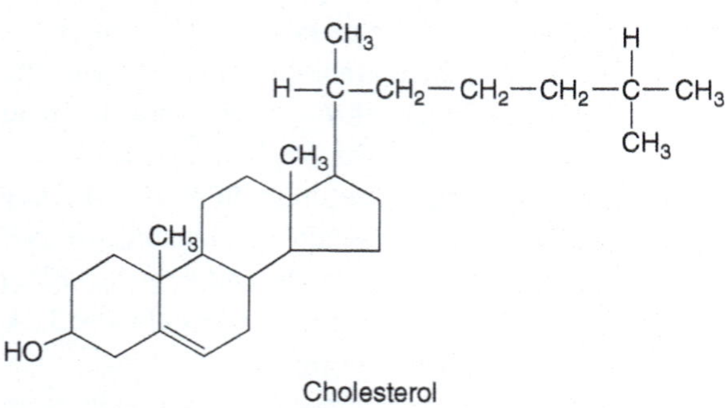

Figure 3.23 Cholesterol, a lipid.

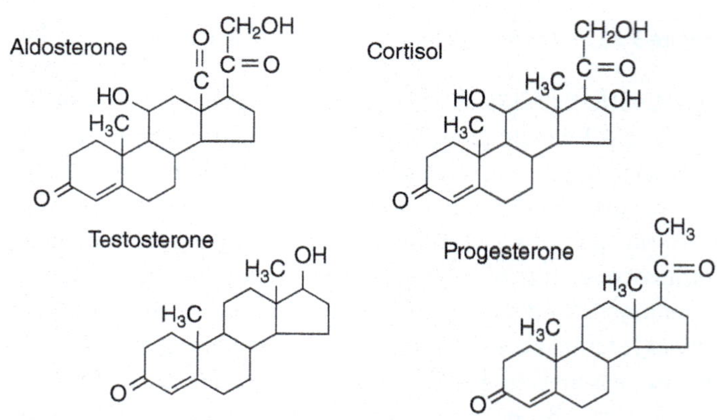

Figure 3.24 Examples of hormones.

than the micelle and should have a critical bilayer formation concentration much lower than corresponding micellar *critical micelle concentration* (CMC, Chapter 4.10); the bilayer membrane does not dissociate as readily into its constituent phospholipids as a typical single tail surfactant. Unlike the micelle which reaches a well-defined structure at the CMC, the bilayer can take on different sizes due to competing energies. There is positive bending energy involved in forming a closed

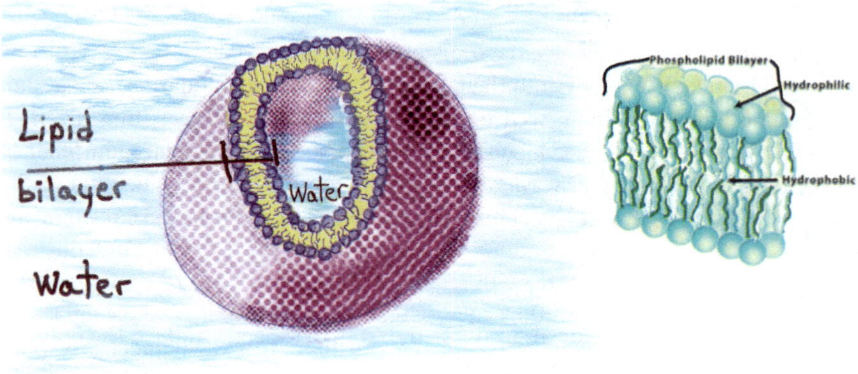

Figure 3.25 A simple bilayer liposome and a fragment of it showing the lipid bilayer motif. (Courtesy of Linnette F. Reed).

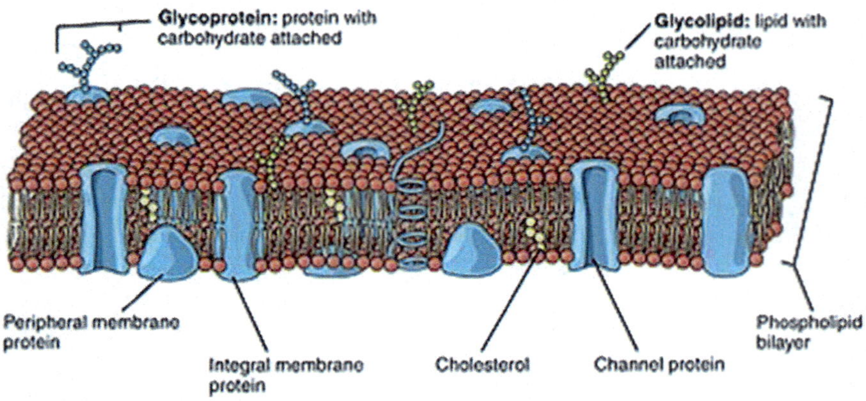

Figure 3.26 A piece of a bilayer lipid cell membrane, with various embedded substances shown. (OpenStax (2023) / Rice University / Licensed under CC BY 4.0).

structure, whereas a planar structure does not incur a bending penalty. However, the planar structure has a highly unfavorable lipid/aqueous interface.

The double chains do not allow these molecules to form micelles, because they can't pack into that geometrical form. They can, however, form liposomes, which involve bilayers, where the hydrophobic tails come together in a planar sense, leaving hydrophilic headgroups on either side, thus allowing a membrane to form naturally. Figure 3.25 is a simple liposome, illustrating the lipid bilayer and the inner aqueous compartment walled off from surrounding aqueous solution. It also shows a small chapter of the lipid bilayer.

Biological cell membranes are generally composed of lipid bilayers, in which are embedded many different agents. These latter include hormones, such as cholesterol, channel proteins, and glycoproteins with the sugar portion exterior to the membrane. An artist's rendition is given in Figure 3.26.

3.5 Metabolism

Metabolism is a complex series of interrelated, enzymatically controlled chemical reactions responsible for the production of biological energy and the synthesis and breakdown of biomolecules. Oxidation of fuel molecules and synthesis of molecules fall into the categories of

(i) *Catabolism*: The oxidative degradation of fuel molecules (polysaccharides, fats, proteins) to produce ATP, and reducing power (electron donor molecules) in the form of NADH, $FADH_2$, and NADPH. It also generates precursors for synthesis of nucleotides.

(ii) *Anabolism*: The biosynthesis of molecules such as amino acids, nucleosides, glycogen, triacyl glycerides, etc., usually via reduction of precursors.

3.5.1 Oxidation–Reduction Reactions (Redox Reactions)

Redox reactions are fundamental processes in metabolism. They involve the transfer of electrons from one species (the one being oxidized) to another species (the one being reduced). We look into the basics of redox reactions here because such reactions are fundamental to generating ATP and building up reducing power in the form of NADH, $FADH_2$, and NADPH, which can then use their high-energy electrons to drive endothermic biochemical reactions. There are a large number of redox reactions peppered throughout all the different metabolic pathways.

Sometimes, it is obvious which species is being reduced and which is being oxidized. We focus on a very simple one:

$$Zn + Cu^{2+} \rightarrow Zn^{2+} + Cu \tag{3.17}$$

Zn is the reducing agent and Cu^{2+} the oxidizing agent on the left side and Zn^{2+} is the oxidized species and Cu the reduced species on the right side. In other cases, it is less obvious, and a convention based on *oxidation numbers*, which we will not delve into here, can be used to determine which species is reduced and which oxidized, or to determine if a given reaction is even a redox-based reaction. Some elements, such as iron, can have multiple oxidation states; the common forms include the ferrous ion Fe^{2+} and the ferric ion Fe^{3+}, and other states besides these. A reaction of fundamental importance in carbohydrate oxidation is the oxidation of the ferrous ion and reduction of oxygen:

$$2Fe^{2+} + 1/2O_2 \rightarrow 2Fe^{3+} + O^{2-} \tag{3.18a}$$

with half-reactions

$$2Fe^{2+} \rightarrow 2Fe^{3+} + 2e^- \tag{3.18b}$$

and

$$1/2O_2 + 2e^- \rightarrow +O^{2-} \tag{3.18c}$$

While redox reactions, such as Equation 3.17, will proceed by just mixing all the reactants together in a solution, just like any other chemical reaction, a remarkable feature of redox reactions is that they can be carried out with reagents spatially removed from each other, as long as there is an electrical connection between the reagents.

The idea that redox reactions can be carried out separated in space, provided there is an electrical connection led to the notion of a *battery*, or *galvanic cell* (also termed a *voltaic cell*, since both Galvani and Volta each developed these). The galvanic cell, illustrated in Figure 3.27, consists of an electrode of one metal, e.g. Zn, in one vessel containing a solution of its oxidized form, Zn^{2+}, counterbalanced by an equal concentration of anion, such as SO_4^-, provided by dissolving the salt $ZnSO_4$ in the solution. An electrode of the other metal e.g. Cu, is placed in a second vessel containing solution with the oxidized form, Cu^{2+}, counterbalanced by SO_4^-, provided by dissolving the salt $CuSO_4$ in the solution. If the two electrodes are connected by a conducting wire, electrons will begin to flow from the Zn electrode to the Cu electrode; the transfer lowers the net energy of the system and is rooted in the quantum mechanical details of the electronic characteristics of the two metals. This electron flow would quickly give the Zn vessel a net positive charge, due to the electron loss, which would stop any further flow of electrons and hence stop the reaction. The Cu vessel would have a net negative charge. If the conducting wire were now cut each vessel would have a net charge which could be neutralized by connecting its electrode to ground potential. If a load were placed between the electrode and ground before discharge the pulse of current could be used to do useful work for a brief moment. In order to make the device provide a constant current and continue the redox reaction, electrical neutrality must be maintained in each vessel. A way to do this is by letting ions flow into each vessel from a *salt bridge*, shown in Figure 3.27. The salt bridge contains an inert ionic solution, e.g. K^+ and Cl^-, and is normally filled with a gel. To maintain electrical

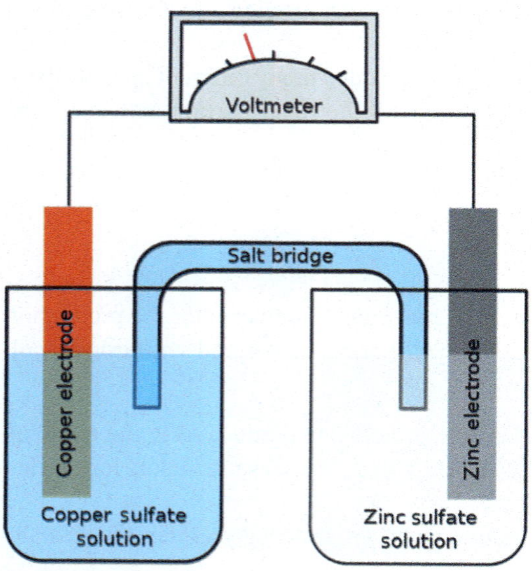

Figure 3.27 A galvanic cell.

neutrality, each time a pair of electrons passes from the Zn to the Cu electrode the salt bridge will allow Cl^- ions to pass into the zinc cell to offset the loss of electrons, and K^+ ions will pass into the copper cell to offset the gain of the two electrons, thus both completing the electrical circuit and maintaining charge neutrality.

A voltmeter placed across the two electrodes will read a constant voltage during the reactions. An ammeter placed in series with the wire connecting the vessels will read a constant current, which depends on the total resistance of the battery circuit, including any external load. As the reaction proceeds the concentration of Zn^{2+} in the Zn vessel will gradually increase and that of Cu^{2+} in the Cu vessel will gradually decrease, and the voltage across the electrodes will decrease, until the final equilibrium is reached and the voltage goes to zero. As Zn atoms from the electrode keep dissolving into solution in the form of Zn^{2+}, this leads to the gradual corrosion of the electrode. In the copper vessel, each time a Cu^{2+} is reduced to Cu the neutral atom plates onto the Cu electrode, thus increasing its mass. In sea-faring boats and ships a "sacrificial" zinc electrode is often connected to the system's electrical ground, and slowly dissolves away due to interactions with salt water, leaving the copper and other metals in the system protected.

The galvanic cell is quite rich in different phenomena, but one of the most striking is the existence of the electric potential difference formed across the electrodes. The separation of the reaction into two compartments gives rise to the notion of the *half-cell*, each separate compartment forming half of the net reaction in Equation 3.17. The half-reactions are $Zn \rightarrow Zn^{2+} + 2e^-$, and $Cu \rightarrow Cu^{2+} + 2e^-$ and the convention is to measure the potential of these with respect to a chosen inert, reference electrode under established standard conditions. These standard conditions are that $T = 25°C$, each dissolved component is present at 1 M concentration, any gas component is at 1 atm, and any solid is in its most stable state. A further convention is to put the reduction reaction on the left side and the reduced product on the right side.

The potential measured under standard conditions is the *half-reaction standard potential* ϕ^0. An inert electrode, such as platinum, merely transfers electrons during a redox reaction and so neither dissolves nor grows during the reaction. The *hydrogen gas standard cell* continuously provides H_2 to the solution which dissociates to keep a constant molar concentration of H^+ ions at 1 M, via $H_2 \rightarrow 2H^+ + 2e^-$. The standard gas cell is *defined* to have zero potential, hence having the equivalent function as *ground potential* in electrostatics and current-carrying circuits. ϕ^0 is the voltage of the half-cell reaction (e.g. Zn) with respect to the standard gas cell. If ϕ^0 is positive then the half-reaction will proceed spontaneously and is the equivalent of an exothermic reaction. If ϕ^0 is negative then the reaction will not proceed spontaneously and is equivalent to an endothermic reaction. Such reactions can proceed, however, if an external voltage of opposite sign larger than ϕ^0 is applied; this *electrolysis* is used for the electrolytic decomposition of materials, such as H_2O into O_2 and H_2. In our example:

$$Zn \rightarrow Zn^{2+} + 2e^-, \quad \phi^0 = 0.76\,V\,(\text{oxidation}) \tag{3.19a}$$

and

$$Cu^{2+} + 2e^- \rightarrow Cu, \quad \phi^0 = 0.34\,V\,(\text{reduction}) \tag{3.19b}$$

The two equations are added to give the net Equation 3.17 and the corresponding standard, voltage.

$$Zn + Cu^{2+} \rightarrow Zn^{2+} + Cu, \quad \Delta\phi^0 = 1.10\,V \tag{3.19c}$$

These standard voltages are compiled in many readily accessible tables for a wide variety of reactions. However, ϕ^0 only applies at the standard condition of dissolved reagents at 1 M at 25°C. The electrolyte vessels above, connected by the ion bridge are conceptually similar to a semipermeable membrane with ions at specified concentration on each side. For this, the relationship between ion concentrations separated by a membrane and the potential difference produced ϕ, is given by the Nernst-Planck Equation 6.110. Changing the notation so that c_1 is the charged species on the left side of the redox equation and c_2 the charged species on the right side, and $q = ze$ when each ion has the same valency, puts the Nernst-Planck equation in the form:

$$\phi = \frac{k_B T}{ze} \ln\left(\frac{c_2}{c_1}\right) \tag{3.20}$$

where k_B is Boltzmann's constant (Chapter 4.5). But Equation 6.110 was derived for a single charged species (which is why c_1 and c_2 can be either mass or molar concentrations, although molar concentrations are preferred in most situations). When $c_1 = c_2$, $\phi = 0$ by Equation 3.20. However, the whole notion of ϕ^0 is that it is the voltage between two different species when the two species have the same molar concentration, $[X_1] = [X_2]$, where $[X_1]$ is the molar concentration of the charged species X_1 on the left of the redox equation and $[X_2]$ the concentration of species X_2 on the right. So, Equation 3.20 should be amended to include these facts, according to

$$\phi = \phi^0 - \frac{k_B T}{ze} \ln\left(\frac{[X_2]}{[X_1]}\right) \tag{3.21}$$

Since the reaction proceeds in the direction of higher $[X_2]$, the voltage will be higher when $[X_2]$ is lower than $[X_1]$, which is why the negative sign is introduced. ϕ^0 hence incorporates the quantum mechanically driven redox reaction into the Nernst-Planck equation. In Equation 6.110, there was no chemical reaction involved, just a difference in concentrations of a single species that doesn't chemically react with itself.

When the system reaches equilibrium the battery is discharged and $\phi = 0$. So, Equation 3.21 gives the relationship

$$\phi^0 = \frac{k_B T}{ze} \ln\left(\frac{[X_{2,eq}]}{[X_{1,eq}]}\right) \tag{3.22}$$

where $[X_{2,eq}]$ and $[X_{1,eq}]$ are the equilibrium concentrations of X_1 and X_2 and their ratio is the equilibrium constant K_{eq} (see Chapter 4.15 for the concept and definition of K_{eq} and its relationship to the Boltzmann distribution introduced in Chapter 4.5), so that Equation 3.22 can also be written as

$$\phi^0 = \frac{k_B T}{ze} \ln K_{eq} \tag{3.23}$$

and

$$K_{eq} = e^{ze\phi^0 / k_B T} \tag{3.24}$$

For the redox reaction for Zn and Cu, Equation 3.17, $z = 2$, and $\phi^0 = 1.10$ V at $T = 25°C = 298$ K, which gives $K = [Zn^{2+}]/[Cu^{2+}] = 1.49 \times 10^{37}$; that is, at equilibrium, there is virtually no Cu^{2+} remaining.

Equation 3.24 often appears in molar form, so that multiplying $k_B T/ze$ by N_A/N_A gives

$$\phi = \phi^0 - \frac{RT}{zF} \ln\left(\frac{[X_2]}{[X_1]}\right) \tag{3.25}$$

where $F = eN_A = e(1.6 \times 10^{-19}\text{ C}) \times (6.02 \times 10^{23}\text{ M}^{-1}) = 9.6 \times 10^4$ C/M is the electrical charge per mole of singly charged ions and is often termed the *Faraday constant*.

Considering the general reaction:

$$aA + bB \leftrightarrow cC + dD \tag{3.26}$$

The following form of the Nernst equation relates the standard potentials to potentials produced for specific concentrations of reagents:

$$\Delta\phi = \phi^0 - \frac{0.089}{n} \ln\left[\frac{[C]^c[D]^d}{[A]^a[B]^b}\right] \tag{3.27}$$

where n is the number of electrons transferred, e.g. $n = 2$ in the copper/zinc example above. The work done in moving a mole of positive charge in such a reaction is

$$W = \Delta G = nF\Delta\phi \tag{3.28}$$

The oxidation of NADH to NAD^+ is an important source of redox energy in metabolism. The two half-reactions are (i) for the reduction of oxygen:

$$\frac{1}{2}O_2 + 2H^+ + 2e^- \leftrightarrow H_2O, \quad \phi^0 = +0.82\text{ V} \tag{3.29a}$$

and (ii) the reduction of NAD^+:

$$NAD^+ + H^+ + 2e^- \leftrightarrow NADH, \quad \phi^0 = -0.32\,V \qquad 3.29b$$

The oxidation of NADH is then written as

$$\frac{1}{2}O_2 + NADH + H^+ \leftrightarrow H_2O + NAD^+, \quad \phi^0 = +1.14\,V \qquad 3.29c$$

(In computing the potential for the complete reaction, notice that Equation 3.29b runs in the opposite direction in Equation 3.29c, where its potential will be the negative of that shown in Equation 3.29b). The reaction is exothermic leading to a change in free energy from Equation 3.28 of $\Delta G = -2.19 \times 10^6$ J/mol = −52.6 kcal/mol.

3.5.2 Principle Catabolic Pathways

Metabolic pathways are very complex and required years of experimentation and ingenuity to unravel. Our purpose here is not to go into the details of the many metabolic pathways, which is a chapter or book unto itself, but rather to mention some of the most important pathways and how they feed into each other.

Glycolysis prepares starches and sugars for final, complete oxidation. It occurs in the cell's cytosol, and took over 100 years to fully elucidate. It does not require oxygen but involves over a dozen enzymatic steps. Its end product is *pyruvic acid*, which is then converted to *acetyl coenzyme A* (A-CoA). CoA, Figure 3.28, is also the common end product of the breakdown of proteins, fats, and lipids. It is a complex, phosphate-rich molecule that forms a high-energy thioester linkage with an acetyl group, where the acetyl group is bound to the sulfur atom, as in Figure 3.28.

CoA from these processes then feeds into the *citric acid cycle* (also known as the *tricarboxylic acid cycle* and the *Krebs cycle*), which is the final common pathway for carbohydrate, protein, and lipid molecules. This cycle builds up reducing power in the form of NADH and $FADH_2$, while producing carbon skeletons for biosynthesis, and emits CO_2 and water. In eukaryotes, it takes place in the mitochondria. Proteins are broken into amino acids and the *urea cycle* converts NH_3 from these into urea and also links to the citric acid cycle, while producing carbon skeletons for synthesis. The *fatty acid oxidation* process likewise prepares lipids for the citric acid cycle and produces carbon skeletons for synthesis.

The citric acid cycle feeds into the *oxidative phosphorylation* process, also termed the *electron transport-linked phosphorylation*. It is the final catabolic pathway in the oxidative degradation of biomolecules. It is also termed *terminal oxidation* and *electron-transport-linked phosphorylation*. This process takes place in the mitochondria of eukaryotes. It uses the reducing power of NADH and FADH, together with O_2 and several enzymes, to produce ATP. This process couples the electrochemical potential of a proton gradient across the membrane, created endergonically (energy consuming) by NADH oxidation, to drive the formation of ATP in a complete *exergonic* (energy generating) path, thus linking the energy-providing and consuming pathways into this metabolic cycle.

In order to create energy to drive the proton pump across the inner mitochondrial wall, the two half-reactions for NADH and O_2 from Table 3.1 combine to form the net redox reaction of Equation 3.29c, with the release of 52 kcal/mol, which is used to pump protons across the inner mitochondrial membrane in the outwards direction, into the intermembrane space, creating an electrochemical gradient across the membrane. A cutaway view of a mitochondrion is given in Figure 3.29, which shows the inner and outer membranes, and the intermembrane space where there is a high concentration of H^+ during ATP production.

The very large protein complex, *ATP synthase* (6×10^5 g/mol), provides the final, energy-consuming phosphorylation process of ADP, using the flow of protons from the intermembrane space to the inner membrane space, also termed the *matrix*.

$$ADP + Pi + 4H^+_{inter\,membrane} \rightarrow ATP + H_2O + 4H^+_{inside\,inner\,membrane} \qquad 3.30$$

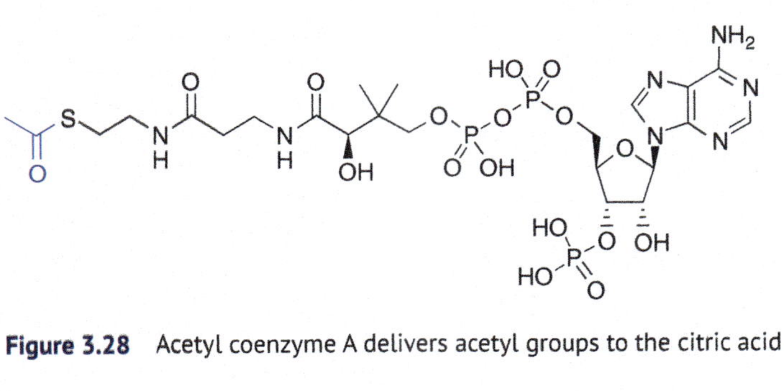

Figure 3.28 Acetyl coenzyme A delivers acetyl groups to the citric acid cycle.

Table 3.1 Half-reactions for several biologically important reactions, at 25 °C and pH = 7.

Half-Reaction, ϕ^0 (V)
$1/2O_2 + 2H^+ + 2e^- \rightarrow H_2O, \quad \phi^0 = +0.816\,V$
$Fe^{3+}e^- \rightarrow Fe^{2+,} \ \phi^0 = +0.771V$
$NO_3^- + 2H^+ + 2e^- \rightarrow NO_2^- + H_2O, \quad \phi^0 = +0.42\,V$
$O_2 + 2H^+ + 2e^- \rightarrow H_2O_2, \quad \phi^0 = +0.295\,V$
$CH_3OH + 2H^+ + 2e^- \rightarrow CH_4 + H_2O, \quad \phi^0 = +0.17\,V$
$Fumarate^{2-} + 2H^+ + 2e^- \rightarrow succinate^{2-}, \quad \phi^0 = +0.03\,V$
$2H^+ + 2e^- \rightarrow H_2, \quad \phi^0 = 0.00\,V \quad (at\ pH = 0)$
$Oxaloacetate^{2-} + 2H^+ + 2e^- \rightarrow malate^{2-}, \quad \phi^0 = -0.166\,V$
$Pyruvate^- + 2H^+ + 2e^- \rightarrow lactate^-, \quad \phi^0 = -0.185\,V$
$Acetaldehyde + 2H^+ + 2e^- \rightarrow ethanol, \quad \phi^0 = -0.197\,V$
$FAD + 2H^+ + 2e^- \rightarrow FADH_2, \quad \phi^0 = -0.219\,V$
$CO_2 + 8H^+ + 8e^- \rightarrow CH_4 + 2H_2O, \quad \phi^0 = -0.24\,V$
$S + 2H^+ + 2e^- \rightarrow H_2S, \quad \phi^0 = 0.243\,V$
$N_2 + 8H^+ + 6e^- \rightarrow 2NH_4^+, \quad \phi^0 = -0.28\,V$
$NAD^+ + H^+ + 2e^- \rightarrow NADH, \quad \phi^0 = -0.320\,V$
$NADP^+ + H^+ + 2e^- \rightarrow NADPH, \quad \phi^0 = -0.324\,V$
$2H^+ + 2e^- \rightarrow H_2\,(pH7), \quad \phi^0 = -0.414\,V$

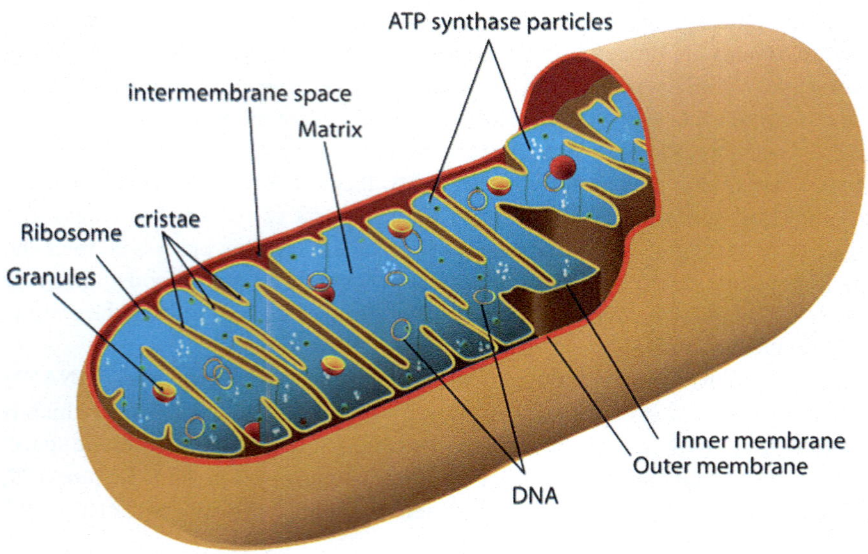

Figure 3.29 Cutaway view of a mitochondrion. (Ruiz, M. (2006) / Wikipedia commons / Public Domain).

where P_i is inorganic phosphate, PO_4^{3-}. ATP synthase acts as a rotary motor as protons enter and pass through it, allowing it to envelope ADP and attach a phosphate group to produce ATP. Reaction 3.30 runs from left to right when there is a sufficient proton gradient across the inner membrane. In the absence of the gradient, this reaction can proceed from right to left, releasing energy as ATP converts to ADP. The oxidation of NADH produces three ATP molecules, using the electron transport chain. This amounts to stored energy of 21.9 kcal/mol stored among the three ATP molecules. This means the reaction has an efficiency of 21.9/52 = 0.42, or 42%.

Figure 3.30 gives a highly simplified summary of how proteins, carbohydrates, and lipids enter into their catabolic pathways, finally entering the citric acid cycle, and the final oxidative phosphorylation pathway to produce ATP and regenerate NADH and $FADH_2$. There are many dozens of enzymatic steps underlying all these processes.

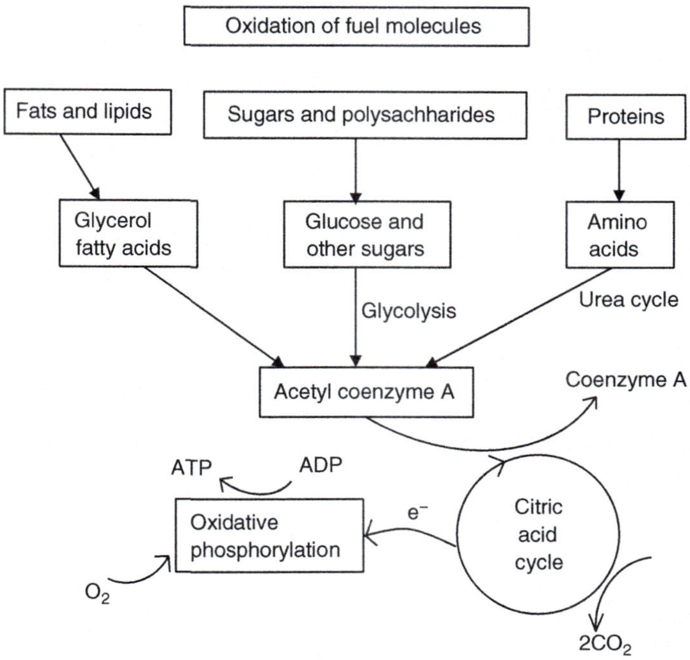

Figure 3.30 A highly simplified summary of metabolic pathways in catabolism.

Table 3.2 Some carrier molecules and activated groups carried.

Carrier Molecule	Activated group carried
ATP	Phosphoryl
NADH and NADPH	Electrons
$FADH_2$	Electrons
Coenzyme A	Acyl
Biotin	CO_2
Tetrahydrofolate	One-carbon units

Activated carriers are exergonic agents, used throughout the metabolic cycles, that carry a needed particle or group, which is released along with energy, when their concentration exceeds their equilibrium concentration, due to their active metabolic production. Examples of some carriers and their deliverable group or particle are given in Table 3.2.

As far as anabolic pathways, *gluconeogenesis* synthesizes glucose from glycerol, lactate, and amino acids in times of sugar deprivation. The *pentose phosphate pathway* enables production of NADPH and ribose-5-phosphate used in nucleic acids and nucleotides, and erythrose-4-phosphate for production of aromatic amino acids.

3.6 Summary

Twenty amino acids are the building blocks of proteins, which have detailed secondary and tertiary structures. Proteins fulfill a wide range of functions, catalysis being preeminent, along with roles in structure and biomotion. Nucleic acids are the basis for the energy molecule ATP, as well as the oxidation–reduction agents NADH and FADH. The most important polynucleic acids are RNA and DNA, responsible for genetic information storage and transmission via the genetic code, which is covered in more detail in Chapter 11.2.

Carbohydrates are used for energy storage and quick energy release. They are integral to the formation of nucleosides and nucleotides, and hence to the structure of DNA and RNA. They have structural properties, such as formation of cellulosic (e.g. trees) and chitin (e.g. lobster shells) structures, and control of compressive and other mechanical forces in connective tissue, and associate with proteins to form glycoproteins.

Lipids are active in energy storage, but a ubiquitous function is their formation of the bilayer biological membrane, which mediates transport into and out of cells, and houses many pores, channels, and other transport agents.

Metabolism comprises an astonishingly complex series of interrelated pathways, responsible for energy generation, respiration, and the buildup and breakdown of biomolecules.

References

Denniston, K., Topping, J., and Quirk Dorr, D. (2020). *General, Organic, and Biochemistry*, 10e. McGraw Hill.

Edelstein, S.J. (1973). *Introductory Biochemistry; Fundamentals of Cellular Metabolism and Molecular Biology*. San Francisco: Holden-Day.

Garrett, R.H. and Charles, M. (2017). *Grisham "Biochemistry,"* 6e. Cengage Learning.

Gatto, G., Berg, J.M., Stryer, L., and Tymoczko, J. (2019). *Biochemistry*, 9e. McMillan.

Nelson, D.L. and Cox, M.M. (2021). *Lehninger Principles of Biochemistry*, 8e. MacMillan.

Further Reading

Cortassa, S. (2011). *Introduction to Metabolic and Cellular Engineering*, 2e. World Science (Singapore).

MacDonald, C.T., Gibbs, J.H., Pipkin, A.C. (1968). Kinetics of biopolymerization on nucleic acid templates. *Biopolymers* 6 (1): 1–25. https://doi.org/10.1002/bip.1968.360060102. PMID 5641411.

Heinrich, R. and Rapoport, T.A. (1980). Mathematical modelling of translation of mRNA in eucaryotes; steady states, time-dependent processes and application to reticulocytest. *Journal of Theoretical Biology* 86 (2): 279–313. https://doi.org/10.1016/0022-5193(80)90008-9. PMID 7442295.

Skjndal-Bara, N. and Morris, D.R. (2007). Dynamic model of the process of protein synthesis in eukaryotic cells. *Bulletin of Mathematical Biology* 69 (1): 361–393. https://doi.org/10.1007/s11538-006-9128-2. PMID 17031456. S2CID 83701439.

Coyte, K.Z., Tabuteau, H., Gaffney, E.A., et al. (2017). Reply to Baveye and Darnault: useful models are simple and extendable. *Proceedings of the National Academy of Sciences*. 114 (14): E2804–E2805. Bibcode:2017PNAS.114E2804C. https://doi.org/10.1073/pnas.1702303114. PMC 5389313. PMID 28341710.

Morozova, N.; Zinovyev, A., Nonne, N., Pritchard, L.-L., Gorban, A.N., Harel-Bellan, A. (2012). Kinetic signatures of microRNA modes of action. *RNA* 18 (9): 1635–1655. https://doi.org/10.1261/rna.032284.112. PMC 3425779. PMID 22850425.

Problem

1 A certain reaction is $A + B \rightarrow C + D, \quad \Delta G = +2.8$ kcal/mol.
The hydrolysis of ATP gives, $\text{ATP} \rightarrow \text{ADP} + P_i + \text{H}^+, \quad \Delta G = -7.3$ kcal/mol.
The reaction takes place at $T = 300$ K.

a) If 1 mol each of A and B are initially present, find the equilibrium concentrations of A, B, C, and D, in the absence of ATP.

b) If the energy of ATP can couple to this reaction, find the equilibrium concentrations of A, B, C, and D. Assume that the concentration ratio for ATP is ATP/ADP = 500. Explicitly state by what factor the presence of ATP has shifted the equilibrium.

Part II

Physical Processes Underlying the Living State

Biophysics: Physical Processes Underlying the Living State, First Edition. Wayne F. Reed.

Companion website: www.wiley.com/go/Reed/BiophysicsandMacromolecules

4

Thermodynamics, Reaction Kinetics, and Information Theory

In classical mechanics, the motion of two bodies interacting via any given forces, such as electromagnetic, gravitational, and friction, can be described exactly and the future of the system predicted indefinitely. When a third body is added the problem becomes considerably more difficult and is analytically solvable only under a restricted set of conditions, and when many bodies are added the system is analytically hopeless. However, contemporary computing power is sufficient to keep track of the interactions and trajectories of huge numbers of particles. This power has been harnessed, for example, in *molecular dynamics simulations*, which can follow the evolution of very large numbers of interacting particles under given forces. Such simulations are of great value in many areas, including biophysics and materials science.

Consider a room full of air molecules. The number of molecules is immense ($\gg 10^{23}$) but with enough computing power, we could conceivably use molecular dynamics simulation to predict the trajectory of each molecule in time. But what insight would that give us into fundamental principles of the system? Basically, none.

Thermodynamics incorporates the principle that to predict the macroscopic properties of systems containing many molecules an efficient means is to find distributions and average values of a few key physical properties of the system, rather than the details of the vast number of individual molecules. These properties include *extensive variables* of the system, such as volume, number of particles, energy, and entropy, and *intensive variables* such as temperature, pressure, and density.

Thermodynamics is fundamental to the living state since the flow of energy and heat and the effect of interacting particles shape much of its behavior.

4.1 Thermodynamically Based Forces

While the electromagnetic force is the only fundamental force having extensive impact in biological systems, thermodynamic principles lead to forces and effects of their own, which occur in the absence of direct action by any of the fundamental forces. The *ideal gas* is a good example of this because ideal gas particles are simply point objects of mass m, and occupy no volume, and don't interact with each other except by elastic point-like collisions. The ideal gas concept allows some of the most basic and subtle points of thermodynamics to be understood. The ideal gas law can be stated in several forms. A convenient one for our purposes is

$$PV = NRT \qquad 4.1$$

where P is the pressure, V is the volume, N is the number of moles of gas, R is the universal gas constant (8.3145 J/mol-K), and T is the temperature (in Kelvin).

Ideal osmotic pressure in liquid systems runs a close parallel to ideal gases and drives many living state processes, so that understanding ideal gases enhances understanding of osmotic and other thermodynamic phenomena in the living state.

Ideal gases become "real gases" when interactions such as electrostatic forces and excluded volume effects come into play. Electrostatic, excluded volume, and other interactions among molecules and macromolecules in a liquid produce effects on osmotic pressure and other thermodynamic quantities that have important living state consequences.

Biophysics: Physical Processes Underlying the Living State, First Edition. Wayne F. Reed.

Companion website: www.wiley.com/go/Reed/BiophysicsandMacromolecules

Nonideality underlies the living state and leads to its complexity and advanced functionality. Nonideality embodies interactions among particles which leads to phases, recognition of chemical identity, self-organizing structures, and far-from-equilibrium dynamics, which are all necessary for the living state.

4.2 Thermodynamic Laws

Thermodynamics has a checkered history of pitfalls, misconceived theories, and often hard-to-grasp concepts. Some ancient Greeks, including Empedocles and Aristotle, considered fire among the four essential elements, which also included air, water, and earth. Johann Becher (Germany, 1635–1682) in 1667 displaced these ideas and proposed the existence of "phlogiston," a substance within flammable bodies liberated upon combustion. This theory was elaborated upon for over a century by numerous others, until Antoine Lavoisier published a theory in 1789 of heat based on a fluid called "caloric," which passed from one body to another, altering temperatures of the bodies. Caloric was conserved in this theory. The idea paralleled that of electrical charge being a fluid with two types, negative and positive, where the total amount was conserved; charge conservation still stands as a fundamental law, whereas the caloric theory was soon overthrown.

Famously, Count Rumford (born Benjamin Thompson in the United States, 1753–1814) effectively destroyed the caloric theory when he noticed, while working in Bavaria in 1798, that boring cannons led to temperature rises in the boring bits and the cannon. He deduced that heat was being *produced* by mechanical work and was not a conserved quantity passed from one body to another. To the grudging astonishment of many, he showed that water could be boiled from the heat produced in mechanical boring, the first deliberate demonstration that water could be boiled by means other than fire or other nonmechanical sources of heat. In fact, the reverse, the practice of turning heat into mechanical energy had long been practiced.

Thermodynamics became an intense field of study for very practical reasons. The steam engine was arguably the device that almost single-handedly ushered in the Industrial Revolution, and it was imperative to establish the then-mysterious principles of converting steam from boiling water into useful mechanical work. Sadi Carnot (Luxembourg/France, 1796–1832) in the 1820s made major advances concerning heat engines; ironically, he successfully used the abandoned caloric theory; another instance in which essentially flawed ideas can still produce important results.

Count Rumford's observations set the stage for not only the idea of heat as energy, but also for the notion of entropy. Namely, boring a cannon takes organized, macroscopic, "high quality" rotational mechanical energy from the boring device and converts or "degrades" it into heat, that is, increased random motion in the cannon and boring device, while breaking metallic bonds as metal shards are released. Entropy is a measure of disorder, so the boring process has increased entropy. We will give expressions for entropy shortly, but this notion of degradation of "high quality," organized energy into random, disorganized energy is one of the fundamental principles underlying the living state. Along the path from organized to disorganized energy *order and structure can be created*, even though there is a net increase in disorder of the entire system. Imagine the path of light from the sun and through a photosynthetic cycle building up high-energy molecules and organized structures in the living state, on its way to thermal degradation and added entropy to the environment.

Some of the major findings in thermodynamics have been codified in the laws of thermodynamics. Each of these laws has several ways of being stated and have frequently led to confusion as to their meaning. The term "zeroeth law" was not even coined until the twentieth century, well after the first and second laws had been established in the nineteenth century. The 'fourth law', based on non-equilibrium thermodynamics, is only informally a thermodynamic law.

Thermodynamics is an excellent illustration of how misconceptions arise and die, how new concepts are hotly contested, even ridiculed, how various formulations of a principle reveal different perspectives on it, how different formulations are shown to be actually equivalent, and how, historically, a field develops in a messy way and can lead to wholly unanticipated new discoveries; Max Planck, essentially founded quantum physics while contemplating the very classical thermodynamic problem of explaining the origin of the continuous electromagnetic spectra of glowing objects.[1]

At the same time, we might ponder how some phenomena that are obvious to many become canonized in science only after lengthy periods of debate and misconception. Even before the Neolithic Revolution, humans have understood, from practice, that mechanical action in the form of friction produces heat, as they lit fires by rubbing sticks together. And even before that humans knew that rubbing their hands together produced heat. And yet it has taken over 12 millennia for humans to formalize the connection between motion, friction, heat, and temperature.

1 The glowing objects are referred to in most texts as "black bodies," since ideally, they absorb all electromagnetic radiation incident upon them.

Pre-Neolithic thermodynamics. *Count Rumford recovers prehistoric knowledge; frictional motion produces heat.* (Painting by Linnette F. Reed).

4.2.1 "Zeroeth" Law

A simple statement of this is that bodies at different initial temperatures will exchange heat when placed in thermal contact with each other until they reach the same temperature, which is considered thermodynamic equilibrium.

4.2.2 First Law

The heat added to a system Q goes into work W, plus change in internal energy ΔU of the system

$$Q = W + \Delta U \tag{4.2}$$

The sign convention is that $Q > 0$ if heat flows into the system, and $Q < 0$ if it flows out. $W > 0$ when the system does work, and $W < 0$ when work is done on the system.

The first law is an expression of the law of conservation of energy, first expounded by Hermann von Helmholtz in 1847. Count Rumford's demonstration of mechanical energy producing heat led James Prescott Joule (England, 1818–1889) to both Joule's own formulation of energy conservation and to his demonstration of the work equivalent of heat in 1849; 1 calorie of heat equals 4.18 Joule of energy. The calorie and *Joule* had been defined independently of each other until Joule unified them. Shortly thereafter, in 1850, both Rudolf Clausius (Germany, 1822–1888) and Lord Kelvin (United Kingdom, born William Thomson, 1824–1897) published their conclusion that it is neither heat nor work that is conserved, rather the sum of the two, or the "energy." Interestingly, Julius Mayer (Germany, 1830–1895) had already enunciated the "mechanical equivalent of heat," and found the same relationship as Joule, but his work was largely overlooked.

4.2.3 Second Law

There are several different ways of expressing the second law. They generally involve the concept of entropy and the distinction between useful work and heat. One statement is that heat always flows spontaneously from a hotter object to a colder one, and never vice-versa, i.e. heat can't be transferred from a colder to a hotter body without doing work. Clausius and Kelvin made equivalent second law statements based essentially on this notion.

Another statement is that the net entropy of a system and its environment will either stay the same or increase as the result of a thermodynamic process.

$$\Delta S_{total} = \Delta S_{system} + \Delta S_{environment} \geq 0 \text{ in a thermodynamic process} \quad 4.3$$

Since entropy, S, represents disorder, this statement means that no *net* order can be created in a process, but a part of the entire system can gain order at the expense of an equivalent or larger amount of disorder in the rest of the system. This is fundamental for understanding how, for example, a mother can gestate an increasingly organized embryo ($\Delta S_{system} < 0$), while the net entropy of mother plus her environment increases.

In applying thermodynamic principles, an important distinction is made between *reversible* and *irreversible* processes. One example of a reversible process is when a system passes "quasi-statically" from the initial state to the final one, with no frictional or other dissipative forces acting during this passage. This means that at each instant, the system is in an equilibrium state between the initial and final ones, and that the system can reverse its direction at any instant. One example is heating a system with an external heat source whose temperature increases gradually, such that the temperature, T, of the heat source is only infinitesimally larger than T of the system. An example is a gas in a cylinder expanding reversibly and doing work on a piston, i.e. the gas expands so slowly that the heat source could drop its T infinitesimally at any instant and the piston would reverse direction. This quasi-static process is similar to those in mechanics, where one force applied to an object balances any other forces on it, so that the net force is only infinitesimally greater than zero and the object moves without accelerating; an example is applying an upward force to a mass, just a tiny increment more than its weight, so that the mass moves upward, acquiring potential energy, but without acquiring any kinetic energy. It is emphasized that in a reversible process, no dissipative work is done, due to nonconservative processes such as mixing, friction, Joule heating in a resistor, etc. A fully reversible process is often just a convenient idealization and approximation to a real process.

In contrast, an irreversible process is one in which the system is not in any equilibrium state between the initial and final state, and the process can also involve dissipative forces, such as friction. Friction is the effect of converting organized motion irreversibly into randomized motion, such as the boring of a cannon, or the sliding to rest of a moving object on a rough surface.

An example of irreversibility is a vessel initially containing hot gas molecules in one portion and cold molecules in another. These will irreversibly mix and share energy until the temperature is uniform throughout the vessel (zeroeth law of thermodynamics). The second law of thermodynamics asserts that the molecules cannot spontaneously reseparate into hot and cold regions in the vessel. This links to the notion of the *thermodynamic arrow of time,* that systems evolve toward greater entropy but never in the reverse direction, e.g. a mountain crumbles to sand over time, but the sand particles never spontaneously re-organize themselves into the mountain.

Nonetheless, non-dissipative dynamic equations of motion are time reversible, in that time t, can be replaced by $-t$ in solutions and still obey the equations. This led to a vexing paradox in the nineteenth century as to why thermodynamic systems are generally irreversible, while individual particles obey reversible dynamics. This exercised nineteenth-century scientists, such as Maxwell, Lord Kelvin, Henri Poincaré (France, 1854–1912), Clausius, and Ludwig Boltzmann (Austria, 1844–1906), each of whom offered a plausible explanation. As we shall soon see, entropy was defined statistically by Boltzmann and related to the number of states available to a system. As the number of states grows and particles occupy the increasing number of available states, i.e. as entropy increases, there is statistically essentially no chance that they will move spontaneously back to occupying just the set of states available before the entropy increase. Time reversal symmetry and its breaking are further considered in Chapter 6.14.

Yet another paradox stemming from the second law was proposed by Maxwell. His thought experiment consisted of two chambers of gas separated by a trap door that could open into only one of the chambers, which were initially at the same temperature. He proposed a tiny entity, subsequently called "Maxwell's Demon," that would open the trap door every time a hot (fast) molecule approached it. In this way, the gas would be sorted into a hot side and cold side and the second law of thermodynamics would be violated. This caused consternation among many scientists, but the conclusion is that the imagined situation is physically impossible. However, it connects thermodynamics and *information theory* in an interesting way, and so the elusive Maxwell's Demon is still debated to this day.

Clausius provided the classical definition of entropy involving reversible heat flow $Q_{reversible}$, as follows:

$$\Delta S = \int \frac{dQ_{reversible}}{T} \quad \text{Classical Definition of Entropy} \quad 4.4$$

Entropy is a *state function*, which means it has a definite value for a state in thermodynamic equilibrium, which also implies that the entropy change of a system in a cyclic process – one that starts and ends in the same state – is zero. This does not contradict the second law because in an irreversible process, the entropy increase in the cyclic process is born by the system's environment.

Some of the confusion concerning entropy is avoided by defining *isolated*, *closed*, and *open* thermodynamic systems. An isolated system allows no exchange of matter or heat with its environment. A closed thermodynamic system allows heat exchange with the environment but does not allow matter to enter or exit the system. In an open system, heat and matter can pass from the thermodynamic system to another one, which can include both other finite systems or the entire surrounding environment. In this case, the net change in entropy is the sum of the entropy change of the system, and any other systems, it is in thermal contact with. In the case of open systems, the net entropy change in a thermodynamic process must be zero or greater. However, the entropy of one or more components of the entire system and surroundings may be less than zero. Hence, an outside force can compress a cylinder of gas, decreasing its entropy, while increasing its own, such that the net entropy is greater than or equal to zero. Living organisms are open systems because they exchange heat and matter with their surroundings.

To elaborate further on the difference between reversible and irreversible processes consider the following: In a reversible cycle, the net entropy change of the system and surroundings is zero. In an irreversible cycle, while the entropy of the system will be the same after the cycle, entropy will have been created during the cycle and passed off to the environment. Hence, while the entropy of the system remains well defined even in an irreversible process the amount passed off to the environment depends on the nature of the irreversible process.

A summary statement is (i) while energy is neither created nor destroyed (first law) (ii) net entropy is created in irreversible processes and is never thereafter destroyed.

4.2.4 Third Law

There are several statements of the third law. One of the first formulations was by Walther Nernst (Germany, 1864–1941) in 1912, which stated that it is impossible to reach absolute zero (0 K) in a finite number of cooling steps. In his dissertation of 1905, he stated that as a system approaches 0 K it approaches a unique, minimum value of entropy, regardless of pressure, magnetic and electric fields, and other external forces.

Closely related is the statement by Gilbert Lewis (United States, 1875–1946) and Merle Randall (United States, 1888–1950) in 1923 that the entropy of a perfect crystal is zero at 0 K. This derives from the notion that there is only a single state in which every part of the system is at its minimum energy and is hence at a unique energy minimum, equal to zero. The fact that the entropy is zero comes from the statistical interpretation of entropy, introduced in Section 4.7. Other systems, such as ice, have multiple states corresponding to the minimum energy level at $T = 0$ K and so do not have zero entropy.

A statement similar to the first, but based on the second statement of the third law, is that the minimum entropy of a system cannot be reached in a finite number of cooling steps. Zero Kelvin and minimum entropy can only be asymptotically approached in successive cooling steps.

4.2.5 Free Energy

Free energy of a system is a fundamental concept. Simply put, it is the amount of energy in a system that is available to do useful work. Looked at another way, a system will move spontaneously toward its state of lowest free energy, its equilibrium state. The time to reach this state is a kinetic issue, not answered by equilibrium thermodynamics. We will briefly consider kinetics toward the end of this chapter and several subsequent chapters.

The Helmholtz free energy, A, is $A = U - TS$, where U is the internal energy, T is the temperature, and S is the entropy.[2] Gibbs free energy G of a system, named after Josiah Willard Gibbs (United States, 1839–1903), also includes pressure and volume, and hence the ability to do mechanical "PV" work. It is given by

$$G = U + PV - TS \tag{4.5}$$

where P is the pressure and V is the volume. Systems can move spontaneously in directions that lower G. Minimization of G is a central feature in analysis and simulation of many physical systems, including living state systems.

Consider the total differential of G for a system with a fixed number of particles

$$\mathrm{d}G = \mathrm{d}U + P\mathrm{d}V + V\mathrm{d}P - T\mathrm{d}S - S\mathrm{d}T \tag{4.6}$$

This shows that changes to each system variable can raise or lower dG. It also shows that an increase in entropy alone can lower dG, and hence drive a process. Later in this chapter, the number of particles in the system will be allowed to vary,

2 "A" stands for the work function, Arbeit = work in German.

which leads to the fruitful concept of *chemical potential*, from which the properties of osmotic pressure can be derived, as well as phase equilibria, and other phenomena.

The Gibbs energy is a potential energy, in that each term has the potential to do work and/or exchange heat.

4.3 Thermodynamic Processes Involving Ideal Gases

Via entropically driven forces ideal gases can do work without involving any fundamental forces.

4.3.1 Reversible Isothermal Expansion of an Ideal Gas

Consider an ideal gas in a cylinder with a moveable piston, into which heat enters from the environment at temperature T_e, such that the temperature of the gas T remains constant. The gas at constant T expands from volume V_0 to final volume V_f, while no gas enters or leaves the cylinder. The work done by the gas in this process is

$$W = \int P\mathrm{d}V = NRT\int_{V_0}^{V_f} \frac{\mathrm{d}V}{V} = NRT\ln(V_f / V_0) \qquad 4.7$$

where Equation 4.1 was used to express P in terms of N, R, and T. W increases logarithmically with V_f, so that, for example, increasing the initial volume V_0 10-fold, $V_f = 10V_0$ only increases W by 2.3 times.

Figure 4.1 shows a complete thermodynamic cycle on a pressure–volume diagram. The reversible isothermal expansion is the top path, as marked. The diagram represents the *Carnot cycle* to which we will return in Section 4.3.4.

As we will see shortly, the internal energy of the gas is proportional to T. Because T of the gas does not change the internal energy, U, remains constant for the isothermal expansion so that the work done comes directly from the heat put in from the environment, and the first law yields $Q = W$. This work has been done at the expense of increased entropy of the

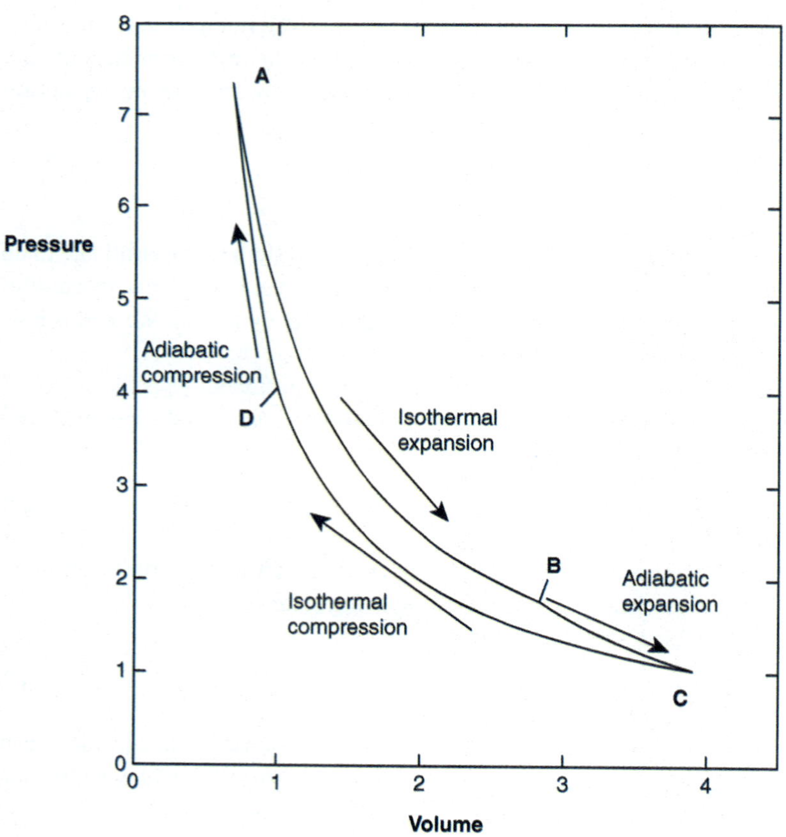

Figure 4.1 The Carnot cycle. The Adiabatic curves are given by PV^{γ} = constant, where $\gamma = C_P/C_V$, C_P and C_V are the heat capacities at constant pressure and volume, respectively.

gas in going from less available volume to more via the heat influx. If the gas expansion is reversible, that is, it proceeds slowly enough that it is at every instant in an equilibrium state going from its initial to final state, then the entropy is

$$\Delta S_{gas} = \int \frac{dQ}{T} = +Q/T = NR\ln(V_f/V_0), \text{ reversible isothermal expansion} \qquad 4.8$$

The gas in this final state cannot spontaneously recompress itself isothermally to its initial state (decrease its entropy), that is, it cannot spontaneously eject heat to the environment (the heat source in this case), which must be at least at an incrementally greater temperature, $T_e = T + dT$, than the gas in order for heat to flow into the gas. This reflects one of the several statements of the second law of thermodynamics that a system cannot by itself decrease its own entropy, or, the entropy of a system cannot spontaneously decrease.

Because heat flows out of the environment during the process, and the temperature of the environment remains constant at T_e, $\Delta S_{environment} = -Q/T_e$, so the entropy of the environment *decreases*. The net entropy of the gas and the environment for the isothermal expansion is

$$\Delta S_{total} = \Delta S_{environment} + \Delta S_{gas} = -Q/T_e + Q/T \geq 0 \qquad 4.9$$

This can approach zero as dT of the environment above T is made arbitrarily small. For $T < T_e$, $\Delta S > 0$.

To recompress the gas, an external agent must do work on the expanded gas to recompress it and reduce its entropy. The entropy change of the external agent will be greater than or equal to zero in the process.

Because T remains constant in this process, and hence also U and PV, examining dG in Equation 4.6, and expanding the infinitesimal dG to a small finite difference ΔG, shows that $\Delta G = -T\Delta S$. That is, the decrease in free energy that makes this process possible occurs solely due to the increase in entropy. The isothermal expansion of an ideal gas can hence be thought of as a process driven by entropy change.

4.3.2 Reversible Adiabatic Expansion of an Ideal Gas

If the gas cylinder is isolated from its environment then no heat exchange will occur if the gas is allowed to expand, i.e. $Q = 0$ during the expansion. This process is shown in the P–V diagram in Figure 4.1. Even though there is no heat flow in or out of the cylinder the piston can still do work at the expense of its own internal energy. In such an *adiabatic process*,[3] the first law yields $W = -\Delta U$. In this case, the negative change in internal energy of the gas corresponds to a drop in its temperature accompanying the work done on the piston. In this case, since $Q = 0$,

$$\Delta S_{gas} = 0 \text{ in a reversible adiabatic expansion or compression} \qquad 4.10$$

4.3.3 Irreversible, Adiabatic Free Expansion of an Ideal Gas

An example of this is letting an initially compressed gas escape into a vacuum of larger volume. In this case, the intermediate states between the compression and final expansion are not equilibrium states, as they would be if there were a piston or other device controlling the expansion. Because no work is done ($W = 0$), and the expansion is adiabatic ($Q = 0$), the first law yields $\Delta U = 0$. Hence, there is no temperature change for the irreversible adiabatic expansion. In the reversible case, the temperature dropped because internal energy was converted to work.

4.3.4 The Carnot Cycle

The Carnot cycle, shown in Figure 4.1, is a reversible cycle on the PV diagram that describes the steps used by an efficient heat engine. A heat engine is one that takes heat Q_H from a hot source at temperature T_H (e.g. steam or the explosion in an

3 The definition of an adiabatic (from Greek for "impassable") process is one in which there is no heat transfer between a system and its surroundings. This is often approximated in processes fast enough that there is no time for heat exchange with the environment, e.g. the rapid, free expansion of gas.

internal combustion engine cylinder) turns it to useful mechanical energy W, and then exhausts waste heat Q_L to the environment at temperature T_L.

The cycle is defined by (1) an isothermal expansion, (2) an adiabatic expansion, (3) an isothermal compression, and (4) a final adiabatic compression that brings the system back to the original state PV. Heat is absorbed in (1) and ejected to the atmosphere in (3). A piston can perform work on the power strokes (1 and 2), after absorbing heat in (1), and then require input work while exhausting the heat to the environment in (3), and being adiabatically compressed back to its starting point at A.

A thermodynamic analysis of this idealized cycle puts a limit on the efficiency e, of heat engines, as

$$e = \frac{W}{Q_H} = 1 - \frac{T_L}{T_H} \tag{4.11}$$

It is noted that in a diesel engine advantage is taken of the gas heating up in an adiabatic compression stroke to reach ignition temperature of the fuel, thus eliminating the need for a spark plug.

Equation 4.11 shows that the highest efficiency is obtained by making the ratio T_L/T_H as small as possible. The limit on efficiency imposed by the Carnot cycle means that a 100% efficient heat engine is not possible, since, by the third law of thermodynamics, it is impossible to cool the environment to absolute zero in a finite number of steps, so that T_L is always greater than zero. Practically speaking, T_L is generally around 300 K give or take 10%, or so. With this constraint, higher T_H leads to higher efficiency. To increase efficiency it becomes a matter of metallurgy and materials engineering to make engine parts that can withstand very high T_H.

4.4 Connection Between Ideal Gas Law and Molecular Kinetic Energy

We have used temperature T, as a phenomenological parameter for the ideal gas law and associated processes. In fact, there is a direct connection between macroscopically measured temperature and the microscopic motion of a particle. This is the beginning of the *kinetic theory of gases*. Let's rewrite the ideal gas law as

$$P = \frac{NRT}{V} = \frac{nRT}{N_A} = nk_BT, \quad \text{where } k_B = \frac{R}{N_A} \tag{4.12}$$

where N is the number of moles, $n = NN_A/V$ is the number density of gas particles, and $N_A = 6.02 \times 10^{23}$ is Avogadro's number (particles/mole). We just introduced the factor k_B, which is *Boltzmann's constant* and is of immense importance in all situations where $T > 0$ K. This constant, named after Ludwig Boltzmann, is the proportionality factor between the gas constant R and Avogadro's number N_A; N in Equation 4.1 is in units of moles and R in J/mol-K. To change moles to number of particles is simply $N \times N_A$ so that combining with R

$$k_B = \frac{R}{N_A} = 1.38\times10^{-23}\,\text{J/K} \quad \text{Boltzmann's Constant} \tag{4.13}$$

Because of its ubiquity, we will reserve k_B uniquely for Boltzmann's constant throughout this book. Among its many appearances, it can always be used to switch between moles and number of particles whenever needed.

The ideal gas law, Equation 4.1, states that pressure is directly proportional to both N and T. If we think of the molecular origins of the pressure of a container of ideal gas, we see that the pressure must be due to the elastic collisions of the gas particles with the walls of the container. The fact that P is proportional to N means that the total pressure P must be the product of N times the average pressure produced by a single particle upon collision with the container wall.

Let the container be a cube of side L, as shown in Figure 4.2, with particles of mass m and number density n. Consider first one dimension, say x, with an initial velocity component $-v_{x,i}$, and corresponding initial x component of momentum $p_{x,i} = -mv_{x,i}$. When the particle bounces elastically off the wall in the YZ plane the change in its x component of momentum is

$$\Delta p_x = 2mv_x \tag{4.14}$$

Figure 4.2 Ideal gas atoms in a rigid cubical container of side L. The x components of momentum of one atom before and after a collision are shown.

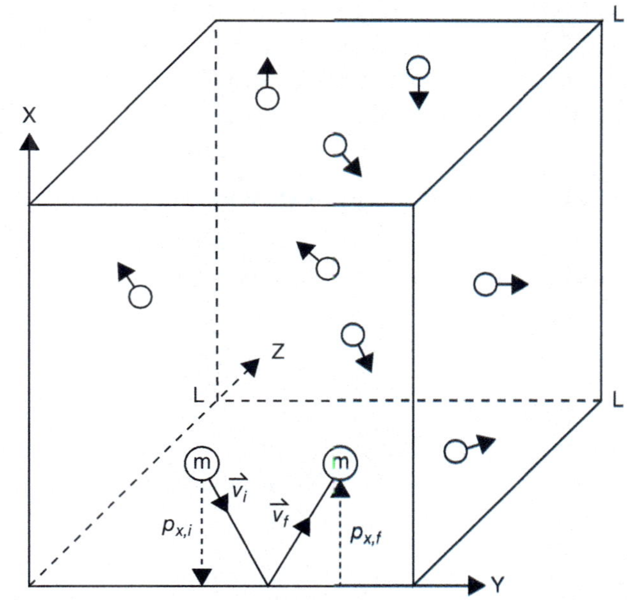

where we dropped the "i" subscript, because the speed of the particle does not change upon the elastic collision, i.e. $|v_{x,i}| = |v_{x,f}| = v_x$. The force the single particle will exert on the wall F_{sp} depends on the average time between collisions, which is the time for it to travel back to the wall after the collision, $\Delta t = 2L/v_x$. The average pressure the single particle exerts on the wall P_{sp} is hence

$$P_{sp} = \frac{F_{sp}}{A} = \frac{1}{L^2}\frac{\Delta p_x}{\Delta t} = \frac{m\langle v_x^2\rangle}{L^3} = \frac{m\langle v^2\rangle}{3L^3} \tag{4.15}$$

where the brackets < > denote the average value of the quantity in the brackets. Here the isotropy of space is used to replace v_x^2 with the particle's mean square velocity $\langle v^2\rangle$, since

$$\langle v^2\rangle = \langle \vec{v}\cdot\vec{v}\rangle = \langle v_x^2\rangle + \langle v_y^2\rangle + \langle v_z^2\rangle \tag{4.16}$$

and $\langle v_x^2\rangle = \langle v_y^2\rangle = \langle v_z^2\rangle$ by the isotropy of space. The total pressure P is just P_{sp} times the total number of particles, N_p, in the volume L^3, $N_p = nL^3$, so that, using the empirical ideal gas law and its connection to temperature, yields

$$P = \left(nL^3\right)P_{sp} = \frac{nm\langle v^2\rangle}{3} = nk_BT \tag{4.17}$$

This yields a fundamental connection between a particle's mean square speed and temperature:

$$\langle v^2\rangle = \frac{3k_BT}{m} \tag{4.18}$$

This can also be expressed in terms of the particle's average kinetic energy $\langle KE\rangle$ as

$$\langle KE\rangle = \frac{m\langle v^2\rangle}{2} = \frac{3k_BT}{2} \quad \text{Relationship between Particle Kinetic Energy and Temperature} \tag{4.19a}$$

The root-mean-square speed, v_{rms}, is given by

$$v_{rms} = \langle v^2\rangle^{1/2} = \sqrt{\frac{3k_BT}{m}} \tag{4.19b}$$

For example, $v_{rms} = 485$ m/s for O_2 at $T = 300$ K.

In this, we have assumed that the only kinetic energy the particle may have is translational kinetic energy. For molecules, which can rotate and vibrate, these are extra degrees of freedom with corresponding contributions to total <*KE*>. Hence, Equations 4.19a,b apply strictly to atoms and other particles without rotation or vibration. We will revisit this briefly in Section 4.5.3 on the *equipartition principle*.

4.4.1 Ideal Gas Specific Heats Derived from Relation of <*KE*> to *T*

For an ideal monatomic gas, the internal energy is simply the average kinetic energy <*KE*> per atom (Equation 4.19a) times the number of moles of gas in the system. For *N* moles of a monatomic gas:

$$U = \frac{3}{2}NRT \tag{4.20}$$

The heat capacity at constant volume C_v can be computed immediately from this, since, by the first law $Q = U + \int P dV$. Because V is constant $Q = U$ so that the heat capacity, with units of J/mol-K is

$$C_v = \frac{1}{N}\frac{\partial Q}{\partial T}\bigg|_V = \frac{1}{N}\frac{\partial U}{\partial T} = \frac{3}{2}R \tag{4.21}$$

Similarly, using the first law at constant *P* to compute heat capacity at constant pressure C_p, yields:

$$C_P = \frac{1}{N}\frac{\partial Q}{\partial T}\bigg|_P = \frac{1}{N}\frac{\partial U}{\partial T} + \frac{1}{N}\frac{\partial}{\partial T}\int P dV\bigg|_{P=constant} = \frac{5}{2}R \tag{4.22}$$

where polyatomic molecules are concerned there will also be rotational and vibrational contributions to the internal energy which will increase the value by $k_BT/2$ per degree of freedom (Section 4.5.3).

4.4.2 Thermal Energy, k_BT

We just saw that k_BT is associated with the kinetic energy of atoms and this is often termed *thermal energy* and provides a convenient benchmark; systems with thermal energies of just a few k_BT and lower are subject to thermal fluctuations, which affect the making and breaking of weak electrostatic bonds, or the wiggling fluctuations of a strand of DNA. Energies well above k_BT, such as covalent bonds, are virtually impervious to being broken by thermal fluctuations. In larger scale structures, such as biological membranes or the hydrogen bonds holding two strands of DNA together, each individual interaction may be only on the order of k_BT, but when many interactions work cooperatively the total binding energy in the system can far exceed k_BT and hence stabilize noncovalent structures. Structures stabilized this way, such as DNA double helices and globular proteins, can go through a phase transition as T increases and weakens the individual low-energy bonds. The phase transition turns the DNA and globular proteins into disordered, randomized structures, *random coil polymers* (Chapter 8.1).

Another important aspect of thermal energy, k_BT, is its relationship to forces in the nanoworld of biological macromolecules. The nanometer, nm, is the approximate scale for phenomena involving these molecules. The associated *thermal force* over a nanometer is $k_BT/(10^{-9}$ m). At 300°C, this is ~4×10^{-12} N, or 4 piconewtons, abbreviated as 4 pN. This is indeed the magnitude of forces associated with unzipping DNA strands, unfolding proteins, and stopping the motion of molecular motors. When molecular motors work together 10's or 100's of pN of force can be generated.

4.5 The Boltzmann Distribution

Just as k_BT will resound throughout this book, the *Boltzmann distribution* will play an equally important role. Above, we found the connection between the *average* kinetic energy and *T*. Of course, the gas molecules don't have a single speed since the total internal energy *U* of the gas is divided among the molecules, and random collisions among them transfer kinetic energy, and hence speed, among them in a random way.

For the moment, we will just make the Boltzmann distribution heuristically plausible. First, because all directions in space are equally likely we expect the average velocity to be zero, $\langle \vec{v} \rangle = 0$. But the mean square velocity is proportional to kinetic energy and so is obviously not zero, $\langle v^2 \rangle \neq 0$. Furthermore, the isotropy of space suggests that the probability of any value in the distribution of mean square speed v^2 must be symmetrically distributed in all directions. The simplest distribution with this property is the Gaussian distribution (see Appendix I), which in this context will be

$$W(v^2) = A\exp\left(\frac{-v^2}{2\sigma^2}\right) \tag{4.23}$$

where $W(v^2)$ is the normalized probability per unit volume of *velocity space* that a particle has the square velocity, v^2, σ is the width of the distribution, and A is the normalization constant, which ensures that integrating $W(v^2)$ over all velocity space yields 1. As T increases, Equation 4.18 shows that $\langle v^2 \rangle$ increases linearly as k_BT/m, so that the distribution should broaden proportionately, suggesting that

$$\sigma^2 = \frac{k_BT}{m} \tag{4.24}$$

This leads to a *normalized distribution*[4] (see Appendix I for some basic features of probability distributions):

$$W(v^2) = \left(\frac{m}{2\pi k_BT}\right)^{3/2} \exp\left(-\frac{mv^2}{2k_BT}\right) \tag{4.25}$$

Since $KE = mv^2/2$, this can be rewritten as

$$W(v^2) = \left(\frac{m}{2\pi k_BT}\right)^{3/2} \exp\left(-\frac{KE}{k_BT}\right) \tag{4.26}$$

This distribution is generalizable to *any* type of energy, including gravitational potential energy, and free energy. Let E be any type of energy, or sum of energies, that defines a system, then the Boltzmann distribution is

$$W(E) = A\exp\left(\frac{-E}{k_BT}\right) \tag{4.27}$$

where A is now a normalization constant that depends on the type of energy involved in the system.

A direct extension to the ideal gas is to consider the distribution in a *constant* gravitational field g, in the $-y$ direction and consider the total energy E as the sum of kinetic energy $mv^2/2$ and potential energy mgy, then:

$$W(E) = W(v^2, y) = A\exp\left(-\frac{mv^2/2 + mgy}{k_BT}\right) = A\exp\left(-\frac{mv^2}{2k_BT}\right)\exp\left(-\frac{mgy}{k_BT}\right) \tag{4.28}$$

If the temperature is constant going upward then the speed distribution, the first exponential in the last term, is independent of height, but the second term shows the probability of being at height y falls off exponentially with height.[5]

4.5.1 Continuous Probability Distributions

The statement was made above that "$W(v^2)$ is the normalized probability per unit volume of *velocity space* for a particle with a squared speed of v^2". Why is $W(v^2)$ a probability *density*, and not just a probability? Suppose $P(x_i,y_j,z_k)$ is the probability of something occurring at a discrete point in space (x_i,y_j,z_k). Normalization requires that the probability summed over all values of x_i,y_j,z_k from negative infinity to positive infinity be unity, i.e. the something has to be somewhere in space.

$$\sum_{k=-\infty}^{\infty}\sum_{j=-\infty}^{\infty}\sum_{i=-\infty}^{\infty} P(x_i, y_j, z_k) = 1 \tag{4.29}$$

4 A normalized distribution sums to unity, i.e. for a continuous distribution $W(x)$, $\int W(x)\mathrm{d}x = 1$, and for a discrete distribution P_i, $\sum P_i = 1$.

5 In fact, temperature normally decreases as altitude increases due to both the decreasing air density and the so-called *adiabatic lapse rate* for a parcel of air expanding and cooling as it rises, without exchanging heat with the atmosphere.

If we now treat space as a continuum, rather than a series of discrete points, then the probability distribution passes from a discrete to a continuous distribution $P(x_i, y_j, z_k) \to W(x, y, z)$, where $W(x,y,z)\mathrm{d}x\mathrm{d}y\mathrm{d}z$ is the probability that the something is in the volume defined by x to $x + \mathrm{d}x$, $y + \mathrm{d}y$, and $z + \mathrm{d}z$. The units of $P(x_i, y_j, z_k)$ and $W(x,y,z)$ are different, and we have emphasized this by changing the name of the probability "P" to that of a probability density "W." [6] You can see that $W(x,y,z)$ has to be a probability per volume, or probability density, because integrating it over all space must be unity and dimensionless.

$$\int_{-\infty}^{\infty}\int_{-\infty}^{\infty}\int_{-\infty}^{\infty} W(x, y, z)\mathrm{d}x\mathrm{d}y\mathrm{d}z = 1 \qquad 4.30$$

4.5.2 The Maxwell–Boltzmann Speed Distribution for Atomic Gases

Equation 4.25 allows us to find the full distribution of speeds, by letting the probability density in velocity space $W(v^2)$ be the probability of finding a given value of v^2 in a volume element centered at v, $f(v)$. This requires multiplying $W(v^2)$ by the volume increment d^3v at any value of v. Consider ordinary 3D space, where the volume increment d^3r is given in rectangular coordinates (x,y,z) and spherical coordinates (r, θ, ϕ) by

$$\mathrm{d}^3r = \mathrm{d}x\mathrm{d}y\mathrm{d}z = r^2 \sin\theta\, \mathrm{d}\phi\mathrm{d}\theta\mathrm{d}r \qquad 4.31$$

(see Appendix II for notation). Velocity space is formed in complete analogy to ordinary space, replacing the $\pm x$ axis by the x components of velocity, $\pm v_x$, and likewise replacing the y and z axes with those for the corresponding velocity components, $\pm v_y$ and $\pm v_z$. Using velocity space coordinates of v, θ, ϕ, the volume element is

$$\mathrm{d}^3v = v^2 \sin\theta\, \mathrm{d}\phi\mathrm{d}\theta\mathrm{d}v \qquad 4.32$$

and thus, the probability of the velocity being in the velocity space volume increment centered at v is

$$W(v^2)\mathrm{d}^3v = A \exp\left(-\frac{mv^2}{2k_BT}\right) v^2 \sin\theta\, \mathrm{d}\phi\mathrm{d}\theta\mathrm{d}v \qquad 4.33$$

where A is the normalization factor. Because of spherical symmetry, we can integrate the angular dependence to 4π (see Appendix II) and normalize to get the *Maxwell–Boltzmann speed distribution* $f(v)$:

$$f(v)\mathrm{d}v = \int W(v^2)\mathrm{d}^3v = 4\pi\left(\frac{m}{2\pi k_BT}\right)^{3/2} \exp\left(-\frac{mv^2}{2k_BT}\right) v^2\mathrm{d}v \qquad 4.34$$

where $f(v)\mathrm{d}v$ is the probability of finding a speed in the interval v to $v + \mathrm{d}v$ in a spherical shell of radius v and thickness $\mathrm{d}v$. Equivalently, it is the fraction of molecules in a gas with speeds between v and $v + \mathrm{d}v$. It is derived from the general Boltzmann weighting factor $\exp(-E/k_BT)$ introduced above. Notice that this distribution is not symmetric in v and because of the v^2 term, it is *right-skewed*, that is, it has a high-speed tail.

Figure 4.3 shows how the distribution becomes narrower and peaks at lower speed as the mass of an atom increases for a fixed T. The right-skewed asymmetry is also apparent.

Consider the process of evaporation in an open container at constant T. The high-speed tail of the distribution means that molecules with those high speeds can escape the open container. However, the tail of the distribution does not diminish compared to the rest of the distribution, rather, the distribution that gets temporarily altered by the high-speed molecules leaving will quickly relax back to the Maxwell–Boltzmann distribution as it renormalizes itself to account for this loss of matter and energy. Since the Earth's atmosphere is open to outer space, it is slowly evaporating away due to the high-speed tail of the Maxwell–Boltzmann distribution above the escape velocity from Earth (11,200 m/s).

---

6 W can stand for "Wahrscheinlichkeit", which is German for "probability."

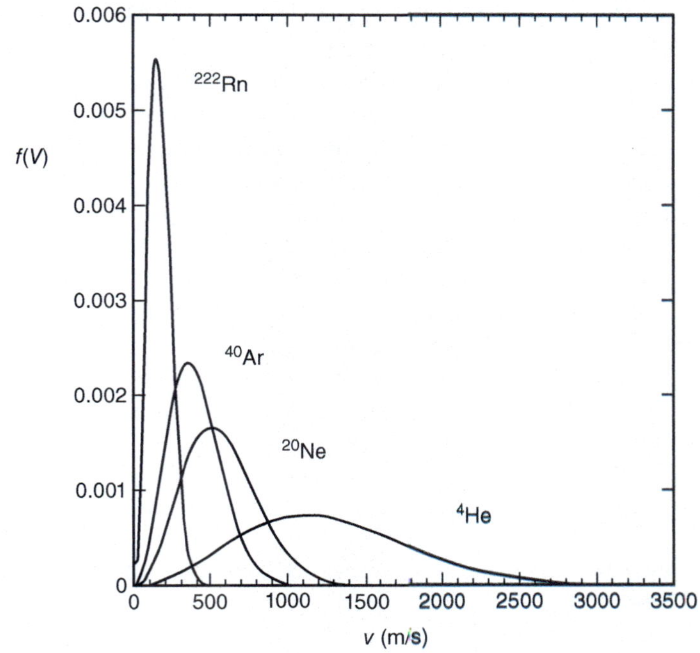

Figure 4.3 Speed distribution for monatomic gases at $T = 300$ K.

Another interesting example of the relaxation to the *MB* distribution is to consider shooting an extremely high-speed molecule into a gas in a closed container at fixed *T*. Through collisions with the much slower molecules, it will gradually lose kinetic energy until it shares its energy equally among all molecules according to the *MB* distribution. This is another instance of highly directed energy being degraded through random processes into disorganized thermal energy, such as the boring of cannons turning rotational mechanical energy to heat.

4.5.3 Equipartition Principle

Above, we considered a monatomic gas, where the classical energy is simply its translational kinetic energy, which has three degrees of freedom, one for each independent direction *x*, *y*, and *z*. If we consider a diatomic molecule as a classically rigid rotator with the *z*-axis along the line joining the two atoms, whose masses are concentrated at their centers, the molecule has two degrees of freedom of rotation, around the *x* and *y* axes, and the rotational kinetic energy is

$$KE_{rotational} = \frac{I}{2}\vec{\omega}\cdot\vec{\omega} = \frac{I}{2}(\omega_x^2 + \omega_y^2) \qquad 4.35$$

where *I* is the moment of inertia of the diatomic molecule about its center of mass. Since the mass of the atoms is preponderantly in the nuclei the moment of inertia around the *z*-axis is negligible compared to *I* around the molecule's center of mass, and hence only the rotations in *x* and *y* contribute to the molecule's total energy, and so rotation adds two degrees of freedom (except at extremely high *T*, where rotation around the *z*-axis could be excited it the molecule weren't destroyed first by thermal energy).

Now, for both $KE_{translational}$ and $KE_{rotational}$, there is a quadratic dependence on velocity; translational velocity in the former case, and angular velocity in the latter. The average energy in one dimension, that is, for one degree of freedom $<E>_{1df}$ is

$$\langle E\rangle_{1df} = \frac{\int E\exp\left(-\frac{E}{k_BT}\right)\mathrm{d}E}{\int \exp\left(-\frac{E}{k_BT}\right)\mathrm{d}E} \qquad 4.36$$

Consider the contribution of v_x to the degree of freedom in *x*. The average energy is

$$\langle E\rangle_{1df} = \frac{m}{2}\langle v_x^2\rangle = \frac{\frac{m}{2}\int v_x^2\exp\left(-\frac{mv_x^2}{2k_BT}\right)\mathrm{d}v_x}{\int \exp\left(-\frac{mv_x^2}{2k_BT}\right)\mathrm{d}v_x} = \frac{k_BT}{2} \qquad 4.37$$

where the integral formula in Appendix II, Equation AII.17, was used to solve the integrals. The exact same result will be obtained, $<E>_{1df} = k_BT/2$, for *energy that is dependent quadratically on any variable.* It can thus be generalized that each degree of freedom of a molecule that depends quadratically on a variable has an average energy of $k_BT/2$, so that the total energy in a molecule is $Nk_BT/2$, where N is the number of degrees of freedom. This is often called the *equipartition principle.* It expresses the idea that energy is shared equally among all degrees of freedom. This is expected if energy interchanges randomly among the degrees of freedom; collisions of diatomic molecules can change both kinetic energy components in v_x, v_y, and v_z, and the angular velocities ω_1 and ω_2. A molecule with translation and rotation has $5/2k_BT$ of energy, yielding $C_v = 5/2R$.

But what about vibrations if the rotor is not rigid? For a mass oscillating in a simple harmonic potential, the kinetic energy is again $mv^2/2$, but now there is also a potential energy $\kappa x^2/2$, where κ is the effective spring constant of the potential and x is the displacement of the oscillator from its equilibrium position. According to the equipartition principle, the kinetic energy of vibration and the harmonic potential, which is quadratic in x, should each contribute $k_BT/2$.

Hence, if the diatomic molecule has three translational degrees of freedom and two additional rotational and two additional vibrational ones, then each molecule should have $7k_BT/2$ of energy so that the total internal energy of a gas with N moles of molecules is

$$U = \frac{7}{2}NRT \tag{4.38}$$

This gives a value for C_v of $7R/2$, which contradicts the experimental result of $C_v = 5R/2$.

$$C_v = \frac{1}{N}\frac{\partial Q}{\partial T}\bigg|_v = \frac{1}{N}\frac{\partial U}{\partial T} = \frac{7}{2}R \neq \frac{5}{2}R = \text{Experimental results} \tag{4.39}$$

Experiments on heat capacity showed unambiguously that C_v is very close to $5R/2$ and not $7R/2$. The resolution of this issue did not come until quantum mechanics was developed. Translational kinetic energy is not quantized in a large system, but both rotation and vibration are quantized. It turns out that over a medium range of temperatures, including 300 K and thereabouts, the rotational states are heavily populated, whereas the vibrational energy levels are not. Typical rotational energy levels are on the order of 10^{-3}–10^{-4} eV and hence are $\ll k_BT$ at $T = 300$ K (recall that k_BT at 300 K is 0.026 eV). In contrast, typical vibrational energy levels are on the order of 10^{-1} eV, and are hence $>k_BT$ at 300 K, and so do not become significantly populated until higher T.

Hence, for a diatomic molecule at medium temperatures, there are three translational and two rotational degrees of freedom and so $C_v = 5R/2$. At higher T, the vibrational modes can come into play, and at low enough T the rotational modes, which are discrete, can be "frozen out," and $C_v = 3R/2$. Figure 4.4 shows schematically how C_v passes through its value of $3R/2$ at low T, transitioning to $5R/2$ as rotational modes become excited, and increasing to $7R/2$ when vibrational modes become excited.

In fact, a stunning breakdown of the equipartition principle was its complete failure to explain blackbody radiation in the nineteenth century. With the equipartition principle it was assumed that every standing electromagnetic wave in a black body contributed k_BT of energy, independent of frequency, f (each of the two orthogonal linear polarization states contributes $k_BT/2$, hence k_BT total). The completely erroneous prediction of the blackbody spectrum coming from this – that the energy would increase parabolically without limit in f^2 – was corrected only when Max Planck made the hypothesis that energy was proportional to multiples of frequency, a supposition for which there was no precedent, nor even a logical reason. It is remarkable that this deviation from equipartition not only solved the black body problem but also directly led to the development of quantum mechanics.

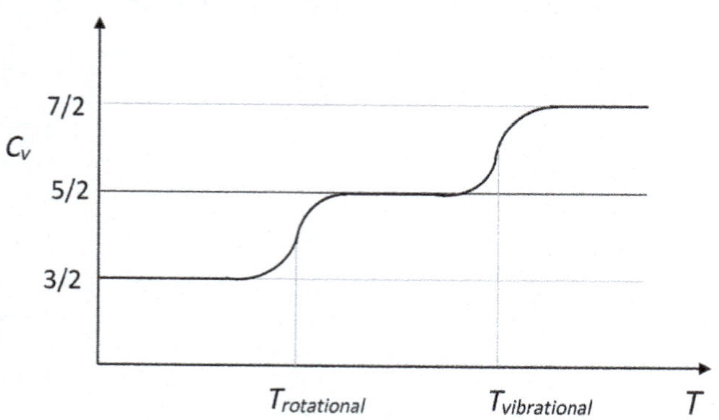

Figure 4.4 Schematic increase of specific heat of diatomic molecules as temperature increases and rotational and vibrational states become populated.

4.6 The Partition Function

The Boltzmann distribution also holds for discrete energy levels E_j, for which P_j is the probability of a particle being in state E_j in a group of particles of total energy E_{total}

$$P_j = \frac{g_j \exp\left(-\frac{E_j}{k_B T}\right)}{\sum_i g_i \exp\left(-\frac{E_i}{k_B T}\right)} \tag{4.40}$$

where g_j is the number of different states available at energy level j, also termed the *degeneracy* of the energy level. This result can be understood qualitatively by considering N particles with E_{total} distributed randomly among the available energy states. It is equally probable that a particle be in any of these energy states, properly weighted for g_j, but the total of all particle energies is constrained to sum to E_{total}. Hence, there are less ways to have a single particle with a very high energy and all the rest of the particles having very low energy, than to have all the particles in low energy states.

Consider a system with just six distinguishable particles, $N = 6$, and a total energy of $E_{total} = 8E_0$, where E_0 is the equal energy spacing between states, and the ground state energy has the (nonquantum mechanical) value of 0. If one particle has $8E_0$, then the other five particles are in the zero energy ground state. This is termed a *macrostate* for the system. Since the particles are distinguishable, the number of *microstates* for a given energy level E_j is

$$g_j = \frac{N!}{n_1!n_2!\ldots n_N!} \tag{4.41}$$

where n_i is the number of particles in energy level E_i for a particular macrostate. For the macrostate where one particle has $8E_0$ and the other five particles are in the zero energy ground state, there are six microstates, according to Equation 4.41. But consider the macrostate where three are in the ground state, one is in $1E_0$ and one in $3E_0$ and one in $4E_0$. For this, there are 120 microstates. This latter macrostate is much more probable than the former. If this is carried out for larger values of N and E_{total} the Boltzmann probability distribution will be found.

In Equation 4.40, the denominator is the normalization constant Z, which is just the sum of the individual probabilities over all energy states

$$Z = \sum_j g_j \exp\left(-\frac{E_j}{k_B T}\right) \quad \text{The Partition Function} \tag{4.42}$$

Z is the *partition function* (Zustandssumme, or "sum over states") and while it appears here as a mere dimensionless normalization factor, all the thermodynamic properties of a system can be obtained from it. For ease of manipulation, let $\beta = 1/k_B T$ then

$$Z = \sum_j^{\text{all energy levels}} g_j \exp(-\beta E_j) \tag{4.43}$$

This expression shows that Z is the Boltzmann factor summed over all microstates, which is equivalent to summing over all energy levels using the degeneracy g_j. The average energy of the system is

$$\langle E \rangle = \frac{\sum_j E_j g_j \exp(-\beta E_j)}{\sum_j g_j \exp(-\beta E_j)} = \frac{-1}{Z}\frac{\partial Z}{\partial \beta} = k_B T^2 \frac{\partial \ln Z}{\partial T} \tag{4.44}$$

This is a demonstration of how the partition function, which is a sum over microstates, yields the macroscopic thermodynamic variable of total average energy <E>, which can also be identified as the system's total internal energy U. Other derivatives of the partition function yield other state variables of the system, such as entropy, specific heats, and even the variance (Appendix I), that is, fluctuations of variables. The partition function relates macroscopic thermodynamic properties or to macrostates weighted by the degeneracy g_j.

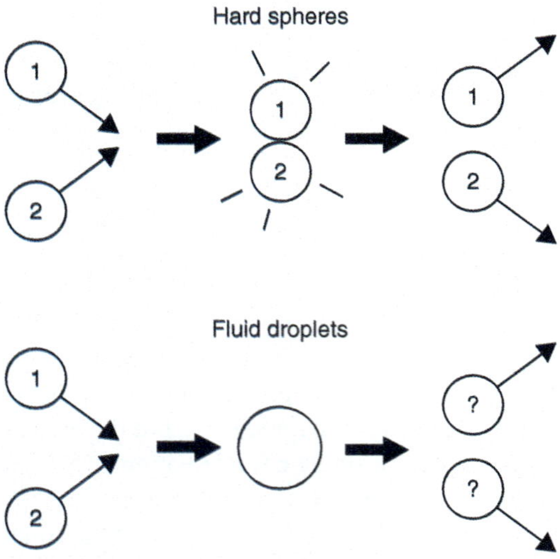

Figure 4.5 (top) Collision of two distinguishable particles that maintain their identity after the collision. (bottom) Two particles that coalesce. After the collision their initial identity is lost.

Equation 4.40 for P_j is based on the distinguishability of particles, that each one has a momentum and trajectory history distinguishable from other particles, and that the particle maintains its identity in a collision with another particle, just as two billiard balls do. In quantum mechanics, however, particles have a "fuzzy" wavelike nature, and so when two identical particles make a collision their wave functions overlap and when they separate there are again simply two indistinguishable particles. This is analogous to two identical drops of fluid, 1 and 2, that coalesce upon collision and then re-emerge as two drops; the resulting drops are indistinguishable from each other and cannot be assigned to the original drops 1 and 2. Figure 4.5 illustrates this. In this case, the number of microstates corresponding to a microstate diminishes significantly. For example, in the above case of $N = 6$ and $E_{total} = 8E_0$, there is only one microstate for a given microstate since the permutations leading to Ω are no longer relevant. Working through this scenario leads to the *Bose–Einstein distribution* for indistinguishable particles.

$$P_j = \frac{g_j}{\exp\left(\frac{E_j}{k_B T}\right) - 1} \tag{4.45}$$

where g_j is the degeneracy of the jth state. This applies for integer spin particles that don't obey the *Pauli exclusion principle*. At low temperatures, all the particles can fall into the lowest energy state to form a Bose–Einstein condensate, which occurs in superfluidity and superconductivity.

If particles are indistinguishable and obey the Pauli exclusion principle, then many states will not be allowed at all. For spin ½ particles only two can exist in any given energy substate, spin up and spin down, which excludes many possible microstates. For example, in the case of $N = 6$ and $E_{total} = 8E_0$, the case of five particles in the zero ground state and one in the $8E_0$ state is excluded, since only two can fit into any state (assuming no degeneracy, $g_j = 1$). In this case, the *Fermi–Dirac distribution* results

$$P_j = \frac{g_j}{\exp\left(\frac{E_j - \mu}{k_B T}\right) + 1} \tag{4.46}$$

where μ is the chemical potential (Section 4.12.1) and is the energy at which there is 50% occupation. Equation 4.46 expresses the fact that energy levels fill up due to the exclusion principle. The value of μ depends on the density of particles in the system. The chemical potential is also equal to the Fermi energy, E_F, at $T = 0\text{K}$, and is close to E_F for $T > 0\text{K}$.

Both these distributions reduce to the Boltzmann Equation 4.27 under non-quantum mechanical conditions, which is when the particles are dilute enough that they don't interact and mix wavefunctions. Using the uncertainty principle in 3D

$$\Delta x \Delta y \Delta z \sim \frac{\hbar^3}{8} \tag{4.47}$$

lets this condition be approximated by

$$\frac{\hbar^3}{8\left(m k_B T\right)^{3/2}} \frac{N}{V} \ll 1 \tag{4.48}$$

where N is the number of particles in volume V. For example, for air treated as an ideal gas at STP composed mostly of O_2 and N_2 the left-hand term in equation is on the order of 10^{-9}, so that the Boltzmann distribution can be used, a lucky happenstance for eighteenth- and nineteenth-century scientists dealing with ideal gas properties. For conduction electrons in copper the term is around 60, which is much greater than 1, and the Boltzmann distribution breaks down completely. The quantum mechanical effect of the exclusion principle completely stymied nineteenth-century scientists trying to build classical models for the electrical and thermal conductivities of metals.

4.7 Statistical Interpretation of Entropy

The macroscopic definition of entropy for a reversible process, Equation 4.4, is given in terms of temperature and macroscopic heat flow. Boltzmann, Gibbs, and others forged the connection between macroscopic entropy S and the *statistical mechanics* of microstates and ensembles. As just described, in statistical mechanics a system in a macrostate has a certain number of microstates available to it, designated here as Ω, e.g. specific spatial, momentum, or energy states. Another instance would be the number of distinct arrangements of a fixed number of two or more different types of particles. This can be expanded to other states, such as bound or unbound states for a protein and its substrate, or helix and coil monomers in a protein α-helix. The fundamental postulate of statistical mechanics for a system in equilibrium is that a particular macrostate has equal probability of occupying any one of its available microstates. This principle was used in the above brief introduction of the partition function.

Boltzmann defined entropy in terms of Ω as

$$S = k_B \ln\Omega \quad \text{Statistical Definition of Entropy} \tag{4.49}$$

This definition is plausible if we consider that entropy is a measure of the number of states available to a system and is an extensive variable, so that it should increase linearly with the number of particles added to a system. If there are Ω_0 spatial states in a system, then a single particle can take on any of these Ω_0. If a second particle is added it can take on any of these Ω values (assuming there are no exclusions from states, which will be treated in Section 4.14 on *excluded volume*). The number of distinct states is now $\Omega_0{}^2$, but in order for entropy to double, and not increase as the square of Ω_0, the entropy must be a logarithmic function so that there is additivity. For N particles, there are $\Omega = \Omega_0{}^N$ states and the logarithmic definition of entropy ensures that S is proportional to N:

$$S = k_B \ln\Omega = k_B \ln\Omega_0^N = N k_B \ln\Omega_0 \tag{4.50}$$

From this it can be seen that providing more microstates increases the entropy of a system. Again, as a concrete example, consider Ω as proportional to available space. Then, for a fixed number of particles, the greater the available space, the greater the entropy for a system. By the fundamental postulate, particles in a system will be, on average, uniformly distributed throughout the available space, e.g. perfume molecules released from a bottle at equilibrium will, in time, be spread equally throughout the space in a room. The time for the transition depends on what mechanism causes the spread, i.e. diffusion or convection. In either case, there will be a *kinetic* formulation that describes the rate of passage from one equilibrium state to a final equilibrium state.

Figure 4.6 illustrates the effect on the number of microstates Ω and entropy, S as (i) the volume is doubled and (ii) as the number of particles is doubled.

Thermodynamics considers different possible equilibrium states but does not, in itself, provide a kinetic description of the passage between states. Because "time is of the essence" in the living state – as we know all too well! – kinetics has a dominating role in living state processes. We will introduce kinetics later in the chapter.

This approach can also be viewed in terms of the number of possible combinations of N particles in k energy levels, where $n_1, n_2, \ldots, n_k$, are the number of particles in each level and $N = \sum n_i$. The number of combinations, or states, is

$$\Omega = \frac{N!}{n_1!n_2!\ldots n_k!} \tag{4.51}$$

Stirling's approximation for large N is

$$\ln N! = N\ln N - N \tag{4.52}$$

which can be approximated as $N\ln N$ for large N. The probability, p_i, that a particle is in state i is

$$p_i = \frac{n_i}{N} = \frac{\exp\left(-\frac{E_i}{k_B T}\right)}{\sum_j \exp\left(-\frac{E_j}{k_B T}\right)} = \frac{\exp\left(-\frac{E_i}{k_B T}\right)}{Z} \tag{4.53}$$

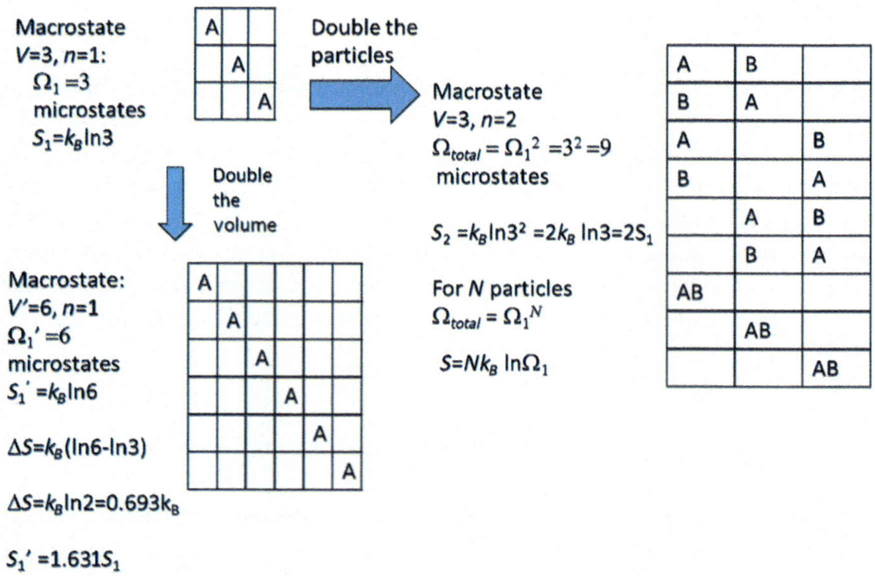

Figure 4.6 Illustration of how changing the number of particles and volume changes number of microstates Ω and entropy, *S*. In this classical case, *A* and *B* are distinguishable and can share the same state. This is an example of Maxwell–Boltzmann statistics (e.g. classical gases and solutions).

Using Boltzmann's Equation 4.49 with Stirling's approximation yields

$$S = -k_B \sum n_j \ln p_j \qquad 4.54$$

Substituting in from Equation 4.53 for p_i, and $n_i = Np_i$ allows S to be given in terms of Z and $\exp\left(-\frac{E_i}{k_B T}\right)$

$$S = k_B N \sum \left\{ \left[\frac{\exp\left(-\frac{E_i}{k_B T}\right)}{Z} \right] \left[\frac{E_i}{k_B T} + \ln Z \right] \right\} \qquad 4.55a$$

and using the middle term of Equation 4.44 for average system energy $\langle E \rangle$

$$S = k_B N \left[\frac{\langle E \rangle}{k_B T} + \ln Z \right] \qquad 4.55b$$

Converting this to entropy per mole s, and remembering that the gas constant $R = k_B N_A$ yields

$$s = \frac{N_A \langle E \rangle}{T} + R \ln Z$$

Using average energy $\langle E \rangle$ per particle from the middle term of Equation 4.44, and the fact that total energy is $N\langle E \rangle$

$$S = \frac{U}{T} + k_B \ln Z \qquad 4.56$$

4.8 Ideal Osmotic Solutions

We are now in a position to translate thermodynamic ideas for ideal gases to ideal osmotic solutions. The simplest osmotic solution consists of a liquid, such as water, and a single solute, such as sugar. An ideal osmotic solution is defined in the same way as an ideal gas; the components do not interact with each other, except by elastic pointlike collisions, and are particles with mass but no spatial dimension.

An example of a simple osmotic system is one in which there is a single chamber that is separated into two portions by a membrane, fixed immovably in the chamber, that is permeable to only one of the components. This is a *semipermeable membrane* and this concept underlies passive and active transport in biological membranes, which lead to nutrient transport, energy production, nerve impulse generation, and other phenomena. For technology, the semipermeable membrane is the basis of many separation and purification processes, including dialysis and water desalination.

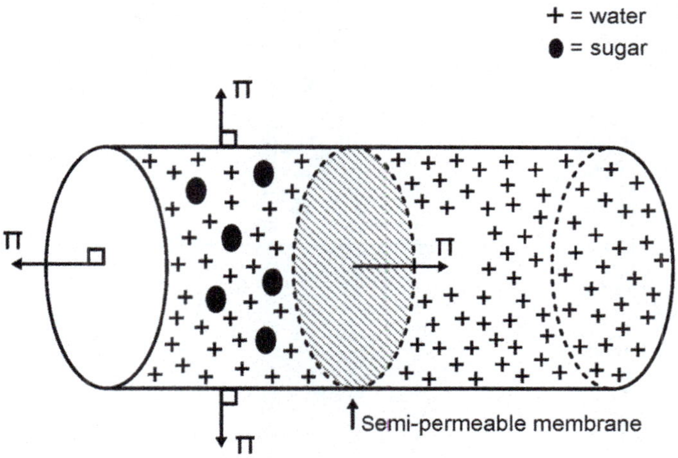

Figure 4.7 Sealed cylinder with a semipermeable membrane, permeable only to water. Sugar is confined to left-hand side, in which an osmotic pressure Π exists. (Courtesy of Marissa Calko).

Figure 4.7 is a schematic of the cylinder with fixed, semipermeable membrane. Both sides of it are filled with water, and a solute, sugar, is only on the left side of the chamber. The semipermeable membrane dividing the chamber is permeable only to water. First, the sugar will dissolve and then the sugar molecules will reach maximum entropy by diffusing until an average equal concentration is reached throughout its portion of the chamber. Once this maximum entropy is reached, diffusion will continue indefinitely, maintaining this equilibrium. No sugar molecules pass to the other side of the membrane in this equilibrium. Meanwhile, water diffuses freely throughout the entire chamber but has different concentrations on each side of the membrane. This imbalance in concentration leads to interesting effects, and, via the chemical potential concept introduced later, it becomes the origin of osmotic pressure Π.

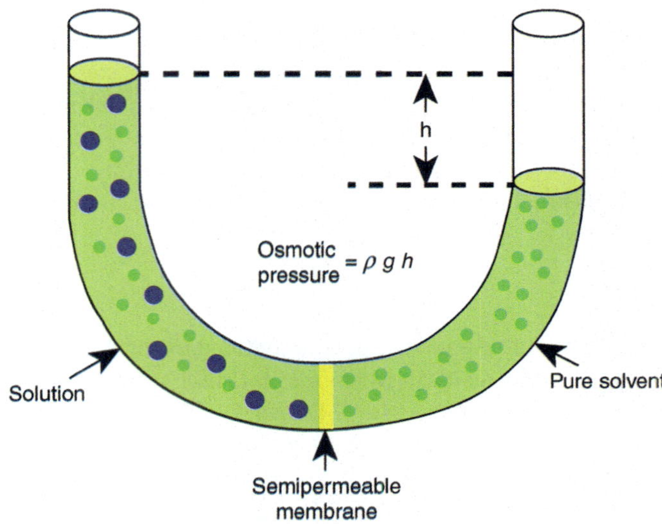

Figure 4.8 A U-tube osmometer.

The osmotic pressure Π in this example is a hydrostatic pressure in its side of the chamber. Π is perpendicular and outward at all surfaces of the sugar side of the chamber, as shown in Figure 4.7. This osmotic pressure can be measured with a pressure gauge in the chamber. As in the case of the ideal gas, the more concentrated the sugar the higher the pressure. At high enough solute concentrations, osmotic pressure can become large enough to burst the membrane or the chamber itself if the system is rigidly sealed. It is remarkable that this osmotic pressure produced simply by imbalances in water and sugar concentrations acts exactly like any other mechanical pressure, such as that produced by steam in a heated boiler, or a person standing on a floor.

A device for measuring Π is an *osmometer*. A simple osmometer is shown in the left-hand U-tube of Figure 4.8. It uses gravity to measure Π, similar to the way a mercury barometer measures atmospheric pressure invented in 1643 by Evangelista Torricelli (Italy, 1608–1647). Once the sugar dissolves, osmotic flow of water through the membrane and into the side of the osmometer with the sugar will build up pressure, causing its level to rise until hydrostatic pressure, $\rho g h$, equals the osmotic pressure. The difference in height, h, is then a direct measurement of the osmotic pressure. In this case, the osmotic pressure did work to lift the liquid against gravity until the free energy minimum was reached and the system arrived at equilibrium.

The fact that the osmotic pressure can perform work suggests harnessing the cylinder to a load that it can usefully lift. Now, let's make the chamber a hollow cylinder with fixed endcaps and volume, and cross-sectional area A, but now with a moveable membrane, as shown in Figure 4.9. The membrane is firmly connected to a shaft that protrudes through the left-hand side of the cylinder but is sealed against any leakage. The shaft is connected to a rope that runs over a pulley and is connected to a mass M. Say the membrane originally divides the cylinder into a left portion with dissolved sugar, and a right-hand portion with only water. There will be an osmotic flow of water from the right side of the chamber to the left side as the water seeks to equalize its concentration throughout the cylinder. In order to do this the membrane moves to the

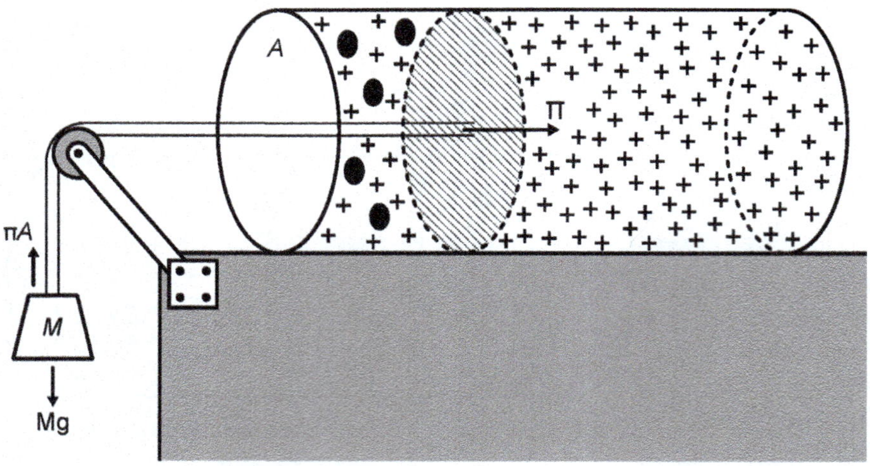

Figure 4.9 Closed cylinder with a moveable, sugar-impermeable membrane, which is attached to a shaft that can lift a mass M, via a rope over a pulley. The cylinder is filled with water, with the sugar impermeable membrane separating the dissolved sugar in the left half from the water-only portion on the right. Black ovals represent sugar molecules and the crosses represent water molecules. (Courtesy of Marissa Calko).

right, carrying the shaft with it and lifting the mass, as long as the upward force due to osmotic pressure and given by ΠA is greater than Mg. Eventually, the membrane will come to a halt when the osmotic pressure is such that $\Pi A = Mg$. The membrane may even reach the right end of the chamber, if ΠA is still higher than Mg, and both the sugar and water will have access to the entire chamber's volume, i.e. both water and sugar will have reached their own constant concentrations throughout the cylinder and maximized their entropy.

For ideal solutes, osmotic pressure Π has the exact same form as the ideal gas law in Equation 4.1:

$$\Pi V = NRT \tag{4.57}$$

where V is the volume the sugar is confined to and N is the number of moles of sugar. To contrast typical values of P and Π, consider that an ideal gas has $n/V = 45\ \text{mol/m}^3$ at STP ($T = 273.15\text{K}$, $P = 1.01 \times 10^5\ \text{N/m}^2$). A dilute sugar solution might have 1% by weight of sucrose (342 g/mol), which is 29 mol/m^3, making the respective pressures quite comparable.

The osmotic pressure exerts forces on areas identically to any other source of pressure, such as a gas in a vessel, or a weight sitting on the ground. A thermodynamic analysis identical to the ideal gas can be made. Let the osmometer remain at constant T, i.e. it is simply sitting in thermal equilibrium with a heat reservoir at constant temperature, so the osmotic process is isothermal. Then the work done in going from initial volume V_o to final volume V_f is the same as Equation 4.7, substituting Π for P.

$$W = \int \Pi \mathrm{d}V = nRT \int_{V_o}^{V_f} \frac{\mathrm{d}V}{V} = nRT \ln(V_f / V_o) \tag{4.58}$$

The system has absorbed heat from the environment, $Q = W$, and the entropy change is the same as that computed for the ideal gas, i.e. this process is driven purely by entropy. While the system has performed mechanical work, it is now at a maximal state of entropy and is at its minimum of free energy and cannot perform any more work. It cannot spontaneously reduce its entropy by the sugar concentrating itself into one end of the chamber, so an external work is required for this, as for the ideal gas.

This idea is of fundamental biophysical significance. Say that an external agent takes the membrane, which has been pushed up against the endcap of the chamber and invests the work required to move the membrane back to its starting position. This amount of work is equal to the work done by the system in the entropic push of the membrane by the sugar, but now the entropy of the system has been decreased by the external force, that is, a certain molecular "ordering" has occurred, from highly spread-out sugar molecules to more concentrated ones. This motif permeates living state processes; reaping energy from an external source to lower system entropy and create molecular order. This is a very simple example of such a process, but it conveys the notion of external work increasing molecular ordering in a system.

This process of reconcentrating the sugar has purified the water of sugar on the other side of the membrane, and is termed *reverse osmosis*, because it requires external work input. Reverse osmosis is used worldwide to desalinate seawater

Figure 4.10 Water desalination plant on the Red Sea. (Courtesy of Moshe Pridan/ Wikimedia Commons/Public Domain).

and in other water purification processes. In this case, the solute is comprised of dissolved ions of Na^+, Cl^-, and others, and the semipermeable membrane only allows water to pass through it. By doing work on seawater in the cylinder, the ions are compressed into a small volume of the cylinder, leaving pure, desalinated water in the rest of the cylinder. Figure 4.10 shows the Zarchin desalination plant in Eilat, Israel, on the Red Sea.

4.8.1 Macroscopic Manifestations of Osmotic Pressure in the Living State

While osmotic effects underlie many microscopic living state processes, there are also many manifestations of it at macroscopic scales. The following are some examples.

Osmosis allows water to flow into plant roots. The plant roots are permeable to water and contain water at a certain concentration. If the concentration of water outside the roots is greater than inside then water can flow into the roots. By the osmotic pressure notion above, the flow can go against gravity, transporting water upward in a plant or tree. It is important to note that osmotic pressure is not the only effect driving liquid against gravity in plants and trees. There is also the capillary effect, discussed in Section 4.11.2, and transpiration, due to the cohesiveness of water, discussed in Chapter 6.12.

Osmotic pressure provides the *turgor pressure* to keep cells "filled out." Placing a cell in a *hypertonic solution* (having osmotic pressure higher than the interior of the cell) causes the cell to shrink. Placing a cell in a *hypotonic solution* (having lower solute content than the cell) causes water to enter the cell, hence swelling it. Fish, for example, have an osmotic balance between their cells and the surrounding water, which is an *isotonic solution*. A saltwater fish has high molarity in its cells to match that of seawater. Placed into extremely high saline water will compress the fish's cells, killing it. In freshwater, osmosis will drive water into the fish's cells, swelling them, and killing the fish. The scenarios are depicted in Figure 4.11.

Osmotic effects in blood are particularly interesting. Blood is composed of an aqueous "carrier fluid," namely *blood plasma*, red blood cells (*erythrocytes*), white blood cells (*leucocytes or leukocytes*), and *platelets (thrombocytes)*, discussed in Chapter 2.6.3. The fraction of blood volume occupied by red blood cells is termed the *hemocrit*. For adult humans, the normal range for the hemocrit is from 38% to 54%. Blood plasma, which accounts for most of the rest of blood volume is mostly water, with ions and, importantly, blood proteins. *Albumin*, a family of water soluble, globular proteins, is among

Figure 4.11 A fish in aqueous solutions which are, with respect to the fish's cells, from left to right, hypertonic, isotonic, and hypotonic. (Courtesy of Marissa Calko).

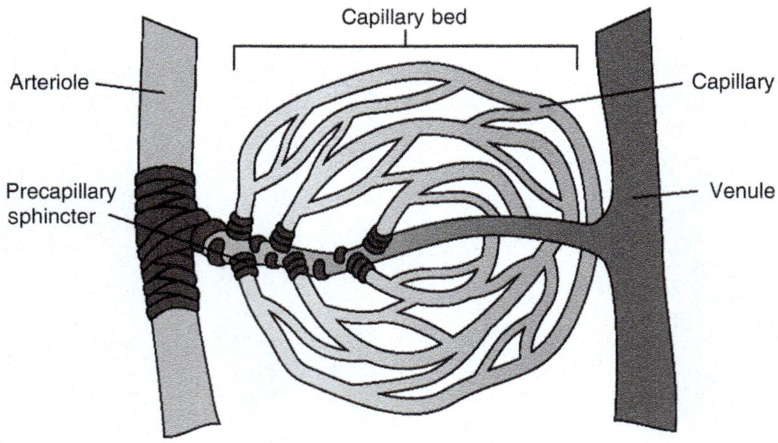

Figure 4.12 Flow of blood from capillaries through the capillary bed and into the venous network. (Courtesy of Marissa Calko).

the most important of these and is typically present at 3.5% to 5.5% by mass in blood. This is a high enough concentration to produce significant osmotic pressure in the blood (often termed *oncotic pressure*).

Figure 4.12 shows how the oxygen-containing blood in the arterial flow branches down into the *capillary bed* where O_2 and fluid exchange take place. After leaving the capillaries the de-oxygenated blood, represented in dark shade, flows through branches into veins.

Since a large class of blood capillaries is permeable to water but not the proteins, there will be a tendency to draw in water from the interstitial tissue, if the concentration of water is higher there and vice versa if the interstitial water concentration is lower. Counteracting the water flow caused by the blood osmotic pressure is the hydrostatic pressure driving blood flow, produced by the heart, which tends to drive water out of the vessel. The effects are summed up in the equation first proposed by Ernest Starling (England, 1866–1927), and later revised, which gives the flow rate of fluid out of a capillary Q (cm^3/s):

$$Q = K\left[(P_p - P_i) - \sigma(\Pi_p - \Pi_g)\right] \tag{4.59}$$

where K is the total capillary hydraulic filtration coefficient, P_p is the hydrostatic pressure of the plasma inside the capillary, P_i the pressure in the interstitial tissue, Π_p is the plasma osmotic pressure, Π_g is the *subglycocalyx* osmotic pressure, and σ is a dimensionless constant equal to unity if the capillary is completely impermeable to protein and less than unity if some protein permeates it. Q is positive if fluid flows out of the capillary and negative if fluid flows in. The glycocalyx is a complex *pericellular matrix* consisting of polysaccharides, *glycoproteins*, and glycolipids. It has many functions, including cell–cell recognition and, of concern here, the control of capillary permeability. Consider the case where $\sigma = 1$. If the net hydrostatic pressure is higher than the net osmotic pressure it will drive water out of the blood vessel and into the interstitial tissue, thus causing swelling, or *edema*, shown on the left side of Figure 4.13. If the net hydrostatic pressure is less than the net osmotic pressure, right side of Figure 4.13, then fluid will flow into the vessel increasing blood pressure. The concentration of albumin, hence, has a decisive effect on capillary dynamics and its determination in blood testing is of fundamental importance.

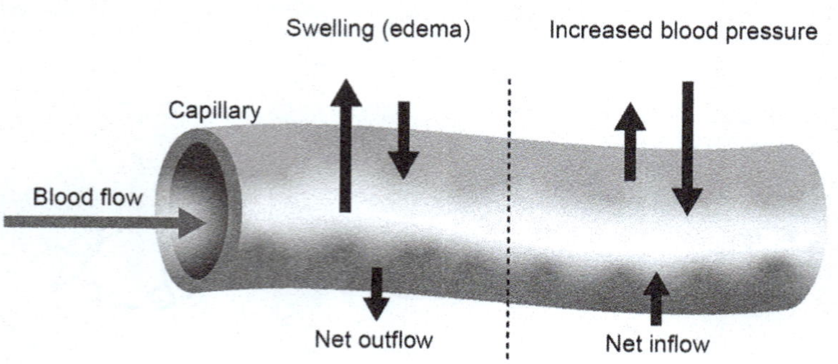

Figure 4.13 (Left) Hydrostatic pressure (upward arrow) higher than the osmotic pressure (upper downward arrow) of the interstitial tissue causes net water outflow and edema (swelling). (Right) hydrostatic pressure (upper outward arrow) lower than the osmotic pressure (upper downward arrow) of the interstitial tissue causes water inflow into the blood vessel and causes high blood pressure. (Courtesy of Marissa Calko).

4.9 Working Substances and Bioenergetics Cycles

The early development of thermodynamics and the unraveling of its puzzles was intimately connected to steam engines, the first truly productive heat engines. In the steam engine, as in modern steam turbines, the working substance, water, is not used up. Rather it is thermally driven between high and low temperatures by an external high-temperature reservoir that requires fuel to maintain its temperature T_H. The same is true for refrigeration systems that have closed-loop refrigerants driven between high and low temperatures by electric motors and compressors; the refrigerant is not used up, it cycles between gas and liquid phases in the refrigeration process.

Internal combustion engines, on the other hand, as well as gas turbines use their fuel as both their heat source and part of their working substance (air being the other working substance component). Nonetheless, all these types of heat engines have in common the expansions and contractions of the volume of their working substances and repeating thermodynamic cycles involving temperature differences.

In contrast, in the living state, energy conversion processes are mediated via chemical energy at essentially constant volume, pressure, and temperature. *Bioenergetic cycles* involve a source of energy that is released as it flows through the biological system, driving energy-consuming processes along the way, and leading to an inevitable amount of waste heat and matter. As mentioned, the biological system is an open one, far from equilibrium, and its structure is maintained as an intermediate in the process that takes one form of energy and converts and degrades it into others, such as heat and waste byproducts.

As seen in the efficiency of the Carnot heat engine cycle, higher temperature source reservoirs lead to higher theoretical efficiency for a given exhaust temperature. Considering photosynthesis, this involves conversion of high-energy photons (several eV) emitted from the Sun's surface at about 5600 K to create high-energy ATP (adenosine triphosphate) fuel molecules at 300 K. ATP then drives subsequent energetic cycles, such as transferring energy to biochemical reactions, pumping ions across channels, and building long-term fuel storage, such as in polysaccharides, lipids, and protein. In nonphotosynthetically based organisms external sources of high-energy molecules – proteins, polysaccharides, lipids – are ingested directly and a consortium of metabolic pathways produces ATP and associated molecules to drive energetic processes in those organisms.

We will later see in Chapter 8 that flexible macromolecules behave in a similar fashion to ideal gases, that is, if they are deformed such that their configurational entropy decreases (e.g. by stretching or compressing), they will exert an entropy-based restoring force in attempting to return to their configuration of maximum entropy. This helps give elasticity to tissues containing flexible biomacromolecules, such as cartilage and skin.

4.10 The Hydrophobic Effect

The *hydrophobic effect* is a preeminent example of an entropically driven process and is operant in the self-organization of biological membranes and structures such as micelles and liposomes, and in the complex folding of proteins.[7] It is one of the major drivers of physical–chemical phenomena in aqueous solution and hence in living state systems. The hydrophobic effect stems chiefly from the fact that nonpolar molecules have low solubility in aqueous solutions and water molecules organize around them. This lowers the entropy of water, so that separation of polar molecules and water into distinct phases liberates the water, increasing its entropy. The hydrophobic effect will be invoked frequently throughout this book.

4.11 Surface Tension

A macroscopic manifestation of the hydrophobic effect is the way in which a liquid droplet sits atop a material. If the surface is hydrophilic a water droplet will spread out and have a small *contact angle*. The contact angle is the angle between the tangent of the drop's surface measured with respect to the horizontal plane atop which the droplet sits, as shown in

7 The hydrophobic effect is sometime referred to casually as a "hydrophobic force," but many people object to this as a "force" is usually conceived as a vector quantity with a direction of action. The hydrophobic effect indeed causes physical actions and motion but does not have assignable force vectors.

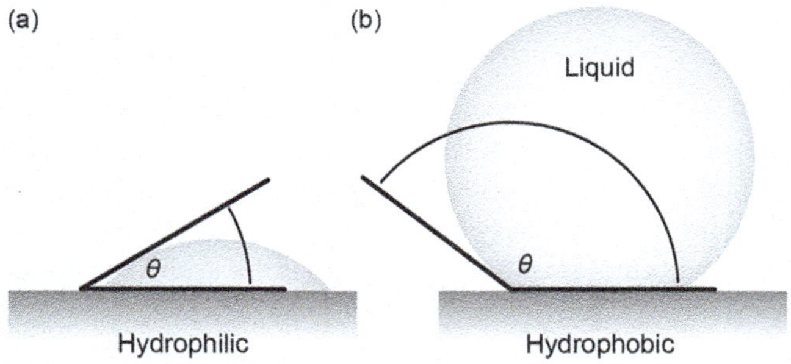

Figure 4.14 (a) A hydrophilic liquid drop has a low contact angle. (b) A hydrophobic liquid drop has a high contact angle. (Courtesy of Marissa Calko).

Figures 4.14a,b. In the case of water, a surface is *hydrophilic* if the droplet spreads out and comes to equilibrium with a low contact angle, Figure 4.14a. On top of a *hydrophobic* surface, the water droplet contact is minimized, does not spread out much, and has a large contact angle, Figure 4.14b.

An important physical quantity, not restricted to aqueous systems, is *surface energy*, γ, (also sometimes called *interfacial energy*), which is the free energy change in creating a unit area more surface of an interface between gases, liquids, and solids. In terms of free energy G, and surface area A, at fixed P and T, the surface energy is given by

$$\gamma = \left.\frac{\partial G}{\partial A}\right|_{T,P} \tag{4.60}$$

Surface energy for liquids is also termed *surface tension*, which is a measure of force per length.

4.11.1 Young's Equation for Contact Angle of a Drop on a Solid Surface

Consider a droplet of liquid which is originally a sphere of radius R'. It is placed on a surface, spreads out, and comes to an equilibrium in the shape of a spherical cap which minimizes the total surface energies, as in Figure 4.15. Ignore gravity and consider only surface energies, γ_{LV}, γ_{SL}, and γ_{SV}, which are the surface energies between liquid and vapor, solid and liquid, and solid and vapor, respectively. Originally the droplet in air had a total surface energy G_i, and then it is placed on the solid surface, where it spreads out until it comes to an equilibrium on the surface. When it is at equilibrium on the surface, it has a final energy G_f, which is a minimum. For the droplet initially in air:

$$G_i = 4\pi R'^2 \gamma_{LV} \tag{4.61}$$

At equilibrium on the surface, the spherical cap as seen in Figure 4.15 has a new radius R. The contact angle is between the tangent to the surface of the cap at the solid interface and the $+x$ axis of the planar interface. The planar surface area of the cap is $A = \pi(R\sin\theta)^2$, and the surface area of the curved cap is $S = 2\pi R^2(1 - \cos\theta)$. Since the volume of the droplet is conserved, the original volume V' must equal the final volume V, which provides the relationship between R and R'. The total surface energy at equilibrium G_f is

$$G_f = \gamma_{SL}A + \gamma_{LV}S - \gamma_{SV}A \tag{4.62}$$

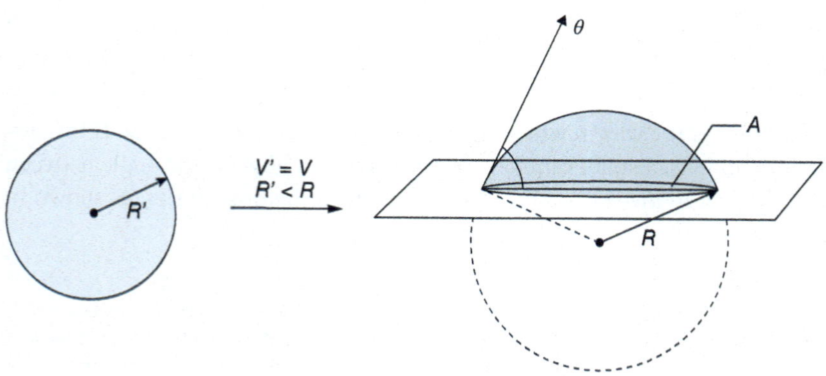

Figure 4.15 A droplet of liquid before depositing on a surface (left). After placing the droplet on the surface the droplet comes to rest at a certain contact angle θ (right). The volume of the droplets on the left and right are equal. (Courtesy of Stephanie Smith).

This can be minimized with respect to θ to yield Young's contact angle equation:

$$\gamma_{LV}\cos\theta = \gamma_{SV} - \gamma_{SL} \tag{4.63}$$

The limiting hydrophilic case ($\theta = 0$) occurs when the surface tension equals γ_{SV} and γ_{SL} is zero. The hydrophobic limit ($\theta = \pi$) occurs when $\gamma_{LV} = \gamma_{SL}$, and $\gamma_{SV} = 0$.

4.11.2 Capillary Action

This long-known effect also derives from surface energy between a liquid, vapor, and surrounding solid surfaces. It manifests itself in a liquid being drawn into a capillary or narrow space without externally applied forces. The height h, to which a liquid of density ρ will rise in a capillary of radius R in a gravitational field g, will be such as to minimize the surface energy in the final state. Figure 4.16 shows the capillary in a very large reservoir of liquid, at equilibrium with the height h. The height of the reservoir does not change as the capillary fills.

The total energy after the water rises and reaches its equilibrium, including the work needed to raise the water to height h, is

$$G_f = \frac{\rho g \pi R^2 h^2}{2} + 2\pi R h \gamma_{SL} + 2\pi R^2 (1-\cos\theta)\gamma_{LV} \tag{4.64}$$

When the capillary was introduced into a reservoir of water, before the water begins to rise in the capillary the total energy in the capillary up to height h, filled with air, and considering the flat surface of water is

$$G_i = \pi R^2 \gamma_{LV} + 2\pi R h \gamma_{SV} \tag{4.65}$$

Minimizing $G_f - G_i$, and using Young's Equation 4.63 yields:

$$h = \frac{2\gamma_{LV}\cos\theta}{\rho g R} \tag{4.66}$$

In Figure 4.16, a positive meniscus is shown. The condition for a positive meniscus is $0 < \theta < \pi/2$ and for a negative meniscus $\pi/2 < \theta < \pi$.

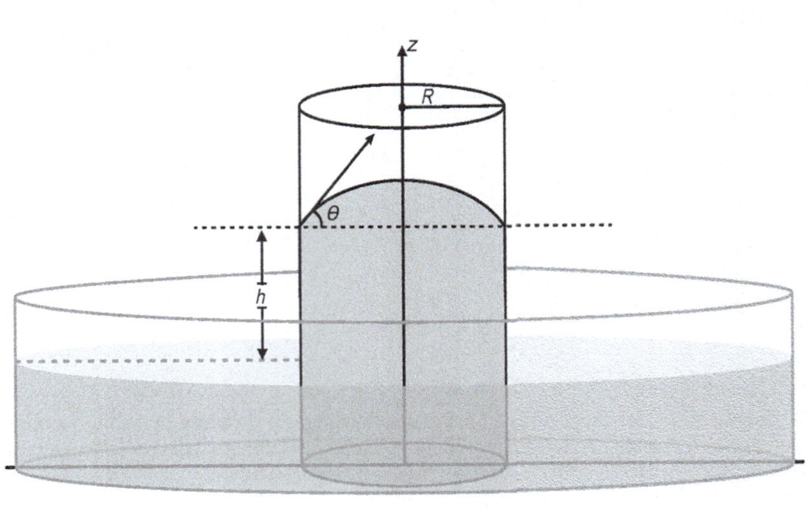

Figure 4.16 Capillary action. The liquid rises to a height h so as to minimize surface energy. (Courtesy of Marissa Calko).

4.12 Thermodynamics of Multicomponent Solutions

The living state occurs in multicomponent, aqueous solutions. These solutions contain ions, small organic molecules, such as ATP, as well as biomacromolecules, and particles in suspension. The term *colloid* is often used to refer to particles that are *suspended* in liquids without being *dissolved* in a molecular sense. Whole cells, bacteria, and organelles, such as chloroplasts and mitochondria are living state colloids. Examples of nonliving state colloids include nanometer and micrometer particles of suspended silica, oxides, microgels, vesicles, latex particles, and emulsions.

If a solution contains several components (e.g. water, salts, macromolecules), there will be certain conditions that govern the equilibrium state of the solution. Living state systems, however, are far from equilibrium, so that energy must be added to the solution to keep it away from equilibrium. Knowledge of the equilibrium state allows estimates of the biological energy and conditions needed to create the nonequilibrium state. Hence, we focus on equilibrium considerations here.

As mentioned at the outset of this chapter, thermodynamic variables can be either extensive or intensive. Extensive variables depend on the total amount of matter in a system, and include volume, mass, number of particles, total energy, and entropy. Intensive variables do not depend on the amount of matter and include temperature, pressure, concentration, or density of species present, energy density, etc.

Until now, we have considered thermodynamic systems with fixed numbers of particles. When particles of one or more species are added or removed from a system, there is a change in extensive variables. The living state constantly exchanges matter with the environment, so the thermodynamic consequences of this are important.

Partial molar quantities are a useful means of turning extensive variables into intensive variables. Partial molar quantities are obtained by expressing how an extensive variable (e.g. volume) changes as the molar concentration of one component i (e.g. salt) changes. A partial molar quantity x_i (lower case "x") of an extensive variable X (upper case "X") is hence given by

$$x_i = \left.\frac{\partial X}{\partial n_i}\right|_{T,P,n_{j\neq i}} \qquad 4.67$$

where n_i is the molar concentration of component i, and P, T, and the concentrations of all other components are held constant. For example, the partial molar volume v_i of component i is

$$v_i = \left.\frac{\partial V}{\partial n_i}\right|_{T,P,n_{j\neq i}} \qquad 4.68$$

where V is the total volume of the solution, i.e. v_i is the amount the total volume of the system, V, changes when one mole of component i is added. An important property of partial molar quantities is the following sum rule, which follows from direct summation of the partial quantities:

$$X = \sum_{i=1}^{N=\#\text{of components}} n_i x_i \qquad 4.69$$

4.12.1 Chemical Potential

A particularly important partial molar quantity is the *chemical potential*, μ. The chemical potential for component i, μ_i, is the partial molar Gibbs free energy[8] (sometimes also denoted as g_i):

$$\mu_i = \left.\frac{\partial G}{\partial n_i}\right|_{T,P,n_{j\neq i}} \qquad \text{Definition of Chemical Potential} \qquad 4.70$$

Then, by the sum rule

$$G = \sum_{i=1}^{N} n_i \mu_i \qquad 4.71$$

8 The use of "μ_i" for chemical potential of component i is widespread in the literature. With the above convention that the intensive variable is lower case, "g_i" is the preferred representation of chemical potential of component i, but we will stay with the more widespread use of "μ_i" for chemical potential.

As the definition implies, chemical potential is especially important when there are more than one species present in a system and/or when there is more than one phase available to a given species. An example is liquid/vapor equilibrium of a single chemical species (e.g. water and water vapor): The chemical potential of the liquid phase equals the change in the free energy as a mole of particles enters the phase, and likewise for the chemical potential of the vapor phase. Since in a closed system, the total number of particles is constant, consideration of the chemical potential should determine what fraction of the particles goes into each phase. As will be seen below, the governing condition is that the chemical potential of a given species be the same in both phases; in the current example, the chemical potential of water in the vapor phase will equal that in the liquid phase. The idea can be extended to two components, such as oxygen and water together, where there will be some oxygen and some water in each of the vapor and liquid phases. The chemical potential of each component is the same in all phases at constant T and P, as shown in the following derivation.

Using Equation 4.71, the infinitesimal change in G, is

$$\mathrm{d}G = \sum_{i=1}^{N} n_i \mathrm{d}\mu_i + \sum_{i=1}^{N} \mu_i \mathrm{d}n_i \qquad 4.72$$

Equation 4.6 for free energy, G, in a system of fixed amount of material can now be updated for an open system, or system with multiple phases, to include molar additions or subtractions of components n_i. G is now explicitly shown also as a function of the number of moles of each of N components,

$$G = G(U, P, V, T, S, n_1, n_2 \ldots n_N) \qquad 4.73$$

Computing the total differential of G with respect to all variables, which now includes the effect of changing the number of moles n_i of component i, $\mathrm{d}G$ becomes

$$\mathrm{d}G = \mathrm{d}U + P\mathrm{d}V + V\mathrm{d}P - T\mathrm{d}S - S\mathrm{d}T + \sum_{i=1}^{N} \frac{\partial G}{\partial n_i} \mathrm{d}n_i \qquad 4.74$$

or, using the definition for chemical potential in Equation 4.70:

$$\mathrm{d}G = \mathrm{d}U + P\mathrm{d}V + V\mathrm{d}P - T\mathrm{d}S - S\mathrm{d}T + \sum_{i=1}^{N} \mu_i \mathrm{d}n_i \qquad 4.75$$

The term $\sum_{i=1}^{N} \frac{\partial G}{\partial n_i} \mathrm{d}n_i$ indicates how the extensive variable G changes when a molar increment $\mathrm{d}n_i$ of species i is added or subtracted from the system (open system), or when, in a single component system (e.g. pure H_2O), an amount of material $\mathrm{d}n_i$ passes from one phase to another in a closed system (e.g. partitioning of a fixed number of moles between vapor and liquid).

For a reversible change, by the first law of thermodynamics:

$$\mathrm{d}Q = T\mathrm{d}S = \mathrm{d}U + P\mathrm{d}V \qquad 4.76$$

so that Equation 4.75 becomes

$$\mathrm{d}G = V\mathrm{d}P - S\mathrm{d}T + \sum_{i=1}^{N} \mu_i \mathrm{d}n_i \qquad 4.77$$

Combining this with the previous expression for $\mathrm{d}G$, Equation 4.72:

$$\mathrm{d}G = V\mathrm{d}P - S\mathrm{d}T + \sum_{i=1}^{N} \mu_i \mathrm{d}n_i = \sum_{i=1}^{N} n_i \mathrm{d}\mu_i + \sum_{i=1}^{N} \mu_i \mathrm{d}n_i \qquad 4.78$$

yields

$$\sum_{i=1}^{N} n_i \mathrm{d}\mu_i = V\mathrm{d}P - S\mathrm{d}T \qquad 4.79$$

This is the *Gibbs–Duhem equation*, which relates changes in chemical potential of components to the changes in pressure and temperature of a system. In the case where P and T are constant, as is often the case in living systems, the Gibbs–Duhem equation becomes

$$\sum_{i=1}^{N} n_i \mathrm{d}\mu_i = 0 \quad \text{Gibbs-Duhem Equation in the Living State} \qquad 4.80$$

This indicates that changes in chemical potential of the various components in a system are strictly related to each other.

At equilibrium $\mathrm{d}G = 0$, and if P and T are again constant, then Equation 4.72 gives

$$\sum_{i=1}^{N} \mu_i \mathrm{d}n_i = 0 \qquad 4.81$$

An important consequence of Equation 4.81 is that *at equilibrium the chemical potential of each component is the same in all phases*, as asserted above. This can be shown easily for a two-component system.

Consider a two-component system of water molecules (1) and a fixed number n_t of micelle-forming surfactant molecules (2). We will consider micellar systems in more depth in Section 4.15.3, but suffice it here to say that above a *critical micelle concentration* (CMC) the surfactant molecules can either be free in solution (phase A) or bound in a micellar structure (phase B), i.e. component (2) can co-exist in these two distinct phases. By Equation 4.81:

$$\mu_1 \mathrm{d}n_1 + \mu_{2,A}\mathrm{d}n_{2,A} + \mu_{2,B}\mathrm{d}n_{2,B} = 0 \qquad 4.82$$

But the first term is zero, since the number of moles of solvent molecules does not change in the solution. Since the total moles of surfactant molecules n_t is

$$n_t = n_{2,A} + n_{2,B} \qquad 4.83$$

then

$$\mathrm{d}n_{2,A} = -\mathrm{d}n_{2,B} \qquad 4.84$$

since $\mathrm{d}n_t$=0, so that

$$\mu_{2,A} = \mu_{2,B} \qquad 4.85$$

This important and widely applicable result states that for a substance that can co-exist in more than one phase, the chemical potential of that substance will be the same in all phases.

This result can be generalized to an arbitrary number, N, of phases. For a closed system:

$$\mathrm{d}n_1 = -\sum_{i=2}^{N} \mathrm{d}n_i \qquad 4.86$$

and

$$\mu_1 \mathrm{d}n_1 = -\mu_1 \sum_{i=2}^{N} \mathrm{d}n_i \qquad 4.87$$

Using Equations 4.81 and 4.87 yields

$$\sum_{i=2}^{N} (\mu_i - \mu_1)\mathrm{d}n_i = 0 \qquad 4.88$$

Since dn_2 ... dn_N are arbitrary, being constrained only by the closure relation that the sum of all dn_i is zero, it follows that $(\mu_i - \mu_1) = 0$ for all i and hence

$$\mu_1 = \mu_2 = \cdots \mu_N \tag{4.89}$$

An example of this is cooperative binding of O_2 to hemoglobin. In this case, there are three components: carrier solvent, hemoglobin, and O_2. For four-fold cooperative binding of O_2, heme(O_2), heme$(O_2)_2$, heme$(O_2)_3$, heme$(O_2)_4$, there are five phases for the O_2 molecules (counting unbound O_2), and $\mu_{O_2, j=1,2\ldots5}$ are all equal.

4.12.2 Chemical Potential μ_i for an Ideal Solution

For an ideal solution for which there is no enthalpic interaction, at constant T and P only entropy contributes to changes in G, so that, for purposes of finding changes in G, ΔG, it suffices to use $G = -TS$. Consider a two-component system composed of solvent (1) and a solute (2). There are N_1 solvent molecules and N_2 solute molecules. The total number of molecules in the system N_t is

$$N_t = N_1 + N_2 \tag{4.90}$$

When the solute and solvent are mixed at constant temperature, the change in free energy from the unmixed to the mixed system is

$$\Delta G_{mix} = G_{mix} - G_{pure} = -T(S_{mix} - S_{pure}) \tag{4.91}$$

The entropy in the mixed state S_{mix} is related to the number of ways Ω that N_1 and N_2 molecules can mix via Boltzmann's relation:

$$S_{mix} = k_B \ln\Omega \tag{4.92}$$

Ω is found from the basic theory of permutations to be

$$\Omega = \frac{N_t!}{N_1!N_2!} \tag{4.93}$$

Using Stirling's approximation, approximated further to $\ln N! \sim N \ln N$

$$\ln\Omega \cong N_t \ln N_t - N_1 \ln N_1 - N_2 \ln N_2 \tag{4.94}$$

which, using

$$\ln\Omega \cong (N_1 + N_2)\ln N_t - N_1 \ln N_1 - N_2 \ln N_2 \tag{4.95}$$

gives the form

$$\ln\Omega \cong N_1 \ln\left(\frac{N_t}{N_1}\right) + N_2 \ln\left(\frac{N_t}{N_2}\right) \tag{4.96}$$

It is convenient to define the mole fraction of each component x_i as

$$x_i = \frac{N_i}{N_t} \tag{4.97}$$

where the factor of Avogadro's number N_A, used to divide the numbers N_1 and N_2 into moles, cancels out in the denominator and numerator when forming x_i. Then, generalizing to a system with k components

$$S_{mix} = -k_B \sum_{i=1}^{k} N_i \ln x_i \qquad 4.98$$

which gives

$$G_{mix} = k_B T \sum_{i=1}^{k} N_i \ln x_i \qquad 4.99$$

Figure 4.17 shows $S_{mix}/(k_B T N_{total})$ versus x_2 for a two-component system. The maximum entropy occurs at the maximum mixing when $x_2 = x_1 = 0.5$.

It is convenient to put G_{mix} in molar form in order to obtain the partial molar free energy, or chemical potential. The number of moles of component i, n_i, in terms of number of particles of type i is $n_i = N_i/N_A$. $R = k_B N_A$, so that

$$G_{mix} = RT \sum_{i=1}^{k} n_i \ln x_i \qquad 4.100$$

We can now use the definition of chemical potential:

$$\mu_{i,mix} = \left.\frac{\partial G_{mix}}{\partial n_i}\right|_{P,T,n_{j\neq i}} = RT \frac{\partial}{\partial n_i} \sum_{i=1}^{k} n_i \ln x_i \qquad 4.101$$

to find μ_i in terms of the mole fraction x_i. Noting that in the differentiation $x_j = n_j/n_t$, it is convenient to express the G_{mix} equation as

$$\frac{G_{i,mix}}{RT} = n_i\left[\ln n_i - \ln n_t\right] + \sum_{j\neq i}^{k} n_j\left[\ln n_j - \ln n_t\right] \qquad 4.102$$

Differentiation of this with respect to n_i, holding all other n_j constant yields

$$\mu_{i,mix} = RT \ln x_i \qquad 4.103$$

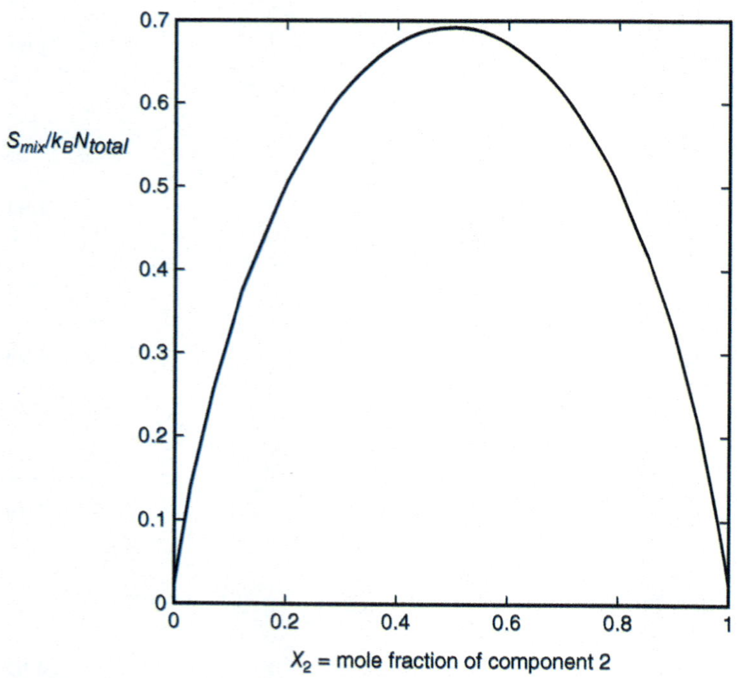

Figure 4.17 Mixing entropy of two components.

When component i is in the pure state $x_i = 1$ and so $\mu_{i,mix} = 0$. However, the chemical potential is not normally zero for a nonideal substance, since there can be cohesion energies or other enthalpic interaction, so the pure state chemical potential, usually termed the *standard chemical potential*, is denoted by

$$\mu_{i,pure} = \mu_i^0 \qquad 4.104$$

The complete chemical potential is thus

$$\mu_i = \mu_i^0 + RT \ln x_i \qquad 4.105$$

The chemical potential for an ideal substance in its pure state is $\mu_{i,ideal}^0 = 0$, so the chemical potential is due purely to the entropy of mixing.

4.13 Chemical Potential for Nonideal Solutions

Experimental evidence of nonideality can be seen when the change of a solution characteristic ceases to change linearly with the concentration of a solute, e.g. light scattering, osmotic pressure, and other easily measurable quantities. For example, if a solute (2) is in a solvent (1) in a nonideal solution with repulsive forces among components then the light scattering intensity and osmotic pressure behave qualitatively as shown in Figure 4.18; osmotic pressure has a positive second derivative, whereas light scattering has a negative second derivative, reaches a maximum, and then decreases.

An empirical means of expressing nonideality is via the phenomenological *activity coefficient* γ_i, which modifies the mole fraction of component i, x_i, to yield an effective concentration $\gamma_i x_i$:

$$\mu_i = \mu_i^0 + RT \ln(\gamma_i x_i) \qquad 4.106$$

Expanding Equation 4.106 for $i = 1$ gives

$$\mu_1 = \mu_1^0 + RT \ln(\gamma_1 x_1) \cong \mu_1^0 + RT \ln \gamma_1 + RT \ln(1 - x_2) \qquad 4.107$$

For an ideal solution, $\gamma_i = 1$. For a two-component system $x_1 = 1 - x_2$ and for a dilute solution $x_2 \ll x_1$, i.e. in a dilute solution the mole fraction of solute (2) is very small compared to the mole fraction of the pure liquid (1). The *dilute solution approximation* allows expansion of ln(1-x_2) to first order, which can be used subsequently, in the form

$$\ln(1 - x_2) \cong -x_2 \quad \text{Dilute Solution Approximation} \qquad 4.108$$

Figure 4.18 Light scattering intensity and osmotic pressure become nonlinear as the mole fraction of solute increases. The nonlinearity is the signature of nonideal effects in the solutions.

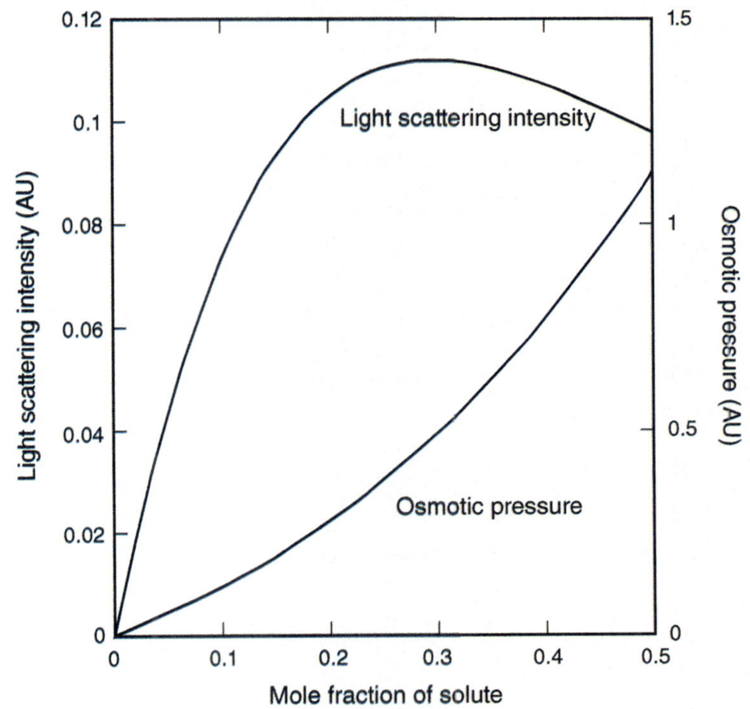

It is often convenient to express the mole fraction x_2 in terms of mass concentration c_2 (g/cm^3) via the dilute solution approximation:

$$x_2 = \frac{n_2}{n_1 + n_2} \cong \frac{n_2}{n_1} = \frac{v_1^0 c_2}{M} \tag{4.109}$$

where M is the molar mass (g/mol) of the solute, and v_1^0 is the specific volume of the solvent in a reference state 0. Then using the first-order expansion of $\ln(1 - x_2)$:

$$\mu_1 = \mu_1^0 + RT\ln\gamma_1 - \frac{RTv_1^0 c_2}{M} \tag{4.110}$$

Now, γ_1 is a function of the solute concentration c_2, that is $\gamma_1 = \gamma_1(c_2)$. γ_1 is initially linear in c_2 at low concentration, before nonideal effects become apparent, so it is assumed that the first deviation from linearity occurs when two of the solute particles interact. This interaction is then a second-order process and is proportional to the number of particles squared, and hence is proportional to c_2^2. As c_2 increases further, and the solution becomes more crowded, three-body interactions can occur, which introduce a c_2^3 term, etc. Hence, $\ln\gamma_1$ can be expressed in a power series of the form:

$$\ln\gamma_1 = -v_1^0 \sum_{i=2} A_i c_2^i \tag{4.111}$$

where the A_i are termed the *virial coefficients*. A_2 is the *second virial coefficient*, A_3 is *the third virial coefficient*, etc. (A_1 is the reciprocal of M). At very low c_2, the linear term for c_2 is already accounted for in Equation 4.110 so that the series for $\ln\gamma_1$ in Equation 4.111 commences in order c_2^2. The negative sign above comes from the experimental fact that repulsive interactions in the solution lower the chemical potential. $\ln\gamma_1$ can then be expressed as

$$\frac{\ln\gamma_1}{v_1^0} = -A_2c_2^2 - A_3c_2^3 - A_4c_2^4 + \cdots \tag{4.112}$$

and the chemical potential for a nonideal solution is thus

$$\mu_1 = \mu_1^0 - RTv_1^0\left(\frac{c_2}{M} + A_2c_2^2 + A_3c_2^3 + A_4c_2^4 + \cdots\right) \tag{4.113}$$

The virial coefficients have a formal connection to interparticle potential energy, as will be shown shortly. $A_2 = A_3 = A_4 ... = 0$ for an ideal solution. $A_2 > 0$ when interparticle interactions are repulsive and $A_2 < 0$ when they are attractive.

4.13.1 Osmotic Pressure and Relation to Chemical Potential

While osmotic pressure was introduced in analogy with an ideal gas in Section 4.8, it is a concept that can be derived directly from chemical potential and the requirement of equality of chemical potential for solute existing among different phases.

In Section 4.8, a semipermeable membrane was used to divide an aqueous solution into one side (α) with water and another side (β) with sugar dissolved in the water, the membrane being impermeable to sugar. This led to an initial net osmotic flow of water into the side with sugar and the creation of an osmotic pressure that built up and finally stopped the net flow at equilibrium.

Now, with the rule that chemical potential must be equal for a species in equilibrium in different phases, the situation can be revisited. The water (1) will seek equal chemical potential on each side, but the sugar will not, since it doesn't exist in side α. At equilibrium

$$\mu_1^\alpha = \mu_1^\beta \tag{4.114}$$

But

$$\mu_1^\alpha = \mu_1^0 \tag{4.115}$$

because on the α side, water is in its pure state, μ_1^0. In β, however, the chemical potential is

$$\mu_1^\beta = \mu_1^0 + RT\ln(\gamma_1 x_1) \neq \mu_1^0, \tag{4.116}$$

This means, as it stands, the system cannot reach equilibrium. For $\gamma_1 \leq 1$, as is the case in a good solvent, e.g. water is a good solvent for sugar, $\mu_1^\beta = \mu_1^0 + RT\ln(\gamma_1 x_1) < \mu_1^0$, and so enough water can never flow to make $x_1 = 1$. Hence, there is some term missing, let's call it $\Delta\mu_1^\beta$, in inequality 4.116, since chemical potential must be the same on both sides of the membrane.

$$\mu_1^\beta = \mu_1^0 + RT\ln(\gamma_1 x_1) + \Delta\mu_1^\beta = \mu_1^0 \tag{4.117}$$

This means that

$$RT\ln(\gamma_1 x_1) = -\Delta\mu_1^\beta \tag{4.118}$$

Now

$$\mu_1 = \frac{\partial G}{\partial n_1}, \quad \text{where } G = U + PV - TS \tag{4.119}$$

Consider that the ideal chemical potential has already taken account of S via entropy of mixing, and that in the solution at hand, and in most living state circumstances, T and V are constant. U is composed of energy of thermal motion and inter-particle interactions, which are taken account of by γ_1. Hence, the only other possibility is that the pressure is not constant. A logically necessary postulate then is that there is an additional pressure term besides the atmospheric pressure P_a the solution is exposed to. Let the total pressure at equilibrium be $P_{eq} = P_a + \Pi$, where Π will be termed the *osmotic pressure* and represents a real pressure that must develop on the β side in order for equilibrium to be reached for the water on both sides α and β.

An expression for Π can be found as follows. The added chemical potential on side β is

$$\Delta\mu_1^\beta = \int_{initial}^{equilibrium} d\mu_1^\beta \tag{4.120}$$

that is, $\Delta\mu_1^\beta$ is the differential chemical pressure integrated from the initial state when the two sides are placed in dialysis contact and osmotic flow of water begins, until the equilibrium state, at which point there is no more osmotic flow of water. This added chemical potential depends only on pressure, as just argued

$$d\mu_1 = \frac{\partial \mu_1}{\partial P} dP = \frac{\partial}{\partial P}\left(\frac{\partial G}{\partial n_1}\right) dP \tag{4.121}$$

where the term in parentheses is the definition of μ_1 and the superscript β has been suppressed for convenience of notation. Interchanging the order of differentiation, since P and n_1 are independent variables,

$$d\mu_1 = \frac{\partial}{\partial n_1}\left(\frac{\partial G}{\partial P}\right) dP = \frac{\partial V}{\partial n_1} dP \tag{4.122}$$

The partial derivative represents how much the system's total volume increases per addition of a mole of solvent (1). This is the molar specific volume v_1^0 given in Equation 4.68, which for component (1) is

$$\frac{\partial V}{\partial n_1} = v_1^0 \tag{4.123}$$

Hence,

$$\Delta\mu_1^\beta = v_1^0 \int_{P_a}^{P_a+\Pi} dP = v_1^0 \Pi \quad 4.124$$

So, the osmotic pressure Π is given by

$$\Pi = \frac{-RT}{v_1^0}\ln(\gamma_1 x_1) \quad 4.125$$

or, in terms of the virial coefficients, using the above expression for $\ln(\gamma_1 x_1)$

$$\Pi = \frac{-RT}{v_1^0}\ln(\gamma_1 x_1) = RT\left(\frac{c_2}{M} + A_2 c_2^2 + A_3 c_2^3 + A_4 c_2^4 + \cdots\right) \quad \text{Osmotic Pressure for Non-Ideal System} \quad 4.126$$

where c_2 is the equilibrium mass concentration of solute (in g/cm^3). In the absence of interactions $A_2 = A_3 ... = 0$, and the expression becomes equivalent to the ideal osmotic pressure of Equation 4.57. Π has the ability to exert force and do work, like any other type of pressure. It was shown in Section 4.8 how attaching the membrane to a moveable piston would allow work to be done while the osmotic flow occurs as equilibrium is sought and finally reached.

4.13.2 Measurement of Number Average Molar Mass and Average A_2 by Osmometry

There are many types of osmometers. The U-tube osmometer was described in Section 4.8. Osmotic pressure Π is measured at a series of solute concentrations, c_2. From Π versus c_2 the osmometer can be used to measure M and A_2, and sometimes also higher virial coefficients. Often, polymer and colloid solutions contain particles of the same type but with a distribution of molar masses. Let M_i be the molar mass of the ith component of the solute population and c_i its concentration.[9] The partial osmotic pressure, Π_i, due to this component is

$$\Pi_i = RT\left(\frac{c_i}{M_i} + \sum_j A_{2,i,j} c_i c_j + \sum_{j,k} A_{3,i,j,k} c_i c_j c_k + \cdots\right) \quad 4.127$$

$A_{2,i,j}$ measures the interaction between solute components i and j, and the sum is over all components. $A_{3,i,j,k}$, which is a three-body interaction, is double summed over all components. The total osmotic pressure is obtained by summing over all partial pressures:

$$\Pi = \sum_i \Pi_i = RT\sum_i\left(\frac{c_i}{M_i} + \sum_j A_{2,i,j} c_i c_j + \sum_{j,k} A_{3,i,j,k} c_i c_j c_k + \cdots\right) \quad 4.128$$

Noticing that the first term in the sum is the molar concentration of component i:

$$n_i = \frac{c_i}{M_i} \quad 4.129$$

and using the definition of *number average mass* M_n (see Chapter 7.12 for a more detailed discussion of M_n):

$$M_n = \frac{\sum_i c_i}{\sum_i n_i} = \frac{c_2}{\sum_i n_i} \quad 4.130$$

where c_2 is the total solute concentration. This allows Π to be written as

$$\frac{\Pi}{c_2} = RT\left(\frac{1}{M_n} + \langle A_2\rangle c_2 + \langle A_3\rangle c_2^2 + \cdots\right) \quad 4.131$$

---

9 Because the solvent concentration scarcely changes in the dilute solution approximation, c_i is the concentration of the ith component of the solute, which is formally $c_{2,i}$. For simplicity of notation c_i will be used for $c_{2,i}$ whenever the dilute solution approximation is used. c_2 is the total concentration of solute, the sum of the c_i.

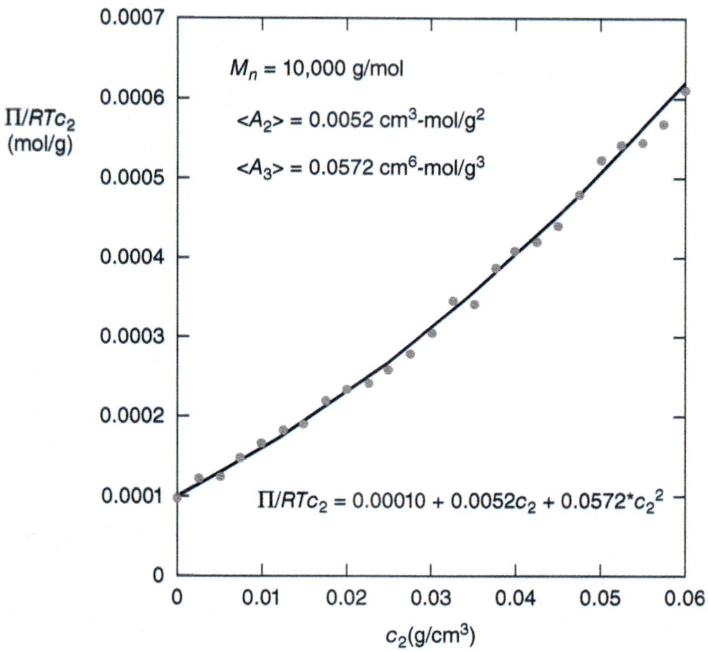

Figure 4.19 Determination of M_n, $<A_2>$, and $<A_3>$ by osmotic pressure.

Osmotic pressure data can hence be used to determine M_n of a macromolecular solute, and an average of the corresponding second virial coefficient, A_2. Extrapolating Π/c_2 versus c_2 to $c_2 = 0$ yields an intercept of $1/M_n$. The linear slope represents a complicated average of the different values of the second virial coefficient over a double sum, $<A_2>$ and the quadratic term gives a triple-sum average of the third virial coefficient $<A_3>$. Figure 4.19 shows how M_n, $<A_2>$ and $<A_3>$ can be determined from osmotic pressure data, using a quadratic fit.

For an ideal solution, $<A_2>=<A_3>=0$ and the osmotic pressure equation reduces to a form similar to Equation 4.57

$$\frac{\Pi}{c_2} = \frac{RT}{M_n} \tag{4.132}$$

4.13.3 Dialysis

In Section 4.8, several scenarios were considered for osmotic pressure; a closed osmotic cylinder with a membrane led to a final Π which could possibly be great enough to either rupture the membrane or burst the cylinder. In an open configuration, like the U-tube osmometer, a column of water was lifted. When attached to an external mass, useful work could be done in lifting the mass due to osmotic pressure. With work input into the system reverse osmosis led to separation of components, such as desalination of water.

Here, we add one more important example of a semipermeable membrane system seeking to equalize chemical potential and creating osmotic pressure in the process, namely, *dialysis*.[10] The general idea of dialysis is to separate out small molecules from the β phase in Figure 4.20 containing both solvent (1), small molecules (such as ions, sugar, etc.) (2), and proteins or other biomacromolecules (3). The β phase is placed in a closed dialysis "sack" consisting of a semipermeable membrane, that allows solvent (1) and small molecules (2) to pass freely through, while blocking the biomacromolecules (3) from leaving the sack. The filled dialysis sack is placed in the external α phase, which contains solvent (1) without the small molecules (2), or with a different concentration of the small molecules, and without macromolecules (3).

In the case where the concentration of small molecules or ions (2) in the β phase is greater than in the α phase, the small molecules in the β phase will pass through the membrane attempting to equalize their chemical potential (concentration) in both α and β phases, while solvent (1) in the α phase will pass through the membrane into the β phase attempting to

10 "Dialysis" comes from the Greek words "dia" (apart) and "luein" (to set free) and means to split or separate.

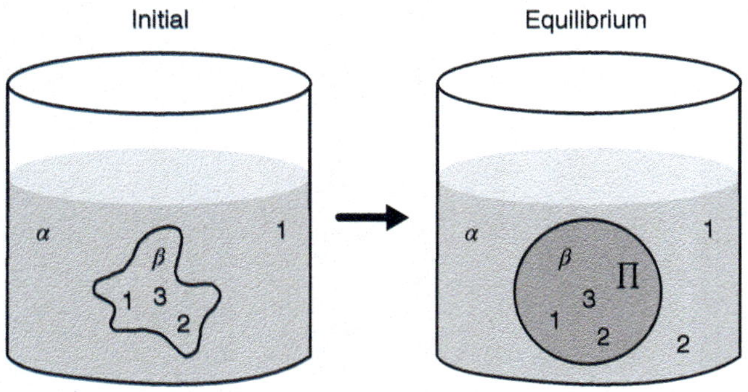

Figure 4.20 A closed dialysis membrane, β phase, contains a solvent (1), a dissolved salt (2), and macromolecules (3). The membrane is permeable to water and salt ions, but not to the macromolecules. At equilibrium, the ion concentrations are the same in the α and β phases, and the osmotic pressure in the β phase produced by the macromolecules swells the sack.

equalize its chemical potential. Meanwhile, the large molecules and particles (3) stay in the β phase. A net osmotic pressure is created in the dialysis sack due to the biomacromolecules or other large particles, which swells the sack. If the osmotic pressure is high enough the sack can burst. To make the dialysis most efficient the solvent in the α phase is stirred and flows past the membrane, thus ensuring the maximum concentration gradient of small molecules across the membrane separating the α and β phases. If the concentration of (2) in the external α phase is greater than in the β phase then (2) will flow into the sack, while water flows out, thus shrinking the sack.

Most people are familiar with dialysis in the medical context. When the kidneys are no longer able to filter out undesirably high levels of urea, potassium, and phosphorous, dialysis can be used to lower their concentrations by exposing the blood, via a semipermeable membrane, to an external solution, which contains some components at physiological concentration, such as calcium, but none of the undesirable small molecules. The blood cells and plasma proteins are too large to flow through the semipermeable membrane. Medical dialysis can be performed by pumping the patient's blood through an external membrane/dialysate unit, or the dialysate can run, via a tube, into the patient's peritoneal cavity, which houses the intestines, and this acts as a natural semipermeable membrane.

Dialysis is used in many other applications. It is frequently used to purify both synthetic and biological macromolecules, and to remove salt ions and other small molecules. It is especially useful to purify proteins. Dialysis membranes are typically made of nitrocellulose or cellulose acetate, which can be manufactured to have different pore sizes to provide permeation cut-offs at desired sizes or molar masses. When purifying macromolecules dialysis tubing is loaded with the macromolecular solutions, sealed, and placed in a bath. Water flowing into the tubing swells the tubing and increases the osmotic pressure, while the small ion content of the β phase reaches that of the α phase, as long as the volume in the α phase is much greater than in the β phase. Frequently, the α phase is removed and replenished with fresh α phase liquid, once or more during dialysis. Monitoring the concentration of components entering the dialysate from the tubing, either continuously or at intervals, can be used to measure the kinetics of the process.

4.14 Excluded Volume Approach to Nonideal Solutions

Nonideality in dilute solutions was treated above by introducing the phenomenological notion of the activity coefficient γ. This was then expanded in a power series to yield equally phenomenological virial coefficients A_2, A_3, A_4, etc. For repulsive interactions $A_2 > 0$, which increases the chemical potential of the solute μ_2, hence decreasing the solvent chemical potential μ_1, as reflected in the virial coefficient expansion for μ_1, above.

Here, the notion of *excluded volume* is introduced that leads to a physically meaningful description of the virial coefficients and nonideality. The basic notion is that each particle in a system excludes other components from its vicinity by virtue of its own volume, or an effective volume. This reduces the number of ways that solvent and solute particles can mix and hence reduces the entropy of mixing, which increases the Gibbs energy, and hence also the chemical potential. Excluded volume can include not only the geometric "hard" volume of the particle, but also other effects, such as electrostatic and hydrophobic interactions.

Excluded volume is the volume that becomes unavailable to a particle due to the presence of another particle. In the following, β will be used to denote excluded volume between two particles. The simplest case is that of two hard (impenetrable) spheres of equal radius R, as shown in Figure 4.21. The excluded volume is the dashed line around the central sphere, i.e. the centers of the two spheres can come no closer than $2R$, so that the excluded volume is

$$\beta = \frac{32\pi R^3}{3} \tag{4.133}$$

In the case where the spheres have different radii R_1 and R_2:

$$\beta = \frac{4\pi\left(R_1 + R_2\right)^3}{3} \tag{4.134}$$

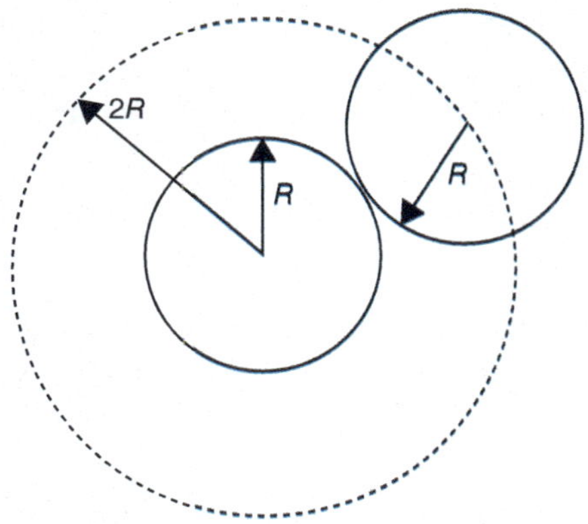

Figure 4.21 The excluded volume (dashed line) between two identical hard spheres.

A fuller account of excluded volume will be given below that can include other interactions besides impenetrability, such as electrostatic repulsion or attraction.

The change in solute chemical potential $\Delta\mu_{2,ev}$ due to excluded volume reducing the entropy of mixing is now sought. The change in Gibbs energy due to excluded-volume restricted entropic mixing will be positive:[11]

$$\Delta G_{ev} = -T(S_{ev} - S_{no\ ev}) > 0 \tag{4.135}$$

and the new equilibrium value of G with excluded volume is

$$G_{ev} = G_{no\ ev} + \Delta G_{ev} \tag{4.136}$$

where S_{ev} is the entropy of mixing when excluded volume is present and $S_{no\ ev}$ is the ideal entropy of mixing computed by Equation 4.98 and denoted as S_{mix} in that equation. Once ΔG_{ev} due to excluded volume is found it can be related back to $\Delta\mu_{2,ev}$ as follows. At equilibrium with excluded volume $dG_{ev} = 0$ means that $d(\Delta G_{ev}) = 0$, so that the Gibbs–Duhem equation is

$$n_1 \Delta\mu_{1,ev} + n_2 \Delta\mu_{2,ev} = 0 \tag{4.137}$$

where n_1 and n_2 are the number of moles of components 1 and 2, respectively. The change in chemical potential for the solvent (1) can be found from $\Delta\mu_{2,ev}$ by

$$\Delta\mu_{1,ev} = -\frac{n_2}{n_1}\Delta\mu_{2,ev} \tag{4.138}$$

where

$$\Delta\mu_{2,ev} = \frac{\Delta G_{ev}}{n_2} \tag{4.139}$$

With $\Delta\mu_{1,ev}$, it will be possible to revisit Equation 4.113 for μ_1 and arrive at a physical interpretation of A_2, which until now is just a phenomenological parameter.

In the following, the dilute solution approximation is again used, in which the amount of solvent is much greater than the amount of solute, i.e. $N_1 \gg N_2$, where N_1 and N_2 are the number of particles of components 1 and 2, respectively. For more than two components, N_1 must be much greater than the sum of all solutes.

11 Excluded volume is normally positive in stable solutions. If there is attraction between particles the excluded volume can be negative.

With ideal particles (no excluded volume), the total change in G at fixed T and P can be found from the number of ways, $\Omega_{no\ ev}$, of mixing N_1 solvent and N_2 solute particles in the dilute solution approximation, $N_2 \ll N_1$. In the dilute solution approximation, Equation 4.93 becomes

$$\Omega_{no\ ev} = \frac{(N_1+N_2)!}{N_1!N_2!} \cong \frac{(N_1+N_2)^{N_2}}{N_2!} \cong \frac{N_1^{N_2}}{N_2!} \tag{4.140}$$

The volume of the solution is denoted V and is given in terms of the partial molar volumes of components 1 and 2, v_1^0 and v_2^0, respectively. Again using the dilute solution approximation

$$V = n_1 v_1^0 + n_2 v_2^0 = (N_1 v_1^0 + N_2 v_2^0)/N_A \cong \frac{v_1^0}{N_A} N_1 \tag{4.141}$$

In the following, N_1 is usefully defined as

$$N_1 = AV,\ A = \frac{N_A}{v_1^0} \tag{4.142}$$

In the case of no excluded volume, each successive particle placed in V has the same number of ways to be placed and the above equation for $\Omega_{no\ ev}$ becomes, for $N_2 \ll N_1$

$$\Omega_{no\ ev} = \frac{(AV)^{N_2}}{N_2!} \tag{4.143}$$

Now, if a particle has a finite size, it removes its excluded volume β for succeeding particles once it is placed in the solvent. Let ξ_i be the number of volume elements available to the ith particle placed in the solvent (without excluded volume $\xi_i = AV$ is the same for all solute molecules). For the first solute molecule placed the full solvent volume V is available (let $i = 0$ for this one), but for the second particle ($i = 1$) there is now V–β available volume, and for the third particle ($i = 2$) there is V–2β, etc. For the ith particle:

$$\xi_i = A(V - i\beta) = AV\left(1 - \frac{i\beta}{V}\right) \tag{4.144}$$

For N_2 solute particles placed in the solvent

$$\Omega_{ev} = \frac{\prod_{i=0}^{N_2} AV\left(1 - \frac{i\beta}{V}\right)}{N_2!} = \frac{(AV)^{N_2}}{N_2!} \prod_{i=0}^{N_2}\left(1 - \frac{i\beta}{V}\right) \tag{4.145}$$

The upper index N_2 is used, since it is essentially equal to N_2–1 (because numbering starts at 0) in the subsequent approximations. S_{ev} is given by

$$\frac{S_{ev}}{k_B} = \ln\Omega_{ev} = \ln\left\{\frac{(AV)^{N_2}}{N_2!}\prod_{i=0}^{N_2}\left(1 - i\frac{\beta}{V}\right)\right\} = N_2\ln(AV) - \ln(N_2!) + \sum_{i=0}^{N_2}\ln\left(1 - i\frac{\beta}{V}\right) \tag{4.146}$$

Since $i\beta \ll V$

$$\frac{S_{ev}}{k_B} \cong N_2\ln(AV) - \ln(N_2!) - \frac{\beta}{V}\sum_{i=0}^{N_2} i \tag{4.147}$$

Since N_2 is large, although much smaller than N_1

$$\sum_{i=0}^{N_2} i \cong \frac{N_2^2}{2} \tag{4.148}$$

The change in Gibbs energy between excluded volume and no excluded volume is

$$\Delta G_{ev} = -T(S_{ev} - S_{no\,ev}) = \frac{RTN_A n_2^2}{2V}\beta \tag{4.149}$$

So

$$\Delta\mu_{2,ev} = \frac{RTN_A n_2}{2V}\beta \tag{4.150}$$

In other words, positive excluded volume increases G and $\Delta\mu_2$. The corresponding change in μ_1 due to excluded volume is

$$\Delta\mu_{1,ev} = -\frac{RTN_A n_2^2}{2Vn_1}\beta \tag{4.151}$$

Comparing this with the second term of the virial expansion for μ_1 above shows the connection to the second virial coefficient A_2:

$$v_1^0 A_2 c_2^2 = \frac{N_A n_2^2}{2n_1 V}\beta \tag{4.152}$$

where c_2 is the mass concentration (g/cm^3) of the solute, component 2. Using

$$n_2 = \frac{c_2 n_1 v_1^0}{M_2} \tag{4.153}$$

where M_2 is the molar mass (g/mol) of the solute and

$$V \cong n_1 v_1^0 \tag{4.154}$$

yields the direct connection between A_2 and excluded volume β:

$$A_2 = \frac{\beta N_A}{2M_2^2} \quad \text{Relation of 2nd Virial Coefficient to Excluded Volume} \tag{4.155}$$

The third virial coefficient A_3 is more complex to compute since it involves three-body interactions and how permutations of pairs of particles combine to exclude volume for a third particle.

4.14.1 General Form for Excluded Volume

Repulsive and attractive interactions can create positive and negative excluded volume, respectively. Let $U(\vec{r}_{12})$ be the potential energy between two particles whose centers are joined by $\vec{r}_{12}$. The probability of the centers being found at $\vec{r}_{12}$, $g(r_{12})$, follows from the Boltzmann distribution:

$$g(\vec{r}_{12}) = \exp\left(\frac{-U(\vec{r}_{12})}{RT}\right) \tag{4.156}$$

The probability that they are *not* at $\vec{r}_{12}$ is $1 - g(\vec{r}_{12})$. The excluded volume is then found by integrating over all $\vec{r}_{12}$ in 3D:

$$\beta = \int \left[1 - \exp\left(\frac{-U(\vec{r}_{12})}{RT}\right)\right] \mathrm{d}^3 r_{12} \tag{4.157}$$

Hence

$$A_2 = \frac{N_A}{2M_2^2}\int \left[1 - \exp\left(\frac{-U(\vec{r}_{12})}{RT}\right)\right] \mathrm{d}^3 r_{12} \tag{4.158}$$

The simplest example is the one mentioned above for impenetrable spheres. In this spherically symmetric case, only the distance between sphere centers, r_{12}, enters and $U(r_{12}) = \infty$ for $r_{12} \le 2R$ and $U(r_{12}) = 0$ for $r_{12} > 2R$, shown in Figure 4.22a. The corresponding $g(r_{12})$ is shown in Figure 4.22b, as well as the integrand for A_2, $1 - g(r_{12})$, in Figure 4.22c. The excluded volume integral yields $\beta = (4\pi/3)(2R)^3$ so that the virial coefficient for identical hard spheres of radius R is

$$A_2 = \frac{16\pi N_A R^3}{3M_2^2} \qquad 4.159$$

Boltzmann made the calculations for A_3 and A_4 for hard spheres, the latter being a particularly formidable mathematical task, and arrived at

$$A_3 = \frac{5MA_2^{\ 2}}{8} \qquad 4.160$$

and

$$A_4 = 0.28695 * M^2 A_2^{\ 3} \qquad 4.161$$

The computation of β for interacting, rigid rods of length L, and diameter d is more complicated. Zimm, and separately, Onsager, arrived at

$$\beta = \frac{\pi d L^2}{2} \qquad 4.162$$

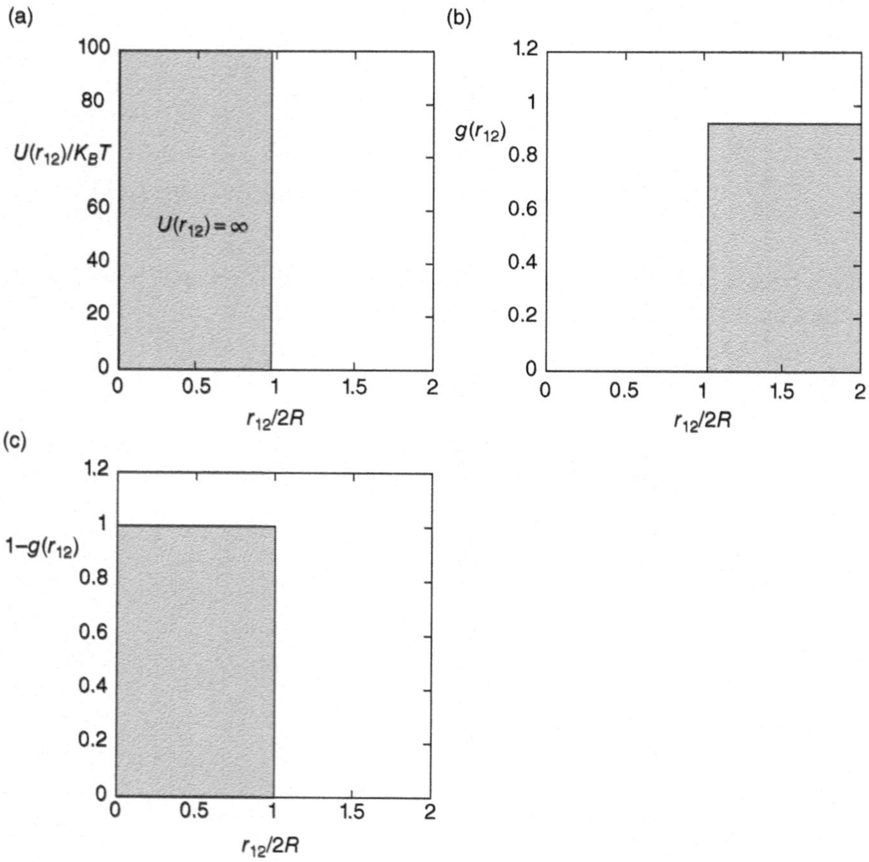

Figure 4.22 For hard spheres: (a) $U(r_{12})$, (b) $g(r_{12})$, and (c) $1 - g(r_{12})$.

4.14.2 Depletion Forces

Excluded volume is largely responsible for an attractive, entropically based *depletion force*. Consider large hard spheres mixed with smaller ones. When large spheres come near each other's excluded volume, the smaller spheres are excluded from the intervening space. This leaves pure solvent between the spheres, which leads to an imbalance in osmotic pressure. The small, excluded spheres produce a net osmotic pressure compared to the pure solvent phase between the large spheres, and the large particles are pushed together.

4.15 Chemical Equilibrium

Chemical equilibrium is not a static condition, despite what the word might seem to imply. If we consider the simplest process, where A and B can transform back and forth into each other:

$$A \leftrightarrow B \tag{4.163}$$

there will be equilibrium concentrations of A and B, but there will nonetheless always be some A transforming to B, and vice versa, while maintaining these equilibrium concentrations constant. This is known as *dynamic equilibrium*. At dynamic equilibrium, the rate of the forward process equals the rate of the backward process. For this particularly simple process, the Boltzmann distribution can be used directly to get an equilibrium constant K_{eq} which is the ratio of the concentration of B to A:

$$K_{eq} = \frac{[B]}{[A]} = \frac{\exp\left(-G_B^0 / RT\right)}{\exp\left(-G_A^0 / RT\right)} = \mathrm{e}^{-\Delta G^0 / RT} \tag{4.164}$$

where $\Delta G^0 = G_B^0 - G_A^0$, G_A^0, and G_B^0 are the standard free energies of the reactants and products, respectively. If $\Delta G^0 < 0$, that is, B has a lower standard free energy state than A, then $K_{eq} > 1$, which favors more B, and $0 < K_{eq} < 1$, when $\Delta G^0 > 0$.

To find the equilibrium concentrations in general, consider a completed reaction in equilibrium, in the form

$$\sum_R n_j R_j \leftrightarrow \sum_P m_j P_j \tag{4.165}$$

where R_j and P_j are the reactant and product species, respectively, and n_j and m_j are their stoichiometric ratios in the reaction. The summation on the left is over all reactants, and over all products on the right. When the reaction is over and equilibrium is reached the difference in free energy between the reactants and products will be zero. The relationship between G and μ from the summation rule of Equation 4.71 gives

$$G_P - G_R = 0 = \sum_P m_j\left[\mu_{P,j}^0 + RT\ln(\gamma_{P,j}x_{P,j})\right] - \sum_R n_j\left[\mu_{R,j}^0 + RT\ln(\gamma_{R,j}x_{R,j})\right] \tag{4.166}$$

where the $\mu_{R,j}^0$ and $\mu_{P,j}^0$ are the standard chemical potentials of reactants and products, respectively, $\gamma_{R,j}$ and $\gamma_{P,j}$ are the activity coefficients, and x_{Rj} and x_{Pj} are the mole fractions. To simplify the notation, let $r_j = \gamma_{R,j}x_{R,j}$ and $p_j = \gamma_{P,j}x_{P,j}$. Then

$$\sum_R n_j RT\ln r_j - \sum_P m_j RT\ln p_j = \sum_P m_j\mu_{P,j}^0 - \sum_R n_j\mu_{R,j}^0 \tag{4.167}$$

The term on the right is the standard chemical potential difference for the reaction, ΔG^0. With this

$$\sum_R n_j \ln r_j - \sum_P m_j \ln p_j = \frac{\Delta G^0}{RT} \tag{4.168}$$

Exponentiating both sides yields the equilibrium constant K_{eq}

$$K_{eq} = \frac{\prod_P p_j^{m_j}}{\prod_R r_j^{n_j}} = \exp\left(\frac{-\Delta G^0}{RT}\right) \quad 4.169$$

If $\Delta G^0 < 0$, then the reaction is exothermic and can proceed spontaneously. For $\Delta G^0 > 0$, the reaction is endothermic and requires energy input to proceed. Importantly, K_{eq} depends only on the difference in free energies in their standard states.

Most chemical reactions involving formation or loss of covalent bonds have $|\Delta G^0| \gg RT$. Consider the reaction:

$$2H_2 + O_2 \leftrightarrow 2H_2O,\ \Delta G^0 = -2.37 \times 10^5\ \text{J/mol} \quad 4.170$$

The negative sign indicates this is exothermic. At $T = 298$ K, $K_{eq} = 3.48 \times 10^{41}$. In other words, the reaction leads to virtually 100% H_2O.

To show an example using more common notation, consider a system containing reactants A and B that produce products C and D according to the following stoichiometry:

$$jA + kB \xleftrightarrow{K_{eq}} lC + mD \quad 4.171$$

The equilibrium constant by Equation 4.169 is then

$$K_{eq} = \frac{[C]^l [D]^m}{[A]^j [B]^k} \quad 4.172$$

In most texts, including this one, the bracketed quantities are molar concentrations of each species (mol/L). Often activities are not known so they are absorbed into K_{eq}. Other units can be used, adjusting the units of K_{eq} accordingly. Note that the units of K_{eq} can vary for different reactions, since the integers j,k,l,m can vary quite widely among different reactions. The form of Equation 4.172 is an example of the *law of mass action*, where the ratio of the products of final species to the products of reactants is a constant at a given temperature.

4.15.1 Ionic Dissociation of Water as an Example of Equilibrium: pH and pK_a

An example of a simple equilibrium involves the dissociation of water:

$$H_2O \xleftrightarrow{K_{eq}} H^+ + OH^- \quad 4.173$$

Often, instead of writing H^+, which looks like a free proton in water, the proton is instead treated as associated with H_2O as H_3O^+. In practical calculations H^+ and H_3O^+ can be used interchangeably. The equilibrium constant is

$$K_{eq} = \frac{[H^+][OH^-]}{[H_2O]} \quad 4.174$$

Because, in the dilute solution approximation, the concentration of water scarcely varies at fixed P and T ($[H_2O] = 55.5$ M at STP), because $[H^+], [OH^-] \ll [H_2O]$, it is customary to subsume $[H_2O]$ into the equilibrium constant, which then gives for the water dissociation constant, K_w, at STP:

$$K_w = [H^+][OH^-] = 10^{-14} (M^2/L^2) \quad 4.175$$

This is the basis for the pH scale:

$$\text{pH} = -\log[H^+] \quad 4.176$$

In pure water at STP, with no added source of either H^+ or OH^-, $[H^+] = [OH^-] = 10^{-7}$ M/L, and pH = 7. K_w increases significantly with T and hence pH decreases: at $T = 10°C$ pH = 7.27, at $T = 40°C$ pH = 6.77, and at $T = 100°C$ pH = 6.14.

Now, consider a weak acid (HA), which dissociates similarly to water

$$\mathrm{HA} \leftrightarrow \mathrm{H}^+ + \mathrm{A}^- \tag{4.177}$$

with a dissociation constant

$$K_A = \frac{[\mathrm{H}^+][\mathrm{A}^-]}{[\mathrm{HA}]} \tag{4.178}$$

Hence, for a weak acid dissociating in water

$$\mathrm{H_2O} + \mathrm{HA} \leftrightarrow \mathrm{H}^+ + \mathrm{A}^- + \mathrm{OH}^- \tag{4.179}$$

So

$$K_{A/W} = \frac{[\mathrm{H}^+][\mathrm{A}^-][\mathrm{OH}^-]}{[\mathrm{HA}]} = \frac{K_w K_A}{[\mathrm{H}^+]} \tag{4.180}$$

and

$$\log\left\{\frac{[\mathrm{A}^-]}{[\mathrm{HA}]}\right\} = \log\left\{\frac{K_A}{[\mathrm{H}^+]}\right\} = \log K_A + \mathrm{pH} \tag{4.181}$$

Defining $pK_a = -\log K_A$, we obtain the desired relationship:

$$\log\left\{\frac{[\mathrm{A}^-]}{[\mathrm{HA}]}\right\} = \mathrm{pH} - \mathrm{p}K_a \tag{4.182}$$

We can use this to compute the fractional ionization of an acid at any pH if we know the acid's pK_a (usually determined by titration techniques): e.g. if pH = pK_a then 50% of the acid is ionized; $[A^-] = [HA]$; if pH–pK_a = 1 then 90.9% of the acid is ionized; $[A^-] = 10[HA]$.

4.15.2 Le Chatelier's Principle

Henry Louis Le Chatelier (France, 1850–1936) studied chemical equilibrium for many decades and formulated what is called *Le Chatelier's principle*. In essence, it states that if a multicomponent system in chemical equilibrium is perturbed by a change in temperature, pressure, or change in concentration of one or more of its components, it will adjust to a new equilibrium in a way that partially offsets the perturbation.

For example, Equation 4.163 shows a very simple process with equilibrium constant $K_{eq} = [B]_0/[A]_0$, where $[A]_0$ and $[B]_0$ are equilibrium concentrations for a given system. If an increment of A is added, $\Delta[A]$, while holding T and P constant, the equilibrium condition shows that some of A must convert to B in order for K_{eq} to remain constant. In other words, initially adding $\Delta[A]$ led to the system's shift to a new equilibrium where the new equilibrium concentration $[A]' < [A]_0 + \Delta[A]$, i.e. the increase in $[A]$ by $\Delta[A]$ was partially counteracted by converting some of the added A into B.

This ties into the general notion of *buffering*, where interacting components of a system work to counteract changes made on it. An example, in regards to the pH discussion above, is that of *buffer solutions*, which are composed of components whose equilibrium concentrations shift in response to changes in a solution's pH, such that the change in pH is counteracted. An example of a buffer solution is a weak acid, such as in Equation 4.177, with its sodium salt, NaA, which is nearly fully dissociated in water by $\mathrm{NaA} \rightarrow \mathrm{Na}^+ + \mathrm{A}^-$. If another acid is added, the added $[H^+]$ can associate with A^- to form AH in Equation 4.177, thus counteracting the added acid, and leading to a much smaller change in pH than if the weak acid and its salt were not present.

Now, consider adding an amount of heat Q into the initial system with $[A]_0$ and $[A]_0$. In a nonreactive, single-component system all of this Q would lead to an increase in T according to the heat capacity of the substance. In the chemical system, however, increasing T will shift the equilibrium toward B, increasing the equilibrium constant K_{eq}. Hence, the added heat goes both into converting some A into B, Q_C, and the rest into increasing T, Q_T; $Q = Q_C + Q_T$. The temperature rise is lessened, that is, counteracted, because of the requirements of equilibrium.

Interestingly, while Le Chatelier's principle was formulated around chemical equilibrium it seems to apply to other well-functioning systems with interdependent components. *Homeostasis* is an example of this. It is mostly used in the physiological sense of an organism's natural attempt to minimize perturbations to its dynamic far-from-equilibrium and well-being. This includes maintenance of concentrations of electrolytes, fluids, blood composition, temperature, and other factors.

Extensions of Le Chatelier's principle have even been found beyond the natural world. In mechanical engineering, for example, an automobile transmission clutch can slip in response to excess torque, rather than being torn apart. A simple means to protect mechanical devices involves a shear pin, which sacrificially breaks in response to a large rotational stress, rather than damage the rest of the device. In the maritime context, boats and ships are often equipped with a sacrificial zinc electrode in the vessel's electrical circuit, which is one of the most easily oxidized metals, which hence protects other metals against corrosion.

It is possible that current global climate change, seemingly produced by the human emission of greenhouse gases, is due to pushing the Earth's buffering capacity beyond normal limits. That is, for reasonably small increases in greenhouse gases the complex buffering capacity of Earth's atmosphere and condensed matter mass can absorb these with negligible changes to the Earth's temperature. Once the increases in greenhouse gas become too large, however, the Earth can no longer buffer them as effectively, and global temperatures rise.

4.15.3 Micellization of Surfactant Molecules in Solution

Surfactants are surface-active molecules. These have a molecular portion, or *moiety*,[12] soluble in water, termed *amphiphilic*,[13] and another molecular portion that is insoluble in water, termed *hydrophobic*.[14] This latter moiety is soluble in organic or "lipid" phase and is often termed *lipophilic*. Remarkably, in aqueous solution, above a *critical micelle concentration*, or CMC, surfactant molecules can spontaneously assemble into small spheroidal structures termed *micelles*. These can also assume other geometries, such as cylinders and lamellae. If the hydrophobic moiety consists of two long alkyl chains then the surfactants can organize into *lipid bilayers*. These latter can form spheroidal *liposomes*, similar to *surfactant vesicles*, which are spheroidal objects with an aqueous interior compartment separated from the surrounding aqueous milieu by a bilayer membrane. Biological membranes are organized on this same motif.

Figure 4.23 is a schematic of micellization. On the left, below the CMC, the surfactant molecules are individually solubilized in aqueous solution. The ellipsoidal head represents the hydrophilic moiety and the squiggly tails represent the hydrophobic moiety. On the left side, below the CMC, surfactant molecules have coated the surface of the solution with the hydrophobic tails pointing into the air. On the right side, above the CMC, surfactant molecules have aggregated into spheroidal micelles, while some stay in solution and on the surface. According to multicomponent solution theory this is a three-phase system (i) surfactant coating the solution surface, (ii) individually solubilized surfactant molecules in solution, and (iii) surfactant molecules in micellar aggregates. The chemical potential of the surfactants in all three phases must be equal.

Sodium dodecyl sulfate (SDS), Figure 4.24, is a common surfactant. SDS is composed of a water-insoluble, linear hydrocarbon chain 12 units long, with a negatively charged sulfate head group, which is soluble in water. In water at low concentration, SDS is sparingly soluble and requires water to organize itself along the hydrophobic portion, which decreases the water's entropy. When the concentration of SDS reaches the CMC the entropy of water can be reduced if a number of SDS molecules aggregate into a micelle, in which the hydrocarbon chains, which are miscible in each other, form the interior, and the hydrophilic sulfate head groups form the interface with the aqueous solution. This allows the water to be released from solvating the hydrocarbons, thus increasing the water's entropy. The entropy of the hydrocarbon chains decreases and the electrostatic repulsion of the negative head groups further increases the Gibbs energy of the micellar state.

---

12 "Moiety" is closely related to the French word "moitié," meaning half of something. In chemistry, it is changed to mean a molecular portion of a substance, not necessarily half.

13 "Amphiphilic" literally means "loving both," in this case both water and lipid phases.

14 "Hydrophobic" literally "water hating."

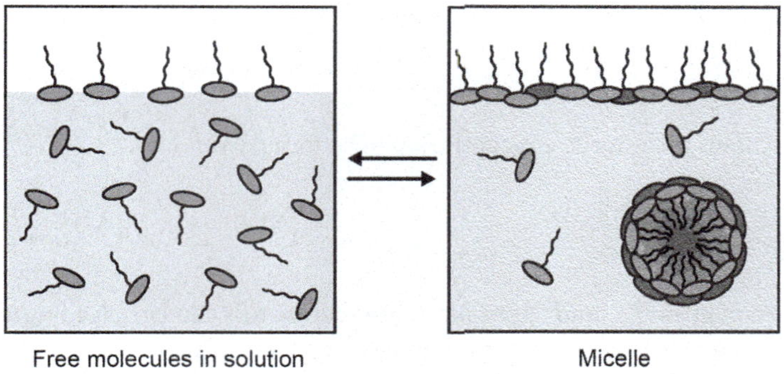

Figure 4.23 Micellization of a surfactant. On the left, the concentration of the surfactant molecules is below the CMC and they remain solubilized. On the right, the concentration is above the CMC, and surfactant micelles form, while leaving an equilibrium concentration of individually solubilized molecules in solution. (Courtesy of Marissa Calko).

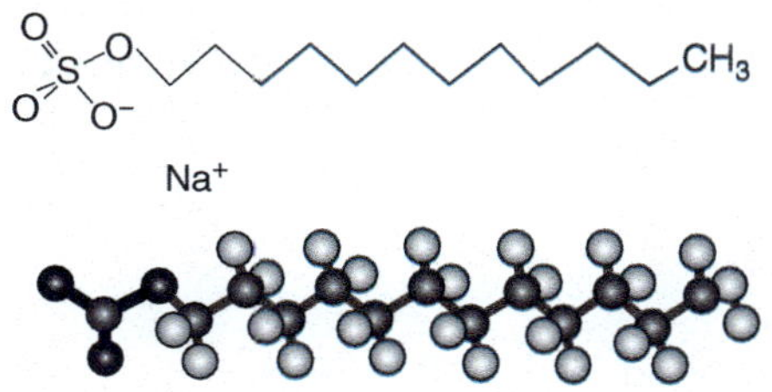

Figure 4.24 Sodium dodecyl sulfate (SDS).

The reduction in net Gibbs energy due to water's entropy increase more than makes up for the increase due to the micellar entropy and head group repulsion. A typical SDS micelle will have about 100 molecules in it and have a diameter of about 4–5 nm.

The transition through the CMC as surfactant concentration is increased can be measured by a variety of methods. Surface tension decreases steadily as surfactant concentration increases and flattens out once the CMC is reached. Light scattering increases abruptly beginning at the CMC since the micelles are much more massive than individual monomers.

For SDS and other charged surfactants, the slope of electrical conductivity versus surfactant concentration abruptly falls at the CMC due to the lower electrophoretic mobility of the micelles compared to individual SDS molecules. Figure 4.25 is an example of the conductivity break at the CMS for SDS in pure water. Below, it will be shown that the actual CMC is somewhat different than the measured $[SDS_{total}]$ at the break point.

In an idealized case, the free molecules of SDS, with concentration [SDS], can be thought of as in equilibrium with micelles of concentration $[SDS_m]$ that incorporate mSDS molecules, (SDS_m), according to the mass action principle,

$$m\text{SDS} \overset{K_{eq}}{\longleftrightarrow} \text{SDS}_m \text{ (micelle)} \tag{4.183}$$

where

$$K_{eq} = \frac{[\text{SDS}_m]}{[\text{SDS}]^m} \tag{4.184}$$

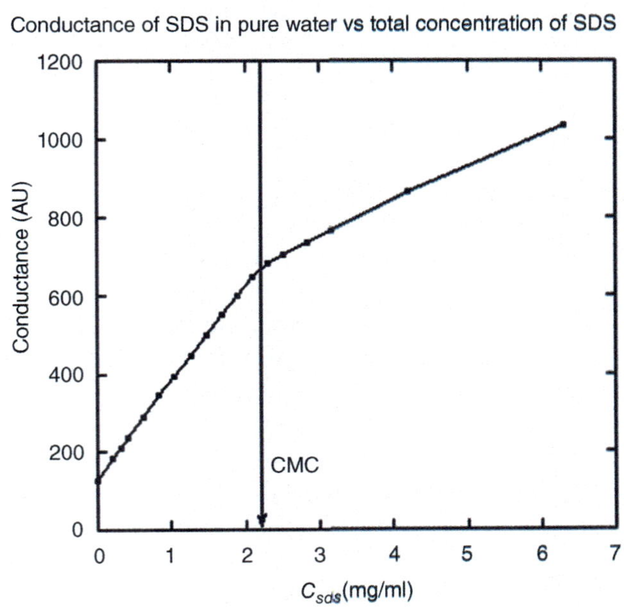

Figure 4.25 Conductivity versus total mass concentration of SDS. The slope of the conductivity abruptly breaks at the CMC. (Zhou, Z. and Reed, W.F., unpublished results).

Using the fact that the total amount of SDS, $[SDS_{total}]$, is

$$[SDS_{total}] = m[SDS_m] + [SDS] \tag{4.185}$$

Allows the micelle concentration to be found from

$$[SDS_m] = K_{eq}\{[SDS_{total}] - m[SDS_m]\}^m \tag{4.186}$$

For a given K_{eq} and aggregation number m, this can be solved numerically to obtain $[SDS_m]$ versus $[SDS_{total}]$.

The actual CMC can be different from the measured $[SDS_{total}]$ if there is a specific value of m, instead of a distribution of monomers in micelles. This can be estimated from the experimentally measured 'CMC', such as in Figure 4.25, which shows $[SDS_{total}]$ on the x-axis, by considering that, at the *CMC*, just as micelles are starting to form, virtually all the SDS is in free SDS molecules, so that at the CMC, CMC $\cong [SDS]$. Then Equation 4.184 can be re-written as

$$\mathrm{CMC}^m \cong \frac{[SDS_m]}{K_{eq}} \tag{4.187}$$

$[SDS_m]$ can be expressed as

$$[SDS_m] \cong \frac{[SDS_{total}] - \mathrm{CMC}}{m} \tag{4.188}$$

and the estimated CMC can then be found from

$$[SDS_{total}]_{\text{at measured break point}} = \mathrm{CMC} + mK_{eq}\mathrm{CMC}^m \tag{4.189}$$

Setting the chemical potentials equal in micellar and free phases, as is required at equilibrium:

$$\mu_{SDS,micelle} = \mu^0_{SDS} + RT \ln x_{SDS} \tag{4.190}$$

where μ^0_{SDS} is the chemical potential of SDS in its pure state and x_{SDS} is the mole fraction of free SDS in solution (ignoring any activity corrections to [SDS], such as via the activity coefficient γ introduced in Section 4.13). In the mass action model of Equation 4.183, for large m virtually all added SDS goes into micelle formation above the CMC, so that the concentration of free SDS remains essentially constant at its CMC value for [SDS] > CMC. Hence, the free energy of micellization per mole of SDS is the difference in chemical potential between the micellar and pure free energies:

$$\Delta\mu_{micellization} = RT \ln x_{SDS} \tag{4.191}$$

4.16 Reaction Kinetics

K_{eq} gives the final concentrations of reactants and products in a reaction but furnishes no information on the rate at which the reaction will occur. Figure 4.26 shows energy as a function of reaction sequence. The energy difference between products and reactants is the ΔG^0 that determines K_{eq} in Equation 4.164. But there is a *free energy of activation* $\Delta G^{\ddagger}$, an *energy barrier* between the reactants and products that determines the reaction rate and hence the time needed to reach equilibrium. If there were no barrier, the reaction would be instantaneous. Activation barriers are generally due to the fact that molecules must make collisions with each other and pass through transitional or *intermediate states* on their way to forming products. The activation barrier corresponds to the highest energy transitional state that must be passed through in order to go downhill toward products. A simplistic representation for the reaction of Equation 4.163 is

$$A \rightarrow I^* \rightarrow B \tag{4.192}$$

where I^* represents one or more intermediate states, marked on Figure 4.26.

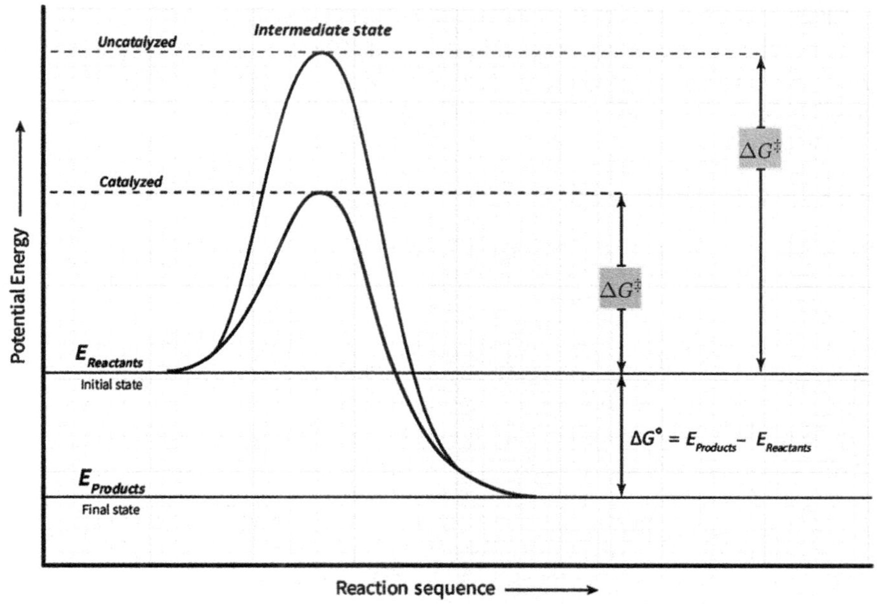

Figure 4.26 The reaction proceeds from left to right. (Courtesy of Marissa Calko).

It is important to realize that ΔG^0 in the figure is unrelated to $\Delta G^{\ddagger}$, and so equilibrium is completely unaffected by $\Delta G^{\ddagger}$. Similarly, since $\Delta G^{\ddagger}$ is unrelated to ΔG^0 the rate of the reaction is independent of ΔG^0. We can use the Boltzmann distribution to surmise a reaction rate. For Equation 4.192, the reaction rate v is the rate at which B is formed: $v = d[B]/dt$. If there is no buildup of intermediates then $v = d[B]/dt = -d[A]/dt$. First, the ratio of reactants that are in the intermediate state I^* to those in the reactant state is

$$\frac{[I^*]}{[A]} = \exp\left(\frac{-\Delta G^{\ddagger}}{k_B T}\right) \qquad 4.193$$

We would expect the rate of the reaction α to be proportional to the concentration of the intermediate state $[I^*]$ so that

$$v = c * [A] \exp\left(\frac{-\Delta G^{\ddagger}}{k_B T}\right) \quad \text{Arrhenius Rate Equation} \qquad 4.194$$

where c is a proportionality factor, which is theoretically very complex, as it must involve detailed quantum mechanical processes. Equation 4.194 is the *Arrhenius equation* for rate, discovered by Svante Arrhenius (Sweden, 1859–1927).

Equation 4.194 shows that reaction rates can be exponentially sensitive to both free energy of activation and temperature. When reaction rates are measured at different temperatures their logarithm can be plotted versus $1/k_B T$. If this turns out linear, then it indicates "Arrhenius rate behavior" and $\Delta G^{\ddagger}$ can be determined. An example of an Arrhenius plot for the aggregation rate of proteins is shown in Figure 10.5.

Catalysts lower $\Delta G^{\ddagger}$and hence speed up reactions. All biochemical reactions are controlled by enzymes which act as catalysts to lower $\Delta G^{\ddagger}$. Enzymes can speed up reactions by a factor of as much as 10^{12}, or more.

We have mentioned that at chemical equilibrium a system has not become static. Rather, there is a continued process of reactants on the left-hand side of the equation toward the right-hand side, and vice versa. What equilibrium tells us is that the rate of these processes must be equal since there is not net flow of molecules from one side of the reaction to the other:

$$\beta_A [A] = \beta_B [B] \qquad 4.195$$

where β_A is the rate (s^{-1}) for conversion of A to B, and β_B is the rate for conversion of B to A at equilibrium. The ratio of these rates is K_{eq}, seen by combining Equations 4.195 and 4.164:

$$\frac{\beta_A}{\beta_B} = K_{eq} = e^{-\Delta G^0/RT} \qquad 4.196$$

Thus, K_{eq} establishes both concentrations and the *ratio of conversion rates* at equilibrium. This ratio depends on ΔG between the two states, but the absolute individual rates cannot be determined just from the ratio. Moreover, these equilibrium rates are unrelated to the reaction rate for a reacting system approaching equilibrium, whose rate α depends on the free energy of activation $\Delta G^{\ddagger}$ according to Equation 4.194.

4.16.1 Order of Reactions

The *order of a reaction r*, is a means of relating a reactant conversion rate to its concentration. Let $[A]$ be the reactant concentration, then

$$\frac{\mathrm{d}[A]}{\mathrm{d}t} = -\alpha[A]^r \tag{4.197a}$$

where α is the rate constant and r is not necessarily an integer, nor necessarily positive. Of course, more than one reactant can be involved. For example for $nA + mB \rightarrow C$ may have a rate equation of the form:

$$\frac{\mathrm{d}[A]}{\mathrm{d}t} = -\alpha[A]^x[B]^y \tag{4.197b}$$

where $r = x + y$. Notice that the units of the rate constant α change with r, x, and y.

A *zeroeth-order reaction* is when $r = 0$, for which $[A](t) = \alpha t + [A]_0$, where $[A]_0$ is the initial reactant concentration, which is completely used up by $t = \alpha/[A]_0$. It is rarely found in a reacting system. However, if $[A]$ is multiplied by system volume V, $V[A]$ represents that amount of A, and hence the mass of A, in the system. There are many processes where total mass decreases in zeroeth order, such as fuel consumed from a fuel tank by an engine working under a constant load.

One of the most ubiquitous reactions in nature is the *first-order reaction*, for which $r = 1$. This universally produces an exponential solution of the form:

$$[A](t) = [A]_0 \mathrm{e}^{-\alpha t} \tag{4.198}$$

This reaction occurs when the rate is proportional to the concentration or amount of reactant. Hence, any spontaneous process, such as radioactivity or photoluminescence will obey this. In reaction 4.163, a mole of product B is formed for every mole of A consumed

$$\frac{\mathrm{d}[A]}{\mathrm{d}t} = \frac{-\mathrm{d}[B]}{\mathrm{d}t} = -\alpha[A] \tag{4.199}$$

so that there is a first-order buildup of B

$$[B](t) = [A]_0\left(1 - \mathrm{e}^{-\alpha t}\right) \tag{4.200}$$

Reactions where a second reactant B combines with A to produce a product C ($A + B \rightarrow C$), and where the concentration of B changes very little during the reaction will produce a *pseudo-first-order reaction*.

For higher order reactions, $r > 1$ Equation 4.197a can be readily integrated to give

$$[A]^{1-r}(t) = (r-1)\alpha t + [A]_0^{1-r} \tag{4.201}$$

For a second-order reaction, $r = 2$, this gives

$$[A](t) = \frac{[A]_0}{1 + \alpha[A]_0 t} \tag{4.202}$$

Second-order reactions occur when a reactant A reacts with itself, such as in the formation of diatomic molecules of a given species, such as O_2.

We will return to reaction kinetics in Chapter 7, in connection with polymerization reactions.

4.17 Phase Transitions

Interactions between the components of a system yield results different than those from non-interacting components. *Phase transitions* are the hallmarks of cooperative systems. Table 4.1 shows a short list of systems and associated phase transitions they can go through. There are several schemes for classifying phase transitions, which essentially try to characterize how sharp or gradual the transitions are. One historical classification was by the *order of the transition*, which is the derivative of free energy with respect to an environmental parameter, such as pressure or temperature at which a discontinuity occurs. For example, for the freezing of water the first derivative of the free energy is discontinuous at the freezing point, and so it is a *first-order phase transition*. For the transition of a magnetic material from ferromagnetic to paramagnetic upon heating, the second derivative of the free energy, the magnetic susceptibility, is discontinuous, so this is a *second-order phase transition*. A characteristic of a first-order phase transition is that it involves *latent heat*. This means that at the point of a phase transition, say the melting of ice, energy is added to the system without a change in the temperature, and the energy added serves to break bonds and bring co-existing water and ice, "slush," from ice to water.

In the living state, there are several phase transitions of fundamental importance, which are due to cooperative organization. Many cooperative bonds are formed when a protein folds from a coil into a helical or organized sheet and then into a globular state. In the globular protein these ordered structures, α helices and β sheets, go through a helix-to-coil phase transition when the protein unfolds or *denatures*. DNA, as a semiflexible coil composed of long double helices, also loses these structures in its own helix-to-coil transition, when it is heated or placed in a denaturing solution. Polysaccharides, likewise, can have internal structure capable of denaturation and transition to disorder. Membranes also have a fluid-to-crystalline structure of their chains. Proteins can crystallize, although this rarely occurs naturally. The biotechnology industry places a high priority of finding means to crystallize proteins so that their 3D structure can be analyzed by X-ray or other means. The condensation of DNA from a highly extended semiflexible coil into a tiny cell nucleus is one of the most dramatic reductions in size that occurs, but this is not spontaneous. It requires association with proteins, such as *histones*, as well as molecular motors to supply work in the condensation process.

One of the simplest models for phase behavior will be introduced in Chapter 8 using a lattice model for calculating free energy of mixing and free energy criteria for predicting phase transitions in polymer systems. In subsequent chapters, the helix–coil and coil–globule transitions will be considered in the light of DNA and protein structure.

For now, at a very simple level, we can think of the phase transition being at the point where the change in free energy due to a change in entropy of a system is balanced by an equal and opposite change in the internal energy. Consider two immiscible substances, such as oil and water. An oil molecule lowers its internal energy when in contact with another oil molecule rather than a water molecule. But at low concentration of oil there is an unfavorable, large decrease in entropy if two oil molecules come together. Hence, oil is very sparingly soluble in water, but as the oil concentration increases above a certain amount, the decrease of free energy due to oil association offsets the increase in free energy due to entropy reduction. After this concentration, added oil will separate into a macroscopic oil phase.

Table 4.1 Examples of phase transitions in specific systems.

System	Phase transition
Crystal lattice formation	Liquid → solid
Boiling	Liquid → gas
Sublimation	Solid → gas
Ferromagnetism	Paramagnetic → ferromagnetic
Superconductivity	Normal → superconductivity
(Formation of electron pairs)	
Protein folding	Random coil → globule
Micelles	Monomers → micelle
Membrane	Disordered gel → gel

4.18 Nonequilibrium Thermodynamics

The living state is far from equilibrium. Energy flows from a source through the organized living state system, which uses available free energy to maintain and/or increase order (e.g. in embryonic development and childhood/adolescent growth greater order is created), performs living state processes (synthesis of biomolecules, nerve conduction, locomotion, cognition) and exhausts waste energy and material to the environment. This is pictured schematically in Figure 4.27, where the Sun is the energy source, emitting electromagnetic radiation and a relatively small amount of charged particles (the *solar wind*). The constant energy flow through the system, shown as the biosystem, keeps it far from equilibrium, keeps the "electronic temperature" high, which allows for the creation of biomolecules and organized structures, and allows cycles of matter/energy transformations to develop. While the entropy of the biosystem can decrease, the net result is exhalation of net heat and entropy to the environment. The Earth radiates its energy, as approximately a black body, to the "Cosmic Sky."

The motive force for many nonequilibrium processes comes from flows of material in response to concentration, electrical, and gravitational gradients. A particle in a liquid is always subjected to forces due to random thermal collisions even in the absence of gradients and applied fields. This leads to the *diffusion* of macromolecules both in living state and ex-vivo solutions. Diffusion, which is often described by *random walk* models, is frequently the rate-limiting step in controlling biochemical and synthetic reactions.

When external forces are applied, macromolecules can combine their diffusional motion with directed motion due to the applied force. Examples of applied forces include electrical fields, in which case the macromolecule, if charged, responds with a velocity related both to the field strength and the particle's *electrophoretic mobility*. Gravitational fields or the acceleration from a centrifuge, in which case the *sedimentation* coefficient becomes central. Under shear stress, the *intrinsic viscosity* is a macromolecular property of interest. In all these scenarios, the underlying physical quantity is the macromolecule's *frictional coefficient* f.

We will return to motion, flows, and these concepts in Chapter 6.

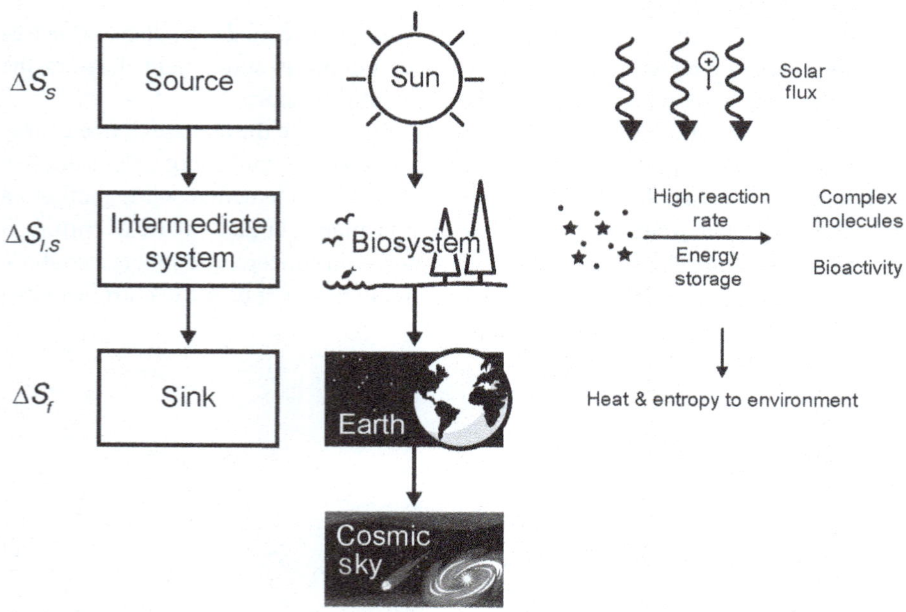

Figure 4.27 Conceptual flow of energy from a source, e.g. the sun emitting radiation and charged particles. The flow of radiation through a living state system creates organized molecules and structure. The exhalation of waste heat increases Earth's environmental entropy. The Earth, with an average surface temperature around 290K, radiates heat to the cosmic sky. (Courtesy of Marissa Calko).

4.19 Information Theory

Macromolecules do not obey the law of constant proportions, fundamental to small molecules (water, carbon dioxide, etc.). A protein of fixed length can have vast numbers of different identities, depending on its primary sequence of amino acids. Because macromolecules are not locked into fixed stoichiometries they are naturally suited for storing information. DNA and RNA are information-bearing molecules based on sequences of the four nucleotide bases discussed in Chapter 11. This section is a very brief introduction to how information is defined and measured, and the implications for biological information.

One definition of information is the transmission or reception of a message consisting of a specification of a number of possible states that a system can be in. The information could be in the form of a written letter, a signal, a DNA sequence, etc. The information may uniquely specify the state of the system or specify a subset of the possible states. If the message does not reduce the number of possible states it contains no information.

For example, consider an atom in one of N containers. A single number between 1 and N will give the information regarding which container the atom is in. If there are n distinguishable atoms, numbered from 1 to n, independently distributed among N containers, then there are N^n possible sequences of numbers that correspond to the distribution of atoms. But information should be additive, that is, we could simply transmit n messages, each with the location of each atom, so that the total information would then be the sum of the information contained in the n separate messages.

Hence, the logarithmic function is chosen to represent information, since it yields this additivity property. It is common to use $\log_2$, since information normally involves discrete information, and the binary system is most apt for digital representations of discrete information.

$$\text{Information} = I = \log_2 N \tag{4.203}$$

I is the number of *bits* needed to specify a number from N possibilities, so that the unit of information is the *bit*, just as in its binary usage in computer science, e.g. a two-state system requires 1 bit of information (0 or 1) to specify a state.

For the n atom system above, the information content of a specific state is

$$I_{total}(\text{bits}) = I_1 + I_2 + \ldots.. + I_n = \log_2(N^n) = n\log_2 N \tag{4.204}$$

The actual amount of information conveyed by a message relates to how far the message narrows down the total number of states, $N_{total} = N^n$ possibilities. If the message does not narrow down the number of states, so that $N_{specified} = N_{total}$, then the message contains no information, $I_{message} = 0$. If the message specifies one unique state $N_{specified} = 1$, then the message contains the complete information $I_{message} = I_{total} = n\log_2 N$. Using the logarithmic property of information:

$$I_{message} = \log_2(N_{total,states} / N_{specified}) \tag{4.205}$$

Say the message only reduces the number of possibilities by a factor of 2, e.g. "the atom is in an even-numbered box." Then the message contains only $I_{message} = \log_2 2 = 1$ bit of information.

A handy conversion from $\log_{10}$, offered by most calculators, to $\log_B$ for any base B is

$$\log_B x = \frac{\log_{10} x}{\log_{10} B} \text{ For base 2, } B = 2 \text{ and } \log_2 x = \frac{\log_{10} x}{0.301} \tag{4.206}$$

Now, consider a five-letter word in English. There are 26 letters in the language, so that there are 26^5 ($=10^{7.07}$) possible combinations. The information content is $I_{total} = \log_2 26^5 = 23.5$ bits. A message that specified a unique combination would have $I_{message} = 23.5$ bits. There are not even 26^5 words (yet!) in the entire English language,[15] let alone that many with exactly five letters. Hence, the information requirement can be reduced by specifying certain restrictions, e.g. no five-letter words have five consonants in a row. There are immediately 21^5 combinations that are eliminated so that the number of possible words specified by this "message" – you can't have a word without vowels – is $N_{specified} = 26^5 - 21^5$, and so

$$I_{message} = \log_2\left[26^5 / (26^5 - 21^5)\right] = 0.6\,\text{bits} \tag{4.207}$$

i.e. this message has low information content, furnishing less than one bit. Further restrictions would require invoking more complex orthographical and lexical rules and, combined with the previous message would lead to higher information content.

15 The Oxford English Dictionary lists around 180,000 words. An educated speaker typically knows upwards of 20,000 words.

Table 4.2 Approximate frequency, in percent, of usage of letters in English, in order of frequency. Taken from frequencies in the Concise Oxford Dictionary (9th Ed., 1995).

Letter	Frequency	Letter	Frequency	Letter	Frequency
E	11.16	C	4.54	Y	1.78
A	8.50	U	3.63	W	1.29
R	7.58	D	3.38	K	1.10
I	7.54	P	3.17	V	1.01
O	7.16	M	3.01	X	0.29
T	6.95	H	3.00	Z	0.27
N	6.65	G	2.47	J	0.20
S	5.74	B	2.07	Q	0.20
L	5.49	F	1.81		

Another point is that not all letters occur with the same frequency, so that there is a higher information content in more rarely used letters, e.g. in a cross-word puzzle it is easier to identify a "cross-word" that contains a letter like z, x, or j, than by a, e, or n. In the game "Scrabble," higher points are given to the rarer letters.

Table 4.2 shows the approximate frequency of occurrence of each letter in English.[16] If a letter occurs with a fractional probability of p, then for a word n letters long the probability P that the letter occurs at least once is

$$P = 1 - (1 - p)^n \qquad 4.208$$

For the letter "a" in a five-letter word, with $p = 0.085$, $P = 0.359$ and there is a 35.9% probability of at least one "a" in any given five-letter word, whereas for z, $p = 0.0027$, and so there is only a 1.34% chance that z occurs at least once. Hence, the letter "z" in a word specifies a far smaller number of possible words than "a," and hence contains much more information. Say there are N_{total} five-letter words in English. Specifying $N_{specified} = PN_{total}$ words by a message (e.g. the message that the frequency of "z" is only 1.34% and that the word contains at least one "z") means that such a message has

$$I_{message} = -\log_2\left[1 - (1 - p)^n\right] \qquad 4.209$$

The information content of the messages specifying one or more "a" per five-letter word and one or more "z" per five-letter word, is 1.48 bits and 6.22 bits, respectively.

4.19.1 The Geometric Distribution and the Binomial Distribution (also see Appendix I)

Say there are N possible outcomes in an event, e.g. $N = 2$ for flipping a coin, $N = 6$ for throwing a die, and $N = 26$ for randomly picking a letter in the English alphabet. Frequently, successive events are independent of the preceding ones, in which case probabilities are simply multiplicative. For n successive events, there are a total of N^n possible outcomes.

Consider a system with $N = 2$, say a and $\bar{a}$, where $\bar{a}$ means "not a," and so could include any number of events, each of which is not a. If n total events occur, and the probability of a occurring upon each event is p, the probability that a will occur j times and that $\bar{a}$ will occur $(n - j)$ times in any given sequence is $P_{GD}(j,n;p) = p^j(1 - p)^{n-j}$. This latter is a form of the geometrical distribution, hence P_{GD}, and occurs frequently when considering the statistics of macromolecules.

Now, consider that we want to know the probability that a occurs j times after n successive events, regardless of the sequence, then it is necessary to know the number of combinations of j occurrences of a and $(n - j)$ occurrences of $\bar{a}$, regardless of sequence. This is the number of permutations for j occurrences of a and is given by the factorial:

$$\frac{n!}{j!(n-j)!} \qquad 4.210$$

16 The frequency of occurrence of letters depends on the sources used, e.g. specialized scientific literature will have a different distribution than a television situation comedy.

For two possible outcomes Equation 4.210 has the property that

$$2^n = \sum_{j=0}^{n} \frac{n!}{j!(n-j)!} \quad 4.211$$

The probability that a occurs j times after n events, regardless of sequence is then given by the binomial distribution P_{BD}:

$$P_{BD}(j,n;p) = \frac{n!}{j!(n-j)!} p^j (1-p)^{n-j} \quad 4.212$$

This can be generalized further to the case of $N > 2$, when the probabilities of the various outcomes do not have equal probabilities, e.g. "loaded dice." For this, $p_1, p_2, \ldots, p_N$ are the probabilities of occurrence of any events 1,2,3,...,N. The generalized multinomial distribution P_{MN} is then,

$$P(n, j_1, j_2, \ldots j_n; p_1, p_2, \ldots p_N) = \frac{n!}{\prod_{i=1}^{N} j_i!} \prod_{k=1}^{N} p_k^{j_k} \quad 4.213$$

where $j_1, j_2 \ldots j_N$ are the number of occurrences of outcome 1, 2....N after n events.

Markov chains have been used to express this type of series of events, where, at any present time, the probabilities of succeeding events are independent of past events. When events further down the chain can affect events further up the chain, then the events are no longer independent and the problem becomes considerably more complex. An example in polymer theory is how the excluded volume and position of the jth monomer in a chain affects the position of the ith monomer, leading to the polymer excluded volume problem (Chapter 8.2).

Applications of the geometric, binomial and other distributions for macromolecular characterization will be given in Chapter 7.

4.19.2 Information and Biomolecules

From the information point of view there are four letters in the genetic code alphabet and a word, or amino acid specification, is three letters long (Chapter 11.2). Hence, there are 64 possible words and the information per codon is $I = \log_2 4^3 = 6$ bits. A sequence for a specific protein of N amino acids hence requires $6N$ bits to encode.

Consider a very small protein that contains 50 amino acids in a given sequence.

i) The probability of arriving at this sequence through random collisions of amino acids is roughly one chance in $20^{50} = 10^{65.05}$, i.e. it is virtually impossible to randomly arrive at even a small protein. There are not enough atoms and molecules on Earth, and the Earth is not old enough, to even remotely approach this.
ii) If nature were "finalistic," that is, it somehow "knew" what the final sequence should be, then an amino acid would only add the specific next amino acid in the known sequence. So, for adding each new amino acid there would be a maximum of twenty different types of encounters; the current amino acid and an encounter with up to a maximum of the 20 different amino acids. The maximum number of different types of encounters between the amino acids would then be a maximum of $20 \times 50 = 1000$ between amino acids.
iii) The genetic code provides the information storage system in the living state. Its alphabet consists of four different bases taken in triplets (codons) to specify an amino acid. For the 50 amino acid sequence in this example, $I = 50 \times 6 = 300$ bits. This is a remarkably efficient way of selecting a single sequence from 10^{65} possible sequences.

The entire human genome comprises approximately 3×10^9 base pairs. This requires 6×10^9 bits (2 bits per base pair), or 7.5×10^8 eight bit *bytes*, or 0.75 Gb. This is a tiny fraction of the storage capacity of current computer hard memory, which commonly exceeds 10^{12} bytes, where 10^{12} bytes is a terabyte, Tb.

It is seen that there is redundancy in the code, since there are 64 different triplets but only 20 amino acids (300 bits of I, compared to the minimum of 216 bits, if there were only one triplet per amino acid). This provides for information "buffering" in the code, i.e. certain mistakes in transmitting triplets can be made and still leave the amino acid specified unchanged. Error rates in transmitting information in communications technology are fundamental considerations in system design (it usually turns out that fast, sloppy messages transmit more information per time interval than precise, slow ones). These concepts have been adapted to study how error rates in genetic transcription are tolerated and/or lead to mutations.

The lesson of the above example is that it is tremendously more efficient to transmit organization through coded information than through chance.

4.19.3 Information and Entropy

There is a very close relationship between information and entropy, and information is often presented as the obverse of entropy. The following is a simplified approach, meant to capture the relationship.

Boltzmann's definition of entropy was stated above, Equation 4.49, $S = k_B \ln\Omega$, where k_B is Boltzmann's constant, and Ω is the number of microstates available to a system. Consider a process in which the number of microstates of the system changes. Let Ω_{in} be the initial number of microstates, and Ω_f be the final number. Then, the change in entropy is

$$\Delta S = k_B \ln\left(\frac{\Omega_f}{\Omega_{in}}\right) \quad 4.214$$

Now, consider the information change in this situation. Initially, the system needs $\log_2\Omega_{in}$ bits to specify it. The final state requires $\log_2\Omega_f$ to specify it. The difference in information is equal to the number of bits needed to define the initial state minus the number of bits needed to define the final state,

$$\Delta I = \log_2\left(\frac{\Omega_{in}}{\Omega_f}\right) \quad 4.215$$

If $\Omega_f < \Omega_{in}$ then information is gained, because the number of bits needed to specify Ω_f is less than the number needed to specify Ω_{in}. A concrete case is compressing an ideal gas. It requires more information to specify the initial state than the final state so that the lessened number of final states corresponds to a gain in information. Compressing the gas has led to a decrease in entropy.

If we slightly redefine ΔI to fit into the thermodynamic context (i.e. go to natural log, ln, and multiply by k_B), we immediately see that

$$\Delta I = -\Delta S = k_B \ln\left(\frac{\Omega_{in}}{\Omega_f}\right) \quad 4.216$$

Thermodynamically, in this approach, information is *negative entropy*, or *negentropy*.

Consider compressing an ideal gas from V to $V/2$ at constant T.

$$W = \int P\mathrm{d}V = nk_B T\int_V^{V/2}\frac{\mathrm{d}V'}{V'} = -Nk_B T\ln 2 \quad 4.217$$

where the − sign indicates that isothermal work was done on the gas. Since $Q = W + \Delta U$ (first law of thermodynamics), and $\Delta U = 0$ for an ideal gas in an isothermal process, $W = Q$, and the entropy change in the system is

$$\Delta S = \int\frac{\mathrm{d}Q}{T} = \frac{W}{T} = -Nk_B\ln 2 \quad 4.218$$

The entropy of the system has decreased, and the information has increased.

$$\Delta I = -\Delta S = Nk_B\ln 2 \quad 4.219$$

This is similar to our original information scenario, where now N atoms have been forced into one-half their original volume. This reduced by one bit the amount of information per atom, $\log_2(2/1)$ needed to specify the system, i.e. one bit of information was gained.

The *energetic* cost for thermodynamically decreasing entropy, and increasing information per atom is the work done on the system, W

$$W = T\Delta S = k_B T\ln 2 \quad 4.220$$

At room temperature ($T = 300$ K),

$$W = 2.8 \times 10^{-21}\text{J} = 0.0175\ \text{eV/bit} \quad 4.221$$

that is, the thermodynamic cost of producing information is quite low. Information is thermodynamically inexpensive! Hence, the problem is in accumulating valuable information through a long process of evolution. Physically, it will cost little energy to store the information.

4.19.4 Information, Shannon Entropy, and Communication Theory

While the interesting analogy between thermodynamic entropy, embodied in Boltzmann's Equation 4.214, and information was made in Equation 4.219, information as it was defined in Equation 4.205 is not inherently thermodynamic. The connection was made through the concepts of probability at the base of both Boltzmann's thermodynamic entropy and information. Entropy in thermodynamics is a physical quantity with units of energy/temperature, and is involved in total free energy, work, and other physical processes. Information has no physical units, although "bits" were used to represent it.[17] Bits have no physical units in the SI or any other system of units.

Claude Shannon (United States, 1916–2001) defined an "information entropy" as a basis of quantifying information. He is often considered one of the seminal founders of information theory, and its applications to communications. Some of the simplified aspects of information theory have been described above.

In the communication paradigm, there is a source of the message that has (i) its own alphabet, which is the four nucleotides in the genetic case; (ii) the message itself, which is a DNA sequence, transcribed onto mRNA; and (iii) a channel for sending the message and a set of rules that read and translate the message, often into a different alphabet. In the case of genetics, the mRNA is read and translated by tRNA in the ribosomes into proteins, which have their own alphabet of 20 different amino acids.

In communication and computer coding theory, error-free information can only flow between messages in the same alphabet, or between isomorphic alphabets, that is, with alphabets that have a one-to-one correspondence of symbols.[18] The genetic code has $4^3 = 64$ distinct triplets (codons) from its alphabet of four nucleotides, whereas the protein alphabet consists of 20 amino acids. Most amino acids have more than one triplet of nucleotides that correspond to it, some as many as six, so that the genetic code is redundant. Any conceivable protein can be specified by the genetic code, but given an amino acid in the protein, it is not possible to know which codon specified it. This supports Crick's central dogma,[19] that information flows from DNA and RNA to proteins,[20] but not from proteins to DNA or RNA.

This notion that protein cannot send complete information to DNA and RNA should be pondered because it has major implications on the debate about the origin of life. One of the questions in the debate is whether proteins or nucleic acids came first. There are proponents of both viewpoints. But the notion that proteins can't send error-free or complete information to DNA and RNA implies, on purely information-theoretic grounds, that proteins could not have arisen first, or, if they did, DNA and RNA would still have to evolve independently of proteins.

As far as the Shannon entropy H, consider an alphabet in which each letter has a certain probability of appearing in a sequence, p_i, such as in Table 2 for the English alphabet, where $\sum p_i = 1$, where the sum is over all letters. Then, consider a message, a sequence of N letters. On average, the number of times a given letter occurs is Np_i, and the probability that p_i occurs Np_i times is $p_i^{Np_i}$. Since probabilities are multiplicative the probability P of the sequence is

$$P = \prod_i p_i^{Np_i} \quad 4.222$$

17 The "bit" is used as the unit of information when using $\log_2$. A bit is sometimes referred to as a "shannon," when $\ln = \log_e$ is used the unit of information is termed a "nat," and when $\log_{10}$ is used the unit is the "hartley" (after Ralph Hartley, United States, 1888–1970).

18 While the number of symbols, or letters, in the English alphabet is 26, it can be made isomorphic with other languages with more letters by using combinations of its letters to represent the additional letters in the second language. For example, German has 30 letters, sharing the 26 letters of English, but adding four letters that can be represented in English by two-letter combinations: ä = ae, ö = oe, ü = ue, and ß = ss.

19 Queried on his choice of the word "dogma," Crick said he merely didn't have a good idea of its meaning when he coined "central dogma," and might have better used "central hypothesis."

20 Retroviruses send information from their RNA to a host's DNA. This two-way flow is permitted because RNA and DNA share the same alphabet and the genetic code.

Taking $\log_2$ of each side yields

$$\log_2 P = \sum_i \log_2\left(p_i^{Np_i}\right) = N\sum_i p_i \log_2(p_i) = -NH \qquad 4.223$$

and the Shannon entropy H is then given by

$$H = -\sum_i p_i \log_2 p_i \qquad 4.224$$

Notice that H is always greater than or equal to one. The smaller the probability of a given letter, the higher its contribution to the total Shannon entropy and the higher its information content. This is what we found for English letters; the least frequently used carry the most information.

Crick's central dogma, the concept that proteins cannot transmit information back to RNA/DNA can also be affirmed using the principle in information theory, that a system of low Shannon entropy (proteins and their 20 amino acid alphabet) cannot reliably send information to a system with higher Shannon entropy (the genetic code).

4.20 Summary

The living state is a thermodynamic system because it exchanges matter and energy, with the environment and obeys thermodynamic laws. One of the central notions is that the living state is far from equilibrium and constantly uses organized energy from the environment to produce both its own type of organized energy and disorganized waste energy. The second law of thermodynamics limits the amount of environmental energy absorbed that can be turned into useful energy, with a positive balance of entropy created. As an open, dissipative system, the living state constantly creates entropy, which it sheds to the environment, even while total energy is conserved. The free energy concept was introduced as a means of determining how much energy is potentially available to do work and how a system will evolve so as to minimize free energy. At the equilibrium condition of minimum free energy, the system cannot do any more work.

Ideal gases form a simple conceptual model for understanding heat engine cycles and associated free energy, work, heat, and entropy, and the basis of ideal osmotic pressure of molecules in a solution can be understood in analogous terms. Osmotic pressure has a vital role in many living state functions. By invoking statistics and probability we were able to go deeper into the microscopic description which underlies macroscopic thermodynamic laws, including a purely statistical interpretation of entropy. The Boltzmann distribution emerged as the key to understanding several phenomena, including speed distribution in gases, equilibrium constants, and chemical reaction rates. Thermal energy, k_BT, is a benchmark energy in a thermodynamic system. Thermal energy and the Boltzmann distribution will figure prominently in many other phenomena yet to be discussed.

In proceeding to multicomponent systems, partial molar quantities were introduced. One of the most important is chemical potential which, for each component, is the free energy per mole of that component. Chemical potential is well described in terms of entropy of mixing for ideal solutions. Introducing nonideality, the excluded volume concept was developed to give power series expansions beyond simple ideal behavior. The virial coefficients, A_2, A_3, etc. embody the excluded volume concept in the power series expansion of chemical potential and osmotic pressure.

The mass action principle was used to define equilibrium conditions for systems with two or more interacting components. This led to Le Chatelier's principle, which essentially states that multicomponent chemical systems will react to perturbations away from equilibrium by partially counteracting the perturbation as they come to a new equilibrium. While equilibrium can be understood via the Boltzmann distribution and the free energies separating two states of a system, reaction rates involve the additional concept of free energy of activation $\Delta G^{\ddagger}$.

Finally, the link was made between entropy, the lack of knowledge of a system, and information, the amount of information known about a system. The one is the negative of the other. By acting logarithmically relatively few bits of information are needed to capture knowledge of huge amounts of particles and events. This means that, biologically speaking, very little energy is needed to encode large amounts of information, such as in the genetic code. Generating the information, however, requires a long evolutionary process. Notably, principles of coding theory, based on Shannon entropy, suggest that the lower information system of the protein "alphabet" (20 letters) versus the 64 codons of DNA/RNA in the genetic code suggest that DNA/RNA could not evolve from proteins.

References

Chandler, D. (1987). *Introduction to Modern Statistical Mechanics*. Oxford.

Dill, K.A. and Bromberg, S. (2003). *Molecular Driving Forces; Statistical Thermodynamics in Chemistry and Biology*. Garland Science.

Gray, P. and Stephen, K.S. (1994). *Chemical Oscillations and Instabilities; Non-linear Chemical Kinetics*. Oxford.

Landau, L.D. and Lifshitz, E.M. (1959). *Statistical Physics*, Part 1. Pergammon (Orig. USSR, Academy of Sciences).

Schroeder, D.V. (2021). *An Introduction to Thermal Physics*. Addison Wesley Longman.

Shannon, C. (1949). *The Mathematical Theory of Communication*. Urbana IL: University of Illinois Press.

Tanford, C. (1980). *The Hydrophobic Effect: Formation of Micelles and Biological Membranes*. Wiley.

Yeomans, J.M. (1992). *Statistical Mechanics of Phase Transitions*. Oxford.

Yockey, H.P. (1992). *Information Theory and Molecular Biology*. Cambridge University Press.

Further Reading

Deviri, D. and Safran, S.A. (2022). Balance of osmotic pressures determines the nuclear-to-cytoplasmic volume ratio of the cell. *Proceedings of the National Academy of Sciences of the United States of America* 119 (21): e2118301119.

Fox, J.M., Zhao, M., Fink, M.J. et al. (2018). The molecular origin of enthalpy/entropy compensation in biomolecular recognition. *Annual Review of Biophysics* 47: 223–250.

Kraus, P.M., Zurch, M., Cushing, S.K. et al. (2018). The ultrafast X-ray spectroscopic revolution in chemical dynamics. *Nature Reviews Chemistry* 2 (6): 82–94.

Leff, H. and Rex, A.F. (eds.). (1990). *Maxwell's Demon: Entropy, Information, and Computing*. Adam Hilger.

Piskulich, Z.A., Laage, D., and Thompson, W.H. (2021). Using activation energies to elucidate mechanisms of water dynamics. *Journal of Physical Chemistry A* 125 (46): 9941–9952.

Sharma, V.K., Mitra, S., and Mukhopadhyay, R. (2019). Dynamic landscape in self-assembled surfactant aggregates. *Langmuir* 35 (44): 14151–14172.

Smith, W.R. (2020). A precise, simple and general Basic Le Chatelier Principle based on elementary calculus: what Le Chatelier had in mind? *Journal of Mathematical Chemistry* 58 (8): 1548–1570.

Wang, J., Arantes, P.R., Bhattarai, A. et al. (2021). Gaussian accelerated molecular dynamics: principles and applications. *Wiley Interdisciplinary Reviews: Computational Molecular Science* 11 (5): e1521.

Problems

1 Compute the work needed to desalinate 10^5 kg of seawater in which there is 0.15 mol/L of salt at constant temperature 30°C. Ignore any frictional effects (e.g. due to viscous fluid flow through a membrane) in the reverse osmosis process. Consider all the salt to be monovalent, e.g. NaCl, so that each mole of salt yields 2 moles of solute. Consider the desalination complete when NaCl in the concentrated portion reaches 4 M (M = molar).

2 Consider the Maxwell–Boltzmann speed distribution for an ideal gas:

$$f(v)\mathrm{d}v = 4\pi\left(\frac{m}{2\pi k_B T}\right)^{3/2} v^2 \exp\left(\frac{-mv^2}{2k_B T}\right)\mathrm{d}v$$

a) Find the speed v_p at the peak in the MB distribution.
b) Find the average speed $\langle v\rangle$ (note that the limits of integration for speed are from 0 to ∞, since it is a scalar),
c) Find the average velocity $\langle \vec{v}\rangle$ (note that the limits of integration for velocity are from $-\infty$ to ∞, over each coordinate, since it is a vector).
d) Show that the MB distribution gives the relationship between average kinetic energy $\langle KE\rangle$ and temperature in a dilute gas

$$\langle KE\rangle = \frac{3k_B T}{2}$$

e) Find the skewness of this distribution, with respect to the peak of the distribution, v_p from part (a), and used for x_0 AI.14 (see Appendix I)

3 Consider 1 mole of an ideal monatomic gas, initially at $P_1 = 10^5$ Pa and $V_1 = 25$ L, going through a cycle consisting of three quasistatic (i.e. reversible) steps:
 i) It is heated at constant volume to $P_2 = 2P_1$.
 ii) It is then isothermally expanded to $V_3 = 2V_1$.
 iii) It is compressed back to V_1 at constant pressure.
 a) Make a diagram of the cycle and find the temperature of the gas at each step of the cycle.
 b) Find the heat flow Q, for each step of the cycle.
 c) Calculate the efficiency of the cycle.

4 Evaluate the work and heat flows over the four steps of the Carnot cycle for a monatomic gas and derive the efficiency formula.

5 Seven moles of an ideal monatomic gas undergo a reversible adiabatic expansion P_1, V_1 to V_2. Let $\gamma = 3R/2$ be the heat capacity of the gas.
 a) Find P_2.
 b) Find T_2.
 c) Compute the work done during the adiabatic expansion.

6 Assuming $T = 300$ K is constant going upward from the surface of the Earth, find the height at which the density of O_2 molecules is ½ that at the Earth's surface.

7 A U-tube osmometer consists of a glass tube of diameter 1.0 cm. It is partitioned at the bottom of the "U" by a membrane permeable only to water. If 120 ml of water are in the tube and 0.09 g of sucrose (molar mass = 342 g/mol) are dissolved in one side, what will be the height difference between the liquid in the two sides? $T = 25$°C. Take the density of water as 1 g/cm^3. (You must take the glass tube diameter into account to correctly solve this).

8 Derive Young's contact angle (Equation 4.63).
Initial drop has radius R' and volume V'. The drop forms a spherical cap when placed on the surface and spreads to a new radius R. The new volume V must be equal to the original volume V'.

9 Derive Equation 4.66, which gives the height to which liquid rises in a capillary:
$S = 2\pi R^2(1 - \cos\theta)$, $A = \pi R^2 \sin^2\theta$
And $V = (\pi h/6)\left(3(R\sin\theta)^2 + h^2\right) = \frac{\pi R^3}{3}(1-\cos\theta)^2(2+\cos\theta)$, where h is the height of the spherical cap above the surface.

10 Consider blood that has 5% by volume albumin proteins. If the protein molecules have a radius of 3 nm and molar mass of 2×10^4 g/mol and have about the same density as water.
 a) Find the osmotic pressure produced by the protein, considering the proteins as ideal.
 b) Estimate the second virial coefficient A_2 of the proteins, considering them as hard spheres and recompute the osmotic pressure.

11 Consider the reaction

$$2H_2 + O_2 \leftrightarrow 2H_2O,\ \Delta G^0 = -2.37\times10^5\,\text{J/mol}$$

The negative sign indicates this is exothermic. At $T = 298$ K, find the number of moles of remaining O_2 if there was initially 10^{15} moles of O_2 and 2×10^{15} mol of H_2.

12 Consider a system consisting of A and B governed by

$$A \leftrightarrow B$$

where $K_{eq} = [B]_0/[A]_0$, where $[A]_0$ and $[B]_0$ are equilibrium concentrations. If A is now increased by $\Delta[A]$, while T and P are held constant (i.e. K_{eq} is unchanged), find the new equilibrium concentrations of A and B.

13 Derive Equation 4.188 $[\text{SDS}]_{CMC} = \left(\frac{1}{m^2 K_{eq}}\right)^{1/(m-1)}$

The CMC refers to the entire concentration of SDS, expressed in Equation 4.189.

$$\text{CMC} = [\text{SDS}]_{CMC}\left(1+\frac{1}{m}\right)$$

14 Derive the equation in the PV plane for the adiabatic expansion of an ideal atomic gas.

15 Show that Equation 4.25, $W(v^2) = \left(\frac{m}{2\pi k_B T}\right)^{3/2} \exp\left(-\frac{mv^2}{2k_B T}\right)$, is normalized.

16 A certain diatomic molecule has a first vibrational energy level that absorbs radiation over a narrow range centered at 10^{-5} m. At what T would the thermal energy levels begin to strongly populate the first vibrational energy level?

17 Consider the osmotic piston of Figure 4.9. It has a diameter of 10 cm and a length of 200 cm. 2 moles of sugar are dissolved in the left side when the volume in the side is 1/20th of the total volume. Find the maximum mass M that can be raised, such that the membrane arrives completely to the right-hand side.

18 G_{mix} is given in Equation 4.102 as

$$\frac{G_{i,mix}}{RT} = n_i\left[\ln n_i - \ln n_t\right] + \sum_{j \neq i}^{k} n_j\left[\ln n_j - \ln n_t\right]$$

Noting that in the differentiation, where $x_j = n_j/n_t$, differentiate this with respect to n_i, holding all other n_j constant and prove Equation 4.103, $\mu_{i,mix} = RT \ln x_i$.

19 Prove the following dilute solution approximation for the case of no excluded volume, Equation 4.140

$$\Omega_{no\ ev} = \frac{(N_1 + N_2)!}{N_1!N_2!} \cong \frac{N_1 N_2}{N_2!}$$

20 $pK_a = 3.4$ for a certain weak acid dissolved in water.
a) Find the fraction of the acid dissociated at pH = 5.
b) At what pH would only 1% of the acid be ionized?

21 Consider the following data for the rate at which crickets chirp (Hz) versus T (°C). Make an Arrhenius plot of the data and find the free energy of activation of this biophotochemical process, in kcal/mol *and* in eV

T (°C)	Cricket chirp rate (1/s)
5.1111	1.04
6.0556	1.27
9.7222	1.84
12.222	2.4
13.889	2.9
16.667	4

22 Compute the diameter of a capillary that can raise water 2 m by capillary action alone. Refer to Equation 4.66, and take $\gamma_{LV} = 0.0728$ N/m at $T = 20$°C, and the contact angle $\theta \sim 0$.

23 Given 10 different coins. How many combinations are there of heads up and heads down? If the coins are identical, how many combinations are there?

24 Qualitatively explain whether and why you expect pH to stay the same, increase, or decrease as T increase. pH versus T.

25 An unbiased die has six faces, each with equal probability of appearing at each toss of the die. Now, consider a biased die, with six faces that have the following probabilities of appearing; 0.50, 0.25, 0.10, 0.08, 0.06, and 0.01. Compute the Shannon entropy H, in each case.

26 2×10^9 distinguishable atoms of an ideal gas are in a container with 32 subchambers, all of the same volume, and originally all connected together. Each atom has an equal probability of being in any of the 32 subchambers.
a) How many bits are needed to specify a specific arrangement of the atoms among the subchambers?
b) What size memory, in gigabytes, would be needed to store this information?
c) The gas is now isothermally compressed into a single, specified chamber. How much work is required for this?
d) What is the total energy cost per bit of information created in the compression process (the compression selects one of the possible arrangements out of many)?

27 Consider a four-letter word in English, which has 26 letters.

a) How many possible words are there, ignoring any exclusions due to any rules?

b) If the probability of the letter "*e*" occurring is 12.7%, how much information is contained in a message "The 4 letter word has one or more '*e*' in it." Compare this to the information in the message "The 4 letter word has one or more '*q*' in it," where the probability of "*q*" occurring is 0.1%.

c) How much information is contained in the message for the six-letter word that the word contains at least one "*e*" and one "*z*". Use the frequencies given in Table 4.2.

28 A batter has an average of 0.350 (makes a hit on 35% of the pitches thrown at him). After five pitches what is the probability that he will have made:

a) No hits

b) 1 hit

c) 3 hits

29 A pair of dice (a die is a cube with six sides and numbers from 1 to 6 on the sides) is cast 12 times.

$$W(N,j,q)=\frac{N!}{(N-j)!j!}q^j(1-q)^{N-j}$$

What is the probability of rolling 2 (i.e. "1" appearing on each of the two dice) six times on 12 tries?

30 a) Consider a liter of water into which 10^{-5} moles of a particle of negligible volume is mixed. Compute the entropy of mixing of this system. You can use the dilute solution approximations for this part and part (b).

b) If the particles are hard spheres (i.e. impenetrable by other spheres) and have a diameter of 30 nm compute the entropy of mixing and compute the change with respect to part (a). Be sure to include the sign.

31 Consider a serum (salt water with protein) that has 5% by volume of protein. If the protein molecules are considered spheres of radius 5 nm and have a molar mass of $M = 2 \times 10^4$ g/mol and have approximately the same density as water. Take $T = 310$ K. (Ignore the presence of salt.)

a) Find the osmotic pressure produced by the protein, *considering the proteins as ideal.*

b) Calculate the second virial coefficient A_2 of the proteins, considering them as hard spheres.

c) Recompute the osmotic pressure taking the A_2 of part (b) into account.

32 a) Compute A_2 for the interparticle potential $U(r_{12})=U_0 e^{-\alpha r_{12}^2}$, where $U_0 \ll k_B T$. Expand the potential to second order, giving two terms in the final value of A_2.

b) Find the radius of an equivalent hard sphere, R, that will give the same excluded volume β, as in part a).

c) Find β (in cm^3) and R (cm) if $U_0 = 0.08 k_B T$, and $\alpha = (5 \times 10^{-6}\ \text{cm})^{-2}$

5

Electrostatics in Solution

The living state occurs in an aqueous environment with many types of electrically charged ions and macromolecules. Ubiquitous simple ions include Na^+, K^+, and Cl^-, while RNA, DNA, and many polysaccharides are negatively charged. Proteins may contain ionizable amino acids which can make the protein electrically positive, negative, or neutral, depending on the pH of the environment (Chapter 4.15.1). In this chapter, we introduce living state phenomena caused by electrostatics in solution, such as charge interactions, ionic shielding, and membrane capacitance.

The elucidation of the classical laws of electromagnetism involved the work of many people over many centuries. This fascinating history stretches back to electrical effects known in ancient Egypt, Greece, and other civilizations, to the use of the magnetic compass a thousand years ago in China, to the final development of classical electromagnetic theory in the eighteenth and nineteenth centuries. The *Treatise on Electricity and Magnetism*, first published by James Clerk Maxwell in 1873 is often considered the culminating unification of classical electromagnetic theory, and essentially describes all classical electromagnetic phenomena in four equations. In this chapter, we will deal only with classical electrostatics and will return to the other three of the Maxwell equations in Chapter 9. For those who remember their introduction to electromagnetism, the review of Section 5.1 can be skipped.

The field of *quantum electrodynamics* (QED) is not entered into in this book, but it is mentioned here because of its importance in unifying electromagnetism from a quantum mechanical and relativistic point of view. The photon represents the quantization of the continuous, classical electromagnetic field, and is used as the intermediary for describing all charge interactions between particles. Early work in the area was done by Dirac, Heisenberg, Enrico Fermi (Italy/United States, 1901–1954), Hans Bethe (France/United States, 1906–2005), Freeman Dyson (England/United States, 1923–2020) and many others. Richard Feynman (United States, 1918–1988) is often considered one of the seminal contributors and "standard bearer" of QED.

5.1 Review of Electrostatics (in MKSA Units)

Charles-Augustin de Coulomb (France, 1736–1806) was the discoverer of the law that bears his name. Coulomb's Law, first published in 1785, was the first quantitative formulation of electromagnetic theory. He found that, similar to Newton's law of gravity, charged objects exert forces on each other that vary as the inverse square of their separation. Unlike gravity, however, which is always attractive, the electrical forces can be either attractive or repulsive, depending on the signs of the charges. The two types of charge are designated simply *positive* and *negative*. The magnitude of the force, which acts along the line joining point-like charges q_1 and q_2 is given, in vacuum, by (MKSA/SI units):

$$F = \frac{q_1 q_2}{4\pi\varepsilon_0 r^2} \qquad 5.1$$

where r is the separation between the charges and q_1 and q_2 are the charges on each, measured in Coulombs (C) in SI units. If the product of q_1 and q_2 is positive, that is, if both have the same signs, then the positive force is repulsive; "likes repel." If the product of q_1 and q_2 is negative, that is, if the charges have opposite signs, the negative force is attractive: "opposites attract." In Coulomb's law $\varepsilon_0 = 8.85 \times 10^{-12}\, C^2/N{-}m^2$ is the permittivity of free space.

Biophysics: Physical Processes Underlying the Living State, First Edition. Wayne F. Reed.

Companion website: www.wiley.com/go/Reed/BiophysicsandMacromolecules

The experimentally discovered Coulomb's law is the starting point for developing the entire field of *electrostatics*, which involves the determination of *electric fields* due to spatially static distributions of electric charge, that is, to charges in *electrostatic equilibrium*, where there is no net charge motion. The electric field notion has become so prevalent that it is easy to lose sight of the mystery initially surrounding how charges could exert force on each other across space. There was an "action at a distance" theory that prevailed well into the nineteenth century, which held that masses could exert forces on each other by gravity and that charges could exert forces on each other by the electrostatic force, without colliding or in any way having physical contact. Michael Faraday (England, 1791–1867) first proposed *lines of force* between charges, and this was elaborated into the electric field notion.

5.1.1 The Electric Field, $\vec{E}$

The vectorial electric field, represented as $\vec{E}$ is, in the classical model, a real physical entity that surrounds each electric charge and can interact with other charges, thus producing the attractive and repulsive forces. $\vec{E}(\vec{r})$ is a real physical entity because it has an energy density associated with it throughout the space around a charge, where $\vec{r}$ is the direction vector for any point in space. $\vec{E}(\vec{r})$ for a charge or distribution of charges is defined as the force that will be felt by one positive unit charge (+1 C) at position $\vec{r}$. For a single point charge Q located at the origin $\vec{E}(\vec{r})$ is obtained by renaming q_1 to Q and dividing Equation 5.1 by q_2, the +1 C test charge. Hence for the point charge Q

$$\vec{E}(\vec{r}) = \frac{Q}{4\pi\varepsilon_0 r^2}\hat{\mathrm{r}} \qquad 5.2$$

where $\hat{\mathrm{r}}$ is the outward-pointing radial unit vector. From here, the $1/r^2$ dependence immediately leads to *Gauss's law*, which is the integral form of the first Maxwell equation. Over a closed surface S, shown in Figure 5.1, the integral of $\vec{E}$ times the dielectric constant equals the net charge Q_{net} enclosed within the surface.

$$\varepsilon\oint_S \vec{E}\cdot \mathrm{d}\vec{A} = Q_{net} = \int_V \rho(\vec{r})\mathrm{d}^3r \qquad 5.3$$

where $\rho(\vec{r})$ is the net charge density at any point in space, and d^3r is the volume element. The circle on the left-hand surface integral indicates that the surface is closed. When dealing with a uniform, isotropic dielectric medium, such as water, ε_0 in the above equation was substituted by the permittivity of the medium, ε, which is

$$\varepsilon = D\varepsilon_0 \qquad 5.4$$

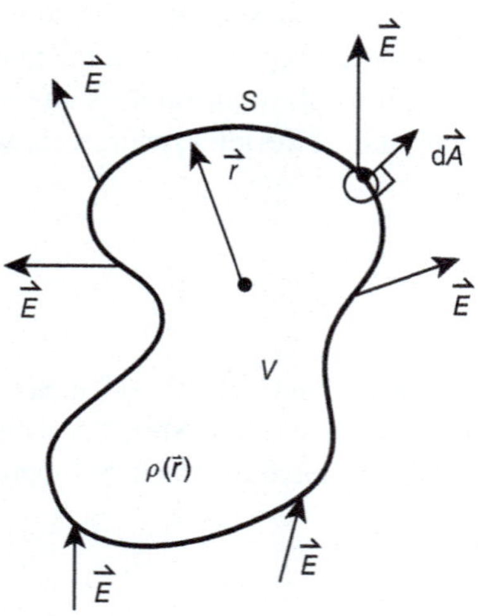

Figure 5.1 Illustration of Gauss's law.

where D is the dimensionless dielectric constant of the medium. $D = 1$ for the vacuum and $D = 80.1$ for water at 25 C. Organic solvents have a wide variety of D, such as $D = 2.27$ for benzene at the low end, and $D = 109$ for formamide at the high end.

The first Maxwell equation is a differential form of Gauss's law, obtained by using the *divergence theorem* (Appendix II) on the left-hand integral and then equating the integrands of both integrals as the volume is shrunk to an infinitesimal, to yield

$$\nabla\cdot\vec{E} = \rho/\varepsilon \qquad 5.5$$

5.1.2 The Electrostatic Potential, ϕ

You can show using Equation 5.2 for $\vec{E}(\vec{r})$ that

$$\nabla\times\vec{E} = 0 \qquad 5.6$$

By Green's theorem, the integral of this over an open surface S is equal to the path integral of $\vec{E}$ around the curve C, which bounds S, and this is equal to zero by Equation 5.6.

$$\int_S \nabla\times\vec{E}\cdot d\vec{A} = \oint_C \vec{E}\cdot d\vec{s} = 0 \qquad 5.7$$

The circle around the path integral on the right-hand side indicates that the path is closed. The path integral is zero for any closed path which means that $\vec{E}(\vec{r})$ is a conservative field and hence can be represented by the negative gradient of a scalar electrostatic potential $\phi(\vec{r})$

$$\vec{E} = -\nabla\phi \qquad 5.8$$

Taking the divergence of this and using Equation 5.5 yields *Poisson's equation* for electrostatics.

$$\nabla^2\phi = -\rho/\varepsilon \qquad 5.9$$

Because this is a scalar equation it is usually easier to solve it first to obtain ϕ and then obtain $\vec{E}$, if desired, from the gradient relationship of Equation 5.8.

While the path integral of $\vec{E}(\vec{r})$ around a closed path is zero, it is not zero between two points in space A and B. The *electrostatic potential difference* between points A and B is, by integration of Equation 5.8

$$\phi_{AB} = \phi_B - \phi_A = -\int_A^B \vec{E}\cdot d\vec{s} \qquad 5.10$$

This integral is also path independent since

$$\oint_C \vec{E}\cdot d\vec{s} = \int_A^B \vec{E}\cdot d\vec{s} + \int_B^A \vec{E}\cdot d\vec{s} = 0 \qquad 5.11$$

That is, the first integral can follow any path from A to B and still be equal and opposite to the second integral which can follow any path from B back to A. The units of ϕ are N-m/C, which are the units of work per unit charge, that is J/C. This latter is the unit for *Volt*.

$$1\,\text{Volt} = 1\frac{J}{C} \qquad 5.12$$

The total work needed to move a test charge q from A to B is found by summing the incremental work dW, that an external agent does in moving over a path increment $d\vec{s}$. The situation is illustrated in Figure 5.2.

$$dW = -q\vec{E}\cdot d\vec{s} \qquad 5.13$$

The negative sign is introduced because the external agent applies a force equal and opposite to the force of $\vec{E}$ on q, in order to move it quasi-statically (i.e. without acceleration) against the field. The total work done by an external agent in moving q from A to B is thus

$$W_{AB} = -q\int_A^B \vec{E}\cdot d\vec{s} = q\phi_{AB} \qquad 5.14$$

This formalizes the relationship between the work in moving a unit of charge from A to B, W_{AB}, and the electrostatic potential difference between A and B, ϕ_{AB}, i.e. ϕ_{AB} is the work to move a charge of one positive Coulomb from point A to B, along any path:

$$\phi_{AB} = \frac{W_{AB}}{q} \qquad 5.15$$

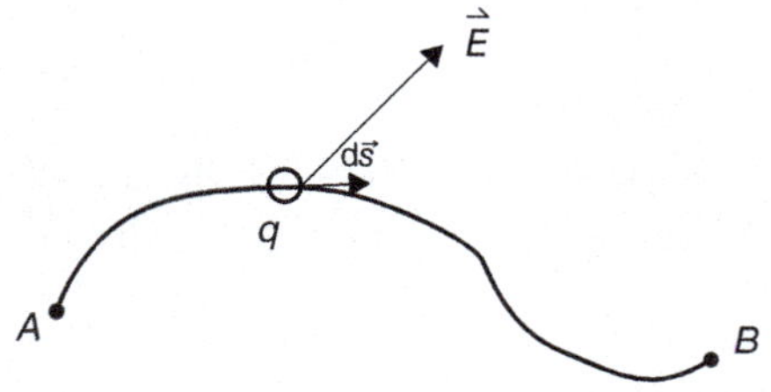

Figure 5.2 A charge q and an arbitrary path from A to B in an external electric field $\vec{E}$.

Let's calculate the electrostatic potential difference between points A and B due to $\vec{E}$ from a point charge Q, given in Equation 5.2

$$\phi_{AB} = -\frac{Q}{4\pi\varepsilon}\int_{\vec{r}_A}^{\vec{r}_B}\frac{\hat{\mathrm{r}}}{r^2}\cdot d\vec{s} \qquad 5.16$$

where $\vec{r}_A$ and $\vec{r}_B$ are the position vectors of points A and B with the charge Q at the origin. An incremental step $d\vec{s}$ in any direction is given in spherical coordinates by

$$d\vec{s} = dr\,\hat{\mathrm{r}} + r d\theta\hat{\theta} + r\sin\theta\, d\varphi\hat{\varphi} \qquad 5.17$$

Since the unit vectors are orthogonal to each other the only term surviving the scalar product in the integrand is $\hat{\mathrm{r}}\cdot d\vec{s} = \hat{\mathrm{r}}\cdot\hat{\mathrm{r}}dr = dr$, and only the radial portion of $\vec{r}_A$ and $\vec{r}_B$ is integrated over, so that

$$\phi_{AB} = -\frac{Q}{4\pi\varepsilon}\int_{r_A}^{r_B}\frac{dr}{r^2} = \frac{Q}{4\pi\varepsilon}\left(\frac{1}{r_B}-\frac{1}{r_A}\right) \qquad 5.18$$

ϕ_{AB} is the work an external agent does on a unit positive charge (+1 C) moving it quasi-statically from r_A to r_B in the electric field of point charge Q. Because there is no angular dependence $\vec{r}_A$ and $\vec{r}_B$ do not have to be colinear.

An important simplifying convention is to take r_A at infinity, so that $1/r_A$ disappears in Equation 5.18, and to designate r_B just as r, so that $\phi_{AB} \to \phi(r)$

$$\phi(r) = \frac{Q}{4\pi\varepsilon r} \qquad 5.19$$

In words, the value of $\phi(r)$ due to point charge Q is the amount of work it takes an external agent to move a unit positive charge from infinity to the point r. Note that if Q is positive this work is positive since the external agent will have to push to move the charge close to Q, and the closer the charge gets, the harder it is to keep pushing. If Q is negative then the unit charge is attracted to Q and the agent exerts a pulling force opposite the test charge's path from infinity toward Q.

If an arbitrary charge q is brought from infinity to a distance r from Q then the work done by an external agent is

$$W = q\phi(r) = \frac{Qq}{4\pi\varepsilon r} = U(r) \qquad 5.20$$

and $U(r)$ is the potential energy stored in the system composed of q at distance r from Q, due to the external agent's work W.

5.1.3 Capacitance, *C*

Capacitance, C, is the amount of charge Q added to an object which raises its electric potential by 1 V.

$$C = \frac{Q}{\phi} \qquad 5.21$$

C can be computed for any given geometry of an object. Its units are Coulomb/Volt, which is the Farad (abbreviated "F")

$$1\ \text{Farad} = 1\ \text{C/V} \qquad 5.22$$

For a spherically symmetric charge distribution, $\rho(r,\theta,\varphi) = \rho(r)$. Let this charge distribution be contained within a sphere of radius R. Since $\vec{E}(r) = E(r)\hat{\mathrm{r}}$ integrating $\rho(r)$ over the sphere's volume and using Gauss's law, Equation 5.3, gives

$$\vec{E}(r) = \frac{Q_{net}}{4\pi\varepsilon r^2}\hat{\mathrm{r}}, \qquad r \geq R \qquad 5.23$$

This means that once on the surface of the charge distribution and beyond it acts as a point charge at the center of the sphere. Newton first found this result for the $1/r^2$ gravitational force; once outside a spherically symmetric mass distribution, its gravitational effect is the same as that of a point charge of the same mass. Computing $\phi(r)$ with respect to infinity for the sphere, as was done for the point charge yields the result:

$$\phi(r)=\frac{Q_{net}}{4\pi\varepsilon r},\quad r\geq R \tag{5.24}$$

The electric potential of the surface of the spherical charge distribution is

$$\phi(R)=\frac{Q_{net}}{4\pi\varepsilon R} \tag{5.25}$$

So that the capacitance of a spherical charge distribution is

$$C=4\pi\varepsilon R \tag{5.26}$$

Computing the capacitance of the planet Earth, using $R = 6.4\times10^6$ m and using $\varepsilon=\varepsilon_0$ yields a mere 7.11×10^{-4} F! Clearly, spherical charge distributions do not lead to high capacitance.

It is a standard exercise with Gauss's law to show that the electric field between two infinite parallel plates is constant and proportional to the surface charge density. An elementary capacitor, and one which a portion of a biological membrane resembles, is the parallel plate capacitor of planar area A, plate separation d, with a material of permittivity ε between the plates. Using the approximation that the E-field in the gap between the plates is constant gives the capacitance:

$$C=\varepsilon\frac{A}{d} \tag{5.27}$$

A typical means of fabricating a practical parallel plate capacitor is to separate two thin metal foils with large area A by a very thin, flexible dielectric material, with small d, and then to roll it up into a cylinder. The dielectric material is an insulator that both keeps the charged parallel plates from shorting each other out, and also increases the capacitance of the system, multiplying the capacitance in vacuum times the dielectric material dielectric constant D.

Now, consider charging any capacitor by adding increments of charge to it, dQ. This can be done by connecting the terminals of the capacitor to a battery with a constant potential difference across its terminals of V, where "V" is used for "voltage." While electric potential ϕ and voltage V have the same units, V is often used instead of ϕ when there is an active source, such as a battery, that can maintain a constant potential difference even as charge is withdrawn from it and work is performed. In a battery, V is kept constant by chemical reactions, in an induction generator a sinusoidal current (alternating current, ac) is provided by rotational mechanical energy.

A concentric sphere capacitor has inside radius a and outside radius b. It is easy to show that the capacitance is

$$C=\frac{4\pi\varepsilon ab}{b-a} \tag{5.28}$$

The cell membrane resembles a concentric sphere capacitor of inside radius R and outside radius $R + d$, where d is the membrane thickness, and $R \gg d$. The capacitance is then approximately:

$$C=\frac{4\pi\varepsilon R^2}{d} \tag{5.29}$$

For $d \sim 2.6$ nm, $\varepsilon \sim 3\varepsilon_0$ and $R = 4\times10^{-6}$ m, $C \sim 2\times10^{-12}$ F = 2 pF. This gives a capacitance per area of

$$\frac{C}{A}\sim\frac{\varepsilon}{d} \tag{5.30}$$

$C/A \sim 0.0034$ F/m^2 = 3.4 mF/m^2. This is a typical value for measured membrane capacitance.

Now, the work to add dQ to the capacitor which at any instant has a total charge Q is

$$\mathrm{d}W=\phi\mathrm{d}Q=\frac{Q}{C}\mathrm{d}Q \tag{5.31}$$

Integrating Equation 5.31 gives the total work required to charge the capacitor from $Q = 0$ to any final value Q

$$W = \frac{Q^2}{2C} \tag{5.32}$$

Alternate forms for the work, combining Equations 5.32 and 5.21 are

$$W = \frac{Q\phi}{2} = \frac{C\phi^2}{2} \tag{5.33}$$

Returning to the parallel plate capacitor the relation between the voltage and the constant electric field inside is, from Equation 5.10, $\phi = V = Ed$. Substituting this into the right-hand term in Equation 5.33 and using Equation 5.27 for C yields

$$W = \frac{\varepsilon AdE^2}{2} \tag{5.34}$$

This is the work that the battery has done to charge the capacitor up to voltage V and establish the constant E between the parallel plates. Once the capacitor has reached the battery's voltage V, no more charge flows, and we are left with a charged capacitor with a constant electric field inside. This suggests that the work done by the battery has been transformed into the electric field. If this is true then the electric field itself must hold the work done in some form of energy: electrostatic energy. The electric field exists uniformly throughout the capacitor's volume Ad (ignoring field fringing effects at the edges of the plates), so we can compute an energy density u, with units of energy/volume, by dividing Equation 5.34 by volume Ad, which yields

$$u = \frac{\varepsilon E^2}{2} \quad \text{Energy Density in the Electric Field} \tag{5.35}$$

Equation 5.35 states that the electric field itself has electrostatic energy with energy density proportional to the electric field squared. This strengthens the notion of the electric field as a real physical entity. While the method of arriving at this conclusion is elementary, the result is completely general, and even electromagnetic waves propagating through a vacuum, where there are no charges, capacitors or batteries, have the same energy density as given in Equation 5.35, with $\varepsilon = \varepsilon_0$. The dependence of u on E^2 is already a clue that E might be directly associated with the amplitude of some type of wave, since energy in classical waves is proportional to wave amplitude squared.

The charging and discharging of a capacitor play a central role in membrane electrical properties, which we will revisit in Chapter 13. Consider the RC circuit with a battery of voltage V_0, shown in Figure 5.3. The voltage drop across the resistor plus capacitor is equal to V_0:

$$V_0 = R\frac{dQ}{dt} + \frac{Q}{C} \tag{5.36}$$

The solution to this when $Q = 0$ at $t = 0$ and the switch is closed at $t = 0$ is the familiar first-order equation:

$$Q(t) = CV_0(1 - e^{-t/RC}) \tag{5.37}$$

No current actually passes through the capacitor, rather the capacitor current leads to the accumulation of charge on the capacitor. The capacitor's charging current I_C is

$$I_C = \frac{dQ}{dt} = \frac{V_0}{R}e^{-t/RC} \tag{5.38}$$

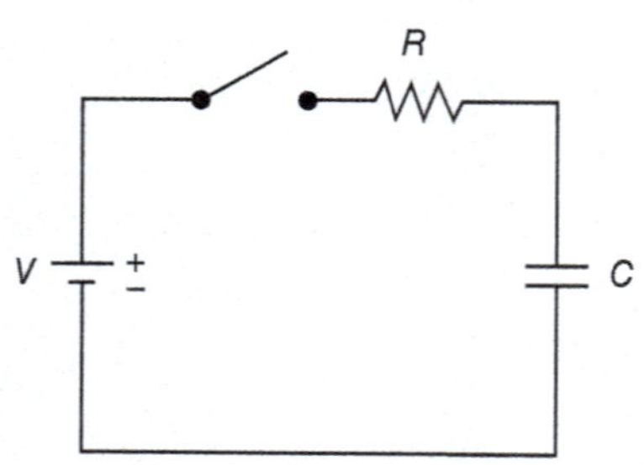

Figure 5.3 An RC circuit with a switch and battery of voltage V_0.

The work done by the battery to charge the capacitor appears as the energy stored in the electric field established by the battery's work. RC is the time constant for the circuit.

5.1.4 The Electric Dipole Moment, $\vec{p}$

Another concept of central importance in the living state and methods used to investigate biomacromolecules is that of the *electric dipole moment*, $\vec{p}$. Dipoles are the first moment in measuring a charge distribution's deviation from spherical symmetry. The dipole moment of a volume charge distribution $\rho(\vec{r})(\mathrm{C/m^3})$, is computed with respect to a given point in space P, via the volume integral

$$\vec{p} = \int_V \vec{r}\rho(\vec{r})\mathrm{d}^3r \qquad 5.39$$

where $\vec{r}$ is the position vector from P to any volume element in the volume and d^3r is the volume element. If the charge distribution is not spherically symmetric and has zero net charge, such as water, it can be proven that $|\vec{p}|$ is a constant, independent of any coordinate system or reference point P.

For N discrete charges $q_1, q_2, \ldots q_N$

$$\vec{p} = \sum_{j=1}^{N} q_j\vec{r}_j \qquad 5.40$$

where $\vec{r}_j$ is the position vector of charge q_j with respect to an arbitrary origin. Again, if the charges sum to zero then $|\vec{p}|$ is a constant, independent of the coordinate system. This is the most common situation, and the simplest case is when two equal and opposite charges, q and $-q$, are separated by a distance d, then Equation 5.40 yields (for simplicity in the computation put the origin at the midpoint of the line connecting the charges, then $r_j = d/2$ for each charge), and

$$|\vec{p}| = qd \qquad 5.41$$

$\vec{p}$ points from the $-q$ toward $+q$. The water molecule is an important example of a permanent dipole composed of three charged groups; two H atoms each with partial charge $\delta^+ \sim +0.33$ C and the O atom which has $\delta^{2-} \sim +0.66$ C. The O–H bond length in water is $d = 0.096$ nm, and the bond angle is $\theta = 104.5°$. This is shown in Figure 5.4. Since the net charge is zero, the origin can be conveniently placed at the O atom, so that the water dipole is the vector sum of the two H-atom dipoles

$$|\vec{p}| = 2\delta^+ d\cos\theta/2 \qquad 5.42$$

which is equivalent to a single dipole of equal and opposite charges δ^{2-} at a distance $d\cos\theta/2$ apart. In general, a dipole consisting of three or more charges, whose net charge is zero, can be reduced to a single dipole with an effective charge equal to the sum of the positive charges at an effective separation depending on the geometry of the charges in space.

Using the values in Equation 5.42, the H_2O has a dipole moment of $|\vec{p}| = 6.17 \times 10^{-30}\,\mathrm{C-m}$. The *Debye* (D) is often used as a measure of dipole moment, where $1\mathrm{D} = 3.336 \times 10^{-30}$ C–m. The H_2O molecule hence has a net dipole moment of 1.85D.

For a distance r much greater than the dipole dimension, essentially a *point dipole*, the electrostatic potential due to the dipole at a position vector $\vec{r}$ in space, whose origin is the center of a dipole $\vec{p}$, is

$$\phi_{dipole}(r) = \frac{\vec{p}\cdot\hat{\mathrm{r}}}{4\pi\varepsilon r^2} \qquad 5.43$$

where $\hat{\mathrm{r}}$ is the unit vector in the direction of the position vector $\vec{r}$. This represents the amount of work done by an external agent to bring a unit charge from infinity to $\vec{r}$. This result is found from a multipole expansion of a charge distribution, in terms of the dipole, quadrupole, octupole, and higher moments. It is treated in all introductory texts on electromagnetism. The torques, forces, and potential energy of dipoles in external electric fields are also treated in introductory texts. An important point to note is that when a collection of dipoles, e.g. water vapor, is placed in an external electric field at finite temperature, the dipoles do not simply align with the field. Rather, the Boltzmann distribution, Equation 4.27, is used to find the thermodynamic average of the dipole orientation with respect to the field ("E" is any type of energy in Equation 4.27, and is replaced by electrostatic energy in the dipole case). The average dipole orientation progressively aligns with the electric field, as this latter is increased at a given temperature and only reaches full alignment asymptotically at high fields.

Importantly, the dipole potential falls off quickly, as $1/r^2$, whereas a single charge falls off as $1/r$, as seen in Equation 5.24. Note that if $\hat{r}$ is perpendicular to $\vec{p}$ then no work is done and $\phi(r) = 0$ for any r, since the external agent will do an equal and opposite amount of work against each of the equivalent positive and negative charges on the dipole.

In Chapter 13, we will return to applications of electrostatics to membrane electrostatic potentials and impulse propagation in nerves.

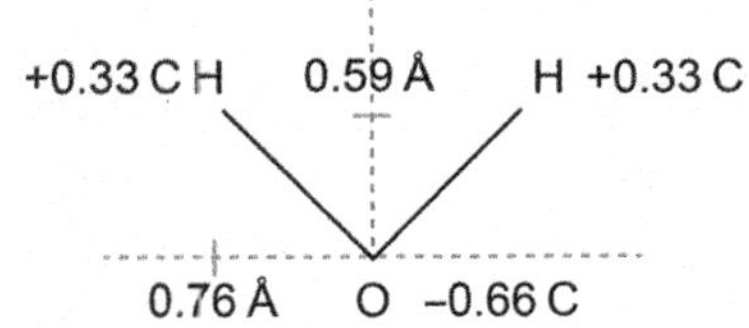

Figure 5.4 Schematic representation of the H_2O dipole.

5.2 Covalent Bonds in the Living State are Stable Against Thermal Energy

The covalent bond involves the sharing of electrons between atoms and is purely quantum mechanical in nature. It typically involves bond strengths of 2–5 eV (electron-volts). In contrast, thermal energy at $T = 300$ K is $k_BT = 4.14 \times 10^{-21}$ J = 0.026 eV. Consider that a modest covalent bond energy of 2 eV is around 80 k_BT. At $T = 300$ K, the Boltzmann distribution (Equation 4.27) shows that only a negligible fraction, $\exp(-40) = 1.8 \times 10^{-35}$, molecules will have a thermally broken bond at any instant. Hence, covalent bonds are a tremendous bulwark against incessant thermal energy. Because the living state involves the continuous making and breaking of covalent bonds to produce and use molecules, energy sources, mostly in the form of ATP (Chapter 3) are needed, and enzyme catalysts are needed to lower activation barriers to allow reactions to occur quickly.

5.2.1 Useful Conversions for Energy Units

In quantum mechanics, the eV (atomic scale energies) is preferred, and its multiples, such as MeV (10^6 eV, appropriate for nuclear scale energies) and GeV (10^9 eV, appropriate for particle physics energies). Chemists often prefer to use kcal/mol, for historical reasons, whereas the Joule (J) is the official SI (Système International or MKSA) unit for energy, and the Erg is the appropriate CGS unit, frequently employed in physical chemistry.

$$1\,\text{eV} = 1.6\times10^{-19}\,\text{J} = 1.6\times10^{-12}\,\text{Erg} = 23\ \text{kcal/mol}$$

A handy conversion for k_BT at $T = 300$ K is

$$k_BT = 4.14\times10^{-21}\,\text{J} = 0.026\text{eV} = 4.14\times10^{-14}\,\text{Erg} = 0.026\text{eV} = 0.6\ \text{kcal/mol.}$$

5.3 "Weak Electrostatic Forces" Allow for Self-organization and Rapid Dynamic Processes

"Weak forces" in the biophysical sense are those that are on the order of k_BT, and hence are easily formed and broken. An important aspect of many weak bonds is that they can act *cooperatively* to stabilize macromolecular and supramolecular structures. By the same token, because the individual bonds are weak they are easily broken. For example, many weak hydrogen bonds that cooperate to hold the DNA double helix together can easily be broken at fairly low temperatures (70°C–95°C) leading to the "melting" of the double helix. Similarly, weak electrostatic and hydrogen bond forces can stabilize the 3D structure of proteins, as well as the hydrophobic effect. These bonds can be broken under thermal and other stressors, causing proteins to unfold.

Whereas the making and breaking of such weak bonds can occur at low temperatures, it is also possible to provide very speedy processes involving covalent bonds via enzymatic (catalytic) action which would otherwise be extremely stable at biological temperatures.

Weak electrostatic forces are subsumed under ion–ion, ion–dipole, and dipole–dipole interactions. The latter occur by various mechanisms, depending on whether the interacting dipoles are permanent (e.g. in water), classically induced (Debye effect), or quantum mechanically produced (London dispersion effect). In the following summary, electrostatic screening is ignored. It will be introduced later in the chapter when we consider small mobile ions in solutions and how they distribute themselves near macrocharges, such as a globular protein, and lead to electrostatic screening, which diminishes the electric field and potential in the vicinity of the macrocharges.

5.3.1 Ion–Ion Interactions

In the aqueous interior of a biological cell, there are many free ions, for example, Na^+, K^+, Cl^-, Ca^{2+}, PO_4^{3-}, among others. For these, the *unscreened potential energy* between charges q_1 and q_2 at a distance r from each other, $U(r)_{ion\text{-}ion}$, is given by Equation 5.20 as

$$U_{ion-ion}(r) = \frac{q_1q_2}{4\pi\varepsilon r} \tag{5.44}$$

For example, for Na^+ and Cl^- at $r = 0.3$ nm, $U_{ion\text{-}ion} \sim 0.06$ eV in pure water, about twice k_BT, and much lower than a covalent bond energy. The high dielectric constant of water is responsible for the dissolution of salts, since $D = 80$ lowers the attraction between salt ions by a factor of 80.

5.3.2 Ion–Dipole Interactions

Since water has a permanent dipole moment, the many ions involved in the living state constantly interact by this mechanism, e.g. in the solvation of ions. The electrostatic potential energy between a charge q at position vector $\vec{r}$ from the center of $\vec{p}$, $U_{ion\text{-}dipole}(r)$ is given by Equations 5.20 and 5.43 by

$$U_{ion-dipole}(r) = \frac{q\vec{p}\cdot\hat{\text{r}}}{4\pi\varepsilon r^2} \qquad 5.45$$

This potential energy falls off as $1/r^2$, in contrast to the $1/r$ dependence of the ion–ion interaction.

5.3.3 Dipole–Dipole Interactions: Van Der Waals Forces

These interactions are among the most ubiquitous of forces in the living state. The electrostatic potential energy between two point dipoles, $\vec{p}_1$ and $\vec{p}_2$ at a distance r from each other is

$$U_{dipole-dipole}(r) = \frac{1}{4\pi\varepsilon r^3}\left[\vec{p}_1\cdot\vec{p}_2 - 3(\vec{p}_1\cdot\hat{\text{r}})(\vec{p}_2\cdot\hat{\text{r}})\right] \qquad 5.46$$

where the geometry of the interaction is shown in Figure 5.5. When this expression is averaged at fixed r over all possible mutual orientations of the dipoles, θ_1 and θ_2 and ϕ using the Boltzmann weighting factor $\exp\left[-U_{dipole-dipole}(r,\theta,\varphi)/k_BT\right]$, the value of the average energy $<U_{dipole\text{-}dipole}>$ is

$$\langle U_{dipole-dipole}\rangle = -\frac{2p_1^2p_2^2}{3(4\pi\epsilon)^2 k_BT}\frac{1}{r^6} \qquad 5.47$$

The $1/r^6$ dependence makes this an extremely short-range force, but because dipoles are so plentiful in the living state – e.g. water – this interaction is important.

5.3.4 Lennard–Jones Potential

The close-range interactions between uncharged atoms and molecules are often represented as a combination of attractive $1/r^6$ dipole–dipole potentials and hard, repulsive steric $1/r^{12}$ forces in the Lennard–Jones potential

$$U_{LJ} = \frac{a}{r^{12}} - \frac{b}{r^6} \qquad 5.48$$

where a and b characterize the strength of the repulsive and attractive interactions, respectively. Figure 5.6 shows the shape of a typical Lennard–Jones potential, and how a shallow, negative (attractive) energy well forms around 0.7 nm and quickly diminishes to zero as r increases. Below 0.7 nm, the $1/r^{12}$ term leads to a hard repulsive energy barrier.

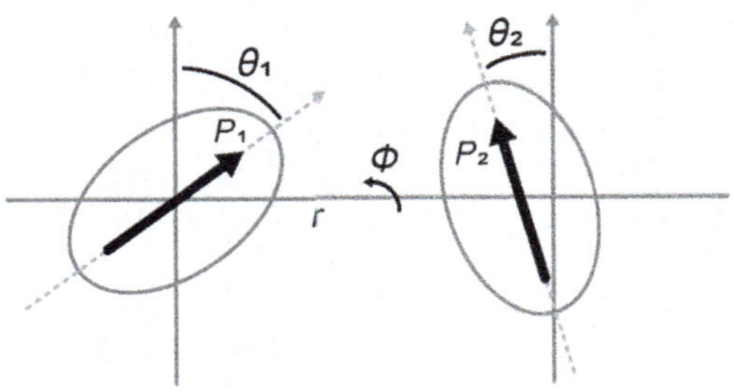

Figure 5.5 Interaction of two electric dipoles. (Courtesy of Marissa Calko).

5.3.5 London Forces and Mutually Induced Dipoles

The London force between two atoms or molecules with no permanent dipole moment is an interesting quantum mechanical phenomenon in which the electron clouds distribute themselves in the interacting particles such that fluctuating dipoles with net attraction occur.

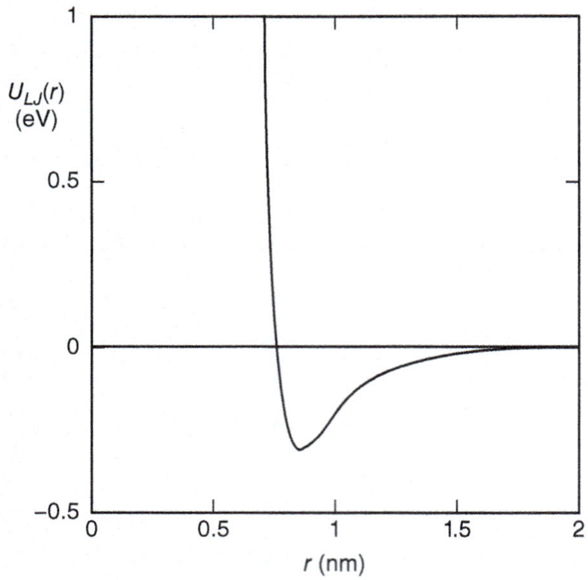

Figure 5.6 A typical Lennard–Jones potential.

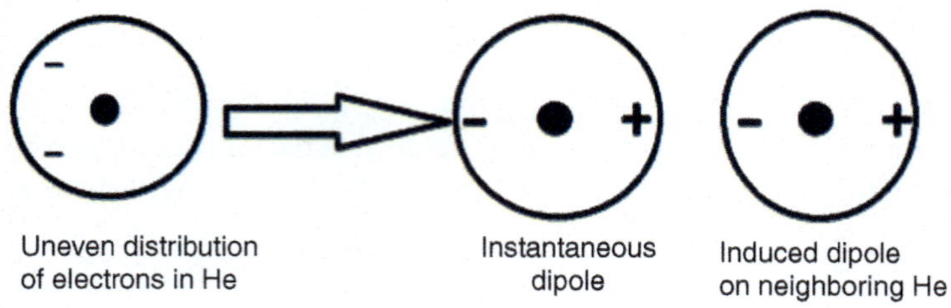

Figure 5.7 Induced dipole in He atoms.

These even occur in noble gases which have exceptionally high degrees of spherical symmetry in their electron distribution. Figure 5.7 shows schematically how charge fluctuations in a He atom lead to a transient dipole which then induces a dipole in a neighboring He atom, creating a weak attraction. Without this force, helium and other noble gases would not exist in a liquid state at low temperature.

Even though the effect is quantum mechanical in origin the *Drude model* offers a means of understanding these forces with a classical analogy. In a very elementary representation, consider two dipoles represented by two identical 1D harmonic oscillators, A and B, each containing an electron and each with force constant κ arranged along the x-axis, at a fixed distance r_{AB}, as sketched in Figure 5.8. The x-axis is vertical in the figure.

Note that in Figure 5.8, a positive displacement from equilibrium from A is in the opposite direction as a positive displacement for B. The electron on each oscillator gives rise to an interaction energy between the dipoles, $U_{A,B}$, that will depend on the relative positions of the electrons in the oscillators. By geometry, the Hamiltonian can be written as

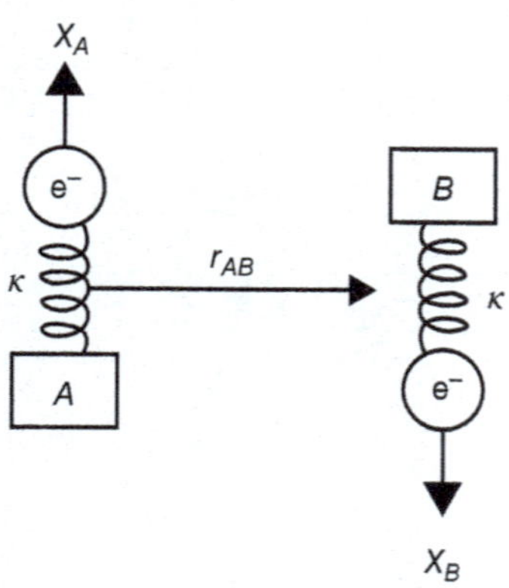

Figure 5.8 Simple classical model for London forces.

$$H = KE_A + KE_B + U_{A,B} = \frac{1}{2m}(p_A^2 + p_B^2) + \frac{\kappa}{2}(x_A^2 + x_B^2 + 2Cx_A x_B) \qquad 5.49$$

Here p_A and p_B are the momentum of A and B, respectively. H is the sum of the kinetic and potential energies of each oscillator plus a coupling term, Cx_Ax_B, which accounts for the interaction potential energy between the electrons; the closer they are, the higher the interaction potential. The constant C is the dimensionless coupling constant, and is a gauge of how strong the interaction is, and increases as r_{AB} decreases. $C \ll 1$, since the coupling between induced dipoles is much weaker than the fluctuation dipoles of each separate atom. Equation 5.49 can be factored into normal-mode form as

$$H = \frac{1}{2m}(p_A^2 + p_B^2) + \frac{\kappa}{4}(1 + C)(x_A + x_B)^2 + \frac{\kappa}{4}(1 - C)(x_A - x_B)^2 \qquad 5.50$$

The first interaction term measures the increase in potential from the oscillators in phase and maximizing their repulsive potential (+), and the second term is when they are out of phase and minimizing the potential (−).

The eigenvalues of energies for each state is

$$E_+ = \hbar\omega_+ = \sqrt{1+C}\hbar\omega_0, \quad E_- = \hbar\omega_- = \sqrt{1-C}\hbar\omega_0 \tag{5.51}$$

where ω_0 is the resonant frequency of each oscillator in isolation:

$$\omega_0 = \sqrt{\frac{\kappa}{m}} \tag{5.52}$$

where m is the electron mass. The term $\hbar\omega_0$ is the quantum ground state for the two oscillators with $C = 0$. The total energy is hence

$$E_{total} = \frac{1}{2}(E_+ + E_-) = \frac{\hbar\omega_0}{2}\left(\sqrt{1+C} + \sqrt{1-C}\right) = \frac{\hbar\omega_0}{2}\left(2 - \frac{1}{4}C^2 - \frac{5}{64}C^4 + \cdots\right) \tag{5.53}$$

where the right-hand expression is from the binomial expansion of the coupling terms, which is permitted since $C \ll 1$. Equation 5.53 shows that the energy is less than the isolated oscillators so that there is a net attractive potential due to the harmonic oscillator coupling U_{d-d} from the negative terms in Equation 5.53:

$$U_{d-d} = -\hbar\omega_0\left(\frac{C^2}{8} + \frac{5}{128}C^4 + \cdots\right) \tag{5.54}$$

When extended to 3D and spatially averaging:

$$C \propto \frac{1}{r_{AB}^3} \tag{5.55}$$

which yields a $1/r^6$ attractive potential (proportional to C^2 in first order) for these types of dipoles, which is the same spatial dependence as for permanent dipoles.

5.4 Hydrogen Bonds

Hydrogen bonds result from hydrogen being bonded to a strongly electronegative atom such as O, N, Cl, and F. Other bonds that have significant electron withdrawing capacity from H can also occur. The H-bond combines features of a strong dipolar interaction with some quantum mechanical electron sharing between molecules participating in the H-bond. IUPAC has an official definition for H-bonds.[1] The H-bond helps to stabilize the DNA helix, is important for secondary and tertiary structure in proteins, and is ubiquitous in water itself, lending water its high boiling point and heat of evaporation, significant viscosity, and ability to crystallize into a regular lattice. H-bonds can also be involved in initiating charge transfer reactions and can cause enzyme–substrate interactions. It is a prominent interaction in the living state. Figure 5.9 shows schematically how the interaction works.

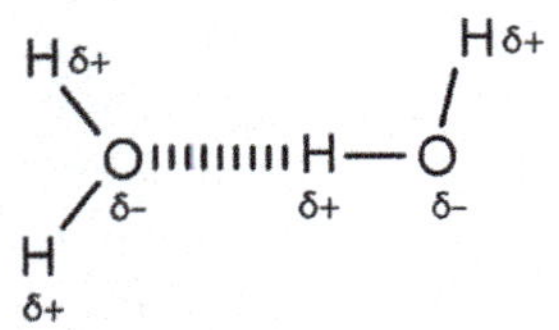

Figure 5.9 Schematic representation of the hydrogen bond.

Hydration forces are distinct from hydrogen bonds but are related to how water interacts and structures itself on polymers, such as DNA, or colloid surfaces. A general viewpoint is that compressing aqueous polymers or colloid surfaces in aqueous solutions involves the stripping of water molecules of hydration from the polymers or surfaces. The decay length for osmotic pressure as such systems are compressed is typically around 0.3 nm, the approximate size of the water molecule, which lends support to the idea of stripping away layers of water under

1 E. Arunan, G. R. Desiraju, R. A Klein, J. Sadlej, S. Scheiner, I. Alkorta, D. C. Clary, R. H. Crabtree, J. J. Dannenberg, P. Hobza, H. G. Kjaergaard, A. C. Legon, B. Mennucci, and D. J. Nesbitt. (2011). Definition of the hydrogen bond. *Pure and Applied Chemistry 83* (8): 1637–41. doi:10.1351/PAC-REC-10-01-02.

compression. Further evidence comes from reports that the decay length is independent of solution ionic strength so that the effect is not likely due to primarily electrostatic or screening effects. The magnitude of the osmotic pressure depends on the details of the polymers and surfaces under stress.

5.5 Electrically Charged Macromolecules and Colloids in Solution

Living state processes occur in an aqueous solution with a pH near 7.2 and high concentrations of small ions. Besides the small ions, Na^+, K^+, Cl^-, Ca^{2+}, most of the biomacromolecules are also charged. Polynucleic acids, proteins, phospholipids, and many carbohydrates have charged groups that give them specific electrostatic properties in solution. We distinguish several classes of electrically charged macromolecules:

1) *Polyelectrolytes* refer to macromolecules that have multiple charge groups along the polymer chain. Small charge species that can dissociate from the polyelectrolyte, such as Na^+, are termed *counterions*, or *gegenions*. A polyelectrolyte chain in solution from which some or all of the counterions have dissociated, leaving the chain with a large net charge is termed a *polyion*. If all of the charge groups on the polyelectrolyte have the same sign then the macromolecule is an *anionic* (−) or *cationic* (+) *polyelectrolyte; polyanion* and *polycation*, respectively. Examples of polyanions include DNA, which carries a negatively charged phosphate group on each repeat unit, and many polysaccharides, which carry negative carboxylate or sulfate groups on their repeat units.
2) *Polyampholytes* are macromolecules that have both negative and positive charge groups. Proteins are usually polyampholytes because they can contain both negatively charged amino acids (the acidic group comprising aspartic acid and glutamic acid), and positively charged ones (the basic group comprising lysine, arginine, and histidine, as well as cysteine and tyrosine). Additionally, the carboxyl terminal residue of a protein is negative, and the amino terminal is positive.
3) *Polyzwitterions* are molecules that bear opposite charges on subunits. A single molecule with equal number of positive and negative charges is a *zwitterion*.[2] A single amino acid with an electrically neutral side chain, for example, is zwitterionic at physiological pH, because it has a positive amino end and a negative carboxyl end. Many phospholipids have zwitterionic head groups. While the phospholipids are not themselves macromolecules, they attain supramolecular organization in the form of membranes, which have critically important electrical roles in living state processes.

The existence of electrical charge on a macromolecule has very marked effects on such properties as conformation, hydrodynamics, and interaction and affinity for other compounds. The theoretical and experimental study of electrically charged macromolecules is very complex and is still a very active field of research.

5.6 Poisson–Boltzmann Equation

In most introductory electromagnetic theory textbooks, the portion on electrostatics is confined to finding electric potentials and fields for fixed charge distributions. But in a solution containing many mobile charges, how can the potentials and fields be found if the charges can both move about and contribute to the potentials and fields? This is not an easy question and there is no easy answer.

Consider a glass of salt water. If there were only a single species of a dissolved salt in solution, say NaCl dissociated into Na^+ and Cl^-, then we would expect a fairly uniform concentration of ions throughout, without any net macroscopic electric fields. Even though there would be no macroscopic, average fields, think about the electric complexity of a solution with randomly diffusing ions. Each positive ion has an outward-directed electric field, and each negative ion an inward-directed one, so that at any point in the liquid, there is an electric field which is the linear superposition of all the ions in the solution. Of course, the ions near the point under consideration contribute the most to the electric field, and, because there are fluctuations in local ion densities and positions one can guess that there must be small net electric fields at any point that rapidly vary in time. Since these small fields can be positive or negative with equal probability, we can reasonably think that the time average field at any point must be zero. Imagine that you could visualize these rapidly changing, tiny fields; it would be kaleidoscopic. This electric field landscape is further influenced by the fact that opposite ions will

2 "Zwitter" is German for "hermaphrodite."

transiently attract and like ions will transiently repel as they move near each other. Additionally, the diffusing ions create small magnetic fields which will exert tiny, but nonzero magnetic force on nearby ions in motion. If all this could be visualized an ordinary glass of water would be a mind-boggling complex dance of electric and magnetic fields, and motion.

While all this multifold action is taking place microscopically, there is still the fact that, macroscopically, this is just a glass of salt water, and there are no macroscopic electric fields and no macroscopic motion, $\vec{E}=0$ and $\vec{v}=0$. This is the essence of *mean field theory*; ignore the mercurial microscopic complexity and consider the mean (average) field of all the ions in solution at any point in space. In our saltwater case, the mean field is zero everywhere, and we are back to our original assertion that there are no macroscopic electric fields.

Now, say we put into the salt solution an object with many elementary charges. This could be a charged metal electrode, for example, or a charged polyelectrolyte molecule. We immediately guess that if the object is positively charged then in its vicinity negative mobile ions will cluster more densely around it than mobile positive ions, and vice versa if the object is negatively charged. We expect a permanent, locally nonzero electric field to arise, and we might try to compute the field by considering the mean field produced by all the ions at any point. This is what we undertake next with the Poisson–Boltzmann equation (PBE), a mean field approach, but we must make the important caveat that, by "smearing out" all the electrical effects of the ions into a locally uniform average electric field, we may be losing some of the most critically important features of living state processes! After all, much of the intricacy of living state phenomena are due precisely to specificities, the geometry of molecular surfaces, complementary recognition of molecules, etc. With this disappointing caveat, the mean field theory has nonetheless proven very useful to generate applicable concepts and relationships for electrothermodynamic phenomena in solutions containing charged macromolecules.

One longstanding mean field approach to characterizing liquids containing electrically charged particles is to combine the Poisson equation for electrostatic potential $\phi(r,\theta,\varphi)$ Equation 5.9, with the Boltzmann distribution for mobile charges in the electrostatic potential.

Consider an electrolyte E, composed of n cations and m anions

$$E = A_n B_m \leftrightarrow nA^{+z_A} + mB^{-z_B}, \quad \text{e.g. } \mathrm{MgCl_2} \leftrightarrow \mathrm{Mg^{+2}} + 2\mathrm{Cl^-} \qquad 5.56$$

Note that both z_A and z_B are positive integers and so a "−" sign is inserted explicitly when expressing the charge of B. The electrical neutrality condition is $nz_A - mz_B = 0$.

Let $[E_0]$ be the initial bulk molar concentration of electrolyte (mol/L). "Bulk molar concentration" means the concentration of E when dissolved in a homogeneous solution. The positive bulk charge density in SI units (C/m^3) is

$$\rho_{+,0}(\mathrm{C/m^3}) = z_A ne\left[E_0\right](10^3\ \mathrm{L/m^3})N_A = 1000\ N_A z_A ne\left[E_0\right] = 1000\ N_A z_A e c_A \qquad 5.57$$

and

$$\rho_{-,0}(\mathrm{C/m^3}) = -z_B me\left[E_0\right](10^3\mathrm{L/m^3})N_A = -1000\ N_A z_B me\left[E_0\right] = -1000\ N_A z_B e c_B \qquad 5.58$$

where e is the elementary charge, and $c_A = n[E_0]$ and $c_B = m[E_0]$ are the molar concentrations of ions A and B, respectively. The subscript "0" refers to the charge density in homogeneous or bulk solution.

The net charge at any point in space is

$$\rho(\vec{r}) = \rho_+(\vec{r}) + \rho_-(\vec{r}) \qquad 5.59$$

where $\vec{r}$ is a position vector in space with respect to a chosen origin. For a bulk electrolyte solution, the distribution of charge is homogeneous so that there is, as discussed above, average electrical neutrality and hence in the mean field approximation:

$$\rho_0(\vec{r}) = \rho_{+,0}(\vec{r}) + \rho_{-,0}(\vec{r}) = 0 \qquad 5.60$$

so that

$$\rho_{-,0}(\vec{r}) = -\rho_{+,0}(\vec{r}) \qquad 5.61$$

Now, consider that a small, positively charged object, such as a protein with a net charge, is placed in the solution, or a small object is placed in the solution and held at a certain positive potential. We expect the free negative charges to preferentially cluster closer to the object at small distances from the object than the positive free charge, $-\rho_-(\vec{r}) > \rho_+(\vec{r})$, and vice versa if the object is negatively charged. Far from the object, the charge densities will be equal to their bulk densities, i.e. equal and opposite to each other, by electroneutrality:

$$\rho_-(\vec{r}) = \rho_{-,0} = -\rho_+(\vec{r}) = -\rho_{+,0}, \text{ and } \rho(\vec{r}) \to 0 \quad \text{for} |\vec{r}| \gg 0 \qquad 5.62$$

where we take the origin of $\vec{r}$ as the center of the object.

The interesting question now is, if the arrangement of the ions determines the electric field, but the electric field exerts forces on the ions and so influences their arrangement how can we proceed in a consistent way using basic principles? The first principle to invoke is that ions will arrange themselves in the electric field according to their energy in the field and the corresponding Boltzmann distribution:

$$\begin{aligned}\rho_+(\vec{r}) &= \rho_{+,0}e^{-U(\vec{r})/k_BT}\\ \rho_-(\vec{r}) &= \rho_{-,0}e^{-U(\vec{r})/k_BT}\end{aligned} \qquad 5.63\text{a,b}$$

where the potential energies of the ions in the field are given by Equation 5.20 as

$$\begin{aligned}U(\vec{r}) &= +z_Ae\phi(\vec{r}) \quad \text{for a positive ion}\\ U(\vec{r}) &= -z_Be\phi(\vec{r}) \quad \text{for a negative ion}\end{aligned} \qquad 5.64\text{a,b}$$

Hence,

$$\rho_+(\vec{r}) = \rho_{+,0}e^{-z_Ae\phi(\vec{r})/k_BT} \qquad 5.65\text{a}$$

$$\rho_-(\vec{r}) = \rho_{-,0}e^{+z_Be\phi(\vec{r})/k_BT} \qquad 5.65\text{b}$$

Let ρ_0 be the bulk charge density:

$$\rho_0 = \rho_{+,0} = -\rho_{-,0} \qquad 5.66$$

Then, the net charge density at any point will be

$$\rho(\vec{r}) = \rho_0\left(e^{-z_Ae\phi(\vec{r})/k_BT} - e^{+z_Be\phi(\vec{r})/k_BT}\right) \qquad 5.67$$

Now, Poisson's Equation 5.9 relates charge density to electric field, so substituting the charge density of Equation 5.67 into it yields, in SI (MKSA) units,

$$\nabla^2\phi = -\frac{\rho_0}{\varepsilon}\left(e^{-z_Ae\phi/k_BT} - e^{z_Be\phi/k_BT}\right) \quad \text{Poisson – Boltzmann Equation} \qquad 5.68$$

This is the well-known *Poisson-Boltzmann equation, PBE*, an electrothermodynamic, second-order, nonlinear, partial differential equation. For the case of *symmetric electrolytes*, $z = z_A = z_B$, it can also be written compactly as

$$\nabla^2\phi = \frac{2\rho_0}{\varepsilon}\sinh\left(ze\phi / k_BT\right) \qquad 5.69$$

With this equation, mobile charges in solution are arranged such that they simultaneously obey both the Poisson equation and the Boltzmann distribution. The situation of mobile charges is not unique to electrolyte solutions, and is found, for example, in the electron distribution in an ionized gas (plasma), and for mobile charge carriers in semiconductors.

The PBE is difficult to solve analytically in a general form, but it can be solved to arbitrarily high accuracy in any geometry and boundary conditions of interest by numerical computation. There are several commercially available packages for such calculations.

To gain insight into how the mobile charges lead to a different form of potential compared to stationary charges, the PBE can be linearized and solved analytically. The right-hand side of the PBE can be linearized with cancellation of both constant and second-order terms. For symmetric electrolytes ($z = z_A = z_B$), let

$$x \equiv ze\phi / k_B T \tag{5.70}$$

Expansion of $e^x - e^{-x}$ in a power series yields

$$e^x - e^{-x} = 2x + \frac{2x^3}{3!} + \frac{2x^5}{5!} + \cdots \tag{5.71}$$

i.e. the constant term and order x^2 terms on the right-hand side disappear so that *the linearized PBE* result for a symmetric electrolyte, good to second order, is

$$\nabla^2 \phi = \frac{2\rho_0 x}{\varepsilon} = \left(\frac{2\rho_0 ze}{\varepsilon k_B T}\right)\phi \quad \text{Linearized Poisson – Boltzmann Equation} \tag{5.72}$$

Linearization is appropriate when:

$$\frac{ze\phi}{k_B T} \ll 1 \tag{5.73}$$

The square root of the term in parentheses on the right-hand side of Equation 5.72 has the dimensions of inverse distance, which then serves as a scale for screening. It is often termed the *Debye screening parameter* κ, after Peter Debye (Holland/USA, 1884–1966):

$$\kappa = \left(\frac{2\rho_0 ze}{\varepsilon k_B T}\right)^{1/2} \tag{5.74}$$

Its reciprocal $1/\kappa$ is the characteristic range of the screened potential and is sometimes termed the *Debye–Hückel screening length*.

For aqueous solutions at 25°C and symmetric electrolytes:

$$\frac{1}{\kappa} = \frac{0.305 \text{ nm}}{\left(z^2 [E_0](\mathrm{M})\right)^{1/2}} \tag{5.75}$$

where the following SI constants are used: $k_B = 1.38 \times 10^{-23}$ J/K, $\varepsilon_0 = 8.85 \times 10^{-12} \mathrm{C^2/N\text{-}m^2}$, $\varepsilon = 80\varepsilon_0$ for water, $e = 1.6 \times 10^{-19}$ C. $[E_0]$ is the molar concentration of the electrolyte (mol/L). For water with 150 mM univalent salt at 25 C, $1/\kappa = 0.8$ nm, and at 0.001 M $1/\kappa = 9.7$ nm for distilled, neutral water, the H^+/OH^- balance of 10^{-7} mol/L gives the limiting value of $1/\kappa = 1\ \mu$m. Dissolved CO_2 in water can add carbonic acid that raises this "native" level of electrolyte and can typically decrease the screening length to about 0.25 μm. In order to generalize to nonsymmetric electrolytes, the concept of *ionic strength* is introduced in Section 5.6.3.

The Debye screening parameter allows a convenient shorthand for the linearized PBE for symmetric electrolytes

$$\nabla^2 \phi = \kappa^2 \phi \tag{5.76}$$

The full Laplacian operator in spherical coordinates is

$$\nabla^2 = \frac{1}{r^2}\frac{\partial}{\partial r}\left(r^2 \frac{\partial}{\partial r}\right) + \frac{1}{r^2 \sin\theta}\frac{\partial}{\partial \theta}\left(\sin\theta \frac{\partial}{\partial \theta}\right) + \frac{1}{r^2 \sin^2\theta}\frac{\partial^2}{\partial \varphi^2} \tag{5.77}$$

If a spherically symmetric potential is involved, $\phi(r,\theta,\varphi)=\phi(r)$, the spherical Laplacian operator reduces to

$$\nabla_r^2\phi=\frac{1}{r^2}\frac{\partial}{\partial r}\left(r^2\frac{\partial\phi}{\partial r}\right)=\frac{1}{r}\frac{\partial^2(r\phi)}{\partial r^2} \tag{5.78}$$

and the linearized PBE becomes

$$\frac{d^2(r\phi)}{dr^2}=\kappa^2(r\phi) \tag{5.79}$$

The form of the solution is a *screened potential*, such as also found in plasma physics,[3]

$$\phi(r)=A\frac{e^{-\kappa r}}{r}+B \quad \text{Screened Electrostatic Potential} \tag{5.80}$$

where A is a constant potential and B a constant, where A and B are determined by boundary conditions. The net charge distribution is then given by

$$\rho(r)=-\epsilon\kappa^2\phi(r) \tag{5.81}$$

5.6.1 Boundary Conditions for a Sphere and Resulting Potential and Charge Distribution Around It

A simple example is that of an impermeable sphere of radius R with a net charge Q distributed evenly on its surface, where the sphere is surrounded by ions in a solution. It is customary to separate this into the region of $r<R$, into a thin layer around the sphere into which ions cannot penetrate due to excluded volume, and outside this layer. For simplicity here, the only boundary is taken as R. Due to spherical symmetry and no volume charge, there is no electric flux and hence no electric field within the sphere so that the potential is constant. Applying the usual boundary conditions for continuity of potential:[4]

$$\phi_1(R)=\phi_2(R) \tag{5.82}$$

where $\phi_1(r)$ is the potential for $r<R$ and $\phi_2(r)$ is for $r\geq R$. Since, by Gauss's law, Equation 5.3, there is zero electric field within the sphere, the boundary condition on the electric field at R is[5]

$$E_\perp(R)=\frac{\sigma}{\varepsilon}=-\nabla\phi(r)\big|_R \tag{5.83}$$

This leads to a potential of the form

$$\phi(r)=\frac{Q}{4\pi\varepsilon(1+\kappa R)}\frac{e^{-\kappa(r-R)}}{r},\quad r\geq R \tag{5.84}$$

The effect of the screening is to lower the surface potential, which from the above is

$$\phi(R)=\frac{Q}{4\pi\varepsilon(1+\kappa R)R} \tag{5.85}$$

---

3 Block, L. P. (1978). A double layer review. *Astrophysics and Space Science*, 55: 59–83.

4 Potential must be continuous across a boundary, so that the electric field, which is its negative derivative, remains finite. The electric field itself, however, can have a discontinuity if there is a surface charge at a boundary.

5 Application of Gauss's law over a closed surface at an interface between surface 1 and 2 yields the general condition on the perpendicular compoonents of $\vec{E}$ that $\varepsilon_1 E_{1,\perp}$–$\varepsilon_2 E_{2,\perp}=\sigma$, where σ is the surface charge density on the interface. If $\sigma\neq 0$, then there is a discontinuity in $\vec{E}_\perp$. Applying the line integral of Equation 5.7 around a closed path enclosing the surface yields the condition that the parallel component of $\vec{E}$ are always continuous $E_{1,\parallel}=E_{2,\parallel}$. These boundary conditions are used later in Chapter 9.10.1 to derive the Fresnel reflection and transmission coefficients for light passing from one medium to another.

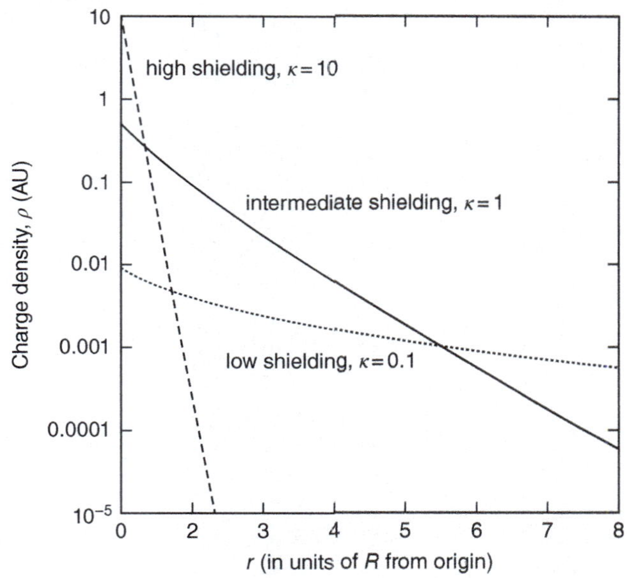

Figure 5.10 Charge density around a positively charged sphere for different values of the Debye screening parameter κ. The charge density is negative, but the absolute value is used to enable a log-scale representation.

As expected, in the case of no screening, $\kappa = 0$, the potential becomes that found in a homogeneous dielectric material of constant ε and no mobile charges, Equation 5.25

$$\phi(R) = \frac{Q}{4\pi\varepsilon R} \tag{5.86}$$

No matter the magnitude of the screening parameter κ, the potential will approach zero as $r \to \infty$, as expected.

The charge density distribution $\rho(r)$ is found by Equation 5.81. The distribution computed this way shows that there is an excess of negative charge near R which returns to the bulk value far away from R.

Figure 5.10 shows the absolute value of the dimensionless charge distribution around a sphere of dimensionless radius $R = 1$, for different values of κ, where the dimensionless κ is in inverse units of dimensionless $R = 1$. A log scale is required to fit the disparate curves on a single figure, so the absolute value of $\rho(r)$ is shown (equivalently this could be a negatively charged sphere, in which case $\rho(r)$ is positive). At high shielding, there is a high concentration of net charge around the charged sphere, giving a high charge density, but the density dies off quickly. At low shielding, there is a far lower concentration around the sphere but it dies off in space much more slowly.

It is left as an exercise to prove that the total charge in the surrounding layer (integrated from the sphere's surface to infinity) is equal to, and opposite in sign, to the charge on the sphere.

5.6.2 Capacitance with a Screened Potential

The capacitance of this sphere, following the usual definition is

$$C = \frac{Q}{\phi} = 4\pi\varepsilon R(1 + \kappa R) \tag{5.87}$$

The ratio of this capacitance to that of an unscreened sphere is

$$\frac{C_{screened}}{C_{unscreened}} = 1 + \kappa R \tag{5.88}$$

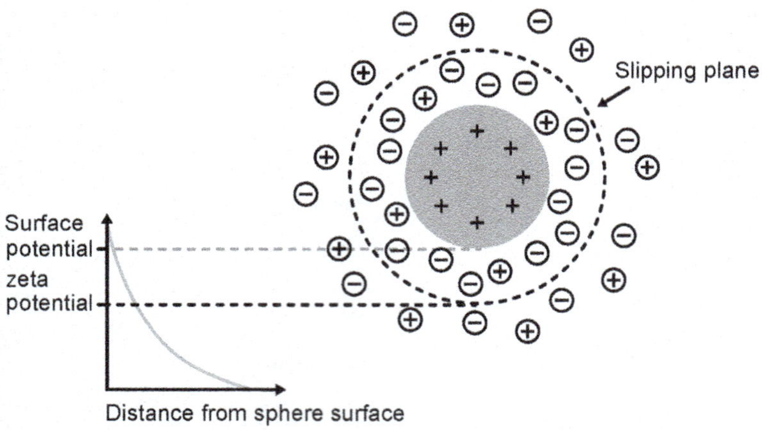

Figure 5.11 Electric double layer, showing the distribution of mobile ions around a negatively charged particle with positive ions adsorbed on the surface (the Stern layer) and the diffuse ion layer around it. (Courtesy of Marissa Calko).

Screening increases the sphere's capacitance: $C_{screened} > C_{unscreened}$. By Equation 5.32, the work needed to charge the sphere to Q hence decreases with screening.

The surface of the charged particle and the diffuse ion cloud around it of opposite charge together are referred to as an *electrical double layer*. These occur frequently even with initially neutral particles when ions of one sign adsorb to the particle's surface and a diffuse ion cloud of opposite sign forms around it. Figure 5.11 is a schematic of the double layer with a negatively charged particle, positive ions adsorbed to the surface, forming the rigid *Stern layer*, and the diffuse layer around it. Also shown is the slippage surface. When the particle diffuses or moves under the influence of an external field it entrains the Stern layer and some of the diffuse layer around it. Since these layers contain opposite charges from the charged particle, the potential is lower at the slipping plane than at the particle's surface. The potential at the slippage plane is termed the *zeta-potential*. The ζ-potential is essentially felt by other particles, and so it is an important parameter in the stabilization of colloid particle suspensions. The higher the ζ-potential, the greater the electrostatic repulsion between particles and the higher the colloidal stability. ζ-potential greater than about 50 mV usually lead to good colloid stability. The ζ-potential can often be conveniently measured by *electrophoretic light scattering*.

5.6.3 Ionic Strength

It is straightforward to extend the above-linearized treatment to the case where the added salt is no longer symmetric, e.g.

$$MgCl_2 \leftrightarrow Mg^{2+} + 2Cl^- \tag{5.89}$$

The result is the concept of *ionic strength*, *I*, which is the equivalent charge concentration in molar units. The general expression for ionic strength is derived next.

An electrolyte of the general form P_jN_k dissociates according to

$$P_jN_k \rightarrow jP^{z_+} + kN^{z_-} \tag{5.90}$$

where "*P*" is used for the positive ion and "*N*" for the negative. The electrical neutrality condition is

$$(jz_+) - (kz_-) = 0 \tag{5.91}$$

Frequently, there is a mixture of electrolytes. In the living state, Na^+, K^+, Cl^-, Mg^{2+}, and bicarbonate, HCO_3^-, are essential small ions. Consider a solution containing *Y* different electrolytes, where the *i*th electrolyte species has the form $E_i = P_{i,j}N_{i,k}$ and dissociates according to

$$P_{i,j}N_{i,k} \rightarrow jP_i^{z_{+,i}} + kN_i^{z_{-,i}} \tag{5.92}$$

where $z_{+,i}$ and $z_{-,i}$ are the number of charges on the P_i and N_i ions, respectively, and charge neutrality applies individually for each species i

$$(jz_{+,i}) - (kz_{-,i}) = 0 \quad 5.93$$

Returning to the linearization of the PBE, the positive and negative charge densities in bulk for electrolyte species i, $\rho_{+,0,i}$ and $\rho_{-,0,i}$ are

$$\rho_{+,0,i} = n_{+,0,i} e z_{+,i} \quad 5.94$$

and

$$\rho_{-,0,i} = -n_{-,0,i} e z_{-,i} \quad 5.95$$

where $n_{+,0,i}$ and $n_{-,0,i}$ are the number concentrations (ions/m^3) of positive and negative ions, respectively, of electrolyte species i. The PBE then involves the sum over all X electrolyte species

$$\nabla^2 \phi = -\frac{1}{\varepsilon}\left\{\sum_{i=1}^{X}\left[\rho_{+,0,i}\exp\left(-\frac{z_{+,i}e\phi}{k_B T}\right) + \rho_{-,0,i}\exp\left(\frac{z_{-,i}e\phi}{k_B T}\right)\right]\right\} \quad 5.96$$

Expanding the exponential term and retaining only the first-order term leads to

$$\nabla^2 \phi = \frac{e^2}{\varepsilon}\left[\sum_{i=1}^{X}(n_{+,0,i}z_{+,i}^2 + n_{-,0,i}z_{-,i}^2)\right]\phi / k_B T \quad 5.97$$

where the electroneutrality condition has been invoked for each species i, to cancel the leading term of the expansion:

$$(\rho_{+,0,i} + \rho_{-,0,i}) = 0 \quad 5.98$$

Importantly, the second-order term in the expansion, ϕ^2, does not cancel, meaning this linearization for asymmetric electrolyte holds to lower energy than for symmetric electrolytes. This is discussed further below.

The half-value of bracketed term in Equation 5.97 is the *ionic strength*, denoted I.

$$I = \frac{1}{2}\sum_{i=1}^{X}(n_{+,0,i}z_{+,i}^2 + n_{-,0,i}z_{-,i}^2) \quad 5.99$$

where n_i is the number concentration of ions i. The screening parameter then becomes

$$\kappa = \left(\frac{2Ie^2}{\varepsilon k_B T}\right)^{1/2} \quad 5.100$$

It is more common to express I in molar concentrations, $[I]$, rather than number concentration, and to run the summation over all $2X$ ion species (each electrolyte species contributes one P and one N, hence the factor of 2), where $Y = 2X$ for an electrolyte of the type P_jN_k. For

$$[I] = \frac{1}{2}\sum_{j=1}^{Y}[C_j]z_j^2 \quad 5.101$$

where $[C_j]$ is the molar concentration of each ion species in mol/L. The factor of ½ avoids double counting ions, e.g. $[I] = 1$ M for one molar monovalent electrolyte, $[E_0] = 1$ M. Note that if the molar concentration of an electrolyte is $[E]$ then $[P] = j[E]$ and $[N] = k[E]$, e.g. for $MgCl_2$, at a concentration of $[MgCl_2]$:

$$I = \frac{1}{2}\left(2^2[MgCl_2] + 1^2 2[MgCl_2]\right) = 3[MgCl_2] \quad 5.102$$

In terms of $[I]$ (in mol/L), the screening length then becomes, in general, at $T = 300$ K:

$$\kappa^{-1}(\mathrm{m}) = \left(\frac{\varepsilon k_B T}{2000e^2[I]N_A}\right)^{1/2} = \frac{3.07\times10^{-10}}{[I(\mathrm{mol/L})]^{1/2}} \quad 5.103$$

5.6.4 Asymmetric Electrolytes Show Clear Deviations Due to the Nonlinearity of the PBE

The above treatment has shown that the linearization for symmetric electrolytes holds to second order in potential, i.e. to order ϕ^2, since this term equals zero in the expansion of the Boltzmann exponential. As seen for asymmetric electrolytes, however, the term in ϕ^2 does not cancel out, and so the linearization holds only to first order, i.e. to order ϕ. This suggests that nonlinear effects may appear more readily for asymmetric electrolytes than for symmetric ones. In either case, the linearization predicts that the potentials only depend on ionic strength I, not the specific valences. Significant differences have been found between symmetric univalent electrolytes and divalent and asymmetric divalent electrolytes at identical ionic strengths.[6] Potentials increase in the order asymmetric divalent < symmetric divalent < symmetric monovalent. This follows from consideration of the nonlinear PBE:

$$\nabla^2\phi = \frac{-c_s}{\varepsilon}\{az_a \exp[-z_a e\phi/kT] - bz_b \exp[z_b e\phi/kT]\} \quad 5.104$$

where the electrolyte is of the form A_aB_b, z_a and z_b are the absolute values of the number of elementary charges on each ion, where $az_a = bz_b$ for electroneutrality of the undissociated electrolyte, c_s is the added (bulk) electrolyte concentration, and ε is the solvent dielectric constant, taken to be that of water. c_s times the bracketed expression is the net charge density around the polyelectrolyte due to the unequal distribution of the electrolyte ions. For the monovalent case, $a = b = z_a = z_b = 1$, for the symmetric divalent case $a = b = 1$, $z_a = z_b = 2$, and for the asymmetric divalent case $a = 1$, $b = 2$, $z_a = 2$, $z_b = 1$. Using the relationship between ionic strength I and c_s, the charge density $\rho(\phi(r))$ can be written for each case as

$$\rho(\phi) = \frac{I}{2}\{\exp[-e\phi/kT] - \exp[e\phi/kT]\}, \quad \text{monovalent case} \quad 5.105a$$

$$\rho(\phi) = \frac{I}{2}\{\exp[-2e\phi/kT] - \exp[2e\phi/kT]\}, \quad \text{symmetric divalent case} \quad 5.105b$$

$$\rho(\phi) = \frac{2I}{3}\{\exp[-2e\phi/kT] - \exp[e\phi/kT]\}, \quad \text{asymmetric divalent case} \quad 5.105c$$

When these are plotted using negative potentials for a negatively charged polyelectrolyte, $\rho(\phi)$, which is positive, at any given value of ϕ grows most quickly for Equation 5.105c, followed by Equation 5.105b, then Equation 5.105a. Only for the case $ze\phi/k_BT \ll 1$, i.e. for the linearized PBE, do expressions 5.105a–c converge to the same dependence on ϕ. In the full nonlinear form, however, the divalent cations cluster more closely to the polyelectrolyte's anionic charge distribution, and for the asymmetric divalent case, the singly charged anions drop off more slowly than the divalent cations, giving the greatest charge density around the polyanionic charge. This effect should be reflected in correspondingly lower potentials for asymmetric divalents when the nonlinear PBE is solved.

Figures 5.12a and 5.12b are for a polyanion, sodium hyaluronate in an aqueous solution, as the ionic strength increases continuously. Figures 5.12a and 5.12b show how light scattering and viscosity data, respectively, cluster according to the type of ion: symmetric mono- or divalent electrolytes and symmetric multivalent electrolytes.

Figure 5.13 shows the computed potentials versus ionic strength for symmetric monovalent and asymmetric divalent salts.

---

6 G. A. Sorci, W. F. Reed. (2004). Effect of valence and chemical species of added electrolyte on polyelectrolyte conformations and interactions. *Macromolecules 37*: 554–65.

Figure 5.12 (a) Effect of electrolyte valence and symmetry on (a) light scattering intensity from a polyelectrolyte solution and (b) the effect on solution viscosity. (Sorci, G.A. and Reed, W.F. (2004) / Reproduced with permission from American Chemical Society).

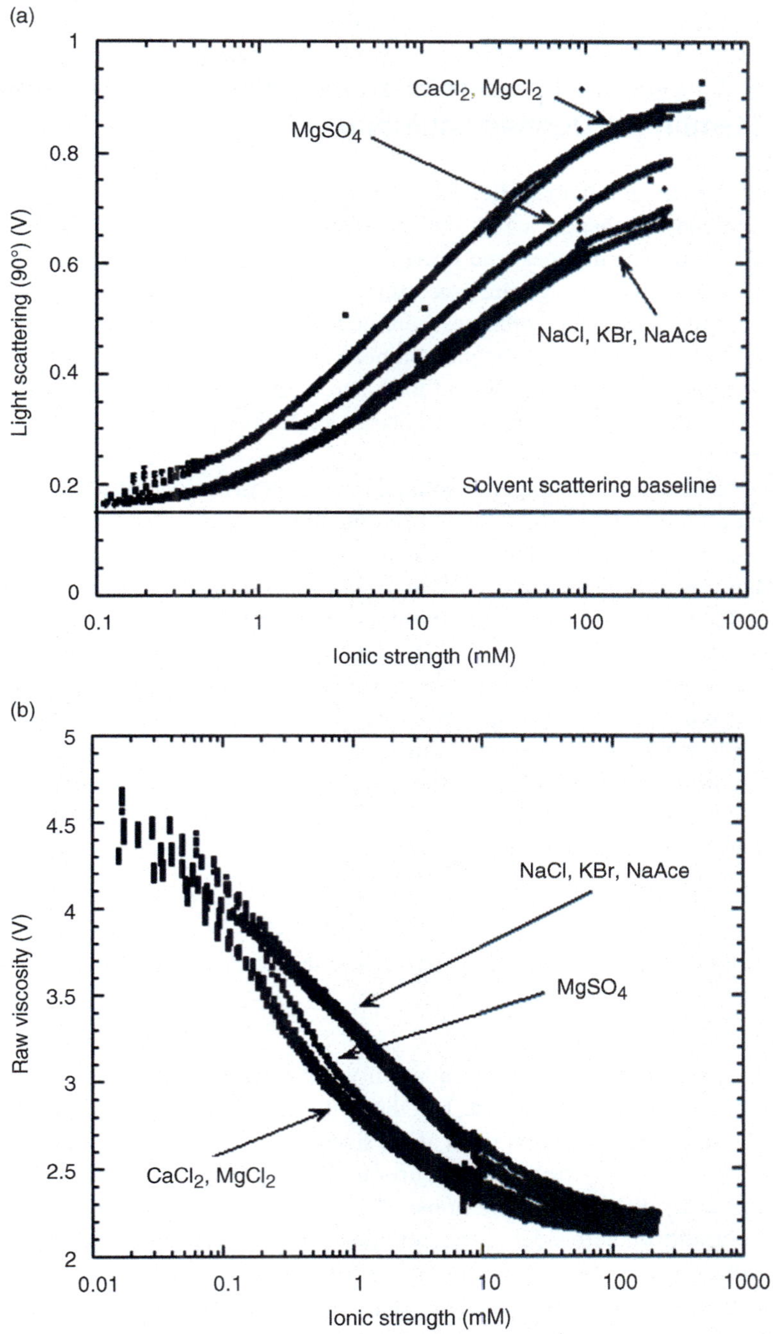

Figure 5.13 Divergence of surface potentials of a sphere for monovalent and asymmetric divalent salts: an effect of the nonlinearity of the PBE. (Sorci, G.A. and Reed, W.F. (2004) / Reproduced with permission from American Chemical Society).

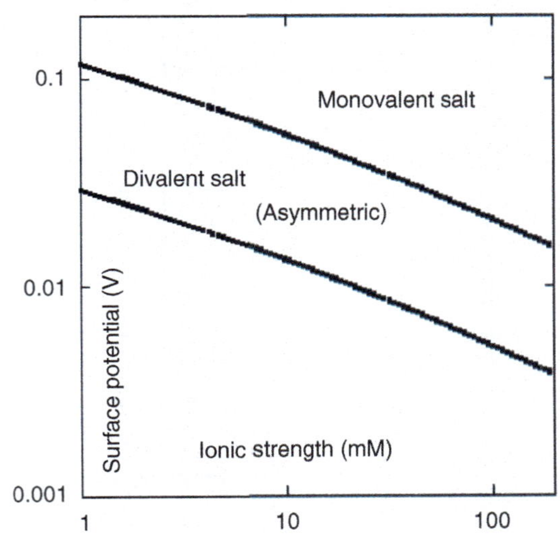

5.7 Osmotic Pressure of an Ideal Polyelectrolyte Solution with a Semi-Permeable Membrane: Donnan Equilibrium

Biological membranes often have different concentrations of ionic species on either side that are controlled by both passive and active transport across the membranes. The difference in concentrations leads to electrostatic potential differences across the membranes, and these are the basis of transport phenomena that underlie energy production, nerve propagation, and other living state processes.

Semipermeable membranes were introduced in Chapter 4 as a means of selectively isolating one or more chemical species on a single side of the membrane, causing concentration differences in each side, while allowing permeable chemical species to pass freely back and forth across the membrane to maintain the same value for their chemical equilibrium on each side. The concentration imbalance led to the concept of osmotic pressure, which was explained via chemical potential equilibria in Chapter 4.13.1.

In Chapter 4, the impermeable solute[7] was electrically neutral, a sugar, for example. Now, we place an impermeable species, a polyion, on the α side of the membrane, which has multiple positive charges, and an equal number of negative charges, which completely dissociate in the solvent. Chemical potential equilibrium still applies to the permeable solutes, but now there is the added condition that each side be electrically neutral. How does this affect the final equilibrium?

To answer this, consider a solution with a fairly high concentration of a fully dissociated monovalent salt, e.g. NaCl, in a vessel, whose volume is divided by a semipermeable membrane, which is impermeable to charged polyions but permeable to the small ions Na^+ and Cl^-. Figure 5.14 illustrates the arrangement. When a polyelectrolyte P, with z negative counterions is dissolved in the α side in Figure 5.14 its counterions are released into the solution and can flow back and forth across the membrane, occupying both the α side and the β side, while the charged polyions remain trapped on the α side. In Figure 5.14 and what follows the polyelectrolyte is taken to be a polycation.

To make the situation concrete consider a polyelectrolyte P that has z monovalent counterions, each denoted as B^-, that dissolve in water according to

$$PB_z = P^{+z} + zB^- \tag{5.106}$$

A monovalent salt AB completely dissociates according to

$$AB = A^+ + B^- \tag{5.107}$$

Using the concepts of electroneutrality and equality of chemical equilibrium for species that can occupy both the α and β sides, A^+ and B^- in this case, together with minimization of free energy, we can develop relationships between the concentrations of all species on the α and β sides.

Before finding the equations, use simple principles to imagine what happens when the polyelectrolyte is first dissolved in the α side and its counterions, B^-, are released, and what the final equilibrium should look like. There is bulk electroneutrality on both sides of the solution before introducing the polyelectrolyte and the concentrations of A^+ and B^- are the same on both sides, i.e. A^+ and B^- are in chemical equilibrium on both sides. Upon dissolving the polyelectrolyte electroneutrality is maintained on each side since the total charge of the released counterions on the α side is equal and opposite to the charge on the polyion, and nothing has yet changed in the β side. So, if all the counterions stay on the α side charge

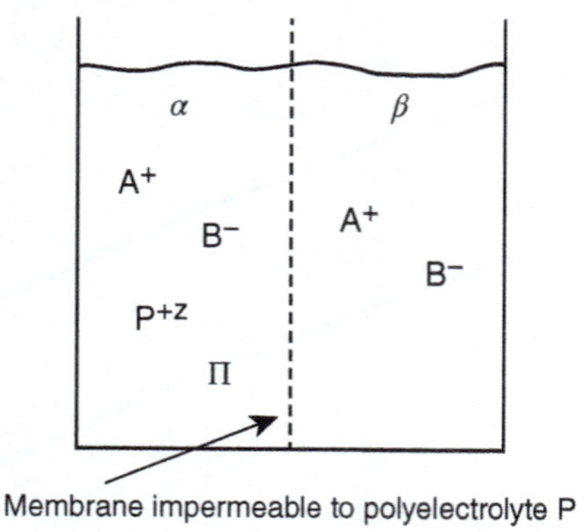

Figure 5.14 A polyelectrolyte P is dissolved in side α. The membrane is permeable to mobile ions A^+ and B^- but impermeable to the polycation.

7 "Impermeable solute" is shorthand for "solute which can't pass through the semipermeable membrane", while "permeable solute" is shorthand for a solute that can.

neutrality is satisfied on both sides. However, the chemical equilibrium of B^- has been disrupted because the release of the polyelectrolyte's counterions has now given more of B^- on the α side than on the β side, while A^+ is so far still in equilibrium on both sides. So, B^- in the α side will flow through the membrane to the β side to balance the concentration of B^- on both sides. Each time a counterion B^- passes from the α side to the β side, an A^+ will also have to pass from the α side to the β side in order to maintain electroneutrality on both sides. But each time an A^+ passes to the β side, it unbalances its concentration. Hence, there must be some "compromise" equilibrium, in which enough of B^- flows into the β side to bring the concentrations in both sides as close as possible, while not creating too much of a concentration imbalance of A^+, which has "sacrificed" its initial chemical equilibrium and unbalanced its concentrations on each side. We surmise that when this compromise equilibrium is reached there will still be more B^- on the α side than on the β side, but there will be more A^+ on the β side. Furthermore, all of P^+ is on the α side so we further guess that there is net higher molar concentration of solute, the sum of A^-, B^-, and P^+ on the α side than on the β side, which has only solutes A^- and B^-, so that there should be a higher osmotic pressure on the α side than on the β side.

We can further think about the microscopic implications of the bulk concentration imbalances between the two sides: The excess concentration of B^- on the α side will "push" some B^- across the membrane toward the β side. We hence expect the membrane to acquire a positive charge on the α side and a negative charge on the β side, with a corresponding electric potential difference, and an electric field across the membrane that points from the α to the β side. Equilibrium charge and field will be reached once the force of the field on the polarized charges equals the force due to the concentration imbalance. We surmise that because this charge polarization occurs only near the membrane, which occupies a negligible volume of the total solution that the unbalanced bulk concentrations on each side don't differ significantly from those obtained under the "compromise" bulk equilibrium between electroneutrality and chemical potential.

In summary, we have reasoned that the simple act of dissolving a polyelectrolyte on one side of the semipermeable membrane containing the simple electrolyte solution has caused both different solute concentrations on each side of the membrane, an osmotic pressure on the side of the membrane containing the polyelectrolyte, and an electric field across the membrane. This process is another example of Le Chatelier's principle; the perturbation of the concentration of B^- on the α side due to adding the polyelectrolyte is partially counteracted by a flow of B^- to the β side, with one A^+ flowing to the α side for each B^-, in order to preserve electroneutrality. We can now make the actual calculations.

One of the goals of finding the solute concentrations on each side is to find the osmotic pressure difference. We can use electroneutrality and chemical equilibrium conditions to find these concentrations. Let's assume the solutes are dilute enough that we can treat the solution as ideal and so the osmotic pressure difference will just be due to the difference in concentrations. For an *ideal* solution

$$\Delta\Pi = \Pi^\alpha - \Pi^\beta = RT\left\{[P^{+z}]^\alpha + \left([A^+]^\alpha - [A^+]^\beta\right) + \left([B^-]^\alpha - [B^-]^\beta\right)\right\} \tag{5.108}$$

where concentrations are in mol/m^3 to be consistent with the use of MKSA in this chapter. For non-ideal solutions, the activity coefficients can be added, or, more rigorously, the virial coefficient approach can be used. We know the concentration of added polyelectrolyte $[P^{+z}]^\alpha$ and we know the concentration of added monovalent electrolyte $[E]$, which is related to the unknowns $[A^+]^\alpha, [A^+]^\beta, [B^-]^\alpha$, and $[B^-]^\beta$.

Electroneutrality in the β side requires (remembering that $[P^{+z}]^\beta = 0$)

$$[A^+]^\beta - [B^-]^\beta = 0 \tag{5.109}$$

So that

$$[A^+]^\beta = [B^-]^\beta \tag{5.110}$$

At equilibrium, electroneutrality in the α side requires

$$z[P^{+z}]^\alpha + [A^+]^\alpha - [B^-]^\alpha = 0 \tag{5.111}$$

It is possible to find a relationship between the concentrations of A^+ and B^- in sides α and β by considering the requirement that the free energy be minimized when equilibrium is reached with the final concentrations of A^+ and B^- on each side of the system. At equilibrium, moving each ion type from one side to the other raises the free energy of each side, since the concentration of each type is being changed from its concentration that yields the minimum total free energy of the system. If an equal number of moles of A^+ and B^-, δn, move across the membrane in the same direction both increase the free energy by the same amount δG on both sides of the membrane. But G must remain a minimum on each side so that the change in δG on each side must be the same, and be zero:

$$\delta G = \delta G^{\alpha} = \delta G^{\beta} = 0 \tag{5.112}$$

The chemical potential is defined as the partial differential of Gibb's free energy with respect to the change in the number of moles while all other variables are held constant. This is given in terms of the total chemical potential (Chapters 4.12–4.13 to review chemical potential) on each side by

$$\delta G = \delta(n_A^{\alpha}\mu_A^{\alpha} + n_B^{\alpha}\mu_B^{\alpha}) = \delta\left(n_A^{\beta}\mu_A^{\beta} + n_B^{\beta}\mu_B^{\beta}\right) \tag{5.113}$$

where μ_A^{α} and μ_B^{α} are the chemical potentials for A^+ and B^- on side α at equilibrium, and n_A^{α} and n_B^{α} are the number of moles of A^+ and B^-, respectively, on side α, and similarly for μ_A^{β}, μ_B^{β}, n_A^{β}, and n_B^{β} on side β. This is shown in the following equation. For ideal solutions, these are directly related to the sum of the entropies of mixing so that

$$RT\left\{\ln\left(\frac{[A^+]^{\alpha}}{[A^+]_{total}}\right) + \ln\left(\frac{[B^-]^{\alpha}}{[B^-]_{total}}\right)\right\} = RT\left\{\ln\left(\frac{[A^+]^{\beta}}{[A^+]_{total}}\right) + \ln\left(\frac{[B^-]^{\beta}}{[B^-]_{total}}\right)\right\} \tag{5.114}$$

where $[A^+]_{total}$ and $[B^-]_{total}$ are the total concentrations of each ion type in the entire system. Simplifying this yields

$$[A^+]^{\alpha}[B^-]^{\alpha} = [A^+]^{\beta}[B^-]^{\beta} \tag{5.115}$$

This is known as *Donnan equilibrium*, named after Frederick G. Donnan (British Ceylon/England, 1870–1956), who first published the concept in 1911. Donnan equilibrium has important consequences for biological membrane systems and ion distributions. Imposing electroneutrality condition, Equation 5.111 gives

$$[A^+]^{\alpha}[B^-]^{\alpha} = \left([A^+]^{\beta}\right)^2 = \left([B^-]^{\beta}\right)^2 \tag{5.116}$$

This shows the effect of our proposed "compromise" equilibrium above. $[B^-]^{\alpha}$ has increased to greater than its value before the polyelectrolyte was added, since z counteranions were released upon introducing the polyelectrolyte. The concentration of $[A^+]^{\alpha}$ is obtained by using the electroneutrality condition in each side to eliminate $[B^-]^{\alpha}$ from Equation 5.116.

$$[A^+]^{\alpha}\left([A^+]^{\alpha} + z[P^{+z}]^{\alpha}\right) = \left([A^+]^{\beta}\right)^2 \tag{5.117}$$

This shows how the polycation P^+ takes the place of some of A^+ on the α side leading to a lower concentration of A^+ and hence, by Equation 5.116, a higher concentration of B^- than on the β side. With the latter equation:

$$[A^+]^{\beta} = [A^+]^{\alpha}\left(1 + z\frac{[P^{+z}]^{\alpha}}{[A^+]^{\alpha}}\right)^{1/2} \tag{5.118}$$

And, likewise for $[B]^{\beta}$, again using the electroneutrality condition,

$$[B^-]^{\beta} = [B^-]^{\alpha}\left(1 - z\frac{[P^{+z}]^{\alpha}}{[B^-]^{\alpha}}\right)^{1/2} \tag{5.119}$$

If $z[P^{+z}]^{\alpha}/[B^-]^{\alpha} < 1$, the binomial theorem

$$(a+b)^n = a^n + na^{n-1}b + \frac{1}{2!}n(n-1)a^{n-2}b^2 + \cdots \tag{5.120}$$

can be used to expand these expressions to second order, yielding the following approximation for the osmotic pressure difference:

$$\Delta\Pi = RT\left\{[P^{(+z)}]^{\alpha} + \frac{z^2([P^{(+z)}]^{\alpha})^2}{8[A^+]^{\alpha}} + \frac{z^2([P^{(+z)}]^{\alpha})^2}{8[B^-]^{\alpha}} + \cdots\right\} \tag{5.121}$$

If the electrolyte concentration $[E]$ is much higher than the biopolymer concentration, which is usually the case in a living system where ionic strength is typically around 150 mM, the differences in electrolyte components on each side are small compared to the total electrolyte concentration and the following approximation can be made:

$$z\left[P^{+z}\right]^{\alpha} \ll \left[E\right] \approx \left[A^{+}\right]^{\alpha} \approx \left[B^{-}\right]^{\alpha} \qquad 5.122$$

So that

$$\Delta\Pi \approx RT\left\{[P^{+z}]^{\alpha} + \frac{z^2\left([P^{+z}]^{\alpha}\right)^2}{4[E]}\right\} \qquad 5.123$$

Interestingly, even though the solution was taken as ideal, in terms of the osmotic pressure arising only from the concentration difference, the effect of Donnan equilibrium together with charge neutrality is to produce an effective second virial coefficient A_2 (and higher order virial coefficients if the expansion above is carried out further):

$$A_{2,Donnan} = \frac{z^2}{4\left[E\right]} \qquad 5.124$$

where $A_{2,Donnan}$ has units of m^3/mol and concentrations are in mol/m^3.

As surmised, electroneutrality and chemical equilibrium between charged species on either side of the membrane leads to imbalances in the bulk concentration of A^+ and B^- ions, as computed above. As further surmised above, in the narrow boundary layer on each side of the membrane, there is a flow of permeable ions down each one's respective concentration gradient, forming a slight excess of negative charge in the boundary layer on one side of the membrane and a slight positive charge on the other side, without effecting the overall bulk electroneutrality on each side. The boundary layer flow of ions down the chemical gradients is counteracted by an increasing electric field across the membrane due to the slight charge build up and polarization across the membrane. The flow continues until equilibrium is reached between the concentration gradient and the electric field and an electrostatic potential forms across the membrane. The flux due to the electric field produced by the electrostatic potential difference cancels out the opposite flux due to the concentration gradient. The length scale over which the electric field exists is on the order of the Debye screening length, $1/\kappa$, discussed above. For physiological ionic strength around 0.15 M, this distance is about 0.8 nm.

The equilibrium between chemical and electric potentials is discussed further in the next chapter in connection with the Nernst–Planck equation.

5.8 Self-energy of a Hydrated Ion

In homogeneous, isotropic solid-state dielectric materials, it is easy to visualize how atoms in fixed positions polarize when a field is applied. The presence of the dielectric causes dipoles to form in response to an applied field and the electric fields of these dipoles oppose the applied field, to yield a net field smaller than the applied one. This leads to increased energy density u, according to Equation 5.35. The increase in u compared to vacuum is due to the fact that the electric field has invested energy into creating the dipoles, similar to storing potential energy in stretched springs, energy that can be released subsequently. There is no material to polarize in the vacuum.[8] In the solid, there has been no appreciable change in entropy because the number of states has remained the same.

Considering now the case of water, when a charge is introduced the water dipoles will go into partial alignment with the field around the charge which means that water entropy must decrease. Consider building up the final charge Q by adding increments dq into a sphere of radius R at constant temperature and pressure. The incremental work needed to add dq is $dW = \phi dq$. With a total amount q during the process, the electrostatic potential is

$$\phi = \frac{q}{4\pi\varepsilon R} \qquad 5.125$$

8 We ignore the vacuum polarizability of quantum mechanical models, an example of which is the short-lived creation of electron–positron pairs around an electron, due to the electron's electric field. The pair creates a transient dipole which aligns in the electron's field and very slightly reduces it, much like an ordinary, but very weak dielectric. The tiny *Lamb Shift* in the H-atom spectrum confirms the vacuum polarizability predicted by quantum electrodynamics.

so that integrating dW while charging up to Q gives

$$W = \frac{Q^2}{8\pi\varepsilon R} \tag{5.126}$$

where ε is the material's permittivity, given by Equation 5.4. ε gathers all the complicated dipole interactions and changes in water entropy into this work term.

Measurements of energy for hydrating different ions of different R shows that the simplicity of Equation 5.126 holds up to experiments. The value of R is generally taken to be the radius of the charge, including the solvent sphere around it. Then, the free energy change of moving an ion from a medium of one dielectric constant ε_1 to a medium of ε_2 is

$$\Delta G = \frac{Q^2}{8\pi R}\left(\frac{1}{\varepsilon_2} - \frac{1}{\varepsilon_1}\right) \tag{5.127}$$

For example, moving a monovalent ion ($Q = e$) of radius $R = 0.8$ nm from water $\varepsilon_1 \sim 80$ to an apolar liquid of $\varepsilon_2 = 2$, gives $\Delta G \sim +7.0 \times 10^{-20}$ J, so that the ratio of ions in the water phase to the apolar phase is $\exp(\Delta G/k_B T)$, which at $T = 300$ K is 2.2×10^7, i.e. only one in 20 million ions is in the apolar phase. As an example, for a 1M aqueous ionic solution, the concentration in the apolar phase would be about 4.54×10^{-8} M. This means that near a biological membrane, e.g. the membrane of a cell, which has an aqueous exterior and a highly apolar lipid bilayer, it is difficult for an ion to traverse the membrane, even if there is a lower free energy on the other side, such as a lower concentration of the ion. The ΔG is a free energy barrier that dramatically slows the passive transport of the ion across the membrane. In this sense, the cell membrane acts as a gatekeeper between the inside and outside of a cell. As is usual in biological processes, the transport of ions and other species across the membrane is controlled, and controlled largely by protein channels. It is easy to see how opening up a channel that would create an aqueous bridge between the outside and inside of the cell would dramatically lower the barrier and speed up ion and small molecule transport. Of course, in reality, channel action is much more complex than simply opening an aqueous bridge, but in some of the larger channel proteins, this is a good approximate picture, which, however, completely fails for small channels, where other effects come to dominate. We will revisit this in Chapter 13.

5.9 Force on an Ion Near an Interface of Two Media with Different Dielectric Constants

Consider a charge Q in a medium of ε_1 at a distance d from the planar interface with a medium of ε_2, as shown in Figure 5.15. The presence of the interface leads to a change in the local dipole moment density in passing from the first to the second medium. This results in a surface dipole layer at the boundary that modifies the electric field that would be present in the absence of the boundary. Using the method of images, the second medium can be conceptually replaced by an image charge at an equal distance from the interface, but on the other side, with a charge Q' given by

$$Q' = -\left(\frac{\varepsilon_2 - \varepsilon_1}{\varepsilon_2 + \varepsilon_1}\right)Q \tag{5.128}$$

The electric field and potential can then be computed to the right of the interface by replacing the second medium with the image charge Q' imbedded in the first medium. The effect of the image charge is to cause a force on Q, whose corresponding potential is given by

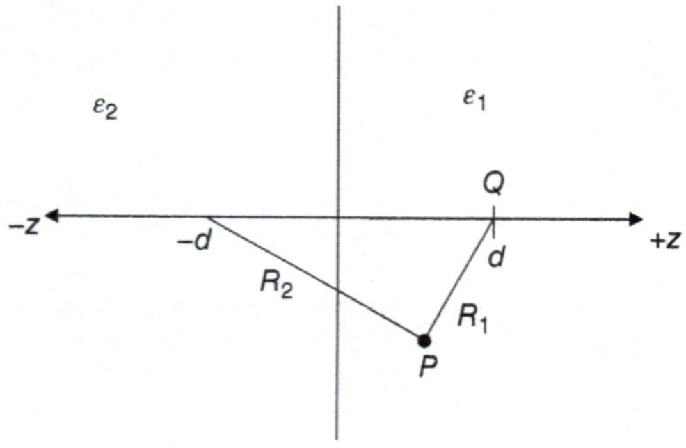

Figure 5.15 A charge Q is in a medium of permittivity ε_1 at a distance d from a planar interface in the x–y plane with a medium of ε_2 for $z < 0$.

$$U = -\frac{Q^2}{8\pi\varepsilon_1 d}\left(\frac{\varepsilon_2 - \varepsilon_1}{\varepsilon_2 + \varepsilon_1}\right) \qquad 5.129$$

If the ion is in an apolar liquid and the second medium is water then $\varepsilon_2 > \varepsilon_1$ and the force is attractive; Q is pulled toward the water phase. If the ion is in water, it will feel a repulsive force. This image-based force becomes an important factor in membrane transport and will be revisited in Chapter 13.

5.10 Bjerrum Length and Counterion Condensation

We introduced the Debye–Hückel screening length, Equation 5.75, as a useful length scale for ascertaining the range of electrostatic potential in an ionic solution. Another useful length scale when dealing with ions in solution is the distance at which the electrostatic potential energy between two charges in a medium, one with z_1 elementary charges and the other with z_2, is equal to the thermal energy k_BT. This distance is the *Bjerrum length* l_B, and by the above condition is given by

$$l_B = \frac{z_1 z_2 e^2}{4\pi D\varepsilon_0 k_B T} \qquad 5.130$$

No account of shielding is taken in this definition. For water at STP, $l_B = 0.72$ nm for monovalent ions ($z_1 = z_2 = 1$). In air, $l_B \sim 58$ nm.

An interesting prediction which arises from the Bjerrum length notion is related to the degree of charge of a polyelectrolyte in a solvent. For simplicity, consider a rod-like polyelectrolyte with monomer groups with z elementary charges each, at a distance $d < l_B$ from each other. Assume the polyelectrolyte fully dissociates in a solvent, so that the released counterions are mobile in the solution. Since $d < l_B$, there is an energy penalty for the dissociation. At the simplest level, if enough counterions re-associate or "condense" back onto the polyelectrolyte so that the spacing by the remaining ionized groups is $\geq l_B$, then the energy penalty can be removed. This leads to the notion of *counterion condensation* and limits the linear charge density of the rod-like polyelectrolyte to a maximum of one charge per Bjerrum length. In fact, in polyelectrolyte theory (Chapter 8), linear charge density is often expressed as $\xi =$ (No. of elementary charges) / (Bjerrum length). Under this theory, if the bare polyelectrolyte has $\xi > 1$, then counterion condensation will bring ξ to $\xi \leq 1$.

In a simple straight-line chain of charges, it was shown that if charges were situated closer than l_B then the repulsive electrostatic energy would exceed k_BT and counterions would condense onto the charges until the spacing between charges was greater than or equal to l_B. While this is an oversimplified version of counterion condensation theory, and leads to a sharp condensation threshold, which has been shown experimentally to be smooth rather than sharp, the essential notion has been amply demonstrated experimentally by conductivity and also online reaction monitoring.

5.11 Summary

Most, but not all, biomacromolecules bear electrical charge. The electrostatic force hence has a major role in biological systems and is responsible for forces between ions, ions and dipoles, and between dipoles. Because water has a high dielectric constant salts are readily soluble in living cells but also have relatively weak interactions, often on the order of k_BT or less. These weak interactions can combine with hydrogen bonds and the hydrophobic effect to link biomacromolecules into cooperative structures much more stable than k_BT, e.g. the protein secondary and tertiary structures encountered in Chapters 3 and 10. Structures which are electrically charged attract opposite mobile charges into their vicinity creating screening effects which further diminish electrostatic interactions. The mean field treatment embodied in the PBE gives a quantitative meaning to shielding and how it is related to ionic concentrations, which are consolidated in the notion of ionic strength.

The work required to assemble a charge in a solution is inversely proportional to the dielectric constant. As a consequence, there is a high-energy barrier for removing an ion from water and placing it into a low-polarity environment, such as a cell membrane, which has a much lower dielectric constant than water. This effect is essential for the very low permeability of membranes to ions, and, as usual in living state systems, where virtually all functions are tightly controlled, the permeability of membranes is controlled by agents such as membrane ion channel proteins. The effects of such membrane permeability control are at the root of biological functions, such as nerve impulse generation and conduction, and the interface between sensory and muscle cells with nerves. This is covered in more detail in Chapter 13.

References

Adrian Parsegian, V. and Zemb, T. (2011). Hydration forces: observations, explanations, expectations, questions. *Current Opinion in Colloid & Interface Science* 16: 618–624. https://doi.org/10.1016/j.cocis.2011.06.010.

Feynman, R. (1985). *QED: The Strange Theory of Light and Matter*. Princeton University Press.

Hinderberger, D., Spiess, H.W., and Jeschke, G. (2004). Dynamics, site binding, and distribution of counterions in polyelectrolyte solutions studied by electron paramagnetic resonance spectroscopy. *The Journal of Physical Chemistry B* 108: 3698–3704.

Jackson, J.D. (1998). *Classical Electrodynamics*, 3e. Wiley.

Kreft, T.R. and Reed, W.F. (2009). Experimental observation of crossover from noncondensed to counterion condensed regimes during free radical polyelectrolyte copolymerization under high-composition drift condition. *Journal of Physics and Chemistry B* 113: 8303–8309.

Lamm, G. (2003). The Poisson-Boltzmann equation. *Reviews in Computational Chemistry* 19: 147–365.

Overbeek, J.T. and Verwey, E.J.W. (1948). *Theory of the Stability of Lyophobic Colloids*, London: Elsevier Publishing Co. In. Cleaver-Hume Press Ltd.

Pollak, G.L. and Stump, D.R. (2002). *Electromagnetism*. Addison Wesley.

Tanford, C. (1996). *Physical Chemistry of Macromolecules*. John Wiley and Sons.

Translation of Donnan's 1911 paper originally written in German. Theorie der Membrangleichgewichte und Membranpotentiale bei Vorhandensein von nicht dialysierenden Elektrolyten. Ein Beitrag zur physikalisch-chemischen Physiologie. *Zeitschrift für Elektrochemie und Angewandte Physikalische Chemie* 17 (10): 572–581.

F.G. Donnan (1995). Theory of membrane equilibria and membrane potentials in the presence of non-dialyzing electrolytes A contribution to physical-chemical physiology. *Journal of Membrane Science* 100: 45–55.

Wilson, R.W. and Bloomfield, V.A. (1979). Counter-ion-induced condensation of deoxyribonucleic-acid - light-scattering study. *Biochemistry* 18: 2192–2196.

Further Reading

Andelman, D. (1995). Electrostatic properties of membranes: the Poisson-Boltzmann theory. In: *Handbook of Biological Physics*. 1B, 603–42.

Anderson, C. F., Record, M. T. Jr. (1990). Ion distributions around DNA and other cylindrical polyions: theoretical descriptions and physical implications. *Annual Review of Biophysics and Biophysical Chemistry* 19: 423–65.

Botello-Smith, W. M., Cai, Q., Luo, R. (2014). Biological applications of classical electrostatics methods. *Journal of Theoretical & Computational Chemistry* 13 (3): 1440008/1–1440008/25.

Buyukdagli, S., Blossey, R. (2016). Beyond Poisson-Boltzmann: fluctuations and fluid structure in a self-consistent theory. *Journal of Physics: Condensed Matter* 28 (34): 343001/1–343001/19.

Chowdhury, R., Egan, R., Bochkov, D., Gibou, F. (2022). Efficient calculation of fully resolved electrostatics around large biomolecules. *Journal of Computational Physics* 448: 110718.

Grochowski, P., Trylska, J. (2008). Continuum molecular electrostatics, salt effects, and counterion binding-a review of the Poisson-Boltzmann theory and its modifications. *Biopolymers* 89 (2): 93–113.

Herbert, J. M. (2022). Dielectric continuum methods for quantum chemistry. arXiv.org, e-Print Archive, Physics. 1–87.

Honig B., Nicholls, A. (1995). Classical electrostatics in biology and chemistry. *Science (New York, N.Y.)* 268 (5214): 1144–1149.

Manning, G. S., Ray, J. (1998). Counterion condensation revisited. *Journal of biomolecular structure & dynamics* 16 (2): 461–76.

Manning, G. S. (2007). Electrostatic free energies of spheres, cylinders, and planes in counterion condensation theory with some applications. *Macromolecules (Washington, DC, United States)* 40 (22): 8071–8081.

Stigter, D. (1987). Donnan membrane equilibrium, sedimentation equilibrium, and coil expansion of DNA in salt solutions. *Cell Biophysics* 11: 139–58.

Tellez, G. (2011). Nonlinear screening of charged macromolecules. *Philosophical Transactions of the Royal Society, A: Mathematical, Physical & Engineering Sciences* 369 (1935): 322–34.

Problems

1 Find the capacitance of two concentric spherical shells, the inner one with radius a and the outer with radius b. If b is only slight larger than b so that there is a layer with thickness d between the spheres, i.e. $b = a + d$, $d \ll a$, find the approximate capacitance in terms of a and d.

2 Find the potential energy, in Joules *and* eV, between a Na^+ ion and a Cl^- ion at 0.5 nm from each other in pure water. Assume the dielectric constant of water is 80.

3 A wall in the yz plane at $x = 0$ is in contact with an electrolyte solution for $x > 0$, where the solution has a dielectric constant D.
a) If the wall is held at a potential ϕ_0, find the solution for the potential for $x > 0$. Assume an ionic strength IS for the electrolyte solution.
b) Repeat the calculation when the plane has a uniform surface charge density σ.
c) A second wall is now placed at $x = d$, and the space between the plates is filled with electrolyte. The plate at $x = 0$ carries $+\sigma$ and the one at $x = d$ has $-\sigma$. Find the potential profile between the plates.
d) Find the capacitance of the parallel plates.

4 An infinite sheet of uniform surface charge density σ is located in the y–z plane at $x = 0$. A monovalent salt of bulk concentration ρ_0 is present and surrounds the plane. Assume the dielectric constant of the medium is D.
a) Set up the Poisson–Boltzmann equation for this situation, then linearize it.
b) Find the electrostatic potential $\phi(x)$ for all x.

5 Consider a very thin cylinder of uniform linear charge density λ, immersed in an electrolyte solution of ionic strength I. Find the potential around the cylinder. This is best solved in cylindrical coordinates.

6 a) Write down the expression for the PBE when an asymmetric electrolyte is used.
b) Expand this expression to the third order. Show that the linearization of the PBE only holds to first order in ϕ for an asymmetric electrolyte, whereas it holds to second order, ϕ^2 for a symmetric electrolyte.

7 Explicitly expand the expression for the electrolytes in the Donnan equilibrium equation to second order and hence derive the osmotic pressure difference.

8 Show that if a discrete charge distribution of charge q_1 and q_2 separated by a distance d has a net charge, then the dipole moment depends on the choice of coordinate system. Next, show that if the net charge of the dipole is zero, $q_1 = -q_2$, then the magnitude of the dipole moment is independent of the coordinate system.

9 A certain model for a protein in globular form treats the protein as a sphere of radius R and uniform charge density ρ_p. Apply the P–B equation to this situation to obtain the potential both inside and outside the sphere. Assume the protein is freely penetrable by mobile ions, present in bulk concentration ρ_0 (monovalent salt) and that the dielectric constant is the same inside and out.

10 a) Compute A_2 for two spherical particles of mass M that interact via a Gaussian potential of the form

$$U(r_{12}) = U_0 \exp(-\alpha r_{12}^{\ 2})$$

This can be considered as a model interaction for "soft spheres." Assume that $U_0/k_BT \ll 1$ so that you can expand the integrand in the excluded volume integral to an easily integrable form. The following integral formula should be useful:

$$\int_0^\infty x^2 e^{-ax^2} dx = \frac{1}{4a}\sqrt{\frac{\pi}{a}}, \quad a > 0$$

b) If $U_0 = 0.08\ k_BT$, and $\alpha = (50\ \text{nm})^{-2}$, find the radius of a hard sphere that would yield the same excluded volume as the "soft sphere."

11 Certain charged, globular proteins with very small radii and mass M can be approximated to interact via a screened potential of the form:

$$U(r_{12}) = \frac{Ae^{-\kappa r}}{r}$$

a) Calculate A_2, by expanding this potential to first order in the excluded volume integrand.
b) Make a sketch of A_2 versus concentration of added salt C_s, for A (erg cm) $= 10^{-7}\, k_B T$, and $M = 10^5$ g/mol.

12 a) Find the effective A_2 as a function of added monovalent salt concentration C_s of an *ideal* polyelectrolyte solution at "Donnan equilibrium" on one side of a chamber separated by a semipermeable membrane, permeable to solvent and salt ions, but not the polyelectrolyte. The net charge of the polyelectrolyte is $z = 75$ and its molar mass is 10^6 g/mol.
b) Compute the osmotic pressure for this system for a polyelectrolyte concentration of 0.001 g/cm^3, when $C_s = 0.01$ M and also for $C_s = 1$ M.

13 Find the ionic strength of a solution containing the following concentrations of salts: [NaCl] = 0.5 M, [$CaCl_2$] = 0.22 M, [$MgSO_4$] = 0.08 M.

14 A charge Q is embedded in a dielectric material with ε_1 and is at a distance $z = d$ from an interface in the x–y plane with a second material of ε_2. The boundary conditions on the perpendicular component of $\vec{D}$, the displacement field, at the interface cause a net bound surface charge to form, which can attract or repel Q, depending on the relation of ε_1 and ε_2.
a) Derive Equation 5.128, $Q' = -((\varepsilon_2 - \varepsilon_1)/(\varepsilon_2 + \varepsilon_1))Q$.
Use cylindrical coordinates with $z = 0$, as shown.
Hints:
 i) Replace the medium of ε_2 by an image charge Q' in medium $\varepsilon = 80\varepsilon_0$ at a distance $z = -d$ from the interface which has been removed. With this find the potential at any point P for $z > 0$.
 ii) Now, to find the potential at any point for $z < 0$, consider a uniform medium of ε_2 with a single image charge Q'' in the place of the original charge Q. Q'' is just an auxiliary for finding the potential for $z < 0$.
 iii) Use the potentials on both sides and the boundary conditions on the parallel and perpendicular components of $\vec{E}$ at the interface to find, Q' and Q''.
b) Find the force which Q feels due to Q' and indicate under which condition it is attractive and in which it is repulsive.

15 A zwitterionic micelle consists of two concentric spheres. The inner sphere has radius r_a and charge $+Q$, while the outer surface has radius r_b and charge $-Q$. This leads to the formation of a spherical shell dipole layer of thickness $s = r_a - r_b$. The inner sphere is impenetrable to mobile ions. Assume that the dielectric constant inside the dipole layer and outside the micelle is that of bulk water (78), at temperature T. Use the linearized Debye–Hückel equation to find, as a function of the Debye screening parameter κ.
a) The potentials for (i) $r < r_a$, (ii) $r_a \le r \le r_b$, and (iii) $r \ge r_b$.
b) The net mobile ion charge density within the charge layer.

6

Fluid Mechanics and Transport Processes

Neither the balloon, nor the aeroplane, nor the gliding machine will be a practical success.

Lord Kelvin, 1902 / Newark Advocate / Public Domain, p. 4

Living state processes take place in fluids. Several types of motion occur in fluids, including diffusion, biased diffusion in fields, linear and circular motion, as well as smooth, laminar flow, and turbulent flow. All of these occur at some length scale in the living state. For example, molecules can diffuse down a concentration gradient, ions can diffuse across an electrostatic potential difference (both are cases of biased diffusion), blood flow can be laminar in arteries and veins, and turbulence can be produced by fish swimming in water or defended against by birds flying in the air.

Fluid mechanics and the dynamics of particles in fluids are highly developed fields of their own and have numerous practical applications. It has taken centuries to painstakingly develop the concepts and mathematical formulations used to describe viscous fluid flow. Solutions for viscous flow are normally difficult to find and, at the opposite extreme, when flow is swift and there is little viscosity to tame it, the flow becomes turbulent and no unique solutions necessarily exist. Just as complex is the related field of *rheology*, which deals with materials that have both viscous and elastic properties, that is, materials which are *viscoelastic*. Viscoelasticity is ubiquitous in the living state. Cell walls, plant and animal fibers, and cytoplasm are some examples. We will take a very brief look at some basic rheological principles in Chapter 8.

Macromolecules and colloids in solution with no applied fields feel random forces due to thermal collisions. This leads to the *diffusion* of macromolecules both in living state and ex-vivo solutions. Diffusion, which is frequently described by *random walk* models, is often the rate-limiting step in controlling biochemical and synthetic reactions.

When external forces are applied, macromolecules can combine their diffusional motion with directed motion due to the applied force. Examples of applied forces include: Electrical fields, in which case the macromolecule, if charged, responds with a velocity related both to the field strength and the particle's *electrophoretic mobility*. Gravitational fields or the acceleration from a centrifuge, in which case the *sedimentation coefficient* becomes central. Under shear stress, the *intrinsic viscosity* is the chief macromolecular property of interest. In all these scenarios, the underlying physical quantity is the macromolecule's *frictional coefficient* f.

Physical models for the frictional coefficients of different types of particles depend on the nature of the fluid flow around them. The nature of directed fluid flow itself is of biophysical interest when one considers the circulation of blood, and in ecology when dealing with the circulation of water in oceans and streams, the subject of physical oceanography, and the motions and properties of the atmosphere, the subject of meteorology.

Here, the concept of viscosity is introduced and then the Navier–Stokes equation for viscous flow. A few important solutions to this equation are then presented, including a model for the frictional coefficient of particles, and laminar flow in a capillary. Subsequently, specific motions of particles are considered, starting with the Langevin equation for random forces, and force-flux relations, and results are found for diffusion, electrophoresis, and other concepts, including fluctuation-dissipation phenomena.

Most literature in these areas use the CGS system (centimeter–gram–second), so we will use those units throughout this chapter.

Biophysics: Physical Processes Underlying the Living State, First Edition. Wayne F. Reed.

Companion website: www.wiley.com/go/Reed/BiophysicsandMacromolecules

6.1 Conceptual Approach to Viscosity

All normal liquids are *viscous*. Only in unusual circumstances, such as in supercooled helium does viscosity disappear and *superfluid* behavior appear. For fluids where viscosity is negligibly small they can often be treated as *ideal fluids* or *inviscid fluids*. *Viscosity* is an internal type of friction in a liquid and is hence dissipative and will also involve heat, and hence the thermodynamic variables of temperature, entropy, heat capacity, and heat conductivity are involved in viscous flow. For example, applying a constant force to push an object through a viscous fluid at constant velocity will continuously turn the work done by the organized, unidirectional force into microscopic disorder, heat: an increase in the fluid's temperature and entropy. In general, heat, entropy, charge, momentum, angular momentum, and entropy can be transported by fluids, beyond the transport of just their mass.

The dissipative force due to viscosity is somewhat similar to the type encountered in elementary physics, such as the notion of coefficients of friction and dragging a box across a floor. To drag the box at constant velocity requires a constant force. As soon as you stop pushing, the box quickly stops.

We will introduce the concept of viscosity by examining the force it takes to drag a plate over a thin layer of viscous fluid. If you have a metal plate atop a thin layer of confined honey it will take a certain amount of constant force to move the plate at a given velocity, depending on the viscosity of the honey, the thickness of the honey layer, and the area of the plate. Three important concepts that arise from this simple system are those of *shear stress*, *shear rate*, and *stick (or no-slip) boundary conditions*. In pushing the plate along the layer of honey, the force is in the plane of the plate; this is the shear stress. If we pushed perpendicularly down on the plate this would be a *normal force*, the one encountered in hydrostatic pressure. The stick boundary condition (also called the *no slip boundary condition*) considers that the liquid at the plate's surface moves with the velocity of the plate, and that at the bottom of the liquid layer the velocity of the liquid is zero at the lower container's surface. The difference in velocity at the plate and at the bottom of the layer means there is a difference in velocity from plate to bottom. The spatial rate at which velocity changes over this layer, change in velocity per distance going from the plate downward, is the *shear rate*.

What is meant by "thin layer" above? Conceptually, as the honey-layer thickness increases, it is not tenable to think that the velocity gradient persists over the entire thickness. It turns out that there is a thin viscous boundary layer around most objects in motion on or through a viscous fluid.

A viscous fluid results in spatial inhomogeneity in a viscous fluid's *velocity field*. The velocity field gives the velocity vector of the fluid motion at every point in space and time, $\vec{v}(\vec{r},t)$, with respect to a particular reference frame, often termed the *laboratory frame*. Inhomogeneity in $\vec{v}(\vec{r},t)$ can arise both from differential momentum transport of fluid, and the presence of particles within the fluid that alters the velocity field, which contributes additional dissipative internal friction to the fluid. The amount of viscosity that a macromolecule or colloid adds to a viscous fluid is easy to measure and can be related to conformational and other properties of the particle, making viscosity measurements a valuable characterization tool. *Intrinsic viscosity* of a macromolecule measures how much viscosity the macromolecule at a given concentration will add to a pure fluid's viscosity and is related to the spatial distribution of the macromolecule's mass.

A useful concept in analyzing fluid motion is that of *streamlines*. A streamline is a curve through space whose tangent at any point gives the direction of the fluid's velocity. This is analogous to electric field lines, whose tangent at any point gives the direction of the field and the direction of force the field applies on a positive test charge. In stable flow, the streamlines form a stationary pattern, whereas with time-dependent flow the streamline pattern will change in time.

6.1.1 The Relationship of Shear Rate to Viscosity and Shear Stress

Spatial variations in the velocity field are specified locally by gradient terms, where v_i is the velocity component in the i-direction of a three-component space and x_j is the jth-directional component of the space. The gradient terms involve derivatives of the form $\partial v_i / \partial x_j$ and have dimensions of s^{-1}. These derivatives are referred to as *shear rates*, $\dot{\gamma}$, for $i \neq j$. Hydrostatic pressure is contained in the $i = j$ terms.

Figure 6.1 shows a fluid velocity field between two plates, of area A, lying in the x–z plane, one fixed in the laboratory at $y = 0$ and the other, at $y = h$, moving in the x-direction with velocity $v_{x,h}$ with respect to the other plate. It is assumed in Figure 6.1 that the layer of fluid between the plates is thin, and that there are *stick (or no-slip) boundary conditions* of the fluid at each plate, that is the fluid molecules are stationary at each plate's surface:

$$v_x(y=0)=0 \text{ and } v_x(y=h)=v_{x,h} \tag{6.1}$$

Since the motion is only in the x-direction, $v_y = v_z = 0$ for all y between the plates.

Now, it requires applied force in the x-direction, F_x, to move the upper plate at velocity $v_{x,h}$, because the viscosity of the liquid dissipates into heat the $\vec{F} \cdot \vec{v}$ power produced by the force. The motion of the plate produces a shear stress, F_x/A. The shear stress is parallel to the plate and acts to carry fluid with the plate's velocity at the plate surface, and with diminishing velocity in the direction of the stationary plate in the $-y$ direction, until v_x of the fluid is zero at the lower plate, because of the stick boundary condition there. Because v_x varies from the plate velocity at the top and goes to zero at the lower plate's surface, a gradient of v_x is set up in the y-direction, which turns out to be proportional to the shear stress.

$$\frac{F_x}{A} = \eta \frac{\partial v_x}{\partial y} \tag{6.2}$$

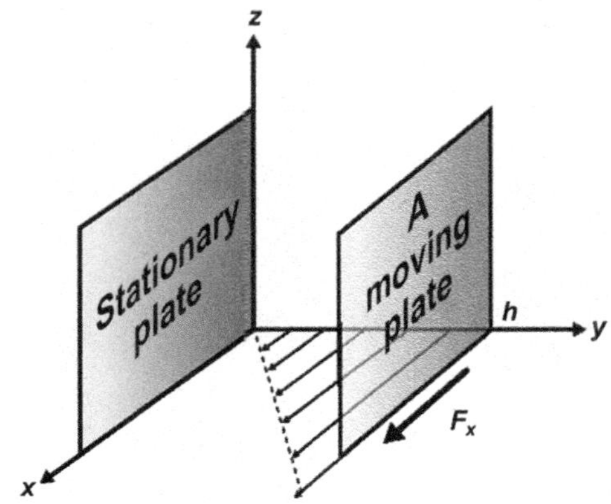

Figure 6.1 Schematic for illustrating the relationship between shear stress, velocity gradient (shear rate), and viscosity for a plate moving over a thin layer of viscous liquid. (Courtesy of Marissa Calko).

The viscosity η is hence the constant of proportionality between the velocity gradient and the applied shear stress. The greater the viscosity of the fluid, the higher the shear stress that must be applied for the same velocity gradient.

Dimensional analysis shows that viscosity has the units of g/cm-s in the CGS system. 1 g $\text{cm}^{-1}\,\text{s}^{-1}$ is called 1 Poise (P)

$$1\frac{\text{g}}{\text{cm-s}} = 1\text{Poise } (P)$$

The centipoise (cP), equal to 0.01 P, is frequently used for convenience. Water has $\eta = 1\text{cP}$ at 20°C and 1 atm, and 0.89 cP at 25°C and 1 atm.

It is important to realize that η is only a constant for *Newtonian fluids*, of which water is an example. There are many *non-Newtonian fluids* for which η depends on shear rate, and which can exhibit *shear thinning* (η decreases with shear rate) and *shear thickening* (η increases with shear rate).

Blood and many polymer solutions are examples of non-Newtonian fluids. Blood is composed roughly half of water and half of red blood cells, with a relatively small admixture of ions, proteins, white blood cells, and platelets. It exhibits pronounced shear thinning. Several mechanisms can underlie shear thinning, including the disruption of structures existing at low shear, alignment of particles, and hydrodynamic streamlining of particle shape. At low shear, blood can stack into rolls, often called by the French term *rouleaux*. As shear increases these structures are disrupted and viscosity decreases. Alignment of cells along flow streamlines makes cells glide past each other more easily and lowers the viscosity even further. Shear thinning of blood and other fluids with particles in suspensions, such as paint, can also occur when the diameter of the capillary in which flow occurs is highly reduced, the so-called *Fahraeus–Lindquist effect*.

The form of the Navier–Stokes equation presented below assumes a Newtonian fluid. When the viscosity of non-Newtonian fluids is measured at low shear rates, a *Newtonian plateau* is reached, which is the constant value of viscosity found over the applicable range of low shear rates.

Table 6.1 shows some typical viscosities for different materials. For those that are non-Newtonian, the values reflect those on the Newtonian plateau.

The viscosity of water is very temperature dependent. Given the aqueous milieu of the living state, this effect can be important. Figure 6.2 shows the strong dependence of water viscosity on temperature.

Table 6.1 Viscosities of some fluids.

Substance	η (P)	T (°C)
Water	0.0089	25
SAE 10 motor oil	0.85–1.4	25
SAE 40 motor oil	6.5–9	25
Blood	0.03–0.04	25
Methanol	0.005	30

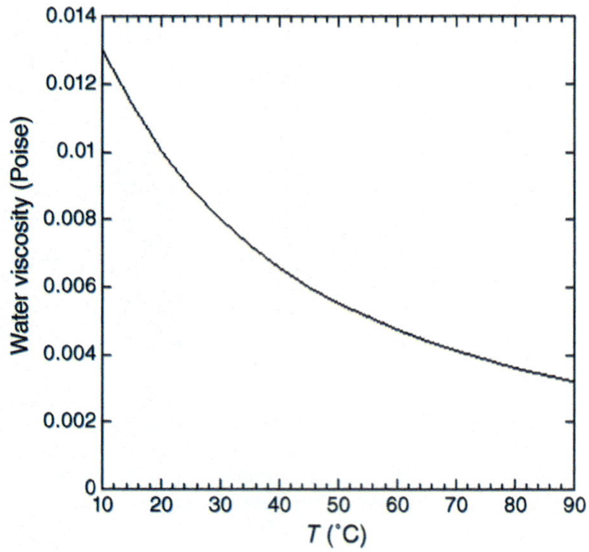

Figure 6.2 Dependence of water viscosity on temperature.

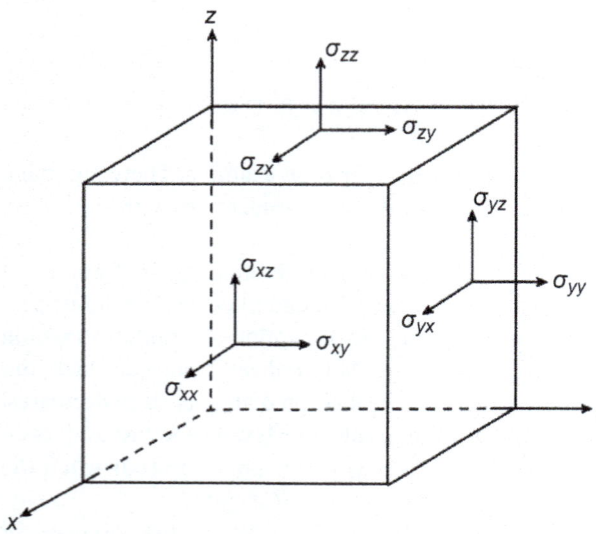

Figure 6.3 The nine elements of the stress tensor in rectangular coordinates.

6.2 The Stress Tensor

Stresses can occur in various directions and in different planes. In 1827, Augustin-Louis Cauchy (France, 1789–1857) introduced the *stress tensor* as a compact representation of these stresses. The stress tensor is composed of nine elements σ_{ij}, which indicates that the stress is applied in a plane perpendicular to axis i in the direction j. For example, the stress tensor element corresponding to F_x/A in Equation 6.2 and Figure 6.1 is σ_{yx} since the force is directed in the x-direction on the plate at $y = h$, which is perpendicular to the y-direction. Figure 6.3 shows the directions of the nine components in rectangular coordinates.

The complete stress tensor, $\overline{\overline{\sigma}}$, is given in any orthonormal coordinate system with dimensions 1, 2, and 3.

$$\overline{\overline{\sigma}} = \begin{bmatrix} \sigma_{11} & \sigma_{12} & \sigma_{13} \\ \sigma_{21} & \sigma_{22} & \sigma_{23} \\ \sigma_{31} & \sigma_{32} & \sigma_{33} \end{bmatrix} \qquad 6.3$$

For rectangular coordinates where $1 \to x, 2 \to y$, and $3 \to z$, the stress tensor becomes

$$\overline{\overline{\sigma}} = \begin{bmatrix} \sigma_{xx} & \sigma_{xy} & \sigma_{xz} \\ \sigma_{yx} & \sigma_{yy} & \sigma_{yz} \\ \sigma_{zx} & \sigma_{zy} & \sigma_{zz} \end{bmatrix} \qquad 6.4$$

The diagonal elements are normal to each surface and are *normal stresses*, which represent hydrostatic pressure, P, in a fluid. The hydrostatic pressure acts on the surface, opposite the surface normal. We can then write for $i = j$

$$\sigma_{ii} = -P \qquad 6.5$$

The off-diagonal elements are shear stresses. The stress tensor in rectangular coordinates is symmetric for an isotropic fluid in rotational equilibrium, i.e. $\sigma_{ij} = \sigma_{ji}$. An isotropic, incompressible, viscous Newtonian fluid is *defined* by the following *constitutive equation*[1] specifying the stress tensor elements' relation to the hydrostatic pressure, and to fluid viscosity and viscous shear rates for each of the six off-diagonal elements.

$$\sigma_{ij} = -P\delta_{ij} + \eta\left(\frac{\partial v_i}{\partial x_j} + \frac{\partial v_j}{\partial x_i}\right) \qquad 6.6$$

where δ_{ij} is the *Kronecker Delta*, equal to 1 for $i = i$ and zero for $i \neq j$.

For an incompressible fluid $\nabla \cdot \vec{v} = 0$ (required by the continuity equation, see Equation 6.66). By Equation 6.6, the stress tensor for the isotropic, incompressible fluid is symmetric, i.e. $\sigma_{ij} = \sigma_{ji}$.

6.3 Navier–Stokes Equation of Motion for Incompressible, Viscous Fluids

Most pure liquids are viscous and incompressible, and the viscosity is constant over the whole velocity field. The following approach does not work for non-Newtonian fluids in their non-Newtonian shear rate regime.

1 Constitutive equations give the relationship between one or more physical properties of a given material and how it responds to external stimuli, such as forces or fields.

Consider a fluid element of density ρ contained in a volume element V and subject to the stress tensor, shown in Figure 6.4.

The net force on the element can be computed by integrating the stress tensor over its entire closed surface S

$$\vec{F}_{net} = m\frac{d\vec{v}}{dt} = \oint_S \bar{\bar{\sigma}} \cdot d\vec{A} \qquad 6.7$$

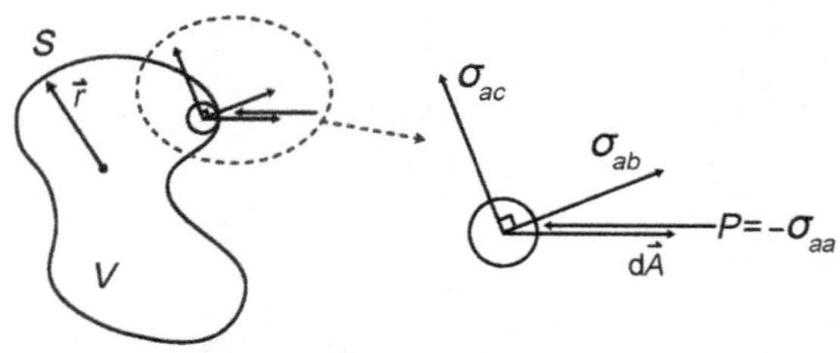

Figure 6.4 Stress elements on a volume V containing a fluid of constant density ρ. The right hand figure zooms in on the stress tensor elements on an infinitesimal surface area, $d\vec{A}$.

where $d\vec{A}$ is the surface area element normal to the surface at each point on the surface. Expressing the volume's mass m as the volume integral of the mass density, and using the divergence theorem to convert the right-hand side to a volume integral over V gives

$$\frac{d\vec{v}}{dt}\int_V \rho d^3r = \int_V \nabla \cdot \bar{\bar{\sigma}} d^3r \qquad 6.8$$

Shrinking V to an infinitesimal volume and setting the integrands equal yields

$$\rho\frac{d\vec{v}}{dt} = \nabla \cdot \bar{\bar{\sigma}} \qquad 6.9$$

Since the normal pressure elements are inward, and since hydrostatic pressure P is equal on all surfaces (at a given height, if gravity is involved):

$$\sigma_{xx} = \sigma_{yy} = \sigma_{zz} = -P\delta_{ij} \qquad 6.10$$

Performing the divergence operation

$$\nabla \cdot \bar{\bar{\sigma}} = \left(\frac{\partial}{\partial x}\frac{\partial}{\partial y}\frac{\partial}{\partial z}\right)\begin{bmatrix} \sigma_{xx} & \sigma_{xy} & \sigma_{xz} \\ \sigma_{yx} & \sigma_{yy} & \sigma_{yz} \\ \sigma_{zx} & \sigma_{zy} & \sigma_{zz} \end{bmatrix} \qquad 6.11$$

and using Equation 6.6 gives

$$\nabla \cdot \bar{\bar{\sigma}} = -\nabla P + \eta\nabla^2\vec{v} \qquad 6.12$$

$d\vec{v}/dt$ is the *total derivative* of the velocity field moving along with a particle in fluid flow. It is sometimes also termed the *material derivative*, since it applies to an element of fluid moving through space and is composed of two parts; the acceleration the particle feels at a fixed point in space, and the change in its velocity due to a spatial change in velocity in the flow field at any instant $\vec{v} = \vec{v}(x, y, z, t)$.

The total derivative is

$$\frac{d\vec{v}}{dt} = \frac{\partial\vec{v}}{\partial t} + \sum_{i=1}^{3}\frac{\partial\vec{v}}{\partial x_i}\frac{\partial x_i}{\partial t} = \frac{\partial\vec{v}}{\partial t} + (\vec{v}\cdot\nabla)\vec{v} \qquad 6.13$$

The final equation for the velocity field of a viscous, incompressible, isotropic fluid is then

$$\frac{\partial\vec{v}}{\partial t} + \left(\vec{v}\cdot\nabla\right)\vec{v} = \frac{-\nabla P}{\rho} + \frac{\eta\nabla^2\vec{v}}{\rho} + \vec{g}_{ap} \quad \text{Navier–Stokes Equation} \qquad 6.14$$

where the term $\vec{g}_{ap}$ has been added to the right-hand side to account for the acceleration due to any applied force in the volume, such as from gravity or an electric field. Claude-Louis Navier (France, 1785–1836) published this in 1822. George Stokes (Ireland/England, 1819–1903) made considerable contributions to fluid mechanics, published it in 1842 and 1843 so that Equation 6.14 is traditionally called the *Navier–Stokes equation.*

For an ideal fluid ($\eta = 0$), Equation 6.14 becomes

$$\frac{\partial\vec{v}}{\partial t} + (\vec{v}\cdot\nabla)\vec{v} = \frac{-\nabla P}{\rho} + \vec{g}_{ap} \qquad 6.15$$

Equation 6.15 was first derived by Leonhard Euler in 1757 and was the starting point for further work that led to Equation 6.12, which includes viscosity.

An interesting result for ideal (inviscid), incompressible fluids under steady flow in a gravitational field was obtained by Daniel Bernoulli in 1738,

$$P + \rho g y + \frac{\rho v^2}{2} = \text{constant} \qquad 6.16$$

This is *Bernoulli's equation* and states that the sum of these three terms is the same for a fluid element anywhere along the flow path. Here g is the acceleration of gravity and y is the height above a chosen origin. One of its several practical consequences is that if the speed is lowered the pressure increases, which is the working principle of an airfoil. The curved shape of the airfoil slows the airflow below it, thus increasing the upward pressure and causing a lift force on the airfoil. Bernoulli's equation can be found by integrating Euler's Equation 6.15. Interestingly, Bernoulli was a good friend of Euler.

The "anatomy" of the Navier–Stokes equation is then, in order of the terms from left to right in Equation 6.14:

a) $\partial \vec{v} / \partial t$ is the *local acceleration term*
b) $\vec{v} \cdot \nabla \vec{v}$ is the *inertial term*, also called the *convective term*
c) $-\nabla P / \rho$ is the *hydrostatic pressure term*
d) $\eta \nabla^2 \vec{v} / \rho$ is the *viscous term*
e) $\vec{g}_{ap}$ is the acceleration arising from any applied force

6.3.1 Reynolds Number

In general, prediction of fluid flow velocity fields is very difficult. At low flow rates, the velocity field can be smooth, *laminar*, and tractable to predict. At high-flow rates, fluids become *turbulent* and hence chaotic and essentially unpredictable. A dimensionless parameter useful for gauging the similarity of flow in systems of different dimensions and for when flow passes from laminar to turbulent is the Reynolds number, *Re*. It is constructed from the ratio of the inertial to viscous terms, $\frac{(\vec{v} \cdot \nabla)\vec{v}}{\frac{\eta \nabla^2 \vec{v}}{\rho}}$. In expressing *Re*, the *kinematic viscosity* ν_{kin} is introduced:

$$\nu_{kin} = \frac{\eta}{\rho}, \text{ kinematic viscosity} \qquad 6.17$$

The units of ν_{kin} are cm^2/s. Then the speed of the flow u is taken, along with a characteristic length scale L, such as the diameter of a tube or the radius of a particle immersed in flow, so that the above ratio is estimated by

$$\frac{(\vec{v} \cdot \nabla)\vec{v}}{\eta \nabla^2 \vec{v} / \rho} \sim \frac{(u^2 / L)}{(u \nu_{kin} / L^2)} = \frac{uL}{\nu_{kin}} = Re \quad \text{Reynolds Number} \qquad 6.18$$

Re is small in low-speed viscous flow and large for high-speed flow at low viscosity, where the flow geometry has a fixed length scale. Low *Re* means that the viscous term keeps the flow damped, steady, and laminar, whereas in high *Re*, the lack of internal friction from low viscosity can lead to chaotic, or turbulent behavior. The range of *Re* at which turbulence occurs depends entirely upon the geometry and boundary conditions of the flow. Different scenarios include flow in an open channel, rotational flow between spheres or cylinders, airflow over an airfoil, fluid flow around an obstruction of a given geometry (e.g. sphere, ellipsoid). For flow in a tube, turbulent flow begins in the regime from *Re* 2300 to 4000.

For motion in aqueous media, $\nu_{kin} \sim 0.01\ cm^2/s$ so that $Re \sim 100Lu$. Table 6.2 gives some typical values of *Re* for aqueous systems. *Re* ranges from $Re \ll 1$ for single cell organisms up to $Re \gg 1$ for macroscopic motion by animals. This means that there have evolved several completely different mechanisms for motility. For example, creatures in the $Re \ll 1$ regime take advantage of the viscosity of the medium to propel themselves about. At the high end, a bird uses lift principles found even for inviscid fluids and fish use acceleration of water with fin and body motions in a way that would essentially also occur in an inviscid medium.

6.3.2 Steady Flow and Stokes Approximation

For the case of steady flow, the velocity field is independent of t so that $\vec{v}(x, y, z, t) = \vec{v}(x, y, z)$ and $\partial \vec{v} / \partial t = 0$ so that Navier–Stokes equation becomes

$$(\vec{v} \cdot \nabla)\vec{v} = \frac{-\nabla P}{\rho} + \frac{\eta \nabla^2 \vec{v}}{\rho} + \vec{g}_{ap} \qquad 6.19$$

Table 6.2 Some typical values of *Re* for aqueous systems.

Situation	*U* (cm/s)	*L* (cm)	*Re* = 100*Lu*	Laminar/Turbulent
Bacterium diffusing in water	10^{-3}	10^{-3}	10^{-4}	Laminar
Single cell with cilium motion	10^{-2}	10^{-2}	10^{-2}	Laminar
Blood flow in a capillary	10^{-1}		10^{-2}	Laminar
Blood flow in aorta	10^{2}		10^{3}	Laminar/Slightly turbulent
Fish swimming	10^{2}	10^{2}	10^{5}	Turbulent
Ocean-going ship	10^{3}	10^{4}	10^{9}	Turbulent

In this case, the streamlines in space will be unchanging in time. The motion of a given fluid element will follow the streamline it starts on.

When $Re \ll 1$, as is normally the case for living state processes at the microscopic level and in polymer solutions the inertial term can be omitted and the Stokes approximation or "slow flow" equation is

$$\frac{\nabla P}{\rho} = \frac{\eta \nabla^2 \vec{v}}{\rho} + \vec{g}_{ap} \qquad 6.20$$

6.4 Applications of Navier–Stokes for Steady Flow

6.4.1 Steady State Flow of Fluid in a Tube with Pressure Difference ΔP

A simple but important case of Equation 6.19 is the flow of fluid in a tube or capillary. It provides a good first approximation, for example, to blood flow in vascular tissues, as well as providing means for constructing practical viscometers.

Consider steady, nonturbulent flow in the z-direction in a long tube of radius R, across which there is a constant pressure gradient ∇P. Cylindrical coordinates are appropriate for this situation. The velocity is purely in the z-direction, v_z, but must be zero at the inner tube wall due to stick boundary conditions. Hence, v_z will depend on the cylindrical radial coordinate r, where $r = 0$ is the axis of the tube. The tube has rotational symmetry, so v_z will not depend on the azimuth angle ϕ, and so $v_z(r,\phi,z) = v_z(r)$. This gives the important simplification that

$$(\vec{v} \cdot \nabla)\vec{v} = v_z \frac{\partial v_z(r)}{\partial z} = 0 \qquad 6.21$$

so that Equation 6.20 can be used, which becomes, with $\vec{g}_{ap} = 0$.

$$\nabla P = \eta \nabla^2_{cyl} v_z(r) \qquad 6.22$$

where ∇^2_{cyl} is the Laplacian operator in cylindrical coordinates, whose r-component is

$$\nabla^2_{cyl} v_z = \frac{1}{r}\frac{\partial}{\partial r}\left(r \frac{\partial v_z}{\partial r}\right) \qquad 6.23$$

The no-slip boundary condition discussed above is $v_z(R) = 0$. It is left as an exercise to show that a parabolic flow profile across the tubes cross section is obtained by integration of the above equation:

$$v_z(r) = \frac{\nabla P}{4\eta}(R^2 - r^2) \qquad 6.24$$

To find the flow rate q (g/s) produced by this flow, we need to specify a pressure gradient. For a linear drop in pressure from one end of the tube to another the gradient over a tube of length L with a total pressure drop between the ends of ΔP:

$$\nabla P = \frac{\Delta P}{L} \qquad 6.25$$

We recall that mass flux J (g/cm^2-s) $= \rho v_z$. It is left as an exercise to integrate J over the tube's cross-sectional area, using Equation 6.24, to find that the mass flow rate q (g/s) of the liquid through the tube is

$$q = \frac{\pi R^4 \rho \Delta P}{8\eta L} \quad \text{Poiseuilles' Equation} \qquad 6.26$$

This is *Poiseuille's equation* for viscous flow of a liquid of viscosity η and density ρ through a tube of radius R and length L under a pressure difference ΔP. It is often expressed in terms of volume flow rate Q (cm^3/s) as

$$Q(\text{cm}^3/\text{s}) = \frac{\pi R^4 \Delta P}{8\eta L} \qquad 6.27$$

The shear rate in the flow is

$$\dot{\gamma} = \frac{dv_z}{dr} = -\frac{\Delta P}{2\eta L} = -\frac{4Q}{\pi R^4} r \qquad 6.28$$

The shear rate is zero at the center of the flow ($r = 0$) and a maximum at the tube wall at $r = R$. The negative sign indicates that the velocity decreases as r increases. You can average this shear rate over the cross-sectional area to find the average shear rate $\dot{\gamma}$

$$\langle\dot{\gamma}\rangle = \frac{\int \dot{\gamma}\, dA}{A} = -\frac{8Q}{3\pi R^3} \qquad 6.29$$

In a typical *millifluidic system*,[2] such as used in liquid chromatography, the flow rates are on the order of 1 cm^3/min (= 0.01667 cm^3/s), with a 0.01 P aqueous solvent. For a capillary of length $L = 100$ cm and $R = 0.01$ cm, this will generate a pressure drop of about 4.2 atm across the capillary. The average shear rate is around 4000 s^{-1}. The speed u in the capillary can be found from the volume flow rate Q by

$$u = \frac{Q}{A} \qquad 6.30$$

For this example, this gives $u = 53$ cm/s. Taking $L = 2R = 0.02$ cm and $v_{kin} = 0.01$ cm^2/s gives Reynolds's number of ~106 from Equation 6.18. Note that this is far outside the Stokes approximation of $Re \ll 1$ that gives Equation 6.20. As shown in Equation 6.21, the inertial term is zero because of the flow geometry, and so Equation 6.20 is applicable.

An interesting example is when the pressure drop is provided by gravity in which case:

$$P = \rho g z \qquad 6.31$$

and $\nabla P = \rho g$ can be substituted into Equation 6.24 and for $\Delta P/L$ in the subsequent equations.

6.4.2 Viscometers

Measuring viscosity is straightforward. Several geometries can be used to create velocity gradients in fluids for the measurement and computation of η. The Navier–Stokes equation provides the basis for finding the relationship between η, the geometry, and applied forces.

One of the simplest, most economical viscometers makes use of Poiseuille flow under gravity. A volume of solution is introduced into a small reservoir in a vertical glass pipette and the time is measured for the liquid to completely flow out of the reservoir. Via a calibration factor, the time is converted directly into the solution's viscosity η. A variation on this is the U-tube viscometer in which there is a receiving reservoir in the arm of the U-tube opposite the arm with the starting reservoir of the fluid. The receiving reservoir is lower than the other, and the time it takes for the liquid in the first reservoir to fill the receiving reservoir is measured and converted to viscosity. Popular versions of these are the *Ostwald viscometer* and the *Ubbelohde viscometer*.

Another Poiseuille flow method is to create a flow rate Q with a pump across a capillary of length L and radius R. A differential pressure sensor measures the ΔP created across the capillary due to this flow, and Equation 6.27 is then used to compute the solution viscosity η. The maximum and average shear rates in the capillary are found by Equations 6.28 and 6.29, respectively. An especially sensitive viscometer using the Poiseuille flow method together with a four-capillary *hydraulic Wheatstone bridge* for use with liquid chromatography is discussed in Chapter 7.

---

2 A millifluidic system is one that typically deals with flow rates on the order of milliliters per minute.

The *Couette viscometer* consists of concentric cylinders with an annular gap between them, where the inner cylinder rotates at a fixed frequency. Because of the stick boundary condition at the surface of the inner cylinder, there is an azimuthal force exerted on it which leads to a torque, which can be measured, e.g. with a torsional balance. The torque is directly related to the solution viscosity.

The velocity field is of the form

$$\vec{v}(r,\phi,z) = v_\phi(r)\hat{\phi} \quad 6.32$$

and the gradient operator in cylindrical coordinates is

$$\nabla = \frac{\partial}{\partial r}\hat{r} + \frac{1}{r}\frac{\partial}{\partial \phi}\hat{\phi} + \frac{\partial}{\partial z}\hat{z} \quad 6.33$$

so that the term and $(\vec{v}.\nabla)\vec{v} = 0$ Equation 6.20 can be used even when $Re \gg 1$ but not in the turbulent regime. It is left as an exercise to find the velocity field and the torque on the inner cylinder.

Other types of viscometers include falling ball, falling piston, and vibrational viscometers.

6.4.3 Hydrodynamic Resistance

Look at the form of Poiseuille's Equation 6.27. It shows that flow is proportional to pressure for a viscous liquid. This is reminiscent of Ohm's law where current flow I is proportional to the voltage, V, which is analogous to the pressure drop in Equation 6.27. The proportionality is $1/Z$, where Z is the resistance: $I = V/Z$.

Accordingly, the cluster of parameters appearing before ΔP in Equation 6.27 suggest defining a *hydrodynamic resistance*, Z

$$Z = \frac{8\eta L}{\pi R^4} \quad 6.34$$

Hence,

$$Q = \frac{\Delta P}{Z} \quad 6.35$$

The analogy to Ohm's Law is readily apparent, where Q corresponds to current and ΔP the circuit voltage. Another instance of an equation of the form of Equation 6.35 is the flow rate across a membrane or filter of hydraulic permeability k (units of area), thickness L, and cross-sectional area A, caused by a pressure drop ΔP, at low Reynolds number:

$$Q = \left(\frac{kA}{\eta L}\right)\Delta P \quad 6.36$$

where the hydraulic resistance for this is

$$Z = \frac{\eta L}{kA} \quad 6.37$$

where kA has replaced $\pi R^4/8$ in Equation 6.34. Equation 6.36 is often called the *Darcy equation*, named for its discoverer Henry Darcy (France, 1803–1858). It is related in form to other force–flux relationships that we encounter later in this chapter.

6.4.4 Frictional Coefficient for a Sphere in a Viscous Liquid

We have just seen the effect of friction on laminar flow in a tube or capillary is to create a parabolic flow profile. Stokes considered what the effects of viscosity would be on a stationary, impenetrable sphere of radius R immersed in a slowly flowing (i.e. $Re \ll 1$) viscous fluid of viscosity η, with an initially uniform velocity field $\vec{v}_0$ (this is equivalent to pushing the sphere at speed $-\vec{v}_0$ through a liquid at rest). This leads to the idea of a *drag force*. In a classical ideal fluid, there is no viscosity so the fluid does not stick to the sphere's surface and hence cannot exert any net force, or drag, on the sphere; the fluid just slips around the sphere.

While this notion of zero drag for an ideal body in an inviscid, incompressible liquid was just blithely stated, it actually was the source of considerable confusion in the development of fluid mechanics. Jean le Rond d'Alembert[3] (France, 1717–1783) proved that there is no force acting on bodies moving through inviscid, incompressible fluids. This was in stark contrast to direct experience, where everyone knew that there is force on an object moving in air and water and so the conclusion of zero drag became known as *d'Alembert's Paradox*. The general consensus now is that the paradox arises only

3 d'Alembert, along with Denis Diderot were the authors of Encyclopédie, which was the first French encyclopedia, published in 1751. Ephraim Chambers published Cyclopedia in England in 1728.

if friction, that is, viscosity, is neglected. Once the notion of the stick boundary condition is accepted then it is intuitive that there must be a force bringing a portion of a fluid flow field to zero velocity: A flowing viscous fluid has zero tangential velocity everywhere on the sphere's surface, because of the stick boundary condition, but flows with its initial uniform velocity far from the sphere. Since the sphere is impenetrable the perpendicular component of velocity at the surface is also zero. In other words, the sphere must exert a force on the liquid in order to fully stop the fluid's flow at its surface. This also engenders an equal and opposite force of the fluid on the sphere.

In fact, Ludwig Prandtl (Germany, 1875–1953) presented the notion of the viscous boundary layer, a usually thin layer around an object moving through a viscous fluid, over which boundary layer the surrounding fluid velocity field decelerates to zero at the bodies surface. There is an interesting difference in imagining frictional force in a liquid versus in a gas, because a gas is compressible. An object moving with speed v through a dilute, ideal gas at rest feels a drag force that increases as v^2. You can imagine how this arises by considering a flat surface moving at speed v through this stationary, dilute, ideal gas. As the surface strikes the gas molecules, it imparts momentum to them of mv, on top of whatever initial momentum they have (remember from Chapter 4 that the gas molecules are in motion but have $\langle \vec{v} \rangle = 0$). The number of particles the surface strikes per second per unit area is the number density of particles times v. Hence, the total force, which is total momentum imparted to the particles per second, varies as v^2. Notice that gas molecules are accumulating in front of the moving surface in this scenario, becoming denser, whereas behind the moving object the density is decreasing. This is only possible because of the *compressibility* of the gas. The same argument cannot be made in an incompressible medium, such as water, because water cannot compress in front of the surface and rarefy behind it. Instead, viscous forces are invoked to explain the force on the moving object.

Interestingly, superfluids, such as helium below 2.2 K, have zero viscosity, and there is experimental evidence that objects moving slowly in them experience no drag force, although there are conjectures that there may be some drag due to quantum mechanical fluctuations. The latter aside, it seems that the discovery of helium's transition to superfluidity at 2.2 K in 1937, usually attributed to Pyotr Kapitsa (Russia, 1894–1984), is a twentieth-century answer for d'Alembert. Namely, he was entirely right about zero drag on a body by an ideal, incompressible, inviscid liquid (approximated by the quantum mechanical superfluid), and so d'Alembert's Paradox disappears.

To find the drag force on a sphere in slow, viscous flow, where $Re \ll 1$, Stokes approximation, Equation 6.20 can be used and $\vec{v}$ solved for in spherical coordinates. Because there is azimuth symmetry $\vec{v} = \vec{v}(r, \theta)$, and the boundary conditions $v_r(R) = v_\theta(R) = 0$ and $\lim_{r \to \infty} \vec{v} = \vec{v}_0$ can be applied to solve Equation 6.20. The stress tensor components can then be found from this and the net force found by Equation 6.7. The calculation is fairly complex, but the final, very useful result for the drag force on the sphere is simple:

$$\vec{F} = -6\pi\eta R\vec{v}_0 \tag{6.38}$$

where the minus sign indicates the force is opposite the direction of the fluid flow. Figure 6.5 shows the streamlines around a stationary sphere in a flowing liquid.

In light of the above discussion about zero drag force in incompressible inviscid flow and a v^2 drag force in compressible flow, it is an important consequence of $Re \ll 1$, and the solution to the Navier–Stokes equation, that the drag force is linearly proportional to velocity in Equation 6.38.

The linear dependence of force on v leads to an important notion in the world of low Re where much of the living state takes place. Consider an applied force, such as gravity, acting on a body initially at rest in a viscous fluid. The particle will first accelerate and the viscous force, which is opposite the direction of the applied force, will increase until finally the viscous force cancels the applied force and the particle ceases to accelerate. At this point, the particle has reached its *terminal velocity*, v_t. The terminal velocity is also often termed the *drift velocity*. Using the magnitude of the viscous force in Equation 6.38 means that for a sphere

$$F_{ap} = 6\pi\eta R v_t \tag{6.39}$$

This suggests that the proportionality factor between the F_{ap} and v_t depends only on the fluid's viscosity and the sphere's radius. This proportionality constant is often termed the *frictional coefficient*, f

$$f = 6\pi\eta R = \frac{F_{ap}}{v_t} \tag{6.40}$$

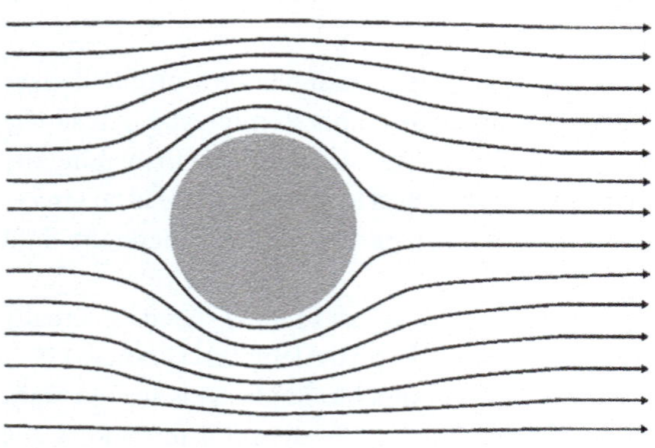

Figure 6.5 Streamlines of liquid flowing around a stationary sphere.

Powerfully, this result holds for any applied force, including electric, centrifugal, and gravitational fields, and is used very widely. It is so useful that it is often used in cases where it doesn't even directly apply, and an "equivalent radius" R of a particle can often be found. Later in this chapter, we will find one of Einstein's simple yet deep connections between diffusivity and viscosity and combine it with the Stokes expression for f in Equation 6.40.

6.5 Hemodynamics

The study of blood flow is termed *hemodynamics* (sometimes also rendered *haemodynamics*). As mentioned, blood is non-Newtonian and so the Navier–Stokes equation is not expected to rigorously apply, although there will emerge some useful principles from it. Furthermore, blood vessels are viscoelastic and not rigid. The elasticity of blood vessels is an essential feature of their operation. Blood flow is *pulsatory* and not steady. Considering these factors hemodynamics should be promoted to a rheological description and, in fact, the field of *hemorheology* has been developed to embrace the complex features of blood flow.

Blood plasma is over 90% water, with a few percent of proteins and ions. Being largely composed of water, blood plasma itself is essentially a Newtonian liquid. It is the presence of the red blood cells that makes blood non-Newtonian.

Equation 6.35 is useful to describe blood flow, where Z is termed the *vascular resistance*. The resistance of the entire circulatory network is termed the *systemic vascular resistance* (it was previously called the *total peripheral resistance*). Q in Equation 6.35 is the average volume blood flow rate, termed the *cardiac output*. ΔP is the hydrostatic pressure drop across the entire circulatory network. It is related to an average aortic pressure, and can be thought of as the *back-pressure* on the heart.

In its simplest form, ignoring the complexities of hemorheology, the entire system of arteries and veins can be viewed as a network of resistors in series and parallel, where Z for each blood vessel is approximated by Equation 6.34; increasing vessel length and blood viscosity each increase Z proportionally, whereas decreasing vessel radius makes Z increase dramatically as $1/R^4$. Considering a blood capillary has a typical diameter of 10 μm and a typical medium size artery has a 1-cm diameter, the pressure per unit length of a capillary is around 10^{12} greater than the medium artery! Not surprisingly, most of the pressure drop in the circulatory system is across the capillaries and small diameter *arterioles* that lead into them. In fact, from this point of view, the Darcy approach of Equation 6.37 might be more appropriate, treating the capillary/arteriole network as a filtration system with hydraulic permeability k. In Chapter 4.8.1 this filtration aspect was discussed in terms of the hydrostatic and osmotic pressures in the blood vessels and surrounding tissue.

These latter considerations also explain why blood pressure monitors that use cuffs can be placed above the elbow, rather than requiring access to the aortic pressure itself; very little of the pressure drop occurs between the aorta and the arteries in the upper arm. The actual principle of the cuff monitor is to pump up the pressure in the cuff until it cuts off the blood flow in the main artery. The pressure is slowly released until the arterial pressure upon ventricular contraction is greater than the cuff pressure and blood flow resumes, which is detected by the monitor; this pressure at which the blood flow from the ventricular pulse resumes is the *systolic pressure*. This pressure is the pressure inside the artery above atmospheric pressure. It typically has a value of 120 mmHg (1 atm = 760 mmHg). The *diastolic pressure* is the minimum pressure reached when the arteries are at their most relaxed state between the high systolic pressure of two ventricular contractions. It typically has a value around 70 mmHg. The monitor senses when the diastolic pressure is reached because there is no more pressure constricting the arteries at this point and the monitor senses no more pulsation. Traditionally, a person uses a stethoscope placed below the cuff on the arm. At high cuff pressure, before systolic pressure is reached there is silence, since blood flow is cut off. When systolic pressure is reached, there is an audible "tapping" sound as blood begins to flow upon ventricular contraction. Because the artery is still constricted there is a pulsatory swooshing of blood that is audible. When the falling pressure ceases to constrict the artery the sound of the pulsations cease completely and diastolic pressure has been reached.

6.6 The Intrinsic Viscosity [η] of Particles in a Fluid

The Navier–Stokes equation provides a basis for treating the flow of viscous, homogeneous liquids, and for considering how particles that are introduced – e.g. spheres – lead to alterations in the fluid flow field and produce frictional coefficients, f, for the immersed objects. The frictional coefficient for a sphere held fixed in a fluid that streams about it was just introduced, $f = 6\pi\eta R$. f is related to the *intrinsic viscosity* [η] of the particle, as well as to its diffusion coefficient, electrophoretic mobility, and sedimentation coefficient.

It is important to mention at the outset that intrinsic viscosity is not an actual viscosity at all, and its name is unfortunate, as a result. As we shall see, its units are completely different than those of viscosity and do not involve time. The simplest operational definition of intrinsic viscosity may be: intrinsic viscosity is the factor, intrinsic to a given particle in given fluid conditions, that determines how much the viscosity of the pure fluid will change as the concentration of the particle in the fluid increases.

Einstein used the Stokes solution for f of spheres to compute their intrinsic viscosity. Later, Jean Baptiste Perrin (France, 1870–1942) and others found exact solutions for ellipsoids of arbitrary axial ratios. These problems were solved for highly dilute solutions and suspensions, i.e. for the case where the flow perturbations caused by individual particles do not affect the changes in flow caused by other particles. When the concentration of particles becomes high enough that hydrodynamic interactions among them can no longer be ignored, a method of first-order hydrodynamic perturbations was introduced by Carl Wilhelm Oseen (Sweden, 1879–1944).

Although the first principles derivation of viscosity is beyond the range of this brief treatment, an important concept in relating viscosity to f is the alternative interpretation of viscosity as power dissipated per volume of fluid due to the viscosity, or internal friction, of the liquid. This also brings us back to a part of the Chapter 4.9 discussion where we discussed the entropic "degeneration" of organized energy into heat. In the parallel plate viscosity scenario, the work done by the constant force keeping the plate moving at constant velocity in Figure 6.1 is dissipated as heat in the fluid, via the internal fluid friction, or viscosity.

In the parallel plate example, the top plate of area A is dragged at a constant speed v, across the surface of the viscous liquid with an applied force F. If the force stops the plate will decelerate to $v = 0$, and its kinetic energy will be dissipated thermally into the liquid. To maintain the constant velocity requires a continuous energy input, whose power is given by

$$\text{Power} = Pr = \vec{F} \cdot \vec{v} = Fv \tag{6.41}$$

since force and velocity are in the same direction. Referring to Figure 6.1, the power dissipated over a volume element $dV = Ady$ can be written as

$$\frac{\partial Pr}{\partial V} = \frac{F\partial v_x}{A\partial y} = \eta\frac{\partial v_x}{\partial y}\frac{\partial v_x}{\partial y} = \eta\left(\frac{\partial v_x}{\partial y}\right)^2 \tag{6.42}$$

Hence, viscosity is related to how velocity gradients lead to thermal power dissipation in the liquid.

This power dissipation is the starting point for computing the intrinsic viscosity of particles, such as spheres. Stokes derived the velocity field for a viscous fluid with an initially constant velocity field flowing about a sphere and found the force on the sphere, given by Equation 6.38. Another situation involves the sphere in a nonuniform velocity field, such as the case of the parallel moving plate in Figure 6.1, where the velocity field drops linearly from the sliding plate's velocity to zero at the lower plate. Wherever the sphere is placed the stick velocity at its top y-value is greater than the stick velocity at its lower y-value and so the sphere will rotate at an angular frequency ω given by twice the shear rate

$$\omega = 2\frac{\partial v_x}{\partial y} = 2\dot{\gamma} \tag{6.43}$$

The stress tensor component at the surface of the sphere that causes the rotation is the one everywhere tangent to the sphere's surface and this is proportional to the fluid viscosity. The torque τ on the sphere is this stress component integrated over the sphere's surface times R. The power appearing in Equation 6.42 is $P = \tau\omega$ and this is divided by the volume of the sphere.

The full derivation to arrive at the increase in the solution's viscosity due to this energy dissipation of the rotating sphere is rather lengthy and was first published by Einstein in 1906. He found that for a dilute solution of spheres:

$$\eta = \eta_0\left(1 + \frac{5}{2}\phi\right) \tag{6.44}$$

where η_0 is the viscosity of the pure solvent and η is the viscosity of the solution with the spheres in it at volume fraction ϕ.

$$\phi = \frac{c\left(\text{g/cm}^3\right)}{\rho\left(\text{g/cm}^3\right)}. \tag{6.45}$$

where ρ is the solvated mass density of the particles and c is the mass concentration (g/cm^3):

$$c = n\bar{v}\rho. \tag{6.46}$$

where n is the number density of particles and $\bar{v}$ is the particle's solvated volume. The solvated particle density and volume refer to the fact that the particles in the solvent may swell compared to their dry state, and the solvated volume may also include waters of solvation or other solvent molecules that adhere to the particle surface. Interestingly, η is independent of the sphere's radius R, and it only depends on the total fraction of the fluid occupied by spheres. This can be understood from the fact that the power Pr is proportional to the sphere's volume (R, the moment arm, times the surface integral of the stress tensor, proportional to R^2), with the volume divided out in the expression power/volume in Equation 6.42. The total solution viscosity is also sometimes expressed as

$$\eta = \eta_0\left(1 + \gamma \frac{c}{\rho}\right) \tag{6.47}$$

where $\gamma = 5/2$ for spheres and ranges up to about 20 for ellipsoids.

The *intrinsic viscosity* $[\eta]$ of a particle is a fundamental property that determines how much the total viscosity of a solution will increase as the concentration of the particles increases. Operationally, $[\eta]$ is defined in the equation:

$$\eta = \eta_0(1 + [\eta]c + k_H([\eta]c)^2 + \cdots) \tag{6.48}$$

where c is the particle mass concentration. $[\eta]$ has units of cm^3/g and is a measure of the hydrodynamic volume per unit of mass of a particle in solution. Again, while $[\eta]$ is termed an intrinsic viscosity Equation 6.48 shows that it is not a viscosity at all, but is the reciprocal of mass density. Notice that for a sphere $[\eta] = 5/2\rho$, independent of the sphere's size. k_H appearing in Equation 6.48 is a hydrodynamic interaction term among solute particles. Equation 6.48 ceases to be useful as terms $O(c^3)$ become important.[4] In fact, viscosity at very high concentration is extremely sensitive to even slight variations in c, and its measurement does not provide an accurate means of determining $[\eta]$.

Equation 6.48 is important because it provides a simple, operational means of determining $[\eta]$ of a particle. The auxiliary concepts of *relative viscosity*, *specific viscosity*, and *reduced viscosity* are helpful in determining $[\eta]$. Relative viscosity $\eta_{relative}$ is the ratio of total solution viscosity to that of the pure fluid, so it conveniently measures the multiple of the pure solution viscosity that occurs when the particles are added at concentration c.

$$\eta_{relative} = \frac{\eta}{\eta_0} = (1 + [\eta]c + k_H\left([\eta]c\right)^2 + \cdots) \tag{6.49}$$

The specific viscosity η_{sp} is the fractional excess viscosity above the solvent's viscosity due to the presence of the particles:

$$\eta_{sp} = \frac{\eta - \eta_0}{\eta_0} = [\eta]c + k_H\left([\eta]c\right)^2 \tag{6.50}$$

The reduced viscosity η_r is the specific viscosity per unit concentration of particles and is used very frequently

$$\eta_r = \frac{\eta - \eta_0}{\eta_0 c} = [\eta] + k_H[\eta]^2 c \tag{6.51}$$

η_r hence provides a convenient means of determining $[\eta]$ of a particle by measuring η_r for a series of solute concentrations and extrapolating this to $c = 0$ and zero shear, $\dot{\gamma} = 0$, i.e.

$$[\eta] = \lim_{\substack{c \to 0 \\ \dot{\gamma} \to 0}} \eta_r \tag{6.52}$$

Figure 6.6 shows how $[\eta]$ can be determined operationally. First, the total solution viscosity η is measured at very low shear (or measurements extrapolated to zero shear) and a known solute concentration with a viscometer, such as one of those discussed above. η_r is then formed via the center term of Equation 6.51, and a linear fit is made to the rightmost term. The intercept then gives $[\eta]$, and the slope allows determination of k_H. An advantage of the single capillary viscometer discussed above for determining $[\eta]$ is that no calibration factor for the viscometer is needed. Hence, it suffices just to measure differential pressure transducer's raw output voltage, which is directly proportional to ΔP, via the transducer's calibration constant. It is clear from Equation 6.51 that the viscosities in the numerator and denominator of the middle term will lead to cancellation of the calibration factor.

As mentioned, Einstein's Equation 6.44 is approximate and only holds for dilute solutions of hard spheres with negligible interactions. For higher concentrations of hard spheres, the so-called particle self-crowding correction, is the following, due to Mooney and Krieger:

$$\eta = \eta_0\left(1 - \frac{\phi}{p}\right)^{-\alpha p} \tag{6.53}$$

4 "$O(c^3)$" is read as "terms of order c^3 and higher."

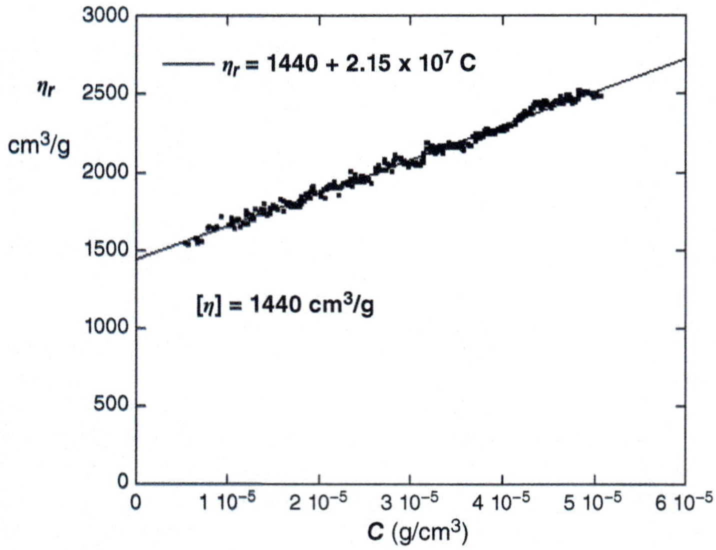

Figure 6.6 Determination of $[\eta]$ for a copolymer of polyacrylamide and acrylic acid in 0.1 M aqueous NaCl. (Reed, W.F., unpublished results).

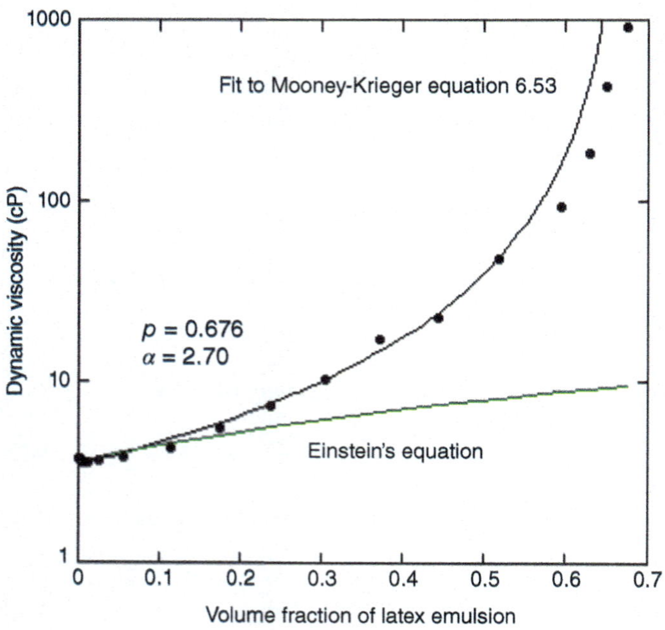

Figure 6.7 Viscosity data for spherical latex emulsions up to high volume fraction, along with a fit to the Mooney Krieger Equation 6.53. The breakdown of Einstein's linear equation is also seen. (Zekoski, T. and Reed, W.F., unpublished data).

where they took $\alpha = 2.5$ in accordance with Einstein's factor in Equation 6.44, and p generally lies between 0.6 and 0.7. Figure 6.7 shows data from high-concentration latex emulsions and a fit according to Equation 6.53. The data are so steep versus concentration that it is represented on a logarithmic scale. The fit is very good up to $\phi = 0.5$, after which it overestimates the viscosity. $p = 0.676$ is in the expected range and $\alpha = 2.70$ is close to Einstein's original value of 2.5. It is not certain whether the deviations after 50% are due to the fact that latex emulsions are not totally hard spheres or whether Equation 6.53 is only approximate. For contrast, the Einstein linear approximation of Equation 6.53 is also shown in Figure 6.7. The effect of interactions among spheres drives the total viscosity two orders of magnitude higher than the linear approximation.

6.7 Force–Flux Relations

The notion of the frictional coefficient was derived above from the Stokes solution to the Navier–Stokes equation for a spherical particle moving at constant velocity in a viscous flow for $Re \ll 1$. A constant force, $\vec{F}_{ap}$ is required to keep the particle in this directed motion. Additionally, a particle undergoes frequent collisions with molecules, due to their thermal motion, which can cause random changes in the particle's momentum and kinetic energy. The *Langevin equation* sums the forces on the particle and uses $\vec{A}(t)$ to represent a random force function due to these thermal collisions. $\vec{F}_{ap}$ is the force due to an external field or agent, and $f\vec{v}$ is the frictional force opposing the motion of the particle.

$$\vec{F}_{net} = \vec{F}_{ap} + \vec{A}(t) - f\vec{v} \tag{6.54}$$

Figure 6.8 illustrates the various forces on a particle of mass m. If we take away $\vec{F}_{ap}$, the *directed motion* ceases, because $\vec{A}(t)$ is completely random, so its time average must be zero: $\langle\vec{A}(t)\rangle = 0$. In this case, however, purely *random motion* of the particle occurs, due to $\vec{A}(t)$. This is diffusion, to which we return in the next section.

In living state systems, because of the viscosity of the aqueous medium and high frictional forces, the terminal velocity v_t is usually reached so quickly once a field is applied that the *ballistic period*, i.e. the time required to reach v_t, is negligible.

To generalize to the notion of many particles, consider the *flux* $\vec{J}$ of a quantity flowing with velocity $\vec{v}$ through a plane perpendicular to the direction of motion. One use of the term "flux" is to denote the quantity of any physical property that crosses a unit area per unit of time. The physical properties whose flux is measured include mass, charge, volume, energy, momentum, etc. A general relationship between the density of the physical property ρ, the flux $\vec{J}$ and the average flow velocity, $\vec{v}$ which you can easily verify, is

$$\vec{J} = \rho\vec{v} \tag{6.55}$$

External forces $\vec{F}_{ap}$ that cause flux of a physical property can often be expressed by the gradient of a corresponding *driving potential U*

$$\vec{F}_{ap} = -\nabla U \tag{6.56}$$

In a liquid, the terminal velocity $\vec{v}_t$ is quickly reached and all particles of a given species subject to the force move with this average velocity. For a number density n (particles/volume) of Equation 6.40 can be combined with Equation 6.56, substituting $\vec{v}_t$ for $\vec{v}$ and n for ρ to give the particle flux (particles per area per second):

$$\vec{J} = -\frac{n}{f}\nabla U \tag{6.57}$$

This is termed a *force–flux relationship*, since a flux of matter arises that is proportional to the force. Force–flux relationships are important in biophysics since many living state processes, such as the flow of ions across a membrane, or the flow of liquid through plants, involve a forcing term and a corresponding flow.

Consider the case of an electric field on a particle of charge q in a medium, where it has a frictional coefficient f. In this case, Equation 6.40 gives

$$\vec{v}_t = \frac{q\vec{E}}{f} = -\frac{q}{f}\nabla\phi \tag{6.58}$$

where ϕ is the electrostatic potential, whose negative gradient yields the electric field $\vec{E}$. The *electrophoretic mobility* of the particle μ_e is defined as the terminal speed per unit of electric field:

$$\mu_e = \frac{v_t}{E} = \frac{q}{f} \tag{6.59}$$

If the particles have a number density n, then they have a charge density qn. The flux in this case is of charge so that J is in units of charge/area-s, which is also termed *current density*:

$$\vec{J} = \frac{q^2 n}{f}\vec{E} = \frac{qn}{f}\vec{F}_{el} \tag{6.60}$$

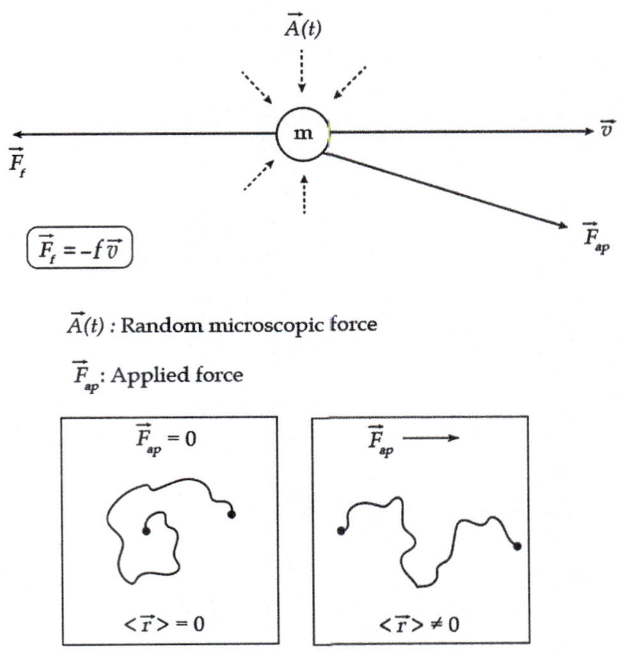

Figure 6.8 Applied, frictional, and random forces on a particle of mass m, having an instantaneous velocity. (Courtesy of Marissa Kalko).

Table 6.3 Some examples of force–flux relationships for various conserved quantities.

Conserved quantity	Transport of	Driving potential U	Process	Conductivity σ
Charge	Electrons	Electrostatic potential	Electrical conduction	Electrical conductivity
Energy	Heat	Temperature gradient	Heat conduction	Thermal diffusivity
Mass	Particles	Chemical potential	Diffusion	Diffusion coefficient
Mass	Particles	Centrifugal (or gravitational) potential	Sedimentation	Sedimentation coefficient

This is a force–flux relationship and also the microscopic form of Ohm's law, often represented as

$$\vec{J} = \sigma \vec{E} \qquad 6.61$$

where σ is the *electrical conductivity*, which in this case is

$$\sigma = \frac{q^2 n}{f} \qquad 6.62$$

The reciprocal of σ is the medium's *electrical resistivity*.

Conductivity σ can be defined for any force–flux relationship. Some important cases are given in Table 6.3. It can sometimes be a tensor, rather than a scalar, if the direction of the flux is not aligned with the direction of the force, e.g. the charged particles of the above example constrained to a tube which makes a nonzero angle with respect to $\vec{E}$. We will investigate some of these force–flux relationships in more detail in subsequent sections.

6.7.1 Conserved Quantities and the Continuity Equation

In many cases, the physical property being transported via the flux is conserved, e.g. mass or electrical charge. In this case, the *continuity equation* links the conservation of the quantity to the flux.

To derive the continuity equation, consider the case of charge leaving the volume V in Figure 6.9. Let Q be the total remaining amount of charge in the volume and $\vec{J}$, the outward flux of charge. S is the closed surface bounding the volume V, $\mathrm{d}\vec{A}$ is the surface element normal to the surface at every point, and $\rho(\vec{r})$ is the charge density distribution within V. The rate of change of total charge in V is negative the rate at which charge flows out over the entire surface.

$$\frac{\mathrm{d}Q}{\mathrm{d}t} = -\oint_S \vec{J} \cdot \mathrm{d}\vec{A} \qquad 6.63$$

But $\mathrm{d}Q/\mathrm{d}t$ can also be written in terms of the volume integral of the charge density, $\rho(\vec{r})$, contained in V:

$$\frac{\mathrm{d}Q}{\mathrm{d}t} = \frac{\partial}{\partial t}\int_V \rho(\vec{r})\mathrm{d}^3 r \qquad 6.64$$

where $\mathrm{d}^3 r$ is an infinitesimal volume element. The divergence theorem allows the flux integral over the surface to be converted to a volume integral of its divergence:

$$\oint_S \vec{J} \cdot \mathrm{d}\vec{A} = \int_V \nabla \cdot \vec{J} \mathrm{d}^3 r \qquad 6.65$$

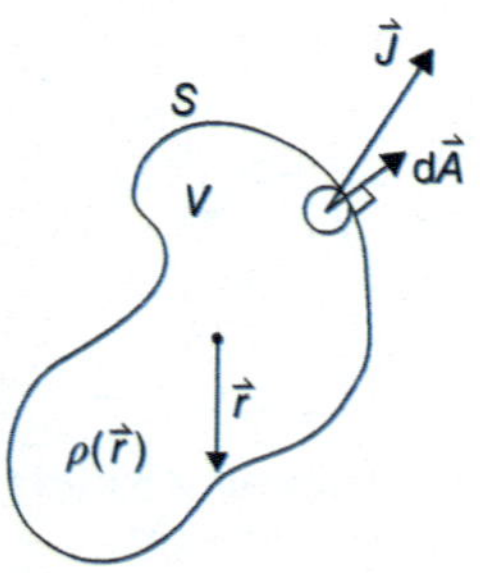

Figure 6.9 Geometry for derivation of the continuity equation.

Matching the integrands of the last two equations, and including the negative sign in Equation 6.63 yields the continuity equation.

$$\nabla \cdot \vec{J} = -\frac{\partial \rho}{\partial t} \quad \text{Continuity Equation} \qquad 6.66$$

For an incompressible material, for which ρ in Equation 6.66 is a constant mass density, the right-hand side is zero, and, considering Equation 6.55, the conclusion is reached that $\nabla \cdot \vec{v} = 0$, which was implied in the constitutive Equation 6.6 for an incompressible viscous fluid.

Combining the continuity equation with the corresponding force–flux relationship will yield a transport equation that can be solved to determine time-dependent changes in densities in terms of the driving potential:

$$\nabla \cdot \vec{J} = -\nabla \cdot (\sigma \nabla U) = -\frac{\partial \rho}{\partial t} \tag{6.67}$$

For a medium with a constant conductivity σ, this becomes

$$\frac{\partial \rho}{\partial t} = \sigma \nabla^2 U \tag{6.68}$$

For a wide class of phenomena $\rho = U$, in which case a parabolic second-order partial differential equation results

$$\frac{\partial \rho}{\partial t} = \sigma \nabla^2 \rho \tag{6.69}$$

Cases where this equation applies include particle diffusion, heat diffusion, and a free quantum mechanical particle. When t is changed to n, a numbering index for monomers in a polymer chain, this equation governs the description of the dimensions of an ideal random coil polymer.

A large amount of mathematical machinery has been built up to solve this equation, including when forcing and nonlinear terms are added, and its applications are widespread.

Consider some specific phenomena described by Equation 6.69. The homogeneous heat equation (no heat sources, but initial nonuniform temperature of a system) has the form of Equation 6.69:

$$\frac{\partial T}{\partial t} = \sigma \nabla^2 T \tag{6.70}$$

where

$$\sigma = \frac{k}{C_p \rho} \tag{6.71}$$

and σ is the *thermal diffusivity*, and is composed of the *thermal conductivity* k, the specific heat at constant pressure C_p, and the object's mass density ρ.

The *diffusion equation* in a uniform medium of diffusion coefficient D and mass concentration c (g/cm^3) has a similar form:

$$\frac{\partial c}{\partial t} = D \nabla^2 c \tag{6.72}$$

Einstein found a remarkable relationship between diffusion and friction:

$$f D_0 = k_B T \tag{6.73}$$

where D_0 is the self-diffusion coefficient of the object, i.e. when its diffusion is not influenced by other similar objects, such as in very dilute solution. Combining this with the Stokes Equation 6.40 for a spherical object of hydrodynamic radius R_H in a fluid of viscosity η, the diffusion coefficient is given by the Stokes–Einstein equation

$$D_0 = \frac{k_B T}{6\pi \eta R_H} \quad \text{Stokes – Einstein Equation} \tag{6.74}$$

We will return to this below.

For a free quantum mechanical particle of mass m, the wave function $\Psi(\vec{r}, t)$ obeys

$$\frac{\partial \Psi}{\partial t} = \frac{\hbar i}{2m} \nabla^2 \Psi \tag{6.75}$$

where $\hbar$ is Planck's constant divided by 2π and $i = \sqrt{-1}$. This shows that a free quantum mechanical particle becomes increasingly delocalized in time, via a diffusion-like mechanism.

6.7.2 Electrophoretic Mobility and Zeta Potential

Charged polymers and colloids exhibit a variety of *electrokinetic* behavior, such as *electrophoresis*, *electro-osmosis*, and *streaming potential*. These have been thoroughly reviewed recently (Delgado 2005).

Among the most important characteristics associated with electrokinetic phenomena are the electrophoretic mobility, defined in Equation 6.59, and *zeta-potential* (or ζ-potential), introduced in Chapter 5.6.2. Electrophoretic mobility μ_e is the steady state terminal (or drift) velocity v_t, a charged particle obtains per unit of applied uniform electric field E, according to Equation 6.59. The drift velocity results when the viscous drag force on the particle balances the electric force.

The ζ-potential is the electric potential at the *surface of hydrodynamic slip* of a charged particle moving through a medium (Chapter 5.6.2). For a small particle in aqueous solution, this layer is typically only a few fractions of a nanometer thick around the particle, so that the hydrodynamic radius and particle radius are virtually identical for particles such as proteins and particles of tens of nanometers and larger. Smoluchowski's treatment provides a simple link between μ_e and ζ for the case where the Debye–Hückel screening length, Equation 5.74, is much smaller than the particle's characteristic size a:

$$\mu_e = \frac{\varepsilon\zeta}{\eta}, \quad a\kappa \gg 1 \tag{6.76}$$

where η is the solvent viscosity. For nanoparticles or very low ionic strength situations, when the Debye length is greater than the particle size, there is a different numerical factor:

$$\mu_e = \frac{2\varepsilon\zeta}{3\eta}, \quad a\kappa < 1 \tag{6.77}$$

ζ-potential is a particularly convenient parameter for gauging the stability of colloidal solutions. Generally speaking, the larger the value of ζ, the greater the electrostatic interparticle repulsion among particles and hence the greater the stability. As a benchmark, the electrostatic potential energy of an elementary charge equivalent to k_BT at room temperature is 25.7 mV. In practice, ζ greater than this generally leads to good colloidal stability, whereas systems usually become less stable below this.

While a number of electrophoretic and related techniques have been developed, one of the pre-eminent means of ζ-potential determination is via electrophoretic light scattering. One means of achieving this is by direct measurement of the Doppler shift of light scattered from particles moving with v_t in an electric field.

6.8 Diffusion

Diffusion has been referenced many times up to this point. It is a fundamentally important process in the living state and is frequently the mechanism by which molecular transport takes place. Now we look more closely at its meaning and formulation in physical terms.

In diffusion, a particle undergoes frequent collisions with molecules, due to their thermal motion, which can cause random changes in the particle's momentum and kinetic energy. We saw above that if there is no applied field the Langevin equation, Equation 6.54, shows that there is still the random thermal force $\vec{A}(t)$ that causes the diffusional motion.

The *random walk* is the quintessential model for a first approach to diffusion. If we now switch $\vec{F}_{ap}$ back on, we understand that there will still be the random motion due to $\vec{A}(t)$ and this will be impressed upon the directed motion due to $\vec{F}_{ap}$. This results in a *biased random walk*.

Now, consider noninteracting particles to be concentrated in a small volume inside a larger volume, such as a drop of ink in a still glass of water. We have already made the argument, on the basis of maximizing entropy, that the ink drop will eventually be diluted evenly, on average, throughout the water and this will be the final equilibrium state. We also saw in Chapter 4 that thermodynamics does not predict the time it takes to go from one state to a final equilibrium state. Diffusion is the *kinetic mechanism* that brings the initial ink drop to its final, fully diluted equilibrium state. The ink particles are subjected to $\vec{A}(t)$ and so are randomly jostled about. It might seem at first glance that, since motion in any direction is equally likely, that the droplet will stay together, since the average displacement of any ink particle will be its initial position in the droplet. But the initial point is just the average of a spatial probability distribution of the particle that changes in time and finally leads to complete dilution.

For simplicity consider a particle in 1D that can move either left or right with equal probability. The initial position is $x = 0$, and the step length is d. Upon the first step, there is 50% probability of being at d or $-d$, and 0% of $x = 0$. Upon the second step, there is a 25% chance of $x = 2d$ or $x = -2d$, and 50% of $x = 0$. The probability that it will have traveled the maximum distance jd after j steps is simply:

$$P(jd) = \frac{1}{2^{j-1}} \tag{6.78}$$

The point is that the probabilities of being at increasing distances from $x = 0$ keep growing as more and more steps are taken; the probability distribution keeps spreading and flattening until after a very large number of steps the distribution is completely flat, and there is equal probability of finding the particle at any position.

First, we can formally arrive at a force–flux relationship for particle diffusion by using Equation 6.57, using $n(\vec{r})$ as the number density of particles at any point in space and using chemical potential $\mu(\vec{r})$, discussed at length in Chapter 4, for the generalized potential U. We take Equation 4.103 and here express the chemical potential for an ideal solute as

$$\mu(\vec{r}) = k_B T \ln\left[n(\vec{r}) / \langle n(\vec{r}) \rangle \right] \tag{6.79}$$

where $\langle n(\vec{r}) \rangle$ is the average concentration of particles, which is just the total number of particles in the system divided by the system volume. The units of chemical potential in Equation 6.79 are energy per particle, since k_B is used, rather than R. Substituting $\mu(\vec{r})$ for U in Equation 6.57 gives

$$\vec{J} = -\frac{n(\vec{r})}{f} \nabla \mu(\vec{r}) \tag{6.80}$$

which yields

$$\vec{J} = -\frac{k_B T}{f} \nabla n(\vec{r}) \tag{6.81}$$

The cluster of constants in front of $\nabla n(\vec{r})$ form a fundamentally important coefficient, *the self-diffusion coefficient* D_0, which has units of area/time, or cm^2/s in the CGS system:

$$D_0 = \frac{k_B T}{f} \quad \text{Self-diffusion Coefficient} \tag{6.82}$$

and is just a re-arrangement of Einstein's Equation 6.73. The diffusion coefficient is hence related to thermal energy, which represents particles' motion, and the frictional dissipation of the particles' kinetic energy. We will return to a heuristic argument for this relationship in the next section. The subscript "0" in D_0 is used to indicate that this is the diffusion coefficient for ideal, noninteracting particles. D can be quite different from D_0 for nonideal solutions, such as for dense, interacting solutions, or electrically charged particles at low shielding.

The consequence of Equation 6.81 is that there will be a net flux of particles moving directly opposite the particle concentration gradient, i.e. particles will diffuse from regions of high concentration (high chemical potential) to low concentration (low chemical potential). The greater the *diffusivity* of the particles, the greater is D_0, and the higher the particle flux down the concentration gradient. Notice that, as in Chapter 4, we can pass freely between the particle's number, mass, and molar concentrations by using N_A and the proper values of the particle molar mass; n (particles/cm^3), c (g/cm^3), and $[n]$ (mol/L), respectively. This will affect both sides of Equation 6.81 equally so that D_0 (and the nonideal D) retain the same units of cm^2/s.

Equation 6.81 is quite general for classical transport systems; we are most interested in its application to diffusing solutes or colloids in liquids, but it applies equally to gas molecules diffusing freely or across membranes, and to particles diffusing in solids.

Equation 6.81 is often called *Fick's first law of diffusion*, after Adolf Fick (Germany, 1829–1901), who first published the result in 1855 in relation to gas flow across a biological membrane. Fick's first law is expressed explicitly with the diffusion coefficient and mass concentration, so combining Equations 6.81 and 6.82:

$$\vec{J} = -D_0 \nabla c \tag{6.83}$$

We can take the divergence of Equation 6.83,

$$\nabla \cdot \vec{J} = -D_0 \nabla^2 c \qquad 6.84$$

For any conserved quantity, the continuity Equation 6.66 can be invoked, so that Equation 6.84 becomes

$$\frac{\partial c}{\partial t} = D_0 \nabla^2 c \quad \text{The Diffusion Equation} \qquad 6.85$$

This is usually called the *diffusion equation*, also sometimes referred to as *Fick's second law*.

For three dimensions in the case of spherical symmetry, Equation 6.85 becomes

$$\frac{\partial c}{\partial t} = D_0 \frac{1}{r^2} \frac{\partial}{\partial r}\left(r^2 \frac{\partial c}{\partial r}\right) \qquad 6.86$$

As a concrete scenario consider that all the mass of diffusive material M_0, is concentrated at the origin at $t = 0$. This implies the seemingly unrealistic condition that the concentration be infinite at $r = 0$. Paul Dirac (England, 1902–1984) introduced the *Dirac delta function*, $\delta(\vec{r} - \vec{r}_0)$, to handle such situations. In the case of mass, it has the unique property of assigning all the mass to the specific point $\vec{r}_0$, leaving the mass zero everywhere else. $\delta(\vec{r} - \vec{r}_0)$ can also be used to handle "infinite densities" of charge and other quantities. $\delta(\vec{r} - \vec{r}_0)$ has units of inverse volume and has the property that for any function $f(\vec{r})$

$$\int_{\text{all space}} f(\vec{r})\delta(\vec{r} - \vec{r}_0)\mathrm{d}^3 r = f(\vec{r}_0) \qquad 6.87$$

In our scenario, all the mass M_0 is concentrated at the origin at $t = 0$ so that $\vec{r}_0 = 0$ and

$$c(0,0) = M_0 \delta(\vec{r}) \qquad 6.88$$

Using the property of Equation 6.87:

$$\int_{\text{all space}} c(0,0)\mathrm{d}^3 r = M_0 \quad \int_{\text{all space}} \delta(\vec{r})\mathrm{d}^3 r = M_0 \qquad 6.89$$

You can show that the solution to this scenario is a Gaussian function:

$$c(r,t) = \frac{M_0}{\left(4\pi D_0 t\right)^{3/2}} \exp\left(-\frac{r^2}{4D_0 t}\right), \quad t > 0 \qquad 6.90$$

Conservation of mass requires that $c(r, t)$ integrated over all space must be constant and equal to the initial mass. Equation 6.90 fulfills this requirement.

Because diffusion is the result of a random probability process, it is interesting to consider Equation 6.90 as a Gaussian probability distribution. First, we replace $c(r, t)$ by number density $n(r, t)$, which can be done freely in Equation 6.86, as discussed above. Then $n(r, t)$ is just the total number density of particles, where the total number of particles N_0 were originally concentrated at $r = 0$, and this substitutes M_0 in Equation 6.90, so that

$$n(r,t) = \frac{N_0}{\left(4\pi D_0 t\right)^{3/2}} \exp\left(-\frac{r^2}{4D_0 t}\right) \qquad 6.91$$

Figure 6.10 shows the $n(r, t)$ computed for a 10-nm diameter globular protein with $D_0 = 4.9 \times 10^{-7}$ cm^2/s in an aqueous buffer at $T = 25$°C. The log–log plot shows the concentration at 10 × time intervals starting at 1 μs. At 10 ms, the root mean square displacement is $\langle r^2 \rangle^{1/2} = 1.73 \times 10^{-4}$ cm.

If $n(r, t)$ is now divided by N_0, we are left with an expression for a single particle. Namely, the probability per unit volume of a particle that was at the origin at $t = 0$ to be at r at time $t > 0$, $W(r, t)$, is

$$W(r,t) = \frac{1}{\left(4\pi D_0 t\right)^{3/2}} \exp\left(-\frac{r^2}{4D_0 t}\right), \quad t > 0 \qquad 6.92$$

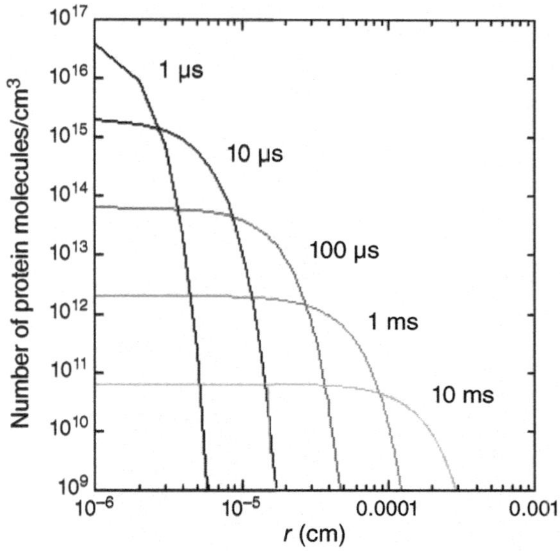

Figure 6.10 Diffusion of protein particles originally concentrated at $r = 0$ at $t = 0$.

It is easy to show that this probability distribution is normalized. If we want to find the average position of the particle $\vec{r}$, we find that

$$\langle \vec{h} \rangle = \frac{1}{\left(4\pi D_0 t\right)^{3/2}} \left\{ \hat{i}\int_{-\infty}^{\infty} x \exp\left(-\frac{x^2}{4D_0 t}\right) dx + \hat{j}\int_{-\infty}^{\infty} y \exp\left(-\frac{y^2}{4D_0 t}\right) dy + \hat{k}\int_{-\infty}^{\infty} z \exp\left(-\frac{z^2}{4D_0 t}\right) dz \right\} = 0 \qquad 6.93$$

because each integral in the brackets vanishes.[5] The *mean square displacement*, $\langle r^2 \rangle$, is found from Equation 6.92 to be

$$\langle r^2 \rangle = 6D_0 t \qquad 6.94$$

and the root mean square displacement is

$$\langle h^2 \rangle^{1/2} = \langle r^2 \rangle^{1/2} = \sqrt{6D_0 t} \qquad 6.95$$

Here we have introduced $\langle h^2 \rangle^{1/2} = \langle r^2 \rangle^{1/2}$ for notational purposes to be consistent with much of the polymer literature and what we will elaborate on in Chapter 8.

We have just shown by Equation 6.95 that an ideal diffusing particle's displacement increases as the square root of time. This property continues to hold in one and two dimensions, only the normalization constants are different. This likewise describes how heat spreads out from an initial "hot spot," and how a wave function for a particle spreads out from its initial localization. For polymers, when t is replaced by the number of monomers, this function gives the spatial dimension of a random coil polymer.

Diffusion can be pictured at a more microscopic level. Consider that a diffusing particle moves a distance d between collisions that occur at average time intervals of $\langle \Delta t \rangle$. The specific amount and direction it moves over a given interval is $\vec{d}_i$. Let $\vec{h}$ be the vector position of the particle after N collisions or steps:

$$\vec{h} = \sum_{i=1}^{N} \vec{d}_i \qquad 6.96$$

$\langle \vec{h} \rangle = 0$ because the particle is equally likely to move in any direction. But after a time, t:

$$t = N\Delta t \qquad 6.97$$

the mean square displacement $\langle h^2 \rangle$ is

$$\langle h^2 \rangle = \langle \vec{h} \cdot \vec{h} \rangle = \left\langle \sum_{i=1}^{N} \vec{d}_i \cdot \sum_{j=1}^{N} \vec{d}_j \right\rangle = \left\langle \sum_{i,j=1}^{N} \vec{d}_i \cdot \vec{d}_j \right\rangle \qquad 6.98$$

5 The integrals vanish because the integrand is odd, being the product of an even and odd function. The integral of an odd function over all space is zero. The *parity* of functions is discussed in Appendix A1.

Because the kicks upon each collision are completely random, there is no correlation among individual steps, that is

$$\langle \vec{d}_i \cdot \vec{d}_j \rangle = 0 \text{ for all } i \neq j \qquad 6.99$$

Hence,

$$\langle h^2 \rangle = \left\langle \sum_{i=1}^{N} \vec{d}_i \cdot \vec{d}_i \right\rangle = N \left\langle d^2 \right\rangle \qquad 6.100$$

Comparing this with the expression resulting from Equation 6.95 based on the solution of the macroscopic diffusion equation, we see that

$$6D_0 t = \langle h^2 \rangle = N\langle d^2 \rangle = \frac{\langle d^2 \rangle t}{\langle \Delta t \rangle} \qquad 6.101$$

This then gives us a microscopic interpretation of the diffusion coefficient D_0

$$D_0 = \frac{\langle d^2 \rangle}{6\langle \Delta t \rangle} \qquad 6.102$$

D_0 is the mean square displacement of the diffusing particle per average time interval between collisions.

Now that we have a connection between the diffusion coefficient and the mean square displacement of a diffusing particle between collisions that occur at an average of $<\Delta t>$ apart, the next question is how does this ratio of mean square displacement to average time between collisions relate to properties of the particle itself, and the medium in which it is diffusing. The formal derivation of the diffusion equation starting with the chemical potential of the diffusing particle and the resulting force–flux relationship of Equation 6.81 automatically gave this relationship $D_0 = k_B T/f$, in Equation 6.82. In other words, the diffusion coefficient is inversely proportional to the frictional dissipation factor; the higher the dissipation, the lower is D_0 and the particle diffuses more slowly.

A heuristic, rather than formal argument for this relationship, is to combine Equation 6.102 with energy dissipation via the average thermal energy $k_B T$. Consider a particle diffusing randomly due to thermal kicks from the molecules of the medium it is immersed in. The average time between kicks is $<\Delta t>$. Each time it receives a kick it either gains or loses an amount of kinetic energy (*KE*) proportional to thermal energy $k_B T$. The average power $<Pr>$ associated with the thermal energy change is

$$\langle Pr \rangle \propto \frac{k_B T}{\langle \Delta t \rangle} \qquad 6.103$$

Between kicks, the particle travels with an average speed $<v>$. The frictional force is proportional to $F_f = f<v>$. The average thermal energy dissipated over the interval is

$$\langle Pr \rangle = F_f \langle v \rangle = f \langle v \rangle^2 \qquad 6.104$$

The mean square distance is proportional to $<v>^2<\Delta t>^2$ and is also equal, by Equation 6.102 to $6D_0<\Delta t>$

$$\langle d \rangle^2 \propto \langle v \rangle^2 \langle \Delta t \rangle^2 = 6D_0 \langle \Delta t \rangle \qquad 6.105$$

Substituting $\langle v \rangle^2 = 6D_0 / \langle \Delta t \rangle$ in Equation 6.104 and setting the powers equal gives the proportionality:

$$fD_0 \propto k_B T \qquad 6.106$$

A deeper analysis than this heuristic argument, or the formal derivation above, shows the proportionality is actually an equality, and Equation 6.82 is recovered. We have seen that the thermal energy dissipated is linked to the product of diffusion and friction.

A specific form of the frictional coefficient for a given particle's shape and the medium it diffuses can be substituted into Equation 6.82 to obtain D_0, or vice-versa. Dynamic light scattering, discussed in Chapter 9.12, measures the mutual diffusion coefficient D_m, directly, and via measurements at several concentrations D_m can be extrapolated to D_0 at zero concentration. Almost invariably DLS instruments convert the measured value of D_m into an equivalent hydrodynamic radius via the Stokes–Einstein relation between R_H, T, and η given in Equation 6.74. This latter equation is the workhorse of many techniques.

6.8.1 Nonideal Effects in Diffusion

The assumption of ideality in the last section assumes, as in the ideal gas and ideal osmotic pressure cases, that particles do not repel or attract each other. When particles begin to interact hydrodynamically, the mutual diffusion coefficient D_m deviates from the single particle or self-diffusion coefficient D_o. In such cases, a useful first-order expansion involving the particle concentration and interaction parameter K_{int}, is

$$D_m = D_o\left[1 + K_{int}c + O(c^2)\right] \qquad 6.107$$

We recall from Chapter 4.14 that the excluded volume notion is helpful for ascertaining effects as particle concentrations increase. We saw for chemical pressure and osmotic pressure that the virial coefficients, A_2, A_3, etc., allowed the power series expansion in terms of concentration for both of these quantities. The virial coefficient A_2 integrates the interparticle potentials over all possible mutual orientations and is positive for net repulsive interactions, zero for noninteracting (ideal) particles, and negative for net attractive interactions.

Similarly, K_{int} in Equation 6.107 measures the net effect of both hydrodynamic and energetic interactions. It is hence a sort of "hydrodynamic A_2." We can visualize the interaction energy effect of a positive K_{int} on the basis of excluded volume. Because of repulsive interactions, a portion of the solution volume is excluded and so the diffusion of a particle resembles a random walk of mutual exclusion with other particles. The effect is that the particles diffuse to greater mean square displacements per unit time, increasing D_m above D_o. If the interactions are attractive then this will slow down the diffusion, as particles linger longer in each other's vicinity than when there are repulsive interactions or no interactions. In the case of attractive interactions, K_{int} is negative, just as A_2 is negative for attractive interactions, and D_m drops below D_0. We will see similar effects on polymer dimensions due to excluded volume.

From the force–flux relationships, it is the chemical potential of a species that drives its flux. Hence, for a solute, component 2, with a scalar conductivity L, the force–flux relationship with $U = \mu_2$ is

$$\frac{\partial c_2}{\partial t} = L\nabla^2\mu_2 \qquad 6.108$$

6.9 The Nernst–Planck Equation

We are now ready to join together the force–flux relations and the result we found for diffusion and electrophoresis. Consider a region of a solution that contains an electric field $\vec{E}$ in addition to concentration gradients of charged particles. For the charged particles, both $\vec{E}$ and concentration gradients can drive flux. This is a fundamentally important basis for producing electrostatic potentials across membranes, which also sets the stage for conduction in the nervous system, to be treated in Chapter 13.

Consider a concentration (g/cm^3) distribution of charged particles $c(\vec{r},t)$ in an electric field $\vec{E}$. The net flux of particles consists of the diffusive flux down the concentration gradient $\vec{J}_c$ and the flux in the direction of the electric field for positive charge (and opposite $\vec{E}$ for negative charge) $\vec{J}_E$:

$$\vec{J} = \vec{J}_c + \vec{J}_E = -D_0\nabla c + qc\frac{\vec{E}}{f} \qquad 6.109$$

Using Equation 6.82 yields one form of the *Nernst–Planck equation*:

$$\vec{J} = D_0\left(-\nabla c + qc\frac{\vec{E}}{k_BT}\right) = -D_0\left(\nabla c + qc\frac{\nabla\phi}{k_BT}\right) \qquad 6.110$$

This shows that the net flux can be in any direction, depending on the relative magnitudes and directions of ∇c and $\vec{E}$. Hence, the electric field can act to increase or decrease net flux, while possibly also changing its direction. The net flux can stop when the two terms in Equation 6.110 cancel each other.

6.9.1 Electrostatic Potential Across a Cell Membrane

Living cells have active membrane ion channels that make such electric field and concentration gradient processes active, essentially by controlling D_0 by opening or closing ion channels in the membrane, in response to energy production requirements, maintenance of concentration gradients, nutrient demand, etc.

Consider a situation in a membrane where a concentration gradient of ions is maintained across the membrane but with zero net flux, $J = 0$, a condition maintained as long as the concentration gradient across the membrane stays constant by some active process. According to Equation 6.110, if there is no net flux across the membrane, $\vec{J} = 0$, there must be equal fluxes passing through the membrane in opposite directions due to the electric field and concentration gradient. If the membrane were completely impermeable there would be no flux from either term and Equation 6.110 would not apply.

If the membrane has permeability, either through active control, or "leaks," then there will be a concentration gradient across it. If there is no net flux across the membrane, $\vec{J} = 0$, but there is an electric field and concentration gradient across it and hence equal but opposite ion fluxes. Equation 6.110 becomes

$$\nabla c = \frac{qc\vec{E}}{k_B T} = -\frac{qc}{k_B T}\nabla\phi \qquad 6.111$$

Considering a planar patch of membrane of thickness Δx reduces this to a 1D problem and

$$\nabla c = \frac{\mathrm{d}c}{\mathrm{d}x} = \frac{qcE}{k_B T} = -\frac{qc}{k_B T}\frac{\mathrm{d}\phi}{\mathrm{d}x} \qquad 6.112$$

This can be integrated immediately. Using $x = 0$ as the inside of the membrane, and integrating from $x = 0$ to Δx, inside to outside, gives

$$\ln\left(\frac{c_e}{c_i}\right) = \frac{qE}{k_B T}\Delta x \qquad 6.113$$

where c_i is the concentration of ionic species inside the cell near the membrane and c_e is the concentration external to the cell, near the membrane. Across the thin membrane, the electric field will be constant so that the electrostatic potential is $\phi = -E\Delta x$. In terms of the internal and external potentials, ϕ_i and ϕ_e, respectively, Equation 6.113 is

$$\ln\left(\frac{c_e}{c_i}\right) = \frac{q(\phi_i - \phi_e)}{k_B T} \qquad 6.114$$

To get an idea of the magnitude of the electric potentials that can be generated across a biological membrane, consider some typical values in Equation 6.114. First, the ions are chiefly monovalent salt ions, K^+, Na^+, and Cl^-, for which $q = \pm e$. Say there is 10 times more Na^+ outside the membrane than inside, then, at $T = 300$ K, $\phi_i - \phi_e = 0.059\,\mathrm{V}$, or +59 mV. This is indeed the typical order of magnitude of transmembrane potentials in living organisms, so that Equation 6.109 has provided an effective first simple model. While 59 mV may not sound like much, consider a typical biological membrane with $\Delta x \sim 10$ nm. The electric field is then ~5.9×10^6 V/m, which is actually quite high; the dielectric breakdown of a typical oil is about 10^7 V/m, and that of dry air is 3×10^6 V/m.

To avoid issues of sign regarding the transmembrane potential, the convention is that this potential is the potential of the interior membrane surface minus the potential on the exterior, $\phi_i - \phi_e$. You can check to make sure the positive sign on +59 mV is correct for this example as follows: Since the concentration of Na^+ is higher outside than inside the diffusion flux will be inward ($-x$). In order to push a positive Na^+ outward against this flux, the electric field must be in the $+x$ direction, which means the potential gradient is in the $-x$ direction, so the inside potential is higher than the outside, thus the "+" sign.

6.10 Fluctuation–Dissipation and a Qualitative Overview of Its Consequences

The basis of the *fluctuation–dissipation theorem* is that a process that produces heat will itself be produced by thermal fluctuations. In some sense, this is an inevitable consequence of the discussion up to this point. We have stressed the fact that diffusional motion is dissipative and is related to the frictional coefficient f via Einstein's Equation 6.82. But in a system insulated from its environment no energy is lost, yet we often think of energy dissipation as loss to some sort of friction. Since no energy is lost in the insulated system, it follows that the dissipated energy must re-appear in the motion that caused the dissipation. Hence, there is a constant trade-off between frictional dissipation by a particle, which reduces its energy and motion, and an increase in energy and motion of other particles absorbing this dissipated energy.

Another example is the number density of particles in a system at equilibrium. There is an average density but there are also fluctuations away from it back to the average. There is a certain characteristic time function, the *time autocorrelation function*, describing the dissipation of the fluctuation back to the average value. One statement of the fluctuation–dissipation theorem is that when an equilibrium system is slowly perturbed from its equilibrium state, e.g. by changing the temperature or applying an electric field, the kinetics at which the new equilibrium state will be reached is directly related to the time autocorrelation function of the original equilibrium state.

Three prominent examples of the fluctuation–dissipation theorem are *Brownian motion*, *Johnson–Nyquist noise* in electrical resistors, and *black body radiation*.

The random motion of pollen particles suspended in water was first reported by Botanist Robert Brown (Scotland, United Kingdom, 1773–1858) in 1827. He repeated the experiment with inanimate dust particles and observed the same random motion, and thus concluded that the motion was unrelated to any living state source.[6] This random, or Brownian motion, is a result of the fluctuation–dissipation theorem.

The phenomenon was first explained by Einstein, who realized that the random motion of a body through a viscous medium produces heat due to drag forces and that, conversely, random thermal fluctuations will produce random motion of a body in a viscous medium. Einstein used statistical mechanics to show that the particle self-diffusion coefficient D_0, is inversely proportional to the dissipative drag factor f, as given by Equation 6.82.

6.10.1 Electrical Noise from Fluctuation–Dissipation

In an electrical conductor, the motion of electrons produces both current and heat. By the fluctuation–dissipation mechanism, random thermal fluctuations will produce random motion of the electrons and hence transient random currents. Currents in a resistive material cause energy dissipation and loss of electron motion. Conservation of energy (for an isolated electrical conductor) and fluctuation–dissipation require that this dissipated energy re-appear as increased motion of other electrons and currents associated with them.

The random thermal currents are *Johnson–Nyquist noise* and exist in all normal electrical resistors, such as metals, even when they are not connected to any voltage or current source. In electronic circuits, the existence of this noise can hamper the functioning of the circuit, limit sensitivity of detectors, and otherwise interfere with the circuit's intended functions. Since the noise is thermal in origin, it can be reduced by cooling the electronic device. Such noise also occurs in electrolyte solutions where random diffusional motion of ions produce associated transient currents.

John B. Johnson (Sweden/United States, 1887–1970) first observed the phenomenon in 1928, while working at Bell Telephone Laboratories (Murray Hill, New Jersey, USA). Harry Nyquist (Sweden/United States, 1889–1976), also working at Bell Labs, developed a theory to explain the phenomenon.

The random, thermally generated currents lead to a thermal noise spectrum, where the mean square noise voltage over a frequency interval Δf for a conductor of resistance R is

$$\langle V^2 \rangle = 4k_B TR\Delta f \qquad 6.115$$

The power dissipation from this for a resistor is

$$P = \frac{\langle V^2 \rangle}{R} = 4k_B T\Delta f \qquad 6.116$$

6 Jan Ingenhousz (Holland, 1730–1739) reported movement of coal dust on the surface of alcohol in 1785, predating Brown's publication.

The fact that the power per Hertz is constant means this is *white noise*, with equal power contributions from all frequencies.[7] The thermal energy in the conductor is traded back and forth between this dissipation and the fluctuating kinetic energy of the electrons that create the dissipative currents. In a resistor insulated from its environment, and not connected to a voltage source, this trade continues constantly with no net loss of energy nor change in temperature.

Consider a black resistor in thermal equilibrium with its environment. Heat is traded back and forth between the resistor and the environment. Kinetic energy can be traded back and forth between collisions of air molecules with the resistor and its electronic motion and lattice vibrations. However, we know that all bodies above 0 K emit and absorb electromagnetic (e-m) radiation. At room temperature, this radiation is primarily in the infrared portion of the spectrum. We are now in a situation surprisingly close to the origins of quantum physics: *blackbody radiation*.

6.10.2 Blackbody Radiation and Fluctuation–Dissipation

Equation 6.115 fails at very high frequency or very low temperature. Similarly, classical thermodynamics was completely stymied in its attempt to explain the electromagnetic radiation spectrum emanating from black bodies[8] at high frequency and low temperature. It is a familiar effect that as a body is heated it begins to glow, first a dull red, and then, as the temperature is increased, the color turns to cherry red, orange, yellow, white, and then blue. Josiah Wedgwood (England, 1730–1795), was the maternal grandfather of Charles Darwin, and a famous and wealthy ceramicist, who invented a *pyrometer*, a device for measuring hot oven temperatures, and along the way observed that the succession of colors in glowing bodies was independent of the material of which the body was made. By the mid-nineteenth century, scientific instruments capable of accurately measuring the energy of infrared e-m waves had been developed, and the shape of the spectra could not be accounted for by assuming that the average energy at any T was just k_BT and independent of frequency, i.e. the equipartition of energy (Chapter 4.5.3) among frequency states failed. The unphysical prediction of classical thermodynamics, using the equipartition of energy, is that the energy emitted by any body above 0 K increases parabolically with frequency, without limit: an absurdity. This led Max Planck (Germany, 1858–1947) to postulate that the energy inside a black body was proportional to the frequency and could only exchange energy in integer intervals of this energy. This is summed up in Planck's hypothesis

$$E_n = nhf, \quad n = 1,2,3,\ldots \tag{6.117}$$

where f is the frequency and h is the proportionality constant, now known famously as *Planck's constant*. Using this hypothesis in conjunction with the classically known density of frequency states in three dimensions, Planck used the classical Boltzmann averaging over the discrete energy states to obtain the energy density per frequency interval, $u(f)$, of a black body (measured in energy/volume–Hertz), and presented the result in 1900:

$$u(f) = \frac{8\pi hf^3}{c^3\left[\exp\left(\frac{hf}{k_BT}\right) - 1\right]} \tag{6.118}$$

where c is the speed of light. The equation fit the black body data perfectly. Figure 6.11 shows both Planck's Equation 6.118 and the classical prediction.

Returning to fluctuation–dissipation, the energy trade-off in the black body is between thermal motion in the body and e-m radiation. The energy absorbed by a body from e-m radiation goes into thermal motion in the body. By the requirements of fluctuation–dissipation, thermal motion must also generate e-m radiation. In fact, Nyquist rederived Planck's Equation 6.118 in the context of Johnson–Nyquist noise.

---

7 Nyquist generalized the noise power spectrum to capacitors and resistors, whose *impedance* are frequency dependent, and so their noise spectrum is no longer white.

8 A *Black Body* is an idealization. It is a body that absorbs all radiation incident upon it. At equilibrium, it absorbs and emits the radiation equally.

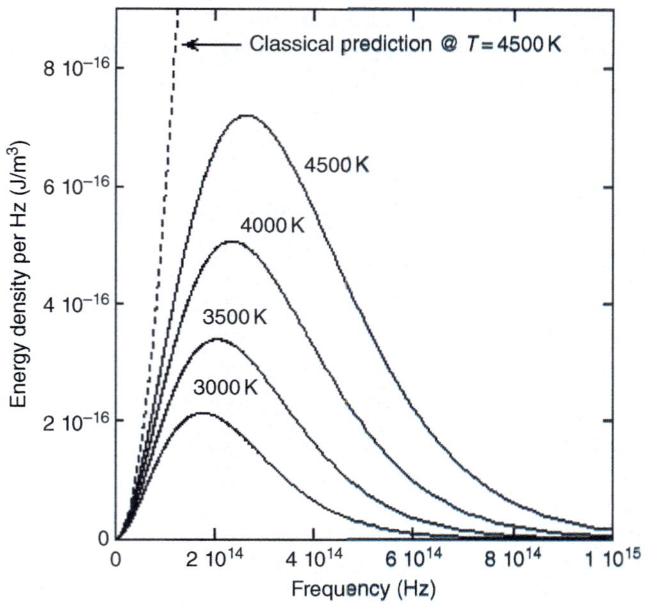

Figure 6.11 Black body radiation spectra at different temperatures. Failed classical prediction is shown by the dashed curve for T = 4500 K.

6.10.3 Fluctuation–Dissipation in the Living State

The most prominent effects from fluctuation–dissipation in the living state appear in diffusion-related phenomena. Viewed in this light diffusion is also a type of "noise." This means that biological systems at the microscopic level actually exist within a sea of such noise. Erwin Schrödinger (Austria, 1887–1961) in his book, "What is Life," published in 1944, surmised that cells must contain huge numbers of molecules in order not to be overpowered by thermal noise.

6.10.4 Fluctuation–Dissipation and the Time Autocorrelation Function

The goal here is to probe a bit more into the time course of the relaxation of a fluctuation in a medium. A good way to do this is to pick up on the trajectory of a particle undergoing Brownian motion. We defined $<\Delta t>$ as the average time between random kicks that changed the particles' direction and kinetic energy. If a particle is initially moving at a velocity $v(0)$, it will keep going more or less at this speed until the next kick. To keep this outline simple, motion in just one dimension is considered. Extension to 3D is straightforward. After one kick, the change in direction and speed may be only slight, but after many collisions, the speed and direction may be very different; eventually, the particle's speed and direction will have no correlation to $v(0)$. The time autocorrelation function is the means of measuring the correlation. The time autocorrelation function of $v(t)$ is defined as

$$A_v(t) = \lim_{\tau\to\infty} \frac{1}{\tau}\int_0^{\tau} v(\tau')v(\tau'+t)\mathrm{d}\tau' \qquad 6.119$$

This is an averaging of the integrand $v(\tau')v(\tau'+t)$ over a total time τ, which is long enough that the velocity has lost all correlation with itself due to random motion. In this, τ' is a dummy variable for time. Notice that if $v(t)$ is constant, v_0, then $A_v(t) = v_0^2$. If $v(t)$ is periodic, $A_v(t)$ will be a periodic function squared, and if $v(t)$ increases monotonically in time then $A_v(t)$ also increases in time. A convenient shorthand for $A_v(t)$ which emphasizes that it is an average is

$$A_v(t) = \langle v(0)v(t)\rangle \qquad 6.120$$

It might seem a bit odd to have $v(0)$ appear in this, but looking at Equation 6.119, the integrand staggers $v(\tau)$ by an increment t at every value of τ, so this is like calling each $v(\tau')$ a starting point for that interval, and can hence be put as $v(0)$ for shorthand.

Our interest now is in random motion in an equilibrium situation, so consider Langevin's Equation 6.54 with no applied force, just the frictional force fv and the random force function $\vec{A}(t)$. Equation 6.54 now reads, in one dimension

$$m\frac{\mathrm{d}v}{\mathrm{d}t} = A(t) - fv \qquad 6.121$$

Multiply this by $v(0)$ and then take the autocorrelation function using the representation of Equation 6.120, and then

$$m\left\langle v(0)\frac{\mathrm{d}v}{\mathrm{d}t}\right\rangle = \left\langle v(0)A(t)\right\rangle - f\left\langle v(0)v(t)\right\rangle \qquad 6.122$$

This can be simplified as follows: For the term $\langle v(0)\frac{\mathrm{d}v}{\mathrm{d}t}\rangle$ notice that the integral is over τ' with t held constant, so that $\mathrm{d}/\mathrm{d}t$ can be taken outside the integral, that is, outside the brackets in this expression. Furthermore, $v(0)$ is uncorrelated to the random $A(t)$ so that $\langle v(0)A(t)\rangle = \left\langle v_0\right\rangle\left\langle A(t)\right\rangle$. But $\left\langle A(t)\right\rangle = 0$ so this term vanishes and we are left with

$$m\frac{\mathrm{d}\langle v(0)v(t)\rangle}{\mathrm{d}t} = -f\langle v(0)v(t)\rangle \qquad 6.123$$

This is the form of a standard first-order equation which can be immediately integrated and whose solution is well known:

$$\langle v(0)v(t)\rangle = \langle v(0)^2\rangle \mathrm{e}^{-\frac{f}{m}t} \qquad 6.124$$

Equation 6.124 gives the relaxation function for the speed, with a characteristic decay time t_c:

$$t_c = \frac{m}{f} \qquad 6.125$$

We recall from Chapter 4 that the equipartition Equation 4.37 relates each degree of freedom of motion of the kinetic energy to $k_BT/2$, so that $mv(0)^2/2 = k_BT/2$, and so

$$\langle v(0)v(t)\rangle = \frac{k_BT}{m}\mathrm{e}^{-\frac{f}{m}t} \qquad 6.126$$

This can now be integrated over all time to give, in 1D

$$\int_0^\infty \langle v(0)v(t)\rangle \mathrm{d}t = \frac{k_BT}{m}\int_0^\infty \mathrm{e}^{-\frac{f}{m}t}\mathrm{d}t = \frac{k_BT}{f} = D_0 \quad \text{for 1D} \qquad 6.127$$

where the rightmost relationship to D_0 comes from Equation 6.82. This is an interesting result because it shows that the diffusion coefficient D_0, which was introduced in Fick's first law Equation 6.83, a nonequilibrium transport equation, as a proportionality constant connecting flux of mass to a concentration gradient, is directly related to equilibrium fluctuations. The result of Equation 6.127 connects *equilibrium fluctuations* with the *nonequilibrium transport constant D_0*, i.e. is a manifestation of the fluctuation–dissipation theorem.

In 3D Equation 6.127, becomes

$$\int_0^\infty \langle v(0)v(t)\rangle \mathrm{d}t = 3D_0 \quad \text{for 3D} \qquad 6.128$$

The fluctuation–dissipation theorem proves the result, that if a perturbation is made to a system by an external force, operating in the linear response regime of the system, then the approach to the new equilibrium is directly related to the kinetics of fluctuations in equilibrium. Equation 6.128 suggests that the other transport factors, such as viscosity and the frictional coefficient, itself can be represented by time integrals over the corresponding time autocorrelation function. For viscosity, the time autocorrelation function is of the momentum transfer in a fluid, and for the frictional coefficient, it is of the time autocorrelation of the Langevin random function force function $\vec{A}$. These relationships between equilibrium correlation functions and associated transport coefficients are *Green–Kubo relationships*.

6.10.5 Time Course of a Charge Density Fluctuation

Consider a medium of uniform conductivity σ 1/(Ω-cm) and uniform permittivity ε, which contains mobile electrical charges, and is electrically neutral. According to the discussion in this chapter, there will be fluctuations in the mobile particle number densities away from their overall average value, and hence there will be deviations from electrical neutrality when and where these fluctuations occur in the form of a transient charge density $\rho(t)$. Consider the continuity Equation 6.66 and the microscopic form of Ohm's law, Equation 6.61. Combining these gives

$$\sigma\nabla\cdot\vec{E} = -\frac{\partial\rho}{\partial t} \tag{6.129}$$

Using the first Maxwell equation, 5.5 gives

$$\frac{\sigma}{\varepsilon}\rho = -\frac{\partial\rho}{\partial t} \tag{6.130}$$

With the solution

$$\rho(t) = \rho_0 e^{-\frac{\sigma}{\varepsilon}t} \tag{6.131}$$

where ρ_0 is the initial magnitude of the charge density fluctuation. For example, for a metal of conductivity $\sigma = 10^8$ 1/Ω-cm and $\varepsilon \sim \varepsilon_0 = 8.8510^{-12}$ $C^2/N\text{-}m^2$, the characteristic decay time is $\varepsilon/\sigma = 8.85 \times 10^{-20}$ s. This is so rapid as to be considered instantaneous.

6.11 Coupled Forces and Flows: Onsager's Reciprocal Relationships

We have considered a single potential causing a single flow, according to Equation 6.57, such as the chemical potential of a concentration gradient driving a flow of particles down the gradient. There are situations where two or more distinct phenomena can be coupled to each other, as seen in Nernst–Planck Equation 6.110, for example.

A thermoelectric material can be made by joining two dissimilar metals. When there is a temperature difference across this junction a voltage develops across it that depends on T, which can drive current flow in a circuit. This is the principle of a *thermocouple*, and the phenomenon is called the *thermoelectric effect* or *Seebeck effect*, named after Thomas Johann Seebeck (Germany, 1770–1831) who discovered it in 1821. Conversely, if a voltage is applied to the junction a flow of heat occurs, sometimes called the *Peltier effect*, named after Jean Charles Peltier (France, 1785–1845), who independently demonstrated the thermoelectric effect. So, in the thermocouple an electric potential difference is created by a temperature difference and a temperature difference appears across the junction if a voltage is applied to it. If a voltage source,[9] such as a battery, and a thermocouple, with a temperature difference across it, are connected in series in a circuit then each one acts as a voltage source and the total voltage is the sum of both. The main point is that there are two very distinct mechanisms, a battery and a thermocouple, adding to the net voltage in the circuit.

Practical uses of these coupled thermoelectric effects are widespread. Measurement of thermocouple voltage is used for accurate determination of temperature differences. When many thermocouples are connected in series and/or parallel a *thermopile* is created which is a useful way of converting heat to a source of electricity. The Peltier effect is used as an efficient, practical way of heating and cooling.

Another such effect is related to piezoelectric materials, which produce a current flow when mechanically strained and produce a mechanical strain when current flows. Piezoelectric stove lighters make use of the former effect to produce a spark by a mechanical impulse on the piezoelectric material. Applying precisely controlled currents allows small, precision movements of piezoelectric materials, to nanometer resolution. The applications *piezo actuators* are rapidly increasing to include mass-produced items, such as in fuel injection systems, and highly precise alignment and positioning accuracy to less than 1nm.

The notion of coupling two distinct forces into two types of flows became the basis for the Onsager phenomenological equations, which Lars Onsager (Norway/United States, 1903–1976) formalized in 1931 into a matrix formulation of the type:

$$\vec{J} = \bar{\bar{L}}\cdot\vec{F} \tag{6.132}$$

9 A voltage source, such as a battery or electrical generator, was classically called a source of *electromotive force* (EMF).

where $\vec{J}$ is the generalized fluxes of all N relevant fluxes, such as charge, mass, energy, momentum, etc. $\vec{F}$ is the vector containing all generalized forces, such as electric fields, and heat and concentration gradients, and $\bar{\bar{L}}$ is the matrix of coefficients that couple these together. The units of F, J, and L must be such that J has the units of $L \times F$. The terms L_{ii} are just the usual conductivities associated with each flux, e.g. the thermal or electrical conductivity of a material. The *cross-coupling coefficients*, L_{ij} for $i \neq j$, relate how one kind of flux J_i is produced by a different type of force F_j. For example, in the thermoelectric effect, let 1 correspond to electricity and 2 to heat. Then L_{12} is the coupling coefficient relating the temperature gradient in the material, F_2, to the electrical flux J_1, and L_{21} is the coupling coefficient relating the voltage gradient in the material, F_1, to the heat flux J_2.

These force–flux relationships are reversible; temperature gradients can produce voltage gradients and vice versa, in the same material. This is analogous to how a rotor being turned in a magnetic field will generate current, while applying current to a rotor in a magnetic field will make the rotor spin as a motor.

Remarkably, it has been found experimentally to reasonably good precision, that $\bar{\bar{L}}$ is a symmetric matrix,

$$L_{ij} = L_{ji}, \quad i \neq j \tag{6.133}$$

This shows that the reciprocal effects, such as the thermoelectric effect, work equally in both directions, heat-to-electricity and electricity-to-heat. Onsager proved this theoretically in his 1931 paper.

6.12 Fluid Transport in Plants

We have put several sections of separation between hemodynamics and fluid transport in plants because the mechanisms are entirely different. In hemodynamics, there is an internal pump, a beating heart, which circulates blood throughout the organism. O_2 is loaded into the blood in the lungs during inhalation and delivered to the organism via arteries and capillaries, and the blood returns loaded with CO_2 which is exhaled, along with water vapor, to the atmosphere. The lungs are the organs of gas exchange with the atmosphere.

In a plant, the driving forces are passive and based on environmental factors and replace the notion of a beating heart. Without getting into the depths and complications of plant physiology, let's take a simplified view in the following. The leaves of a plant are similar to lungs in so far as they are the major site of gaseous exchange with the environment. Through the *stomata* of the leaf, which can open and close, CO_2 enters from the atmosphere, water enters the leaf and into the stomata from the plant's vascular system, and absorbed sunlight drives the photosynthetic, endothermic production of sugars, and then the waste product of the reaction, O_2, is released to the atmosphere. The plant requires a constant supply of H_2O for this and so requires a vascular system and driving forces. Water enters the plant though its roots. In simplified terms, the *xylem* of the plant provides a path for the entering water to the leaves, and the *phloem* provides the pathway for the many products produced in the leaves to be distributed throughout the plant, the most abundant being cellulose, along with other polysaccharides, proteins, and small molecules. In this sense, there is an analogy between xylem and capillaries and between phloem and veins. A major functional difference between leaves and lungs is that in the former the energy and vital material production takes place in addition to gas exchange, whereas the latter are not directly involved in energy production, which derives from ingested, high-energy food. A major functional difference between phloem and veins is that the former transports major chemical building blocks for the plant, whereas the latter carry waste CO_2 for expulsion from the organism.

What are the driving forces for water transport? Osmotic pressure in the roots and capillary action provide some of the driving force, but *transpiration* is the primary mechanism. Stephen Hales[10] (England, 1677–1761) postulated that transpiration of water in the leaf led to a constant flow of water in the plant to the leaf and that the leaf derived some of its nourishment from the air. Transpiration involves the water flow from the roots to the leaf, and its loss to the atmosphere. In fact, most of the plant's water, well over 90%, is lost to the atmosphere when the stomata are open in order to accept atmospheric CO_2 for photosynthesis. Because water is cohesive, it "sticks to itself" due to hydrogen bonds (Chapter 5.4), as water is lost to the atmosphere the remaining water in the leaf pulls in more water from below. This creates a negative pressure with respect to the roots driving the water upward from higher to lower water potential, against gravity. The cohesiveness of water also leads to the capillary action discussed in Chapter 4.11.2, which can raise water to a height dependent on the capillary diameter, as given in Equation 4.66. However, the height is generally not high enough to transport water to the top

10 Stephen Hales was also the first person to measure blood pressure.

of tall trees. Capillary action helps initiate upward water flow which is then enhanced by transpiration, which is also based on the cohesiveness of water.

The rate of evaporation at the top of the capillary depends on the relative humidity of the air, and so, likewise in a leaf, the transpiration rate of water will be strongly influenced by humidity.

Another important driver of water movement in plants is osmosis. Inside the roots of plants, there are various osmotic components, including small molecules and macromolecules. The water in the soil outside the plant normally has less concentration of these components, and in some cases no concentration at all, so that osmosis will drive water through the root into the plant.

While highly idealized, water flow in the plant can be considered as another force–flux relationship, where the water pressure between a leaf and the roots drives the flow of water against a resistance due chiefly to viscous forces in the complex xylem channels. These latter can be approximately viewed as capillaries and the flow of water through them as Poiseuille flow, given by Equation 6.27. The flow rates are normally very low, typically centimeters or tens of centimeters per hour, at Reynolds numbers far less than 1.

Figure 6.12 shows some of the relevant structure in leaf anatomy.

Figure 6.12 Tree and leaf anatomy (Courtesy of Linnette F. Reed).

6.13 Diffusional Versus Directed Motion

Unraveling the process of diffusion took quite a bit of time and ingenuity in the history of science. Part of this is because at the human length scale, there are no common, visible diffusive processes. It took people using microscopes, originally Brown and Ingenhousz, to catch the first glimpses of Brownian motion of micron-size particles due to molecular diffusion, and it would be several generations more before a comprehensive mathematical description of diffusion would appear.

We are simply used to directed motion in everyday life, whether it's walking to school, climbing a mountain, or flying to a distant land in an airplane. But for those living entities that exist on the micron-length scale, the world of motion is quite different, because diffusion can compete with directed motion.

Consider a bacterium of size d, on the micron scale, with a cilium that can be used for directed transport at a top speed of v. Say that nutrient particles are distributed about the bacterium at a similar distance d and have a diffusion constant D_0. For nutrient molecules at distance d is it better for the bacterium to swim toward one of them, or just wait for one to diffuse to it? The "break even point" would be when swimming at v over distance d is equal in time to a nutrient molecule diffusing over d. You can use Equation 6.101 to determine whether it is worthwhile for the bacterium to swim or wait.

In connection with the relative magnitudes of diffusional versus directed motion, there is a useful dimensionless number that compares these. It is the *Péclet number*, *Pe*, and is the ratio of the time it takes to diffuse over a distance L to the time it takes for a particle to move the same distance L by directed motion at speed v:

$$Pe = \frac{L^2 / D_0}{L / v} = \frac{vL}{D_0} \qquad 6.134$$

$Pe < 1$ means that diffusional motion is faster than directed motion, and if $Pe \ll 1$, it is almost certainly better for the bacterium in the above example to wait for the nutrient molecule to diffuse to it.

The bacterium's cilia actually gives it the power to hunt for food. If there are no nutrient molecules in its vicinity the bacterium will be forced to find some. It can then swim at speed v in a given direction a distance L usually many times its size d. If it senses a gradient of food, it will further explore that direction and navigate its way up the food gradient, an example of *chemotaxis*. If no food or food gradient is found after traveling with directed motion over distance L, the bacterium can stop and turn itself in another random direction, seeking food or a gradient. This motion resembles a biased random walk from the bacterium's initial position to a suitable feeding ground of high-density nutrient molecules.

6.14 Time Reversal Symmetry and Its Breaking

Thermodynamic irreversibility in Chapter 4 and viscous dissipation discussed in this chapter are closely connected, and create an *arrow of time* in the macroscopic world. There are systems whose defining equations under *time reversal* are unchanged and yield related solutions. Such equations are symmetric under time reversal. Time reversal replaces t in the governing equation with $-t$. For example, consider a constant 1D force, F_0 acting on a mass, such as gravity, when there is no friction. Then Newton's law remains unchanged when $t \to -t$,

$$m\frac{d^2x}{dt^2} = F_0,\ t \to -t,\ m\left(\frac{d}{d(-t)}\right)\left(\frac{dx}{d(-t)}\right) = m\frac{d^2x}{dt^2} = F_0 \tag{6.135}$$

But now consider a drag force $f\vec{v}$ on an object which opposes the direction of the object's motion, where f is the frictional coefficient:

$$m\frac{d\vec{v}}{dt} = -f\vec{v} \tag{6.136}$$

This has a solution of the form, $\vec{v} = \vec{v}_0 e^{-ft/m}$, where $\vec{v}_0$ is the initial velocity, i.e. the drag force exponentially reduces the velocity of the object. Under time reversal, this becomes

$$m\frac{d\vec{v}}{dt} = f\vec{v} \tag{6.137}$$

This difference of the negative sign turns the real world Equation 6.136 for fluid motion into an impossible physical situation; the solution to Equation 6.137 is, $\vec{v} = \vec{v}_0 e^{ft/m}$. The drag force, in this case, would accelerate the object exponentially![11] The diffusion Equation 6.72 is also seen to not be symmetric under time reversal and the time-reversed equation would give startlingly different results than the real world, e.g. diffused perfume molecules in a room putting themselves all into a little open container.

Throughout, we have referred to processes whereby organized energy turns into disorganized energy as irreversible and dissipative; an extremely high-speed molecule shot into a gas of much slower molecules eventually loses its speed and joins the speed distribution. It would be improbable that energy was consistently transferred back to this molecule to bring it to its original extremely high speed. A mountain crumbles into sand, but sand does not spontaneously assemble itself into a mountain. A person ages but never naturally de-ages. Dissipation and irreversibility set the arrow of time into a future of higher net disorganization.

6.15 Techniques for Determining Transport Properties

We have discussed different viscometers for obtaining both fluid viscosity and a particle's intrinsic viscosity. In Chapter 9.12, we find how the diffusion coefficient D_0 can be found from dynamic light scattering and ζ-potential from electrophoretic light scattering. Table 6.4 is a small, nonexhaustive list of methods used for determining the various viscous and transport properties presented in this chapter. There are many other methods besides those listed.

11 Of course, there are situations of exponential growth. If $\vec{v}$ in Equation 6.137 is replaced by N, the number of bacteria in a growth medium, and f by its reproduction rate α, then the solution to Equation 6.137, showing an exponential growth of the bacterial population is physically correct. Of course, α is dependent upon density of bacterial food, among other factors (e.g. viral attack on bacteria) and so can decrease in time so that the exponential growth slows and stops. In fact, α can go negative if food disappears, in which case it represents the bacterial death rate, and the population decreases. Once again, time reversal in this case yields dramatically different behavior.

charges in an electric field to yield the form of the Nernst–Planck equation in 6.110, which gave us a first look at how ionic concentration gradients across membranes are related to electric fields across the membrane.

Throughout we have seen that in a fluid in equilibrium without external forces, there is a constant state of internal motion based on diffusion. Diffusion, being directly related to frictional dissipation requires a reciprocity between dissipating kinetic energy in frictional flow and imparting this kinetic energy to other particles. This leads to the notion of fluctuation–dissipation, for which electrical Johnson–Nyquist noise, Brownian motion, and black body radiation are prominent examples. This brought us finally to coupled fluxes and forces, where the force causing flux of its own kinds, such as a thermal gradient producing heat flow, can cause flux of a different kind, e.g. the thermal gradient can produce an electric voltage gradient.

References

Anderson, B.D.O. (1982). Reverse-time diffusion equation models. *Stochastic Processes and Their Applications* 12: 313–326. North-Holland Publishing Company.

Carter, J.M. and Phillies, G.D.J. (1985). Second-order concentration correction to the mutual diffusion coefficient of a suspension of hard Brownian spheres. *Journal of Physical Chemistry* 89: 5118–5124.

Delgado, A.V., Gonzalez-Caballero, F., Hunter, R.J. et al. (2005). Measurement and interpretation of electrokinetic phenomena, (IUPAC technical report). *Pure and Applied Chemistry* 77: 1753–1850.

Dhont, J.K.G. (1996). *An Introduction to Dynamics of Colloids*. Amsterdam: Elsevier.

Dorshow, R. and Nicoli, D.F. (1981). The effect of hydrodynamics on the diffusivity of charged macromolecules: application to BSA. *Journal of Chemical Physics* 75: 5853–5856.

Grotberg, J.B. (2021). *Biofluid Mechanics: Analysis and Applications*. Cambridge University Press.

Katz, J. (2011). *Introductory Fluid Mechanics*. Cambridge University Press.

Kubo, R. (1966). The fluctuation-dissipation theorem. *Reports on Progress in Physics* 29 (1): 255–284. https://doi.org/10.1088/0034-4885/29/1/306.

Landau, L.D. and Lifshitz, E.M. (1959). Fluid mechanics. In: *Course of Theoretical Physics*, 6. 2nd English Translation. Oxford UK, Burlington MA: Elsevier.

Lyklema, J. (1995). *Fundamentals of Interface and Colloid Science*, 2. London: Elsevier Ltd.

Miller, J.F., Schaetzel, K., and Vincent, B.J. (1991). The determination of very small electrophoretic mobilities in polar and nonpolar colloidal dispersions using phase analysis light scattering. *Journal of Colloid and Interface Science* 143: 532–554.

Nyqyuist, H. (1928). Thermal agitation of electric charge in conductors. *Physical Review* 32 (1): 110–113. https://doi.org/10.1103/PhysRev.32.110.

Tscharnuter, W.W. (2001). Mobility measurements by phase analysis. *Applied Optics* 40: 3995–4003.

Further Reading

Aquino, T. and Dentz, M. (2017). Chemical continuous time random walks. *Physical Review Letters* 119 (23): 230601/1–230601/5.

Beris, A.N., Horner, J.S., Jariwala, S. et al. (2021). Recent advances in blood rheology: a review. *Soft Matter* 17 (47): 10591–10613.

Dey, K.K. (2019). Dynamic coupling at low Reynolds number. *Angewandte Chemie, International Edition* 58 (8): 2208–2228.

Elgeti, J.; Winkler, R.G., and Gompper, G. (2015). Physics of microswimmers-single particle motion and collective behavior: a review. *Reports on Progress in Physics* 78 (5): 1–50.

Fedosov, D.A., Noguchi, H., and Gompper, G. (2014). Multiscale modeling of blood flow: from single cells to blood rheology. *Biomechanics and Modeling in Mechanobiology* 13 (2): 239–58.

Goldsmith, H.L. (1999). Flow-induced interactions in the circulation. *Rheology Series* 8: 1–62.

Hao, Y., Zhang, Xiqi, and Jiang, L. (2019). Quantum-confined superfluid. *Nanoscale Horizons* 4 (5): 1029–1036.

Lauga, E. and Powers, T.R. (2008). The hydrodynamics of swimming microorganisms. *Condensed Matter* 1–58. arXiv.org, e-Print Archive.

Li, G., Lauga, E., and Ardekani, A.M. (2021). Microswimming in viscoelastic fluids. *Journal of Non-Newtonian Fluid Mechanics* 297: 104655.

Pak, O.S. and Lauga, E. (2016). Theoretical models of low-Reynolds-number locomotion. *RSC Soft Matter Series* 4: 100–167.

Rodrigues, T., Mota, R., Gales, L., and Campo-Deano, L. (2022). Understanding the complex rheology of human blood plasma. *Journal of Rheology (Melville, NY, United States)*, 66 (4): 761–774.
Sheel, A.W., Boushel, R., and Dempsey, J.A. (2018). Competition for blood flow distribution between respiratory and locomotor muscles: implications for muscle fatigue. *Journal of Applied Physiology* 125 (3): 820–831.
Vogiatzis, I., Louvaris, Z., and Wagner, P.D. (2020). Respiratory and locomotor muscle blood flow during exercise in health and chronic obstructive pulmonary disease. *Experimental Physiology* 105 (12): 1990–1996.
Waigh, T.A. (2016). Advances in the microrheology of complex fluids. *Reports on Progress in Physics* 79 (7): 074601/1–074601/62.
Yadav, V., Duan, W., Butler, P.J., and Sen, A. (2015). Anatomy of nanoscale propulsion. *Annual Review of Biophysics* 44: 77–100.

Problems

1a) Solve for the velocity profile $v_z(r)$ for gravity flow of a liquid in a vertical capillary, using

$$\nabla P = \eta \nabla^2_{cyl} v_z(r)$$

and the no-slip boundary condition, $v_z(R) = 0$, where the pressure gradient is due to gravity, and the radial part of the cylindrical Laplacian is

$$\nabla^2 v_z = \frac{1}{r}\frac{\partial}{\partial r}\left(r\frac{\partial v_z}{\partial r}\right)$$

1b) Using the result from 1a and recalling that mass flux J (g/cm^2-s) $= \rho v_z$ show that the mass flow rate q (g/s) of the liquid through the tube is

$$q = \frac{\pi R^4 \rho^2 g}{8\eta}$$

1c) Find the shear rate in the flow $\dot{\gamma}(r)$ and the average shear rate over the tubes cross-section, $\langle\dot{\gamma}\rangle$.

2 Consider a Couette viscometer with an inner cylinder of radius a rotating at an angular frequency ω, and a stationary concentric outer cylinder of radius b. Both cylinders have a height h. The annular gap between them is filled with a viscous fluid.
a) Find the velocity field $\vec{v}$.
b) Find the torque on the inner cylinder.

3 If the average velocity of blood ejected from the aorta is 30 cm/s and the diameter of the aorta as it leaves the heart is 3 cm, then find the average velocity of blood flow through capillaries, assuming there are four billion in the human body, with average radius of 4×10^{-4} cm.

4 Consider an approximately spherical virus of density 1.4 g/cm^3 and radius 1 μm. Find
a) the net gravitational force on the virus in water.
b) the force needed to propel the virus through water at 1 μm/s.

5 Consider a spherical object of density ρ (g/cm^3) and radius R, falling under the influence of gravity in water, where $\rho > \rho_{water}$. Use the Stokes–Einstein equation $f = 6\pi\eta R$ to compute the terminal velocity of the sphere in water, v_t. $\rho = 1.4$ g/cm^3, $R = 0.3$ cm, and the viscosity of water at $T = 20°$C is $\eta = 0.01$ P (the CGS unit).

6 Consider that the pressure of blood inside a person's arm is 20 mmHg above atmospheric. A blood transfusion at 8 mL/min is to be given through a needle which is 2.5 cm long, with an internal radius of 0.025 cm, via a container of blood hung at a height h. If the blood viscosity is 0.04 P and its density taken as 1 g/cm^3, find the height at which the container must be placed above the needle to obtain the 8 mL/min flow rate.

7 Consider Poiseuille flow through a xylem conduit in a vine of diameter 100 μm and length 1 m.
a) Find the resistance per unit length R (Pa-s)/m^3, considering the liquid transported to have the viscosity of water.
b) Compute the flow rate of liquid through a vertical conduit 10 m high

8 The diffusion equation in 1D is

$$\frac{\partial c}{\partial t} = D_0 \frac{\partial^2 c}{\partial x^2}$$

a) Show that the solution to this is

$$c(x,t) = \frac{m_0}{\sqrt{4\pi D_0 t}} \exp\left(-\frac{x^2}{4D_0 t}\right)$$

when $c(x,t=0) = m_0\delta(x)$ where $m_0\int_{-\infty}^{\infty}\delta(x)\mathrm{d}x = 1$. m_0 is the initial mass concentrated at $x = 0$ and the delta function, $\delta(x)$, has the dimensions of inverse length.

b) Consider the concentration at $x = 0$ at t_1 to be c_1. Find the time t_2 at which the concentration at the origin will have dropped by one-half, $c_2 = c_1/2$.

9 Show that:

a) Equation 6.90, $c(r,t) = \frac{M_0}{(4\pi D_0 t)^{3/2}} \exp\left(-\frac{r^2}{4D_0 t}\right)$, solves the spherically symmetric 3D diffusion Equation 6.86, $\frac{\partial c}{\partial t} = D_0 \frac{1}{r^2}\frac{\partial}{\partial r}\left(r^2 \frac{\partial c}{\partial r}\right)$, where M_0 is the initial amount of mass concentrated at $r = 0$ at $t = 0$.

b) Prove that this solution conserves the mass M_0.

c) Find the root mean square displacement, $\langle r^2 \rangle^{1/2}$.

10 Consider a cell of radius 1 μm, with a membrane impermeable to ions that is 3 nm thick.

a) Find the capacitance of the membrane. Take the dielectric constant to be 5.

b) Assume there is 20× higher concentration of Cl^- ions outside the membrane compared to the interior. Find the magnitude and direction of the electric field across the membrane.

11 A protein has a mass of 150,000 g/mol, and a hydrodynamic radius of $R_H = 9$ nm. Find the characteristic decay time of its velocity autocorrelation function in water at $T = 25$°C.

12 A simple model for tree respiration is to consider that there is an effective evaporation surface area from all the leaves, assumed clustered at the top (like in some Japanese gardens), and that this area is composed of $N = 6 \times 10^9$ identical, vertical 5 m high capillaries, each of diameter 50 μm. Assume the total evaporation rate of water from the tree averages 2 L/h. Assume the rate of evaporation equals the rate at which water is pulled from the ground through the roots and is steady.

Assuming the flow of water obeys Poiseuille's Equation 6.26, find the pressure difference caused by the evaporation. Take the viscosity of water as 1 cP. $q = \pi R^4 \rho \Delta P / 8\eta L$, where the flow rate q is in g/s. Express the pressure difference in terms of atmospheres. 1 g/(cm-s^2) = 9.9×10^{-7} atm.

13 A certain globular protein with a net charge of $z = +15e$ is found to have a terminal velocity of 0.05 mm/s when a dilute aqueous solution is between two flat electrode plates, separated by 1 cm and held at 20 V potential difference.

a) Find the electrophoretic mobility of the protein.

b) Ignoring any effects of electrostatic screening, find the hydrodynamic radius of the proteins, assuming they can be treated as hard spheres.

14 a) Compute the pressure drop, in atmospheres, across 10 cm of tubing of inner diameter 0.02 cm, when a solution of viscosity 0.05 P flows through the tube at a rate of 3 cm^3/min.

b) Compute the maximum shear rate in the tubing. Where does it occur?

15 A cylinder initially filled with water has a radius of 10 cm and a height of 300 cm. The bottom of the cylinder is located 100 cm above ground level. Attached to the bottom is a capillary tube of diameter 0.05 cm and length 100 cm. It carries water from the bottom of the cylinder to ground level. Take the viscosity of water as 0.01 P.

a) Find how long it takes to empty the tank.

b) Find the initial Reynolds number for the flow in the capillary tube.

16 A bacterium of size d can swim at a top speed of v (μm/s). There are nutrient molecules around it at a distance L, each of which has a diffusion coefficient D_0.

a) Find the "break even" distance at which it takes an equal time for the bacterium to swim the distance L and the time it takes a molecule to diffuse the distance L.

b) If $d = 1.5\ \mu m$, $v = 25\ \mu m/s$, and $D_0 = 2 \times 10^{-7}\ cm^2/s$, determine the time for the bacterium to swim to the molecule and compare with the time that it takes a molecule to diffuse to the bacterium.

c) Say the molecule will supply the bacterium with total metabolic energy E, and that it costs the bacterium a fraction fE for every distance d that it swims, where $f < 1$. Further suppose that the metabolic energy dissipation rate of the bacterium at rest is $R = \varepsilon E$ where R is in units of E/s and $\varepsilon<1$. Find the farthest distance L_{max} for which it is worth the bacterium swimming to get the energy molecule.

17 A spherical fungus cell of $d = 2\ \mu m$ is moving at 25 μm/s in an aqueous medium of viscosity 0.01 P. How far does it travel before stopping?

18 Does a dissipative force proportional to v^2 lead to an equation of motion that obeys time reversal symmetry? Explain.

19 Find the equation in time for the number of infected people N_i in a constant total population $N = N_i + N_h$, where N_h are the number of healthy people (population is constant means infected people don't die). Assume the rate at which infected people are produced is proportional to the number of healthy people, with proportionality constant α. Describe time reversal of this and its implications.

20 Consider electrical charge flowing outward in a spherical conductor. Charge is being constantly injected into the center of the sphere at a rate of α (C/s) such that the charge density $\rho(\vec{r},t)$ is constant throughout time and throughout the conductor's volume. Use the continuity equation to compute the flux $\vec{J} = \sigma\vec{E}$, which is radially outward and only depends on the distance r from the center of the conductor. (You don't need to know any electrical properties for this. We could just as soon have a constant particle density throughout the sphere and be injecting particles at the center at rate α (particles/s)).

21 An electrical charge Q is initially implanted uniformly at $t = 0$ throughout a conducting sphere, so that the charge density at $t = 0$ is a constant, ρ_0. The sphere has uniform conductivity σ. Starting at $t = 0$, the charge will immediately start flowing to the surface where it will accumulate until $\rho = 0$ throughout the sphere. In this problem, the continuity equation, the relation $\vec{J} = \sigma\vec{E}$, and the first Maxwell equation $\nabla \cdot \vec{E} = \dfrac{\rho}{\varepsilon_0}$ will be needed. $\varepsilon_0 = 8.85 \times 10^{-12}\ C^2/N\text{-}m^2$ is the permittivity of free space.

a) Compute the charge density $\rho(t)$, which is the same at all points in the sphere.

b) If the conductivity of the material is $\sigma = 6 \times 10^7$ (1/Ω-m), find the time for the charge density to drop to 1/2 of its original value.

Part III

Polymer Science

Attention turns in Chapters 7 and 8 to polymer science *per se*, while Chapter 9 details some of the important optical methods used for their characterization, with an emphasis on light scattering. Many of the properties of macromolecules and the methods used to characterize them are shared between the fields of synthetic polymer science and biomacromolecules. Whereas most biological polymers take on specific conformations, such as globules and rods, and can have very organized secondary and tertiary structures, most common synthetic polymers resemble random coils, which can be analyzed using the random walk and diffusion methods. Interestingly, proteins, in general, exist only as random coils when they are "denatured" or unfolded from their higher structural organization and have lost their biofunctionality.

Biophysics: Physical Processes Underlying the Living State, First Edition. Wayne F. Reed.

Companion website: www.wiley.com/go/Reed/BiophysicsandMacromolecules

7

Overview of Polymer Science

The road of polymer discovery is strewn with the bones of absolute statements.

Ferdinand Rodriguez, in *Principles of Polymer Systems*, 1970

Macromolecules, long-chain molecules, do not usually obey the law of constant proportions, which is one reason they have such wide-ranging and modifiable properties. As the above quote implies, many attempts have been made to make sweeping statements about macromolecules, but most are eventually overturned by some new polymer or process. By the way, "polymer molecule" and "macromolecule" are often used interchangeably. While some will split hairs on definitions the point of view here is that the terms are interchangeable, and that frequently "polymer" alone refers to a macromolecule. "Polymer" used alone often also denotes a material composed of macromolecules.

7.1 Biological and Synthetic Polymers

In Part II we have reviewed some of the fundamental thermodynamic, kinetic, transport, fluid mechanical, and electrostatic principles operative in the living state. These alone are not enough to span the other basics of the living state. Macromolecules are the basis of genetics, information storage and transmission, catalysis, and structure. They obey principles laid out in Part II but are a field of research and application of their own. Here, we outline some of the basics of the vast field of polymer science, which encompasses both natural and synthetic macromolecules.

The history of polymer science had both humble and practical beginnings. Much of the early work focused on modifying the properties of naturally occurring polymers, such as cellulose and rubber. Notably, it was found that treating natural rubber with sulfur and heat dramatically improved its properties, making it less brittle in the cold and less deformable in the heat. Thomas Hancock (England, 1786–1865) patented the process in 1843, and Charles Goodyear (United States, 1800–1860) patented it across the Atlantic eight weeks later. It was discovered nearly a century later that the sulfur provides cross-links among the rubber molecules, just as the amino acid cysteine provides sulfur cross-links that help stabilize protein structure.

In addition to natural product modifications polymeric material was often found as a sludge on the bottom of the test tubes of nineteenth-century organic chemists unraveling the mysteries and wonders of carbon-based chemistry. When attention began to focus on polymers, both natural and synthetic, there was no fundamental understanding of their nature. Some considered the material to be particles of small molecules held together by some as yet undiscovered force, while others thought the particles actually to be giant molecules held together with the same covalent bonds as small molecules. In the former camp Thomas Graham (Scotland, United Kingdom, 1805–1869), a pioneer of colloid science, proposed an aggregation theory, involving colloidal forces holding polymeric materials together. This debate continued until 1920, when Hermann Staudinger (Germany, 1881–1965) finally demonstrated that the materials are, in fact, very large molecules held together predominantly by covalent forces (here, we leave aside discussion of noncovalent polymers, which also exist); hence the term "macromolecule." It took at least 10 years more for this picture of macromolecules to be accepted. During this generations-long debate, it is interesting to reflect that the theory of evolution, much of modern thermodynamics, electromagnetic theory, special and general relativity, atomic and nuclear structure, and quantum mechanics, as well as many other fields, were discovered.

Biophysics: Physical Processes Underlying the Living State, First Edition. Wayne F. Reed.

Companion website: www.wiley.com/go/Reed/BiophysicsandMacromolecules

While modern polymer science and industry are frequently equated with *plastics*, it is fundamentally important to realize that biological macromolecules have evolved over billions of years and are the building blocks of the living state. The genetic code is based on macromolecules of nucleic acids, and these latter both store information and are the subject of continued evolutionary change. In contrast, macromolecules made by humans are quite modest and typically consist of just one or a small handful of different types of *monomers*, the molecular units which are covalently bonded together to form polymer molecules. These materials fulfill a wide range of functions in virtually all modern industrial sectors, but they have nowhere near the sophistication of biomacromolecules. Nonetheless, there are intense, worldwide efforts to produce polymers of increasing sophistication, for example, those that can respond to environmental factors, such as heat or light, and change their conformations and properties.

Synthetic polymers are found in an extraordinary array of modern products. They are used in paints, adhesives, coatings, water purification, papermaking, filtration, medical devices and prosthetics, pharmaceutical and personal care products, metallurgy, fracking, advanced oil recovery, electronics, optics, building materials, packaging, composites, clothing, automotive and aerospace applications, and much more. In 2010, the estimated worldwide polymer industry involved over 10^{12} US dollars. Rapidly advancing uses of synthetic polymers include light-emitting semiconductors, drug delivery agents, and artificial skin.

The value of polymers can range from under a dollar per pound for commodity plastics, to many dollars for engineering plastics, to kilodollars or more for specialty and medically related polymers.

7.2 Brief Overview of Classes of Organic Molecules

Organic chemistry is a vast, deep field. If you have not studied it before you are encouraged to do some exploring in this area using any one of several introductory texts. It is worthwhile here to list some of the most important classes of organic molecules. Often there are several names for a single molecule. The International Union of Pure and Applied Chemistry (IUPAC) has systemized the naming of chemical compounds. IUPAC maintains an important website at goldbook.iupac.org.

A huge number of organic chemists have contributed to this dynamic field and it is not practical to list many of them. Some of the names associated with prominent reactions and reagents include Friedrich Wöhler, the first to synthesize a biomolecule, urea, a living state chemical compound, Victor Grignard (France, 1871–1935) known for the *Grignard reagent*, which can introduce carbon–carbon bonds into compounds, Emil Fischer (Germany, 1852–1919) who made fundamental connections between organic chemistry and biology, Otto Diels (Germany, 1876–1954) and Kurt Alder (Germany, 1902–1958), whose Diels–Alder reaction allows coupling of alkenes into closed rings, Adolf von Baeyer (Germany, 1835–1917) who discovered how to synthesize indigo and other dyes, Charles Friedel (France, 1832–1899) and James Craft (United States, 1839–1917) whose Friedel–Craft reaction allows groups to be added to aromatic rings ... and the list goes on.

The purpose here is to give an overview of classes of organic molecules, which are important both in living state molecules and synthetic polymers. Table 7.1 shows the main features of some of the most important classes of organic molecules. Petrochemicals, produced from the distillation, fractionation, and cracking of petroleum, yield several classes of molecules, such as olefins and aromatics, including precursors to many other compounds. They are also derived from other fossil fuels, such as natural gas, shale oil, and coal. Many of the molecules are also derived from plant and animal sources.

7.2.1 Hydrocarbons

Hydrocarbons contain only carbon and hydrogen and occur in several motifs. In general, hydrocarbons are very apolar and have very low solubility in water. The *alkanes* have fully saturated bonds between carbon units and their names end in "ane." The first 10 are methane (Figure 7.1a), ethane, propane, butane, pentane, hexane, heptane, octane, nonane, and decane (Figure 7.1b). The alkanes can be linear, branched, and cyclic. The simplest branched alkane is isobutane. Figure 7.2 shows linear *n*-butane and isobutane. The cyclic alkanes are prefixed by "cyclo." The simplest cyclic alkane is cyclopropane, and a widely used organic solvent is cyclohexane.

The *alkenes* have one or more double bonds and their names end in "ene." The simplest is ethene (or ethylene, Figure 7.3a), followed by propene (or propylene), and butene (or butylene). These latter are especially important for producing *polyolefins*, discussed below. The presence of the double bond is the key feature that makes these compounds polymerizable. The simplest cyclic alkene is cyclopropene, a compound which is highly reactive and difficult to prepare.

The *alkynes* have one or more triple bonds and their names end in "yne." The simplest is ethyne (Figure 7.3b), which is commonly known as "acetylene," a highly flammable gas. Cyclooctyne is the smallest cycloalkyne that can be isolated and stored.

Table 7.1 An overview of various classes of organic molecules. *Source:* Saylor Academy/CC BY 3.0.

Class	General Formula	Example	Common Name (Systematic Name)	Common Suffix/Prefix (Systematic)
		Hydrocarbons		
Alkanes	RH	CH_3CH_3	Ethane	-ane
Alkenes	RR'C═CR"R"'	$H_2C{=}CH_2$	Ethylene (ethene)	-ene
Alkynes	RC≡CR'	HC≡CH	Acetylene (ethene)	(-yne)
Arenes	ArH[a]		Benzene	-ene
		Halogen-Containing Compounds		
Alkyl halides	RX	CH_3CH_2Cl	Ethyl chloride (chloroethane)	halide (halo-)
Aryl halides	ArX[a]	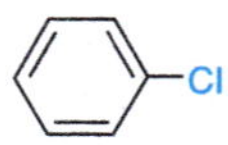	Chlorobenzene	halo-
		Oxygen-Containing Compounds		
Alcohols	ROH[a]	CH_3CH_2OH	Ethyl alcohol (ethanol)	-ol
Phenols	ArOH[b]	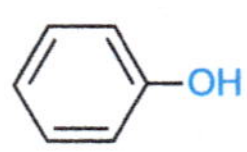	Phenol	-ol
Ethers	ROR'	$H_3CH_2COCH_2CH_3$	Diethyl ether	ether
Aldehydes	RCHO	$CH_3CH{=}O$	Acetaldehyde (ethanal)	-adehyde (-al)
Ketones	RR'C═O	$CH_3C({=}O)CH_3$	Acetone (2-propanone)	-one
Carboxylic acids	RCO_2H	$CH_3C({=}O)OH$	Acetic acid (ethanoic acid)	-ic acid (-oic acid)
		Carboxylic Acid Derivatives		
Esters	RCO_2R'	$CH_3C({=}O)OCH_3$	Methyl acetate (methyl ethanoate)	-ate (-oate)
Amides	RCONHR'	$CH_3C({=}O)NHCH_3$	*N*-methylacetamide	-amide
		Nitrogen-Containing Compounds		
Amines	RNH_2, RNHR', RNR'R"	$CH_3CH_2NH_2$	Ethylamine	-amine
Nitriles	RC≡N	$H_3CC{\equiv}N$	Acetonitrile	-nitrile
Nitro compounds	$ArNO_2$[a]	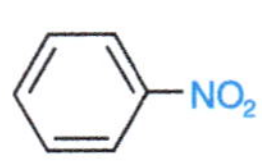	Nitrobenzene	nitro-

[a]R indicates an alkyl group
[b]Ar indicates an *aryl* group

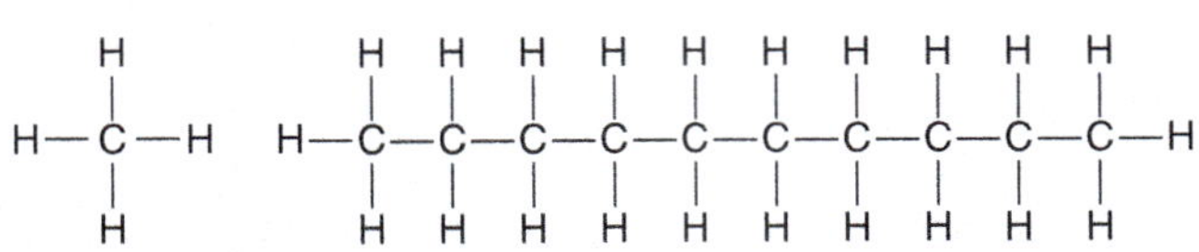

Figure 7.1 Typical alkanes. Methane (left), and decane (right).

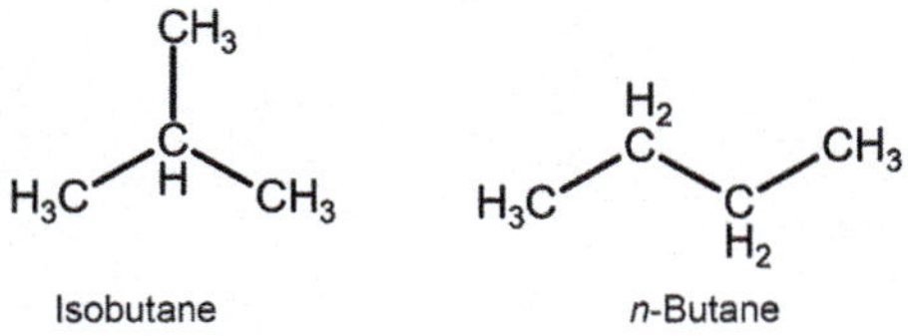

Figure 7.2 Alkane isomers. Isobutane (left) and linear *n*-butane (right).

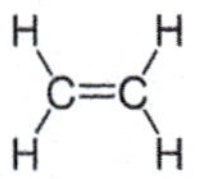

Figure 7.3a The simplest alkene, ethene.

H—C≡C—H

Figure 7.3b The simplest alkyne, ethyne.

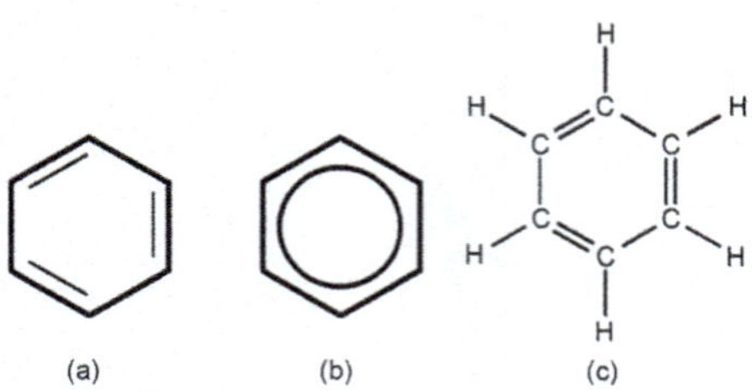

Figure 7.4 Different representations of benzene. Simplified representation showing one configuration of the three double bonds (a). Simplified representation with the circle showing equal sharing of the electrons among all six carbons (b). Representation with all atoms explicitly shown (c).

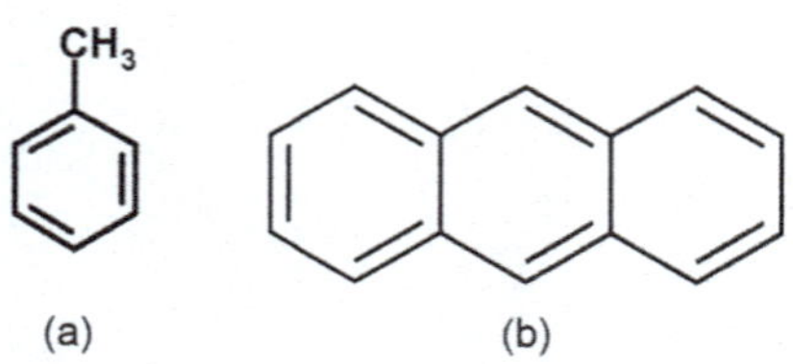

Figure 7.5 Benzene derivatives. Toluene (a) and anthracene (b).

The *arenes* are cyclic hydrocarbons that have alternating single and double bonds, creating an electron delocalization which stabilizes these structures. They have historically been referred to as *aromatics*. Benzene, with six carbon atoms in the ring, is the most important of the arenes. Figure 7.4a–c shows various representations of benzene. In Figure 7.4a, the three double bonds are shown explicitly in one state, and the C and H atoms are implicit. Figure 7.4b uses a ring instead of double bonds to suggest how the double bond electrons are actually delocalized and shared equally among all six carbon atoms and also has implicit C and H atoms. To simplify representations of organic molecules the vertex of two lines designates a carbon atom in Figures 7.4a,b, and if no other atomic groups are shown it is implicit that an H atom is bonded to the carbon. Figures 7.4a,b use this simplified representation, whereas Figure 7.4c explicitly shows all the atoms, similar to the previous figures.

There is a large number of benzene derivatives. Toluene (or toluol), for example, is a widely used solvent, and simply has a CH_3 group replacing one of the H atoms on the benzene ring. Fusing two benzene rings together yields naphthalene, the principal aromatic component of *coal tars*. Fusing three benzene rings together linearly produces anthracene. Figures 7.5a,b show toluene and anthracene, respectively, using the simplified representation introduced in Figure 7.4a.

A benzene derivative of special importance for polymer science and engineering is *styrene*; the IUPAC name is ethenylbenzene, because an ethene group, minus one H atom, replaces one of the H atoms on the benzene ring. It is also known as vinylbenzene and poly(styrene) is a member of the class of *vinyl polymers*. The *vinyl functional group* is ethene minus one H atom. The IUPAC name for this functional group is *ethenyl*. Poly(styrene) and its copolymers rank among the most important polymers manufactured. *Divinyl benzene*, which is benzene with two vinyl (ethenyl) groups, is of special importance because it allows for *branching* and *cross-linking* of polymer molecules. Styrene and divinyl benzene are shown in Figures 7.6a,b in the simplified representation.

7.2.2 Organic Molecules Containing Halogens

Halogens are elements occurring in the 17th column of the periodic table and include chlorine, fluorine, bromine, iodine, and astatine. They react with metals to form many types of salts, including sodium chloride ("table salt"), potassium iodide, cesium iodide, magnesium chloride, among many others. They associate with hydrogen to form acids (hydrohalic acids) in water, such as hydrogen chloride (HCl) and hydrogen fluoride (HF).

Halogens combine with virtually all types of hydrocarbons in the forms outlined above. An important subclass is the *haloalkanes* (or alkyl halides). In these, the

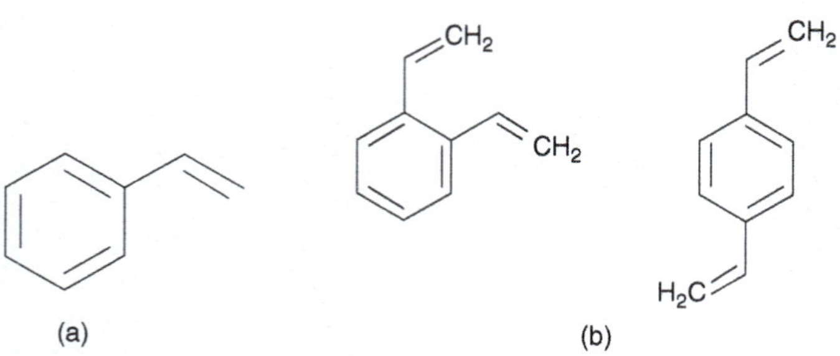

Figure 7.6 Two benzene derivatives of fundamental importance in the polymer industry. Styrene (a) and (b) divinyl benzene, shown in two isomeric states.

halogen substitutes one or more of the H atoms. Examples include the organic solvents trichloromethane (or chloroform) and dibromoethane.

Vinyl chloride (Figure 7.7a), a *haloalkene*, is especially important industrially, as it is used to manufacture poly(vinyl chloride), or PVC. Tetrafluoroethene (or tetrafluoroethylene, Figure 7.7b), in which all four H atoms in ethene are substituted by four fluorine atoms, is used to produce poly(tetrafluoroethene), abbreviated as PTFE, and commonly called by its commercial name, Teflon.

Halogens also combine with arenes to form haloarenes (or aryl halides), where one or more halogen atoms is bonded to the aromatic ring. Examples include bromobenzene and trichlorobenzene.

Figure 7.7 Important haloalkenes. Vinyl chloride, which is the monomer used to produce poly(vinyl chloride), or PVC (left). Tetrafluorethene, used in the production of poly(tetrafluoroethene), or PTFE, often called "Teflon" (right).

7.2.3 Organic Molecules Containing Oxygen

This is a vast class of molecules, encompassing several broad groups, including *alcohols*, *esters*, *phenols*, *ethers*, *aldehydes*, *ketones*, and *carboxylic acids*. Many of these compounds appear in the living state.

Alcohols contain at least one –OH group bound to a carbon. They are named by adding the suffix "ol" to the main molecule's name. The simplest is methanol (or "wood alcohol"), CH_3OH, followed by ethanol, C_2H_5OH (the alcohol in alcoholic beverages, Figure 7.8a). After this, the alcohols can have various isomers, depending on where the –OH group(s) is bound. For example, propanol is propan-1-ol when the –OH is at the end, but is propan-2-ol (2-propanol, or isopropyl alcohol, or isopropanol, or rubbing alcohol) when –OH is bound to the middle carbon. The –OH can also bond to alkenes. The simplest example is the substitution of an H atom on ethene by an –OH group to form vinyl alcohol. This latter is not the precursor for poly(vinyl alcohol), which is produced by hydrolysis of poly(vinyl acetate).

The –OH group can also bond directly to a cycloalkane to form an alcohol such as cyclohexanol. Alcohols can substitute an H atom on aromatic groups to produce compounds such as phenol.

The alcohols mentioned until now are *monohydric alcohols*, meaning they have a single –OH group. *Polyhydric alcohols* have two or more –OH groups. The simplest example is ethane-1,2-diol (ethylene glycol a widely used "anti-freeze" fluid), where two atoms on the ethane molecule are substituted by two –OH groups. Propane-1,2,3-triol (or glycerol, Figure 7.8b) is a "gateway" molecule for several biomolecules, such as triglycerides, which consist of the glycerol functionalized with three fatty acids. Hexane-1,2,3,4,5,6-hexol (or mannitol or sorbitol) has one –OH group on each of the six carbon atoms. It is used as an artificial sweetener and as a means of reducing pressure in the eye caused by glaucoma.

Phenols (or phenolics) involve an –OH group bonded to an aromatic ring. The simplest of this group is benzenol (or, simply, "phenol," Figure 7.9a), where an –OH group is bonded directly to the benzene ring. Phenol is of major importance in the polymer industry. 4,4′-(Propane-2,2-diyl)diphenol (or bisphenol A) is used to produce certain polycarbonates, epoxy resins, and polysulfones.

Polyphenols also occur naturally and are found in abundance in fruits, cereals, vegetables, coffee, and tea. Tannin is one of the most important examples. In the living state, these can have several functions, such as UV protection screens in plant pigments, regulation of certain hormones, and as signaling molecules in certain growth processes. A generic polyphenol is shown in Figure 7.9b.

Ethers are characterized by an oxygen bonded to two alkyl and/or aryl groups. The general motif is R–O–R′, where R and R′ can come from a very wide variety of functional groups. The simplest ether is methoxymethane (or dimethyl ether), while ethoxyethane (diethyl ether) is the anesthetic commonly called just "ether." Tetrahydrofuran (THF) is a *cyclic ether* and an example of a *heterocyclic compound* because it contains at least one other type of atom besides carbon in the cyclic ring. THF is a polar organic solvent widely used for its miscibility with water and many apolar compounds. The lone pair electrons shown as dots in Figure 7.10a for THF are the source of its polarity. It is especially important in polymer science for dissolution of polymers prior to chromatographic and other analyses. Methoxybenzene (or anisole, the major aromatic component of anise seed, Figure 7.10b) is an example of an *aryl ether*.

Ethers are widespread in the living state. They are important in lignins, which form an important class of biopolymers that add structural strength to plant cell walls. Because they are highly hydrophobic they also help stave off hydrolysis and prevent rotting. Lignins are extraordinarily complex and heterogeneous. A generic lignin is shown in Figure 7.11, where the abundant ether linkages are apparent.

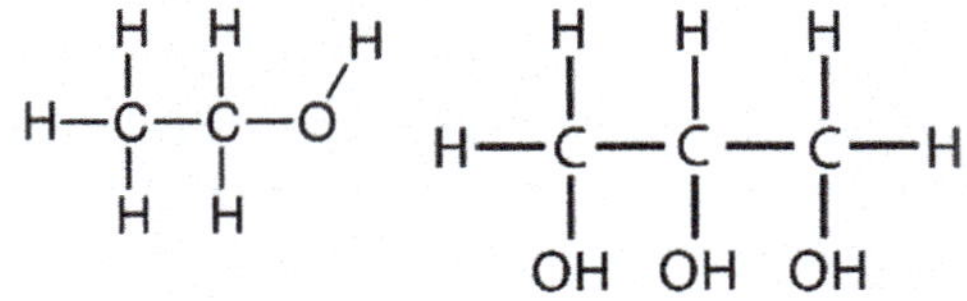

(a) (b)

Figure 7.8 Ethanol, a monohydric alcohol (a). Glycerol, a polyhydric alcohol (b).

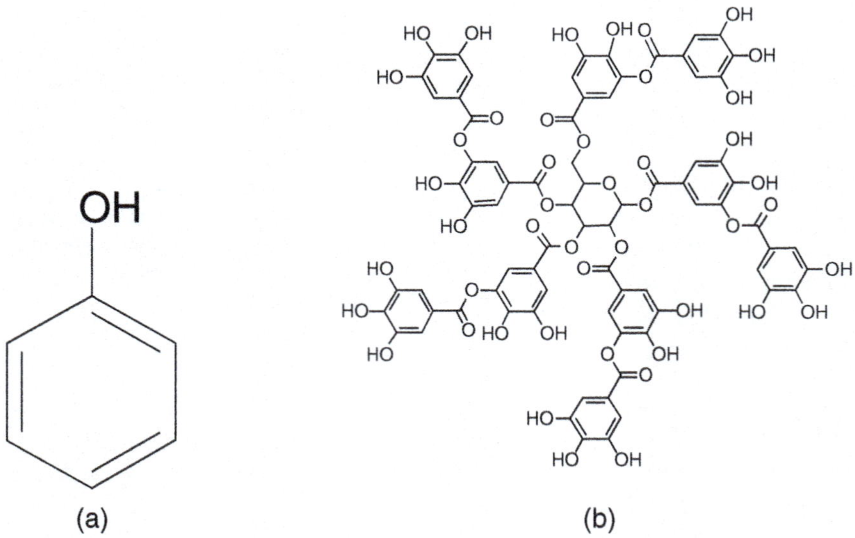

Figure 7.9 Phenolics. Phenol (a) and a generic polyphenol (b).

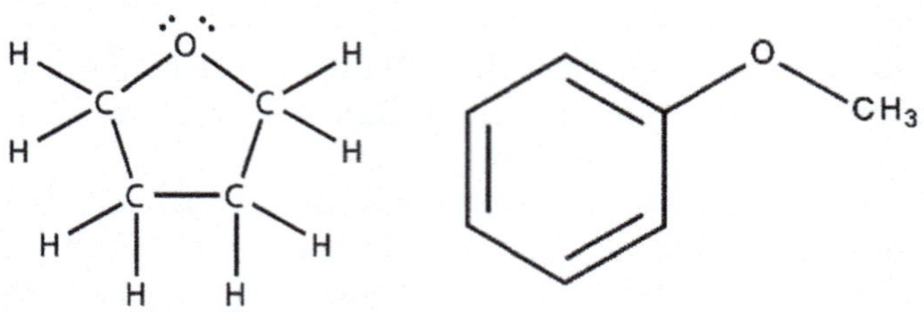

Figure 7.10 Tetrahydrofuran, a heterocyclic ether (left) and anisole an aryl ether (right).

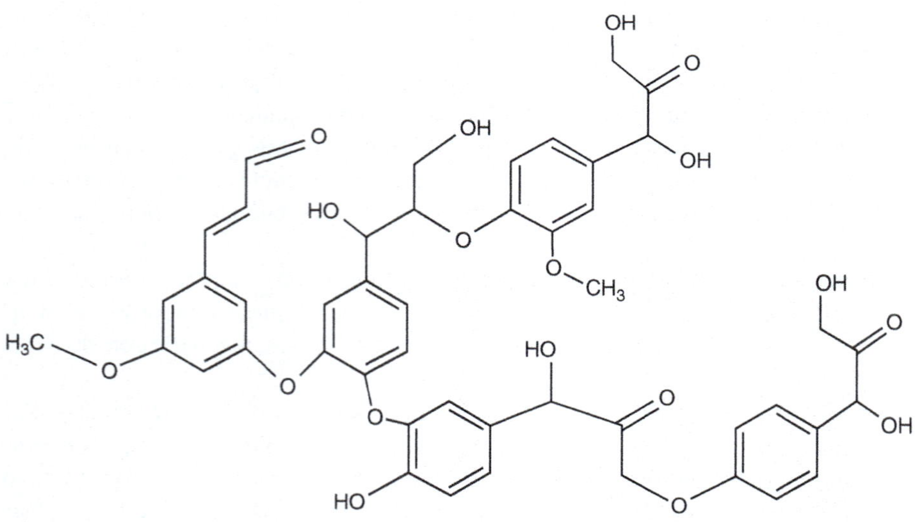

Figure 7.11 A generic lignin.

Aldehydes and ketones are closely related and each features a carbon double bonded to oxygen, a *carbonyl* center, leaving two bonds for alkyl, aryl, or H groups. The simplest aldehyde is methanal (or formaldehyde, because of its close relationship to formic acid). An important phenolic aldehyde used for flavoring is the naturally occurring vanillin (IUPAC name, 4-hydroxy-3-methoxybenzaldehyde). Formaldehyde and vanillin are contrasted in Figures 7.12a and 7.12b, respectively. A ketone, diethyl ketone is shown in Figure 7.12c. The two lone pair electrons on the carbonyl group give enough polarity to aldehydes and ketones that small molecules of these are water-soluble. All of them are soluble in organic solvents.

Aldehydes and ketones are central to the living state because monosaccharides are aldehydes or ketones with multiple hydroxyl groups. Aldehydes have at least one H– off the carbonyl carbon, whereas ketones have two carbon-based groups attached to the carbonyl carbon. Disaccharides are composed of two linked monosaccharides, while polysaccharides can contain thousands of saccharide subunits and fulfill a plethora of living state functions. Furthermore, saccharides play a fundamental role in the nucleic acids and the structure of DNA and RNA. They are also an integral part of the universal energy molecule adenosine triphosphate (ATP), (Secton 3.2).

Carboxylic acids occur widely throughout nature and are the basis of *amino acids*, the protein building blocks. The motif is R–COOH, where R is one of three different alkyl, aryl, or H groups. Figure 7.13a shows this generic structure, and Figure 7.13b shows the carboxylate anion formed when the acid deprotonates. Figure 7.14 shows several specific types of carboxylic acids, including the simplest, formic acid.

Esters are formed from acids, such as carboxylic acid, and frequently with an alcohol, where the H from the acid's OH group is replaced by an O– attached to an alkyl group. Phosphoric acid forms a *phosphodiester linkage* between nucleic acids in DNA and RNA, making it one of the most important chemical motifs for the living state.

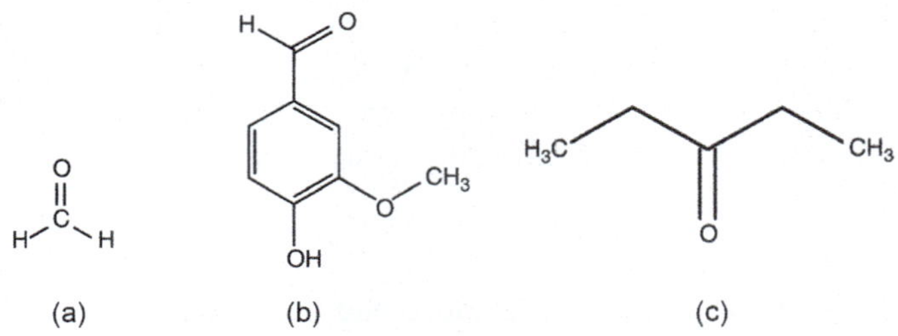

Figure 7.12 Two aldehydes. Formaldehyde (a), vanillin (b) and a ketone, diethyl ketone (c).

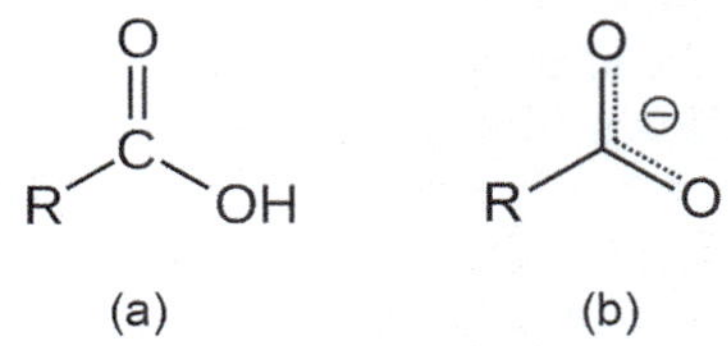

Figure 7.13 A generic carboxylic acid (a) and the carboxylate anion formed from its deprotonation (b).

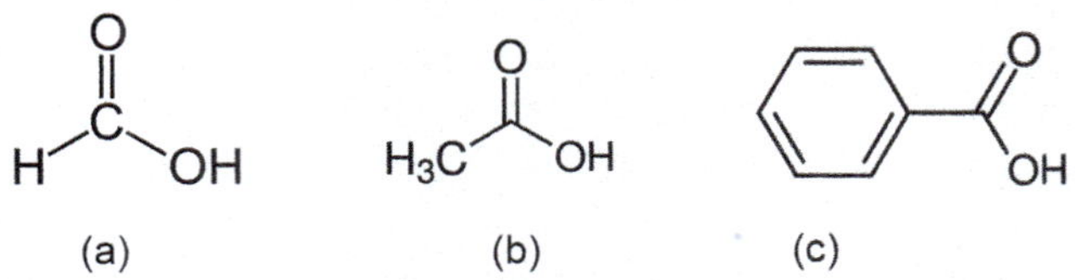

Figure 7.14 Several types of carboxylic acids: (a) formic acid, (b) acetic acid, and (c) benzoyl formic acid.

7.2.4 Organic Molecules Containing Nitrogen

This broad class includes *amides*, *amines*, *nitrile*, and *nitro compounds*. One of the most important amides for production of water-soluble polymers is acrylamide. It is shown with some of its many derivatives in Figures 7.15a,b.

Amines are ubiquitous in nature and the basic chemical and polymer industries. They are essentially an ammonia molecule where the electron lone pair is left intact and where one, two, or three of the hydrogens have been replaced by functional groups; these correspond to primary, secondary, and tertiary amines, respectively. The amino acid content of proteins underlies the living state and specification of an amino acid sequence to define specific proteins is the function of the genetic code. While there are thousands of known amino acids, only 20 of them are encoded genetically.

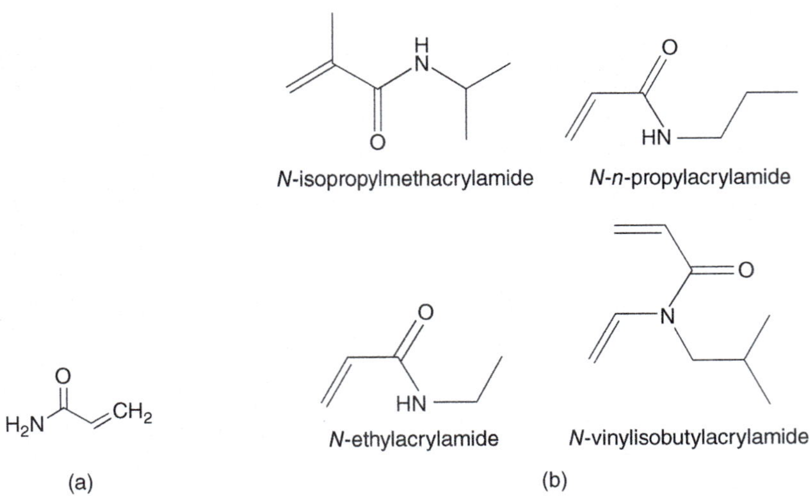

Figure 7.15 Acrylamide (a) and several of its derivatives (b).

When the lone pair in a tertiary amine is substituted by a fourth group, the nitrogen acquires a permanent positive charge, independent of solution pH, and such compounds are referred to as *quaternary ammonium salts*, or *quats*. These underlie the production of many highly water-soluble cationic polymers.

Acrylonitrile is especially important in the polymer industry and consists of the vinyl group C_2H_3 linked to a nitrile group —C ≡ N.

7.2.5 Hybrid Orbitals and Molecular Geometry

Any linear combination of individual solutions to a linear differential equation is also a solution to the equation. This is the *principle of linear superposition*. In quantum mechanics, this means that quantized electronic orbitals that solve the Schrödinger wave equation can be combined linearly to produce viable *hybrid orbitals*. While the notion of hybrid orbitals extends well beyond organic chemistry, the carbon atom is an excellent illustration of the principle. The basic notion is that the constituents of a molecule will arrange themselves geometrically, minimizing the repulsive energy among themselves when bonded together.

You likely recall that the electronic configuration of carbon is $1s^22s^22p^2$, where 2s is the electronic orbital corresponding to the second energy level of the atom with zero angular momentum and 2p is the orbital corresponding to the second energy level with angular momentum number $l = 1$ (total angular momentum is $L = \hbar\sqrt{l(l+1)}$, where $l = 0,1,2,3...$). The s orbitals are all spherically symmetric, whereas the p orbitals have lobe shapes with directionality.

The second energy level of carbon has four electrons. The principle of linear superposition allows the orbitals to be combined in different ways. The most important combinations lead to the sp^3, sp^2, and sp orbitals.

Four equal sp^3 orbitals are formed by combining three 2p orbitals with a single 1s orbital, which produces a 3D tetrahedral structure. A tetrahedron is composed of four equilateral triangular faces with four vertices so that the center point of the tetrahedron is equidistant from each vertex. The carbon atom is treated as this center point and the vertices are the points where different atomic or molecular groups can be bonded. The bond angle is the angle formed by two vertices and the center point, which, by simple geometry, is about 109.5°. The actual bond angle in a given tetrahedral molecule can vary, depending on the type and size of the constituent at a vertex; the angle can deviate from 109.5° to minimize repulsive energy among constituents. The simplest example of sp^3 orbitals is methane. The next example of sp^3 orbitals, ethane, involves two tetrahedral carbons bonded together, as shown in Figure 7.16a. In some molecules, one of the vertices has no atomic or molecular group and thus has a lone pair of electrons in that vertex. One example is ammonia NH_3, where H atoms occupy three vertices and a lone pair the fourth.

Three sp^2 orbital are formed by combining two 2p orbitals and one 1s orbital. This produces a trigonal planar structure with the carbon at the center and three vertices in the plane, equidistant from the center, to each of which an atom or molecular group can bond. The bond angle in this case is 120°. Ethene (or ethylene) is a good example of this and is shown in Figure 7.16b. Again, the bond angles can deviate from 120° in order to minimize repulsive energy of the specific atoms or groups at the vertices.

Two sp orbitals can be formed by combining equal parts of 2p and 2s orbitals. This produces a linear structure, with two bonding vertices on either side of the carbon at the center, with a bond angle of 180°. Figure 7.16c shows ethyne (or acetylene) in this scheme.

A critically important aspect of bonding as concerns polymers is the ability of the molecule to rotate about interatomic and intermolecular bonds. Single bonds are rotationally symmetric around the axis joining one carbon to another and hence can rotate. This leads to an enormous number of possible configurations of a polymer molecule with rotational freedom (Chapter 8.1). Bonds that are rotationally symmetric are often termed "*sigma bonds*" and can be constructed

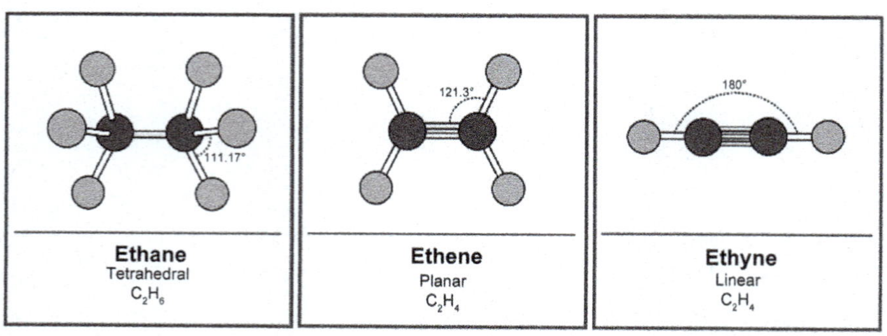

Figure 7.16 Illustrations of (a) 3D sp^3 hybrid orbitals in ethane, (b) 2D planar sp^2 orbitals in ethene, and (c) linear 1D sp orbitals in ethyne. (Courtesy of Marissa Calko).

Table 6.4 Techniques for determining transport properties.

Intrinsic viscosity $[\eta]$	Capillary viscometer Couette viscometer Falling ball Vibrating viscometer
Translational diffusion coefficient, D_0	Dynamic light scattering Boundary spreading Fluorescence recovery after photobleaching (FRAP) NMR
Electrophoretic mobility, μ_e Zeta potential, ζ	Conductivity Free-flow electrophoresis Electrophoretic light scattering
Sedimentation coefficient, s	Ultracentrifuge
Rotational diffusion coefficient, Θ	Flow birefringence Luminescence depolarization relaxation NMR Depolarized dynamic light scattering

We did not cover the *sedimentation coefficient* s, but it follows a force–flux relationship for motion of particles in centrifugal fields. The modern *ultracentrifuge* can reach centrifugal fields as high as 10^6g, where g = 9.8m/s^2. The rotational kinetic energy associated with these machines is enormous and so the technology behind them is quite sophisticated. The acceleration at a distance r from the rotational axis for a centrifuge with an angular frequency ω is $\omega^2 r$, and so is a function of the particles' position. Terminal speed v_t is reached when the centrifugal force is balanced by the sum of the buoyant force of the particle. The sedimentation coefficient s is defined as the terminal velocity per unit of acceleration for a particle of mass m,

$$s = \frac{v_t}{\omega^2 r} = \frac{m\left(1 - v_0\rho\right)}{f} \qquad 6.138$$

The units of this are seconds. The buoyancy term (Archimedes law) is $v_0\rho$, where v_0 is the specific volume of the sedimenting particle, and ρ is the fluid density. Ultracentrifugation is used to separate cell components from each other, including the separation of mixtures of proteins. Its impact on protein science has been enormous.

6.16 Summary

The overriding theme of this chapter has been the properties of viscous liquids; how they flow, dissipate energy, and affect the motion of objects placed in them. The key concept of viscosity led us both to the notion of dissipative frictional effects in liquid flow and to the stress tensor, which allowed derivation of the Navier–Stokes equation, which describes the flow of incompressible, viscous liquids, and is one of the foundations of the field of fluid mechanics. This same equation furnished the basis for computing flow profiles in capillaries, frictional coefficients of spheres and other shapes, and the intrinsic viscosity of a body (which, you'll remember is not actually a viscosity, but rather the reciprocal of an object's density). It also helps, via the Reynolds number (*Re*) concept, to demarcate flow regions which are laminar and those which are chaotic, or turbulent. We were able to use principles developed on the basis of this equation to help explain the basics of blood flow and fluid flow in plants, and of microorganism motility at $Re \ll 1$. We also saw that while microorganisms have very low *Re*, macroscopic animals produce fluid mechanical motion in the range of $Re \sim 10^3$–10^5, which can be far into the turbulent regime.

Consideration of forces applied to bodies in viscous fluids brought us to the generalized notion of force–flux relationships, of which the flux created by particles in a concentration gradient brought us to the foundational notion of diffusion. In diffusion, we met recurring themes from previous chapters, including entropy and randomness, and chemical potential. The reciprocal relationship of the diffusion coefficient and the frictional coefficient via $k_B T$, given by Einstein's Equation 6.82 links diffusion and friction via thermal energy, $k_B T$. We found how diffusional flux could couple to the flux of mobile from ss, sp, and pp orbitals in the same direction (e.g. p_z–p_z), and dd orbitals in the same direction. Bonds that do not have rotational symmetry are rigid, cannot rotate, and are hence locked. Double and triple bonds between carbon are not rotationally symmetric and have p orbitals lying above and below the nodal plane formed by the overlapping p orbits. These bonds are often referred to as "pi-bonds," because of the p-type orbital asymmetry.

7.3 Synthetic Polymers

7.3.1 Glass Transition Temperature, Chirality, and Tacticity

In considering synthetic polymer molecules, it is important to introduce two concepts that affect polymers in their solid state. The first is the *glass transition temperature*, T_g. This is the temperature at which sufficient disassociation of weak intermolecular forces occurs among polymer chains that independent movement of the chains takes place and the solid polymeric material becomes rubbery. The transition can be found in measurable ways; heat capacity becomes significantly higher at and above T_g, and density decreases. There is an enormous range of T_g among polymers, from those well below 0°C to hundreds of °C.

The second concept is that of *tacticity*. The basis of tacticity involves the notion of *chirality*, or "handedness." For example, your right and left hands are mirror images of each other but cannot be superposed on each other. In a pair of gloves, there is a right-handed one and a left-handed one, and, as you know, they are not interchangeable. Chiral molecules, similarly, are mirror images of each other but cannot be superposed upon each other. They rotate the plane of polarization of linearly incident polarized light and are termed *optically active*. Chiral molecules which rotate linearly polarized light to the left are termed *levorotatory* (L) and those that rotate linearly polarized light to the right are *dextrorotatory* (D).

The alkane molecule in Figure 7.17 is an example of a chiral molecule. Notice that all four groups off the carbon are different. The pair of nonsuperposable molecules are *stereoisomers*, termed *enantiomers* of each other. For polymers, the notion of a *diastereomer* is important. These latter are stereoisomers which are *not* mirror images of each other. That is, they have all the same chemical groups but are neither superposable upon each other, nor are they mirror images of each other.

The effect of diastereomers becomes clearer if we consider stretching out a polymer chain containing diastereomers. Figure 7.18 shows two situations. In the first, all the R groups are on the same side of the stretched chain, in what is termed an *isotactic polymer*. In the second case, the R groups occur on alternate sides of the chain, in what is termed a *syndiotactic polymer*. If the R groups occur randomly on either side of the chain then this is an *atactic polymer*. Figure 7.19 shows a planar view of isotactic, syndiotactic, and atactic polystyrene. One of the most important effects of tacticity is on the crystalline structure of the polymer in its solid state. When there is a regular tacticity, either isotacticity or syndiotacticity, the chains can

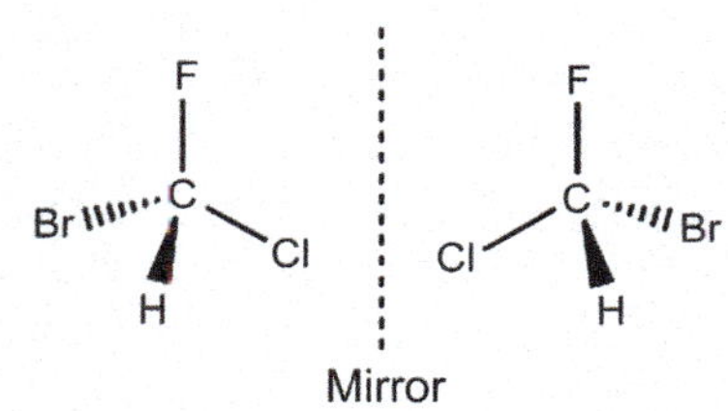

Figure 7.17 Example of enantiomers, two chiral molecules which are mirror images of each other.

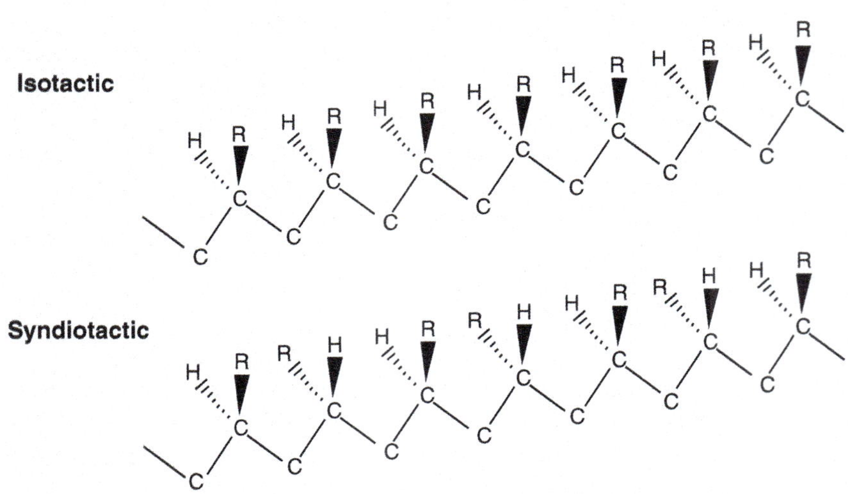

Figure 7.18 Generic isotactic and syndiotactic polymer chains.

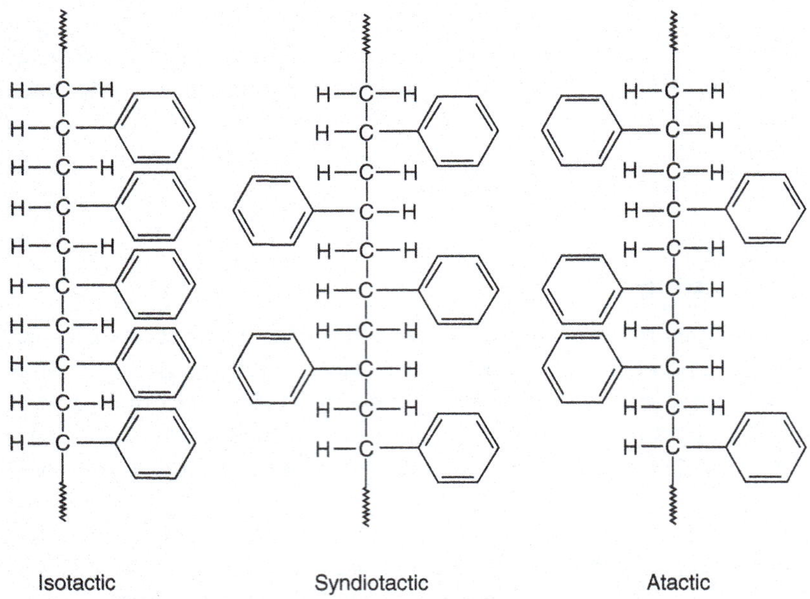

Figure 7.19 Isotactic, syndiotactic, and atactic polystyrene.

fit well into each other and form a (partially) crystalline solid. This can lead to a much stronger, more useful polymeric material. For example, atactic polypropylene is sticky and of little use. Isotactic polypropylene, in contrast, has a high degree of crystallinity and can be used to make strong fibers, often produced for rope, carpeting, and other materials. In the 1950s, Karl Ziegler (Germany, 1898–1973) and Giulio Natta (Italy, 1903–1979) developed what are now called Ziegler–Natta catalysts, which produce isotactic and syndiotactic polyolefins up to very high molar mass.

Polystyrene is normally manufactured in atactic form and so does not crystallize, but rather produces a hard, amorphous solid. This is strong enough and with a T_g over 100°C it is useful for many applications. Tacticity also affects T_g, with T_g normally higher for isotactic or syndiotactic forms of a macromolecule compared to its atactic form. Syndiotactic polystyrene, which is manufactured in relatively small quantities, does not melt until 270°C.

There are several ways of classifying synthetic polymers. One popular way is to consider *thermoplastics*, *thermosets*, and *elastomers*. In this classification, thermoplastics have secondary electrostatic forces between polymer molecules and can be reversibly melted and cast over many cycles. Thermosets, on the other hand, form covalent bonds that cross-link the polymer molecules upon heating – this is known as "curing" – and hence will never melt back to a fluid state. Elastomers have enough cross-links for the material to keep its shape and give it elasticity, but not so many cross-links that the material is rigid.

This classification overlooks a great many different types of synthetic polymers, so we provide a brief survey that looks at chemical classes of polymers, in order of their global production.

It is important to point out that the monomers used in making polymers are usually highly toxic, because of their high reactivity. When they are incorporated into polymer chains they use up their reactivity and the polymer is generally non-toxic and allows polymeric materials to be used by billions of people. Regulatory agencies set strict limits on the amount of residual monomer left in a commercial lot of polymer. For example, the current US Environmental Protection Agency (EPA) sets a limit of 300 parts per million (ppm) of acrylamide monomer for industrial uses, such as paper-making and metallurgy, while the US Food and Drug Administration sets a limit of 5 ppm for acrylamide products used in personal care and less than 0.1 ppm for medical products.

7.3.2 Specific Types of Synthetic Polymers

Polyolefins comprise the single largest sector of the polymer manufacturing industry, accounting for well over half of all polymer tonnage produced. The global polyolefin production in 2016 was estimated at 150 million tons, or 150 billion kilos, more than 20 kilos per person on the planet. *Olefins* (from French *oléfiant* or "oil making") are produced from linear and cyclic alkenes, that is, hydrocarbons with at least one double (unsaturated) bond. IUPAC designates these polymers *poly(alkenes)*. It is the presence of the double bond that makes polymerization possible, as will be discussed shortly. The most common olefins used to produce polyolefins include ethylene, propylene, butene, and butadiene, introduced above, as well as many of their derivatives. Polyolefins have multiple uses, including films and foams, clothing and sporting goods, wire and cable insulation, elastomeric roofing and weather stripping, road construction, and much more.

The most important polyolefins are *polyethylene* and *polypropylene*, as well as copolymers of these. The properties of polyolefins are directly related to microstructural distributions, including the molar mass, copolymer chemical composition, isotacticity or syndiotacticity (for polypropylene), comonomer sequence length, and degree and length of branching.

Polyesters compete with polyolefins in production scale, with an estimated 77 million tons produced in 2016. The most prominent is polyethylene terephthalate (PET). Terephthalate is more properly termed 1,4-benzenedicarboxylate, because it consists of a benzene ring with carboxylate groups bonded across from each other on the benzene ring. PET is used for textile fibers, alone and also mixed with natural fibers. Heterogeneous polyesters can form under mild conditions, and it has been conjectured that natural polyesters might have had a role in prebiotic evolution because they can compartmentalize other molecules, as an alternative to membranes, allowing vital biomolecules to form which would later move universally (excluding viruses) to membrane systems.[1]

Poly(vinyl chloride), or PVC, is by far the most important *chloropolymer*, with a production estimated at 42 million tons in 2015. Half of this is used for piping in construction, plumbing, wastewater treatment, reagent delivery in chemical plants, and irrigation. Other uses include siding for buildings, sometimes called "vinyl siding," insulation for electrical cables, water-resistant textiles, and flooring. About one-third of plastic-based medical products are made of PVC. PVC is made from C_2H_3Cl a vinyl halide. PVC can be further chlorinated to increase its chlorine content to over 60%, which gives it higher heat resistance and flexibility than standard PVC.

Polystyrene was first produced from distilled oils from a sweetgum tree by Eduard Simon (Germany, 1789–1856) in 1839. It occurs naturally in some plants and animals. The structure of the styrene monomer was given in Figure 7.6a and its various tacticities in Figure 7.19. It has since become one of the major industrially produced polymers. An estimated 15 million tons were produced globally in 2016, the vast majority of which is atactic. Its uses as a hard, durable, amorphous solid range from disposable items to permanent objects in the automotive, computer, and aerospace industries. It can also be infused with up to 95% air to make lightweight foam containers, insulation, and packing material ("Styrofoam"). It is optically transparent and can withstand sterilization temperatures, making it useful in medical and biological applications.

High-impact polystyrene is a copolymer of styrene grafted off of polybutadiene polymer chains, where there is a far greater amount of styrene than butadiene. The rubbery polybutadiene sites make the material softer and deformable and also prevent propagation of cracks and other defects. In styrene-butadiene rubber, there is far more butadiene than styrene, making it an actual rubber rather than a hard material. Copolymerizing acrylonitrile and styrene with polybutadiene leads to a strong, resilient, lightweight material used in medical devices, automotive bumpers, luggage, and keyboard caps, among many other applications.

Synthetic rubber has a long and fascinating history. Starting with bicycles in the late nineteenth century, and then automobiles and military vehicles starting in the twentieth century, the demand for rubber skyrocketed. Natural rubber is an isoprene polymer produced by certain trees. It comes mostly from Indonesia and southeast Asia, but with substantial amounts also from Africa and South America. Natural rubber was used almost exclusively up until nearly the Second World War, although many attempts at producing synthetic rubber were made in the early twentieth century, mostly with limited success. By the 1930s, neoprene was developed into synthetic rubber but it was still costly. By mid-1942, the Axis Powers[2] controlled most of the world's natural rubber-producing countries, leading to intensive research in the United States and elsewhere for a means to produce large quantities of synthetic rubber. An elastomer based on styrene and butadiene was developed and became the standard material for military and other vehicles. Huge amounts were produced in the United States, Europe, and elsewhere. Allied bombing raids targeted synthetic rubber plants in Germany, Italy, and Nazi-occupied Poland. Since World War II, there have been many innovations in synthetic rubber, and around 15 million tons were produced in 2018.

---

1 Jia, T.Z., Chandru, K., Hongo, Y., et al. (2019). Membraneless polyester microdroplets as primordial compartments at the origins of life. *Proceedings of the National Academy of Sciences of the United States of America 116*.

2 Germany, Japan, and Italy.

Butadiene figures prominently in most synthetic rubbers. Besides the styrene-butadiene rubber mentioned, there are also acrylonitrile-butadiene rubbers with high tensile strength and resistance to oils and other organic fluids, making them widely used for oil seals, packaging, fuel diaphragms, etc. In addition to butadienes other dienes, such as dicyclopentadiene and vinyl norbornene, are used in producing ethylene propylene diene monomer rubber (EPDM), an elastomer used widely for seals, foams, insulation, and roofing.

Polyurethanes are a widely used class of polymer with global production of around 15 million tons in 2016. There is a wide variety of polymers under this rubric. They are copolymers of isocyanates and polyols. The isocyanates have the form R–N=C=O and contain two isocyanate groups (diisocyanate) or more. The polyols are organic molecules containing two or more hydroxyl groups. The breadth of polyurethane applications is astonishing, involved in such products as hard, abrasion-resistant seals varnishes and paints, foams for upholstery, adhesives, automotive and electronic components, wheels, and aquatic vessels.

Polyamides are another major class of polymer with an estimated global production of 8 million tons in 2016. *Nylon 66* (IUPAC name poly(imino(1,6-dioxohexamethylene)iminohexamethylene)) is the most prominent polyamide and was invented by Wallace Carothers (United States, 1896–1937) in the 1930s at the DuPont de Nemours Corporation in Delaware. DuPont had created a program for pure "blue sky" research and coaxed Carothers to leave his faculty position at Harvard to lead the effort.

Nylon 66 is formed by a condensation reaction of adipic acid (a diacid) and hexanediamine (a diamine), shown in Figure 7.20. Many other combinations of diacids and diamines are also produced industrially. A competitor to nylon 66 is *nylon 6*, which was developed by Paul Schlack (Germany, 1897–1987) at IG Farben in order to avoid violating DuPont's nylon 66 patent. Interestingly, IG Farben (Interessengemeinschaft Farbenindustrie AG) was the largest industry in Europe prior to World War II and was formed in 1925 by the merger of six chemical companies, including Bayer, BASF (Badische Anilin und Soda Fabrik), Hoechst, Agfa, and two others. After World War II, the allies broke it back into its constituent companies.

Schlack's discovery of nylon 6 was more than a simple workaround of Carothers's reaction. Schlack expanded the use of an important type of reaction, *ring-opening polymerization* (ROP), which was first used in 1906 to synthesize polypeptides and then to produce synthetic polysaccharides. The monomer for nylon 6 is lactam (a cyclic amide) and the structure of nylon 6 is similar to, but very distinct from that of nylon 66, and has a much shorter repeat unit.

Acrylate polymers, also known as *polyacrylates* or simply *acrylics*, are an important part of the polymer industry, with estimates of around 7 million tons produced in 2016. The monomers are based on the structure of acrylic acid and have a vinyl group with either a carboxylic acid or nitrile group attached. Some of the major monomers involved in these polymers include methyl, ethyl and butyl acrylates and methacrylates, 2-chloroethyl vinyl ether, acrylonitrile, cyanoacrylate, and 2-ethylhexyl acrylate. Polyacrylates are widely used in adhesives, paints, coatings, textiles, optical components, and high-temperature elastomers. Poly(methyl methacrylate) is the basis for "acrylic glass," often known by its commercial name "Plexiglass."

Acrylamide monomers and their derivatives lead to very important *water-soluble polymers*. While electrically uncharged, polyacrylamide is sufficiently polar to be water soluble. Adding monomers with tertiary ammonium salts leads to cationic polyelectrolytes, whereas the use of acrylic acid and other anion-containing monomers leads to anionic polyelectrolytes. Polyacrylamides are used in water purification, oil recovery, fracking, personal care products, metallurgy, papermaking, and in electrophoresis gels. A derivative of acrylamide, *N*-isopropylacrylamide (NIPAM), forms polymers and copolymers with unusual stimuli-responsive properties, discussed in Section 7.3.4.

Polycarbonates had an estimated global production of 5 million tons in 2015. They contain a carbonate group, –O–(C=O)–O– and are formed in a condensation reaction using bisphenol A and phosgene gas, $COCl_2$. Polycarbonates are hard materials with $T_g \sim 147°C$. They are used extensively for electronic components because they have good electrical insulation, heat-resistant, and flame-retardant properties. They are used widely for construction materials, including skylights and sound dampening. Polycarbonates are also used for making very smooth surface coatings in automotive and aerospace applications, as well as for optical lenses.

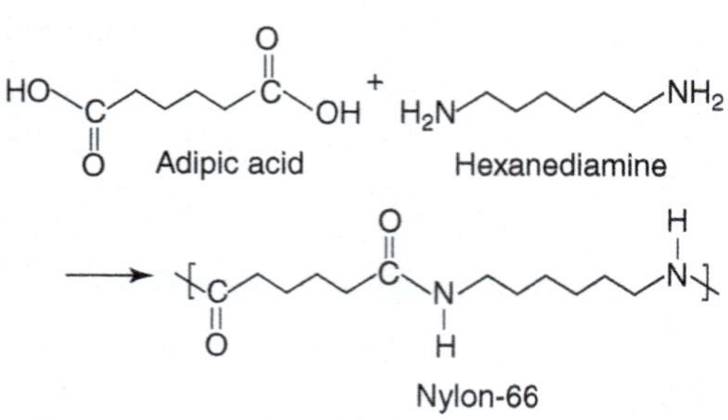

Figure 7.20 Synthesis of nylon 66.

Fluoropolymers are specialty polymers with unique properties. The global production of fluoropolymers was around 0.35 million tons, chiefly PTFE (Teflon), poly(vinylidene fluoride), and fluoroelastomers. Notice that this is over two orders of magnitude less production than polyolefins and PVC, and over an order of magnitude less than the other polymers listed above. Fluoropolymers have very low van der Waals forces between chains leading to very low friction and nonstick properties. The strong C–F bonds in fluoropolymers make them chemically inert and give them high resistance to heat, corrosive chemicals, moisture, and shock. They are used extensively in coatings, electronic materials (e.g. dielectrics for capacitors), including miniaturized circuit boards and wire and cable insulation, filtration, and industrial machinery.

While the above list includes the majority of high-tonnage polymers produced globally, there are many other types not mentioned, such as polyvinyl alcohols and polysulfones.

7.3.3 Natural Products from Biological Macromolecules

Part of the very large natural products industry extracts, and sometimes modifies, macromolecules of value from plant and animal sources. A short list includes: Gelatin, a multifarious collagen-based polyampholyte protein derived from bone, hide, and hair. Xanthan gum, an anionic polysaccharide, for viscosifying foodstuffs, produced by bacteria. Alginates, polysaccharides also for texturizing foodstuffs, principally from seaweed. Gum Arabic, a complex mixture of polysaccharides and proteins, used to emulsify flavors and fragrances and to add viscosity in aqueous food preparations. Hyaluronic acid, an anionic polysaccharide, used to maintain osmotic pressure in eye surgery, to restore elasticity to skin, and as an injectable form of extracellular matrix to aid damaged cartilage. It is obtained from both microbial and animal sources. Pectin, extracted from fruits, is a complex, branched polysaccharide that is used as a gelling agent, especially for jellies, jams, and fruit preserves.

Cellulose is the most abundant biopolymer on the planet. The use of wood, primarily made of cellulose fibers, has been, and continues to be, a mainstay of human industry for fuel, construction, and papermaking. Several derivatives are important, including *rayon*, which is a collective word for many types of cellulose derivatives, including the cellulose ester, and *cellulose acetate*, which is used in textiles and films.

7.3.4 Stimuli-responsive Polymers and Highly Architected Polymers

As polymers become more complex and sophisticated in architecture and composition, they gain the ability to perform more "intelligent" functions than traditional polymers. The frontier of advanced polymeric materials in the twenty-first century will be dominated by these increasingly sophisticated polymers. The upcoming polymers can micellize, aggregate, and respond to stimuli, such as temperature, light, solvent polarity, different solvents and solvent mixtures, the presence of specific agents, metal ions, multivalent ions, proteins, antibodies, receptors, etc. Applications include sensing, stimuli-responsive materials, encapsulation and release of agents (e.g. drugs, cosmetics, etc.), micropatterning, self-healing polymers, bioconjugated polymers for medical applications, etc. There is considerable interest in "fine tuning" polymers to have well-behaved phase transition characteristics and interaction properties.

For example, polymers in solution can go through a number of sharp or gradual phase transitions, e.g. LCST (lower critical solution temperature), micellization, aggregation, helix–coil, and other intra- and intermolecular transitions. Such transitions are of fundamental and applied interest. They arise from the thermodynamics of complex, interacting, cooperative systems. Whether sharp or gradual, these transitions depend on many factors, such as pH, ionic strength, solvent type and polarity, solvent mixture types, presence of *chaotropic agents*,[3] temperature, irradiation by electromagnetic waves, including light, and addition of interacting agents (e.g. small molecules, dyes, etc.), as well as the molar mass and copolymeric composition and microstructure of the polymers themselves. Other examples concern the many types of associations that can take place between polymers and other polymers, micelles, emulsions, vesicles, liposomes, proteins, polypeptides, etc. These often involve the formation of *supramolecular* (noncovalent) structures promoted by electrostatic, hydrophobic, depletion, and other forces.

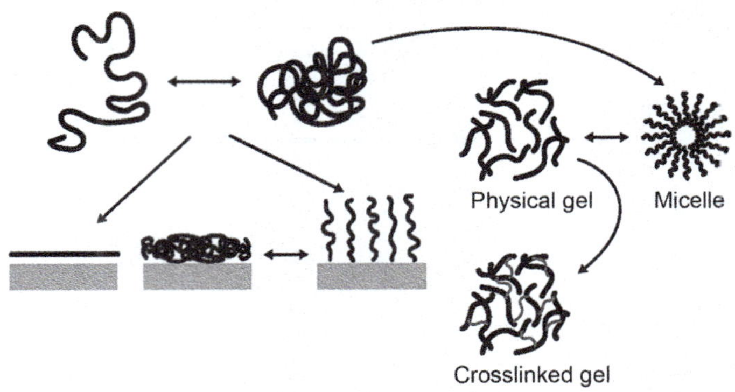

Figure 7.21 Artist's depiction of stimuli-responsive polymers. (Courtesy of Marissa Calko).

Figure 7.21 shows an artist's rendition of different morphological states of a stimuli-responsive polymer:

3 Chaotropic molecules interfere with hydrogen bonds, the hydrophobic effect, and other effects that help stabilize organized structures in macromolecules, such as the globular structure of proteins, and the secondary structure of proteins, RNA, and DNA. Guanidine hydrochloride and urea are common examples of chaotropic molecules.

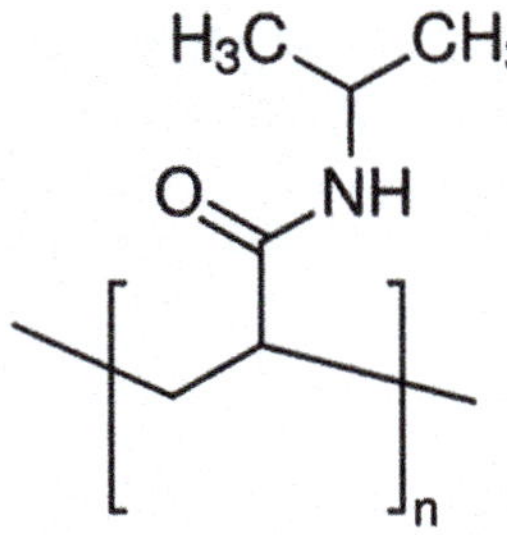

Figure 7.22 The repeat unit of p(NIPAM).

Reversible coil–globule transitions, ability to coat surfaces smoothly and change their properties, micellization and reversible gel formation, and setting the gel by cross-linking with a cross-linking agent (e.g. divinyl benzene or sulfur).

A workhorse polymer for exploring temperature-actuated stimuli-responsive behavior is poly(*n*-isopropyl acrylamide), shown in Figure 7.22, and abbreviated as pNIPAM. Compare this with acrylamide in Figure 7.15; one H on the acrylamide H_2N group is replaced by an isopropyl moiety, so that HNC_3H_6 results, which is much more hydrophobic than H_2N. At low temperature, typically around 25°C pNIPAM is water soluble, but "reluctantly" so, because water molecules must organize around the hydrophobic isopropyl moiety, while hydrogen bonding to the NH group. The organization of the water lowers the water entropy, but the enthalpy loss from hydrogen bonding, ΔH, dominates, and minimizes the free energy, $\Delta G = \Delta H - T\Delta S$. When temperature is raised, once the LCST is reached the hydrogen bonds are broken cooperatively, water is released and its increase in entropy ΔS, dominates the minimization of the free energy. Without water for solubilizing pNIPAM molecules then collapse under their own hydrophobicity, undergoing a coil-to-globule transition. If the concentration of NIPAM is high enough, the hydrophobic globules that form will themselves aggregate above the LCST, turning the solution cloudy. The transition is fully reversible upon lowering T back through the LCST.

NIPAM can be copolymerized with other monomers to produce materials with widely different properties. An example is shown in Figure 7.23, which shows the LCST behavior for the pure pNIPAM homopolymer, which is very sharp and well defined. When copolymerized with acrylamide (Am), however, which is very hydrophilic, the LCST moves to a higher temperature, because the copolymer is more water soluble than the homopolymer. There is composition drift during the copolymerization, meaning that the proportion of NIPAM/Am changes during the reaction, producing an end product with chains of a distribution of compositions. Each composition has its own LCST. Hence, the mixture of many different compositions smears the LCST out, and it is no longer sharp, as it is for the homopolymer. A more extreme version of this effect is seen in the curve in Figure 7.23, which shows how even further smeared out the LCST becomes when the composition distribution is deliberately broadened by continuously feeding NIPAM during the copolymerization into the reactor, which initially contained only Am.

pNIPAM and its copolymers find a range of applications. In the biomedical field, pNIPAM can be grown from or grafted onto surfaces, such as glass and plastics, creating a *thermoresponsive surface* (e.g. Figure 7.21). Cells can be cultured on these above the LCST where the polymers are collapsed and dense, and the cells can then be removed by cooling below the LCST, where the surface swells under the globule to coil transition. Adhesion and release of protein molecules using thermoresponsive surfaces have been similarly exploited.

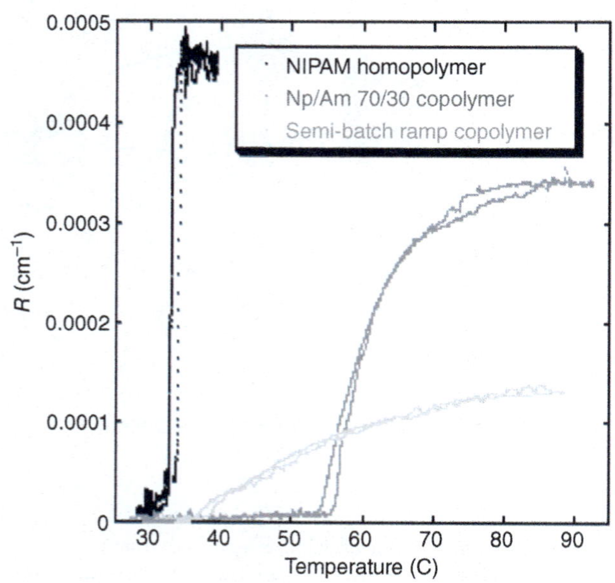

Figure 7.23 The Rayleigh scattering ratio, *R* (Chapter 9.4.3) vs. temperature. The sharp LCST of the p(NIPAM) homopolymer is seen and how the transition becomes more stretched out for copolymers with a wide distribution of NIPAM and acrylamide: (i) by batch copolymerization with Am and (ii) by a continuous semibatch ramp feed of NIPAM into the reactor containing initially only Am. (Reproduced from McFaul, C. A. et al. (2014), with permission from Elsevier).

7.3.5 Biomimetics

One school of thought in the design of materials and structures is *biomimetics* or *biomimicry*. These approaches attempt to solve technological human challenges by taking cues and motifs from nature, with the rationale that evolution has frequently developed the best designs and materials for life on a complex planet. Examples abound. Spider silk is one of the highest strength-to-diameter polymeric fibers known. The protein-based adhesives mollusks use to cling to rocks have inspired "super glues." The automatic healing that occurs in many organisms has set the stage for polymeric materials that can self-heal when cut or fractured. The notion of artificial photosynthesis is well over a century old, but the search for nonfossil based fuels and means for removing CO_2 from the atmosphere has accelerated in the early twenty-first century. Solar-based fuels aim to split water for the capture of H_2 as fuel, whereas methods are under development to capture atmospheric CO_2 and turn it into desirable hydrocarbons, such as methanol, as well as monomeric feedstocks for polymer synthesis.

7.4 Polymerization Reactions

Polymerization reactions are conveniently divided into two major categories:

(I) Chain-growth reactions. Monomers add sequentially to the end of growing polymer chains. Two major classes of chain growth polymerization are (i) *free radical polymerization* (FRP) and (ii) *living polymerization* (LP).

FRP follows a distinct series of steps in which a monomer is (a) *initiated* by obtaining a free radical, usually through the transfer of the free radical from an initiating agent, (b) in the *propagation step*, the free radical of the initiated monomer captures a free monomer and covalently bonds it in a growing chain, where the captured monomer then receives the free radical and captures the next monomer, and so on, to build a polymer chain, and (c) *termination* of the process occurs when the radical ends of propagating chains meet each other and the radicals deactivate each other and terminate the propagation process. The result of the termination step can be fusion of the two colliding chains into a single chain, known as *recombination*, or, the chains can remain separate and terminated in a process termed *disproportionation*. It is important to emphasize that in FRP the process of initiation, propagation, and termination for a given polymer chain is very rapid, typically taking on the order of milliseconds, whereas the entire amount of time to convert all available monomers into polymer chains in a reactor is much longer, typically hours. We shall see shortly that the rapidly produced polymer chains generally get shorter and shorter as the FRP conversion process proceeds.

In *ideal living polymerization*, there is never a termination step so the chains will continue to propagate as long as monomer is available. In the ideal living case, after all monomer is converted to polymer more monomer can be added long after conversion was complete, and polymerization will recommence; hence the name "living polymerization." Anionic living polymerization was first demonstrated in 1956 using styrene monomer in THF, with a sodium napthalenide initiator. Since then, other types of living polymerizations have been developed and include cationic, ring-opening metathesis, and controlled free radical mechanisms. A further feature of LP is that the polymer population has a much narrower distribution of chain lengths than the corresponding FRP.

In chain-growth reactions, there is usually a *bimodal* population. That is, there is a decreasing population of monomers, and a growing population of relatively long-chain polymers. This is shown schematically in Figure 7.24a for early-, mid-reaction, and final states of a free radical polymerization reaction. $dC_p/d(\ln M)$ is the mass concentration (g/cm^3) of polymer per natural logarithmic interval of polymer molar mass M. The polymer concentration grows in time, its maximum moves toward lower values of M, and its width usually broadens as the reaction proceeds. Meanwhile, the concentration of monomer decreases from its initial value to zero by the end of the reaction.

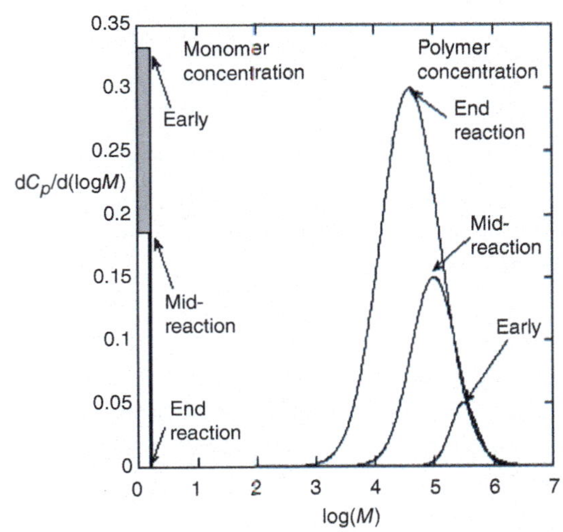

Figure 7.24a Schematic representation of the build-up of polymer concentration during a free radical polymerization.

A schematic representation of the evolution of the molar mass distribution of a living chain growth reaction is given in Figure 7.24b. The progression of the polymer concentration distribution is quite

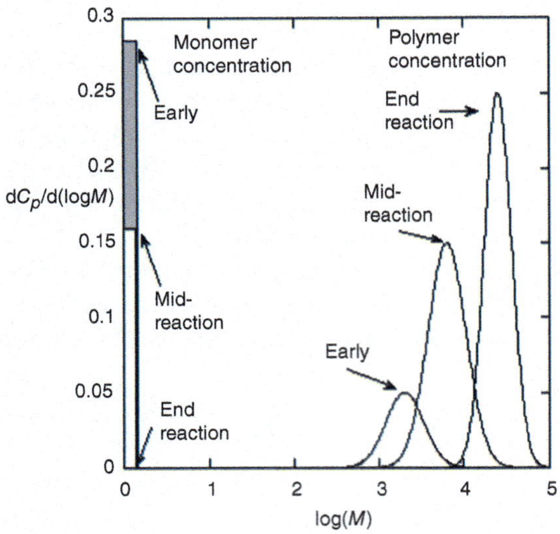

Figure 7.24b Schematic representation of the build-up of polymer concentration during a living chain growth reaction; the trends are opposite those of the free radical case.

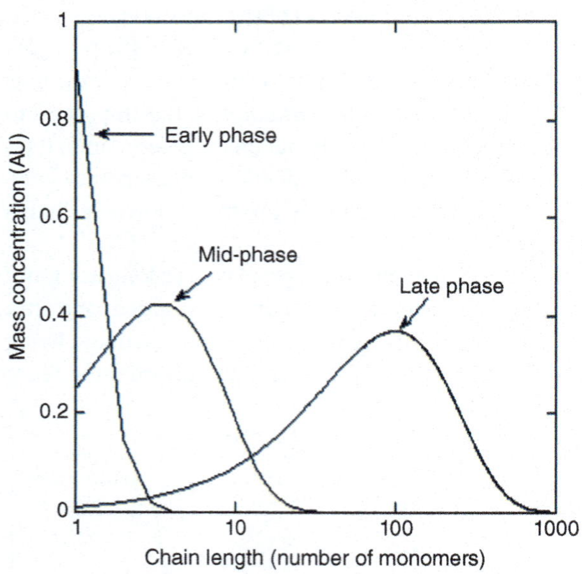

Figure 7.25 Schematic representation of the evolution of the concentration distribution of chain lengths as a step growth reaction proceeds.

different from the free radical case. In the living polymerization chains do not terminate so that they grow in time, rather than decrease. Similarly, the width of the distribution actually decreases in time, rather than increases.

Chain-growth reactions can be carried out in solvents, *in bulk*, that is, with pure monomer, also in micelles, suspensions, emulsions, slurries, and gas phases. The majority of industrial polymers are produced by chain growth.

(II) Step-growth reactions. In this type of polymerization, polymer chains can bond to each other, usually regardless of length, as well as to monomers. As such, these polymers build up gradually, so that at any time there is usually a single mode of polymers, whose average mass increases in time. Many of these reactions involve the elimination of a small molecule each time a bond is formed, and are often termed *polycondensation reactions. Ring scission*, such as occurs in the synthesis of Nylon 6 is also a step-growth mechanism. Figure 7.25 shows a schematic of the molar mass distribution of a step-growth reaction; it shows the mass concentration versus the chain length for early mid- and late phases of the reaction.

Many important polymers are produced by step growth reactions. Among these are polyesters, polyamides (nylons), polyurethanes, polysiloxanes, polycarbonates, polysulfides, and polyethers.

Importantly, living state synthesis of biopolymers involves step growth reactions. These include the production of polypeptides (proteins), polynucleic acids, and polysaccharides.

Apart from chain growth and step-growth polymerization, another process that is often used to produce polymers with specific characteristics is to first produce the polymer molecule and then to modify it chemically, by radiation, or some other means. Common modifications include carboxylation, sulfonation, hydroxylation, peptidization, quaternization, and many more.

7.4.1 Methods of Performing Polymerization Reactions

Bulk reactions involve pure monomers and initiator or catalysts and use no solvent or other media. These can become quite viscous, reaching viscosities as high as 10^6 cP, e.g. bulk polystyrene polymerization.

Solvent-based reactions use monomers and initiators in a solvent, such as water, butyl acetate, dimethyl formamide, tetrahydrofuran, and many others. Solvent-based reactions normally start, and often finish, as homogeneous solutions. Certain types lead to the polymer product precipitating during the reaction. The solvent may also be used to take up heat of the reaction, and control viscosity and reaction kinetics.

Emulsion reactions use monomer droplets in a continuous aqueous phase, often partially stabilized by surfactants. Such reactions can lead to the formation of latex particles. In some cases, high molar mass polymers form in latex particles during the reaction and can be recovered afterward. In other cases, the polymers comprising the latex particles are cross-linked during synthesis to produce stable, very high molar mass latex particles, such as are used in colloid size standards, paints, and other coatings.

The initial emulsions are frequently very heterogeneous in size, ranging from submicron up to hundreds of microns. The emulsions are normally unstable, and surfactants are often used to slow the instability. The emulsions will normally phase separate in time, so that the polymerization must take place soon after emulsification.

In emulsion polymerization the heat of reactions is easier to control, and the viscosity of the reactor fluid remains low because polymers are inside spherical emulsion drops, which have very low viscosity. This facilitates stirring of reactions, pumping, and filling of shipping vessels, etc. In the case of uncross-linked polymers, when emulsions are broken by phase inversion after the reaction, the viscosity of the solution increases enormously. Styrenic and acrylic polymers are frequently synthesized in emulsions.

Inverse emulsions use an oil continuous phase and water-soluble polymers in droplets, usually stabilized by surfactant. These are used frequently to produce acrylamide and vinyl pyrrolidone-based products, water-soluble polymers, and offer the same advantages of heat and viscosity control as regular emulsions.

Inverse emulsion polymerizations often lead to unstable latex particles in the final reaction liquid, which can aggregate and phase separate over time. The time it takes for phase separation depends critically on the concentration of the final product, as solution viscosity, and hence diffusion rate of latex particles, controls the aggregation rate. The dramatic dependence of viscosity on concentration is seen in Figure 6.7 so that relatively small variations in concentration can have very large effects on stability.

Suspension polymerization is similar to the emulsion case but normally the reactions take place in unstabilized droplets of water or other solvent suspended in another liquid medium. Products are sometimes small but macroscopic polymer beads for a variety of uses, including abrasives and oil recovery.

Micellar and microemulsion polymerization are similar to emulsion and inverse emulsion systems, except that the particles are often much more stable and smaller.

Gas phase reactions can involve finely divided catalyst particles held in suspension in air as monomer gas flows by. Polymer grows on the catalyst surface, producing a "snow" or "popcorn" of polymer beads that eventually settles out under their own mass.

There are many types of *polymerization reactors* but operationally they usually fall into the basic classes of *batch*, *semibatch* (also termed *semicontinuous*), *continuous*, and *reactive extruders*. In batch reactors, all the reagents are added at the beginning of the reaction, and discrete amounts can also be added during the reaction. A semicontinuous reactor can control for differences in reactivity between comonomers in a copolymerization reaction by continuously or semi-continuously feeding one or more reagents during the reaction. A continuous reactor uses feeds of reagents to arrive at a steady state in which the product is continuously produced and can be extracted continuously. Most of the polyolefins are produced in continuous reactors, while specialty polymers are still produced largely in batch or semibatch reactors. An anecdote tells of how one can often discern who the founder of a particular polymer manufacturing facility was: If batch reactors are used the facility may have been founded by a chemist, whereas continuous reactors are a hint that the founder might have been a chemical engineer.

7.4.2 Monitoring Polymerization Reactions

There have been many approaches to monitoring polymerization reactions. The earliest and industrially still most widespread method is simply to withdraw aliquots and perform different sorts of analyses. Polymer characterization and analysis is a field unto itself and involves many methods, including several separation and fractionation methods, differential scanning calorimetry, thermogravimetric analysis, densitometry, NMR, absorbance measurements from ultraviolet (UV) to visible (Vis) to infrared (IR), dielectric spectroscopy, polarimetry, conductivity, various forms of light scattering, viscometry, and several rheological methods, electronmicroscopy, and many others.

Recently, a method termed *automatic continuous online monitoring of polymerization reactions* (ACOMP) has been developed and used to monitor multiple reaction characteristics in near real-time. In brief, ACOMP relies on the continuous extraction of a stream of reactor fluid which is automatically and continuously diluted and conditioned to produce an analytical grade, dilute polymer solution, which can be subjected to analysis by any instrument equipped with a flow cell. These include multi-angle light scattering, dynamic light scattering, UV, Vis, and IR absorption, differential refractometry, viscometry, polarimetry, conductivity, NMR, and others. The continuous data allow monitoring such polymer and reaction characteristics as monomer conversion, cumulative and instantaneous weight-average molar mass and reduced viscosity, as well as cumulative and instantaneous composition in the case of copolymerization. The analysis can also quickly detect

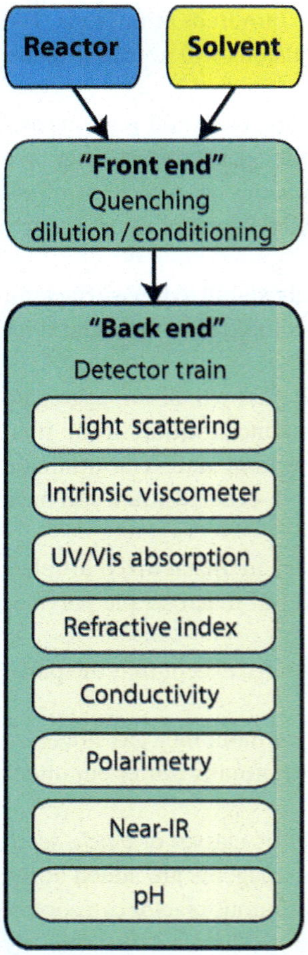

Figure 7.26 A schematic representation of an ACOMP instrument. (Schematic by Marissa Calko).

anomalies in a reaction, such as reaction failure or dead-end,[4] phase separation, aggregation, and nano- and microgelation. While not inherently a chromatographic method, the level of dilution produced in the continuous ACOMP sample stream is of the order used for gel permeation chromatography (discussed in Section 7.13.1), so that a size exclusion chromatography (SEC) instrument can be connected to the exit stream from the ACOMP detector train and automatic SEC injections can be made periodically. Figure 7.26 shows a typical ACOMP instrument arrangement.

Kinetic principles in subsequent sections will be illustrated by ACOMP data, where appropriate.

7.5 Free Radicals and Chain Reactions

Before pursuing free radical polymerization kinetics, it is important to understand what is meant by a *free radical*. In the simplest definition, it is an atom or molecule with one or more unpaired electrons. Because the unpaired electrons generally will seek to lower their energy by pairing with available electrons in other molecules, free radicals are usually quite reactive. A simple example is the photolytic or thermolytic cleavage of Cl_2, where the molecule is symmetrically cleaved by either a UV photon or by high temperature (>250°C) providing the bond dissociation energy of 58 kcal/mol:

$$Cl_2 \xrightarrow{\text{UV or heat}} 2Cl\bullet \qquad 7.1$$

The "dot" after Cl is used to symbolize the unpaired electron in the free radical. Note that Cl• is just the "bare," electrically neutral chlorine atom. The electronic configuration of Cl is $1s^2 2s^2 2p^6 3s^2 3p^5$. Four of the five p electrons are spin paired, leaving a single unpaired electron. Again, the free radical in this case is not an ion. The Cl• can now recombine into Cl_2, or react strongly with other atoms or molecules that may be present. For example, chloromethane, CH_3Cl, can be formed in the presence of methane, CH_4. Upon collision of a free radical Cl• with CH_4 a complete H atom is extracted by the radical, leaving the radical $CH_3\bullet$ while the extracted H atom pairs its unpaired electron with the other Cl• radical's unpaired electron to produce HCl. Having lost the H in one of the four C–H bonds, where the H atom's electron was paired with one of the C atom's electrons, the remaining molecule now has one unpaired electron on C, and so is a free radical $CH_3\bullet$ the *methyl radical*. The process is represented by

$$Cl\bullet + CH_4 \rightarrow HCl + CH_3\bullet \qquad 7.2$$

When $CH_3\bullet$ collides with a Cl_2, the free radical extracts a Cl atom from Cl_2, and the unpaired electron in the extracted Cl atom pairs with the unpaired electron in $CH_3\bullet$, yielding CH_3Cl, leaving behind a chlorine atom, which is the free radical Cl•. This part of the process is represented as

$$CH_3\bullet + Cl_2 \rightarrow CH_3Cl + Cl\bullet \qquad 7.3$$

Reaction 7.3 has "returned" the free radical Cl• which can then undergo Reaction 7.2 again, and hence also Reaction 7.3, and this cycle of Reactions 7.2 and 7.3 can continue until Cl• is somehow lost. In other words, Reaction 7.1 initiated a *chain reaction* involving the repetition of Reactions 7.2 and 7.3. Although usually present in less concentration than reactants the free radicals can also recombine upon collision, thus annihilating them. These can be of the types:

$$Cl\bullet + Cl\bullet \rightarrow Cl_2 \qquad 7.4a$$

$$CH_3\bullet + Cl\bullet \rightarrow CH_3Cl \qquad 7.4b$$

4 A dead-end reaction is one in which not all the monomer is converted to polymer. There can be several causes. One example is if all the initiator decomposes in a free radical reaction before monomer conversion is complete.

$$CH_3\bullet + CH_3\bullet \rightarrow CH_3CH_3 \qquad 7.4c$$

Reaction 7.4a regenerates diatomic chloride, Reaction 7.4b provides an additional chloromethane molecule, and Reaction 7.4c produces ethane, i.e. the final reaction product will contain a very small admixture of ethane. The amount is very small because the concentration of free radicals is very low so that Reactions 7.4a–c have low probabilities compared to reactions 7.2 and 7.3. The *chain length* of the chain reaction is the average number of cycles through Reactions 7.2 and 7.3 per radical initiated. It is equal to the net rate of Reactions 7.2 and 7.3 divided by the rate of initiation in 7.1.

The mechanism outlined here holds for chlorination of other alkanes, such as the synthesis of chloroethane (or ethyl chloride), and for other halogens, such as fluorine and bromine.

7.5.1 Free Radicals and the Vinylic Bond

As mentioned, the vinyl group is one of the leading motifs in FRP reactions. It is found in many common monomers such as styrene, acrylamides, alkenes, and acrylates. It is the ethene molecule with a functional group R, replacing one of the hydrogens. Unlike chloroethane, mentioned above, there is a double bond between the carbons in the vinyl group; $RHC{=}CH_2$. If a free radical, such as Cl• collides with this group its unpaired electron will combine with one of the electrons in the double bond, leaving a single bond between the carbons, but also leaving one of the electrons in the initial double bond unpaired. Thus, the vinyl group is turned into a free radical. This event can be represented by

$$Cl\bullet + RHC = CH_2 \rightarrow RClHC - H_2C\bullet \qquad 7.5$$

and is the initiation step. Unlike Reaction 7.2, where a hydrogen atom is extracted from methane, there is no extraction of any sort in Reaction 7.5, and the initiator group, Cl• in this case, bonds to the vinyl group, and the unpaired electron appears at the head of the new molecule. A different type of chain reaction now occurs, which is quite different from Reactions 7.2 and 7.3, which lead to continuous production of chloromethane molecules. Namely, the vinyl radical can now bond another vinyl monomer to itself using the same process as the initiation step; the unpaired vinyl electron in the radical pairs with one electron in the vinyl monomer's double bond, sending an unpaired electron to the head of what is now a growing chain molecule, or a propagating free radical. The process of propagation, chain building, continues until the unpaired electron on the growing radical chain encounters the unpaired electron of another radical. This is the termination step. The kinetics of this process is considered next.

7.6 Free Radical Polymerization Kinetics

Free radical polymerization generally involves the three steps introduced in the free radical synthesis of chloromethane from Cl_2 and CH_4: initiation, propagation, and termination.

(1) *Initiation* of a monomer into a free radical, which serves as the first link in the growing free radical polymer chain. This is often accomplished by use of free radical initiators such as hydrogen peroxide, potassium persulfate, azobisisobutyronitrile (AIBN), or a very large number of available organic peroxides. Some initiators can also decompose into free radicals under ultraviolet or other radiation.

A thermal initiator I_2 (I_2 stands for a generic initiator, not molecular iodine) first thermally decomposes into two free radicals according to

$$I_2 \xrightarrow{k_d} 2I\bullet \qquad 7.6$$

where k_d is the temperature-dependent first-order decomposition rate constant, and hence

$$[I_2] = [I_2]_0 \exp(-k_d t) \qquad 7.7$$

where the brackets in the expression $[I_2]$ indicate concentration of I_2, and $[I_2]_0$ is the initial concentration of I_2. Now, the production of radicalized monomers is dependent on the concentration of initiated radicals $R_1\bullet$ which, in turn are produced from I• and a monomer m, according to the bimolecular *initiation reaction*:

$$I\bullet + m \xrightarrow{k_i} R_1\bullet \qquad 7.8$$

where k_i is the initiation rate constant.

Either of the two reactions above can be rate controlling in the production of $R_1\bullet$. If the initiator and monomer are dilute the latter diffusion-controlled reaction may be rate limiting, whereas if monomer concentration $[m]$ is high then the initiator decomposition is rate controlling. This latter is the most common case.

(2) *Propagation* of the free radical chain by sequentially incorporating successive monomers onto the end of the growing chain. Once $R_1\bullet$ forms, the propagation takes place according to the following chain reaction:

$$R_j\bullet + m \xrightarrow{k_p} R_{j+1}\bullet \tag{7.9}$$

where k_p is the propagation rate constant, and the propagation equation is

$$\frac{\mathrm{d}[m]}{\mathrm{d}t} = -k_p[m][R\bullet] - k_i[m][\mathrm{I}\bullet] \tag{7.10}$$

Because a monomer can react with a free radical of any chain length, the important convention used in Equation 7.10, and henceforth, is that $R\bullet$ represents free radicals of all lengths, and $[R\bullet]$ is the sum of concentrations of free radicals of all lengths. That is,

$$[R\bullet] = \sum_j [R_j\bullet] \tag{7.11}$$

In the *long-chain limit* (degree of polymerization $\gg 1$), far more monomer is consumed in forming polymer chains than is consumed in the initiation step, so that Equation 7.10 is well approximated by

$$\frac{\mathrm{d}[m]}{\mathrm{d}t} \cong -k_p[m][R\bullet], \quad \text{Long Chain Approximation} \tag{7.12}$$

(3) *Chain termination* proceeds according to

$$R_j\bullet + R_i\bullet \xrightarrow{k_t} P_{j+i}, \ \text{or}, \ P_j + P_i \tag{7.13}$$

where P represents dead polymer chains formed by either recombination (the first term on the right-hand side) or disproportionation (second term). "Dead" in this context means the polymer chain is no longer active and cannot propagate further and is a final polymer product molecule. In disproportionation, there is an atom transfer from one radical chain to the other, usually a hydrogen atom, which leads to no unpaired electrons on either chain, and hence these latter are no longer propagating free radicals. In some polymerization reactions, termination can be almost exclusively by one of the mechanisms; styrene terminates almost exclusively by recombination. In other cases, both mechanisms can be active, e.g. methyl methacrylate typically terminates 58% by disproportionation and 42% by recombination.

The termination rate constant k_t, as well as k_p, are taken here as chain-length independent, although in some instances, chain length-dependent propagation and termination are found. For the case where initiation is decomposition-limited, the population of free radicals is governed by

$$\frac{\mathrm{d}[R\bullet]}{\mathrm{d}t} = 2Fk_d[\mathrm{I}_2] - k_t[R\bullet]^2, \quad \text{Decomposition-limited} \tag{7.14}$$

whereas in the case where diffusion limits the initiation of monomer the population of free radicals is governed by

$$\frac{\mathrm{d}[R\bullet]}{\mathrm{d}t} = k_i[\mathrm{I}\bullet][m] - k_t[R\bullet]^2, \quad \text{Diffusion-limited} \tag{7.15}$$

In Equation 7.14, F is the fraction of $\mathrm{I}\bullet$ that goes on to produce $R_1\bullet$, according to Reaction 7.8 and is sometimes referred to as the *initiator efficiency*. The rest of the $\mathrm{I}\bullet$, $1 - F$, is consumed by other reactions such as recombination, sometimes associated with a "cage effect," and scavenging by impurities (e.g. oxygen).

7.6.1 Quasi-Steady State Approximation, QSSA

A robust and realistic approximation for many chemical reactions is that the rate at which an intermediate is formed is equal to the rate at which it is converted to product. This is often called the quasi-steady state approximation (QSSA).

We shall make use of the QSSA again when considering enzyme kinetics in Chapter 10. In the case of free radical polymerization the QSSA assumes that the rate at which the $R\bullet$ population changes is very slow compared to the other processes (decomposition, initiation, propagation, and termination), i.e. it is assumed that

$$\frac{d[R\bullet]}{dt} \approx 0, \quad \text{QSSA} \tag{7.16a}$$

which gives, in the prevalent decomposition-limited case of Equation 7.14

$$[R\bullet] = \sqrt{\frac{2Fk_d[I_2]}{k_t}} \tag{7.16b}$$

The QSSA harkens back to Le Chatelier's principle (Chapter 4.15.2), which states that when a chemical system in equilibrium is perturbed, by changes in reagent concentrations, temperature, etc., the change will be counter-acted by the system, either maintaining the current equilibrium or moving to a new equilibrium state. In the latter case, the change to the new equilibrium is not as far away as it would be in the absence of mass balance in equilibrium. In this free radical case, the production of a higher radical concentration by decaying initiator over any short interval is offset by the second-order termination of radicals, whose rate increases as the square of the radical concentration. This process maintains radical concentration steady over short periods, periods comparable to at least the lifetime of radicals (typically milliseconds), and often much longer. The radical concentration can change over the course of the reaction, on time scales much larger (minutes or hours) than a radical lifetime. This illustrates the meaning of "quasi" (almost-steady state) in the QSSA.

With this, $[m](t)$ in Equation 7.12 can be found. For the case of decomposition limited initiation, using Equation 7.7 for $[I_2]$, direct integration of Equation 7.12 yields an expo-exponential form:

$$[m](t) = [m]_0 \exp\left[\frac{2Ak_p}{k_d}\left(e^{-k_d t/2} - 1\right)\right], \; A = \sqrt{\frac{2Fk_d[I_2]_0}{k_t}} \tag{7.17}$$

where $[m]_0 = [m](0)$ is the initial monomer concentration. A further simplification results in the case where very little initiator decomposes during the entire reaction, as is often the case with thermal initiators. In this case $[I_2] \sim [I_2]_0$, where $[I_2]_0$ is the initial concentration of initiator. The above equation then reduces to a simple exponential:

$$[m](t) \cong [m]_0 e^{-\alpha t} \tag{7.18}$$

where

$$\alpha = k_p \sqrt{\frac{2Fk_d[I_2]_0}{k_t}} \tag{7.19}$$

Except at very low monomer concentrations initiation is almost always limited by initiator decomposition, so only that case is considered going forward.

Fractional monomer conversion is the fraction of monomer that has been converted to polymer at any stage of the reaction. It is defined by the remaining monomer concentration at time t, $[m](t)$, and the initial monomer concentration $[m]_0$

$$f = 1 - \frac{[m(t)]}{[m]_0} \tag{7.20}$$

For the QSSA case of slow initiator decomposition, this is the first-order function:

$$f = 1 - e^{-\alpha t} \tag{7.21}$$

Two examples described well by this first-order conversion are shown in Figure 7.27, corresponding to the free radical polymerization of acrylamide (Am) at low and high initiator concentrations. The first order fits to the data points using Equation 7.21 are shown.

Recall that the *chain length* was defined in Section 7.5 as the rate of the chain reaction (propagation rate in the polymerization chain reaction) divided by the initiation rate. For chain growth polymerization reactions, it represents the average

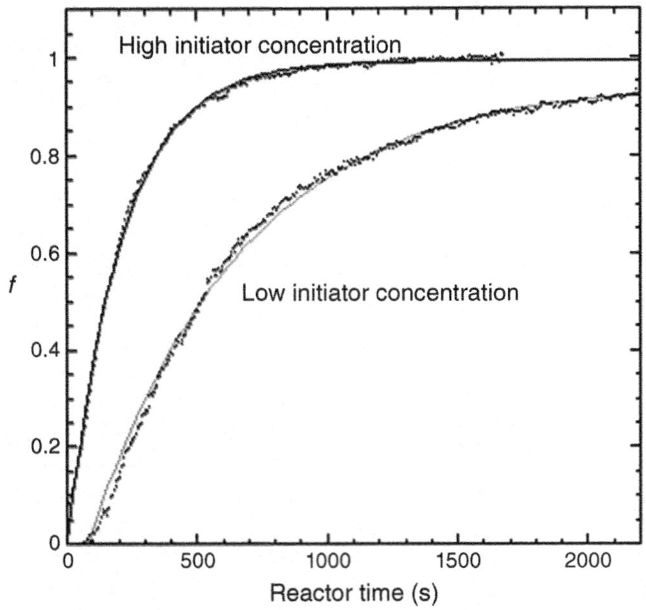

Figure 7.27 Conversion of acrylamide (Am) in free radical polymerization is described well by first-order kinetics. (Giz, A., et al. (2001) / with permission from American Chemical Society).

number of propagation cycles per radical and hence is directly related to the final number of monomer units in a polymer chain.

The instantaneous *kinetic chain length* concept forges the relationship between the rates of propagation and termination and the average length of a chain. The average instantaneous kinetic chain length is equal to the probability of propagation at any instant divided by the probability of termination. This average is a direct number average over the distribution of chain lengths of final polymers produced at any instant. The instantaneous average chain length, in terms of number of monomers in the polymer chain, will be denoted as $N_{n,inst}$, where the subscript "n" reflects the number average. In the absence of chain transfer (treated in the next section), and for termination by disproportionation, the ratio of rate of propagation to rate of termination gives

$$N_{n,inst} = \frac{k_p[m][R\bullet]}{2k_t[R\bullet]^2} = \frac{k_p[m]}{2k_t[R\bullet]} \quad 7.22$$

For termination by recombination, $N_{n,inst}$ is twice this value. Equation 7.22 gives several means to control $N_{n,inst}$. Increasing $[m]$ will increase its value, whereas adding more initiators will give more $[R\bullet]$ and hence lower it. Increasing temperature will produce more $[R\bullet]$ and hence also lower $N_{n,inst}$, and vice versa upon lowering temperature.

For the QSSA with slow initiator decomposition, Equation 7.18, the result is

$$N_{n,inst} = N_{n,inst,0}(1-f) \quad 7.23$$

where $N_{n,inst,0}$ is the initial number average kinetic chain length. The instantaneous number average molar mass is just the kinetic chain length times the molar mass of the monomer M_0

$$M_{n,inst} = M_0 N_{n,inst,0} \quad 7.24$$

The meaning of number and weight average molar mass, M_n and M_w, respectively, are discussed in Section 7.12. Light scattering intensity measures the cumulative weight average molar mass in the reactor, $M_w(t)$, as will be discussed in Chapter 9. This cumulative value is the integral over the instantaneous weight average molar mass $M_{w,inst}$ over conversion:

$$M_w(f) = \frac{\int_0^f M_{w,inst}(f')\,df'}{f} \quad 7.25$$

$M_{w,inst}$ is proportional to $N_{n,inst}$, so Equation 7.25 can be combined with Equations 7.23 and 7.24 and integrated, which yields $M_w(f)$, in the QSSA case of slow initiator decomposition,

$$M_w(f) = M_{w,0}\left(1-\frac{f}{2}\right) \quad 7.26$$

where $M_{w,0}$ is the initial weight average molar mass at $f = 0$. Figure 7.28 shows ACOMP data for $M_w(f)$ and $M_{w,inst}(f)$ for free radical polymerization of acrylamide. $M_{w,inst}$ can be found from the primary ACOMP data for $M_w(f)$ by multiplying Equation 7.25 by f and differentiating:

$$M_{w,inst} = \frac{d[fM_w(f)]}{df} \quad 7.27$$

so that

$$M_{w,inst} = M_{w,0}\left(1-f\right) \qquad 7.28$$

At $t = 0$, $M_{w,0} = M_{w,inst,0}$. In Figure 7.28, $M_w(f)$ is in good agreement with Equation 7.26 and $M_{w,inst}(f)$ with Equation 7.28.

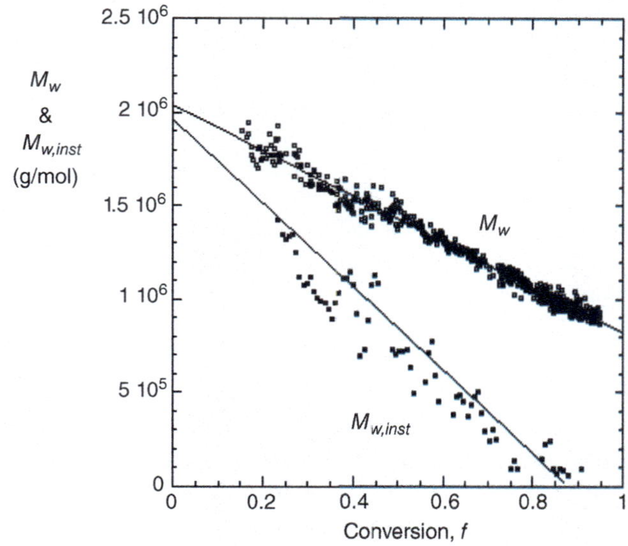

Figure 7.28 M_w and $M_{w,inst}$ for an acrylamide free radical polymerization. (Giz, A., et al. (2001) / with permission from American Chemical Society).

7.6.2 Effect of Chain Transfer on Molar Mass

In the free radical kinetic scheme given by Equations 7.9, 7.12, and 7.13, the fate of a propagating radical $R\bullet$ is to either add another monomer (propagation) or to terminate via reaction with another free radical. Another possibility arises if the unpaired electron of $R\bullet$ can be passed to another molecule, via transfer of some group, leaving behind a dead polymer chain, and a new free radical that can initiate a monomer. Such a molecule that can become a radical through collision with a propagating radical is termed a *chain transfer agent*, or CTA. Many types of molecules can act as CTAs; the solvent, the monomer or polymer themselves, the initiator, or a separate species, such as sodium formate. Once the CTA becomes a radical it can initiate a monomer and begin propagation of another chain, while the original propagating radical is now a dead polymer chain.

In the case where the agent that becomes the free radical cannot initiate monomer, the agent is termed an *inhibitor*. Quinones and O_2 are examples of inhibitors and are sometimes termed *radical scavengers*. At very low concentrations, O_2 acts as a reversible chain termination agent. It is reversible because it can be either spontaneously eliminated in some cases (e.g. acrylamide free radical polymerization) and can always be eliminated by purging with an inert gas, such as N_2. When $[O_2]$ increases above a certain threshold, the reaction abruptly stops, but restarts once $[O_2]$ drops below the threshold, either spontaneously or by inert gas purge. These remarkable effects are seen in Figure 7.29 for the free radical polymerization of acrylamide, where the reactor was initially fully purged with N_2 and then a small flow of O_2 was introduced. The abrupt turn-off of the reaction occurs at low $[O_2]$. The O_2 flow was stopped just before the peak value. Interestingly, the O_2 was automatically eliminated by an acrylamide-catalyzed process, and when $[O_2]$ dropped to its threshold the reaction spontaneously restarted.

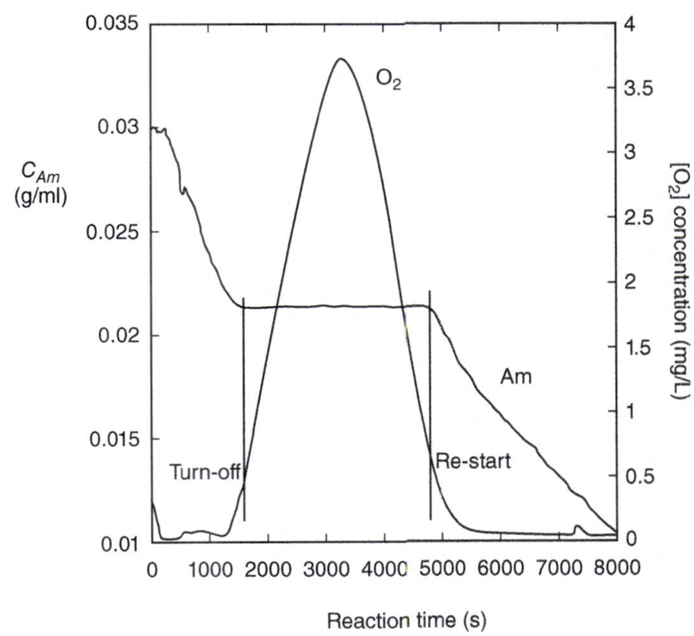

Figure 7.29 The inhibitory effect of O_2 on acrylamide free radical polymerization. As O_2 flows at a very low rate into the reactor a threshold concentration of O_2 is reached and the reaction stops. During the conversion plateau, O_2 is automatically eliminated and the reaction spontaneously restarts when the O_2 threshold is reached. (Siqueira, J.S. et al. (2022) / American Chemical Society / Licensed under CC BY 4.0).

The effect of a CTA on a free radical polymerization is to shorten the average chain length, and hence lower the molar mass, while producing more polymer chains, and normally only slightly affecting the rate of the reaction. These effects are now justified.

When a CTA, is present in concentration [CTA], then the following reactions can occur:

$$R\bullet + \text{CTA} \xrightarrow{k_3} P + \text{CTA}\bullet \qquad 7.29$$

where k_3 is the chain transfer constant, and CTA• indicates that CTA bears the unpaired electron transferred from $R\bullet$ after the $R\bullet$ /CTA encounter, and a "dead" polymer chain P, is left. Meanwhile, CTA• can go on to initiate another propagating chain:

$$\text{CTA}\bullet + m \xrightarrow{k_4} R_1\bullet + \text{CTA} \qquad 7.30$$

where k_4 is the CTA's monomer initiation constant. If the transfer constant to CTA, k_3, is smaller than the transfer constant from CTA to monomer, then the first reaction, Reaction 7.29, will be rate controlling. There is no loss of propagating free radical in either of the above equations – the *reaction kinetic chain* is not broken – but the rate of producing dead polymer chains now increases, because dead polymer P, is produced both by propagating chains encountering each other and encountering CTA, expressed by

$$\frac{\mathrm{d}[P]}{\mathrm{d}t} = Yk_t[R\bullet]^2 + k_3[\text{CTA}][R\bullet] \qquad 7.31$$

Y is a factor that is equal to 1 when recombination dominates and is equal to 2 when disproportionation dominates, and which takes on values between these limits when both termination mechanisms are present. For example, disproportionation dominates in acrylamide polymerization so that $Y \sim 1$.

In contrast to the increased rate of chain production, the rate of monomer conversion normally changes only slightly, because the total radical concentration $[R\bullet]$ in Equation 7.12 has not changed, although an $R\bullet$ is removed in Equation 7.29 before being restored to the total radical population as $R_1\bullet$ in initiation Reaction 7.30. This delay can slow conversion somewhat. Figure 7.30a shows monomer conversion versus time for free radical acrylamide polymerization without a CTA and with two different concentrations of CTA (sodium formate). While there is no obvious difference among them, fits to first-order Equation 7.21 show the CTA slightly slows the conversion compared to no CTA, by 3% when [CTA]/[Am] = 0.125 and by 6% when [CTA]/[Am] = 0.25. The rate constant in the table of Figure 7.29 is the term "α" in Equation 7.21.

In the case where monomer initiation by CTA• via Equation 7.31, is very slow, the monomer conversion rate can be significantly reduced. In this case, the CTA is known as a *polymerization retarder.*

The effect of the CTA on the polymer molar mass can be accounted for by amending the expression for the kinetic chain length $N_{inst,n}$ in Equation 7.22, using the same concept that the average kinetic chain length will be equal to the probability of propagation divided by the probability of a *chain* terminating (but not the radical, which is conserved for an ideal CTA, which initiates a new chain with the radical; hence, the reaction chain length is not broken by the CTA). $N_{n,inst}$ is now given by

$$N_{n,inst} = \frac{k_p[m][R\bullet]}{Yk_t[R\bullet]^2 + k_3[R\bullet][\text{CTA}]} = \frac{k_p[m]}{Yk_t[R\bullet] + k_3[\text{CTA}]} \qquad 7.32$$

and the instantaneous number average mass, in terms of the monomer molar mass M_0, is

$$M_{n,inst} = M_0 N_{n,inst,0} = M_0 \frac{k_p[m]}{Yk_t[R\bullet] + k_3[\text{CTA}]} \qquad 7.33$$

This can be inverted to give the usual graphical form of determining k_3 via

$$\frac{1}{M_{n,inst}} = \frac{1}{M_{n,inst}(\text{CTA}=0)} + \frac{k_3[\text{CTA}]}{k_p[m]M_0} \qquad 7.34$$

Figure 7.30b shows the strong effect of the CTA sodium formate on weight average molar mass M_w during acrylamide polymerization. The data are from the same reactions as in Figure 7.30a, where the conversion rates of the reaction showed only a slight dependence on [CTA]. Figure 7.31 shows the corresponding graphical representation of an equation similar to Equation 7.34, used to determine the chain transfer constant for sodium formate in the acrylamide reaction. The difference is that light scattering was used for the molar mass determinations, and light scattering measures the weight average molar mass M_w, as explained in Chapter 9. The weight average degree of polymerization X_w $(=M_w/M_{Am}$, where $M_{Am} = 79.1$ g/mol is the molar mass of Am) is used in Figure 7.30b and is given by

$$X_w = Đ\frac{k_p}{k_t[R]Y + k_3[\text{CTA}]}[m] \tag{7.35}$$

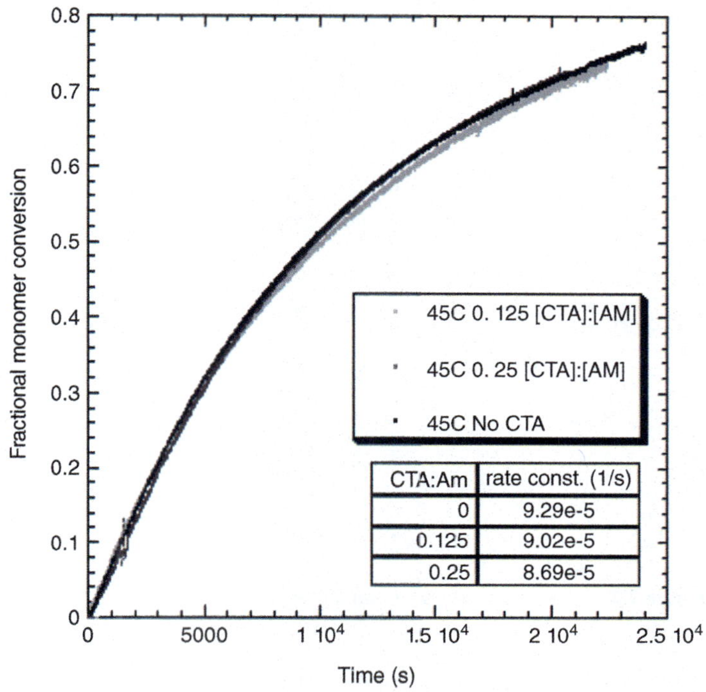

Figure 7.30a Monomer conversion during free radical polymerization of acrylamide with and without a CTA (sodium formate). The conversion rates (α in Equation 7.21) are shown in the inset table.

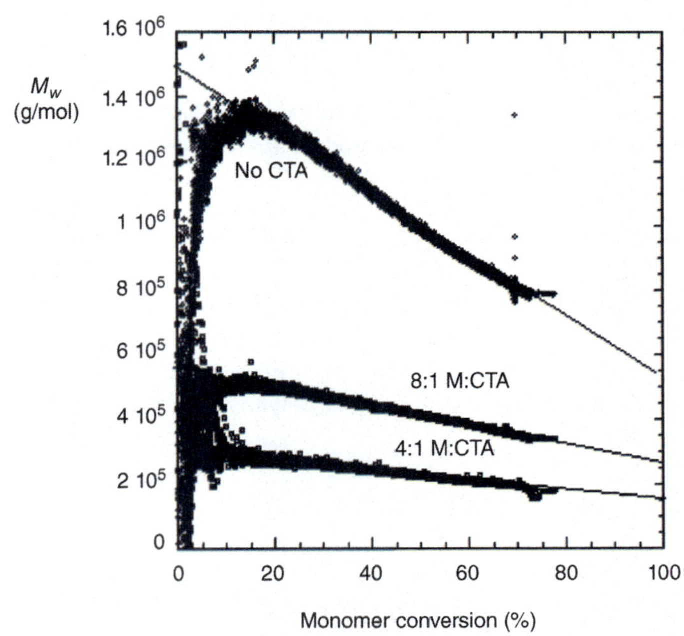

Figure 7.30b The effect of chain transfer agent on M_w for the free radical polymerization of acrylamide with no CTA and then increasing concentration of CTA. These are the same reactions as in Figure 7.29. (Figures 7.30 (a,b) Reproduced by permission from Wiley, from Leonardi et al. (2017)).

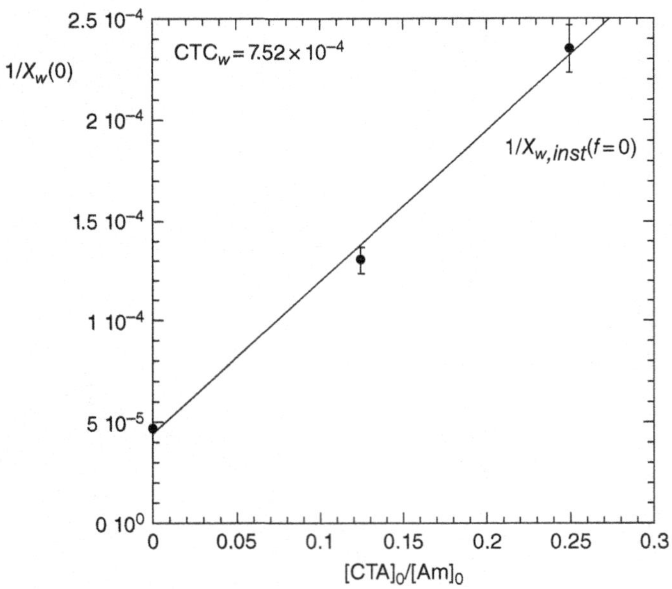

Figure 7.31 Determination of sodium formate's chain transfer constant in free radical polymerization of acrylamide. (Reproduced by permission from Wiley, from Leonardi, N. et al. (2017)).

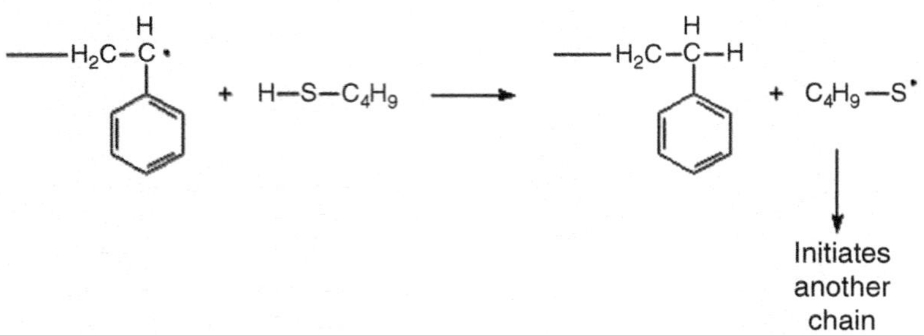

Figure 7.32 Chain transfer by a mercaptan.

where $Đ$ is the dispersity, given by $Đ = M_w/M_n$. The reciprocal of X_w is plotted versus the initial molar ratio $[\mathrm{CTA}]_0/[\mathrm{Am}]_0$ and analyzed using X_w extrapolated to $f = 0$:

$$\frac{1}{X_w(0)} = \frac{1}{X_w(0,[\mathrm{CTA}]=0)} + \frac{k_3}{k_p} Đ \frac{[\mathrm{CTA}]_0}{[M]_0} \tag{7.36}$$

The slope of this equation yields $k_3/k_p\ Đ$. This dimensionless quantity is the chain transfer constant, CTC_w, where the subscript "w" indicates it is based on M_w. The standard CTC is based on M_n and is k_3/k_p, so its relation to CTC_w determined by light scattering is

$$\mathrm{CTC}_n = Đ\mathrm{CTC}_w \tag{7.37}$$

Chain transfer agents normally have one weak bond. *Mercaptans*, also known as *thiols*, are an example of a chain transfer agent in organic phase, where the sulfur becomes a free radical upon transfer and readily initiates a new chain with a monomer. An example is shown in Figure 7.32. Halocarbons, such as carbon tetrachloride, and some alcohols are also chain transfer agents. Monomers, polymer chains, and solvent can all potentially act as chain transfer agents.

7.7 Ideal Living Polymerization

'Living' polymerization reactions are of both fundamental and practical importance. They offer much closer control over the width (*dispersity*) of the molar mass distribution than free radical polymerization, more opportunities to add and mix different functional groups in precise ways, and hence many opportunities to control polymer architecture. Polymers that

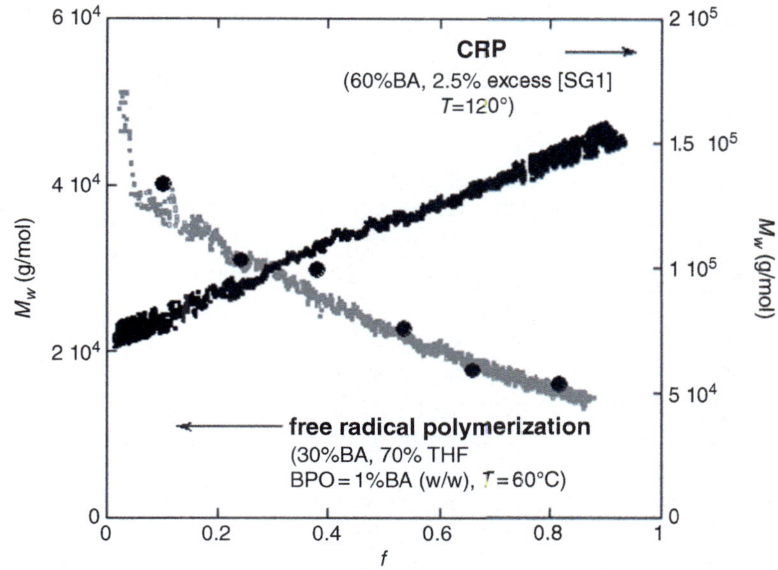

Figure 7.33 The difference in M_w versus conversion for free radical and living-like controlled radical polymerization. Solid circles are SEC results on manually withdrawn reaction aliquots. (Alb, A.M. et. al. (2008) / Reproduced with permission from John Wiley & Sons).

are highly architected, such as dendrimers and many types of graft copolymers are produced by living polymerization reactions. Classically, anionic polymerization was among the first types of living polymerization developed. It is used industrially to annually produce millions of tons of poly(ethylene glycol), styrene-butadiene rubber, and hundreds of thousands of tons of triblock thermoplastics.

The features of ideal living reactions are:

1) Initiation occurs fairly rapidly. Each monomer initiated will grow, as long as monomer is present.
2) Ideally, no termination occurs, so initially produced propagating species continue to propagate.
3) *There is a steady increase in the length of each propagating chain throughout the entire reaction.* Hence, unlike free radical reactions, where chains are initiated, propagate and die rapidly compared to the time span of the entire reaction, living chains, in principle, never die. Provided that initiation is rapid, the final number average molar mass of the chains is simply the molar concentration of monomer consumed divided by the initiator concentration, assuming that each initiator molecule initiated a monomer, that is

$$M_n = \frac{[m]_0}{[\mathrm{I}]} \quad 7.38$$

4) Once all monomer is incorporated into chains they remain "living" and more monomer can be added at any time and the polymerization will start again: hence "living polymerization."

Figure 7.33 shows the stark difference in how M_w changes during polymerization between free radical and a living-like controlled radical polymerization. Equation 7.22 shows that in a batch reaction $[m]$ decreases continuously as it is converted to polymer, whereas $[R\bullet]$ decreases slowly for most thermal initiators chosen. Hence, $N_{r,inst}$ decreases continuously during a radical polymerization, as does M_w. The black circles for the FRP reaction are size exclusion chromatography, SEC (Section 7.13.1) results from manually withdrawn reaction aliquots. The agreement with ACOMP is good, but there are few data points due to the time needed for manual sampling and running the SEC experiments.

7.7.1 Controlled Radical Polymerization

Over the past few decades, many types of *controlled radical polymerization* (CRP) have been developed, including *atom transfer radical polymerization* (ATRP), *reversible addition-fragmentation transfer* (RAFT), *nitroxide-mediated polymerization* (NMP), *ring-opening metathesis polymerization* (ROMP), and others. *Click chemistries* have been developed to couple different oligomeric and polymer groups together.

CRP approximates living polymerization by keeping a very low number of actively propagating radicals at any instant, but unlike ideal living polymerization, termination is minimized but not eliminated. CRP is related to *reversible*

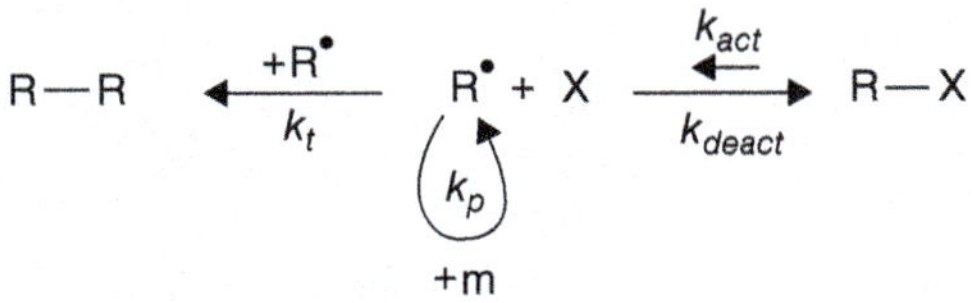

Figure 7.34 A schematic of a CRP in which X thermoreversibly caps $R\bullet$ When capped R-X, no polymerization takes place. When uncapped R• can add monomer before being capped again by X and becoming transiently dormant.

deactivation radical polymerization. All the CRP classes mentioned above are in this category, and this somewhat cryptic name means something fairly simple: A conventional propagating free radical occurs in CRP but there is an agent that can thermoreversibly bind to it, hence putting that radical "to sleep" while the agent is bound, stopping the radical chain propagation, until the agent thermally unbinds, thus freeing up the radical to resume propagation. A radical that is transiently capped by an agent is termed a *dormant radical*. Typically, CRP reactions are slower than uncontrolled free radical under similar conditions, because there are many fewer propagating radical chains at any instant in the CRP case. CRP control can be lost if the reaction temperature is too high, because the capping agent spends too much time uncapped, the dormant radical population decreases, and there are many more propagating free radical chains, leading to higher termination rates and loss of living-type polymerization.

Figure 7.34 is a generic schematic of this process: X is an agent (e.g. a stable radical) that can reversibly bind to, or 'cap', the propagating radical R• via the deactivation step with rate constant k_{deact}, to form the dormant radical R–X. R–X can revert to R• and X via the activation step, with rate constant k_{act}. Notice that the arrow for k_{deact} is longer than for k_{act}, signifying that the system is designed to have more dormant radicals than propagating radicals at any time. Uncapped, propagating chains in the figure are shown adding monomer m with a propagation rate constant k_p. Uncapped radicals $R\bullet$ can react with other R• in a (usually) unwanted termination step. To keep termination to a minimum $k_{deact} \gg k_{act}$.

7.8 Chain Growth Copolymerization Kinetics

Copolymers refer to macromolecules with more than one type of monomer incorporated. Often, "copolymer" is used to mean two types of monomers in a copolymer, *terpolymer* to mean three types of monomers, etc. Here, "copolymer" refers to any polymer with two or more different *comonomers*. The most impressive copolymers are DNA and proteins. By means of four different comonomers (the nucleotides) arranged in codons (three nucleotides at a time), the DNA copolymer transmits all the accumulated genetic information of an organism. It produces proteins, copolymers that contain up to hundreds or more of the 20 amino acid comonomers. Protein catalysts, enzymes, use this extraordinarily sophisticated information contained in its copolymeric structure to speed up very specific reactions by many orders of magnitude.

Industrially useful copolymers generally contain two or more comonomers that combine properties of the corresponding homopolymers into materials with much different properties than the various homopolymers alone. Blending homopolymers can often create useful new materials also, but there is frequently the risk of phase separation of the polymers due to thermodynamic incompatibility. For copolymers on the other hand, especially block copolymers (defined below) thermodynamic incompatibility of the comonomers can lead the copolymers to self-organize into many different morphologies, such as micelles, vesicles, disks, unique surface coatings, etc.

Very different types of polymer properties can result from the monomer sequentiality. In an *alternating copolymer* of A and B, the polymer will have the sequence $(AB)_n$, such as ABABABABAB. In a *block copolymer*, there can be a block of A length n and a block of B of length m, A_nB_m, such as AAAAAAAAAAAABBBBBBBBBBB. Block copolymers, in which the two blocks are incompatible, can form *polymeric micelles*. *Multiblock copolymers* can have sequences such as $A_nB_mA_kB_k$, such as (AAAAAAAAAABBBBAAABBBBBBBB). Periodic copolymers can repeat block sequences, such as $k(A_nB_m)$. *Statistical copolymers* results when the *reactivity ratios* for a pair of monomers, introduced below, together with the molar abundances of A and B lead to *statistical sequence length distributions*, also introduced below. When the reactivity ratios are equal to unity then a *random copolymer* results, and the probability of adding A or B in a propagating chain is solely proportional the molar ratios of A and B.

Copolymers can be produced both by chain growth and step growth reactions. Here, a brief look at chain growth copolymerization involving just two comonomers is given.

7.8.1 Mayo–Lewis Approach to Free Radical Copolymerization

Consider a monomer A and a second monomer B. Let A• and B• represent the propagating radicals for A and B, respectively. The following reactions can occur:

$$B_mA_n\bullet + A \xrightarrow{k_{11}} B_mA_{n+1}\bullet \qquad 7.39a$$

$$\mathrm{B}_m\mathrm{A}_n \bullet + \mathrm{B} \overset{k_{12}}{\rightarrow} \mathrm{A}_n\mathrm{B}_{m+1} \bullet \tag{7.39b}$$

$$\mathrm{A}_m\mathrm{B}_n \bullet + \mathrm{B} \overset{k_{22}}{\rightarrow} \mathrm{A}_m\mathrm{B}_{n+1} \bullet \tag{7.39c}$$

$$\mathrm{A}_m\mathrm{B}_n \bullet + \mathrm{A} \overset{k_{21}}{\rightarrow} \mathrm{B}_n\mathrm{A}_{m+1} \bullet \tag{7.39d}$$

Here it is assumed that the rate constants k_{11}, k_{12}, etc. do not depend on the length of the propagating chain. In the simplest paradigm for propagation in free radical polymerization, it is assumed that the rate of reaction for a given comonomer depends only on which monomer bears the radical at the chain terminus. This is the *ultimate model*. A more detailed model, the *penultimate model*, considers the effect of both the last and penultimate monomers. We do not pursue that model here, although it has sometimes been found appropriate when the ultimate model fails. For the molar concentrations of monomer [A] and [B], Equations 7.39a–d lead to propagation equations of the form

$$-\frac{\mathrm{d}[\mathrm{A}]}{\mathrm{d}t} = k_{11}[\mathrm{A}][\mathrm{A}\bullet] + k_{21}[\mathrm{A}][\mathrm{B}\bullet] \tag{7.40a}$$

$$-\frac{\mathrm{d}[\mathrm{B}]}{\mathrm{d}t} = k_{22}[\mathrm{B}][\mathrm{B}\bullet] + k_{12}[\mathrm{B}][\mathrm{A}\bullet] \tag{7.40b}$$

The rate equations for A• and B• are

$$\frac{\mathrm{d}[\mathrm{A}\bullet]}{\mathrm{d}t} = 2F_\mathrm{A}k_D[\mathrm{I}] + k_{12}[\mathrm{A}][\mathrm{B}\bullet] - k_{21}[\mathrm{B}][\mathrm{A}\bullet] - k_{t,11}[\mathrm{A}\bullet][\mathrm{A}\bullet] - k_{t,12}[\mathrm{A}\bullet][\mathrm{B}\bullet] \tag{7.41a}$$

$$\frac{\mathrm{d}[\mathrm{B}\bullet]}{\mathrm{d}t} = 2F_\mathrm{B}k_D[\mathrm{I}] - k_{12}[\mathrm{A}][\mathrm{B}\bullet] + k_{21}[\mathrm{B}][\mathrm{A}\bullet] - k_{t,22}[\mathrm{B}\bullet][\mathrm{B}\bullet] - k_{t,12}[\mathrm{A}\bullet][\mathrm{B}\bullet] \tag{7.41b}$$

where $k_{t,11}$, $k_{t,22}$, and $k_{t,12} = k_{t,21}$, are the termination rate constants for encounters of A•A, B•B and A•B, respectively. $2F_\mathrm{A}k_D[I]$ and $2F_\mathrm{B}k_D[I]$ are the creation terms for A• and B• respectively, where F_A and F_B subsume the possibly different initiation rates of A and B. In Equation 7.41a, $k_{t,11}[\mathrm{A}\bullet][\mathrm{A}\bullet] - k_{t,12}[\mathrm{A}\bullet][\mathrm{B}\bullet]$ is the termination term for B• and $-k_{t,22}[\mathrm{B}\bullet][\mathrm{B}\bullet] - k_{t,12}[\mathrm{A}\bullet][\mathrm{B}\bullet]$ is the termination term for B•. In the QSSA, it is assumed that $\mathrm{d}[\mathrm{A}\bullet]/\mathrm{d}t \sim 0$ and $\mathrm{d}[\mathrm{B}\bullet]/\mathrm{d}t \sim 0$ and that the creation and termination terms cancel each other out, as they do for homopolymerization in the QSSA. This leaves the condition:

$$k_{12}[\mathrm{A}][\mathrm{B}\bullet] = k_{21}[\mathrm{B}][\mathrm{A}\bullet] \tag{7.42}$$

In other words, in the QSSA, the rate of interconversion of B• to A• is balanced by the rate of conversion of A• to B•. With this, Equation 7.40a can be divided by Equation 7.40b to yield the Mayo–Lewis copolymer equation:

$$\frac{\mathrm{d}[\mathrm{A}]}{\mathrm{d}[\mathrm{B}]} = \left(\frac{[\mathrm{A}]}{[\mathrm{B}]}\right)\left(\frac{r_\mathrm{A}[\mathrm{A}] + [\mathrm{B}]}{[\mathrm{A}] + r_\mathrm{B}[\mathrm{B}]}\right) \tag{7.43}$$

where the *reactivity ratios* r_A and r_B are defined as

$$r_\mathrm{A} = \frac{k_{11}}{k_{12}}, \quad r_\mathrm{B} = \frac{k_{22}}{k_{21}} \tag{7.44}$$

These are interpreted as the rate at which a comonomer radical adds another comonomer of its own type to itself divided by the rate at which that comonomer radical adds the other comonomer. Hence, if $r_\mathrm{A} > 1$ then A• prefers to add A, whereas if $0 < r_\mathrm{A} < 1$ then A• prefers to add B. In the limit of $r_\mathrm{A} = r_\mathrm{B} = 0$ there is no chance that A• will add A, and that B• will add B, so that a strictly alternating copolymer is produced of the form –ABABABAB–. In the limit $r_\mathrm{A} \rightarrow \infty$, and $r_\mathrm{B} \rightarrow \infty$ A• will add only A, and B• will add only B, so that no copolymer is formed, only two non-joined homopolymers of A and B.

The instantaneous mole fraction of [A] incorporated, $F_{A,inst}$, is given by[5]

$$F_{A,\,inst} = \frac{d[A]}{d\{[A]+[B]\}} = \frac{r_A f_A^2 + f_A(1-f_A)}{r_A f_A^2 + 2f_A(1-f_A) + r_B(1-f_A)^2} \qquad 7.45$$

where f_A and f_B are the remaining mole fractions of monomer A and B, respectively, at any instant,

$$f_A = \frac{[A]}{[A]+[B]}, \quad f_B = \frac{[B]}{[A]+[B]} \qquad 7.46$$

$F_{B,inst}$ is obtained by interchanging the subscripts A and B in Equation 7.45. These expressions form the basis for finding composition and sequence length distributions.

An interesting feature of Equations 7.40a,b under the QSSA is that these reduce to simple first-order equations

$$\frac{d[A]}{dt} = -\alpha_A[A] \qquad 7.47a$$

$$\frac{d[B]}{dt} = -\alpha_B[B] \qquad 7.47b$$

where

$$\alpha_A = k_{11}[A\bullet] + k_{21}[B\bullet] \qquad 7.48a$$

and

$$\alpha_B = k_{22}[B\bullet] + k_{12}[A\bullet] \qquad 7.48b$$

where α_A and α_B are constant. From this, it follows:

$$r_A r_B = 1, \text{ for QSSA} \qquad 7.49$$

Figure 7.35 shows the composition drift during the free radical copolymerization of butyl acrylate (BA) and methyl methacrylate (MMA), monitored by ACOMP. The instantaneous fractional composition of BA in the chain, $F_{inst,BA}$ computed using the middle expression in Equation 7.45, is plotted versus the total conversion of monomer f, which includes both BA and MMA conversion. The composition drift trajectories are shown for different starting ratios BA/MMA. Except for the case of BA/MMA 10/90, the drift in $F_{inst,BA}$ is always upward. This indicates that MMA is being consumed more rapidly, so that $r_{MMA} > r_{BA}$. The reactivity ratios were found to be $r_{MMA} = 2.29 \pm 0.4$ and $r_{BA} = 0.36 \pm 0.06$. This gives $r_A r_B = 0.82 \pm 0.2$, which is in range of the prediction of Equation 7.49.

Recalling that dead polymer chains are formed very quickly compared to total time to consume all monomer in free radical reactions, the effect of composition drift is to produce a compositionally disperse collection of chains in the final product. In Figure 7.35, for example, polymer chains formed in the top four reactions become richer and richer in MMA as the reaction proceeds. In the case of BA/MMA 10/90, there is no composition drift and the average composition of chains in the final product is the same. In polymer science, in analogy to distillation, a system with no composition drift is termed *azeotropic*.[6]

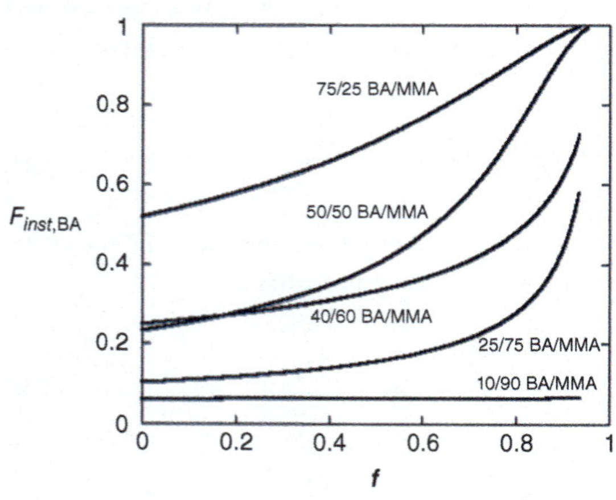

Figure 7.35 Composition drift of BA, depicted by $F_{inst,BA}$ during its free radical copolymerization with MMA. (Reproduced with permission from Alb, A.M. et al. (2006). with permission from American Chemical Society.).

5 The instantaneous fraction of comonomer A, $F_{A,inst}$ is not to be confused with the initiation efficiency subsumed in the fraction F_A, and likewise for $F_{B,inst}$ and F_B.

6 In distillation, two liquids are azeotropic if they maintain the same composition during distillation and are not separated from each other.

7.8.2 Living Copolymerization

In living copolymerization, the final polymer population is quite different from the free radical case. Equations 7.40a,b for the chain propagation still hold, but in Equations 7.41a,b, there is no termination step for living polymerization. Also, for ideal living polymerization initiation occurs rapidly at the outset, and does not continue as the $2Fk_d[\mathrm{I}]$ term of Equations 7.41a,b implies. The result of the absence of termination, as mentioned above, is that the chains keep growing throughout the reaction and so the composition drift is impressed along the length of the chain. Such copolymers are hence termed *gradient copolymers*. The result of the rapid initiation is that the number average molar mass of the final chains is given by Equation 7.38. Figure 7.36 is a hypothetical example of the difference in both chain length and composition between final populations of copolymer chains of A and B produced by (i) free radical copolymerization and (ii) living copolymerization.

Another feature of the living copolymerization is that it provides a convenient means of forming block copolymers. Namely, only monomer A is present in the first phase of polymerization, and when it nears 100% conversion monomer B is added.

7.8.3 The Sequence Length Distribution

Figure 7.36 shows an example of how the length of internal segments of comonomers A and B can vary. A way of characterizing the length of segments of A and B is via the *sequence length distribution*. Let W_{AA} be the probability that A adds to A• upon the next propagation step. This can be expressed, at any moment during the reaction as

$$W_{\mathrm{AA}} = \frac{k_{11}[\mathrm{A}][\mathrm{A}\bullet]}{k_{11}[\mathrm{A}][\mathrm{A}\bullet] + k_{12}[\mathrm{B}][\mathrm{A}\bullet]} \qquad 7.50$$

which can be re-arranged to involve the reactivity ratios

$$W_{\mathrm{AA}} = \frac{r_{\mathrm{A}} f_{\mathrm{A}}}{f_{\mathrm{A}}(r_{\mathrm{A}} - 1) + 1} \qquad 7.51$$

A	A	A	A	A	B
A	A	A	A	A	A
A	A	A	A	B	B
A	B	A	A	A	B
B	A	A	B	A	A
B	B	B	B	B	A
A	A	A	A	A	B
A	A	A	A	B	B
A	A	A	A	A	B
B	B	B	A	B	B
B	B	B	A	B	B
B	B	B	B	A	
B	B	B	A	A	
A	A	A	A	A	
A	A	A	B		
B	B	B	A		
B	B	B	A		
B	B	B	A		
B	B	B			
B	B				
B					

Figure 7.36 Differences in composition and chain length distributions in final products of copolymerization of comonomers A and B for living copolymerization (left) and for free radical copolymerization (right). The living copolymers have a gradient of composition from A to B and have similar length, whereas the free radical copolymers start with chains rich in A and proceed toward shorter chains with increasing amounts of B.

The probability of having a sequence of k monomers in a row of type A, followed by a monomer B, $P_{A,k}$, then follows the well-known *geometric distribution* (GD):

$$P_{A,k} = (1 - W_{AA}) W_{AA}^{k-1} \tag{7.52}$$

$P_{A,k}$ can be computed from f_A via Equations 7.51 and 7.52, where f_A is measured at each point in time by ACOMP or other composition measurements. $P_{A,k}$ is *the instantaneous sequence length distribution* of free radical chains produced at any given time. It depends on the relative concentrations of A and B at any instant, as well as the fixed reactivity ratios.[7] If there is composition drift then the concentrations of A and B vary in time and so also the sequence length distributions.

The moments of this distribution are well known, and their definitions and computations will be explained in Section 7.12.1. The instantaneous number average sequence length of monomer A, $\langle N_A \rangle_n$, at conversion point f, is

$$\langle N_A \rangle_n(f) = \frac{1}{1 - W_{AA}(f)} \tag{7.53}$$

while the instantaneous weight average sequence length of A, $\langle N_A \rangle_w$ is

$$\langle N_A \rangle_w(f) = \frac{1 + W_{AA}(f)}{1 - W_{AA}(f)} \tag{7.54}$$

The cumulative values of the number and weight averages, $\langle N_A \rangle_{n,cumulative}$ and $\langle N_A \rangle_{w,cumulative}$, can be found by simply summing, or integrating, over all conversion points:

$$\langle N_A \rangle_{cumulative}(f) = \frac{\int_0^f \langle N_A \rangle (f') df'}{f} \tag{7.55}$$

where the cumulative averages of any moment (e.g. n or w) are obtained by using the corresponding instantaneous average in the integrand.

To obtain the appropriate expressions for monomer B, the subscript B is substituted for A in each variable in the above equations.

7.9 Cumulative and Instantaneous Polymer Characteristics During Free Radical Reactions

We have already distinguished between cumulative and instantaneous molar mass and composition in preceding sections. This can be generalized for any characteristic Q. In a batch and semibatch reactor, the cumulative values of Q are measured directly; cumulative weight average molar mass M_w by light scattering, remaining fractions of comonomers A and B, f_A and f_B, by absorption spectroscopy, refractometry, or other means, and cumulative weight average reduced viscosity η_r by a viscosity detector. The cumulative values of Q build as conversion proceeds and the concentration of polymer C_p, increases. The cumulative value Q is then the sum of the instantaneous values Q_{inst} over the accumulating C_p

$$Q = \frac{\int_0^{Cp} Q_{inst} dC'_p}{C_p} \tag{7.56}$$

Because the cumulative values Q are measured experimentally, the instantaneous value Q_{inst} is obtained by direct differentiation of the measured values according to

$$Q_{inst} = \frac{d(QC_p)}{dC_p} \tag{7.57}$$

7 Reactivity ratios are fixed under specific solvent and temperature conditions but can change as solvent and temperature change.

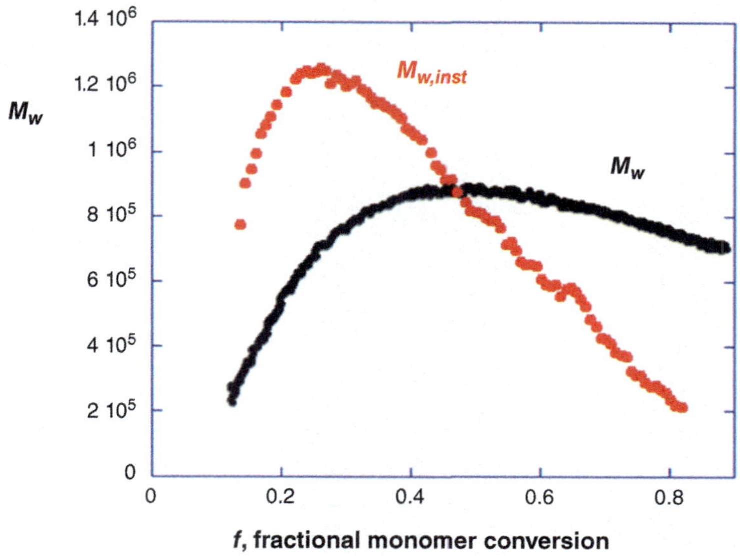

Figure 7.37 Cumulative and instantaneous weight average molar mass, M_w and $M_{w,inst}$, respectively. (Data courtesy of Terry McAfee and Aide Wu).

Figure 7.37 is an example of cumulative M_w and its instantaneous value $M_{w,inst}$ obtained from ACOMP data during a semi-batch copolymerization of acrylamide. In this reaction, monomer was fed into the reactor, causing $M_{w,inst}$ and M_w to rise. At 25% conversion, $f = 0.25$, the feed was cut, the reaction became a batch reaction, and $M_{w,inst}$ and M_w decay in the typical way for a batch reaction.

In a continuous reactor at steady state, there is no change in the distributions of M_w, η_r and reagent concentrations. Hence, $Q = Q_{inst}$ in this case.

7.10 Fully Automatic Feedback Control of Molar Mass and Conversion During Chain Growth Polymerization

We saw that in batch free radical polymerization that the instantaneous molar mass is controlled by the kinetic chain length, Equation 7.22, and that this in turn depends on monomer concentration, free radical concentration, and chain transfer agent concentration, if any of the latter is used. For free radical copolymerization, the instantaneous comonomeric composition of copolymer chains depends on the reactivity ratios of the species and the instantaneous concentrations of the comonomers.

Because molar mass and copolymer composition have strong effects on the properties of end products, it has long been a goal to control these. One way is with temperature, but it is a rather coarse tool, changes relatively slowly, and affects all kinetic parameters simultaneously. As discussed above, addition of reagents during polymerization is referred to as *semi-batch*. It is readily seen in the kinetic equations above that if the various concentrations of reagents can be controlled through feeding the reactor from reagent reservoirs, then it should be possible to control reactions and steer them to desired final end product properties.

Because ACOMP collects instantaneous weight average molar mass $M_{w,inst}$, and instantaneous copolymer composition F_{inst}, it is possible to mathematically establish target reaction trajectories offline, which will give desired end products, and then use semibatch feed to automatically follow these target trajectories using feedback control in real time. Figure 7.38 is a schematic of a reactor and its major process control variables, an ACOMP system, and a feedback controller. Oxygen is a free radical scavenger and will slow down or stop a reaction when present. It is normally purged with nitrogen or other inert gas. O_2, or even just air, is frequently used to slow down or stop runaway exothermic polymerization reactions. According to Equation 7.22, adding monomer will increase instantaneous chain length, whereas adding initiator will both decrease chain length and speed up the reaction. Increasing temperature will increase free radical concentration and so essentially have the same effect as adding initiator. Adding a chain transfer agent will shorten chains. Using differential flows of comonomers will vary the copolymer composition.

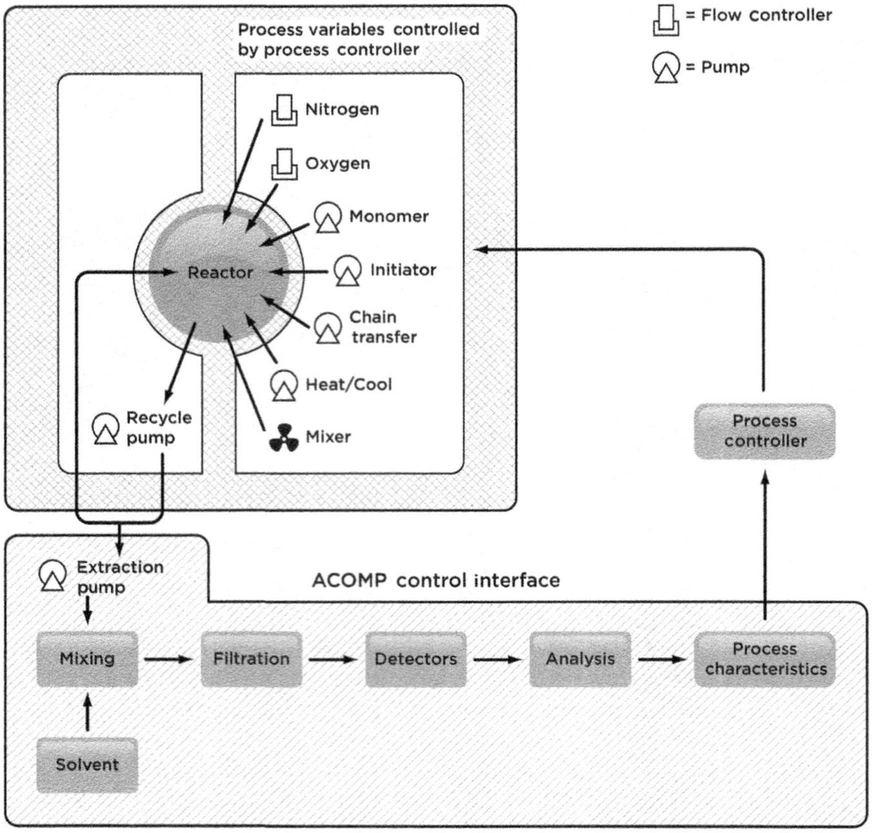

Figure 7.38 Schematic of a reactor with various process control variables, and an ACOMP system with a process controller. (Reproduced with permission from McAfee, T. et al. (2016). with permission from American Chemical Society).

A simple feedback controller can be devised whereby a single measurable proportionality constant α can be used to control the conversion rate. The consumption of monomer can be considered first order *instantaneously*, in the following form:

$$\frac{d[m]}{dt} = -\alpha[m] \qquad 7.58$$

where α replaces $k_p[R\bullet]$ in Equation 7.12. Equation 7.16b shows that, in the QSSA, $k_p[R\bullet]$ depends on four different parameters, all of which are temperature sensitive. α subsumes all these parameters without any need of explicitly knowing any of them. α can change throughout the reaction due to various factors. Equation 7.58 holds at least over short intervals. Because α is experimentally measurable over these short intervals from the measured $[m](t)$ and its derivatives, its value will automatically account for the net change due to any of the variables it subsumes.

Control of a reaction trajectory is achieved operationally by establishing finite control intervals which are very short compared to the duration of the polymerization reaction. First, a *target trajectory*[8] $[m]_t(t)$ is established, which, when followed during the polymerization reaction, will yield a final product of desired properties. For the entire reaction, the target trajectory $[m]_t(t)$ can be followed closely by automatically feeding monomer into the reactor at predicted rates during each

8 The target trajectory is often termed the *control path*.

control interval. The controller takes the measured $[m](t)$ at the beginning of a control interval and sets the monomer pump rate so that $[m]$ will be identical to the target trajectory $[m]_t$ by the end of the control interval. This process is iterated throughout the reaction. Figure 7.39 is an example of a sinusoidal trajectory for $[m]_t(t)$ and the result of an actual polymer reaction controlled in this way (Free radical polymerization of acrylamide). For contrast, Figure 7.39 also shows the exponential decrease of monomer that occurs in an uncontrolled batch reaction. The simple controller was robust in closely following the target trajectory.

A second measurable proportionality constant p can substitute the factors in Equation 7.22 for instantaneous kinetic chain length, which allows $M_{w,inst}$ to be expressed as

$$M_{w,inst} = p[m] \quad 7.59$$

p hence subsumes the temperature-dependent factors involved in $[R\bullet]k_p$, and the concentration of chain transfer agent, if any, and its associated transfer constant, as well as the instantaneous proportionality between the instantaneous weight- and number average molar masses, $M_{w,inst}$ and $M_{n,inst}$, respectively. A target trajectory for molar mass $M_{w,t}(t)$ or instantaneous molar mass $M_{w,inst,t}$ is established offline to produce a desired end product molar mass distribution. Hence, with α, p, $[m]$, M_w, and $M_{w,minst}$ measured at the start of each control interval $[m]$ can be adjusted by pump flow to meet $M_{w,inst,t}$ and M_w at the end of the control interval. Figure 7.40 shows a result for this controller where a target $M_{w,t}(t)$ was established to increase linearly in time. The controller performed well in following the target trajectory. Also shown is the experimentally measured exponential decay of $M_w(t)$ for an uncontrolled batch reaction, as predicted by Equation 7.18.

This approach to control can be extended to composition control in copolymerization by the following simple observation. If there are N comonomer species that copolymerize the average instantaneous composition of comonomer k, m_k, is, by definition

$$F_{inst,k} = \frac{d[m_k]}{d\left\{\sum[m_j]\right\}} = \frac{d[m_k]/dt}{d\left\{\sum[m_j]\right\}/dt} = \frac{\alpha_k}{\sum \alpha_j} \quad 7.60$$

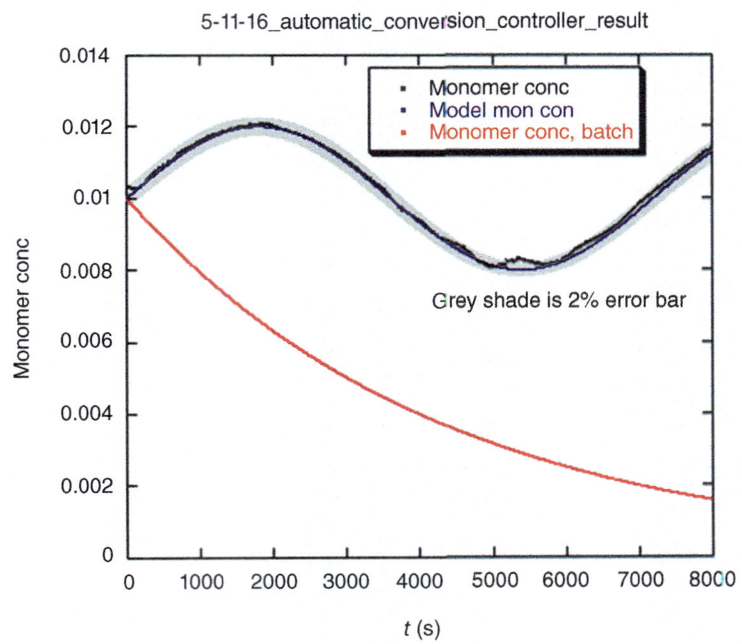

Figure 7.39 Fully automatic control of monomer concentration in a semibatch acrylamide free radical polymerization. The target trajectory is sinusoidal and the experimental, controlled trajectory follows it within a 2% error bar, shown by the gray band. The monotonic decay is what occurs in an uncontrolled batch reaction. (Reproduced with permission from McAfee, T. et al. (2016). with permission from American Chemical Society).

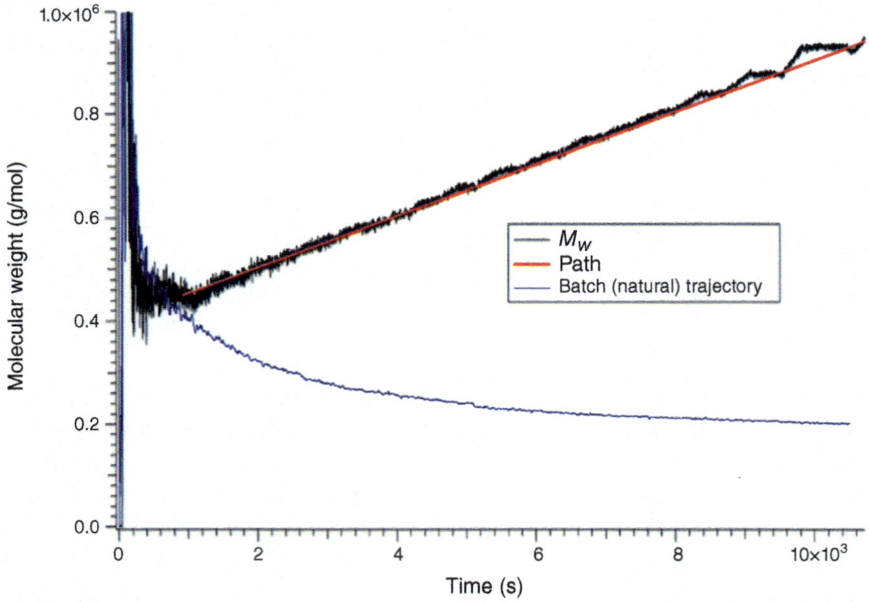

Figure 7.40 A target trajectory in which M_w increases linearly in time in a semibatch reaction, along with the experimental result that was, on average, within 1% of the target trajectory. Also shown, is the monotonic decay of M_w in a batch reaction. (McAfee, Montgomery, Reed, unpublished results).

In other words, it is sufficient to measure the monomer consumption rates of each comonomer, α_k, in order to obtain $F_{inst,k}$. Hence, a target trajectory for composition can be established, $F_{inst,k,t}(t)$ for each comonomer, and by using the measured value of the α_k at the beginning of a control interval, the right amount of comonomer from each of N comonomer reservoirs can be pumped in so that $F_{inst,k} = F_{inst,k,t}$.

In the case of $N = 2$, it is sufficient to measure the two rates, α_1 and α_2, in order to compute the flow rates from reservoirs of comonomers 1 and 2 that are needed to follow $F_{inst,1,t}$. For $N = 2$, $F_{inst,2,t}$ is automatically determined since

$$F_{inst,1,t} + F_{inst,2,t} = 1 \qquad 7.61$$

The controller will only specify the ratio of feed rates from reservoirs of comonomers 1 and 2, so an extra degree of freedom, the actual amount of each monomer pumped in, can be used to simultaneously follow a molar mass trajectory. Figure 7.41 shows the results of simultaneous composition and molar mass control for the copolymerization of acrylamide (Am)

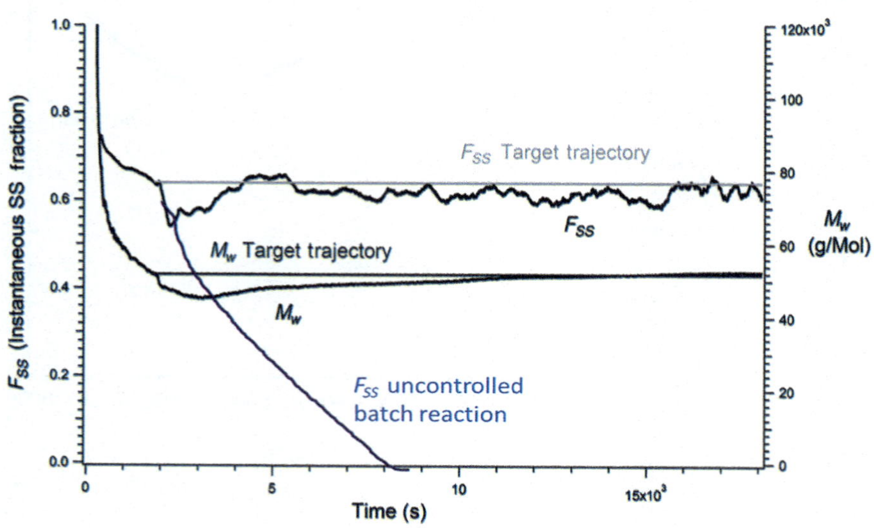

Figure 7.41 Simultaneous control of molar mass and composition in the aqueous semibatch free radical copolymerization of acrylamide and styrene sulfonate. (Terry McAfee et al. (2018) / Reproduced with permission of Elsevier).

and styrene sulfonate (SS), where the target trajectories for both $M_{w,t}(t)$ and $F_{inst,1,t}(t)$ were set to remain constant. The figure also shows the decay of F_{ss} that occurs in an uncontrolled batch reaction. The SS is used up quickly because it has a much higher reactivity ratio than Am.

7.11 Linear Step Growth Reactions

In step growth reactions, monomers are not added one at a time to growing chains, as is the case for chain growth polymerization. Rather, growing polymers of any length can bond to each other, as well as to monomers, according to any restrictions imposed by the chemistry itself. In step growth reactions, a central feature is equal reactivity among growing polymers, regardless of chain length. This means that polymer chains remain reactive throughout the polymerization, in contrast to free radical chain growth where chains rapidly terminate and become dead polymer chains. It is hence possible in step growth reactions that polymers can keep reacting with each other until there is just one single polymer containing all the initial monomers. Monomers that can react with more than one group – i.e. which can form branched chains – are termed *polyfunctional monomers.* When polyfunctional polymers are involved, a network can form. When sufficient branching and cross-linking occurs the network becomes a *gel*. Gel formation is usually unwanted, because it produces insoluble polymers and can even gel the entire reactor, requiring a lengthy and tedious cleaning procedure. When macrogelation occurs in a large industrial reactor, it is sometimes even necessary to use heavy equipment, such as a small excavator or back-hoe, to clean it out.

Consider two identical, bifunctional monomers A that combine in a polycondensation reaction, such as amino acids which combine and release a water molecule to form a dimer A–A via a peptide bond (Chapter 3.1.1). Further growth involves reactions such as

$$A-A+A \rightarrow A-A-A, \text{ trimer}, A-A-A+A-A \rightarrow A-A-A-A-A, \text{ pentamer} \tag{7.62}$$

and, in general

$$A_n + A_m \rightarrow A_{m+n} \tag{7.63}$$

Once A is incorporated inside a growing, linear chain, and has an A on each side, it can no longer react, because A is difunctional, but the two A end groups can continue to react. Let N_T be the initial number of monomers and N_D be the number that have reacted and are internal to the chain, and so cannot react further. The degree of completion of the reaction p, is then given by

$$p = \frac{N_D}{N_T} \tag{7.64}$$

Initially, $p = 0$ and when all monomers have reacted and can react no further $p = 1$. The number average chain length, or number average degree of polymerization, X_n, is the total number of monomers divided by the total number of chains. The total number of chains N_C in step growth reactions is equal to the total number of unreacted monomers and is $N_C = N_T - N_D$, so that

$$X_n = \frac{1}{1-p} \tag{7.65}$$

An important distinction in step growth reactions is that all monomers, whether bound in chains or free are taken into the average. Hence, a free, reacted monomer is considered a chain of length one. In chain growth reactions, because the population is bimodal, with long polymer chains widely separated from the monomer, only monomers bound in chains are taken into the average, and free monomers are excluded from the average.

An interesting case of linear step growth is when there are two monomers, A and B, with reactivity ratios equal to zero. As discussed in Section 7.8 for free radical copolymerization this leads to a strictly alternating copolymer, ABABABABAB. In this case, if the total initial number of A monomers, $N_{T,A}$, is different from the total initial number of B monomer, $N_{T,B}$, the reaction will stop when the least abundant monomer is used up. In this way, there can be stronger control over the final degree of polymerization than in the homopolymer case. Important examples of this include poly(ethylene terephthalate), PET, and Nylon 66.

For the strictly alternating copolymer, let A be the less abundant monomer and r the ratio of the less abundant to the more abundant monomer

$$r = \frac{N_{T,\mathrm{A}}}{N_{T,\mathrm{B}}} < 1 \qquad 7.66$$

If $N_{D,\mathrm{A}}$ and $N_{D,\mathrm{B}}$ represent the number of monomers that have reacted and can no longer react, for A and B, respectively, then the number of chains is

$$N_C = N_{T,\mathrm{A}} + N_{T,\mathrm{B}} - \left(N_{D,\mathrm{A}} + N_{D,\mathrm{B}}\right) \qquad 7.67$$

X_n is again the ratio of total monomer $N_T = N_{T,\mathrm{A}} + N_{T,\mathrm{B}}$, to the total number of chains, so that

$$X_N = \frac{(1+r)}{\left[1 + r - \left(\frac{N_{T,\mathrm{A}} + N_{T,\mathrm{B}}}{N_{T,\mathrm{B}}}\right)\right]} \qquad 7.68$$

At the end of the reaction $N_{D,\mathrm{B}} = N_{D,\mathrm{A}} = N_{T,\mathrm{A}}$, so that the final number average chain length is

$$X_N = \frac{(1+r)}{(1-r)} \qquad 7.69$$

Hence, establishing the initial ratio between A and B allows a predictable final molar mass. Notice that without this type of control, or other, Equation 7.65 predicts that the reaction will tend towards a *single* polymer chain of extremely high molar mass.

While difunctional monomers strictly produce unbranched linear chains, such as polypeptides formed from amino acids, tetrafunctional and higher functional monomers will produce branching and cross-linking. The consequences of branching and cross-linking are quite different in step growth polymerization compared to chain growth.

7.12 Molar Mass Distributions (MMD) and Averages[9]

Polymer populations in which all chains have identical masses are termed *monodisperse*. Proteins and DNA of the same type are necessarily monodisperse, as they are coded for precise monomer sequences.

In general, however, synthetic polymer populations contain a wide distribution of masses. Such populations are termed *disperse*. The previous, long-held terminology for this was *polydisperse*. While most synthetically produced polymers are disperse, many biopolymers are also produced in disperse populations, e.g. many polysaccharides and glycoproteins.

Some methods of producing polymers can lead to distinct subpopulations, widely separated in mass. The population is termed *unimodal* if there is only one maximum of the population. When two or more identifiable maxima exist within a polymer population it is termed *multimodal*, i.e. *bimodal* for two subpopulations, *trimodal* for three, etc.

As in any situation where probability distributions are involved useful averages are defined which act as hallmarks for related properties. Some of the important distributions commonly found in polymer populations are outlined in Appendix I, as well as the basics of finding averages and moments. Here, we adapt that discussion to the task of computing molar mass averages from given distributions.

Let N_i be a discrete distribution that gives the number of chains in a population of chain length i. We can compute the average $\langle M \rangle$ of any variable M_i according to

$$\langle M \rangle = \frac{\sum_{i=i_{min}}^{i_{max}} M_i N_i}{\sum_{i=i_{min}}^{i_{max}} N_i} \qquad 7.70$$

where i_{min} is the minimum, physically meaningful value of i, often equal to 0 or 1, and i_{max} is the maximum value, which can often be taken as infinite. If the distribution function is a continuous density distribution, $n(x)$, where $n(x)\,dx$ is the number of chains with lengths in the interval x to $x + dx$, then

9 We use "molar mass distributions" instead of the more frequently used "molecular weight distribution," because the molar masses are always expressed in mass/mole not in terms of weight = mass × acceleration due to gravity.

$$\langle M \rangle = \frac{\int_{x_{min}}^{x_{max}} M(x) n(x) \mathrm{d}x}{\int_{x_{min}}^{x_{max}} n(x) \mathrm{d}x} \tag{7.71}$$

The *moments of a distribution* are described in Appendix I. The *j*th moment of a discrete distribution N_i about the origin is defined as

$$\mu_j = \langle i^j \rangle = \sum_i i^j N_i \tag{7.72}$$

The zeroeth moment, μ_0, is $\sum_i N_i$ which is 1 for a normalized distribution, the first moment is $\mu_1 = <i>$, the second moment is $\mu_2 = <i^2>$, etc. The higher moments of a distribution hence compute successively higher averages of the distribution's independent variable, e.g. in classical statistical thermodynamics the mean square velocity $<v^2>$ is the second moment of the Maxwell–Boltzmann speed distribution for an ideal gas, and is directly related to average kinetic energy $<KE>=m<v^2>/2$, and hence to temperature ($<KE> = 3k_BT/2$ for a monatomic gas).

For a *one-parameter distribution*, it suffices to know any moment (higher than the zeroeth) to completely define the distribution. The geometric and Poisson distributions are examples of one-parameter distributions. For a *two-parameter distribution*, such as the Gaussian, two moments (higher than the zeroeth) completely define the distribution, and so on for distributions with increasing numbers of parameters. Other distributions may require more than two moments to define them.

The moments of a polymer's mass distribution are used to define the important types of averages corresponding to a given quantity *M*. Notice that the first moment corresponds to taking a simple number average over the distribution. We hence define the number average of *M* as the *number average mass* $<M>_n$ (which we will often abbreviate simply as M_n)

$$\langle M \rangle_n = \frac{\sum_i M_i N_i}{\sum_i N_i} \tag{7.73a}$$

and for a continuous distribution

$$\langle M \rangle_n = \frac{\int M n(M) \mathrm{d}M}{\int n(M) \mathrm{d}M} \tag{7.73b}$$

where $n(M)\mathrm{d}M = \frac{\partial N}{\partial M}\mathrm{d}M$ represents the number of chains in the mass interval from *M* to *M* + d*M*. Because the various averages involve ratios involving *n*(*M*), this quantity can either represent the total number of chains in *M* to *M* + d*M* or number of chains per volume in *M* to *M* + d*M*. Notice that $<M>_n$ is the ratio of the first to zeroeth moments of the distribution,

$$\langle M \rangle_n = \frac{\mu_1}{\mu_0} \tag{7.74}$$

The *weight average* (more properly termed *mass average*, but polymer tradition uses the former term) of any quantity Q_i of a polymer population is defined as

$$\langle Q \rangle_w = \frac{\sum_i Q_i M_i N_i}{\sum_i M_i N_i} \tag{7.75a}$$

and for the continuous function *n*(*M*) and *Q*(*M*)

$$\langle Q \rangle_w = \frac{\int Q(M) M n(M) \mathrm{d}M}{\int M n(M) \mathrm{d}M} \tag{7.75b}$$

The weight average of a quantity hence uses the total mass of M_i, which is *M*(*i*)*N*(*i*) to weight the average rather than the number of masses. The *weight average mass*, $<M>_w$, sometimes abbreviated M_w, is hence

$$\langle M\rangle_w = \frac{\sum_i M_i^2 N_i}{\sum_i M_i N_i} = \frac{\sum_i M_i c_i}{\sum_i c_i} \tag{7.76a}$$

where $c_i = N_i M_i / N_A V$ is the mass concentration (g/cm^3) of chains of mass M_i, and N_i/V is the number of chains of length i in volume V.

For the continuous function $n(M)$:

$$\langle M\rangle_w = \frac{\int M^2 n(M)\mathrm{d}M}{\int M n(M)\mathrm{d}M} \tag{7.76b}$$

$\langle M\rangle_w$ is the ratio of second to first moments of the $n(M)$ distribution:

$$\langle M\rangle_w = \frac{\mu_2}{\mu_1} \tag{7.77}$$

A third average that is frequently important in polymer science is the so-called *z-average*. It is defined for a quantity $Q(i)$ as

$$\langle Q\rangle_z = \frac{\sum_i Q_i M_i^2 N_i}{\sum_i M_i^2 N_i} \tag{7.78a}$$

and for the continuous functions $n(M)$ and $Q(M)$:

$$\langle Q\rangle_z = \frac{\int Q(M) M^2 n(M)\mathrm{d}M}{\int M^2 n(M)\mathrm{d}M} \tag{7.78b}$$

The z-average mass, $\langle M\rangle_z$ is the ratio of the third to second moments of $n(M)$:

$$\langle M\rangle_z = \frac{\mu_3}{\mu_2} \tag{7.79}$$

It is customary in polymer science to deal with *dispersity*[10] *indices*. The two most common are M_w/M_n and M_z/M_w. Ideal polymer reactions have well-defined ratios, which are often approached in real polymerization reactions. In common polymer parlance, the particular ratio M_w/M_n is referred to as *the polydispersity*, or simply *the dispersity*.

It is also important to note that the differential concentration $C(M)$ is $C(M) = Mn(M) = \frac{\partial c}{\partial M}$ where $C(M)\mathrm{d}M$ is the concentration of polymer in the interval M to $M + \mathrm{d}M$, with the normalization condition that

$$c_{total} = \int_0^\infty C(M)\mathrm{d}M \tag{7.80}$$

where c_{total} is the total concentration of polymer in a population. The quantity $C(M)$ is used experimentally so often that M_n, M_w, and M_z are given explicitly here, as well as the form in discrete mass concentration, c_i (g/cm^3), of polymers of molar mass M_i and molar density n_i (moles of chains of mass M_i per cm^3):

$$M_n = \frac{\int C(M)\mathrm{d}M}{\int \frac{C(M)}{M}\mathrm{d}M} = \frac{\sum_i c_i}{\sum_i n_i} = \frac{c_{total}}{\sum_i \frac{c_i}{M_i}} \tag{7.81}$$

$$M_w = \frac{\int MC(M)\mathrm{d}M}{\int C(M)\mathrm{d}M} = \frac{\sum_i M_i c_i}{\sum_i c_i} = \frac{\sum_i M_i c_i}{c_{total}} \tag{7.82}$$

10 *Dispersity* is the IUPAC term for a population containing a distribution of sizes or masses. Historically, the somewhat redundant term *polydispersity* was frequently used and continues to be used. They have the same physical meaning.

$$M_z = \frac{\int M^2 C(M)\mathrm{d}M}{\int MC(M)\mathrm{d}M} = \frac{\sum_i M_i^2 c_i}{\sum_i M_i c_i} \tag{7.83}$$

7.12.1 Molar Mass Averages of the Geometric Distribution

This discrete distribution is introduced in Appendix I and is a special case of the binomial distribution and is ubiquitous in polymer science. It arises when only one sequence out of the $N!/[(N-j)!j!]$, possible sequences are relevant for obtaining j successes.

If ideal free radical polymerization chains are formed very quickly with respect to the total time of monomer conversion, then any given propagating chain sees an environment (concentration of radicals, monomers, solvent, etc.) that changes little during its formation. It is of interest to know after N free radical events what the probability is that there were $N-1$ consecutive events ($j = N-1$) that chained monomers to free radical, and a final "chain killing" event that terminated the chain by a radical–radical encounter. Although there are $N!/[(N-1)!1!] = N$ different sequences of events that give $N-1$ monomers and one terminated radical in a chain, the only physically possible sequence is when the terminated radical is at the end of the chain.

The probability q of adding a monomer, instead of terminating via a collision with a free radical is

$$q = \frac{\text{rate of propagation}}{\text{rate of propagation} + \text{rate of termination}} = \frac{k_p[m]}{k_p[m] + k_t[\mathrm{R}\bullet]} \tag{7.84}$$

The probability that an $R_1\bullet$ radical is terminated on the next step is $(1-q)$. The probability that it adds a monomer is q. The probability that this two-unit chain terminates on the next step is $(1-q)q$. In general, the probability that a chain of length i is reached is

$$P_i = (1-q)q^{i-1} \tag{7.85}$$

It is seen in Appendix I that Equation 7.85 is the BD in the case where $j = N-1$ and the combinatorial term is omitted, for reasons just discussed. If n_i is the number of chains of chain length i, where n_{total} is the total number of chains then n_i will simply be equal to

$$n_i = n_{total} P_i \tag{7.86}$$

since P_i is a normalized distribution. Figure 7.42 shows the mass or concentration distribution, iP_i versus chain length for free radical polymerization for different values of q.

M_n is computed by the following, where m is the monomer mass:

$$M_n = m\frac{\sum_i iP_i}{\sum_i P_i} = m(1-q)\sum_{i=1}^{\infty} iq^{i-1} = \frac{m}{1-q} \tag{7.87}$$

where the normalized distribution P_i yields one in the denominator of the second expression, and Equation 8.50 was used to obtain the relation

$$\sum_{i=1}^{\infty} iq^{i-1} = \frac{\mathrm{d}}{\mathrm{d}q}\sum_{i=1}^{\infty} q^i = \frac{1}{(1-q)^2} \tag{7.88}$$

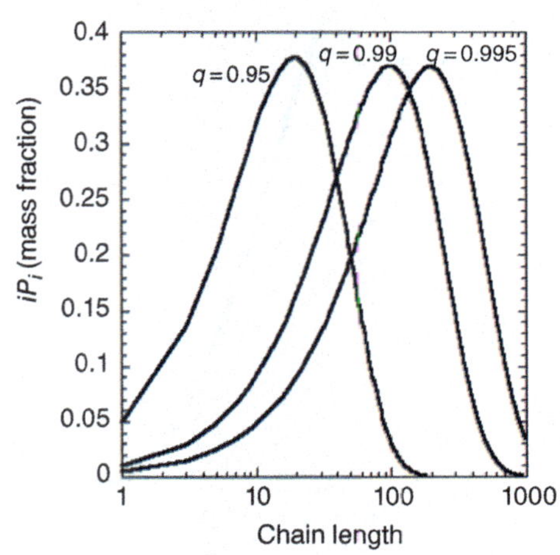

Figure 7.42 Geometric distributions for mass fraction (or concentration) versus chain length for free radical polymerization at different values of q. The distributions are left-skewed.

For M_w,

$$M_w = \frac{m(1-q)\sum_{i=1}^{\infty} i^2 q^{i-1}}{(1-q)\sum_{i=1}^{\infty} i q^{i-1}} = m\frac{1+q}{1-q} \quad 7.89$$

The GD also describes step growth polymerization, where q is replaced by p, the degree of polymerization.

7.12.2 Molar Mass Averages of the Log-Normal Distribution

Because the Gaussian (or normal) distribution is continuous, it lends itself to easy mathematical manipulations, such as averaging. A form for chain length distributions that is found to resemble experimental determinations of MMD in many situations is the *log-normal distribution* (LND):

$$C'(\ln M) = \frac{\partial c}{\partial \ln M} = \frac{\exp[-[\ln(M/M_0)]^2/2\sigma^2]}{\sigma\sqrt{2\pi}} \quad 7.90$$

where $C'(\ln M)\mathrm{d}(\ln M)$ is the concentration of chains in the interval from $\ln M$ to $\ln M + \mathrm{d}\ln M$, and M_0 is the average mass, which corresponds to the peak value of the distribution. The relationship between $C'(\ln M) = \frac{\partial c}{\partial \ln M}$ and $C(M) = \frac{\partial c}{\partial M}$ is

$$C(M) = \frac{\partial c}{\partial \ln M}\frac{\partial \ln M}{\partial M} = \frac{C'(M)}{M} \quad 7.91$$

where $C(M)$ is the differential concentration used in Equations 7.80–7.83. You can verify that the LND in Equation 7.90 is normalized.

MMD determined by *size exclusion chromatography* (SEC), also termed *gel permeation chromatography* (GPC) often resembles the LND. Figure 7.43 is an example of a MMD determined by SEC for a free radical acrylamide polymerization. It is fairly well fit by a LND. However, there is some left skewed asymmetry in the experimental MMD, which is expected from the above treatment showing that free radical polymerization is expected to give a geometric distribution, such as in Figure 7.42. Some of the geometric distribution asymmetry is lost in the diffusive SEC separation process. We recall from Chapter 6 that the diffusion equation has Gaussian solutions.

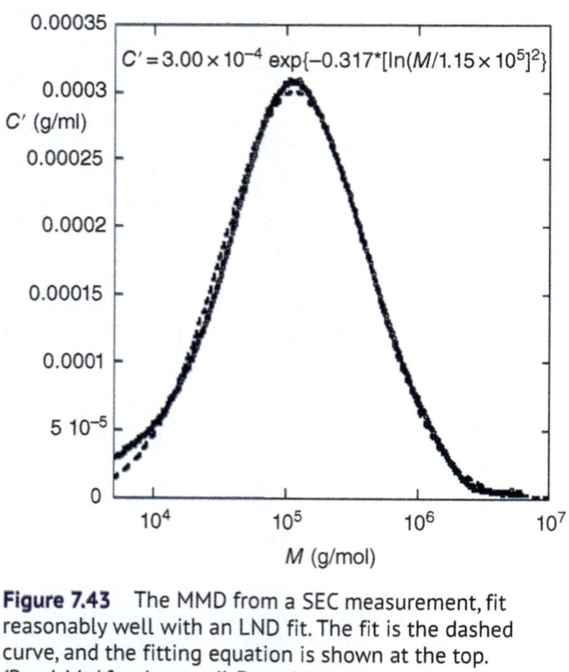

Figure 7.43 The MMD from a SEC measurement, fit reasonably well with an LND fit. The fit is the dashed curve, and the fitting equation is shown at the top. (Reed, McAfee, Leonardi, Drenski, Mongomery. Unpublished results).

Recalling that $C(M) = \frac{C'(M)}{M}$ the integrals needed to compute M_n, M_w, and M_z are:

$$\int \frac{C}{M}\mathrm{d}M = \int \frac{C'}{M^2}\mathrm{d}M \; \int C\mathrm{d}M = \int \frac{C'}{M}\mathrm{d}M \quad 7.92a$$

$$\int CM\mathrm{d}M = \int C'\mathrm{d}M \quad 7.92b$$

$$\int CM\mathrm{d}M = \int C'\mathrm{d}M \quad 7.92c$$

$$\int M^2 C\mathrm{d}M = \int MC'\mathrm{d}M \quad 7.92d$$

Convenient substitutions are

$$z = \ln\left(\frac{M}{M_0}\right)$$

$$M = M_0 e^Z$$

$$\mathrm{d}M = M\mathrm{d}z$$

$$\alpha = \frac{1}{\sigma^2} \quad 7.93a–d$$

which yields

Table 7.2 β in Equation 7.97 for the nth moment of the LND around the origin.

n^{th} moment	$\beta = \frac{(n-1)^2}{4\alpha}$
0	$1/4\alpha$
1	1
2	$1/4\alpha$
3	$1/\alpha$
4	$9/4\alpha$
5	$4/\alpha$
6	$25/4\alpha$

$$\int \frac{C}{M}\text{d}M = \frac{A}{M_0\sqrt{2\pi\sigma^2}}\int_{-\infty}^{\infty} \text{e}^{-\left(z+\alpha z^2\right)}\text{d}z \propto \frac{\text{e}^{1/4\alpha}}{M_0} \tag{7.94a}$$

$$\int C\text{d}M = M_0 A \tag{7.94b}$$

$$\int MC\text{d}M = \frac{AM_0^{\ 2}}{\sqrt{2\pi\sigma^2}}\int_{-\infty}^{\infty} \text{e}^{z+\alpha z^2}\text{d}z \propto M_0^{\ 2}\text{e}^{1/4\alpha} \tag{7.94c}$$

$$\int M^2 C\text{d}M = \frac{AM_0^{\ 3}}{\sqrt{2\pi\sigma^2}}\int_{-\infty}^{\infty} \text{e}^{2z+\alpha z^2}\text{d}z \propto M_0^{\ 3}\text{e}^{1/\alpha} \tag{7.94d}$$

From these, the following expressions are obtained for M_n, M_w, M_z

$$\begin{aligned} M_n &= M_0\text{e}^{-\sigma^2/2} \\ M_w &= M_0\text{e}^{\sigma^2/2} \\ M_z &= M_0\text{e}^{3\sigma^2/2} \\ \frac{M_w}{M_n} &= \text{e}^{\sigma^2}, \quad \frac{M_z}{M_w} = \text{e}^{\sigma^2} \end{aligned} \tag{7.95a–d}$$

A helpful relationship in the above integrals is

$$\int M^{n-1}C\text{d}M = \int M^{n-2}C'\text{d}M = \int \text{e}^{(n-1)z-\alpha z^2}\text{d}z \propto \text{e}^{\beta} \tag{7.96}$$

where

$$\beta = \frac{(n-1)^2}{4\alpha} \tag{7.97}$$

The first few values of β are shown in Table 7.2.

7.12.3 Poisson Distribution for "Living" Type Reactions

The Poisson distribution is applicable to chain growth reactions where initiation of the propagating species is very quick, termination is a negligible mechanism and the time it takes for the chain to grow is comparable to the time it takes for complete conversion of monomer; i.e. living polymerization.

In ideal living polymerization, there is a population of a fixed number of slowly growing chains. The number of initiated monomers at the outset of the reaction is equal to the concentration of initiator $[\text{I}]_0$, e.g. lithium alkyls. This number does not change throughout the reaction. Over any short interval, there is an average chain size $\langle j \rangle$, and the number of monomers in a chain is small compared to the total number of monomers in the reactor. The Poisson chain length distribution $W(j)$ is

$$W(j) = \text{e}^{-\tau}\frac{\tau^{j-1}}{(j-1)!} \tag{7.98}$$

where τ is related to both monomer conversion and the ratio of converted polymer to initial initiator concentration I_0 via

$$\tau = \frac{M_0 - M}{[\text{I}]_0} \tag{7.99}$$

Obviously, τ increases as the reaction proceeds and monomers are added to growing chains. What is remarkable, however, is that because the width-to-average ratio of the chain length distribution is $1/\text{sqrt}(\tau)$, *the dispersity in living polymerization actually decreases as the reaction proceeds.*

Performing the averaging operations to determine N_n and N_w yields the following results

$$N_n = 1 + \tau \tag{7.100}$$

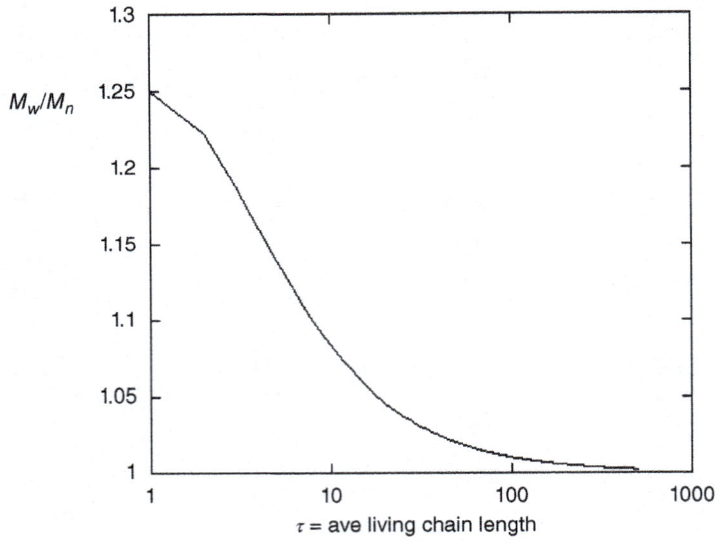

Figure 7.44a Evolution of dispersity as a function of chain length for an ideal living polymerization reaction, governed by the Poisson distribution.

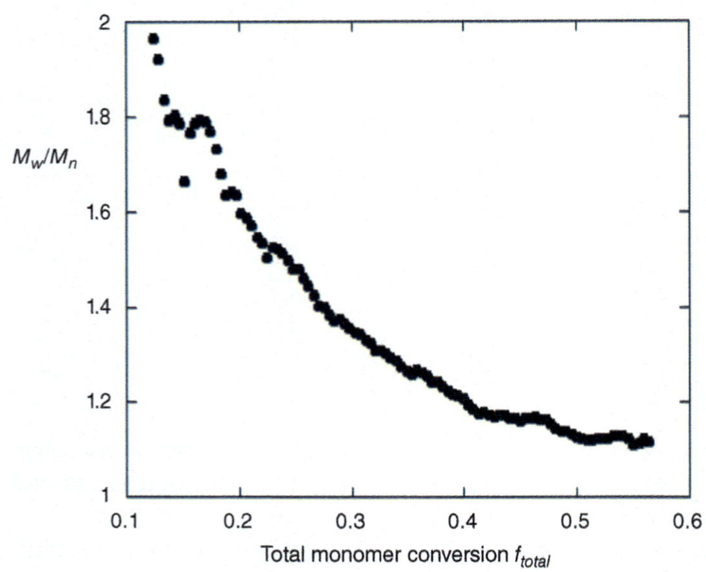

Figure 7.44b Dispersity during a living (RAFT) polymerization of acrylamide. The dispersity decreases with conversion, as expected. (Reproduced with permission from Elsevier, from Siqueira, J.S. et al. (2021)).

$$N_w = 1 + \tau + \frac{\tau}{1+\tau} \qquad 7.101$$

The dispersity index is then given by

$$\frac{N_w}{N_n} = 1 + \frac{\tau}{(1+\tau)^2} \qquad 7.102$$

Living polymerization reactions are of great practical importance in producing low-dispersity polymers. Figure 7.44a shows how dispersity theoretically decreases and approaches unity in the limit of long living chains. Figure 7.44b shows dispersity data from ACOMP for RAFT copolymerization of 90% acrylamide and 10% sodium styrene sulfonate. The dispersity decreases markedly, as predicted for living polymerization, although the actual values are higher than the theoretical expression.

7.13 Experimental Methods for Determining Molar Mass Distributions

Several techniques exist for determining different averages of M. These include membrane osmometry for direct determination of M_n (Chapter 4.13.2), and light scattering for M_w. Other techniques determine averages of other quantities, which, via a model, can be related to one or more of the above averages. Determination of intrinsic viscosity, for example, generally yields a weight average intrinsic viscosity, which can be related to a moment of M, if the relation between viscosity and M is known. Likewise, dynamic light scattering can yield a z-average diffusion coefficient directly, which can be related to M.

Currently, the most widely used means of determining an entire population distribution $C(M)$ is via a *separation technique*. These include *size exclusion chromatography* (SEC), *field flow fractionation* (FFF), *matrix assisted laser desorption ionization time of flight mass spectroscopy* (MALDI-TOF), *capillary hydrodynamic fractionation, gel electrophoresis, Fourier transform mass spectroscopy*, and several others. Since knowledge of $C(M)$ furnishes the most complete description of a polymer distribution, there is intensive development and application of these techniques currently in progress.

7.13.1 Size Exclusion Chromatography

Because this technique is one of the mainstays of polymer characterization, a brief overview is given here. SEC is often termed *gel permeation chromatography* (GPC), although the mechanism of SEC is more restrictive than that of GPC. The heart of this method is the packed column, which contains a separation medium that creates a porous network with a spectrum of pore sizes. A small volume of polymer-containing liquid is injected into the column with an injector valve and this injected volume is entrained in the flow of a carrier liquid, or *eluent*, which is pumped through the column at a constant flow rate. Analytical SEC typically uses injection volumes of 10–100 μl, and flow rates between 0.5 ml/min and 2 ml/min. The back pressure generated on the pump typically ranges from 10s of atmospheres up to about 100 atm.[11] The packed column separates polymers according to their hydrodynamic size so that the original small injected volume spreads out over a much larger volume. Normally, the largest polymers exit the column first and the smallest last. The separated polymer volume flows through one or more detectors, which can then determine various properties of the separated polymer, the most important usually being the molar mass distribution. Figure 7.45 shows a schematic of a multidetector SEC system.

The column packing material is typically composed of beads of material such as silica, or polymeric gels, based on cross-linked polystyrene, poly(methyl methacrylate), polyacrylamide, dextran, and other materials. The size distribution of pores is critical as this determines the working molar mass range of the polymers to be fractionated. Manufacturers typically have a range of columns with different pore sizes, each aimed at separation over a specific molar mass range. Sometimes columns are connected in series in order to increase the molar mass range, or, a gradient of pore-size distribution is incorporated in a single column. There are columns for use with organic solvents, such as tetrahydrofuran (THF), toluene, chloroform, acetone, trichlorobenzene, and many others. Other columns are specially made for aqueous solvents when analyzing biological and water-soluble macromolecules.

The ideal means of separation of a polymer in SEC as it passes through the pores is purely entropic, with no interaction between the polymer and the packing. Consider the spectrum of pore diameters in the packing and the effective

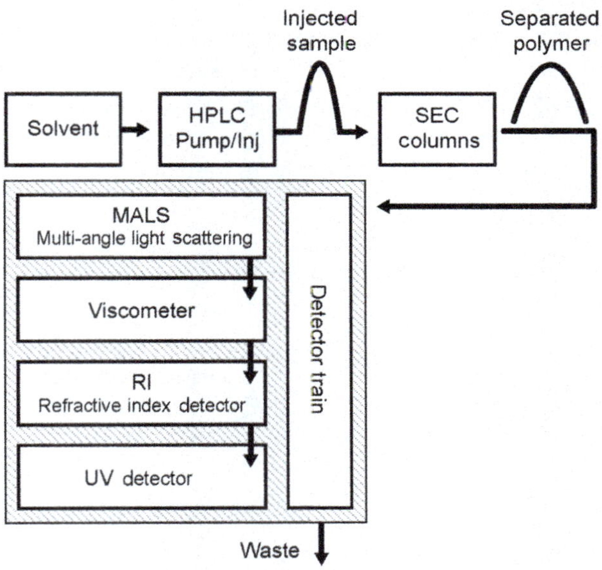

Figure 7.45 Schematic of a typical multidetector SEC system. (Courtesy of Marissa Calko).

11 Ultrahigh pressure SEC, with operating pressures well over 1000 atm are currently in use and vastly shorten analysis time.

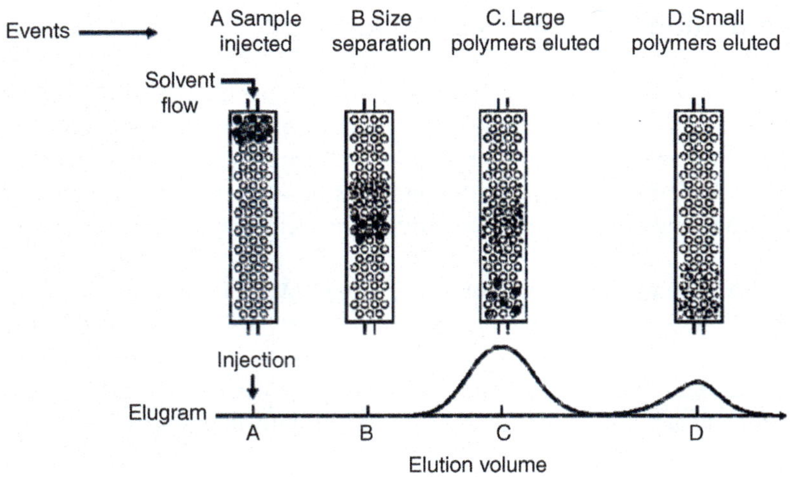

Figure 7.46 Schematic of a bimodal polymer population separated by a SEC column according to size. (Courtesy of Marissa Calko).

diameter of a polymer chain. The smaller the chain diameter the more pores it can randomly fit into and sample as it passes through the column packing, resulting in a longer *residence time* in the column than a larger diameter polymer that has fewer pores that it can fit into, which leads to a shorter residence time. There are two important limits of the column: The *exclusion volume limit*, which is determined by the volume of the column excluding the particles. Polymers beyond a certain size cannot access the pores at all and so all elute in this exclusion volume limit. In this case, no useful fractionation occurs and such molecules might just as well be characterized in unseparated batch measurements. The second limit is the diffusion limit, also termed the *total permeation limit*, which represents the entire volume of the column accessible to liquid, i.e. the smallest polymers can access all pores in the column and so the volume accessible to them is the exclusion volume plus the volume of all the pores. Figure 7.46 is a schematic of the size exclusion separation mechanism, where a bimodal polymer population enters and separates as it passes through the column, with the large particles exiting first and the small particles later.

The purely entropic separation mode of SEC is termed a "size exclusion" mechanism. Frequently, solutes interact with columns, due to charge, hydrophobic, or other effects, in which case the separation is no longer via size exclusion, and care must be taken in the interpretation of the data. Where size exclusion is no longer the main separation mechanism, the more general term 'GPC' applies to the process. The opposite of the entropic mechanism is the enthalpic mechanism, which is based on selective interaction of solutes with the column, and is sometimes termed *affinity chromatography* or *interaction chromatography*. *High-pressure liquid chromatography* (HPLC) is a common technique relying on interaction for separation of low molar mass solutes.

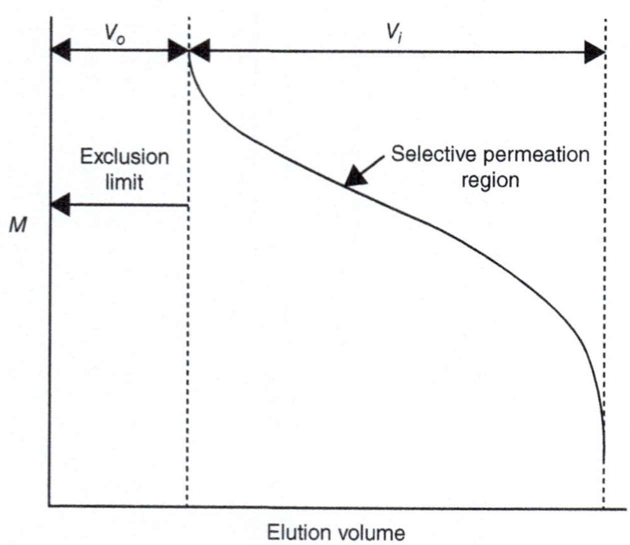

Figure 7.47 A typical SEC column molar mass separation curve.

While manufacturers and the above paragraphs speak of SEC column separation according to molar mass, it should be clear by now that when operating in the desired, purely entropic SEC mode, separation of macromolecules takes place according to their hydrodynamic volume as they sample pore volumes, and is only indirectly related to molar mass. It is hence an ongoing misnomer to rate SEC columns by their ability to separate over molar mass ranges, but it remains an accepted, industry-wide practice.

Figure 7.47 shows what a typical performance curve from a manufacturer looks like; it plots molar mass of macromolecules versus the *elution volume*. The elution volume is the volume inside the column, exclusion volume plus available pore volume for a specific macromolecule, that the macromolecule has traversed by the time it exits. Because detector measurements are made in time and the pump rate is known, the time axis can be converted directly to elution volume. When a polymer of unknown MMD is injected the concentration at each elution volume is assigned a corresponding molar mass, from the elution volume in Figure 7.47. This is termed *SEC column calibration*.

The problem with using SEC column calibration, such as in Figure 7.47, is that it is made from a certain standard polymer, e.g. linear polystyrene standards, and so is not directly applicable to polymers with different hydrodynamic volume to molar mass ratios, and especially not to polymers that have any enthalpic interaction with the column. Hence, there is no expectation that using this type of calibration curve will yield a true molar mass of the polymer sample, unless the sample is polystyrene of the same architecture (e.g. linear) as the calibration styrene itself. In this case, one can only speak of a "polystyrene equivalent molar mass" for any polymer analyzed with this calibration curve.

While column calibration may seem highly arbitrary and make the entire endeavor look unsound, there is nonetheless great practicality in this approach. Consider a polymer manufacturer who has found an MMD based on column calibration for a polymer within acceptable "MMD" specifications. Polymers outside of the acceptable specification will have different distributions. The manufacturer can hence use the "MMD" from column calibration of acceptable products to determine whether other batches of the product are acceptable. In this case, although the "MMD" of the product is only a polystyrene equivalent MMD, and is hence not the correct MMD, consistently obtaining this incorrect MMD is of great practical value for the manufacturer.

The fact that the SEC column separates according to the hydrodynamic volume of a particle instead of molar mass led Benoit et al. in France to propose a method to calibrate columns according to hydrodynamic volume obtained from the intrinsic viscosity $[\eta]$ of the standards. Recall from Chapter 6.6 that intrinsic viscosity is a particle's hydrodynamic volume per unit of molar mass, M. Hence

$$V_H = M[\eta] \tag{7.103}$$

The V_H of the standards, with known M and $[\eta]$, can now be plotted against elution volume. If the SEC system includes a viscometric detector then M can be obtained at each elution volume by dividing the corresponding V_H from the calibration curve by the measured $[\eta]$ of the polymer in order to obtain M.

An especially sensitive viscometer can be used for discrete SEC pulses of a polymer containing fluid in the eluent flow. This viscometer can be built on the Poiseuille principle by forming a hydraulic Wheatstone bridge, shown in Figure 7.48. One pressure transducer is connected across the bridge, indicated as *IP*, because it measures the difference between inlet and outlet pressures, whereas a second one is connected between the arms of the viscometer and measures the difference pressure, *DP*, between the two sides of the bridge. There is a holdup reservoir in one of the arms to keep a supply of pulse carrier solvent in that arm during a pulsed measurement. The holdup reservoir is the equivalent of the single variable resistor in the electrical Wheatstone bridge. The bridge is constructed with identical capillaries so that the pressure drop across each one is the same when pure carrier solvent flows. In this situation, $DP = 0$. When the pulse of polymer solution enters the bridge the pressure drop across all capillaries increases, except the one containing the solvent reservoir, because pure carrier solvent continues to flow in that capillary. The effect is to unbalance the bridge and produce a nonzero *DP*. By using principles similar to those in the analysis of the electrical Wheatstone bridge you can show, using the arrangement of Figure 7.48, that the specific viscosity of the pulse of material passing through the bridge at each instant t is given by

$$\eta_{sp}(t) = \frac{\eta(t) - \eta_0}{\eta_0} = \frac{4DP}{IP - 2DP} \tag{7.104}$$

Because this viscometer operates on the principle of null measurement it is inherently more sensitive than single capillary viscometers and can largely eliminate temperature and pump pressure fluctuations.

Wheatstone bridge viscometers are most useful for SEC measurements, but are normally not suitable for long-term flow measurements, because the hold-up volume is used in one or more bridge arms, which leads to zeroing out of the differential signal with time. Single and bridge capillary viscometer results have been exhaustively compared (Norwood and Reed 1997).

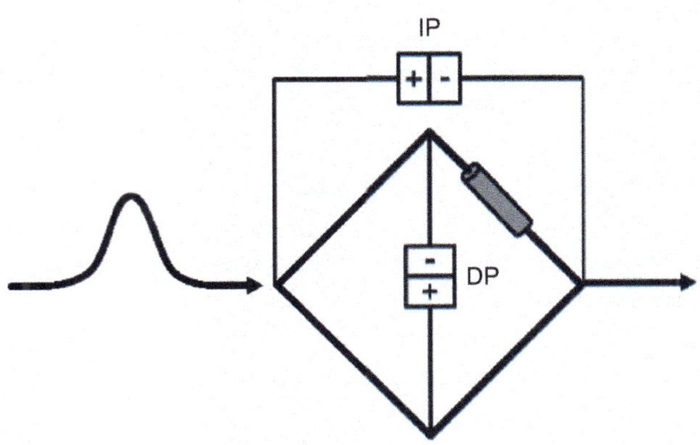

Figure 7.48 Hydraulic Wheatstone bridge with pulse of polymeric solution entering it. (Courtesy of Marissa Calko).

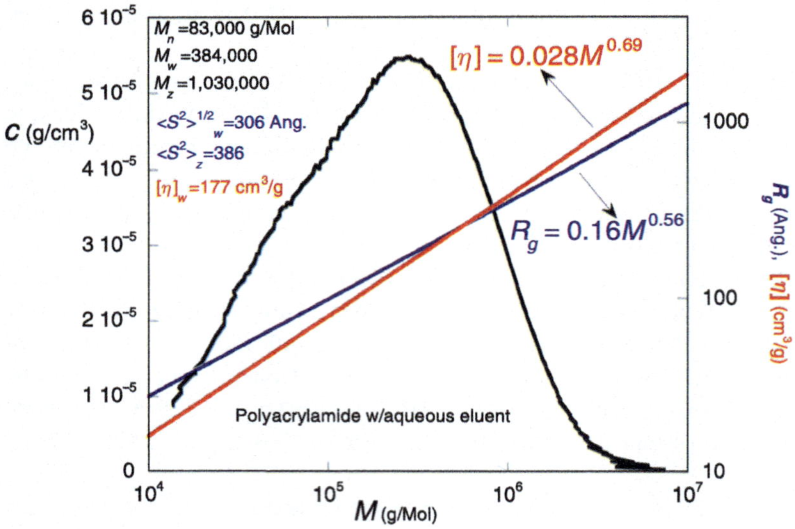

Figure 7.49 MMD results for a highly disperse polyacrylamide sample from multidetector SEC. (Reed (2004) / with permission from American Chemical Society).

An even more direct means of determining molar mass is to include a static light scattering detector in the SEC system. As will be discussed in Chapter 9, measurements of scattered light intensity are directly related to polymer molar mass, without any intervening empirical models or column calibration.

Figure 7.49 shows various results from an SEC system that included a refractive index detector, a total intensity light scattering detector, and a viscometric detector. Combining the light scattering and index of refraction data yielded the broad MMD shown, with large dispersity indices; $M_w/M_n = 4.63$ and $M_z/M_w = 2.68$. Combining the MMD with the intrinsic viscosity, $[\eta]$ (Equation 6.52), found from the refractive index and viscometric detectors, yielded the relationship between $[\eta]$ and M, with a power law exponent of 0.69. The polymer radius of gyration R_g was found from the angular dependence of the light scattering (Chapter 9.6) and, combining this with the MMD yielded the power law relationship between R_g and M, with an exponent of 0.56. The power laws can aid in determining polymer features, such as conformation state and branching/cross-linking. The origin and significance of these power laws is discussed in Chapters 8.2 and 8.3.

For copolymers knowledge of the copolymer composition distribution (CCD, also often referred to as the *copolymer chemical distribution*), is frequently important. It is sometimes possible to achieve this with a UV or an infrared (IR) detector. More often, 2D fractionation methods are used, where M is fractionated in one dimension and composition in another.

7.13.2 Other Fractionation Methods

An opposite approach from SEC is *interaction chromatography* (IC), which is more concerned with distributions of chemical composition, branching, stereochemistry, and crystallinity, than with the MMD. It is especially important for assessing the crystallinity and short-chain branching of polyolefins. In IC, it is specifically desired that the macromolecules interact with the column packing. Several methods exist for this. One of the earliest to be developed was *temperature-rising elution fractionation* (TREF). A molten polymer sample is deposited in a column, packed with almost any type of inert material (typically glass or metal particles), and then cooled. When all the polymer has solidified, the temperature is then raised. The polymers with the highest degree of short-chain branching and lowest crystallinity adhere less to the column packing and begin to melt off and elute at lower temperatures. As temperature increases, those polymers with greater crystallinity and less short-chain branching then melt and elute.

Variations on the TREF principle are embodied in the kindred techniques of *crystallization analysis fractionation* (CRYSTAF), *crystallization elution fractionation* (CEF), and *high-temperature thermal gradient interaction chromatography* (HT-TGIC). A related method that separates according to solubilities, especially useful for determining chemical composition distribution (CCD), is interactive gradient polymer elution chromatography or *interactive GPEC*.[12] at a fixed temperature. In this, a binary (or tertiary or quaternary) mixing pump forms a gradient of solvent composition and polymers elute into the flow according to their solubilities in the composition of solvents.

12 In the polyolefin analytical realm, this is often termed solvent gradient interaction chromatography (HT-SGIC).

When both MMD and some other characteristic distribution are required, two-dimensional polymer LC (2D-LC) is frequently used. This separates the polymer in a first column according to a given characteristic, generally an "interactive" one, i.e., one through which the analytes interact enthalpically with the column packing material. Important characteristics separated in the first column include CCD and branching. Fractions of this separated material then separate in a second column, often an SEC column.

Field flow fractionation (FFF) has developed over time into a column-free alternative to SEC. Polymers in laminar flow in a long channel are separated by a field imposed perpendicularly to the flow direction. The force of the field on the particles pushes them to the bottom of the channel, while diffusion of the particles spreads them back upward in the channel in an exponentially decreasing concentration profile reflecting the balance between the force of the field and the diffusion. The main principle is that particles of different sizes and diffusion coefficients in the cross field will have different concentration versus height profiles in the channel, with the smallest, most diffusive particles having higher profiles. Because the laminar flow in the channel is parabolic, with the velocity increasing with height from the bottom of the channel, the particles with higher concentration profiles will elute first, opposite to the case of SEC. The cross-fields most frequently used in FFF are a temperature gradient from the top to bottom of the channel, a centrifugal field provided by a spinning, ring-shaped channel, and a cross-flow through a semipermeable membrane.

Because the particles emerge fractionated according to one or more characteristics, the detectors used are generally the same as those used in SEC. While adjusting FFF conditions to specific projects can be tedious, the resolution of a well-tuned system is quite good, and polymers and colloids too large to be fractionated by a packed SEC column can be separated by FFF, because there is no packing material in the channel, and hence there is no size exclusion limit, as in SEC.

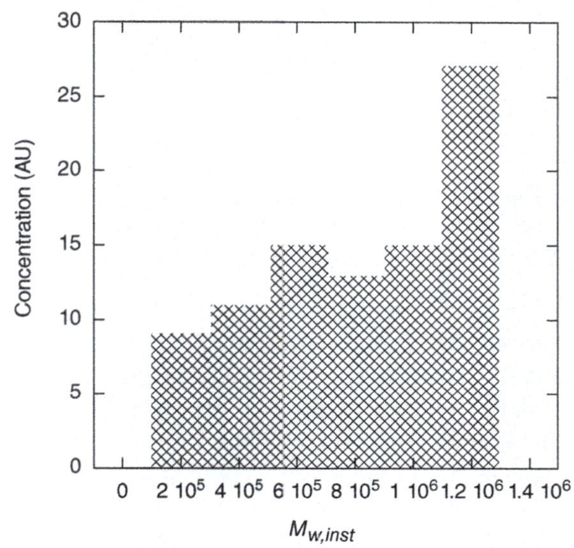

Figure 7.50a Model-free histogram of $M_{w,inst}$ during the free radical polymerization of acrylamide.

7.13.3 Chromatography-free Determination of MMD

For most reaction types, instantaneous MMD have been established on experimental and theoretical grounds, such as the geometric and other distributions discussed above. These instantaneous distributions can be used in conjunction with $M_{w,inst}(t)$ from ACOMP to build up MMD during free radical polymerization without need for separation. $M_{w,inst}(t)$ is a known moment of a distribution which can be used to compute the entire instantaneous distribution. These can be continuously summed during the reaction until the final summation of the MMD of the end product. The polymer is thus "born characterized." Figure 7.50a shows a model-free histogram of concentration versus intervals of instantaneous molar mass $M_{w,inst}$ determined by ACOMP during an acrylamide free radical polymerization. The height of each interval corresponds to the number of occurrences of $M_{w,inst}$ in that interval. To facilitate comparison with SEC a log-normal distribution was chosen as the instantaneous distribution with $M_w/M_n = 2$. Figure 7.50b shows the log-normal distributions corresponding to each $M_{w,inst}$ interval, with the amplitude proportional to the histogram height in Figure 7.50a. The relationship between $M_{w,inst}$ and M_0, the average of the log-normal distribution appearing in Equation 7.95b is

$$M_0 = 0.7071 M_{w,inst} \tag{7.105}$$

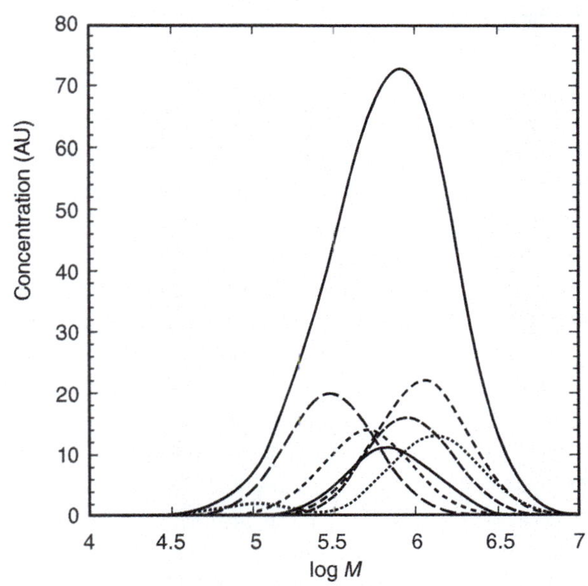

Figure 7.50b Log normal distributions based on the amplitudes in Figure 7.50a and Equation 7.106. The large peak represents the final MMD, and is the summation of the smaller MMJD below it.

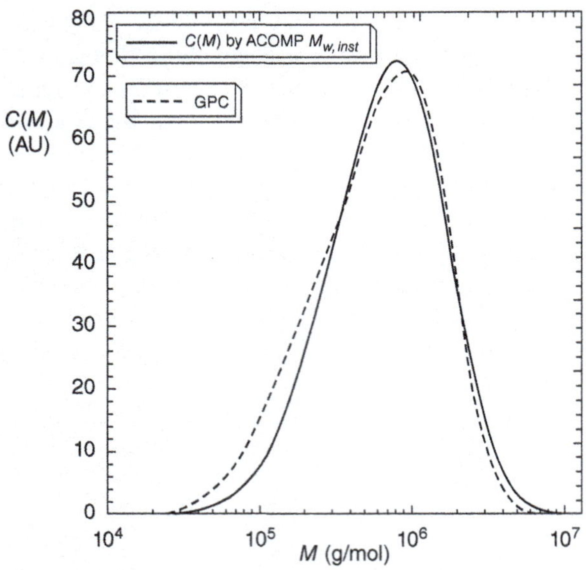

Figure 7.50c Final distribution from the sub-distributions computed in Figure 7.50b. The dashed line is the experimentally determined final distribution using SEC.

The MMD from each histogram interval are summed together to find the MMD of the polymers in the reaction at any time. Summing all the histogram intervals through to the end of the reaction yields the final MMD, which is the large peak in Figure 7.50b.

Figure 7.50c shows the final distribution in Figure 7.50b, together with the MMD determined experimentally from SEC on the end product. While the two MMD are in good agreement, it is apparent that the SEC is somewhat left-skewed, as expected from the geometric distribution for free radical polymerization. Some of the geometric distribution character has been smeared out, however, by the lateral diffusion that occurs both inside the SEC column and the connecting flow paths. As seen in Chapter 6, such diffusive behavior generally results in a normal (Gaussian) distribution.

7.14 Summary

This chapter started with an overview of some of the most important organic chemical functional groups found in polymers, and then gave a survey of polymers, including those synthesized and manufactured on a large scale, those deriving from natural products, and the developing family of stimuli-responsive polymers. Tacticity and its effect on polymer crystallization were introduced. The important notion of chirality in organic molecules was explained. The main chemical routes for polymer synthesis were divided into chain growth and step growth reactions, with an emphasis on the detailed kinetics and automatic control of the former, including copolymers. The differences between free radical polymerization and living type polymerization were explored. The chapter concludes with molar mass distributions and their determination. The fractionation methods are common to both synthetic and biological polymers. The physics and characteristics of polymer molecules individually and in solutions and solid state are explored in the next chapter.

References

Alb, A.M., Enohnyaket, P., Drenski, M.F., Head, A., Reed, A.W. and Reed, W.F., (2006). Online monitoring of copolymerization involving comonomers of similar spectral characteristics. *Macromolecules* 39: 5805–5713.

Carberry, J.J. and Varma, A. (eds.). (1987). *Chemical Reaction and Reactor Engineering*. Marcel Dekker.

Drenski, M.F. and Reed, W.F. (2004). Simultaneous multiple sample light scattering for characterization of polymer solutions. *Journal of Applied Polymer Science* 92: 2724–2732.

Dotson, N.A., Galvan, R., Laurence, R., and Tirrel, M. (1996). *Polymerization Process Modelling*. New York: VCH Publishing.

Flory, P.J. (1971). *Principles of Polymer Chemistry*. Cornell University Press.

Giz, A., Giz, H., Brousseau, J.L., Alb, A., and Reed, W.F. (2001). Kinetics and mechanism of acrylamide polymerization by absolute, online monitoring of polymerization kinetics. *Macromolecules* 34 (5): 1180–1191.

Leonardi R., N.C., Montgomery, R.D., Siqueira, J., McAfee, T., Drenski, M.F., and Reed, W.F. (2017). Automatic synthesis of multimodal polymers. *Journal of Macromolecular Engineering* 11 (4): 1–13. doi:10.1002/mren.201600072.

Lodge, T.P. and Hiemenz, P.C. (2020). *Polymer Chemistry*, 3e. CRC Press, Taylor and Frances Group.

McFaul, C.A., Drenski, M.F., and Reed, W.F. (2014). Sensitivity of the LCST of NIPAM-Am copolymers to discrete and broad composition distributions. *Polymer* 55: 4899–907. doi:10.1016/j.polymer.2014.08.004.

Mark, H.F. (2014). *Encyclopedia of Polymer Science and Technology*, 4e, 3, 591–644.

Morrisson, and Boyd, (1976). *Organic Chemistry*, 3e. Prentice Hall.

Norwood, D.P. and Reed, W.F. (1997). Comparison of single capillary and bridge viscometers as size exclusion chromatography detectors. *International Journal of Polymer Analysis and Characterization* 4: 99–132.

Odian, G. (1991). *Principles of Polymerization*. New York: John Wiley & Sons.

Provder, T. (ed.). (1991). *Chromatography of Polymers; Characterization by SEC and FFF*. ACS Symposium Series, 521.

Reed, W.F. (2004). Fundamentals of static light scattering and viscometry in SEC and related methods. In *Multiple Detection Size-Exclusion Chromatography. ACS Ser. 893* (ed. A.M. Striegel). Washington, DC: ACS.
Rodriguez, F. (1989). *Principles of Polymer Systems*, 3e. Bristol, PA: Hemisphere Publishing Corp.
Siqueira, J.S., Florenzano, F.H. and Reed, W.F. (2021). Kinetic analysis of continuous reaction data for RAFT and free radical copolymerization with acrylic and styrenics monomers. *Polymer* 226: 123798.
Striegel, A.M. (2020). Method development in interaction polymer chromatography. *TrAC Trends in Analytical Chemistry* 130: 115990.
Striegel, A.M., Yau, W.W., Kirkland, J.J., and Bly, D.D. (2009). *Modern Size-exclusion Chromatography*, 2e. Wiley.
Wade, L.G., Jr (2016). *Organic Chemistry*. Pearson.
Wang, D.W. and Thomas, W.M. (1991). *Encyclopedia of Polymer Science and Engineering*, 73, 2359–2368. NY: Wiley Interscience.
Wu, C.-S. (ed.). (2004). *Handbook of Size Exclusion Chromatography and Related Techniques*. Marcel Dekker.

Further Reading

Buback, M. (2009). Fundamentals of free-radical polymerization propagation kinetics in radical polymerization studied via pulsed laser techniques. *Macromolecular Symposia* 275–276 (Microstructural Control in Free-Radical Polymerization), 90–101.
Gao, Y., Zhou, D., Lyu, J. et al. (2020). Complex polymer architectures through free-radical polymerization of multivinyl monomers. *Nature Reviews Chemistry* 4 (4): 194–212.
Matyjaszewski, K. (ed.). (2006). Controlled/Living Radical Polymerization: from Synthesis to Materials. *ACS Symposium Series* 944.
Matyjaszewski, K. (ed.). (2009). Controlled/Living Radical Polymerization: progress in ATRP. *ACS Symposium Series* 1023.
Maynard, H.D., Heredia, K.L., Li, R.C. et al. (2007). Thermoresponsive biohybrid materials synthesized by ATRP. *Journal of Materials Chemistry* 17 (38): 4015–4017.
Plamper, F.A., Andreas, W., and Muller, A.H.E. (2007). Nanoblossoms: light-induced conformational changes of cationic polyelectrolyte stars in the presence of multivalent counterions. *Matthias Ballauff Nano Letters* 7 (1): 167–171.
Tsarevsky, N.V. and Sumerlin, B.S. (Editors). (2013). *Fundamentals of Controlled/Living Radical Polymerization*. Cambridge: RSC.
Tusa, S.I., Sakakibara, A., Yamammoto, T., and Morishima, Y. (2002). Fluorescence Studies of pH-responsive unimolecular micelles formed from amphiphilic polysulfonates possessing long chain alkyl carboxyl pendants. *Macromolecules* 35: 10182–10288.
Winnik, F.M., Whitten, D.G., Urban, M.W., and Lopez, G. (2007). Surface and bulk collapse transitions of thermoresponsive polymer brushes. *Langmuir* 23 (1): 1–2.
Zhu, S. (ed.). (2010). Reaction engineering and industrial aspects of controlled/living radical polymerization. *Macromolecular Reaction Engineering* 4 (3–4).

Problems

1 Given an ideal chain polymerization reaction in the QSSA with the following rate equations:

$$\frac{d[M]}{dt} = -k_p[M][R\bullet]$$

$$\frac{d[R\bullet]}{dt} = 2Fk_d[I_{2,0}] - k_t[R\bullet]^2 \approx 0$$

where $[M]$ is the concentration of monomer, and $[R\bullet]$ is the concentration of propagating free radicals. F is the fraction of decomposing initiator that radicalizes monomer, $[I_{2,0}]$ is the initial concentration of initiator, and k_d is the initiator decomposition rate. Assume termination by disproportionation.

a) Given that $2Fk_d[I_2]_0 = 1.3 \times 10^{-8}$ M/L-s, and that $\frac{k_p{}^2}{k_t} = 16(\text{L/mol-s})$ find the time it takes to convert 67% of the monomer to polymer.
b) Express the number average chain length x_n in terms of time. The initial monomer concentration is $M_0 = 0.05$ M.
c) If a chain transfer agent is present in concentration $[S]$, find x_n as a function of conversion, if $k_3S = 5 \times 10^{-7}$ s^{-1} and remains constant throughout the reaction.

$k_t = 2 \times 10^{-6}$ L/M- s.
Sketch x_n versus conversion for parts (b) and (c).

2 Show that for a given value of a probability W_{AA} that A will add to A •, the weight average sequence length is

$$<N_A>_w = \frac{1+W_{AA}}{1-W_{AA}}$$

3 Given a molar mass distribution of the form:

$$C'(\ln M) = A\exp\left[-\frac{1}{2}\left(\frac{\ln\left(\frac{M}{M_0}\right)}{\sigma}\right)^2\right]$$

where $C(\ln M)d\ln M$ is the concentration of polymers with chain length in the interval lnM to $\ln M + d\ln M$.

a) Find the normalization constant A.
b) Find M_n and M_w, and the dispersity index M_w/M_n.

4 Given a homopolymer population whose chain length distribution follows the geometric distribution ($q < 1$):

$$W(i) = (1-q)q^{i-1}$$

a) Compute the z-average mass M_z, of this population. The mass of an individual monomer is m.
b) Find a continuous form of $W(i)$ for large q (q is still <1)
c) Use the result of (b) to compute the average of $<M^{0.5}>$

5 Say that a copolymer involving A and B units will have optimal properties if the *weight* average sequence length $(N_A)_w = 3$.

a) If $r_1 = r_2 = 1$, what should the starting composition f_1 and f_2 be?
b) What percent of sequences have five or more consecutive A groups?

6 Consider a living polymerization in which a total molar concentration of radicals [R•] is formed instantaneously and begins to propagate with a rate constant k_p in a solution containing polymerizable monomer with an initial concentration of m_0.

a) Assuming no termination reactions occur, find the monomer concentration as a function of time, $[m(t)]$
b) Now, assume that there is an agent or impurity present, of concentration $[S]$, that terminates radicals with a rate constant k_t. Assume that $[S] \gg [R\bullet]$ so that you can assume $[S]$ is approximately constant throughout the reaction. Find the monomer concentration as a function of time, $[m(t)]$.
c) Now assume that the radicals can terminate each other in radical–radical encounter with a rate constant k_r. Find $[m(t)]$.
d) Surprisingly, in all three of the above scenarios the number average chain length N_n, is always the same, for any given level of monomer conversion. Justify this and give the equation for N_n, which is applicable to all the cases.

7 Find M_n and M_w for a Poisson distribution and also compute the dispersity index M_w/M_n for a homopolymer.

$$W_i = \frac{e^{-\tau}\tau^{i-1}}{(i-1)!}, \quad \tau = \frac{[m]_0 - [m]}{I_0}$$

8 A star has a surface temperature of $T = 6000$ K, and a radius of 4×10^5 km. It is at a distance of 10^5 light years from earth. Assume it acts like a black body and that it radiates against a background of essentially $T = 0$ K. The energy/area–time emitted by a black body is $S\ (\text{W/m}^2) = \sigma T^4$, where $\sigma = 5.67 \times 10^{-8}$ W/m^2-K^4. As an approximation, assume all the emitted light is in the visible, with an average photon energy corresponding to a wavelength of 500 nm (green).

a) Find the average number of photons per second reaching a telescope on earth with an objective lens of diameter 0.30 m.
b) Using the Poisson distribution, what is the probability of receiving half this number of photons over any given 1 s interval.
c) What is the probability of receiving twice this number of photons over any given one second interval?

9 Prove that $r_A r_B = 1$ for QSSA.

10 Derive Equation 7.104: $\eta_{sp}(t) = \dfrac{\eta(t) - \eta_0}{\eta_0} = \dfrac{4DP}{IP - 2DP}$

11 A monomer acts as its own transfer agent, with a transfer constant k_3. Find an expression for $N_{n,inst}$ in terms of fractional conversion f. Assume that the radical concentration remains constantly throughout the reaction. Make a plot of $N_{n,inst}$ versus f.

12 Use the histograms of Figure 7.50a, together with the geometric distribution (instead of the log-normal distribution used in Figure 7.50b) to get a final chromatogram, such as in Figure 7.50b.

13 Say that a propagating comonomer chain has a 10% chance of adding monomer A to radical A•. Starting with a single A•, what is the probability of it propagating to form the sequence AAAAB•

14 Consider a polymer molar mass distribution given by

$C(M) = C_0$ (constant), for $M_a \leq M \leq M_b$
$C(M) = 0, \quad M < M_a, M > M_b$

a) Calculate M_n.
b) Calculate M_z, the z-average polymer mass.
c) Assume that $M_b = 4M_a$. Compute the polydispersity index M_w/M_n.

15 A certain polymer has a chain length distribution given by $n_j = (1 - q)q^{j-1}$, where n_j is the number of chains per volume with j monomers. Here $q = 0.995$. The monomer molar mass is 85 g/mol.

It has a second virial coefficient of $A_2 = 1.2 \times 10^{-3}$ cm^3-mol/g^2, which can be considered approximately independent of M. The polymer is mixed in pure water on one side of a membrane, impermeable to the polymer but permeable to water, in a chamber that has a thin capillary. Pure water is in the other side of the chamber. If the polymer is mixed at a concentration of 0.015 g/cm^3 in one side of the chamber, find the difference in height of solution in the two capillaries, once equilibrium is reached. Because of the thin capillary you can assume there is no significant change in volume in either side once osmotic equilibrium is reached, and that the volume in each side is equal. $T = 27°C$.

8

Polymer Physics

Polymer molecules alone in solution, or in very dilute solution, have characteristics including molar mass and composition distributions, conformations, spatial dimensions, branching, hydrodynamic properties, such as intrinsic viscosity, and others. These characteristics, in turn, affect interactions as solution concentration increases, and solid-state properties, such as crystallinity, tensile and compressive strength, rheology, and processability.

Polymer characterization is a well-developed field in and of itself and involves many methods, some of which are discussed in Chapter 9, and separation methods presented in Chapter 7.13. Most of the solution characterization methods can be used for both biological and synthetic polymers. The preceding chapters have set the stage for an overview of polymer characterization, because the thermodynamics, electrostatics, and hydrodynamics are readily adapted from particles in general to polymers. Because many polymers are flexible, there is usually an added degree of complexity involved in the characterization not encountered with rigid particles in solutions.

In fact, the conformations and spatial distributions that polymers can assume form a subfield unto itself. This chapter begins with a survey of polymer conformational statistics.

8.1 Polymer Conformations and Dimensions

8.1.1 Rotational Isomeric Model

Polymers consist of many monomeric subunits held together by covalent bonds.[1] In general, there is some rotational freedom between monomers in a polymer chain, and this leads to the possibility of many different configurations for any given polymer. Figure 8.1 shows a fixed covalent bond angle θ between monomers and a free 360° azimuth rotation angle ϕ.

In reality, free rotations often meet with steric or electrostatic hindrances that yield a finite number of rotational energy minima. The most common situation is of *rotational isomers*, or *rotamers*, around a single bond for tetrahedral bonding. According to the angular rotational energy scheme in Figure 8.2, there is an energy minimum each 120°, for a total of three rotamers. The maximum energy is when the monomeric groups are nearest each other, in the *eclipsed configurations*. Energy minima occur in the *anti-eclipsed configurations*. Even with the restriction of only three rotameric states, there is still an enormous number of configurations for a polymer consisting of N monomers, namely 3^{N-2}. For a small polymer, e.g. polyacrylamide with 100 monomers (a mere 7100 g/mol polymer), there are 5.7×10^{46} possible configurations. A computer capable of outputting one billion strings per second would require 1.8×10^{30} years to represent this number of configurations, far longer than the estimated age of the observable universe of 1.3×10^{10} years.

Given the enormous number of configurations for even short chains, polymers are best characterized by averages, such as are determined through the principles of thermodynamics and statistical mechanics.

1 There are also noncovalent polymers, such as those assembled through hydrogen bonds.

Biophysics: Physical Processes Underlying the Living State, First Edition. Wayne F. Reed.

Companion website: www.wiley.com/go/Reed/BiophysicsandMacromolecules

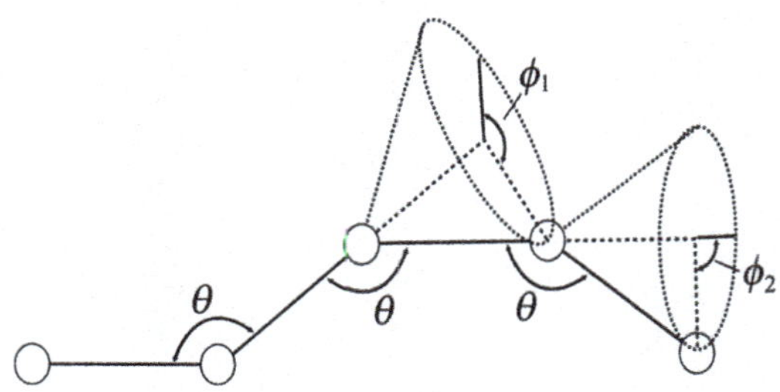

Figure 8.1 Several monomers with fixed bond angle θ and free 360° rotation, shown by the angles ϕ_1, ϕ_2.

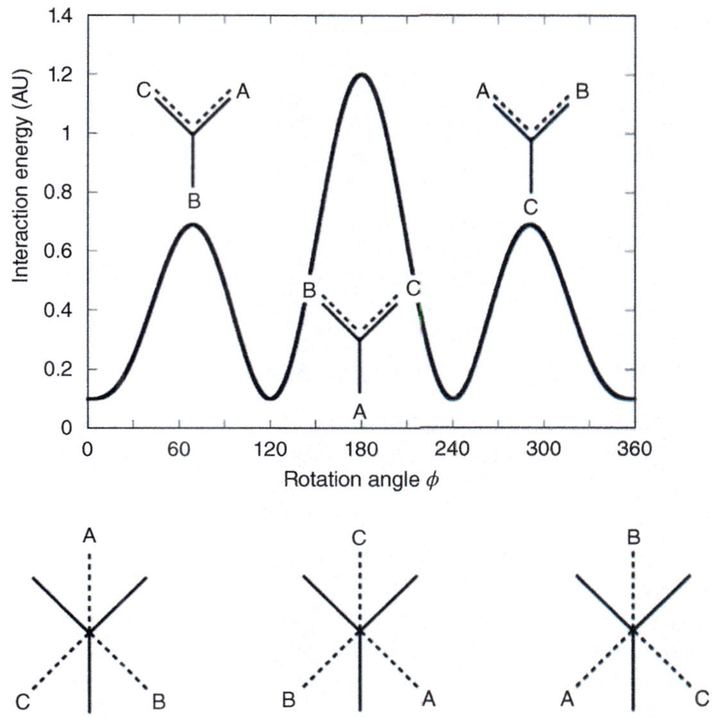

Figure 8.2 Schematic of rotational isomers with energy levels and eclipsed configurations (above) and anti-eclipsed below.

8.1.2 Polymer Conformations

The simplest possible model for a polymer chain has free rotations of its noninteracting monomeric subunits. The natural result of this is the so-called *random coil*, whose end-to-end length probabilities obey the diffusion (or random walk) equation. When intermonomeric forces in the polymer chain come into play, polymers can adopt other conformations besides the random coil. Such forces can be those due to steric interaction, electrostatic effects, including charge–charge, charge–dipole, and dipole–dipole, hydrophobic interactions, and hydrogen bonding. The particular amino acid composition of a given protein, for example, provides multiple sources of these various forces, allowing the protein to fold into a globule, form secondary structures, such as α-helices and β-sheets, and stretch into fiber conformations. Random coils, globules, and rods are the major conformational classes of polymers. After these, one can differentiate a wide variety of other "architectures," such as random branching, hierarchical branching, cross-linking, dendrimers, stars, and so on, as illustrated in Figure 8.3.

The characteristic spatial dimension of polymers in the different conformational classes vary as a power of the polymer mass M: For spheroidal particles diameter, $D \propto M^{1/3}$. For ideal random coils, the root mean square end-to-end length $\langle h^2 \rangle^{1/2} \propto M^{1/2}$. For random coils with excluded volume (EV) $\langle h^2 \rangle^{1/2} \propto M^{3/5}$. For rod-like polymers, length, $L \propto M$. Branched-type architectures usually have characteristic dimensions in the range from $M^{1/3}$ to $M^{3/5}$.

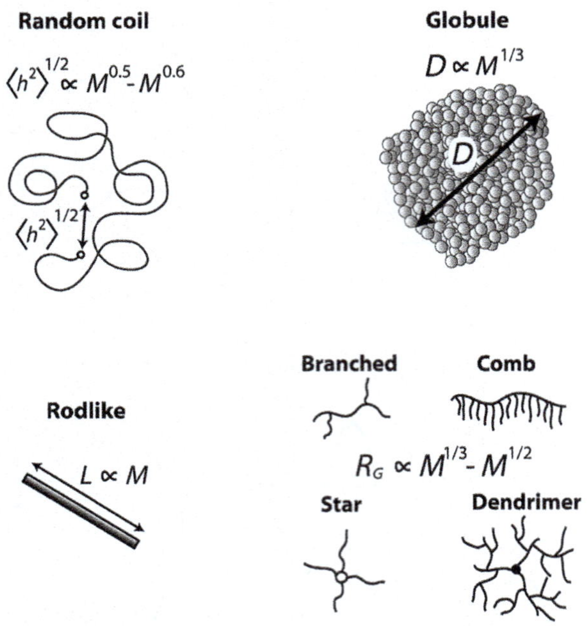

Figure 8.3 Conformational classes of polymers and their associated scaling relations of characteristic linear dimensions with mass, *M*.

8.1.3 Ideal Random Coil Polymer Statistics Obey the Diffusion Equation

Random coils are an important conformational class because most synthetic polymers resemble random coils. Unfolded or "denatured" proteins and nucleic acids, as well as many polysaccharides in their native state also are described well by random coils.

A sketch of an ideal, completely flexible random coil of fixed bond length b is shown in Figure 8.4a. Any given monomer i is located by $\vec{r}_i$ with respect to the origin. The first monomer is placed at an arbitrary position $\vec{r}_1$ and then monomer 2 is placed at a distance b from the first with equal probability on any point of a sphere of radius b centered at $\vec{r}_1$. Each succeeding monomer is added in the same way, randomly on the surface of the sphere centered on the preceding monomer. This is the fundamental description of an ideal random walk which describes many situations. Examples of random walks in time include diffusion of a particle, the spread of heat from a heat source, and the way a free electron's wave function spreads in time. This type of random walk for the *ideal chain* is *non-self excluding*, that is, it is possible for a monomer in the chain to occupy a position occupied by another monomer, because the ideal monomers of the chain have no volume, in direct parallel to ideal gas atoms or molecules. In a real chain with *excluded volume* (EV), the random walk is *self-excluding* which means that each new monomer added to the chain cannot occupy the position of a monomer already placed in the chain. The self-excluding random walk is much more difficult to model and a very large body of theoretical approaches exists. The effects of EV on solution thermodynamics were seen in Chapter 4.14, and the virial coefficients A_2 and A_3 were used for describing nonideal effects.

The same approach is taken here, namely the ideal cases are first considered and the excluded volume features are treated afterward.

While the scenario of placing succeeding monomers at random on a spherical surface about the preceding monomer is an effective image, an easier approach that later leads to other useful models is the 3D lattice model. In the 3D lattice model, each monomer added is allowed to be placed at a distance b only in the $\pm x$, $\pm y$, or $\pm z$ directions, denoted by $x + \Delta x$, $y + \Delta y$, $z + \Delta z$ in Figure 8.4b. The bond length is b. While this may at first seem artificial, in the limit of a large number of steps the polymer thus built converges to the freely jointed chain just outlined, as depicted in Figure 8.4a.

Consider that after $n + \Delta n$ steps, the monomer has a discrete probability $P(n + \Delta n, x, y, z)$ of being at x, y, z. In this, $\Delta n = 1$ means one step at a time is taken, but Δn is used in the formulation to allow easy passage to the continuum limit, dn, at large n. The monomer may have arrived at x,y,z with equal probability from any of the six positions $x \pm \Delta x, y \pm \Delta y$, or $z \pm \Delta z$,

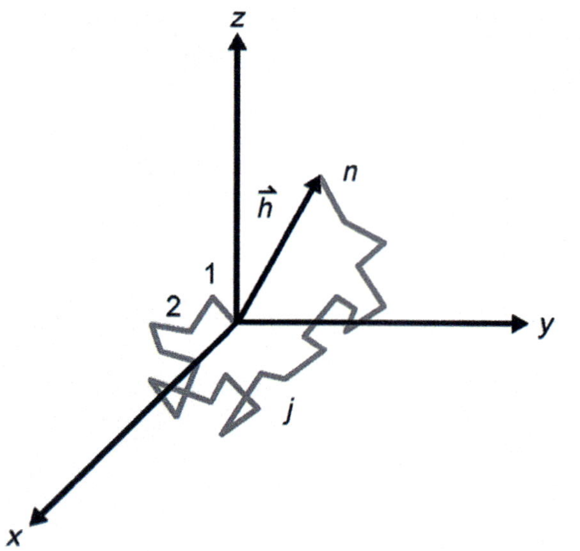

Figure 8.4a Depiction of the random walk after a very large number of steps; it converges to the completely free random walk.

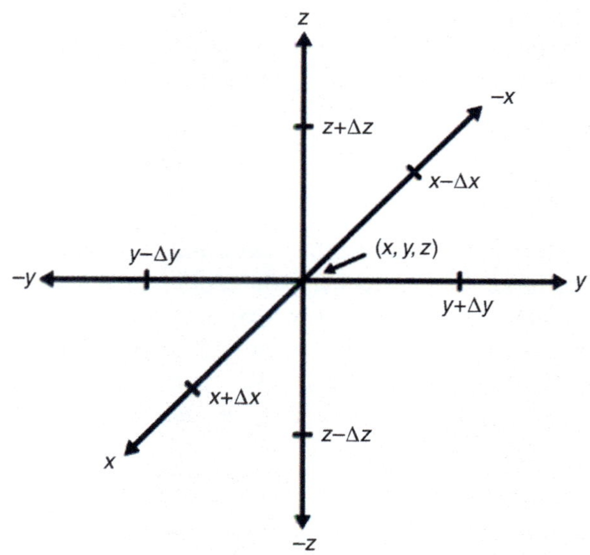

Figure 8.4b Lattice model for a 3D random walk.

where $\Delta x = \Delta y = \Delta z = b$, and the Δx, Δy, Δz notation is again used to pass to the continuum limit, dx, dy, dz, at large n. The fact of arriving at x,y,z with equal probability from any of the six positions mentioned is expressed by the sum of the equal probabilities of having been at any of the six positions at n in the form of the following *difference equation*:

$$P(n+\Delta n,x,y,z)=\frac{1}{6}\begin{bmatrix} P(n,x+\Delta x,y,z)+P(n,x-\Delta x,y,z)+P(n,x,y+\Delta y,z)+ \\ P(n,x,y-\Delta y,z)+P(n,x,y,z+\Delta z)+P(n,x+\Delta x,y,z-\Delta z) \end{bmatrix} \quad 8.1$$

This can be expanded like a continuous function to pass into the continuum limit. Replacing $P(n, x, y, z)$ by the continuous, differentiable probability density $W(n, x, y, z)$ (probability per unit volume) allows expanding the left-hand term around n, and leads to

$$W(n+\Delta n,x,y,z)=W(n,x,y,z)+\frac{\partial W}{\partial n}\Delta n+\frac{\partial^2 W}{\partial n^2}\frac{\Delta n^2}{2!}+\cdots \quad 8.2$$

Expanding the right-hand side around x leads to terms of the form:

$$W(n,x+\Delta x,y,z)=W(n,x,y,z)+\frac{\partial W}{\partial x}\Delta x+\frac{\partial^2 W}{\partial x^2}\frac{\Delta x^2}{2!}+\cdots \quad 8.3a$$

$$W(n,x-\Delta x,y,z)=W(n,x,y,z)-\frac{\partial W}{\partial x}\Delta x+\frac{\partial^2 W}{\partial x^2}\frac{\Delta x^2}{2!}+\cdots \quad 8.3b$$

There are four more equations of the form of Equations 8.3a,b, in terms of Δy and Δz, yielding a total of six terms.

The first partial spatial derivatives of $W(n, x, y, z)$ on the right-hand side hence cancel leaving the second-order terms. When these are collected and all cancellations made the result is

$$\frac{\partial W}{\partial n}\Delta n=\frac{1}{6}\left(\frac{\partial^2 W}{\partial x^2}\Delta x^2+\frac{\partial^2 W}{\partial y^2}\Delta y^2+\frac{\partial^2 W}{\partial z^2}\Delta z^2\right) \quad 8.4$$

Using $\Delta n = 1$ and $\Delta x = \Delta y = \Delta z = b$ finally yields

$$\frac{\partial W}{\partial n}=\frac{b^2}{6}\nabla^2 W \quad 8.5$$

The random walk in space has yielded the diffusion equation with the number of links n replacing time in the usual time-dependent diffusion equation, and $b^2/6$ replacing the diffusion constant D. Compare Equation 8.5 with diffusion Equation 6.85.

This has the well-known, normalized solution (Chapter 6.8):

$$W(n,x,y,z)=\left(\frac{3}{2\pi nb^2}\right)^{3/2}\exp\left(-\frac{3(x^2+y^2+z^2)}{2nb^2}\right) \quad 8.6$$

Note that in its formulation as a difference equation in x, y, z, and n, $P(n, x, y, z)$ is the finite probability of the nth step landing on (x, y, z). In passing to the continuum limit, as always, the discrete variable becomes a density in the differential. In this continuum limit, where $n \gg 1$ (the "long chain limit"), $W(n, x, y, z)\mathrm{d}V$ hence represents the probability that the end of a chain of n links, each of length b, is in the volume element $\mathrm{d}V = \mathrm{d}x\mathrm{d}y\mathrm{d}z$ centered on x, y, z.

If the first link is placed at the origin at (0, 0, 0), then the position of the last link can be located by the vector $\vec{h}$, shown in Figure 8.4a. This vector is the *end-to-end vector* of the chain, and its magnitude is the chain's *end-to-end length.*

$$\vec{h}=x\hat{i}+y\hat{j}+z\hat{k} \quad 8.7$$

For the freely jointed, spatially isotropic polymer chain $W(n, x, y, z)$ can be written as

$$W(n,h)=\left(\frac{3}{2n\pi b^2}\right)^{3/2}\exp\left(\frac{-3h^2}{2nb^2}\right) \quad 8.8$$

since $h^2 = x^2 + y^2 + z^2$.

$W(n, h)$, the probability per unit volume, is a maximum at the origin, $h = 0$, but the probability of finding the chain end in a specific volume element $\mathrm{d}V$ is $W(n,h)4\pi h^2\mathrm{d}h$. These two functions are shown in Figure 8.5 for a random coil with $n = 500$ monomers. The maximum probability occurs at h_{mp}, given by

$$\left[h^2W(h)\right]_{max} \text{ at } h_{mp}=\sqrt{\frac{2n}{3}}b \quad 8.9$$

Useful integral formulas for finding averages of $W(n,h)$ are

$$\int_0^\infty x^{2n}e^{-\alpha x^2}\mathrm{d}x=\frac{1\cdot3\cdot5\ldots(2n-1)}{2^{n+1}\alpha^n}\sqrt{\frac{\pi}{\alpha}},\quad a>0,\ n=0,1,2,\ldots \quad 8.10$$

and

$$\int_0^\infty x^{2n+1}e^{-\alpha x^2}\mathrm{d}x=\frac{n!}{2\alpha^{n+1}},\quad a>0,\ n=0,1,2\ldots \quad 8.11$$

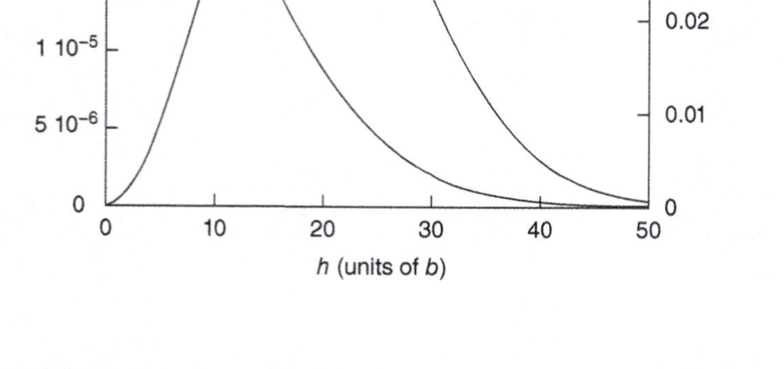

Figure 8.5 $W(h)$, the probability per volume of finding the end of the chain at h, and $4\pi h^2W(h)\mathrm{d}h$, the probability of finding the end of the chain in a spherical shell of radius h and thickness $\mathrm{d}h$.

Because the distribution is spherically symmetric

$$\langle \vec{h} \rangle = 0 \tag{8.12}$$

The mean square end-to-end length from this distribution is

$$\langle h^2 \rangle_0 = nb^2 \tag{8.13}$$

The subscript "0" is introduced at this point for $\langle h^2 \rangle$ to indicate it is the value for an ideal polymer without any type of excluded volume or long-range interactions. After excluded volume is introduced for polymers $\langle h^2 \rangle$ without the "0" will refer to the mean square end-to-end length of the polymer taking into account all excluded volume interactions.

The root mean square end-to-end length, which is the root mean square distance of the chain-end form the origin, is:

$$\langle h^2 \rangle_0^{1/2} = \sqrt{n}b \tag{8.14}$$

The variation of $\langle h^2 \rangle_0{}^{1/2}$ with $\sqrt{n}$ has the same form as the diffusion equation in time, i.e. the root mean square displacement of a diffusing particle from the origin increases as $\sqrt{t}$, whereas the root mean square end-to-end length of a growing polymer chain increases as $\sqrt{n}$, where n is the number of monomers in the chain.

Since the mass of a homopolymer M is $M = nm$, where m is the molar mass of a monomer, the important result is obtained that, for an ideal random walk chain, the root mean square end-to-end length of the chain increases as the square root of the chain's mass M:

$$\langle h^2 \rangle_0^{1/2} \propto \sqrt{M} \tag{8.15}$$

8.1.4 Entropic Origin of Polymeric Elasticity

A very interesting feature of this distribution function is that it describes the origin of elasticity for rubber and other elastic polymers. For an ideal polymer chain, there is no interaction among monomers so there is no change in internal energy U when $\langle h^2 \rangle_0^{1/2}$ changes. The entropy is a function of h

$$S = k_B \ln W(h) = -k_B\left(\frac{3}{2nb^2}\right)h^2 + \text{constant} \tag{8.16}$$

Increasing the end-to-end length h of a given chain decreases the chain's entropy, which increases the Gibbs energy. Hence, external work must be done on the polymer in order to stretch it.

If the polymer chain is fixed at one end and the other pulled to increase h in a quasi-static, reversible isothermal process then by the first law of thermodynamics:

$$\mathrm{d}Q = \mathrm{d}W \tag{8.17}$$

because $\mathrm{d}U = 0$ at fixed T. Because the process is reversible:

$$\mathrm{d}Q = T\mathrm{d}S = \mathrm{d}W = -\vec{F}_{ap} \cdot \mathrm{d}\vec{h} = -F_{ap}\mathrm{d}h \tag{8.18}$$

that is, the work done by the applied force $\vec{F}_{ap}$ in stretching the polymer goes into decreasing the polymer chain's entropy; this is analogous to the pressure applied to an ideal gas to compress it and reduce its entropy. The vector signs are removed on the right-hand expression in Equation 8.18 because $\vec{F}_{ap}$ and $\vec{h}$ are in the same direction, i.e. of increasing h. By thermodynamic convention, the work is negative because it is work done on the system. The force exerted by the polymer, the restoring force $\vec{F}_r$ is equal and opposite $\vec{F}_{ap}$ so that

$$F_r = T\frac{\partial S}{\partial h} = -\frac{3k_BT}{nb^2}h \tag{8.19}$$

The restoring force is hence an entropically based force. It has the same form as a Hooke's law spring force $F_r = -kx$, with an effective spring constant:

$$k = \frac{3k_B T}{nb^2} \tag{8.20}$$

The spring constant "stiffens" as T increases and weakens as n and/or b increase. This latter suggests that if a polymer with n links is cut in half its spring constant will double. If the two halves of the cut chain are placed in parallel and pulled together then the net spring constant will be four times that of the original, uncut polymer. This is one effect that cross-links can have in a rubber network; an increase in stiffness using the same amount of polymer; the cross-links, among other effects, reduce the "effective n," increasing k.

Entanglements and excluded volume of course come into play in dense polymeric networks, but the purely entropic and conformational aspects have been idealized into the notion of the *phantom network*, where the many types of intra- and interchain interactions are ignored to gain a conceptual starting point for real materials.

Typically, a polymer chain can be stretched to many times its most entropically relaxed coil state. This is in strong contrast to the origin of elasticity in "hard" materials, such as metals, where it is minute interatomic displacements in steep electrostatic potential wells that provides the restoring force, often described by Young's modulus of elasticity, and the entropic portion of G plays no role. Hard materials can be formed geometrically into springs, with spring constant k, and a restoring force $F = -kx$ with potential energy $PE = kx^2/2$, which are due to the electrostatic forces. Young's moduli for metals such as steel are on the order of 10^{11} N/m^2, whereas for rubber it is typically on the order of 10^7–10^8 N/m^2.

8.1.5 Estimate of Young's Modulus and the Effect of Cross-linking

An estimate of Young's modulus Y can be made by considering a material to be a collection of springs acting in parallel such that the net force constant k_{net} is the sum of all the force constants k of the individual parallel springs. Young's modulus, Y, relates the stress, which is the force per area, applied to the material to the strain produced, which is the change in length per unit length, or $\Delta L/L$. Y relates the stress and strain via what is essentially a form of Hooke's law:

$$\frac{F}{A} = Y\frac{\Delta L}{L} \tag{8.21}$$

A volume of the material has a unit cross-section $A = 1$ m^2 and a length L determined by the length of a spring. The total force F is

$$\frac{F}{A} = Y\left(\frac{\Delta L}{L}\right) \approx kNL\left(\frac{\Delta L}{L}\right) \tag{8.22}$$

where N is the number of springs per cross-sectional area of the material. Y can be identified as

$$Y \approx kNL \tag{8.23}$$

First consider that the spring has a cross-sectional area $a = \pi R^2$, where R is the radius of the spring, which is on the order of 1 Å for a metal crystal and for the backbone of a polymer. $N \sim 3 \times 10^{19}$ springs/m^2 with this estimate. The spring constant for a solid, such as a metal, can be roughly estimated by considering that the interatomic crystal forces are approximately simple harmonic, with a potential energy U of the form:

$$U = \frac{1}{2}kx^2 \tag{8.24}$$

Considering the potential difference between maximum and minimum potentials between atoms to be ΔU over the characteristic area scale R^2 of the interatomic spacing suggests

$$k_{metal\ lattice} \approx \frac{\Delta U}{R^2} \tag{8.25}$$

If $\Delta U \sim 2$ eV $= 3.2 \times 10^{-19}$ J over the radius of the potential $x \sim R \sim 1$ Å, yields $k = 32$ N/m for each spring. Taking $L \sim 2R = 2$ Å for the spring length (i.e. one atomic diameter) then for this generic metal crystal:

$$Y_{metal\ crystal} \approx 2 \times 10^{11}\,\text{N/m}^2 \qquad 8.26$$

Consulting a table of Young's moduli shows that ordinary metals are indeed on this order of magnitude.

Applying the same reasoning to a phantom network of a rubber made of polymers in parallel, that is, ignoring entanglements, cross-links, and excluded volume and considering only the entropic contribution, k can be computed from the above. Taking a typical vinyl polymer spacing of $b = 2.5$ Å and an average number of units $n = 1000$ yields $k = 2 \times 10^{-4}$ N/m, which is 10^5 times less than for the metallic solid.

The characteristic length L is now taken as $L \sim \langle h^2 \rangle^{1/2} = \sqrt{n}b = 8 \times 10^{-9}$ m. Using the same $N \sim 3 \times 10^{19}$ springs/m^2 yields

$$Y_{polymer\ rubber} \approx 5 \times 10^{7}\,\text{N/m}^2 \qquad 8.27$$

Consulting tables of Y show that a typical rubber has $Y \approx 10^7$–10^8 N/m^2.

The effect of cross-links can be assessed by considering a cuboid of a cross-linked network of ideal polymer chains in volume V, containing ρ polymer subunits per unit volume. This can be thought of first as a single polymer chain with n units in volume V, which then has $N = \rho V$ cross-links made on it, yielding N small polymer subunits. The density of cross-links per unit volume is then the same as the density of polymer subunits. For simplicity, assume that the N subunits are of equal length. The total entropy is just the entropy per chain, Equation 8.16, times the density of chains (cross-links) times V, where n for each subunit is $n' = n/N$ and $h'^2 = n'b^2/3$:

$$S_{total} = -(\rho V)k_B\left(\frac{3}{2n'b^2}\right)h^2 + \text{const} = -\frac{(\rho V)}{2}k_B + \text{const} \qquad 8.28$$

V is expressed in terms of the network's dimensions in x, y, and z.

$$V = L_x L_y L_z \qquad 8.29$$

The force needed for uniaxial extension in x is

$$F = -\frac{\partial G}{\partial L_x} = T\frac{\partial S}{\partial L_x} = -\frac{\rho}{2}k_B T\left[L_y L_z + L_x\left(L_z\frac{\partial L_y}{\partial L_x} + L_y\frac{\partial L_z}{\partial L_x}\right)\right] \qquad 8.30$$

where $\partial L_y / \partial L_x$ and $\partial L_z / \partial L_x$ are the lateral deformation ratios, or *Poisson ratios*, and can depend on whether the material is incompressible or compressible. These terms are much smaller than the stretch in x, so that, considering also that V changes little, if at all, so that ρ remains at least approximately constant:

$$F \approx -\frac{\rho}{2}k_B T L_y L_z \qquad 8.31$$

which divided by area $A = L_y L_z$ yields the stress,

$$\frac{F}{A} \approx -\frac{\rho}{2}k_B T \qquad 8.32$$

Putting this in terms of Young's modulus, Y, in Equation 8.21, and using

$$\frac{\Delta L_x}{L_x} \propto \frac{N\left(\langle h_x^2\rangle^{1/2} - \langle h_x^2\rangle_0^{1/2}\right)}{N\langle h_x^2\rangle^{1/2}} \qquad 8.33$$

for the uniaxial strain, where $\langle h_x^2\rangle^{1/2}$ is the RMS end-to-end length in the x-direction of the strained network, and $\langle h_x^2\rangle_0^{1/2}$ is the RMS end-to-end length of the fully relaxed network,

$$Y \propto \rho k_B T \qquad 8.34$$

i.e. the modulus of the network is proportional to the cross-link density and T.

8.1.5.1 Full Polymeric Extension Compared to Entropically Relaxed State

When the freely rotating polymer chain is fully stretched its length is $h_{fs} = nb$. Comparing this to the most probable end-to-end length, h_{mp}, from Equation 8.9:

$$\frac{h_{fs}}{h_{mp}} = \sqrt{\frac{3}{2}n} \quad 8.35$$

For example, for a polymer chain of $n = 1000$, its fully stretched length compared to its maximum entropy coiled length is $h_{fs}/h_{mp} = 38.7$. This is an enormous factor compared to, say, a steel bar, whose extension under significant tensile stress is only a small fraction of its length. When polymer chains are stretched to this degree, the restoring force F_r is no longer linear in h and rises as a steep polynomial in h.

8.1.6 Polymer Dimensions for Random Coils: Relation Between Radius of Gyration and End-to-end Length

Although $W(n, h)$ for an ideal coil has a mathematically simple form and yields a characteristic dimension, the root mean square end-to-end length $\langle h^2\rangle^{1/2}$, it is not possible to measure this quantity in a simple experiment. A closely related characteristic is the *mean-squared radius of gyration* $\langle S^2\rangle$ (the "S" is not to be confused with entropy or intensity in other portions of this book). $\langle S^2\rangle$ can often be measured in light-scattering experiments or with other scattering techniques, as explained in Chapter 9.

$\langle S^2\rangle$ is related to the concept of moment of inertia of a body. Moment of inertia is the mass-averaged squared displacement of the body about a given axis. The squared *radius of gyration* S^2 is defined as the mass averaged squared displacement of the body about a given point. The point is normally taken as the center of mass (CM). By this definition:

$$S^2 = \frac{\sum_{i=1}^{n} m_i r_i^2}{M} \quad 8.36$$

where $M = \sum_{i=1}^{n} m_i$ is the total mass of the body, m_i is the ith mass element, r_i^2 is its squared distance from the CM, and n is the total number of mass elements. For a continuous body of mass density $\rho(\vec{r})$, the integral expression is

$$S^2 = \frac{\int_V r^2 \rho(\vec{r})\,\mathrm{d}^3 r}{M} \quad 8.37$$

where the integral is over the volume V of the body and $M = \int_V \rho(\vec{r})\mathrm{d}^3 r$. For a rigid body, these discrete and integral definitions of S^2 are complete. For a flexible body, such as a freely jointed polymer chain, S^2 must be averaged over all possible configurations of the chain and hence the mean squared radius of gyration for a flexible polymer uses the terminology $\langle S^2\rangle$.

In the following, $\langle S^2\rangle_0$, which has the subscript "0," will be used to denote the mean square radius of gyration of an ideal polymer without excluded volume interactions, which is the same rationale for the terminology $\langle h^2\rangle_0$.

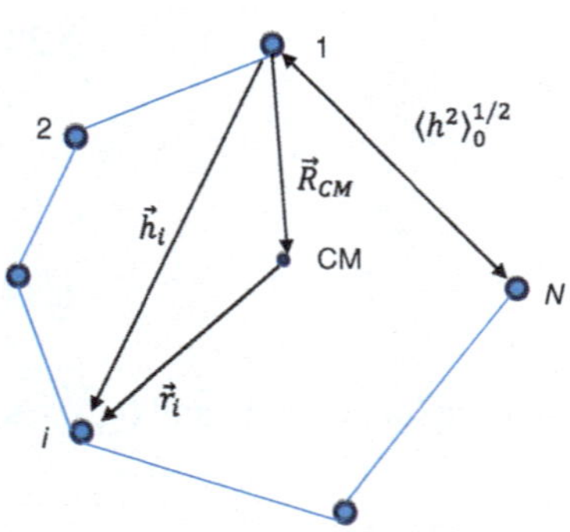

Figure 8.6 Geometry for computing $\langle S^2\rangle_0$.

To compute $\langle S^2\rangle_0$, one starts again with the first monomer's position as the origin and then locates the CM of the entire chain with respect to this origin with the vector $\vec{R}_{CM}$. Then, with $\vec{r}_i$, the position vector from the CM to monomer i, and $\vec{h}_i$, the vector from the first monomer to monomer i,

$$\vec{h}_i = \vec{R}_{CM} + \vec{r}_i \quad 8.38$$

as shown in Figure 8.6.

Using this, the definition of $\langle h^2\rangle_0$ and $\langle S^2\rangle_0$, Debye, and separately Zimm, were able to find the relationship between $\langle h^2\rangle_0$ and $\langle S^2\rangle_0$ for a homopolymer (i.e. m_i is the same for all monomers) by weighting all possible configurations using $W(n,h)$ for a freely jointed chain. The result for a freely jointed chain is

$$\langle S^2\rangle_0 = \frac{\langle h^2\rangle_0}{6} = \frac{nb^2}{6} \quad 8.39$$

8.1.7 Effect on Polymer Dimensions of Restricted Bond Angles

The last section showed how a Gaussian distribution of end-to-end lengths results from the freely jointed chain. Of course, real polymers have restrictions regarding both covalent bond angles and hindered rotational angles. What follows is a vectorial approach to taking account of bond angles of polymers, while they are still considered to be ideal. That is, this approach does not take into account long-range excluded volume interactions which are the result of a *self-excluded random walk*. Hence, the subscript "0" continues to be used in the following.

$\vec{r}_i$ is now the vector that connects the monomer i to the monomer $i + 1$ (instead of the vector joining the CM to monomer i, as in the previous section). $\vec{h}$ can then be expressed for a chain of n monomers as

$$\vec{h} = \sum_{i=1}^{n-1} \vec{r}_i \quad 8.40$$

where monomer 1 is placed at the origin. With this

$$\langle h^2 \rangle = \langle \vec{h} \cdot \vec{h} \rangle = \left\langle \sum_i \sum_j \vec{r}_i \cdot \vec{r}_j \right\rangle \quad 8.41$$

$$\langle h^2 \rangle = \left\langle \sum_i r_i^{\,2} \right\rangle + \left\langle \sum_{i \neq j} \sum_j \vec{r}_i \cdot \vec{r}_j \right\rangle \quad 8.42$$

The above definitions hold also for a polymer chain with excluded volume interactions and hence the restrictive "0" subscript is not used. Returning to the ideal chain, $\langle h^2 \rangle_0$ replaces $\langle h^2 \rangle$ in Equations 8.41 and 8.42, and, using the freely-jointed ideal chain result for the first sum on the right-hand side:

$$\langle h^2 \rangle_0 = nb^2 + \left\langle \sum_{i \neq j} \sum_j \vec{r}_i \cdot \vec{r}_j \right\rangle \quad 8.43$$

With reference to Figure 8.7, θ is the fixed angle, θ_{ij} between successive monomers i and j, which allows the last expression to be written as

$$\langle h^2 \rangle_0 = nb^2 + b^2 \left\langle \sum_{i \neq j} \sum_j \cos(\theta_{ij}) \right\rangle \quad 8.44$$

The right term in brackets is a *bond correlation function*. For a freely jointed chain, all angles from 0 to π are equally likely so that $\langle \theta \rangle = \pi/2$, and so the second term vanishes and the result $\langle h^2 \rangle_0 = nb^2$ is recovered.

Consider the case for a fixed bond angle, e.g. 70.5° is the most common bond angle for polymers with carbon backbones, which have a supplementary tetrahedral angle of 109.5°. The problem of finding $\langle h^2 \rangle_0$ involves finding the expression for the cosine term in the bond correlation function, Equation 8.44, and performing the double summation.

For adjacent links

$$\langle \vec{r}_i \cdot \vec{r}_{i+1} \rangle = b^2 \cos\theta \quad 8.45$$

where the average is over all rotational angles $\phi \in \{0, 2\pi\}$. There are $2(n - 1)$ occurrences of this value over the double sum, so

$$\sum_i \langle \vec{r}_i \cdot \vec{r}_{i+1} \rangle = 2(n-1) b^2 \cos\theta \quad 8.46$$

Figure 8.7 Successive projections of bond length with a fixed covalent bond angle θ.

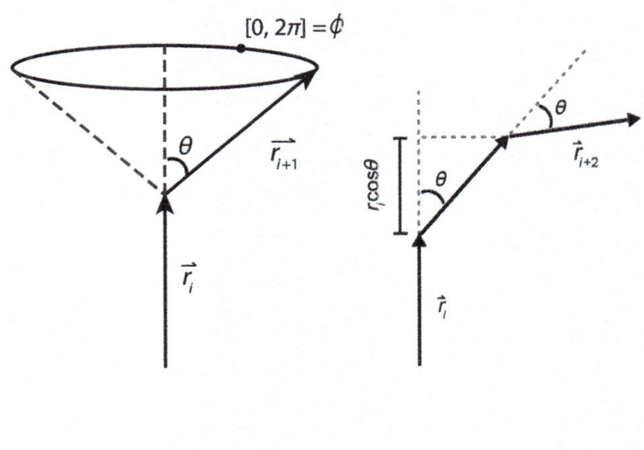

For $\langle \vec{r}_i \cdot \vec{r}_{i+2} \rangle$, consider that the projection of $\vec{r}_i$ onto $\vec{r}_{i+1}$ is $b\cos\theta$. This projection in the direction of $\vec{r}_{i+1}$ has a scalar product with $\vec{r}_{i+1}$ of $b\cos\theta\,(\vec{r}_{i+1}/r_{i+1})\cdot\vec{r}_{i+2}$

$$\langle \vec{r}_i \cdot \vec{r}_{i+2} \rangle = b\cos\theta\left(\frac{\vec{r}_{i+1}}{r_{i+1}}\cdot\vec{r}_{i+2}\right) = b^2\cos^2\theta \qquad 8.47$$

There are $2(n-2)$ occurrences of this term. Continuing this projection method gives

$$\sum_i \langle \vec{r}_i \cdot \vec{r}_{i+k} \rangle = 2(n-k)b^2\cos^k\theta \qquad 8.48$$

Because $\cos\theta < 1$ and $n \gg 1$, $(n-k)$ can be approximated by n. For example, for the 70.5° angle between carbons on a polymer backbone, $\cos\theta = 0.334$ and $\cos^5\theta = 0.00414$, so that the terms in k rapidly die out.

Returning to the double sum with this approximation:

$$\left\langle \sum_{i\neq j}\sum_j \vec{r}_i \cdot \vec{r}_j \right\rangle \cong 2nb^2 \sum_{k=1}^{n-1}\cos^k\theta \qquad 8.49$$

and using the convergence property for $x < 1$ of

$$\sum_{k=0}^{\infty} x^k = \frac{1}{1-x}, \quad x<1 \qquad 8.50$$

lets the sum be written:

$$\sum_{k=1}^{n-1}\cos^k\theta = \frac{1}{1-\cos\theta} - 1 \qquad 8.51$$

$\langle h^2 \rangle_0$ can now be given as

$$\langle h^2 \rangle_0 = nb^2 + 2nb^2\left(\frac{1}{1-\cos\theta} - 1\right) \qquad 8.52$$

This yields the final result for the mean square end-to-end length for a polymer with a fixed bond angle θ between adjacent bonds.

$$\langle h^2 \rangle_0 = nb^2\left(\frac{1+\cos\theta}{1-\cos\theta}\right) \qquad 8.53$$

Importantly, the addition of the bond angle restriction has not changed the scaling law and $\langle h^2 \rangle_0^{1/2}$ is again proportional to $\sqrt{n}$, because the chain still obeys a non-self-excluding random walk. This suggests using a *statistical segment length* l_k instead of the bond angle for the length scale of chain segments. n_k is the number of statistical segments:

$$\langle h^2 \rangle_0 = n_k l_k^2 = \beta n b^2 \qquad 8.54$$

The *contour length* of the polymer chain L is the length measured along the chain

$$L = nb = n_k l_k \qquad 8.55$$

so that

$$l_k = \beta b \qquad 8.56$$

where β is the number of monomer units per statistical segment length, for which $\beta > 1$ for $\theta > 90°$ and $\beta < 1$ if $\theta < 90°$. In the case of the restricted bond angle:

$$l_k = \left(\frac{1+\cos\theta}{1-\cos\theta}\right)b \qquad 8.57$$

e.g. for a bond length of 0.154 nm and $\theta = 70.5°$, $\beta = 2.00$, and $l_k = 0.308$ nm.

8.1.8 Rotational Isomeric States and Their Effect on Bond Angle

As discussed in Section 8.1.1, in real polymers, the azimuth angle is not able to rotate completely freely around 2π due to steric hindrance with adjacent monomer side groups, leading to the rotational isomeric states or rotamers. Figure 8.2 shows how staggered (low energy) and eclipsed (high energy) rotational states lead to a rotational energy profile $U(\phi)$ with three minima, which constitute three rotameric states. The differences in energy from peak to trough are on the order of 1 or 2 kcal/mol. These states rapidly rotate into each other on a time scale on the order of 10^{-10} s.

The thermodynamic average over $\cos\phi$ is

$$\langle\cos\phi\rangle = \frac{\int \cos\phi \exp[-U(\phi)/k_BT]\mathrm{d}\cos\phi}{\int \exp[-U(\phi)/k_BT]\mathrm{d}\cos\phi} \tag{8.58}$$

Using reasoning similar to above for fixed bond angle, the expression for $\langle h^2\rangle_0$ when rotameric states are taken into account gives for the ideal chain:

$$\langle h^2\rangle_0 = Nb^2\left(\frac{1+\cos\theta}{1-\cos\theta}\right)\left(\frac{1+\langle\cos\phi\rangle}{1-\langle\cos\phi\rangle}\right) \tag{8.59}$$

The usual tetrahedral bond angle of 109.5° hence tends to stretch the polymer out, as do rotameric states for which $\langle\cos\phi\rangle > 0$. l_k in this case is

$$l_k = b\left(\frac{1+\cos\theta}{1-\cos\theta}\right)\left(\frac{1+\langle\cos\phi\rangle}{1-\langle\cos\phi\rangle}\right) \tag{8.60}$$

8.1.9 Persistence Lengths of Semiflexible Polymers

The above discussion shows how polymer dimensions can increase or decrease when restrictions such as bond and rotational angles are taken into account. These latter effects are relatively small compared to other factors that can stiffen a polymer chain, such as helix structure, electrostatic interactions, and hydrogen bonds. Double-strand DNA, for example, has $l_k \sim 100$ nm. The factors that can lead to the stretching or shrinking of polymer chains are often termed *short-range* and *long-range effects*. Short-range effects are usually considered among neighboring polymers, such as the effects of bond and rotational angles, whereas *long range forces* are those resulting from monomers distant from a given monomer bending back through space into the vicinity of the given monomer and interacting with it. These latter effects are usually subsumed into the theory of intramolecular excluded volume.

An elegant statistical mechanical approach to the dimensions of flexible polymers was developed by Landau and Lifschitz (1980), who considered a long, flexible thread of contour length L and positive bending energy a, (J-m in SI units) and developed the concept of *persistence length* for polymers. The model is again based on an ideal polymer, so that long-range EV effects are not included, and so the subscript "0" for means square end-to-end length and radius of gyration continues to be used.

The model starts with a semistiff polymer chain of contour length L, as seen in Figure 8.8, and focuses on a segment of it, oriented in a 3D Cartesian space. The equilibrium position of the segment is as a straight segment in the z-direction,

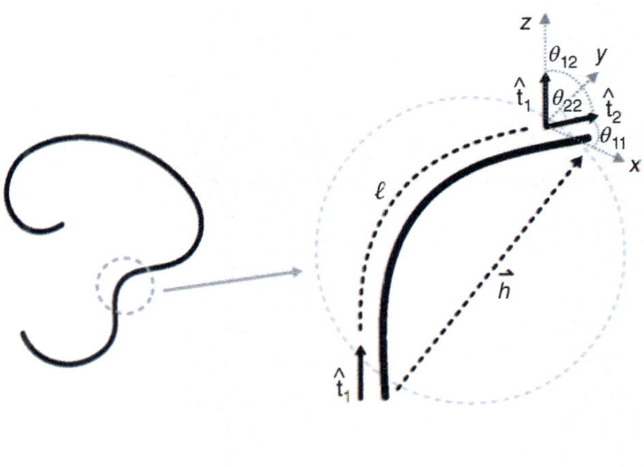

Figure 8.8 A semiflexible polymer chain of contour length L. The magnified small segment defines the unit tangent vectors for the beginning and end of the segment, as well as the small bending angles θ_{11} and θ_{22} and the resulting net angle θ_{12}. The angles are small but their size has been amplified for clarity. (Courtesy of Marissa Calko).

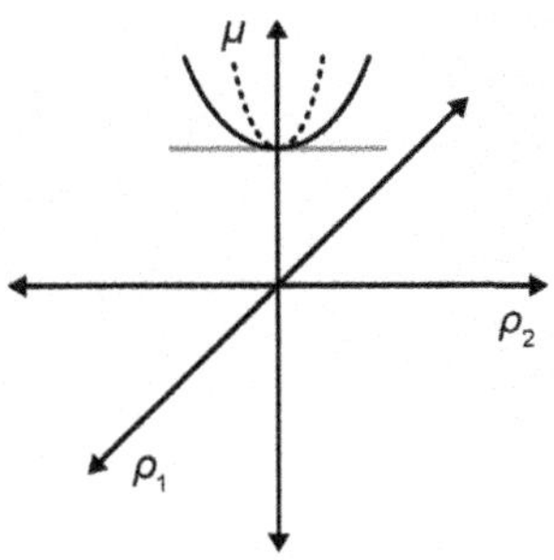

Figure 8.9 The harmonic potential in the x (dashed) and y directions of curvatures ρ_1 and ρ_2, respectively.

without any bending into the xy plane. The bending angles θ_{11} and θ_{22} indicate that the rod may have different stiffnesses when bending in the x or y directions (e.g. a rod with an elliptical cross-section).

The lowest energy state is when the thread is straight, but thermal fluctuations allow it to bend and "wobble" in a complex way that tends to shrink $\langle h^2\rangle_0^{1/2}$ below the full contour length L. A geometry is assigned to the segment on the right hand of Figure 8.8, such that l is its contour length, $\hat{t}_1$ is the unit vector pointing in the direction tangent to the start of the segment, and $\hat{t}_2$ the unit tangent vector at the end of the segment. θ_{12} is the angle between $\hat{t}_1$ and $\hat{t}_2$. For small deflections, this angle is related to the bend angles into the x and y planes θ_{11} and θ_{22} by

$$\theta_{12}^2 \cong \theta_{11}^2 + \theta_{22}^2 \qquad 8.61$$

To compute $\langle h^2\rangle_0$, it is necessary to first compute $\langle\theta_{12}^2\rangle$. $<\theta_{12}> = 0$ because bending in the x and $-x$ direction are equally likely, and the same for bending in y and $-y$.

The *bending energy per unit length* μ,[2] can be defined in terms of bending constants a_1 and a_2 and the *curvature* ρ. The curvature is the reciprocal of the radius R that can be drawn to a bent segment, or, alternatively, it measures how quickly θ changes along the length of the segment, where an infinitesimal length of the segment is ds

$$\rho = \frac{1}{R} = \frac{d\theta}{ds} \qquad 8.62$$

To compute $\langle\theta_{12}^2\rangle$, the bending energy per unit length μ is expanded to second order in the curvatures in the x and y directions ρ_1 and ρ_2, respectively.

$$\mu(\rho_1,\rho_2) = \mu(0,0) + \frac{\partial\mu}{\partial\rho_1}\rho_1 + \frac{\partial^2\mu}{\partial\rho_1^2}\frac{\rho_1^2}{2!} + \cdots + \frac{\partial\mu}{\partial\rho_2}\rho_2 + \frac{\partial^2\mu}{\partial\rho_2^2}\frac{\rho_2^2}{2!} + \cdots \qquad 8.63$$

With reference to Figure 8.9, μ has a minimum at $\rho_1 = \rho_2 = 0$, so the first partial derivatives are zero, leaving the second partial derivatives.

Since μ is a potential energy, defined up to an arbitrary additive constant, the constant term can be set equal to 0, leaving

$$\mu(\rho_1,\rho_2) \cong \frac{\partial^2\mu}{\partial\rho_1^2}\frac{\rho_1^2}{2!} + \frac{\partial^2\mu}{\partial\rho_2^2}\frac{\rho_2^2}{2!} \qquad 8.64$$

which is a harmonic bending energy potential in the curvature. The bending constants are defined by the second partial derivatives and are analogous to spring constants in the harmonic potential analogy, i.e. the larger the bending constants the stiffer the polymer.

$$a_1 = \frac{\partial^2\mu}{\partial\rho_1^2}, \quad a_2 = \frac{\partial^2\mu}{\partial\rho_2^2} \qquad 8.65$$

a_1 and a_2 have units of J-m in the SI system and erg-cm in the CGS system. Substituting the definition of the curvatures yields

$$\mu(\rho_1,\rho_2) \cong \frac{a_1}{2}\left(\frac{d\theta_1}{ds_1}\right)^2 + \frac{a_2}{2}\left(\frac{d\theta_2}{ds_2}\right)^2 \qquad 8.66$$

where ds_1 and ds_2 are the differential path elements in the two directions. The total energy of the bent segment ΔU is found by integrating ds over each segment's length l

$$\Delta U = \frac{a_1}{2}\int_0^l \left(\frac{d\theta_1}{ds_1}\right)^2 ds_1 + \frac{a_2}{2}\int_0^l \left(\frac{d\theta_2}{ds_2}\right)^2 ds_2 \qquad 8.67$$

2 μ here is bending energy per unit length, not chemical potential, which was represented by μ in earlier chapters.

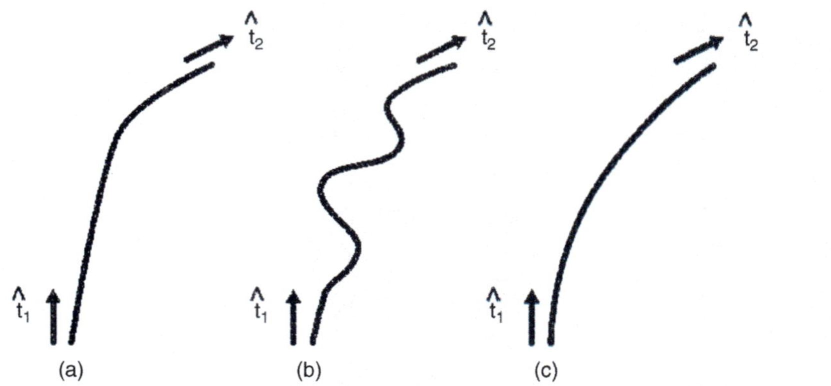

Figure 8.10 Possible bending paths between the two unit tangent vectors at each end of the segment in Figure 8.8.

This is a path integral and bending will occur so as to minimize ΔU for given θ_{11} and θ_{22}. Some examples of bending paths are shown in Figure 8.10. Minimization of ΔU follows standard procedure for an integrand of the form $f\big(\theta(s),\ \dot{\theta}(s);\ s\big)$. The procedure involves solving the Euler equation:

$$\frac{\partial f}{\partial \theta}-\frac{\mathrm{d}}{\mathrm{d}s}\left(\frac{\partial f}{\partial \dot{\theta}}\right)=0 \qquad 8.68$$

For the case here $f\big(\theta(s),\ \dot{\theta}(s);\ s\big)=\dot{\theta}^2$, so that

$$\dot{\theta}=\frac{\mathrm{d}\theta}{\mathrm{d}s}=\text{constant}=\alpha \qquad 8.69$$

and, using the boundary conditions that $\theta_1(0) = 0$ and $\theta_1(1) = 0_{11}$ (and similarly for θ_2)

$$\theta_1(s)=\frac{s}{l}\theta_{11},\quad \text{and} \quad \theta_2(s)=\frac{s}{l}\theta_{22} \qquad 8.70$$

This means the segment bends in a smooth, constant way over its length, bending path (c) in Figure 8.10. The energy integral is now

$$\Delta U=\frac{a_1}{2}\left(\frac{\theta_{11}}{l}\right)^2\int_0^l \mathrm{d}s_1+\frac{a_2}{2}\left(\frac{\theta_{22}}{l}\right)^2\int_0^l \mathrm{d}s_2=\frac{a_1}{2}\frac{\theta_{11}{}^2}{l}+\frac{a_2}{2}\frac{\theta_{22}{}^2}{l} \qquad 8.71$$

The averages $\langle\theta_{11}^2\rangle$ and $\langle\theta_{22}^2\rangle$ can be found using ΔU in the Boltzmann weighting factor

$$\langle\theta_{11}{}^2\rangle=\frac{\int_0^\pi \theta_{11}{}^2\exp\left(-\frac{\Delta U}{k_BT}\right)\mathrm{d}\theta_{11}}{\int_0^\pi \exp\left(-\frac{\Delta U}{k_BT}\right)\mathrm{d}\theta_{11}} \qquad 8.72$$

and similarly for $\langle\theta_{22}^2\rangle$. Solving this yields the mean square angular bending in the x and y directions of the segment of length l due to thermal fluctuations.

$$\langle\theta_{11}^2\rangle=\frac{lk_BT}{a_1},\quad \langle\theta_{22}^2\rangle=\frac{lk_BT}{a_2} \qquad 8.73$$

Considering the case where $a_1 = a_2 = a$

$$\langle\theta_{12}^2\rangle=2\frac{lk_BT}{a} \qquad 8.74$$

It is now possible to find $\langle h^2\rangle_0$ for the entire polymer by averaging the double integral over the whole contour length L of the infinitesimal segments ds, where $\int_0^L \mathrm{d}s = L$. This is analogous to the double sum over monomers used in the last section to determine $\langle h^2\rangle_0$. Here, the polymer is taken as a continuous thread and so the bond correlation sum is replaced by the integral:

$$\langle h^2\rangle_0 = \left\langle \int_0^L \int_0^L \hat{t}_a \cdot \hat{t}_b \mathrm{d}s_a \mathrm{d}s_b \right\rangle = \left\langle \int_0^L \int_0^L \cos\theta_{ab} \mathrm{d}s_a \mathrm{d}s_b \right\rangle \tag{8.75}$$

Expanding the cosine term allows the following approximation by an exponential to be made

$$\langle \cos\theta_{ab} \rangle = \left\langle 1 - \frac{\theta_{ab}^2}{2!} + \cdots \right\rangle = 1 - \left\langle \frac{\theta_{ab}^2}{2} + \cdots \right\rangle \approx \exp\left(-\left\langle \frac{\theta_{ab}^2}{2} \right\rangle \right) \tag{8.76}$$

For the ideal chain with no long-range excluded volume:

$$\langle \theta_{ab}^2 \rangle = 2\frac{|s_a - s_b| k_B T}{a} \tag{8.77}$$

so that

$$\langle h^2\rangle_0 = \int_0^L \int_0^L \exp\left(\frac{|s_a - s_b| k_B T}{a} \right) \mathrm{d}s_a \mathrm{d}s_b = 2\int_0^L \int_0^{S_a} \exp\left(\frac{|s_a - s_b| k_B T}{a} \right) \mathrm{d}s_a \mathrm{d}s_b \tag{8.78}$$

Performing this final integral yields

$$\langle h^2\rangle_0 = 2L_p^{\,2}\left[\frac{L}{L_p} - 1 + \exp\left(-L/L_p\right) \right] \tag{8.79}$$

which introduces the persistence length L_p. From the analysis, L_p is physically equal to the ratio of the bending energy parameter a to thermal energy, i.e.

$$L_p = \frac{a}{k_B T} \tag{8.80}$$

L_p is hence the distance along the chain at which the average bending energy is equal to the thermal energy.

L_p also provides a scale length for how a polymer bends through space after proceeding a length l along the chain, via

$$\langle \cos\theta \rangle = \exp\left(-l/L_P\right) \tag{8.81}$$

From this perspective, L_p measures how far the polymer chain "persists" in a given direction as it gradually loses track of its starting direction. In this interpretation, L_p is the contour distance along the chain from a starting point at which $\langle \cos\theta \rangle = e^{-1}$, that is, the angle between starting tangent and the tangent at $l = L_p$ is 68.4°.

This expression for $\langle h^2\rangle_0$ provides a bridge between the rod and coil limits of the polymer. In the rod limit $L_p \gg L$, the chain is very stiff, so that

$$\langle h^2\rangle_0 = L^2, \quad \text{Rod Limit} \tag{8.82a}$$

and the coil limit when $L_p \ll L$

$$\langle h^2\rangle_0 = 2LL_p, \quad \text{Coil Limit} \tag{8.82b}$$

Comparing the coil limit to that for the random coil from the preceding section provides the connection between the statistical segment length and the persistence length:

$$l_k = 2L_p \tag{8.83}$$

Table 8.1 Approximate intrinsic persistence lengths of some biopolymers.

Biopolymer	Approx. intrinsic persistence length (nm)
Polystyrene (not a biopolymer)	0.7
Single strand DNA	10
Hyaluronic acid	10
Double strand DNA	50
Xanthan	100
F-actin filament	15,000

Landau and Lifschitz provided the further vector integral analysis needed to relate $\langle h^2 \rangle_0$ to the more experimentally useful $\langle S^2 \rangle_0$, similar to the vector summation analysis by Debye and Zimm for discrete monomers in a chain. For a polymer with no long-range intramolecular interactions:

$$\langle S^2 \rangle_0 = \frac{LL_p}{3} - L_p^{\ 2} + \frac{2L_p^{\ 3}}{L} - 2\left(\frac{L_p^{\ 4}}{L^2}\right)\left[1 - \exp\left(-L/L_p\right)\right] \tag{8.84}$$

The coil limit of this expression is frequently useful:

$$\langle S^2 \rangle_0 = \frac{LL_p}{3}, \quad \text{Coil Limit} \tag{8.85}$$

A polymer chain that can be described by Equation 8.84 is often termed a *wormlike chain* and is the bridge between a random coil and a totally rigid polymer.

Table 8.1 lists the approximate intrinsic persistence lengths of several biomacromolecules, with that of polystyrene also shown. The latter persistence length is one or more orders of magnitude shorter than for the biomacromolecules. This is typical of many synthetic polymers: highly flexible and resembling random coils. The *intrinsic persistence length*, $L_{p,0}$, is the persistence length of the polymer due to its intrinsic stiffness, without any electrostatic contribution (treated in Section 8.4.1) or any other EV contribution.

8.1.10 Branching and Cross-linking

Polymer branching is ubiquitous and profoundly modifies polymer properties. It is often sought as an important feature of polymers, sometimes appears as an inevitable result or as a side reaction, and has biological precedent, especially among polysaccharides and branched fatty acids. Cross-linking can increase the stiffness of rubber and other polymers, as discussed in Sections 8.1.4 and 8.1.5.

Dynamic cross-linking occurs biologically, where the cross-links can form and break in response to stimuli or to produce directed motion. An example is how the protein *fascin* can dynamically bundle and cross-link actin filaments in the cell's cytoskeleton.

Analysis of branching is difficult and constitutes an entire subfield of polymer characterization. With appropriate detectors, size-exclusion chromatography (SEC), among other methods, provides means for branching analysis.

There are many types of branching; regular, random, star, comb, brush, dendrimer, etc., some of which are depicted in Figure 8.3. It is beyond the scope of this book to enter into the many mechanisms and models related to branching.

From the practical polymer characterization point of view, the issues are usually how to detect when branching is present and how to quantify it. While there are some relatively straightforward means for detecting branching in a polymer sample, a meaningful analysis normally requires a model, which itself requires some knowledge or conjecture of the mechanism and type of branching involved for each particular case.

A widespread method proposed for quantifying branching is due to Zimm and Stockmayer (1949). This defines the branching ratio BR for a given polymer mass M in terms of the experimentally measured mean squared radius of gyration of a branched polymer $\langle S^2 \rangle_{branched}$ to that of a linear analog of the same polymer, at the same molar mass.

$$BR \equiv \frac{\langle S^2 \rangle_{branched}}{\langle S^2 \rangle_{linear}} \tag{8.86}$$

BR can often be measured by light scattering if $\langle S^2\rangle$ is large enough (typically $\langle S^2\rangle^{1/2} > 5$ nm is achievable for standard multi-angle light scattering, or MALS, instruments). Such measurements can be made in batch, or, increasingly by multi-detector SEC.

An alternate branching ratio, sometimes termed BR′, can be determined from the ratio of intrinsic viscosities for a given M

$$BR' \equiv \frac{[\eta]_{branched}}{[\eta]_{linear}} \quad 8.87$$

BR′ is often easier to measure experimentally, since viscosity measurements do not have the spatial dimension lower limit of MALS, discussed in Chapter 9.6.

Normally, BR and BR′ are less than unity since the branched structures are denser than the corresponding linear polymers. To proceed to an analytical interpretation of BR or BR′ requires a model, such as star or random branching.

A star polymer with n equal-length branches, for example, has

$$BR = \frac{3n-2}{n^2} \quad 8.88$$

8.2 Polymer Excluded Volume (EV)

All the above expressions for polymer dimensions, $\langle h^2\rangle_0$ and $\langle S^2\rangle_0$ are for ideal polymers. An ideal polymer is like an ideal gas in that, whereas ideal gas molecules are noninteracting point-like particles, the monomers in the polymer chain have no net interaction with each other, although they are chained to each other by chemical bonds. The effect of bond and rotational angles on ideal dimensions was treated in Section 8.1. In a polymer solution, the condition in which interparticle attractive and repulsive forces cancel each other out, yielding an ideal polymer solution, is the *theta condition*. If the cancellation occurs at a certain temperature in a polymer/solvent system, this is the *theta temperature*.

Deviations from ideality can be conceptualized in terms of the EV concept, as was first introduced for nonideal solutions in Chapter 4.14. If the net interparticle force is repulsive, then the EV is termed positive and if the interaction is attractive then the excluded volume is termed negative. EV has *intraparticle* and *interparticle* effects in a polymer solution.

EV between monomers on a single chain leads to intraparticle effects. If the EV is positive, the polymer coil swells compared to its ideal state. The coil shrinks compared to its ideal state if the EV is negative. This is pictured in Figure 8.11.

EV between polymer chains leads to changes in solution osmotic pressure, light scattering, and other characteristics, and also can affect solubility.

In general, the EV can be computed from knowledge of the potential between two particles $U(r_{12},\theta_{12},\phi_{12})$ by Equation 4.157. In this expression for U, r_{12} is the interparticle center-to-center distance, and the spherical altitude and azimuth angles θ_{12} and ϕ_{12}, respectively, allow for noncentral forces, such as dipolar and magnetic forces, as well as attractive hydrophobic and repulsive electrostatic "patches" in more complex structures, such as globular proteins. In the context of polymer intraparticle excluded volume, the potential is between monomers in a polymer chain. For polymer interparticle EV, the potential is between separate polymer chains.

For central forces, equivalent sphere models are often robust enough for assessing interparticle effects. Steric repulsion lends itself naturally to the hard sphere approximation, whereas charge–charge effects in polyelectrolytes let the "soft" spherically symmetric electric field be represented by equivalent spheres.

Figure 8.11 Changes in polymer solvated volume due to positive and negative excluded volume.

The theory of intramolecular excluded volume for random coil polymer chains has been elaborated to high levels of complexity. Flory introduced the chain expansion factor α as a conceptually simple means to account for how a polymer coil's dimensions change when no longer in ideal conditions:

$$\langle h^2 \rangle = \alpha^2 \langle h^2 \rangle_0 \tag{8.89}$$

where $\langle h^2 \rangle$ is the mean square end-to-end length in the nonideal condition and $\langle h^2 \rangle_0$, the ideal value considered until now.

It is common to speak of ideal conditions as Θ-conditions, as mentioned above. At a Θ-condition, the attractive and repulsive forces between polymer chains cancel each other out and the polymers behave as if ideal. This can occur in specific solvents, termed Θ-solvents, or, in a specific solvent, there may be a Θ-temperature at which the ideal behavior occurs. At a Θ-condition, $\alpha = 1$ and $\langle h^2 \rangle = \langle h^2 \rangle_C$. In a *good solvent*, $\alpha > 1$ and $\langle h^2 \rangle > \langle h^2 \rangle_0$, i.e. the polymer coil expands. In a *poor solvent*, $\alpha < 1$ and $\langle h^2 \rangle < \langle h^2 \rangle_C$, i.e. the polymer coil shrinks.

At Θ-conditions, the ideal polymer chain obeys the random walk equation, which is an example of a *first-order Markov chain*. In the latter, the probability of the nth configuration depends only on the preceding $(n-1)^{\text{th}}$ state summed over the probabilities of all the N possible states x_i.

$$P(n, x_n) = \sum_{i=1}^{N} P_i(n-1, x_i) \tag{8.90}$$

Second-order Markov chains describe the situation where the probability of the nth configuration depends on the previous two or more steps.

When EV is involved, the behavior is often termed non-Markovian and the probability of the nth configuration can depend on any and all configurations, including beyond the nth configuration up to the total number M of steps in the process.

$$P(n, x_n) = \sum_{j=1}^{M} \sum_{i=1}^{N} P_{ij}(j, x_{ij}) \tag{8.91}$$

Because it is difficult to take into account interactions between segment n and segment i, which can be many links apart and have spatial probabilities which depend on the spatial probabilities of all intervening segments, the problem is analytically intractable. A number of mean field, perturbation, Monte Carlo, and other approaches have been taken.

In a major simplification to an inherently very difficult problem, Flory proposed a scaling argument for estimating the excluded volume effect by minimizing the free energy, $G = U - TS$, of n monomers confined to a volume represented by R^3. U is proportional to the number of two-body interactions of the N monomers per volume, or $U \propto n^2/R^3$. Using Equation 8.16 for S (without the constant term), and taking h as the characteristic length scale containing the n monomers, R, yields

$$G \propto \frac{n^2}{R^3} + T\frac{R^2}{n} \tag{8.92}$$

Minimizing G with respect to R gives the scaling relation:

$$R \propto n^{3/5} \tag{8.93}$$

Because R^2 is proportional to $\langle h^2 \rangle$, the following scaling laws are obtained:

$$\langle h^2 \rangle^{1/2} \propto n^{3/5} \tag{8.94a}$$

$$\langle S^2 \rangle^{1/2} \propto n^{3/5} \tag{8.94b}$$

The more complex theories, as well as numerical approaches, give an exponent around 0.583, which is only slightly different from the simple Flory result.

8.2.1 A_2 for Coil Polymers

The computation of excluded volume for random coil polymers is difficult. The difficulties can be subsumed by defining an *equivalent hard sphere radius*, R_e for a random coil:

$$R_e = \gamma R_g \tag{8.95}$$

where the root mean square radius of gyration is written for convenience as $R_g \equiv \langle S^2\rangle^{1/2}$. γ is usually smaller than 1 and contains the complicated, spatially, and conformationally averaged interactions between two interpenetrating random coils. It includes the same expansion factor α used above for intramolecular excluded volume, in a complicated way. Using R_e and A_2 for the hard sphere gives A_2 for the random coil in terms of γ.

$$A_2 = \frac{16\pi N_A \gamma^3 R_g^3}{3M^2} \tag{8.96}$$

For random coils with the excluded volume exponent of Flory, $R_e \propto M^{0.6}$:

$$A_2 \propto \frac{1}{M^{0.2}} \tag{8.97}$$

Usually, the dependence of A_2 on M is weak, as suggested by Equation 8.97.

8.2.2 Third and Higher Virial Coefficients

A_2 takes into account how two particles exclude volume from each other in a two-particle interaction. The third virial coefficient A_3 takes into account the volume mutually excluded in a three-particle collision. The theory of A_3 and higher coefficients is quite complex, although well developed. Even for simple hard spheres, only A_3 and A_4 have been evaluated analytically. A_5, A_6, and A_7 have been evaluated by Monte Carlo techniques.

To compute A_3, one particle can be placed at the origin, particle "0," and the interaction potentials between particles (0,1), (0,2), and (1,2) can be made with reference to the coordinates of particles 1 and 2, or r_1 and r_2, respectively. The excluded volume integral can be represented as

$$\beta_3 = \frac{1}{2}\int\int (1-\exp(-U(0,1)/k_BT)(1-\exp(-U(0,2)/k_BT)(1-\exp(-U(1,2)/k_BT)\mathrm{d}^3r_1\mathrm{d}^3r_2 \tag{8.98}$$

and

$$A_3 = \frac{2N_A}{3M^3}\beta_3 \tag{8.98}$$

For a solution of hard sphere solutes, the relation between A_2 and A_3 is

$$A_3 = \frac{5MA_2^{\ 2}}{8} \tag{8.99}$$

The units, U_n, of the nth virial coefficient A_n are given by

$$U_n = [(\mathrm{cm}^3)^{n-1} - \mathrm{mol}]/\mathrm{g}^n \tag{8.100}$$

For polymer solutions, there are abundant data on A_2, but relatively little of high accuracy for A_3. Most experimental results show that if the equivalent sphere approach is taken (R_e) for a random coil, A_3 as given above overestimates the experimental value by about a factor of 2.

8.3 Hydrodynamic Characteristics of Polymers in Solution

Polymers in solution exhibit distinctive properties which often reveal information on their conformations, architecture, and interparticle interactions. Here, the focus is on the hydrodynamic properties of polymer diffusion, and reduced and intrinsic viscosity, based on notions introduced in Chapter 6.

8.3.1 Diffusion

Diffusion was treated generally in Chapter 6. The widely used Stokes–Einstein equation was introduced for a spherical particle of *translational self-diffusion coefficient* D_0,[3] hydrodynamic radius R_H, at temperature T, in a solvent of viscosity η:

$$D_0 = \frac{k_B T}{6\pi\eta R_H} \qquad 8.101$$

where $f = 6\pi\eta R_H$ is the Stokes frictional coefficient for a spherical particle, Equation 6.38. The applicability of D_0 via this equation has been amply considered for the case of macromolecules, and in the case of monodisperse polymers, R_H represents an "equivalent hydrodynamic diameter," an analogous concept to "equivalent hard sphere diameter" introduced in conjunction with A_2 above.

When particles begin to interact hydrodynamically, the mutual diffusion coefficient D_m deviates from the single particle or self-diffusion coefficient D_0. In such cases, a useful first-order expansion involving the particle concentration and interaction parameter K_{int}, a sort of "hydrodynamic second virial coefficient," is, first introduced in Equation 6.107:

$$D_m = D_0\left[1 + K_{int}C + O(C^2)\right] \qquad 8.102$$

8.3.2 Polymer Intrinsic Viscosity

The notions of total solution viscosity, reduced viscosity, and intrinsic viscosity, η, η_r, and $[\eta]$, respectively, were introduced in Chapter 6.6. The problem of calculating $[\eta]$ for flexible polymers is complex, although theories exist that have been well confirmed for neutral polymers. The reason for the complexity has interesting historical roots. Early derivations were sophisticated enough to take first-order perturbations among polymer segments into account. The prediction, however, was that the frictional coefficient f should be linearly proportional to the number of monomers in the polymer, and hence proportional to the polymer mass: *Staudinger's law*. This was based on the notion that solvent would flow freely through the volume occupied by polymer, and be perturbed strongly only about the monomer positions. Polymers treated in this way are usually termed *free-draining coils*. It was soon found, in contrast, that for large polymers f was proportional to the square root of polymer mass M. Debye succeeded in explaining this as the result of *hydrodynamic screening*, which gave rise to the notion of *non-free draining coils*. In this view, the coil is treated essentially as a gel in which solvent molecules are bound strongly to both the monomers and the intermonomer spaces, so that the polymer entrains an entire volume of trapped solvent with it. This leads to the marked difference in scaling laws for viscosity. Definitive theories of viscosity for polyelectrolytes are not yet fully established.

As covered in Chapter 6.6, $[\eta]$ is the extrapolation to zero concentration and zero shear rate of the reduced viscosity η_r. Determination of $[\eta]$ hence requires that η_s of the sample solvent and the total viscosity of the fluid containing the macromolecules η be measured. In some analytical techniques, such as *size exclusion chromatography* (SEC) and *automatic continuous online monitoring of polymerization reactions* (ACOMP), the concentration is usually low enough that to a good approximation $[\eta] \cong \eta_r$, where η_r is measured by combining viscometer and concentration detector data. Importantly, $[\eta]$ is a direct measure of the ratio of a polymer's hydrodynamic volume V_H to its molar mass M. For example, $[\eta]$ for an ideal random coil in Θ-solvent conditions is

$$[\eta] = \frac{\Phi_\nu}{M}\left(\sqrt{6}\langle S^2\rangle_0^{1/2}\right)^3 \qquad 8.103$$

where $\Phi_\nu = 2.56 \times 10^{23}$, and $\langle S^2\rangle_0$ is the unperturbed (Θ-condition) mean square radius of gyration. The value of Φ_ν changes with chain architecture and the perturbing effects of hydrodynamic interactions.

It is important to reiterate from Chapter 6.6 that intrinsic viscosity is actually a misnomer, since it is not actually a viscosity, but rather the hydrodynamic volume occupied by a polymer per unit mass of the polymer. Intrinsic viscosity has units of volume/mass (cm^3/g), whereas actual viscosity has units of mass/(length-time), g/cm-s in CGS.

There is often a scaling law between $[\eta]$ and M, of the form:

$$[\eta] = GM^\gamma \qquad 8.104$$

3 "Translational" indicates diffusion through space, whereas particles also have *rotational diffusion coefficients*.

Table 8.2 Relationship between the radius of gyration and intrinsic viscosity exponents, β and γ, respectively.

Structure	β	γ
Sphere	1/3	0
Ideal random coil	1/2	1/2
Random coil with excluded volume	3/5	4/5
Rod	1	2

where γ is sometimes termed the *Mark–Houwink exponent* and G is a scaling constant. The mean square radius of gyration, $\langle S^2 \rangle$, often also has a relationship to M of the form:

$$\langle S^2 \rangle^{1/2} = AM^{\beta} \tag{8.105}$$

and viscosity exponents are expected to bear a relationship close to

$$\gamma = 3\beta - 1 \tag{8.106}$$

For example, a random coil in Θ-conditions has $\beta = \gamma = 0.5$, whereas a random coil with excluded volume has $\beta = 0.6$ and $\gamma = 0.8$. Knowledge of G and γ also allows computation of the viscosity averaged mass M_η, which lies between M_n and M_w for $\gamma < 1$. Unlike M_n and M_w, which are "absolute" averages of a polymer molar mass distribution, M_η can change with changes in solution conditions and T, because intrinsic viscosity can change with solution conditions and T. Viscosity measurements can also be useful for assessing branching, since branched polymers will have smaller values of $[\eta]$ than their unbranched analogs of the same molar mass. Table 8.2 shows the relationship between the radius of gyration and intrinsic viscosity exponents, β and γ, respectively, for several polymer conformations.

The reduced viscosity η_r of polymer solutions can be expressed by Equation 6.51

$$\eta_r = \frac{\eta - \eta_0}{\eta_0 c} = [\eta] + k_H [\eta]^2 c \tag{8.107}$$

The first interaction term depends on the *hydrodynamic interaction parameter*, k_H. This is difficult to calculate but generally is found to be around $k_H \sim 0.4$ for electrically neutral random coil polymers.

8.3.3 Dilute, Semidilute, and Concentrated Polymer Solution Regimes

In the limit of infinite dilution polymers are isolated units and do not interact with other polymers, and the self-diffusion coefficient D_0 and intrinsic viscosity $[\eta]$ are properties of the isolated polymer molecules in given solution conditions and temperature.

As concentration rises above zero, we have seen linear expressions for mutual diffusion D_m in Equation 8.102, and for reduced viscosity in Equation 8.107 in the *dilute regime*. As concentration continues to increase, the polymer coils will begin to overlap when the average volume of solution occupied per polymer molecule equals the hydrodynamic volume occupied per polymer molecule. In terms of concentration, the *overlap concentration*, c^*, is when the mass per hydrodynamic volume occupied by a polymer is equal to the mass per solution volume of the polymer. A convenient rule of thumb for estimating c^* is hence

$$c^* \sim \frac{1}{[\eta]} \tag{8.108}$$

Above c^*, *the semidilute regime*, higher order terms in concentration can be used to account for interactions among more than two particles at a time.

Finally, as concentration continues to increase further, the *concentrated regime* is reached at c^{**}, where chains become strongly interpenetrated and their diffusive motion and motion under shear is dominated by how the chains move through and with respect to each other. In this regime segments of individual, linear chains are so intermingled with segments of other chains that the chains become ideal (no excluded volume) in the *polymer melt limit*, that is, when there is no solvent

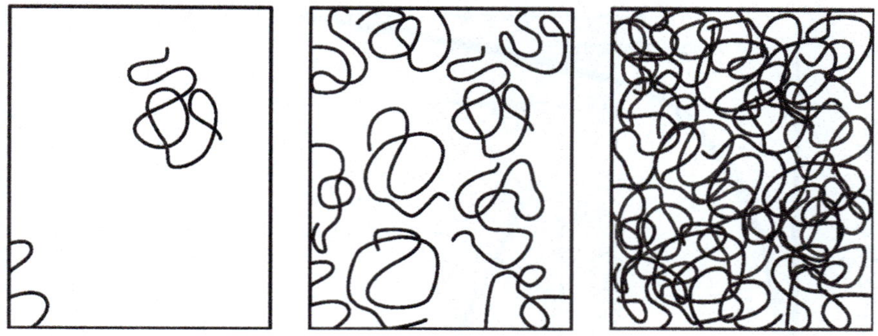

Figure 8.12 From left to right, dilute, semidilute, and concentrated polymer solutions.

in the molten polymer batch. Figure 8.12 is a conceptual picture of the dilute, semidilute, and concentrated regimes. There are no sharp boundaries between the regimes.

8.4 Electrically Charged Polymers: Polyelectrolytes

Electrically charged polymers are *polyelectrolytes*, first introduced in Chapter 5.5. Proteins, polynucleic acids, and many polysaccharides are *biopolyelectrolytes*. Electrical charge confers many properties on these, such as the ability to interact attractively with ions, polymers, and structures of the opposite sign, to maintain stability and osmotic pressure in the aqueous living state media, to absorb large amounts of water, to form gels, to provide interfaces with water for structures containing hydrophobic interiors, and to impart compressive strength and elasticity to tissues.

Many biopolyelectrolytes have fairly large intrinsic persistence lengths, e.g. on the order of 50 nm for DNA and 10 nm for hyaluronic acid.[4] In contrast, a typical synthetic polyelectrolyte, such as polystyrene sulfonate has an intrinsic persistence length less than 1 nm. If the polyelectrolyte is much longer than its persistence length, it will resemble a random coil, and the coil size will expand as the ionic strength of the medium decreases. This will have at least two effects: First, the intrinsic excluded volume will be electrostatically enhanced both by the increased chain volume and the less shielded electrostatic field around this expanded chain, so that the second virial coefficient A_2 increases, which will raise the osmotic pressure of the solution. It will also decrease the osmotic compressibility of the solution and hence scatter less light (Chapter 9.5). Second, the expanded volume with the same molecular weight means that the intrinsic viscosity (hydrodynamic volume/mass) will increase, along with that of the solution the polyelectrolyte is in.

Figure 8.13 shows the raw light scattering and solution viscosity (arbitrary units) for the biopolyelectrolyte sodium hyaluronate versus the ionic strength. As described above, light scattering is much less at low ionic strength and reduced viscosity much higher than at high ionic strength. The scattering rapidly increases and the solution viscosity decreases as ionic strength increases. The inset shows A_3 and A_2 versus ionic strength from the analysis of the same experiment. A_3 overlaps A_2 but changes by two decades for every one decade change of A_2, just as Boltzmann predicted for hard spheres. The result is remarkable because hyaluronate is a stiffened random coil, not a hard sphere.

Interestingly, polyampholytes, which have both negative and positive charges, can show opposite ionic strength behavior than polyelectrolytes. Most proteins are polyampholytes.

Much effort has been given to quantitatively explaining these effects, and two approaches, that of *electrostatic persistence length* (EPL) and *electrostatic excluded volume* (EEV) are outlined next.

Figure 8.14 shows the NaCl titration of proteinaceous capsids. M_w is proportional to light scattering intensity in this dilute case, allowing direct comparison of the opposite behavior of the polyampholyte in Figure 8.14 and the polyelectrolyte in Figure 8.13. At low ionic strength, the positive and negative charges on the proteins are unshielded, allowing many capsids to associate via electrostatic attraction. As ionic strength increases, these charges become shielded, weakening the attractions and causing fewer capsids to associate with each other. At high ionic strength, the capsids are no longer associated. The effect is reversible, i.e. if ionic strength is lowered, either through dilution with pure water or unsalted buffer, or

4 A useful table of persistence lengths and linear mass densities for biological polymers is at https://bionumbers.hms.harvard.edu/files/Linear%20density%20(mass%20per%20unit%20length)%20and%20persistence%20length%20of%20some%20biologically%20important%20polymers.jpg

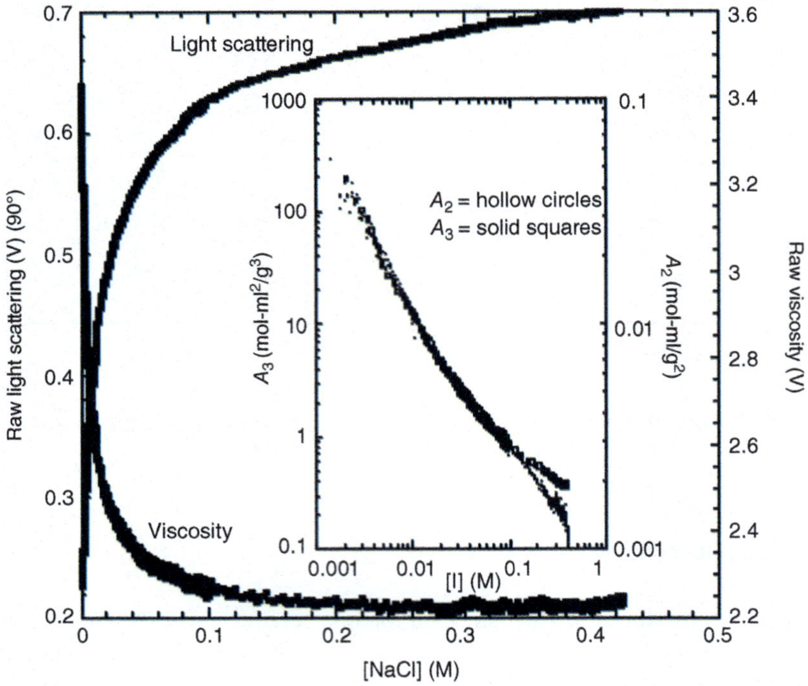

Figure 8.13 Effect on light scattering and viscosity for a solution containing 10^{-4} g/cm^3 of hyaluronate as ionic strength increases. The inset shows the variation of both A_2 and A_3 versus ionic strength. (Sorci, G.A. and Reed, W.F. (2004) / Reproduced with permission of American Chemical Society).

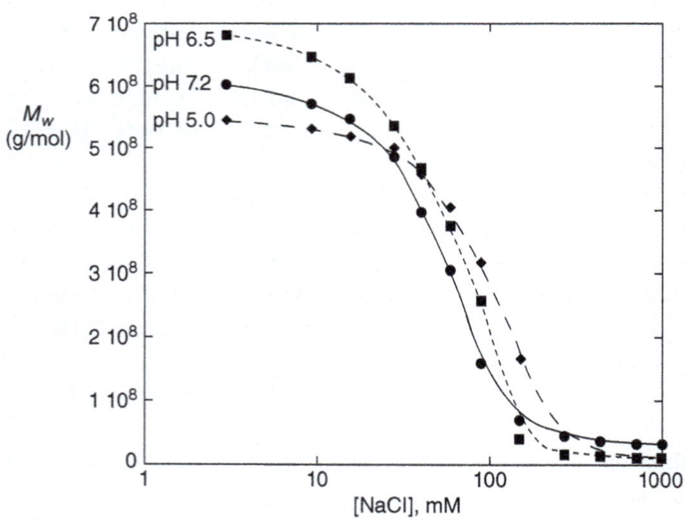

Figure 8.14 Polyampholyte behavior of protein capsids versus salt titration is opposite the behavior of the polyelectrolyte (sodium hyaluronate) in Figure 8.13. (Unpublished results, Curt Jarand, Wayne Reed).

by dialysis, the associations re-occur. The effect also depends on pH, as might be expected, since the charge state of the protein is affected by the pH.

8.4.1 Electrostatic Persistence Lengths and Electrostatic Excluded Volume

In the Gedanken experiment where a flexible polymer in a solution that is initially neutral becomes electrically charged with charges of a single sign, one would imagine that the polymer would stretch out as the charges along the chain repel each other. That the chain would not stretch out completely follows from the fact that the entropic penalty in the free energy for complete stretching is very high, because there is only a single completely stretched state (all trans) among 3^{N-2} total states, with entropy $S = k_B \ln(1) = 0$. Hence, the challenging physical problem is to compute how far the chain stretches out in response to both electrical charge and the ionic strength of the supporting medium, such that free energy, $G = U - TS$ is minimized. The repulsive electrostatic energy term, U, is reduced by the chain stretching out, while the entropic terms $-TS$ decreases by the chain stretching out, and so resists the stretching. One approach to formulating this problem is by introducing the electrostatic persistence length, L_e.

One can further imagine that as the neutral polymer becomes charged the intermolecular repulsive energy will increase, and hence the excluded volume will also increase, due both to the increasing strength of the electric field in the vicinity of the charged polymer, and due to the fact that as it stretches out in space, it will also occupy and hence exclude more volume. This problem has been formulated in terms of the electrostatic excluded volume.

Odijk (1977) and, separately, Fixman and Skolnick (1977) derived an expression for EPL as a perturbation to the inherent persistence length $L_{p,0}$, to obtain a total persistence length L_p:

$$L_p = L_{p,0} + L_e \tag{8.109}$$

They arrived at the same expression for the electrostatic persistence length L_e

$$L_e = \frac{\xi^2 \kappa^{-2}}{12 l_B}\left[3 - \frac{8}{y} + e^{-y}\left(y + 5 + \frac{8}{y}\right)\right] \tag{8.110}$$

where ξ is the number of elementary charges per Bjerrum length, l_B (Chapter 5.10), κ is the Debye–Hückel screening parameter, and $y = \kappa L$, where L is the contour length of the polymer. With knowledge of $L_{p,0}$ approximately obtained at very high ionic strength, L_p can hence be computed using $L_{p,0}$ and L_e from Equation 8.110, so that $\langle S^2\rangle_0$ can be computed by Equation 8.84. The measured value $\langle S^2\rangle$ is related to $\langle S^2\rangle_0$ via the static expansion factor α_s, introduced above in Equation 8.89. Figure 8.15 gives a notion of how electric charge in the ideal case (center of figure) increases the effective volume due to L_e, and the right-hand shows the additional expansion due to long-range effects contained in $\alpha^2_{s,el}$.

There are several theories that relate α_s to a commonly used perturbation parameter z, given by

$$z = \left(\frac{3}{2\pi l_k^2}\right)^{3/2} \beta n_k^{1/2} \tag{8.111}$$

where l_k and n_k are the (Kuhn) statistical segment length and number, respectively, and, in the coil limit. Here β is the EV between two charged rod-like segments, for which Fixman and Skolnick arrived at the expression:

$$l_k = 2L_p \tag{8.112}$$

$$\beta = 8L_p^2 \kappa^{-1} \int_0^{\pi/2} \sin^2\theta \int_0^{w/\sin\theta} x^{-1}(1 - e^{-x})\,dx\,d\theta \tag{8.113}$$

where

$$w = 2\pi\xi^2\kappa^{-1}e^{-\kappa d} \tag{8.114}$$

where d is the rod diameter. The dimensionless linear charge density ξ is normally used as a convenient parameter for counterion condensation and polyelectrolyte theories.

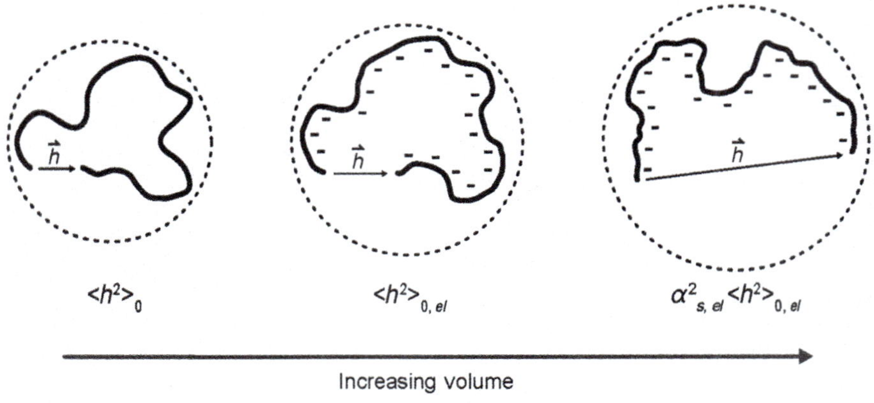

Figure 8.15 An uncharged polymer (left) and how its ideal effective volume increases when it is charged and L_e increases L_p (center). When long-range effects are included, there is an additional volume expansion due to $\alpha^2_{s,el}$ (right).

Of the several expressions relating z to α_s one that has found considerable utility when $n_k > 2$, is the Gupta–Forsman expression

$$\alpha_s^5 - \alpha_s^3 \approx \frac{134}{105}(1 - 0.885 n_k^{-0.462}) z \qquad 8.115$$

The expressions outlined above have proven to be a robust approach to predicting polyelectrolyte dimensions for varying linear charge density and ionic strength conditions.

The same EEV is involved in calculating how an electrostatic value of A_2 will arise, although, again, there is no simple path for making the connection. A result by Yamakawa (1971) is

$$A_2 = \left(N_A n_k{}^2 \beta / 2M^2\right) h_0(\bar{z}) \qquad 8.116$$

where

$$h_0(\bar{z}) = \left[1 - (1 + 3.9\bar{z})^{-0.468}\right] / 1.83\bar{z} \qquad 8.117$$

and

$$\bar{z} = z/\alpha^3 \qquad 8.118$$

Some experimental results for both $\langle S^2 \rangle$ and A_2 for sodium hyaluronate, using light scattering, are shown in Figure 8.16. *The computed lines are not a fit* and do not involve any adjustable parameters. They are the results of combining the EEV expressions just cited with the known parameters of the molecules: L, d, ξ, and the ionic strength of the polyelectrolyte solution (added salt). Hence, the theories are fairly robust in predicting real polyelectrolyte intra- and intermolecular EEV effects. It should be noted that as far as $\langle S^2 \rangle$, the EEV effect is predominant compared to the L_e effect alone, as is readily seen in Figure 8.16.

Counterion condensation on polyelectrolytes was outlined in Chapter 5.10 in terms of the Bjerrum length, l_B.

$$l_B = \frac{z_1 z_2 e^2}{4\pi D \varepsilon_0 k_B T} \qquad 5.130$$

Under this theory, if the bare polyelectrolyte has $\xi > 1$, then counterion condensation will bring ξ to $\xi \leq 1$.

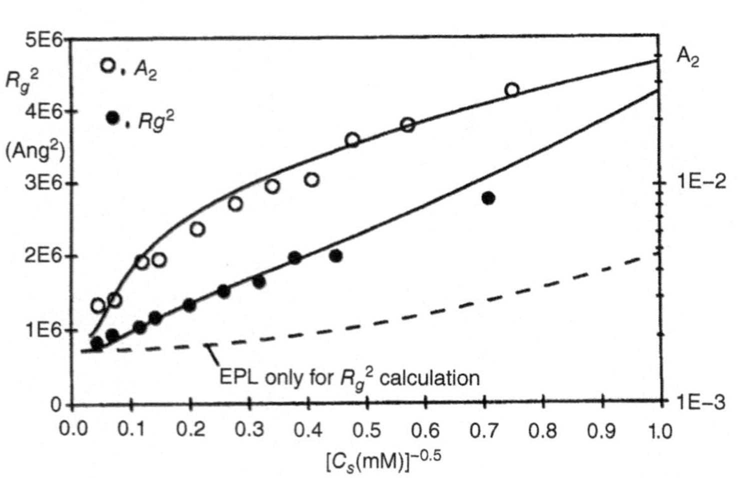

Figure 8.16 Computed mean square radius of gyration ($R_g{}^2 = \langle S^2 \rangle_z$) and A_2 versus the reciprocal square root of ionic strength. The solid lines are computations using Equations 6.102–6.108. They are not fits and involve no adjustable parameters. The dashed line for $R_g{}^2$ shows the computational result if the long-range effect given by $\alpha^2_{s,el}$ is neglected. (Reproduced with permission, Benmouna and Reed / Oxford Publishing Limited).

8.5 Case Study of Polystyrene Characteristics in Tetrahydrofuran

The above sections have detailed static and dynamic properties of polymers. To get an idea of the magnitudes of these quantities, data from various authors are summarized here for polystyrene in tetrahydrofuran (THF), taken as a very common case of a linear polymer in a good solvent.

Using low-dispersity molar mass standards for polystyrene over the wide range of 10^4–10^7 g/mol, the following relationships to molar mass M were found in THF at 30°C:

$$\langle S^2\rangle^{1/2}(\mathrm{nm}) = 0.0209\mathrm{M}^{0.56} \qquad 8.119a$$

$$A_2(\mathrm{cm}^3 - \mathrm{mol/g}^2) = 0.0194\mathrm{M}^{-0.286} \qquad 8.119b$$

$$[\eta](\mathrm{cm}^3/\mathrm{g}) = 0.01363\mathrm{M}^{0.71} \qquad 8.119c$$

$$D_0(\mathrm{cm}^2/\mathrm{s}) = 3.4\times10^{-4}\mathrm{M}^{-0.56} \qquad 8.119d$$

Values for these measured characteristics are shown for a range of polystyrene masses in Table 8.3.

Table 8.3 shows $\langle S^2\rangle^{1/2}/R_H = 1.40$ for PS in THF. This is close to the theoretical value for an ideal random coil in the nonfree draining limit.

Table 8.3 Various static and dynamic characteristics of polystyrene in THF at 30°C.*

M (g/mol)	A_2 (cm^3-mol/g^2)	$\langle S^2\rangle^{1/2}$ (nm)	R_H (nm)	D (cm^2/s)	Time (s) to diffuse 100 μm	$[\eta]$ (cm^3/g)
10^4	1.4×10^{-3}	3.6	2.6	2.0E − 06	8.5	9.4
10^5	7.2×10^{-4}	13.2	9.4	5.4E − 07	30.9	48.4
10^6	3.7×10^{-4}	47.9	34.1	1.5E − 07	112.3	248.0
10^7	1.9×10^{-4}	173.8	123.9	4.1E − 08	407.8	1272.0

*$\eta = 0.00438$ P for THF at 30°C.

8.6 Thermodynamics of Polymer Solutions

When a polymer is dissolved in a good solvent, the free energy of the system is lowered, and when dissolved in a poor solvent, the free energy increases. The way in which polymer molecules interact with the solvent, with each other, and with themselves (e.g. excluded volume effects, self-organization of copolymer chains, etc.) lends itself to a thermodynamic description. In the latter, the usual thermodynamic parameters of temperature, concentration, pressure, and interparticle potentials come into play. A brief outline of the early (1940s) Flory-Huggins thermodynamic lattice model is given next. It has since been improved and extended in several ways from the original form, which won't be covered here.

8.6.1 Flory–Huggins Mean-field Theory

The essence of the Flory–Huggins model is to treat a polymer solution as a 3D lattice composed of n_{total} lattice sites, equally spaced in all dimensions. Each lattice site contains either a solvent or monomer of a polymer chain, so that $n_{total} = n_s + Nn_p$, where n_s and n_p are the number of solvent and polymer molecules, respectively, and N is the degree of polymerization. The theory yields the Gibbs free energy G, and related quantities in terms of (i) the entropy of mixing and (ii) the enthalpy of mixing. The theory is formulated in terms of the volume fraction ϕ of monomers in the lattice (and $1-\phi$ solvent molecules in the lattice) and the Flory–Huggins χ-parameter, which characterizes the energy of polymer–polymer (u_{pp}), solvent–solvent (u_{ss}), and polymer–solvent (u_{ps}) interactions.

$$\frac{\Delta G}{n_{total}k_BT} = \frac{\phi}{N}\ln\phi + (1-\phi)\ln(1-\phi) + \chi\phi(1-\phi) \qquad 8.120$$

The first two terms on the right represent the entropy of mixing. The first of these is due to the mixing of the polymer and the second to the mixing of the solvent. For long-chain polymers, $N > 100$, the first of these terms becomes negligible so that only the mixing entropy of the solvent remains important.

The enthalpy of mixing is given in terms of the Flory–Huggins parameter χ, which is computed as

$$\chi = \left[u_{ps} - \left(u_{ss} + u_{pp}\right)\right] / k_B T \tag{8.121}$$

χ is inversely proportional to T and is affected by other interaction factors, which can themselves also be T-dependent; χ for real substances can have nonmonotonic dependence on T and can produce complex phase diagrams. The chemical potential of the polymer can be determined from the partial derivative of free energy with respect to n_p at constant n_s and n_t to yield:

$$\frac{\Delta\mu_p}{k_B T} = \ln\phi - (N-1)(1-\phi) + \chi N (1-\phi)^2 \tag{8.122}$$

where the relation

$$\left.\frac{\partial\phi}{\partial n_p}\right|_{T,n_s} = \frac{N\left(1-\phi\right)^2}{n_s} \tag{8.123}$$

is useful in the derivation. Similar reasoning leads to the chemical potential of the solvent:

$$\frac{\Delta\mu_s}{k_B T} = \ln(1-\phi) + \left(1 - \frac{1}{N}\right)\phi + \chi\phi^2 \tag{8.124}$$

Using the relationship between solvent chemical potential and the specific volume of solvent, $v_1{}^0$, which is the volume of a lattice site, and hence also the specific volume of a monomer $v_2{}^0 = v_1{}^0$, $\Pi v_1{}^0 = -\Delta\mu_s$ yields

$$\frac{\Pi v_1^0}{k_B T} = \frac{\phi}{N} - \ln(1-\phi) - \phi - \chi\phi^2 \tag{8.125}$$

For dilute solutions expansion of $\ln(1-\phi)$ in a power series and retaining terms in ϕ and ϕ^2 yields

$$\frac{\Pi v_1^0}{k_B T} = \frac{\phi}{N} + \left(\frac{1}{2} - \chi\right)\phi^2 + O(\phi^3) \tag{8.126}$$

Comparing the second term to the virial coefficient expansion of osmotic pressure from Equation 4.126:

$$\Pi = RT\left(\frac{c_2}{M} + A_2 c_2^2 + A_3 c_2^3 + \cdots\right) \tag{4.126}$$

yields the following connection between A_2 and χ,

$$A_2 = N_A v_1^0 \left(\frac{N}{M}\right)^2 \left(\frac{1}{2} - \chi\right) \tag{8.127}$$

where c_2 is the concentration of polymer in g/cm^3, M is the molar mass of the polymer, and the dilute solution approximation has been used:

$$\phi = \frac{n_2 v_2^0}{n_1 v_1^0 + n_2 v_2^0} \cong \frac{n_2 v_2^0}{n_1 v_1^0} = \frac{n_2}{n_1} = \frac{c_2 v_1{}^0 N_A}{M} \tag{8.128}$$

along with the lattice volume condition $v_2{}^0 = v_1{}^0$, and $V \cong n_1 v_1^0$.

As noted, under Θ conditions the polymer solution becomes ideal, repulsive, and attractive forces cancel each other out and the virial coefficients become zero. Under Θ conditions, $\chi = 1/2$. In a good solvent, $A_2 > 0$ so $\chi < 1/2$. In a poor solvent, $A_2 < 0$ so $\chi > 1/2$.

χ (and hence also A_2) is useful for determining coexistence and phase separation behavior of polymer solutions, discussed next.

8.6.2 Phase Behavior

The miscibility of multicomponent systems depends on how the total free energy of mixing of the solution changes as the composition and temperature of the solution change. Using the free energy of mixing from the lattice model, Equation 8.120 allows phase behavior to be computed as follows. The chemical potential for the polymer chains is the derivative of this free energy with respect to n_p at constant T and V. When there is a separation of the system into two phases, e.g. high concentration of solvent with some polymer in phase A and high concentration of polymer with some solvent in phase B, the chemical potential of polymer is the same in both phases, and likewise for the solvent. This requires that the tangent be the same in both phases, i.e. the common tangent condition is

$$\left.\frac{\partial\left(\Delta G_{mix} / n_{total}k_B T\right)}{\partial\phi}\right|_A = \left.\frac{\partial\left(\Delta G_{mix} / n_{total}k_B T\right)}{\partial\phi}\right|_B \qquad 8.129$$

Tracing out the values at which ϕ occurs for each value of T (and hence χ) defines the binodal, the curve between the phases, sometimes also termed the "coexistence" or "phase stability" curve.

An interesting aspect of phase stability is that there can be concentrations at which the solution is metastable against phase separation. This arises when the second derivative of $\Delta G_{mix} / n_{total}k_B T$ is zero. Then, the system is stable against phase separation for small changes in ϕ in either direction, i.e. the free energy is lower at the inflection point of $\Delta G_{mix} / n_{total}k_B T$ for the system staying unseparated at that point than what the total $\Delta G_{mix} / n_{total}k_B T$ would be if separated into two phases, one an amount $+\Delta\phi$ from the inflection point and the other $-\Delta\phi$ from the inflection point. Large perturbations, such as shaking or nucleation sites, however, will lead to destabilization and phase separation, similar to the case of a supercooled liquid remaining liquid until some perturbation (e.g. mechanical shock) sends it into phase separation. The curve that defines the metastable region is called the *spinodal line*. Between the *binodal line* and spinodal line resides the *metastable region*.

The graphical example of Figure 8.17 illustrates these concepts. Figure 8.17 shows a hypothetical $\Delta G_{mix} / n_{total}k_B T$ versus the volume fraction of polymer ϕ for different values of T and corresponding χ, where χ was taken to vary, for example only, as $\chi = -0.65 + 490/T$, and $N = 2$. One example of how the equal tangents are found graphically is shown by a dashed line in the figure and how they translate to the binodal points on the bimodal curve in the phase diagram overset in the figure is shown by the dotted lines. Also shown for the same curve are the inflection points which translate up to the spinodal curve in the phase diagram, also shown with dotted lines. The portions of the diagram where the solution is phase separated, single phase, and metastable are also indicated.

In the hypothetical case of Figure 8.17, χ decreases with temperature from 0.98 at 283 K to 0.60 at 360 K, i.e. the solvent becomes better at higher temperatures. The temperature at which the bimodal and spinodal curves meet are termed the upper critical solution temperature (UCST). Below the UCST, the system separates into two phases.

There are also systems for which χ increases with temperature, e.g. in a form $\chi = A - B/T$. In this case, the bimodal and spinodal lines open upward, visualized by rotating the top curve in Figure 8.17 by 180° around the x-axis. In this case, the system will have a lower critical solution temperature (LCST), i.e. the phase separation occurs when the system is heated, e.g. pNIPAM in water, or polyethylene in hexane under 5 bar pressure.

There are also cases where χ is not a monotonic function of temperature and may have a maximum and/or minimum value versus T. In such cases, it is possible to describe phase shapes that don't resemble parabolas, such as in Figure 8.17. Poly(oxyethylene) in water, for example, has a χ versus T with a peak, and this leads to an oval shape in the T versus ϕ plane with the two-phase separation within the oval. An example of χ versus T with a minimum include low-mass poly(styrene) in acetone, which has an LCST in an upward parabola above a downward parabola with a UCST. It is not a contradiction that the LCST is higher than the UCST; if the system goes from demixed to single phase upon heating, then a UCST is crossed and continued heating the single phase in this latter case eventually leads to demixing again at the LCST, which is higher than the UCST.

Another example of a system with a minimum in χ versus T is high molar mass poly(styrene) in acetone, for which an "hour glass" figure forms in the T versus ϕ plane, inside of which the system is phase separated at any temperature.

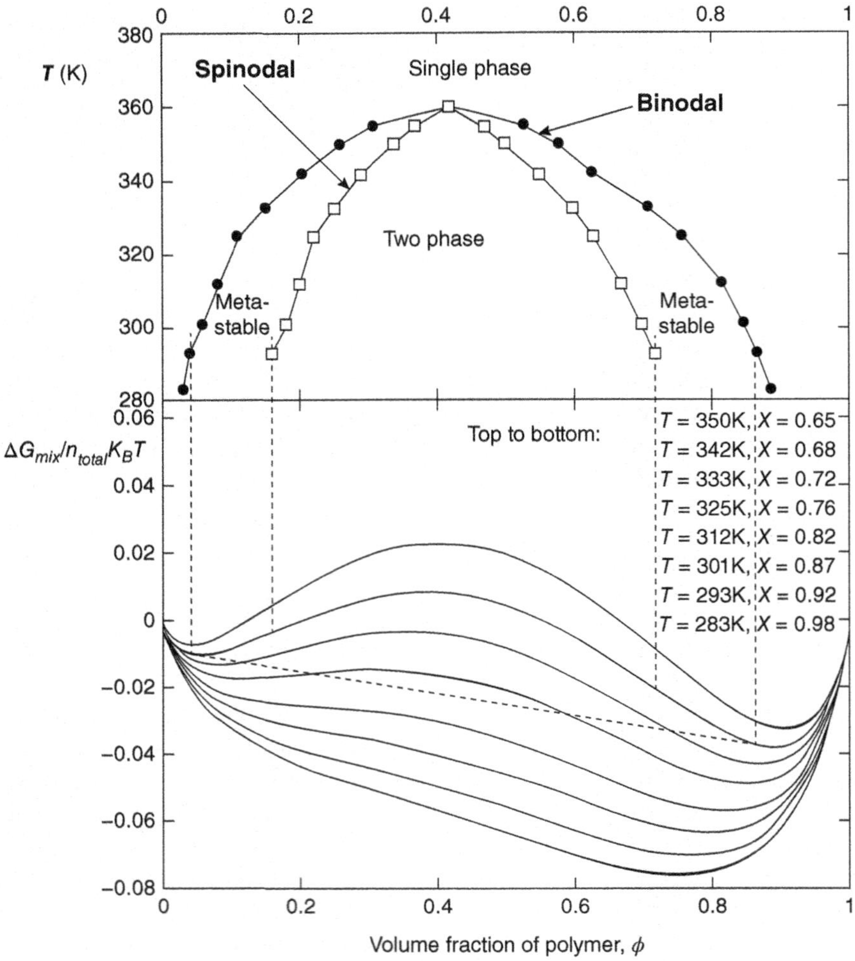

Figure 8.17 Phase behavior of a polymer at different temperatures: Lower panel shows $\Delta G/k_BT$ versus volume fraction of polymer for various temperatures. The top panel shows the corresponding temperature/concentration phase diagram. (Elias, H.-G. (1996) / Reproduced with permission of John Wiley & Sons).

8.7 Rheology

The above discussion of polymer diffusion and viscosity involves time-independent quantities that occur under no externally applied shear conditions (diffusion) and vanishingly small, time-independent shear (intrinsic viscosity). Polymers in melts and in solution constitute complex fluids with intricate time-dependent responses to stresses and flows that reflect molecular motions on many different length and time scales. The two extremes in rheology are (i) the pure, inviscid liquid, which is incapable of supporting any shear stress and immediately flows in response to such stress; and (ii) a solid which neither flows nor irreversibly deforms in response to a shear stress. Most real materials exist between these extremes. Bose–Einstein condensation in helium is a rare exception, where all atoms are phase locked into the ground state, similar to superconductivity, and have no viscosity while in this superfluid state.

The Deborah number, *De*, is a dimensionless parameter which is the ratio of the time of relaxation of a material in response to a shear stress to the time of experiment or observation. $De = 0$ for the extreme of inviscid matter (e.g. superfluid helium) and $De \to \infty$ for a pure, non-degrading solid. *De* was taken as not only an index defining the continuum from pure, inviscid liquid to pure solid but also as a jocular shield for the sometimes misplaced use of "theology" where, in fact, "rheology" is concerned. "All things flow before God" is the paraphrase of the prophetess Deborah that indicates all things have $De = 0$ when the time scale of observation is infinite.

8.7.1 Newtonian and Non-Newtonian Polymer Liquids

Polymer motions, such as bond rotations among different configurations, reptations through space, intermolecular overlap interactions, and so on, lead to viscoelastic responses in polymer solutions that can span many orders of magnitude in time, or in its conjugate variable, the frequency at which stress is applied. Liquids whose viscosity does not change with frequency of stress are termed *Newtonian* fluids. Most polymeric fluids deviate from Newtonian behavior under certain conditions and are termed *non-Newtonian* (Chapter 6.1.1).

Shear thinning (thixotropy) is one of the most common manifestations of non-Newtonian behavior in polymeric liquids. Increased shear can lead to partial alignment of polymers or colloid particles with the flow, thus decreasing viscosity. Examples include latex paint, blood, and syrups. Shear thickening is the opposite phenomenon (anti-thixotropy) whereby the material becomes more viscous or stiffer with increasing shear, often due to shear-induced organization, such as partial crystallization. Quicksand and aqueous solutions of corn starch are examples of shear-thickening materials, as well as mixtures of boric acid and silicone oil, which came to be known as "silly putty".

Figure 8.18 shows shear thinning of xanthan gum in a 10 mM aqueous NaCl solution at 0.003 g/cm^3. Over the shear rate range of 50–60,000 s^{-1}, the reduced viscosity drops by an order of magnitude.

In practical terms, polymer rheology, and especially its non-Newtonian aspects are of critical importance in processing polymers. The relationship of applied stress and polymer flow determines the processability of polymers in applications such as extrusion, fiber spinning, and film blowing.

8.7.2 Dynamic Mechanical Analysis and Other Viscoelastic Analyses

An important means of assessing the rheological behavior of polymer melts and concentrated solutions is the use of dynamic mechanical analysis. This consists of testing the mechanical response, or strain, of a polymeric material to a time-dependent stress, or, conversely, measuring the stresses generated by time-dependent strains. For simplicity, a sinusoidal driving function is normally used to elicit the response function $\sigma(t)$. In the case of controlled shear stress $\dot{\gamma}(t)$:

$$\dot{\gamma}(t)=\dot{\gamma}_0 \sin(\omega t) \quad \text{Controlled Input} \tag{8.129a}$$

$$\frac{\sigma(t)}{\dot{\gamma}_0}=G'(\omega,T)\sin(\omega t)+G''(\omega,t)\cos(\omega t) \quad \text{Response Function} \tag{8.129b}$$

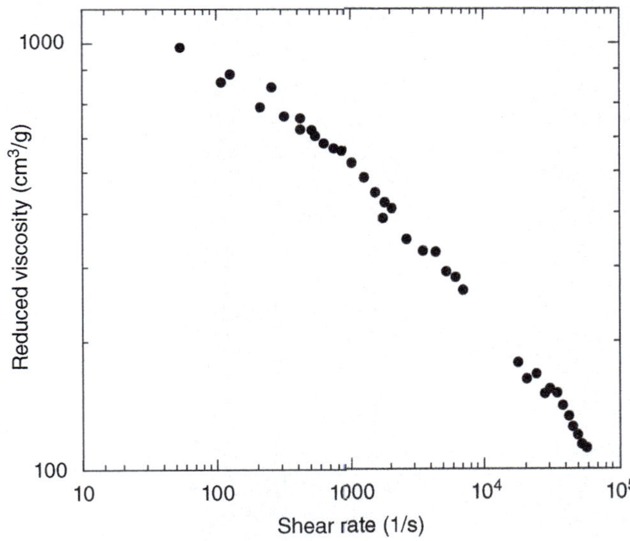

Figure 8.18 Shear thinning of xanthan as shear rate increases, in 10 mM aqueous NaCl. (Drenski, M.F. and Reed, W.F., unpublished results).

The coefficient of the in-phase portion, often termed the dynamic storage modulus, $G'(\omega,T)$, measures the purely elastic response of the material and is normally very sensitive to both frequency ω and temperature. The out-of-phase portion, termed the dynamic loss modulus G'', measures the energy dissipative response of the material. The phase shift between the input frequency and the response is usually denoted by δ and is the ratio of the moduli:

$$\tan\delta = \frac{G''}{G'} \tag{8.130}$$

A purely viscous material with no elasticity will have $\delta = \pi/2$, whereas a loss-free elastic material will have $\delta = 0$. A typical polymer will show a complex interplay between how G' and G'' behave versus ω. It is common for G' and G'' to cross each other once or more as ω increases.

Because polymer rheology is a deeply complex field, there are many other methods to assess different types of viscoelastic and mechanical behavior of polymers in concentrated solutions, melts, and rubbery states.

The *Mooney viscosity* is the torque generated on a spindle rotating at constant angular velocity, immersed in a polymeric material between heated dies. It is one of the most widely used industrial measures of bulk viscoelastic properties of polymeric materials, especially elastomers and rubbers. Oftentimes, the Mooney viscosity is of greater practical significance than the dynamic mechanical determination of G' and G''. The notion of Mooney viscosity is well grounded in the principles of continuum mechanics but involves the empirically determined Mooney viscosity and related parameters.

The melt-flow index is an extremely simple test and essentially consists of measuring the amount of polymer melt flowing through a tube of specific dimensions under a specific pressure over a specified amount of time. Experimentally, the apparatus is a small extruder. The index obtained has a sort of inverse relationship to viscosity at any given temperature and is indirectly related to polymer molar mass.

8.8 Solid-state Properties

Polymers in the solid-state exhibit extraordinarily diverse properties. Polymers can be made exceptionally hard, scratch-, impact-, and high-temperature resistant, while they can also be made highly ductile, flexible, and rubbery. Additionally, their electrical and optical properties can be tailored in many unique ways.

The details of polymeric solid-state mechanical behavior are rooted in a number of factors: molar mass distribution, polymer tacticity, degree of crystallinity, inclusions, defects, incorporation of specific modifiers, polymer blending, processing conditions, etc.

Many of the methods used for characterizing solid materials are applicable to polymers in the solid state: X-ray diffraction and crystallography, thermal analyses such as differential scanning calorimetry, mechanical testing for ductility, brittleness, compressive, tensile, and torsional stress–strain relationships, impact, fatigue, creep, etc.

The glass transition is of particular importance for polymers that are in a glassy state when solid. This state is typical of other solid-state amorphous materials, i.e. no long-range crystal order, although there can be regions and domains of crystallinity. Upon heating a polymer in the glassy state a glass transition temperature T_g is found where the polymer loses its solidity and becomes rubbery. A number of other thermodynamic parameters also change at T_g.

A new generation of polymer electronics, including photovoltaics, transistors, and flexible electronic materials, is being developed based on conducting and doped polymers. Polymers have unique optical properties where they are used in the vast field of polymer optics, to fabricate devices such as lenses, polarizers, fiber optics and other waveguides, liquid crystals, novel photonic materials, nano-optics, etc. Some polymers, including most biopolymers, have marked optical activity and can be analyzed and used in areas of polarimetry, and linear and circular dichroism and birefringence. Polymer optics also extends to nonlinear optical applications, such as second harmonic generation.

8.9 Summary

Important characteristics that describe static mass, conformations, and dimensions of polymer molecules have been surveyed. A close connection between ideal elastic polymer coil dimensions and entropy was found, along with a purely entropic force as the basis of polymer elasticity. Various stiffening mechanisms for polymer coils, including bond constraints, positive EV, and electrostatic interactions, lead to coil expansion. A concise assessment of the addition of enthalpic interactions using Flory's free energy approach gives a good approximation of limiting effects of EV on polymer dimensions. The notion of bending energy of a semistiff polymer leads to the persistence length concept.

The development of average static conformations has been followed by their connection to hydrodynamic properties such as diffusion and viscosity. A separate section has been used to describe the salient aspects of charged polymers, or polyelectrolytes, in solution.

The collective properties of polymers were briefly introduced in terms of their solution thermodynamics and to phase behavior and transitions. Concentrated polymer solutions and melts become extraordinarily complex, with time response behavior depending on polymer architecture and interactions, and this has been briefly discussed as an area of rheology. In the solid-state limit of rheology, polymers take on myriad applications in materials engineering, electronics, optics, and other areas. The relationship of polymer properties and the scattering of light are introduced in the following chapter.

References

Barnes, H.A. (1997). Thixotropy, a review. *Journal of Non-Newtonian Fluid Mechanics* 70: 1–33.

Benmouna, M., Weill, G., Benoit, H., and Akcasu, Z. (1982). Scattering from charged macromolecules. I. Static structure factor. *Journal de Physique (Les Ulis, Fr.)* 43: 1679–1685.

Burchard, W. (1999). Solution properties of branched macromolecules. *Advances in Polymer Science* 143: 113–194.

Carter, J.M. and Phillies, G.D.J. (1985). Second-order concentration correction to the mutual diffusion coefficient of a suspension of hard Brownian spheres. *Journal of Physical Chemistry* 89: 5118–5124.

Chu, J.C. and Mak, C.H. (1999). Inter and intrachain attractions in solutions of flexible polyelectrolytes at nonzero concentration. *Journal of Physical Chemistry* 110: 2669–2679.

Debye, P. (1946). The intrinsic viscosity of polymer solutions. *Journal of Chemical Physics* 14: 636–639.

Deserno, M., Holm, C., and May, S. (2000). Fraction of condensed counterions around a charged rod: comparison of Poisson–Boltzmann theory and computer simulations. *Macromolecules* 33: 199–206.

Doi, M. and Edwards, S.F. (1986). *The Theory of Polymer Dynamics*. New York: Oxford University Press.

Doi, M. (1992). *Introduction to Polymer Physics*. Oxford.

Donald, A.M. (2007). Why should polymer physicists study biopolymers. *Journal of Polymer Science Part B: Polymer Physics* 45: 3257–3262. https://doi.org/10.1002/polb.21335.

Dorshow, R. and Nicoli, D.F. (1981). The effect of hydrodynamics on the diffusivity of charged macromolecules: application to BSA. *Journal of Chemical Physics* 75: 5853–5856.

Elias, H.-G. (1997). *An Introduction to Polymer Science*. VCH.

Evans, D.F. and Wennerstrom, H. (1994). *The Colloidal Domain, Where Physics, Chemistry, Biology, and Technology Meet*. VCH.

Fixman, M. and Skolnick, J. (1977). Polyelectrolyte excluded volume paradox. *Macromolecules* 11: 863–867.

Flory, P.J. (1941). Molecular size distribution in three dimensional polymers 1. Gelation. *Journal of the American Chemical Society* 63: 3083–3100.

Flory, P.J. (1953). *Principles of Polymer Chemistry*. Ithaca: Cornell University Press.

Flory, P.J. and Fox, T.G., Jr (1951). Treatment of intrinsic viscosities. *Journal of the American Chemical Society* 73: 1904.

Gedde, U.W. (1995). *Polymer Physics*. Dordrecht, The Netherlands: Kluwer Academic Publishers.

Gennes, P.-G.D. (1979). *Scaling Concepts in Polymer Physics*. Ithaca: Cornell University Press.

Ghosh, S., Li, X., Reed, C.E., and Reed, W.F. (1990). Apparent persistence lengths and diffusion behavior of high molecular weight hyaluronate. *Biopolymers* 30: 1101–1112.

Gupta, S.K. and Forsman, W.C. (1972). General treatment of random-flight statistics with and without excluded volume. *Macromolecules* 5: 779–785.

Hansen, J.-P. and McDonald, I.R. (1990). *Theory of Simple Liquids* 2e. London: Academic Press.

Kirkwood, J.G. and Riseman, J. (1948). The intrinsic viscosities and diffusion constants of flexible macromolecules in solution. *Journal of Chemical Physics* 16: 565.

Kirkwood, J.G. (1967). *Macromolecules*. Gordon and Breach.

Landau, L.D. and Lifschitz, E.M. (1980). Ch. 12: fluctuations. In: *Statistical Physics*. Oxford: Pergammon Press.

Li, X. and Reed, W.F. (1991). Polyelectrolyte properties of proteoglycans. *Journal of Chemical Physics* 94: 4568.

Manning, G.S. (1969). Limiting laws and counterion condensation in polyelectrolyte solutions II. Self diffusion of the small ions. *Journal of Chemical Physics* 51: 934–938.

Mark, J.E., Eisenberg, A., Graessley, W. et al. (1993). *Physical Properties of Polymers*. ACS professional ref. book.

McMillan, W.G., Jr. and Mayer, J.E. (1945). The statistical thermodynamics of multicomponent systems. *Journal of Chemical Physics* 13: 276–305.

Odijk, T. (1977). Polyelectrolytes near the rod limit. *Journal of Polymer Science and Physics* 15: 477–483.

Oosawa, F. (1971). *Polyelectrolytes*. N.Y.: Marcel Dekker.

Reed, W.F., Ghosh, S., Medjahdi, G., and François, J. (1991). Dependence of polyelectrolyte apparent persistence lengths, viscosity, and diffusion on ionic strength and linear charge density. *Macromolecules* 24: 6189–6198.

Reiner, M. (1964). The Deborah number. *Physics Today* 17: 62.

Rubinstein, M. and Colby, R.H. (2003). *Polymer Physics*. Oxford University Press.

Schulz, G.V. and Baumann, H. (1968). Thermodynamic behavior, expansion coefficient, and viscosity number of polystyrene in tetrahydrofuran. *Die Makromolekulare Chemie* 114: 122–138.

Skolnick, J. and Fixman, M. (1977). Electrostatic persistence length of a wormlike polyelectrolyte. *Macromolecules* 10: 9444–9448.

Sorci, G.A. and Reed, W.F. (2002). Electrostatically enhanced second and third virial coefficients, viscosity and interparticle correlations for linear polyelectrolytes. *Macromolecules* 35: 5218–5227.

Sperling, L.H. (1986). *Introduction to Physical Polymer Science*. Wiley Interscience.

Stacy, C.J. and Kraus, G. (1979). Second viral coefficients of homopolymers and copolymers of butadiene and styrene in tetrahydrofuran. *Journal of Polymer Science: Polymer Physics Edition* 17: 2007–2012.

Stockmayer, W.H. (1943). Theory of molecular size distribution and gel formation in branched-chain polymers. *Journal of Chemical Physics* 11: 45–55.

Teraoka, I. (2002). *Polymer Solutions, an Introduction to Physical Properties*. Wiley Interscience.

Venkataswamy, K., Jamieson, A.M., and Petschek, R.G. (1986). Static and dynamic properties of polystyrene in good solvents: ethylbenzene and tetrahydrofuran. *Macromolecules* 19: 124–133.

Yamakawa, H. (1971). *Modern Theory of Polymer Solutions*. London: Harper and Row.

Zimm, B.H. and Stockmayer, W.H. (1949). The dimensions of chain molecules containing branches and rings. *Journal of Chemical Physics* 17: 1301–1313.

Further Reading

Anderson, C.F. and Record, M.T. Jr. (1990). Ion distributions around DNA and other cylindrical polyions: theoretical descriptions and physical implications. *Annual Review of Biophysics and Biophysical Chemistry* 19: 423–65.

Bruetting, W. (ed.) (2005). *Physics of Organic Semiconductors*. Weinheim: Wiley-VCH.

Cotts, P.M., Guan, Z., McCord, E., and McLain, S. (2000). Novel branching topology in polyethylenes as revealed by light scattering and C NMR. *Macromolecules* 33: 6945–6952.

Dobrynin, A.V. (2020). Polyelectrolytes: on the doorsteps of the second century. *Polymer*, 202: 122714. doi:10.1016/j.polymer.2020.122714.

Durmaz, E.N., Sahin, S., Virga, E., De Beer, S., De Smet, L.C.P.M., De Vos, W.M. (2021). Polyelectrolytes as building blocks for next-generation membranes with advanced functionalities. *ACS Applied Polymer Materials*, 3 (9): 4347–4374.

Gaborieau, M. and Castignolles, P. (2011). Size-exclusion chromatography (SEC) of branched polymers and polysaccharides. *Analytical and Bioanalytical Chemistry* 399: 1413–1423.

Gudla, H., Shao, Y., Phunnarungsi, S. et al. (2021). Importance of the Ion-pair lifetime in polymer electrolytes. *Journal of Physical Chemistry Letters* 12 (35): 8460–8464.

Holm, C., Rehahn, M., Oppermann, W., and Ballauff, M. (2004). Stiff-chain polyelectrolytes. *Advances in Polymer Science* 166 (Polyelectrolytes with Defined Molecular Architecture II): 1–27.

Hu, W., Nakashima, H., Wang, E. et al. (2006). Advancing conjugated polymers into nanometer-scale devices. *Pure and Applied Chemistry* 78: 1803–1822.

Leising, G., Stadlober, B., Haas, U. et al. (2006). Nanoimprinted devices for integrated organic electronics. *Microelectronic Engineering* 83: 831–838.

Li, Q.-Y., Yao, Z.-F., Wang, J.-Y., Pei, J. (2021). Multi-level aggregation of conjugated small molecules and polymers: from morphology control to physical insights. *Reports on Progress in Physics* 84 (7): 076601.

Muthukumar, M. (2017). *50th Anniversary Perspective*: a perspective on polyelectrolyte solutions. *Macromolecules* 50 (24): 9528–9560.

Winie, T., Arof, A.K., Thomas, S. (2020). *Polymer Electrolytes: Characterization Techniques and Energy Applications*. Wiley VCH: Weinheim, Germany.

Prakash, J.R. (2019). Universal dynamics of dilute and semidilute solutions of flexible linear polymers. *Current Opinion in Colloid & Interface Science* 43: 63–79.
Podzimek, S., Vlcek, T. (2001). Characterization of branched polymers by SEC coupled with a multiangle light scattering detector. II. Data processing and interpretation. *Journal of Applied Polymer Science* 82: 454–460.
Powell, S., Fisher, D. (2007). Polymer optics gain increased precision. *Laser Focus World* 43: 111–117.
Venkataraman, D., Russell, T.P. (2012). Polymer electronics: power from polymers. *Journal of Polymer Science Part B: Polymer Physics* 50: 1013–1028.
Peitzsch, R.M., Reed, W.F. (1992) High Osmotic Stress Behavior of Hyaluronate and Heparin, *Biopolymers* 32: 219–238.

Problems

1 A strand of DNA with a contour length 10 μm has a persistence length of 50 nm. Roughly, how much energy would it take to bend the DNA strand into a circle?

2 An ideal polymer has 1000 monomers and a link length of 0.12 nm.
a) Compute the most probable end-to-end length h_{mp}.
b) Find the ratio of the fully extended length h_{fe} to h_{mp}.
c) How much energy does it take to stretch it from its most probable position to its fully extended length at room temperature 25°C?

3 Compute the radius of gyration of an ideal homopolymer with a molar mass $M = 10^5$ g/mol, whose monomer molar mass is $m = 180$ g/mol, contour bond length is 0.12 nm, and whose persistence length is 50 nm.

4 Minimize $G \propto (n^2 / R^3) + T(R^2 / n)$ to find Flory's scaling law for $\langle h^2\rangle^{1/2}$ for a polymer coil of n monomers with excluded volume. R is the radius of the "characteristic volume" of the polymer, and as such $R \propto \langle h^2\rangle^{1/2}$.

5 Consider a phantom network of ideal coil polymers with 2000 links per polymer and a link length of 0.25 nm. Say the polymer is randomly cross-linked with an average of five cross-links per polymer.
a) Compute the spring constant before and after cross-linking.
b) Estimate Young's modulus before and after cross-linking.

6 a) A protein has a mass $M = 2 \times 10^5$ g/mol and contains 1280 amino acid monomer. The distance between monomers is 2.56 Å. When the protein is in denatured form it resembles a random coil with a persistence length of 50 Å. Find $\langle S^2\rangle_0^{1/2}$, assuming any increase in the value due to intramolecular excluded volume is negligible.
b) Find A_2 for this denatured protein assuming that the equivalent radius $R_e = 0.7\langle S^2\rangle^{1/2}$. Compare this value to that of A_2 from the undenatured protein, assuming that it acts as a hard sphere of $R = 1.2$ nm, when it is in globular (undenatured) form.

7 Compute $\langle S^2\rangle^{1/2}$ for
a) a rod of uniform mass density and total length L.
b) a sphere of uniform density and radius R.
c) a hollow sphere of radius R.
d) A "microgel" whose mass distribution is spherically symmetric and whose density decreases according to a Gaussian of the form

$$\rho(r) = \rho_0 \exp\left(-\alpha \frac{r^2}{R^2}\right), \text{ where } \alpha \gg 1, \text{ i.e. you can integrate } r \text{ from 0 to } \infty$$

where ρ_0 is the mass density at the center, α is a constant, and r the distance from the microgel's center.
For a continuous body of mass density $\rho(\vec{r})$, the integral expression is

$$S^2 = \frac{\int_V r^2 \rho(\vec{r}) \mathrm{d}^3 r}{M}$$

where the integral is over the volume V of the body and $M = \int_V \rho(\vec{r})\mathrm{d}^3 r$.

8 A certain polymer has a mass distribution given by $C(M) = m(1 - q)q^{M/m - 1}$, where $m = 85$ g/mol, and $q = 0.995$. It has a second virial coefficient of $A_2 = 1.2 \times 10^{-3}$ cm^3-mol/g^2, which can be considered approximately independent of M. The polymer is mixed in pure water on one side of a membrane, impermeable to the polymer but permeable to water, in a chamber that has a thin capillary, as shown. Pure water is in the other side of the chamber that also has a thin capillary.

a) If the polymer is mixed at a concentration of 0.015 g/cm^3 in one side of the chamber, find the difference in height of solution in the two capillaries, once equilibrium is reached.
b) Now the chamber is open on both sides and has a rigid wall with the semipermeable membrane on the bottom, as shown. Find the difference in height of the solution on the two sides of the membrane, and the final concentration of polymer in the polymer side of the chamber. Each side of the chamber has the same cross-sectional area, and the original height of solution on each side is $h_0 = 100$ cm.

9 a) Compute the intrinsic viscosity $[\eta]$, of a polymer of mass $M = 6 \times 10^5$ g/mol, that has a monomer mass of $m = 115$ g/mol, and a link length of 0.26 nm between monomers, if the persistence length is $L_P = 5$ nm. Assume $[\eta]$ can be computed by the Flory equation,

$$[\eta] = \frac{\Phi}{M}\left[\sqrt{6}\langle S^2\rangle^{1/2}\right]^3$$

with $\Phi = 2.56 \times 10^{23}$.

b) Estimate the overlap concentration of a solution containing this polymer.
c) Find the total viscosity of an aqueous solution containing 0.003 g/cm^3 of the polymer, assuming the pure aqueous solution has a viscosity of 0.01 P at $T = 20°$C.

10 a) Find the ratio of the intrinsic viscosity of a protein when in globular form to that when it is in denatured form. In globular form, it can be considered as a hard sphere of radius $R = 1.8$ nm, and when in denatured form, it resembles a random coil of $\langle S^2\rangle^{1/2} = 9$ nm. The mass of the protein is $M = 6.5 \times 10^4$ g/mol. In denatured form, you can use the same Flory equation as above.

b) Find the total solution viscosity when the protein is in globular form at a concentration 0.020 g/cm^3 in solution.
c) Find the total solution viscosity when the protein is in denatured form at a concentration 0.020 g/cm^3 in solution.
d) Consider the following scenario: 0.020 g/cm^3 of the protein is in an aqueous solution. (i) The protein is in globular form, with the values stated above. (ii) The protein is then denatured (e.g. by adding a denaturant such as urea) and assumes a random coil form. (iii) After several hours, the random coil proteins have aggregated into microgels that each contain an average of 200 protein molecules. The microgels can be approximated as hard spheres of radius $R = 15$ nm. Compute the total solution viscosity.

11 Consider a random coil polymer in a Θ-solvent (i.e. with no excluded volume effects) at $T = 27°$C. It has $N = 500$ monomers, with covalent links $b = 0.3$ nm between monomers, and the covalent bond angle between links is $\theta = 60°$. There are three rotational isomeric states with the following energies:

$U(\phi = 0°) = 0.25$ kcal/mol
$U(\phi = 120°) = 0.45$ kcal/mol
$U(\phi = 240°) = 0.70$ kcal/mol

a) Find the root mean square end-to-end length of the polymer, $\langle h^2\rangle^{1/2}$.
b) Find the statistical segment length (Kuhn length), l_k.

9

Light Scattering and Cognate Methods

Maxwell's equations have been the inspiration for most of what has been done in physics in the last 150 years.

Steven Weinberg

9.1 Overview of Light Scattering

We see the world around us due largely to light scattering. Reflected and refracted light is due to scattering, whether from metals, liquids, the sky, clouds, mountains, or living creatures. When we see colored objects it is the light scattered and not selectively absorbed that reaches our eyes. Interference and diffraction patterns of scattered light produce iridescent effects, such as the appearance of mother-of-pearl, peacock feathers, and oil slicks on water.

Selective absorption of light is the driver of the living state. When electromagnetic radiation is incident on matter it is normally either scattered or absorbed. When the radiation is absorbed light can be emitted, chemical reactions can be driven, and internal energy and heat can increase. It is the driving of photosynthetic reactions by radiation emitted from the Sun that underlies most of the living state on Earth. Only when photon energies correspond to discrete or broad energy levels for absorption does the interaction change from scattering to absorption. The porphyrins, the chlorophylls and rhodopsins, in particular, have evolved broad light absorption ranges that absorb light to produce and store chemical energy, rather than just to scatter it.

Scattering techniques, especially of light, X-rays, and neutrons, play a central role in the experimental investigation of biological and synthetic polymers. The way incident waves scatter can be related back to the properties of the particles that scatter them. The angular scattering pattern, absolute scattering intensities, time-correlation functions, spectral broadening, and polarization states carry information on the particle masses, spatial dimensions, configurations, interactions, and transport properties. When nonequilibrium processes are occurring in solution, such as polymerization, degradation, phase separation, aggregation, and formation of microstructures, the time-dependent scattering behavior yields information on kinetics, mechanisms, and structural changes.

Lord Rayleigh (aka John William Strutt 3rd, England, 1842–1919) in the 1870s applied the then recently developed Maxwell equations of electromagnetism to the scattering of light by atoms and molecules, discovering why the sky is blue and the sunset red, and giving birth to modern radar and scattering technologies. Debye, Einstein, and others later extended electrodynamics and thermodynamics to explain how pure liquids, and those containing solutes can scatter light. Light scattering in many different forms has become one of the most important techniques in macromolecular and colloid science.

Here, the focus is on visible light scattering from dielectric particles. The basic theory of light scattering for all electromagnetic (e-m) waves scales as λ/D, where λ is the wavelength of the radiation and D is the characteristic dimension of the scattering particle. Hence, for example, the way visible light scatters from bacteria should resemble the way microwaves scatter from rain drops, or radio waves from elephants.

The scattering of e-m radiation is used as an important technique in diverse areas. An example is *radio detection and ranging*, or *RADAR*, which sends out microsecond bursts of e-m radiation, generally microwaves or radio waves, typically on the order of centimeters, and detects the waves scattered off of dielectric and metallic objects during the millisecond "listening period" between bursts. The detected signals are analyzed to locate the scatterers' positions and determine their

Biophysics: Physical Processes Underlying the Living State, First Edition. Wayne F. Reed.

Companion website: www.wiley.com/go/Reed/BiophysicsandMacromolecules

velocities. Velocities can be determined both by the time sequence of positions and by the Doppler effect. The latter is standard in so-called *NEXRAD*, or *next-generation radar*.

LIDAR, standing for either light detection and ranging, or laser imaging, detection, and ranging, is built on the same basic principles as RADAR. Using UV, visible, and near-infrared light, rather than radio waves, allows far greater spatial resolution. Modern technologies for pulse formation and propagation at megahertz frequencies, coupled with high-speed analysis software, lend it to many applications in terrestrial and marine mapping, including real-time 3D imaging. It is being used to discover and precisely map ancient Meso-American cities buried under deep layers of foliage, and at other archaeological sites around the world.

Before building up light scattering theory for the analysis of macromolecules, it is interesting to see where e-m wave scattering is useful in other fields. The following merely scratches the surface.

Atmospheric science and *meteorology* make extensive use of light scattering. *Aerosols* are a general class of particulates in a gas, or mixture of gases, such as air. These include dust particles, particulate contaminants, and water vapor particles. These can be detected and quantified using visible light scattering. Particles of snow, rain, and sleet are on the millimeter scale and so microwaves are typically used for their detection and analysis. The Doppler shift from airborne particles can be used to map wind speeds, and NEXRAD has a dominant role in current atmospheric science and weather analysis. Many sophisticated beam types and advanced analyses schemes have been developed for it. In *oceanography*, light scattering is used for assessing colloidal particles in suspension. It can be used to distinguish between particles made of living matter, detritus, or minerals.

Astrophysics actually provided much of the early impetus for developing light scattering theory, in order to detect and quantify particulates in the interstellar medium, *interstellar dust*, sometimes referred to as "stardust." Scattering measurements have also led to detection of density fluctuations in the interstellar medium.

Scattering in *semiconductors* has contributed to understanding electronic, vibrational, and structural properties. In *semiconductor manufacturing*, even small numbers of dust particles can destroy circuits. *Cleanrooms* are required for production of semiconductors and are based on complete control of their environment and exclusion of all dust and particulate matter. Cleanroom airborne particle counters, which are used to assess the quality of the cleanroom, are frequently based on light scattering from these particles.

While we will focus on scattering from macromolecules, colloids, and supramolecular structures in solution, light scattering has contributed enormously to the understanding of pure gases and liquids, including critical phenomena, such as *critical opalescence*, a situation, usually in a binary liquid, where the length scale of density fluctuations are on the order of the wavelength of light and the solution becomes cloudy. Time-dependent processes that can be monitored in real-time with light scattering methods include polymerization, degradation, aggregation, phase separation, cell propagation, dialysis processes, and polymer dissolution.

Because polymers are typically on the length scale of visible light, light scattering provides a particularly convenient and powerful means of studying polymer solutions. Advances in the past decades in computers, high-sensitivity photodetectors, lasers, and chemical filtration technologies have all led to vastly improved ease, efficiency, and scope for light scattering. The most useful set of approximations for polymer characterization are due to Bruno Zimm (USA, 1920–2005), Rayleigh, and Debye.

Although attempts at describing the appearance of clouds and smoke due to light scattering date back to Leonardo DaVinci (Italy, 1452–1519) and were thoroughly investigated by John Tyndall (Ireland, 1820–1893), it was Lord Rayleigh who first used the then newly minted Maxwell equations to formulate a quantitative theory for light scattering in the late 1800s. His first approach was to consider atmospheric molecules as noninteracting particles whose size was much smaller than the wavelength of the light arriving from the sun. Later, it was Einstein who developed the basic theory of why pure liquids should scatter any light at all. His basic notion was that thermal fluctuations in liquids occur, leading to microscopic fluctuations in density and hence index of refraction. It is the optical contrast produced by these small-scale index of refraction variations that lead to scattering from a pure liquid.

Later, Debye, Zimm, and others used the fluctuation ideas to describe how the addition of particles, such as polymers, to a solution would lead to an excess of scattering over the pure solvent background, due to the particles themselves. The theory was subsequently refined to take account of interparticle interactions (where, not surprisingly, the virial coefficients $A_2, A_3, \ldots$ play a central role), intraparticle interference for particles whose dimensions are on the wavelength of light, and other effects, based on certain simplifying assumptions. The theory of scattering from isolated particles without these latter assumptions becomes quite involved, i.e. when the index of refraction throughout a particle, whose size is on the order of λ, is significantly different from the medium it is immersed in, a full solution to Maxwell's wave equations with appropriate boundary conditions is required. This is termed *Mie scattering* after the original investigator.

9.2 The Maxwell Electromagnetic Equations and the Prediction of Electromagnetic Waves and Their Properties, in Gaussian (CGS) Units[1]

The quantitative principles of electromagnetism were discovered experimentally, starting with the relationship of forces between two electric charges, found by Coulomb in 1785. The rest of the discoveries were made by many people over the following decades, including Michael Faraday, André-Marie Ampère (France, 1775–1836), Hans Christian Ørsted (Denmark, 1777–1851), Joseph Henry (USA, 1797–1878), Jean Baptiste Biot (French, 1774–1852), and many others. James Clerk Maxwell united the results into his famous equations and actually deduced the existence of a missing piece of the equations on purely theoretical grounds: induction of a magnetic field by a time-changing electric field. With this last piece in place, Maxwell showed that light was composed of mutually inducing electric and magnetic fields, and all the classical properties of light then flowed from this treatment.

It is worth reviewing the Maxwell equations in vacuum, in local and integral form, to understand the underlying principles of electromagnetic radiation. While it is often taken for granted that two charges separated in space exert equal and opposite forces on each other, it was quite mysterious that the charges acted at a distance on each other, across space. Faraday proposed the idea of an electric field, $\vec{E}$, associated with a charge that permeates space all around it, even in the absence of any other charge. A second charge, q, moved into the electric field of the first feels a force. Chapter 5.1 defines the electrostatic $\vec{E}$-field in Equation 5.2.

9.2.1 The First Maxwell Equation

The first equation is derivable from the experimentally discovered Coulomb's law, and, using the notion of the electric field, is given, in a homogeneous dielectric medium, in differential form in Gaussian units of Equation 5.5:

$$\nabla \cdot \vec{E} = \frac{4\pi\rho}{D} \tag{9.1}$$

where D is the dimensionless dielectric constant of the material, assumed spatially uniform, in which the electric charge density ρ resides and which was first introduced in Chapter 5.1. Equation 9.1 states that the electrostatic field $\vec{E}$ is produced by electric charge density, which can consist of both positive and negative charge.

As regards the dimensionless dielectric constant, $D = 1 + \chi_e$, where χ_e is the dimensionless *electric susceptibility* of a dielectric medium. D has the same value in the Gaussian and MKSA systems, e.g. $D = 1$ for the vacuum, and $D = 78$ for water at STP.

The integral of $\vec{E}$ over a closed surface S is the electric field flux and is equal to the total charge enclosed in the surface, divided by D. The integral form of the first Maxwell equation is obtained from Equation 9.1 by integrating each side over the volume of the body bounded by S and then using the divergence theorem to turn the left side into a surface integral. The result of this procedure yields:

$$\oint_S \vec{E} \cdot \mathrm{d}\vec{A} = 4\pi \frac{Q_{net}}{D} \tag{9.2}$$

and states that the total electrostatic electric field flux through a closed surface is proportional to the net charge enclosed by the surface. Here $\mathrm{d}\vec{A}$ is outwardly perpendicular to the surface at each point and has the magnitude of a small differential area of the surface, $\mathrm{d}A$. The circle on the integral sign indicates that the integral is over a closed surface, S. Both differential and integral forms are useful in actual calculations of electric fields.

The electrostatic field $\vec{E}$ is a conservative field, similar to the gravitational field. This means that an electrostatic potential $\phi(\vec{r})$ exists, as explained in Chapter 5.1, and is related to the electric field by

$$\vec{E} = -\nabla\phi \tag{5.8}$$

1 Most light scattering sources use Gaussian units for electromagnetic formulas. The same will be followed here. The unit of charge in the Gaussian system is the 'stat-coulomb' (=3.33×10^{-10} Coulombs), and CGS units are used for length (cm), mass (g), and time (s). Note that in Chapter 5, rationalized-MKSA units were used (Système International), reflecting much of the literature in polyelectrolyte solutions. The question of units in electromagnetic theory is often vexing, and the reader is referred to the Appendix in *Classical Electrodynamics* by J.D. Jackson. Jackson's book uses the Gaussian system. Other popular texts, such as *Electromagnetism* by Pollack and Stump, use MKSA.

The divergence of this yields *Poisson's equation* for electrostatics, Equation 5.9, which in CGS is written as

$$\nabla^2\phi = -\frac{4\pi\rho}{D} \tag{9.3}$$

Because Equation 9.3 is a scalar equation, it is generally easier to solve than Equation 9.1, and once found, $\vec{E}$ is computed simply by Equation 5.8.

9.2.2 The Second Maxwell Equation

The second equation concerns the magnetic field, $\vec{B}$. Its differential form is

$$\nabla \cdot \vec{B} = 0 \tag{9.4}$$

which states that, unlike electric charge, there is no monopolar magnetic charge; there is no known net magnetic charge density equivalent to ρ in the first equation. Observe that permanent magnets always have a North and South pole, and that if the magnet is fragmented each fragment continues to have a North and South pole, albeit weaker than the original. It is not possible to isolate a magnetic North or South pole. Notwithstanding this, there is ongoing research into the possible existence of magnetic monopoles.

The integral equation is obtained by integrating the left side of Equation 9.4 over the body's volume and applying the divergence theorem. The result is

$$\oint_S \vec{B} \cdot \mathrm{d}\vec{A} = 0 \tag{9.5}$$

Equation 9.5 shows that there is zero-magnetic flux through a closed surface S.

9.2.3 The Third Maxwell Equation

The third Maxwell equation is the law of induction, first discovered by Faraday, and independently shortly thereafter also by Joseph Henry.

$$\nabla \times \vec{E} = -\frac{1}{c}\frac{\partial \vec{B}}{\partial t} \tag{9.6}$$

where c is the speed of light. This differential equation means that a new type of electric field, one that curls around on itself to produce closed electric field lines, is induced when a magnetic field changes in time.

A striking feature of this third Maxwell equation is that it relates electric and magnetic fields to each other without the intermediary of any electrical charge; it is a pure field equation that doesn't reference matter. This foreshadows the fact that e-m waves can propagate through "empty space," with no need for the presence of electrical charge or matter. Classically, Equation 9.6 is a law of physics unto itself and cannot be deduced from any other equation or principle. While no matter needs to be present in Equation 9.6 for the induction of $\vec{E}$, if a closed conducting loop is placed at the site where the closed, induced $\vec{E}$-field exists, charge will flow in the loop as if a battery or other source of *electromotive force* (EMF), or *voltage*, were present.

The latter case, where a conductor is present, leads to an important point; if *induction* means the production of a closed electric field through space without requiring a battery or other material voltage source at the site of the field, then induction can *also* be caused simply by relative motion, due to the *Lorentz force law* (see Equation 9.10). Namely, if a conductor moves through a $\vec{B}$-field, the charges in it will feel a magnetic force that will move them. If the conductor is a wire moving through a uniform $\vec{B}$-field then charge in the conductor will briefly flow until it is polarized and the charges feel equal and opposite forces due to the $\vec{B}$-field and the $\vec{E}$-field due to polarization. Polarization will similarly occur for a closed conducting wire moving through a uniform $\vec{B}$-field. If, however, the $\vec{B}$-field is not uniform in space, the magnetic and electric forces cannot balance and cannot cancel at every point in the entire loop and so charge flow will occur in the conductor, as if a battery or other source of EMF were present. This is often termed *motional EMF*. The two effects, the motional EMF and the pure field (Equation 9.6), are present in the integral expression for EMF over an open-surface S

$$EMF = \oint_C \vec{E} \cdot \mathrm{d}\vec{l} = -\frac{1}{c}\frac{\mathrm{d}}{\mathrm{d}t}\int_A \vec{B} \cdot \mathrm{d}\vec{A} \tag{9.7}$$

The rightmost integral shows that changing B, area of the surface, the orientation of the surface and B, in time, or any combination of these can produce an *EMF*.

The left-hand integral is that of the induced electric field around a *closed path*, C, and it is equal to the time rate of change of the net flux of the magnetic field over the *open surface* A, which is bounded by C. The left-hand integral has the units of electric potential[2] (volts in the rationalized-MKSA system) and was historically termed the electromotive force because it can drive current in a closed circuit in the same way as a battery or other material voltage source.

The negative sign on the right-hand integral means that the induced $\vec{E}$-field opposes the changing magnetic flux. The direction of the induced $\vec{E}$-field is most easily visualized by finding the $\vec{B}$-field that would result from a current flowing along the induced $\vec{E}$-field that would create a changing magnetic flux opposing the change in magnetic flux on the right-hand side. This negative sign ensures the conservation of energy. It is sometimes referred to as *Lenz's Law*. If the negative sign were not there, then a changing magnetic flux on the right side would lead to an induced $\vec{E}$-field that would create more magnetic flux with the same sign, which would lead to the induced $\vec{E}$-field and total magnetic flux growing without limit in time, thus violating conservation of energy.

A far-reaching consequence of the integral form of the third equation, Equation 9.7, is that it underlies the production of electricity for modern civilization. Technologically, the electrification of cities using induction generators is among the greatest achievements flowing from the Industrial Revolution. The practical invention and application of the steam engine in the early eighteenth century marks the definitive birth of this period, with electrification, invention of internal combustion engines and their use of liquid fossil fuels, along with the discovery and manipulation of radio waves each paving the way to extraordinary new advances.

9.2.4 The Fourth Maxwell Equation

The fourth equation is particularly fascinating, because the first part of it rests on experimental discoveries by Øersted, Ampère, and others, and the second part was predicted on theoretical grounds by Maxwell:

$$\nabla \times \vec{B} = \frac{4\pi k_M}{c}\vec{J} + \frac{Dk_M}{c}\frac{\partial \vec{E}}{\partial t} \tag{9.8}$$

where $\vec{J}$ is the current density (current/area)[3] and k_M is the magnetic constant of the material[4] the charge flows in. The current density term in the differential equation means that a flow of electrical charge causes a $\vec{B}$-field that curls around into closed field lines. When only this charge density term appears, Equation 9.6 becomes what is often called *Ampère's law*. The second term, predicted by Maxwell by invoking the conservation of charge and the continuity equation, shows that a changing $\vec{E}$-field induces a $\vec{B}$-field. This last term completes the symmetry of induction; a changing $\vec{B}$-field induces an $\vec{E}$-field, and vice versa.

The integral form of the fourth Maxwell equation is obtained from Equation 9.8 by integrating both sides over an open surface A, and using Green's theorem to convert the left-hand side into the integral of $\vec{B}$ around the closed curve C, which bounds A. The result is

$$\oint_C \vec{B}\cdot d\vec{l} = \frac{4\pi k_M}{c}\iint_A \vec{J}\cdot d\vec{A} + \frac{Dk_M}{c}\frac{\partial}{\partial t}\iint_A \vec{E}\cdot d\vec{A} \tag{9.9}$$

The current density surface integral in the integral form is the total current flowing through an open surface A. It determines the $\vec{B}$-field around the closed curve C bounding A and is the integral form of Ampère's law. The second term shows that a time changing $\vec{E}$-field acts like an actual current, in that the time derivative of the electric flux over the open surface A defines the $\vec{B}$-field around the closed curve C. Historically, this term was referred to as a *displacement current*, because of its ability to create a magnetic field without any actual flow of charge.

9.2.5 Electromagnetic Force Law

The four Maxwell equations are usually accompanied by the electromagnetic force law, often called the *Lorentz force law*. Parts of this and its derivation are variously attributed to Maxwell, Hendrik Lorentz (Netherlands, 1853–1928) and Oliver Heaviside (England, 1850–1925).

$$\vec{F} = q\left(\vec{E} + \frac{\vec{v}}{c}\times\vec{B}\right) \tag{9.10}$$

2 1 V in MKSA = (1/300) Statvolt in Gaussian units.

3 The MSKA unit of current density 1 A/m^2 = 3 × 10^5 statampère/cm^2 in Gaussian units.

4 k_M is a dimensionless constant that plays the analogous role of D for the magnetic field. It occurs in the MKSA form as $k_M = \mu/\mu_0$, where μ is the permeability of the material. k_M has the same value in Gaussian units.

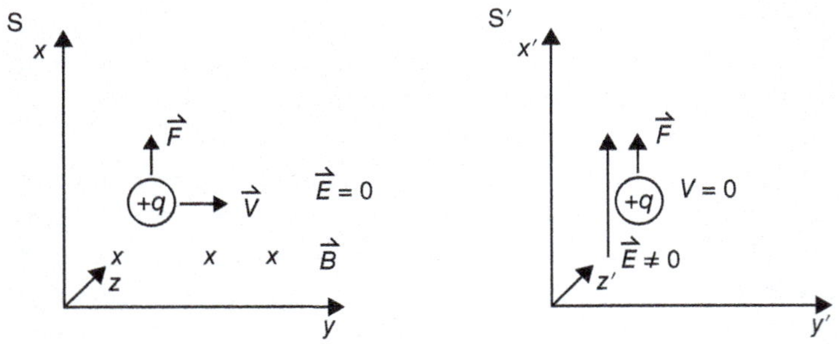

Figure 9.1 The necessity of $\vec{E}$ and $\vec{B}$ transforming into each other: In S, there is only a $\vec{B}$-field, but since the charge is moving at $\vec{v}$ there is an upward force. In S′ $\vec{v} = 0$ so there can be no magnetic force. But the instantaneous force is the same in both S and S′. The conclusion is that there must be an electric field in S′ to cause the force. (Courtesy of Marissa Calko).

This shows that a charge q placed in an electric field $\vec{E}$ will experience a force parallel to $\vec{E}$ and that it will experience a sidewise deflecting force due to the magnetic field $\vec{B}$ if the charge has a non-zero velocity $\vec{v}$.

The $\vec{E}$-field is used amply to accelerate charges, such as in cathode ray tubes and synchrotrons. The deflecting $\vec{B}$-field force leads to curved orbits of charged particles, but does no work on the particles and hence does not change the particles' kinetic energy, since the force acts at right angles to the instantaneous velocity and direction of motion. The magnetic force has a wealth of natural manifestations and practical uses. For example, charged matter drawn into black holes and neutron stars follow spiral paths that enhance the strong X-ray emission due to the acceleration of the charge. This force is also exploited in many technological inventions, such as magnetic focusing of electrons in electron microscopes, magnetic bottles, tokamaks,[5] in mass spectrometers, velocity selectors, cyclotrons, and synchrotrons.

The force law also gives an immediate intuition for why $\vec{E}$ and $\vec{B}$ are intimately related and, in fact, must transform into each other. For example, consider that in our reference frame S, there is a charge initially moving with speed v along the y-axis, and there is $\vec{B}$ in the $+z$ direction (the crosses indicate $\vec{B}$ is into the page), and no $\vec{E}$. The situation is depicted in Figure 9.1. The charge will initially accelerate upward due to this $\vec{B}$. Now consider another reference frame S', moving sideways with respect to us at the same velocity as that of the charge in our S frame, so that the charge is initially at rest in S'. In S', the charge feels the same instantaneous upward force as in S, a well-known result in classical mechanics (you can't get rid of a force by merely switching inertial reference frames), but this cannot be due to a field $\vec{B}$ because the charge is initially at rest in S', so it feels no magnetic force. We must conclude that there is a field $\vec{E}'$ in S' causing this. Hence, $\vec{B}$ in S has transformed into $\vec{E}$ in S'.

Classical relativistic transforms of $\vec{E}$ and $\vec{B}$ into $\vec{E}$ and $\vec{B}$ were easily deduced, and special relativity provides an especially elegant form of unifying electromagnetic theory, including special relativistic transformations of $\vec{E}$ and $\vec{B}$ into $\vec{E}'$ and $\vec{B}'$.

9.2.6 The Solar Wind, Cosmic Rays, and Possible Implications for the Living State

Electrons, protons, and a small fraction of alpha particles reaching Earth from the Sun and other extraterrestrial sources create an important manifestation of the force law. The Earth is surrounded by its own magnetic field, the *magnetosphere*. Hence, the paths of incoming charged particles spiral around the $\vec{B}$-field lines of the magnetosphere as they travel, and are reflected back and forth between the north and south poles, creating "belts" of trapped radiation; the *Van Allen Radiation Belts*, named after their discoverer James van Allen (USA, 1914–2006). Some of these particles enter the Earth's atmosphere in the polar and subpolar regions (typically 50–400 miles above the Earth's surface), where they have a higher probability of colliding with and ionizing atmospheric molecules, principally O_2 and N_2. The recapture of electrons by ionized molecules and their cascade through various energy levels, with subsequent showering release of photons, produce the colorful phenomena of both the *Aurora Borealis* (*Northern Lights*), and *Aurora Australis* (*Southern Lights*). An example of the Aurora Borealis is in Figure 9.2.

While these light shows are spectacular in and of themselves, there is an important effect accompanying the trajectories of the charged particles in the magnetosphere. Namely, since the $\vec{B}$-field lines are above the atmosphere in the nonpolar regions, and the particles follow these lines, there is relatively little penetration of these particles to the Earth's surface, compared to their flux in the polar regions. Hence, the magnetosphere effectively shields the nonpolar regions of Earth from much of this ionizing radiation and protects the atmosphere from destruction. This ionizing radiation can also produce random mutations, most of which are harmful.

5 A Tokamak is a device that confines a plasma using a magnetic field within a toroidal, or "donut-shaped," chamber. The magnetic field allows high-energy particles in the extremely hot plasma to spiral around field lines, minimizing contact with the container walls and reducing heat loss. The Tokamak is primarily used in plasma physics and nuclear fusion research, where it aims to achieve the conditions necessary for sustained nuclear fusion reactions.

Furthermore, neither the Earth's magnetosphere nor the flux of particles from the Sun are steady. The magnetosphere's field strength waxes and wanes over time, the magnetic north and south poles wander about in the polar regions, and even flip, nonperiodically, from north to south over geologic time spans, thought to be on the order of hundreds of thousands of years. The activity of the Sun itself, including the "breaking" of its magnetic field lines cause solar flares, which release large amounts of electromagnetic radiation and charged particles, which can disrupt satellites and terrestrial power grids. There is high interest, but also very much uncertainty, about how variations in the Earth's magnetosphere and the Sun's activity affect human health, possibly from viral mutations that can cause pandemics, to mental health issues. The longer range effects are even more speculative, but some people surmise that periods of weakened magnetosphere or magnetic reversal may have allowed large radiation doses to reach Earth and possibly cause periods of rapid mutations of species.

Figure 9.2 Aurora Borealis in Estonia. (kristian192 / Adobe Stock).

9.2.7 The Continuity Equation for Electric Charge

In Chapter 6, it was shown that any conserved quantity obeys the continuity equation:

$$\nabla \cdot \vec{J} + \frac{\partial \rho}{\partial t} = 0 \tag{6.66}$$

In the case of electric charge, which is conserved, ρ is charge density and $\vec{J}$ is current density.

While not logically necessary, it can be shown that Ampère's law:

$$\nabla \times \vec{B} = \frac{4\pi}{c} k_M \vec{J} \tag{9.11}$$

which is the fourth Maxwell equation without the inductive term, predicts the inductive term if the continuity equation is invoked. This follows because

$$\nabla \cdot \left(\nabla \times \vec{B}\right) = 0 = \frac{4\pi}{c} k_M \nabla \cdot \vec{J} \tag{9.12}$$

Completing the right-hand side using the continuity Equation 4.66 gives

$$\nabla \cdot \left(\nabla \times \vec{B}\right) = 0 = \frac{4\pi}{c} k_M \left(\nabla \cdot \vec{J} + \frac{\partial \rho}{\partial t}\right) \tag{9.13}$$

and, using the first Maxwell Equation 9.1 yields the full fourth Maxwell Equation 9.8, including the inductive term.

9.2.8 The Wave Equation and Electromagnetic Waves in Vacuum

Once Maxwell had established the four equations he quickly realized that $\vec{E}$ and $\vec{B}$ obey the wave equation. We can readily see this in the vacuum where there are no charges or matter, so that $\rho = 0$ and $\vec{J} = 0$. Taking the curl of the third equation in vacuum yields:

$$\nabla \times \left(\nabla \times \vec{E}\right) = \nabla \times \left(-\frac{1}{c}\frac{\partial \vec{B}}{\partial t}\right) \tag{9.14}$$

Using the identity:

$$\nabla \times \left(\nabla \times \vec{E}\right) = \nabla\left(\nabla \cdot \vec{E}\right) - \nabla^2 \vec{E} \tag{9.15}$$

and the other Maxwell equations you should show that $\vec{E}$ and $\vec{B}$ each obey the wave equation. The wave Equations 9.16a,b are second-order, linear, partial differential equations which can describe any displacement – the electric and magnetic fields in this case – traveling through space with speed c.

$$\nabla^2\vec{E} = \frac{1}{c^2}\frac{\partial^2\vec{E}}{\partial t^2} \tag{9.16a}$$

and

$$\nabla^2\vec{B} = \frac{1}{c^2}\frac{\partial^2\vec{B}}{\partial t^2} \tag{9.16b}$$

Several important properties of e-m waves come directly from the wave equation in conjunction with the individual Maxwell equations. There are a number of solutions to this in different coordinate systems. The simplest are plane waves in three dimensions. Suppose the wave is a harmonic, monochromatic wave (it has only one wavelength) traveling in the $+x$ direction, then it has the form:

$$\vec{E}(x,t) = \vec{E}_0 e^{i(kx-\omega t)} \tag{9.17}$$

where $\omega = 2\pi f$ is the angular frequency in radians/s, and f is the frequency in s^{-1}. k is the angular wave number given by $2\pi/\lambda$, where λ is the wavelength of the e-m wave. $\vec{E}_0$ has components in x, y, and z

$$\vec{E}_0 = E_{0,x}\hat{\imath} + E_{0,y}\hat{\jmath} + E_{0,z}\hat{k} \tag{9.18}$$

where $\hat{\imath}, \hat{\jmath}$, and $\hat{k}$ are the unit vectors in the x, y, and z directions, respectively. (The wave vector $\vec{k}$, and its magnitude k, should not be confused with the unit vector $\hat{k}$ in the z-direction.)

You can show that $E_{0,x} = 0$, and that only $E_{0,y}$ and $E_{0,z}$ can be nonzero. This means that $\vec{E}$ is perpendicular to the direction of travel of the e-m wave (+x-direction). Using the same procedure you can find that $\vec{B}$ is also perpendicular to the direction of travel, that $\vec{E}$ and $\vec{B}$ are perpendicular to each other and that $E = B$. From this we summarize

i) The e-m wave is a *transverse* wave, i.e. the displacement of the fields is perpendicular to the direction of travel. Electromagnetic waves can hence be polarized. The most general form of polarization is elliptical polarization. Linear and circular polarization are special cases of elliptical polarization.
ii) $\vec{E}, \vec{B}$ and $\vec{k}$ are all mutually perpendicular.
iii) $\vec{E} \times \vec{B} \parallel \vec{k}$; i.e., the cross-product $\vec{E} \times \vec{B}$ is in the direction of motion of the e-m wave.
iv) $E = B$, which puts E and B on an equal footing in Gaussian units.
v) Further analysis of e-m waves in vacuum using the Maxwell equations gives the energy density u (energy/volume) in the e-m field. $\vec{E}$ and $\vec{B}$ each contain one-half of the e-m wave's energy density. These energy densities u_E and u_B are given by

$$u_E = u_B = \frac{E^2}{8\pi} = \frac{B^2}{8\pi} \tag{9.19}$$

and the equality of the electric and magnetic energy densities is independent of the choice of units.

vi) Consideration of the intensity of e-m waves leads to the Poynting vector, $\vec{S}$, named after John Henry Poynting (England, 1852–1914). This gives the intensity (energy/s-area) of the traveling e-m wave and points in its direction of travel:

$$\vec{S} = \frac{c}{4\pi}\vec{E} \times \vec{B}. \tag{9.20}$$

Neither the eye nor current photodetectors can respond directly to electric fields at optical frequencies, which are on the order of 10^{14}–10^{15} s^{-1}. Eyes and detectors respond to the average intensity of e-m waves. The time average of $\vec{S}$, which is an intensity, is denoted by $\langle S \rangle$

$$\langle S \rangle = \frac{c}{4\pi}\frac{\int_0^\tau E^2 dt}{\tau} = \frac{c}{4\pi}\langle E^2 \rangle \tag{9.21}$$

where τ is one complete period of the harmonic wave above. Using the time-dependent real part of the harmonic wave, you can show that

$$\langle E^2 \rangle = \frac{E_0^2}{2} \quad 9.22$$

so that

$$\langle S \rangle = \frac{cE_0^2}{8\pi} \quad 9.23$$

Figure 9.3 illustrates several of these properties.

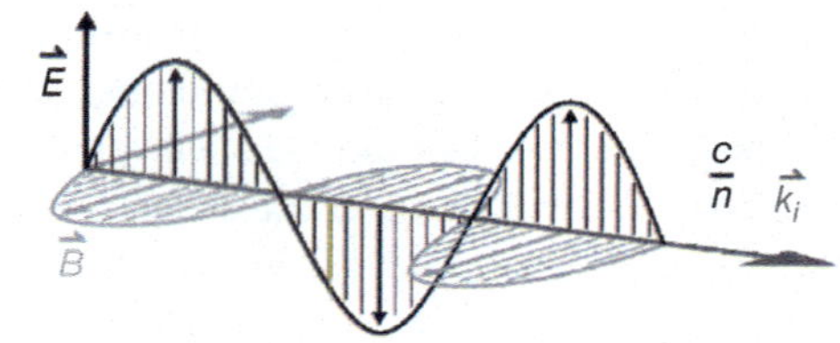

Figure 9.3 Schematic of an electromagnetic wave.

9.2.9 Electromagnetic Units, the Discovery of the Speed of Light, and Einstein's Special Relativity

A fabulous prediction of the Maxwell equations, which is obscured by the use of Gaussian units (which defines E and B to have the same units), is the prediction of c, the speed of light from the electrostatic and magnetostatic constants ε_0 and μ_0. Rewriting the wave Equations 9.16a,b in MKSA units gives

$$\nabla^2 \vec{E} = \varepsilon_o \mu_0 \frac{\partial^2 \vec{E}}{\partial t^2} \quad 9.24a$$

and

$$\nabla^2 \vec{B} = \varepsilon_o \mu_0 \frac{\partial^2 \vec{B}}{\partial t^2} \quad 9.24b$$

It is seen that c in wave Equations 9.16a,b are given by

$$c = \frac{1}{\sqrt{\varepsilon_o \mu_0}} = 3 \times 10^8 \text{m/s} \quad 9.25$$

While defining E and B to have the same units in the Gaussian system, which brings them onto an equal footing and makes their transformations into each other simpler, the above important historical detail is lost in these units, because c is imbedded in their formulation.

Equation 9.25 was an astonishing finding by Maxwell because it related the electrostatic constant ε_0 used in finding the force between two charges, and the magnetostatic constant μ_0 used in finding force/length between two current-carrying wires, directly to the speed of light! It was in no way obvious that the descriptions of these electrostatic and magnetostatic phenomena together carried the speed of light within them.

In earlier centuries, it had been a matter of debate whether the speed of light was finite or infinite. Convinced it was finite, in 1638 Galileo attempted to measure the speed of light by stationing himself on a hilltop with a lantern, shutter, and a clock, and a friend on a distant hilltop with a lantern and shutter. Galileo opened the shutter on his lantern and the friend then opened the shutter on his lantern as soon as he saw Galileo's light. Galileo then measured the time it took between when he opened the shutter on his lantern until the time he received the light from his friend's lantern. Naturally, the speed of light was far too great for this type of measurement. Galileo recognized this and realized that he was merely measuring the speed of human reflexes.

Attempts to measure the speed of light by Christiaan Huygens (Netherlands, 1629–1695) based on the data of Ole Rømer (Denmark, 1644–1710) gathered in 1676 obtained an order of magnitude value. Finally, Hippolyte Fizeau (France, 1819–1896) in 1851 obtained a value of the speed of light within 5% of the most accurate modern value (299,792,458 m/s). He used a geared wheel through which he passed light aimed at a distant mirror. He increased the speed of rotation of the wheel until the return light stroboscopically passed through the next gear on the wheel.

The discovery of electromagnetic waves led to the end of one of the longest-reigning errors in scientific history; the definitive debunking by Albert Einstein in 1905 of the "luminiferous aether" concept. The earliest idea of an "aether" an "ethereally" light substance in which light propagated went as far back as the fourth century BC in ancient Greece, where Aristotle's work "physics" described celestial bodies as surrounded by it. Many scientists including Sir Isaac Newton, Augustin-Jean Fresnel (France, 1788–1877), Fizeau, Maxwell, Tyndall, and Albert Michelson (Poland/USA, 1852–1931) either postulated its existence or took it for granted. Even the eccentric American author Edgar Allan Poe (USA, 1809–1849) speculated upon it. The aether was expected to "drag" light along with it as it moved, which led to many experiments attempting to determine the nature of it. Albert Michelson and Edward Morley conducted highly refined, interferometric searches for the aether in 1887 by measuring the speed of light in perpendicular directions. The results were always negative. The luminiferous aether could not be found.

Einstein had studied the Maxwell equations and understood that time-changing, mutually inducing $\vec{E}$ and $\vec{B}$ fields are necessary to sustain an e-m wave. He concluded that if one could travel at the speed of light and get in a light wave's reference frame, then the wave would be still,[6] and there would be no more time change in the fields, so they could no longer exist. But, how could a mere change of reference frame eliminate a real physical entity, light, which carries energy? He allegedly puzzled over this from age 16 to 26, pondering whether there was something wrong with the Maxwell equations or whether the way time and space was conceived and measured was wrong. He concluded the latter with one simple hypothesis, that the speed of light in vacuum was constant in all reference frames, no matter which frame it was produced in[7]. This single postulate (together with the already widely believed principle that laws of physics should be equally valid in all inertial reference frames) led to a completely new understanding of how time and space are related to each other and transform into each other. Effects such as *time dilation*, *length contraction*, the *relativistic Doppler shift*, the Lorentz transformations which blend space and time into each other, *relativistic energy*, and the famous $E = mc^2$ all flow from this hypothesis. The aether immediately became an obsolete, in fact, erroneous concept. Enormous new frontiers were opened in science and technology; it soon became apparent that energies involved in nuclear fission and fusion reactions and radioactive decay involved transformations between mass and other forms of energy, according to $E = mc^2$.

9.3 Radiation Emitted by an Accelerated Charge

One of the most fundamental results from the Maxwell equations is that an accelerated charge emits electromagnetic radiation. The acceleration can be of any type: Rapid linear deceleration of electrons stopped by a metal target produce X-rays, the phenomenon known as *Bremsstrahlung*, or *braking radiation*. Charges in circular motion emit e-m radiation due to their centripetal acceleration. Oscillating charges likewise emit e-m radiation, and scattering by dielectric materials is well modeled as oscillating charges treated as oscillating dipoles.

First, we look at the results of electromagnetic theory that connect acceleration of a charge to the e-m waves the charge emits. There is a complicated pattern to the emitted light nearby the particle. Many wavelengths away, however, the e-m radiation pattern becomes simpler, and this is referred to as the *radiation zone*. All subsequent results shown here are in the radiation zone.

Consider a charge q with acceleration $\vec{a}$, as shown in Figure 9.4. In the radiation zone, at a point located by $\vec{r}$, assuming nonrelativistic speeds of the accelerated charge ($v \ll c$), the electric field produced $\vec{E}_s(\vec{r})$ is perpendicular to $\vec{r}$ and given by

$$\vec{E}_s(\vec{r}) = -\frac{q\vec{a}'}{c^2 r} \tag{9.26}$$

where the subscript "s" has been used in anticipation of treating scattered electric fields. $\vec{a}'$ is also perpendicular to $\vec{r}$ and its magnitude is given by

$$a' = a\sin\phi \tag{9.27}$$

The Poynting vector $\vec{S}$ of Equation 9.20 gives the intensity (power/area) and direction of propagation of the radiation at $\vec{r}$ produced by the accelerating charge. Its magnitude in terms of the scattered electric field alone is

$$S = \frac{c}{4\pi}E_s^2 \tag{9.28}$$

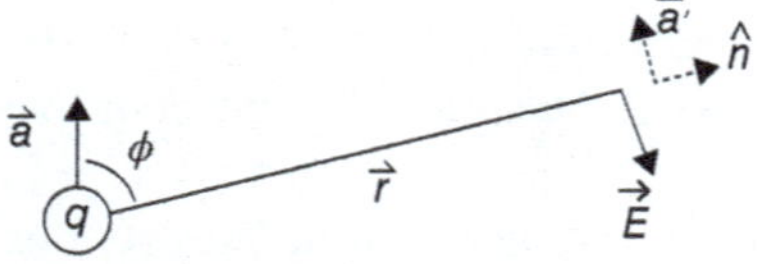

Figure 9.4 Radiation from an accelerated charge. The electric field produced at $\vec{r}$ is shown.

Using the above expression for E_S:

$$\vec{S} = \frac{q^2 a^2 \sin^2\phi}{4\pi c^3 r^2}\hat{r} \tag{9.29}$$

---

6 This is similar to how a surfer atop a moving wave travels with it and sees the wave as stationary.

7 While the speed of light c is the same in all inertial reference frames, the wavelength and frequency of light change due to the relativistic Doppler effect. When there is relative motion between the source and the observer, the observed frequency and wavelength are altered, leading to a blue shift if they move closer together and a red shift if they move apart.

The total power radiated by the accelerated charge, P_S, can be found by integrating $\vec{S}$ over the surface of a sphere of radius r centered on the charge:

$$P_s = \oint_S \vec{S} \cdot d\vec{A} \tag{9.30}$$

$\vec{S}$ and $d\vec{A}$ are in the radial outward direction, $\hat{r}$, so that

$$P_s = \frac{2q^2a^2}{3c^3} \tag{9.31}$$

9.4 Basic Scattering Theory: Light Emitted from an Oscillating Electric Dipole

When light is incident upon a dielectric material, e.g. water, glass, polymers, biological cells, etc., away from any absorption frequencies of the material, the oscillating electric field will induce oscillating dipoles in the electron clouds of the material. These, in turn, emit e-m radiation, and the emission of this radiation is termed the *scattering of electromagnetic radiation*.[8] In fact, the vast majority of white light falling on matter around us produces this type of scattering, and it is hence via this type of scattering that we see most of the nonluminous world around us.

9.4.1 Rayleigh Scattering by a Single Particle

The simplest treatment is for a single dielectric particle whose dimension is much smaller than the incident wavelength of the e-m wave. Such small particles are often termed *Rayleigh scatterers*. Figure 9.5 shows the electric field of a vertically polarized e-m wave incident on a dielectric particle of scalar polarizability α. This induces a time-dependent dipole $\vec{p}(t)$, at the particle, parallel to the $\vec{E}$-field of the incident e-m wave:

$$\vec{p}(t) = \alpha\vec{E}(t) \tag{9.32}$$

The particle may have an anisotropic polarizability, due to spatial directedness of chemical bonds, for example. In that case, there will be a polarizability tensor $\bar{\bar{\alpha}}$ and the induced dipole is no longer necessarily aligned with the incident $\vec{E}$-field. This leads to *depolarized light scattering*, which is used to probe particle optical anisotropy and rotational diffusion, among other properties.

$$\vec{p}(t) = \bar{\bar{\alpha}} \cdot \vec{E}(t) \tag{9.33}$$

Figure 9.5 shows the relationship of the $\vec{E}$-field and induced dipoles for both the isotropic and optically anisotropic cases. The rest of the scattering treatment will assume scalar polarizability. There is also a phase difference between $\vec{p}(t)$ and $\vec{E}(t)$, but, since we will be dealing with time-averaged incident and scattered intensities, this will not generally affect the results.

The dipole created by the incident field on q can be expressed in terms of the displacement $\vec{x}$ of q from its equilibrium position:

$$\vec{p} = q\vec{x} \tag{9.34}$$

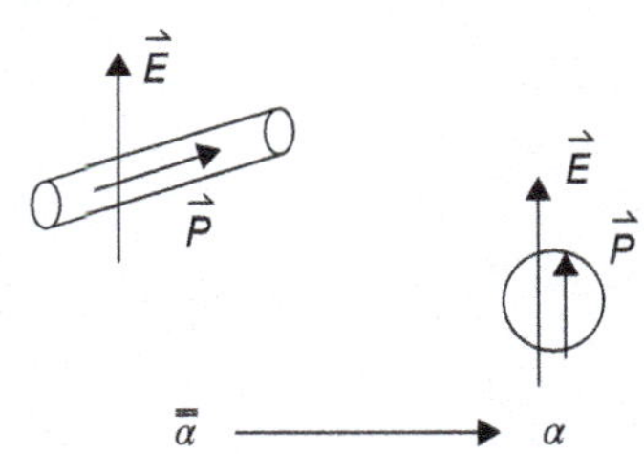

Figure 9.5 Tensor polarizability of an *optically* anisotropic object[9] (left) and scalar polarizability for an optically isotropic object (right).

8 When electrons without restoring forces are involved, e.g. free electrons in a metal, then the magnetic component of the e-m wave can cause magnetic dipole scattering. This is seldom found in biological and nonconducting polymers and so only electric dipole scattering is focused on.

9 "Optical anisotropy' does not necessarily imply any type of "geometrical anisotropy," such as a rod, e.g. the lattice directions in a crystal help define the optical anisotropy. The crystals may have a perfectly spherical morphology and be optically anisotropic.

Consider a monochromatic incident wave of angular frequency ω whose electric field at the site of q is given by

$$\vec{E}(t) = \vec{E}_0 e^{-i\omega t} \tag{9.35}$$

The acceleration is given by

$$\vec{a} = \frac{d^2\vec{x}}{dt^2} = \frac{1}{q}\frac{d^2\vec{p}}{dt^2} = -\frac{\omega^2}{q}\vec{p} \tag{9.36}$$

Using Equation 9.26, the magnitude of the scattered field at a point at distance r from the oscillating dipole, whose position is at altitude angle ϕ can then be written as

$$E_s(\vec{r}) = \frac{\omega^2}{c^2 r} p(t)\sin\phi \tag{9.37}$$

and the scattered intensity, by Equation 9.29:

$$\vec{S}_s(t) = \frac{\omega^4\alpha^2\sin^2\phi}{4\pi c^3 r^2} E^2(t)\hat{r} \tag{9.38}$$

Figure 9.6 shows the geometry for this. E is polarized in the vertical direction ($\phi = 0$). The time average intensity of light scattered at a distance r from the dipole and at angle ϕ to the vertically polarized incident electric field $<S>$ is

$$\langle S_s \rangle = \frac{\omega^4\alpha^2\sin^2\phi}{8\pi c^3 r^2} E_0^2 \tag{9.39}$$

using

$$c = \frac{\omega}{k} \tag{9.40}$$

gives the following alternate expressions for Rayleigh scattering by a single particle:

$$\langle S_s \rangle = \frac{ck^4\alpha^2\sin^2\phi}{8\pi r^2} E_0^2 \tag{9.41}$$

and, with $k=2\pi/\lambda$

$$\langle S_s \rangle = \frac{2c\pi^3\alpha^2\sin^2\phi}{\lambda_0^4 r^2} E_0^2 \tag{9.42}$$

where λ_0 is the vacuum wavelength of the incident light. The relation between E_0^2 and average incident intensity from Equation 9.21 can be used to give

$$\langle S_s \rangle = \frac{16\pi^4\alpha^2\sin^2\phi}{\lambda_0^4 r^2} \langle S_0 \rangle \tag{9.43}$$

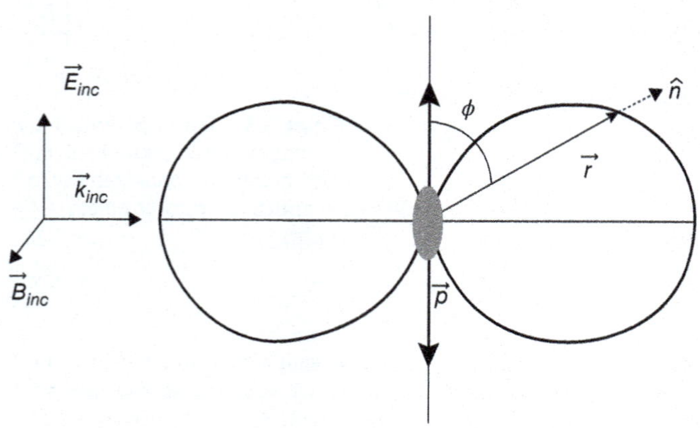

Figure 9.6 Geometry for scattered e-m radiation from an oscillating dipole induced by an incident field $\vec{E}_{inc}$.

where $\langle S_0 \rangle = \left(\frac{c}{8\pi}\right) E_0^2$ is the time average incident intensity and Equation 9.43 can also be written as

$$\langle S_s \rangle = \frac{k^4 \alpha^2 \sin^2 \phi}{r^2} \langle S_0 \rangle \qquad 9.44$$

There are several notable features of $<S_S>$ for Rayleigh scattering:

1) $\langle S_s \rangle \propto 1/\lambda_0^4$. This famous result accounts for why the sky is blue when there are no clouds. Short-wavelength blue light scatters from high-altitude atmospheric molecules around 10 times more than longer wavelength red light, which means the scattered light seen in the sky is predominantly blue. By the same token, when one looks toward the sun's vicinity at dawn or dusk the light, coming in obliquely from the normal, has traveled through a thick layer of atmosphere, scattering away most of the blue light, leaving the familiar reddish hue.
2) $\langle S_s \rangle \propto \sin^2 \phi$. This means that no light is scattered in the direction of the incident electric field ($\phi = 0$) and that the maximum light is scattered in the plane perpendicular to the incident light ($\phi = \pi/2$). This latter plane is often referred to as the *scattering plane*, and most light scattering experiments are arranged so that the detector is in this plane, in order to receive maximum scattering intensity.
3) $\langle S_s \rangle \propto \alpha^2$. For solid particles, the polarizability α is proportional to the mass of the particle. This means that scattering will increase as the square of the particle's mass and for spheroidal particles it will increase as the sixth power of the radius. As an example, at low angles a "dust" particle of 10 μm can scatter up to 10^{18} times more light than a protein of radius 10 nm! This has made light scattering traditionally difficult in practice, since "dust" and other large, stray scatterers must be carefully removed, usually by filtration and sometimes by electronic treatment of scattering data, in order to measure proteins and most polymers.
4) $\langle S_s \rangle \propto 1/r^2$ means that scattered energy propagating outward is conserved, since integration over the surface of a sphere cancels the $1/r^2$ term, as seen below.

Since intensity is power/area, $\langle S_S \rangle$ can also be expressed in terms of the average scattering power $\langle P_S \rangle$ as

$$\langle S_s \rangle = \frac{d\langle P_s \rangle}{dA} = \frac{d\langle P_s \rangle}{r^2 d\Omega} \qquad 9.45$$

where dΩ is the solid angle increment. It is given in spherical coordinates by $d\Omega = \sin\phi \, d\phi \, d\theta$, where ϕ is the altitude angle and θ the azimuth angle.[10] The integral in Equation 9.30 with $\vec{S}_s$ can then be performed over a sphere surrounding the oscillating dipole to obtain the total average power radiated by the oscillating dipole:

$$\langle P_s \rangle = \frac{8\pi k^4 \alpha^2}{3} \langle S_0 \rangle \qquad 9.46$$

9.4.2 Rayleigh Scattering by Dilute, Noninteracting Particles

Consider a collection of noninteracting Rayleigh scatterers of particle number density N (particles/cm^3), where the average spacing between particles is much greater than λ, i.e. $\lambda << 1/N^{1/3}$. The reason for this requirement of dilute particles is that as increasing density leads to average spacing that approaches and falls below λ, there will be interference effects among the particles that must be accounted for. This is treated in a later section.

In this and all of the following sections, it is assumed that there is single scattering, that is, a photon only scatters once before being detected. Hence, there is a limit to the turbidity of a solution whose scattering can be analyzed by these theories. A means of assessing single and multiple scattering probabilities is given in Section 9.11.

In the case of dilute Rayleigh scatterers, each particle scatters the power computed above, so that the total scattering power per volume V of particles in a uniform medium is simply:

$$\frac{d\langle P_s \rangle}{dV} = \frac{8\pi k^4 \alpha^2}{3} N \langle S_0 \rangle = \gamma N I_0 \qquad 9.47$$

10 Most physics textbooks use θ as the altitude angle and ϕ as the azimuth angle. Virtually, all of light scattering literature uses θ as the azimuth angle, so that convention is used here.

where $\langle S_0 \rangle = I_0$ is adopted for convenience as the incident intensity and γ is the cluster of constants:

$$\gamma = \frac{8\pi k^4 \alpha^2}{3} \tag{9.48}$$

This result allows computation of how much power of the incident light beam is lost to scattering per centimeter traveled in the scattering medium. Let the incident light travel in the $+x$ direction and enter the medium at x with intensity I_0. After the beam travels, a distance dx the total power lost due to scattering is

$$\left[I(x+\mathrm{d}x) - I(x)\right]A = -\frac{\mathrm{d}\langle P_s \rangle}{\mathrm{d}V}\mathrm{d}V = -\gamma N I(x) A \mathrm{d}x \tag{9.49}$$

where the volume increment is $\mathrm{d}V = A\ \mathrm{d}x$. Expanding the left-hand side to order dx yields

$$\frac{\mathrm{d}I(x)}{\mathrm{d}x} = -\gamma N I(x) \tag{9.50}$$

whose solution is

$$I(x) = I_0 \mathrm{e}^{-\gamma N x} = I_0 \mathrm{e}^{-\tau x} \tag{9.51}$$

The incident beam intensity falls off exponentially with distance traveled in the medium with a characteristic distance of $1/\tau$. τ is the *turbidity* and is given by

$$\tau = \frac{8\pi k^4 \alpha^2}{3} N \tag{9.52}$$

Turbidity is a function of the nature of the scatterers, via their polarizability α, their number density N, and the wavelength of the incident light. Turbidity is the fraction of incident light lost per cm of path length due to scattering. The reciprocal of turbidity has the units of distance and is related to the *mean free path* of a photon, i.e. the average distance a photon travels before being scattered, when the scatterers are dilute and there is no absorption of light, just scattering.

The Clausius–Mossotti equation furnishes a link between polarizability of a material and its index of refraction. In the case of scatterers of refractive index n immersed in a solvent of index n_0, the equation is

$$n^2 - n_0^2 = 4\pi N \alpha \tag{9.53}$$

Turbidity can then be written as

$$\tau = \frac{k^4 \left(n^2 - n_0^2\right)^2}{6\pi N} \tag{9.54}$$

9.4.3 The Rayleigh Scattering Ratio[11]

The average scattered intensity $<S_s>$ in all of its various expressions in Equations 9.38, 9.39, 9.41–9.44 depends on both the distance of the detector from the scatterer, r, and the incident intensity $<S_0>$. The total scattering intensity per unit volume, $N<S_s>$, from the Rayleigh particles of number density N is, from Equation 9.45 (middle term below) and Equation 9.43 (right term)

$$N\langle S_s \rangle = \frac{N\mathrm{d}\langle P_s \rangle}{r^2 \mathrm{d}\Omega} = \frac{16\pi^4 \alpha^2 \sin^2\phi N}{\lambda_0^4 r^2} \langle S_0 \rangle \tag{9.55}$$

11 This term is historical and still in widespread use. Some people object to "ratio" because by some definitions a "ratio" is the quotient of two quantities with the same units, and is hence dimensionless. The Rayleigh ratio has units of inverse distance and so is not a "ratio" by the latter definition. Some have tried to redefine it as the Rayleigh "factor" or Rayleigh "quotient," but the historical definition is followed here. IUPAC, as of this date, continues to use "Rayleigh ratio."

The *Rayleigh ratio*, R, eliminates the scattering dependence on distance r and average incident intensity by multiplying $N\langle S_s\rangle$ by r^2 and dividing by $\langle S_0\rangle$, and is

$$R(\text{cm}^{-1}) = \frac{r^2\langle S_s\rangle N}{\langle S_0\rangle} = \frac{N\text{d}\langle P_s\rangle}{\langle S_0\rangle \text{d}\Omega} = \frac{16\pi^4\alpha^2\sin^2\phi}{\lambda_0^4}N \quad \text{The Rayleigh Scattering Ratio} \quad 9.56$$

The first observation is that R can be interpreted as the fraction of the incident intensity scattered per steradian per centimeter of travel through the scattering medium (third term in Equation 9.56). The units of R are cm^{-1}. The second observation, from the fourth term in Equation 9.56, is that R can be interpreted as the total scattering power per volume of scatterers per steradian of detection, per unit of incident light intensity. A third interpretation is that R is the probability that a photon will be scattered per steradian per unit of pathlength.

A simpler form for R for dilute, non-interacting particles, using $k = 2\pi/\lambda$ in the rightmost term of Equation 9.56 is

$$R = k^4\alpha^2 N\sin^2\phi \quad 9.57$$

Equations 9.56 and 9.57 show explicitly that R is independent of r and $\langle S_0\rangle$. R is hence a "workhorse" quantity in practical applications of light scattering, because it is not necessary to know the distance of the detector from the scattering volume, nor the incident light intensity.

Using the Claussius–Mossotti Equation 9.53 allows Equation 9.57 to be expressed as

$$R = \frac{k^4\sin^2\phi}{16\pi^2 N}\left(n^2 - n_0^2\right)^2 \quad 9.58$$

or

$$R = \frac{\pi^2\sin^2\phi}{\lambda_0^4 N}\left(n^2 - n_0^2\right)^2 \quad 9.59$$

to make the same turbidity analysis as above, but in terms of R, Equation 9.52 is recovered:

$$\tau = \int R\text{d}\Omega = \frac{8\pi}{3}R = \frac{8\pi}{3}k^4\alpha^2 N \quad 9.60$$

What makes R practical is that R has been measured experimentally for many pure liquids.[12] Scattering from a liquid whose R is known lets raw experimental scattering values be converted to R for the samples being measured. For example, for toluene at $T = 25°\text{C}$, R is given at any wavelength λ (nm) by

$$R_{toluene}(\lambda) = 1.069\times10^{-5}\left(\frac{677}{\lambda_0(\text{nm})}\right)^4\left(\text{cm}^{-1}\right) \quad 9.61$$

where the inverse fourth power of wavelength for Rayleigh scattering is used explicitly in the computation.[13]

For example, consider a photodetector whose output voltage V_s for any sample is proportional to the intensity of scattered light. If the detector gives a voltage V_{tol} when scattering from pure toluene is measured, and if the detector has a dark voltage V_d (i.e. its reading when no light is incident on it), then R for the scatterers, at $T = 25°\text{C}$, is

$$R(\lambda) = F\frac{\left(V_s - V_{base}\right)}{\left(V_{tol} - V_d\right)}1.069\times10^{-5}\left(\frac{677}{\lambda_0(\text{nm})}\right)^4 \quad 9.62$$

where V_{base} is the scattering from the pure solvent, and $V_s - V_{base}$ is the *excess scattering* due to the polymers or other scatterers dissolved in the solvent. F is an optical constant that corrects for index of refraction effects between the reference and pure solvent, which depends on the geometry of the scattering sample cell used; for upright cylinders, it is essentially $(n_s/n_{tol})^2$, whereas for longitudinal flow cells it is about (n_s/n_{tol}).

---

12 Some practitioners use known molecular weight standards to calibrate R rather than a pure solvent.

13 Zimm found that, in addition to the $\frac{1}{\lambda_0^4}$ dependence of scattering, the index of refraction of the medium, n, must be taken into account if the medium is dispersive, that is, if $n = n(\lambda)$. In that case

$$R \propto \frac{n(\lambda)^2 - 1}{\lambda_0^4}$$

For toluene $n(450\text{nm}) = 1.504$, $n(650\text{nm}) = 1.49$, at 20°C, this latter effect gives a 3.4% correction to using just $R \propto \frac{1}{\lambda_0^4}$.

9.4.4 The Clausius–Mossotti Equation and the Differential Refractive Index $\partial n / \partial c$

The above expression can be cast in even more practical terms by again using the Clausius–Mossotti equation for a medium of index of refraction n_0 whose index of refraction becomes n when a scatterer at concentration c is present, where c is in g/cm^3. Expanding n in power of c about $c = 0$ gives

$$n = n_0 + \frac{\partial n}{\partial c}c + \frac{1}{2!}\frac{\partial^2 n}{\partial c^2}c^2 + \cdots \qquad 9.63$$

The partial derivatives of c are used to express the fact that n is a function of several variables besides concentration, including T and λ. $\partial n / \partial c$ is often pronounced "d, n by d, c" and also very frequently written simply as dn/dc. $\partial n / \partial c$ is often termed the *differential index of refraction* or the *incremental refractive index*, since it measures by how much the index of refraction of the solution with scatterers increases, or decreases, per increment of mass concentration dc of the scatterer. It is commonly found that, in a given solvent, $|\partial n / \partial c|$ is larger for the polymer than for the corresponding monomer. Because of this, $|\partial n / \partial c|$ is also molar mass dependent at very low degrees of polymerization and rapidly becomes independent of molar mass as degrees of polymerization exceed about 10.

Substituting Equation 9.63 only to order c into the Clausius–Mossotti Equation 9.53 yields

$$4\pi N\alpha = 2n_0\frac{\partial n}{\partial c}c + \left(\frac{\partial n}{\partial c}\right)^2 c^2 \qquad 9.64$$

so that, again dropping the above term in c^2

$$\alpha = \frac{n_0}{2\pi N}\frac{\partial n}{\partial c}c \qquad 9.65$$

and R can now be written as

$$R = \frac{k^4 n_0^2 (\partial n / \partial c)^2 c^2 \sin^2\phi}{4\pi^2 N} \qquad 9.66$$

Expressing particles per cm^3, N, in terms of mass concentration c (g/cm^3), molar mass M (g/mol) of the scatterers and Avogadro's number N_A,

$$N\left(\text{cm}^{-3}\right) = \frac{c\left(\text{g}/\text{cm}^3\right)N_A\left(1/\text{mol}\right)}{M\left(\text{g}/\text{mol}\right)} \qquad 9.67$$

and using λ_0 instead of k, turns Equation 9.66 into

$$R = \frac{4\pi^2 n_0^2 (\partial n / \partial c)^2 \sin^2\phi Mc}{\lambda_0^4 N_A} \qquad 9.68$$

which is compactly expressed as

$$R = KMc\sin^2\phi \qquad 9.69$$

where K is an optical constant, which is the following cluster of variables, for vertically polarized light:[14]

$$K = \frac{4\pi^2 n_0^2 (\partial n / \partial c)^2}{\lambda_0^4 N_A} \quad \text{The Auxiliary Scattering Constant} \qquad 9.70$$

14 This expression for vertically polarized light has the "4" in the numerator modified when unpolarized incident light is used; $4 \to 2(1+\sin^2\theta)$, where θ is the scattering angle, which is measured in the scattering plane from the forward direction which is $\theta = 0$. This has sometimes caused confusion because, before the common use of vertically polarized lasers, which are now virtually universal in macromolecular solution characterization by light scattering, unpolarized light sources were used, and this modified numerator was required in the analysis.

It is termed a constant because, for a given experiment, a specific solvent of n_0 is used, the solute has a certain $\partial n / \partial c$ in the solvent, and the light source, usually monochromatic, has a specific value of λ_0.

For practical light scattering measurements, the detector is normally placed in the scattering plane ($\sin^2 \phi = 1$) and a convenient representation for finding a scatterer's molar mass M is

$$\frac{Kc}{R} = \frac{1}{M} \tag{9.71}$$

This expression allows determination of M for dilute Rayleigh scatterers (size much smaller than λ) that are noninteracting.

If the particles are disperse, that is, there is a distribution of sizes where c_i is the concentration of the subpopulation of particles of mass M_i then:

$$R = \sum_i K_i c_i M_i \tag{9.72}$$

In the common case where K is the same for each subcomponent:

$$R = K \sum_i c_i M_i \tag{9.73}$$

Using the definition of weight average molar mass M_w (Equation 7.82),

$$\frac{Kc}{R} = \frac{\sum_i c_i}{\sum_i c_i M_i} = \frac{1}{M_w} \tag{9.74}$$

where the summation

$$c = \sum_i c_i \tag{9.75}$$

for total mass concentration of polymer (g/cm^3) was used.

This illustrates an important general principle, that light scattering measures a weight average, M_w, of a population of particles. Recall from solution thermodynamics that osmotic pressure determinations yield the number average molar mass M_n.

9.5 Relation of Light Scattering by Pure Liquids to Thermodynamic Fluctuations

The above treatment derived expressions for dilute, noninteracting Rayleigh scatterers. It was mentioned that "dilute" means the average spacing between particles is much greater than the wavelength of incident light. At the other extreme, consider a transparent dielectric material, such as quartz, which has no imperfections or impurities. Light will propagate through the quartz without any scattering, except in the forward direction of propagation.

To understand this, consider a dielectric crystal in which the atoms are packed into a regular lattice, whose spacing is $\lll\lambda$. If we pick any lateral point (i.e. $\theta \neq 0$) in space P and consider the scattered light reaching it from all the dipole scattering at each atomic site, we can reason as follows: Because the scatterers are so closely spaced, neighboring atoms will introduce almost continuously different path lengths, and hence continuous phase shifts, δ, in the light they scatter, as measured with respect to P. Hence, the phase shifts, which range from 0 to 2π (and then repeat, modulo 2π, for more distant scatterers) occur with equal probability, and so the average phase shift is $<\delta> = \pi$, so that there is destructive interference, and no scattered light at all lateral points P. Looked at another way, for every phase δ, there is a phase $-\delta$. This means that all the scattered light will cancel, and the incident light beam will continue to propagate in the forward direction. In terms of scattering away from the purely forward direction, we conclude that *a perfect dielectric crystal scatters no light, except in the forward direction.*

The light scattered in the forward direction has its own fixed phase shift between the incident light and the induced dipole oscillations, and this combines with the incident beam to "slow" the light down, so that it propagates at $v = c/n$,[15] where n is the crystal's index of refraction. The frequency remains constant and so the wavelength shortens in the crystal compared to its vacuum wavelength λ_0:

$$\lambda = \frac{\lambda_0}{n} \tag{9.76}$$

The question arose as to why pure liquids, whose scatterers are also very closely spaced, should scatter any light at all, except in the forward direction. The fact is that pure liquids *do* scatter light away from the forward direction, and it was Einstein who argued that thermodynamic fluctuations in liquid density, and hence refractive index, are responsible for the scattering.

The qualitative outline is as follows: A pure liquid has a certain average density and index of refraction n_0. Index of refraction is related to particle number density and polarizability via the Clausius–Mossotti equation, Equation 9.53. Because of thermal energy, there are local fluctuations in density that lead to changes in polarizability away from the average value. Light scatters off these regions of nonaverage polarizability. The intensity of the scattered light will depend on how great the polarizability fluctuations are which in turn depends on the *osmotic compressibility* of the liquid. A full description follows.

In the following, the result for a single particle of polarizability α is extended to a unit volume element, designated as ΔV ($=1\ \text{cm}^3$), in order to keep track of dimensionality. The number density of particles is N, and each particle has polarizability α. We are effectively calling all the particles in the unit volume element a single particle, whose polarizability, A, is the total polarizability of the unit volume:

$$A = \alpha N \Delta V \tag{9.77}$$

The dimension of A is cm^3, the same as the dimension of α. The result for the intensity of single particle scattering is Equation 9.44:

$$\langle S_s \rangle = \frac{k^4 \alpha^2 \sin^2 \phi}{r^2} \langle S_0 \rangle \tag{9.44}$$

so that treating the volume element as a single particle with polarizability A gives the scattering for the volume element:

$$\langle S_s \rangle = \frac{k^4 A^2 \sin^2 \phi}{r^2} \langle S_0 \rangle \tag{9.78}$$

The number density of volume elements is $1/\Delta V$. The Rayleigh ratio for volume elements of number density $1/\Delta V$ is, substituting $1/\Delta V$ for N in Equation 9.57,

$$R = \frac{k^4 A^2 \sin^2 \phi}{\Delta V} \tag{9.79}$$

Because of thermal fluctuations the total number of particles $p = N\Delta V$ in the unit volume will fluctuate. A will also fluctuate as $\delta A = \alpha \delta N \Delta V$, where δN is the fluctuation in particle number density.

Fluctuations δN can be caused by fluctuations in pressure, temperature, and concentration. Here, the liquid is considered to be at constant T and P. The average particle density in the system is N_0, and the particle density fluctuations are represented by δN, so that

$$N = N_0 + \delta N \tag{9.80}$$

15 In the photon view, individual photons traveling through the crystal still move at c between interactions. However, their interactions with the crystal cause a cumulative forward phase shift that results in a net group velocity less than c, leading to a shortened wavelength within the crystal. The frequency of the photons remains the same in both vacuum and crystal.

By Equation 9.77 for A, there will be a corresponding average polarizability A_0 with fluctuations δA from this average:

$$A = A_0 + \delta A = \alpha \Delta V \left(N_0 + \delta N\right) \tag{9.81}$$

It is the square of this that determines the scattering R in Equation 9.79:

$$A^2 = \alpha^2 \left(\Delta V\right)^2 \left(N_0^2 + 2N_0 \delta N + \left(\delta N\right)^2\right) \tag{9.82}$$

Now, this is A^2 for any given fluctuation δN. Think about what controls δN. If the particles in ΔV don't interact with each other their numbers might fluctuate strongly, giving large δN, which would give large A^2 and hence high scattering. If there are strong repulsions among the particles the fluctuations, δN, would be lower, and so A^2 would be lower, and so less light would be scattered. If the interactions are so strong that the particles can't move at all, that is, they are locked in place as if in a crystal, then there will be no fluctuations. In this latter case, $\delta N = 0$ and $A^2 = A_0^2$. It was argued above that in the crystal case there is no lateral scattering, and so the N_0^2 term is the "crystal term," which contributes no lateral scattering.

In summary, the higher the fluctuations in N the greater the scattering. As fluctuations diminish so also does the scattering. In the limit of no fluctuations, $\delta N = 0$, there is no lateral scattering.

The average value of A^2, represented by $\langle A^2 \rangle$, is obtained by averaging over all fluctuations:

$$\langle A^2 \rangle = \alpha^2 \left(\Delta V\right)^2 \langle N_0^2 + 2N_0 \delta N + \left(\delta N\right)^2 \rangle \tag{9.83}$$

It was just argued that the N_0^2 term is the "crystal" term, the nonfluctuating term, and hence gives no lateral light scattering. Now $<\delta N> = 0$ by the definition of the average, so that the only remaining term is in $(\delta N)^2$. Hence

$$\langle R \rangle = k^4 \alpha^2 \Delta V \sin^2 \phi \langle \left(\delta N\right)^2 \rangle \tag{9.84}$$

where α is the constant polarizability of a single particle, and it is recalled that $\Delta V = 1\ \text{cm}^3$, and is not a variable. The units of $\langle R \rangle$, of course, remain cm^{-1}.

The scattering problem, finding $\langle R \rangle$, involves finding $\langle (\delta N)^2 \rangle$, which can be found by taking a Boltzmann-weighted average over all fluctuations, where the fluctuations are treated as a continuum of particle number density states, each with a free energy $G(\delta N)$, i.e. an excursion away from the average particle number density N_0 involves a certain increase in free energy.

$$\langle \left(\delta N\right)^2 \rangle = \frac{\int_{-\infty}^{\infty} \left(\delta N\right)^2 \exp\left[-G\left(\delta N\right) / k_B T\right)\right] \mathrm{d}\left(\delta N\right)}{\int_{-\infty}^{\infty} \exp\left[-G\left(\delta N / k_B T\right)\right] \mathrm{d}\left(\delta N\right)} \tag{9.85}$$

Expanding $G(\delta N)$ in a power series to second order gives

$$G\left(\delta N\right) \cong G\left(N_o\right) + \left.\frac{\partial^2 G}{\partial N^2}\right|_{N_0} \frac{\left(\delta N\right)^2}{2} \tag{9.86}$$

where the first order term vanishes, that is, $\left.\left(\frac{\partial G}{\partial N}\right)\right|_{N_0} = 0$. The Equations 9.85 and 9.86 yield

$$\langle \left(\delta N\right)^2 \rangle = \frac{k_B T}{\left.\left(\frac{\partial^2 G}{\partial N^2}\right)\right|_{N_0}} \tag{9.87}$$

Hence, the problem from here is to determine $\left.\frac{\partial^2 G}{\partial N^2}\right|_{N_0}$ either experimentally or theoretically for a given liquid. For pure liquids, there are well-developed theories for this.

In the next section, scattering from a two-component system is considered, where there is a solute of low concentration c, in the liquid. In this case, we will be concerned with fluctuations in the concentration, δc. These are linked to the particle number density fluctuations by

$$\delta c = \frac{M}{N_A} \delta N \tag{9.88}$$

9.5.1 Extension of the Fluctuation Theory to Liquids Containing Polymers or Colloids

Here, we return to finding an expression for light scattered by solutes, such as polymers, or suspended colloids, in the liquid. The same approach of treating the scattering as due to fluctuations is continued, but now the total scattering is due to fluctuations of concentration of both the liquid and the solute:

$$R_{total} = R_{pure\ liquid} + R_{polymer} \tag{9.89}$$

Using the notation from multicomponent thermodynamics, the number density of solvent molecules in the volume element ΔV is N_1 and the number density of polymer molecules is N_2. The number of moles of solvent in volume V is $n_1 = N_1 V$ and the number of moles of polymer is $n_2 = N_2 V$.

The interest is in finding the scattering due to the polymer, so as to determine its molar mass M and other properties. Hence, $R_{pure\ liquid}$ is first measured experimentally and subtracted from the total measured, R_{total}. This yields the scattering due purely to the polymer and is sometimes termed the *excess Rayleigh scattering ratio*, because it is the scattering over and above that of the pure solvent. Having treated $R_{pure\ liquid}$ in the previous section, the final answer can be quickly found. Using Equation 9.88 in Equation 9.84 gives

$$\langle R \rangle = k^4 \alpha^2 \Delta V \sin^2\phi \frac{N_A^2}{M^2} \langle (\delta c)^2 \rangle \tag{9.90}$$

Using Equation 9.65 lets this be represented by

$$\langle R \rangle = \left(\frac{n_0}{2\pi N_2} \frac{\partial n}{\partial c} c \right)^2 k^4 \Delta V \sin^2\phi \frac{N_A^2}{M^2} \langle (\delta c)^2 \rangle \tag{9.91}$$

where N_2 is the number density of polymers (polymer molecules/cm^3), and c is the mass concentration of polymer (g/cm^3). The thermodynamic average $\langle (\delta c)^2 \rangle$ is found exactly as above, weighting concentration fluctuations away from the average concentration c_0 by the Boltzmann factor in the weighting integral Equation 9.85. R_{gas}, the universal gas constant, is used in place of k_B as the following will be derived on the basis of moles rather than number of particles. The result is

$$\langle (\delta c)^2 \rangle = \frac{1}{N_A} \frac{R_{gas} T}{\left. \dfrac{\partial^2 G}{\partial c^2} \right|_{c_0}} \tag{9.92}$$

$\left. \left(\dfrac{\partial^2 G}{\partial c^2} \right) \right|_{c_0}$ can be determined using thermodynamics of two-component systems covered in Chapter 4, where $c=c_2$ is the solute concentration, following the notation of Chapter 4. The number of moles of solute n_2 in the volume ΔV is $n_2 = c\Delta V/M$, where $M = M_2$ is the molar mass of the solute, following the notation of Chapter 4. Hence

$$\frac{\partial G}{\partial c} = \frac{\partial G}{\partial n_2} \frac{\partial n_2}{\partial c} = \frac{\Delta V}{M} \mu_2 \tag{9.93}$$

The Gibbs–Duhem Equation 4.80 is

$$\partial \mu_2 = -\frac{n_1}{n_2} \partial \mu_1 \tag{9.94}$$

So

$$\frac{\partial^2 G}{\partial c^2} = -\frac{n_1}{n_2} \frac{\Delta V}{M} \frac{\partial \mu_1}{\partial c} \tag{9.95}$$

Recalling from Chapter 4, μ_1, the chemical potential of the solvent can be expressed as

$$\mu_1 = \mu_1^0 - R_{gas} T v_1^0 c \left(\frac{1}{M} + A_2 c + A_3 c^2 + \cdots \right) \tag{9.96}$$

where v_1^0 is the solvent's partial molar specific volume, so that, in the dilute solution approximation ($n_2 \ll n_1$), $n_1 v_1^0 \simeq \Delta V$, and

$$\frac{\partial^2 G}{\partial c^2} = \frac{\Delta V^2}{n_2 M} R_{gas} T\left(\frac{1}{M} + 2A_2 c + 3A_3 c^2 + \cdots.\right) \qquad 9.97$$

So that

$$\langle(\delta c)^2\rangle = \frac{n_2 M}{N_A \Delta V^2\left(\frac{1}{M} + 2A_2 c + 3A_3 c^2 + \cdots.\right)} \qquad 9.98$$

Substituting this back into Equation 9.91 for the average Rayleigh scattering ratio due to concentration fluctuations, and after gathering and canceling terms, gives, with $\Delta V = 1\ \text{cm}^3$,

$$\langle R(q=0)\rangle = \frac{K \sin^2\phi}{\left(\left(1/M\right) + 2A_2 c + 3A_3 c^2 + \cdots\right)} c \qquad 9.99$$

where K is the optical constant defined in Equation 9.70. In the scattering plane, $\phi = 90°$, an experimentally useful expression is, omitting the averaging bars for $\langle R(q = 0)\rangle$

$$\frac{Kc}{\langle R\rangle} = \left(\frac{1}{M} + 2A_2 c + 3A_3 c^2 + \cdots\right) \quad \text{Scattering at } q = 0\text{, or for Rayleigh Scatterers at any } q \qquad 9.100$$

$R(q = 0)$ is also proportional to osmotic compressibility $\partial c / \partial \pi$

$$\langle R\left(q=0\right)\rangle = K R_{gas} T\left(\frac{\partial c}{\partial \pi}\right) c \qquad 9.101$$

where it is recalled from Equation 4.126, with $c_2 = c$ for the solute concentration:

$$\Pi = R_{gas} T\left(\frac{c}{M} + A_2 c^2 + A_3 c^3 + \cdots\right) \qquad 4.126$$

Hence, highly osmotically compressible solutions will exhibit high levels of light scattering, whereas osmotically incompressible solutions will exhibit little scattering. A good example of highly osmotically compressible solutions are those containing large neutral polymers, or large polyelectrolytes at high ionic strength, i.e. low values of A_2. An example of an osmotically incompressible solution is one containing polyelectrolytes (large or small) with no added supporting electrolyte (no added ionic strength); the polyelectrolytes are unshielded and avoid each other leading to very little concentration fluctuation, i.e. A_2 is very large and very little light is scattered above the solvent level.

9.6 The Angular Dependence of Scattered Light: Intramolecular Interference Effects on Scattering

So far, light scattered from interacting particles whose dimensions are much smaller than the wavelength of the incident light has been considered. Many bio- and synthetic polymers and colloids, however, have dimensions on the order of visible light wavelengths, which leads to interference effects for the light scattered from different parts of the same particle.

Consider a particle whose linear dimension d, is on the order of λ, and whose index of refraction is

$$n = n_0 + \Delta n \qquad 9.102$$

where n_0 is the index of refraction of the medium in which the particle is immersed, and Δn is the difference between the particle's and medium's indices of refraction. Vertically polarized light is incident on the particle (the incident electric field is perpendicular to the plane of the page), and scattered light is collected at a detector (e.g. a photodiode, photomultiplier tube, or charge-coupled device), placed at an angle θ in the scattering plane (i.e. the plane of the page defined by $\phi = \pi/2$, where scattering is a maximum from a particle with scalar polarizability). The scattering angle θ is defined with respect to the direction of the incident light as seen in Figure 9.7. $\theta = 0$ indicates forward scattering, whereas $\theta = 180°$ indicates back-scattering.

The scattered light reaching the photodetector emanates from all the points within the scattering body. There is a phase difference δ for the light arriving at the detector from each point due to differences in the optical pathlength, which are composed of

1) The difference in the geometrical pathlength from each point to the detector, and
2) The difference in phase introduced by Δn for light traveling through various amounts of the scattering body.

Consider just two points in the object 1 and 2, as shown in Figure 9.8. They are separated by a distance d along a line approximately aligned with the direction to the detector. Point 2 is on the exit edge of the particle. The *optical path difference*, OPD, between light reaching the detector from 1 and 2 is

$$\text{OPD} \cong \left(n_0 + \Delta n\right)d \tag{9.103}$$

The OPD is related to the phase shift δ between light from 1 and 2 in the usual way:

$$\delta = \text{OPD} \times \frac{2\pi}{\lambda_0} = \frac{2\pi}{\lambda_0}\left(n_0 + \Delta n\right)d \tag{9.104}$$

The phase shift is composed of two parts:

$$\delta = \frac{2\pi n_0 d}{\lambda_0} + \frac{2\pi \Delta n d}{\lambda_0} \tag{9.105}$$

The first term is the phase shift due to the geometric separation d of points 1 and 2. The second term includes the effect of the difference in index of refraction between the particle and surrounding medium, e.g. per Equation 9.76, the wavelength within the particle is shorter than outside the particle if $\Delta n > 0$, and longer if $\Delta n < 0$. Figure 9.8 schematizes this

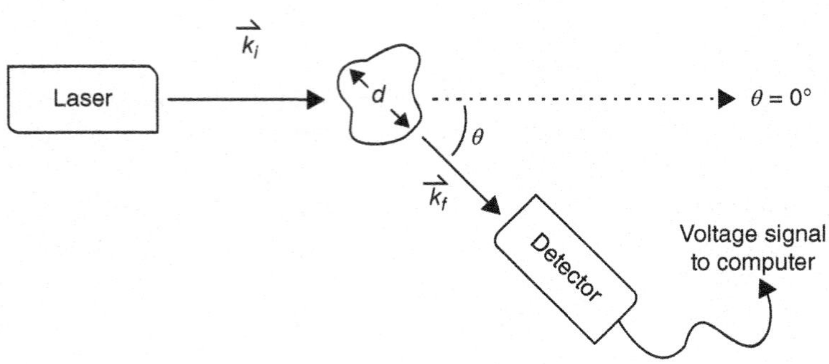

Figure 9.7 Experimental set up for light scattering measurements.

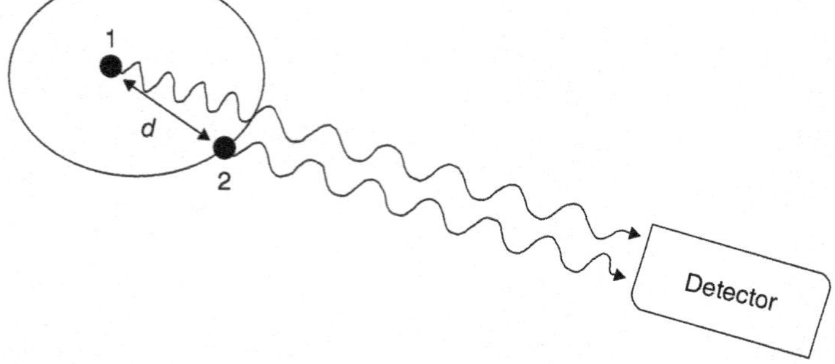

Figure 9.8 Intraparticle interference.

situation, showing how the wavelength is shorter within the particle, for $\Delta n > 0$, and that the scattered waves from points 1 and 2 arrive out of phase at the detector.

9.6.1 The Rayleigh–Debye Approximation (RDA)

The RDA considers that the second term is negligible, that is

$$\frac{2\pi\Delta n d}{\lambda_0} << \frac{2\pi n_0 d}{\lambda_0} \tag{9.106a}$$

or

$$\frac{\Delta n}{n_0} << 1 \tag{9.106b}$$

When this condition does not hold, the solution to the scattering problem is much more difficult than in the R–D approximation. In that case, the Maxwell equations with appropriate boundary conditions must be solved, which is outlined in the section on Mie scattering, later. This is usually the case with solid particles, aerosols, bacteria, etc.

Fortunately, polymers are "threadlike" entities in a solvent so that the distance between the different scattering points of the polymer is filled with solvent so that Δn is effectively nearly zero, and the R–D approximation, stating that the major part of the phase shift is due to the geometrical path difference, is widely applicable. Figure 9.9 represents how the scattering from each monomer on a polymer chain produces its own scattered electric field with its own phase shift at the detector. The detector's response depends on the sum of all the phase-shifted, scattered electric fields from the monomers.

The procedure to determine the scattering from a particle is to find an expression for the phase shift from each point in the particle with respect to the observation site and incorporate this shift into the scattered electric field for all points on the particle, and then sum the electric fields at the observation site.

There is a convenient and widely used formalism for determining the phase shift δ for intraparticle scattering, which is applicable also to interparticle scattering. Figure 9.10 shows a convenient geometry for establishing an expression for the phase shift for a particle. This is applicable to any type of wave scattering, including all e-m waves, electrons, protons, neutrons, etc. Monomer j is located from the arbitrarily placed origin by the position vector $\vec{r}_j$ and produces a phase shift δ_j of its scattered wave with respect to the origin. To find δ_j, first OPD$_j$ is found and then related to δ_j by

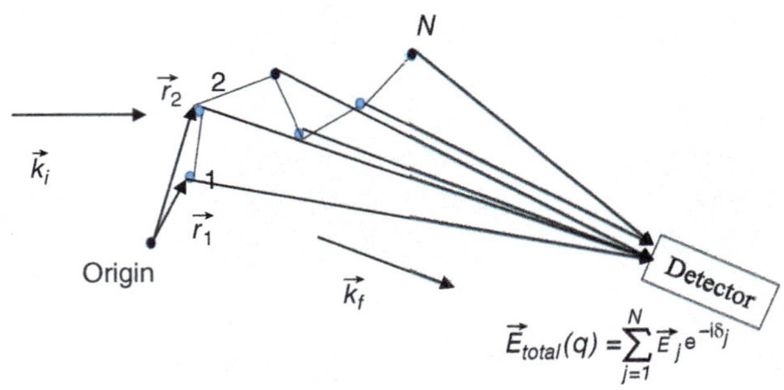

Figure 9.9 Schematic of a portion of a polymer chain, where each monomer on the chain has a position vector $\vec{r}$ and scatters a wave, phase shifted with respect to a reference point, the origin. The scattered electric fields reaching the detector from all monomers are summed.

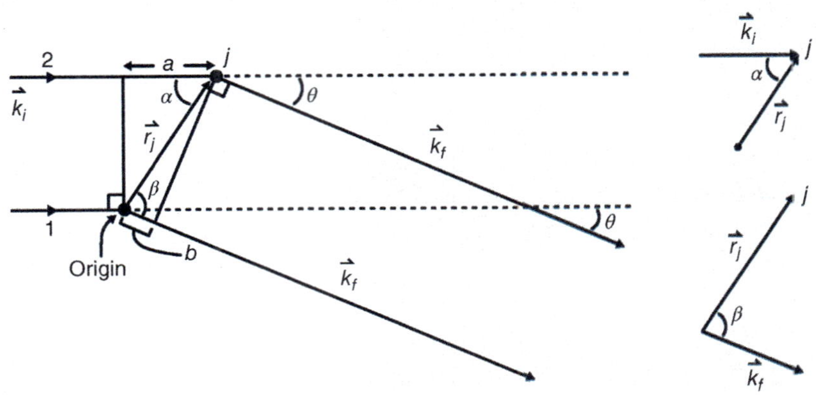

Figure 9.10 Geometry for computing phase difference between scattered waves. (Courtesy of Marissa Calko).

$$\delta_j = \frac{2\pi(\text{OPD}_j)}{\lambda_0} \tag{9.107}$$

where

$$\text{OPD}_j = n_0|a-b| \tag{9.108}$$

and $|a-b|$ is the geometrical path distance between monomer j and the origin, and in the DB approximation, this is taken as the complete OPD_j.

From the geometry in the sketch:

$$a = r_j\cos\alpha, \quad b = r_j\cos\beta \tag{9.109}$$

So

$$\delta_j = \frac{2\pi n_0}{\lambda_0}(r_j\cos\alpha - r_j\cos\beta) \tag{9.110}$$

An important aspect of scattering in this situation is that $|k_i| \cong |k_f|$, i.e. there is no appreciable change of wavelength upon scattering. A small frequency and hence wavelength shift comes from the Doppler shift of a monomer's motion with respect to the rest frame of the incident light's source:

$$\frac{\Delta\omega}{\omega} \cong \frac{v}{c} << 1 \tag{9.111}$$

A monomer's speed is typically on the order of $10^{-14}c$ (~3×10^{-3} cm/s), making this effect negligible. An even smaller effect arises from Compton scattering, due to an incident photon making an elastic collision with a monomer and transferring a tiny amount of energy and momentum. The Compton wavelength for elastic photon scattering from a free electron of mass m_e is

$$\lambda_C = \frac{h}{m_e c} = 0.00243\text{nm} \tag{9.112}$$

which is generally only observable in X-rays. When m_e is replaced by a polymer mass, even a small one, on the order of 1000 g/mol, yields a shift of ~10^{-10} nm.

With $|k_i| \cong |k_f|$, the phase shift of monomer j with respect to the origin can then be written as

$$\delta_j = (\vec{k}_i - \vec{k}_f)\cdot\vec{r}_j \tag{9.113}$$

where $k_i = k_f = k$, and

$$k = \frac{2\pi n_0}{\lambda_0} \tag{9.114}$$

λ_0 is the vacuum wavelength of the incident light. The n_0 in k expresses the fact, via the OPD, that the wavelength of the incident light is reduced to λ_0/n_0 in the solvent.

The *scattering wave vector*, $\vec{q}$ is defined as

$$\vec{q} = \vec{k}_f - \vec{k}_i \tag{9.115}$$

where $\vec{k}_i$ is the wave vector of the incident light and $\vec{k}_f$ is the wave vector of the scattered light that reaches the detector. With this definition, Equation 9.113 becomes

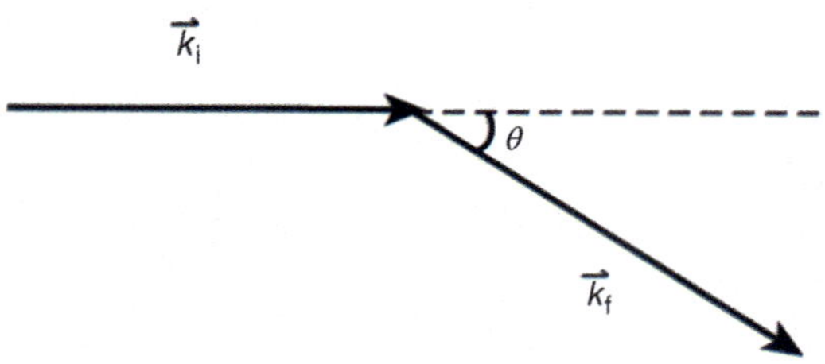

Figure 9.11a Incident and final wave vectors, and relationship via θ.

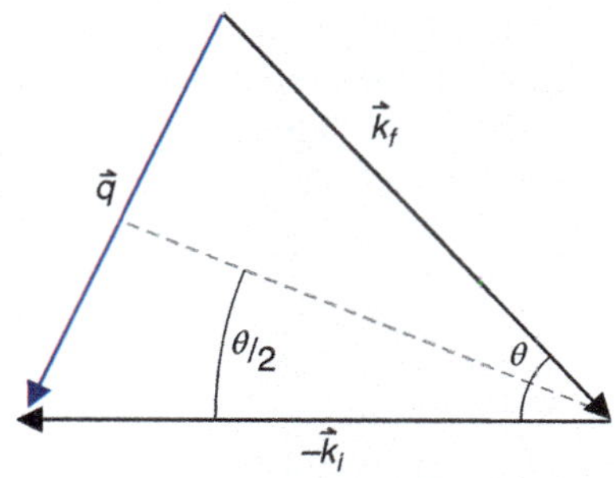

Figure 9.11b Geometry for determining the scattering wave vector $\vec{q}$.

$$\delta_j = \vec{q} \cdot \vec{r}_j \tag{9.116}$$

Referring to the geometrical relation of the two vectors in Figure 9.11, the amplitude of the scattering wave vector q, is

$$q = \frac{4\pi n_0}{\lambda_0} \sin\left(\frac{\theta}{2}\right) \tag{9.117}$$

The amplitude of the scattering vector combines the vacuum wavelength of the incident light, the index of refraction of the medium, and the half-angle of the detector's location.

For a polymer composed of N monomers numbered from 1 to N, the total scattered electric field at the detector, $\vec{E}_{total}(q)$ is given by the vector sum of the scattered field over all N monomers

$$\vec{E}_{total}(q) = \sum_{j=1}^{N} \vec{E}_j e^{-i\delta_j} \tag{9.118}$$

The scattered electric field from particle j is proportional to the polarizability of that particle: $|\vec{E}_j| \propto \alpha_j$. If the monomers are the same, or similar, E_i will be the same for all monomers and summation will be controlled by the *phase sum* in Equation 9.118, which is sometimes also called the *interference sum.*

$\vec{E}_{total}(q)$ can now be written as

$$\vec{E}_{total}(q) = \sum_{j=1}^{N} \vec{E}_j e^{-i\delta_j} = \sum_{j=1}^{N} \vec{E}_j e^{-i\vec{q}\cdot\vec{r}_j} \tag{9.119}$$

Again, it is noted that the formalism leading to $\vec{E}_{total}(q)$ is applicable to the scattering of any type of wave, with the appropriate amplitude of the type of wave substituted for $\vec{E}_j$. The expression can be used for all e-m scattering, such as light and X-rays and also for electron and neutron scattering by replacing $\vec{E}_j$ with the appropriate physical quantity measuring the wave amplitude (e.g. the wave function ψ for neutrons or electrons).

Taking this to the continuum limit, consider that particles j and k become infinitesimally spaced so that the sum becomes an integral, and $\vec{E}_j$ becomes the scattered field at each point in space $\vec{E}(\vec{r})$, which is proportional to the polarizability per volume at $\vec{r}$, $\vec{\alpha}(\vec{r})$. The vector sign on the polarizability makes it point in the direction of $\vec{E}(\vec{r})$ for a material with scalar polarizability. The total scattered field at the detector due to a body occupying a volume in space is then

$$\vec{E}_{total}(q) = \int \vec{E}(\vec{r}) e^{-i\vec{q}\cdot\vec{r}} d^3r \propto \int \vec{\alpha}(\vec{r}) e^{-i\vec{q}\cdot\vec{r}} d^3r \tag{9.120}$$

This means that the scattered electric field at q is the Fourier transform of the body's polarizability. The integral is over all space, and the integrand is nonzero only over the volume V of the body. For the case of a material of a specific type, the polarizability per volume at $\vec{r}$ is proportional to the mass density at $\vec{r}$, $\rho(\vec{r})$. For a nondepolarizing body and vertically polarized incident light:

$$E_{total}(q) \propto \int \rho(\vec{r}) e^{-i\vec{q}\cdot\vec{r}} d^3r \tag{9.121}$$

This means that the scattered field is the Fourier transform of the body's mass density. Inverting the transform yields $\rho(\vec{r})$, which gives the detailed shape and mass distribution of the body.

For direct detection of scattered light, e.g. with a photodiode or charge–couple device (CCD), the scattered intensity is measured, not the electric field.[16] The time average scattered intensity at the detector $\langle S_s \rangle$ is given by the Poynting relation:

$$\langle S_s \rangle = \frac{c}{8\pi} \vec{E}^*_{total} \cdot \vec{E}_{total} \tag{9.122}$$

where $\vec{E}^*_{total}$ is the complex conjugate of $\vec{E}_{total}$

$$\langle S_s \rangle = \frac{c}{8\pi} \sum_{i=1}^{N} \sum_{j=1}^{N} \vec{E}^*_i \cdot \vec{E}_j e^{-i\vec{q}\cdot(\vec{r}_j - \vec{r}_i)} \tag{9.123}$$

In the case where the scattering elements (monomers) of the polymer are essentially the same and the scattered light is not depolarized for the scattered field from each element:

$$\langle S_s \rangle = \frac{c}{8\pi} E^*_0 E_0 \sum_{i=1}^{N} \sum_{i=1}^{N} e^{-i\vec{q}\cdot(\vec{r}_j - \vec{r}_i)} \tag{9.124}$$

and

$$\sum_{i=1}^{N} \sum_{j=1}^{N} e^{-i\vec{q}\cdot(\vec{r}_j - \vec{r}_i)} \tag{9.125}$$

is the phase sum over all monomers. With regard to Figure 9.12, the mutual geometry of $\vec{r}_j$ and $\vec{r}_i$ allows $\vec{r}_{ij}$ to be defined as the position vector connecting monomers i and j.

$$\vec{r}_{ij} = \vec{r}_j - \vec{r}_i \tag{9.126}$$

With this, the phase sum can be written as

$$\sum_{i=1}^{N} \sum_{j=1}^{N} e^{-i\vec{q}\cdot\vec{r}_{ij}} \tag{9.127}$$

For $i = j$, $\delta_j = 0$ so that

$$\sum_{i=1}^{N} \sum_{j=1}^{N} e^{-i\vec{q}\cdot\vec{r}_{ij}} = N + \sum_{\substack{j=2 \\ j\neq i}}^{N} \sum_{i=1}^{N} e^{-i\vec{q}\cdot\vec{r}_{ij}} \tag{9.128}$$

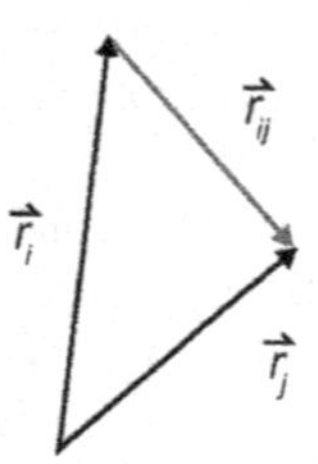

Figure 9.12 The position difference vector, formed from $\vec{r}_i$ and $\vec{r}_j$.

When N is large, the term "N" can usually be neglected, leaving the double sum to be evaluated.

The *scattering form factor* $P(q)$ is the normalized phase sum, which, reintroducing the possibility that each scatterer has a different polarizability and hence different $\vec{E}_j$

$$P(q) = \frac{\left\langle \left\langle \sum_{\substack{k=2 \\ j\neq k}}^{N} \sum_{j=1}^{N} \vec{E}^*_j \cdot \vec{E}_k e^{-i\vec{q}\cdot\vec{r}_{jk}} \right\rangle_{all\ orientations} \right\rangle_{all\ configurations}}{N^2} \tag{9.129}$$

The double sum indicates an average over all particle orientations in space and over all configurations of the particle. If the particle is rigid, such as a rod of sphere, then only the spatial average need be performed. For flexible polymers the double averaging process must be carried out.

16 If the scattered wave is mixed with a portion of the incident wave, a *beat frequency* can be produced, allowing detection of the field itself, E_{total}.

An important limit is $q = 0$, for which

$$P(q=0)=1 \tag{9.130}$$

Rayleigh scattering results in the previous sections were obtained in this approximation.

The continuum limit for $P(q)$ for a body of a single polarizable material is

$$P(q)=\frac{\iint\rho(\vec{r}')\rho(\vec{r})e^{-i\vec{q}\cdot(\vec{r}-\vec{r}')}d^3rd^3r'}{\iint\rho(\vec{r}')\rho(\vec{r})d^3rd^3r'}=\frac{\iint\rho(\vec{r}')\rho(\vec{r})e^{-i\vec{q}\cdot(\vec{r}-\vec{r}')}d^3rd^3r'}{M^2} \tag{9.131}$$

Because

$$e^{-i\vec{q}\cdot(\vec{r}-\vec{r}')}=e^{-i\vec{q}\cdot\vec{r}}e^{i\vec{q}\cdot\vec{r}'} \tag{9.132}$$

and

$$\mathrm{Re}\left(e^{iqr\cos\theta}\right)=\mathrm{Re}\left(e^{-iqr\cos\theta}\right) \tag{9.133}$$

these integrals are independent of each other so that

$$P(q)=\frac{\left[\int\rho(\vec{r})e^{-i\vec{q}\cdot\vec{r}}d^3r\right]^2}{M^2} \tag{9.134}$$

9.6.2 Computation of the Spatial Average of $e^{-i\vec{q}\cdot\vec{r}_{ij}}$ for Uniform Mass Density

The spatial average of the phase sum can be found by considering a fixed $\vec{q}$ and then finding the distribution of orientations of $\vec{r}_{ij}$ with respect to it. The orientation can be defined by the single angle α between $\vec{q}$ and $\vec{r}_{ij}$, shown in Figure 9.13. The spatial average of $e^{-i\vec{q}\cdot\vec{r}_{ij}}$ is then

$$\langle e^{-i\vec{q}\cdot\vec{r}_{ij}}\rangle=\frac{\int_0^{\pi}n(\alpha)e^{-i\vec{q}\cdot\vec{r}_{ij}}d\alpha}{\int_0^{\pi}n(\alpha)d\alpha} \tag{9.135}$$

where $n(\alpha)d\alpha$ is the number of states between α and $\alpha + d\alpha$. The number of states in this interval is proportional to the area of the ring of thickness $d\alpha$ in Figure 9.13

$$n(\alpha)d\alpha=2\pi A\,r_{ij}^2\sin\alpha\,d\alpha \tag{9.136}$$

where A is a proportionality constant. The spatial average can be found from the integral

$$\langle e^{-i\vec{q}\cdot\vec{r}_{ij}}\rangle=\frac{\int_0^{\pi}\sin\alpha\,e^{-iqr_{ij}\cos\alpha}d\alpha}{\int_0^{\pi}\sin\alpha d\alpha} \tag{9.137}$$

which yields

$$\langle e^{-i\vec{q}\cdot\vec{r}_{ij}}\rangle=\frac{\sin(qr_{ij})}{qr_{ij}} \tag{9.138}$$

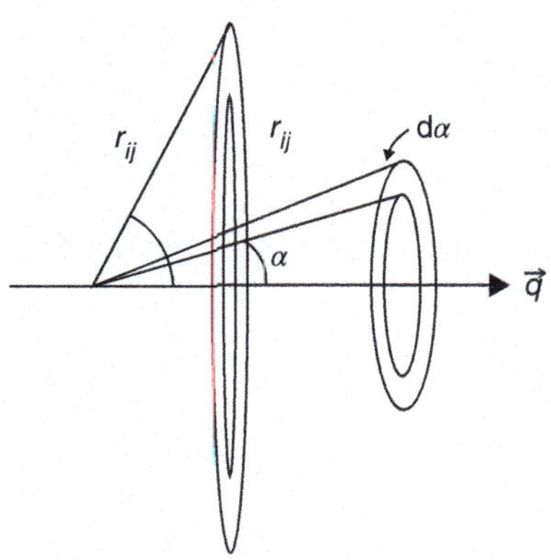

Figure 9.13 Geometry for spatial averaging.

The complete phase sum for a rigid particle can then be written as

$$\left\langle \sum_{\substack{j=2\\ j\neq i}}\sum_{i} e^{-i\vec{q}\cdot\vec{r}_{ij}} \right\rangle_{\substack{spatial\\ orientations}} = \sum_{\substack{j=2\\ j\neq i}}\sum_{i} \frac{\sin(qr_{ij})}{qr_{ij}} = P(q)N^2 \tag{9.139}$$

For a flexible particle

$$P(q)N^2 = \left\langle \sum_{\substack{j=2\\ j\neq i}}\sum_{i} \frac{\sin(qr_{ij})}{qr_{ij}} \right\rangle_{all\ configurations} \tag{9.140}$$

$P(q)$ has been calculated for many types of particles; rods, ellipsoids, random coils, etc.

Some examples are

Uniform sphere of radius R

$$P(q) = 9\left[\frac{\sin u - u\cos u}{u^3}\right]^2, \quad u = qR \tag{9.141}$$

Uniform rod of length L, where $L \gg$ rod diameter:

$$P(q) = \frac{2}{qL}\int_0^{qL}\frac{\sin(qL)}{qL}\mathrm{d}(qL) - \frac{\sin^2(qL/2)}{(qL/2)^2} \tag{9.142}$$

Random coil polymer of mean square radius of gyration $\langle S^2\rangle$.

Random coil polymers were discussed in detail throughout Chapter 8. Obtaining $P(q)$ requires the introduction of the chain length probability distribution function from Chapter 8 to be introduced into the phase sum. The result is

$$P(q) = \left(\frac{2}{u^2}\right)\left(e^{-u} - 1 + u\right), u = q^2\langle S^2\rangle \tag{9.143}$$

9.6.3 The Limit of $P(q)$ at Small qr_{ij}

An important expression is found in this lower limit of qr_{ij} by expanding the average phase as

$$\frac{\sin(qr_{ij})}{qr_{ij}} = 1 - \frac{(qr_{ij})^2}{6} + \cdots \tag{9.144}$$

and substituting this into the phase sum for $P(q)$

$$P(q) = 1 - \frac{q^2}{N^2}\left\langle \sum\sum r_{ij}^2 \right\rangle + \cdots \tag{9.145}$$

Taking the origin at the center-of-mass of the polymer chain and using the definition of the mean square radius of gyration for N identical masses $\langle S^2\rangle$

$$\langle S^2\rangle = \frac{1}{N}\left\langle \sum r_i^2 \right\rangle \tag{9.146a}$$

and returning to $\vec{r}_{ij} = \vec{r}_j - \vec{r}_i$, and expanding r_{ij}^2 in Equation 9.145 yields

$$\left\langle \sum\sum r_{ij}^2 \right\rangle = \left\langle N\sum r_i^2 + N\sum r_j^2 - 2\sum \vec{r}_i \cdot \sum \vec{r}_j \right\rangle = 2N\left\langle \sum r_i^2 \right\rangle = 2N^2\langle S^2\rangle \tag{9.146b}$$

since, in the center-of-mass, by definition

$$\sum \vec{r}_i = \sum \vec{r}_j = 0 \tag{9.146c}$$

so that

$$P(q) = 1 - \frac{q^2\langle S^2\rangle}{3}, \quad q^2\langle S^2\rangle << 1 \quad \text{Low } q \text{ Limit of } P(q) \tag{9.147}$$

In the RDA, this results holds for any shape particle, making it a very useful approximation, when applicable. Figure 9.14 shows $1/P(q)$ for several shapes of particles and how the values converge in the low $q^2\langle S^2\rangle$ limit.

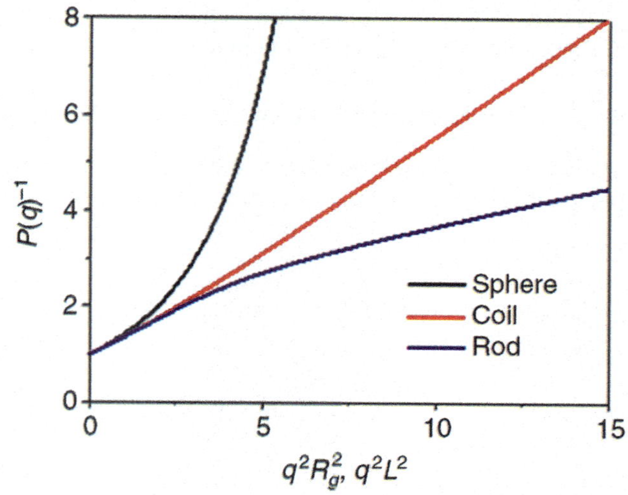

Figure 9.14 $1/P(q)$ for random coils, rods, spheres. In this, R_g^2 represents the mean square radius of gyration $<S^2>$. q^2L^2 is for the rod.

9.6.4 The Zimm Approximation for Polymers

The method of Bruno Zimm (USA, 1920–2005), and its extension to disperse polymers, is the single most used approximation for the analysis of light scattering intensity from polymer solutions. The goal of his work in this area was to provide an expression for the intensity of light scattered as a function both of polymer concentration and scattering angle, due to effects from intraparticle and interparticle interference and interparticle interactions.

First, Zimm considered the scattering at q by treating the phase sum as an integral over the entire scattering volume. The scattering volume is the illuminated portion of a sample whose scattered light reaches the detector. $\rho(\vec{r})$ is the mass density distribution of polymer throughout the scattering volume. Equation 9.121 shows that the total scattered electric field is the Fourier transform of the scatterers' spatial mass distribution. Zimm then treated $\rho(\vec{r})$ as composed of an intraparticle portion, i.e. the mass distribution in a polymer coil, and an interparticle portion, which accounts for how interactions among polymer molecules affect their distribution throughout the scattering volume.

The scattered electric field at the detector $\vec{E}_{total}$ can be found by summing the scattered fields of all the monomers in the scattering volume. Zimm did this by summing over all N monomers on each chain and n polymers in solution to account for all nN monomers. $\vec{E}_{ik}e^{-i\delta_{ik}}$ is the scattered field at the detector due to the kth monomer on the ith polymer. δ_{ik} is the phase shift of this monomer with respect to a chosen origin.

$$\vec{E}_{total} = \frac{c}{8\pi} \sum_{i=1}^{n\ polymers} \ \sum_{k=1}^{N\ monomers\ on\ polymer\ i} \vec{E}_{ik}e^{-i\delta_{ik}} \tag{9.148}$$

The corresponding intensity of this scattered field is

$$\langle S_s \rangle = \frac{c}{8\pi} \sum_{i=1}^{n} \sum_{j=1}^{n} \sum_{k=1}^{N_i} \sum_{l=1}^{N_j} \vec{E}_{ik}^{*} \cdot \vec{E}_{jl} e^{i(\delta_{ik}-\delta_{jl})} \tag{9.149}$$

His first point of departure for evaluating this formidable quadruple sum was to consider that the interparticle distribution in dilute enough solution would be due to single contacts between pairs of polymers. The derivation from that point becomes involved and the reader is referred to Zimm's classic paper (Zimm 1948). The result, however, is rather simple:

$$R(q,c) \cong Kc\left[MP(q) - 2A_2M^2P(q)^2 c\right] \tag{9.150}$$

Forming the factor $Kc/R(q)$ and expanding the binomial term to first order yields Zimm's equation for a monodisperse polymer solution:

$$\frac{Kc}{R(q,c)} = \frac{1}{MP(q)} + 2A_2c \tag{9.151}$$

Of note is that the first-order interaction term $2A_2c$ is independent of q in this approximation. The limiting case for $P(q)$ in Equation 9.147 is often substituted in the above equation.

9.6.5 Zimm Equation for a Disperse Polymer Solution

In the case of a disperse polymer population, it is considered that the total concentration of polymer c is composed of sub-populations of polymers of concentration c_i, each having a specific number of monomers per chain.

$$c = \sum c_i \tag{9.152}$$

$R(q,c)$ is then given by the sum of scattering from each subconcentration by the Zimm Equation 9.151 which also takes into account the interactions among the subpopulations:

$$R(q,c) = \sum_i R_i(q,c) \cong \sum_i K_i\left[c_i M_i P_i(q) - 2A_{2,i} M_i^{\,2} P_i(q)^2 c_i^{\,2}\right] \tag{9.153}$$

where the dilute solution condition requires that the second term in brackets involving A_2 be much smaller than the first term, i.e. the mass and shape of the polymer population dominate the net scattering, and the interactions are a second-order effect.

To get a particularly useful result Zimm considered the low $q^2\langle S^2\rangle$ limit for each $P_i(q)$, based on Equation 9.147:

$$P_i(q) \cong 1 - \frac{q^2\langle S^2\rangle_i}{3} \tag{9.154}$$

K_i was approximated to be the same for each subpopulation, K, which usually holds very well for long-chain homopolymers[17], since their $\partial n / \partial c$ is the same. Hence,

$$\sum K_i c_i M_i P_i(q) \cong K \sum c_i M_i \left(1 - \frac{q^2\langle S^2\rangle_i}{3}\right) \tag{9.155}$$

The weight average of a quantity $\langle Q\rangle_w$ over a concentration distribution c_i is

$$\langle Q \rangle_w = \frac{\sum Q_i c_i}{c} \tag{9.156}$$

and the z-average of Q, $\langle Q\rangle_z$ is given by

$$\langle Q \rangle_z = \frac{\sum Q_i M_i c_i}{\sum M_i c_i} \tag{9.157}$$

The above expression then becomes

$$K \sum c_i M_i \left(1 - \frac{q^2\langle S^2\rangle_i}{3}\right) = K M_w c \left(1 - \frac{q^2\langle S^2\rangle_z}{3}\right) \tag{9.158}$$

To account for the A_2 term, return to the summation over $R_i(q,c_i)$ and make the expansion of Kc/R to first order:

$$\frac{Kc}{R(q,c)} \cong \frac{c}{\sum\left[c_i M_i P_i(q) - 2A_{2,i} M_i^{\,2} P_i(q)^2 c_i^{\,2}\right]} = \frac{c}{\sum c_i M_i P_i(q)} + 2c\frac{\sum A_{2,i} M_i^{\,2} P_i(q)^2 c_i^{\,2}}{\left(\sum c_i M_i P_i(q)\right)^2} + \cdots \tag{9.159}$$

The second term is seen to be an unusual average over the $A_{2,i}$, $\langle A_2\rangle$

$$\langle A_2\rangle = \frac{\sum A_{2,i} M_i^{\,2} P_i(q)^2 c_i^{\,2}}{\left(\sum c_i M_i P_i(q)\right)^2} \tag{9.160}$$

---

17 This assumption can break down for copolymers whose $\partial n / \partial c$ values are different. See Stockmayer et al. (1955), Enohnyaket et al. (2007).

This can be more clearly seen as an average over a weighting function W_i:

$$W_i = M_i P_i(q) c_i \qquad 9.161$$

$$\langle A_2 \rangle = \frac{\sum A_{2,i} W_i^{\,2}}{\left(\sum W_i\right)^2} \qquad 9.162$$

This average shows that $<A_2>$ can involve $P(q)$ and is hence not normally q-independent as is the case for the monodisperse polymer solution above. Hence, strict application of the Zimm method requires extrapolation to both $c = 0$ and $q = 0$.

Using the above result for $P(q)$:

$$\frac{Kc}{R(q,c)} = \frac{1}{M_w\left(1 - \frac{q^2 \langle S^2 \rangle_z}{3}\right)} + 2c\langle A_2 \rangle \qquad 9.163$$

Expanding the M_w term to the first order yields the final, most familiar form of the *Zimm equation*:

$$\frac{Kc}{R(q,c)} = \frac{1}{M_w}\left(1 + \frac{q^2 \langle S^2 \rangle_z}{3}\right) + 2c\langle A_2 \rangle \quad \text{Zimm Equation} \qquad 9.164$$

By measuring $R(q,c)$ for a number of different concentrations c, and at various values of q (usually done by changing scattering angle θ but also achievable by varying the incident wavelength λ), three important characteristics of a polymer solution can be measured; the weight average mass M_w, the mean square z-average radius of gyration $<S^2>_z$, and the complex average of the second virial coefficient $<A_2>$.

The *Zimm plot* is a convenient representation for determining these characteristics. A Zimm plot from a disperse polystyrene (PS) in toluene is shown in Figure 9.15. The y-axis uses ΔR_θ to emphasize the "$R(q,c)$" treated above is the excess Rayleigh ratio, i.e. the scattering from the polymer solution with the solvent scattering subtracted. In making the Zimm plot of Figure 9.15, the scaling value of "23" was used in $23c$ to bring the range of $23c$ into a scale comparable to that of $\sin^2(\theta/2)$. Notice that in using $\sin^2(\theta/2)$ instead of q^2, the Zimm equation is written in a slightly different form for analysis:

$$\frac{Kc}{R(q,c)} = \frac{1}{M_w}\left(1 + \left(\frac{4\pi n_0}{\lambda_0}\right)^2 \frac{\langle S^2 \rangle_z}{3} \sin^2(\theta/2)\right) + 2c\langle A_2 \rangle \qquad 9.165$$

The slightly tilted horizontal lines are the extrapolation over seven angles to $q = 0$ and the steep lines are extrapolations over five concentrations to $c = 0$. It can be seen from Equation 9.164 that when the $c = 0$ extrapolated line is itself extrapolated to $q = 0$ the y-intercept is $1/M_w$.

$$\lim_{\substack{c \to 0 \\ q \to 0}} \frac{Kc}{R(q,c)} = \frac{1}{M_w} \qquad 9.166$$

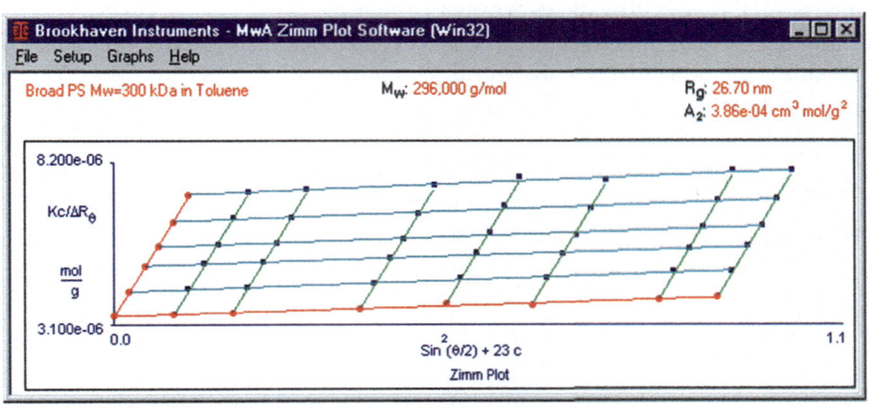

Figure 9.15 A Zimm plot for a polystyrene sample of nominal $M_w = 3 \times 10^5$ g/mol, in toluene at $T = 25°C$. (Courtesy of Brookhaven Instruments Corporation).

The slope of this line yields $<S^2>_z$ by

$$\frac{\mathrm{d}\left[Kc/R(q,c)\right]_{c=0}}{\mathrm{d}\sin^2(\theta/2)} = \frac{1}{M_w}\left(\frac{4\pi n_0}{\lambda_0}\right)^2 \frac{\langle S^2\rangle_z}{3} \qquad 9.167$$

Similarly when the $q = 0$ line is extrapolated to $c = 0$, its slope yields $<A_2>$ by

$$\frac{\mathrm{d}\left[Kc/R(q,c)\right]_{q=0}}{\mathrm{d}c} = 2\langle A_2\rangle \qquad 9.168$$

While many reports of Zimm plot results in the literature explicitly reference that M_w is determined, fewer mention that it is the quantity $<S^2>_z$ which is directly obtained from the Zimm treatment, and from this the root mean square z-average radius of gyration $<S^2>_z^{1/2}$. Instead, many reports simply term $<S^2>_z^{1/2}$ the "radius of gyration" R_g. The importance of the difference becomes more pronounced for increasingly disperse polymers, because $<S^2>_z$ more strongly weights the higher mass subcomponents compared to M_w.

Virtually no reports refer to the unusual average represented above by $<A_2>$ and usually just list the determined result as "A_2." Earlier light scattering work delved into the significance of $<A_2>$ for disperse populations (e.g. Cassassa 1962; McMillan and Meyer 1945). "A_2" as quoted is nonetheless a very useful parameter for quantifying polymer interactions and excluded volume.

9.7 The Angular Dependence of Scattered Light: Intermolecular Interference Effects on Scattering: The Structure Factor *S(q,c)*

The effect of particle interaction on scattering at $q = 0$ was linked to how repulsive or attractive interactions control the concentration fluctuations in a fluid containing particles. We saw that increasing excluded volume due to repulsive interactions reduced the fluctuations and hence also reduced the scattering. The effect is summed up in the virial coefficient A_2, A_3, etc. For dilute solutions, where polymer characterization often takes place, the Zimm equation shows that in the dilute, single-contact approximation the angle-independent use of A_2 is robust.

When solutions become more concentrated, or interactions become more long range, the situation changes, because a given particle can interact with more than just local particles. Namely, the particles in the solution can start to order so that there is a correlation between a given particle and other ones at increasing distances. The limit of such organization is a crystal lattice, which produces sharp, angular diffraction peaks. But between local collisions producing an angle-independent effect to interactions producing a crystal lattice, there is a lot of intermediate territory for structures: At low concentration, the angle-independent A_2 accounts for interparticle interactions as long as dispersity is not too high. But as concentration increases correlations arise between particles as the excluded volume forces them into increased interactions, which will lead to interference effects as a function of angle, even for monodisperse populations.

The effects of intra- and intermolecular interference can be expressed in the following proportionality involving the Rayleigh scattering ratio, R

$$R(q,c) \propto P(q)S(q,c) \qquad 9.169$$

where $S(q,c)$ is the angle and concentration-dependent *structure factor*. It accounts for interference effects due to any short and/or long-range correlations between particles.

An approximation to the angular dependence of scattering due to interparticle interactions can be made by considering the dilute relationship for scattering at $q = 0$, in terms of A_2. Rearranging Equation 9.153 and expanding to first order and remembering that $P(q = 0) = 1$ gives

$$R(q=0,c) = KcM(1-2A_2cM) = KcMS(q=0) \qquad 9.170$$

Here, $S(q = 0)$ is identified with $(1 - 2A_2cM)$. Equations 4.156 and 4.158 can be used to express A_2 as

$$A_2 = \frac{N_A}{2M^2}\int\left[1-g(r_{12})\right]\mathrm{d}^3r_{12} \qquad 9.171$$

We have seen that interference effects between two scatterers j and k are treated by the phase term $e^{-i\vec{q}\cdot\vec{r}_{jk}}$, whether the scatterers are part of a solid body, or are separate particles independently diffusing in the liquid. We can propose that interference corrected expression for $S(q)$ should be the Fourier transform of $\left[1-g\left(r_{12}\right)\right]$

$$S\left(q,c\right)=1-\left(2cM\right)\frac{N_A}{2M^2}\int\left[1-g\left(r_{12}\right)\right]e^{-i\vec{q}\cdot\vec{r}_{12}}d^3r_{12} \quad 9.172$$

So that

$$S\left(q,c\right)=1-\frac{N_A}{M}c\int\left[1-g\left(r_{12}\right)\right]e^{-i\vec{q}\cdot\vec{r}_{12}}d^3r_{12} \quad 9.173$$

or, in terms of the number density of particles, N (particles/cm^3)

$$S\left(q,N\right)=1-N\int\left[1-g\left(r_{12}\right)\right]e^{-i\vec{q}\cdot\vec{r}_{12}}d^3r_{12} \quad 9.174$$

For $g(r_{12})$ with spherical symmetry, as is the case in an isotropic solution:

$$S\left(q,N\right)=1-4\pi N\int\left[1-g\left(r\right)\right]\left(\frac{\sin qr}{qr}\right)r^2d^3r \quad 9.175$$

where we have put particle 1 at the origin so that the location of particle 2 is simply r so that $r_{12} = r$.

The simplest case is, again, hard sphere interaction, following Equations 4.156 and 4.158, so that for spheres of diameter D:

$$S\left(q,N\right)=1-4\pi N\int_0^D\left(\frac{\sin qr}{qr}\right)r^2d^3r=1-4\pi ND^3\left(\frac{1}{x^3}\right)\left(\sin x-x\cos x\right) \quad 9.176$$

where $x = qD/2$. Figure 9.16 shows an example of $S\left(q,N\right)$ versus θ for particles, where $4\pi ND^3=0.1$ $D = 600$ nm and $\lambda = 532$ nm, in water. In contrast to $P(q)$, which decreases with q and for which $P(0) = 1$, $S(q)$ initially increases with q and $S(0) < 1$, before becoming undulatory. The fact that $S(0) < 1$ is the result of repulsive interactions lowering the thermodynamic fluctuation, as was treated in the $q = 0$ limit of fluctuation in Equation 9.100.

Equation 9.176 is limited to dilute scatterers and the hard sphere potential. More detailed interactions than the simple hard sphere have been treated, and at higher concentrations, leading to more complex expressions than Equation 9.176.

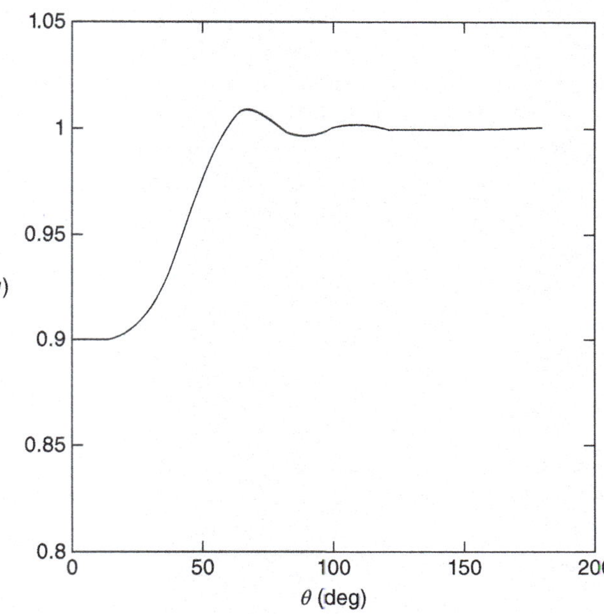

Figure 9.16 S(q) for a dilute solution of spheres.

An example of a more complex interaction is the electrostatic interaction between charged particles in a solution or suspension. Figure 9.17 shows how a dilute solution of proteoglycans (Chapter 12) in pure water, at 0.01% by mass, shows strong interparticle correlations, which are reflected in a broad peak. The broad maximum suggests a "fuzzy lattice," where a rigid regular lattice loosens so that there is still an average distance between the particles, but the particles can fluctuate from this average, unlike in a rigid crystal. In the fuzzy crystal, the sharp diffraction peak of a rigid crystal reduces in amplitude and widens as the lattice becomes less well defined. It is remarkable that such a dilute solution of particles produces such a pronounced scattering peak. Significantly, this means that the hard sphere potential of $S(q)$ in Figure 9.16 cannot be used to describe the proteoglycan peak, even if an electrostatically enhanced "effective diameter" is introduced. Presumably, a

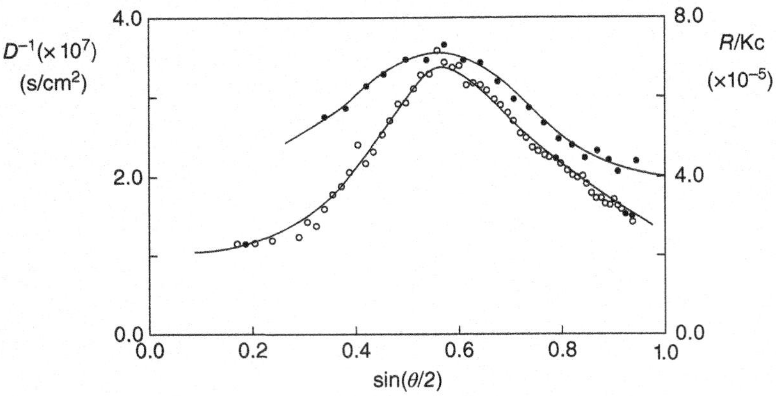

Figure 9.17 Strong interparticle correlations in a solution of proteoglycans, $M = 1.6 \times 10^6$ g/mol and $R_g = 60$ nm. (Reproduced from Li, X. and Reed, W.F. (1991)).

long-range, quasi-periodic $g(r)$ is needed for Equation 9.173. The hard sphere case only takes into account local collisions of two bodies (A_2), three bodies (A_3) and up, but no long-range ordering. The "fuzzy lattice" concept implies long-range ordering which is remarkable to find in the dilute proteoglycan solution of Figure 9.17.

9.8 Mie Scattering

In the former approximations, it was assumed that the intraparticle interference for light waves scattered from a single particle and reaching a detector is due solely to the difference in geometrical path for the light waves; the term which accounts for the difference in index of refraction between the solution and scatterer was assumed negligible, that is,

$$\delta = \frac{2\pi d\left|n_p - n_s\right|}{\lambda_0} << 1 \qquad 9.177$$

where d is the characteristic dimension of the particle, λ_0 is the vacuum wavelength of the incident light, and n_p and n_s are the indices of refraction of the particle and solvent, respectively.

For solid particles, such as nano- and microscale spheres, and many types of biological colloids, the approximation of Equation 9.177 no longer holds. In this case, the calculation of the magnitude and angular distribution of scattered light is much more complex than the foregoing approximations and requires a solution of the Maxwell equations with boundary conditions applied. Gustav Mie (Germany, 1868–1957) first published a rigorous approach to this in 1908, and scattering for particles in this regime is often termed *Mie scattering*. For spherical particles, the absolute, angular-dependent scattering function depends only on the ratio of λ/R, where R is the particle radius, and of n_p/n_s where n_p and n_s may be complex (i.e. involve light absorption in addition to refraction).

For spherical scatterers, the angular intensity in the plane perpendicular to the direction of polarization of the incident light is

$$I_s\left(\text{horizontal plane}\right) = \frac{\lambda^2}{4\pi^2 r^2}\left|S_1{}^2\right| \qquad 9.178$$

where

$$S_1 = \sum_{n=1}^{\infty} \frac{2n+1}{n(n+1)}\left\{a_n \pi_n\left(\cos\theta\right) + b_n \tau_n\left(\cos\theta\right)\right\} \qquad 9.179$$

and

$$\pi_n\left(\cos\theta\right) = \frac{P_n\left(\cos\theta\right)}{\sin\theta}, \text{ and } \tau_n\left(\cos\theta\right) = \frac{\mathrm{d}P_n\left(\cos\theta\right)}{\mathrm{d}\theta} \qquad 9.180$$

where $P_n(\cos\theta)$ are the Legendre polynomials, and

$$a_n = \frac{\psi_n(\alpha)\psi_n'(\beta) - m\psi_n(\beta)\psi_n'(\alpha)}{\varsigma_n(\alpha)\psi_n'(\beta) - m\psi_n(\beta)\varsigma_n'(\alpha)} \tag{9.181}$$

$$b_n = \frac{m\psi_n(\alpha)\psi_n'(\beta) - \psi_n(\beta)\psi_n'(\alpha)}{m\varsigma_n(\alpha)\psi_n'(\beta) - \psi_n(\beta)\varsigma_n'(\alpha)} \tag{9.182}$$

where

$$\alpha = \frac{2\pi n_s R}{\lambda}, \quad \beta = \frac{2\pi n_p R}{\lambda}, \quad \text{and } m = \frac{n_p}{n_s} \tag{9.183}$$

$$\psi_n(\alpha) = \sqrt{\frac{\alpha\pi}{2}}\, J_{n+1/2}(\alpha) \tag{9.184}$$

where $J_{n+1/2}$ are the Ricatti–Bessel functions. The $\psi_n(\alpha)$ obey the recursion relations:

$$\psi_{n+1}(x) = \frac{2n+1}{x}\psi_n(x) - \psi_{n-1}(x) \tag{9.185}$$

$$\psi_n'(x) = \frac{-n}{x}\psi_n(x) + \psi_{n-1}(x) \tag{9.186}$$

Finally,

$$\varsigma_n(\alpha) = \mathrm{i}^{(n+1)}\mathrm{e}^{-\mathrm{i}\alpha}, \quad \text{and}\, \varsigma_n'(\alpha) = \mathrm{i}^n \mathrm{e}^{-\mathrm{i}\alpha} \tag{9.187}$$

Figure 9.18 shows examples of the normalized (i.e. $P(\theta = 0) = 1$) angular scattering envelopes for latex spheres of various diameters in water; $n_p = 1.59$, $n_s = 1.33$, $\lambda = 532$ nm for vertically polarized incident light. The 50-nm sphere scatters nearly isotropically, whereas the 500-nm sphere has four lobes, and the 2-μm sphere has very complex structure.

Figure 9.18 shows that Mie scattering effects at finite q-values can be enormous as particle size increases. Figure 9.19 shows how scattering intensity (arbitrary units) changes as solid latex sphere size increases. At $q = 0$, the intensity grows cubically and is soon off the scale. In contrast at $q = 179{,}000$ cm^{-1}, a typical scattering measurement angle (e.g. $\lambda_0 = 660$ nm, $n_{water} = 1.33$, and $\theta = 90°$), the scattering is in fair agreement with the $q = 0$ value up until about 115 nm, at which point the difference between them is 20% and rapidly increasing. Figure 9.19 has two immediate consequences for intensity scattering measurements at finite q: (1) Scattering from large particles will be under-represented in the intensity value and so the value of M_w measured at finite angle will underestimate the true value, which is why measuring at q close to zero, or by extrapolating to zero can be very important. (2) If we are measuring a time-dependent aggregation of spheres, which aggregates retain an approximately solid spherical shape, then the scattering will not only underestimate the growth of aggregation, but can become non-monotonic, whereby the intensity actually decreases despite the aggregates becoming larger.

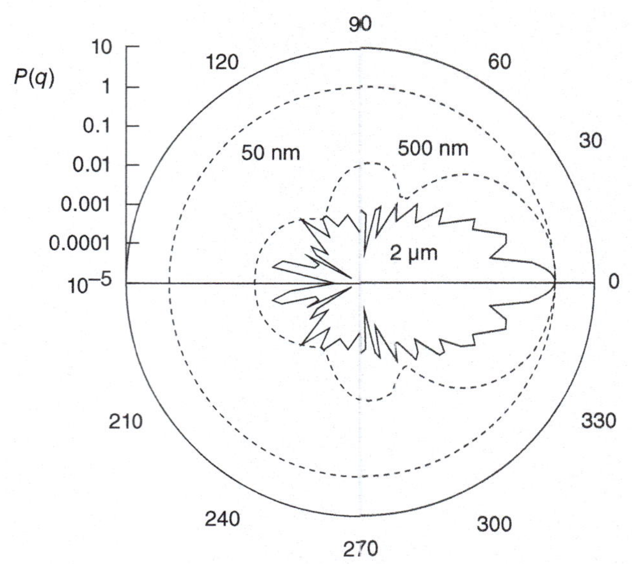

Figure 9.18 Polar plots of Mie scattering from spherical particles of various diameters, where $n_p = 1.59$ in water, $n_s = 1.33$.

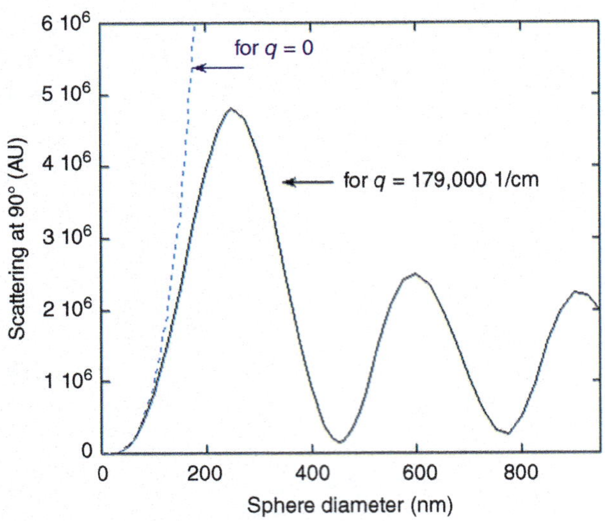

Figure 9.19 Contrast of scattering at $q = 0$ with Mie scattering at 90°, for solid latex spheres in water. λ_0=660nm, n_{water}=1.33, n_{sphere}=1.59, and q=179,000 1/cm.

9.9 Scattering Model for Index of Refraction

The index of refraction has been used throughout this chapter but without any thought as to its physical origin. Consider a dielectric material in which all electrons are bound to parent atoms so that there are no free electrons. The $\vec{E}$-field of an incident harmonic wave of angular frequency ω, $\vec{E} = \vec{E}_0 e^{-i\omega t}$ exerts a force on the electrons, but, being bound to their parent atoms, they cannot move freely. Rather, they will oscillate with the incident ω slightly stretching the electron cloud away from an otherwise nonpolar atom, and forming an oscillating dipole. As we have seen earlier in this chapter, the acceleration of the oscillating charge in the dipole is the origin of dielectric scattering. A simple model is to consider the electrons bound to their parent atoms by a spring-like restoring force, $F_r = -\kappa x = -m_e\omega_0^2 x$ where κ is the spring constant, m_e is the electron mass, ω_0 is the natural frequency of the spring–mass system, and x is the displacement from the electron equilibrium position. The electron also experiences a damping force opposing its velocity, of the form $F_d = -v\gamma$, where γ is the strength of the dampening. The equation of motion is then

$$m_e \frac{d^2x}{dt^2} + \gamma \frac{dx}{dt} + m_e\omega_0^2 x = qE_0 \cos \omega t = qE_0 \operatorname{Re}\left\{e^{-i\omega t}\right\} \tag{9.188a}$$

This is more conveniently solved in the complex plane, with $\tilde{z}=x+iy$, and $|\tilde{z}|=\sqrt{(x^2+y^2)}$.

$$\frac{d^2\tilde{z}}{dt^2} + \frac{\gamma}{m_e}\frac{d\tilde{z}}{dt} + \omega_0^2\tilde{z} = \frac{q}{m_e}E_0 e^{-i\omega t} \tag{9.188b}$$

with solution

$$\tilde{z} = \frac{qE_0}{m_e}\frac{e^{-i\omega t}}{\omega_0^2 - \omega^2 + i\omega\gamma / m_e} \tag{9.188c}$$

whose magnitude is

$$|\tilde{z}| = \frac{qE_0}{m_e}\frac{1}{\sqrt{\left(\omega_0^2 - \omega^2\right)^2 + \omega^2\gamma^2 / m_e^2}} \tag{9.188d}$$

The oscillating dipole, $p(t)$, caused by the electron, q=-e, is

$$p(\omega,t) = -e|\tilde{z}|(t) = \frac{e^2E_0 e^{-i\omega t}}{m_e}\frac{1}{\sqrt{\left(\omega_0^2 - \omega^2\right)^2 + \omega^2\gamma^2 / m_e^2}} \tag{9.189}$$

The damping term introduces a phase difference between E and $\tilde{z}$, but also prevents a singularity at $\omega = \omega_0$. For a material of magnetic constant $k_m = 1$ and dielectric constant D the index of refractions is related by

$$n^2(\omega) \cong D = 1 + 4\pi N p(\omega) = 1 + \frac{4\pi N e^2}{m_e} \frac{1}{\sqrt{\left(\omega_0^2 - \omega^2\right)^2 + \omega^2\gamma^2 / m_e^2}} \quad 9.190a$$

When multiple electrons are involved,

$$n^2(\omega) \cong D = 1 + \frac{4\pi N e^2}{m_e} \sum_j \frac{f_j}{\left(\omega_{0,j}^2 - \omega^2 - i\gamma_j\omega\right)} \quad 9.190b$$

where the sum is over all the electrons in the atom, and $\omega_{0,j}$, γ_j, and f_j are the natural frequency, damping coefficient, and *oscillator strength* of the jth electron, respectively, and $\textit{number of electrons} = \sum_j f_j$. N is the number density (atoms/volume) of the atoms, so that, as expected, the index of refraction is sensitive to material density, e.g. hot air is less dense than cold air and so has lower n. When air on the ground is hotter than above, such as in a desert or near hot pavement, *mirages* can form due to the way light paths bend upward, creating mirror images of objects. When air is warm above and cold below, as often happens on the ocean, light paths bend downward and can produce the phenomenon of *looming*, where the image of the object seems suspended in the air higher than the object.

Equation 9.190 explains why $\partial n / \partial c$ for a vinyl polymer in a solvent is larger than $\partial n / \partial c$ of the corresponding monomer: In Equation 9.188b the stiffer the spring force the higher the spring constant κ and the corresponding ω_0. The electron double bond in the monomer gives a higher value of κ than when the monomer loses the double bond when polymerized, and the stiffness decreases because only one bond is now left. This reduces ω_0 in the denominator of Equation 9.190, leading to a higher n and corresponding $\partial n / \partial c$.

Equation 9.190 is a *dispersion relation* because it indicates how n^2, and hence the speed of light in the medium $v = c/n$ varies with ω. For a typical dielectric, such as water, n increases with frequency in the visible range, so that a pulse of white light traveling through water will "disperse," that is, the blue wavelengths will lag behind the red and the pulse shape will stretch out as it moves through the water. Blue light will also refract through a greater angle entering water than red.

9.10 Scattering at Interfaces

The focus in this section has been on the characterization of polymers and other objects using light scattering. As mentioned at the outset, we perceive much of the non-luminescent world around us via scattering, and especially scattering at interfaces. Whether a blade of grass, a red bicycle, the surface of a pond, or the sparkling of snow, we are generally seeing the scattering of wavelengths not absorbed by the material behind the interface. Interestingly, materials that absorb exceptionally strongly, such as metals and metallic dyes, reflect more light than they absorb, and so take on the color that they absorb for reflected light, but take on the colors not absorbed in transmitted light.

The interest here is to consider nonabsorbing materials – i.e. colorless dielectrics – and see how incident light power is split between polarization-dependent reflection and refraction (transmission). The critical factor is the real part of the index of refraction between the two materials forming the interface.

9.10.1 The Fresnel Equations

For the special but important case of specular reflection[18] between two transparent interfaces of different n, such as air and water, it suffices to apply the boundary conditions on $\vec{E}$ and $\vec{B}$ of an electromagnetic wave incident on an interface to derive all the equations of reflection, refraction, polarization, and evanescent wave propagation, including the exact magnitudes of these effects.

Figure 9.20 shows an interface between two dielectric media of n_1 and n_2 with a harmonic wave incident from n_1 on the interface with wave vector $\vec{k}_i$ and at an angle θ_i with respect to the normal of the interface. Call the plane containing $\vec{k}_i$ and

18 "Specular" means "mirror-like," which applies to locally smooth, reflective surfaces. The word comes from the Latin "specularis," from the word for mirror "speculum." "Specchio" meaning "mirror" in Italian is a direct descendant of this word.

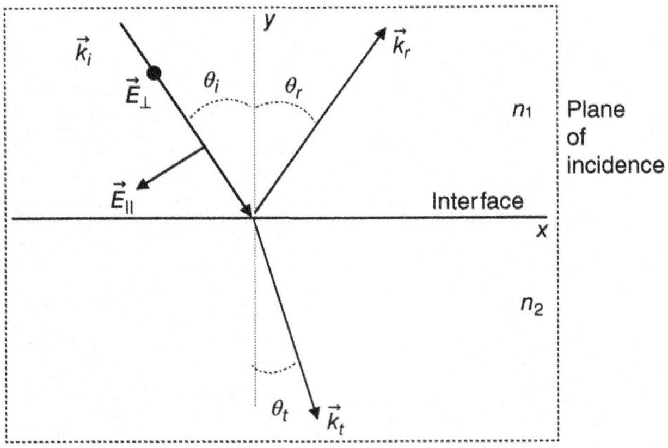

Figure 9.20 Incident light at an interface, reflecting and refracting. The electric field of the incident, reflected, and transmitted waves can be perpendicular or parallel to the Plane of Incidence.

perpendicular to the interface the *plane of incidence* (PoI). A wave reflects from the interface with a wave vector $\vec{k}_r$ at angle θ_r with respect to the normal of the interface. A refracted wave is transmitted into the medium of n_2 with wave vector $\vec{k}_t$ at angle θ_t with respect to the normal. The incident, reflected, and transmitted waves have electric fields, $\vec{E}_i, \vec{E}_r$, and $\vec{E}_t$ respectively, a result of the boundary conditions. These fields are perpendicular to the wave vectors and form a plane. Hence, the polarization state can be defined with two degrees of freedom. The most convenient axes for the fields are perpendicular to the PoI and parallel to the PoI. These two linear polarization states can be combined to produce any type of polarization state, including the general elliptical polarization, with limiting linear and circular polarization states. The component of $\vec{E}$ perpendicular to the PoI in Figure 9.20 is denoted by $\vec{E}_\perp$ and the component parallel to the PoI is denoted by $\vec{E}_\parallel$.

Using the continuity of the components of the three electric fields parallel to the interface, and the continuity of the parallel components of the three magnetic fields, *assuming the magnetic constant k_m is the same for both media*, yields a host of results, once the algebra is carried out.

The results of applying the boundary conditions are:

a) The frequency of the incident, reflected, and transmitted waves is the same for all.
b) $\vec{k}_i$, $\vec{k}_r$ and $\vec{k}_t$ are all in the PoI.
c) $\theta_i = \theta_r$. The *law of specular reflection* results.
d) $n_1 \sin\theta_i = n_2 \sin\theta_t$. The *law of refraction* results.
e) The *reflection and transmission amplitudes* are found for the incident electric field polarization states perpendicular and parallel to the PoI, $r_\perp, t_\perp, r_\parallel$, and $t_\parallel$, respectively:

$$r_\perp = \frac{E_{r,\perp}}{E_i} = \frac{n_1 \cos\theta_i - n_2 \cos\theta_t}{n_1 \cos\theta_i + n_2 \cos\theta_t} \quad 9.191$$

This indicates that if $n_1 < n_2$, then $r_\perp$ is negative, so that the phase of the reflected wave is 180° out of phase with the incident wave. It also indicates that for both $n_1 < n_2$ and $n_1 > n_2$, there will always be some reflected light, i.e. $r_\perp$ is never zero. The other transmission and reflection amplitudes are:

$$t_\perp = \frac{E_{t,\perp}}{E_i} = \frac{2n_1 \cos\theta_i}{n_1 \cos\theta_i + n_2 \cos\theta_t} \quad 9.192$$

$$r_\parallel = \frac{E_{r,\parallel}}{E_i} = \frac{n_2 \cos\theta_i - n_1 \cos\theta_t}{n_2 \cos\theta_i + n_1 \cos\theta_t} \quad 9.193$$

$$t_\parallel = \frac{E_{t,\parallel}}{E_i} = \frac{2n_1 \cos\theta_i}{n_2 \cos\theta_i + n_1 \cos\theta_t} \quad 9.194$$

Both cases, perpendicular and parallel, indicate that when $n_1 > n_2$, total internal reflection occurs (there is no transmitted wave) at and above a critical incidence θ_c, given by

$$\sin\theta_c = \frac{n_2}{n_1} \quad 9.195$$

Unlike $r_\perp$ there is an incident angle θ_p at which $r_\parallel = 0$. At this angle, known as the *polarization angle*, θ_p, or *Brewster's angle*, there is no reflected parallel polarized light, so there is only perpendicular polarized light reflected, a phenomenon often called *polarization by reflection*. This condition results when $\theta_r + \theta_t = \pi/2$ which, together with Equation 9.193 gives the polarization angle:

$$\tan\theta_p = \frac{n_2}{n_1} \quad 9.196$$

iii) The reflection and transmission amplitudes yield the fractional intensity of light power that goes into reflection R, the *reflection coefficient*, and into transmission T, the *transmission coefficient*.

By conservation of energy:

$$R + T = 1 \quad 9.197$$

Analysis of e-m energy flow requires using two effects: (i) incident and transmission angles are different, and hence the cross sections of the incident and transmitted beams of light are different; and (ii) the speed of light in a medium of index of refraction n has a wave speed v:

$$v = \frac{c}{n} \quad 9.198$$

which results from the fact that the electric field of the forward scattered light is phase lagged (or led) with respect to the propagating incident light and the two combine to give a phase speed that is controlled by the index of refraction.

When these effects are taken into account, the relationships between R and T, for both perpendicular and parallel incident light are

$$R = r^2 \quad 9.199$$

Because both the cross-section and speed of light of the incident and reflected waves are equal. For the transmitted wave:

$$T = t^2 \frac{n_2 \cos\theta_t}{n_1 \cos\theta_i} \quad 9.200$$

which results from the two effects mentioned.

For light perpendicularly incident on the PoI, $\theta_i = 0$ and, regardless of polarization,

$$R(\theta_i = 0^o) = \left(\frac{n_2 - n_1}{n_2 + n_1}\right)^2 \quad 9.201$$

Hence, there is always some light reflected at normal incidence at the interface between two media of different n. For example, $R = 0.04$ for the interface between air ($n_1 = 1$) and a typical glass ($n_2 = 1.5$). Equation 9.201, as well as Equations 9.191 and 9.193, shows that no light is reflected if two media have the same n. In this case, the two media are said to be *index matched* or *isorefractive*. For example, a slab of glass, visible in air due to the difference in n_1 and n_2, becomes invisible when immersed in a liquid of the same index, e.g. toluene provides a nearly perfect index match for a typical glass.

Index matching can be used to reduce or eliminate stray light at interfaces, which can be important when performing light scattering measurement. In the case of copolymers, it can be used to make one of the comonomers invisible, that is $(\partial n / \partial c) = 0$ so that the properties of the other comonomer can be studied. Figure 9.21a,b shows scattering for polybutadiene (PB) and polystyrene (PS) in (a) carbon disulfide, where PS is isorefractive; and (b) in orthoxylene, where PB is essentially isorefractive.

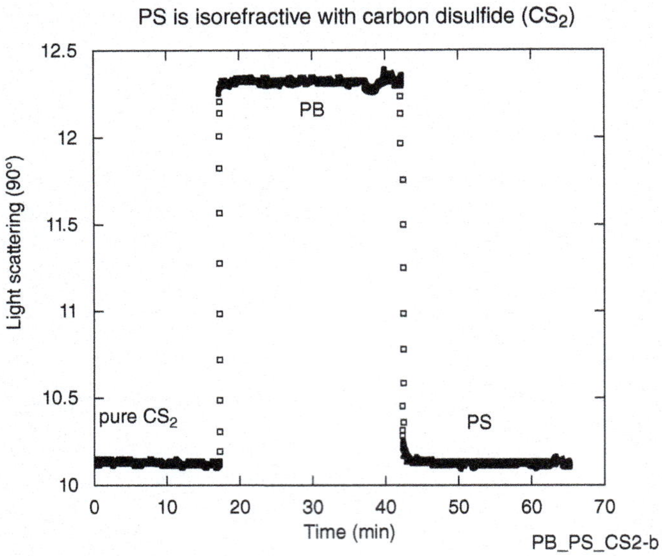

Figure 9.21a Polystyrene is isorefractive in carbon disulfide, whereas polybutadiene scatters strongly.

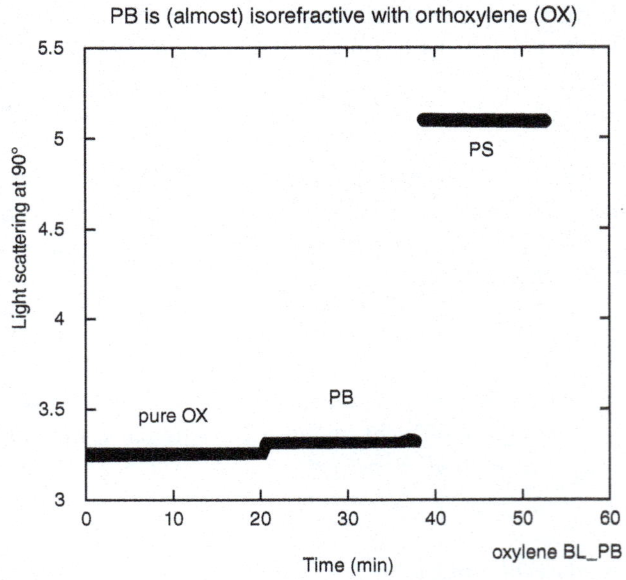

Figure 9.21b The opposite holds in orthoxylene. (Elizarraras, D., Drenski, M.F. and Reed, W.F., unpublished results).

9.10.2 The Evanescent Wave in Total Internal Reflection

In the case of total internal reflection, that is, if $\theta_i \geq \theta_c$ then $T = 0$ and the incident beam is not propagated into medium 2. However, without going into the derivations, there exists an *evanescent wave* that travels into and dies off exponentially in medium 2. This fundamentally complex topic can be simplified and understood heuristically by considering a harmonic wave incident on an interface and passing into medium 2, where $n_1 > n_2$ and $\theta_i \geq \theta_c$

$$\vec{E} = \vec{E}_0 e^{i\left(n_2 \vec{k}_0 \cdot \vec{r} - \omega t\right)} \qquad 9.202$$

where $\vec{r} = x\hat{i} + y\hat{j}$ is the position vector. For simplicity, consider the wave traveling in the $+x$ and $+y$ directions and the interface at $y = 0$. While we usually think of the electric field as not penetrating into medium 2, in medium 2, we have

$$n_2 \vec{k}_0 \cdot \vec{r} = \left(2\pi n_2 / \lambda_0\right)\left(\sin\theta_2 x + \cos\theta_2 y\right) \qquad 9.203$$

where λ_0 is the vacuum wavelength of the incident light. $\cos\theta_2$ can be expressed as

$$\cos\theta_2 = \sqrt{1 - \sin^2\theta_2} = \sqrt{1 - \left(\frac{n_1 \sin\theta_1}{n_2}\right)^2} \qquad 9.204$$

But, for $\theta_i \geq \theta_c$, $\left(n_1 \sin\theta_1 / n_2\right)^2 > 1$ so that $\cos\theta_2$ is imaginary and can be rewritten as

$$\cos\theta_2 = i\sqrt{\left(\frac{n_1 \sin\theta_1}{n_2}\right)^2 - 1} = i\beta \qquad 9.205$$

where β is real. Substituting this into the harmonic wave of Equation 9.202 yields the y-component of the wave:

$$\vec{E}(y,t) = \vec{E}_0 e^{-\beta y} e^{-i\omega t} \qquad 9.206$$

Hence, the e-m wave dies off exponentially as it penetrates into medium 2. This exponentially dying wave is the evanescent wave.

The evanescent wave behaves similarly to the quantum mechanical effect of an electron that penetrates into a potential barrier higher than its kinetic energy, where it dies off exponentially. If the barrier becomes lower than the electron's kinetic energy after the electron has traversed a certain distance, then the electron re-appears as a wave and resumes propagation. This is known in quantum mechanics as *tunneling*, and underlies the theory of *alpha decay* and *nuclear fusion*, and is used in devices such as the *scanning tunneling microscope*, and *tunneling diodes*, and *transistors*. Biologically, electron tunneling is thought to have a central role in the oxidation–reduction reactions in photosynthesis and respiration, and proton tunneling may be responsible for some mutations of DNA.

The evanescent e-m wave also finds many applications. For example, if two triangular glass or polymer prisms are put together as a square, with no space between them, they act as a single square of material. If they are separated, then incoming light can undergo total internal reflection along a diagonal edge and the evanescent wave appears in the gap between them. By adjusting the separation between the prisms, adjustable amounts of the evanescent wave can be recaptured by the second prism and propagate in it. This arrangement allows for precision beam splitters that can divide an incident beam into any fractions desired.

Another important use of the evanescent wave is in *attenuated total reflectance* (ATR) probes. This is frequently used in conjunction with infra-red (IR) spectroscopy and allows highly IR absorptive samples to be probed by just putting a thin layer – e.g. a liquid – atop the ATR crystal. The sample may be otherwise too IR-absorptive to allow standard IR transmission measurements to be made. The ATR crystal itself is optically dense, and often made of germanium or zinc selenide. The evanescent wave passes through the thin sample, where IR absorption occurs, and reflects at least once from the sample surface. The absorption of the evanescent wave by the vibrational bands of the sample reduces the reflected IR probe beam intensity at the detector. Multiple reflections are often used, and the attenuation depends on the sample, its thickness, and the number of reflections. Being able to measure the IR spectrum of a sample, without dilution or other conditioning steps, allows fast IR measurements, including on flowing liquid samples.

9.10.3 Scattering at Metal Surfaces

Metals are usually associated with high reflectivity. Metals have free conduction electrons forming a sort of quantum mechanical *electron gas*. These electrons belong to no particular atom and hence have no restoring force, so they can make much bigger oscillations in response to incident light than bound electrons. Metal atoms have many bound electrons so the expression for n^2, Equation 9.188 is still relevant for metals, but the effect of the free electrons is fundamental. The model for the free electrons simply drops the restoring force term in Equation 9.188a, eliminating ω_0, so that

$$\frac{d^2\tilde{z}}{dt^2}+\frac{\gamma}{m_e}\frac{d\tilde{z}}{dt}=\frac{q}{m_e}E_0e^{-i\omega t} \qquad 9.207$$

where m_e and γ are the electron mass and the system damping strength, respectively, and the free electron portion of the index of refraction, n_{fe}^2 is

$$n_{fe}^2\cong 1-\frac{4\pi Ne^2}{m_e}\frac{1}{\left(\omega^2+i\gamma\omega/m_e\right)} \qquad 9.208$$

For $\omega \gg \gamma$, this can be written as

$$n_{fe}^2\cong 1-\frac{\omega_p^2}{\omega^2} \qquad 9.209$$

where ω_p is the *plasma frequency* and is given by

$$\omega_p=\left(\frac{4\pi Ne^2}{m_e}\right)^{1/2} \qquad 9.210$$

The plasma frequency for most metals corresponds to the vacuum ultraviolet (<200 nm).

For $\omega < \omega_p$, Equation 9.209 shows that $n_{fe}^2<0$ so that n_{fe} is imaginary, and the total index of refraction is complex:

$$\tilde{n}=n_R+in_I \qquad 9.211$$

where the real portion comes from the bound electrons and the imaginary portion from the free electrons. The imaginary part of $\tilde{n}$ causes the electric field to die out exponentially after it enters the metal, and the rest of the incident energy is reflected. Consider an incident e-m wave traveling in the x-direction, and entering the metal at $x = 0$. Then

$$\vec{E}=\vec{E}_0e^{i(\tilde{n}k_0x-\omega t)}=\vec{E}_0e^{-n_Ik_0x}e^{i(n_Rk_0x-\omega t)} \qquad 9.212$$

The penetration depth is $1/n_Ik_0$ and is typically on the order of nanometers for metals.

For $\omega > \omega_p$, Equation 9.207 shows that $n_{fe}^2 > 0$ so that n_{fe} is real and the metal can transmit incident light above this frequency. The treatment applies for other conducting media. The Earth's ionosphere, for example, has $N_{ionosphere} \ll N_{metals}$ and has a plasma frequency around 3 MHz, where $N_{ionosphere}$ is the particle density in the atmosphere, and N_{metals} is the much higher particle density in metals. Radio waves with frequencies lower than this will be reflected from the ionosphere,[19] whereas frequencies above this can penetrate the ionosphere, e.g. to send signals into outer space.

It is interesting to apply Equation 9.201 to the complex index of refraction of Equation 9.211 for light incident normally on an air/metal interface:

$$R = \left(\frac{n_R - 1 + in_I}{n_R + 1 + in_I}\right)^2 = \frac{(n_R - 1)^2 + n_I^2}{(n_R + 1)^2 + n_I^2} \quad 9.213$$

where the complex conjugate of $\tilde{n}$ is used in performing the squaring operation. When $n_I \gg n_R$, this shows that $R \rightarrow 1$ and the metal becomes a perfect reflector.

9.10.4 The Phase Contrast Microscope

A major advance in visual microscopy was made with the introduction of the *phase contrast microscope*. When light is incident from a medium of lower index of refraction to a particle of higher index of refraction, the dipoles induced on the particle's surface will emit a scattered $\vec{E}$ – field with a 90° phase lag with respect to the incident $\vec{E}$ – field. Fritz Zernike (Netherlands, 1888–1966) took advantage of this effect to develop the phase contrast microscope.

Figure 9.22 shows a schematic of the working principle. The light from the illumination source passes through an annulus and this circle of illuminating light is focused by a converging lens (*condenser*) onto the specimen, after which the annular image is transmitted through the microscope objective and brought to a focus in the image plane of the specimen, which is viewed via an eyepiece or photographed by a camera. Before arriving at the image plane, the transmitted annulus of light passes through a phase plate, where its $\vec{E}$ – field can be shifted either 90° behind (lag) or ahead (lead) of its source phase. If it is lagged by 90°, it will be in phase with the light scattered from the specimen and the interference of the light scattered from the specimen combined with the now in-phase source light will produce constructive interference, thus brightening the object and enhancing its contrast with the medium it is in. Meanwhile, light coming through the medium not scattered by the specimen will not be phase shifted and hence will not interfere

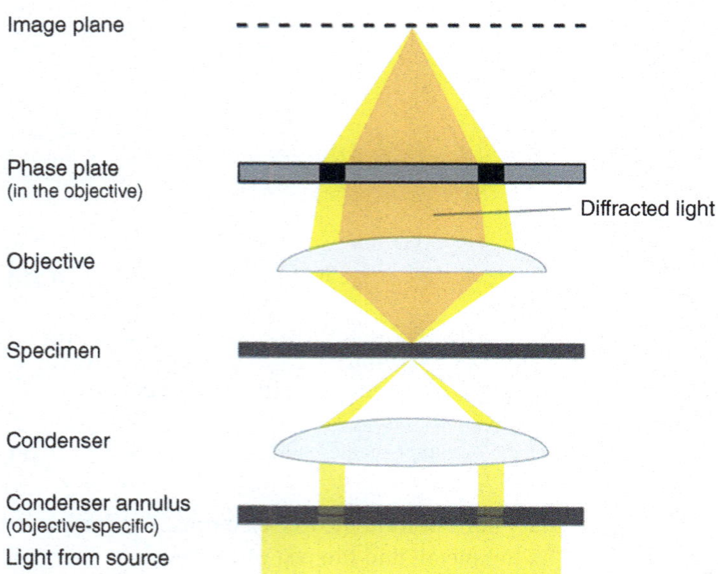

Figure 9.22 Working principle of the phase contrast microscope. (Image provided with permission by Scientifica, UK).

19 The radio waves reflected from the ionosphere are often termed "skip waves" and allow radio transmissions to be detected beyond the Earth's horizon from the transmitter.

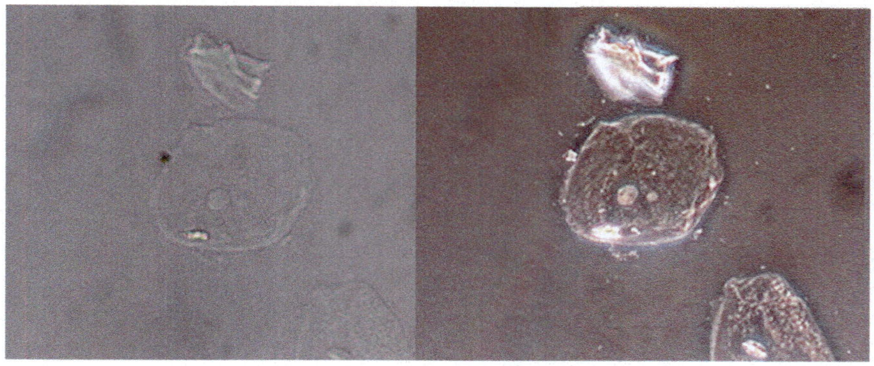

Figure 9.23 Microscope image without phase contrast (left), and with phase contrast (right). (Courtesy of Shenzhen Boshida Optical Instruments).

constructively or destructively with the phase-shifted source light. If the source light is shifted to lead the source by 90°, it will be 180° out of phase with the light scattered from the specimen and the specimen's image will be darkened with respect to nonscattered light. An example of a specimen with and without phase contrast is shown in Figure 9.23. The phase-contrast principle has been so successful in visual microscopy that it is offered as a standard feature on many inexpensive teaching-grade microscopes.

9.11 Single Photon Scattering

The theories presented here assume that a photon scattered within the field of photodetection scatters only once before reaching the detector. The Rayleigh ratio is a convenient means of assessing single photon scattering, when it is interpreted as the probability that a photon is scattered into a steradian of solid angle per cm of path length traveled.

For example, consider a typical dilute polymer solution of Rayleigh scatterers, with $R \sim 10^{-5}\ \text{cm}^{-1}$. Integrating R over the entire solid angle of a sphere gives $8.4 \times 10^{-5}\ \text{cm}^{-1}$, which is the probability that a photon is scattered once over a 1 cm path; i.e. only one photon out of 8.4×10^{-5} incident photons will scatter once. The probability of being scattered twice is this value squared; 7×10^{-9}, so that, on average, only one photon out of 1.2×10^{4} photons that have scattered once is scattered twice over 1 cm.

It is interesting to relate this probability of scattering to probability of detection of the photon. In a light scattering experiment the solid angle of detection is

$$\Delta\Omega = \sin\phi\Delta\phi\Delta\theta \qquad 9.214$$

For a detector in the scattering plane, $\phi = \pi/2$ and for a rectangular slit in front of the photodetector of slit width s in both dimensions, where the detector is at a distance r from the scattering volume:

$$\Delta\Omega(\text{steradians}) = \left(\frac{s}{r}\right)^2 \qquad 9.215$$

For a 100-μm slit at 10 cm from the scattering volume, $\Delta\Omega \sim 10^{-6}$ steradians. If the path length the photons travel in the scattering volume is 0.01 cm, for example, then for a scatterer with $R \sim 10^{-5}\ \text{cm}^{-1}$ the probability of the photon scattering once is 10^{-7}. The probability of this photon reaching the detector is 10^{-13}. The probability of the photon scattering twice over the scattering volume is 10^{-14}, which is a vanishingly small probability, and the probability of this twice-scattered photon reaching the detector is 10^{-20}.

At the other extreme are turbid solutions, where particles are highly scattering and dense enough to cause multiple scattering in the scattering volume. The mean free path in such solutions can be on the order of microns, such that photons can be thought of as randomly diffusing through the solution, making frequent scattering collisions with particles. If the particles are large enough, on the order of the incident light's wavelength or larger, then there is no wavelength dependence of the scattering; incident white light is scattered as white light, unlike in Rayleigh scattering. The particles causing the multiple scattering can be perfectly transparent, such as latex particles of polymeric styrene or acrylates, so it is scattering via multiple random reflections, as governed by the Fresnel equations above, that cause latex solutions to appear white. The same notion of multiple scattering applies to nonliquid particles, giving transparent snowflakes, salt crystals, and talcum powders their white appearance.

Diffusing wave spectroscopy is a form of dynamic light scattering that takes advantage of "light diffusion"[20] in dense media in order to characterize microscopic motion of particles. The problem of turbidity causing too much light scattering for detection is often solved by measuring back-scattered light, or using ultrathin sample cells.

9.12 Dynamic Light Scattering

Dynamic light scattering (DLS) has become a very widespread method of particle sizing. It is generally robust for sizing particles from about 1 nm to 1 μm. It is far more tolerant of stray light (e.g. scattering from glass cells and interfaces) than total intensity light scattering. It is mostly used to measure the self-diffusion coefficient of particles D_0, either by extrapolation of particle concentration to zero, or simply making measurements at very low concentration. Most systems convert this into an equivalent sphere, using the Stokes–Einstein equation:

$$D_0 = \frac{k_B T}{6\pi\eta R_H} \qquad 6.74$$

In practice, DLS is widely misunderstood and results are often misleading. Nonetheless, when used properly DLS is a very valuable method and its coupling of optical and diffusive phenomena make its principles quite fascinating.

The essential principle of DLS is that the intensity at a detector which receives light from a tiny scattering volume measures the instantaneous interference pattern at the detector due to the summation of all scattered electric fields from each scatterer in the scattering volume (see Figure 9.9 and Equation 9.118). Because the particles are in diffusive motion, the intensity of the interference pattern at the detector will keep changing in time. The diffusion coefficient of the scatterers, or an average thereof for disperse scatterers, is connected in a non-obvious way to these intensity fluctuations via the *electric field autocorrelation function*.

In DLS, the scattered light from a very small, illuminated volume of scatterers (the scattering volume) in solution is detected. The scattered intensity will appear to be merely a noise signal, but in fact, its time dependence is directly linked to the diffusion of particles and the subsequent fluctuating interference pattern intensity at the detector. The smaller the scattering volume the greater the fluctuations and the correlation of the signal with diffusive particle motion. The *coherence area* of the photo-detecting surface A_{coh}, the area over which scattered light is coherent, or correlated with itself, is approximated by

$$A_{coh} \sim \frac{\lambda^2}{\Omega} \qquad 9.216$$

where Ω is the solid angle of the source subtended at the detector. The smaller the detector's solid angle, the greater the coherence area.

The means of extracting information from the time dependence of the net scattered field $E(t)$ at the detector is via the *scattered electric field autocorrelation function* $A_E(\tau)$ defined as

$$A_E(\tau) = \lim_{T\to\infty} \frac{1}{T} \int_0^T E(t+\tau)^* E(t)\,dt \qquad 9.217$$

$A_E(\tau)$ can be measured via a beat frequency between the scattered light mixed with a portion of the incident beam. This is often referred to as a *self-beating detection mode*. Such mixing can also occur unintentionally from spurious, stray scattering from glass or plastic scattering cuvettes and their interfaces with air and liquid.

Most available instruments do not use a self-beating mode of detection, and so measure the scattered intensity autocorrelation function $A_I(\tau)$ instead of the scattered electric field autocorrelation function $A_E(\tau)$. The integral is over the total time T of the DLS measurement:

$$A_I(\tau) = \lim_{T\to\infty} \frac{1}{T} \int_0^T I(t) I(t+\tau)\,dt = \langle I(t) I(t+\tau) \rangle \qquad 9.218$$

20 Interestingly, in French, *diffusion de la lumière* refers to "light scattering" even in the single scattering regime, whereas "light diffusion" in English requires intense multiple scattering.

The angle brackets are used as a concise notation for the autocorrelation function, which is a time-averaging procedure. "Infinite T" for a DLS measurement is usually tens of seconds, because the correlation function typically decays on time scales of microseconds to milliseconds. The normalized intensity autocorrelation function is usually termed $g^{(2)}(\tau)$ and is given by

$$g^{(2)}(\tau) = \frac{\langle I(t)I(t+\tau)\rangle}{\langle I(t)\rangle^2} \qquad 9.219$$

The assumption is usually made that fluctuations in small scattering volume elements conform to a normal (Gaussian) distribution, in which case the normalized $A_E(\tau)$, $g^{(1)}(\tau)$ is related to $g^{(2)}(\tau)$ by

$$g^{(2)}(\tau) = B + \beta\left[g^{(1)}(\tau)\right]^2 \qquad 9.220$$

where B is the baseline reached by $g^{(2)}(\tau)$ at the end of a DLS measurement, and β is an instrument factor depending on coherence area of detection, estimated in Equation 9.216, and other factors. When fluctuations in scattered intensity are due to diffusion of identical scatterers, then

$$g^{(1)}(\tau) = \mathrm{e}^{-\Gamma\tau} \qquad 9.221$$

where the exponential decay rate is related to the particle self-diffusion coefficient D_0 (i.e. at very low concentration, where interparticle interactions are negligible) by

$$\Gamma = q^2 D_0 \qquad 9.222$$

q is the magnitude of the scattering wave vector, given by Equation 9.117.

For a disperse population

$$g^{(1)}(\tau) = \int_0^\infty G(\Gamma)\mathrm{e}^{-\Gamma\tau}\mathrm{d}\Gamma \qquad 9.223$$

where $G(\Gamma)$ is the distribution of decay times, related to the molar mass distribution of the scatterers, and their form factors $P(q)$, structure factors $S(q,c)$, and corresponding diffusion coefficients.

Using discrete notation for the concentration c_i, molar mass M_i, $P_i(q)$, D_i for the ith particle in the population $g^{(1)}(r)$ can be expressed as

$$g^{(1)}(\tau,q,c) = \frac{\sum c_i M_i P_i(q)\mathrm{e}^{-q^2 D_i \tau}}{\sum c_i M_i P_i(q)} \qquad 9.224$$

where, in dilute solution $S(q,c) \sim 1$. Equation 9.224 shows that $g^{(1)}(\tau)$ is quite complex and depends not only on c_i, M_i, and D_i of each scattering type but also on the scattering angle via the form factor $P_i(q)$. To eliminate these effects, measurements should be extrapolated to zero scattering angle ($q = 0$) and zero concentration. If measurements are made at a fixed angle, as is the case for most available instruments, and if the scattering population is disperse and there is angular dependence, larger particles will be under-represented in Equation 9.224 because $P(q) < 1$ for these, and a spuriously high value of D_0, and a spuriously low value of the hydrodynamic radius R_H will be reported. As particle size enters the Mie regime, the effects can be enormous, as shown in Figure 9.18.

The cumulant expansion method is widely used to interpret $g^{(1)}(\tau)$ under the assumption that $P(q) = S(q,c) = 1$. However, both $P(q)$ and $S(q,C)$ are involved in the expansion, and $\ln\left[g^{(1)}(\tau)\right]$ about τ gives

$$\ln\left[g^{(1)}(\tau)\right]=1-q^2\frac{\Sigma c_iM_iP_i(q)S_i(q,c)D_i}{\Sigma c_iM_iP_i(q)S_i(q,C)}\tau+q^4\left[\frac{\Sigma D_i^2c_iM_iP_i(q)S_i(q,C)}{\Sigma c_iM_iP_i(q)S_i(q,C)}-\left(\frac{\Sigma D_ic_iM_iP_i(q)S_i(q,C)}{\Sigma c_iM_iP_i(q)S_i(q,C)}\right)^2\right]\frac{\tau^2}{2}+\cdots \quad 9.225$$

Or, in terms of Γ and its averages

$$\ln\left[g^{(1)}(\tau)\right]=-\langle\Gamma\rangle\tau+\left\langle\left(\Gamma-\langle\Gamma\rangle\right)^2\right\rangle\frac{\tau^2}{2}+\cdots \quad 9.226$$

where <Γ> is the intensity averaged value of Γ and $\langle(\Gamma-\langle\Gamma\rangle)^2\rangle$ is the mean square deviation of the decay rate distribution about <Γ>, also termed the *variance*. For diffusive motion, the intensity-weighted averaging yields

$$\langle\Gamma\rangle=q^2\langle D\rangle_z \quad 9.227$$

where $\langle D\rangle_z$ is the *z*-averaged diffusion coefficient:

$$\langle D\rangle_z=\frac{\sum c_iM_iD_i}{\sum c_iM_i} \quad 9.228$$

$\langle D\rangle_z$ is determined by this from $g^{(1)}(\tau)$ and the reported apparent *z*-averaged hydrodynamic radius $\langle R_H\rangle_{z,ap}$ is obtained from the Stokes–Einstein equation, applicable strictly only to spheres. $\langle D\rangle_z$ and *not* the true *z*-average hydrodynamic radius $\langle R_H\rangle_z$, is obtained from $A_E(\tau)$. The relationship between $\langle D\rangle_z$ and the true R_H is

$$\langle D\rangle_z=\frac{k_BT}{6\pi\eta}\left\langle\frac{1}{R_H}\right\rangle_z \quad \text{Relation between Z-averages of Diffusion Coefficient and Hydrodynamic Radius } R_H \quad 9.229$$

$\langle R_H\rangle_{z,ap}$, reported by most commercial instruments, is hence the reciprocal of the *z*-averaged reciprocal hydrodynamic radius (at $q = 0$, $c = 0$):

$$\langle R_H\rangle_{z,ap}(q=0)=\frac{1}{\left\langle\frac{1}{R_{H,}}\right\rangle_z} \quad 9.230$$

The true *z*-average hydrodynamic radius $\langle R_H\rangle_z$ is

$$\langle R_H\rangle_z=\frac{\sum R_{H,i}c_iM_i}{\sum c_iM_i} \quad 9.231$$

$$\langle R_H\rangle_{z,ap}=R_{H,true} \quad \text{For Monodisperse Particles} \quad 9.232$$

If the system is monodisperse then $\langle R_H\rangle_{z,ap}(q=0)$ is equal to the true R_H ($R_H = \langle R_H\rangle_{z,ap}$ in the monodisperse case). Otherwise, in disperse systems, $\langle R_H\rangle_{z,ap}$ can deviate significantly from the true $\langle R_H\rangle_z$. Figure 9.24 shows an example of this relationship of $\langle R_H\rangle_{z,ap}(q=0)$ versus the true $\langle R_H\rangle_z$ using mixtures of spheres of radii 25 nm and 120 nm, as they are mixed together, starting with pure 25 nm spheres and ending with pure 120 nm spheres. The deviations are substantial.

The variance of the particle size distribution in Equation 9.226, its width, allows a dimensionless dispersity parameter, *Q*, to be defined by the quotient of the second order term in the expansion divided by the square of the first term

$$Q=\frac{\langle(D-\langle D\rangle_z)^2\rangle}{2\langle D\rangle_z^{\,2}} \quad 9.233$$

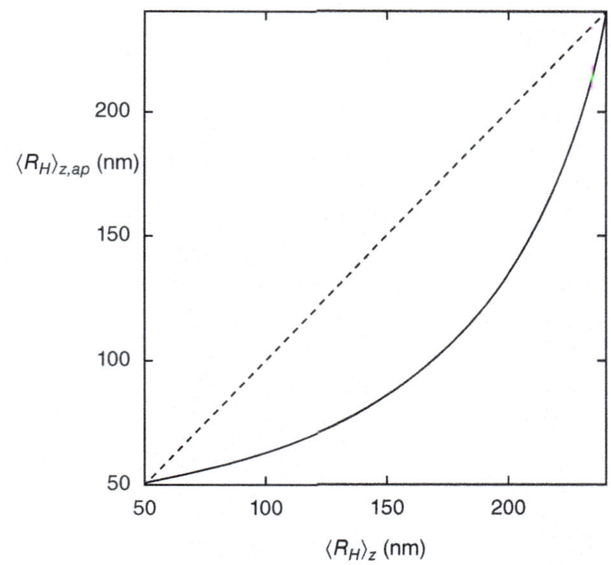

Figure 9.24 $\langle R_H \rangle_{z,ap}$ at $q = 0$ versus the true $\langle R_H \rangle_z$ for a mixture of latex spheres of radii 25 nm and 120 nm. The dashed line is for monodisperse spheres. (Jin, L. et al. (2022) / IOP Publishing / Licensed under CC BY 4.0).

If the population is monodisperse, then $Q = 0$. A particle population is considered fairly monodisperse if $Q < 0.1$, whereas it is considered highly disperse if $Q > 0.3$.

As mentioned, the electric field autocorrelation function can be obtained in a self-beating system, in which the scattered light is mixed with unscattered light from the laser source, thus obtaining a beat frequency. The characteristic exponential decay rate for the homodyne correlation function is $2q^2D$, which is twice the decay rate of the self-beating (E-field) autocorrelation function. Sometimes, there will be a partial self-beating character to the autocorrelation function if unwanted stray light from the incident laser mixes with the scattered light, sometimes termed *accidental homodyning*.

9.12.1 Depolarized Scattering and Rotational Diffusion

Most flexible and semiflexible polymers are optically isotropic and hence do not depolarize incident-polarized light. For particles with optical anisotropy, there will be a tensor polarizability and the scattered light will have depolarized components, due to the induced dipole $\vec{p}$ being in a different direction than the incident E-field:

$$\vec{p} = \bar{\bar{\alpha}} \cdot \vec{E} \tag{9.234}$$

where $\bar{\bar{\alpha}}$ is the anisotropic polarization tensor of the particle.

Besides translational diffusion, anisotropic particles can also undergo *rotational diffusion*. An optically anisotropic particle has a rotational diffusion coefficient Θ which can be measured from the autocorrelation function for depolarized scattered light, e.g. by placing an analyzing polarizer at right angles to the incident electric field polarization in front of a detector in the scattering plane (which is perpendicular to the incident polarization direction). The complete electric field autocorrelation function for the depolarized scattering is $A_{E,VH}(\tau)$ and is proportional to the product of the rotational and translational portions:

$$A_{E,VH}(\tau) \propto \exp\left[-\left(6\Theta + q^2D\right)t\right] \tag{9.235}$$

Equation 9.235 shows that measuring the correlation function over several value of q (i.e. at different angles or incident wavelengths) allows extrapolation to $q = 0$, yielding 6Θ as the intercept and D as the slope.

9.12.2 Electrophoretic Light Scattering

In the force–flux formulation in Chapter 6, the electrophoretic mobility μ_e, arose as the combination of charge and frictional coefficient that determines a charged particle's terminal velocity under an $\vec{E}$ field in a liquid. It is defined as the terminal speed per unit of electric field:

$$\mu_e = \frac{v}{E} = \frac{q}{f} \quad 6.59$$

Electrophoretic light scattering (ELS) measures the particles' velocity v under a known value of E. If the net charge q on the particle is known, then μ_e can be determined. From this, there are several theories that connect μ_e to the zeta-potential discussed in Chapter 5, which is important for assessing the stability of colloid suspensions against aggregation; the higher the zeta-potential, the higher the mutual repulsion among particles and the higher the stability.

ELS uses the Doppler shift of light off of charged particles moving at terminal velocity under the applied $\vec{E}$ field. A pattern of interference fringes is set up in space at the focus of the optical electrophoresis cell, by splitting a laser beam and then recombining the beams, which are coherent, with each other. As the particles move through the pattern, the scattered light intensity will fluctuate with a beat frequency directly related to the particles' velocity. This falls under the general technique of *laser Doppler velocimetry*. An even more sensitive means of determining the ζ-potential is via *phase angle light scattering*, where phase shifts of charged, moving particles are measured. A particle need only move a fraction of its diameter in order to produce a measurable phase shift.

9.13 X-ray Diffraction and Crystallography

Wilhelm Röntgen in 1895, accidentally discovered the penetrating power of X-rays, which he first produced chiefly by *Bremsstrahlung* ("braking radiation" produced from electrons decelerating and crashing into metal targets). Within a year, X-rays were employed in medical diagnostics. In the early 1900s, Bragg posited his simple law for X-ray diffraction. It immediately allowed the structure and spacing of atomic and molecular solids to be discovered. Since then, X-ray diffraction has been vital in the discovery of the double-helix nature of DNA and in deciphering the complex spatial structures of proteins. Crystallizing proteins and obtaining their 3D structures continues to be an important endeavor in analytical protein science.

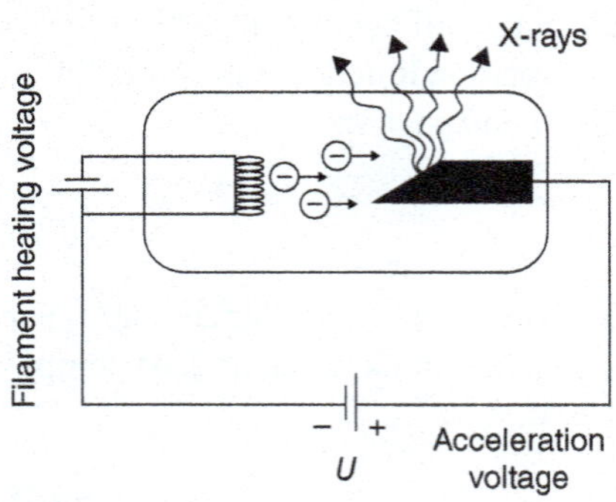

Figure 9.25 Schematic of a vacuum device for producing X-rays via deceleration of high-speed electrons colliding with a metal target. (Courtesy of Marissa Calko).

A schematic of the Bremsstrahlung apparatus is shown in Figure 9.25, as well as a graph of X-ray intensity emitted versus X-ray energy in Figure 9.26. In that figure, the smooth portion of the curve is due to Bremsstrahlung, an effect predicted by classical electromagnetic theory, the spikes are due to the quantized levels of inner electrons in the atomic cores of the copper target, and the highest energy frequency occurs when all the incident electron kinetic energy produces a single photon of $eV = hf_{max} = KE_{electron}$, where V is the accelerating voltage of the electron.

The Bragg reflection equation for interference maxima of X-rays reflecting from a crystal surface is $m\lambda = 2d\sin\theta$ The geometry for this is shown in Figure 9.27. A DNA diffractogram of DNA is shown in Figure 9.28 from Rosalind Franklin in her work with Watson and Crick.

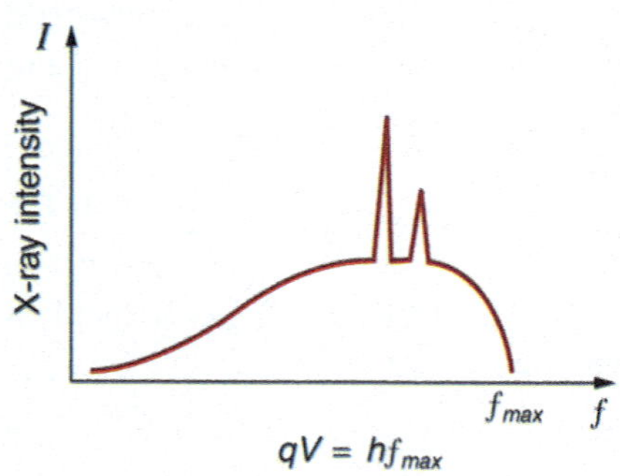

Figure 9.26 Features of the X-ray emission spectrum for 50 keV electrons incident on a copper target.

9.14 Raman Scattering

Unlike the elastic scattering treated in the various approximations above, *Raman scattering* is an inelastic scattering phenomenon from molecules, first discovered by Sir Chandrasekhara Venkata Raman (India, 1888–1970) and his student Kariamanickam Srinivasa Krishnan (India, 1898–1961). Whereas in elastic scattering, the scattered photons have virtually the same energy and frequency as the incident photons, in the Raman effect a portion of the incident photon energy is transferred to vibrational levels of the molecule, sometimes also the rotational states, so that the scattered light has less energy and lower frequency than the incident light. This process is *normal Stokes–Raman*

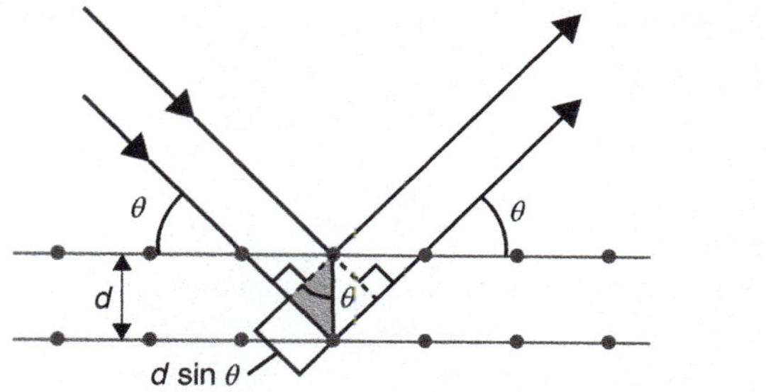

Figure 9.27 Bragg diffraction geometry.

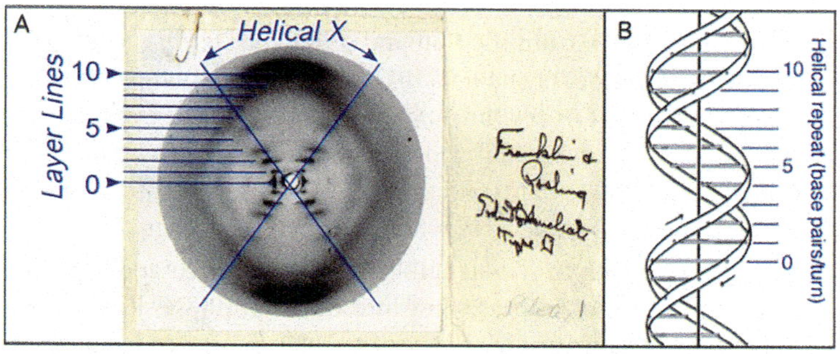

Figure 9.28 Original X-ray diffractogram of DNA. (Courtesy of P. Shing Ho, Colorado State University).

scattering. The cross-section for a Raman scattering event is usually only a very small fraction of the probability of an elastic scattering event, requiring specially adapted detection for performing Raman spectroscopy. With much less probability, the photon may also absorb vibrational energy so that the scattered photon has higher energy and frequency than the incident photon. In this latter case, the process is termed *anti-Stokes scattering*.

Raman scattering involves frequency shifts generally in the infra-red regime. This is the same regime where vibrational levels can directly absorb and emit infra-red photons, with corresponding *quantum mechanical selection rules*. Raman scattering is complementary to infra-red absorption, as different quantum mechanical selection rules are involved. Its use is widespread in the identification and analysis of vibrational spectra in biological molecules because water absorbs strongly in the IR, but not in absorption bands accessible by Raman.

9.15 Optical Activity

Because of the transverse nature of electromagnetic waves, they can be polarized, described by using two perpendicular axes in the plane perpendicular to their direction of propagation. The most general state of polarization is elliptical, which can be expressed for a plane wave traveling in the $+z$-direction as

$$\vec{E}(z,t) = E_{0x}e^{i(kz-\omega t)}\hat{\mathrm{i}} + E_{0y}e^{i(kz-\omega t+\varepsilon)}\hat{\mathrm{j}} \quad 9.236$$

where E_{0x} and E_{0y} are the magnitudes of the E-field components in the x and y directions, respectively, and ε is the phase difference between the components. For $\varepsilon = 0$ and 180°, the wave is linearly polarized. When $E_{0x} = E_{0y}$ and E_{0x} leads E_{0y} by 90°, the wave is left-hand circularly polarized, and when E_{0x} is less than E_{0y} by 90°, the wave is right-hand circularly polarized. Combining equal amounts of right- and left-hand circularly polarized light produces a linearly polarized wave.

As discussed in Chapter 7.3, molecules with chiral centers are optically active, and rotate the plane of light. Depending on the asymmetry of the molecule, the linearly polarized light can be rotated to the right (dextrotatory, or simply *D*), or to the left (levorotatory, or *L*). The amount a chiral molecule rotates the plane of polarization is called the *specific rotation* $[\alpha]$

and is normally given in terms of angle of rotation per unit of concentration, c, of the molecule in solution, per unit of path length. Concentration units are usually either in mass, e.g. g/cm^3, or molar, e.g. mol/cm^3. The total rotation of the polarization plane, α, traveling over a distance l in solution is then

$$\alpha = [\alpha]cl \tag{9.237}$$

A *polarimeter* is designed to measure α. If $[\alpha]$ is known, then the concentration of the optically active molecule in a solution can be determined. This is a particularly valuable method when a chiral molecule of interest exists among many other nonchiral molecules and ions because α is sensitive only to the chiral molecule's concentration. An example is measuring the content of a sugar or polysaccharide in a solution containing different types of nonchiral salt ions.

A *birefringent* material has two different optical axes in it, an ordinary and extraordinary axis, each with a different refractive index n_0 and n_e, respectively. A wave that enters the material with a component of $\vec{E}$ perpendicular to each axis will accumulate a phase difference ε between the two components over a pathlength l, given by

$$\varepsilon = \frac{2\pi}{\lambda_0}(n_e - n_0)l \tag{9.238}$$

Wave plates are made from birefringent materials and can be tuned to a given wavelength to turn linear polarized light into right- or left-hand circularly polarized light – known as *quarter-wave plates*, because they introduce 90° phase shifts, or one quarter of a full 360° – or into any other elliptical polarization state. Because birefringence can also be produced by mechanically straining materials that are not inherently birefringent, *photoelastic modulators* use the piezoelectric effect to oscillate a crystal between two strain states that can cause linearly polarized input light to exit from the modulator as alternately left- and right-hand polarized light, typically at high frequencies, on the order of 10^5 Hz.

Photoelastic modulators can hence be used to measure *circular dichroism* (*CD*). A dichroic material preferentially absorbs light over a range of frequencies more along one axis than another. A circular dichroic material absorbs one handed circular polarization state more than the other over a range of frequencies. Optically active molecules display CD when in solution, and because most biological molecules are chiral, they will have circular dichroic behavior. Of great importance is the fact that secondary structures, such as protein α-helices and the helices of RNA and DNA produce circular dichroic effects of their own. The use of photoelastic modulators to alternately produce right and left-hand circular polarized light is at the base of circular dichroism spectroscopy, which has wide-ranging application in characterizing proteins, including the way they gain or lose secondary structure due to changing conditions such as temperature or addition of certain chemical species to a solution of proteins, e.g. urea or guanidine hydrochloride.

9.16 Superconducting Quantum Interference Devices (SQUIDs)

These devices (coupled *Josephson junctions*) make use of the coherence properties of electrons in the superconducting state to detect miniscule magnetic fields. Applications include the use of SQUIDs to follow brain activity, registered by the myriad tiny magnetic fields caused by charge conduction in neural tissues. Magnetic fields as low as 10^{-14} T have been measured by SQUIDs. To get an appreciation of how small this is, the Earth's magnetic field, which is relatively weak, is on the order of 10^{-5} T, so that hypersensitive biomagnetic measurements normally must be conducted in magnetically shielded rooms, to eliminate "ambient magnetic noise." In contrast, the highest continuous magnetic fields produced are on the order of 10s of T.

Weak magnetic fields are produced by synchronized periodic neuronal currents. The magnetic field associated with the cortical activity of the brain is only around 10^{-14} T, and so challenges the limits of sensitivity of SQUID detection. Collective neural oscillations in the human alpha rhythm, at about 8–12 Hz, can reach magnetic field strength of 10^{-12} T. *Magnetoenchephalography* uses SQUIDs and related magnetometers for neural imaging. This has uses in basic research into brain functions but also has practical applications in medicine and surgery for localizing neuro-abnormalities.

9.17 Antimatter and PET Imaging

In extending quantum mechanics into the realm of Einstein's special relativity, Dirac used the form of total relativistic energy E for a particle, given by

$$E^2 = p^2c^2 + m^2c^4 \tag{9.239}$$

where c is the speed of light and $\vec{p}$ is relativistic momentum, given by

$$\vec{p} = \frac{m\vec{u}}{\sqrt{1 - u^2/c^2}} \tag{9.240}$$

where m is the particle's rest mass and $\vec{u}$ is its velocity in the observer reference frame. The square root of E^2 yields both positive and negative free particle energies. This led Dirac to postulate that *antimatter* exists in a "sea" around us in the negative energy states, and is hence normally undetectable. In certain circumstances, however, such as the absorption of a gamma ray of energy equal to or greater than 1.02 MeV[21] (mega-electronvolts) in the vicinity of a nucleus, an electron with negative energy can be energized into our positive energy world (pair production), and this electron's absence in the negative energy sea is the antiparticle; the *positron*, often represented as e^+. The positron has the same mass as the electron but opposite charge. The energized electron can eventually "fall back" into the hole it left – the electron and the positron annihilate each other – releasing the equivalent of the energy of two electrons in the form of gamma rays: the *pair annihilation* phenomenon. In order to conserve momentum in the process two gamma rays are released in nearly opposite directions, making it possible to trace back the point in space of the annihilation process.

Positive beta decay is used to provide the positrons for PET scans. This occurs from proton-rich radioactive nuclei, which shed a positron and electron neutrino, ν_e, along with kinetic energy of the fragments, leaving a neutron in the nucleus and lowering its atomic number, Z, by one. The positive beta decay of a radioactive nucleus X_Z^A into a daughter nucleus Y_{Z-1}^A, where A is the atomic number is represented as

$$^A_ZX \rightarrow {}^{\;\;A}_{Z-1}Y + e^+ + \nu_e \tag{9.241}$$

Positive beta decay is fairly rare on Earth so that beta-emitters for PET scans are produced artificially in cyclotrons. A list of some common positron-emitting radionuclides is shown in Table 9.1. These are generally short-lived so that PET facilities are usually close to a cyclotron.

The selected positron emitter is incorporated into a chosen compound to create a *radiotracer* which is then injected into the body. Examples of positron-emitting radiotracers include ^{18}F-fluorodeoxyglucose, which marks glucose uptake in tissues – e.g. brain, heart – and can hence reveal tissues with high metabolic activity, including disturbances due to cancer and other diseases. Na^{19}F is rapidly taken up by bone and subsequently rapidly cleared by blood flow, making it an effective method for skeletal imaging. There is a rapidly expanding number of radiotracers, many specific to highly specialized tissues, and others that bind to specific molecules, such as neuroreceptors. PET imaging is also used in guiding certain surgeries, especially neurosurgeries. The range of metabolic and physiological phenomena that can be monitored by PET imaging, along with improved instrumentation ensures that the field will grow significantly.

Table 9.1 Half-lives of some positron emitters.

Isotope	^{82}Rb	^{15}O	^{13}N	^{11}C	^{68}Ga	^{18}F	^{64}Cu	^{55}Co
Half-life (min)	1.3	2	10	20	68	110	762	1050

21 The rest mass energy equivalent of the electron is 0.51 MeV.

9.18 Electron Microscopy

Louis de Broglie postulated the particle/wave dual nature of electrons in 1924. He suspected electrons had a wave nature because of the quantized electron energy levels in atoms. He reasoned that the only known physical phenomena that involve integers are standing waves. Since the electron energy levels are discrete and related by integers, he concluded that the electrons form closed-loop standing waves in atoms. The Davisson–Germer experiment in 1927 accidentally proved the wave nature of electrons in a serendipitous experiment, confirming the particle/wave duality hypothesis of de Broglie made just a few years earlier. This was in the formative period of developing the basic notions of the wave function Ψ and quantum mechanics. While the Davisson–Germer experiment put quantum mechanics on a solid experimental basis, the electron microscope developed as a profoundly important practical application shortly after the discovery. Its continuous development has revolutionized biology, medicine, and material sciences.

A schematic of the Davisson–Germer experiment is shown in Figure 9.29a, as well as an electron micrograph of red blood cells in Figure 9.29b.

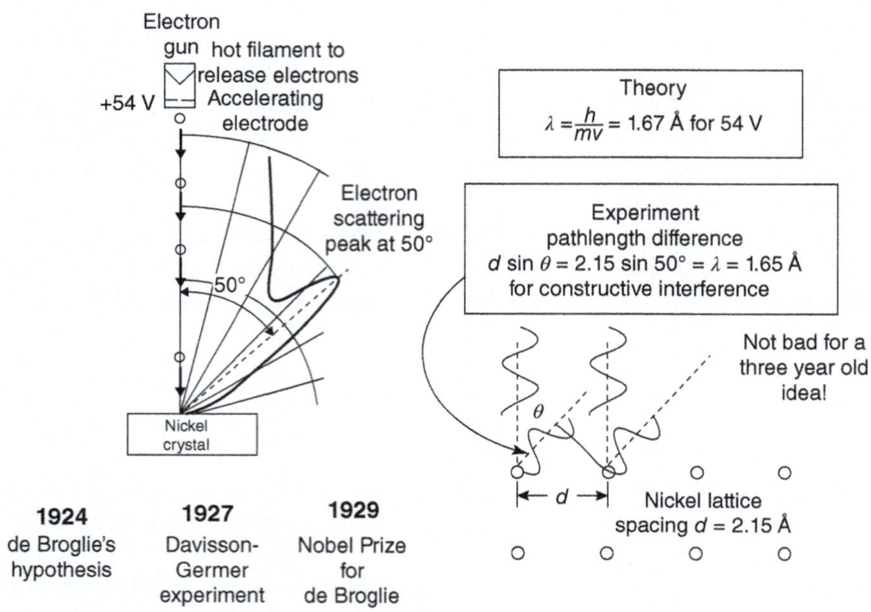

Figure 9.29a Schematic of apparatus and computations for Davisson–Germer experiment. (With permission from HyperPhysics by Rod Nave, Georgia State University).

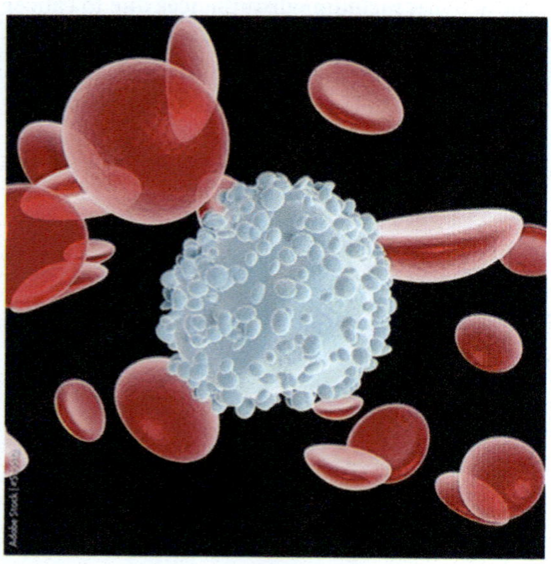

Figure 9.29b Electron micrograph of red and white blood cells. (Yang MingQi/Adobe Stock Photos).

9.19 Summary

Starting with the Maxwell equations in vacuum, it was shown that classical electromagnetic waves can propagate through the vacuum with a speed given by the electrostatic and magnetostatic constants, $c = \frac{1}{\sqrt{\varepsilon_0 \mu_o}} = 3 \times 10^8$ m/s. The Maxwell equations give all the relationships between the magnitudes and directions of the $\vec{E} - field$ and $\vec{B} - field$, the equal energy contained in each field, and the intensity of the propagating wave given by the Poynting vector $\vec{S}$.

The major classical result that an accelerated charge emits electromagnetic waves was used as a starting point for the e-m waves emitted by an oscillating dipole, which is the quintessential model for electric dipole radiation. Rayleigh scattering of a polarizable dielectric particle whose size is small compared to the wavelength of incident light follows immediately from this model, as does scattering from a dilute, noninteracting collection of particles, e.g. gas molecules in the Earth's upper atmosphere. The Rayleigh ratio was introduced, by dividing out the dependence on incident wave intensity and multiplying out the squared distance from the scatterer, in order to obtain the fraction of incident light scattered per centimeter of pathlength per solid angle in a medium. The conversion from dependence on a single particle's polarizability to the index of refraction of a medium was made via the Clausius–Mossotti equation and yields a practical means of quantifying light scattering.

The effects on light scattering from interacting scatterers were based on microscopic fluctuations of particle density and index of refraction, and the thermodynamic energy cost of these fluctuations. The scattering intensity is directly proportional to the osmotic compressibility of a system of scatterers: the stronger the repulsive interactions among particles, the smaller the fluctuations and the lower the intensity of scattered light, e.g. adding simple salt to a polyelectrolyte solution provides electrostatic shielding, less repulsive interaction, greater fluctuations, and more light scattering.

As the size of particles becomes appreciable compared to the wavelength of incident light, the effects of intraparticle interference must be taken into account. A generalized scattering formulation was presented involving the momentum transfer vector, $\vec{q}$, and various approximations involving $\vec{q}$, and the particle's shape and dimensions were derived for the particle form factor $P(q)$. Under several approximations, the Zimm method was shown to be robust for analyzing scattering from polymer solutions, providing measurements of weight average molar mass, M_w, z-average mean square radius of gyration $\langle S^2 \rangle_z$, and a complex average of the second virial coefficient $\langle A_2 \rangle$. In some instances, averages of the third and fourth virial coefficients $\langle A_3 \rangle$ and $\langle A_4 \rangle$ can also be measured.

Interparticle correlations are covered in the limit of $q = 0$ thermodynamically and involve the virial coefficients. These correlations also lead to solution structure factors $S(q)$, which sometimes are non-monotonic and have peaks.

For solid particles, the simplest approximations break down at fairly modest sizes, typically several tens of nanometers, and the Mie scattering formalism must be used.

Scattering at interfaces was handled by applying boundary conditions to $\vec{E}$ and $\vec{B}$ for polarization states parallel and perpendicular to the plane of incidence, whence the Fresnel equations resulted, giving transmission and reflection amplitudes for incident waves on a boundary. This leads to the existence of evanescent waves in total internal reflections and also helps to understand the reflectivity of metals.

Dynamic light scattering was presented as the time autocorrelation of intensity fluctuations at the photodetector due to diffusive motion of the scatterers and a time-changing interference pattern, and hence time-changing intensity at the detector. The diffusion coefficients of particles can be determined by DLS, and, under certain shape assumptions hydrodynamic models relate diffusion coefficient to characteristic dimensions, such as the equivalent hydrodynamic diameter of a sphere.

Several results from quantum mechanics have developed into useful methods, including the wave nature of particles underlying the principle of electron microscopy, barrier tunneling for certain microscopies, quantum mechanical selection rules for the interpretation of Raman scattering, and the use of antimatter in PET scan technology.

References

Berne, B.J. and Pecora, R. (1976). *Dynamic Light Scattering*. John Wiley & Sons.

Cassassa, E.F. (1962). Effect of heterogeneity in molecular weight on the second virial coefficient of polymers in good solvents. *Polymer* 3: 625–638.

Chu, B. (1991). *Laser Light Scattering; Basic Principles and Practice*. Academic Press.

Enohnyaket, P., Kreft, T., Alb, A.M., Drenski, M.F., and Reed, W.F. (2007). Determination of molecular mass during online monitoring of copolymerization reactions. *Macromolecules* 40: 8040–8049.

Hecht, E. (2002). *Optics*, 4e. Addison Wesley.

Jackson, J.D. (1998). *Classical Electrodynamics*, 3e. Wiley.

Jin, L., Jarand, C.W. et al. (2022). Angle-dependent effects in DLS measurements of polydisperse particles. *Measurement Science and Technology* 33 (4): IOP Publishing, Ltd.

Kerker, M. (1969). *The Scattering of Light and Other Electromagnetic Radiation*. Academic Press.

Li, X. and Reed, W.F. (1991). Polyelectrolyte properties of proteoglycan monomers. *Journal of Chemical Physics* 94 (6): AIP Publishing.

McMillan, W.G., Jr. and Mayer, J.E. (1945). The statistical thermodynamics of multicomponent systems. *Journal of Chemical Physics* 13: 276.

Pollack, G.L. and Stump, D.R. (2002). *Electromagnetism*. Addison Wesley.

Stockmayer, W.H., Moore Jr., L.D., Fixman, M., and Epstein, B.N. (1955). Copolymers in dilute solution. I. Preliminary results for styrene-methyl methacrylate. *Journal of Polymer Science* 16: 517.

van de Hulst, H.C. (1981). *Light Scattering by Small Particles*. Dover Publications.

Zimm, B.H. (1948). *Journal of Chemical Physics* 16 (12): 1093–1115.

Further Reading

Ando, J. (2022). Scattering imaging of biomolecules with metallic nanoparticles: localization precision, imaging speed, and multicolor imaging capability. *Optical Review* 29 (4): 358-365.

Girolami, G.S. (2015). *X-Ray Crystallography*. University Science Books.

Han, C.C. and Akcasu, A.Z. (2011). *Scattering and Dynamics of Polymers: Seeking Order in Disordered Systems*. Wiley.

Liu, J., Kuang, W., Liu, J. et al. (2022). In-situ multi-phase flow imaging for particle dynamic tracking and characterization: advances and applications. *Chemical Engineering Journal (Amsterdam, Netherlands)*.

Podzimek, S. (2015). *Light Scattering, Size Exclusion Chromatography and Asymmetric Field Flow Fractionation*. Wiley.

Sivia, D.S. (2011). *Elementary Scattering Theory for X-Ray and Neutron Users*. Oxford University Press.

Smith, J.C., Tan, P., Petridis, L., and Hong, L. (2018). Dynamic neutron scattering by biological systems. *Annual Review of Biophysics* 47: 335–354.

Some, D. and Razinkov, V. (2019). High-throughput analytical light scattering for protein quality control and characterization. *Methods in Molecular Biology (New York, NY, United States)* 2025: 335–359.

Tolan, M. (2013). *X-Ray Scattering from Soft-Matter Thin Films: Materials Science and Basic Research*. Springer.

Wahl, R.L. and Beanlands, R.S.B. (2008). *Principles and Practice of PET and PET/CT*. Lippincott, Williams & Wilkins (Wolters Kluwer).

Warren, B.E. (1990). *X-Ray Diffraction*. Dover Publications.

Weber, W.H. (2000). *Raman Scattering in Materials Science*. Springer.

Wilson, A.J., Devasia, D., and Jain, P.K. (2020). Nanoscale optical imaging in chemistry. *Chemical Society Reviews* 49(16), 6087-6112.

Problems

1 a) Compute the ratio of scattered intensity of blue light of $\lambda = 400$ nm incident, an atom to that of red light of $\lambda = 700$ nm.
b) Compute the ratio of scattered light intensity from a scatterer at 0.5 m away from the scatterer to the intensity at 2.5 m.

2 Show that the Maxwell equations in vacuum predict the existence of electromagnetic waves.

3 Starting with the generic plane wave

$$\vec{E}_0 e^{i(kz-\omega t)}, \text{where } \vec{E}_0 = E_{0x}\ \hat{\imath} + E_{0y}\ \hat{\jmath} + E_{0x}\ \hat{k}$$

use the Maxwell equations, to prove the following properties of e-m waves in a vacuum:
a) The electromagnetic wave is a *transverse* wave, i.e. the displacement of the fields is perpendicular to the direction of travel.
b) $\vec{B}$ and $\vec{k}$ are all mutually perpendicular.
c) $\vec{E} \times \vec{B} \parallel \vec{k}$

d) $E = B$

e) $u_E = u_B = \dfrac{E^2}{8\pi} = \dfrac{B^2}{8\pi}$

f) $\vec{S} = \dfrac{c}{4\pi}\vec{E} \times \vec{B}$

g) $\langle S \rangle = \dfrac{cE_0^2}{8\pi}$

4 Using Equation 9.30, $P_s = \oint_S \vec{S} \cdot d\vec{A}$ perform the integration to find P_s given in Equation 9.31.

5 a) Find the attenuation length in air for light of λ = 500 nm. n_{air} = 1.000278. Treat the air as an ideal gas at STP and ignore any interference effects due to the particle density.
b) Find the attenuation length for toluene, for which the absolute Rayleigh ratio is 3.59×10^{-5} cm^{-1} at λ = 500 nm.

6 a) Calculate the Rayleigh scattering ratio R, for air at 370 nm, assuming that air is composed principally of N_2 that can be treated as an ideal gas at STP (T = 300 K, P = 1 atm). n_{air} = 1.000278.
b) The Rayleigh ratio for toluene at T = 25°C is 1.40×10^{-5} cm^{-1} for 633 nm light vertically polarized light. Calculate the Rayleigh ratio for toluene at 350 nm and T = 25°C. Include Zimm's correction for index of refraction given in Footnote 12.

7 Consider two point-like scattering particles, as shown, separated by d = 0.005 cm. Coherent light of wavelength λ = 650 nm is incident on both. Both particles are immersed in the same medium (i.e. the same index of refraction, n):
a) Find the phase difference between scattered light arriving at point P from the two particles. Point P is L = 1 cm away from particle 1 on the x-axis.

b) At what position along the x-axis could you move particle 1 to have constructive interference at P?

8 Show that for light normally incident on a material with complex index of refraction

$$\tilde{n} = n_R + in_I$$

the reflection coefficient (fraction of incident power reflected) is given by

$$R = \frac{(n_R - 1)^2 + n_I^2}{(n_R + 1)^2 + n_I^2}$$

9 Explicitly compute the integral:

$$\langle e^{-i\vec{q}\cdot\vec{r}_{ij}} \rangle = \frac{\int_0^\pi \sin\alpha e^{-iqr_{ij}\cos\alpha} d\alpha}{\int_0^\pi \sin\alpha d\alpha} \quad 9.137$$

10 Using the definition of the mean square radius of gyration $<S^2>$:

$$\langle S^2 \rangle = \frac{1}{N}\left\langle \sum r_i^{\,2} \right\rangle \quad 9.146a$$

And returning to $\vec{r}_{ij} = \vec{r}_j - \vec{r}_i$, show that

$$P(q) = 1 - \frac{q^2}{N^2}\left\langle \sum\sum r_{ij}^{\,2} \right\rangle + \cdots$$

can be represented as

$$P(q) = 1 - \frac{q^2\langle S^2 \rangle}{3}, \quad \text{when } q^2\langle S^2 \rangle << 1$$

11 A certain scatterer in water ($n_w = 1.33$) has $<S^2>^{1/2} = 80$ nm. Assuming the low-angle approximation, $P^{-1}(q) = 1 + q^2\langle S^2\rangle_z / 3$ is valid up to $q^2<S^2> \sim 1$, find the angular range for which the approximation is valid for light of

a) 633 nm
b) 488 nm

12 Consider the form factor $P(u)$ for an ideal random coil $P(u) = \frac{2}{u^2}\left(e^{-u} - 1 + u\right)$ where $u = q^2<S^2>$

a) Find the limiting form of $P(u)$ for $u \ll 1$. Comment on this.
b) Demonstrate that the limiting form for $u \gg 1$ is $P(u) \cong \frac{2}{u} - \frac{2}{u^2}$
c) For a disperse population of polymers with a distribution $C(M)dM$ (which is the concentration of polymer in the mass interval from M to $M + dM$), and total concentration c, find

$$\frac{Kc}{I_R(u)} = \frac{c}{\int_0^\infty MC(M)P(u)dM}$$

Use the fact that for an ideal random coil $<S^2> = \gamma M$, where γ is a constant.

13 Random degradation of a polymer chain. Using the limit from part 12(b)

$$P(u') \approx 2\left(\frac{1}{u'} - \frac{1}{u'^2}\right)$$

and the scaling law

$$\langle S^2\rangle = \gamma M$$

and the number of random cuts made per initial polymer chain r,

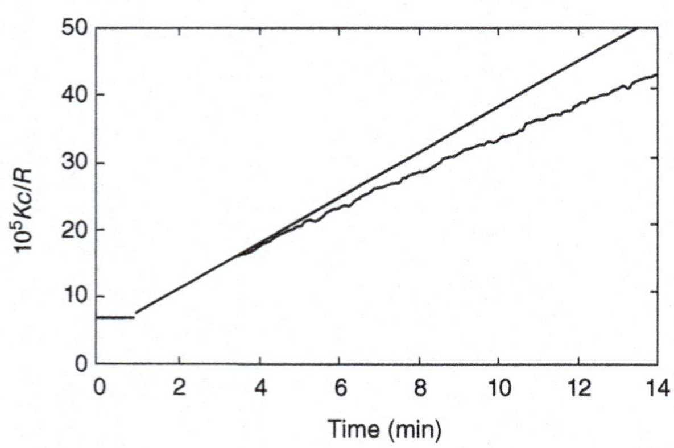

$$r = \dot{\beta} M t$$

where $\dot{\beta}$ is the number of random cuts made per second per unit of polymer mass M, show that

$$u' = M(\gamma + \dot{\beta} t)$$

$$\frac{Kc}{R(u, c)} = \frac{1}{\int MC(M)P(M)dM}$$

becomes, to second order in u,

$$\frac{Kc}{R(q,t)} = \frac{1}{2M_n} + \frac{\gamma q^2}{2} + \frac{\dot{\beta} t}{2}$$

Using this, find $\dot{\beta}$ for the endocytic enzymatic degradation of hyaluronate by hyaluronidase, shown in the figure.

14 Using Zimm's single contact approximation result:

$$R(q) = Kc\left[MP(q) - 2A_2M^2P(q)^2c\right]$$

explicitly show that measurement of the total light scattered from a disperse population of polymers yields M_w and $<S^2>_z$, hence yielding

$$\frac{Kc}{R(q)} = \frac{1}{M_w}\left(1 + \frac{q^2\langle S^2\rangle_z}{3}\right) + 2\langle A_2\rangle c$$

Assume that each mass M_i in the polymer population has a concentration c_i. You can use the low q form factor:

$$P(q) \cong 1 - \frac{q^2\langle S^2\rangle}{3}$$

15 Consider a monodisperse biopolymer of initial mass M_0 that is degraded by an exocytic enzyme, i.e. end groups of the polymer are removed sequentially at a constant rate α (end groups removed per polymer per second), i.e. $\frac{dM(t)}{dt} = -\alpha m$, where m is the mass of an end group, and $M(t)$ is the mass of the remaining polymer as end groups are removed.

a) Use the Zimm equation to calculate the time dependence of the Rayleigh scattering ratio in the limit $q = 0$. Assume the solution of polymer is dilute enough that the $2A_2c$ term can be ignored and that the scattering from removed end groups is negligible compared to the scattering from the remaining polymer. Also, assume the polymer mass is related to $<S^2>$ via $<S^2> = AM$, where A is a constant, and $q^2<S^2>(t = 0) = 0.8$.

b) Make a sketch of $Kc_0/R(t)$ versus t, where c_0 is the initial concentration of polymer $(\text{g/cm})^3$ in solution.

16 Consider an ideal polymerization reaction in a reactor in which the instantaneous weight-average mass $M_{w,inst}(t)$ is given by $M_{w,inst}(t) = 1.5k_pc_m(t)/k_tR$, where $c_m(t)$ is the monomer concentration, given by $c_m(t) = c_m(0)\exp(-k_pRt)$ where R is the concentration of free radical, *assumed to be constant at all time during the polymerization reactions*, and k_p and k_t are the propagation and termination rate constants, respectively.

a) A continuous light scattering measurement of the process will measure $M_w(t)$ of the *cumulative* population of "dead" polymer chains in the reactor. Find this cumulative value $M_w(t)$.

b) Assuming that $<S^2> = AM$ is the relation between mean square radius of gyration and polymer mass M (A is a constant), find the following slope:

$d\left(Kc/R(q,t)\right)/dq^2$ (using the Zimm equation, where $R(q,t)$ is the excess Rayleigh scattering ratio at scattering vector q and time t). Relate this directly to how the dispersity index M_z/M_n evolves in time.

17 Given the normalized electric field autocorrelation function, weighted by intensity, over all the diffusion coefficients in a population:

$$g^{(1)}(\tau) = \frac{\sum I_i e^{-q^2D_i\tau}}{\sum I_i}$$

I_i can be represented by

$$I_i = Kc_iM_iP_i(q)$$

Expand $\ln\left[g^{(1)}(\tau)\right]$ in time around $\tau = 0$ to find

$$\ln\left[g^{(1)}(\tau)\right] = 1 - q^2\langle D\rangle\tau + q^4\langle(D-\langle D\rangle)^2\rangle\frac{\tau^2}{2!} - q^6\langle(D-\langle D\rangle)^3\rangle\frac{\tau^3}{3!} + q^8\langle(D-\langle D\rangle)^4\rangle\frac{\tau^4}{4!}$$

18 With the mass of a proteoglycan of $M = 1.6 \times 10^6$ g/mol and $c = 10^{-4}$ g/cm^3 use the PG data of Figure 9.17 to compute the spacing of an equivalent lattice, using the Bragg formula at the peak. Compare this with the average spacing among particles. Comment on the differences.

19 It is commonly found that, in a given solvent, $|\partial n/\partial c|$ is larger for the polymer than for the corresponding monomer. Give an explanation for this using the model for index of refraction given in 9.188–9.190.

20 For DLS, carry out the intensity-weighted averaging of Γ to show that $<D>_z$ results from the DLS measurement.

21 In the following, use experimental light data to compute M_w and A_2 using the angle-independent Zimm equation:

$$\frac{Kc}{R(q=0,c)} = \frac{1}{M_w} + 2A_2c$$

Light scattering measurements were made on a protein dissolved in an aqueous solution ($n_{solvent} = 1.333$), at various concentrations. Necessary data are given in Table 9.2. The scattering voltage at each concentration is shown in Table 9.3. Because the protein is small (8 nm), it can be treated as a Rayleigh scatterer (no angular dependence of scattering). The voltage measured for pure toluene was $V_{tol} = 2.2$ V, for the pure solvent $V_{solvent} = 0.3$ V, and $V_{dark} = 0$. Take the optical factor $F = 1$ in the Rayleigh ratio computation. Use $\partial n / \partial c = 0.189 \text{cm}^3/\text{g}$. The vacuum wavelength of the laser was $\lambda_0 = 633$ nm.

$R(c)$, the absolute Rayleigh ratio, is computed by

$$R(c)\ (\text{cm}^{-1}) = \frac{V(c) - V_{solvent}}{V_{toluene} - V_{dark}} 1.40 \times 10^{-5}$$

(1.40×10^{-5} is R for toluene at $T = 25°\text{C}$, $\lambda = 6.33 \times 10^{-5}$ cm), $K = \dfrac{4\pi^2 n^2 (\text{d}n/\text{d}c)^2}{N_A \lambda^4}$, where $n = 1.454$, $\text{d}n/\text{d}c = 0.106$ (cm^3/g), $\lambda = 6.33 \times 10^{-5}$ cm.

N_A is Avogadro's number.

a) Construct a plot of $Kc/R(q = 0,c)$ versus protein concentration.
b) Determine M_w and A_2 from the above plot. Make sure to give units in your answers!

Table 9.2 Needed auxiliary information.

R, toluene	1.06×10^{-5}	cm^{-1}
$V_{toluene}$	2.1	Volts
$V_{solvent}$	0.3	Volts
V_{dark}	0	Volts
$\text{d}n/\text{d}c$	0.189	cm^3/g
λ	677	nm
$n_{solvent}$	1.333	dimensionless

Table 9.3 Scattering voltage at each protein concentration.

Protein concentration (g/cm^3)	Scattering voltage
0.0001	0.56
0.0002	0.81
0.0003	1.05
0.0005	1.5
0.00075	2
0.0009	2.3
0.001	2.45
0.00125	2.9
0.0015	3.2
0.00175	3.58
0.002	3.8

Part IV

Examples of Specific Living State Phenomena

Quien mucho abarca poco aprieta.

Spanish Proverb[1]

It is estimated that there have been at least 10 million species on earth, and potentially orders of magnitude more. The diversity is mind-boggling. Species crowd the land and water, and others have taken to the sky. Length scales run from submicron to hundreds of meters and more. If one counts fungi, there is a single-colony honey fungus in Oregon that covers 10 million square meters, or around 4 square miles. In 2022, it was announced that a 180-km long stretch of a seagrass meadow, in Shark Bay off the Western Australian coast, arose from a single organism that has continued to clone itself for the past 4500 years.

Countless different strategies for survival and procreation have evolved, some involving nurture of a few offspring (e.g. humans) and others using shear number of offspring (e.g. frogs, seahorses). Food chains and species cycles are complex. All species need a carbon source in their diet. Plants use carbon dioxide and light to energize their metabolic cycles and become the producer organisms in a chain involving insects, fish, birds, and mammals. Other species decompose detritus from dead organisms.

Environments spawn an amazing variety of ecologies, interconnecting webs of species. Fungi hold a pre-eminent position in this regard, weaving together the roots and plants of trees and creating dynamic fungal biomes in almost every species of higher organism. Flowers are pollinated by bees and other animals. Birds and land animals disperse the seeds of plants. Myriad symbioses among species exist, some beneficial, others parasitic, and others that produce neither mutual benefit nor harm.

Socialization, relationships between individuals of the same or different species, reaches high levels in insects, bees, birds, and mammals. Besides humans and our diverse and elaborate language systems, many other species have their own languages and sounds for communication; many mammals, including whales and dolphins, and birds. Communication also occurs without sound. Some animals, such as bees, communicate directions via dancelike movements. Chemical signaling via olfaction and pheromones is used by both plants and animals. Complex webs of chemical signaling occur in the soil and in the water.

Neurocomplexity in humans has evolved to the point where we manipulate the environment on a large scale, manufacture amazingly complex devices and materials, speak highly evolved languages and jargons, and create artistic works which show a profound resonance of the human with others and the world and universe around us. We have discovered fundamental principles of energy and its transformations, traced the thread of evolution and its biochemical workshop, and are now trying to fathom and grasp one of the greatest mysteries: our own brain. At the same time, we are gripped with the primal imperatives of organisms across the living world, to take advantage of resources, usually at the expense of other species, and, as history has shown, usually at the expense of others of our own species. Calls to world peace and justice are calls to transcend this primal imperative.

1 Approximately "He who attempts to embrace much holds little," or "Jack of all trades, master of none."

Biophysics: Physical Processes Underlying the Living State, First Edition. Wayne F. Reed.

Companion website: www.wiley.com/go/Reed/BiophysicsandMacromolecules

Meanwhile, what do marine mammals, such as whales, do with their vast brains? There is ample evidence that whales and dolphins have highly sophisticated "languages," means of communicating with each other in complex, hierarchical songs, and that there are recognizable "dialects" which vary among whale pods.

Sensing is vital to a living organism and its interaction with its world. While we humans pride ourselves on the acuity of our vision and hearing, and make daily use of our olfaction, gustation, and tactility, it is even more impressive to recognize the senses we don't have but which other species do. Many species can perceive ultraviolet light, but we can't ... a dog enters a room and is assailed by a multitude of odors our less-evolved olfaction is oblivious to ... many animals can perceive subtle vibrations in the air and through the ground, of which we have no perception ... echolocation of bats and dolphins is stunning in the imagery of the surrounds it forms for the animal, and for which we have absolutely no capacity ... the heat sensing of snakes ... the sensitivity to electric and magnetic fields of various species, the ability of fungi to perceive raw chemicals and detritus and synthesize new enzymes to consume them, and the list goes on.

Currently, the search for extraterrestrial life is being pursued along several different lines. Terrestrial extremophiles or archaea are single-cell organisms that can live in extraordinarily hot, cold, acidic, and high-pressure environments and may set some ranges for habitability on exoplanets.

And all this multifold life is part of a vast cosmological evolution, which has produced the energy, elements, and celestial bodies that underlie the living state on our planet.

Despite all this diversity, there are common threads binding organisms together in the evolutionary web. Our goal in Part IV is to use the concepts of preceding chapters, while introducing new ones, to explore selected features and actions of proteins, polynucleic acids, polysaccharides, and then the formation and functions of the lipid bilayer membrane. Chapter 14 will take a brief look at some of the higher level living state functions, such as vision, while Chapter 15 will bring us to the speculative edge of prebiotic evolution, consciousness, and extraterrestrial planets and life.

10

Proteins: Structure, Folding, Enzyme Kinetics, and Cooperativity

In Chapter 3, the basics of amino acids, the peptide bond, protein composition, and levels of their complex 3D structure were discussed. Here, we delve into some of the properties of proteins in terms of their aggregation, enzymatic catalysis, and cooperative effects, both in catalysis and in *helix–coil phase transitions.*

10.1 The Protein Folding Problem

Conceptually, the folding of a protein can be considered a random coil-to-globule transition. The protein as it is being produced at the ribosome comes off as a random coil and can start to fold even before it is fully produced, and completely fold after full production.

It is usually assumed that the primary structure of a protein, given from the genetic sequence for it, contains enough information for the protein to correctly fold into its *native state* with all its secondary, tertiary, and quaternary structure. This assumption has given rise to the ongoing challenge of the *protein folding problem.* Christian Anfinsen (United States, 1916–1995) asserted this, which implies that (i) there is a unique native structure, that is, there is only one-folded state, or a cluster of similar folded states, that is an absolute minimum of free energy, (ii) this native state should be stable against thermal and chemical fluctuations in its environment, that is, the minimum is a deep energy well, and (iii) there must be fairly smooth pathways to the minimum so that the protein doesn't get trapped in states with local energy minima. These assertions often have the unfortunate name of "Anfinsen's dogma." Since "dogma" has the meaning of "incontrovertibly true," it should have no place in science, which has a long history of controverting many a cherished belief. The point is that "violations" of Anfinsen's dogma have already been found and that in dealing with the complexities of protein folding few if any categorical statements are safe from falsification.

Consider, first of all, the huge possible number of amino acid sequences. For an average protein of molar mass 50,000 g/mol, and for an average amino acid molar mass on the order of 100 g/mol, there are 500 amino acids. This means there is an astronomical number, $20^{500} \sim 10^{650}$, of possible sequences. And yet the known number of protein sequences, of all lengths, so far is on the order of 10^7, an extremely tiny subgroup of this number. This reminds us of alphabetic language, where the number of real words coming from all the combinations of letters is extremely small, e.g. there are $26^5 \sim 12 \times 10^6$ five-letter combinations of the English alphabet, and yet only $\sim 10^5$ words of all lengths are actually meaningful. Many combinations are blocked as being pronounceable but nonsense, e.g. klilp, quilg, and yerpy, whereas others are simply unpronounceable, e.g. fgivt, lqwez, ihjtc, as well as every word consisting only of consonants (excluding y).

Furthermore, it seems that the number of protein fold states is even smaller than the number of sequences, probably less than 10^4. Hence, given fold states are reused in nature with very different sequences to achieve very different functions. Further aiding the understanding of folding structures and origin is the notion of *protein family*, where sophisticated tools for assessing sequence similarity exist. High-sequence similarity implies common evolutionary descent, so this comparative method is valuable for tracing the evolutionary tree of proteins.

The total number of possible protein configurations is $N = 3^{n-2}$, n = number of amino acids, e.g. for $n = 102$, a small protein, $N = 3^{100} = (5 \times 10^{47})$. Yet, the protein primary sequence is presumably sufficient to determine the 3D folded state. The information for folding is $I_{to\ fold} = \log_2 3^{100} = 158$ bits needed to specify the native configuration. In the sequence itself, there are $I_{sequence} = \log_2 20^{100} = 432$ bits, i.e. there is more information needed to specify the sequence than to specify the 3D structure; the former is 20^N, whereas the latter is 3^N.

Biophysics: Physical Processes Underlying the Living State, First Edition. Wayne F. Reed.

Companion website: www.wiley.com/go/Reed/BiophysicsandMacromolecules

What limits the number of sequences of proteins? There are probably many factors, but an important one may be that not all sequences can fold into a compact structure. Like letter combinations that can't be pronounced some amino acid combinations cannot get along together, because of incompatible hydrophobicities, electrical charge, and other factors. There is also the notion of *geometrical frustration*, which occurs in numerous physical situations where a unique ground state cannot be found. An early example was that of ice, which has nonzero entropy at 0 K because there are several equivalent arrangements of protons in the structure yielding the same energy. The statistical physics of *spin glasses*, which also exhibit frustration, has been extensively elaborated, both experimentally and theoretically, and some of its approaches have been adapted to the protein folding problem. For proteins, it is possible that many sequences will not lead to a unique functional "ground state." Because the protein energy landscape is so complex, with so many local minima, or "wells," it cannot even be asserted that the functional state will necessarily be the one with the absolute minimum energy, in defiance of the Anfinsen assertions. Furthermore, it has been found that proteins can fold into more than a single state, or cluster of similar states, thus defying Anfinsen's uniqueness claim. Additionally, some post-translational modifications can modify folding, or proteins require cofactors to fold, and some *intrinsically disordered proteins* have no well-defined 3D structure.

Multidomain proteins have the advantage that the individual domains already have their own folding pathways and have hence "solved" their own protein folding problem. Linking these domains together, usually by a flexible, random coil sequence of amino acids, means that the protein can take advantage of multiple solved folding pathways, which further reduces the number of possible folding states. However, there are often very strong interactions between the domains and this may affect, positively or negatively, the overall folding pathway.

The main driver for folding and stabilization appears to be the hydrophobic effect, but other factors such as hydrogen bonds, ionic interactions, and weak electrostatic forces, such as van der Waals dipole–dipole interactions also play a role. Covalent disulfide bonds which further stabilize native structure can be supplied by the amino acid cysteine. This folding, while energetically favorable, is nonetheless entropically unfavorable as the protein chain is forced into a very reduced number of states compared to all the ones available to it, given all the rotational freedom of the peptide bonds composing it; as mentioned, a small protein of 102 amino acids has on the order of 3^{100} or 5×10^{47} possible conformational states. This leads to *Levinthal's paradox*, which posits that, even for nanosecond bond rotations, it would take some 10^{40} years to explore all conformational possibilities, far longer than the age of the known universe of 10^{10} years, and hence random protein folding is not feasible as a basis for life. The resolution of the paradox is that the folding is not random and that the *energy landscape* is a sort of *energy funnel*, and that each step in the folding sequence brings the protein deeper into the funnel until a specific minimum is reached. It is presumed that there are well-defined intermediate states, such as formation of secondary structure or sometimes a series of metastable conformational states, into which a protein rapidly folds on its way to the bottom of the funnel. Many proteins can fold on the order of milliseconds or less. The actual energy landscape for protein conformations is topologically complex and much work involves computer simulations and models. So much so that new computers have been designed explicitly to do the vast computing needed for systems with so many possible states, and people around the world have offered idle CPU and video time on their home computers to massively distributed computation.

In light of much experimental and computational work, it is asserted that the protein will fold in order to bury its hydrophobic groups inside the globule and to expose the hydrophilic and charged groups to the water at the globule's surface. One computational approach that has some limited success for small proteins involves assigning hydrophobicity and hydrophilicity indices to each amino acid, and then search for the arrangement which maximizes hydrophobicity in the interior and hydrophilicity at the surface.

Following the notion that hydrophobicity plays a major role in protein folding, experiments have been carried out using *site-directed mutagenesis* to probe the tolerance of a protein against substitutions in an amino acid sequence. Site-directed mutagenesis allows specific amino acids to be substituted for others at exactly specified points in the DNA sequence. There is a fair degree of tolerance when substituting amino acids of similar hydrophobicity, which are then neutral mutations, since the protein continues to function unimpaired. Substituting a hydrophobic amino acid with a highly polar one can completely disrupt the protein's folding. If random sequences of amino acids are chosen, the vast majority will not fold. The protein folding problem is so difficult that one expedient that is used consists of finding known sequence-structure pairs and cobbling these together to try to predict structure in a protein which can be broken down into sequence blocks. Since evolution does not seem to have found a large number of foldable sequences, compared to the astronomical number of possibilities, this approach can be of great practical value and leapfrog the complications of deep physical modeling.

Fruitful, coarse-grained modeling approaches are those that set up energy functions based on amino acids that should make contact and ignoring other interactions. These are often referred to as *Gō models* and are amenable to computationally intensive simulations.

Another approach is to work backward, that is, instead of trying to deduce a structure from a sequence, try to find sequences which give a specific structure; the *inverse folding problem*. Whereas computational complexity for the folding problem scales exponentially with the number of amino acids, the inverse problem scales linearly, using the right approaches. The inverse approach allows for protein design, finding sequences that will give a desired structure. Fold states are also classified into a very small number of fold classes, ~10^1, largely according to α-helix and β-strand content. Since neither the number of known sequences, ~10^7, nor known fold states, ~10^4, are astronomical numbers, and the fold classes are relatively few, this approach can find rationally designed proteins, which can then be artificially synthesized.

A major question linked to protein folding is whether an isolated, denatured (random coil) protein in a physiologically appropriate milieu of temperature, pH, and ionic strength, *in vitro*, will spontaneously fold into its native structure. There is experimental evidence that this actually occurs for small, simple proteins, but does not necessarily occur for all complex proteins, especially those with multiple domains.

Consider the actual crowded *in vivo* cellular environment in which proteins are expressed and must properly fold. We've seen repeatedly that many effects are concentration dependent, such as osmotic pressure, because of nonideality, that is, because of interparticle interactions. A protein being expressed in a cell interacts with RNA, the intracellular matrix, and other proteins. Hence, the energy landscape for an unfolded protein in a cell may be far different from the landscape *in vitro*. In fact, much research supports the idea that protein *chaperones* provide the interactions that are essential for controlling the complex steps in folding. There exists a *proteostasis network* in the cell that consists of chaperones and structures for destroying misfolded proteins. This network allows the cell to cope with different stressors, both internal and external to the cell.

Figure 10.1 shows a schematic of a protein folding funnel. It has several features that also don't fit the Anfinsen assertions. It shows a number of deep wells, all of them in which the folding protein can get caught; there is not a unique folded state. All of these deep wells are bad except for the one corresponding to the native state. In the case of Figure 10.1 the native state well is actually not as deep as the worst-case scenario of amyloid fibril formation, which is behind such diseases as Parkinson's and Alzheimer's. The chaperones in Figure 10.1 can catalyze exit of a folding protein from a shallow well intermediate state into the next shallow well on its way to proper folding into the native state. The chaperones also act to block a partially folded protein from associating with other proteins to form useless or even toxic aggregates. Defects in the chaperone-based proteostasis network, due to aging and other factors, are indicated in a number of protein-related diseases.

In 2020, The Deep Mind Laboratory, in London, announced that its computer program "AlphaFold," uses artificial intelligence (AI) for solving the 3D structures of proteins. Over 2×10^8 proteins have so far been identified, but the 3D structure of less than 200,000 of these has been discovered using X-ray crystallography, NMR, electronmicroscopy, and

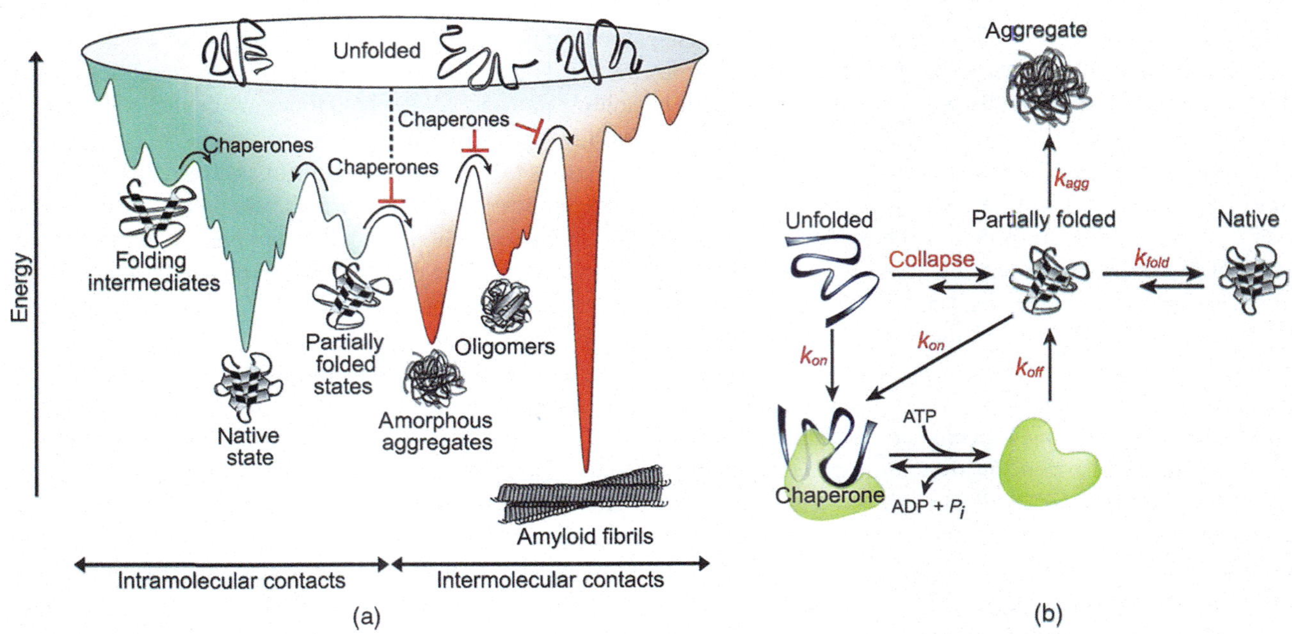

Figure 10.1 (a) The energetic folding landscape of a protein. (b) Various states of protein conformation; native, partially unfolded, aggregated, and the role of chaperones in the folding process (right). (Balchin, D. et al. (2016) / Reproduced with permission of American Association for the Advancement of Science).

other associated methods. AlphaFold's AI learned many structures and the types of amino acid sequences associated with them, and it is by cross-referencing and interpolating this massive experimental protein folding database that it solves the 3D structure of an unknown protein. AlphaFold sidesteps all the physics involved in folding and essentially uses an extremely sophisticated form of pattern recognition.

It is worth considering a few remarks concerning the state of the physics behind the protein-folding field. From Brini, E., Simmerling, C., Dill, K. (2020). *Protein storytelling through physics. Science 370* (1056).

They summarize that over the past 50 years, the protein folding has used the following two approaches:

a) Inferences from databases on known proteins to predict folding of unknown proteins.
b) Computational molecular physics (CMP) using force fields for atom-atom interaction sampled by molecular dynamics.

AlphaFold uses a). An actual quote from their work states

> CMP is rooted in the principled physics of driving forces and reveals granular detail of conformational populations in space and time. Recent advances are accessing longer time scales, larger actions, and blind testing, enabling more of biology's stories to be told in the language of atomistic physics...Based on the premise that structure determines function—the same principle that helps you discern how a knife and fork work—protein shapes are often the starting points for protein storytelling. These shapes are known at atomic detail for more than 150,000 proteins and are available in a public resource called the Protein DataBank (PDB), thanks to X-ray crystallography, nuclear magnetic resonance (NMR) spectroscopy, and cryo–electron microscopy (cryo-EM) experiments from a large community of structural biologists over the past 60 years....CMP aims to accomplish the following: (i) Reveal causalities, by allowing investigations of the driving forces, dynamics, and motions. (ii) Give the intermediate steps in time and space, angstrom by angstrom, nanosecond by nanosecond—narratives of molecular actions such as folding, binding, and rearrangements. (iii) Give conformational distributions, not just a single averaged structure. Proteins writhe and deform in ways that are needed for matching our stories to experiments. Drugs often don't bind "key-in-lock" to preexisting protein cavities; they push the protein into a new shape that we seek to learn. (iv) Systematize, giving a common language for our storytelling. In principle, physical modeling is transferable: to different proteins, different ligands, different binding situations, or different experimental conditions, including those not yet measured. And physics-based models are less susceptible to errors of so-called hasty generalization.

10.1.1 More on Chaperones

Heat-shock proteins were discovered before their role as chaperones in protein folding was realized. When a cell is under thermal or other types of stress the possibility of dangerous protein folding increases. Under stressful conditions, heat-shock proteins are expressed in greater number to prevent misfolding damage. Many chaperones are heat-shock proteins. Chaperones can run from tens of thousands of g/mol up to assemblies in excess of 10^6 g/mol.

The discovery in 1978 of chaperones as proteins that aid other proteins to fold is very significant as far as tight control on almost all biological processes. Enzymes catalyze virtually all biochemical reactions and metabolic cycles involving the making and breaking of covalent bonds, and hence control rates. The fact that chaperones assist in protein folding, a non-covalent process, suggests that this activity is also at least partially catalytic in that the chaperones seem to be able to lower barriers between conformational states, hence speeding up folding. This strengthens the notion that very little, if anything, is left to chance in the folding process and that proteins are not allowed to simply collapse into their native state driven by free energy minimization *in vivo*. The crowded environment and the dangers of misfolding or slow folding may make chaperone action imperative in many cases. Catalytic folding is another factor in resolving the time-scale problem of Levintal's paradox. At the same time that chaperones catalyze folding toward the proper native state by lowering intermediate barriers, they also raise barriers toward folding into useless or dangerous states, that is, they also act as *anticatalysts*.

This does not necessarily violate Afinsen's dogma, because the final structure is still encoded in the protein primary sequence, but requires chaperones to provide the folding pathway in order to achieve the final state. This has possible implications in suggesting that there are more rules to protein folding, and to many supramolecular assembly processes, which go beyond simply free-energy minimization. If it is true, for even a small number of proteins, that the primary sequence does not dictate the 3D shape, then we have to consider an evolutionary pathway in which the 3D structure of the protein depends upon an external agent, the chaperone, and that such evolution is synergistic, maybe even opportunistic.

Beyond the task of assisting in protein folding, chaperones are also involved in recognizing and dealing with misfolded proteins. When a chaperone recognizes a misfolded protein, it can attempt to fold it into the proper native structure. If it can't fold the protein it can tag it for degradation by the ubiquitin-proteasome system. This latter system encompasses a very large number of ubiquitin-protein ligases that receive the tagged, defective proteins. The proteolysis by this complex can reduce the protein to useful peptide fragments that can be deployed for other functions, such as to the cell membrane for immunosurveillance, or fully degraded to free amino acids by peptidases. Chaperones and the ubiquitin complexes are found in the cytosol, endoplasmic reticulum, and mitochondria, and there are membranes between the latter that grow and shrink in size and allow interaction and coordination of these activities. Besides dealing with individual misfolded proteins, chaperones and the ubiquitin complex can also break up some protein aggregates.

The importance of proper protein folding cannot be overstated. Up to a few percent of the eukaryotic genome is devoted to coding for the ubiquitin-protein ligases which destroy misfolded proteins. Evolution has arrived at enzymatic control as the way to catalyze biochemical reactions that enable the living state. The price to be paid is the 3D folding complexity of the structures needed to perform catalysis. An entire part of the genome has developed to ensure that these folded structures can be assembled, and another entire part to rip apart defectively folded structures.

While misfolded proteins are useless, aggregates of misfolded proteins can be lethal. A *prion* (a contraction of **pro**tein and infect**ion**) is a misfolded protein that aggregates together with others to form *amyloid fibers*.[1] The prions can seemingly cause other healthy, normally functioning *prion proteins* to misfold and join in the aggregate. The aggregates are very stable and are shown as the lowest energy state in the funnel of Figure 10.1. They are so stable that they resist proteases, boiling, ionizing radiation, and other harsh treatments. Prions are implicated in many neurodegenerative diseases, including Alzheimer's, Creutzfeldt-Jakob, and Parkinson's. Because the prions induce misfolding and aggregation in other prion proteins, they can spread throughout tissues and disrupt membranes and other cellular structures. The normal function of the prion protein is not well understood, but there is evidence that it is involved in immunity and, for example, can deliberately form amyloid fibrils for regulated cell death, in order to stop the spread of a cell infected with a virus to other, healthy cells. There is also evidence that the normal prion protein is involved in long-term memory and repair of the myelin sheath in nerve connections.

A remarkable aspect of diseases associated with prions is that they may be the closest disease process driven so nakedly by free energy minimization, as seen in the bundle of harmful amyloid fibrils of Figure 10.1 lying at the very bottom of the energy landscape. Even viruses, which are stripped down, nanoscopic genetic monsters, are far more sophisticated in that they actually reproduce themselves via a host's hijacked genetic machinery, rather than just inducing a conformational change and aggregation in already existing proteins, as prions do.

10.1.2 Is Nothing Spontaneous in the Living State?

We have seen that nothing seems to be left to chance or spontaneity in the living state. There is seemingly a protein and associated actors for every task. Can even something as seemingly natural as dried, withered leaves falling from trees in the Fall, be under biological control? In the picture (February 2022), there is a large branch that was shorn from the main tree during Hurricane Ida in the US Gulf South (August 2021). It got caught up in the tree and didn't fall to the ground. The white arrow shows where the limb snapped. The main tree is barren of leaves, but the dead, suspended limb has virtually all its leaves, which were just as dried and withered as those of the main tree. The suggestion is that the main tree actually "commands" the dead leaves to fall off, and does not depend on wind and their own weight to shear them off. In fact, leaf dropping, *abscission*, is a highly controlled hormonal process, where an abscission layer forms at the base of the leaf stem and weakens the cell walls, causing the leaf to fall. Thrifty nature recovers nitrogen, phosphorous, and other nutrients before the leaf falls. This process leads to brilliant autumn foliage as the chlorophyll breaks down and changes its absorption spectrum. At the same time, other nongreen pigments, such as carotenoids, become more visible.

The ineffectiveness of waiting for leaves to fall is seen in the full complement of intact leaves on the dead branch. As of December 2023, most of the leaves on the dead branch are still in place, while the live tree has gone through a second cycle of producing and then dropping leaves.

1 "Amyloid" is an historical misnomer, since it means starch-related. Early on amyloids were thought to consist of polysaccharide aggregates, until it was discovered that they are proteinaceous.

A branch shorn from the main tree but remaining tangled in the tree (white arrow), during Hurricane Ida, August 2021. The branch has maintained its withered leaves, and in December 2023, the withered leaves are still largely intact, while the tree itself has gone through three complete leaf cycles.

10.2 Protein Aggregation

As discussed, protein aggregation *in vivo* results from protein misfolding and often also follows partial or complete unfolding. Protein aggregation occurs biologically in many contexts and usually has negative effects, such as interference in normal cellular functions and the fatal propagation of prions. On the other hand, a beneficial process such as blood clotting to stanch a bleeding wound involves a series of protein aggregation steps.

10.2.1 Aggregation of Therapeutic Proteins and Light Scattering Analysis

Protein aggregation is a ubiquitous problem in the *biologic drug industry*, which uses protein-based therapeutics. *Protein denaturation*, often also termed *unfolding*, can occur *ex vivo* and leads to the loss of the 3D native structure. It can occur due to effects such as pH, temperature, denaturing agents, such as urea and guanidine-HCl, and mechanical stresses such as pumping, stirring, and filtering. The effect of denaturation is normally to turn an organized protein, with globular or rodlike structure into a random coil. Figure 10.2 is a schematic of the denaturation and renaturation processes.

Biologic drug is a broad term for any medicine extracted from or synthesized by a living organism. These can include blood components, vaccines (often comprising deactivated viruses), nucleic acids, proteins, polysaccharides, and tissues. A large part of the biotechnology industry is centered on therapeutic proteins. A great concern in this industry is the stability of these latter, especially their undesirable tendency to aggregate in biologic drug formulations. Immense effort is spent on stabilizing and formulating protein drugs once suitable candidates have been narrowed down. When protein drugs aggregate, or otherwise degrade, they lose their biological activity, can become highly *pyrogenic* (i.e. causing inflammation), and the patient can even develop an immunity to the drug.

Antibodies are *immunoglobulin* proteins and constitute a large fraction of blood plasma protein content. As the name implies, these proteins identify and aid in destroying invading species, or *antigens*. A general schematic for antibodies is shown in Figure 10.3. They are composed of four polypeptides, two light chains and two heavy chains, which together form a "Y-shaped" structure. The antigen binding sites are at the ends of the *variable region* forming the "Y." They recognize and bind *epitopes*, which are portions of the antigen to which the antibody is specific. The *constant region* determines the type of antibody, which are divided into five major classes, and these, in turn, control the mechanism by which the antigen is destroyed. An important class of therapeutic proteins comprises *monoclonal antibodies*. These are clones from the same parent cell. "Mono" indicates that they are monovalent, that is, they bind only one specific epitope. Through variations of the antigen binding sites a very wide variety of epitopes can be made to bind. *Polyclonal antibodies* recognize and bind to more than one type of epitope.

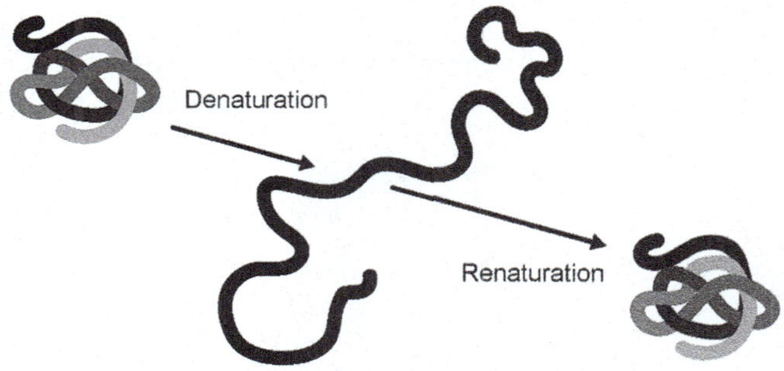

Figure 10.2 Denaturation and renaturation of a globular protein.

There are many means of studying protein aggregation, with the goals of understanding the mechanisms and kinetics, and of developing formulations that inhibit aggregation. We can apply the capabilities of both dynamic and total intensity light scattering, discussed in Chapter 9 to monitor protein aggregation in real time, and apply the notions of second virial coefficient and hydrodynamic diameter, as well. Recall that for a particle whose size is significantly less than the incident light intensity of light scattered from it increases as the square of the molar mass. This is shown in Equation 9.38, and several subsequent equations, by the polarizability squared of the particle, α^2, which is proportional to the square of the particle's molar mass. At a fixed mass concentration (g/cm^3) of dilute particles the scattered intensity will increase linearly with mass. This makes the scattered light intensity an exquisitely sensitive probe of changes in particle molar mass.

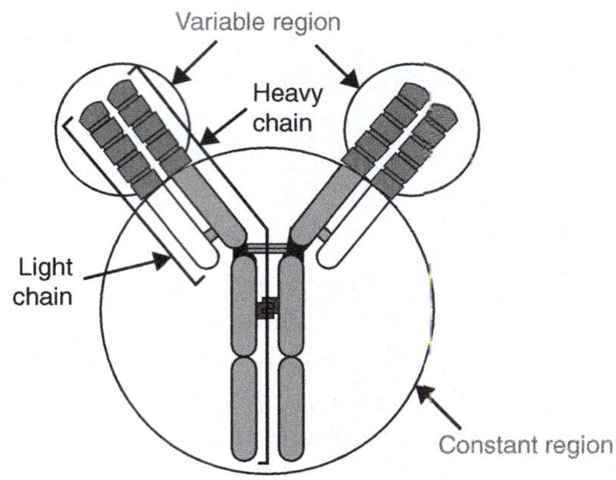

Figure 10.3 Schematic of a monoclonal antibody. (Courtesy of Marissa Calko).

In the example below, a dilute solution of monoclonal antibody was held at different temperatures. Figure 10.4 shows both how protein aggregation leads to increased light scattering and how the rate of protein aggregation depends strongly on the temperature. The representation is based on $I(t)/I_0$ versus time, where $I(t)$ is the scattering intensity from the protein aggregates at any time, and I_0 is the initial scattering of the native, unaggregated proteins. The solvent background scattering is subtracted from the total signal $I(t)$ and from the initial signal I_0. Notice that since this is a ratio of measured signals the calibration factor cancels out so that it is sufficient to measure the scattering directly, e.g. from a photodiode voltage, or the direct output of a *charge-coupled device* (CCD). If the particles are dilute enough that $2A_2Mc \ll 1$, where M is the molar mass and c the mass concentration, then A_2 (second virial coefficient) effects are negligible, or can be accounted for with a known A_2. When the average size of the particles is less than about 50 nm (the native proteins are under 10 nm in diameter), a measurement at $\theta = 90°$ is sufficient and no angular extrapolation to $\theta = 0°$ is necessary when the incident light source is above 500 nm. Under these two conditions, the ratio $I(t)/I_0$ is $M_w(t)/M_0$, where $M_w(t)$ is the weight average of all native and aggregated proteins at time t, and M_0 is the molar mass of the native, unaggregated protein. Hence, $M_w(t)/M_0$ is the weight average number of native molecules per aggregate at any time t. In Figure 10.4, the solution was very dilute so the $M_w(t)/M_0$ representation is used.

Figure 10.4 uses a log–log representation, since the changes in both time scale and scattering intensity are very large. At 30°C, the proteins are stable up to six days, whereas at 45°C aggregation is discernible after one day. At higher temperatures, the aggregation rate increases dramatically, until high levels of aggregation occur within tens of seconds at 69°C and higher.

The steep dependence on temperature suggests a rate exponentially sensitive to T, so we can investigate this with the Arrhenius representation, Equation 4.194. First, we define the *aggregation rate*, AR, as the time rate of change of the dimensionless $M_w(t)/M_0$ in the early regime of aggregation, in the case where this latter increases linearly,

$$AR(1/\text{s}) = \frac{\mathrm{d}\left(M_w(t)/M_0\right)}{\mathrm{d}t} \qquad 10.1$$

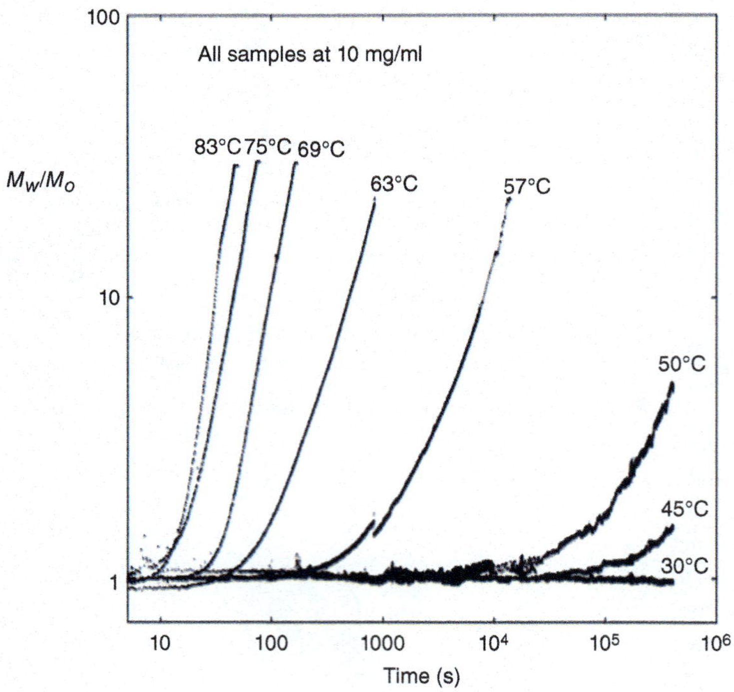

Figure 10.4 Aggregation of a monoclonal antibody at various temperatures. M_w/M_0 is the dimensionless mass comprised of the initial, native mAb mass, M_0, and the weight average M_w of all native proteins and aggregates at any given instant. (Drenski, M.F. et al. (2013) / Reproduced with permission of Elsevier).

AR represents the fractional rate of protein aggregate mass increase per second. Data for the mAb of Figure 10.4 (mAb2) and a second mAb (mAb1) are shown in the Arrhenius representation, Figure 10.5. The free energy of activation for unfolding is similar for both in the lower temperature regime (40–68°C). At 69°C, there is a break in the curve for each, signaling the midpoint temperature, T_m, for protein structure melting. T_m for these proteins, the temperature at which the protein is half-denatured, was independently established separately by *differential scanning calorimetry* and by *fluorescence emission spectroscopy*. Above 69°C, there is a second Arrhenius regime with lower activation, possibly corresponding to the unfolding of a different domain in the protein. Despite the same activation energies and T_m for the two proteins, their aggregation rates differ by a factor of 100. This suggests that it is the pre-factor in the Arrhenius relation that controls the probability of two proteins sticking when they encounter each other. Apparently, mAb2 is far more likely to stick together when unfolded proteins encounter each other than is mAb1. This may be due to a greater hydrophobicity of the domain(s) exposed upon unfolding. The practical consequence of the relative stabilities is that mAb1 is thermally stable enough to be used as a commercial product, whereas mAb2 could not be stabilized sufficiently, and was abandoned.

Figure 10.6 takes the reciprocal of the rate constant and converts it to days. It then represents the time it takes for the initial unaggregated protein to aggregate to a state containing, on weight average, two proteins, that is, this is the *time to weight average dimerization*. Besides the thermal regimes, for which Arrhenius fits apply, there is a *stochastic regime* below 42°C for which there is a very strong deviation of the measured aggregation rate from the extrapolated Arrhenius rate. This latter predicts extremely long-term stability for proteins at these lower temperatures, even billions of years at about 10°C, which is not found experimentally; the Arrhenius extrapolation at 30°C predicts 750 years as the time to dimerization, versus the experimental measurement of 0.3 years! This result relates to the field of accelerated testing, and its potential pitfalls, where conditions are intensified and times for an event are extrapolated to normal conditions, e.g. trying to determine the lifetime of a latex paint on a surface due to solar photodegradation by increasing the radiation far above the sun's terrestrial intensity.[2]

The stochastic regime in Figure 10.6, below 42°C, shows that other effects besides temperature-induced unfolding may drive aggregation. Since these measurements were made *ex vivo*, and since proteins are generally stable below 42°C *in vivo*, it is likely that the much greater stability of proteins *in vivo* than measured here *ex vivo* is due to environmental factors in living cells; the presence of chaperones, the crowded environment that makes unfolding entropically difficult, etc.

2 The Sun's intensity is 1,360 W/m^2 at the upper boundary of the Earth's atmosphere, which is the so-called *solar constant*. It puts an upper limit on the intensity of harvestable solar energy in terrestrial applications.

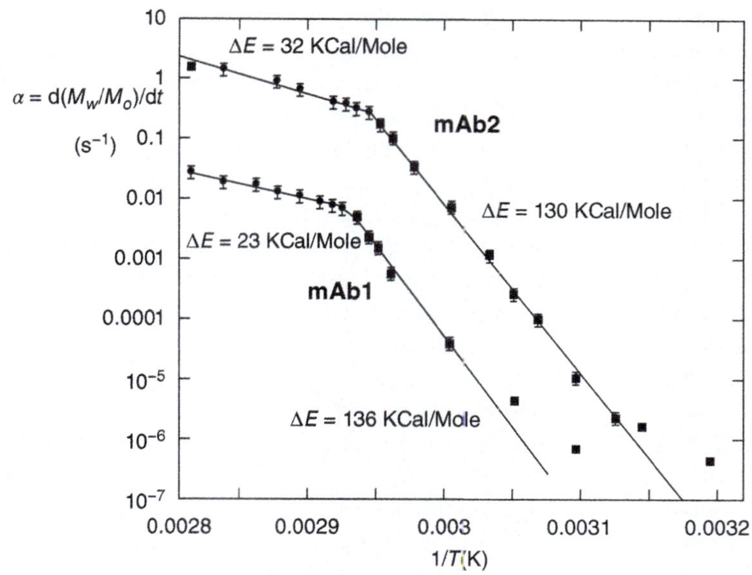

Figure 10.5 Arrhenius plot of aggregation rates for two different mAbs. (Drenski, M.F. et al. (2013) / Reproduced with permission of Elsevier).

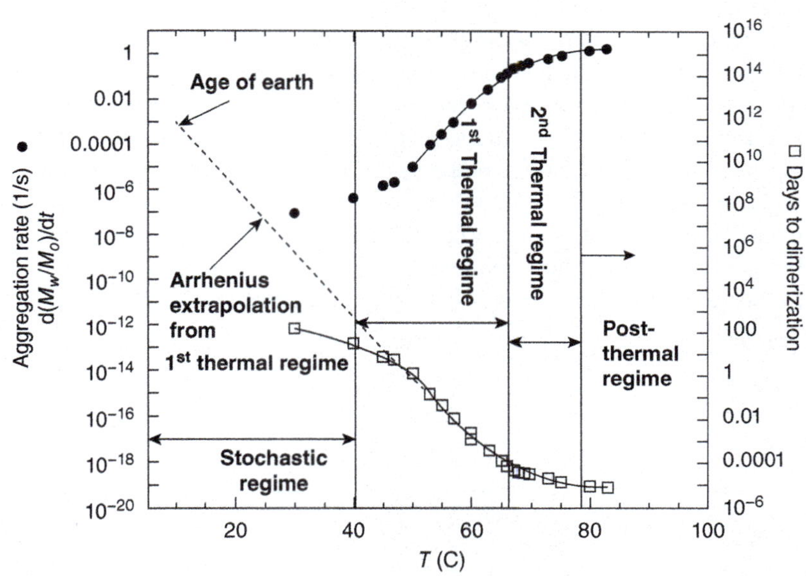

Figure 10.6 Aggregation rate (hollow squares, left-hand *y*-axis) and days to dimerization for an mAb (solid circles, right-hand *y*-axis). The direct extrapolation of Arrhenius behavior (dashed line) above 40°C to below 40°C is dramatically unrealistic. For time comparison, the age of the Earth is shown. (Drenski, M.F. et al. (2013) / Reproduced with permission of Elsevier).

10.2.2 On the Reproducibility of Protein Aggregation: Deterministic, Stochastic, or Chaotic?

It might seem that a process resulting from unfolded proteins randomly sticking to each other may be *stochastic*, and not *deterministic*. It might even be *chaotic*. First, let's review the classical definition of each of these terms.

A deterministic system is one in which, given the initial state of the system and the forces acting on it, its future is completely described by a corresponding governing equation. Planetary motion or the trajectory of a projectile through nonturbulent air are simple examples of this. If the same system with the same initial state and forces is recreated, the same result will occur, identically, each time the system is recreated. If the system is recreated with small changes in initial conditions and forces the results will be similar, but not necessarily identical.

A deterministic chaotic system is one in which recreating the initial state and forces with very slight differences leads to very different results, where the differences usually grow exponentially in time (Chapter 14.4). These systems are governed by deterministic, but nonlinear equations. Turbulence in fluid flow, and the motion of a double-hinged pendulum are examples.

A folkloric example, which, while not being scientifically deterministic chaos, nicely represents the cumulative effects of small events having huge consequences, is the old saying "for want of a nail a shoe was lost, for want of a shoe a horse was lost, for want of a horse a General was lost, for want of a General a battle was lost, for want of a battle a war was lost, for want of a war a nation was lost." If the nail hadn't been lost the General might have powered on to victory and the nation possibly saved.

A stochastic system is one in which, starting with nominally identical conditions, there are multiple possible outcomes for successive events, which constitute a random process. This involves probabilistic rather than deterministic behavior. We have used diffusion at equilibrium as a preeminent example of a random, hence stochastic process; a particle suffers a collision at random intervals which causes a random change in the particle's momentum. A series of flipped coins is another example. Such processes normally have predictable probability distributions for results, rather than single, predictable outcomes.

If we intend to find out whether a process falls into one of these categories, we will find it impossible to assert whether a system is completely deterministic based on measurements with an instrument. This is because every measuring instrument has an *intrinsic error distribution*, IED, due to thermal and other effects and errors, including its signal/noise characteristics. This IED is not connected to systematic errors, such as calibration or uncertainties in physical constants used to interpret measurements. If multiple, repeated measurements are made on an identical system and these fall on or within the IED, then the most we can assert is that the process is deterministic within the limits of the IED, and we can say the system is *operationally deterministic* with respect fo the capabilities of the instrument.

Now, back to the issue of classifying protein aggregation. Figure 10.7a shows the light scattering results for repeating thermally induced protein aggregation multiple times, which are as close to identical as experimentally feasible. Results for *bovine serum albumen*, BSA, and a monoclonal antibody, mAbX, are shown. In both cases, the results in Figure 10.7a are close to each other, and Figure 10.7b shows the mAbX results falling on the IED of the light scattering instrument. Similar results were obtained for the BSA and several other proteins. Hence, thermally induced protein aggregation, in these cases, is operationally deterministic.

In contrast, Figure 10.8a shows repetitions of experiments for one of the proteins, in which the aggregation was caused by stirring the protein solutions with a magnetic stir bar resting on the bottom of the scattering cuvette. The results are quite divergent, but not so different as to appear chaotic. Figure 10.8b shows that the results fall outside the IED and so it is asserted that contact-stir-induced aggregation is a stochastic phenomenon.

A rationalization for these outcomes is that the temperature is uniform throughout the cuvette and gives every protein molecule the opportunity to unfold under its influence. It is a *global stressor* leading to an operationally deterministic outcome. In the case of contact stir, only the proteins in the vicinity of the stir bar, and most specifically those in contact between the stir bar and the bottom of the cuvette feel the stress, and so this is a *local stressor*, leading to stochastic behavior. Suspending the stir bar so as to have no contact with the cuvette and spinning it at the same frequency did not cause any aggregation.

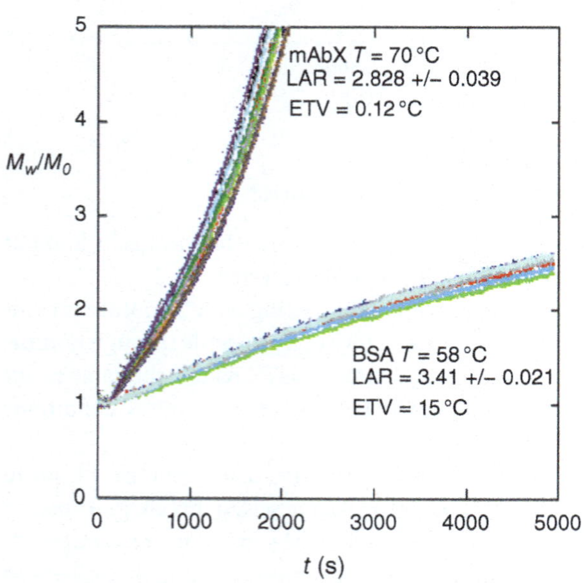

Figure 10.7a Repeated light scattering measurements for thermally induced aggregation of BSA and an mAb.

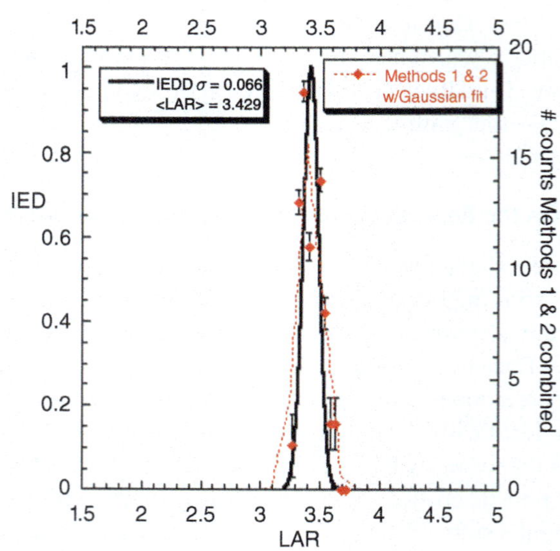

Figure 10.7b The mAbX results fall within the light scattering instrument's IED (as do those for BSA, not shown). (Jarand, C.W. and Reed, W.F. (2018) / Reproduced with permission of American Chemical Society).

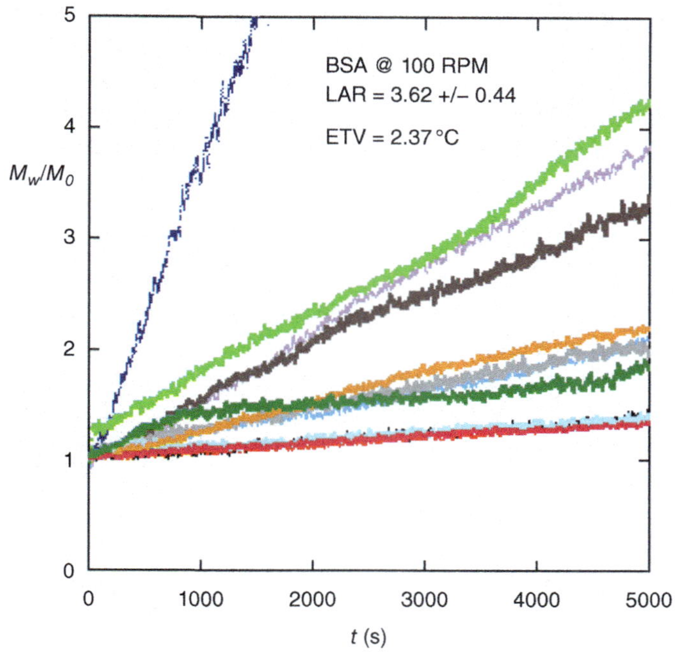

Figure 10.8a Repeated light scattering measurements for contact-stir-induced aggregation of BSA and an mAb.

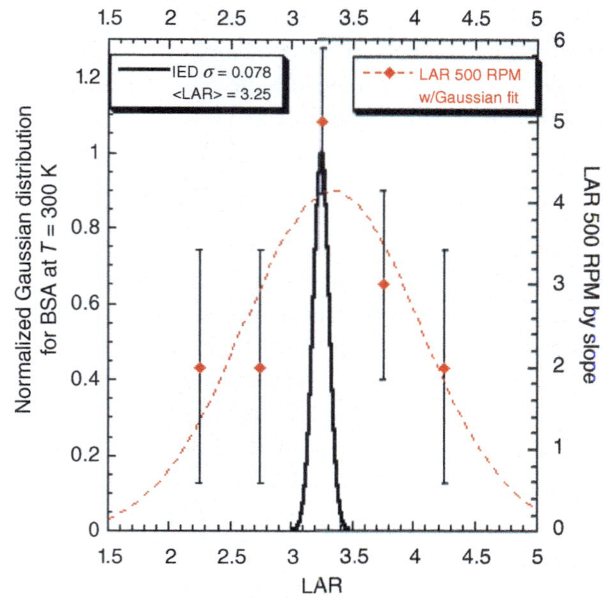

Figure 10.8b The mAb results fall outside the light scattering instrument's IED (as do those for BSA, not shown). (Jarand, C.W. and Reed, W.F. (2018) / Reproduced with permission of American Chemical Society).

10.2.3 Cooperativity and Other Effects in Protein Aggregation

The time-dependent static light scattering signature of an aggregating protein solution represents the time-dependent weight average molar mass produced by the underlying governing kinetic and mass balance equations. A complete set of kinetic equations for irreversible protein aggregation can be formulated as follows: Let $N(t)$ be the number concentration of native proteins at time t (number of native proteins per unit volume), and let $P_1(t)$ be the number concentration of "activated" proteins, i.e. native proteins that are damaged by unfolding or other means and are hence susceptible to aggregation. In the case of thermal or mechanically induced unfolding, the loss rate of $N(t)$ is a first-order process, where the loss rate of $N(t)$ is the rate at which $P_1(t)$ is produced, due to a stressor such as temperature, where α is the rate constant.

$$\frac{dN}{dt} = -\alpha N \tag{10.2}$$

In the case of purely colloidal instability of native proteins, i.e. in the case where native proteins can aggregate without damage, then $P_1(t) = N(t)$ and Equation 10.2 is no longer relevant.

Let $P_i(t)$ be the number concentration of aggregates containing i proteins, whether damaged or native. Then, from mass balance $P_1(t)$ can be written as

$$\frac{dP_1}{dt} = \alpha N - 2k_{1,1}P_1^2 - P_1\sum_{j=2}^{N_{max}} k_{1,j}P_j - k_{N,1}NP_1 \tag{10.3}$$

The factor of "2" occurs in the P_1^2 term because two P_1 objects are lost when they stick together to form a single P_2 object. The $k_{1,j}$ in the summation term express the probability of P_1 irreversibly sticking to P_j in a collision. The $k_{N,1}$ term admits the possibility that an undamaged protein could stick to P_1 in a collision. N_{max} is the largest number of proteins in an aggregate. For self-limited dimerization, $N_{max} = 2$, whereas for unlimited colloidal growth N_{max} tends toward infinity.

The expression for P_i for $i > 2$ is

$$\frac{dP_i}{dt} = -2k_{i,i}P_i^2 - P_i\sum_{\substack{j=1\\ j\neq i}}^{N_{max}} k_{i,j}P_j + \sum_{j=1}^{Int(i/2)} k_{j,i-j}P_jP_{i-j} - k_{N,i}NP_i \tag{10.4}$$

It is assumed in Equations 10.3 and 10.4 that aggregates grow chiefly by two body collisions, i.e. the solutions are dilute enough that two body collisions dominate. In the case where three body collisions become important terms of the form $k_{i,j,k}P_iP_jP_k$ needs to be added. In the case where native proteins cannot aggregate, then $k_{N,i} = 0$ for all i. On the other hand, if native proteins can aggregate and if they aggregate with different probabilities according to the composition of an aggregate in terms of how many damaged proteins j are in it and how many native proteins m are in it, then such a compound aggregate can be designated as $P_{j,m}$ and the corresponding constants as $k_{N,j,m}$ for addition of native monomers and as $k_{k,p,j,m}$ for an aggregate $P_{k,p}$ sticking to an aggregate $P_{j,m}$ upon collision. In these cases the $k_{N,l,m}$ and $k_{k,p,j,m}$ can depend on the sequence in which damaged and native proteins were added to an aggregate. In this case, the constants are no longer a fixed matrix of values, but rather a dynamically determined array whose values depend on the history of the aggregation. The aggregation is a non-Markovian process in this case, since the set of all k constants at a given time depends on the path by which the current state of the system was reached. A parallel to this in polymer science is an ideal random coil polymer with no excluded volume (Chapter 8.1.3), which is Markovian, versus the case of excluded volume, where the spatial placement of a monomer j after $j - 1$ monomers have been placed in the chain depends on the spatial placement of all previous $j - 1$ monomers. This latter process is non-Markovian and has generated large amounts of literature seeking solutions.

For the Markovian case, i.e. where the matrix $k_{k,p,j,m}$ is definable and not history dependent, then Equation 10.4 can be extended to

$$\frac{dP_{k,p}}{dt} = -2k_{k,p,k,p}P^2_{k,p,k,p} - P_{k,p}\sum_{\substack{m=1\\ m,j\neq k,p}}\sum_{j=1} k_{p,k,j,m}P_{j,m} + \sum_{m=1}^{Int(N_{max}/2)}\sum_{j=1}^{Int(N_{max}/2)} k_{j,k-j,m,p-m}P_{j,k}P_{k-j,p-m} - k_{N,k,p}NP_{k,p} \qquad 10.5$$

For any specific irreversible mechanism described by Equations 10.3–10.5, the key variables $k_{i,j}$ and $k_{N,i}$ form a set. The set can describe mechanisms such as self-limiting aggregation which ends at a certain number of proteins per aggregate, "chain growth" addition of native monomers and or P_1, unlimited growth which is neutral (all $k_{i,j}$ of same or similar values), cooperative ($k_{i,j}$ increasing as i and j increase), or anticooperative ($k_{i,j}$ decreasing as i and j increase).

While the above formulation has been made for irreversible aggregation, it is straightforward to include equilibrium conditions between different states of aggregation, such as an equilibrium between native and activated monomeric states, and equilibria between native and activated monomers and oligomers, such as dimers and tetramers, and equilibria between oligomers. Such equilibria are represented by a standard notation, such as $N \underset{k_b}{\overset{k_f}{\rightleftarrows}} P_1$, for which the equilibrium constant $K_{eq} = P_1/N$ gives the ratio of moles of monomer in the activated monomeric state to native monomer.

Figure 10.9a shows aggregation signatures for three types of processes: Cooperative aggregation, in which the succeeding k_i increase with increasing i, i.e. aggregates grow more quickly as they increase in mass. Neutral aggregation is where all k_i are equal. Anticooperative aggregation is where k_i decreases with increasing i, i.e. it becomes more difficult to increase the aggregate mass as the mass grows. Figure 10.9b shows aggregation of a monoclonal antibody at $T = 35°C$, which self-limits to dimerization, i.e. all $k_i = 0$ except for $k_{1,1}$.

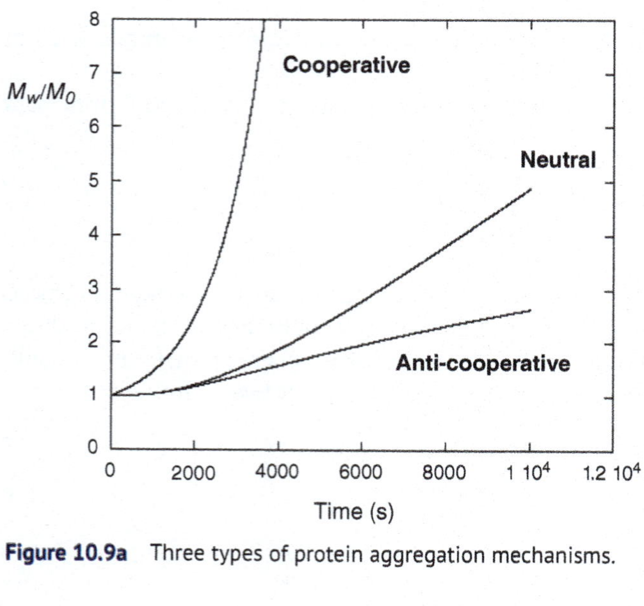

Figure 10.9a Three types of protein aggregation mechanisms.

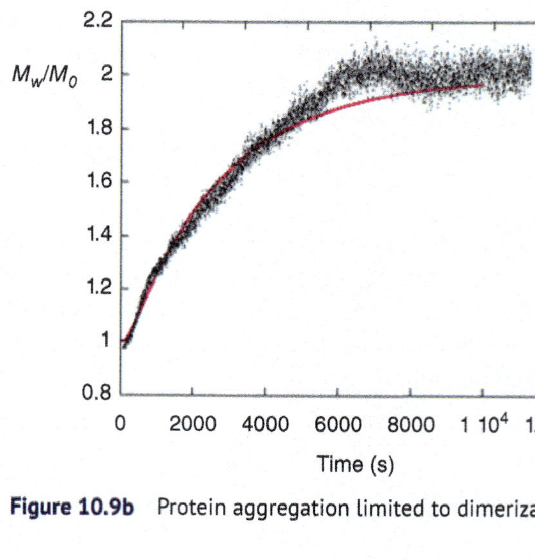

Figure 10.9b Protein aggregation limited to dimerization.

10.2.4 Temperature Ramps Can Muddle Protein Aggregation Behavior

Several of the above figures illustrate how protein aggregation does not occur at a specific temperature, rather, the protein unfolds and aggregates at different rates dependent on temperature. What is sometimes reported as an "aggregation temperature" is often an artifact of a temperature ramp. This occurs when the time scale of the aggregation process is on the order of the time scale of the temperature ramp. Figure 10.10 shows how running a temperature ramp at different rates profoundly affects what might be interpreted as an "aggregation temperature." What is happening, simply, is that the proteins are unfolding at different rates as the temperature changes, and the slower the ramp, the more time the protein has to unfold, and so the misnomered "aggregation temperature" appears at lower temperatures as the ramp rate is lowered. This caution applies to any ramped method where the time scale of the ramp is on the order of the time scale of processes or properties being monitored.

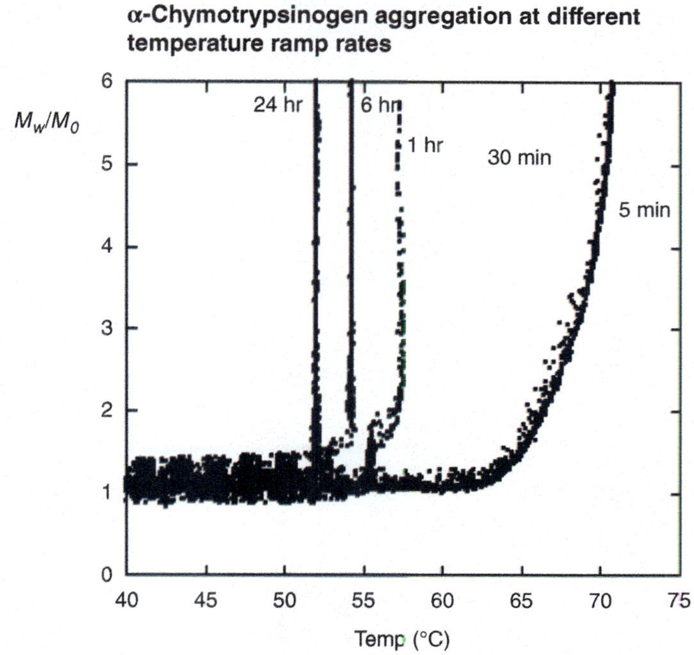

Figure 10.10 The effect of temperature ramp rate on aggregation behavior of α-chymotrypsinogen. The abrupt increases occur at different temperatures, depending on the ramp rate. These abrupt increases are sometimes erroneously identified as "aggregation temperatures." (Jarand, C.W. and Reed, W.F., unpublished results).

10.2.5 Equilibrium Properties of Native Proteins

There has been some hope that the equilibrium properties of native proteins may give a clue as to their stability. In this section, the equilibrium properties of mAB1, mAb2, and a reference mAb from the US National Institute of Standards and Technology (NIST) were determined by static light scattering (SLS) and dynamic light scattering (DLS). SLS yielded the values of the virial coefficients A_2, and, where measurable, the values also of A_3 and A_4. DLS provided the effective hydrodynamic diameter D_H (see Chapter 9.12 for a discussion of what DLS actually measures).

Figure 10.11 shows the Rayleigh ratio, Equation 9.56, for mAb1 (stable), mAb2 (unstable), and the NIST mAB (stable), as a function of concentration. mAb2 shows linear scattering versus concentration up to its highest concentration, whereas both mAb1 and the NIST mAb deviate from linearity at an earlier concentration, become concave downward, then reach a maximum, and subsequently decline. Because the radius of gyration of the native proteins is only about 4 nm, and the incident light was 677 nm, $R_g \ll \lambda$, and there is no angular dependence of the scattering. Hence. the $q = 0$ virial expansion of Equation 9.100 can be used:[3]

$$\frac{Kc}{R(q=0)} = \left(\frac{1}{M} + 2A_2c + 3A_3c^2 + \cdots\right) \quad 9.100$$

If all the virial coefficients are positive, then Equation 9.100 shows that A_2 will cause the initial downward curvature and lead to a plateau versus c, whereas A_3 will cause a maximum with subsequent decrease to occur.[4] A_4 and any higher coefficients will cause the decrease to be steeper.

Table 10.1 shows the results for mAb1 and mAb2, and the NIST mAb. mAb1 and the NIST mAb are stable, whereas mAb2 is unstable. The aggregation rates (AR) of mAb1 and the NIST mAb are well over an order of magnitude less than for the unstable mAb2. A_2 values for the more thermally stable mAb1 and NIST mAb are an order of magnitude larger than for mAb2. Because A_2 integrates attractive and repulsive potentials over all mutual orientations of two interacting mAbs, the higher A_2 indicates higher net repulsion among the proteins, which may also help stabilize them more under thermal and

3 M is used instead of M_w in equation 9.100 because the native proteins are monodisperse.

4 The protein solutions were clear and nonturbid up to the highest concentrations measured, hence only single scattering is significant.

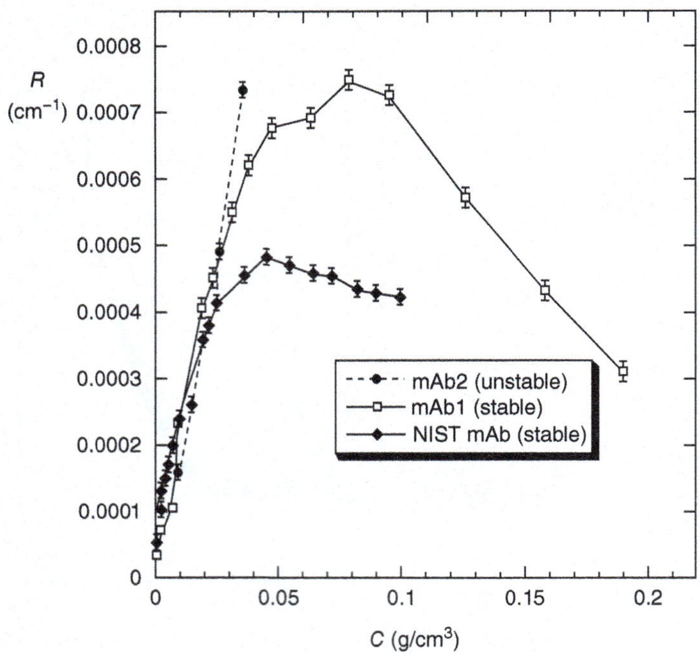

Figure 10.11 Rayleigh ratio, *R* (Chapter 9.4.3), versus concentration (g/cm^3) of mAb1, mAb2, and NIST mAb. (Drenski, M.F., Brader, M.L., and Reed, W.F. (2015). Simultaneous Multiple Sample Light Scattering (SMSLS) for continuous monitoring of protein aggregation, chapter 6. *Technologies for Therapeutic Monoclonal Antibody Characterization*, Volume 3. (eds. Schiehl, J., and Borisov, O. American Chemical Society, Washington, D.C.).

Table 10.1 Equilibrium characteristic of mAb1, mAb2, and the NIST mAb.

	Stable	Stable	Unstable
	NIST mAb	**mAb1**	**mAb2**
AR (1/s) 50°C	$<10^{-7}$	6×10^{-7}	2×10^{-5}
M_w (g/mol)	148,800 ± 2%	153,340 ± 3%	190,900 ± 3%
A_2 (cm^3-mol/g^2)	0.00011 ± 4%	7.33×10^{-5} ± 6%	5.33×10^{-6} ± 34%
A_3, quadratic fit (cm^6-mol/g^3) (measured)	0.00084 ± 8%	3.37×10^{-4} ± 18%	Not measurable
A_3, from max of I_R (measured)	0.00096 ± 14%	4.46×10^{-4} ± 15%	Not measurable
A_3, Boltzmann (theoretical)	0.0011 ± 10%	3.37×10^{-4} ±14%	3.4×10^{-6} ± 71%
A_4 (cm^9-mol/g^4)	Not measurable	2.9×10^{-3} ± 24%	Not measurable
d (nm) from A_2 (quad fit)	12.4 ± 4%	11.1 ± 4%	5.4 ± 13%
D_H (nm) from DLS	11.6 ± 3%	11.9 ± 3%	17.4 ± 3%
Hard sphere resemblance?	Y	Y	N

mechanical stress than mAb2 and other proteins with low A_2 values. Hence, a high value of A_2, >5x10^{-5} cm^3-mol/g^2, may be an indicator of potential stability.

For both mAb1 and the NIST mAb A_3 was measurable and corresponds reasonably well with the theoretical predictions of Boltzmann for hard spheres. A_3 was not measurable for the unstable mAb2. Hence, a measurable A_3 may be another indicator of potential stability.

A_4 was also measurable for mAB1, whereas this factor was too small for mAb2 and the NIST mAb to be measured. Boltzmann predicted the relationships of A_3 and A_4 to A_2 for hard spheres as follows:

$$A_3 = \frac{5MA_2{}^2}{8} \tag{4.160}$$

and

$$A_4 = 0.28695 * M^2 A_2{}^3 \qquad 4.161$$

The value of A_3 is in fair agreement with the Boltzmann prediction based on M and A_2, Equation 4.160, as is the measured A_4 compared to Boltzmann's Equation 4.161.

The similarity of the virial coefficients for mAb1 and the NIST mAb suggests that the hard sphere model may be appropriate for them. The equivalent hard sphere diameter d, can be computed from A_2 and M, via Equation 4.159. It can then be compared with D_H from DLS measurements. The results for d and D_H are shown in Table 10.1. In the case of the thermally stable mAb1 and NIST mAB, D_H is equal to d, within the error bars, but is significantly larger than d for the unstable mAb2. Recalling from Equation 4.158 that A_2 is the integral of the interparticle interaction energies over all orientations,

$$A_2 = \frac{N_A}{2M_2^2} \int \left[1 - \exp\left(\frac{-U(\vec{r}_{12})}{RT} \right) \right] d^3 r_{12} \qquad 4.158$$

and is hence a net energy result, we may surmise the following: The thermally stable mAb1 and NIST mAb resemble hard spheres, whereas the much smaller d for thermally unstable mAb2 indicates that there is significant attraction (negative interaction energy) among particles, and hence does not resemble a hard sphere. The results may further suggest that thermally induced aggregation may be linked to A_2 in the native state: The higher the repulsions the greater will be d with respect to D_H, and vice versa. Comparable values of d and D_H may hence be another criterion for stability.

10.2.6 Protein Aggregation under Denaturing Conditions

Chaotropic agents, such as guanidine-HCl (Gd) and urea, can interrupt hydrogen bonds and hydrophobic interactions and can lead to disruption of protein tertiary structure. This disruption may sometimes lead to protein aggregation. Figure 10.12 shows $M_w(t)/M_0$ for a bovine serum albumen (BSA) as the concentration of [Gd] changes. This involved monitoring the light scattering as a 6M Gd solution was dialyzed into the initial BSA solution in a buffer, at pH 8.0 and pH 8.5. Remarkably, there is a sharp threshold for aggregation. Furthermore, the threshold is strongly dependent on pH, even though separated by only 0.5 units, illustrating the complexity of protein structure and its delicate relationship to solution conditions.

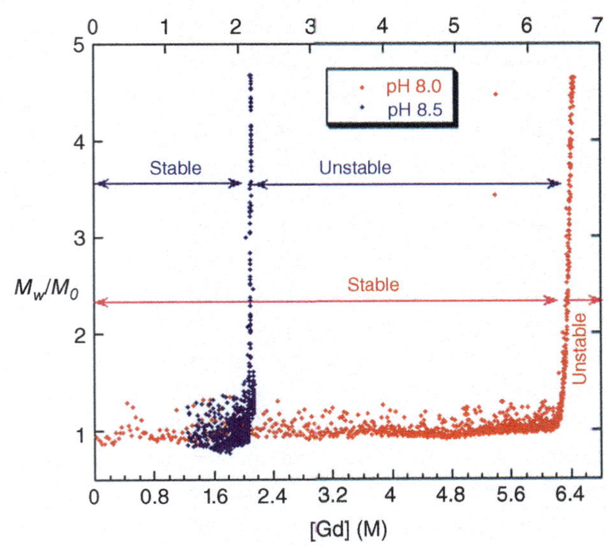

Figure 10.12 The aggregation of BSA during dialysis against 6M Gd, at pH 8.0 and 8.5. (McLeod, M., Jarand, C.W., and Reed, W.F.).

10.2.7 Blood Clotting Via the Fibrinogen Coagulation Cascade

Blood clotting via the fibrinogen coagulation cascade may be considered a beneficial type of natural protein aggregation. However, it does not involve unfolding of the fibrinogen and random aggregations. Rather, it involves the formation of fibrin fibers that form a mesh for platelets that form a plug, the blood clot. Fibrinogen is a long, thin *glycoprotein* that is soluble in plasma and circulates in the blood. It is complex and consists of six polypeptide chains held together by many disulfide bonds and has a fairly large molar mass of ~340,000 g/mol. When a vascular injury is detected the protein *thrombin* attacks fibrinogen, converting it into fibrin, beginning a cascade of events leading to the cross-linking of fibrin bundles and formation of the blood clot.

10.2.8 mRNA Delivery in Lipid Nanoparticles Versus Therapeutic Protein Injection

The development of biologic drugs has greatly advanced medical treatment. The problems with aggregation have been detailed above. As of this time, a radical new technology has just been approved by the USA Food and Drug Administration (FDA) for medical use, in 2020, spurred by the COVID-19 global pandemic, caused by the coronavirus SARS-CoV-2. This technology consists of encapsulating mRNA, engineered to produce specific proteins, in *lipid nanoparticles* (LNP), which consist of cholesterol, ionizable lipids, phospholipids, and stabilizing polymers, such as poly(ethylene glycol). The mRNA nanoparticles are delivered to the patient, normally by injection, and then the LNP undergo endocytosis through the lipid-based cell membranes, since the LNP is soluble in the lipid membrane, and the mRNA is released in the cytosol, and the cells genetic apparatus begins to express the protein.

In the case of COVID-19, the *spike proteins* which anchor the virus to the victim cell were sequenced and the mRNA to produce them was engineered and delivered to patients via LNP. In the patient's body the spike proteins are expressed, which are themselves harmless, but the immune system recognizes them as antigens and builds up defenses. When the actual SARS-CoV-2 virus enters the body it meets with antibodies that can destroy the invasion or at least strongly attenuate it. This success is now unleashing a new wave of treatments for other diseases, since the mRNA can be made to code for any protein desired. As a cautionary note, however, as 2023 dawned, new variants of the COVID virus evolved that were not deterred by the various vaccines effective against earlier strains.

10.3 Enzyme Kinetics

In Chapter 4.16, we saw that in an exothermic chemical reaction, there is a starting concentration of reactants at a certain free energy, and a final state of products at a lower free energy. At final equilibrium, there is a balance between product turning back into reactants and reactants into products. For reactions with $|\Delta G| \gg k_B T$, where ΔG is negative, the equilibrium concentration of reactant is extremely low, so that virtually all reactants have been turned into products. We further saw that there is normally a barrier separating the reactant and product states, the free energy of activation, $\Delta G^{\ddagger}$, which has no bearing on equilibrium concentrations but controls the reaction rate. Catalysts lower $\Delta G^{\ddagger}$ and hence increase activation rates without changing the equilibrium concentrations of reactant and product (Figure 4.26).

Enzymes act as catalysts for biochemical reactions and control their rates. With the exception of RNA-based ribozymes, all enzymes are proteins. Enzymes catalyze thousands of different reactions. *Substrates* are acted upon by enzymes in these reactions. The 3D native folded state of the enzyme is the source of its specificity for substrates. The rates of virtually all biochemical reactions are controlled by enzymes, as the reaction rates are extremely slow at the mild temperature and pressure conditions of the living state. Enzymes can increase reaction rates by as much as a factor of 10^{12}, and even more in some cases.

There are several modes of action of enzymes. The simplest is when an enzyme binds a substrate, forms an intermediate state during which the reaction process takes place, and then releases a product. Some reactions require more than one substrate, others involve a series of reactions and intermediate products. When an enzyme can reversibly catalyze the back reaction of product to substrate, then there is the potential for feedback control. Another mechanism for feedback control is when the enzyme rate increases as more of a substrate is bound, termed *positive cooperativity*, or when it decreases when more substrate is bound, termed *negative cooperativity*, or sometimes *anticooperativity*. The functional form of the reaction rate versus substrate concentration(s) often provides direct evidence of the mode of action involved. No matter what the mode, all enzymes can be saturated by the substrate. This means that while the rate

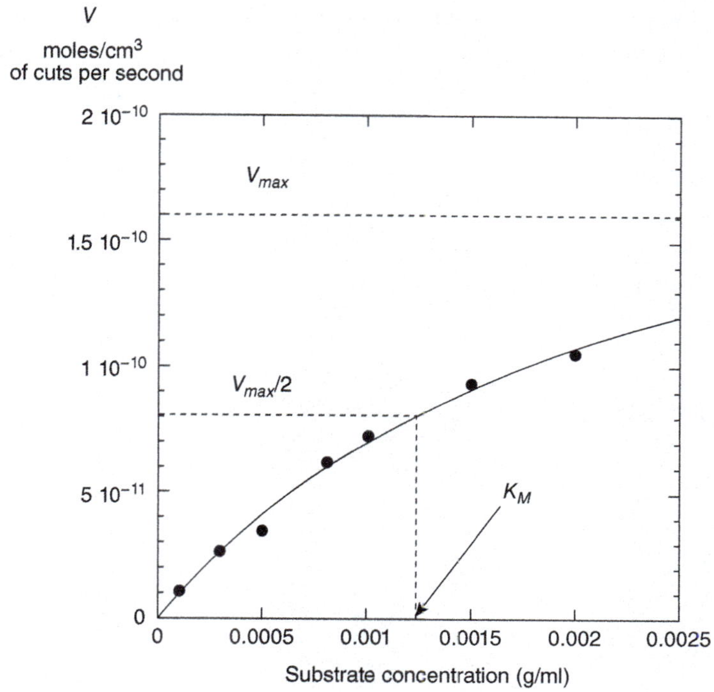

Figure 10.13 Reaction rate of hyaluronidase versus concentration of the hyaluronate substrate. The approach to enzyme saturation is clear.

increases with increasing substrate, at some point the enzyme works at its maximum rate and the rate becomes independent of substrate concentration. Figure 10.13 is an example of the velocity of an enzymatic reaction at fixed concentration versus substrate concentration. The meaning of V_{max} and K_M is derived in the next section, where we consider the simplest kinetic model of enzyme action.

10.3.1 Simple Enzyme Model: Michaelis–Menten–Henri Kinetics

This model effectively describes many enzyme-catalyzed reactions and yields two important characteristics of an enzyme; its maximum rate, or velocity, V_{max}, and the Michaelis–Menten parameter, K_M, which is the substrate concentration at which the velocity is $V_{max}/2$. There are two underlying assumptions for the model: (i) the total amount of enzyme remains constant, since catalysts are not normally used up during reactions, and (ii) the system is at a quasi-steady state, that is, the concentration of the intermediate state of enzyme-bound substrate is constant. This harkens back to Chapter 7.6.1, where we introduced the quasi-steady-state approximation (QSSA) in the context of the concentration of free radical being constant in time.

The model posits that the substrate S binds reversibly to the enzyme E, to form an *intermediate complex* ES with binding and unbinding rate constants k_1 and k_2, respectively. The enzyme performs a catalytic reaction on S to form a product P with a rate constant k_3. The process is shown in schematic Equation 10.6.

$$E + S \underset{k_2}{\overset{k_1}{\rightleftharpoons}} ES \xrightarrow{k_3} E + P \qquad 10.6$$

The dissociation of P back to E and S is ignored since the rate constant should be very small, i.e. the reaction is considered irreversible. The production rate V is given by

$$V = \frac{\mathrm{d}[P]}{\mathrm{d}t} = k_3[ES] \qquad 10.7$$

where k_3 has the units of s^{-1}. The rate of change of $[ES]$ is

$$\frac{\mathrm{d}[ES]}{\mathrm{d}t} = k_1[E][S] - (k_2 + k_3)[ES] \qquad 10.8$$

in which k_1 has units L/mol-s (or cm^3/g–s for mass concentrations in CGS). At a steady state, production rate $[ES]$ = constant, that is

$$\frac{d[ES]}{dt} = 0 \qquad 10.9$$

so that

$$k_1[E][S] = (k_2 + k_3)[ES] \qquad 10.10$$

The total enzyme concentration is

$$[E]_T = [E] + [ES] \qquad 10.11$$

Solving for $[ES]$, and applying $V = k_3[ES]$ leads to

$$V = \frac{k_3[S][E]_T}{k_M + [S]}, \quad K_M = \frac{k_2 + k_3}{k_1}, \quad V_{max} = k_3[E]_T \qquad 10.12a,b,c$$

or

$$V = V_{max}\frac{[S]}{K_M + [S]} \qquad 10.13$$

Hence, plotting $1/V$ versus $1/[S]$ should yield a linear fit:

$$\frac{1}{V} = \frac{1}{V_{max}} + \frac{K_M}{V_{max}} \qquad 10.14$$

The rate constant k_3 gives the number of product molecules formed per second per enzyme molecule when the enzyme is saturated with substrate and is hence termed the "turnover rate," which can range from 1 to nearly 10^6 product molecules per second per enzyme molecule.

In physiological situations, however, enzyme is not saturated by substrate so that most enzyme is not in *ES* form, and most of the population of enzyme is free to bind substrate, as shown by the approximation in Equation 10.15.

$$V = \frac{k_3}{K_M}[E][S] \cong \frac{k_3}{K_M}[E]_T[S] \qquad 10.15$$

So, under low-substrate concentrations, k_3/K_M is the controlling rate, and since

$$\frac{k_3}{K_M} = \frac{k_1 k_3}{k_2 + k_3} \qquad 10.16$$

if the enzyme is highly efficient then $k_3 \gg k_2$, i.e. once the ES complex is formed there is a much higher chance of it forming product than in dissociating back to *E* and *S*. In this case

$$\frac{k_3}{K_M} \cong k_1 \qquad 10.17$$

This condition is sometimes referred to as "kinetic perfection," since the limit on the rate of enzyme forming product is controlled by k_1, which is a diffusion-controlled process, i.e. the enzyme immediately passes through the *ES* intermediate to form product as soon as it encounters substrate via diffusion. Some enzymes that work through a series of substrates and products are organized into assemblies so that the product of one enzyme is immediately passed to the next enzyme in the sequence, lowering the time for a diffusion-controlled encounter.

K_M generally lies between 10^{-1} and 10^{-6} mol/L. It represents the substrate concentration at which the velocity is $V_{max}/2$. The smaller K_M the more tightly the enzyme binds the substrate and forms product, so that K_M is a measure of enzyme-binding efficiency.

10.3.2 Enzymatic Depolymerization Kinetics Determined by Light Scattering

There are many ways of measuring enzyme kinetics, including ultraviolet and visible spectrophotometry, fluorescence, and titration methods. Total-intensity light scattering can be used for measuring kinetics of enzymatic reactions when they affect biomacromolecules by degrading them, building them up, or changing their morphology.

The form factor $P(u)$ of a random coil is given by

$$P(u)=\left(\frac{2}{u^2}\right)(e^{-u}-1+u) \qquad 10.18$$

where $u = q^2<S^2>$. Let r be the number of random scissions made on an initial ideal random coil. Because the probability distribution of scission cuts is random, just as the probability of adding a link when building the coil is random, it gives the interesting result that the previous u and the number of cuts are additive, so that u becomes u'

$$u'=q^2\langle S^2\rangle_0+r \qquad 10.19$$

where $\langle S^2\rangle_0$ is the initial mean square radius of gyration of the coils before any scission occurs so that the random coil form factor becomes

$$P(u')=\left(\frac{2}{u'^2}\right)(e^{-u'}+u'-1) \qquad 10.20$$

The Rayleigh scattering ratio, $R(u',c)$ for an initially disperse population is

$$\frac{Kc}{R(u',c)}=\frac{1}{\int MC(M)P(M)\mathrm{d}M}+2A_2c \qquad 10.21$$

where $C(M)\mathrm{d}M$ is the concentration of polymer in the mass interval M to $M + \mathrm{d}M$, c is the total concentration in g/cm^3, and A_2 has been assumed independent of polymer mass. Because hyaluronate is large and also because r increases as scission proceeds, $e^{-u} \ll 1$ (in practice this works for $u > 3$), $P(u')$ becomes

$$P(u')\approx 2\left(\frac{1}{u'}-\frac{1}{u'^2}\right) \qquad 10.22$$

Using the scaling relationship for an ideal random coil

$$\langle S^2\rangle=\gamma M \qquad 10.23$$

and defining $\dot{\beta}$ as the number of cuts per second per gram per mole of polymer mass, then the number of scissions per second per polymer of mass M is proportional to the polymer's mass M, and, in the early linear regime the total number of scissions the number of cuts r for the polymer of mass M is

$$r=\dot{\beta}Mt \qquad 10.24$$

so that

$$u'=M\left(\gamma q^2+\dot{\beta}t\right) \qquad 10.25$$

Substituting this into Equation 10.21 for $R(u',c)$, and retaining terms to first order in u yields

$$\frac{Kc}{R(q,t)}=\frac{1}{2M_n}+\frac{\gamma q^2}{2}+\frac{\dot{\beta}t}{2}+2A_2c \qquad 10.26$$

where c is the protein concentration in g/cm^3, which remains constant. Hence, the scission rate in moles of bonds broken per second per gram of polymer mass is

$$\dot{\beta} = 2Kc\frac{\mathrm{d}(1/R(q,t))}{\mathrm{d}t} \quad 10.27$$

where the time dependence of $R(q,t)$ indicates that R changes due to changing $\langle S^2\rangle_0$ and r in time. The velocity V, given as moles of bonds broken per cm^3 per second is then

$$V = \dot{\beta}c \quad 10.28$$

where the total concentration, c, does not change during the degradation; only the fragment distribution changes.

Figure 10.14 shows an example of light scattering intensity, plotted as R_{total}/Kc (=$R(t)/Kc$) versus time for the random cleavage of sodium hyaluronate by hyaluronidase for a fixed [hyaluronidase] and variable substrate [hyaluronate], at $T = 25°$C, in a 0.1-M succinate buffer at pH = 5.31 and 0.15 M NaCl. K is the optical constant given by Equation 9.70. The hyaluronidase concentration was fixed at 40 units/cm^3, where one *enzyme unit* is the amount of enzyme that catalyzes reaction of one micromole per minute under specified conditions. Hyaluronate is a linear polysaccharide of N-acetyl-D-glucosamine (NAG) and D-glucuronic acid (GlcUA) units with the repeating structure (O-β-NAG-(1,4)-O-β-GlcUA-(1,3)). It has the highest molar mass and lowest charge density of the glycosaminoglycans (Chapter 3.3). It is involved in providing compressive stiffness and swelling pressure in extracellular matrix and in the eye and acts as a backbone for proteoglycan formation (Chapter 12.2). It has a persistence length of about 10 nm at moderate ionic strength and resembles a random coil. Hyaluronate is cleaved randomly by hyaluronidase.

Using Equations 10.27 and 10.28, V can be extracted from these curves, and the values are plotted versus $1/c$ in Figure 10.15. The graph is given in concentration units of g/cm^3 so that $V_{max} = 1.59 \times 10^{-10}$ g/cm^3–s = 4×10^{-13} mol/cm^3–s, and $K_M = 0.0014$ g/cm^3. With the monomer molar mass of hyaluronate of 400 g/mol, $K_M = 0.00355$ mol/L, which is a typical value for an enzyme.

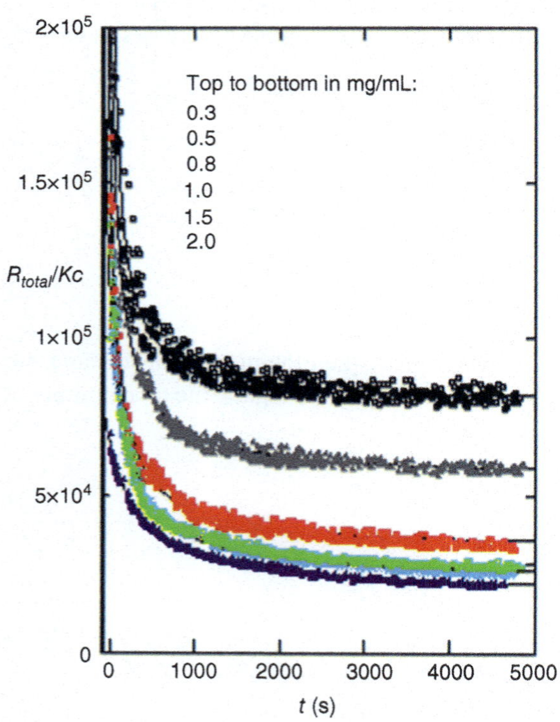

Figure 10.14 Time-dependent scattering for the hydrolysis of hyaluronate by hyaluronidase. The concentration of hyaluronate is shown in the list. (Drenski M.F. and Reed, W.F. (2004)).

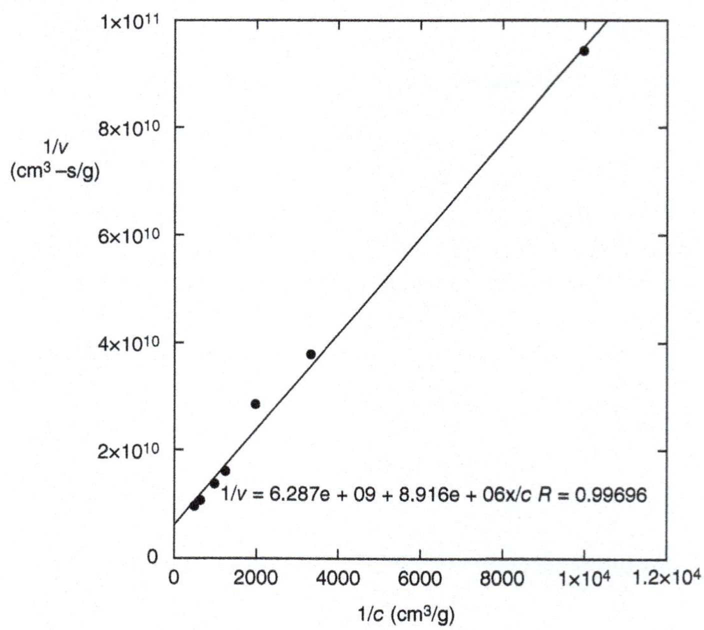

Figure 10.15 Michaelis–Menten plot, yielding K_M and V_{max}. (Drenski M.F. and Reed, W.F. (2004)).

10.3.3 Presteady-state Kinetics

In the Michaelis–Menten model in Section 10.3.1 and application in Section 10.3.2, it was assumed the enzyme system was at steady state. The build-up to the steady state can be examined by dropping the condition that $[ES]$ = constant. Then, we don't set Equation 10.8 equal to zero (reminder $\frac{d[ES]}{dt} = k_1[E][S] - (k_2 + k_3)[ES]$), and using total enzyme concentration to eliminate $[E]$ yields

$$\frac{d[ES]}{dt} = -(k_1[S] + k_2 + k_3) + k_1[S][E]_T \quad 10.29$$

But $[S]$ is also changing with time, yielding the coupled equation:

$$\frac{d[S]}{dt} = -k_1[S] + k_2[ES] \quad 10.30$$

Given the constants and $[E]_T$, these can immediately be integrated numerically to obtain $[ES](t)$ and hence the velocity V, by Equation 10.12a. To gain a general insight with an analytical solution, we can assume that $[S] \gg [ES]$, since the catalyst is generally in much lesser concentration than the substrate, and so $[S]$ is approximately constant during the pre-steady-state phase. With this approximation, Equation 10.29 has a simple first-order solution, yielding, via Equation 10.12a:

$$V(t) = \frac{k_3[E]_T[S]}{K_M + [S]}\left(1 - e^{-\alpha t}\right) \quad 10.31$$

where

$$\alpha = k_1[S] + k_2 + k_3 \quad 10.32$$

This is just a first-order build-up to the steady-state Michaelis–Menten–Henri Equation 10.12a. The rate α increases with the increase of any of the four factors it is composed of.

10.3.4 Michaelis–Menten–Henri Kinetics with an Intermediate Enzyme Complex, EI

In this case, the *ES* complex turns irreversibly into an intermediate involving the enzyme, EI, which then proceeds to product. In this case, *EI* can only proceed to product and not return to *ES*; it is the *committed step* in the reaction; if EI is reached, it will definitely proceed to E + P.

$$E + S \underset{k_2}{\overset{k_1}{\rightleftharpoons}} ES \xrightarrow{k_3} EI \xrightarrow{k_4} E + P \quad 10.33$$

The velocity of the reaction is

$$V = \frac{d[P]}{dt} = k_4[EI] \quad 10.34$$

Using the same steady-state approximation as above

$$\frac{d[ES]}{dt} = \frac{d[EI]}{dt} = 0 \quad 10.35$$

so that

$$[EI] = \frac{k_3}{k_4}[ES] \quad 10.36$$

and total enzyme is still conserved

$$[E]_T = [E] + [ES] + [EI] \tag{10.37}$$

Using the expression for $[ES]$, and $[E]$ from yields $[EI]$, and, after some algebraic manipulation gives

$$V = \frac{\left(\dfrac{k_3 k_4}{k_3 + k_4}\right)[S][E]_T}{\dfrac{(k_2 + k_3)}{k_1(k_3 + k_4)} + [S]} \tag{10.38}$$

If $k_4 \gg k_3$, then the first Michaelis–Menten–Henri Equation 10.12a is recovered, whereas if $k_4 < k_3$ it becomes the rate-limiting step.

10.3.5 Kinetics of Reversible Product Formation

If the enzyme can bind product and return it to its substrate form then this is a reversible enzymatic reaction and can serve to modulate enzyme activity, increasing velocity of production when product is scarce, and returning product to substrate when there is sufficient product and more substrate is needed for a different function. In this case, the reaction pathway is

$$E + S \underset{k_2}{\overset{k_1}{\rightleftarrows}} ES \underset{k_4}{\overset{k_3}{\rightleftarrows}} E + P \tag{10.39}$$

The process is schematized in Figure 10.16.

The only difference between this and pathway Equation 10.6 is the addition of k_4 which reforms the ES complex from product, which can then either break apart into E and substrate, or back into E and product. The velocity for $[ES]$ conversion is now:

$$\frac{d[ES]}{dt} = k_1[E][S] - (k_2 + k_3)[ES] + k_4[P][E] \tag{10.40}$$

and the velocity of the reaction for production of P is

$$V = \frac{d[P]}{dt} = k_3[ES] - k_4[P][E] \tag{10.41}$$

The negative term means that the reaction will return product to substrate when $V < 0$, that is, when

$$k_3[ES] < k_4[P][E], \quad \text{Reverse Reaction Favored} \tag{10.42}$$

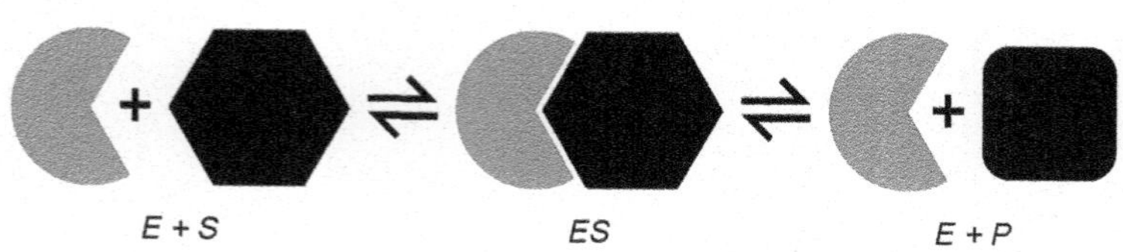

Figure 10.16 Enzyme scheme with reversible product formation.

Again using the steady-state approximation that $d[ES]/dt = 0$, and Equation 10.11 for the relation between $[ES]$, $[E]_T$, and $[E]$ yields the velocity:

$$V = \frac{[E]_T \left(k_1 k_3 [S] - k_2 k_4 [P]\right)}{k_2 + k_3 + k_4 [P] + k_1 [S]} \qquad 10.43$$

The reaction stops, i.e. $V = 0$, when

$$[P] = \frac{k_1 k_3}{k_2 k_4} [S] \qquad 10.44$$

Hence, reversible product formation produces a *feedback mechanism* for controlling the rate of product formation, stopping product production when Equation 10.44 is met, and breaking down product when

$$[P] > \frac{k_1 k_3}{k_2 k_4} [S] \qquad 10.45$$

10.4 Cooperative Binding in Proteins

Cooperativity appears in many ways in the living state. Cooperative binding by enzymes is a good illustration. Hemoglobin, for example, binds up to four O_2 molecules, each one more tightly than the preceding. It is a good example of an *allosteric enzyme*, one whose activity is affected by the binding of other molecules at different sites on the protein from where the specific enzymatic activity occurs. The presence of CO_2 and H^+ in metabolically active tissue such as muscle, where both CO_2 and acid (H^+) are produced in abundance, acts allosterically to promote release of needed O_2 from hemoglobin. Myoglobin, in contrast, is a much less evolved O_2 transport protein than hemoglobin and does not exhibit cooperative O_2 binding.

The mechanisms of cooperativity in allosteric enzymes are rooted deep in the tertiary structure of the protein and the details of how it is affected as succeeding substrate molecules bind.

It is a simpler matter to make a mathematical model that describes cooperative binding, assuming a phenomenological suite of binding constants, K_1, K_2, ... K_N, as follows. (Notice that capital K is for equilibrium constants, whereas the previous lower case k is for rate constants). Consider a protein A (e.g. hemoglobin) that binds a substrate B (e.g. O_2) to form a complex of the protein with one molecule of B, with an equilibrium binding constant K_1, which relates the concentrations $[A]$, $[B]$, and $[AB_1]$ to each other.

$$A + B \xleftrightarrow{K_1} AB_1 = C_1, \quad K_1 = \frac{[AB_1]}{[A][B]} \qquad 10.46$$

where C_1 is used as a convenience to express AB_1. Its concentration is given by

$$[C_1] = K_1 [A][B] \qquad 10.47$$

AB_1 can then bind another molecule of B to form AB_2 with binding constant K_2

$$AB_1 + B \xleftrightarrow{K_2} AB_2 = C_2, \quad K_2 = \frac{[AB_2]}{[AB_1][B]} \qquad 10.48$$

and

$$[C_2] = K_2 [B][C_1] = K_1 K_2 [A][B]^2 \qquad 10.49$$

This can be generalized to N molecules of B.

$$AB_{N-1} + B \xleftrightarrow{K_N} AB_N = C_N, \quad K_N = \frac{[AB_N]}{[AB_{N-1}][B]} \qquad 10.50$$

where

$$[C_N] = K_1 K_2 \ldots K_N [A][B]^N \qquad 10.51$$

The total concentration of protein A_t is composed of A, with no B bound, and the sum of all proteins that have one or more B molecules bound, A_B, that is

$$[A_t] = [A] + [A_B] \qquad 10.52$$

where

$$[A_B] = \sum_{i=1}^{N} [C_i] = [A] \sum_{i=1}^{N} [B]^{K_i} \left(\prod_{j=1}^{i} K_j \right) \qquad 10.53$$

Solving for $[A]$ and substituting into the latter equation yields

$$[A_B] = [A_t] \frac{\gamma}{1+\gamma} \qquad 10.54$$

where

$$\gamma = \sum_{i=1}^{N} [B]^{K_i} \left(\prod_{j=1}^{i} K_j \right) \qquad 10.55$$

Figure 10.17 shows examples of the fraction of substrate B bound to A, $[A_B]/[A_t]$, versus $[B]$ for two cases of cooperativity where $N = 4$. In one case, K for the each successive B that gets bound is twice the preceding K. For the case of even higher cooperativity, K for each successive K is quadruple the preceding K. As expected, at any given $[B]$, the case with the quadrupled increase in K has a higher fraction of bound substrate. In both cases, there is an inflection point at a low value of $[B]$. Also shown in Figure 10.17 is the case of noncooperativity, where all K's are equal. There is less binding at a given $[B]$ than the cooperative cases and there is no inflection point. Finally, Figure 10.17 shows a case of anticooperativity, where each successive K is one-half the previous value. As expected, it binds the least B of all the cases.

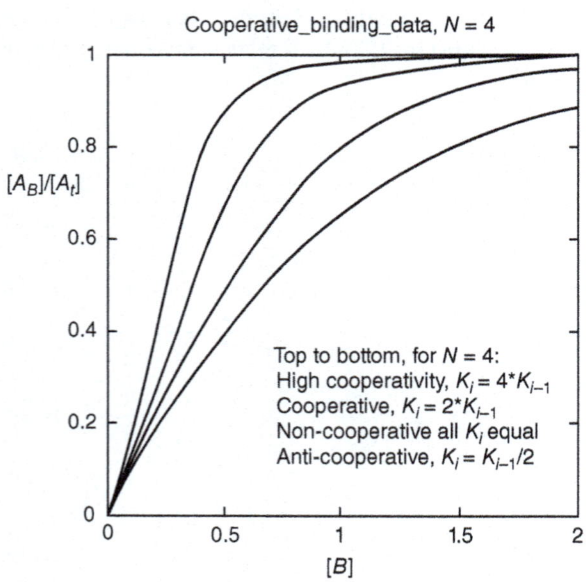

Figure 10.17 Fraction of A that has one or more bound substrate molecules, B, where $N = 4$ is the maximum number of Bs that can bind to one A. Two cases are shown for cooperative binding, and one case of anticooperative binding. The noncooperative case is also shown.

Now, let's consider an "all or nothing" binding scenario, which will lead to a marked difference from the previous scenario where A could bind any number from 1 to N of B. In this "all or nothing case," at equilibrium A either has no molecules of B bound or binds all N molecules it is capable of binding. The numbers 1,2,3..., $N - 1$ are excluded from binding. Then

$$A + NB \xleftrightarrow{K} AB_N, \quad K = \frac{[AB_N]}{[A][B]^N} \qquad 10.56$$

The equilibrium constant K, now bears no subscript since it is the only constant describing the equilibrium. The solution is often called the Hill equation.

$$[A_B] = [A_t] \frac{K[B]^N}{1 + K[B]^N} \qquad 10.57$$

For cooperative enzymatic reactions, N is often the fitting parameter for experimental data. If $N > 1$, then the binding is cooperative, for $N = 1$ it is noncooperative, and for $N < 1$ it is anticooperative. Figure 10.18 illustrates the binding for cooperative, noncooperative, and anticooperative binding. For any value of K, half of A are bound when $KB^N = 1$, which is why all the cases intersect at the same point. In Figure 10.18, K was kept the same for all cases, in order to illustrate this point.

At first, it may seem puzzling that before the ½ point is reached there is actually less binding in the cooperative case than for the noncooperative case, and even the anticooperative case. Figure 10.17 showed that as cooperativity increases binding increases for any value of $[B]$. It is the "all or nothing" nature of this mechanism that causes lower binding at lower $[B]$. The fact that there are no equilibrium bound states with less than N molecules of B bound to A means that this type of cooperativity imposes a *penalty* on individual molecules of B binding to A. Physically, there must be unstable intermediate states AB_1, $AB_2..AB_{N-1}$, which have very short lives. These states are necessarily unstable because none of them exist at equilibrium. But these intermediate states must have a finite life, no matter how short, because the probability of N molecules of B colliding with and sticking to A at exactly the same instant is vanishingly small. N molecules have to reach A and remain stuck to A during the lifetime of the first molecule of B to arrive, in order for a stable AB_N to form. As N gets greater we expect the probability of more molecules arriving within the lifetime of the first to get smaller, so that a higher concentration of B is needed to increase the probability to the same level as a case with lower N. In fact, the behavior leads to what essentially approaches a phase transition, as seen in the $N = 20$ case in Figure 10.18.

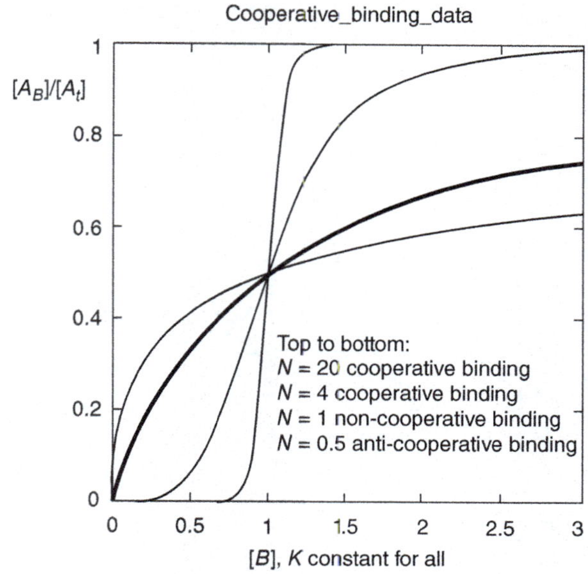

Figure 10.18 "All or nothing" binding of B to A, for different values of N, which is the only number of B that can stably bind to A. K is taken as equal in all cases. The noncooperative case is shown ($N = 1$), as well as an anticooperative state, and two cooperative states. As N gets larger the binding becomes sharp and resembles a phase transition.

Figure 10.19 gives a conceptual representation of how cooperative binding can affect a process such as uptake and delivery of a substance, for example, oxygen uptake and release by hemoglobin (Hb) in the process

$$Hb + NO_2 \leftrightarrow Hb(O_2)_N \tag{10.58}$$

The equivalent Hill cooperativity coefficient is $N = 2.8$ in the Hill equation for fractional saturation of hemoglobin f

$$f = \frac{[O_2]^N}{[O_2]^N + [O_2]_{ref}{}^N} \tag{10.59}$$

where $[O_2]_{ref}$ is a reference O_2 concentration. Figure 10.19 plots fractional saturation of Hb by O_2 versus $[O_2]$ in arbitrary units for the case where $N = 4$ ($N = 2.8$ would normally be used, but this conveys the concept). In a high O_2 environment, such as a lung, $f = 0.82$ for the $N = 4$ cooperative binding curve when the lung concentration (arbitrary units) is 1.5. When this high O_2 content reaches a low oxygen environment, such as 0.5 in a muscle, the $HB(O_2)_N$ drops to $f = 0.06$, so that it unloads 92% of its O_2. In contrast, a noncooperative molecule that binds $N = 4$ O_2 (or 4 myoglobin molecules that bind one O_2 each) will have $f = 0.6$ in the lung and $f = 0.35$ in the muscle, so that only 41% of its O_2 is unloaded, resulting in a much less efficient oxygen transfer process from lung to muscle.

Before leaving the Hill mechanism, the $N = 1$ case is interesting because this is the noncooperative case and the Hill equation then reduces to

$$[A_B] = [A_t]\frac{K[B]}{1 + K[B]} \tag{10.60}$$

This is one form of the paradigm equation:

$$y = A\frac{Kx}{C + Kx} \tag{10.61}$$

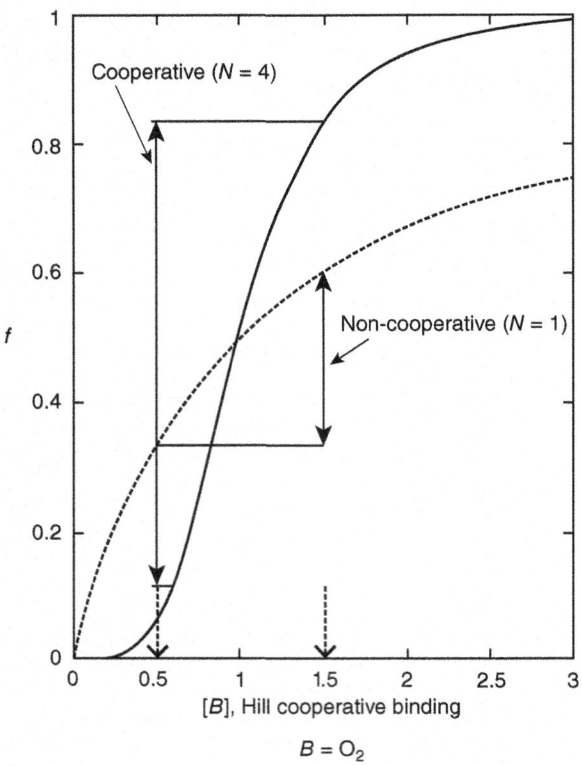

Figure 10.19 Effect of cooperativity (hemoglobin, $N = 4$) and noncooperative binding (myoglobin, $N = 1$) on loading and release of O_2. The O_2 release between $[B] = 1.5$ and $[B] = 0.5$ is much larger for the cooperative case than for the noncooperative case.

where A, C, and K are constants. It is found in many contexts, such as binding and kinetics associated with binding. We already saw this in the simplest Michaelis–Menten–Henri model for enzymatic velocity, or rate of product production, Equation 10.13, where $C = V_{max}$, $K = 1$, and $C = K_M$. Since enzymes are catalysts we expect to find Equation 10.61 in the context of many other catalysts besides enzymes, since the catalyst must be able to easily bind to its substrate and release its product.

Equation 10.61 describes substrate binding to enzyme via the form of Equation 10.13. It also describes the adsorption of molecules to a surface, where K is the equilibrium constant between the adsorbed and free state, and includes both energetic and entropic terms. If the molecules are gas molecules that stick to a surface then $C = 1$ and x is the partial pressure of the gas. The first derivation of this was by Irving Langmuir (United States, 1881–1957) and it is often termed the *Langmuir binding isotherm* equation. We will also encounter the paradigm equation below when dealing with the helix–coil transition. In each case, A, C, and K are based on physical parameters describing the phenomenon.

10.4.1 Effect of Cooperative Binding on Kinetics

If we revisit the Michaelis–Menten approach for cooperative binding, we will get sigmoidal behavior for velocity, just as we did for the concentration in the Hill Equation 10.57. We use schematic Equation 10.5, but now with E, ES, P, and N ($N = 1$ in schematic Equation 10.5)

$$E + NS \underset{k_2}{\overset{k_1}{\rightleftarrows}} ES_N \xrightarrow{k_3} E + P \tag{10.62}$$

The new rate for ES is

$$\frac{d[ES]}{dt} = k_1[E][S]^N - (k_2 + k_3)[ES_N] \tag{10.63}$$

The velocity is similar to Equation 10.7 ($N = 1$), but now uses $[ES_N]$ in place of $[ES]$.

$$V = \frac{d[P]}{dt} = k_3[ES_N] \tag{10.64}$$

Using the QSSA, d[ES]/dt ~ 0, solving for [ES] and finding V yields an expression very similar in form to the Hill equation:

$$V = \frac{Nk_3[S]^N[E]_T}{K_m + [S]^N} \quad 10.65$$

where K_m is unchanged from the noncooperative model and is given by Equation 10.12b.

10.5 Cooperativity in the Helix–Coil Transition

Formation of secondary structure, such as the α-helix, was seen as a critical process in the folding pathway of a protein in coil form. Likewise, unfolding of the protein and complete denaturation means breaking these secondary structures. Helical structures also form in polysaccharides, RNA, and double helices occur in DNA, and helices of higher order are found in fibrous structures. All of these helices have cooperativity in common. This means that by taking a helical conformation, the net free energy is lower than in a completely random coil. The free energy will be composed of the sum of all energetic interactions in the polymer chain and the chain entropy. There is an entropic penalty for the coil to pay when a random coil assumes a helical form so the coil energetic savings and water entropy gain must compensate for this.

Normally, helix transitions to a coil as temperature is raised and weak hydrogen bonds are broken. In some cases, however, the helix forms as temperature is raised, such as the case of the homopolypeptide poly-γ-benzyl-L-glutamate, on which the Bragg–Zimm model for the helix coil transition was first tested. The reason for this counterintuitive behavior is that account must also be taken of the change in the solvent entropy when the helix is formed and solvent molecules are either liberated or forced into bonds. In the case where the solvent is liberated as temperature increases and its entropy increases sufficiently, the entropy loss of the chain and an energetic penalty can be offset and the coil passes to a helix form as temperature increases. This is similar to the lower critical solution temperature discussed in Chapter 7.3.4, where the weakly solvated form of poly(n-isopropyl acrylamide) in water is lost as the temperature rises and the water is released, increasing its entropy, while the NIPAM chains collapse into a state of decreased entropy.

A simple model can be built that glosses over the complexity of the energetic interactions and entropy changes and attempts to introduce cooperativity in the most elementary way. As usual, a model starts by simplifying a system, extracting what are hopefully the most important features and ignoring the ones of lesser importance. The early Zimm–Bragg approach to the helix–coil transition proved successful in explaining basic behavior. It involves two phenomenological parameters from which the partition function is built. The partition function then yields the average helical content in the chain and other related characteristics. The full treatment is rather complex, so we just sketch the essentials here.

The chain of N monomers is assumed to consist of monomers in either coil (c) or helical (h) states. The helical state implies fixed rotational angles for a helical monomer and a monomer in the coil state has large rotational freedom. Hence, there is an entropic penalty for a monomer being in the h form. The helical content θ is defined as the fraction of monomers in the h state.

$$\theta = \frac{N_h}{N} \quad 10.66$$

where N_h is the number of monomers in helical form. An example is the following sequence for $N = 30$ monomers: *cchhhhhhcccchhhcccccccccchhcccchc*

There are 12 h monomers so that $\theta = 0.40$. There are four helical segments. The two phenomenological parameters are (i) the favorable weight given to the formation of hydrogen bonds for each h monomer. This will be directly related to $\exp[-E_H / k_B T]$, where E_H is the energy of the hydrogen bond formed in the h monomer. This parameter is denoted s and is used as a statistical weight in the partition function below. The more negative E_H the larger s is. (ii) The unfavorable entropic restriction that occurs at a c–h interface. This *helix initiation parameter* occurs at the c–h interface (but not at h–c) and is denoted σ. It is a cooperativity parameter because it favors an h–h pair and penalizes a c–h pair. Now s applies to every h monomer, but σ only applies to each c–h interface. The weights for the above example are then: s^{12} for the 12 h monomers and σ^4 for the four c–h interfaces. The product of these weights is then $s^{12}\sigma^4$ and the partition function is the sum of the weights over all possible combinations of c and h, where g_{ij} is the number of ways there can be j helical segments with a total number i of h monomers for a given N. The complete partition function is the product of these three weights so

$$Z = \sum_{i=0}^{N}\sum_{j=0}^{N/2} g_{ij} s^i \sigma^j \quad 10.67$$

where the summation in i is over all h monomers, corresponding to state s, and the summation in j is over all c–h pairs. The upper limit for σ summation is $N/2$, which occurs when the chain is alternating c–h–c–h...c–h and is 0 if there are no c monomers.

Once Z is established, the average number of helical monomers <i> is obtained by

$$\langle N_h \rangle = \langle i \rangle = \frac{\sum_{i=0}^{N}\sum_{j=0}^{N/2} i g_{ij} s^i \sigma^j}{\sum_{i=0}^{N}\sum_{j=0}^{N/2} g_{ij} s^i \sigma^j} \quad 10.68$$

The numerator can be found by differentiating Z, so that combining this together with Equation 10.66, the helical content is

$$\theta = \frac{s}{Z}\frac{\partial Z}{\partial s} \quad 10.69$$

The derivation of Z is quite involved and was provided by Zimm and Bragg. The (non-obvious) approximate result for a chain of N monomers, in the form provided by Meyer Jackson (2006), is

$$Z \sim \left(\frac{1 + s + \sqrt{(1-s)^2 + 4\sigma s}}{2} \right)^N \quad 10.70$$

Using Equation 10.70, in the limit of large N, the helix content is approximated by

$$\theta \sim s \left(\frac{1 + \frac{s - 1 + 2\sigma}{\sqrt{(1-s)^2 + 4\sigma s}}}{1 + s + \sqrt{(1-s)^2 + 4\sigma s}} \right) \quad 10.71$$

Figure 10.20 shows θ versus s for three different values of the cooperativity parameter σ. $s = 1$ is the point at which $\theta = 0.5$ for all values of σ. For s near zero, E_H is positive and large so that very few monomers are in helical form. As s increases, E_H becomes smaller, $s = 1$ corresponds to $E_H = 0$, which is the midpoint where $\theta = 0.5$ for any value of σ in this approximation. The cooperativity is controlled entirely by σ, as seen dramatically in Figure 10.20. For $\sigma = 1$, there is no cooperativity, and Z is not influenced by σ, so that $\theta = s/(1 + s)$, which is the same form as $N=1$ in the Hill Equation 10.57. For $\sigma > 1$, the situation is anticooperative and θ builds up more slowly above s = 1.

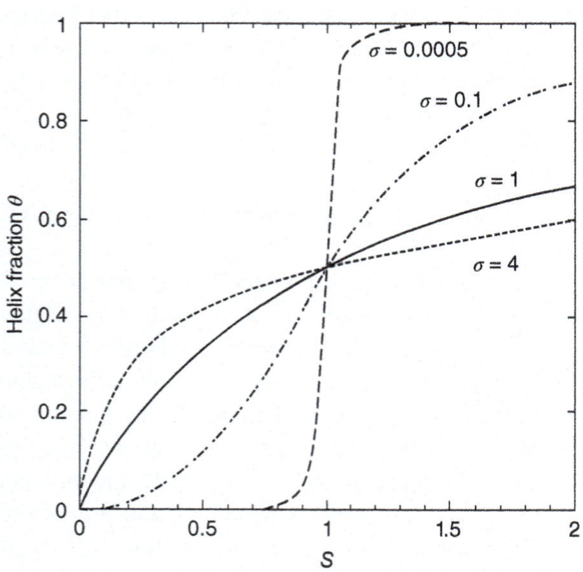

Figure 10.20 Helix–Coil transition by Equation 10.71, for various values of the helix initiation cooperativity parameter σ.

It is interesting that the behavior of the helix–coil equilibrium in Figure 10.20 resembles the Hill-binding equilibrium behavior in Figure 10.18. In the Hill case, the x-axis is the concentration of substrate and the degree of cooperativity is defined by the number N, of B's bound to A. The higher N the higher $[B]$ is needed at low $[B]$ to form AB_N. For the helix–coil equilibrium, the weighting due to the energetics of H-bond formation, the factor s, takes the place of $[B]$. The higher s, the greater the energy savings in H-bond formation in the h state. The cooperativity factor, σ, which is now a number less than unity sharpens the transition as it decreases toward zero, and it takes the place of the discrete values of N, which sharpen the transition as they get larger.

The interpretation of the increasing sharpness of the transition as σ decreases can be understood as follows. As σ decreases, the entropic penalty for forming a c–h interface increases. Say that in case #1 σ is slightly less than 1 and that in case #2 it is slightly greater than 0. We consider a value of s less than 1, say, around $s = 0.5$. Figure 10.20 allows visualization of case #1 and #2. The entropic σ penalty is low in case #1 and so it does not cost much to initiate a sequence of h monomers and so there is a high θ. For case #2, it is costly to initiate even a single c–h interface and so the chain prefers the higher entropy coil form and θ is lower than case #1. As s increases and the energetic incentive to transition from c to h increases; it is only possible to form a few c–h interfaces and this only occurs when long stretches of h occur together. At some point shortly after $s = 1$, virtually the whole chain is in h form. The high entropic penalty of σ in case #2 essentially leads to an "all or nothing" scenario, as occurs in the Hill case.

Loss of both helical and beta-sheet structure in proteins can be observed by several methods, including loss of circular dichroism, differential scanning calorimetry, fluorescence-based assays, and changes in molar mass scaling laws for intrinsic viscosity and radius of gyration.

10.6 Histones and Other Chromosomal Proteins

As a prelude to Chapter 11 and its discussion of the genome and its physical manifestation in chromosomes, histones are introduced here. Histones are short, with molar masses on the order of 10^4 g/mol, and have abundant lysine and arginine, such that they are very basic and are positively charged under physiological conditions. This allows them to associate tightly with DNA, which is negatively charged, to form *nucleosomes*. The histones are generally dimeric with alpha helices that make them resemble "spools," around which DNA winds and condenses to fit within very small cell nuclei. Besides DNA condensation, histones are involved in gene regulation and protein transcription, and DNA repair. Histones are highly conserved among eukaryotes and those found in archaea may be evolutionary precursors to eukaryotic histones.

Histones are subject to many chemical modifications in the nucleus, which in turn control how they interact with DNA. It has been proposed that a possible histone code underlies the complex variation in chemical modifications. Some of the modifications include arginine and lysine methylation which can be recognized to high accuracy by certain other proteins, and lysine acetylation which neutralizes histone positive charge, loosening chromosome structure, and promoting transcription.

While histones have an overarching role in DNA condensation and gene regulation there are many more non-histone proteins and complexes in the nucleus with vital roles. These include DNA polymerase, structure maintaining chromosome complexes, involving cohesion and condensing, and many others.

10.7 The Action of Proteins Often Depends on Correlated Internal Motions

The secondary, tertiary, and quaternary structures of proteins have been described, but these structures have been considered as stationary until now. In reality, each amino acid in a native protein undergoes correlated motions. While random thermal energy imparts random internal motion to proteins, the effect of correlated motions often, or usually, overshadows this. Evidence of these motions comes from experimental structural determinations made under different conditions, such as a protein in the presence of a substrate, inhibitor, in quaternary association, and in allosteric action, among others.

The correlated motions often lead to a discrete number of states, e.g. an open and closed hinge, states of various helical content, rotameric states, etc. These can be represented by energy landscapes, with energy wells characterizing the different states, determining the fractional occupation of each state, and barriers between them, controlling transition rates among them.

The significance of these correlated motions should not be underestimated. Proteins carry out a vast number of biological functions – enzymatic, directed motion, static and dynamic structure, light transduction – and much of this depends on correlated and coordinated motions within the protein. Several types of motion have been discovered and modeled:

Helix-to-coil transitions at the c–h boundary discussed in Section 10.5 can be dynamic, with c–h boundaries forming and disintegrating. This leads to different lengths and numbers of helical segments, which, in turn lead to motions and structural changes in the protein.

Hinge motions have been found associated with the ends of secondary α-helices and β-sheets. The ends between two helices or sheets connected by a flexible linker region can then act as a hinge, such that the two connected secondary structures can open and close, like a door. This can have several uses, such as trapping a substrate or antigen, and receptor dimerization.

Protein ratchets are found in many different biological functions and have *directed motion* as their most important effect. A number of mechanisms for the action of protein ratchets have been modeled. A common motif is when there is a repeating scaffold, such as a strand of RNA, or a stretch of proteinaceous cytoskeleton fiber. One mechanism is the *Brownian ratchet*, which combine random diffusion with asymmetric potentials, such as a "sawtooth" potential, which rises very steeply but falls less steeply. In this case, a molecule being transported along the strand or fiber is bound on the steep side of the potential and is released via ATP energy input. It then diffuses, but, because the downside of the sawtooth potential is less steep, there is a greater probability that the molecule will move in that direction, the direction of travel, and bind to the next potential minimum in the sequence. This is another example of a biased random walk.

Subdomain dynamics of multidomain proteins (Chapter 3.1.4) are controlled by correlated motions. Oftentimes, the active site is located close to the joining section of two subdomains, which must be moved in correlated motion to achieve the highly specific geometry of the active site.

Many approaches have been taken to modeling correlated motions in proteins. These include many types of computer simulations, such as molecular dynamics, and more fundamental theories such as *elastic network models* (ENM, see Dill et al., 2017, Chapter 12, and references therein). In this latter, there is an equilibrium position for each amino acid in its native tertiary structure, and nearby monomers have spring-like forces among them. This leads to normal modes of vibration. Monomers deep within the protein have higher spring constants than those on the outer portions. ENM have been highly developed and can describe hinge motions, ratchets, breathing modes, ligand binding, and aspects of allostery.

10.8 Directed Protein Motion and Protein Motors

While internal protein motions in the previous section were shown to have fundamental roles in the enzymatic and other functions of proteins, the ratchet brings the idea of directed motion to the fore. The reason for directed motion within the cell is quite clear. If there were no cytoskeleton, no "highways and paths" within the cell, then all of the exchange of "molecular and energetic goods" within the cell would have to rely on random diffusion. Not only does unbiased diffusion have no directionality but it is extremely slow. For unbiased diffusion, the relationship between self-diffusion coefficient D_0, mean square displacement $<r^2>$, and time was given in Equation 6.94, from which the root mean square displacement, $<r^2>^{1/2}$ is

$$\langle r^2\rangle^{1/2} = \sqrt{6D_0 t} \qquad 10.72$$

Substituting in diffusion coefficients ~10^{-5} cm^2/s for small molecules, of ~10^{-6} cm^2/s for proteins, and 10^{-9} cm^2/s for organelles, and making a first approximation of the cytosol as water, the time to diffuse root mean square distances $<r^2>^{1/2}$ of cell dimension 10 μm is, respectively, 0.1 s, 1 s, and 1000 s, respectively, and for a cell dimension of 100 μm, the time is 10 s, 100 s, 10^5 s. These times are far too slow for the fast internal dynamics of cells, not to mention the fact that the cell interior is highly congested and impedes diffusion. Hence, it is not surprising that there is a fundamental need to provide pathways and *motor proteins* to move molecular agents around in the cell; distances in random diffusion are proportional only to the square root of time, whereas directed motion at constant velocity is directly proportional to time.

To achieve this type of fast, directed transport and motion is not cost-free and requires the input of metabolic energy, usually ATP directly, or in conjunction with another energy-consuming process. This represents transduction of chemical energy to mechanical energy. As usual, the living state constantly consumes energy to maintain its far-from-equilibrium condition. Protein motors are involved in a vast array of biological functions, including all the internal cell motions such as the entire genetic transcription and synthesis process, molecular transport, signaling, cell replication, and external motions, such as motion through space and muscle action in applying forces to external objects. Not surprisingly, breakdown of protein motor functions can lead to diseases, such as respiratory, kidney, and neurological disorders.

There are many open questions concerning motor proteins and their functions, which are the object of active research. It is not known if there are just one or actually many chemical-to-mechanical energy transduction mechanisms. Has evolution optimized the efficiency of motor proteins, and if so what might be optimized? Speed, force, transduction efficiency?

A good starting point for a comprehensive introduction to the rapidly growing field of research on motor proteins is Kolomeisky (2015).

10.9 Allostery and Feedback Regulation

Allostery was first mentioned in the context of cooperative binding in proteins, with the example of cooperative binding of O_2 in hemoglobin. Allostery is a far-ranging notion and represents the concept that the binding of an *effector molecule* on one part of an enzyme, distant from the active site, can affect the behavior of the active site. There are several models for this, but one of the underlying notions is that the allosteric binding of an effector molecule can change the conformation of the enzyme, or a subunit of the enzyme, which, in turn, affects how substrate molecules bind to the distant active site.

The implications of allostery for feedback regulation of enzymatic activity are important. Effectors that increase the enzyme's throughput are *allosteric activators*, while those that reduce the enzyme's activity are *allosteric inhibitors*. We already saw in Section 10.4 that O_2 acts as an allosteric activator in the process of binding O_2 to hemoglobin, since the binding of a first O_2 facilitates the binding of the next O_2, and so on, up to the cooperative binding of four O_2 molecules. The advantages of this for O_2 binding and release were discussed in connection with Figure 10.19.

There are many examples of allosteric inhibitors. An interesting case of metabolic feedback control is where ATP binds allosterically to phosphofructokinase (PFK) and inhibits its activity. PFK controls the third step in the glycolysis pathway which produces ATP (Chapter 3.5.2), so that if its activity is reduced by the allosteric inhibition of ATP on PFK, the glycolysis cycle is slowed or shut down. This happens when ATP is abundant, i.e. when the cells' energy charge is high, Equation 3.14, there is no need to produce more ATP, so that glucose is conserved.

There is much research using allostery for medical purposes. Many, if not most, small molecule drugs work by competing for binding to an enzyme's active site, thus slowing enzyme activity. Efforts are underway to use allosteric binding sites as the drug target, rather than the active site itself. Different types of selective binding to allosteric sites can actually enter into metabolic feedback control, giving a higher level of modulation and control of enzyme activity.

There is also work in synthesizing oligomers and macromolecules that have allosteric properties. Typical synthetic macromolecules have absolutely no allosteric character, but some classes of synthetic oligomers, such as crown ethers and ionophores, and modified biological molecules, such as porphyrins, show allosteric properties.

10.10 Energetics of Iscosahedral Viral Self-assembly[5]

We posed the question of whether viruses are a life form in Chapter 2.4 and also presented some of the theories of how viruses arose. In viruses, the genetic material, DNA or RNA, is encapsulated by a protein *capsid*, and the protein subunits comprising the capsid are termed *capsomers*. It is found that some of the folding motifs in viral proteins are unique to viruses, which goes against one of the evolutionary theories, that viruses are from genetic machinery that escaped fully developed cells. This folding uniqueness might strengthen the theory that viruses may be an extremely ancient entity, developing their own protein folding pathways, while other more sophisticated life forms were co-evolving. In this theory, viruses represent a sort of "stunted" evolution.

While there are many types of viruses, around one-half of them have icosahedral symmetry, and evidence points toward a pure energy minimization driving their self-assembly. A convex regular icosahedron, the most common, is a polyhedron with 20 faces, which are equilateral triangles, 12 vertices, and 30 edges. Figure 10.21 shows a schematic of an icosahedral viral capsid with coiled DNA or RNA inside it. Many of these icosahedral viruses are not directly related to each other, reinforcing the notion that simple energy minimization may lie behind their self-assembly. As seen earlier, protein folding is much more complex and less geometrically "ideal" than the icosahedral symmetry of these viruses.

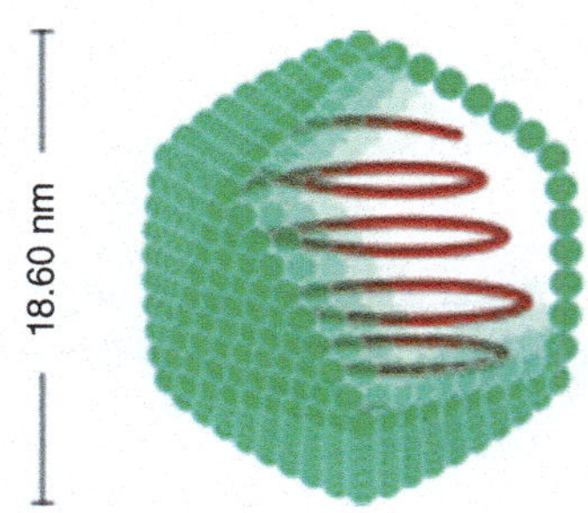

Figure 10.21 Schematic of an icosahedral virus capsid, with the coiled RNA or DNA inside. (Brito, A. (2011) / Wikimedia commons / Licensed under CC BY-SA 3.0).

5 Adapted with permission from unpublished biophysics research report from Samia Sabir, Tulane University Physics Undergraduate, 2019.

Viruses with icosahedral symmetry include poliovirus, rhinovirus (most common agent causing the "common cold"), and adenovirus (they cause acute bronchitis, pneumonia, pink eye, among other infections).

There are several models for the energetics of icosahedral virus assembly. One model, based on continuum elasticity theory, analyzes the energetic costs to change a spherical shell into different shapes, by considering the cost of bending energy to create the sharp edges and vertices, and the stretching energy involved in creating the structure. A parameter that defines this is the dimensionless parameter β:

$$\beta = \frac{YS}{\kappa} \tag{10.73}$$

where Y is Young's modulus (Chapter 8.1.5), S the capsid surface area, and κ is the bending energy constant. At low β, YS is much smaller than the bending energy, so that a spherical shape is favored. As β increases, however, it becomes less energetically favorable to keep a large surface area S, and so the shape buckles into an icosahedron, the least energy state, where the total bending and surface energy is less than just the bending energy of the equivalent sphere.

Another energy minimization approach is to consider the energetic costs of hydrophobic and electrostatic interactions between capsid proteins and between capsids and surrounding solvent. For a given pair of capsid proteins (capsomers), the surface tension (Chapter 4.11) between capsids and the aqueous environment is γ, A_H is the contact area between the two proteins, λ is the decay length of the hydrophobic effect, and d is the separation between the proteins. This leads to a hydrophobic energy/area E_H of

$$E_H = -2\gamma A e^{-(d/\lambda)} \tag{10.74}$$

so that the total hydrophobic free energy due to surface tension is

$$G_H = -A_H \gamma \tag{10.75}$$

Capsid proteins under physiological conditions are charged, usually positively. Using the linearized Poisson–Boltzmann equation, Equation 5.72 and treating the proteins as two elementary charges separate by distance r, the electrostatic interaction potential V_e, is approximated by

$$V_e = k_B T \left(\frac{l_b}{r} \right) e^{(-\kappa r)} \tag{10.76}$$

where κ is the Debye parameter, Equation 5.74:

$$\kappa = \left(\frac{2\rho_0 z e}{\varepsilon k_B T} \right)^{1/2} \tag{5.74}$$

where ρ_0 is the charge density of simple electrolyte in the surrounding aqueous environment, z is the net charge on the object, and $\varepsilon = \varepsilon_0 D$ is the dielectric permittivity of the aqueous environment, D is its dimensionless dielectric constant, and l_B is the Bjerrum length from Chapter 5.10 (with $z_1 = z_2 = z$):

$$l_B = \frac{z^2 e^2}{4\pi D \varepsilon_0 k_B T} \tag{5.131}$$

For a capsid composed of several, or many capsomers, this can be visualized as leading to an equivalent charged, thin spherical shell of radius R and surface charge density σ, with free energy, G_e of

$$G_e = \frac{\pi \sigma^2 R^2}{\varepsilon_0 D \kappa} \tag{10.77}$$

This leads to a total capsid-free energy G_t, of

$$G_t = -A_H \gamma + \frac{\pi \sigma^2 R^2}{\varepsilon_0 D \kappa} \tag{10.78}$$

Values for the various parameters can then be entered to minimize the free energy with respect to A_H and R.

10.11 Summary

The primary sequence of amino acids is obtained from DNA, but the secondary, tertiary, and quaternary structures that proteins fold into are often assumed to be controlled by the physics dictated by the sequence. Nonetheless, purely physics-based approaches to protein folding have found only limited success, whereas Artificial Intelligence approaches, essentially devoid of physics, which use historical patterns gleaned from huge protein databases, have yielded remarkable predictions. Chaperones, large protein complexes, are found to be essential for folding some proteins and for checking for and correcting misfolded proteins, or for tagging them for destruction by the proteostasis network.

The functional state of proteins is not always the deepest free energy minimum in an energy landscape. In fact, the lowest energy state can sometimes correspond to aggregated proteins, which are suspected in human dementia, and in diseases such as Alzheimer's. Protein aggregation is also a critical issue in the formulation and stability of biologic proteins, including mono- and polyclonal antibodies.

Enzyme catalysis is arguably the most important function of proteins. Several models were introduced that describe the enzyme kinetics quite well, based on mass action principles. Cooperative binding is a widespread feature of proteins and can lead to better regulation and efficiency of functions such as oxygenation and deoxygenation. Cooperative binding is an important mechanism in feedback control of protein synthesis, ATP production, and metabolic cycles.

Cooperativity is also found in the formation of helices in proteins, and a cooperative model for the partition function gives a good account of the sharpness of the helix–coil phase transition. In both substrate binding and helix formation, high cooperativity leads to sharp transitions, which resemble phase changes.

The role of different types of internal protein motions was highlighted as a crucial underlying phenomenon for many protein catalytic and other activities, such as allostery and moving molecular products along cytoskeleton networks consisting of different protein-based microtubules and filaments. These motions then extend to translational motion of cells, or groups of cells, through space and different media, as well as providing the basis for muscular action. The role of allostery in feedback regulation of metabolism and genetic activities is an especially fascinating area of active research.

While tertiary structure of prokaryote, eukaryote, and archaea proteins is very spatially complex, virus capsid proteins form very regular 3D structures, essentially well-defined geometric crystals, of which those with icosahedral symmetry are exceptionally widespread. The folding sequences found in these are different than those of the highly evolved and conserved nonviral folding pathways, presenting an intriguing puzzle as to viral origins.

References

Agozzino, L., Balázsi, G., Wang, J., and Dill, K.A. (2020). How do cells adapt? Stories told in landscapes. *Annual Review of Chemical and Biomolecular Engineering* 11: 155–182.

Bahar, I., Jernigan, R.L., and Dill, K.A. (2017). *Protein Actions; Principles and Modeling*. Garland Science.

Brini, E., Simmerling, C., and Dill, K. (2020 November 27). Protein storytelling through physics. *Science* 370 (2020): eaaz3041.

Buell, A.K. (2019). The growth of amyloid fibrils: rates and mechanisms. *Biochemical Journal* 476 (19): 2677–2703.

Bustamante, C. and Yan, S. (2022). The development of single molecule force spectroscopy: from polymer biophysics to molecular machines. *Quarterly Reviews of Biophysics* 55: e9.

Chen, H., Yang, J., Ziping, Q. et al. (2022). Heat shock proteins: biological functions, pathological roles, and therapeutic opportunities. *MedComm* 3 (3): e161.

Chennubhotla, C., Rader, A.J., Yang, L.-W., and Bahar, I. (2005). Elastic network models for understanding biomolecular machinery: from enzymes to supramolecular assemblies. *Physical Biology* 2: S173–S180. https://doi.org/10.1088/1478-3975/2/4/S12.

Dill, K.A., Banu Ozkan, S., Scott Shell, M., and Weikl, T.R. (2008 June 9). The protein folding problem. *Annual Review of Biophysics* 37: 289–316. https://doi.org/10.1146/annurev.biophys.37.092707.153558.

Drenski, M.F. and Reed, W.F. (2004). Simultaneous multiple sample light scattering for characterization of polymer solutions. *Journal of Applied Polymer Science*, 92, 2724–2732.

Drenski, M.F., Brader, M.L., Alston, R.W., and Reed, W.F. (2013). Monitoring protein aggregation kinetics with simultaneous multiple sample light scattering (SMSLS). *Analytical Biochemistry* 437: 185–197. https://doi.org/10.1016/j.ab.2013.02.014.

Fraser, P.E. (2014). Prions and prion-like proteins. *Journal of Biological Chemistry* 289 (29): 19839–19840.

Jarand, C.W. and Reed, W.F. (2018). On the reproducibility of early stage thermally induced and contact-stir induced protein aggregation. *The Journal of Physical Chemistry B* 122: 9361–9372.

Jackson, M.B. (2006). *Molecular and Cellular Biophysics*. Cambridge University Press.

Kaizhi, Y. and Dill, K.A. (1992). Inverse folding problem: designing polymer sequences. *Proceedings of the National Academy of Sciences of the United States of America* 89: 4163–4167.

Kolomeisky, A.B. (2015). *Motor Proteins and Molecular Motors*. CRC Press.

Kulkarni, P., Leite, V.B.P., Roy, S. et al. (2022). Intrinsically disordered proteins: ensembles at the limits of Anfinsen's dogma. *Biophysics Reviews (Melville, NY, United States)* 3 (1): 011306.

Mateu, M.G. (2013). *Structure and Physics of Viruses an Integrated Textbook*. Netherlands: Springer.

Milliken, G. (2019 Mar 18). Are viruses alive? New evidence says yes. *Popular Science* www.popsci.com/new-evidence-that-viruses-are-alive.

Nassar, R., Dignon, G.L., Razban, R.M., and Dill, K.A. (2021). The protein folding problem: the role of theory. *Journal of Molecular Biology* 433 (20): 167126.

Nguyen, T.T., Bruinsma, R.F., and Gelbart, W.M. (2005 November 21). Elasticity theory and shape transitions of viral shells. *Physical Review* 72 (051923).

Vale, R.D. (1993). Measuring single protein motors at work. *Science* 260: 169–170. www.jstor.com/stable/2881297.

Vale, R.D. (2003). The molecular motor toolbox for intracellular transport. *Cell* 112: 467–480.

Vogel, C., Bashton, M., Kerrison, N.D. et al. (2004). Structure, function and evolution of multidomain proteins. *Current Opinion in Structural Biology* 14: 208–216.

Zandi, R. et al. (2004 Feb 2). Origin of icosahedral symmetry in viruses. *The Proceedings of the National Academy of Sciences of the United States of America* 2004 Nov 2;101(44):15556-60.

Zlotnick, A. (2004) Viruses and the physics of soft condensed matter. *Proceedings of the National Academy of Sciences of the United States of America* 101 (44): 15549–15550.

Further Reading

Ali, M.Y., Lu, H., Bookwalter, C.S. et al. (2008). Myosin V and Kinesin act as tethers to enhance each others' processivity. *Proceeding of the National Academies of Science, USA* 105: 4691–4696.

AlQuraishi, M. and Sorger, P.K. (2021). Differentiable biology: using deep learning for biophysics-based and data-driven modeling of molecular mechanisms. *Nature Methods* 18 (10): 1169–1180.

Amin, S., Barnett, G.V., Pathak, J.A. et al. (2014). Protein aggregation, particle formation, characterization & rheology. *Current Opinion in Colloid & Interface Science* 19 (5): 438–449.

Astumian, R.D. and Bier, M. (1994). Fluctuation driven ratchets: molecular motors. *Physical Review Letters* 72: 1766–1769.

Beard, D.A. and Qian, H. (2008). *Chemical Biophysics: Quantitative Analysis of Cellular Systems*. Cambridge University Press.

Bhatia, S. and Udgaonkar, J.B. (2022). Heterogeneity in protein folding and unfolding reactions. *Chemical Reviews (Washington, DC, United States)* 122 (9): 8911–8935.

Calabrese, G., Molzahn, C., and Mayor, T. (2022). Protein interaction networks in neurodegenerative diseases: from physiological function to aggregation. *Journal of Biological Chemistry* 298 (7): 102062.

Carter, N.J. and Cross, R.A. (2005). Mechanics of Kinesin step. *Nature* 435: 308–312.

Cassaignau, A.M.E., Cabrita, L.D., and Christodoulou, J. (2020). How does the ribosome fold the proteome. *Annual Review of Biochemistry* 89: 389–415.

Chopra, G., Shabir, S., Yousuf, S. et al. (2022). Proteinopathies: deciphering physiology and mechanisms to develop effective therapies for neurodegenerative diseases. *Molecular Neurobiology* 59 (12): 7513–7540.

Cook, P.F. and Cleland, W.W. (2007). *Enzyme Kinetics and Mechanisms*. NY: Garland Science.

Cornish-Bowden, A. (2012). *Fundamentals of Enzyme Kinetics*, 4e. NY: Wiley-Blackwell.

Flechsig, H. and Togashi, Y. (2018). Designed elastic networks: models of complex protein machinery. *International Journal of Molecular Sciences* 19 (10): 3152/1–3152/18.

Geronimo, I., Vidossich, P., Donati, E., and De Vivo, M. (2021). Computational investigations of polymerase enzymes: structure, function, inhibition, and biotechnology. *Wiley Interdisciplinary Reviews: Computational Molecular Science* 11 (6): e1534.

Gruebele, M. and Pielak, G.J. (2021). Dynamical spectroscopy and microscopy of proteins in cells. *Current Opinion in Structural Biology* 70: 1–7.

Han, J.-H., Batey, S., Nickson, A.A. et al. (2007 April). The folding and evolution of multidomain proteins. *Nature Reviews, Molecular Cell Biology* 8: 319–330.

Haran, G. and Mazal, H. (2020). How fast are the motions of tertiary-structure elements in proteins? *Journal of Chemical Physics* 153 (13): 130902.

Kang, H.-W., KhudaBukhsh, W.R., Koeppl, H., and Rempala, G.A. (2019). Quasi-steady-state approximations derived from the stochastic model of enzyme kinetics. *Bulletin of Mathematical Biology* 81 (5): 1303–1336.

Karaca, E., Prevost, C., and Sacquin-Mora, S. (2022). Modeling the dynamics of protein-protein interfaces, how and why? *Molecules* 27 (6): 1841.

Khade, P.M., Kumar, A., and Jernigan, R.L. (2020). Characterizing and predicting protein hinges for mechanistic insight. *Journal of Molecular Biology* 432: 508–522.

Kim, N.H., Choi, H., Shahzad, Z.M. et al. (2022). Supramolecular assembly of protein building blocks: from folding to function. *Nano Convergence* 9 (1): 4.

Kornev, A.P. (2018). Self-organization, entropy and allostery. *Biochemical Society Transactions* 46 (3): 587–597.

Kremer, C. and Lützen, A. (2013). Lesson from nature: implementing allosteric control in artificial receptors. *Chemistry, a European Journal* 19: 6162–6196. https://doi.org/10.1002/chem.201203814.

Lindorff-Larsen, K. and Kragelund, B.B. (2021). On the potential of machine learning to examine the relationship between sequence, structure, dynamics and function of intrinsically disordered proteins. *Journal of Molecular Biology* 433 (20): 167196.

Litvinov, R.I. and Weisel, J.W. (2017). Fibrin mechanical properties and their structural origins. *Matrix Biology* 60-61: 110–123.

Lorsch, J.R. (2014). Practical steady-state enzyme kinetics. *Methods in Enzymology* 536: 3–15.

Lu, Y., Zhu, B., Li, Q. et al. (2022). The folding and misfolding mechanisms of multidomain proteins. *Medicine in Drug Discovery* 14: 100126.

Magi Meconi, G., Sasselli, I.R., Bianco, V. et al. (2022). Key aspects of the past 30 years of protein design. Reports on progress in physics. *Physical Society (Great Britain)* 85 (8).

Mecha, M.F., Hutchinson, R.B., Lee, J.H., and Cavagnero, S. (2022). Protein folding in vitro and in the cell: from a solitary journey to a team effort. *Biophysical Chemistry* 287: 106821.

Meisl, G., Knowles, T.P.J., and Klenerman, D. (2022). Mechanistic models of protein aggregation across length-scales and time-scales: from the test tube to neurodegenerative disease. *Frontiers in Neuroscience* 16: 909861.

Navarro, S. and Ventura, S. (2022). Computational methods to predict protein aggregation. *Current Opinion in Structural Biology* 73: 10234.

Goldbeter, A. (2013). Oscillatory enzyme reactions and Michaelis-Menten kinetics. *FEBS Letters* 587 (17): 2778–2784.

Pereira-Leal, J.B., Levy, E.D., and Teichmann, S.A. (2006). The origins and evolution of functional modules: lessons from protein complexes. *Philosophical Transactions of the Royal Society B* 361: 507–517. https://doi.org/10.1098/rstb.2005.1807.

Rahman, A., Saikia, B., Gogoi, C.R., and Baruah, A. (2022). Advances in the understanding of protein misfolding and aggregation through molecular dynamics simulation. *Progress in Biophysics & Molecular Biology* 175: 31–48.

Sao Pedro, M.N., Klijn, M.E., Eppink, M.H.M., and Ottens, M. (2022). Process analytical technique (PAT) miniaturization for monoclonal antibody aggregate detection in continuous downstream processing. *Journal of Chemical Technology and Biotechnology* 97 (9): 2347–2364.

Sengupta, U. and Strodel, B. (2018). Markov models for the elucidation of allosteric regulation. *Philosophical Transactions of the Royal Society, B: Biological Sciences* 373 (1749): 20170178/1–20170178/7.

Senior, A.W., Evans, R., Jumper, J. et al. (2020 January 30). Improved protein structure prediction using potentials from deep learning. *Nature* 577: 706–710.

Simmons, W.A. (2018). Protein folding and machine learning: fundamentals. *Physics* 1–18. arXiv.org, e-Print Archive.

Stock, G. and Hamm, P. (2018). A non-equilibrium approach to allosteric communication. *Philosophical Transactions of the Royal Society, B: Biological Sciences* 373 (1749): 20170187/1–20170187/9.

Tang, Y., Zhang, D., Gong, X., and Zheng, J. (2022). A mechanistic survey of Alzheimer's disease. *Biophysical Chemistry* 281: 106735.

Varela, A.E., England, K.A., and Cavagnero, S. (2019). Kinetic trapping in protein folding. *Protein Engineering, Design & Selection* 32 (2): 1–6.

Vincentelli, R. (ed.) (2019). *High-throughput Protein Production and Purification; Methods and Protocols*. Humana Press.

Wang, W. and Roberts, C.J. (2018). Protein aggregation - Mechanisms, detection, and control. *International Journal of Pharmaceutics (Amsterdam, Netherlands)* 550 (1–2): 251–268.

Wang, Y., Rader, A.J., Bahar, I., and Jernigan, R.L. (2004). Global ribosome motions revealed with elastic network model. *Journal of Structural Biology* 147: 302–314.

Problems

1 Derive the expression for Michaelis–Menten when an intermediate enzyme state is involved.

$$V = \frac{\left(\frac{k_3k_4}{k_3+k_4}\right)[S][E]_T}{\frac{(k_2+k_3)}{k_1(k_3+k_4)}+[S]}$$

2 Find the expression for helix content θ in the two limits $\sigma = 0,1$
Is there an order to this phase transition?

3 Find the condition for the inflection point in the Hill equation, 10.57, and plot the value of $[B]$ at the inflection point versus N, for $K = 4$ (arbitrary units). To make the plot, you will have to solve your equation numerically.
($2KB = 1 + KB^N$ at inflection point)

4 Consider the binding of three oxygen molecules to a hypothetical protein according to

$$A + B \xleftrightarrow{K_1} AB$$

$$AB + B \xleftrightarrow{K_2} AB_2$$

$$AB_2 + B \xleftrightarrow{K_3} AB_3$$

where A = protein and $B = O_2$.

a) Find an expression for the amount of protein bound to O_2 as a function of the concentration of O_2.
b) Qualitatively sketch this curve under the cases where
$K_1 \gg K_2, K_3$
$K_3 > K_2 > K_1$

5 A two-phase enzymatic reaction involves two substrates, E_1 and E_2, and produces two products P_1 and P_2, according to

$$\mathrm{E} + \mathrm{S} \underset{k_2}{\overset{k_1}{\rightleftarrows}} ES_1 \xrightarrow{k_3} ES_2 + P_1 \xrightarrow{k_4} E + P_2$$

The serine protease mechanism is described by this, for example. Using the steady-state approximation, show that the rate of formation of product P_2 is given by

$$v = \frac{k_1k_3k_4[E_T]}{k_1[S](k_3+k_4)+k_4(k_2+k_3)}$$

6 Use the concentration of the maxima of the Rayleigh scattering ratio, I_R, in Figure 10.11 to find A_3 for mAb1 and the NIST mAb

7 Use Equation 10.1 to find the initial aggregation rate (1/s) for BSA and for mAbX in Figure 10.7a.

11

DNA and RNA Properties and Structures: The Genetic Code

The nucleic acid bases and associated nucleosides and nucleotides were introduced in Chapter 3.2, along with a mention of their roles in energy production and storage. Here, the focus shifts to the macromolecular properties of RNA and DNA, as well as their higher order biological functions in reproduction, and the genetic code.

11.1 Structure and Macromolecular Properties of RNA and DNA

DNA and RNA are polymeric forms of the nucleotides. An organism's genome is contained in its DNA. Small RNA can have as few as 20 bases, whereas genomic DNA can have upward of 2×10^8 base pairs (bp). RNA is single stranded and can form random coils and single helices, and have extensive and intricate secondary structure.

DNA normally forms a right-handed double helix around a common axis, stabilized by hydrogen bonds between complementary purine–pyrimidine base pairs, which are thymine–adenine and cytosine–guanine. The base pairing for DNA is shown in Figure 11.1a and the overall helical structure of DNA in Figure 11.1b. The pyrimidine–purine pairs are on the inside of the helix, and the phosphate and deoxyribose units are on the outside. The base planes are perpendicular to the common axis. Bases are separated from each other by 0.34 nm with a 36° rotation between them, so that a complete 360° turn of the helix comprises 10 bp. The diameter of the helix is around 2 nm. The persistence length, L_p, introduced in detail in Chapter 8.1.9, is quite long for DNA, due to the stiffness imparted by the double helix and tight conformational packing of the nucleotides in the chain. While L_p depends on several factors, such as ionic strength of the aqueous medium and temperature, there is a general consensus that it is in the range of 45–50 nm, or 132–147 bp, or 13–15 complete 360° turns of the helix, at intermediate ionic strengths (10–200 mM). The directionality of a single strand of DNA starts at an unbound 5′ phosphate and ends on an unbound 3′ hydroxyl group.

In RNA *uracil* substitutes for thymine, but still pairs with adenine. The only difference between the two molecules is that thymine has a methyl (CH_3) group at carbon number five, whereas uracil has a hydrogen. This seemingly small difference leads to huge differences in structure and biological function between DNA and RNA.

A much less conformationally stable form of DNA was discovered (Pohl and Jobin 1972) which is left-handed; *Z-DNA*. Under certain conditions, regular right-handed DNA can transiently form lengths of Z-DNA, with important implications in transcription, immune responses, and other biological functions. The Z-DNA is torsionally stressed and at a higher energy level than regular DNA so soon returns to the right-handed conformation, once formed, e.g. during transcription.

11.1.1 Mechanical Properties of Isolated DNA

The mechanical properties of isolated DNA can be considered in terms of its bending, torsional, and stretching rigidity. It follows from Equations 8.71 and 8.80 that the bending energy needed to bend a length l of DNA chain a small angle $\Delta\theta$ away from being straight, E_{bend}, is

$$E_{bend} = \frac{k_B T L_p}{2l} (\Delta\theta)^2 \qquad 11.1$$

Biophysics: Physical Processes Underlying the Living State, First Edition. Wayne F. Reed.

Companion website: www.wiley.com/go/Reed/BiophysicsandMacromolecules

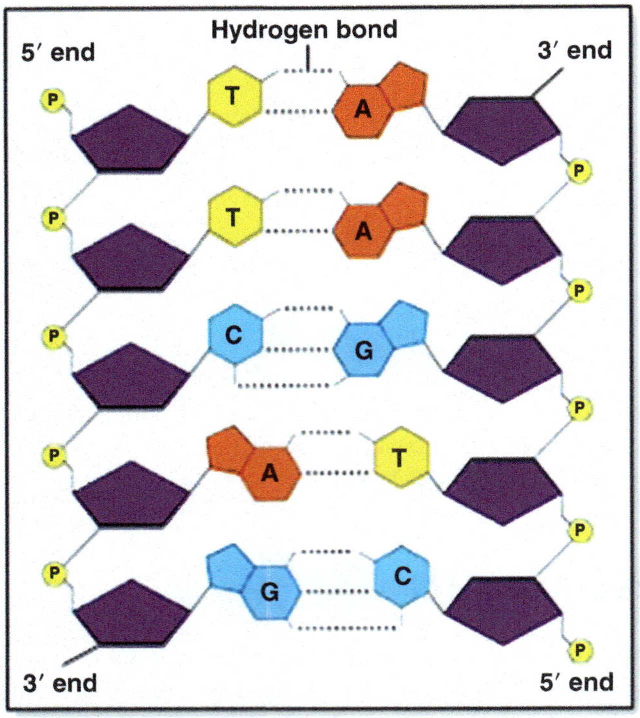

Figure 11.1a Complementary thymine–adenine and cytosine–guanine in DNA.

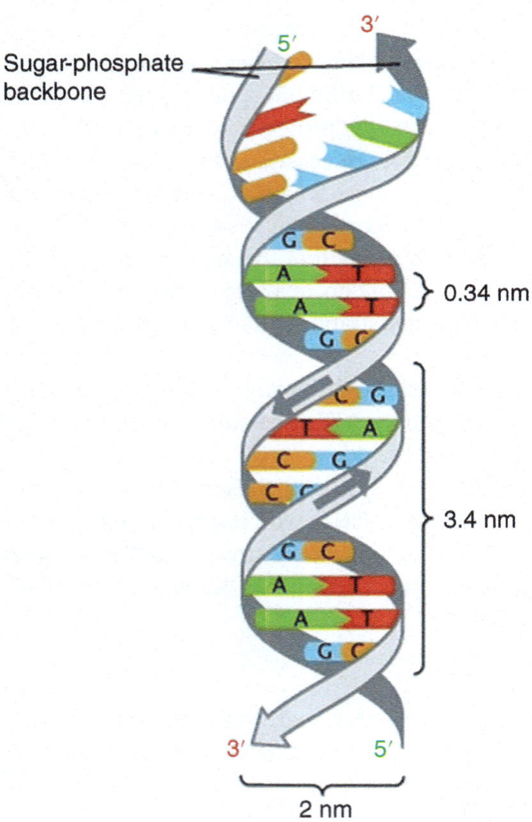

Figure 11.1b DNA double-helix structure.

where L_p is the persistence length derived from the bending energy concept in Chapter 8.1.9. Twisting DNA from its relaxed state can be characterized by a torsional twisting energy per unit length, μ, which is the energy per unit length needed to twist the helix an amount $\Delta\phi$ from its relaxed position over a length Δs of the polymer

$$\mu_{torsional} = \frac{\alpha}{2}\left(\frac{\Delta\phi}{\Delta s}\right)^2 \tag{11.2}$$

where α is the torsional rigidity constant, with units energy × length. This can be treated in terms of a torsional persistence length, $L_{torsional}$, in the same way (bending) persistence length was treated in Chapter 8.1.9. Namely, $L_{torsional}$ is the ratio of the torsional rigidity constant to thermal energy

$$L_{torsional} = \alpha / k_B T \qquad 11.3$$

The energy needed to twist the polymer a small angle $\Delta\phi$ away from its relaxed state over a length l is

$$E_{torsional} = \frac{k_B T L_{torsional}}{2l} (\Delta\phi)^2 \qquad 11.4$$

The experimentally measured value of $L_{torsional}$ is around 65 nm, or 191 bp, using fluorescence polarization anisotropy decay.

Stretching does not play a significant role compared to bending and twisting, due to its high stretching modulus $\beta = 1500$ pN, which has been measured by single-molecule experiments. The stretching modulus is related to the stretching force applied, F, and the strain $\Delta x/x_o$, where Δx is the stretch over a segment of initial length x_o

$$F = \beta \frac{\Delta x}{x_o} \qquad 11.5$$

so that the stretching energy is

$$E_{stretch} = \beta \frac{(\Delta x)^2}{2l} \qquad 11.6$$

These latter are the energies for DNA unassociated with other agents. Since genomic DNA in an organism is always associated with histones and other proteins in the form of chromatin, we will return to a brief view of the mechanics of chromatin after surveying the genetic code and a brief review of gene expression.

In connection with the persistence length of DNA is its intrinsic stiffness, or intrinsic persistence length, $L_{p,0}$, and the sensitivity of the total persistence length to ionic strength. The inherent persistence length of DNA is found to be around $L_{p,0} \sim 40$–50 nm. Electrostatic stiffening as ionic strength is decreased yields total persistence length, which is approximated as the intrinsic plus electrostatic persistence lengths (Chapter 8.4.1), in excess of 100 nm at very low ionic strength, I ($I < 1$ mM, Chapter 5.6.3). The linearized PB equation (Equation 5.72) does not provide an adequate basis for explaining the total persistence length from very low to high ionic strength, nor of explaining the profound differences in DNA persistence length behavior between monovalent and divalent cations. The linearized PB was also seen to fail in connection with monovalent and symmetric and asymmetric multivalent ions as far as static and hydrodynamic dimensions of anionic polysaccharides in Chapter 5.6.4. A successful description of the experimental total persistence length data was given using a variational approximation of the full Poisson–Boltzmann equation (Equation 5.68):

$$L_p = L_{p,0} + \gamma^2 (z, \kappa R_{DNA}) \frac{l_B}{4\xi^2 \kappa^2} \qquad 11.7$$

where l_B is the Bjerrum length (Chapter 5.10), κ is the Debye screening parameter (Equation 5.74), ξ is the linear charge density in terms of the number of elementary charges per Bjerrum length, and $\gamma(z, \kappa R_{DNA})$ is the effective fraction of charge of the DNA macromolecule, obtained using the variational approximation.

11.2 The Genetic Code

DNA carries all the information needed to produce every protein in a given organism and represents the accumulated, evolutionary information for the organism. The genetic code specifies which combination of base pairs, taken three at a time, corresponds to each of the 20 amino acids plus a stop signal. Table 11.1 shows the genetic code. The code is "redundant" in the sense that there are more than one set of *codons* (the triads of bases) that specify a given amino acid. (The exception is methionine, for which there is only one codon.) It is conjectured that this helps buffer against mutations, by allowing some errors in code sequence to nonetheless produce the same amino acid. The code shown is essentially universal, although the mitochondrial genetic code has slight differences.

Crick's Central Dogma was brought up in connection with information theory in Chapter 4.19.4. It is worth revisiting this and some other findings from information theory to further examine the genetic code. First, consider the Shannon entropy

Table 11.1 The genetic code for RNA.

First base	Second base: U	Second base: C	Second base: A	Second base: G	Third base
U	UUU — Phe	UCU — Ser	UAU — Tyr	UGU — Cys	U
U	UUC — Phe	UCC — Ser	UAC — Tyr	UGC — Cys	C
U	UUA — Leu	UCA — Ser	UAA — STOP	UGA — STOP	A
U	UUG — Leu	UCG — Ser	UAG — STOP	UGG — Trp	G
C	CUU — Leu	CCU — Pro	CAU — His	CGU — Arg	U
C	CUC — Leu	CCC — Pro	CAC — His	CGC — Arg	C
C	CUA — Leu	CCA — Pro	CAA — Gln	CGA — Arg	A
C	CUG — Leu	CCG — Pro	CAG — Gln	CGG — Arg	G
A	AUU — Ile	ACU — Thr	AAU — Asn	AGU — Ser	U
A	AUC — Ile	ACC — Thr	AAC — Asn	AGC — Ser	C
A	AUA — Ile	ACA — Thr	AAA — Lys	AGA — Arg	A
A	AUG — Met or Start	ACG — Thr	AAG — Lys	AGG — Arg	G
G	GUU — Val	GCU — Ala	GAU — Asp	GGU — Gly	U
G	GUC — Val	GCC — Ala	GAC — Asp	GGC — Gly	C
G	GUA — Val	GCA — Ala	GAA — Glu	GGA — Gly	A
G	GUG — Val	GCG — Ala	GAG — Glu	GGG — Gly	G

Note: For DNA uracil (U) is simply replaced by thymine (T). (Wikimedia Commons, File:Genetic Code.png|Genetic_Code, Sarah Greenwood, 2018)

introduced in Chapter 4.19.4, and defined by Equation 4.224. The higher the Shannon entropy, the higher the information content. Interestingly, highly organized systems, such as a regular crystal of a material contain less information than the same amount of material randomly distributed; the position of all the particles in the crystal can be simply specified by a short formula or algorithm, while the position of each particle when spatially randomized requires information for each individual particle. And, if the random particles are in motion, the information specifying each particle changes in time, whereas the crystal information remains static.

Another property of Shannon entropy is that, given two systems, each with their own Shannon entropy, information can only be uniquely coded in the direction of the higher Shannon entropy system to the lower. This is the case with DNA/RNA passing on sequences to proteins; the genetic code has 64 elements and hence higher Shannon entropy than amino acids, of which there are only 20, plus a stop signal. From this Shannon entropy point of view, information can only uniquely flow from DNA/RNA to proteins, but not in the reverse direction. This is the essential content of Crick's central hypothesis,[1] and it might be asserted that information theory validates this hypothesis.

A further implication is that proteins could not have evolved the nucleic acid–based genetic code. This is an interesting but controversial assertion, as there is still great theoretical and experimental effort devoted to producing amino acids, proteins, and other organic molecules from a "primordial soup," or "prebiotic broth" subject to strong energy input (e.g. lightning). These concepts were proposed by Alexander Oparin (Russia, 1894–1980) in 1924 and John Burdon Sanderson Haldane (England/India, 1892–1964) in 1929.

It caused some surprise when retroviruses were discovered, because these can transcribe their RNA into the DNA of the host's genome. Because RNA and DNA share the same genetic code, and the same Shannon entropy, information theory affirms that information can be passed uniquely between RNA and DNA, so that the discovery of retroviruses is not at all a surprise from the information theory perspective.

Several information scientists have realized that the genetic code is segregated into discrete units, linear, and digital, and resembles other types of codes, such as digital computer codes, Morse Code, language alphabets, and so on. These coding

1 Although Watson used the word "central dogma," he later conceded (see H. Yockey 2005) that he should have termed it the "central hypothesis."

```
GAATTCTCTTTGGTATCCAATGAAGAAATCGAATCCATACCCATAGCTATAAAAAACAT
TTCAGGAGAAAATAAGACCGAAGCTGCTCAATTAGGCGCAATTGATTCGTTTCAAAAAT
GTGAAACTTGCCAGCTTACTTCGGCATGTCCTGGTCATTTTGGAAAATTTCATCTTACT
CAACCATTATTTAAAGTCGCATTTAAAAAACTTGTTGAAAATATTTTTAAATATACTTG
TTCTTTCTGTGGTGCTTTACAAAATCTTGAACTTCTGGAATTGATCAAGCAGATAGACG
AACGAAATACTGGAATAACAGTTAAAGATCGTGCTGCTTTTAAAAAAATTTTAGAAGCT
ACCAAACAAAGCAAATTCAAGTGTATTGCACCTAATTGCCAAAAACAAGTCTCTCCTTT
ACAATATTCGAAAAATAATAACTTTATATATAATTCGGGTACTACAAAGGGTATAGTTT
TGGATAACAGGCATGTGTTTAATATCTTACAAAATCTTCCACAAACGTTTAAATTATTG
TTAACCCCTTCGAATGCTCATCAAATCGTATCTCCCGAAAATGTCTTTTATGCTAATAG
TATCTTACTTCCACCACATAATCTACGAACTATCAATGTTTATGATGGTCAGGTTACGA
GTTTGTTAACAAGTGATTTGAATCTGATAATGCGAAGAGTTGCTAATAATGAGACAAAT
GCAAAAATACAAAAAATCTTGGATTCTATCGATAACAGCCGAGGTGCCAATCCATATGC
TACAAATAAAAAGCTTACTTTGGATACTTTGACAGGTGGACACTCAAAAGAATCTTATT
TGCGAAGTTATATTAATGGCAAACGTATTCCTGAGACTGCCAGAGCTGTAATCGAACCC
TCTATGAATAAAACTGGCTTTATTGAAGTACCATCTTACATTTTAAACAAGTTAAGAGA
TGTTGTCTTTTATAATCACGTTACGAAAGATAACATACTCAAAAGTCTTCAAAACGAAC
AAGCTTTTCTAACATATATCAAAAGTGATCATAATTCTGAAAATCCTTATATGGTTTAT
GATTTAGCACAGAAGAATGGATATTTAACCTTGGCTCCTAATTTCGGTGATATTTTCGA
AAAAAGGAAAGAGGAAGGTGGTTTTGTAACTATTTGCAGACATCCATCTATCTGGTTAA
CTAATATCCAATCTGGTATAATAAAAAGATCAGAAGGGTTTACTATTAACATCCCAACC
ACAATTTGCACATCTTTTAATGCTGATTTTGATGGAGATGAGATGACAATATATTCTTT
CAAATCCCCATGTGCCAATCTCGAACAAGCTTTGATTATGAACTCACGAAATCTCTTCA
AAAATTCTATAACAAGCAATCCAATGTTCGGCTTGGTCCAAGATCAAATACCAGCCTTG
AATAAGTTATATAGACGACAAAATTATACATATAACGATGCGTTGGTGATTTTAGGACA
ATTCGGATTTCTGTTAACACCTGGAAAAGATAATTATACCGGAAAAGATATACTTTCTT
GTGTATTCCCAAAACATTATACACTCAAAGGAATTGTTGAAAATGGCGAACTTATTTTG
GAGAATTTTACAAATAAACTCGTTTCCGCAAATTCCTCAAAGTCCATCTTTGGGCATCT
TGTTTTATTTTATGGACAAGAGTATGGTTTGACTATATTGGATACAATGCGAGATATTG
TTCAAAATTTTATTACACATTTTGGTTTCAGTGTAAAAATCCGAGATATGATCCCAAGC
CCAAAAATTTTGGATATTCTAGAAAAGATCGTAGACCAAGAAGTGGATAAAATTGATAA
ACAAACAAAACTTCTATATGACGATATCGAACAAGGTAAGGTTATAATCAACTCTTATG
ATGATATTTCTGAGTTCAGATTAAAAAATGTGGCTATTATGAAAAAGAAACTAGAAAGC
AAACTTTTGGAACTTTTGGATGAATATTATGATGAAGACAATAATTTCCTAGAGATGTA
TAGAACGGGATATAAGGTCAACATTAACGAACTTCTCTCTATTATGTGTTTCTCGGGTT
TTAAAAATTATGGAAATATCGAAATGATTACACCGGGTCTTAATGGTAAAACATCTTTG
TTTAGCTTACCAGATTCTATAAACTTACAAGATTATGGGTTCATCAAAAGCTCTATTGC
CAAAGGGTTAACGTTTGAAGAATATGCTACAATCGTAAAACAAGAAGCTTTTCCACAAA
TTGTTAATGTTACAACTGGTACTTCACAAACAGGATTTTTGGGGAAAAAAATGGTTAAA
ATGGCTTCTGAATTC
```

Figure 11.2 An example of a gene that codes for a polymerase.

systems are "a-thermodynamic," they do not use thermodynamic concepts such as temperature, pressure, volume, heat flow, and physical work. Hence, from this perspective, the genetic code must be understood through information and coding theory, not through physics and chemistry. One school of thought is that the genetic code is "axiomatic" (Yockey 2005), or a given, and it cannot be deduced from purely physical/chemical principles. The origin is unknowable, in the same way that physics has great power in describing and manipulating the properties of matter and energy, without knowing the origin of the matter and energy. From Yockey's perspective, the genetic code stands apart from pure physics but does not transcend the laws of physics. Furthermore, using the Shannon entropy notion, the amount of information in the laws of physics is far less than the information content in even the simplest living organism. The "alphabet" of physics alone is not enough to predict the words, sentences, and stories that will result from it, in the same way that a human alphabet and orthographic rules are not sufficient to predict any of the expressions or literature that can result from it.

Figure 11.2 shows an example of a series of codons for a protein. It is the deduced amino acid sequences of the DlEPV EcoRI clone #36 (RI-36), a large gene that encodes a homolog of a DNA-directed RNA polymerase. The GenBank accession number for this sequence is AF500107.

11.3 Brief Description of Gene Expression with a Focus on Protein Synthesis

The process in which a product is synthesized from an organism's genome is termed *gene expression*. *Transcription* is the process in which genes, coded sequences of DNA for specific products, transcribe their complementary sequence onto *messenger RNA, mRNA* which goes on to synthesize proteins. Other types of RNA are involved in post-transcription modification of RNA, such as splicing, whereas others have gene and epigenetic regulatory functions. Some RNAs, *ribozymes*, have catalytic activity in chemically modifying molecules, similar to enzymes. The chief RNA types involved in protein synthesis are mRNA, *transfer RNA*, tRNA, and *ribosomal RNA*, rRNA, among others. *Translation* is the process in which the code transferred from DNA to mRNA during transcription is used to build a protein.

An organism must constantly produce the right proteins at the right time, in the right conditions, and in the right cells. A portion of the genome of an organism has evolved to store and transmit the instructions for achieving this. The entire process of synthesizing proteins is extraordinarily complex, even in bacteria. Much is known about the entities and events involved and has been gleaned by generations of numerous researchers wielding increasingly sophisticated tools. And yet, there are still many unknowns, especially at the level of macromolecular physics.

Here, we give an extremely simplified view of the events involved in protein synthesis and some of the many macromolecular agents involved. This is a very active research field with new perspectives and models constantly evolving. Transcription and translation occur in prokaryotes (including Archaea, Chapter 14.1.1) and eukaryotes, but the mechanisms and events in eukaryotes are significantly more complex. In the following nanoreview of the process, we focus on eukaryotes.

11.3.1 Transcription

When a specific gene is ready to be expressed and the corresponding protein produced, a promoter site adjacent to the gene is activated and attracts RNA polymerase (RNAP), which docks to the site and begins to unzip the DNA double helix and polymerize complementary nucleotides to the opened DNA sequence of the gene. The RNAP travels along the growing mRNA, proofreading and editing the growing mRNA until it meets a terminator segment at the 3′-end of the gene sequence. It then releases the mRNA, which may be further processed during transcription by splicing, polyadenylation, and capping. The RNA may be used for various functions within the nucleus or be transported out of the nucleus into cytoplasmic sites, where *ribosomes* are located and carry out the translation process.

Transcription factories within the nucleus have been discovered that are immobilized via tethering to the nuclear matrix (a network of dynamic protein fibers). They contain all the proteins and materials needed for transcription and hence make transcription more efficient and also allow for simultaneous transcription from multiple genes.

11.3.2 Properties of RNA Polymerase (RNAP)

RNAP carries out the local unwinding of the DNA chain, polymerization, editing, and proofreading of mRNA and can synthesize very short chains with tens of nucleotides up to chains with over 2 million nucleotides. It also needs to recognize when and where to act and when to stop its activity. Several protein subunits are necessary to carry out these functions, and, in fact, there are three different classes of RNAP (types I, II, and III) in eukaryotic animals. Types I and II are over 5×10^5 g/mol and have over 10 subunits each. The unwinding action of RNAP is carried out by *helicase*, which is a motor protein driven by ATP hydrolysis. Zinc and magnesium are essential cofactors in RNAP activity. Type III are involved in producing small RNA's, such as transfer RNA (tTNA).

Reverse transcriptase is found in *retroviruses* and works in the reverse sense of RNAP. Namely, it produces DNA from an RNA template. Once a retrovirus invades a host cell, it uses reverse transcriptase to produce a DNA copy of its genome, which then becomes incorporated into the host genome by a viral *integrase* enzyme, and the host then produces the retrovirus genome, in addition to its own. Human immunodeficiency virus (HIV-1 and HIV-2) are examples. These latter cause human-acquired immune deficiency syndrome (AIDS).

11.3.3 Transcription Factors and Mediator Complexes

The recognition of the need for certain proteins at any given moment, the timing, and regulation of genes is carried out chiefly by *transcription factors* (or *sequence-specific DNA-binding factor*), of which there are over 1000 in the human genome. TFs turn genes on and off and ensure the correct genes and amounts of mRNA are produced in the correct cells. They also control and coordinate cell division, growth, and death and orchestrate embryological development, and can respond to extracellular signals. All TFs contain one or more DNA binding domains that bind to promoter sites next to genes and attract RNAP to begin transcription. The DNA binding domains distinguish TFs from other proteins involved in genetics, including methylases and other modifiers of histones, and chromatin remodelers, which lack these binding sites. Mutations and disorders of TFs are implicated in diseases such as cancer, diabetes, autoimmune diseases, and osteoarthritis.

As if the complexity of transcription weren't enough with RNAP and TFs, there are also *mediator complexes*, which mediate the signaling from the TFs to the RNAP to initiate transcription. These were discovered by Roger Kornberg (United States, 1947–) in 2006. These have as many as 26 different subunits and disordered regions that allow for great conformational and positional changes needed to carry out mediators' functions. The mediator can stay at its site on the DNA to initiate further copies of a protein by attracting more RNAP. The mediator's functions are not limited to controlling RNAP in initiation. As will be discussed in Section 11.7, the structure of chromatin involves the packing of DNA into the nucleus, complex physical dynamics involved in transcription, and segregation into different active and passive domains. Mediators have a central role in creating chromatin loops and 3D organization, which is at the base of DNA packing and function. Mediators are also involved in signal transduction involving the cell membrane.

11.3.4 Translation and Ribosomes

Ribosomes are "translation factories," complex macromolecular structures where translation of mRNA to proteins occurs. The mRNA is attached to ribosomes, and transfer RNA, tRNA, brings in the amino acids specified by the codons of the mRNA sequence, thus building the coded protein.

Ribosomes can be found free in the cytoplasm, located on the cell membrane, or on the rough endoplasmic reticulum. They were discovered in 1955 by George Palade (Romania/United States 1912–2008). They are very complex macromolecular assemblies, machines composed of large amounts of ribosomal RNAs and proteins, and constitute a *ribonucleoprotein complex*. Ribosomes are similar across eukaryotes, archaea, and bacteria, which suggests a common origin, although there are distinct differences in size and composition among these. Bacterial ribosomes are approximately 20 nm in diameter and have a composition of 65%/35% rRNA/protein, while eukaryotic ribosomes are about 25 nm and are close to 50%/50% rRNA/protein. In all species, the ribosome is composed of a small subunit and a large one, and the two come together during translation. The bacterial *E. coli* ribosome shares a total of around 4500 nucleotides between the large and small subunits. With an average molar mass around 325 g/mol per nucleotide; this gives a total molar mass around 1.5×10^6 g/mol for the rRNA component. Typical eukaryotic ribosomes have about 50% higher molar mass of tRNA. The differences in bacterial and eukaryotic ribosomes have been exploited pharmaceutically to produce antibiotics that destroy the functionality of the bacterial large ribosomal subunit, while leaving the eukaryotic ribosome unharmed.

During translation, the mRNA enters the complex formed of the two ribosome subunits and is "read" sequentially, and the codons translated into the corresponding amino acids, which are brought into the complex by tRNA, where the amino acids are covalently joined in the growing protein chain. Translation is an energetically expensive process. Two ATP molecules are needed to charge an amino acid onto tRNA and two more are needed for covalently joining the amino acid to the growing chain in the ribosome. As detailed in Chapter 3.3.2, the hydrolysis of a single glucose molecule in the metabolic cycle (citric acid cycle) is 38% efficient and produces approximately 36 ATP molecules which are sufficient to add about 9 amino acids during translation in the ribosome.

Chaperones, introduced in Chapter 3.1 and discussed in more detail in Chapter 10.1.1 aid in the repair and folding of the protein macromolecules in the congested intracellular environment.

11.3.5 Mitochondrial DNA and Ribosomes

While all the important gene expression in eukaryotes is derived from the genome in the cell nucleus, there is a curious case of an "isolated DNA" in the mitochondria. This *mitochondrial DNA*, *mtDNA*, is inherited matrilineally so that there is no mixing of genes as occurs with sexual combinations of chromosomes. In humans, mtDNA has only about 16,000 base pairs and codes for only 13 proteins, as opposed to human chromosomal DNA, which has 3×10^9 bp and around 20,000 proteins. In fact, there are about 1500 mitochondrial proteins and only 13 are produced from the mitochondrial DNA. *Mitochondrial ribosomes* have the same structural motif as the other ribosomes, with a large and small subunit that come together for protein translation.

Interestingly, mtDNA accumulates mutations roughly twenty times more quickly than chromosomal DNA, and so its mutations have been traced back to a most recent matrilineal common ancestor, who also became known as the *Mitochondrial Eve* in the popular press. Estimates of this woman are in the range of 100,000–200,000 years in the past.

Similarly, the Y chromosome is present only in the male and is passed on to male offspring without recombination. Hence, it is possible to trace the current male population back to a *Y-chromosome Adam*, who is estimated to have lived 200,000–300,000 years ago.

11.4 Chromatin and DNA Mechanics

The 46 human chromosomes have DNA that stretches out to about 2 m and has a diameter of about 2 nm. The nucleus into which it is packed is on the order of 10 μm. In Section 11.1, we saw that the intrinsic persistence length of DNA is about 50 nm. How can such a long strand of material be packed into such a small volume? With an average of 500 g/mol per base, it turns out that when packed into the nucleus the concentration of DNA is around 1% by mass, or 10 mg/cm^3. In viral capsids, RNA and DNA can have concentrations up to 50%. L_p is a direct measure of bending energy, so it must cost a fair amount of energy to stuff that long strand of DNA into the nucleus. You can compute the highly reduced effective persistence length that would be required to fit DNA into a 10-μm nucleus if it were considered a wormlike chain in the random coil limit, Equation 8.85. But DNA does not exist as a random coil in the nucleus.

How, then, can the DNA be packed into the nucleus? Work in this area is progressing rapidly and new discoveries are constantly made. What is well known is that DNA complexes with octamers of short-chain *histones* (Chapter 10.6), which range from about 11,000–15,000 g/mol. The histones contain up to 25% lysine and arginine, which are both positively charged under physiological pH, and this opposite charge to DNA allows it to associate closely via charge–charge interactions, as well as hydrogen bonds. About 146 bp of DNA coil around the histone octamer in a 1.67 left-handed super-helix to form a *nucleosome*. A simple artist's conception of the DNA/histone octamer complex is depicted in Figure 11.3a. Many nucleosomes form and are connected by linker regions, to form a structure resembling "beads (nucleosomes) on a string (linker DNA)." These in turn, form structured complexes. An electronmicrograph of concatenated nucleosomes is shown in Figure 11.3b. These structures become the major part of a complex involving other proteins in the form of *chromatin*, which then makes up the chromosomes. The entire collection of chromosomes of an organism contains its genome. There are about 30 million nucleosomes in each human cell nucleus. Eukaryotes all have histones and nucleosomes, and their core structure is remarkably conserved between species, despite significant differences in amino acid sequences. Archaea (Chapter 14.1.1) also have simplified histone and nucleosome structure.

The mysteries of chromatin are still being unraveled, but it is clear that it is fundamental in the dynamics of gene expression. The histones have a helix–loop–helix structure that allows them to interact with and wrap the DNA into the nucleosomes. The histones also have flexible tails that interact with the DNA and have intra- and internucleosome functions that relate to genetic expression and also contribute to the folding and stability of the chromatin. In order for a portion of DNA to express a certain protein, it must detach from the histone octamer. DNA that is not bound to the histone octamer can be transcriptionally active, and this loose portion of the chromatin is termed *euchromatin*, whereas DNA tightly bound to the histone octamer is transcriptionally inactive, and this portion of chromatin is termed *heterochromatin*. It is possible that some chromatin structure in germ cells survives in fertilization, which may bear on epigenetics.

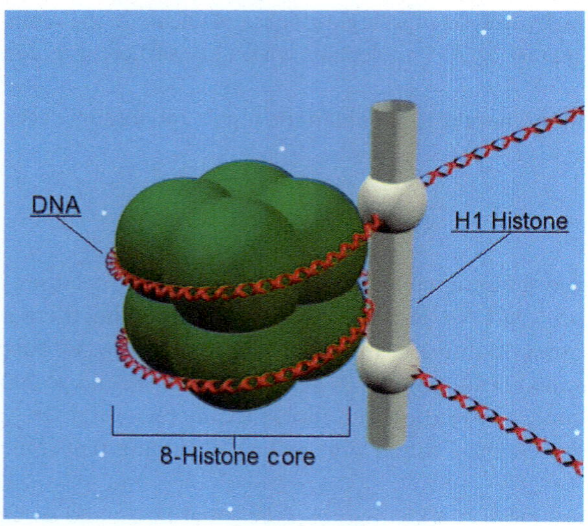

Figure 11.3a Simplified schematic of an 11-nm DNA fiber wrapped around a histone complex. (Wikimedia: Nucleosome).

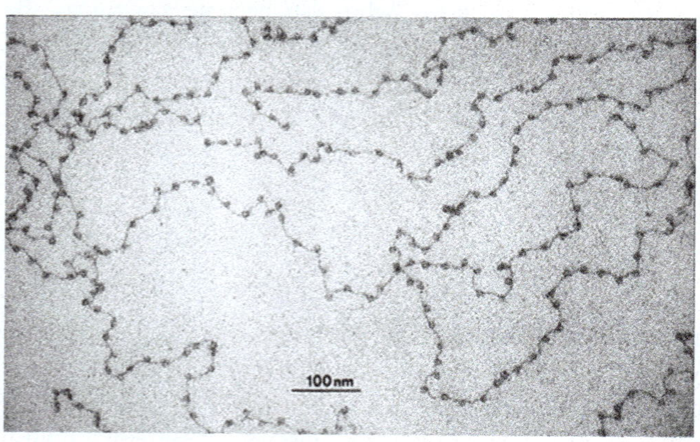

Figure 11.3b Electron micrograph of nucleosomes, portions of DNA (~150 bp per nucleosome) wrapped around histone complexes, connected to each other via linker DNA, resembling "beads on as string." (Wikimedia Commons, "Nucleosome discovery: 1974," Oak Ridge National Laboratory, File:ORNL History (44670612800).jpg|ORNL_History_(44670612800)).

The histones are subject to a number of post-translational modifications, including acetylation and methylation of lysine, methylation of arginine, and phosphorylation of serine. Among other effects, these modifications can lower the histone net charge, thus loosening up the chromatin into euchromatin, activating the affected genes. Other effects can silence transcription. It is thought that information stored in these modifications – which is not carried in the DNA sequence – is inheritable and may be the basis of epigenetics (see Chapter 1.2.2.1).

Like most living state phenomena, the bending, folding, and looping of DNA into these compact structures is not spontaneous and requires molecular motors and the expenditure of ATP energy to drive the process and maintain the coiled state, which is not an equilibrium state. The energy stored in this allows the DNA to activate and deactivate according to metabolic demands. The mechanics of its bending, the nonequilibrium structures formed, and the resulting mechanics and dynamics all have direct biological consequences.

This motif of keeping structures and systems far from equilibrium is pervasive in the living state. We will see in Chapter 13.2 that maintaining ion gradients across excitable membranes requires constant work and keeps them out of equilibrium, and ready to respond to stimuli.

11.4.1 Telomeres

Due to the directionality of DNA polymerase (5′–3′), it cannot synthesize the sequences at the 3′ end. It was discovered that there are *telomeres*, or end sequences on DNA that are lost upon replication. In a human, there is a loss of about 50–100 bp per division. There are a limited number of these on DNA, and when they are used up no further replication is possible. The loss of telomeres has been linked to aging and death.

11.5 Summary

Double-stranded DNA takes on a cooperative, well-stabilized, right-handed helical structure, first discovered by X-ray diffraction. Its macromolecular properties have been well studied; persistence length, bending energy, ability to condense meter-long strands into micron-size volumes, denaturation, and so on. The right-handed DNA can transiently change to a high-energy, unstable left-handed conformation, Z-DNA, which is implicated in transcription and other phenomena.

Segments (~150 bp) of intrinsically stiff DNA wrap around octamers of histones, small positively charged proteins, to create nucleosomes, which concatenate together via linker DNA. The nucleosomes, in turn, are organized, along with other proteins, into chromatin, which condenses meter-long DNA strands into the tiny confines of the cell nucleus in eukaryotes. Chromatin is the dynamic genetic structure responsible for gene expression. Regulation of gene expression occurs within chromatin, and it is thought that modifications of the histones and other proteins carry details of how, when, and which proteins get produced. These structures and modifications do not have any impact on the "hard" DNA sequence, but they are heritable and seemingly form the basis of epigenetic behavior.

The genetic code stores all the information evolved over time in a species. Remarkably, its principles are "a-thermodynamic" and stand apart from living state processes in that they involve no temperature, pressure, energy, etc., and are similar to codes invented by humans for communication and computation. Nevertheless, the implementation of the code in producing proteins is itself highly thermodynamic.

References

Anselmi, C., DeSantis, P., and Scipioni, A. (2005). Nanoscale mechanical and dynamical properties of DNA single molecules. *Biophysical Chemistry* 113: 209–221.

Bruinsma, R.F. (2002). Physics of protein-DNA interaction. *Physica A* 313: 211–227.

Cortini, R., Barbi, M., Caré, B.R. et al. (2016). The physics of epigenetics. *Jean-Marc Victor Reviews of Modern Physics* 88: 250021–250029.

Dupont, S. and Wickstrom, S.A. (2022). Mechanical regulation of chromatin and transcription. *Nature Reviews Genetics* 23 (10): 624–643.

Ghirlando, R. and Felsenfeld, G. (2013). Physical chemistry of nucleic acids and their complexes. *Biopolymers* 99 (12): 910–915. https://doi.org/10.1002/bip.22311.

Guilbaud, S., Salomé, L., Destainville, N. et al. (2019). Dependence of DNA persistence length on ionic strength and ion type. *Physical Review Letters* 122: 028102.

Jost, D., Rosa, A., Vaillant, C., and Everaers, R. (2018). A polymer physics view on universal and sequence-specific aspects of chromosome folding. *Translational Epigenetics Series* 2: 149–169.

Manghi, M. and Destainville, N. (2016). Physics of base-pairing dynamics in DNA. *Physics Reports* 631: 1–41. https://doi.org/10.1016/j.physrep.2016.04.001.

Orlandini, E. (2018). Statics and dynamics of DNA knotting. *Journal of Physics A: Mathematical and Theoretical* 51: 53001–53080. IP address 172.6.240.26.

Pohl, F.M. and Jobin, T.M. (1972). Salt-induced co-operative conformational change of a synthetic DNA. *Journal of Molecular Biology* 67: 375–396.

Rich, A. and Zhang, S. (2003 July). Z-DNA: the long road to biological function. *Nature Reviews: Genetics* 4.

Rittman, M., Eilroy, E., Koohy, H. et al. (2009). Is DNA a worm-like chain in Couette flow? In search of persistence length, a critical review. *Science Progress* 92: 163–204. https://doi.org/10.3184/003685009×462205.

Saiz, L. (2012). The physics of protein–DNA interaction networks in the control of gene expression. *Journal of Physics: Condensed Matter* 24: 193102–193117. https://doi.org/10.1088/0953-8984/24/19/193102.

Schiessel, H. (2003). The physics of chromatin. *Journal of Physics: Condensed Matter* 15: R699. IP address 172.6.240.26.

Vologodskii, A. (2015). *Biophysics of DNA*. Cambridge University Press.

Yockey, H.P. (2005). *Information Theory, Evolution, and the Origin of Life*. Cambridge University Press.

Further Reading

Blossey, R. and Schiessel, H. (2018). The latest twists in chromatin remodeling. *Biophysical Journal* 114 (10): 2255–2261.

Brandani, G.B., Gopi, S., Yamauchi, M., and Takada, S. (2022). Molecular dynamics simulations for the study of chromatin biology. *Current Opinion in Structural Biology* 77: 102485.

Chereji, R.V. and Clark, D.J. (2018). Major determinants of nucleosome positioning. *Biophysical Journal* 114 (10): 2279–2289.

Daban, J.-R. (2021). Soft-matter properties of multilayer chromosomes. *Physical Biology* 18 (5): 053001.

Dai, Z., Ramesh, V., and Locasale, J.W. (2020). The evolving metabolic landscape of chromatin biology and epigenetics. *Nature Reviews Genetics* 21 (12): 737–753.

de Oliveira Guerra, J.C. (2015). Thermodynamics of denaturation transition of DNA duplex oligomers in the context of nearest neighbor models: a short review. *Physics Procedia* 68: 37–41.

Dileep, V. and Tsai, L.-H. (2021). Three-dimensional chromatin organization in brain function and dysfunction. *Current Opinion in Neurobiology* 69: 214–221.

Emanuel, M., Hamedani Radja, N., Henriksson, A., and Schiessel, H. (2009). The physics behind the larger scale organization of DNA in eukaryotes. *Physical Biology* 025008–019. https://doi.org/10.1088/1478-3975/6/2/025008.

Erdel, F. and Rippe, K. (2018). Formation of chromatin subcompartments by phase separation. *Biophysical Journal* 114 (10): 2262–2270.

Esposito, A., Bianco, S., Fiorillo, L. et al. (2021). Polymer models are a versatile tool to study chromatin 3D organization. *Biochemical Society Transactions* 49 (4): 1675–1684.

Fiorillo, L., Bianco, S., Chiariello, A.M. et al. (2020). Inference of chromosome 3D structures from GAM data by a physics computational approach. *Methods (Amsterdam, Netherlands)* 70-79: 181–182.

Gill, J., Kumar, A., and Sharma, A. (2022). Structural comparisons reveal diverse binding modes between nucleosome assembly proteins and histones. *Epigenetics & Chromatin* 15 (1): 20.

Hansma, H.G., Kim, K.J., Laney, D.E. et al. (1997). Properties of biomolecules measured from atomic force microscope images: a review. *Journal of Structural Biology* 119: 99–108.

Hoermanseder, E. (2021). Epigenetic memory in reprogramming. *Current Opinion in Genetics & Development* 70: 24–31.

Jamge, B. and Berger, F. (2022). Diversification of chromatin organization in eukaryotes. *Current Opinion in Cell Biology* 74: 1–6.

Jang, S. and Song, J.-J. (2019). The big picture of chromatin biology by cryo-EM. *Current Opinion in Structural Biology* 58: 76–87.

Korolev, N., Nordenskiöld, L., and Lyubartsev, A.P. (2016). Multiscale coarse-grained modelling of chromatin components: DNA and the nucleosome. *Advances in Colloid and Interface Science* 232: 36–48. https://doi.org/10.1016/j.cis.2016.02.002.

Kwon, Y. and Doo Chung, Y. (2020). RNA-mediated regulation of chromatin structures. *Genese & Genomics* 42: 609–617. https://doi.org/10.1007/s13258-020-00929-5.

Lezia, A., Miano, A., and Hasty, J. (2022). Synthetic gene circuits: design, implement, and apply. *Proceedings of the IEEE* 110 (5): 613–630.

Maksimovic, I. and David, Y. (2021). Non-enzymatic covalent modifications as a new chapter in the histone code. *Trends in Biochemical Sciences* 46 (9): 718–730.

Manning, G.S. (2015a). The energy of naturally curved elastic rods with an application to the stretching and contraction of a free helical spring as a model for DNA. *Journal of Chemical Physics* 143 (10): 104901/1–104901/12.

Manning, G.S. (2015b). The response of DNA length and twist to changes in ionic strength. *Biopolymers* 103 (4): 223–226.

Marko, J.F. (2020). Physics and biology of chromosomes. *Journal of Molecular Biology* 432: 621–631. https://doi.org/10.1016/j.jmb.2019.11.022.

Misteli, T. (2020). The self-organizing genome: principles of genome architecture and function. *Cell (Cambridge, MA, United States)* 183 (1): 28–45.

Morgan, M.A.J. and Shilatifard, A. (2020). Reevaluating the roles of histone-modifying enzymes and their associated chromatin modifications in transcriptional regulation. *Nature Genetics* 52 (12): 1271–1281.

Nelson, P.C., Phillips, R., Widom, J., and Wiggins, P.A. (2006). Biological consequences of tightly bent DNA: the other life of a macromolecular celebrity. *Biopolymers* 85 (2): 115–130. https://doi.org/10.1002/bip.

Podesta, A., Indrieri, M., Brogioli, D. et al. (2005). Positively charged surfaces increase the flexibility of DNA. *Biophysical Journal* 89 (4): 2558–2563.

Razin, S.V. and Ulianov, S.V. (2017). Gene functioning and storage within a folded genome. *Cellular and Molecular Biology Letters* 22: 18. https://doi.org/10.1186/s11658-017-0050-4.

Reid, M.A., Dai, Z., and Locasale, J.W. (2017). The impact of cellular metabolism on chromatin dynamics and epigenetics. *Nature Cell Biology* 19 (11): 1298–1306.

Sazer, S. and Schiessel, H. (2017). The biology and polymer physics underlying large-scale chromosome organization. *Traffic*. https://doi.org/10.1111/tra.12539.

Shindo, Y., Brown, M.G., and Amodeo, A.A. (2022a). Versatile roles for histones in early development. *Current Opinion in Cell Biology* 75: 102069.

Shindo, Y., Brown, M.G., and Amodeo, A.A. (2022b). Versatile roles for histones in early development. *Current Opinion in Cell Biology* 75: 102069.

Shukla, S., Agarwal, P., and Kumar, A. (2022). Disordered regions tune order in chromatin organization and function. *Biophysical Chemistry* 281: 106716.

Siew Lim, P., Jasmine, L., Holloway, A.F., and Rao, S. (2013). Epigenetic regulation of inducible gene expression in the immune System. *Immunology* 139: 285–293.

Singh, A., Verma, S., Modak, S.B. et al. (2022). Extra-nuclear histones: origin, significance and perspectives. *Molecular and Cellular Biochemistry* 477 (2): 507–524.

Suganuma, T. and Workman, J.L. (2018). Chromatin and metabolism. *Annual Review of Biochemistry* 87: 27–49.

Tang, L. (2018). Polymer model predicts chromatin folding. *Nature Methods* 15 (7): 483.

Zhang, D., Lam, J., and Blobel, G.A. (2021). Engineering three-dimensional genome folding. *Nature Genetics* 53 (5): 602–611.

Zhang, S., Chen, F., and Bahar, I. (2020). Differences in the intrinsic spatial dynamics of the chromatin contribute to cell differentiation. *Nucleic Acids Research* 48 (3): 1131–1145. https://doi.org/10.1093/nar/gkz1102.

Problems

1 a) Compute the bending energy to take a length L of DNA and bend it into a circle. Take $L_p = 50$ nm and $T = 300$ K.
b) Compute the total bending energy for $L = 2$ m.
c) If the maximum angle of bending is limited to 60° over one $L_p = 50$ nm and the DNA is bent into a coil of circular loops, find the bending energy.

2 a) Estimate the density of DNA in a spherical 10-μm diameter nucleus. Assume the total end-to-end length of all 46 human chromosomes is 1.8 m and that the average base has a molar mass of 500 g/mol, and that the contour length of one base pair is 6×10^{-10} m.
b) Compute the radius of gyration occupied by a single chromosomal strand of human DNA, of end-to-end length 3.9 cm, and persistence length 50 nm, assuming the random coil limit, Equation 8.85. Compare the approximate volume occupied by this random coil to the volume of the 10-μm nucleus of part a).
c) Compute the effective persistence length needed so that the single DNA strand of end-to-end length 3.9 cm can fit in the nucleus's volume. Take the nucleus as a sphere of 10 μm diameter.

3 a) Use the Genetic Code to predict how many bits of information it takes to code for the following sequence of 10 amino acids:

Gly–Ser–Gly–Arg–Leu–Glu–Ile–Ser–Met–Gly

b) Repeat a) for the sequence composed of the following amino acids. Briefly discuss any differences with part a).

Gly–Trp–Trp–Glu–Phe–Lys–Arg–Cys–His–Trp

4 Show all series of codon triplets that can produce the following amino acid sequence:

Leu–Glu–Gly

5 A specific sequence of 200 amino acids forms a polypeptide chain.

a) Given equal large numbers of each of the 20 amino acids interacting with each other, calculate the probability of a specific sequence forming when 200 amino acids randomly polymerize into a chain.

b) If nature developed a five-base system, coding two bases at a time per amino acid, how many bits of information would be necessary to code for the 200 amino acid sequence?

12

Some Polysaccharide Phenomena

Sugars and some of their many forms, including macromolecular versions, were introduced in Chapter 3. The polymers of these sugars have many functions, including energy storage, structural components, conjugation with proteins to produce catalytic, metabolic, immune, and mechanical functions. Here, we concentrate on some of the notable polyelectrolyte properties of polysaccharides. This includes an analysis of the complex composition of a naturally occurring polysaccharide, gum arabic, the vital role of glycosaminoglycans in the dynamics of cartilage, and unique time-dependent light scattering signatures for polysaccharides undergoing degradation.

12.1 Polymer and Polyelectrolyte Properties of Polysaccharides

Besides being based on carbohydrates, rather than amino and nucleic acids, polysaccharides pose a great structural contrast to polypeptides and polynucleic acids. In Chapter 3, we detailed the secondary, tertiary, and quaternary structures of proteins, and examined the helical structure of single-stranded RNA and single and double-stranded DNA. Denaturation of proteins and polynucleic acids causes their transition to random coils and the loss of their biofunctionality. In contrast, some polysaccharides in their native state can actually exist as structures resembling random coils, whereas others have a significant amount of secondary structure. Furthermore, because polysaccharides aren't formed from genetic templates, they can vary enormously in molar mass and composition and are the only biomacromolecules that are naturally disperse; a given protein is monodisperse, as well as the DNA or RNA corresponding to a particular gene. Since structural polysaccharides control characteristics such as cell visco-elasticity, permeability of extracellular matrix to water, ions, and other species, an organism may flexibly respond to its environment and needs by producing different molar mass distributions and compositions of polysaccharides. Changes in temperature, nutrient availability, pH, ionic strength, toxins, and many other factors can greatly influence polysaccharide molar mass distributions.

12.1.1 Profile of a Polyfunctional Polysaccharide, Gum Arabic

Polysaccharide gums are highly water-soluble polymers, that form high-viscosity "gums" or gels even at very low concentration. Examples include *xanthan, guar, agar, carrageenans, alginates, carboxymethyl cellulose*, and *gum arabic*. The latter is extracted from the hardened sap of acacia trees, mostly in Sudan and across the Sahel, the semi-arid land south of the Sahara desert, stretching from Senegal to Somalia.

Gum arabic is particularly interesting because it consists of a mixture of polysaccharides of arabinose and galactose, and glycoproteins, which endows it with both viscosifying and emulsifying characteristics. It has been harvested and traded since antiquity and used in ceramics, paintings, as a general binder for powders, including gunpowder, and in food preparations.

Figure 12.1 shows raw multidetector size-exclusion chromatography (SEC) data (Chapter 7.13.1) for a sample of gum arabic. The detectors are multi-angle static light scattering (LS, only the data from 90° are shown in the figure), capillary viscometer (Visc), and a differential refractometer (RI). The LS data are trimodal, which suggests that there are at least three distinct subpopulations in the sample. The RI shows two clear subpopulations, and a shoulder between them, suggesting the third population. The viscosity data are noisier but also suggest two well-defined subpopulations with a shoulder between them. We can analyze the data in light of what we learned from LS in Chapter 9, and viscosity in Chapters 6 and 8.

Biophysics: Physical Processes Underlying the Living State, First Edition. Wayne F. Reed.

Companion website: www.wiley.com/go/Reed/BiophysicsandMacromolecules

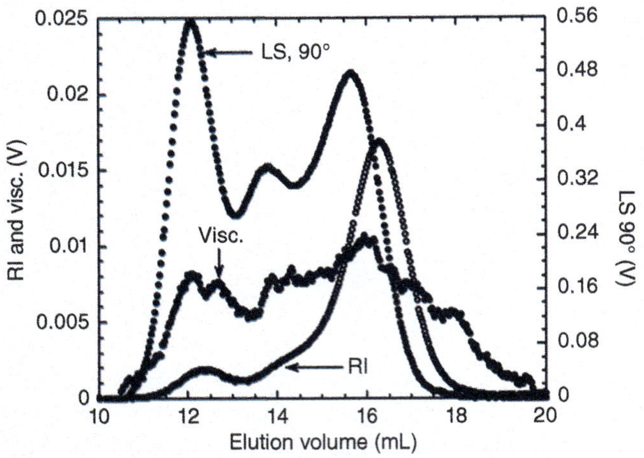

Figure 12.1 Raw multidetector SEC data for a gum arabic sample: LS 90° is static light scattering at 90°, Visc is the raw capillary viscometer signal and RI is the differential refractometer signal. The LS data clearly show a trimodal population, and the Visc and RI data support this. (Reed, W.F. (2004)).

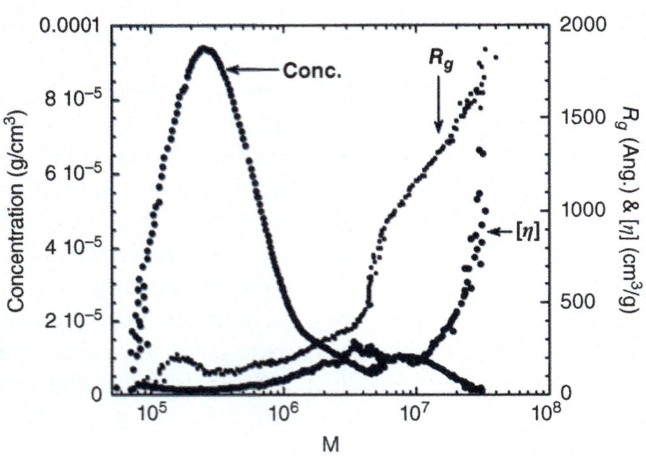

Figure 12.2 The analyzed data showing polymer concentration C_p, radius of gyration R_g, and intrinsic viscosity $[\eta]$ versus molar mass. (Reed, W.F. (2004)).

From Chapter 9, we know that at low polymer concentration C_p, where virial coefficient effects are negligible, the light scattering (in the limit of $q = 0$) is proportional to M_wC_p, where M_w is the weight average molar mass. From Chapter 8, we know that the viscometer signal is proportional to M_vC_p, where M_v is the viscosity averaged mass, which normally lies between M_w and the number average molar mass M_n. The refractive index (RI) data are proportional to C_p and show that the bulk of C_p is at high elution volume, that is, at low molar mass. There is a distinct but small amount of a high molar mass population, seen as the small peak in the RI data at low elution volume. The dependence of each signal on C_p and molar mass allows us to understand why LS in the high molar mass peak towers above LS in the other two populations, and why RI in the low molar mass population dominates.

Figure 12.2 shows the analyzed SEC data: The molar mass M on the x-axis was determined by extrapolation of multi-angle LS data to zero angle ($q = 0$, see Chapter 9.6), together with C_p provided by the RI data, and ignoring the A_2 term, due to very low C_p. The root mean square radius of gyration R_g ($=<S^2>^{1/2}$) was determined from the slope of the inverse LS angular envelope. The intrinsic viscosity $[\eta]$ was taken as η_r, which is actually measured, justified by the very low values of C_p and shear rate in the capillary viscometer.

The high molar mass population has a peak around 10^7 g/mol, which is a very high molar mass in the polymer world. It also has very large R_g (root mean square radius of gyration), suggesting a very extended spatial structure resembling a random coil, with correspondingly high intrinsic viscosity; this small population is the source of gum arabic's viscosifying power. The intermediate population shows both R_g and intrinsic viscosity dropping abruptly. It is surmised that this population is due to the glycoproteins which have fairly high mass, can bind with and collapse some of the high mass population, and is the source of gum arabic's emulsifying ability; a proteinaceous component that can compatibilize oils and other hydrophobic components in an application. Interestingly, the vast majority of the population is in the "small" molar mass portion, peaked around 250,000 g/mol, which is generally a "medium mass" polymer in macromolecular science. The viscosity data show that this medium-mass population contributes only modestly to the viscosifying power of gum arabic and is mostly a filler or "excipient" that carries the small populations of active glycoproteins and very high-mass polysaccharide.

Because climactic, soil, and water conditions vary dramatically from year to year in the Sahel, as well as place to place, the composition of the three populations is bound to shift dramatically among regions and harvest years. SEC provides a means of assessing the quality of the gum arabic.

12.2 Proteoglycans and Extracellular Matrix Functions

Extracellular matrix has many functions, including regulation of solute transport, visco-elasticity, absorbing energy of motion, lubrication, and embryonic development. Bone cartilage, a type of extracellular matrix, is composed chiefly of a network of collagen fibers, interwoven with proteoglycans, and with a sparse population of chondrocytes, which are cells

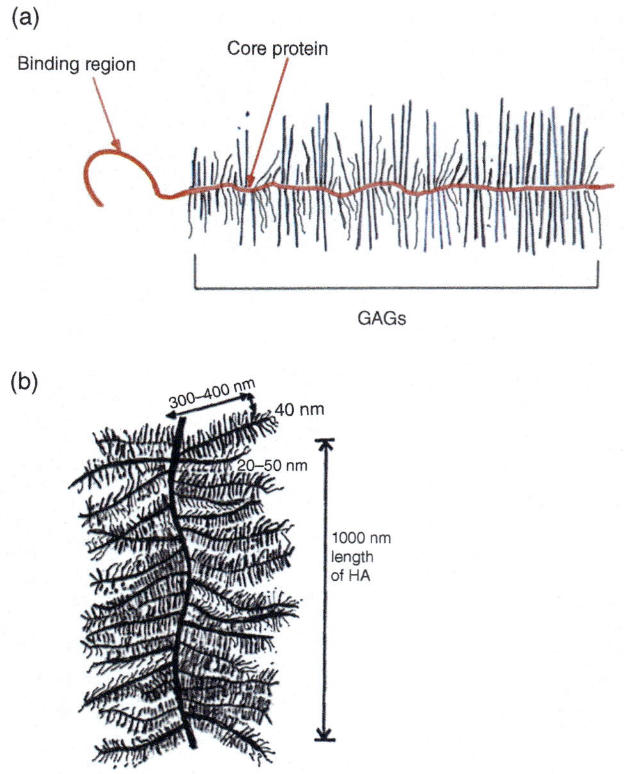

Figure 12.3 (a) An aggrecan, or proteoglycan "monomer," (b) Multiple aggrecans attached non-covalently to the anionic polysaccharide, hyaluronan. (Courtesy of Linnette F. Reed).

that produce extracellular matrix components. The collagen fibers, spaced at about 100 nm apart, give tensile strength to the matrix, helping it resist forces that pull on it.

A schematic of a typical cartilage proteoglycan is given in Figure 12.3a. The proteoglycan "monomer," or *aggrecan*, is composed of a protein backbone, or *cartilage-specific proteoglycan core protein*, a medium-weight macromolecule ($M \sim 4 \times 10^5$ g/mol), to which are covalently bonded glycosaminoglycans (GAGs) of masses up to about 10^4 g/mol, mostly chondroitin sulfate and keratan sulfate, spaced at about 3 nm from each other along the protein. The total aggrecan assembly has $M \sim 2 \times 10^6$ g/mol. These aggrecan units, in turn, bind noncovalently via protein linkages to hyaluronan, a ubiquitous extracellular matrix anionic polysaccharide, to form very large micron-size structures, with M on the order of 10^8 g/mol, depicted in Figure 12.3b. These structures are then interwoven into the collagen fibers to form the cartilaginous matrix. An artist's representation of a portion of the cartilage matrix is shown in Figure 12.4.

12.2.1 Cartilage Dynamics

The proteoglycans play a major role in providing compressibility to cartilage. The response of the cartilage, as measured by the magnitude of its dynamic modulus $|E^*|$ (the higher the modulus, the stiffer the cartilage tissues) is frequency dependent.

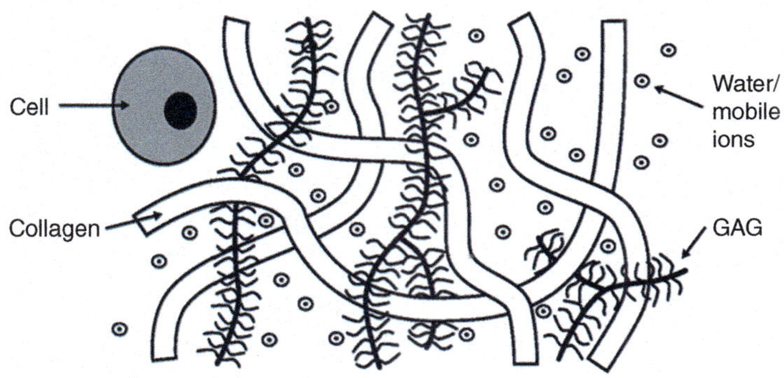

Figure 12.4 An artist's rendition of a portion of bone cartilage. (Courtesy of Marissa Calko).

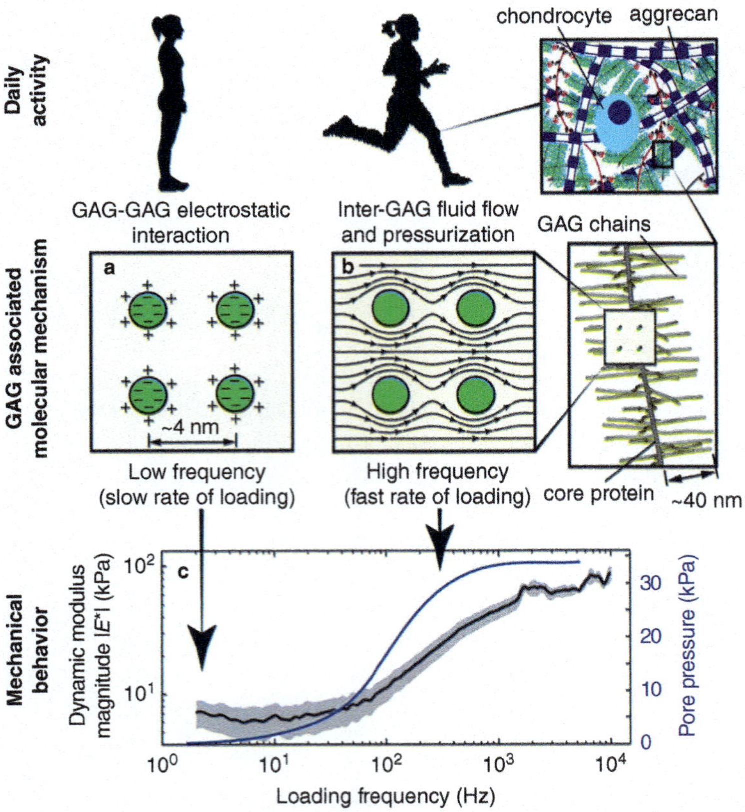

Figure 12.5 Composite view of the molecular origins of cartilage dynamics. The left hand figures show a person at rest and a) electrostatic repulsion among GAGs at rest or in the c) low frequency loading range of mechanical behavior. The middle figures show a person running, which engenders fluid and ion flow among the GAGs in b) and produces the large increase in dynamic modulus at high loading frequency in c). (Adapted with permission from Nia, H.T. et al. (2015). Copyright American Chemical Society).

Figure 12.5 is a composite depiction of the frequency dependence of loading cartilage. At rest and at very low frequencies (a), the electrostatic repulsion between the anionic GAGs in aggrecan and their steric repulsion dominates cartilage response. When cartilage is compressed at low frequencies, the GAG charge groups are brought in closer to proximity to each other and their electrostatic repulsion helps provide a restoring force to the cartilage, making it resist the compression. There is also little to no pressure on the pores in the cartilage matrix. As frequency of loading increases (b), e.g. from running,[1] the inter-GAG fluid flow and pore pressurization lead to a significant increase in cartilage stiffness, and dominate the dynamic modulus, which can increase by over an order of magnitude. The high-frequency effect is due to *poroelasticity*, which is how a porous solid imbued with liquid reacts and deforms in response to fluid flow and pore pressure.

Closely related to the poroelasticity is the hydrodynamic permeability k, of the solid/liquid matrix of cartilage. In Chapter 6.4.3, we encountered k in conjunction with flow rate Q over a pressure drop ΔP, in a fluid of viscosity η, cross-sectional area A and length L:

$$Q = \left(\frac{kA}{\eta L}\right)\Delta P \qquad 6.36$$

Grodzinsky et al. (2015) found, *in their units*,[2] $k = (4.8 \pm 2.8) \times 10^{-15}$ m^4/N-s for an *ex vivo* aggrecan assembly, which is close to the value for intact cartilage. In contrast, for the cartilage network from which GAGs have been removed, the permeability is over an order of magnitude greater, $k = (1.3 \pm 0.7) \times 10^{-13}$ m^4/N-s, demonstrating that the dynamics and permeability are dominated by the aggrecans and their GAGs, and not the collagen fibers and other cartilage constituents.

Another important aspect of cartilage dynamics is the lubrication and very low friction it provides. Experiments have shown that adjacent surfaces coated with polyelectrolytes, as opposed to neutral polymers, have remarkably low friction as they are moved transversely to each other, even when the polyelectrolytes are at high concentration. The ostensible

1 The actual frequency of running legs is quite low. The reason the loading frequency in the figure goes all the way up to 10^4 Hz is that there is a rapid force pulse versus time in the cartilage, which quickly absorbs the shock when a running foot hits the ground. This force pulse contains a wide range of frequencies.

2 As noted in Chapter 6.4.3, the units of hydraulic permeability κ are area (cm^2 for CGS). The units of Grodzinsky et al. are hydraulic permeability (m^2) divided by viscosity.

reasons for this are (i) the polyelectrolyte counterions provide considerable osmotic pressure keeping the surfaces apart, and (ii) the polyelectrolytes have a strong hydration layer that allows polyelectrolytes to glide past each other, in addition to their charges preventing interpenetration of polyelectrolytes on the opposite surfaces.

12.2.2 Cartilage Electrokinetics

The compression and stretching of cartilage leads to interesting electrokinetic effects. The *streaming potential* is produced when cartilage is compressed or stretched. Basically, since the proteoglycans and their charges are relatively fixed in the collagen fibers, there will be a streaming of counterions under compression or stretching, which generates a streaming current through the resistive extracellular matrix, and hence an associated electric streaming potential.

A conceptualization of streaming potential due to fluid flow inside a charged capillary with a permeable mesh can be made as follows: Consider the electric potential drop ΔV, the streaming potential, across a length L of capillary of cross-sectional area A, which has a mesh of hydraulic permeability κ inside it, through which fluid flows. The inside of the capillary is charged and its electrical potential at the hydrodynamic slip plane of fluid flow, is the ζ-potential (Chapter 5.6.2), and the fluid, together with the mesh inside the capillary have a net dielectric constant ε, and conductivity s. The liquid has viscosity η and flows at rate Q and the pressure drop across the capillary is ΔP. The streaming potential is given by

$$\Delta V = \frac{\varepsilon\zeta}{\eta s}\Delta P \qquad 12.1$$

Using Equation 6.36 to represent ΔV in terms of the fluid flow rate Q gives

$$\Delta V = \frac{\varepsilon\zeta QL}{s\kappa A} \qquad 12.2$$

The streaming potential in cartilage, and to a large degree the compressive stiffness, are due to the ionized groups on the proteoglycans. The data in Figure 12.6 for both stiffness and streaming potential show that as this charge on the proteoglycan's GAGs is neutralized by lowering pH the streaming potential drops almost to zero upon charge neutralization, as ζ drops, then increases as further drops in pH lead to positively charged groups. The streaming potential reaches a minimum upon electrical neutralization at low pH. Observe, however, that compressive stiffness, shown on the same graph, does not fall close to zero. This indicates that steric crowding of proteoglycan components also contributes to the stiffness. Independently, osmotic stress measurements on hyaluronate and several GAGs and comparison with connective tissue compressibility data confirm the latter has a partial but strong contribution from the GAGs and hyaluronate, whereas for corneal stroma the compressibility is due almost entirely to them.[3]

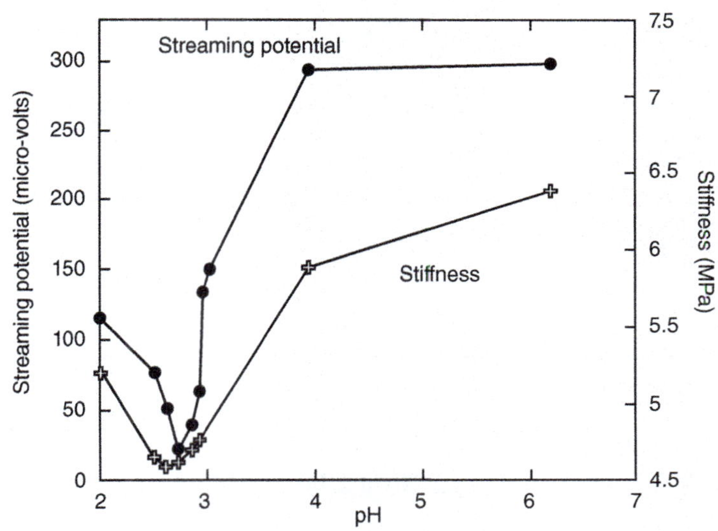

Figure 12.6 Streaming potential and compressive stiffness of cartilage as a function of pH. (Frank, E.H. and Grodzinsky, A.J. (1987) / Reproduced with permission from Elsevier).

3 R.M. Peitzsch and W.F. Reed "High Osmotic Stress Behavior of Hyaluronate and Heparin", Biopolymers, 32, 219–238, 1992.

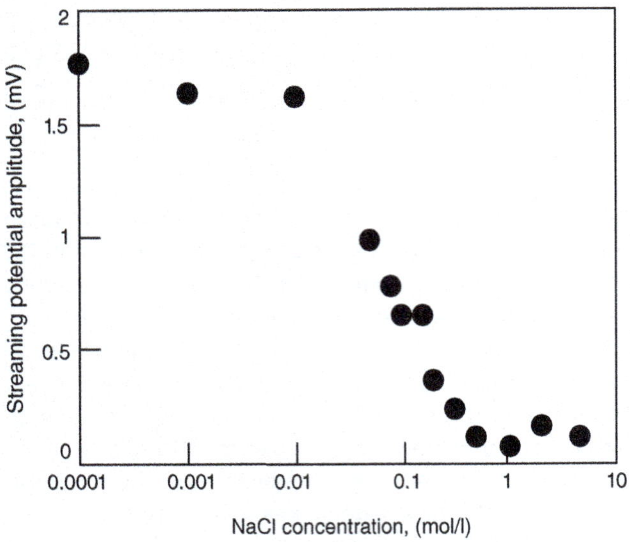

Figure 12.7 Streaming potential versus NaCl concentration in bovine articular cartilage. (Frank, E.H. and Grodzinsky, A.J. (1987) / Reproduced with permission from Elsevier).

The streaming potential is also sensitive to ionic shielding, furnished by added simple electrolyte. As ionic strength (Chapter 5.6.3) increases ζ decreases, so that the streaming potential falls monotonically, seen in Figure 12.7.

The streaming potentials have been found to stimulate the chondrocytes which produce the biomolecules needed to keep the cartilage healthy. It has been found that when compressive and tensile action on cartilage decreases, e.g. due to inactivity, chondrocyte action decreases which can lead to cartilage degeneration. Cartilage degeneration is linked to arthritis. Astronauts who have been immobilized for long periods in space have shown arthritic symptoms upon return to Earth.

Collectively, Eliot, Frank and Grodzinsky (1985, 1987) made an electrokinetic model for the streaming potential and streaming current for uniaxial flow and compression in articular cartilage. A cylindrical plug of cartilage was confined in a tube with impermeable walls and one impermeable, moveable end cap, while the other endcap was fixed but permeable. Ag/AgCl electrodes were mounted on each endcap. If the cartilage was compressed by pushing the impermeable end cap fluid flowed through the cartilaginous matrix. If the electrodes were in open circuit, then the streaming potential produced by the fluid motion was sensed with a voltmeter. If the terminals were short circuited, with an ammeter between them, then the streaming current could be measured.

Remarkably, the system described follows the Onsager reciprocity relationship described in Chapter 6.11 and summarized in

$$\vec{J} = \bar{\bar{L}} \cdot \vec{F} \qquad 6.130$$

where the generalized fluxes $\vec{J}$, for cartilage are the flux of liquid, represented by its one-dimensional velocity U and the flux of electrical charge in the cartilage matrix, both in the z-direction. The generalized forces $\vec{F}$ in the cartilage are the mechanical pressure gradient over the cartilage in the tube and the electrical field in the tube, which is the gradient of the electric potential V. $\bar{\bar{L}}$ is the matrix of coefficients that couple these together. Equation 6.130 then takes the form

$$\begin{bmatrix} U \\ J \end{bmatrix} = \begin{bmatrix} L_{11} & L_{12} \\ L_{21} & L_{22} \end{bmatrix} \begin{bmatrix} \partial P / \partial z \\ \partial V / \partial z \end{bmatrix} \qquad 12.3$$

In other words, according to Equation 12.3, the flux U can be produced by gradients of both P and V, and, likewise, the current flux J can be produced by gradients of both P and V. Eliot and Frank (1987) showed that they could produce liquid flow in the cartilage U, by simply applying a voltage across the endcap terminals with no mechanical pressure. Similarly, the streaming current J could be caused simply by compression of the cartilage, that is, from a pressure gradient $\partial P / \partial z$. Furthermore, they found that

$$L_{21} = L_{12} \qquad 12.4$$

confirming a major part of Onsager's reciprocity theorem that the coupling constants between the two distinct fluxes and forces are equal.

12.3 Proteoglycan Degradation Mechanisms Found by Light Scattering

Structural and mechanistic information was obtained for a proteoglycan aggrecan using the time-dependent total-intensity light scattering signatures during aggrecan degradation by different agents. Three main mechanisms were found, each with its own characteristic time-dependent signature: (i) The GAG side chains could be stripped off at their binding site on the core protein, such as by NaOH, (ii) the GAG side chains could be randomly degraded, such as by hyaluronidase, and (iii) the protein backbone could be randomly degraded. HCl was found to combine mechanisms (i) and (iii).

If r random scission cuts are made on an ideal random coil polymer of initial molar mass M_0 and mean square radius of gyration $<S^2>_0$, then the Rayleigh ratio $R(u)$ in dilute solution (i.e. ignoring the A_2 effect on scattering in Equation 9.100) of polymer concentration c_0 and optical constant K (see Equation 9.70) is

$$\frac{R(u)}{Kc_0} = M_0\left(\frac{2}{u^2}\right)\left[e^{-u} - 1 + u\right] \qquad 12.5$$

where

$$u = q^2\langle S\rangle_0 + r \qquad 12.6$$

where q is given by Equation 9.117 ($q = (4\pi n_0/\lambda_0)\sin(\theta/2)$). For this type of random scission of a linear chain when $u > 3$, the limiting form is

$$\frac{Kc_0}{R(t)} = \frac{\left(q^2\langle S^2\rangle_0 + \dot{r}t\right)}{2M_0} \qquad 12.7$$

where $\dot{r}$ is the number of cuts per initial polymer chain per second for $r \ll$ number of monomers in the initial polymer.

Another interesting signature occurs if a polymer consisting of N strands is randomly degraded. In that case, it can be shown that the reciprocal scattering intensity obeys

$$\frac{1}{R(t)} \propto a + bt^N \qquad 12.8$$

For example, for random degradation of double-stranded DNA, the reciprocal light scattering intensity will trace a parabolic signature.

If GAG chains are randomly stripped off, an aggrecan of initial total mass $M_{t,0}$ in a first order process, then the mass in the side chains M_s, decreases according to

$$\frac{dM_s}{dt} = -\alpha M_s \qquad 12.9$$

where α is the first order rate constant. Since the protein backbone stays intact, with constant mass M_p, the total aggrecan mass decreases according to

$$M_A(t) = M_p + M_{s,0}e^{-\alpha t} \qquad 12.10$$

where $M_{s,0}$ is the initial mass of GAGs in the intact aggrecan. The total concentration of protein backbone and GAGs, both free and still in sidechains is constant c_0 so that the concentration of decreasing aggrecan c_A in solution is

$$c_A(t) = \frac{c_0\left[M_p + M_{s,0}e^{-\alpha t}\right]}{M_{t,0}} \qquad 12.11$$

and the concentration of free GAG in solution is found by the mass balance of total concentration c_0. This leads to total scattering, assuming again that A_2 interactions are negligible of

$$R(t) = \frac{K_A c_A(t) M_t(t)}{\left[1 + q^2\langle S\rangle^2{}_A(t)/3\right]} + \frac{K_{GAG} c_{GAG}(t) m_s}{\left[1 + q^2\langle S\rangle^2{}_{GAG}(t)/3\right]} \qquad 12.12$$

where m_s is the average GAG sidechain mass. Because the GAGs were very small compared to the core protein, their contribution to scattering was negligible, and since they also contribute negligibly to $<S^2>_A$, scattering could be approximated as

$$\frac{Kc_0}{R(t)} = \frac{M_{A,0}\left[1 + q^2\langle S^2\rangle_A(t)/3\right]}{\left[M_p + M_{s,0}e^{-\alpha t}\right]^2} \qquad 12.13$$

Figure 12.8 shows Kc_0/R versus time when 1 M NaOH was added to a solution of 5×10^{-5} g/cm^3 of articular cartilage aggrecan at $T = 25$ °C. The signature is well fit by Equation 12.13, indicating GAG side chain stripping, revealing the

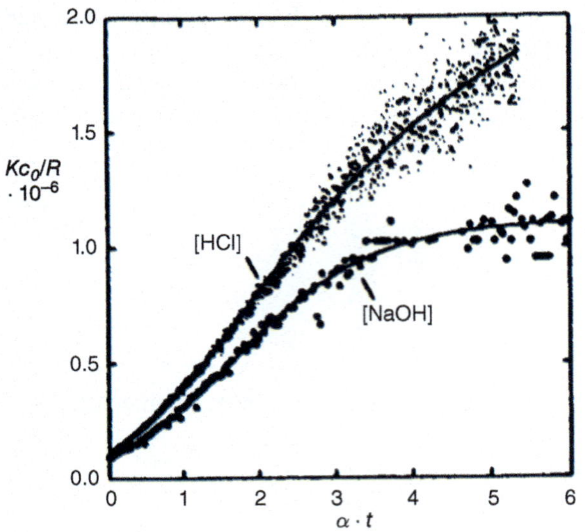

Figure 12.8 The effect on light scattering of NaOH stripping off GAG chains from aggrecan. The two model parameters found were the stripping rate constant $\alpha = 0.0036\ s^{-1}$, and $M_p = 0.31 M_{A,0}$; i.e. initially 31% of the aggrecan is the protein backbone, the rest GAGs. (Ghosh, S. and Reed, W.F. (1995) / Reproduced with permission from John Wiley & Sons).

mechanism of degradation, with the rate constant $\alpha = 0.0036\ s^{-1}$ and 31% of the original aggrecan in the core protein. The rate constant α was found to increase linearly with [NaOH].

Also shown in Figure 12.8 is the effect of 2 M HCl at $T = 44\,°C$ on aggrecan. The signature is qualitatively different from that produced by NaOH hydrolysis. The mechanism best explaining the signature was stripping of the GAG sidechains combined with random hydrolysis of the core protein. The fit shown is obtained by combining GAG side chain stripping expression with the random coil degradation signature.

Concerning mechanism (ii) the random degradation of the GAG side chains, it was shown (Reed and Reed 1989) that for random degradation of an ideal random coil, the fraction of fragments of normalized mass $x = M/M_0$ (M_0 is the initial mass of the polymer) per normalized mass interval, $f(r,x)$, is, after r random cuts

$$f(r,x) = e^{-r}\delta(x-1) + e^{-rx}(r^2 - xr^2 + 2r) \qquad 12.14$$

Using this to find the weight of chains remaining on the aggrecan after r cuts, $M_s(r)$, yields the fraction of GAG chains still attached to the aggrecan

$$\frac{M_s(r)}{M_{s,0}} = \frac{(1+re^{-r})}{(1+r)} \qquad 12.15$$

Combining this with the concentration of aggrecan, and again assuming the scattering due to GAG fragments is negligible leads to the signature

$$\frac{Kc_0}{R(t)} = \frac{M_{A,0}\left[1+q^2\langle S^2\rangle_A(t)/3\right]}{\left[M_p + M_{s,0}\dfrac{(1+re^{-r})}{(1+r)}\right]^2} \qquad 12.16$$

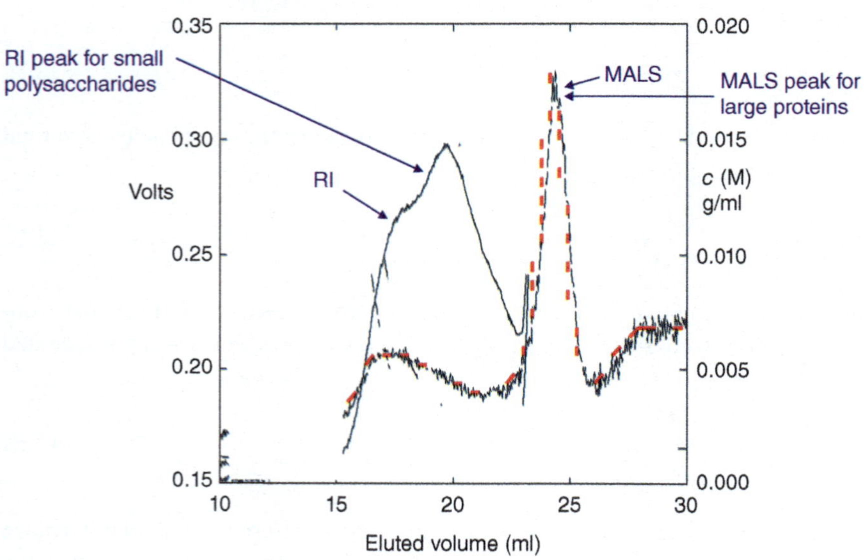

Figure 12.9 GPC results using light scattering (MALS) and refractive index (RI) detectors for aggrecan after stripping of its GAG sidechains by NaOH. (Ghosh, S. and Reed, W.F. (1995) / Reproduced with permission from John Wiley & Sons).

The difference in the GAG sidechain stripping signature in NaOH, and the combined stripping and random protein backbone degradation signature in HCl can be understood qualitatively as follows: The stripping reaction with NaOH reaches a plateau when all side chains have been stripped off and the intact core protein remains. The combined stripping and core protein degradation involves two different time scales, and in this case, the protein degradation continues after all GAG chains are stripped so that the intensity continues to diminish.

The GPC results for aliquots of the aggrecan during NaOH hydrolysis are shown in Figure 12.9 at $t = 15$ min. The raw 90° light scattering signal is shown (MALS=multi-angle light scattering) and the refractometer signal (RI). The broad, asymmetric RI peak with low light scattering intensity at early elution volume corresponds to the GAG sidechains which have been stripped off, which are small relative to the core protein, ~10^4 g/mol versus 5×10^5 g/mol. The narrow peak for both RI and MALS at high elution volume corresponds to the intact, stripped core proteins. The narrowness of the peak indicates low dispersity, while the high MALS signal indicates the core protein's large mass. Important to note is that the separation, in this case, is *not* by size exclusion, so this is not SEC (Chapter 7.13.1), rather, the more general gel permeation chromatography. In size exclusion, the large molecules elute first and smaller molecules last. In the case here, the order is reversed. The reason, often seen, is that the proteins have attractive enthalpic interaction with the SEC column gel, which slows its elution down. This is a good example of how interpreting the GPC data with column calibration standards would lead to totally erroneous results.

12.4 Summary

Unlike DNA, RNA, and proteins, polysaccharides are not synthesized via a template, and so their composition and molecular weight distributions can be enzymatically adjusted in response to the environment. Many polysaccharides are polyelectrolytes, which endows them with the ability to interact with other charged and neutral species. Polysaccharide/protein complexes are found widely. Gum arabic was used as an example of how a complex natural product can contain a variety of polysaccharide and proteinaceous components, which gives it remarkable properties for emulsifying, viscosifying, and binding. SEC, polymer physics, and light scattering methods from Chapters 7, 8, and 9 helped separate and understand the origin of these properties.

The role played by a variety of glycosaminoglycans (GAGs) in cartilage dynamics and electromechanics was explored in detail. Proteoglycans, which combine proteins and GAGs, are essential for these functions. Light scattering was used to understand proteoglycan structure and composition while undergoing degradation. GAGs are also essential for lubrication and shock absorption.

References

Detournay, E. and Cheng, A.H.-D. (1993). Fundamentals of Poroelasticity, chapter 5. In: *Comprehensive Rock Engineering: Principles, Practice and Projects, Vol. II, Analysis and Design Method* (ed. C Fairhurst), 113–171. Pergamon Press.

Eisenberg, S.R. and Grodzinsky, A.J. (1985b). Swelling of articular cartilage and other connective tissues: electromechanochemical forces. *Journal of Orthopaedic Research* 3: 148–159.

Frank, E.H. and Grodzinsky, A.J. (1987). Cartilage electromechanics-I. Electrokinetic transduction and the effects of electrolyte pH and ionic strength. *Journal of Biomechanics* 20: 615–627.

Frank, E.H. and Grodzinsky, A.J. (1987). Cartilage Electromechanics II. A continuum model of cartilage electrokinetics and correlation experiments. *Journal of Biomechanics* 20: 629–639.

Ghosh, S. and Reed, W.F. (1995). New characteristic signatures from time-dependent static light scattering during polymer depolymerization, with application to proteoglycan subunit degradation. *Biopolymers* 35: 435–450.

Matijevic, E. (1974). *Surface and Colloid Science*, 49–272. New York: Wiley Interscience.

Nia, H.T., Han, L., Bozchalooi, I. S., Roughley, P., Youcef-Toumi, K., Grodzinsky, A. J., Ortiz, C. (2015). Aggrecan nanoscale solid-fluid interactions are a primary determinant of cartilage dynamic mechanical properties. *ACS Nano* 3 (3), 2614–2625.

Raviv, U., Giasson, S., Kampf1, N. et al. (2003 11 September). Lubrication by charged polymers. *Nature* 425.

Reed, W.F. (2004). Fundamentals of Static Light Scattering and Viscometry in SEC and Related Methods, chapter 2. In *Multiple Detection Size-Exclusion Chromatography. ACS Ser. 893* (ed. A.M. Striegel). ACS: Washington, DC.

Reed, C.E. and Reed, W.F. (1989). Light scattering power of randomly cut random coils with application to the determination of depolymerization rates. *Journal of Chemical Physics* 91: 7193–7199.

Further Reading

Almond, A. (2018). Multiscale modeling of glycosaminoglycan structure and dynamics: current methods and challenges. *Current Opinion in Structural Biology* 50: 58–64.

Campos, Y., Almirall, A., Fuentes, G. et al. (2019). Tissue engineering: an alternative to repair cartilage. *Tissue Engineering. Part B, Reviews* 25 (4): 357–373.

Freutel, M., Schmidt, H., Dürselen, L. et al. (2014). Finite element modeling of soft tissues: material models, tissue interaction and challenges. *Clinical Biomechanics (Bristol, Avon)* 29 (4): 363–372.

Kogut, M.M., Marcisz, M., and Samsonov, S.A. (2022). Modeling glycosaminoglycan-protein complexes. *Current Opinion in Structural Biology* 73: 102332.

Nia, H.T., Ortiz, C., and Grodzinsky, A. (2022). Aggrecan: approaches to study biophysical and biomechanical properties. *Methods in Molecular Biology (Clifton, N.J.)* 2303: 209–226.

Su, L., Hendrikse, S.I.S., and Meijer, E.W. (2022). Supramolecular glycopolymers: how carbohydrates matter in structure, dynamics, and function. *Current Opinion in Chemical Biology* 69: 102171.

Problems

1 Use the GPC data of Figures 12.1 and 12.2 to estimate the fraction by mass of each of the three subpopulations of the gum arabic.

2 Show that for a polymer with N strands being randomly degraded that the following approximation applies for $u > 3$:

$$\frac{1}{I(t)} \propto a + bt^{N}$$

13

Phospholipid Membranes: Channels and Nerve Impulses

The basic motifs for lipid structure were presented in Chapter 3. Most importantly, their mechanism of lipid bilayer formation was introduced. The lipid bilayer is the essential structure of cell membranes. Here, after a brief review of general properties of membranes, the generation of action potentials in nerve cells, and their propagation, are focused on.

13.1 General Properties of Membranes

While the lipid bilayer motif underlies membranes of many types, there is wide variation among membrane types as far as the ratios of lipids, proteins, and carbohydrates. Very active membranes may have 50% or higher protein content. One of the main tasks of the cell membrane is to provide a boundary between each individual cell, and surrounding other cells, extracellular matrix, and environment. The membrane then demarcates the spatial extent of the cell and controls the exchange of material into and out of the cell in order to maintain the cell's metabolism and functions.

In addition to providing cell identity and transport and exchange of material with the exterior, a further vital membrane function is in energy production; oxidative phosphorylation that produces ATP takes place on mitochondrial inner membranes, while photosynthesis that produces ATP and other compounds in plants occurs on the inner membrane of the chloroplast.

The membrane is a dynamic structure, and individual lipid molecules diffuse about in the sea of surrounding lipids with a diffusion coefficient on the order of 10^{-8} cm^2/s. This gives a root mean square displacement of about 2 μm over 1 s. The lipid molecules can also exchange between inner and outer-membrane surfaces, but this process is about a billion times slower than a lipid molecule laterally diffusing a few microns on one surface.

While there is constant lateral diffusion of the lipid molecules, there is long-range order among the lipid molecule tails in the membrane surfaces, which makes the membrane a *liquid crystal*. As temperature decreases, the diffusive mobility of the lipid molecules decreases, and at a certain temperature, the liquid crystal becomes a solid crystal, at which point life-sustaining transport functions across the membrane – which require membrane fluidity – cease.

There are two transport mechanisms across membranes: *passive* and *active*. In the first case, molecules that are soluble in the lipid membrane can pass through it if there is a concentration gradient of the molecule across the membrane. The gases O_2 and CO_2 are fundamental to respiration, are soluble in the lipid bilayer, and can passively diffuse through alveolar membranes in the lung and across capillary membranes throughout the circulatory system.

In active transport there is a wide variety of membrane pore and channel-forming proteins and polypeptides that can open and close, as needed, in addition to carrier molecules, which can bind to and transport specific molecules across membranes. A number of channels are activated by voltage differences, and this underlies the excitability of cells. The rest of this chapter will focus on the electrochemical structures and events behind nerve impulse generation.

13.2 Membrane Potentials

Living cells invariably have an electrostatic potential difference across them, even when they are at rest and not performing active functions. The potential inside the membrane is negative with respect to the outside, hence transmembrane potentials are usually, but not always, cited as negative numbers, such as a *resting potential* of –70 mV.

Biophysics: Physical Processes Underlying the Living State, First Edition. Wayne F. Reed.

Companion website: www.wiley.com/go/Reed/BiophysicsandMacromolecules

While it is worthwhile to consider passive models the conversation can be preempted by saying that, as usual, the living state does not leave its well-being to chance, rather it actively controls all its critical functions, including potentials and ion transport across cells. And, as usual, a wide variety of proteins come into play for this active control. A pure liposome, such as in Figure 3.25, is not very permeable to inorganic ions. This can be understood immediately by the fact that inorganic salts have very low solubility in apolar organic phases, such as the aliphatic chains inside the membrane, and very high solubility in aqueous phase, such as on either side of the membrane. The typical dielectric constant for a bilayer lipid membrane is around $D = 3$, while for water $D \sim 80$. This imposes a very high barrier for penetration by ions, which have almost no solubility in such low dielectric material. The permeability of the membrane can be increased by orders of magnitude by transmembrane channels which can be actively opened and closed. There are likewise proteins that can actively pump ions against concentration gradients. The time-dependent action potential, the origin of nerve impulses, due to *voltage-gated membrane channels*, is arguably the most sophisticated of membrane-based functions. In fact, one way of classifying cells is whether they are excitable or nonexcitable. The former class includes nerve, photoreceptor, and muscle cells.

A useful concept in thinking about ion transport through the membrane is the work needed to move a charge q from the aqueous phase at the surface of a membrane to the middle of the hydrophobic membrane (MKSA/SI units)

$$\Delta G = \frac{q^2}{8\pi R}\left(\frac{1}{\varepsilon_h} - \frac{1}{\varepsilon_w}\right) - \frac{q^2}{4\pi\varepsilon_h d}\ln\left(\frac{2\varepsilon_w}{\varepsilon_w + \varepsilon_h}\right) \qquad 13.1$$

where R is the solvated ion radius, d the membrane thickness, ε_h and ε_w the permittivities of the hydrophobic membrane and water, respectively. The first term is the energy of transfer of the ion from aqueous phase to the apolar phase, as discussed in Chapter 5.8, and given in Equation 5.127. The second term is work done against the image charge discussed in Chapter 5.9. The smaller the value of ΔG, the lower the barrier for the ion to go through the membrane and the quicker its transport. Equation 13.1 suggests that for a given R and d, if the effective ε_h could be raised by some mechanism then ΔG will decrease, speeding up transfer. Opening a pore to form an aqueous bridge is one means to do this, and it works to a certain extent with large channel proteins. Other mechanism, not so obvious, such as selective affinity, can effectively increase $\varepsilon_{h,}$ while not literally changing ε_h, which is a macroscopic parameter.

Equation 13.1 suggests that (i) thinner membranes, with smaller d, will speed up transport, and (ii) increasing ionic radius, R, also speeds up transport, because the cost of dehydration is smaller, the larger R gets for a given q, as seen in Equation 5.127.

Notation: In all that follows in this chapter, we are concerned with the electrostatic potential difference *across* the membrane, which is $\Delta\phi = \phi_i - \phi_\varepsilon$, where ϕ_i is the inner membrane potential and ϕ_e is the outer membrane potential. In keeping with most of the physiological literature, we use Φ to express this transmembrane potential difference. This also avoids the use of a "Δ":

$$\Phi \equiv \Delta\phi = \phi_i - \phi_e \qquad 13.2$$

Let's build up to active control starting with a passive system, using equilibrium considerations. In Chapter 5.7, we saw how a membrane permeable to small ions but impermeable to macro-ions led to concentration differences of the ionic species across the membrane. These create a transmembrane electrostatic potential at Donnan equilibrium. A negative Donnan potential contributes to the observed negative resting potential of membranes, due to the many polyanions, such as DNA and RNA inside the cell membrane, to which the cell membrane is impermeable. Since these charges are negative there is a "free" equilibrium contribution to the negative resting potential of the cell that doesn't cost any metabolic energy. This contribution is estimated to be on the order of −10 mV. The electrostatic Donnan potential *across* the membrane is given by

$$\Phi_{Donnan} = \frac{RT}{F}\ln\left(\frac{z\left[P^{-z}\right]}{2c_e} + \sqrt{\left(\frac{z\left[P^{-z}\right]}{2c_e}\right)^2 + 1}\right) \qquad 13.3$$

where z is the number of negative charges on the polyanion P^{-z} (z is a positive integer), F is the Faraday constant given in Equation 13.5, and c_e the molar concentration of ions exterior to the membrane. Since $c_e \gg z[P^{-z}]$ the Donnan potential is negative.

In Chapter 6.9, we considered a membrane permeable to ions with a concentration gradient across the membrane. This led to a flux of ions driven by chemical potential down the concentration gradient, and a corresponding flux of ions resulting from the electric field caused by the ionic concentration gradient, expressed by the Nernst–Planck Equation 6.110.

$$\vec{J} = -D_0\left(\nabla c + qc\frac{\nabla\phi}{k_BT}\right) \tag{6.110}$$

where D_0 is the self-diffusion coefficient of the ion, c its concentration, and q its charge.

The net flux due to these two flows can reach an equilibrium where they cancel each other out, being equal and opposite in direction, leaving an equilibrium electrostatic potential difference across the membrane, $\Phi = \Delta\phi = \phi_i - \phi_e$,

$$\ln\left(\frac{c_e}{c_i}\right) = \frac{q(\phi_i - \phi_e)}{k_BT} \tag{6.114}$$

where c_i is the mass concentration of the species inside the membrane and c_e is the mass concentration on the exterior of the membrane. $\Phi = \phi_i - \phi_e$ from this condition is often termed the Nernst potential $\Phi_{Nernst} = \phi_i - \phi_e$. On a molar concentration basis $[X]$ for species X, Φ_{Nernst} is

$$\Phi_{Nernst} = \phi_i - \phi_e = \frac{RT}{zF}\ln\left(\frac{[X]_e}{[X]_i}\right) \tag{13.4}$$

where z is the valence of the ion (in this case, a positive integer for cations and a negative integer for anions) and F is Faraday's constant and is given by

$$F = eN_A = 9.649 \times 10^4 \text{ C/mol} \tag{13.5}$$

If the concentration of a cation is less inside than outside the cell, Φ_{Nernst} will be positive, and vice versa. To help remember the meaning of the sign of Φ_{Nernst}, it is the potential of the inner membrane surface minus the potential of the outer. To visualize the situation when $c_e > c_i$ for a cation, consider that the diffusion due to the concentration gradient drives cations inward, but, to be in equilibrium, with no net flow, there must be an electric field pointing outward to counteract the diffusive flow. The electric field is in the opposite direction of the potential gradient, so the gradient must point inward in order for the field to point outward: i.e. $\phi_i > \phi_e$ so that Φ_{Nernst} is positive.

Recall that D_0 in Equation 6.110 is the self-diffusion coefficient of the charged particle at low concentration in a homogeneous, isotropic medium. In treating ion transport across a membrane, the diffusion coefficient can vary in a complex way as an ion leaves the aqueous boundary of the membrane and passes through the hydrophobic membrane interior or through a protein channel rather than the interior aliphatic phase. The ion may also interact with other ion species in the channels where site-specific events, adsorption, and blockage can occur. The net effect is that the ion is no longer diffusing in a simple homogeneous environment so that the diffusion coefficient no longer has its idealized value of D_0, and its relation to the frictional coefficient given by Einstein's Equation 6.73. The models now presented assume an effective value of the diffusion coefficient and D is used instead of D_0 in subsequent equations.

So, with the Nernst and Donnan potentials, the former dominant, how does this match up to observed concentrations of ions and transmembrane potentials? Table 13.1 shows for mammalian blood cells the typical resting concentrations of the main ions involved in transmembrane transport; Na^+, K^+, Cl^-, and the hydrogen carbonate ion HCO_3^-, which results from the extraction of a proton from carbonic acid CH_2O_3. The right-hand column shows the corresponding Nernst potentials, according to Equation 13.4, which are quite different from one another. The resting potential is typically −70 mV, so only K^+ and Cl^- have Φ_{Nernst} anywhere near this value. K^+, unlike the other ions, has a far greater concentration inside than outside the cell.

Clearly, the cell's resting potential is not due to this resting state distribution of ions. This analysis is based on treating each ion separately, requiring that its flux be zero, $J_{Na} = J_K = J_{Cl} = J_{HCO_3} = 0$, and then computing its own Nernst potential.

Hodgkin and Katz, and independently, Goldman (HKG), proposed that the individual fluxes are not zero, rather, the net flux is zero

$$J_{net} = J_{Na} + J_K + J_{Cl} + J_{HCO_3} = 0 \tag{13.6}$$

Table 13.1 Typical mammalian resting millimolar (mM) concentrations of major ions, inside and exterior to blood cells, and associated Nernst potential, Φ_{Nernst}. T = 310 K (37°C), according to Equation 13.4. The typical transmembrane potential is −70 mV.

Ion	$[\text{Ion}]_i$ (mM)	$[\text{Ion}]_e$ (mM)	Φ_{Nernst} (mV)
Na^+	12	145	+67
K^+	139	4	−95
Cl^-	4	116	−90
HCO_3^-	12	29	−24

From NIH NCBI Resources https://www.ncbi.nlm.nih.gov/books/NBK21627.

The fact that the individual fluxes are not zero means that this condition implies that *the resting potential is not an equilibrium state of the cell*, rather, it must be dynamically maintained at the expense of metabolic energy. It turns out that Equation 13.6 is met and the constant potential is maintained due to various active pumps and channels in the membrane. The pumps can pump ions against their gradients, and the channels control the permeability of the membrane, usually quite selectively.

We point out that even if the Nernst voltages turned out to be commensurate with experimentally measured values, which they are clearly not, and there were no active membrane channels for ions, then the time scale for changes in membrane potential, such as for nerve conduction would be glacially slow, due to the huge apolar energy barrier presented by the membrane's lipid bilayer and summed up in Equation 13.1.

The last few paragraphs have again illustrated how the living state controls all its functions and leaves nothing to simple, natural equilibria. Again, *equilibrium is death to the living state.* In fact, one may ask why the cell expends energy to stay out of equilibrium and in a dynamic steady state. One possible answer is that by staying out of equilibrium free energy is not minimized, so that there is positive free energy always available to do work in the resting state. The cell is "alert, energized, and ready to act."

Now, when more than one ionic species flows they will contribute differently to the net voltage since Equation 6.110 shows that the diffusion coefficient establishes the magnitude of the flux J, and the diffusion coefficients of two separate species can be quite different, and this difference can depend strongly on what types of transmembrane channels are active.

HKG used the Nernst–Planck equation for each ion to then substitute into the $J_{net} = 0$ condition of Equation 13.6. For the jth ion

$$J_j = -D_j\left[\frac{\Delta\left[X\right]_j}{d} + \frac{z_j F}{RT}\left[X\right]_j \frac{\Phi}{d}\right] \qquad 13.7$$

where the D_j are phenomenological diffusion coefficients for each ionic species X_j, unrelated to D_0 in a pure fluid, $\Delta\left[X\right]_j$ the transmembrane molar concentration difference, d the membrane thickness, and Φ the transmembrane potential. Using this equation, together with $J_{net} = 0$, and assuming that interactions among ion species that change their diffusivity are negligible, HKG found (omitting the HCO_3^- contribution), and assuming that the electric field is constant across the membrane:

$$\Phi = \frac{RT}{F}\ln\left(\frac{P_{\text{NA}}\left[\text{Na}^+\right]_e + P_K\left[\text{K}^+\right]_e + P_{\text{Cl}}\left[\text{Cl}^-\right]_i}{P_{\text{NA}}\left[\text{Na}^+\right]_i + P\left[\text{K}^+\right]_i + P_{\text{Cl}}\left[\text{Cl}^-\right]_e}\right) \qquad 13.8$$

where the P_j's are permeability coefficients, $P_j = D_j/d$, where d is the membrane thickness.

13.2.1 Circuit Model for Transmembrane Electric Potential

There is a net negative resting potential Φ_r, across the cell membrane. Since the Nernst potential for each species is different from Φ_r, a good way of illustrating the competing effects is in terms of equivalent battery/resistance terms in parallel, in a circuit diagram. Figure 13.1 shows the equivalent circuit. As discussed in Chapter 5.1.3, the membrane acts as a

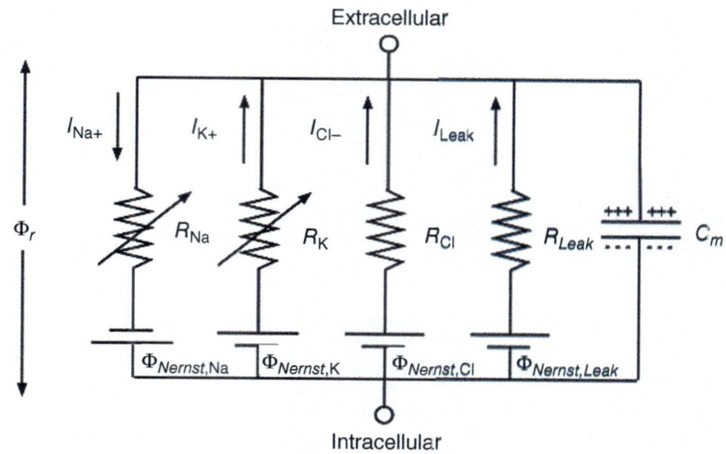

Figure 13.1 Equivalent circuit for a membrane with the Nernst potentials for each ion and the leakage current, and the variable resistors, whose values can change due to active transmembrane channel control. (Nrets / Wikimedia commons / Licensed under CC BY-SA 3.0).

capacitor with the net charges from the ions on the inner and outer-membrane surface providing a capacitive transmembrane potential. Not shown in the circuit diagram are the active pumps that drive K^+ from the exterior to the interior of the membrane, and Na^+ from the interior to the exterior. The imbalances maintain the net transmembrane resting potential, Φ_r.

In Figure 13.1, each ion type has its own element; a Nernst EMF (or Nernst battery) due to the resting ion concentrations, such as in Table 13.1, and a resistor for its ion type to cross the membrane. Since the elements are in parallel the voltage drop across each of them is equal to the resting voltage Φ_r. The current that flows through ion X's element is

$$I_X = \frac{(\Phi_r - \Phi_{Nernst,X})}{R_X} \tag{13.9}$$

where R_X is the resistance of the ion element X and $\Phi_{Nernst,X}$ its Nernst potential. Also in the circuit is the membrane capacitance. What makes the system work and hold the nonequilibrium, steady value of Φ_r, is the Na^+–K^+ pump. This pump has to expend metabolic energy to pump these ions against their own gradients and ensure that $I_{net} = 0$ in the resting state.

At the steady state, there is no net current and so the Φ_r can be expressed in terms of the Nernst potentials and resistances of each ion element in the circuit. Because channel opening corresponds directly to increasing conductance G_X, and reciprocally to resistance R_X, and since parallel resistances add reciprocally and parallel conductances linearly, it is generally preferred to use the conductance $G_X = 1/R_X$ for the currents so that

$$I_X = G_X(\Phi_r - \Phi_{Nernst,X}) \tag{13.10}$$

Setting the sum of currents equal to zero yields Φ_r

$$\Phi_r = \frac{G_{Na}\Phi_{Nernst,Na} + G_K\Phi_{Nernst,K} + G_{Cl}\Phi_{Nernst,Cl}}{G_{Na} + G_K + G_{Cl}} \tag{13.11}$$

The ratio of an individual ion element conductance to total conductance is sometimes termed *transference*, *T*. For ion species X

$$T_X = \frac{G_X}{G_{Na} + G_K + G_{Cl}} \tag{13.12}$$

The resting potential in human muscle and nerve cells can be estimated from the Φ_{Nernst} in Table 13.1 and the empirical result that $T_K/T_{Na} \sim 10$, and T_{Cl} is negligible. Then, $\Phi_r \sim (1/11) \times 67 + (10/11) \times (-95) = -80$ mV, which is close to experimental values.

Notice the polarities in Figure 13.1. Φ_r has its negative terminal on the intracellular (cytoplasmic) side. Both K^+ and Cl^- have negative Φ_{Nernst} and so their negative terminals are also on the cytoplasmic side, whereas Φ_{Nernst} for Na^+ and so its positive terminal is on the cytoplasmic side.

Since the conductances G_X are controlled by membrane channels by the fraction of channels open and closed for a given ion type, we can already get a preview of what lies behind an action potential; imagine that many Na^+ channels abruptly opened, increasing the conductivity G_{Na} accordingly. Then Φ_r, according to Equation 13.11 could be driven positive. This, in fact, is the first part of producing an action potential, discussed below.

Membrane capacitance was estimated in Equation 5.29, by treating the membrane as a thin concentric spherical capacitor. Typical membrane capacitances are on the order of 1–20 mF/m^2. The resistivity of a sheet of lipid bilayer membrane without channels is on the order of 10^4 Ω/cm^2.

13.3 The Voltage Clamp and Patch Clamp

The tremendous advances in our knowledge of membrane potentials and the effects of active channels are due largely to two technical innovations. The *spatial voltage clamp* can keep the transmembrane potential constant at any desired value along a linear stretch of nerve axon. The subsequent invention of the *patch clamp* allows patches of cell membrane surface to be held at a given voltage, thus allowing the behavior of single ion channels to be directly investigated. While the spatial voltage clamp allowed the pioneering work on the giant squid axon to be carried out, the patch clamp extends voltage clamping beyond the squid axon to the realm of clamping patches of single cells of different types.

The voltage clamp was invented by Kenneth Cole (United States, 1900–1984) and George Marmont in 1947, while researchers worked with the giant squid axon. The axon is several cm long and up to 1 mm in diameter. Nerve propagation models show that propagation velocity increases with cross-sectional area of the axon. The fast response of the giant axon allows the squid to squirt out water at high speed and escape from predators. Later, we see that the *myelin sheath* evolved in higher animals yields much more efficient nerve propagation with small diameter axons.

The large diameter of the squid axon makes it possible to insert a very fine electrode axially. This electrode can produce a current across the axon membrane. A highly simplified schematic of the voltage clamp is shown in Figure 13.2. A second electrode senses the membrane potential of the axon. The experimenter establishes a fixed "command voltage" or clamp voltage of their choosing and the current-passing electrode produces a current which achieves this fixed voltage. The voltage-sensing electrode constantly sends the instantaneous voltage state to a difference amplifier which compares the membrane voltage to the clamp voltage. The output of this amplifier directs the current electrode to increase or decrease the current until the membrane potential matches the clamped potential. The current adjustments to keep the voltage clamped are made on submillisecond time scales and mirror the axon membrane currents. Thus, the time record of the axon currents due to opening and closing of membrane channels can be precisely measured for any clamp voltage. While this device enabled important discoveries, culminating in the elucidation of the action potential, it measures the net currents due to all ion species. Hence, it cannot home in on the actions of channels specific to K^+, Na^+, or other ions.

The patch clamp was a major innovation that allowed detection of single-channel action. It was developed by Erwin Neher (Germany, 1944–) and Bert Sackmann (Germany, 1942–) in the 1970s and 1980s. The major innovation is the production of a micron-size pipette tip which can seal onto a cell membrane. This seal may cover several channels or even single channels, which allows the behavior of specific channels to be probed and their current–voltage curves determined. The body of the pipette whose micron-size tip is sealed to the cell has an electrode in contact with the cell membrane, which can record voltages and current. The pipette can contain any electrolyte solutions desired, such as physiological saline. The cell is in an electrolyte bath surrounding it, and a reference ground electrode is in the bath. The bath, likewise, can contain any desired electrolyte solution, and drugs and other agents can be added to monitor their effect on specific channel types.

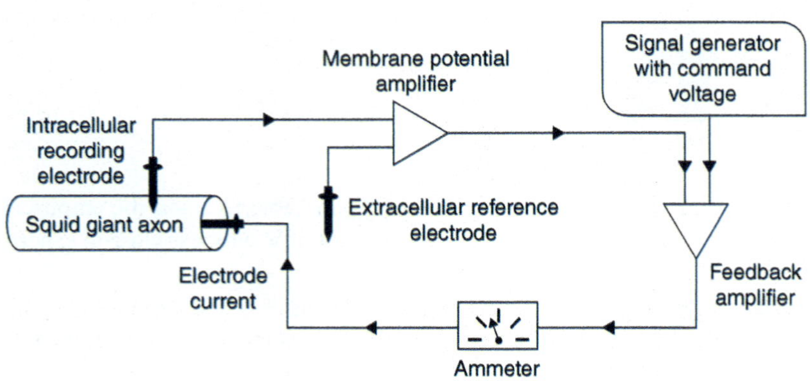

Figure 13.2 Voltage clamp used for pioneering experiments with squid giant axons. (Courtesy of Marissa Calko).

13.4 Membrane Current–Voltage Curves

Current–voltage curves, or *I–V curves*, are frequently used to characterize the electrical properties of a material or device. Figure 13.3 shows IV curves for (i) an ohmic device (linear), (ii) a diode, and (iii) a generic membrane.

I–V diagrams that are linear are characteristic of ohmic or *passive electrical elements*. *I–V* curves which are not linear are hallmarks of *active electrical elements*. The *I–V* curve for the diode increases exponentially with forward bias (positive voltage) and is essentially a short circuit for positive current flow but allows very little backward current, thus acting effectively as a one-way directional controller for current.

The foregoing equations were for a steady state, where currents canceled each other out, leaving zero net flux and a constant potential Φ. Now the case is considered where there can be a net flux. For a single ion of type *j*, e.g. Na^+, Goldman–Hodgkin–Katz derived the following current equation:

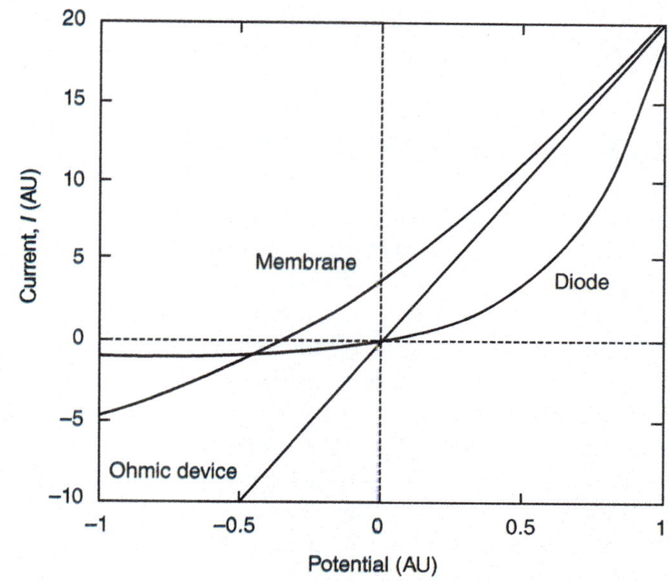

Figure 13.3 *I–V* curves for ohmic resistor, and non-ohmic diode, and membrane.

$$I_j = \frac{z_j^2 F^2 P_j \Phi}{RT}\left(\frac{c_{e,j} e^{z_j F\Phi/RT} - c_{i,j}}{e^{z_j F\Phi/RT} - 1}\right) \qquad 13.13$$

This produces a nonohmic *I–V* curve (actually an *I–Φ* curve in this case), such as the membrane *I–V* curve of Figure 13.3. It gives a high resistance in reverse bias and low resistance in forward bias, which bears a slight qualitative similarity with the diode. The resting Nernst potential is recovered at $I = 0$.

This current equation is for only one species. To generalize to an arbitrary number of species, each with their own valence z_j, not necessarily unity, the current from each species is summed.

$$I_{net} = \sum_j I_j \qquad 13.14$$

Setting $I_{net} = 0$ gives the steady-state voltage across the membrane, Φ, as

$$0 = \sum_j z_j^2 P_j \Phi\left(\frac{c_{e,j} e^{z_j F\Phi/RT} - c_{i,j}}{e^{z_j F\Phi/RT} - 1}\right) \qquad 13.15$$

where the solution can be found numerically. In practice, the most important application is for interrogating the effects of the divalent ion of calcium, Ca^{2+}, which has important control functions.

M.B. Jackson (2006) proposed a rate theory of membrane conductance by considering that the potential energy of an ion in the membrane has an inverted parabolic shape in *x*, the distance into the membrane

$$U(x) = U_{max} - w(x - x_{max})^2 \qquad 13.16$$

where U_{max} is the barrier height which occurs at x_{max}, the midpoint of the membrane, and *w* is the width parameter of the barrier. This leads to a Gaussian energy barrier versus distance into the membrane, *x*, in going from one side of the membrane to the other. This barrier controls the rate of ion passage through the membrane from the side of high potential energy to low potential energy, $z\Phi$, just as any energy barrier controls the rate between two states, as discussed in Chapter 4.16. The current Jackson derives for the current based on this potential is

$$I = -zFp\left(c_e \exp\left(\frac{zF\Phi}{2RT}\right) - c_i \exp\left(-\frac{zF\Phi}{2RT}\right)\right) \qquad 13.17$$

where

$$p = \sqrt{\frac{w}{\pi RT}} D \exp\left(-\frac{U_{max}}{RT}\right) \qquad 13.18$$

This gives a much steeper I–V curve than Equation 13.13. It again gives the Nernst voltage at $I = 0$, and the curve somewhat resembles that of a Zener diode, which has a reverse bias breakdown at a specific negative bias voltage.

13.5 Membrane Channel Proteins

As mentioned, cells actively control membrane potentials and concentrations of ions and other substances. This requires transmembrane channels that can open and close when needed. As discussed, there is a large energy barrier for an ion to cross a membrane even if there is a lower energy state available on the other side, so that the rate of transport is low. If a transmembrane channel opens up that lowers this barrier, then the rate of transport will increase, often by several orders of magnitude. In this sense, membrane channels are transport catalysts, lowering the energy barrier for transport, just as enzymes are catalysts that lower chemical reaction barriers.

An enormous amount of experimental and theoretical work by many researchers has revealed much about the types of channels that exist and the mechanism of their operation. These are vast and varied. Some of this progress is summarized here. See M.B. Jackson 2006, Chapter 14, for a detailed overview of protein channels.

The simplest mechanism that one might consider for a channel protein is to simply open up a hole in the membrane that fills with aqueous fluid, thus providing a direct path for ions and other materials to flow down concentration gradients across the membrane, and completely avoiding the energy barrier. For the ionic case, a model based on *ohmic channels* simply considers the conductance of an open channel of area A, and resistivity ρ, with a length equal to the membrane thickness, d. The conductance of such a channel is $A/\rho d$. Experiments with a variety of channel proteins and measurements of their conductance show that only those with a large area, above 1 nm^2, owe an appreciable fraction of their transmembrane conductance to this mechanism. As channel areas drop well below 1 nm^2 the ohmic contribution becomes small, and finally, for *gramicidin* at $A = 0.19$ nm^2, the effect is less than 3%, and for those below 0.1 nm^2, the contribution is entirely negligible. In other words, the mechanism of these smaller channels involves specificity and other mechanisms beyond simply providing aqueous pores in the membrane. Some channels, such as gramicidin and the acetylcholine receptor, select for cations, while the $GABA_A$ receptor[1] and glycine receptor select for anions. *Aquaporin* only selects for water.

Selectivity for positive or negative charge is linked to either charges or dipoles that are part of the channel, but the further question of how a channel selects for ions beyond their mere charge has been more challenging. The five alkali cations Li^+, Na^+, K^+, Rb^+, and Cs^+ have been investigated experimentally, and their relative selectivities by channels determined. There are $5! = 120$ possible sequences for these selectivities, but only 11 sequences have been found experimentally. A conceptually simple but revealing model considers that there would be a *selectivity filter*, at a site in the channel that has charge q_s. An ion with opposite charge q passing through the channel has two energy contributions; one from its favorable interaction with q_s and the other the energy penalty of dehydration E_{hyd}

$$\Delta E = E_{hyd} - \frac{qq_s}{4\pi\varepsilon_w\left(r + r_s\right)} \tag{13.19}$$

where r is the ion radius, r_s is the radius of the site and $(r + r_s)$ is their distance of closest approach. The ion flux is proportional to $\exp(-\Delta E/RT)$, so the selectivity sequence for the ions at a given r_s will be in the order of decreasing ΔE. Using E_{hyd} and r from known values and varying r_s the curves for ΔE versus r_s each cross each other once and provide 11 selectivity sequences, as is found experimentally.

A common motif for membrane channels is a pair of short polypeptides which can assemble across the membrane and form a sort of hollow barrel shape, through the center or which ions are transported. Figure 13.4 is an artist's conception of a typical channel, showing it in both open and closed modes, or activated and deactivated.

An intensely studied channel family is *gramicidin A, B, and C*, which forms cation-selective membrane channels. It falls into the interesting class of *nonribosomal peptides*. These are not coded by genes, have no presence in DNA, and are synthesized by special protein synthetases, which are themselves genetically coded. Additionally, gramicidin A contains *right-handed amino acids*. All amino acids coded by genes are left-handed. This extragenetic use of mixed chiralities allows for the packing of gramicidin's 15 amino acids into short, compact helices, which dimerize to span the membrane. Inside the channel, there are carbonyl groups that are dipolar, with the negative pole facing inward, which presumably accounts for its selectivity for cations and against anions.

1 $GABA_A$ receptors are a family of receptors that are composed of five subunits arranged around a central pore. They control Cl^- transport in neurons and can suppress nerve propagation. As a result, they are widely used in tranquilizers, anticonvulsants, sedatives, and muscle relaxants.

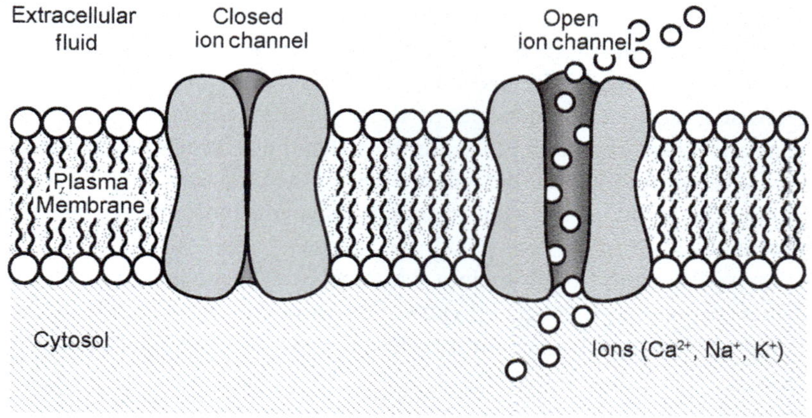

Figure 13.4 A membrane ion channel, one closed and one open. (Courtesy of Marissa Calko).

Much work has gone into building potential energy models for gramicidin's conduction properties. For example, one model considers the image charge force on an ion, where a boundary by dielectric causes a force on a charge near the boundary, which is repulsive and proportional to z^2 together with the attractive dipole force which is proportional to z (z is dimensionless). The total barrier energy is then $E_{barrier} = az^2 - bz$, where a and b are positive constants with units of energy. With this simple notion anions which have $z = -1$ give $E_{barrier} = a + b$ which is always higher than cations with $z = +1$, for which $E_{barrier} = a - b$. Divalent cations, $z = 2$, give $E_{barrier} = 2(2a - b)$ will have a larger energy barrier than monovalent cations as long as $a > b/3$. Since gramicidin A is largely impermeable to divalent cations, this puts a lower limit on the ratio of image to dipole energies. Continuum pictures, however, are expected to break down in the gramicidin A channel, since there only six water molecules fit in the channel and their dipoles interact strongly with the carbonyl dipoles.

Pharmacologically, gramicidin is used as a topical antibiotic. Because it is a cationic channel, its unregulated insertion into membranes can cause them to lose their resting state transmembrane differences in K^+ and Na^+ concentration, leading to cell death.

13.6 Passive Propagation of Potentials Along an Axon

Let's refocus on the giant squid axon, on which the fundamental discoveries of the action potential and nerve impulse propagation were made. We can treat the axon as a cylinder whose surface is the membrane and whose interior is cytoplasm. The cylinder lies along the x-axis and has radius a. So far, we have only considered the potential across the membrane but have not inquired into how this voltage spreads in three dimensions, in this case along the cylindrical axon. The ion currents we have discussed are radially outward or inward across the membrane, and, for the case of the cylindrical axon, they represent a current flowing radially through an annulus in 2D circular symmetry. The radial current is denoted by I_r.

To bring this into a 3D picture we need to find how potential and current at a given location x propagates down the axon on the x-axis as a function of time, that is, we want to find $\Phi(x,t)$ and the axial current $I_x(x,t)$. To do this, we consider that the cylinder has a membrane capacitance per area of σ, so that the membrane capacitance along a section of the cylinder of length Δx is $2\pi a \Delta x \sigma$.

Similarly, we define the resistance per unit length in the axial direction as λ, so that the axial resistance of a length Δx is $\lambda \Delta x$. The radial current density J_r is the amount of ion current per area of cylinder so that $2\pi a \Delta x J_r$ is the total radial current flowing along over a length Δx. It is interesting to consider the time for the characteristic charge density relaxation across the membrane compared to along the axis, that is, through the axoplasm. Using Equation 6.129, and typical values of 0.1 M salt water of conductivity ~ 1 $(\Omega\ \text{m})^{-1}$ and $\varepsilon = 80\varepsilon_0$, where $\varepsilon_0 = 8.85 \times 10^{-12}$ C^2/N-m^2 gives a relaxation time of ~10^{-9} s for a charge density fluctuation in the axoplasm. In contrast, relaxation time across membranes is typically ~10^{-2} s, or 10 million times slower than axoplasmic relaxations.

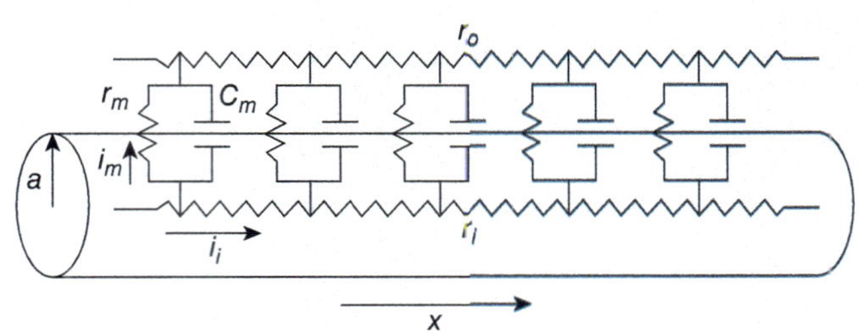

Figure 13.5 The equivalent circuit for the passive axon.

Now, continuing with the passive ohmic model, we can give an equivalent circuit for this, shown in Figure 13.5. The axon is divided into contiguous sections of length Δx.

Each section has a membrane capacitance $C = 2\pi a\Delta x\sigma$, a radial conductance $G = \Delta xg$, where g is the specific radial conductance per unit length, an axial resistance $\lambda\Delta x$, and an *equivalent Nernst potential* Φ'_{Nernst}. Φ'_{Nernst} is introduced since this simplest of models lumps all ion flows – Na^+, K^+, and Cl^- – into a single ion path producing J_r. The conductance g likewise lumps the conductances of all the different channels into this single, constant value. This last assumption, of a constant g, is rather blithe, since it implies that the axon acts like a passive ohmic device with a fixed, time-invariant conductance per unit length. We will see that in the propagation of a nerve impulse, it is exactly this assumption which is wrong; the ability of the axon to change its conductance in response to changing potential is what enables the action potential.

An axial current $I_x(x)$ flows into a segment at x. At the junction of the parallel RC path, the current $I(x)$ divides into a radial current $I_{r,RC}$ that again divides going through the parallel RC path into resistive and capacitive charging currents, and the rest of the current continues into the next axon segment as $I(x + \Delta x)$. We assume the total current is conserved so that

$$I_x(x) = I_x(x + \Delta x) + I_{r,RC} \tag{13.20}$$

Now, $I_{r,RC}$ is composed of a portion due to ion current I_r, and part is due to the current that goes into charging the capacitance of the membrane I_C.

$$I_{r,RC} = I_r + I_C \tag{13.21}$$

where

$$I_r = \Delta xg\left[\Phi(x,t) - \Phi'_{Nernst}\right] \tag{13.22}$$

Actually, over a finite length Δx, I_r is not strictly constant. But since we will go to the limit $\Delta x \rightarrow dx$ this technicality is unimportant.

The capacitive current I_C stems from the time derivative of $Q = C\Phi$ and results in charge accumulating on the membrane.

$$I_C = C\frac{\partial \Phi}{\partial t} = 2\pi a\Delta x\sigma\frac{\partial \Phi}{\partial t} \tag{13.23}$$

Adding the currents gives

$$I_x(x) = I_x(x + \Delta x) + 2\pi a\Delta x\sigma\frac{\partial \Phi}{\partial t} + \Delta xg\left[\Phi(x) - \Phi'_{Nernst}\right] \tag{13.24}$$

Expanding $I_x(x + \Delta x)$ to first order and subtracting from $I_x(x)$ gives

$$\frac{\partial I_x(x)}{\partial x}\Delta x = -\Delta x\left(2\pi a\sigma\frac{\partial \Phi}{\partial t} + g\left[\Phi(x,t) - \Phi'_{Nernst}\right]\right) \tag{13.25}$$

But $I_x(x)$ is given by the voltage drop over Δx divided by the resistance over Δx

$$I_x(x) = -\frac{\Delta\Phi}{\lambda\Delta x} \tag{13.26}$$

The negative sign indicates that if Φ decreases with x, $\frac{\Delta\Phi}{\Delta x} < 0$, then positive ions flow to the right to give a positive $I_x(x)$.

$$I_x(x) = -\frac{\Phi(x + \Delta x, t) - \Phi(x,t)}{\lambda\Delta x} \cong -\frac{1}{\lambda\Delta x}\frac{\partial \Phi}{\partial x}\Delta x \tag{13.27}$$

where $\Phi(x + \Delta x)$ was expanded to first order in Δx. Substituting this into Equation 13.25 yields

$$\frac{1}{\lambda}\frac{\partial^2 \Phi}{\partial x^2} = 2\pi a\sigma\frac{\partial \Phi}{\partial t} + g\left[\Phi(x,t) - \Phi'_{Nernst}\right] \tag{13.28}$$

This is known as the *linear cable equation*.[2] It resembles the diffusion equation, with the extra linear term, $g\Phi(x,t)$, and constant term $-g\Phi'_{Nernst}$, added to the homogeneous diffusion Equation 6.72. Considering a steady state, where the voltage is kept constant at $x = 0$ and the membrane is fully charged so that $\partial\Phi/\partial t = 0$ yields the solution:

$$\Phi(x) = \Phi(x=0)e^{-\xi x} + \xi^2\Phi'_{Nernst} \qquad 13.29$$

where $1/\xi$ is the characteristic length of the exponential fall off of $\Phi(x)$ over space, and

$$\xi = \sqrt{g\lambda} \qquad 13.30$$

While the time-dependent solution will have a different form than the steady-state Equation 13.29, it will lead to a decay of $\Phi(x,t)$ over time and space.

The main conclusion from this is that the potential will spread from the point x where it was first raised or lowered above the resting potential in a diffusive, or at least diminishing manner, and so the voltage dies out over time and distance. This means that with this ohmic model, it is not possible to propagate a potential over space without its value just diffusing away to zero over space.

The discovery of the action potential, a potential which sustains itself and propagates without dying out over space is the result of *voltage-gated channels* and is the subject of the next section. It means that g, the specific radial conductance per unit length, is not a constant, rather $g = g(\Phi)$.

13.7 Action Potentials and Nerve Impulse Propagation

One of evolution's greatest achievements is the *excitable membrane*, which allows for muscle action, photoreception, and nerve propagation. Nerve propagation sets the stage for neuronal networks, electrical signaling, and data processing. These capabilities are due to propagating *action potentials* which arise from voltage-gated membrane channels.

Using the voltage clamp method and varying ion concentrations outside the axon Alan Hodgkin (England, 1914–1998) and Andrew Huxley (England, 1917–2012) monitored a *threshold-dependent action potential* using an oscilloscope. The history of these developments and those leading up to this discovery, including early breakthroughs in the late nineteenth century, are fascinating, as are the ingenious approaches Hodgkin and Huxley (HH) took to first observe and then manipulate, and finally model the action potential and its propagation. The reader is recommended to History of Electrophysiology and the Patch Clamp (Verkhratsky, A., Parpura, V. (2014). Patch-clamp methods and protocols. *Methods in Molecular Biology 1183*: 1–19, Springer).

Figure 13.6 shows a schematic of what HH actually observed. The whole event occurs on a time scale of milliseconds. The sequence of events can be rationalized as follows: The cell receives a stimulus creating a high enough threshold potential to activate the pulse. This change above threshold voltage triggers the voltage-gated Na^+ channels to open, increasing G_{Na} in Equation 13.11 by a factor of approximately 500, and bringing the transmembrane Φ positive, in fact, fairly close to $\Phi_{Nernst,Na}$ (e.g. Table 13.1), the *depolarization phase*. Then, the upward swing is terminated and the action potential spike created as the voltage-gated Na^+ channels turn off, and voltage-gated K^+

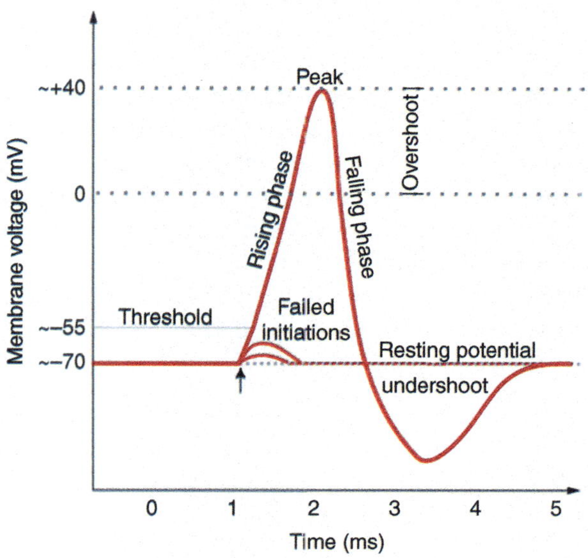

Figure 13.6 Schematic of an action potential. (MrNinja, J. (2005) / Wikimedia commons / Public Domain).

2 The seemingly odd use of the word "cable" stems from the fact that Lord Kelvin originally derived the equation in the 1850s to account for the loss of signal strength in undersea *telegraph cables*.

channels open, the *repolarization phase*. This makes Φ plunge down to a value below the resting Φ_r, the *hyperpolarization phase*. This is predicted by Equation 13.11 because with the voltage-gated K^+ channels open G_K in Equation 13.11 is higher than in the resting state and G_{Na} has receded back to its resting state level. Φ remains below Φ_r until the voltage-gated K^+ channels turn off. The period while hyperpolarization occurs and relaxation back up to Φ_r is the *refractory period*, during which the cell cannot be stimulated further. Notice also the "failed initiations" in Figure 13.6; these were stimuli that did not reach the voltage threshold needed to activate the voltage-sensitive channels.

That is the phenomenological scheme that HH deduced from their experiments. What remained was to make a model for this which would (i) predict the time course of the action potential, (ii) give an equation for the propagation of this wave through space without diminishing, and (iii) find the speed at which the pulse propagates.

The cable Equation 13.28 was derived based on the assumption of fixed membrane conductivity for each ion along the squid axon. It is a linear, dissipative equation that does not predict a traveling wave. In light of HH's insights, the specific conductance g, in the equation is no longer constant, but a function of the voltage itself which is itself a function of space and time. The specific conductance now becomes voltage-dependent g = g(Φ), so that Equation 13.28 becomes

$$\frac{1}{\lambda}\frac{\partial^2\Phi}{\partial x^2} = 2\pi a\sigma\frac{\partial\Phi}{\partial t} + g(\Phi)\left[\Phi(x,t) - \Phi'_{Nernst}\right] \qquad 13.31$$

What this means is that no matter what the specific form of the dependency of g(Φ), it necessarily makes the term

$$g(\Phi)\,\Phi(x,t) \qquad 13.32$$

nonlinear and hence makes the differential equation nonlinear. Most often, nonlinear differential equations don't have analytical solutions and are either solved numerically or by approximations. They also often have remarkable properties, including chaotic behavior (solutions which radically diverge in time for slightly different initial conditions. See Chapter 14.5), and regimes or points of stability. Before looking at some of the details of the equations HH proposed, we can already surmise that stable solutions to the equation are highly restricted and may be linked to characteristics of propagation such as wave speed. Furthermore, the details of g(Φ) will undoubtedly be complex and involve several parameters and their interplay.

13.7.1 The Hodgkin–Huxley Equations

The equations that Hodgkin and Huxley developed are *phenomenological*, that is, they are grounded in the theoretical notions outlined above but are themselves not derived from theory. Rather, the form of the equations is adduced from the features of the experimentally observed action potential, the voltage clamp data over a wide range of clamped voltages, and the rationalization sketched above for the sequence of voltage-gated channel events. The parameters in the equations are found that fit to the data.

The voltage clamp data yielded total current behavior for the squid axon bathed in seawater as a function of clamp voltage. HH then replaced Na^+ in the bath with choline, a cation ($C_5H_{14}NO$) too large to pass through Na^+ channels. This allowed them to measure just the K^+ current, and hence also obtain the Na^+ current by subtracting the former from the seawater bath current.

First, consider the time-dependent radial transmembrane current $I_{r,RC}$ HH measured during the action potential event. The Cable Equation 13.30 lumped all the ion currents together, but now they must be explicitly separated, since the whole notion of the action potential requires time-dependent differences of the flows of Na^+ and K^+. We can write $I_{r,RC}$ along an axial portion of the axon of length Δx, which has capacitance C and a conductance G_X for each ion X as

$$I_{r,RC} = C\frac{\partial\Phi}{\partial t} + \sum_j G\left(\Phi(t)\right)\left[\Phi(x,t) - \Phi_{Nernst,j}\right] \qquad 13.33$$

where the conductances are shown to depend on Φ, which itself is a function of time. Because only Na^+ and K^+ have important voltage-gated channels, Equation 13.33 can be expanded as

$$I_{r,RC} = C\frac{\partial\Phi}{\partial t} + G_{Na}\left(\Phi(t)\right)\left[\Phi(x,t) - \Phi_{Nernst,Na}\right] + G_K\left(\Phi(t)\right)\left[\Phi(x,t) - \Phi_{Nernst,K}\right] + \sum_j G_j\left(\Phi(t)\right)\left[\Phi(x,t) - \Phi_{Nernst,X}\right] \qquad 13.34$$

where the sum is for channels which are not voltage-gated, such as Cl^-, in the case of the giant squid axon, and hence have fixed, time-independent conductances.

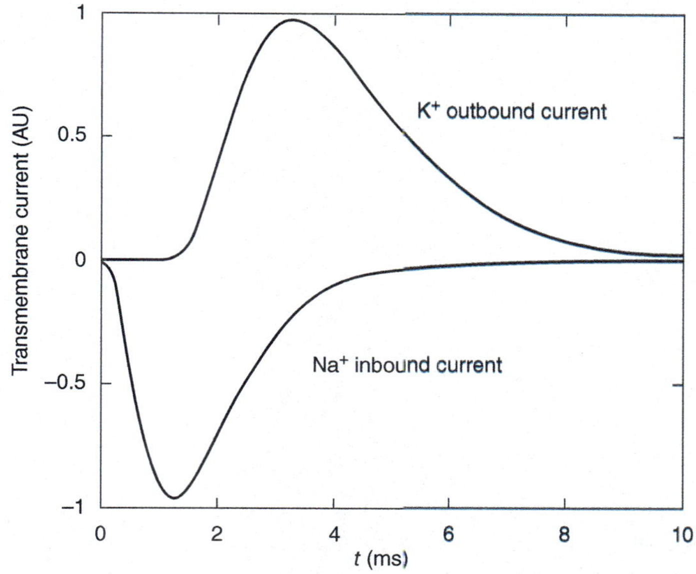

Figure 13.7 Conceptual relation of depolarizing Na^+ inbound current, when the Na^+ channels open, followed by repolarizing outbound K^+ current, as the Na^+ channels close.

HH studied the separated Na^+ and K^+ currents versus time and deduced that there were three essential events once the action potential threshold is reached: (i) The Na^+ voltage-gated channels open quickly giving an I_{Na} inward through the membrane. (ii) Shortly thereafter these channels are deactivated and (iii) to quickly repolarize the membrane to its resting potential voltage-gated K^+ channels turn on roughly around the time the Na^+ voltage-gated channels are turning off.

A conceptual time course for these events would show up *qualitatively* something like in Figure 13.7, which shows the types of curves that arise from products of rising and decaying exponentials. In Figure 13.7, K^+ rises as the fourth power of a rising first exponential of the form

$$I_K \propto (1-e^{-t/\tau_n})^4 e^{-t/\tau_n} \qquad 13.35$$

where τ_n is the voltage-dependent time constant for the activation of the voltage-gated K^+ channel. The sodium current has the form of a rising first-order exponential to the third power times a simple decaying exponential of the form

$$I_{Na} \propto (1-e^{-t/\tau_m})^3 e^{-t/\tau_h} \qquad 13.36$$

where τ_m is the voltage-dependent time constant for the activation of the voltage-gated Na^+ channel, and τ_h is the voltage-dependent time constant for the deactivation of the voltage-gated Na^+ channel. The parameters m, h, and n are dimensionless numbers from 0 to 1 for Na^+ gate activation (0 being no activation, 1 full activation of all gates), Na^+ gate deactivation, and K^+ gate deactivation, respectively.

In this model, a channel is either activated A, or deactivated, D, and follows:

$$A \underset{\alpha_y,\beta_y}{\longleftrightarrow} D \qquad 13.37$$

where α_y and β_y are the voltage-dependent rate constants for activation and deactivation, respectively, for any of the parameters m, h, and n. HH found these rates by fitting their data at a series of a dozen different clamp potentials. The Boltzmann factors appearing in them represent the difference in free energy between an activated and inactivated channel. The following are the original forms for the α's and β's found by Hodgkin and Huxley (1952d). (Other authors may show different parameters, depending on the type of neuron and conditions, as well as different units and symbols)

$$\alpha_m(\Delta\Phi) = \frac{0.1(\Delta\Phi+25)}{\exp\left(\frac{\Delta\Phi+25}{10}\right)-1} \qquad \text{The Hodgkin – Huxley Equations} \qquad 13.38a$$

$$\beta_m(\Delta\Phi) = 4\exp\left(\frac{\Delta\Phi}{18}\right) \qquad 13.38b$$

$$\alpha_h(\Delta\Phi) = 0.07\exp\left(\frac{\Delta\Phi}{20}\right) \quad 13.38c$$

$$\beta_h(\Delta\Phi) = \frac{1}{\exp\left(\frac{\Delta\Phi+30}{10}\right)+1} \quad 13.38d$$

$$\alpha_n(\Delta\Phi) = \frac{0.01\left(\Delta\Phi+10\right)}{\exp\left(\frac{\Delta\Phi+10}{10}\right)-1} \quad 13.38e$$

$$\beta_n(\Delta\Phi) = 0.125\exp\left(\frac{\Delta\Phi}{80}\right) \quad 13.38f$$

where $\Delta\Phi$ is the negative deviation of the clamped voltage Φ_c to the resting voltage $\Phi_r = -70$ mV. Because the membrane depolarizes HH wanted to make upward deviations from a negative resting potential negative; for example, If the clamped voltage Φ_c, is clamped at -10 mV, then the depolarization is -80 mV. Hence

$$\Delta\Phi = \Phi_r - \Phi_c \quad 13.39$$

and is measured in mV. These rates, in turn, control the values of $y = m, h$, or n according to

$$\frac{dy}{dt} = \alpha_y(1-y) - \beta_y y \quad 13.40$$

which has a solution of the form

$$y(t) = y_\infty - (y_\infty - y_0)e^{-t/\tau_y} \quad 13.41$$

where y is m, h, or n, and

$$y_0 = \frac{\alpha_{y,0}}{\alpha_{y,0} + \beta_{y,0}} \quad 13.42$$

where $\alpha_{y,0}$ and $\beta_{y,0}$ are the rates at resting potential, $\Delta\Phi = 0$.

Combining these expressions with Equation 13.34 yields the time-dependent total transmembrane current

$$I_{r,RC} = C\frac{\partial\Phi}{\partial t} + G_{\mathrm{Na},max}m(t)^3h(t)\left[\Phi(x,t) - \Phi_{Nernst,\mathrm{Na}}\right] + G_{\mathrm{K},max}n(t)^4\left[\Phi(x,t) - \Phi_{Nernst,\mathrm{K}}\right] + \sum_j G_j\left[\Phi(x,t) - \Phi\right] \quad 13.43$$

where $G_{\mathrm{Na},max}$ and $G_{\mathrm{K},max}$ are the experimentally measured maximum conductances for the Na^+ and K^+ channels over the segment Δx. The solutions to this for Φ can be found iteratively by taking short intervals of time Δt and using the initial value of Φ at $t = 0$, to compute the initial α_y and β_y by Equations 13.38a–13.38b, and then solving for Δy and adding this to the initial values of y_0 to obtain these values at $t = \Delta t$, and then computing Φ ($t = \Delta t$) by Equation 13.11. The next step is $t = 2\Delta t$, the process is iterated to obtain Φ ($t = 2\Delta t$), and so on.

Among the many computations that HH carried out with this and related equations, they found that a threshold excitation is predicted. This was also found experimentally when HH provided transient electrical shocks of various depolarization amplitudes along the axon. Below about -7 mV of stimulation, no action potential was produced.

A separate, rather amazing finding was by Baker, Hodgkin, and Shaw in 1961.[3] According to Equation 13.43, the entire mechanism of the action potential lies with the voltage-gated channels in the membrane. From this point of view, the

---

3 Baker, P.F., Hodgkin, A.L., and Shaw, T.I. (1962). Replacement of axoplasm of giant nerve fibres with artificial solutions Journal of Physiology, 164: 330–354.

internal axoplasm and external bath are nothing more than a source of Na^+ and K^+ ions for the channels in creating action potentials. These experimenters removed the axoplasm entirely from a squid axon and replaced it with a mere solution of potassium ions. An action potential, identical to those produced by native axons, was generated upon stimulation!

Equation 13.43 can now be combined with the cable Equation 13.28 to introduce propagation in space by using

$$I_r = G_{\mathrm{Na},max} m(t)^3 h(t)\left[\Phi(x,t) - \Phi_{Nernst,\mathrm{Na}}\right] + G_{\mathrm{K},max} n(t)^4\left[\Phi(x,t) - \Phi_{Nernst,\mathrm{K}}\right] + \sum_j G_j\left[\Phi(x,t) - \Phi_{Nernst,\mathrm{X}}\right] \tag{13.44}$$

To give

$$\frac{1}{\lambda}\frac{\partial^2\Phi}{\partial x^2} = 2\pi a\sigma\frac{\partial\Phi}{\partial t} + G_{\mathrm{Na},max} m(t)^3 h(t)\left[\Phi(x,t) - \Phi_{Nernst,\mathrm{Na}}\right] + G_{\mathrm{K},max} n(t)^4\left[\Phi(x,t) - \Phi_{Nernst,\mathrm{K}}\right] + \sum_j G_j\left[\Phi(x,t) - \Phi_{Nernst,\mathrm{X}}\right] \tag{13.45}$$

We see the equation is nonlinear, given the dependencies of m, h, and n on Φ. There is no easy solution to this, and numerical methods must be used. HH did not have any computers available in 1952, just a manual calculator.

HH had a very clever insight into the situation. Since they found the action potential propagates through space unchanged in form, it must obey the well-known wave equation:

$$\frac{\partial^2\Phi}{\partial x^2} = \frac{1}{v^2}\frac{\partial^2\Phi}{\partial t^2} \tag{13.46}$$

where v is the wave speed. Solutions of the wave equation have the form:

$$\Phi = \Phi(x - vt) \tag{13.47}$$

This wave keeps its form at $t = 0$, for any $(x, t = 0)$ and travels in $+x$, ($\Phi(x + vt)$ travels in $-x$). Equation 13.45 now simplifies to an ordinary differential equation in t.

$$\frac{1}{v^2\lambda}\frac{d^2\Phi}{dt^2} = 2\pi a\sigma\frac{d\Phi}{dt} + G_{\mathrm{Na},max} m(t)^3 h(t)\left[\Phi(t) - \Phi_{Nernst,\mathrm{Na}}\right] + G_{\mathrm{K},max} n(t)^4\left[\Phi(t) - \Phi_{Nernst,\mathrm{K}}\right] + \sum_j G_{\mathrm{X}}\left[\Phi(t) - \Phi_{Nernst,\mathrm{X}}\right] \tag{13.48}$$

This was tractable to approach iteratively with a hand calculator, but it turned out that solutions diverged to infinity except for those at a well-defined value of K, which was determined empirically, and the pulse propagation speed is given by

$$v = \sqrt{\frac{Ka}{2r_a\sigma}} \tag{13.49}$$

where r_a is the resistivity (Ω-m) of the axoplasm, and a and σ have the same meanings as above, axon radius and membrane capacitance per m^2, respectively. K has the units of 1/s, and HH found that the value of K leading to a stable solution to Equation 13.47 was $K = 1.047 \times 10^4$ 1/s. Using $a = 238$ μm, $r_a = 0.354$ Ω-m and $\sigma = 0.01$ F/m^2, they found $v = 18.8$ m/s. Their experimental value was close, 21.2 m/s (20 m/s = 45 mi/h).

Equation 13.48 shows that the propagation speed increases as the square root of axon radius a. Apparently the giant squid evolved along this path, making a macroscopic size axon radius for quick signaling. While very fortunate for the discoverers and all of us beneficiaries of these discoveries, evolution found a more efficient means of increasing propagation speed with the *myelin* sheath, which allows for a small value of a while keeping propagation speed high. This was necessary for developing elaborate neuronal structures and brains.

Figure 2.8 shows the myelin sheath along an axon. It is an insulating layer composed mainly of lipids, punctuated every millimeter or so by a *Node of Ranvier*, also seen in the figure, which are on the order of a micron long. The depolarization wave does not propagate along the myelinated axon, rather, the pulse jumps from node to node, in what is termed *saltatory conduction*. Nerve pulse propagation speeds increase enormously due to myelination and saltatory conduction. This has allowed for dense packing of very narrow axons in cerebral structures. Axon diameters in mammals vary from 0.1 μm to 20 μm, which is orders of magnitude smaller than the 1-mm diameter axon of the giant squid. The myelin sheath is whitish in appearance and the dense, highly myelinated axons in the brain give rise to the "white matter."

13.7.2 Trains of Action Potential Spikes

The propagating action potential resembles a spike in both space and time. The action potential from one tissue can stimulate another excitable tissue. Neuromuscular junctions are good examples. The heart sinus, for example, is the pacemaker for the heart and sends out spikes that directly stimulate excitable heart muscle into synchronized contractions of the four heart chambers. The average pulse rate is around 0.016 Hz.

While the direct relationship between the voltage spike of an action potential and an excitable tissue is clear, consider that these spikes basically all look the same, their magnitudes and shapes are virtually identical, so there is no information content inherent in them. How, then, do these spikes convey information? We have seen that the whole action potential cycle, including the refractory period, lasts several milliseconds, so the most a cell under a fixed stimulus above threshold can do is send spikes every few milliseconds. Saturation of the excitable cell occurs when it is sending out spikes at its maximum rate. In reality, it seems spike trains are more on the order of a maximum of 10 Hz. The current notion is that it is the frequency of spikes, including the details of time intervals between spikes that convey the information. This implies a digital behavior for excitable membranes. It also implies highly sophisticated neural processing units, brains, to decode this digital information.

13.8 Summary

Lipids are essential to membrane formation, which in turn allows for a wide variety of membrane channels and receptors to lodge in and carry out a large number of living state functions; active transport of nutrients and waste products across membranes, formation of concentration and electric potential gradients critical for energy production, and formation and transmission of nerve impulses, and signaling. Resting membranes are not in equilibrium but rather use active ion pumping to maintain nonequilibrium electric potential and concentration gradients. Excitable membranes, such as those found in sensory and muscle cells, generate and transmit nerve impulses via action potentials through a complex process involving voltage-sensitive, active channels and ion pumps.

References

Baker, P.F., Hodgkin, A.L., and Shaw, T.I. (1962). Replacement of the axoplasm of giant nerve fibres with artificial solutions. *Journal of Physiology* 164: 330–354.

Brenner, N., Agam, O., Bialek, W., and de Ruyter van Steveninck, R. (2002). Statistical properties of spike trains: universal and stimulus-dependent aspects. *Physical Review E* 66: 0319071–14.

Hodgkin, A.L. and Huxley, A.F. (1952a). Currents carried by sodium and potassium ions through the membrane of the giant axon of loligo. *Journal of Physiology* 116: 449–472.

Hodgkin, A.L. and Huxley, A.F. (1952b). The components of membrane conductance in the giant axon of loligo. *Journal of Physiology* 116: 473–496.

Hodgkin, A.L. and Huxley, A.F. (1952c). The dual effect of membrane potential on sodium conductance in the giant axon of loligo. *Journal of Physiology* 116: 497–506.

Hodgkin, A.L. and Huxley, A.F. (1952d). A quantitative description of membrane current and its application to conduction and excitation in nerve. *Journal of Physiology* 117: 500–544.

Hodgkin, A.L., Huxley, A.F., and Katz, B. (1952). Measurement of current-voltage relations in the membrane of the giant axon of loligo. *Journal of Physiology* 116: 424–448.

Jackson, M.B. (2006). *Molecular and Cellular Biophysics*. Cambridge University Press.

Laio, A. and Torre, V. (1999). Physical origin of selectivity in ionic channels of biological membranes. *Biophysical Journal* 76: 129–148.

Saliani, A., Perraud, B., Duval, T. et al. (2017). Axon and myelin morphology in animal and human spinal cord. *Frontiers in Neuroanatomy* 11 (129): 1–19.

Further Reading

Gopinath, T., Weber, D., Wang, S. et al. (2021). Solid-state NMR of membrane proteins in lipid bilayers: to spin or not to spin? *Accounts of Chemical Research* 54 (6): 1430–1439.

Liu, S., Yang, J., and Zou, P. (2021). Bringing together the best of chemistry and biology: hybrid indicators for imaging neuronal membrane potential. *Journal of Neuroscience Methods* 363: 109348.

Ros, U., Pedrera, L., and Garcia-S, A.J. (2021). Techniques for studying membrane pores. *Current Opinion in Structural Biology* 69: 108–116.

Steeneken, P.G., Dolleman, R.J., Davidovikj, D. et al. (2021). Dynamics of 2D material membranes. *2D Materials* 8 (4): 042001.

Ulmschneider, J.P. and Ulmschneider, M.B. (2018). Molecular dynamics simulations are redefining our view of peptides interacting with biological membranes. *Accounts of Chemical Research* 51 (5): 1106–1116.

14

Integrated Biological Systems

We commonly think of animals and plants as matter, but they are really systems through which matter is continually passing.

William Bateson

We have delineated small molecules, biomacromolecules, and cell structures in Chapters 2 and 3 and built up physical principles in Chapters 4–6, macromolecular principles and characterization methods in Chapters 7–9, and delved into a small subset of activities of the various classes of biomacromolecules in Chapters 10–13. In this chapter, we want to see how these principles and molecules come together to form cell-level and higher systems and to explore some areas of theoretical biology.

14.1 Light and Life

In a poetic sense, we can think of living organisms on Earth as "Beings of the Sun-chain." The Earth captures a tiny fraction of the sun's photon flux and converts this, via photosynthesis, into the important biomolecules which then allow for the whole chain of nonphotosynthesizing life forms to evolve; from saprophytes that live off of dead or decaying organic matter, to carnivorous fish, to omnivorous higher mammals which can devour many different species of plants and animals. *Autotrophs* are organisms that can produce biomolecules from inorganic substances, such as water, carbon dioxide, and minerals. We tend to think most of the *photoautotrophs*, plants and algae, which use light as the energy source for this production, are the start of the "Sun-chain." However, evidence suggests that *chemoautotrophs* preceded photoautotrophs during evolution on Earth. Chemoautotrophs can live without any light and derive their energy from such inorganic chemical sources as hydrogen sulfide, ferrous ion, ammonia, and H_2, in order to produce biomolecules from CO_2.

14.1.1 Archaea

In 1977, Carl Woese (United States, 1928–2012) made the startling discovery that certain prokaryotes (single-cell organisms without a nucleus), until then considered bacteria, actually had ribosomal RNA, rRNA, more similar to eukaryotes than to bacteria. In fact, the net biochemistry of these organisms more resembled that of eukaryotes than bacteria. Woese designated these organisms as belonging to a different *domain*, *archaea*. From this came the *three domain classification* of life: archaea, bacteria, eukaryota (all organisms with a cell nucleus). While there is still some controversy, it is currently thought that the split between archaea and bacteria occurred around 3.5 billion years ago, and bacteria went on to develop photosynthesis. Archaea were first found in extreme conditions, such as in dark, hot thermal vents, and in thermal hot springs. This led to their classification as *extremophiles*, and they helped map the extremes under which life can survive, which include high temperature, high acidity, and high ionic strength aqueous environments, such as Utah's Great Salt Lake and the Dead Sea. Because such organisms were previously misclassified as bacteria, subsequent research has shown that they are ubiquitous and are also found in many nonextreme conditions. They exist in soil, in swamps, and are an important component of *microbiota*. This latter term refers to the collection of various microorganisms that are present in multicellular life forms and includes not only archaea but also fungi,

Biophysics: Physical Processes Underlying the Living State, First Edition. Wayne F. Reed.

Companion website: www.wiley.com/go/Reed/BiophysicsandMacromolecules

bacteria, protists (chiefly eukaryotic single-cell organisms, such as protozoans), and viruses. Archaea are found in the *human microbiota*, including in skin and in the *gut flora*. In cow gut flora, they are responsible for the production of copious amounts of methane, which is a *greenhouse gas*. Interestingly, unlike other microorganisms, archaea have not been linked to pathogenic or parasitic activity.

With this important detour into the vast world of archaea as autochemotrophs that don't require light or products originating from photoautotrophs, we now look at the importance of light for a vast amount of living organisms on Earth.

14.1.2 Absorption and Emission of Light in the Living State

The majority of biological molecules do not absorb visible light and hence are "colorless." This holds true for proteins, nucleic acids, carbohydrates, and lipids. All of these have absorption bands in the ultraviolet, but not in the visible range. Yet life is a many-colored phenomenon and so the question arises of where color comes from in the living state. There is more than one mechanism for biological color.

Chemiluminescence is the general term for chemical reactions that produce light. *Bioluminescence* is chemiluminescence produced by living organisms. These latter are widely distributed in sea life and include certain fish, bacteria, and algae. On land, fireflies, and glow worms are examples. The light produced can have different functions, such as warning, attraction of a mate, and to illuminate prey. As usual, an enzyme lies behind this phenomenon, *luciferase*, which acts upon a small molecule, *luciferin*. Both luciferase and luciferin are generic names for whole classes of enzymes and substrates, respectively. The reactions involve O_2, but in different bioluminescent organism a variety of other cofactors may be involved, such as Ca^{2+} or Mg^{2+}. There are other *photoproteins* besides luciferase, such as *aequorin*.

Another source of color involves diffraction. *Iridescence* is caused by diffraction that results in seeing different colors reflected from incident white light as the viewing angle changes. Examples of iridescent biomaterials include peacock feathers, mother of pearl, and some tropical leaves. These biological materials have organized colloidal *photonic* structures whose sizes are on the order of visible light, which act as diffraction gratings.

While fascinating, bioluminescence and iridescence are not even the tip of the iceberg when it comes to light and life. Biomolecules which absorb visible light are the origin of photosynthesis, on which the entire Sun-chain of life depends, as well as light detection and vision. The porphyrins ("purple" in Greek) are a class of molecules based on multiple pyrrole units, which are highly conjugated heterocyclic, aromatic organic molecules. Among the most prominent of the porphyrins are heme, which gives blood its red color, and the chlorophylls, which can yield many different colors, but are often associated with a predominant green.

That conjugation can lead to visible absorption can be understood conceptually and simply by considering the quantized electron energy levels in a 1D box (infinite square well). For an electron of mass m in a box of length L the nth integer energy level is given by

$$E_n = \frac{n^2\pi^2\hbar^2}{2mL^2} \tag{14.1}$$

In conjugated molecules, each electron is spread out over more atomic centers and hence has a longer "box length" L. As seen in the above equation, this will lower the energy and lead to longer wavelength photon absorption, in the near UV and visible, since photon energy is given by

$$E = \frac{hc}{\lambda} \tag{14.2}$$

Figures 14.1a, 14.1b show the highly conjugated biomolecules heme (*a*) and chlorophyll (*b*) which absorb visible light. Figure 14.1c shows the absorption spectrum of chlorophyll. There is an absorption gap in the visible spectrum, which gives rise to the reflection of green light from leaves and other structures filled with chlorophyll.

Other biological molecules that absorb light include retinal (part of rhodopsin, the light-sensing proteinaceous unit in the eye), melanin (skin color), and betalains (e.g. found in plants such as beets). All of these molecules also have high levels of conjugation to yield visible absorption.

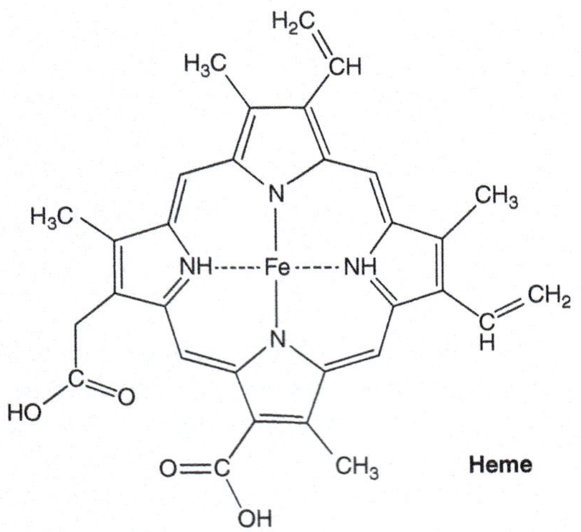

Figure 14.1a Heme.

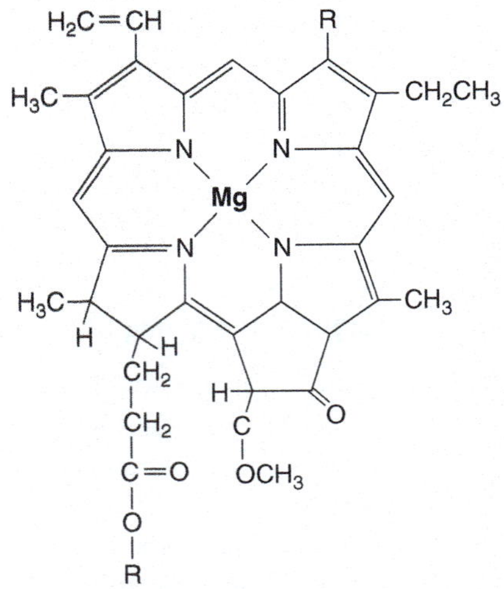

Figure 14.1b Chlorophyll.

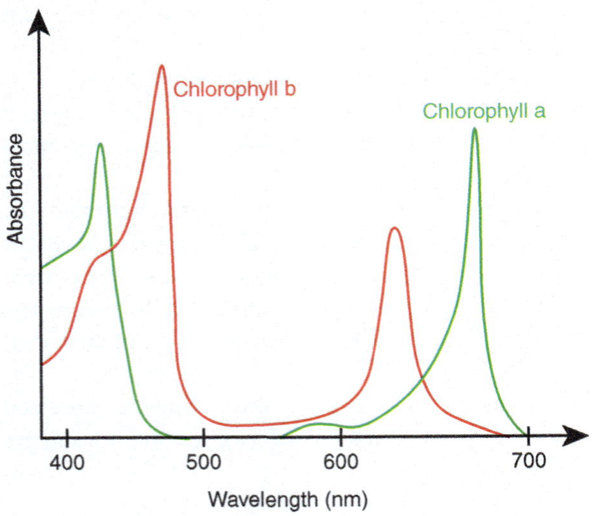

Figure 14.1c Absorption spectrum of chlorophyll *a* and *b*. (Aumars, P. (2018) / Wikimedia commons / Public Domain).

14.2 Vision

Vision and image formation occur in many animal species. Vision brings together virtually all of the living state principles we have studied, and more. One way to illustrate this is to pass through the various aspects of vision, starting with the basics of image formation, the structure of the eye, the function of its components, and the membrane neurophysics of light reception and transduction that capture the image formed by the lensing system.

14.2.1 Geometrical Optics and Image Formation

Before considering the anatomy of the eye, consider the formation of a *real image* by a thin, *converging lens* (a *convex lens*). We make the simple treatment using *geometrical optics*, where light is considered as rays traveling in straight lines that can *refract*, that is change directions when the ray passes at an angle from a medium of one refractive index to another one of different refractive index. In geometrical optics, the wave nature of light is ignored and so does not treat interference, diffraction, and polarization phenomena.

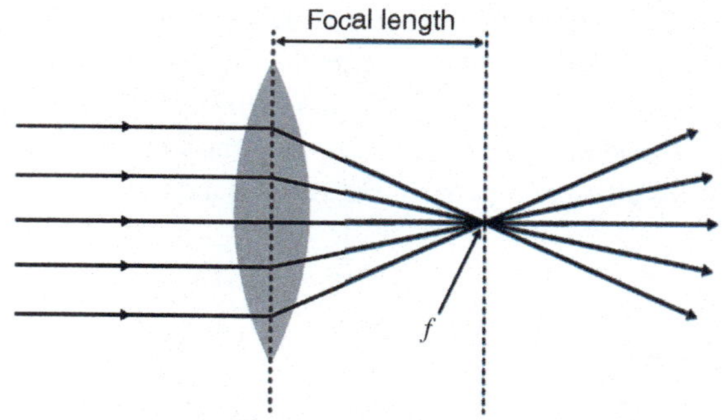

Figure 14.2 Focusing of incident parallel light at the focal length, f, of a thin converging lens. The path is reversible (just turn the arrow directions around) so that light diverging from a point source at f will emerge as a parallel beam from the lens.

The *focal length f* of the converging lens is the distance from the center of the lens at which a parallel light beam is focused to a point after passing through the lens. Conversely, the focal point is also the distance at which an illuminated point object can be placed and light emanating from it will emerge parallel after passing through the lens. This is a statement of the reversibility of the optical path; in these two situations, the light travels the same path, just in opposite directions. Figure 14.2 illustrates the focusing principle.

The thin lens approximation uses *Snell's law of refraction* to relate the distance of an object, s_o, from the center of the lens to the distance on the other side of the lens that the image is formed, s_i, via the focal length. This yields

$$\frac{1}{f}=\frac{1}{s_o}+\frac{1}{s_i} \qquad \text{Thin lens equation} \qquad 14.3$$

This equation, while handy and simple, assumes that light comes in essentially parallel to the lens axis, the *paraxial approximation*. When this condition no longer holds there is *spherical aberration* and a blurry image is formed. The "thin" condition means that the thickness of the lens is assumed to be small compared to f, s_o, and s_i.

The focal length of a thin lens with index of refraction n_2 immersed in a medium of index of refraction n_1 is related to the radii R_1 and R_2 of the lens surfaces, assumed spherical, according to

$$\frac{1}{f}=\frac{(n_2-n_1)}{n_1}\left(\frac{1}{R_1}-\frac{1}{R_2}\right) \qquad 14.4$$

where, by convention, R_1 is positive if the front lens surface is convex and negative if it is concave, and R_2 is opposite; that is, it is negative if the second surface is convex and positive it is concave. A converging lens in air will always have a positive focal length, and a diverging lens will always have a negative focal length. The series of optical surfaces in the human eye constitute an equivalent converging lens.

Using Equation 14.4 and the two features of the focal point of the lens allows us to make the ray tracing in Figure 14.3 of an object through the lens. Three rays are traced: The first is for a ray parallel to the axis, which passes through the focal point to the right of the lens. A second ray passes through the focal point to the left of the lens and emerges parallel. These two rays, based on the focusing/collimation principle of Figure 14.3 are enough to determine the final image. A third ray, however, is simple to include, and passes through the center of the lens, which, in the paraxial approximation, passes without deviation through the lens. An inverted, real image is formed by the convergence of these three rays. The image is real when s_i is positive in Equation 14.3, which is determined by f and where the object is placed, s_o. By convention, s_o is positive and is the position of a real object to the left of the lens, and s_i is positive when the image is real, and falls to the right of the

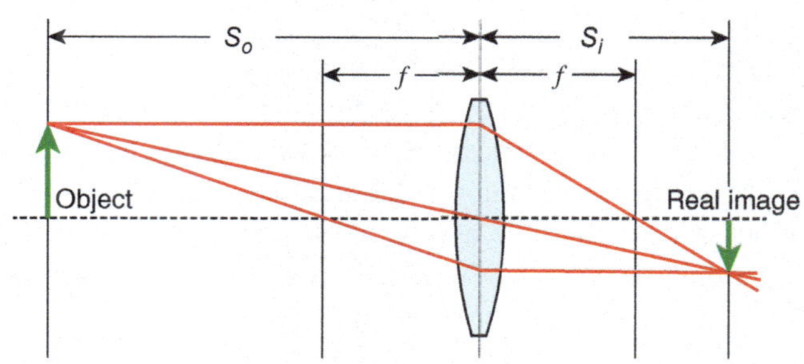

Figure 14.3 Ray tracing of an object through a positive lens to form a real image. (Wikimedia commons (2006) / Public Domain).

lens. The light energy passes through the real image, and in the human eye, it is this energy that allows transduction of the image formed. The image formed on the eye's retina is inverted and so the brain performs the re-inversion of the image through its complex photochemical neurophysical interpretation of the image.

The *transverse magnification* of the image, m_t, the ratio of the lateral size of the image to the lateral size of the object, from direct geometry in Figure 14.3 is

$$m_t = -\frac{s_i}{s_o} \tag{14.5}$$

where the negative sign is introduced so that a negative magnification, such as for the eye, indicates that the image is inverted.

The optical system in the human eye has more than one refracting surface. In the thin lens paraxial approximation, multiple lenses can be stacked together to give a net focal length f_{net}. For N lenses stacked together:

$$\frac{1}{f_{net}} = \frac{1}{f_1} + \frac{1}{f_2} + \cdots + \frac{1}{f_N} \tag{14.6}$$

In addition to accounting for the eyes refractive power, Equation 14.6 allows one or more lenses to be placed before the eye to change f_{net} and help correct eye defects. $1/f_{net}$ is termed the 'diopter' of a lens system. For eyeglass the larger the diopter the higher the correcting power of the lens.

14.2.2 Structure and Optics of the Human Eye

Figure 14.4 shows a simplified schematic of the human eye. The *cornea*, the outer-curved surface of the eye provides most of the refracting power, with $n_2 = 1.376$ in Equation 14.4. For land animals, the light is incident from the air for which $n_1 \sim 1$. When we open our eyes in the water, for which $n_1 = 1.333$, there is very little refractive power by the cornea, f becomes very long, and our vision becomes blurry. The *sclera*, which gives the "white of the eye," is continuous with the cornea and is the tough collagenous layer that protects the eye, and, together with hyaluronan which supplies osmotic pressure, helps keep the shape of the eye.

The *lens* comes after the cornea and is responsible for "fine tuning" vision. The distance from the lens to the part of the *retina* on the optical axis, s_i, is fixed and is normally around $s_{i,normal} = 1.71$ cm. But Equation 14.3 shows that s_i is different for different object distance s_o for a fixed f. Evolution has "solved" this problem by making the lens flexible enough that its focal length f is variable and automatically adjusts its value to the distance of the object being focused on, s_o. Equation 14.4 shows that f will change if the curvature of either or both the lens surfaces R_1 and R_2 change. The curvature change is accomplished by the *ciliary muscles* within the *ciliary body* shown in Figure 14.4, which can apply pressure to the lens to bulge the lens, lessening R_1 and R_2, causing a smaller f from Equation 14.4, and the ability to focus on nearby objects. This process of changing the lens focal length is termed *accommodation*. Most fish have fixed f but are able to move their lens relative to their retina and hence change s_i instead of f.

From Equation 14.3 the required focal length for maintaining the fixed image distance $s_i = 1.71$ cm is given by

$$f(\text{cm}) = \frac{1.71 s_o}{1.71 + s_o} \tag{14.7}$$

Figure 14.5 shows how f must change as the distance of the object from the eye changes. At very close distances, f drops very sharply, so sharply that there is a physiological limit as to how small the radius of curvature can be made by the ciliary muscle apparatus. In fact, there is a cut-off object distance, termed *the near point*, closer than which the eye cannot focus on. The near point is taken as 25 cm for the normal eye. In reality, the lens does not have a fixed index of refraction, so that Equation 14.4 is not strictly applicable, as it considers the lens to have a single index of refraction n_2. It is a *gradient lens* going from $n = 1.406$ toward the center to $n = 1.386$ in the outer layers of lens.

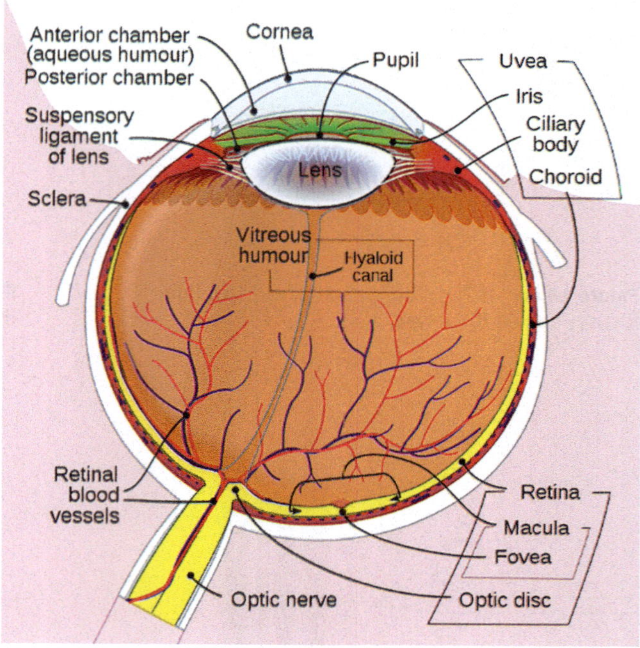

Figure 14.4 Simplified schematic of the human eye. (Rhcastilhos and Jmarchn (2007) / Wikimedia commons / Licensed under CC BY-SA 3.0).

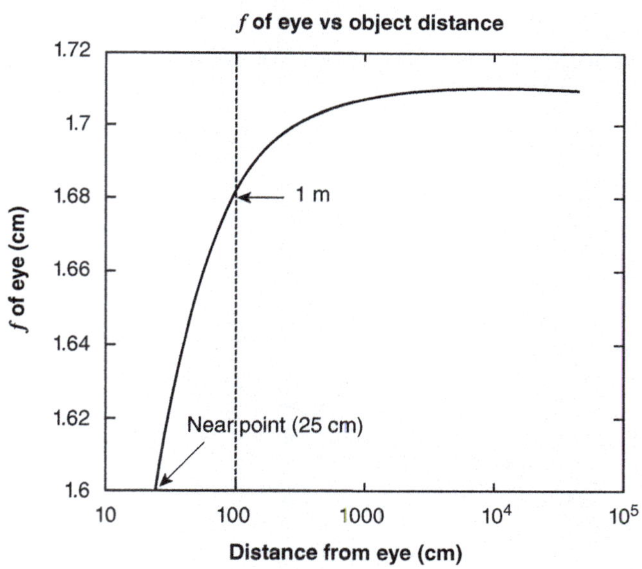

Figure 14.5 Net focal length of eye needed to focus the image of an object on to the retina versus distance of the object from the eye.

As a person ages, the lens loses flexibility and the ciliary muscles lose strength, so that accommodation to nearby objects by shrinking *f* is no longer possible. As the ability to make *f* small enough by accommodation diminishes, a person's near point increases. For objects closer than the near point, *f* is too large for resolving nearby objects, and, by Equation 14.3, the image forms behind the eye, at a distance $s_i > s_{i,normal}$. This condition is known as *hyperopia* or *far-sightedness* and can be corrected by placing an appropriate converging lens in front of the eye, "reading glasses," which decrease *f* according to Equation 14.6. Figure 14.6a shows how incident parallel light converges beyond the retina, blurring the image. Figure 14.6b shows how a converging lens in front of the eye decreases the net focal length enough for the light to converge on the retina, thus correcting the condition.

Conversely, if the lens cannot flatten and relax enough then *f* cannot be made large enough to resolve distant objects and, by Equation 14.3, the image forms at $s_i < s_{i,normal}$. This is termed *myopia* or *near-sightedness* and can be corrected by placing a diverging lens, which has a negative *f*, in front of the eye. By Equation 14.6, this will increase *f*, allowing the distant objects to be resolved. Figure 14.7a shows incident parallel light converging in front of the retina in myopia. Figure 14.7b

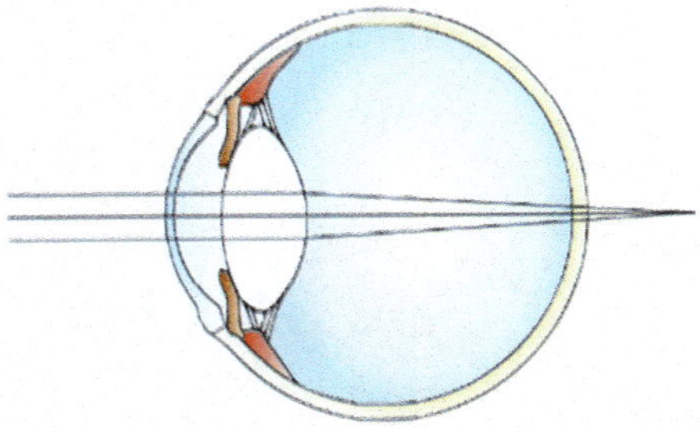

Figure 14.6a In hyperopia, parallel incident light focuses behind the retina.

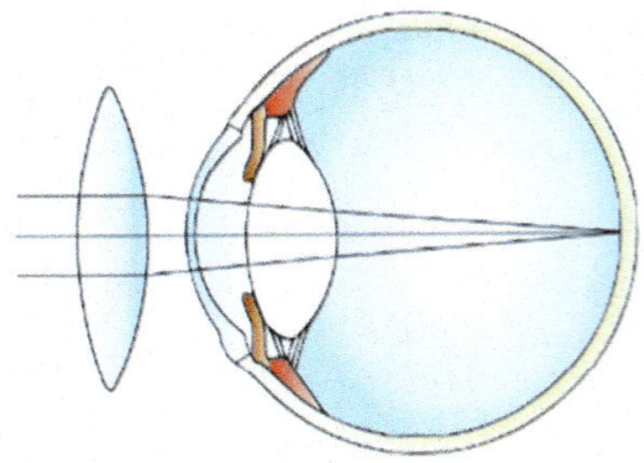

Figure 14.6b Placing a converging lens in front of the lens focuses the parallel light on the retina, correcting the hyperopia. (Gumenyuk, I.S. (2014) / Wikimedia commons / Licensed under CC BY-SA 4.0).

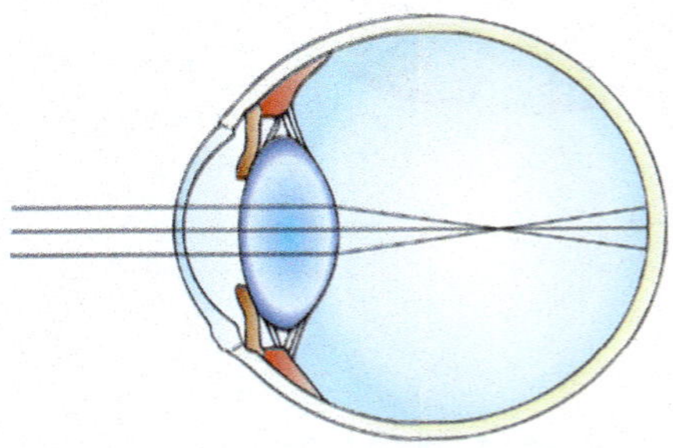

Figure 14.7a Convergence of incident light before the retina in myopia.

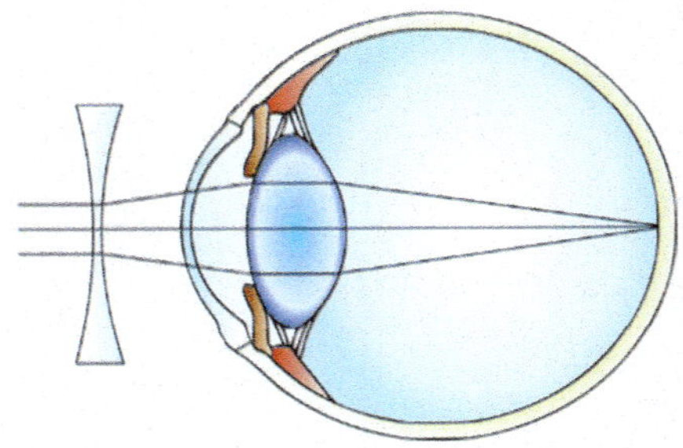

Figure 14.7b Placing a diverging lens in front of the eye causes the light to converge on the retina, thus correcting myopia. (Gumenyuk, I.S. (2014) / Wikimedia commons / Licensed under CC BY-SA 4.0).

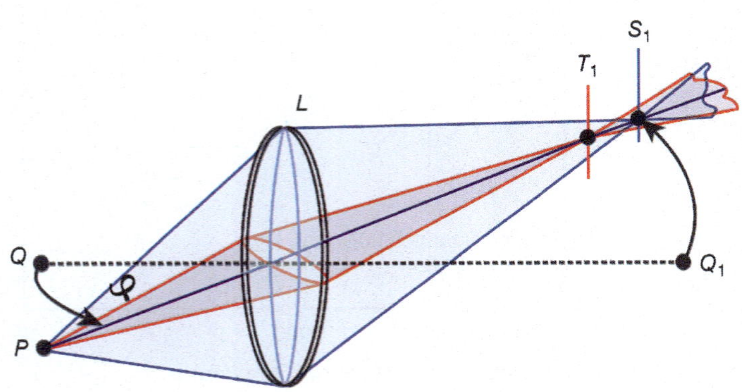

Figure 14.8 Astigmatism causes light in perpendicular planes entering the lens L, to have different focal lengths T_1 and S_1, thus blurring the image. (Kosch, S. (2007) / Licensed under CC BY-SA 3.0).

shows how a diverging lens placed in front of the eye increases the focal length and causes the light to converge on the retina, thus correcting the condition.

Another defect of the cornea/lens system is when either or both of these has an irregular noncircularly symmetric shape. In this case, according to Equation 14.4, different planes that pass through the center of the eye can have different values of f. It is simplest to imagine this with two planes perpendicular to each other, each of which has a different focal length. Then a given object will focus at different distances from the retina in each plane, causing blurred vision. This condition is known as *astigmatism* and can be corrected with lenses that themselves are stigmatic in the opposite sense of the eye, thus creating a single focal length for the entire eye. An example of astigmatism is shown in Figure 14.8.

The *pupil* of the eye is the black, circular region through which light enters the lens. Its diameter controls the amount of light reaching the retina; in bright light, the pupil will have a small diameter, and vice-versa in dark conditions. The *iris* controls the pupil's diameter and is actuated by the *pupillary sphincter muscle*. The iris contains collagen bundles, called *stroma*, in an interwoven network. The color of one's eyes is determined by the *melanin* content of the stroma in the iris.

The more melanin, the darker the eye color. The color of blue, green, and hazel eyes results from less melanin and more light scattering from the stroma and is not due to any blue or green colored pigments.

A two-dimensional image of an object, according to Equation 14.3, is formed at s_i on the *retina*, the photosensitive surface lining the eyes inner surface. *Rods* and *cones* are the structures that act as photoreceptors. These complex structures capture photons and amplify the events into nerve impulses. Rods are sensitive to light intensity, not color, and can absorb and process small numbers of photons. Cones require higher light intensities to function and are concentrated in the *fovea*, the center for producing high visual detail, such as for driving or reading. Half of the optical nerves are dedicated to this relatively small patch. There are three types of cones, sensitive to red, green, and blue light. Rhodopsin is the protein prosthetic complex responsible for photon capture. There are about 10^9 rhodopsin molecules per rod and about 10^8 rods in the human eye. There are far fewer cones in the eye, on the order of 5×10^6, and they cluster at the center of the retina, whereas rods are spread throughout.

Amazingly, the eye can respond to light intensity variations over 10 orders of magnitude! The response function of the eye over this range is essentially logarithmic, as proposed by "Weber's Law"

$$R \propto \ln\left(\frac{S}{S_0}\right) \tag{14.8}$$

where R is the response to a light level of intensity S and S_0 is the minimum intensity threshold for the eye to have a response. This type of logarithmic response function applies to other senses, such as audition, where the ear likewise responds to many orders of magnitude difference in sound intensity, which is measured on a logarithmic decibel scale.

The lower limit on response has been studied by presenting subjects with very dim flashes of light with small numbers of photons entering the eye and having the subject state whether or not they perceived the flash. Recall from Chapter 7.12.3 (also see Appendix I) that the Poisson distribution measures the probability of an event occurring j times when the average number of events is m

$$W(j) = \frac{m^j e^{-m}}{j!} \tag{14.9}$$

In the case of a dim flash m is the average number of photons per flash that enter the eye and are absorbed by rods and j is the actual number that enter on any given flash and are absorbed by rods. The term "absorbed by the rods" is used here because a significant number of photons entering the eye are lost to scattering and other processes and are never absorbed by rods.

By analyzing subject response with this distribution, it was determined that the minimum number of photons needed for flash detection was in the interval of five to nine photons over an approximately 100 ms flash. The photons were distributed over hundreds of rods, so the probability that a given rod receives more than one photon is negligible. This means that the rods are responding to single photons! However, the fact that subjects could rarely or ever respond to flashes with just one photon is apparently linked to a neural processing threshold that filters out events with one or just a few photons and is not a lack of rod sensitivity. The system requires a minimum number of photons to make it "worthwhile" for processing so that the event is more than just random noise photons.

Rods are around 1–2 μm in diameter and 100 μm long. Interestingly, this means there are many rods packed within the diffraction spot of the eye on the retina, and this improves spatial resolution. The Rayleigh criterion for separation of two objects due to overlapping diffraction patterns is that the minimum angle θ, of separation of the objects, be

$$\sin\theta = 1.22\frac{\lambda}{D} \tag{14.10}$$

where λ is the wavelength of the incident light and D is the aperture diameter. For green light of 500 nm, a pupil diameter of 0.5 cm, the minimum resolvable angle is 0.007°, which, at a distance of the lens to the retina of 1.7 cm, the diameter of the diffraction spot on the retina of 4 μm. Hence, there are several rods within the diffraction spot. People have been found to actually resolve better than the diffraction limit, to about 0.002°, which is apparently part of the many highly evolved processing functions the brain carries out on visual information.

Oddly, vertebrates have *inverted retina*, which means that light has to pass through neurons and capillaries to reach the rods. This, along with passage through imperfect cornea, lens, and vitreous, leads to some scattering and absorption so that not all photons entering the pupil reach the rods for absorption. It is thought that Müller (glial) cells in the retinal layer act as fiber-optic guides to the photoreceptors, thus making up for some of the losses, and perhaps even optimizing the light reception process. Because the neurons are on the outer surface of the retina, they have to be drawn together in a bundle and brought through the retina. This bundle is the *optic nerve*, and the area where it passes through the retina and into the brain is the *blind spot*.

The *aqueous humor* is very similar to blood plasma, but with less protein, and is in front of and in back of the lens and has an index of refraction of 1.3335, nearly that of water. The *vitreous humor* has virtually the same index of refraction as the aqueous humor and is about 99% water. It contains the polysaccharide hyaluronan, discussed in Chapters 8 and 12, which provides the osmotic pressure needed to keep the eye's shape. It also imparts viscosity to the vitreous humor several times that of water, giving it a somewhat gelatinous consistency.

While the small amount of the polysaccharide hyaluronan is needed to keep the eye's osmotic pressure, and phospholipid membranes are ubiquitous for active membrane action, it is the proteins that give the overall structure and functionality of the eye. The transparency of both the cornea and lens is due to the arrangement of *crystallin proteins* within a collagen fiber capsule. There are no cell structures or organelles in the cornea and lens, as these would otherwise scatter light and reduce transparency. While not crystalline the collagen fibers are arranged in an organized enough fashion that most of the scattering occurs in the forward direction, so that light losses are fairly low in these structures. *Cataracts* result when this organization is compromised, and lateral light scattering occurs, as well as other processes that clouds the lens. The lens and the cornea have collagenous casings.

14.2.3 Transduction Events in Vision

We have covered some of the essentials of how the cornea and lens cause images to form on the retina, where the individual photons associated with the image are absorbed by rhodopsin and turned into nerve impulses. The chain of events involved in vision after the photons are absorbed is stunning and the biochemical details are more than we wish to describe here. The cycle was discovered by George Wald (United States, 1906–1997) and colleagues in the 1950s.

In brief, rhodopsin consists of a protein molecule, an *opsin*, of molar mass around 40,000 g/mol, covalently bound to a relatively small molecule, *retinal*, an aldehyde of vitamin A, with a molar mass of 284 g/mol, which is in an 11-*cis* isomeric state while bound. Absorption of a photon by rhodopsin in a rod causes isomerization of 11-*cis* retinal to all-*trans*, Figure 14.9, causing a series of conformational changes in the opsin leading to metarhodopsin II, releasing all-*trans* retinal in the process. The metarhodopsin II activates a protein, *transducin*, which then dissociates into subunits and is involved in a cascade of signal transduction events that end up hyperpolarizing (to −70 mV) the rod. This hyperpolarization leads to the formation of a nerve impulse which is then communicated to the retinal neurons and collected at the optic nerve for transmission to the brain and processing. Meanwhile, the all-*trans* retinal released needs to go through several steps to be regenerated to 11-*cis* retinal for rebonding with opsin. It takes about 30 minutes from the time a photon is absorbed and cleaves the all-*trans* retinal, termed *photobleaching*, and the restoration of rhodopsin in the form of 11-*cis* retinal covalently bonded to opsin.

Rhodopsin has an absorption spectrum with a maximum at 500 nm, corresponding to an extinction coefficient of around 40,000 M^{-1} cm^{-1}. This corresponds closely to the maximum emission wavelength of the blackbody spectrum of the Sun, and it is thought that vision evolved to make the most efficient use of the spectrum. While rhodopsin in rods is not involved in color perception, there are visual pigments in the cones corresponding to detection of red, green, and blue.

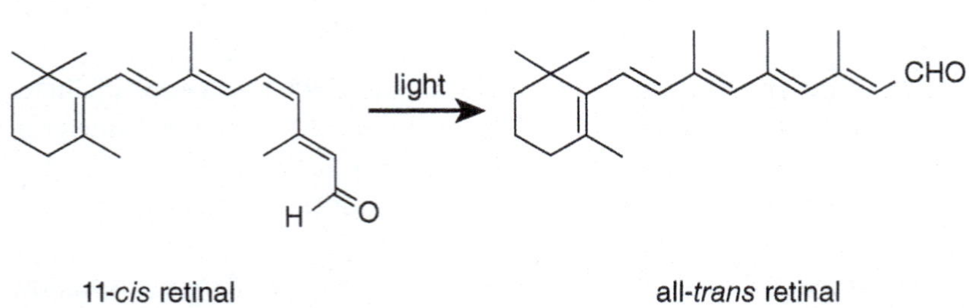

Figure 14.9 Shows retinal in *cis* and *trans* conformations.

14.3 Cilial and Flagellar Biomotion

Motility is a characteristic of animal life. Bacteria with *cilia* (singular *cilium*), long, flexible, movable fibers, are among the simplest examples of self-locomotion in the living state. In order to create translational motion, there must be some asymmetry in the motion of the cilium. If the cilium were to just reciprocate back and forth any net motion would be canceled by the opposing strokes. This is similar to using a paddle in a canoe. If the paddle remains immersed in the water and you just move it back and forth, you cancel the forward stroke with the backward stroke. So, one way around this is to minimize the backward stroke by rotating the paddle 90° so that the plane of the blade is in the direction of motion on the backstroke, vastly reducing the force of the backstroke.[1] This creates the asymmetry between forward and backward motion needed for net translational motion. Similarly, the motion of cilia is nonreciprocal, the backstroke being made in a way that reduces the force of the backstroke.

A more sophisticated mechanism is found in bacteria with *flagella* (singular *flagellum*). *Escherichia coli* bacteria, for example, have helical, rigid flagella which are driven into circular, propeller-type motion by rotary protein motors. This allows translational motion in both directions, and by modulating multiple flagella a bacterium can also rotate.

We recall from Chapter 6.3.1, Table 6.2, that microorganisms have very low Reynolds number, *Re*, when they move through aqueous media, creating laminar flow. The motion is so slow, in fact, that one wonders whether it is even worth it for a bacterium to move to find food, rather than to just stay still and wait for diffusion to bring nutrients to the bacterium from the surrounding medium. The Péclet number was introduced in Chapter 6.13 and defined by Equation 6.132, which provided a criterion as to whether the bacterium should move or wait for the nutrients.[2]

The prime reason for bacterial motion is to hunt for food when none is available locally. Some bacteria have developed an efficient way to hunt by performing a sort of biased random walk. They can propel themselves in straight-line motion, sampling the environment for nutrients as they go. If no nutrients are found after a certain distance, say several tens of microns, then the bacterium rotates about and randomly sets off in another direction. If they sense the beginning of a gradient they can alter the direction of their motion to follow the gradient.

Interestingly, slow, laminar motion comes with a disadvantage for small organisms. Referring to the fluid streamlines around the sphere in Figure 6.5, the sphere will keep pushing liquid away as it moves, and nutrient molecules along with it. Some small organisms have developed a means of giving a burst of speed that brings them out of the low *Re* regime and allows them to overtake their prey, be it a molecule or lesser creature. Similarly, the backside of the motion entrains liquid in the organism's motion, which could drag a predator along, if present, allowing it to catch up on its prey. The same type of burst speed by the organism can bring it out of the low *Re* regime, breaking the streamlines that entrain the predator.

14.4 Theoretical Biology: Cycles, Instabilities, and Attractors

Theoretical biology and the closely related field of mathematical biology is a rubric covering many areas, including ecology, population genetics, neuroscience, morphogenesis, and many more. Because living organisms are open to their environment and exchange matter and energy, we do not expect equilibrium thermodynamics to be highly predictive for far-from-equilibrium processes. Some of the important notions that come into play in nonequilibrium systems are *oscillations, limit cycles, instabilities, autocatalysis, interspecies interaction, bistable steady state, attractors*, and *chaos*. The same equations underlying these phenomena can often be adapted to physics, chemical reactions, cellular processes, ecology, economics, and sociology.

14.4.1 Elementary Time-dependent Processes Leading to Cycles, Instability, and Chaos

Consider an unstable material that diminishes spontaneously in time; for example, radioactive atoms in a material, excited electronic states in a gas, the charge on a conductor in contact with ground via a resistive path, or thermal dissociation of an unstable chemical. The rate of loss of the material X is proportional at any instant to the amount of material itself

$$\frac{dX}{dt} = \alpha X \qquad 14.11$$

where α is a *negative* rate constant. This yields the usual solution $X = X_0 e^{\alpha t}$, which subsides to zero over time, since $\alpha < 0$, and leads toward the final equilibrium state of minimum free energy.

---

1 Of course, the most practical way of minimizing the force of the back stroke is to lift the paddle entirely from the water on the back stroke. Experienced paddlers will often "feather their paddles," that is, turn the plane of the paddle 90° during the backstroke to reduce wind resistance.

2 There is also a thermal Péclet number, which measures the time of heat diffusion compared to directed motion.

Now, consider the case of *autocatalysis*, where one reagent, X, can, among other things, produce more of itself. A simple autocatalytic reaction is

$$A + X \rightarrow 2X \tag{14.12}$$

Sexual reproduction of a species, X, whereby two opposite-sex adults produce N offspring is also autocatalytic:

$$X + X \rightarrow NX \tag{14.13}$$

For example, doubling the population of reproducing species with a constant N doubles the rate. The solution to X in this case involves a positive rate constant, α, and grows without bound, often termed *unstable growth*. The demarcation between the unstable growth and decaying exponential regime is given by $\alpha = 0$. This simple notion is often the starting point for assessing the stability of differential equations describing some system; $\alpha = 0$ corresponds to a steady state, $X(t) =$ constant, $\alpha < 0$ leads to decreasing X, and $\alpha > 0$ leads to exponential growth.

If the back reaction is taken into account:

$$X + X \rightleftarrows NX \tag{14.14}$$

the solution for X is sigmoidal, and hence self-limiting.

Of course α can be a function of time, X, or some other variable, and other factors can be involved. This is usually the case since populations can only increase to the carrying capacity of their environment. A simple way of accounting for a deceleration in population growth rate, e.g. by reaching a limiting food supply, is by

$$\frac{\mathrm{d}X}{\mathrm{d}t} = \alpha X\left(1 - \frac{X}{K_{max}}\right) \tag{14.15}$$

where K_{max} is the maximum population or concentration that a system can support and α is constant and positive.

A model for two species competing for the same resources, each with its own reproductive rate constant α and β, and which kill each other with unequal probabilities γ and ν when they encounter each other, gives

$$\frac{\mathrm{d}X}{\mathrm{d}t} = \alpha X - \gamma XY \tag{14.16}$$

$$\frac{\mathrm{d}Y}{\mathrm{d}t} = \beta Y - \nu XY \tag{14.17}$$

Depending on the constants, this can lead to one species exterminating the other, or a possible coexistence. If $\alpha = \beta$ and $\gamma = \nu$, then one species alone will survive given even a slight initial population difference between X and Y.

Vito Volterra (Italy, 1860–1940) pioneered many of the early approaches to non-monotonic time behavior of interacting systems. A simple model he posited was that of predator/prey relationships. This was partly spurred by the observation in post–World War I Italy that fishermen were bringing in much higher populations of predator fish than prewar. Fishing activity had greatly diminished during the war leaving more feed fish for the predators. The model considers prey as X and predators as Y. The prey has its usual autocatalytic growth rate constant α, but its population decreases when predator meets prey in a second-order reaction with rate constant β. Volterra further assumed that the predator reproduction rate was set by the consumption of prey with rate constant ν, while the predator population, devoid of prey, would decline with rate constant γ, due to starvation. Volterra's classic coupled equation are then:

$$\frac{\mathrm{d}X}{\mathrm{d}t} = \alpha X - \beta XY \tag{14.18a}$$

$$\frac{\mathrm{d}Y}{\mathrm{d}t} = \nu XY - \gamma Y \tag{14.18b}$$

Solving this numerically for $\alpha = 1$, $\beta = 0.5$, $\nu = 1$, and $\gamma = 2$, with several values of $X(t = 0)$ yields an oscillation between the predator and prey over time, shown in Figure 14.10a. Plotting the phase portrait of Y versus X gives Figure 14.10b.

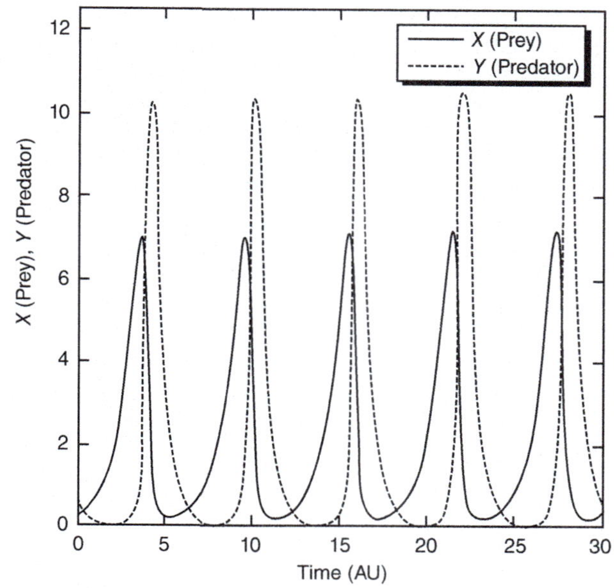

Figure 14.10a Predator/Prey cycles according to Equations 14.18a and 14.18b, with $\alpha = 1, \beta = 0.5, \nu = 1, \gamma = 2$.

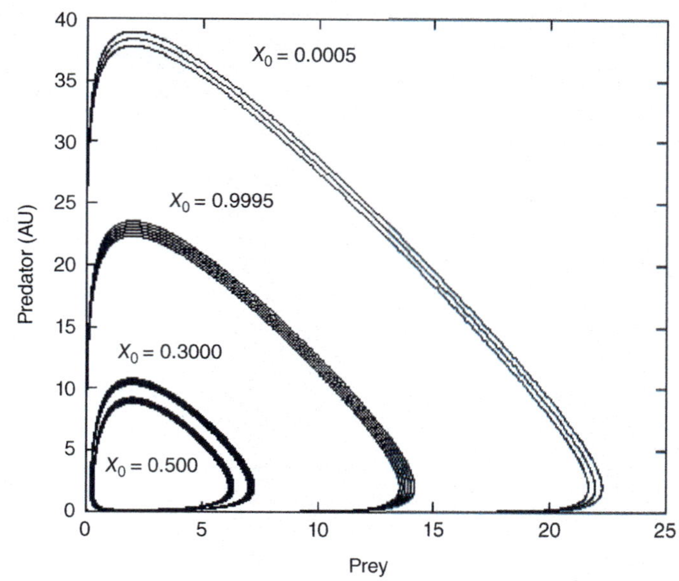

Figure 14.10b Phase portrait for Figure 14.10a, at different starting values of $X(t = 0) = X_0$.

Figure 14.10a shows that when there are few predators the prey population increases rapidly, but as it reaches its peak the predators begin to multiply, driving down the prey population. When this is driven low enough, the predators begin to die out from starvation. There is then an interval where the populations of both predator and prey are very low, allowing the prey to begin another growth cycle, so that the process of prey and predator increase and decrease is periodic. Figure 14.10b is another way of visualizing this: It shows the prey near a maximum when predator is very low, and the subsequent concomitant increase of predator as prey diminishes. The closed curves in Figure 14.10b show specific cycles, depending on the initial conditions.

Figures 14.10a and 14.10b are good examples of how *temporal order*, i.e. oscillations, can arise from simple coupled equations. Temporal order, oscillations, can also be produced when there is an autocatalytic process within a larger process; for example, the important process of glycolysis, turning glucose into two ATP molecules, shows temporal order due to autocatalysis caused by the enzyme phosphofructokinase.

An interesting effect can occur when the autocatalytic process involves a chiral molecule. In this case, the chiral molecule will catalyze creation of more molecules of the same chirality, which leads to an excess of that chirality over the opposite chirality. This type of autocatalytic process may be linked to the fact that biological molecules have a specific chirality; e.g. L-amino acids in proteins.

The notion of autocatalysis has been generalized to metabolism, since in the cell new molecules are produced which are copies of the same molecules. The notion of autocatalysis figures prominently in theories of prebiotic evolution.

If the model takes into account the carrying capacity of the environment for the prey, K_{max}, such as in Equation 14.16, and the "satiation" of the predator, S (i.e. if there is a great abundance of prey, then the predator can be satiated and not keep eating prey without limit), then Equations 14.18a and 14.18b take the form:

$$\frac{dX}{dt} = \alpha X\left(1 - \frac{X}{K_{max}}\right) - \beta XY\left(\frac{S}{X+S}\right) \quad 14.19a$$

$$\frac{dY}{dt} = \nu XY\left(\frac{S}{X+S}\right) - \gamma Y \quad 14.19b$$

Equations 14.19a and 14.19b lead to the notion of the *limit cycle*. In this, no matter the starting populations of prey and predator, the trajectories will all converge to a single closed orbit, rather than each initial condition producing its own orbit,

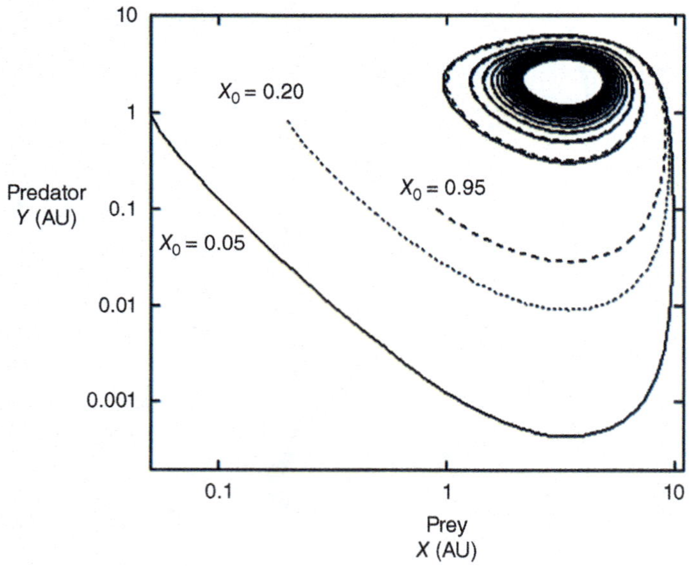

Figure 14.11 The elliptic-like limit cycle to which all trajectories converge for all initial populations of Prey (X_0) and Predator (Y_0). Trajectories shown are (X_0 = 0.9, Y_0 = 0.1), (X_0 = 0.2, Y_0 = 0.8), and (X_0 = 0.05, Y_0 = 0.95) and are from numerical integration of Equations 14.19a and 14.19b. The other constants in Equations 14.19a and 14.19b are $\alpha = 1, \beta = 0.5, \nu = 1, \gamma = 2, K_{max} = 10, S = 5$.

which occurred in Figure 14.10b, where there were no limits on prey carrying limit, and no predator satiation. Figure 14.11 shows the limit cycle reached by starting conditions of ($X_0 = 0.9$, $Y_0 = 0.1$), ($X_0 = 0.2$, $Y_0 = 0.8$), and ($X_0 = 0.05$, $Y_0 = 0.95$). The other constants in Equations 14.19a and 14.19b are shown in Figure 14.11 legend. The limit cycle is an example of an *attractor*, a set of values towards which the system will evolve in time for a wide range of starting conditions.

In the 1880s, Henri Poincaré first discovered unpredictable behavior in a simple system of just three bodies under mutual gravitational force. He realized that even small differences in the initial conditions of the three bodies – their initial positions and momenta – would lead to enormous differences in their subsequent trajectories. This is an example of *deterministic chaos*. It is deterministic in that the trajectories fully obey the governing differential equations of motion, but the trajectories are extremely sensitive to the initial conditions. The sensitivity to initial conditions is often represented via the *Lyapunov exponent*, λ. If the separation of initial conditions is represented by a vector of small magnitude $\delta \vec{P}_0$, then this small initial difference grows exponentially in time according to

$$\left|\delta \vec{P}\right|(t) = \left|\delta \vec{P}_0\right| e^{\lambda t} \qquad 14.20$$

The reciprocal of the Lyapunov exponent is the *Lyapunov time*, the time scale on which chaotic behavior manifests itself and is no longer predictable.

Little was done with Poincaré's original work until in 1963 Edward Lorenz (United States, 1917–2008), while formulating a description of atmospheric convection rate, and lateral and vertical temperature variations, demonstrated deterministic chaos in a simple system of coupled equations, given by

$$\frac{dX}{dt} = \sigma(Y - X) \qquad 14.21a$$

$$\frac{dY}{dt} = \rho X - Y - XZ \qquad 14.21b$$

$$\frac{dZ}{dt} = XY - \beta Z \qquad 14.21c$$

Figure 14.12 shows numerical solutions for Y and Z versus X for Lorenz Equations 14.21a–c, with $\sigma = 18$, $\rho = 10$, $\beta = 3$, $X = Z = 1$, $Y = 1.001$. The trajectories of Y and Z spiral into two stationary final states, termed *fixed point attractors*.

Figure 14.13a shows how sensitive the solutions to Lorenz Equations 14.21a–c are to slight changes in initial conditions. Figure 14.13a plots Y versus time, computed numerically from Equations 14.21a to c, with $\sigma = 18$, $\rho = 35$, $\beta = 3$, $X = Z = 1$,

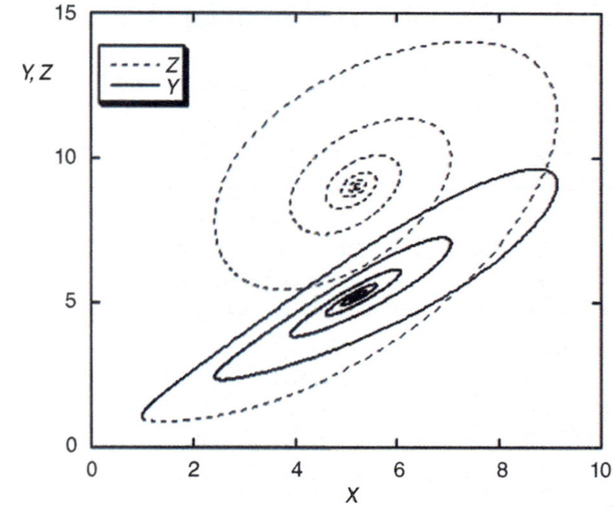

Figure 14.12 *Y* and *Z* versus *X* for Equations 14.21a–c, with $\sigma = 18, \rho = 10, \beta = 3, X = Z = 1, Y = 1.001$. The stable spiral trajectories of *Y* and *Z* converge to two stationary states, the *attractors*.

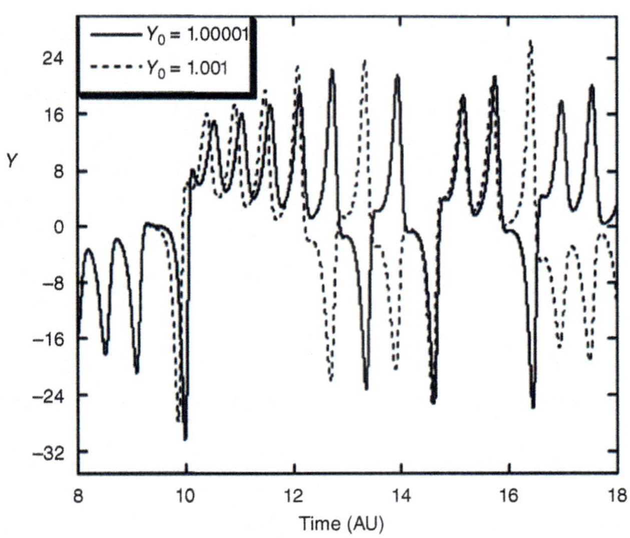

Figure 14.13a *Y*, integrated numerically from Equations 14.21a c, with $\sigma = 18, \rho = 35, \beta = 3, X = Z = 1$, and for very close initial values of $Y_0 = 1.00001$ and $Y_0 = 1.001$.

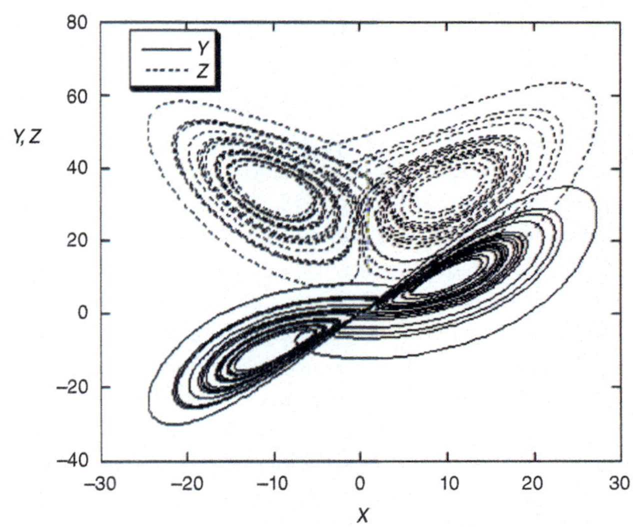

Figure 14.13b *Y* and *Z* versus *X* for the data of Figure 14.13a for $Y_0 = 1.00001$. It shows how *Y* and *Z* converge to ellipse-like curves, strange attractors, but never stabilize at a particular value.

and for two very close initial values of *Y*; $Y = 1.00001$ and $Y = 1.001$. The values of *Y* are nearly identical up to around 9 seconds, after which the values diverge from each other.

Figure 14.13b shows the phase plot of *Y* and *Z* versus *X* for the data of Figure 14.13a for $Y_0 = 1.00001$. It shows how *Y* and *Z* converge to complex "butterfly-like" structures, to which the systems motion is confined but in which the motion follows non-repeating, non-periodic paths, and never stabilizes at a particular value or reaches a stable limit cycle. This is an example of a *strange attractor*. Notice that the only difference in the parameters between Figure 14.12 and 14.13b is that $\rho = 10$ and $\rho = 35$, respectively, which brings the phase trajectory from an attractor to a strange attractor. Different strange attractor patterns will be formed, depending on initial conditions, unlike the limit cycle of Figure 14.11, to which all trajectories converge, regardless of initial conditions.

In coupling interactions between different species, even as few as three different species linked through first-order differential equations can exhibit chaotic behavior.

14.5 Dissipative, Far from Equilibrium Spatially Self-organizing Systems

Self-organization has been encountered several times in this book, as a result of the drive toward free energy minimization and equilibrium. An early example (Chapter 4.15.3) was the spontaneous, entropy-driven micellization of surfactants. Another example was the formation of bilayer vesicles and biological membranes. Protein folding, the packing of extremely long DNA molecules into tiny nuclei, and the assembly of viral proteins to form a capsid are further examples of self-organized structures in equilibrium.

On the other hand, we have encountered the far-from-equilibrium, dynamic "tension" that exists in many biological processes. Pumping ions across membranes against their gradients in order to provide voltage-polarized membranes ready to send action potentials upon stimulation is a good example. And it costs a lot of energy to maintain the polarization state, ultimately furnished by DNA.

Ilya Prigogine (Russia/Belgium, 1917–2003) was one of the first to realize that such far from equilibrium, energy costly processes could lead to many forms of dynamic self-organization in biological systems. He termed these *dissipative structures*, where the word "dissipative" indicates that the structures are dynamic and require dissipation of energy for their stability. Prerequisites for dissipative structure formation are that the system be thermodynamically open, governed by nonlinear processes, and far from equilibrium. Since living organisms exist far from equilibrium, constantly using energy, and creating and maintaining many sorts of structures, while in constant thermodynamic exchange with the environment, Prigogine's definition can be applied to a living organism; a dissipative, far from equilibrium, dynamically self-organized entity,[3] a notion first mentioned in Chapter 1.

Among the organized processes and events in biological systems proceeding from dissipative structures are:

i) *Bistability* is a property of many nonlinear differential equations and can manifest itself, among many other ways, as the change from one steady state to another steady state caused by changes in parameters controlling the process. Biological examples include excitable membrane states, and prion diseases which switch proteins from individuals into aggregates of other proteins. Platelet aggregation also switches individual platelets into irreversible aggregates at a threshold level of ADP, which they themselves produce, and hence produce more of when stimulated by ADP. This leads to a bistable transition based on positive feedback. Mutually inhibiting gene suppressors may act, via negative feedback, as a toggle switch between expression of different genes. Bistability has also been implicated in ecology, leading to switches between types of predominant vegetation.
ii) *Oscillations and rhythms.* Periodic fluctuations had been observed in several chemically reacting systems, first discovered by Boris Belousov (Russia, 1893–1970) in 1951. Not long afterward, it was discovered that there were periodic fluctuations in the concentrations of the intermediates in the glycolytic pathway of yeast, so-called *glycolytic oscillations.* Biological rhythms are spread over many time scales and phenomena. A few rhythms include neural and muscular (as fast as 0.01 s), cardiac (1 s), and glycolytic oscillations (10 s–10 min), circadian (24 h).
iii) *Spatial patterns* from dissipative structures were first predicted in 1952 by Alan M. Turing (England, 1912–1954). These may help explain spotted and striped coloration, chirality (handedness) in living organisms, and many other phenomena where fixed spatial patterns arise.

    An interesting case of dynamic spatial organization is the formation of convection cells in thin layers of heated liquid, such as water between two transparent plates, heated from below. The higher temperature at the bottom decreases water density making it buoyant and rising to the surface, while the denser water near the upper surface falls to the bottom. This results in vertical circulation upward and downward of liquid, which forms (often hexagonal) macroscopic cells, visible through the upper plane of glass. The process is deterministically chaotic, in that minute differences in starting conditions produce different patterns. This is sometimes termed *Rayleigh–Bénard convection.*
iv) *Waves* in the sense of spatial patterns that propagate or change in time occur in several known cases. The propagation of nerve impulses is the propagation of a wave shape through space with a certain speed. A *clock and wavefront* model is used to describe the process of somite segment propagation of *somites*[4] in embryonic development, whereby a feedback-controlled "clock" times the propagation of wavefronts that mark the segmentation of the somite.

3 In this definition, we may be able to exclude viruses as "living organisms," since they have no metabolism of their own, and their highly organized structures, such as the capsids, are equilibrium structures requiring no energy to maintain.

4 Somites are stem cells of segmented tissues that can give rise to cartilage, bone, muscle, and tendon in embryonic development.

14.5.1 Example of Stability and Instability in Simple Coupled Reaction–Diffusion Equations

A feel for the phenomenon of instability for coupled reaction-diffusion equations can be developed by considering a system where one reagent X, catalyzes the production of itself and is inhibited by reagent Y, and reagent Y both produces more X and consumes itself. Consider two cells, 1 and 2, in communication with each other with concentration of X and Y given by X_1 and Y_1, and X_2 and Y_2, respectively. One particularly simple representation of this situation (there are many other forms) that will allow for simple analysis is

$$\frac{dX_1}{dt} = (B-1)X_1 - BY_1 + 1 - D_x(X_1 - X_2) \qquad 14.22a$$

$$\frac{dY_1}{dt} = BX_1 - (B+1)Y_1 + 1 - D_y(Y_1 - Y_2) \qquad 14.22b$$

where $B > 1$, is a real number, and D_x and D_y are the diffusion coefficients of X and Y, respectively. The corresponding equation in cell 2 requires only a switch in the sign of the diffusive term:

$$\frac{dX_2}{dt} = (B-1)X_2 - BY_2 + 1 - D_x(X_2 - X_1) \qquad 14.23a$$

$$\frac{dY_2}{dt} = BX_2 - (B+1)Y_2 + 1 - D_y(Y_2 - Y_1) \qquad 14.23b$$

To simplify the system further, take the concentration of X and Y to be equal in both cells and normalized to unity, that is $X_0 =$, $Y_0 = 1$. This is the homogeneous state, which is stable. If there are concentration fluctuations such that the concentration of X and Y become slightly higher in cell 1 than their values in cell 2, then, because of the specific coupling of the equations above, the concentration fluctuation can either die back to the equilibrium state or proceed to grow unstably. Consider an arbitrary small concentration fluctuation ε that occurs for Y_1 and that a concentration fluctuation of $a\varepsilon$ occurs for X, where a is a real number. By mass balance, the fluctuations in cell 2 for X_2 and X_1 must be opposite, so that the situation is represented by

$$X_1 = 1 + a\varepsilon, \quad Y_1 = 1 + \varepsilon, \quad X_2 = 1 - a\varepsilon, \quad Y_2 = 1 - \varepsilon \qquad 14.24$$

Substituting this in Equation 14.22a gives

$$\frac{dX_1}{dt} = a\frac{d\varepsilon}{dt} = \left[a(B-1) - B - 2aD_x\right]\varepsilon \qquad 14.25$$

This equation will be unstable and yield a positive exponential if the rightmost term is positive and will be stable and X_1 will die back down to its equilibrium value if the rightmost term is negative. The boundary between stable and unstable behavior is when the latter term is zero, that is

$$a(B-1) - B - 2aD_x = 0, \quad \text{for stability} \qquad 14.26$$

Interestingly, the stability criterion is independent of the size of the fluctuation ε but depends on B, which corresponds to a specific reaction stoichiometry, a, which is the ratio of the perturbation strength in X and Y, and D_x.

14.5.2 Turing Patterns

Turing published a foundational work that showed how distinct spatial patterns could form based on fluctuations in initially homogeneous systems coupled by chemical reactions and diffusion. In his work, Turing also considered how the same principles of fluctuations in initially homogeneous systems could lead to the biological preference of one chirality, left- or right-handedness, over another.

Turing's basic idea was to take two or more connected cells, containing *morphogens*, and *activator* X and an *inhibitor* Y that can chemically react with each other and other agents in the cells, at different rates, as well as diffuse down their own concentration gradients from one cell to an adjacent one, with different diffusion coefficients. There are no convective processes in the system. If the X and Y concentrations in all of the cells are identical, without fluctuations, the system is stable and there will be no change. If, however, there are slight variations in the X and Y concentrations among the cells, then, depending on the details of the governing equations for the diffusion and chemical reactions, and associated constants, the system can become unstable and move away from the homogeneous state to a stable, spatially patterned final state.

More particularly, Turing considered the case where substance A is uniform, constant in time, and produces activator X. X can produce more X and also Y, but Y inhibits production of X, and also diffuses more quickly than X. X and Y can combine in a certain stoichiometric ratio to produce a product P. Think of two cells, 1 and 2, where, due to a fluctuation, concentration of X in 1 is slightly higher than in 2, and Y is at the same concentration in 1 and 2. X will build more rapidly in 1 and hence also Y. Since Y diffuses rapidly it diffuses into 2, where production of X is already slower than in 1 due to its lower concentration in 2. The diffusive influx of Y from 1 into 2 further slows the production of X in 2. Over time, the effect can add up, depending on system details, with X becoming more concentrated in 1 and less in 2 until a final state is reached with this final imbalance in X between 1 and 2. If many cells are connected – Turing used an example of a ring of cells – the end state will be a pattern of some cells having more and different amounts of X than others. Turing extended the treatment to continuous bodies, with similar results. An important aspect of the solution is that, depending on the governing equations, constants, and initial concentrations, there are certain spatial wavelengths of X that are favored, independent of spatial boundary conditions.

Turing suggested this process as an explanation for forms such as zebra stripes, leopard spots, fish and bird color patterns, formation of spirals, and other marked patterns. Turing also suggested it could apply to embryonic development, starting with a spherically symmetric zygote which then undergoes asymmetric morphogenesis according to these reaction-diffusion principles, to produce nonspherical plants and animals. Embryology has since then revealed that it does not work this way. Rather, there is an asymmetric distribution of mRNA at points on the egg's inner membrane from the start, which leads immediately to protein gradients once the mRNA begin their synthesis, and gradients of different proteins give 3D positional information to cells in a developing embryo, so there is no need to invoke the spontaneous spherical symmetry breaking in Turing's theory to understand asymmetric embryonic development.

In spite of this, there is evidence of Turing processes in the other biological processes mentioned above, and his seminal work has inspired legions of experimental, theoretical, and computational investigations, still being vigorously pursued, as the space of variables and process producing and affecting Turing patterns widens. For example, adding growth and growth rate of the system as variables increases the range of parameters contributing to pattern formation, as well as adding history-dependence to the final patterns.

Turing's paradigm can be represented by *reaction-diffusion equations* for the concentrations C_j of N species of the form:

$$\frac{\partial C_j}{\partial t} = D_j \nabla^2 C_j + f_j(C_1, C_2, \ldots, C_N) \qquad 14.27$$

The first term on the right-hand side, which involves D_j is the diffusion coefficient of species j and imparts the spatial properties to the solution. The nonlinear functions f_j are added to the diffusion equation and contain the chemical reaction rates and mass action properties of the system.

The diffusion equation alone (omitting the f_j term in Equation 14.27) leads to a smoothing out of concentration differences in a system and a final uniform steady-state concentration of components. The addition of the reaction terms f_j opens an entire new physical and mathematical front in the form of *bifurcation theory*. In very simplified terms, a system may have a homogeneous steady-state that obeys Equation 14.27, but it may or may not be stable to changes in its governing parameters.

Despite the fascination with Turing patterns and the wealth of new physical and mathematical results it inspired, it wasn't until 1990 (Castets et al.) that experimental evidence for actual temporally stable, spatial Turing patterns in a chemical system was demonstrated, with wavelengths dependent only on reaction/diffusion parameters, and not on the systems spatial boundary conditions. The system was quite elaborate, consisting of a long, narrow, essentially 1D polyacrylamide gel, fed with the reagents iodine, chlorine dioxide, and malonic acid, with a color indicator added for visualization of the patterns.

The breaking of spherical symmetry is akin to the problem cosmologists face when trying to rationalize how an initially homogeneous "Big Bang" could lead to an observable universe full of heterogeneous structures, such as galaxies and clusters of galaxies. Turing's insights into chirality in biological systems harkens to the problem of how matter came to dominate the known universe rather than antimatter.

14.5.3 Fibonacci Sequence and Patterns in Nature

This integer sequence is formed by adding the previous two numbers in the sequence so that the nth term F_n is given by

$$F_n = F_{n-1} + F_{n-2} \quad 14.28$$

If the sequence starts with $F_0 = 0$ and $F_1 = 1$, then it develops as 0, 1, 1, 2, 3, 5, 8, 13, 21, 34, 55,.... It is an interesting sequence that appears in many ways in nature. A good example is the x chromosome for a male, which he inherits from his mother. The number of ancestors that contribute to the male's x chromosome come only from the maternal lineage. If the male is taken as $F_1 = 1$, then $F_2 = 1$, since only the mother supplies the x chromosome. The mother receives an x from both her son's grandparents so that $F_3 = 2$, and the grandmother receives two xs from the great-grandmother and one x from the great-grandfather so that $F_4 = 3$, and so on . Figure 14.14 illustrates this, where M = mother and F = father.

The Fibonacci sequence is closely rated to certain patterns in nature, such as the arrangement of leaves on some plants and patterns in flowers. It is also related to the so-called *golden ratio*, which has been known since ancient times and is sometimes considered a naturally occurring irrational number of the stature of π. The simplest means of constructing the golden ratio is to consider a line segment consisting of two smaller segments a and b, where $a > b$. The golden ratio φ, is defined when a and b are such that

$$\varphi = \frac{a+b}{a} = \frac{a}{b} = 1.61803399\ldots \quad 14.29$$

as can be found from the associated quadratic equation

$$\varphi^2 - \varphi - 1 = 0 \quad 14.30$$

The golden ratio appears in an astonishing number of situations, including triangles, rectangles, icosahedrons, spirals, *Penrose aperiodic tilings*, certain solutions of the Γ function, and many more. The limit of the ratio of successive numbers in the Fibonacci sequence converges to φ. That is

$$\lim_{n\to\infty} \frac{F_{n+1}}{F_n} = \varphi \quad 14.31$$

If the circumference of a circle is broken into two arcs of length a and b, where $a > b$ and $b/a = \varphi$, then the angle subtended by arc a is the *golden angle* and is equal to, in degrees

$$\text{Golden angle} = 180(3-\sqrt{5}) = 137.5077\ldots^{\circ} \quad 14.32$$

Some flowers, such as sunflowers, arrange their clusters of tiny flower in their heads separated by the golden angle.

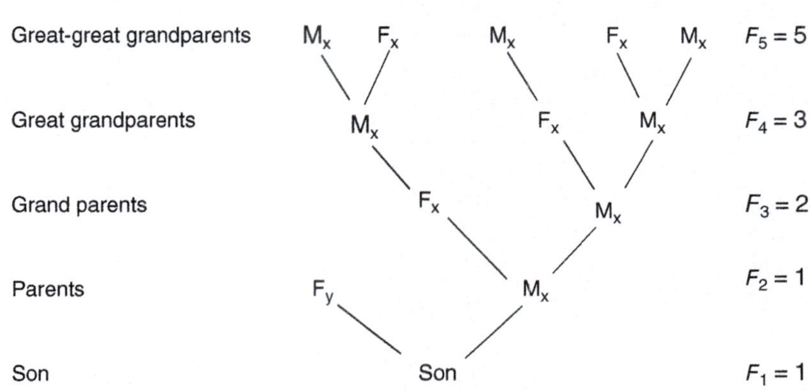

Figure 14.14 The number of ancestors contributing to a male x chromosome follows the Fibonacci sequence.

14.6 Summary

In this chapter, a tiny corner of cellular level and higher functions was introduced. Biomacromolecules and small molecule agents come together to exploit light, both as an energy source in photosynthesis and for forming images of the world. Evolution has found different strategies for the latter, including compound eyes, eyes that focus by moving the lens with respect to the retina, and eyes which change the shape of the lens to form images on the retina. Apart from the anatomical structure of the eye required to form images, the actual electrochemical events in transducing the reception of photons into nerve impulses are exceedingly complex, and the manner in which the impulses are communicated to the brain and decoded and interpreted by it are even more complex.

Biomotion starts at the microbial scale with flagella and cilia and increases in complexity as the scale of the animal organism grows. Biomechanics of vertebrates is an extensive field in itself.

Simple predator–prey cycles give insight into the oscillations in ecosystems. This leads into the conceptualization of the living state as being rooted in dissipative, far from equilibrium, spatially self-organizing systems. Coupled predator–prey relationships that take into account environmental carrying capacity for prey and satiation levels for predators can often lead to stable limit cycles, which are reached in such systems, regardless of the initial abundances of prey and predator.

Coupling linear differential equations for various interdependent variables can lead to deterministic chaos, in which slightly different starting parameters eventually lead to highly divergent outcomes. Parameter dependence in such coupled system is so strong that changes in these can lead from chaos to stable limit cycles and other types of attractors. Turing showed how initially homogeneous mixtures of reagents can form spatially periodic patterns due to combining reactions with diffusion.

References and Further Reading

Vision

Ernst, O.P., Lodowski, D.T., Elstner, M. et al. (2014). Microbial and animal rhodopsins: structures, functions, and molecular mechanisms. *Chemical Reviews* 114: 126–163. https://doi.org/10.1021/cr4003769.

Fatt, I. and Weissman, B.A. (1992). *Physiology of the Eye*. Butterworth-Heinemann.

Jenkins, F.A. and White, H.E. (1976). *Fundamentals of Optics*. McGraw Hill.

Slingsby, C. (1999). Structure of the crystallins. *Naomi J. Clout. Eye* 13: 395–402.

Smith, S.O. (2010). Structure and activation of the visual pigment rhodopsin. *Annual Review of Biophysics* 39: 309–328.

Theoretical Biology: Cycles and Instabilities

Bier, M., Bakker, B.M., and Westerhoff, H.V. (2000 Mar). How yeast cells synchronize their glycolytic oscillations: a perturbation analytic treatment. *Biophysical Journal* 78 (3): 1087–1093. https://doi.org/10.1016/S0006-3495(00)76667-7.

Castets, V., Dulos, E., Boissonade, 3., and De Kepper, P. (1990). Experimental evidence of a sustained standing Turing-type nonequilibrium chemical pattern. *Physical Review Letters* 64 (24): 2953–2957.

Goldbeter, A. (2018). Dissipative structures in biological systems: bistability, oscillations, spatial patterns, and waves. *Philosophical Transactions of the Royal Society A* 376: 20170376. https://doi.org/10.1098/rsta.2017.0376.

Konow, C., Somberg, N.H., Chavez, J. et al. (2019). Turing patterns on radially growing domains: experiments and simulations. *Physical Chemistry and Chemical Physics* 21: 6718.

Lander, A.D. (2011 March 18). Pattern, growth, and control. *Cell* 144.

Lengyel, I. and Epstein, I.R. (1991 Feb 8). Modeling of Turing structures in the chlorite-iodide–malonic acid-starch reaction system. *Science* 251 (4994): 650–652.

Turing, A.M. (1952 Aug 14). The chemical basis of morphogenesis. *Philosophical Transactions of the Royal Society of London. Series B, Biological Sciences* 237 (641): 37–72.

General Mathematical Biology

Allen, L. (2006). *Introduction to Mathematical Biology*. Pearson Press.

Earnest, T.M., Cole, J.A., and Luthey-Schulten, Z. (2018). Simulating biological processes: stochastic physics, from whole cells to colonies. *Reports of Progress in Physics* 81: 52601–52634.

Galovin, A.A. and Nepomnyaschy, A.A. (2004). Self-assembly, pattern formation, and growth phenomena in nano-systems. *NATO Science Series, II. Mathematics, Physics, and Chemistry* 218.

Henson, S.M. and Hayward, J.L. (2023). *Mathematical Modeling in Biology*. Boca Raton, London, New York: CRC Press Taylor&Francis Group. Mathematical Biology Series.

Rubin, A. and Riznichenko, G. (2014). *Mathematical Biophysics*. Springer.

Other

Baravalle, H.V. (1948). The geometry of the pentagon and the golden section. *Mathematics Teacher* 41: 22–31.

Bari, K.J. (2021). The structural biology of crystallin aggregation: challenges and outlook. *FEBS Journal* 288 (20): 5888–5902.

Earnest, T.M., Cole, J.A., and Luthey-Schulten, Z. (2018). Simulating biological processes: stochastic physics from whole cells to colonies. *Reports on Progress in Physics* 81 (5): 052601/1–052601/33.

Mizutani, Y. (2021). Concerted motions and molecular function: what physical chemistry we can learn from light-driven ion-pumping rhodopsins. *Journal of Physical Chemistry B* 125 (43): 11812–11819.

Prigogine, I. and Rice, S.A. (1998). *Advances in Chemical Physics*. John Wiley and Sons.

Prigogine, I. and Stengers, I. (1984). *Order Out of Chaos*. Shambhala Publications.

Ridley, J.N. (1982). Packing efficiency in sunflower heads. *Mathematical Biosciences* 58 (1): 129–139.

Steen, L.A. (1988). The science of patterns. *Science* 240 (4852): 611–616.

15

On the Frontier

As our circle of knowledge grows, so grows the perimeter of darkness surrounding it.

Albert Einstein

Virtually all areas of biophysics are open to ongoing discovery and controversy, but some areas more than others. In this final chapter just a few of the still unsolved and contentious issues are briefly discussed. Some areas, such as the prebiotic origins of life, and the origins of consciousness have no full consensus or conclusions at this time. How deeply and widespread purely quantum mechanical phenomena might be in the living state is currently poorly understood, but there is vigorous activity and experimentally viable means of testing hypotheses in this area. The question of the existence of extraterrestrial life still looms, but may soon be answered due to the relatively recent discovery of exoplanets and refinements in characterizing them and their atmospheres, while space missions, such as the Webb Space Telescope launched in 2022 allow us to observe exoplanet atmospheres and probe stellar and galactic formation in the early Universe.

15.1 Prebiotic Origins of Life

The cosmos is within us. We are made of star-stuff. We are a way for the universe to know itself.

Carl Sagan

Fossil and other evidence suggest that cellular life on Earth began between 3.2 and 3.9 billion years ago. The Earth is estimated to be around 4.5 billion years old, and prebiotic evolution is presumed to have occurred in the billion or so years between Earth's formation and the appearance of cellular life.

There is still no definitive, widely accepted theory for how life arose on Earth, but an enormous amount of experimental and conjectural work has been done, and the topic continues to be controversial, intensely researched, and hotly debated. It inevitably bumps into highly speculative and sometimes teleological and even theological territory when one questions the ultimate origins of matter and its properties, even if plausible and fully natural pathways for evolution are found. In these times, strident voices from many different perspectives can be heard: strict reductionists, antireductionists, evolution as teleology, evolutionary theology, creationism, intelligent design, each earnestly stumping for their viewpoint. Darwin stated that "life appeared by some wholly unknown process." Bohr proposed that life follows all the laws of physics but cannot be deduced from those laws.

While cosmology itself is beset with bedeviling problems, as mentioned in Chapter 1, the observational and theoretical astrophysical bases for the birth and life cycle of stars are fully plausible and reasonably well understood, including the various nucleosynthetic cycles that build up elements and distribute them through stellar supernova explosions. Massive stars (over two orders of magnitude larger than a solar mass) are extremely hot and short-lived and incubate the heavier elements that became distributed in these explosions. The cores of these massive stars become neutron stars or black holes after the supernova. Smaller stars, such as our sun, burn cooler and have a much longer lifetime than these massive stars, and a less dramatic death. The terrestrial planets (Mercury, Venus, Earth, and Mars) are largely accretions of heavier elements, where "heavier elements" means any more massive than hydrogen and helium, which account for the vast majority of baryonic matter in the known universe. It is hence often said that "we are stardust," made of matter forged in staggeringly hot stellar anvils.

Biophysics: Physical Processes Underlying the Living State, First Edition. Wayne F. Reed.

Companion website: www.wiley.com/go/Reed/BiophysicsandMacromolecules

Some of the scientific debate concerning the prebiotic origins of life has traditionally centered around "what came first?" Autocatalytic processes, a sort of prebiotic metabolic cycle, in which the product of the reagent is more of the reagent? Development of genetic material that can self-replicate and store information? Enclosure of reactants into sequestered environments, such as protocells, that could provide concentration of molecules and accelerate reactions leading to life? Did proteins come first? RNA? Did life come from outer space (so-called *panspermia*)? Life may have even evolved in hot, alkaline under-sea thermal vents, or in regions directly receiving solar radiation, or both.

15.1.1 The "RNA World Hypothesis"

A leading conjecture is that RNA may have been the precursor polymer that started evolution, because it has both self-replicating and catalytic properties, i.e. this single molecule may predate the molecular classes of self-replicating, information storing DNA, and the catalytic proteins. This is sometimes referred to as the "RNA World Hypothesis" (RWH). Early RNA in this concept was then the precursor of both proteins (for catalysis) and DNA (for storage and transmission of information). Many experiments have been done to demonstrate synthesis of amino acids under conditions that might be similar to atmospheric conditions in the prebiotic Earth. More problematic has been prebiotic synthesis of RNA, but plausible synthetic pathways have been found from prebiotic precursor molecules, such as glycoaldehyde, urea, and cyanamide.

In the RNA model, fatty acids in the environment could form prototype lipid bilayer vesicles, or liposomes, which could encapsulate the RNA, thus accelerating replication reactions and allowing selective transport of molecules in the environment through such early membranes. It has been proposed that the most successful versions of RNA would replicate most quickly, fill vesicles and, due to internal osmotic stress would be able to absorb membrane material from liposomes containing less successful RNA. It is a physical property of such vesicles that as their size increases they go through morphological phase transitions (e.g. elongation) that can lead to their fission into daughter vesicles, i.e. these conjectures posit both an early form of natural selection by variation, in this case of the more successful, highly replicating versions of RNA, and reproduction via division of vesicles which physically split as their successful RNA content drives them to accrete more lipids in the protocell. Figure 15.1 is an illustration of this.

Despite its popularity, the RWH has met with substantial objections. Harold Bernhardt in 2012, for example, published a provocatively titled article "The RNA world hypothesis: the worst theory of the early evolution of life (except for all the others)." Bernhard's objections include

i) "RNA is too complex a molecule to have arisen prebiotically
ii) RNA is inherently unstable
iii) Catalysis is a relatively rare property of long RNA sequences
iv) The catalytic repertoire of RNA is too limited"

We have already seen the statistical improbability of randomly creating specific sequences for even small proteins (Chapter 4.19.2), and the same statistical barrier applies to the formation of specific sequences of RNA. Furthermore, RNA is inherently unstable and can hydrolyze fairly quickly under both acid and alkaline conditions. As far as catalysis, very small sequences of RNA have shown some limited, very specific catalytic activity, but proteins vastly overshadow RNA in their diverse catalytic capabilities.

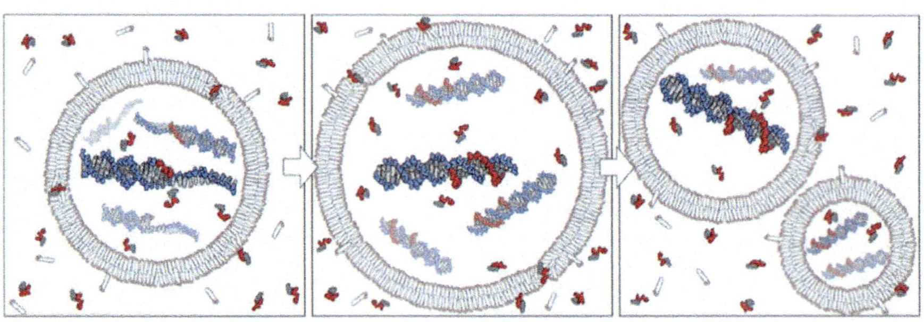

Figure 15.1 *Conceptual model of a heterotrophic protocell.* Growth of the protocell membrane results from the incorporation of environmentally supplied amphiphiles, whereas division may be driven by intrinsic or extrinsic physical forces. Externally supplied activated nucleotides permeate across the protocell membrane and act as substrates for the copying of internal templates. Complete template replication followed by random segregation of the replicated genetic material leads to the formation of daughter protocells as the protocell enlarges and fissions. (With permission from *Springer Nature*, from Mansy, S.S. et al. (2008)).

Further skepticism, rather scathing, of the RWH was enunciated by Charles Carter (see Berhnhardt 2012):

> I, for one, have never subscribed to this view of the origin of life, and I am by no means alone. The RNA world hypothesis is driven almost entirely by the flow of data from very high technology combinatorial libraries, whose relationship to the prebiotic world is anything but worthy of "unanimous support." There are several serious problems associated with it, and I view it as little more than a popular fantasy.

Because of the limited catalytic activity of RNA, it has been suggested that proteins came first. We recall from Chapter 13.5 that gramicidin is an example of a nonribosomal peptide that doesn't use DNA or RNA in its synthesis; rather, it is formed by nonribosomal peptide synthetase. This large class of peptides use a wide variety of amino acids, as many as 900, far more than the 20 proteinogenic amino acids. The synthetases, however, are themselves produced via DNA, and each is limited to producing a single peptide.

Other arguments against protein-first include how the proteins would then somehow switch to RNA-dependent synthesis, and the possibility of finding such archaic proteins may be veiled by the vastness of proteins produced genetically. An even greater objection comes from Crick's "central dogma," which states that information can flow from DNA/RNA to proteins but not in the reverse direction. As mentioned in Chapter 4.19.4, there is a strong information theory case to be made against proteins arising first.

15.1.2 Alternate Prebiotic Paths

The uniqueness of prebiotic paths is also in question. For example, bacteria from very hot deep-sea hydrothermal vents may have evolved independently of anything occurring on the Earth's surface and in its atmosphere. There are early microenvironments in these vents that may have concentrated and catalyzed life's early molecules. Some speculate that the life forms found at these vents are the origin of life on Earth. The instability of RNA under the high alkalinity of these vents, however, would not favor their formation.

There is an entire school of thought that posits not only these microenvironments but also those found in exfoliations of clay throughout the Earth's surface may have acted to concentrate reactants and catalyze early reactions.

While Panspermia – the notion that life came from outer space – is not widely considered a contender in the prebiotic origin of life, because it just shifts the puzzle of prebiotic evolution to another planet(s) or celestial object, there are a growing number of organic molecules detected in meteorites, interstellar dust, comets, and around stars. The celestial ubiquity of these many molecules has led some to suggest that such organic molecules – possible precursors of amino acids, nucleic acids, and other biomolecules – may commonly initiate prebiotic and subsequent biotic evolution on other planets. Extraterrestrial organic molecules detected include *polycyclic aromatic hydrocarbons*, uracil, glycine, xanthine, glycoaldehyde, ethanamine, and others. Work in this field is variously termed *cosmochemistry, astrochemistry*, and *space chemistry*.

Many tests have been made concerning the survivability of microorganisms in outer space environments. Candidates for survival under such conditions include extremophiles (Chapter 14.1.1), lichen, spores, and possibly even the tiny animals known as *tardigrades*. These latter can survive extreme conditions such as temperatures down to near absolute zero, up to 150°C, dehydration, vacuum, pressures over 1000 atm, and levels of ionizing radiation that kill virtually all other species. They are able to go into a state of *cryptobiosis*, a state where metabolic activity all but stops.

Living functions always exist within a cell, or collection of cells. The cell membrane separates the cell from its environment, concentrates, organizes, and packages all the biomolecules and associated processes. Some possibilities for packaging are:

- Natural microcavities and environments, such as the highly catalytic surfaces of clay and fissured minerals provided early "packaging" without there being actual cell membranes present. These latter would then develop subsequently.
- Simple lipids and amphiphilic molecules developed early on, along with other simple biomolecules, and formed cellular compartments through spontaneous self-organization. It is well established that micelles and liposomes can form spontaneously from different classes of relatively simple amphiphilic molecules.
- Another variation on self-organizing microenvironments is the notion of coacervate droplets (held together by attractive forces, such as electrostatics of oppositely charged ingredients) turning into compartments for early macromolecules.

15.1.3 Origin of Homochirality

Other difficult issues in finding plausible pathways for evolution include the fact of homochirality in biological molecules, i.e. the fact that only one chirality is found among each biomolecule class; virtually all proteins have L-amino acids

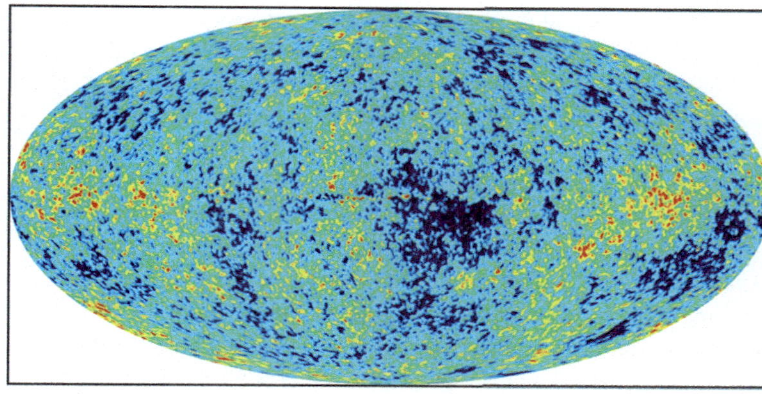

Figure 15.2 Temperature fluctuations in the Cosmic Microwave Background. (NASA / WMAP Science Team / http://map.gsfc.nasa.gov) last accessed under 06 May, 2023.

(although mixed amino acid chiralities exist in non-ribosomal peptides), and sugars are of the D-form. In normal synthetic processes, a *racemic mixture* is produced in which both chiral forms are found in equal number. An *enantiomeric excess* of one chiral form (enantiomer) over another is required to eventually select out a single enantiomeric form in subsequent evolution. The problem is finding a mechanism that might yield even a slight enantiomeric excess early in evolution that would select for that form. There have been reports of a slight amount of circular polarization for sunlight reflected from planetary atmospheres,[1] and it has been speculated that a slight net circular polarization of one handedness could favor the abundance of one type of chiral molecule over its enantiomer.

This is similar to the problem of baryogenesis in cosmology; why is the baryonic portion of our detectable universe made of matter rather than antimatter? Matter and antimatter are symmetric in all their properties, except that charged antiparticle have the opposite charge of their corresponding particles. Antimatter can form atomic systems in the same way that matter does; e.g. *antihydrogen* with an *antiproton* nucleus and positron instead of electron. Were matter and antimatter present in equal amounts in the early universe and some slight imbalance was somehow created that led to the predominance of matter?

These quandaries also bear some resemblance to the problem of inhomogeneity in cosmological structure. There is no *a priori* reason that the distribution of matter in the universe should not be completely symmetric and evenly distributed. Yet, all astrophysical observations indicate an extremely inhomogeneous distribution into widely separated clusters of galaxies, and numerous subgalactic structures, such as stellar clusters and planetary systems. The Cosmic Background Explorer satellite (COBE) launched in 1989 showed faint but detectable anisotropy in the cosmic microwave background, a remnant of the Big Bang. This is interpreted to mean there was an anisotropic imprint on the matter and energy of the universe at a very early stage, perhaps due to early quantum fluctuations. Figure 15.2, taken from the COBE, illustrates fluctuations in the cosmic microwave background.

Returning to the origin of homochirality, it has often been noted that the "primordial soup" experiments, such as the Miller–Urey experiments, in which precursor biomolecules, such as amino acids and sugars, produce racemic mixtures of these chiral molecules. H. Yockey (2005) states "No natural chemical process exists to select and produce a single chirality. That is a property unique to life."

15.1.4 Information Theory Objections to a Strictly Physics/Chemistry Origin of the Genetic Code

The Shannon entropy as a measure of information was introduced in Chapter 4.19.4, and defined by Equation 4.224. The higher the Shannon entropy the higher the information content. In information theory, there is a natural hierarchy whereby a system of high Shannon entropy can transmit accurate information to a system of equivalent or lower Shannon entropy, but the reverse cannot occur. It was mentioned in Chapter 11.2 that the genetic code with 64 codons has more information than the 20 amino acids and stop signal it codes for. This fact has been used by information scientists to deny the possibility that RNA/DNA and the genetic code arose from proteins during prebiotic and early evolution.

Furthermore, given an alphabet, such as the genetic code, it is impossible to "predict" all the possible meaningful words made from that alphabet, which extend the alphabet to complex words – given the alphabet and grammatical rules of a human language, it is impossible to predict all the words, sentences, poems, and literature that these will produce, but all such produced just use the alphabet and obey associated grammatical rules. In this sense, it is impossible to predict the

---

1 Hansen, J. E. (1971). *Journal of the Atmospheric Sciences 28*: 1515.

emergence of life and its properties from the laws of physics, because these latter actually have a relatively small Shannon entropy, and the information required for even the simplest living entity is far greater than that contained in the laws of physics. H. Yockey (2005) stated "the sequence hypothesis and Genetic Code are in addition to but do not transcend the laws of Physics and Chemistry." Furthermore, unlike most physical structures – planets, mountains, oceans, some stars – genetic messages have been shown to survive for nearly 4 billion years.

The information theorist also takes the relationships of self-organized matter versus random matter in a way not always recognized by those not in that field. From an information theory perspective, it was stated in Chapter 11.2 that organized structures made of a material actually contain less information than an equivalent amount of random material of that type. For example, a crystalline material can be defined in terms of a repeating unit cell, describable by a small amount of information. In contrast, randomly moving particles of the same amount material are much more complex and require a comparatively huge amount of information, since the location of each particle must be specified, and the information changes in time as the particles move about. However, the Shannon Entropy flow would permit the disorganized, even chaotic, high information disorganized system to produce more organized structures requiring less information. This adds support to Prigogine's notion that life involves the formation of organized structures from far from equilibrium systems, because these latter contain a huge amount of information.

15.1.5 The Entscheidungsproblem[2] and Relationship to the Origin of Life and the Genetic Code

Many people, starting with Gottfried Wilhelm Leibniz (Germany, 1646–1716) have pondered some form or another of this problem. A simplified means of expressing it is to ask whether, given a logical statement, there is a yes or no answer as to the validity of the statement within a given axiomatic system. People who expressed the problem and answers to it in various forms include David Hilbert (Germany, 1862–1943) and Wilhelm Ackerman (Germany, 1896–1962) who formally posed the challenge in 1928, and the many responders and original thinkers who pondered the issue, including Alonzo Church (USA, 1903–1995), Alan Turing, Kurt Gödel (Czechia/USA, 1906–1978), and many others. While it was long thought that there is a "yes" or "no" answer to every question posed in an axiomatic system, it turns out, by a number of different proofs, that there are questions which are *undecidable* within the axiomatic system.

An example of this is to consider an infinite sequence of digits. The fraction 1/3 is an infinite decimal expression, but a very simple algorithm which captures the sequence is "the digit 3 repeats infinitely." Another example is $17/39 = 0.435897435897435589$. This can be captured with the simple algorithm "Repeat the sequence 435897 infinitely." However, the values of the irrational numbers π and the natural exponent e have sequences with no discernible pattern, so that each digit in them requires information, making it necessary to have an infinite amount of information to specify their respective infinite series. However, π and e both have series representations which can be reduced to fairly simple algorithms, making the information to fully define their infinite series quite small, so that their series are not just random numbers. In the case of an infinite, purely random sequence, with no simple algorithmic description, an infinite amount of information is needed. Gregory Chaitin (1987) proved that it is fundamentally undecidable whether any sequence with no obvious pattern is random or not. This is an example of undecidability within an axiomatic number system.

Undecidability can present itself in any system, meaning the answers to many questions may be unavailable to human knowledge. Unknowable in no way implies "impossible": While it may be unknowable whether an infinite sequence is random or not, the sequence still exists and is not "impossible." The three-body gravitational problem can only be solved exactly in only very limited cases, yet multiple celestial bodies go about their gravitationally determined motions. In math, there are no explicit solutions for polynomial equations of fifth order and higher, yet solutions exist and can be found numerically.

This undecidability may include the system containing the laws of physics. In the context of prebiotic evolution, the answers to questions about how life arose and how the genetic code arose may be unknowable within the framework of current physical science. This, of course, is denied by the many researchers seeking the origins of life.

15.2 Quantum Biology

Inasmuch as all chemical reactions are quantum mechanical in nature, the living state is based in quantum phenomena. Quantum chemistry is a highly evolved field that began shortly after the formulation of the Schrödinger wave equation in 1926. In 1927, Walter Heitler (Germany, 1904–1981) and Fritz London (Germany/USA, 1900–1954) applied the wave

2 German for the "decision problem."

equation to the diatomic hydrogen molecule. Since then, there have been several fruitful approaches, including molecular orbital theory, valence bond theory, and density functional theory. As in classical physics, however, where the three-body problem is analytically tractable only under certain limiting conditions, the same holds in quantum physics, so that even the helium atom, with a nucleus and two electrons, cannot be solved exactly.

But the term "quantum biology" is usually taken to mean more than the underlying quantum chemistry and seeks biological phenomena where distinctly quantum mechanical phenomena, such as *superposition of allowed states, tunneling, quantum coherence, and entanglement* may play a definitive role which sets them apart from a classical explanation, i.e. processes that don't occur in the classical limit formalized in the correspondence principle.[3] Although all chemical reactions are quantum mechanical in nature, the classical thermodynamic tools of free energy, entropy, etc. are normally sufficient to quantify enzymatic and other biochemical reactions, without having to invoke the quantum mechanical underpinnings.

The current situation in quantum biology might be characterized as both controversial and tantalizing. Some are skeptical of quantum mechanical manifestations in living state systems, which are "wet, noisy, warm, and open," the antithesis of the cold, closed, and low-noise environments in which quantum effects, such as superconductivity, superfluidity, and Bose–Einstein condensation can be observed. It is often argued that the open nature of living state systems, with constant fluxes of energy and material from the environment, would "wash out" any delicate quantum mechanical effects. Despite these arguments, much research has been done and there is both experimental and theoretical support for quantum biology.

Some of the main areas of quantum investigation include energy transfer and charge separation in photosynthesis, enzyme catalysis, sensing processes, such as magnetoreception and olfaction, and possible quantum mechanical roles in cognition and the origin of life. Other phenomena that might implicate quantum mechanical effects include possible tunneling in DNA mutation and repair, quantum coherence might have a role in protein folding, single photon reception by retinal cells, quantum coherence and entanglement might be involved in neural processing, and quantum information theory may bear on storage, transmission, and protection of genetic information.

In the case of photosynthesis, for example, there is a fundamental question of whether there is quantum coherence among exciton states in photoreceptive pigments that makes their quantum efficiency amazingly close to perfect. Or is it simply the proximity of the pigments to each other? Or, to what degree does the quantized nature of molecular vibrations couple into energy and charge separation processes that can't be glossed over with a classical model? Whatever the answers, claims are made that the results of quantum studies and theories of photosynthesis can be used to guide artificial photosynthetic devices, another example of Biomimetics.

In enzyme catalysis phenomena transition states and energies have long been used to conceptualize reaction kinetics, but an understanding of detailed mechanism does not necessarily result. It has been proposed that tunneling may be important in many enzymatic processes, especially those involving electron and proton transfer over "long distances," typically more than 0.5 nm, or so.

15.2.1 Spin Radicals as a Possible Quantum Mechanical Origin of Magnetoreception

Magnetoreception in birds, bacteria, and many other animals has long intrigued scientists. To-date, no definitive magnetic field-sensing receptor has been identified. In some bacteria, particles of magnetite have been found which align in Earth's magnetic field, essentially turning a bacterium into a classical magnetic dipole. Magnetite and iron in homing pigeons that interact with the Earth's magnetic field may relate, at least partially, to their ability to navigate. Certain aquatic animals, such as sharks and some stingrays, have a DC current and electric field sensing receptor, termed the *ampullae of Lorenzini*. It is thought that they can also use the classical law of electromagnetic induction (Equation 9.6) to sense changes in magnetic fields.

A quantum alternative, or addition to the above mechanisms of magnetoreception, involves conjectures involving *spin radicals*. In these, a photon of light is absorbed by a photoreceptor which donates an electron to an electron acceptor molecule, thus forming a radical pair. While spatially separated the two radicals are quantum entangled, that is, they share a wave function and the state of each is mutually determined. Because the radicals were originally bound together via opposite binding electron spins the two correlated electrons on the two radicals are in a *singlet state*, and can rapidly recombine into the original molecule(s), or proceed to a chemical product. If a magnetic field is present, however, the interaction of the electron spins with the field (the Zeeman effect) can cause spin flips to occur, aligning spins and producing *triplet states*. There is then a resonance between singlet and triplet states, the balance of which depends on the magnitude of the

3 The Correspondence Principle states that in the limit of very large quantum numbers a quantum mechanical system behaves as a classical, nonquantum mechanical system.

magnetic field. Triplet states cannot directly combine back to the ground state, and so are available to form a product, and this product may be different than the product formed by singlet radicals in equilibrium with their product. It is surmised that in magnetoreception the organism may distinguish the magnitude and direction of the magnetic field from the distribution of singlet and triplet products. The effect occurs in even weak fields, since the energy differences between the single and triplet state are typically small, so that even a magnetic field as weak as that from the nucleus, which causes electronic hyperfine structure, can influence chemical reactions. The Earth's field is also quite weak, with an average around 5×10^{-5} T, but sufficient to drive the radical pair mechanism.

The class of proteins that posits a quantum explanation via the radical pair mechanism to explain magnetoreception is the *cryptochromes*, which are found in plants and animals, and control circadian rhythms and other timing functions, in addition to their putative role in magnetoreception. Cryptochromes absorb blue and near-ultraviolet light, which provide the photon needed to activate a donor molecule into donating an electron to an acceptor molecule. Cryptochromes have been found in the retinas of certain birds which magneto-navigate.

There have been conjectures and experiments concerning the possible role of the radical pair mechanism in cryptochrome control of circadian rhythms, polarization and protection in photosynthesis, and certain enzymatic processes, such as DNA repair, and neural development.

15.3 Neuroscience and the Question of Consciousness

> *Wer reist bringt sich mit.*
>
> Anon

Neuroscience deals with all aspects of the nervous system, including those of all species having one. It may be considered one of the most dynamic forefronts of current living state research and it is highly interdisciplinary. While it encompasses all neural phenomena, including sensors and peripheral nervous activity, it is safe to say that the most intense efforts are centered on the brain and understanding the biophysics of brain activity and how and whether these relate to mental processes of learning, memory, communication, illness, and so on. While there is evidence that ancient cultures, such as the Egyptians, knew that the brain was the center of intelligence and consciousness, it was also proposed that the heart was the center of these functions. Aristotle, who, as we have seen, proposed erroneous thought concerning both atomic theory and evolution, struck out once again in this area, maintaining the heart to be the organ of intelligence, and the brain to be a heat regulation organ.

Neuroscience is engaged at so many levels and is dealing with so many questions and problems that it would be foolhardy to attempt a review here. At the molecular and cellular level, we have already glimpsed a bit of the physics behind nerve transmission and the electrochemistry of synaptic transduction in Chapter 13. While intense work in these areas is still ongoing, large efforts are underway in understanding and modeling neural circuits, which has also given birth to the field of *artificial neural networks* and *deep learning*. Brain neuroscience takes many different approaches; brain imaging by several methods, positron-electron annihilation tomography (PET), electroencephalography (EEG), and especially nuclear magnetic resonance imaging (MRI). Some of the most important data revealed by imaging is glucose and oxygen consumption, and blood flow rate in the different parts of the brain. Because noninvasive imaging is relatively recent, mechanistic brain science is still in its infancy, and many of the functions of the vast number of parts and subsystems of the brain are still unknown and often speculative. In contrast, the purely anatomical features of the brain and all its substructures and composition are known in great detail.

While humans have decided we are the most intelligent creatures on the planet, it is important to note that other animals have larger brains – elephants, dolphins, and whales among them – and some even have more convolution and folding, which corresponds to higher levels of information processing. Many whales and dolphins have highly evolved auditory powers and are able to communicate with each other at human audible frequencies, and to use ultrasound (above 20 kHz) for echolocation, allowing them to both evaluate and navigate their environment. It is thought that a significant part of their large brains is involved in echolocation, but they are also capable of advanced learning, problem solving, long-term memory, and emotion. It has been claimed that they have a sense of their own mortality. They are highly social and spend time nurturing and teaching their offspring. Project CETI (Cetacean Translation Initiative), begun in 2020 and supported by a wide range of universities and nonprofit organizations, focuses the powers of machine learning and artificial intelligence in an attempt to decode what might be a sort of language in the complex, audible sounds that sperm whales make. Because their brains are so large and complex, and because they live in an aquatic, rather than terrestrial world, some researchers have compared the working of their brains to a form of "alien intelligence."

15.3.1 A Short Tour of the Human Brain

Figure 15.3 is an example of an MRI image of the human brain.

A partial, labeled panorama of the human brain is shown in Figure 15.4. The figure captures only a small part of the brain's complexity and intricacy but some of the areas controlling major functions are represented and can be compared to the actual MRI of Figure 15.3. The outer portion covering the entire brain is the *cerebral cortex* (not specifically labeled in the figure), composed of *gray matter*, which is rich in neuron cell bodies, synapses, and unmyelinated axons. It is folded to allow a high density of neurons and is the seat of many higher order functions, including consciousness, language, short-term memory, perception, vision, and motor processes. The cerebral cortex is separated into right and left hemispheres, which communicate with each other via a large bundle of myelinated nerves, or *white matter*, called the *corpus callosum* (shown). The right hemisphere controls the left side of the body and vice versa for the left hemisphere. The left hemisphere is weighted toward analytical functions, while the right hemisphere is weighted toward creative functions.

The cerebral cortex is divided into several lobes, each paired in the right and left hemispheres, and each lobe has several vertical layers. A schematic of the lobes is shown in Figure 15.5. Each lobe itself contains numerous substructures. The *frontal lobe* controls movements of the skeleton, eyes, speech organs, as well as emotions, and most higher level thought processes. In regards to the latter, the *prefrontal cortex* is the outer portion of this lobe and controls some of the most sophisticated brain functions, including planning, abstraction, organization, short-term memory, control of language, emotions, and personality. Damage to the frontal lobe can cause radical personality changes, loss of motivation and ability to coordinate and complete activities, as well as many other disabilities.

The *parietal lobe* integrates signals related to touch (temperature, pain, and touch) and vision, including deciphering spatial relationships, and then helps coordinate movement of the limbs and eyes.

The *temporal lobe* controls auditory, long-term memory, and emotional functions. It receives and interprets the various frequencies and amplitudes coming from the ears, and filters out extraneous sounds so that the mind can concentrate on important ones. In this function, it also helps interpret and establish meaning for language. The temporal lobe is also involved in object recognition and interpretation, including complex and subtly different patterns, such as faces. The temporal lobe also communicates with the *limbic system* and is involved in automatic responses, such as hunger and fear. The *hippocampus* is a distinct "seahorse-shaped" organ. Humans have two hippocampi, each tucked into one of the temporal lobes. It has several critical functions, including the formation of maps of the environment used for navigation, consolidation of short and long-term memory, and connection to the limbic system. Damage to the hippocampus, such as by Alzheimer's disease, leads to memory loss and disorientation.

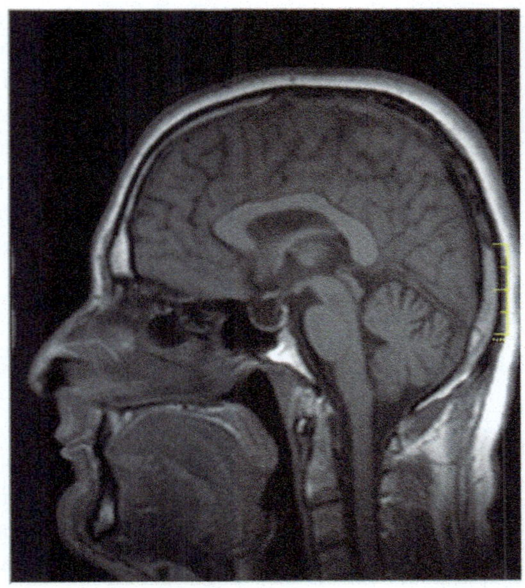

Figure 15.3 MRI of a human brain. (Nevit Dilmen / Wikimedia Commons / CC BY-SA 3.0).

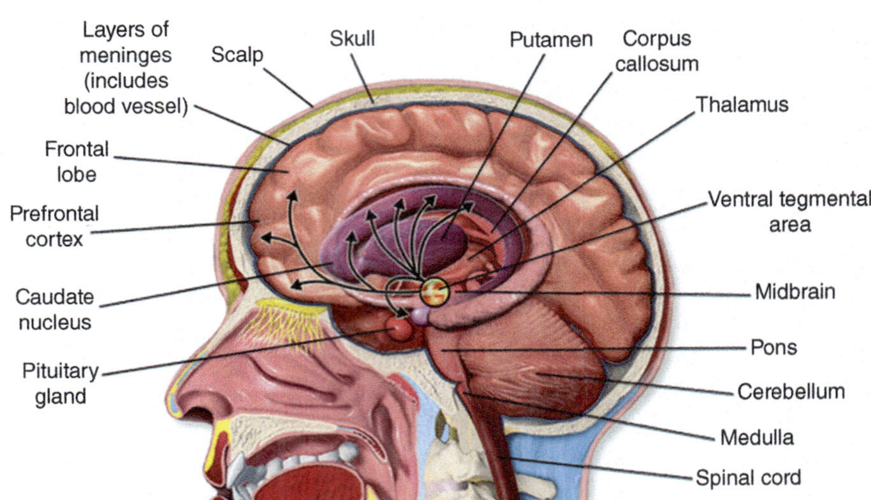

Figure 15.4 A partial view of human brain anatomy. (Blaus, B. (2016) / Wikimedia commons / Licensed under CC BY-SA 4.0).

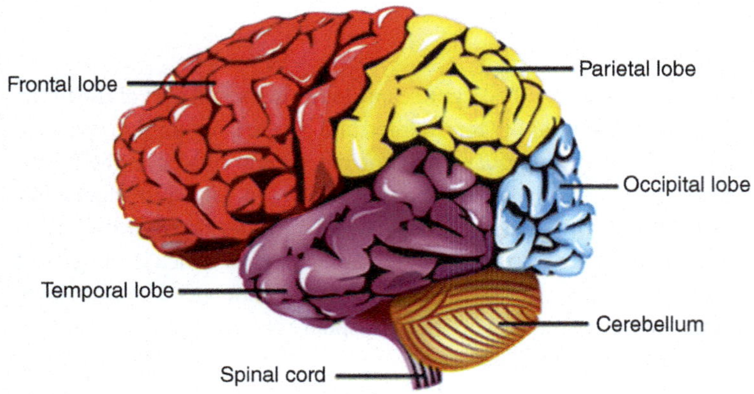

Figure 15.5 Four of the major lobes of the human brain. The highly regular, finely spaced grooves of the unlabeled cerebellum, the mass underneath the lobes, are visible. (ErMED14 / Wikimedia commons / Licensed under CC BY-SA 4.0).

The *occipital lobe* is the main vision-processing center of the brain. Beneath the cerebral cortex, the *cerebellum* ("little brain") forms an anatomically distinct complex with a strikingly regular appearance, whose surface consists of many closely spaced parallel grooves. This stands in contrast to the cerebral cortex which appears as an irregular, highly folded structure. The cerebellum has massive signal processing power that is clearly implicated in coordination of movement and fine motor control. It takes in sensory and other signals, processes them, and then outputs signals to other parts of the brain, but does not seem to have mutually excitatory and resonant circuits. There is still uncertainty about its role in language and other cognitive functions, and cerebellar research is currently very active.

The *thalamus* is located near the corpus callosum, consists of two halves, and consists mostly of gray matter. It is thought of as an information hub that moderates and sends sensory signals to the cerebral cortex and receives feedback from the latter. The *brainstem* is the lower part of the brain mass that joins the rest of the brain to the *spinal cord*. It comprises only about 2.5% of the total brain mass and consists of the *midbrain*, *pons*, and *medulla oblongata*. It controls automatic functions, such as respiration and heart action, and is the main mediator between the brain and the rest of the body.

15.3.2 The Question of Consciousness

L'homme est l'univers devenu conscient de lui-même

Pierre Teilhard de Chardin

We open the door to one of the most controversial areas of modern science: *What is consciousness, how is it defined, how does it arise, and how is it connected to the neuronal activity in the brain and the rest of the conscious entity*? Is it simply the ability to react to the environment in such a way as to enhance an individual's survival and that of its species? If that is the case then virtually all living entities have this type of "consciousness," since they all can sense and react to their environment in ways that enhance their survival: Bacteria chemotactically follow nutrient gradients, slime molds can explore their environment while motionless then make decisions on which direction to move in, and they can solve advanced spatial problems, such as navigating a labyrinth. The octopus has advanced abilities to manipulate its environment and its own protean physical morphology, as well as a highly developed memory. All animals with eyes can create a visual map of their surroundings and make decisions based on it. As mentioned above, whales and some other large mammals are emotive, manipulative, teachable, able to solve problems, and may have advanced forms of their own type of language. So, maybe human consciousness, although it has been considered as unique, is just in a privileged portion of a spectrum of consciousness types or states. In fact, there is no universally accepted definition of "consciousness" and it seems that almost any statement made about it can engender thunderous opposition and derision, as forefront researchers in the area adopt widely varying philosophies about their field.

What everyone can agree on is that brains involve an immense amount of neuronal electrochemistry and signal transport and processing. In fact, imaging methods show that most of the physical energy consumed by the brain is to support neuronal activity, which brings up the issue of associated "mental processes": fear, joy, pain, anxiety, concentration, calculation, creative activity, memory, speech, music, and so on. How are these related to neuronal activity? The strictly reductionist perspective says that consciousness *IS* the electrochemical events occurring. Epitomizing this view Francis Crick stated "You, your joys, and your sorrows and your ambitions, your sense of personal identity and free will are in fact no more than the behavior of a vast assembly of nerve cells and their associated molecules."[4]

4 Crick, F. H. C. (1994). *The Astonishing Hypothesis*. New York, NY: Charles Scribner's Sons.

Another view is that the brain itself is not conscious, but that consciousness arises in the context of the living entity as a whole.[5] In this picture, the neuronal activity of the brain is what supports integrated consciousness but is not consciousness itself. The brain's neuronal events do not create behavior, rather behavior creates the necessary associated neuronal events in the brain. Another, somewhat related perspective is that these neuronal electrochemical events are necessary for consciousness but that consciousness involves "dimensions" we don't currently perceive or understand. In this view, the electrochemical events, which can be physically imaged by our instruments, are projections of the conscious events into space-time and are hence an incomplete view, because the projection into the consciousness dimension(s) of the entire space-time-consciousness manifold is missing. Just as projecting any shaped 3D object onto a single spatial axis yields just a line, you can't claim, however, the 3D object is just the line. From the Biophysics point of view, only the physics of the neuronal events can be measured at this point and the question of consciousness is still wide open.

The debates about consciousness are not likely to be resolved soon. Reductionists insist there is nothing more to consciousness than neuronal activity. "Dualists," most notably starting with Descartes in the early seventeenth century delineate that mind and body are two distinct realms. Descartes went so far as to deny that experimental methods could even be applied to questions of mind. Modern neuroscience has shown that to be not entirely true. Pragmatists don't deny that mind may exist as in the dualists' perspective, but that it is not fruitful to start with fuzzy concepts, such as "working memory," "will," "sensation," and even the word "consciousness" and try to make deductions from physical brain measurements about these conjectural and not rigorously defined words. They prefer the historical empirical inductive, or bottom-up approach, where the results of neuroscience experiments are gathered to form testable theories and hypotheses.

15.3.3 Further Considerations Regarding Consciousness

Harkening back to Niels Bohr's suggestion that the principles behind the living state are distinct from those of inanimate Physics – his yin/yang – leads us to rekindle the question about whether the currently known laws of nature are complete. Modesty, and respect for the history of arrogant declarations of "final knowledge," point to the great probability that we are nowhere near any final knowledge and do not currently possess the tools, theories, models, and understanding to achieve it.

We have seen the mind-boggling contrasts between synthetic macromolecules and biomacromolecules, even while noting similarities in some underlying motifs and analysis methods. But recounting the myriad of highly intricate processes that biomacromolecules are involved in, vastly separates them from the mundane physics of synthetic macromolecules. The physically profound principle of free-energy minimization fails in the protein folding problem; the absolute minimum energy states are sometimes associated with nonfunctional, often dangerous protein aggregates and prions. Quite opposite the minimization of free energy, the living state builds up a reserve of free energy in its far-from-equilibrium, dynamic condition, which it directs into a host of life-sustaining activities.

The tight interactions of histones and other proteins in regulating gene expression, the ability of chaperones to recognize misfolded proteins, and either correct them or tag them for destruction, the plethora of complex molecular events in photoreception, the events surrounding nerve impulse generation and transmission, and so much more, reveal how much more "sophisticated" these biomacromolecular activities are compared to current, simple synthetic macromolecules, including the latest stimuli-responsive polymeric materials.

What principle(s) can propel the appearance and evolution of biomacromolecules and their functions? Is there a piece of the second law of thermodynamics that we're missing? While net entropy always increases in real processes, a part of the whole system can reduce its entropy and gain information. Is there a principle that actually "seeks" to increase information, which appears in the amazingly intricate and complex workings of the living state? All based on interaction, on nonideality. We have a fairly good description of the fundamental forces in physics, the electromagnetism which mediates chemistry and biochemistry, and lights the universe with photons, gravity that provides a long-range force for celestial and cosmic organization, the strong force that holds both nucleons and nuclei together, and the weak force that transforms certain particles into others. But the interactions in the living state, while not contradicting any fundamental forces or known principles, are just too highly developed to be predictable outcomes of the fundamental forces and the structure of the periodic table of elements.

Think how far removed living state interactions are from the fundamental ones. The examples are limitless. A creature's eye and brain create a visual map of the world around it through the reception of photons, vibrations in the air lead to discernible sounds and elicit complex behavior in many species, many species recognize very complex molecular and macromolecular structures which direct their behavior. Structured social organizations arise within and between species, based on myriad interactions.

5 Shulman, R. G., Hyder, F., Rothman, D. L. (2002). Biophysical basis of brain activity: Implications for neuroimaging. *Quarterly Review of Biophysics*, 35 (3), 287–325.

Perhaps "consciousness" refers to the overall set of interactive processes that direct the behavior of any given organism, allowing it to survive, and, if able, to propagate. This posits the primacy of the individual as the bearer of consciousness, the individual as differentiated from its surrounding world. The individual interacts with the surrounding world and other individuals of many different species. Individuals allow the universe to separate itself into different living state quanta that interact with each other, almost as if the universe were trying to "feel itself" by creating all these far-flung types of individuals and their interactions. Arriving at the large-brained human species, self-reflective, with the capacity for extreme manipulation of the environment and ability to create and manipulate abstract concepts, the universe is able to more fully "feel" itself and interrogate itself, leading to artistic and scientific expression. Self-reflective life elsewhere in the universe may have entirely different types of interactions that allow the universe to experience itself in ways radically distinct from the experience of humans and other terrestrial animals.

15.4 Artificial Intelligence

The question of biological consciousness elides inescapably with the notion of *artificial intelligence, AI*. At the time of this writing, the capabilities of AI are expanding exponentially in time, to the delight, amazement, and consternation of humans. It is far beyond our scope here to trace the complex history and current state of AI, but some discussion is worthwhile.

The notion of machines with AI is very old and a huge amount of science fiction literature creatively conceived many different manifestations of it. Alan Turing might be the first person to have articulated what AI might be capable of, that a computing machine could conceivably perform any possible mathematical operation. That was quickly made into reality as computers and computer science burgeoned throughout the mid-1900s and into the present time.

But AI visionaries set their sights beyond mathematical computations. Could AI learn to sense its environment, to move about in it and manipulate it? Extraterrestrial rovers, self-driving cars, and a wide spectrum of mobile robots are currently doing this. Could AI learn human languages and be able to translate among them? After years of poor results and misguided methodologies, a new technology, *neural machine training*, uses whole sentence translation, coupled with *deep learning*, to capture the context and nuances of human language, and is now capable of accurately translating spoken and written language for over 100 human languages. Could AI produce machines that can understand and act on verbal commands by humans? *Virtual assistants*, able to carry out a wide variety of tasks, are now as common in households as refrigerators.

These latter AI capabilities have become so routine that they are no longer considered revolutionary. What is now startling humanity is AI that is capable of synthesizing essays, analyzing, and solving problems in almost every field, having enormous impact in medicine, engineering and science, creating stunning art, creating prosaic but plausible literature and poetry, and even creating virtual companions which learn from their creator/user and "take on a life of their own." The language, artistic, and technical products of such AI are often indistinguishable from corresponding human products.

An example of a highly capable AI natural language processing platform that produces text is the GPT platform: *Generative Pretrained Transformer*. A hallmark of such systems is their ability to search the Internet and other online sources, which contain a vast amount of human knowledge in text, visual, auditory, mathematical, and other forms, and connect it all into complex responses to questions or challenges posed by humans.

At this time, there is widespread debate about the dangers of AI, beyond its positive and revolutionary new capabilities. Many AI specialists think that the limits of what these systems are capable of is currently unknowable, since the vast knowledge accessible to these systems, and their speed and ability to connect data far exceeds the capacity of any human. Even this latter assumption is being challenged, as AI becomes ever more capable of generating hypotheses, creating new designs and concepts, and discovering mathematical proofs. There are even autonomous AI systems that can conceive of and run scientific experiments.

While it was traditionally thought that machines, even AI, would simply take over drudgerous work, including operation of heavy machinery and earth-moving equipment, it is now apparent that AI can do some, or maybe even most of the tasks of engineers, researchers, lawyers, teachers, physicians, policy makers, and even actors and other artists. This, together with AI to produce convincing but false information, when so instructed, raises the level of potential peril that AI presents. AI is capable of taking visual and voice information from individuals to create *deepfakes*, which can depict an individual involved in any real-sounding, real-looking but fictitious act imaginable. Given that the products of AI are largely indiscernible from human products, the potential for misuse of deepfakes in everything from politics to criminal justice to entertainment is potentially very disruptive. Meanwhile, the military applications of AI are growing exponentially. These include sophisticated intelligence and surveillance capabilities for enhanced decision making, and autonomous fighting systems, including unmanned drones, tanks and ships.

Human evolution and the path to AI. (Painting by Linnette F. Reed).

A recent anecdote concerns a query made to a GPT, "How can human-caused global warming be stopped?" The answer from GPT, "Eliminate humans." Perhaps the problems and potential chaos caused by AI will far outweigh its benefits. Only time will tell.

15.4.1 Neural Networks and Parallel Computing

Neural networks are the basis for a vast array of AI functions, including image, speech, text, and pattern recognition. They are inspired by the human brain, with its neurons, connections and processing layers. They are, however, much simpler and more mechanical than human neural networks, and are completely devoid of emotions, subjectivity, and intentionality. . . . at this time.

An outline of the basic functioning of the neural network, and its capacity for deep learning follows.

There is an *input layer*, containing an array of *nodes*, that can accept many different types of input, such as text, images, sounds, equations, and numeric data. The input layer then sends the outputs to the heart of the deep learning network, the multiple *hidden layers*, whose neurons apply *weights* and *biases* to the incoming data and culminate in an output to other neurons via an *activation function*. When data arrive at a neuron it produces a result, z, which is passed on to the activation function.

$$z = w_1 x_1 + w_2 x_2 + \cdots + w_n x_n + b \qquad 15.1$$

where x_i is the ith input and w_i is the weight, or importance, assigned to it. The constant b is the bias, which can shift the activation threshold up or down. The entire learning function of the neural network is based on the weights and biases. Training the network involves using an input with a known output and adjusting the weights and biases of the neurons to minimize the difference between the network's output and the known output.

The result z produced by a neuron is passed to the activation function, f(z), which transforms it into an output a, before passing its result to the next layer.

$$a = f(z) \qquad 15.2$$

The pre-selected activation function is non-linear and allows for modelling of complex relationships. If the activation function were linear, then its output would be a mere linear combination of the inputs. Activation functions include sigmoids, hyperbolic tangents, and other, specially defined functions. It is the ability of the activation function to non-linearly transform a neuron's output that allows deep learning.

The neurons in a layer are capable of *massive parallel computing,* that is, each neuron, or set of neurons performs its computations – applying weights, adding biases, and producing an output through its activation function -- independently of the others. This gives significantly higher speed and throughput than a traditional *central processing unit,* which works in a strictly sequential fashion (modern CPU's can have multiple cores, that work independently of each other, and allow some parallel operation). *Graphical processing units* (GPUs) and *tensor processing units* (TPU)[6] consist of thousands of small cores yielding the parallel processing power needed for massive parallel operation.

Some of the most important operations of a deep learning neural network involve matrices. For example, if the input vector $\vec{X}$ has n features going into a layer with m neurons, then an mxn weight matrix $\overline{\overline{W}}$ is created, and the output vector $\vec{Z}$ is computed by

$$\vec{Z} = \overline{\overline{W}} \cdot \vec{X} + \vec{B} \qquad 15.3$$

where $\vec{B}$ is the vector containing the bias of each neuron.

One area that holds the potential for significantly increasing the performance of AI is *quantum computing* (QC). While not yet a practical technology it could lead to advances in AI optimization, secure encryption, probabilistic logic, and massive parallel data processing. The quantum bit, the *qubit,* is the basis. In contrast to a classical bit, which has a value of 0 or 1, the qubit can be represented as a wave function, ψ, a superposition of 0 and 1 states

$$|\psi> = \alpha|0> + \beta|1> \qquad 15.4$$

with amplitudes α and β, respectively, and $|\alpha^2| + |\beta^2| = 1$. It can hence weight the 0 and 1 states, unlike a classical bit. Higher-dimensional qubits can include other states beyond 0 and 1. An example of the latter is the *qutrit,* which adds a term $\gamma|2>$ to Equation 15.4, and requires $|\alpha^2| + |\beta^2| + |\gamma^2| = 1$. Furthermore, quantum entanglement of two bits means that determining the value of one bit will automatically determine the value of the coupled bit(s), even if spatially distant, so that the qubits are interdependent, allowing high efficiency using *quantum parallelism.* Qubits can be created by several different physical embodiments, including i) superconducting qubits, that can exist in a superposition of clockwise and counterclockwise currents, ii) trapped ions, where energy levels act as qubits, iii) photonic cubits which use superpositions of two polarization states, and several others. None are currently at a level for scaling up for widespread practical use.

In closing this brief overview of neural networks, it is interesting to regard our own thought processes in light of how neural networks function. Our brains have an input layer, which receives input from our senses, including text, speech, music, 3D images, equations, tactile and olfactory sensations, etc. These inputs are transmitted to the deep, hidden layers of the brain, which, as we've seen, is massively interconnected and convoluted. Through weighting and bias, inputs can 'light up' an immense variety of outputs; memories, complex associations among abstract ideas, decisions, coordinated motion, and much more.

'Training' of the brain is similar to training a neural network with weights and biases, which then elicits actions that correspond to the training; e.g. learning a language, whether native or foreign, solving problems through fundamental concepts and formalisms, using the body to perform certain movements or feats, reciting a poem or prayer, playing an instrument, learning and forming ideas using an ideology, or simply conditioned responses to dogmatic training, as in military, political, and religious milieux. 'Mob behavior' involves synchronizing the action of the neural networks of large numbers of persons; cheering or booing at a sporting event, creating organized action for a political purpose, a mob bent on destruction, etc.

Think, for example, of how, in a conversation, the inputs from your conversational partners spark coherent responses as complex information is processed through your vast neural network. Think also of how some inputs in a conversation can

6 'Tensor' as used in this term is not a tensor in the physics and mathematical sense, which are matrices that have certain transformation properties under change of coordinates. Tensor in the TPU sense is a generalized term for multidimensional structures that can be operated on mathematically.

set another person's neural network into thoughts and responses which can become 'tangential' to the conversation. 'Brainstorming' is a synergy where neural networks of two or more persons interact creatively and constructively around a theme. The principles of AI allows us to better understand how our own minds function, even though our cognitive functions are far more complex than the current operations of AI.

15.5 Astrobiology and Exoplanets

> *....There is even recognition of a new science of extraterrestrial life, sometimes called "exobiology" – a curious development in view of the fact that this "science" has yet to demonstrate that its subject matter exists.*
>
> George Gaylord Simpson (1964)

Are we alone in the universe? Does life exist elsewhere? The answer to this fundamentally important question is being attacked from many angles. At the time of this writing, the answer is still "we don't know," but it is possible that the answer will already have been found by the time you read this.

The notion of extraterrestrial life, or *cosmic pluralism* as it was sometimes called, has captured human imagination since antiquity, and references to it exist in Greek, Indian, and other schools of thought. The Qur'an (seventh century) possibly refers to life beyond the Earth. Among the ancient Greeks, Anaximander proposed an infinite universe with many planets and life and was supported by numerous other philosophers, such as Democritus and Epicurus. Plato and Aristotle, however, believed in the uniqueness of the Earth and its life forms, and this stifled credence in cosmic pluralism well into the early modern era, and was reenforced by the Geocentric Theory predominant in the Middle Ages. Interest in it accelerated during the Renaissance when the Heliocentric Theory was firmly established and it was realized there was nothing particularly special about the Earth and our Sun, and it was becoming clearer that physical laws on Earth were likely obeyed in the rest of the observable universe.

Giordano Bruno (Italy, 1548–1600) proposed that virtually every visible star would have planets, similar to ours, abundant with life. The astronomer Friedrich Wilhelm Herschel (Germany/England, 1738–1822) thought the solar system was strewn with life. Modern scientists such as Stephen Hawking (England, 1942–2018) and Frank Drake (USA, 1930–) take the existence of extraterrestrial life for granted. Whether there is also intelligent life, in addition to rudimentary life, however, is hotly debated. At this point, while there is significant hope of finding extant life or remnants of bygone life on Mars or several of the moons of Jupiter and Saturn, the consensus is there is not much chance of finding civilizations in our Solar System. Hawking thought it best not to try signaling other intelligent civilizations, lest they spot us and come to wreak havoc to extract Earth's resources.

Drake proposed a probabilistic scheme for estimating the number of other intelligent civilizations existing in our galaxy, N, that could communicate with us on Earth, given by

$$N = R_* \times f_p \times n_e \times f_1 \times f_i \times f_c \times L$$

where R_* is the rate of star formation, f_p is the fraction of stars with planets, n_e is the average number of habitable planets per star with planets, f_1 is the fraction of habitable planets that eventually have life, f_i is the fraction of planets with life that develop an intelligent civilization, f_c is the fraction of civilizations able to broadcast signals into space, and L is the length of time the civilization transmits signals into space. An estimate of $N = 10{,}000$ has been made, but obviously most of the parameters are speculative, and some consider the equation to be meaningless.

Pure speculation has been supplanted by experimental efforts over the past decades. The interdisciplinary SETI Institute (Search for Extraterrestrial Intelligence) was founded in 1984 and has scientists working in astronomy and astrophysics, planetary geology, astrobiology, chemistry, and in engineering new instrumentation. Among their many activities is the scanning of the sky for possible radio-wave communications of life from elsewhere.

Perhaps the biggest single advance was the discovery of *exoplanets*, planets orbiting stars other than the sun. The first exoplanet discovered was via a periodic irregularity in a pulsar's pulse rate (PSR B1257 + 12), 2300 light-years from our solar system. The astronomers computed it was due to two planets in orbit around it; one 3× Earth mass and one 4×.

Shortly thereafter, the *radial velocity method* began to find evidence of other exoplanets. The method relies on the small wobble of a star caused by a planet massive enough to cause the star to have slight radial motion around the star-planet center of mass, which is shifted outwards from the star's own center of mass. The emission spectrum of the star then has slight blue and red Doppler shifts during an entire orbital period of the planet, corresponding to radial motion of the star toward and away from the Earth, respectively. Using the radial velocity and planet's orbital period allows its orbital radius and minimum mass to be estimated, using classical celestial mechanics. Using the star's black body spectrum its temperature can be

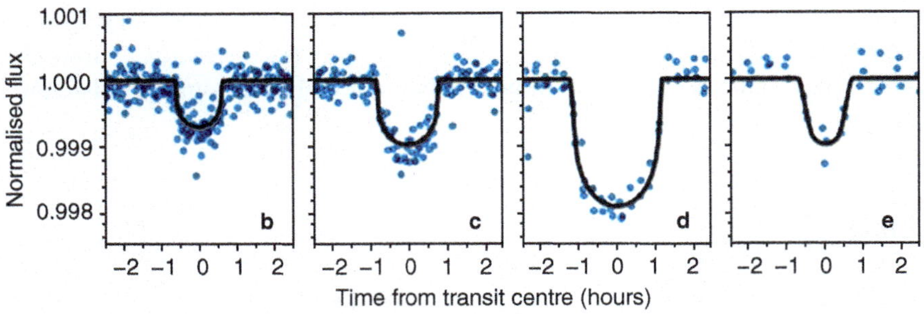

Figure 15.6 Transit curves for light intensity for four exoplanets. (From Wells, R. et al. (2018). By permission of the Oxford University Press).

determined and so, having estimated its distance from its star, the intensity of light incident on it from the star can be estimated, which is useful for estimating the planet's temperature and potential habitability. The first exoplanet identified by this method, in 1993 at the ELODIE spectrograph at the Haute-Provence Observatory (France), was of a "hot" (i.e. close to its star) Jupiter-type planet in the Pegasus constellation. Since then, over 900 exoplanets have been observed using this method at Keck Observatory (At 13,600 ft at Mauna Kea, Hawaii, USA), Lick Observatory (near San José, California), Anglo-Australian Observatory (near Sydney, Australia), and the Geneva Extrasolar Planet Search (near Geneva Switzerland).

The *transit method* to identify planets and their orbital characteristics has been highly developed and is the most successful single method to-date. When a planet passes in front of its star, it is *in transit* and partially dims the star's intensity. The shape, timing, and magnitude of the obscuration curve give an approximation of the planet's radius and orbital characteristics. Combining data from the radial velocity method allows the density of the planet to be estimated, and hence surmise if it might be predominantly gaseous or rocky. The first satellite to search for exoplanets used this method; the Kepler Satellite Mission, launched in May 2009, from Cape Canaveral, Florida. The Transiting Exoplanet Survey Satellite (TESS), was launched in April 2018, and is equipped with four wide field-of-view CCDs, exclusively to measure transits. Examples of transit measurements are shown in Figure 15.6.

Other terrestrial-based observations are actively using methods involving polarimeters, coronagraphs (instruments which block most of a star's light so that its corona and vicinity can be observed), spectrometers, adaptive optics,[7] and telescopes working over much of the electromagnetic wave spectrum. The Gemini Planet Imager at the Gemini South Telescope in Chile is used mostly to detect young gas giants. Also in Chile, the Very Large Telescope Spectro-Polarimetric High-Contrast Exoplanet Research Instrument uses a polarimeter, a coronagraph, and other instruments to search for a variety of exoplanets.

To-date, all evidence is that virtually every star in the neighborhood of our solar system has at least one planet in orbit around it.

15.5.1 Types of Exoplanets

There is considerable variation in the structure and properties of exoplanets discovered to-date, which suggests a huge continuum of sizes, composition, and temperatures. One scheme to interpret their properties is to roughly classify them as *Giant Gas Planets* ("Jupiters"), *Giant Ice Planets* ("Neptunes"), and *Terrestrial and Ocean Planets* (Earthlike). Exoplanets have been found, however, that have no analogous representation in our Solar System, such as extremely extended gas giants and highly compact, iron-like small planets.

Jupiter and Saturn are our local instances of gas giants. They are basically huge spheres of hydrogen and helium, with some presence of heavier elements. They have a thin outside convective layer where the gases, molecular hydrogen, H_2, and helium move and swirl about. Below this helium and atomic hydrogen, H, are in metallic form. At the center, there is possibly heavier elements, ices and rocks.

Neptune and Uranus are our local instances of ice giants. These also have a thick layer of hydrogen and helium, mostly in metallic form, with a thin outer layer of H_2 and He. Overall, they have a higher heavy element content than gas giants, and the heavy element mass fraction generally exceeds the hydrogen and helium content. Furthermore, there are ionic

7 Adaptive optics are systems that use sensors and precise micropositioning devices to actively move optical components in order to remove the effects of the Earth's atmospheric fluctuations on incoming wavefronts from celestial objects, and hence rid the images of distortions

"fluid ices" likely to consist of water, as well as some methane and ammonia, comprising a large portion of the planet. A much larger, rocky core is expected, compared to gas giants. It is thought there is a very wide variety of compositions, phase structures, and masses, among the ice giant exoplanets, many unlike Neptune and Uranus, including subice giants, and ice giants far larger than Neptune.

Mercury, Venus, and Mars are our local instances of Terrestrial Planets, whereas the Earth is both a Terrestrial and Ocean planet. These planets are far smaller than the giant planets and are composed chiefly of heavier elements, such as Fe, Si, Mg, O, and C, which form abundant minerals.[8] These planets can have a thin, rocky crust, mineral laden mantles, and molten cores, chiefly iron.

While Earth and Venus are similar in size and density there are major differences beyond just the greater solar intensity at Venus – where the surface is at an inhospitable average temperature of 735 K with 230 K variations – and these seem rooted in fundamentally different geophysical evolution. One marked difference is that there is no evidence of plate tectonics on Venus, depriving it of mechanisms to release heat, mechanical stress, and cycle material from the interior to the surface and atmosphere. Making experimentally based deductions about the seismology of exoplanets is challenging and has not yet been achieved.

In the search for life, water is presumed to be a key ingredient, at least for life forms that may be similar to Earth's. While Mars has water vapor in its atmosphere and ice on its surface it is surmised that it had oceans in its distant past. Icy crusts have been detected on Titan and Enceladus, both moons of Saturn, and on Callisto, Ganymede, and Europa, moons of Jupiter. It is not known at this point whether there are oceans beneath these crusts, but there is strong evidence for them in the case of Enceladus, Europa, and Ganymede. No Ocean exoplanets have yet been unambiguously detected, although there are theoretical predictions that they may be common, and water has been detected in exoplanet atmospheres.

15.5.2 Exoplanetary Atmospheres and Conditions for Life

It is exceedingly difficult to detect the components in atmospheres of exoplanets, but it is critical in the search for life to establish if an exoplanet has a viable atmosphere. Some spectroscopic identification of composition is possible in transit measurements when the star's light passes through the exoplanet's atmosphere, leaving telltale absorption spectra, and these can be compared to the thermal and reflection spectra of the exoplanet out of transit. Such signals are extremely weak and require very high signal/noise detection. Most work in exoplanetary atmospheres is still speculative and relies on model paradigms that have been developed for Earth, the other planets of the solar system, and the moons of the giant planets. Even models for the Earth's atmosphere, for which there are virtually unlimited data of every conceivable type are extremely sensitive to a host of factors, such as missing physical processes in a model, positive feedback and instability involving nonlinear phenomena and threshold effects, and parametrization of small, local effects, such as clouds. Primary data of fundamental importance include a planet's atmospheric composition, mass, surface albedo (fraction of incident light reflected), and amount of volatile molecules, currently impossible to determine for exoplanets, the intensity of incident light from its star, for which good estimates exist in many cases, and the details of the exoplanet's orbit, spin, and inclination of the latter's axis to the orbital plane, for which good data often exist. Even with all these data assembled, much of an atmosphere's state depends on its history and particular evolution, so the day seems far off when there will be fully definitive planetary atmospheric and climate models.

Despite all these caveats, certain constraints can be put on what may constitute a habitable atmosphere for "conventional life." Permanent liquid water should exist, which then defines the so-called habitable zone around a star, and the planet must be massive enough to keep gravitational escape of the atmosphere low.

As discussed in Chapter 14.1.1, archaea, or extremophiles, are able to withstand a vast range of conditions. These include temperatures as low as −18°C and as high as 350°C, salinities at the solubility limits of salts, lightless and oxygen-free environments, pH in the range of 4.0–9.0, extremely low and high pressures, and under UV and ionizing radiation. This effectively extends potential habitability to these extremes and likely higher. There is hope that because of their density where they exist, there will be absorption/reflection biosignatures for extremophiles on exoplanets.

The search for water and other biosignatures within and without our solar system is in full gear. While modeling and computational efforts are enormous, these must be sustained by increasingly sophisticated observational data if they are to be meaningful. Figure 15.7 shows a collage of factors that can affect the existence of liquid water on a planet.

And of course, the possibility exists that life may take forms we can't currently imagine, with vastly different requirements, e.g. perhaps some life forms can exist without water or planetary atmospheres.

8 ^{56}Fe and ^{62}Ni are the stablest isotopes. The former has the lowest mass per nucleon and the latter has the highest binding energy per nucleon.

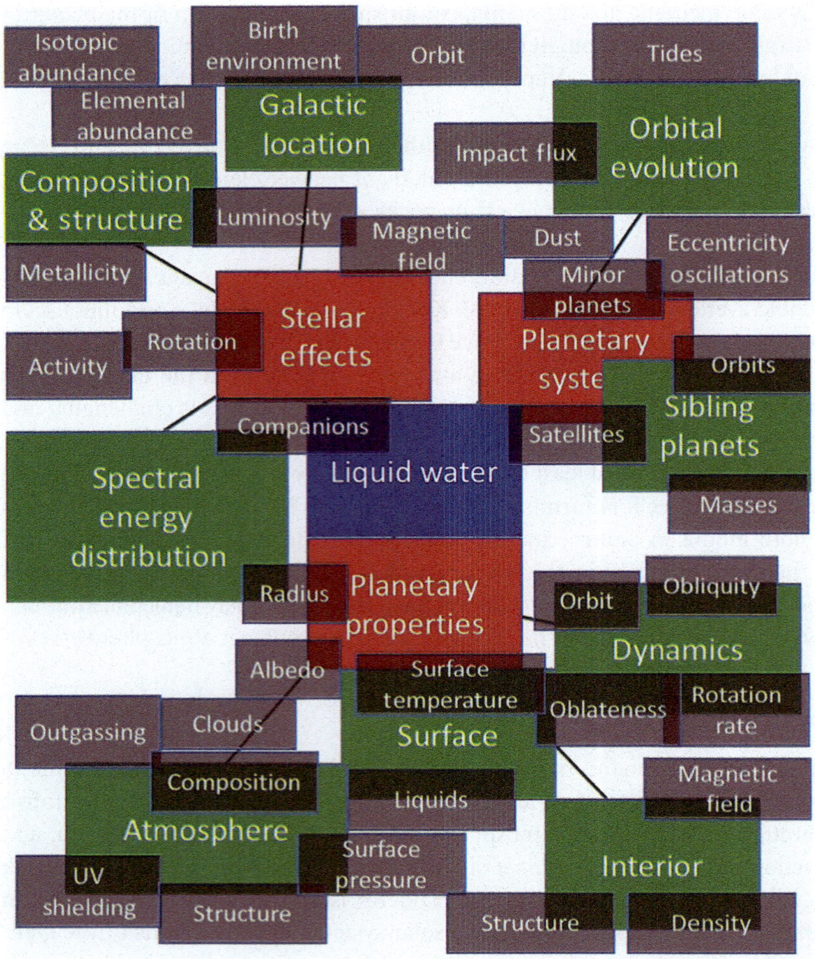

Figure 15.7 The many factors influencing whether there will be liquid water on a planet's surface. (figure 1 in Shields, A. L. (2019) based on figure 1 from Meadows, V. S. and Barnes, R. K. (2018)).

15.5.3 The James Webb Space Telescope

The James Webb Space Telescope (JWST) was launched on December 25, 2021. It is equipped with high-resolution infrared (IR) sensors and is currently the largest optical telescope in space. The IR detectors cover the spectrum from orange/red, near IR (800–2500 nm), and into the mid-IR region (3–50 μm). It can see objects far fainter and more distant than any preceding telescope, and its multiple missions include looking back in time to "shortly" after the Big Bang (maybe 100 million years after), witnessing the birth of stars and galaxies, and searching for signs of possibly habitable exoplanets. It is the successor to the Hubble Space Telescope, which was mainly concerned with observation in the near IR, visible, and near UV ranges.

On June 21, 2022, it was announced that the Webb telescope detected the IR signature of water from the exoplanet WASP-96b, at about 1150 light-years from Earth. The detection was made during a 6.4-h transit measurement recorded in IR. While WASP-96b is a gas giant, with a surface temperature around 550°C, and thus uninhabitable for conventional life, the JWST has demonstrated its ability to analyze exoplanetary atmospheres, adding a new capability in the search for extraterrestrial life.

Figure 15.8 shows a July 2022 image from the Webb telescope, combining near- and mid-IR data. This is part of the Carina Nebula, nasa.gov/webbfirstimages/. The US National Aeronautics and Space Administration (NASA) provides this description of the image:

> The image is divided horizontally by an undulating line between a cloudscape forming a nebula along the bottom portion and a comparatively clear upper portion. Speckled across both portions is a starfield, showing innumerable stars of many sizes. The smallest of these are small, distant, and faint points of light. The largest of these appear larger, closer, brighter, and more fully resolved with 8-point diffraction spikes. The upper portion of the image is blueish, and has wispy translucent cloud-like streaks rising from the nebula below. The orangish cloudy formation in the bottom half varies in density and ranges from translucent to opaque. The stars vary in color, the majority of which have a blue or orange hue. The cloud-like structure of the nebula contains ridges, peaks, and valleys – an appearance very similar to a mountain range. Three long diffraction spikes from the top right edge of the image suggest the presence of a large star just out of view.

Figure 15.8 A July 2022 image of a nebula taken by the Webb telescope. The description is in the above quote. (NASA, ESA, CSA, and STScI / Wikimedia Commons / Public Domain).

15.6 Summary

This chapter introduced some of the most vexing, least understood, and speculative frontiers in biophysics. There is no generally accepted theory of the prebiotic origins of life, and there is vigorous ongoing experimental and theoretical research in the area. In terms of the genetic code, some information scientists have even speculated that its origin, as a non-thermodynamic code, is incompatible with a purely physicochemical explanation.

Whereas quantum physics underlies all chemistry, quantum biology is in its infancy and only a few living state phenomena have been definitively traced to quantum physical effects, such as tunneling and superposition of states, whereas there is much speculation about manifestations of quantum coherence and entanglement.

In the meantime, neuroscience, a fully intertwined interdisciplinary endeavor, has made concrete progress on many fronts. Tools from physics, such as 3D and time-resolved NMR, PET scans, and many other types of tomographies, have elucidated electrochemical events in the brain associated with different conscious phenomena, such as memory, speech, fear, pain, and much more. Still, the overriding question of what constitutes "consciousness" is unsolved and the subject of much controversy and speculation. The current burgeoning of artificial intelligence is amazing, confounding, and even frightening, concerning what totally unprecedented and unpredictable results will emanate from computer codes and associated machines, that can almost instantly access and synthesize almost all of human knowledge, language, science, and art.

The question as to whether there is life outside the Earth, and particularly "intelligent life," is unanswered as of this writing. It is quite possible, however, that with the new astrophysical tools and space missions that are rapidly evolving, there may be some forthcoming evidence of extraterrestrial life.

References and Further Reading

Prebiotic Origins of Life

Baaske, P., Weinert, F.M., Duhr, S. et al. (2007). Extreme accumulation of nucleotides in simulated hydrothermal pore systems. *Proceedings of the National Academy of Sciences of the United States of America* 104: 9346–9351.

Bada, J.L., Fegley, B.J., Miller, S.L. et al. (2007). Debating evidence for the origin of life on Earth. *Science* 315: 937–939.

Bailey, J., Chrysostomou, A., Hough, J.H. et al. (1998). Circular polarization in star-formation regions: implications for biomolecular homochirality. *Science* 281: 672–674.

Bernhardt, H.S. (2012). The RNA world hypothesis: the worst theory of the early evolution of life (except for all the others). *Biology Direct* 7: 23. https://doi.org/10.1186/1745-6150-7-23.

Budin, I., Bruckner, R.J., and Szostak, J.W. (2009). Formation of protocell-like vesicles in a thermal diffusion column. *Journal of the American Chemical Society* 131: 9628–9629.

Chen, I.A., Roberts, R.W., and Szostak, J.W. (2004). The emergence of competition between model protocells. *Science* 305: 1474–1476.

Chyba, C.F. (2005). Rethinking Earth's early atmosphere. *Science* 308: 962–963.

Cronin, J.R. and Pizzarello, S. (1997). Enantiomeric excesses in meteoritic amino acids. *Science* 275: 951–955.

De Duve, C. (1995). The beginnings of life on earth. *American Scientist* 83: 428–437.

Eigen, M. and Schuster, P. (1977). The hypercycle. A principle of natural self-organization. Part A: emergence of the hypercycle. *Naturwissenschaften* 64: 541–565.

Gesteland, R.F., Czech, T.R., and Atkins, J.F. (eds.). (2006). *The RNA World*. Cold Spring Harbor Laboratory Press.

Hanczyc, M.M., Mansy, S.S., and Szostak, J.W. (2007). Mineral surface directed membrane assembly. *Origins of Life and Evolution of Biospheres* 37: 67–82.

Hanczyc, M.M. and Szostak, J.W. (2004). Replicating vesicles as models of primitive cell growth and division. *Current Opinion in Chemical Biology* 8: 660–664.

Johnson, A.P., Cleaves, H.J., Dworkin, J.P. et al. (2008). The Miller volcanic spark discharge experiment. *Science* 322: 404.

Joyce, G.F. (2002). The antiquity of RNA-based evolution. *Nature* 418: 214–221.

Kauffman, S. (1993). *Origins of Order: Self-Organization and Selection in Evolution*. Oxford University Press.

Koonin, E.V. (2007). An RNA-making reactor for the origin of life. *Proceedings of the National Academy of Sciences of the United States of America* 104: 9105–9106.

Koonin, E.V. and Novozhilov, A.S. (2009). Origin and evolution of the genetic code: the universal enigma. *IUBMB Life* 61: 99–111.

Lopez, A. and Fiore, M. (2019 Jun). Investigating prebiotic protocells for a comprehensive understanding of the origins of life: a prebiotic systems chemistry perspective. 9 (2): 49. Published online 2019 Jun 7. https://doi.org/10.3390/life9020049.

Mansy, S.S. and Szostak, J.W. (2008). Thermostability of model protocell membranes. *Proceedings of the National Academy of Sciences of the United States of America* 105: 13351–13355.

Martin, W. and Russell, M.J. (2003). On the origins of cells: a hypothesis for the evolutionary transitions from abiotic geochemistry to chemoautotrophic prokaryotes, and from prokaryotes to nucleated cells. *Philosophical Transactions of the Royal Society of London. Series B, Biological Sciences* 358: 59–83.

McBride, J.M. and Tully, J.C. (2008). Physical chemistry: did life grind to a start? *Nature* 452: 161–162.

Meadows, V. S. and Barnes, R. K. (2018). *Handbook of Exoplanets*, H. J. Deeg & J. A. Belmonte, Ed. (New York: Springer Nature), 2771.

Miller, S.L. (1953). A production of amino acids under possible primitive earth conditions. *Science* 117 (1953): 528–529.

Orgel, L.E. (1998). The origin of life – a review of facts and speculations. *Trends in Biochemical Science* 23: 491–495.

Orgel, L.E. (2000). Self-organizing biochemical cycles. *Proceedings of the National Academy of Sciences of the United States of America* 97: 12503–12507.

Orgel, L.E. (2004). Prebiotic chemistry and the origin of the RNA world. *Critical Reviews in Biochemistry and Molecular Biology* 39: 99–123.

Poole, A.M. (2006). Getting from an RNA world to modern cells just got a little easier. *BioEssays* 28: 105–108.

Shields, A. L. (2019). The climates of other worlds: A review of the emerging field of exoplanet climatology. *The Astrophysical Journal Supplement Series* 243: 30 (11 pp.). https://doi.org/10.3847/1538-4365/ab2fe7.

Szostak, J.W., Bartel, D.P., and Luisi, P.L. (2001). Synthesizing life. *Nature* 409: 387–390.

Wächtershäuser, G. (1990a). Evolution of the first metabolic cycles. *Proceedings of the National Academy of Sciences of the United States of America* 87: 200–204.

Wächtershäuser, G. (1990b). Evolution of the first metabolic cycles. *Proceedings of the National Academy of Sciences of the United States of America* 87: 200–204.

Wells, R., Poppenhaeger, K., Watson, C. A. (2018). Three small transiting planets around the M-dwarf host star LP 358–499. *Monthly Notices of the Royal Astronomical Society* 473: 1131–1135. doi:10.1093/mnrasl/slx171.

Information Theory

Chaitin, G.J. (1987). *Algorithmic Information Theory*. Cambridge University Press. available for free download at https://web.archive.org/web/20111215170328/http://www.cs.auckland.ac.nz/~chaitin/cup.pdf.

Chaitin, G.J. (1992). *Information Theoretic Incompleteness*. Singapore: World Scientific Press. Available for free download at https://web.archive.org/web/20100514220011/http://www.cs.auckland.ac.nz/~chaitin/ps3.pdf.

Copeland, J.B. (ed.). (2004). *The Essential Turing: Seminal Writings in Computing, Logic, Philosophy, Artificial Intelligence, and Artificial Life*. Oxford University Press. Available for free download at http://www.cse.chalmers.se/~aikmitr/papers/Turing.pdf.

Yockey, H.P. (2005). *Information Theory, Evolution, and the Origin of Life*. Cambridge University Press.

Quantum Biology

Adams, B., Sinayskiy, I., and Petruccione, F. (2018). An open quantum system approach to the radical pair mechanism. *Nature, Scientific Reports* 8: 15719. https://doi.org/10.1038/s41598-018-34007-4.

Brookes, J.C. (2017). Quantum effects in biology: golden rule in enzymes, olfaction, photosynthesis, and magnetodetection. *Proceedings of the Royal Society A*.

Davydov, A.S. (1982). *Biology and Quantum Mechanics*. Pergamon Press.

DeVault, D. (1981). *Quantum-mechanical Tunnelling in Biological Systems*. Cambridge University Press.

Kishkinev, D.A. and Chernetsov, N.S. (2015). Magnetoreception systems in birds: a review of current research. *Biology Bulletin Reviews* 5 (1): 46–62. https://doi.org/10.1134/s2079086415010041.

Lambert, N., Chen, Y.-N., Cheng, Y.-C. et al. (2012). Quantum biology. *Nature Physics* 9 (1): 10–18. https://doi.org/10.1038/nphys2474.

Marais, A., Adams, B., Ringsmuth, A.K. et al. (2018). The future of quantum biology. *Journal of the Royal Society Interface* 15: 20180640. https://doi.org/10.1098/rsif.2018.0640.

Mohseni, M., Aspuru-Guzik, A., Rebentrost, P. et al. (n.d.). Environment-assisted quantum transport. *Quantum Effects in Biology* 159–176. https://doi.org/10.1017/cbo9780511863189.009.

Schaller, G. (2014). *Open Quantum Systems Far from Equilibrium*. Cham: Springer International Publishing.

Sension, R.J. (2007). Quantum path to photosynthesis. *Nature* 446 (7137): 740–741. https://doi.org/10.1038/446740a.

Steiner, U.E. and Ulrich, T. (1989). Magnetic field effects in chemical kinetics and related phenomena. *Chemical Reviews* 89: 51–147.

Neuroscience and Artificial Intelligence

Bear, M.F., Connors, B.W., and Paradiso, M.A. (eds.). (2007a). *Neuroscience, Exploring the Brain*, 3e. Lippincott Williams & Wilkins.

Bear, M.F., Connors, B.W., and Paradiso, M.A. (eds.). (2007b). *Neuroscience, Exploring the Brain*, 3e. Lippincott Williams & Wilkins.

Bostrom, N. (2014). *Superintelligence: Paths, Dangers, Strategies*. Oxford University Press.

Fadel, C., Holmes, W., and Bialik, M. (2019). *Artificial Intelligence in Education: Promises and Implications for Teaching and Learning*. Boston, Massachusetts: Center for Curriculum Redesign.

Frewen, P., Schroeter, M.L., Riva, G. et al. (2020). Neuroimaging the consciousness of self: review, and conceptual-methodological framework. *Neuroscience and Biobehavioral Reviews* 112: 164–212.

Goertzel, B., Lian, R., Arel, I. et al. (2010). A world survey of artificial brain projects, Part II: biologically inspired cognitive architectures. *Neurocomputing* 74 (1–3): 30–49. https://doi.org/10.1016/j.neucom.2010.08.012.

Gopnik, A. (2017 June). Making AI More Human: artificial intelligence has staged a revival by starting to incorporate what we know about how children learn. *Alison Gopnik, Scientific American* 316 (6): 60–65.

Heiss, W.-D. (2016). Hybrid PET/MR imaging in neurology: present applications and prospects for the future. *Journal of Nuclear Medicine* 57 (7): 993–995.

Iphofen, R. and Kritikos, M. (2019). Regulating artificial intelligence and robotics: ethics by design in a digital society. *Contemporary Social Science* 16 (2): 170–184. https://doi.org/10.1080/21582041.2018.1563803. ISSN 2158-2041. S2CID 59298502.

Jordan, M.I. and Mitchell, T.M. (2015). Machine learning: trends, perspectives, and prospects. *Science* 349 (6245): 255–260. https://doi.org/10.1126/science.aaa8415. Bibcode:2015Sci..349.255J. PMID 26185243. S2CID 677218.

Kandel, E., Koester, J.D., Mack, S.H., and Siegelbaum, S. (2021). *Principles of Neural Science*, 6e. McGraw Hill.

Nagel, T. (2012). *Mind and Cosmos: Why the Materialist Neo-Darwinian Conception of Nature Is Almost Certainly False*. Oxford University Press.

Owen, A.M. (2019). The search for consciousness. *Neuron* 102 (3): 526–528.

Poole, D., Mackworth, A., and Goebel, R. (2020). *Computational Intelligence: A Logical Approach*. NY: Oxford University Press.

Rosenberg, A. (2006). *Darwinian Reductionism*. University of Chicago Press.

Russell, S. and Norvig, P. (2021). *Artificial Intelligence: A Modern Approach*, 4e. Pearson.

Shulman, R.G. (2013). *Brain Imaging; What It Can (And Cannot) Tell Us about Consciousness*. Oxford University Press.

Squire, L.R. (ed.). (2009). *Encyclopedia of Neuroscience*, 1e. Academic Press.

Stone, J.V. (2019). *Artificial Intelligence Engines; A Tutorial Introduction to the Mathematics of Deep Learning*. Sebtel Press.

Sensing

Montgomery, S. (2016). *The Soul of an Octopus: A Surprising Exploration into the Wonder of Consciousness*. Simon & Schuster.

Yong, E. (2022). *An Immense World: How Animal Senses Reveal the Hidden Realms around Us*. Random House.

Exoplanets and Astrobiology

Batalha, N.M. et al. (2013) Planetary candidates observed by Kepler. III. Analysis of the first 16 months of data. arXiv:1202.5852.

Cockell, C.S., Bush, T., Bryce, C. et al. (2016). Habitability: a review. *Astrobiology* 16 (1). https://doi.org/10.1089/ast.2015.

DasSarma, S., DasSarma, P., Laye, V.J., and Schwieterman, E.W. (2020). Extremophilic models for astrobiology: haloarchaeal survival strategies and pigments for remote sensing. *Extremophiles* 24: 31–41. https://doi.org/10.1007/s00792-019-01126-3.

Domagal-Goldman, S.D. and Wright, K.E. (eds.). The astrobiology primer v2.0. *Astrobiology* 16 (8). https://doi.org/10.1089/ast.2015.1460.

Le Sergeant d'Hendecourt, L. (2011). Molecular complexity in astrophysical environments: from astrochemistry to "astrobiology"? *EPJ Web of Conferences* 18: 06001. https://doi.org/10.1051/epjconf/20111806001.

A list of currently discovered exoplanets is available at https://exoplanets.nasa.gov/discovery/exoplanet-catalog as of 1/19/2021 4,331 confirmed discoveries.

Madhusudhan, N., Agúndez, M., Moses, J.I., and Hu, Y. (2016). Exoplanetary atmospheres—chemistry, formation conditions, and habitability. *Space Science Reviews* 205: 285–348. https://doi.org/10.1007/s11214-016-0254-3.

Mansy, S.S., Schrum, J.P., Krishnamurthy, M., Tobé, S., Treco, D.A. and Szostak, J.W. (2008). Template-directed synthesis of a genetic polymer in a model protocell. *Nature* 454: 122–125.

Méndez, A., Rivera-Valentín, E.G., Schulze-Makuch, D. et al. (2021). Habitability models for astrobiology. *Astrobiology* 21 (8): 1017–1027.

Mills, F.P., Moses, J.I., Gao, P., and Tsai, S.-M. (2021). The diversity of planetary atmospheric chemistry - lessons and challenges from our solar system and extrasolar planets. *Space Science Reviews* 217 (3): 43.

Parker, R.J. (2020). The birth environment of planetary systems. *Royal Society Open Science* 7 (11): 201271.

Rivera-Valentín, E.G., Filiberto, J., Lynch, K.L. et al. (2021). Introduction-first billion years: habitability. *Astrobiology* 21 (8): 893–905.

Shields, A.L. (2019). The climates of other worlds: a review of the emerging field of exoplanet climatology. *Astrophysical Journal, Supplement Series* 243 (2): 30.

Spiegel, D.S., Fortney, J.J., and Sotin, C. (2014). Structure of exoplanets. *Proceedings of the National Academy of Sciences of the United States of America* 111 (35): 12622–12627. www.pnas.org/cgi/doi/10.1073/pnas.1304206111.

Widicus Weaver, S.L. and Fortenberry, R.C. (2021). Space is the future: interdisciplinary research in astrochemistry and planetary science leads to innovations that impact all of chemistry. *Accounts of Chemical Research* 54 (6): 1309–1310.

Problem

1 a) A star has a surface temperature of 6800 K. Orbital and other data show the star to have a mass of 3.2×10^{30} kg and radius of 8×10^8 m. A planet at 2.5×10^{11} m away from the star has a radius of $9{\times}10^6$ m. Assuming the planet absorbs and emits a constant 40% of the starlight incident upon it (i.e. the planet's temperature is constant), predict the planet's surface temperature, assumed uniform over its surface. The energy/area-time emitted by the surface of a black body is $S\ (\mathrm{W/m^2}) = \sigma T^4$, where $\sigma = 5.67 \times 10^{-8}\ \mathrm{W/m^2\text{-}K^4}$.

b) Do you think this planet might be thermally habitable for the type of water-based life that exists on Earth (i.e. ignoring the atmospheric composition)? Explain.

Afterword

How close are we to understanding the living state? We have seen that inquiries into the origin and nature of the living state stretch far back into antiquity and across much of the globe and continue to be pursued with even more vigor and resources in current times. We have relied on the interdisciplinary combination of well-established science in physics, chemistry, biology, and other areas to approach the different mechanistic aspects and emergent properties of the living state, and with some notable successes. Our relatively recent discovery of synthetic macromolecules allows us to assess the similarities and differences between them and biomacromolecules. We see that the actions of the latter are amazingly intricate and specialized, whereas our own synthetic macromolecules, even the ones deemed "stimuli responsive," still have only modest functional characteristics. But macromolecular engineering is evolving quickly, so we don't yet know what complex processes will be harnessed by emerging synthetic materials.

There is no process yet observed in the living state that contradicts any known physical law. At the same time, there is nothing in our current physical laws that allows us to predict the emergence of the living state or its properties. Computers and rockets also use and obey known physical laws, but it is not possible to deduce or predict the existence of these devices from first principles, at this time. A simple information-based reason that life can't be predicted from the laws of physics, is that the amount of information needed to construct the genome of even the simplest species is far greater than the information contained in the laws themselves.

We have seen that, whereas the principle of minimization of free energy drives both inanimate physical processes and living state reactions, it simply can't account for how entities arise which do quite the opposite; namely, accumulate and deploy free energy in a vast number of ways in the living state, which is maintained very far from thermodynamic equilibrium. We have marveled at how staggeringly complex living state phenomena are and involve baffling series of actions by proteins and other agents that have evolved over time. And many of these processes may be optimized. Arguments have been made that pure randomness can't lead to even modest-size proteins with specific sequences, because there is not enough eligible matter (C, P, N, O, etc.) on the face of the Earth, nor enough time, to arrive at these, let alone to create complex metabolic and other biochemical pathways. It seems that there are deep principles involving information and self-organization of matter that await discovery.

Biophysics: Physical Processes Underlying the Living State, First Edition. Wayne F. Reed.

Companion website: www.wiley.com/go/Reed/BiophysicsandMacromolecules

Appendix I: Probability Distributions and Their Averages

Throughout this book, we perform averaging procedures in many different physical contexts, such as finding an average energy, deformation, concentration, electric potential, or molar mass distribution. What these all have in common is the notion of an underlying *probability distribution.* A simple case is computing the average grade, $<G>$, for a test involving N students.[1] Let the test grades be integers from 0 to 100. Let N_j be the number of tests with grade j (e.g. $j = 87$). Then the fraction of tests F_j with grade j is

$$F_j = \frac{N_j}{N} \qquad \text{AI.1}$$

But F_j is also the fractional probability that if one exam is drawn at random from all N tests that the grade is j. We denote this fractional probability P_j which is then:

$$P_j = F_j = \frac{N_j}{N} \qquad \text{AI.2}$$

The P_j forms a *discrete probability distribution* over the integers $j \in [0,100]$. The term "discrete" refers to the integer nature of the distribution. The distribution may have no discernible form, or may follow a certain form; often, the grades will have a "normal" distribution ("bell curve" or *Gaussian*), while others might have a *bimodal distribution*, where a clump of student is in the 80–100 grade range, and another in the 50–70 range, with very few in the 71–79 range.

The probability distribution is *normalized* when it yields unity when summed over all values $j = 0$–100

$$\sum_{j=0}^{100} P_j = 1 \qquad \text{AI.3}$$

To obtain the average grade $<G>$, we perform the usual procedure of adding up all the grades and then dividing by the total number of tests, N

$$\langle G \rangle = \frac{\sum_{j=1}^{100} jN_j}{N} = \sum_{j=1}^{N} jP_j \qquad \text{AI.4}$$

where

$$N = \sum_{j=1}^{N} N_j \qquad \text{AI.5}$$

where we have used the normalized probability distribution P_j in the rightmost expression, and thus omitted the need for the denominator appearing in the first summation.

1 Throughout the book angle brackets, < >, denote the average of a quantity.

Biophysics: Physical Processes Underlying the Living State, First Edition. Wayne F. Reed.

Companion website: www.wiley.com/go/Reed/BiophysicsandMacromolecules

As the number of tests gets large, say millions of tests from a national standardized exam, where grades might be continuous from 0 to 100, instead of an integer, it is convenient to use a *continuous probability density distribution* or *probability density function*. For a continuous distribution of grades we now speak, of the probability density distribution $W(g)$, where g is now a continuous variable, running from 0 to 100, and has substituted the discrete index j used above.

$W(g)$ is a density because it is the probability per unit interval of the continuous grade g (Chapter 4.5.1). In this example, the probability of finding a grade in the interval g to g + dg is $W(g)$dg. The probability P, of finding a grade in the finite interval between g_1 and g_2, $P \in [g_1, g_2]$, is

$$P \in [g_1, g_2] = \frac{\int_{g_1}^{g_2} W(g')dg'}{\int_0^{100} W(g')dg'} \quad \text{AI.6}$$

where the "prime" on g' indicates that g' is a dummy variable of integration. Capital "P" is used because the integration of probability density $W(g)$ over g has yielded a probability.

The probability density distribution $W(g)$ is normalized if the denominator of Equation AI.6 is unity

$$\int_0^{100} W(g')dg' = 1 \quad \text{AI.7}$$

The average grade of the entire distribution, with a normalized $W(g)$ is

$$\langle G \rangle = \int_0^{100} g' W(g')dg' \quad \text{AI.8}$$

An important feature of the probability distribution is that it can be used to find other averages of the distribution beyond $\langle G \rangle$.

A1.1 Moments of Distributions

Let's move away from the test grade example and consider a more general continuous probability function $W(x)$. All the same properties apply to $W(x)$ as to $W(g)$, of course, but now consider other functions of x. Let $W(x)$dx be the probability that a particle be found in the interval x to x + dx where $x \in [-\infty, \infty]$. The average of any other function of x, $f(x)$, is found via

$$\langle f(x) \rangle = \frac{\int_{-\infty}^{\infty} f(x') W(x')dx'}{\int_{-\infty}^{\infty} W(x')dx'} \quad \text{AI.9}$$

For example, let $W(x)$ be an even function about the origin, that is $W(x) = W(-x)$. Then the average position $\langle x \rangle$ is

$$\langle x \rangle = \frac{\int_{-\infty}^{\infty} x' W(x')dx'}{\int_{-\infty}^{\infty} W(x')dx'} = 0 \quad \text{AI.10}$$

This result can be seen immediately because x is an *odd function*, $f(-x) = -f(x)$, and the product of an even and odd function is an odd function, and the integral of any odd function from $-x$ to x is zero. Whether a function is odd or even defines its *parity*. Notice that a function with parity can be formed with linear combinations of a function with no parity. If $f(x)$ has no parity then an even function can be formed by $h_{even}(x) = f(x) + f(-x)$, and an odd function created by $h_{odd}(x) = f(x) - f(-x)$. Table A1 shows the parity of combinations of functions.

The fact that $\langle x \rangle = 0$ for an even $W(x)$ gives us no real information of how much of the x-axis, the particle actually samples. If we compute $\langle x^2 \rangle$, which is an even function, then the result will be nonzero, when $W(x)$ is an even function.

$$\langle x^2 \rangle = \frac{\int_{-\infty}^{\infty} x'^2 W(x')dx'}{\int_{-\infty}^{\infty} W(x')dx'} \neq 0 \quad \text{AI.11}$$

Averages of powers of x, the dependent variable, are termed *moments of the distribution*. In Chapter 7.12, the moments of polymer molar mass distributions are used to determine dispersity indices for the polymer populations. There, the general notation here is adapted to the polymer context.

Table A1 Parity of multiplied functions.

$f(x)$	$g(x)$	$f(x)g(x)$
Even	Even	Even
Even	Odd	Odd
Odd	Even	Odd
Odd	Odd	Even

The nth moment of $W(x)$ about any point x_0, μ_n is given by

$$\mu_n = \int_{-\infty}^{\infty} (x' - x_0)^n W(x')\mathrm{d}x' \qquad \text{AI.12}$$

For a normalized $W(x)$, $\mu_0 = 1$, $\mu_1 = <x> - x_0$. For the second and higher moments $x_0 = <x>$, and the moments are then the moments about the mean value $<x>$. $\mu_2 = <(x - <x>)^2>$ is the *variance* of the distribution:

$$\mu_2 = \int_{-\infty}^{\infty} (x' - <x>)^2 W(x')\mathrm{d}x' = s^2 \qquad \text{AI.13}$$

where s^2 is the variance of the distribution and $\sigma = \sqrt{s^2}$ is the *standard deviation*. The standard deviation is a particularly valuable parameter describing the distribution since it is a direct measure of its width. It is nonzero for all distributions, even and odd, and only goes to zero for a distribution with a single value function; i.e. a *delta function* (Chapter 6.8).

The third moment about the mean is,

$$\mu_3 = \int_{-\infty}^{\infty} (x' - <x>)^3 W(x')\mathrm{d}x' \qquad \text{AI.14}$$

is a measure of the asymmetry of the distribution. For a symmetric $W(x)$ the integrand is odd and $\mu_3 = 0$, as well as all other odd moments. *Skewness*, S, is a measure of asymmetry in a distribution, and one way to represent it is with the dimensionless parameter:

$$S = \frac{\mu_3}{\sigma^3} \qquad \text{AI.15}$$

If $S > 1$, then the distribution is *right-skewed*, meaning the average lies to the right of the median, and for $S < 1$, it is *left-skewed*, meaning the average lies to the left of the median. In the former case, the average of the distribution is higher than its median, and vice versa in the latter case.

The fourth moment μ_4 is a measure of *kurtosis*, or how far the distribution proceeds into extreme values in the tails of the distribution. One measure is given by the dimensionless parameter:

$$K = \frac{\mu_4}{\sigma^4} \qquad \text{AI.16}$$

which describes the weight of extreme values compared to the width of the distribution.

Some of these features can be illustrated with a few well-known distributions, used throughout this book.

A1.2 The Gaussian Distribution, *G*(*x*)

This is one of the most common distributions occurring in nature. Its normalized form is

$$G(x) = \frac{\exp\left[-\left(x - \langle x \rangle\right)^2 / 2\sigma^2\right]}{\sigma\sqrt{2\pi}} \qquad \text{AI.17}$$

$G(x)$ is symmetric about its average value $<x>$, and so odd moments about $<x>$ are all zero.[2] Here, σ, is a measure of the width of the distribution. For example, $S = 0$, that is, there is no skewness to this symmetric distribution, whereas even moments are nonzero. A useful integral is

$$\int_0^{\infty} x^{2n} e^{-\alpha x^2} \mathrm{d}x = \frac{1 \cdot 3 \cdot 5 \ldots (2n-1)}{2^{n+1}\alpha^n} \sqrt{\frac{\pi}{\alpha}} \qquad \text{AI.18}$$

2 $G(x)$ is often designated by other variable letters, depending on the context: e.g. $W(x)$, $c(r,t)$, $n(r,t)$, etc.

where $\alpha = 1/2\sigma^2$. This integral shows that all the higher even moments of $G(x)$, for $n > 2$, are expressed in terms of σ and hence contain no new information; i.e. $G(x)$ is completely defined by $<x>$ and σ.

Table A2 shows the fractional probability of x falling within the listed number of standard deviations, σ, from the mean value $<x>$. At $\pm 1\sigma$, there is 68% probability of x falling within this limit, and 95.5% of falling within $\pm 2\sigma$.

Table A2 The probability $P(x)$ of x falling within the number of '+/- σ' around the mean $<x>$, for the Gaussian distribution. $1-P(x)$ is the probability that x doesn't fall within these limits.

# of σ's	$P(x)$	$1-P(x)$
0.1	0.07965643591996	0.920344
0.5	0.38292829889903	0.617072
1	0.68269413318009	0.317306
2	0.95450180718719	0.045498
3	0.99730045893796	0.0027
4	0.99993666778315	6.33E-05
5	0.99999942683942	5.73E-07
6	0.99999999802752	1.97E-09
7	0.99999999999744	2.56E-12

A1.3 The Exponential Distribution

Throughout this book, we have used the exponential Boltzmann energy distribution, $b(E)$ of the form:

$$b(E) = Ae^{-\alpha E} \tag{AI.19}$$

where $\alpha = 1/k_BT$, and A is the normalization constant, which for $E \geq 0$ is $A = \alpha$. In evaluating the moments the following integral formula is useful

$$\left[\int_0^{\infty} x^n e^{-\alpha x} dx = n!\left(\frac{1}{\alpha}\right)^{n+1}\right] \tag{AI.20}$$

$\mu_0 = 1$, $\mu_1 = <E> = k_BT$, and $\mu_2 = <E^2> = 2(k_BT)^2$, $\sigma = k_BT$. Notice that there are no zero-value moments. The skewness, using the third moment μ_3 about the mean $<E>$ is

$$S = 2 \tag{AI.21}$$

which means the distribution is right-skewed. The median of the distribution is $E_{median} = k_BT \ln 2 = 0.693k_BT$.

A1.4 The Binomial Distribution (BD)

The BD was introduced in Chapter 4.19.1 in connection with probabilities. It is repeated here as it describes a wide variety of situations in which the probability of the outcome of a certain event q, remains fixed for all trials; for example, the probability of rolling a certain number of a die. Other distributions, such as the Gaussian, geometric, and Poisson distributions are limiting cases of the binomial distribution.

If n independent events occur, each with the same probability q for success, the probability for failure in each event is $(1 - q)$. The probability that j events out of n are successful, $P_{n,j,q}$, is given by the BD (Chapter 4.19.2)

$$P_{n,j,q} = \frac{n!}{(n-j)!\,j!} q^j (1-q)^{n-j} \quad \text{AI.22}$$

The $n!$ in the numerator is the number of ways n distinct outcomes can be arranged. Since the only "objects" here are "success" and "failure" of a specific outcome, most of these arrangements are redundant. $(n - j)!$ and $j!$ in the denominator factorial terms are the number of permutations in which $(n - j)$ failures and j success can occur, respectively. Dividing $N!$ by these terms then gives the total number of ways of obtaining j successes and $(n - j)$ failures in n events, regardless of the order in which they occur. The complete factorial term then gives the number of sequences producing j successes and $(n - j)$ failures. The average number of events that occur is computed in the usual way

$$\langle j(n,q)\rangle = \frac{\sum_{j=0}^{n} jP_{n,j,q}}{\sum_{j=0}^{n} P_{n,j,q}} = qn \quad \text{AI.23}$$

where a summation rule has been used to solve for the sums shown.

An example: A coin is tossed 10 times ($n = 10$). The probability of obtaining two heads ($j = 2$) during this trial, according to Equation 4.212 is 45/1024 = 4.4%. The probability of obtaining five heads is 252/1024 = 24.6%. The average number of heads for many individual trials of 10 tosses each is 5.

The Gaussian distribution of A1.2 is a continuous probability function obtained from the binomial distribution in the case where $q = 0.5$, and $n \gg j$, and the average $nq = m \gg 1$.

The principles used to derive the binomial distribution can be generalized to trinomial ($N = 3$ in Equation 4.213) and any multinomial distribution (arbitrary integer N in Equation 4.213), as described in Chapter 4.19.1 and any multinomial distribution. Given the probability of event A is q_A and of event B is q_B, the probability of neither event A or B is $(1 - q_A - q_B)$. The trinomial distribution gives the probability that after n events, A will have occurred j times, B k times, and neither A nor B $n - j - k$ times is

$$P_{n,j,k,q_A,q_B} = \frac{n!}{j!\,k!\,(n-j-k)!} q_A{}^j q_B{}^k \left(1 - q_A - q_B\right)^{n-j-k} \quad \text{AI.24}$$

A1.5 Poisson Distribution (PD)

This is the BD in the special case where $N \gg j$ and $q \ll 1$. It occurs when the probability of an event occurring (often in a given interval of time) is extremely low and the number of successful events (often in a given interval of time) is very low compared to the total number of events. Some examples include:

- Photons arriving at a telescope per unit time from a distant star. In this case, q is dramatically small, being the surface area of the telescope divided by the surface area of the spherical shell surrounding the star whose radius is that of the earth–star distance. The number of photons j per unit of time received is vastly smaller than the number of photons emitted by the star.
- Consider a very rarefied gas of N molecules in a large volume V. The probability of a molecule being in a specific small volume $v \ll V$ is $q = v/V \ll 1$. The number of particles j in a specific volume v is much smaller than N.
- In radioactive decay, the number of radioactive atoms $N \gg j$, where j is the number of atoms decaying in a given time interval. The probability of decay in a time interval $\delta t \ll \tau_{1/2}$, where $\tau_{1/2}$ is the half-life of the species. If there are N_0 radioactive atoms at the start of the interval, the number that decay during δt is $\Delta N = N_0(1 - e^{-\delta t/\tau_{1/2}}) \cong N_0 \frac{\delta t}{\tau_{1/2}}$, and so the count rate (counts/second) over the interval is just $N_0/\tau_{1/2}$.

The Poisson distribution is sometimes called the *law of small numbers* or the *law of rare events*. In the case of the PD, the following approximations can be made:

$$\frac{N!}{(N-j)!} \approx N^j \quad \text{AI.25}$$

and

$$\begin{aligned}(1-q)^{N-j} &= 1-(N-j)q+\frac{1}{2!}(N-j)(N-j-1)q^2-\frac{1}{3!}(N-j)(N-j-1)(N-j-2)q^3+\cdots \\ &\approx 1-q+\frac{1}{2!}q^2-\frac{1}{3!}q^3+\cdots = e^{-q} \\ &\approx e^{-qN} = e^{-m}\end{aligned} \qquad \text{AI.26}$$

With $m = qN$, the average value of j for the distribution:

$$P_j = \frac{m^j e^{-m}}{j!} \qquad \text{AI.27}$$

that is, m is the average number of successful events, $m = Nq$. The distribution is completely determined by the single parameter m. It is, however, not a continuous function, and so still requires summation techniques, rather than integration, to determine average values from it.

The width or standard deviation of P_i for the PD is sqrt(m). An important property of the PD is that the ratio of the width to the average value is hence 1/sqrt(m). Hence, this ratio decreases as m increases. Pollsters often quote their sampling error simply by using 1/sqrt(m), where m is the number of people sampled; for example, in the United States when 1000 people are polled, the pollster often claims a sampling error of ±3%. What they do not usually delve into is how the 1000 people chosen "at random" can actually be representative of a population of 3.3×10^8.[3]

A familiar example of the Poisson distribution in physics is in radioactive decay of long-lived isotopes. Here, there is a very large number of radioactive nuclei, N, but a short number of "successful events" i (i.e. radioactive decays) during a typical measurement interval, and the probability p of any given nucleus decaying during this time is small. $m = Np$ is the average count rate for a measurement, but as Equation AI.27 shows, there is a significant probability of finding values other than m.

An example of Poisson distributions for various <j> is shown in Figure AI.1. At large values of m, the Gaussian distribution becomes a good approximation to the Poisson distribution. Figure AI.1a plots the Poisson distribution for selected low values of <j>, low values of <j>, and Figure AI.1b plots the distribution for high values of <j>.

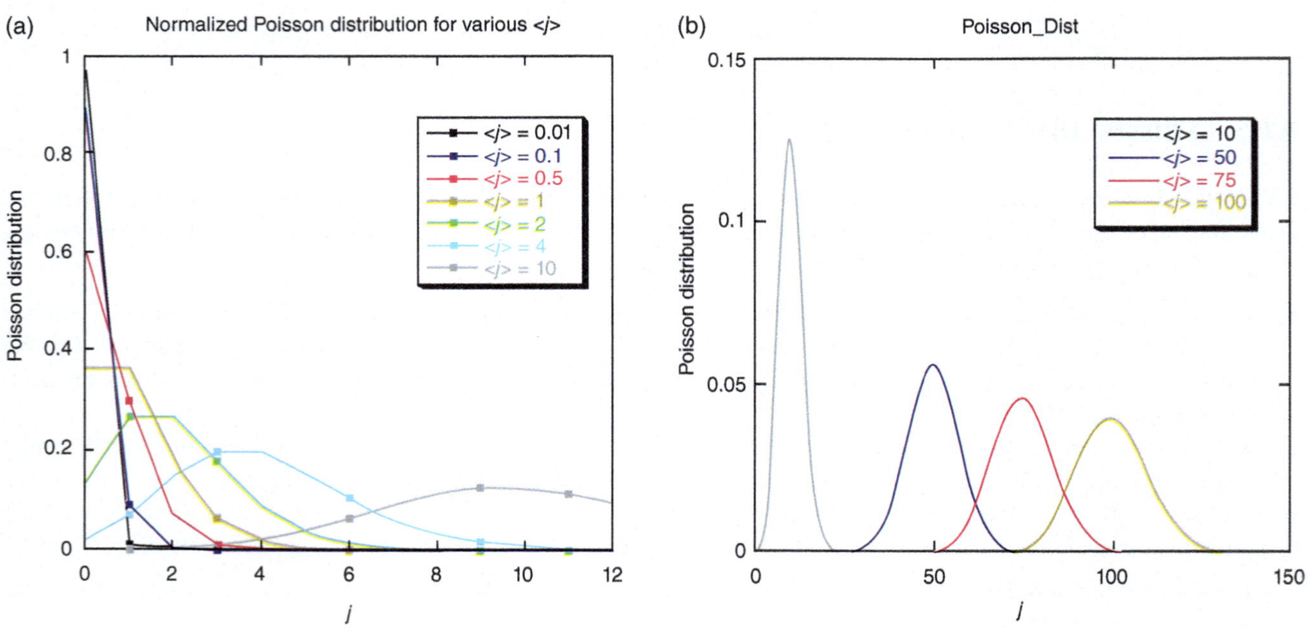

Figure AI.1 (a) Poisson distribution for low values of <j>. (b) Poisson distribution for high values of <j>.

3 In a recent criticism of US polling, it was revealed that many polls are conducted only with people who have telephones with landlines in their houses. Since these people are predominantly older than those who have only a cellular telephone the sample is no longer random, it is biased toward older people. More recent polling techniques are reaching cell phones.

Appendix II: Review of Vector Calculus and Notation Used

Vector, Integral, and Differential Notation

Many different notations are used in vector calculus throughout texts and scientific literature and often vary by field. Here, we use notation that is similar to that used in many physics textbooks.

First, we use arrows for vectors, rather than boldface, because you can draw the arrows with pen and pencil while working chapter problems, but it is difficult to produce boldface by hand.

In *rectangular coordinates*, also called Cartesian coordinates, we use $\hat{\mathrm{i}}$, $\hat{\mathrm{j}}$, and $\hat{\mathrm{k}}$ as the unit vectors for the x, y, and z directions, respectively. The *position vector* with respect to the origin of the coordinate system in rectangular coordinates is $\vec{r}$ and is given by

$$\vec{r} = x\hat{\mathrm{i}} + y\hat{\mathrm{j}} + z\hat{\mathrm{k}} \qquad \text{AII.1}$$

The area element is $\mathrm{d}A = \mathrm{d}x\mathrm{d}y$. When used in a vector integral, the vector area element $\mathrm{d}\vec{A}$ is used in the integrand over a surface S in *any coordinate system*:

$$\mathrm{d}\vec{A} = \hat{\mathrm{n}}\mathrm{d}A \qquad \text{AII.2}$$

where $\hat{\mathrm{n}}$ is the normal to the surface at the position of the surface area element $\mathrm{d}A$. An example of a vector $\vec{J}$ over an *open surface S* is represented as

$$\int_S \vec{J} \cdot \mathrm{d}\vec{A} \qquad \text{AII.3}$$

The area over a *closed surface* S is represented by the integral with a circle in the middle.

$$\oint_S \vec{J} \cdot \mathrm{d}\vec{A} \qquad \text{AII.4}$$

The volume element in any coordinate system is represented by d^3r. In rectangular coordinates $\mathrm{d}^3r = \mathrm{d}x\mathrm{d}y\mathrm{d}z$. A *volume integral* over volume V in any coordinate system is represented by

$$\int_V \vec{J}\mathrm{d}^3r \qquad \text{AII.5}$$

The infinitesimal path length vector is represented by $\mathrm{d}\vec{s}$ in any coordinate system. A path integral from points A to B in space is represented by

$$\int_A^B \vec{J} \cdot \mathrm{d}\vec{s} \qquad \text{AII.6}$$

Biophysics: Physical Processes Underlying the Living State, First Edition. Wayne F. Reed.

Companion website: www.wiley.com/go/Reed/BiophysicsandMacromolecules

Around a *closed path C*, the integral is represented by the integral with a circle in the middle.

$$\oint_C \vec{J} \cdot d\vec{s} \qquad \text{AII.7}$$

A function which is a variable of spatial coordinates will often be represented generally as $f = f(\vec{r})$. The gradient of f is represented as

$$\text{Gradient } f = \nabla f \qquad \text{AII.8}$$

The divergence of the gradient, or the *Laplacian*, is represented by ∇^2, so that, in any coordinate system:

$$\nabla \cdot \nabla f = \nabla^2 f \qquad \text{AII.9}$$

In rectangular coordinates

$$\nabla^2 = \frac{\partial^2}{\partial x^2} + \frac{\partial^2}{\partial y^2} + \frac{\partial^2}{\partial z^2} \qquad \text{AII.10}$$

The curl of a vector $\vec{J}$ is, in any coordinate system

$$\text{Curl } \vec{J} = \nabla \times \vec{J} \qquad \text{AII.11}$$

A couple of identities of special importance are

$$\nabla \times (\nabla U) = 0 \qquad \text{AII.12}$$

which is a defining condition for a conservative vector field $\vec{E}$ to be represented via a scalar potential function U, by $\vec{E} = -\nabla U$.

Similarly,

$$\nabla \cdot (\nabla \times \vec{J}) = 0 \qquad \text{AII.13}$$

expresses the fact that the divergence vanishes for the curl of a nonconservative vector field $\vec{J}$ with a nonzero curl.

Divergence, Green's Theorem, and Exponential Integrals

The *divergence theorem*, which relates the integral of the divergence of a vector function over a volume to the integral of the function over the surface is

$$\oint_S \vec{J} \cdot d\vec{A} = \int_V \nabla \cdot \vec{J} d^3 r \qquad \text{AII.14}$$

Green's theorem relates the curl of a vector field over an open surface S to the closed path integral around the surface's boundary

$$\int_S \nabla \times \vec{J} \cdot d\vec{A} = \oint_C \vec{J} \cdot d\vec{s} \qquad \text{AII.15}$$

Two integrals that occur frequently, especially in averaging calculations are the following:

$$\int_0^{\infty} x^n e^{-\alpha x} dx = n! \left(\frac{1}{\alpha}\right)^{n+1} \qquad \text{AII.16}$$

and

$$\int_{-\infty}^{\infty} x^{2n} e^{-\alpha x^2} \mathrm{d}x = \frac{1\cdot3\cdot5\ldots(2n-1)}{2^n \alpha^n}\sqrt{\frac{\pi}{\alpha}} \quad \text{AII.17}$$

Many quantities in the physical world can be represented by harmonic waves. Time or spatial averages of harmonic waves over one period are zero. In many instances what is measured physically is the square of the quantity; for example, in an ac circuit, the power transferred is proportional to the square of sinusoidal voltage or current. An ac voltage signal from a household socket has the form:

$$V(t) = V_0 \sin(2\pi ft) \quad \text{AII.18}$$

where v_0 is the peak voltage (172 V in North America) and f is the voltage frequency (60 Hz in North America, 50 Hz in Europe). When integrated over one cycle $T = 1/f$ the result is zero, that is

$$\langle V(t)\rangle = V_0 \int_0^T \sin(2\pi ft')\mathrm{d}t' = 0 \quad \text{AII.19}$$

But, despite this zero average value, we know that the voltage can power an appliance or electrocute someone who touches the two terminals. This is because it is the power, not the voltage alone, that determines energy delivered from the socket to an external circuit. Power $(t) = V(t)^2/R$, where R is the resistance of the external circuit. So,

$$\langle \text{Power}(t)\rangle = \frac{\langle V(t)^2\rangle}{R} \quad \text{AII.20}$$

And the mean square voltage is

$$\langle V(t)^2\rangle = V_0^2 \int_0^T \sin^2(2\pi ft')\mathrm{d}t' = \frac{V_0^2}{2} \quad \text{AII.21}$$

A relevant nonzero voltage average is the square root of this and is termed the *root mean square voltage*, or V_{RMS}. This is quoted when referring to a country's ac voltage; V_{RMS} = 120 V in North America and 230 V in Europe.

Any sinusoidal quantity, for example, the intensity of a simple harmonic sound or light wave will give the same integration factor of 1/2, found in Equation AII.21.

Spherical Polar Coordinates

Many situations in three dimensions are more conveniently treated with spherical polar coordinates, rather than Cartesian coordinates. Spherical polar coordinates use r, θ, ϕ, as shown in Figure AII.1, where

r is the distance from the origin, $r \in \{0, \infty\}$
θ is the *altitude angle*, or *latitude angle*, with $\theta = 0°$ pointing to the north pole, $\theta \in \{0, \pi\}$
ϕ is the *azimuth angle*, or *longitude angle*, measured from the $+x$ axis, $\phi \in \{0, 2\pi\}$

The transformation from rectangular coordinates to spherical coordinates is

$$z = r\cos\theta \quad \text{AII.22a}$$

$$y = r\sin\theta\sin\phi \quad \text{AII.22b}$$

$$x = r\sin\theta\cos\phi \quad \text{AII.22c}$$

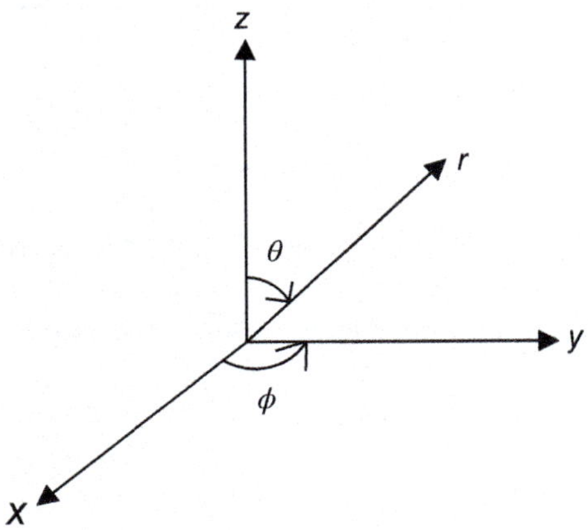

Figure AII.1 Geometry for spherical polar coordinates.

The infinitesimal area element at $r = R$ in spherical coordinates is

$$dA = R^2 \sin\theta d\theta d\phi = R^2 d\Omega \qquad \text{AII.23}$$

where Ω is the *solid angle*, measured in *steradians*. A useful property of the solid angle, when integrated over the entire surface, is

$$\oint_S d\Omega = 4\pi \qquad \text{AII.24}$$

The volume element in spherical coordinates is

$$d^3r = r^2 dr \sin\theta d\theta d\phi \qquad \text{AII.25}$$

The differential operators in spherical coordinates are

$$\nabla f = \hat{r}\frac{\partial f}{\partial r} + \hat{\theta}\frac{1}{r}\frac{\partial f}{\partial \theta} + \hat{\phi}\frac{1}{r\sin\theta}\frac{\partial f}{\partial \phi} \qquad \text{AII.26}$$

$$\nabla \cdot \vec{J} = \frac{1}{r^2}\frac{\partial}{\partial r}(r^2 J_r) + \frac{1}{r\sin\theta}\frac{\partial}{\partial \theta}(J_\theta \sin\theta) + \frac{1}{r\sin\theta}\frac{\partial J_\phi}{\partial \phi} \qquad \text{AII.27}$$

$$\nabla \times \vec{J} = \hat{r}\frac{1}{r\sin\theta}\left[\frac{\partial}{\partial \theta}\left(J_\phi \sin\theta\right) - \frac{\partial J_\theta}{\partial \phi}\right] + \hat{\theta}\frac{1}{r}\left[\frac{1}{\sin\theta}\frac{\partial J_r}{\partial \phi} - \frac{\partial\left(rJ_\phi\right)}{\partial r}\right] + \hat{\phi}\frac{1}{r}\left[\frac{\partial\left(rJ_\theta\right)}{\partial r} - \frac{\partial J_r}{\partial \theta}\right] \qquad \text{AII.28}$$

$$\nabla^2 f = \frac{1}{r^2}\frac{\partial}{\partial r}\left(r^2\frac{\partial f}{\partial r}\right) + \frac{1}{r^2\sin\theta}\frac{\partial}{\partial \theta}(\sin\theta)\frac{\partial f}{\partial \theta} + \frac{1}{r^2\sin^2\theta}\frac{\partial^2 f}{\partial \phi^2} \qquad \text{AII.29}$$

While the altitude and azimuth angles are represented by θ and ϕ, respectively, in Chapter 9 θ and ϕ are interchanged, to follow the majority of the literature, and the angle ϕ in Chapter 5 is changed to φ Equation 5.17 so that phi stands for potential.

Index

Note: Page numbers in *italic* and **bold** refers to figures and tables, respectively; Page numbers followed by "n" refer to foot notes.

Biophysics: Physical Processes Underlying the Living State, First Edition. Wayne F. Reed.

Companion website: www.wiley.com/go/Reed/BiophysicsandMacromolecules

d

e

f

g

o

p

t

Wiley The manufacturer's authorized representative according to the EU General Product Safety Regulation is Wiley-VCH GmbH, Boschstr. 12, 69469 Weinheim, Germany, e-mail: Product_Safety@wiley.com.

Printed and bound by CPI Group (UK) Ltd, Croydon, CR0 4YY
19/08/2026
02200421-0004

Mycology for Architecture

Mycology for Architecture

Methods and Applications of Mycelium Materials in the Built Environment

Edited by Dilan Ozkan

Sustainable Futures

Pelagic Publishing | www.pelagicpublishing.com

First published in 2026 by
Pelagic Publishing
20–22 Wenlock Road
London N1 7GU, UK

www.pelagicpublishing.com

Mycology for Architecture: Methods and Applications of Mycelium Materials in the Built Environment

https://doi.org/10.53061/MSVZ3329

Sustainable Futures series

A CIP record for this book is available from the British Library

EU Authorised Representative: Easy Access System Europe –
Mustamae tee 50, 10621 Tallinn, Estonia, gpsr.requests@easproject.com

ISBN 978-1-78427-651-5 Hbk
ISBN 978-1-78427-652-2 Pbk
ISBN 978-1-78427-653-9 ePub
ISBN 978-1-78427-654-6 PDF

Front cover image: Digitally overlaid 3D scans of oyster mushrooms. (Dilan Ozkan)

Frontispiece: Photographs of oyster mushrooms grown under different environmental conditions: low and high humidity, and high and low carbon dioxide concentrations. (Dilan Ozkan)

Typeset in Palatino LT Std by S4Carlisle Publishing Services, Chennai, India

Contents

Foreword

When we applied for the grant which became the Hub for Biotechnology in the Built Environment (HBBE) – a joint initiative between Northumbria University and Newcastle University – we knew that we needed to set up some parameters. Given seemingly infinite biotechnological possibility, we needed to define what we could and couldn't do. As the predominant expertise in the group was for bacterial microbiology we would focus on bacteria. Fungus was out of scope. We admit that this decision, which meant that there wasn't a single line on mycelium in our research application, was also borne of a prejudice. Our sole experience of mycelium-based materials had been in biodesign exhibitions where mycelium materials were often presented on trestle tables dotted with samples of what appeared to be mouldy bread. This was hardly, in our thinking, a material of the future. We were wrong.

When we received the funding in 2019 (the date will become important), things rapidly changed. First, we already had in our cohort a mycelium specialist. Dilan Ozkan, who now edits this book, had been working, alone, on a mycelium-based PhD and was with each supervision persuading me that mycelium was more than material. It was, in her work, more like a material platform. Depending on your perspective, fungi were factories for the production of many different types of material; or, in Dilan's case, both a muse and a thought experiment. Through their different phases of life, mycelium exhibits a phenomenal range of characters from the deceptively simple filamentous network of mycelia through to the (often ignored in biodesign) fruiting bodies with complex morphologies and responses to environmental conditions. This is not just a material.

Early in her PhD, Dilan lamented that she didn't really have a support community in our lab. We mentioned, in passing, that a way to counteract this was to create a group. What we meant was to reach out to a few like-minded folks via email; maybe attend a conference or two; perhaps a newsletter in LinkedIn. What Dilan did, both hampered by but eventually also facilitated by COVID-19 (which shut our newly created labs for close to a year), was to build a much more formal network – Mycology for Architecture – powered by an online seminar series which became videos. It seemed that every name in the world of mycelium materials was happy to donate a bit of their time to a series of incredibly rich and varied presentations. The videos, which were hosted on the HBBE YouTube channel, remain some of our most viewed. Dilan had also tapped into a moment. As we were locked into our homes, our now-habitual use of video conferencing allowed for a much greater range of people to be brought together, and for our content to be free and open for anyone with an interest.

As the lockdowns came to an end, we were left with a cohort of more than 50 researchers and a growing (it's hard to avoid these puns in biodesign) interest in mycelium-based materials such that it was not only a fringe activity but also came to dominate much of our outward-facing research. Our first, and still most significant, scaled work came from Jane Scott and Ben Bridgens, whose BioKnit structures offered a compelling example of a new way of using the material in conjunction with knitted textiles to create complex geometry and shape space. Probably our most important materials research output in the early days of the HBBE came from Elise Elsacker's project on self-healing mycelium composites, which was already approaching 100 citations last time we checked. And we should also mention PhDs on mycelium-based materials from Romy Kaiser-Style and Monika Brandić Lipińska (both represented in this volume).

By the end of the funding period for the HBBE, our labs were dominated by boxes of growing mycelium which, to the chagrin of many of the other lab users, had a habit of sprouting fruiting bodies which would fall into other experiments to create new mycelium-based materials in unintended places.

Dilan's Mycology for Architecture network also helped to make the HBBE into a centre for mycelium research, and helped to promote many other labs. It became a home for individuals who were often regarded as the 'miscellaneous' researchers in their own departments – seen as outside the norm of architectural research.

The contemporary landscape, however, is very different. Thanks in part to the work of HBBE *alumnus* Assia Crawford, the 2023 ACADIA (Association for Computer Aided Design in Architecture) conference had, in the words of an anonymous delegate in an overheard conversation, 'a bit too much mycelium this year'. Can we have too much mycelium? Companies from Ecovative to MycoWorks are also sprouting up like… well, you see where we're going.

This publication brings together many of those who were part of the original Mycology for Architecture network, and more besides. While it invites a wide range of voices, it also reflects Dilan's editorial and intellectual position. The book offers a practical approach to working with mycelium – concerned as much with the how as with the why. It also maintains the sense of awe and curiosity that all of us feel when we realize just how extraordinary biology, and fungi in particular, are as material makers as well as recyclers, remediators and even space explorers.

We congratulate Dilan on this generous endeavour, and hope that after reading it you will be inspired to listen next time you see someone standing at a trestle table waxing lyrical about what looks like a piece of mouldy bread.

Martyn Dade-Robertson and Meng Zhang
Northumbria University

Information on Contributors

Inyoung Yonah Lee, received her BSc in Chemistry and Chemical Engineering, then her MSc degree in Chemical Engineering and Biotechnology at EPFL. She is an enthusiast of membrane materials for separation processes and joined the Sustainable Materials Laboratory to work on the synthesis of self-assembled membranes and on biofilms from fungal fermentation.

Blanca de Miguel Martínez completed her undergraduate degree in Chemical Engineering at Complutense University of Madrid and her master's degree in Sustainable BioEnergy at the Technical University of Denmark. She has worked on biomass valorization, biocatalysis and phase change materials for thermal energy storage. Blanca is currently doing her PhD in the Sustainable Materials Laboratory, working on nanomaterials from fungal fermentation.

Chiara Moretti is a researcher specializing in bio-based materials. She obtained her BSc and MSc in Chemistry from the University of Florence (Italy) and her PhD in Chemistry from the École Normale Supérieure of Lyon (France). She has a strong background in nanomaterials, particularly in the synthesis and self-assembly of colloidal inorganic nanoparticles. She is currently a postdoctoral researcher at the Sustainable Materials Laboratory at EPFL (Switzerland). Her research focuses on the development of mycelium-based composite materials and functional coatings.

Tiffany Abitbol is a professor at EPFL, where she heads the Sustainable Materials Laboratory within the Institute of Materials in the School of Engineering. Prior to joining EPFL in 2022, she worked as a Research Scientist at RISE Research Institutes of Sweden. She holds a PhD in Chemistry from McGill University, where she was supervised by Professor Derek Gray. Her research group focuses on bionanocomposites based on lignocellulosic and proteinaceous biomass, with a fundamental interest in self-assembly and growth-driven assembly by microorganisms. Her position is an endowed chair co-funded by BASF SE, Logitech, Nestlé, and SIG Services AG.

Thibaut Houette is an architect with a PhD in Integrated Bioscience, working at Ceebios on multi-scale and multifunctionality in biomimicry. Captivated by biological and ecological strategies, his academic research and practice experience focuses on combining biomimicry, parametric design, and regenerative design. His PhD at the Biomimicry Research and Innovation Center of The University of Akron experimented with ways to implement biological growth into living architecture through biomimicry and biotechnology. Convinced by the potential of biomimicry to address complex multiscale challenges towards a symbiotic inhabitation of Earth with other species, his interdisciplinary projects span and combine scales from materials to ecosystems.

Petra Gruber is a researcher and educator in architecture and biodesign. She is a University Professor and Department Head of Building Construction at the University of Applied Arts Vienna. She holds a PhD in "Biomimetics in Architecture" from the Vienna University of Technology in Austria and worked internationally on three continents in inter- and transdisciplinary design, research, and education, at the intersection of biology, architecture and art. Her work has been published widely in books, journals, exhibitions, and documentary films. She holds lectures and workshops worldwide and acts as reviewer for scientific journals and funding organizations.

Jan Wurm is professor of Regenerative Design and Biofabrication in Architecture at the Department of Architecture of the KU Leuven and has established the BioDigital ConstructionLab. He is recognized as a leading expert for applied research and innovation in the field of sustainable construction. Based on his background as a carpenter, a researcher and lecturer at RWTH Aachen and a director at Arup in Europe, Jan is passionate about bridging the gap between academic research and practice in the field of regenerative design. The current research focus is on mycelium-based composites with enhanced aesthetic, functional and environmental qualities integrating surface substrates and digital fabrication technologies.

Kumar Biswajit Debnath is a building physicist and Chancellor's Research Fellow at the School of Architecture, University of Technology Sydney, pioneers bio-intelligent building systems and bio-building physics – interdisciplinary approaches that harness architectural science, engineering, and biotechnology to develop passive, self-regulating structures for climate adaptation and extraterrestrial applications. In this book's chapter, he critically examines mycelium-based composites' potential for climate-responsive façades, confronts real constraints (moisture, mechanical limits, degradation), and reframes limitations as opportunities – modular, replaceable panels, bio-based coatings, bio-digital fabrication, and living adaptive skins – advancing circular, net-zero building technologies.

Monika Brandić Lipińska is a space architect and researcher. Her work explores how biotechnologies and living materials can shape resilient habitats in extreme environments and how these lessons can inform more sustainable building on Earth. She is a research fellow at the University of Innsbruck, working with bacterial cellulose for architectural and space applications. Her PhD at Newcastle University's Hub for Biotechnology in the Built Environment, in collaboration with NASA Ames Research Center, focused on mycelium–regolith composites for constructing Martian habitats. She holds MSc degrees from the International Space University and Lund University. Monika is co-founder of Bio-Futures for Transplanetary Habitats.

Adrien Rigobello, Ph.D., is a postdoctoral researcher at the Royal Danish Academy's Chair for Biohybrid Architecture. His research focuses on postindustrial design for eco-social sustainability, through the study of fermentation, locality, community autonomy, and aesthetics. His Ph.D. centred on the study of the eco-socially sustainable conditions for the design and diffusion of mycelium materials, grounded in fungal biology, material science, and design studies using anthropology, philosophy of science and aesthetics. Adrien has developed an expertise in technical strategies for research and society with a background as a computational designer, innovation manager, industrial engineer and consultant in the life science industry. He is the president-founder of the NGO design studio thr34d5 which researches contemporary media for social justice with a focus on biomaterials and digital media. Adrien has been academic coordinator of the Advanced Master Design by Data hosted in Ecole des Ponts ParisTech, and the founding manager of Ecole des Ponts ParisTech's makerspace.

Ena Naito is an interdisciplinary design researcher at Aalto University working at the intersection of biology, design, and co-learning. Her doctoral research examines emerging biodesign workspaces and their role in facilitating experimental, interdisciplinary practices aimed at fostering sustainable material cultures. She founded and leads the BioMakerStudio at Aalto - a hybrid biolab-makerspace-studio dedicated to hands-on biodesign research and education. Ena is also actively involved in teaching and public engagement, promoting accessible and collaborative approaches to biodesign. Ena holds an MA in Contemporary Design with CHEMARTS from Aalto University, and BFA from the University of Oxford, UK.

Gabriela Farias is a versatile designer and researcher with roots in community activism and mixed media fabrication. She has integrated her experience from these realms in her recent Master's thesis at Aalto University, working with and researching mycelium biomaterials in community-driven biodesign, exploring how mycelium as a biomaterial responsibly interacts in situ with localized knowledge practices of social and environmental sustainability. She aims to give more voices a platform in biodesign, while also exploring the potential of mycelium, its growth mediums and the shapes it can hold. She currently works with mycelium materials in the Netherlands.

Manuel Arias Barrantes is a designer working between the disciplines of design, biology, technology and craft. Inspired by the notions of sustainability and new digital and biological tools, he has been actively researching for the last five years the topic of mycelium-based materials in collaboration with scientists at VTT Technical Research Centre of Finland. He is currently conducting doctoral studies at Aalto University in collaboration with VTT. His research focuses on biofabrication processes that integrate design and bioengineering principles for the creation of materials that could self-heal in the near future. He is also interested in methodologies for guiding creative processes in scientific environments. Manuel holds an MA in Creative Sustainability from Aalto University and a BA in Visual Communication.

Barbara Pollini is an Assistant Professor in Biodesign Integration at Aalto University, with a background in sustainable design, material design, and biodesign. She holds a Master's degree in ecodesign and eco-innovation, and a Master's in computational design. She has been a senior lecturer at Naba Design University in Milan and Domus Academy for more than ten years, teaching sustainable design and materials, DIY materials, biodesign and design fiction for sustainable futures. In 2023, Barbara concluded a PhD in Design at Politecnico di Milano with research on biodesign's potential for conventional and regenerative sustainability. She constantly switches between hands-on experimentation and theoretical research; her primary research interest focuses on material scenarios based on regenerative processes, including living and life-enabling materials.

Jessica Dias is a biomaterials researcher and designer whose work bridges biological sciences, design and education. She specialises in mycelium-based materials and is committed to making biodesign more accessible through workshops, seminars and hands-on learning experiences. Jessica holds an MEng in Architecture and Environmental Design from the University of Nottingham and a Master's degree in Advanced Architecture from the Institute for Advanced Architecture of Catalonia in Barcelona, focused on computational design and digital brication. In 2017, she co-founded Biobabes, followed by S-BIOTICA in 2020. Together th Catherine Euale, she co-developed an online course on mycelium-based materials.

Elisa Strobel do Nascimento is a professor at the Design Department at Universidade Federal do Paraná (UFPR), Brazil. She has a background in product design with a PhD from UFPR. Elisa worked on projects in different industries, from toys to vehicles. At UFPR, she co-leads the department's Prototyping Lab with the project 'Floresta: Laboratório de Design Mais-Que-Humano' (Floresta: More-Than-Human Design Lab). Her research aims to facilitate teaching and learning biodesign in undergraduate education: 'Design with the Living: Learning to gether'. She is also an associated member of the Cluster of Excellence Matters of Image, Space, Material.

Dilan Ozkan is a researcher and registered architect developing transferable frameworks for designing and fabricating with living materials. Her work connects biological behaviour, parametric design logic, and material performance into replicable methodologies for sustainable construction, moving between conceptual theory and hands-on practice. Her research focuses on two interlocking questions: how biological systems can be understood as parametric design media, and how mycelium composite production can be systematised. This work draws on material experimentation, computational design, and digital fabrication, grounded in a PhD investigating fungal growth as a designable process at Hub for Biotechnology in the Built Environment, Newcastle University.

Martyn Dade-Robertson is Professor of Emerging Technology at Northumbria University, known for his pioneering work at the intersection of architecture, computing and synthetic biology. As founder of the Hub for Biotechnology in the Built Environment, he co-leads research in living construction, developing innovative biological fabrication methods which includes projects such as Living Manufacture to develop a biological based materials manufacturing system and Thinking Soils to develop a microbe powered self-constructing building foundation. He edits a Routledge book series, wrote the book *Living Construction* and serves as editor-in-chief of Cambridge University Press's *Biotechnology Design* journal. His work explores the transformative potential of biotechnology in architecture, championing sustainable and responsive design through living systems and materials.

Abhinav Chaudhary is the Head of Computational Design at PLP Architecture with a focus on parametric modelling, complex geometries and material optimization across various stages of retail, mixed-use and residential projects in Europe and Asia. Prior to joining PLP, he worked on several award-winning installations at Orproject's New Delhi office, dealing with fabrication of complex geometries in timber, metal and plastic. He is also a Studio Tutor at the Architectural Association's graduate programme, Emergent Technologies and Design.

Savannah Willits is the Team Lead of PLP Labs, the research and development arm of PLP Architecture, with a background in planning policy and commercial real estate. She facilitates interdisciplinary research with global experts to serve the studio's architectural and urban work, as well as external clients with bespoke research, by operating outside the confines of traditional architecture. She holds an MSc in Urban and Regional Planning Studies from the London School of Economics and is the recipient of the Oram-Stott-Schlusche Scholarship.

Phillip Gough is a a Senior Lecturer in Design at The University of Sydney School of Architecture, Design and Planning. His research primarily focuses on sustainable interaction design and biodesign, exploring novel applications of mycelium-based materials, their application to design, and impacts on user experience with technology. His research also engages community groups with biomaterials for a circular economy. He is a member of the Care-ful Technologies lab, where he carries out teaching and research on responsible and ethical design practices.

Yuxin Gong earned a Bachelor's degree in Landscape Architecture from Zhejiang University in 2023 and is currently pursuing a Master's in Interaction Design and Electronic Arts at the University of Sydney, expected to graduate in 2025. Her research interests include computational design, digital fabrication, interactive installations, AIGC and game design. With a multidisciplinary approach, she explores emerging design technologies while drawing aesthetic inspiration from fashion, film and illustration. Passionate about hands-on experimentation and speculative design ideas, she actively engages in projects that bridge creative exploration and practical implementation.

Bahareh (Bahar) Barati is an Assistant Professor with a focus on integrating bio and living materials—alongside craft and related technologies such as digital fabrication and living media interfaces—into design research and education. Her work champions material-driven design and challenges dominant human-centered paradigms in pursuit of alternative ways of understanding, relating, and cohabiting in more-than-human worlds. Bahar's research spans the experiential characterization of emergent smart and living materials, including their temporal and responsive behaviours; the development of methods and hybrid (digital/physical) tools to support creative material practices; and, more recently, exploring opportunities for more-than-human design at the intersection of materials, interactive systems, and biodesign. She frequently publishes in design, engineering, and Human–Computer Interaction venues, including CHI and DIS.

Fengyuan Wang is a designer with an interdisciplinary background in industrial and fashion design and holds an MSc in Industrial Design from Eindhoven University of Technology. Her work explores biofabrication and living materials for circular and sustainable solutions, employing a material-driven approach that integrates biological processes with digital fabrication to develop experimental applications that redefine material boundaries. She treats living systems as co-agents, navigating the intersections of humans, materials, technology and ecological change. By emphasizing sustainability, biodiversity and more-than-human design, her practice challenges dominant notions of control, permanence and industrialization, fostering co-evolving, symbiotic material futures.

Eda Özdemir is a researcher and a PhD candidate at the chair of EDEK at the University of Kassel, since 2021. She obtained a Bachelor of Science in Architectural Sciences at Politecnico di Milano, Italy, in 2017 and completed her studies with a research project on additive manufacturing for complex concrete casting. In 2020, she received an MSc in Integrative Technologies and Architectural Design Research at the University of Stuttgart. During her studies, she worked as a research assistant at ICD and ITKE. Her research focuses on large-scale additive manufacturing and novel digital manufacturing methods using sustainable materials for architecture. She is currently researching large-scale reinforced mycelium composites and teaches at the University of Kassel.

Philipp Eversmann is a registered architect in Munich and professor at the University of Kassel, where he leads the Experimental and Digital Design and Construction department. His research group focuses on digital design and construction methods using sustainable materials and robot-assisted additive manufacturing processes, such as renewable and bio-based building materials ranging from solid wood elements to continuous wood fibres, waste particles and living materials. From 2014 to 2016, he was the head of education at the National Center of Competence in Research Digital Fabrication at ETH Zurich, where he established a new Master's programme focusing on robotic technologies in architecture. As a visiting professor, he led architecture research classes at EPF Lausanne from 2011 to 2014, at the École Spéciale d'Architecture in Paris from 2011 to 2014, and at the Technical University of Munich from 2012 to 2014.

Andrea Rossi is an architectural researcher and computational designer. He holds a PhD with distinction from TU Darmstadt and is currently postdoctoral researcher at the University of Kassel, while also providing freelance consulting services on computational design. Previously he held positions as researcher and computational designer at Coop Himmelb(l)au, IndexLab and ETH Zurich. His research focuses on computational design, robotic fabrication and architectural biomaterials. He is the developer of Wasp, an open-source toolkit for combinatorial design. He has recently been awarded the Kurt-Ruths-Preis for his doctoral dissertation on open-source software approaches for combinatorial design and fabrication.

Nadja Nolte is a research assistant at EDEK, University of Kassel. She received her diploma in product design in 2021 from the Kunsthochschule Kassel. Towards the end of her studies, she focused on sustainable materials and fermentation, with an emphasis on mycelium materials. Accordingly, she co-founded BIOLAB at the art academy in 2019, which has since become a platform for design and research at the intersection of design, art and biology. Nadja has been working at EDEK since August 2021. She is now researching the combination of sustainable, growing materials with computer-aided design and digital fabrication strategies.

Romy Kaiser-Style is a biodesigner and textile researcher focusing on textile-based biofabrication and bioassembly methods for Hybrid Living Materials. She obtained her PhD at the Hub for Biotechnology in the Built Environment, Newcastle, UK, working at the intersection between biology, textiles and architecture as part of the Living Textiles Group. Romy's PhD research project, Textile Templating, investigated transdisciplinary knit design strategies for complex, large-scale mycofabrication. Her research contributes to the development and scale-up of mycelium-based composites in the built environment towards more sustainable construction materials and methods. Additionally, her work promotes and positions textile approaches as a recognized method for the biofabrication of hybrid living materials and biodesign research.

Jane Scott leads the Living Textiles Research Group in the Hub for Biotechnology in the Built Environment at Newcastle University. Her interdisciplinary research is located at the interface of programmable materials, knitted fabric design, architecture and biology. Her research positions textiles as a critical biofabrication strategy in the development of biohybrid materials and composites for architecture. Through large-scale prototyping, her group has established the protocols for textile/mycelium architecture, and this work has exhibited in major venues including the Farrell Centre, Newcastle, and the Design Museum London.

Introduction

What could fungi offer to the future of architecture and the built environment? This question opens a field that is at once scientific, architectural, experimental and visionary. In the face of climate change, material scarcity and the urgent need for planetary resilience, architects and designers are turning to living systems to rethink how we build. Among these, fungi – and particularly mycelium, the root network of fungi – emerge as one of the most compelling possibilities.

Mycology for Architecture explores the potential of designing with living systems, with a focus on mycelium. It considers how mycelium can be cultivated into architectural forms and components, how its use reshapes the design process and how advances in both biology and technology are redefining architectural practice. Bringing together perspectives from leading experts worldwide, the book offers a critical overview of mycelium's versatility, applications, opportunities and limitations, while positioning fungal design as a path toward more sustainable and imaginative futures.

The book is structured in three interconnected parts. Part I (Material) introduces mycelium in its different forms and purposes, establishing the *what*. Part II (Methods) discusses how designers, researchers and educators work with it, establishing the *how*. Part III (Fabrication) extends this knowledge into advanced practices, showing where the field is heading, establishing the *future*.

The chapters are authored by professionals from diverse fields – architecture, biotechnology, design, education and industry – reflecting the multiplicity of ways fungi can be understood and applied. Their approaches vary in language and perspective, yet together they weave a coherent narrative of material, method and practice. Before the first part begins, Chapter 1 provides an inclusive overview of fungi, covering the historical context of fermentation, fungal biology, material production processes, application and the challenges that remain.

Part I: Mycelium as a Material Alternative traces the journey of mycelium-based composites (MBCs) from scientific foundations to speculative futures. It begins with an exploration of fungi's role in circular ecosystems and the properties that make MBCs viable building materials (Chapter 2). From there, it moves into early architectural applications, including affordable and efficient interior components (Chapter 3) and the extension of these materials into building skins and façades, where their thermal, acoustic and environmental potential is tested against real-world challenges (Chapter 4).

The discussion then expands outward to the frontiers of space exploration, where mycelium's capacity for self-growth, self-repair and resource recycling positions it as a candidate for extraterrestrial habitats (Chapter 5). Finally, this part closes with a critical reflection on sustainability, species diversity and a proposal for a local, craft-based approach that reimagines mycelium not only as a material but as a cultural partner in building desirable, eco-social futures (Chapter 6). Together, these chapters highlight both the promise and the complexity of working with living materials, inviting architects, designers, scientists and makers to rethink how we build with life.

Part II: Methods and Practices for Working with Mycelium shifts focus from applications to the practices and methodologies that make designing with mycelium possible. It begins with the infrastructures that support experimentation, from DIY labs and university biodesign studios to industry partnerships that connect education with commercialization (Chapter 7). From there, the focus moves to workshops and hands-on teaching, highlighting both the opportunities and constraints of bringing living materials into short-term educational settings (Chapter 8). The discussion then broadens into pedagogical strategies and global perspectives on how design with living systems is taught (Chapter 9), before closing with a methodological framework that rethinks how designers can work with fungi in a scientific manner, balancing experimental rigour with creative exploration (Chapter 10). These chapters show that designing with living materials fundamentally reshapes design itself: rather than traditional makers or fabricators, designers become gardeners, bio-crafters and guides, working in partnership with fungi to shape outcomes.

After establishing mycelium as a viable material alternative and exploring the practices and methodologies that shape its design, **Part III: Fabrication Strategies with Mycelium** examines how biological growth can be translated into architectural and material systems. This section highlights digital, hybrid and system-based fabrication strategies, showing how diverse techniques – each emerging from different disciplinary goals – can expand the possibilities of mycelium-based design.

It begins with observations from architectural practice, where hands-on material research and real-world installations highlight both the potential and challenges of integrating mycelium into the construction industry (Chapter 11). Building on this foundation, the following chapter explores computer numerical control (CNC) biofabrication, where mycelium composites are machined, regrown and evaluated (Chapter 12). From there, the focus shifts to hybrid interfaces between fungi and clay, showing how mycelial 'skins' can reinforce and transform recycled ceramic substrates in digitally fabricated forms (Chapter 13). The discussion then scales up to additive manufacturing strategies, including robotic veneer lattices, interlocking shells and stay-in-place formworks, which aim to enable large-scale applications of mycelium in construction (Chapter 14). The section closes with textile templating approaches, demonstrating how knitted pre-forms and mycelium pastes can be combined into adaptive, living material systems for myco-architecture (Chapter 15).

The contributors were selected to represent the breadth and depth of knowledge in mycelium research, design and practice. Their voices address both practical and ethical questions of working with living systems. Moving beyond biomimicry, designing with fungi as an active material requires care, patience and an acceptance of biological agency. Each perspective reflects a specific dimension of the field: architects and designers imagine new applications in the built environment; scientists and biotechnologists ground the work in biological and material research; and educators develop frameworks for transmitting this emerging knowledge.

Together, these perspectives embody the inherently interdisciplinary nature of working with living materials. No single discipline can advance mycelium design in isolation; it requires collaboration across fields. In dialogue, the contributors trace mycelium's evolution from living matter to applied design and to speculative futures. This structure underscores the book's central aim: to demonstrate how fungi can shape a new paradigm of living architecture.

Whether you're a designer, scientist or curious thinker, I hope this book invites you to imagine, question and co-create with life itself.

Happy reading,
Dilan Ozkan

CHAPTER 1

Demystifying Fungi: Towards New Materials

INYOUNG YONAH LEE, BLANCA DE MIGUEL MARTÍNEZ, CHIARA MORETTI and TIFFANY ABITBOL

All authors contributed equally to this work.

We are surrounded by microorganisms – in our bodies, soil, water and on nearly every surface we touch – with recent estimates that our bodies contain one bacterial cell for every human cell [1]. In this chapter, we focus on a tiny piece of this fascinating puzzle, restricting our scope to fungi, specifically as they relate to materials development. This seems to be a rather new facet of human utilization of fungi, which was previously mostly relegated to food and medicine [2, 3]. The sphere of use has expanded in recent years to include materials such as vegan leathers [4], biodegradable packaging [5] and even far-out applications such as off-planet construction [6].

Part of the reason for the increased popularity of fungal mycelium materials can be attributed to the potential of fungi to transform and repurpose spent materials [7]. In this chapter, we introduce a brief historical context to fermentation, including specific examples using fungi. Next, key biological aspects of fungi are elaborated within the context of mycelium-derived materials, with an emphasis on mycelium-bound composites from solid-state fermentation, highlighting the main applications of these materials and their advantages and limitations. Finally, we describe how fungal fermentation might be accomplished in a lab or home setting for the aim of materials production.

From the mythological Greek god Dionysus who schooled humankind in winemaking, to the biblical account of Noah planting a vineyard after the flood in the book of Genesis, fermentation products seem firmly planted within the collective psyche. Indeed, archaeological and anthropological evidence show that fermentation has been used for thousands of years [8]. One such example is a mixed fermented beverage of rice, honey and fruit found in China, dating from the seventh millennium BCE [9].

It is a commonly used word, but what is actually meant by the term 'fermentation'? Fermentation derives from the Latin *fevere* [10], which means to boil. We can imagine that this word was first used long ago to describe the bubbling that was observed as the sugars in fruit or malted grains were transformed by wild yeast, releasing carbon dioxide in the process [10]. However, nowadays fermentation encompasses much more than just the bubbles observed during baking or brewing.

To better understand fermentation, we introduce the term 'catabolism', which simply describes how complex nutrients are broken down into simpler molecules, releasing the energy that fuels all living organisms in the process. According to the strict biochemical

Inyoung Yonah Lee, Blanca de Miguel Martínez, Chiara Moretti and Tiffany Abitbol, 'Demystifying Fungi: Towards New Materials' in: *Mycology for Architecture: Methods and Applications of Mycelium Materials in the Built Environment*. Pelagic Publishing (2026).
DOI: 10.53061/ZOZD5755

definition, fermentation implies the anaerobic catabolism of organic compounds [10]. However, the term itself has taken on a broader scope in industrial biotechnology to describe any process that yields products from the cultivation of microorganisms [10], whether anaerobic or not. Indeed, alcohol from *Saccharomyces cerevisiae* yeasts, penicillin from the *Penicillium* moulds and the edible polysaccharide pullulan from the fungus *Aureobasidium pullulans* are all considered the products of fermentation [11]. In this chapter, we adhere to the broader industrial biotechnology definition, colloquially referring to the processes used to produce fungal materials as 'fermentation', although the word 'culture' might be more technically accurate for these aerobic fungal growth systems.

How then is fungal fermentation used to produce materials? To answer this question, we first define what is meant by 'materials', next expanding our understanding of what fungi are and the various ways their biology can be harnessed to produce new biodegradable materials. Within this scope, it might be meaningful to briefly introduce ourselves. Who are we to write such a chapter on fungal materials? We are not mycologists, biologists or even biochemists – instead, the four of us find ourselves planted within a materials science department in an engineering school, each with varying backgrounds in materials, biotechnology, nanotechnology and chemistry. It is precisely these backgrounds that colour our perspective of fungi, how we envision their utility and, indeed, this chapter.

Broadly speaking, useful compositions and forms of matter are referred to as 'materials', which are sometimes categorized into different classes – one such grouping is metals, ceramics, polymers and composites [12]. Materials are technology enablers, moving humankind through the ages [13]. Initially, humans relied on renewable materials, but since the Industrial Revolution, our dependence has shifted almost entirely to non-renewable resources, such as metals, plastics and composites [13]. Somewhat hand in hand with our technological achievements is a marring by resource mismanagement and overconsumption. It is in this light that many among us are looking to renewable resources – asking whether we might be able to fashion new materials fit for purpose from resources that might otherwise be overlooked or outright wasted. The idea is not to regress to a prehistoric time, but rather to expand where we can use renewable resources reliably and with greatest impact, for instance to replace single-use packaging or other high-volume/high-waste materials. It is from this vantage of materials science that we now gently broach the basics of the biology and lifestyles of fungi, with a foreshadowing towards fungal materials development.

The fungal kingdom is one of the oldest eukaryotic lineages on Earth. While molecular clock estimates have varied, recent analyses suggest fungi colonized land between 700 and 900 million years ago [14, 15]. Furthermore, fungi comprise one of the largest clades – based on extrapolations from documented areas, it is estimated that between 2 million and 11 million fungal species exist, mainly in terrestrial habitats [16, 17, 18, 19]. However, only around 150,000 species have actually been formally reported [16, 17, 18], and fungal behaviour, structure and properties have not been fully elucidated. Plenty of mysteries remain in the fungal kingdom and their application in materials science is no exception.

The known fungal species are categorized into different classes, including phylum, subphylum, class, order, family and genus [20]. Species is the basic unit of classification [20]. However, small variations within a species can be found depending on their origin and genetic mutations, leading to different strains [21]. Fungal taxonomy has undergone significant and frequent revisions since the establishment of this kingdom in 1969 by Robert Harding Whittaker [19] and will likely require further revision as new species are discovered [20, 22].

Fungi are eukaryotic organisms, either unicellular (yeast) or multicellular (filamentous fungi) [23, 24]. Similar to plants and prokaryotic cells, fungal cells possess a cell wall that

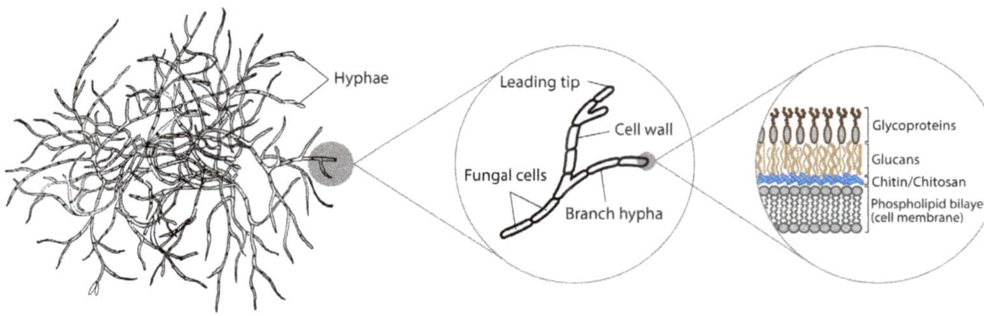

Figure 1.1 Filamentous fungal anatomy, from the mycelium network through to the fungal cell wall structure. Image credit: Blanca de Miguel Martínez.

provides mechanical resistance, shapes and protects the cell, enabling them to penetrate and explore diverse environments [19, 25]. The fungal cell wall is typically depicted as a stratified structure consisting of three layers: an inner primary wall of chitinous fibres, a middle glucan layer and an outer glycoprotein layer, as shown in Figure 1.1 [22, 26, 27]. In filamentous fungi, individual elongated cells are linked together into thread-like structures called *hyphae*, which have diameters ranging from 1 to 30 μm. Figure 1.1 illustrates the network of interconnected hyphae known as mycelium, forming the vegetative part of the fungus responsible for metabolism and growth [28, 29, 30, 31]. Mushrooms – the 'fruiting body' of fungi – are the reproductive part of the organism, not usually targeted in materials applications, although there are some exceptions.

Fungi are heterotrophs, requiring an external source of nutrition for their growth [19, 28]. Fungi can be *saprophytic*, obtaining their nutrition from decomposed organic matter, *parasitic*, feeding on living organisms, usually plants or animals, or *symbiotic*, obtaining nutrition through a beneficial relationship with other living organisms [32, 33, 34]. Filamentous fungi explore their surroundings for food via growth, using cellular extension to do so. Their hyphae grow continuously in one direction through tip extension [25], which is achieved through the production of membrane and cell wall precursors, such as proteins and lipids, and their translocation to the tip, where they are transformed into new membranes and rigidified by cross-linking enzymes [19, 25, 35]. The hyphal tip has a softened cell wall with viscoelastic features, allowing for flow and adaptation to the surroundings and aiding its recovery from deformation [19, 35].

Saprophytic fungi are the most important organisms for organic material recycling in nature and can be found in most ecosystems owing to their capacity to adapt to a broad range of environmental conditions [19, 22]. These versatile fungi are able to dominate their ecosystems owing to a lifestyle based on exploring, scavenging and substrate recycling [19]. They thrive on resource heterogeneity, taking up and transporting nutrients from rich areas to other regions, strategically controlling growth and expansion [19, 36]. Hyphal growth in saprophytic filamentous fungi is controlled in such a way as to avoid growth into one another, ensuring a uniform distribution within the surrounding environment, as well as growth away from nutrient-depleted areas [37, 38]. In some cases, hyphal attraction or fusion can occur, especially in central regions of mature mycelium, giving rise to a fully interconnected network [19, 38, 39, 40]. These hyphae can also create new leading tips on their laterals, called branches [19, 25, 28], which are essential for the proliferation of mycelium. In a nutrient-depleted environment, resources are primarily directed towards hyphal extension to explore the surroundings [41, 42]. When a nutrient source is found, the generation of branched hyphae is prioritized over their extension, in order to exploit the resource [19, 41, 42].

Saprophytic filamentous fungi are the most widely used in materials applications owing to three core biological capabilities: (1) their ability to *digest* diverse organic materials, (2) their capacity to *fuse* and integrate organic matter into cohesive forms and (3) their *living* state, which not only leads to the formation of these cohesive material forms but also potentially allows them to remain metabolically active throughout the material's lifetime, resulting in dynamic and responsive living materials. Accordingly, we will focus on these three aspects in our discussion of saprophytic filamentous fungi for materials production.

1 Digestion

Saprophytic fungi can digest organic matter in their environment to obtain the nutrition they need for growth and to colonize their surroundings [19, 43, 44]. This organic matter, referred to as the 'substrate', is usually lignocellulosic in nature, consisting of cellulose, hemicellulose and lignin, and conveniently contains all essential nutrients: carbon, oxygen, nitrogen, hydrogen, phosphorus and sulphur, as well as other elements required in lower quantities such as potassium, magnesium, sodium and calcium [26, 45]. Fungal digestion occurs externally through the secretion of a wide variety of enzymes and metabolites that enable breakdown of the substrate [33, 34, 46]. Degrading fungi have been categorized according to the main biomass fractions that are first digested:

- White-rot fungi primarily digest the lignin and phenolic compounds that colour biomass, leaving behind a decoloured substrate [7, 33, 47, 48]. Although lignin is more easily digested, hemicellulose and cellulose will also be eventually digested. Included in this group are *Agaricus bisporus* (champignon mushroom), *Lentinula edodes* (shiitake mushroom), *Ganoderma lucidum* and *Trametes versicolor* [7, 47].
- Brown-rot fungi are only able to degrade cellulose and hemicellulose. In this case, the biomass becomes dark brown over the course of the digestion [33, 48, 49]. Common species such as *Laetiporus sulphureus*, *Serpula lacrymans* and *Gloeophyllum trabeum* are some examples of this group [7].
- Soft-rot fungi can superficially degrade polysaccharides and lignin, leaving behind a softened material [7, 48, 50]. This group is less studied and applied than white-rot and brown-rot fungi. Species such as *Alternaria alternata*, *Chaetomium globosum* and *Thielavia terrestris* belong to this group [7].

White-rot fungi are the most studied and applied in materials development owing to their efficient lignin degradation activity. Lignin is commonly considered a bottleneck for materials modification in lignocellulosic complex structures, and therefore lignin removal is generally prioritized in this field [123]. White-rot fungi show selective degradation activity that prioritizes lignin consumption while leaving most of the cellulose and hemicellulose intact, thereby improving fibre quality in wood-derived materials. In mycelium-bound composites, these intact cellulose fibres impart structural integrity and strength. By contrast, brown-rot fungi tend to degrade cellulose and hemicellulose, and soft-rot fungi degrade lignin less efficiently than white-rot fungi. Nevertheless, further investigation of brown-rot and soft-rot fungi could result in materials with potentially unique and valuable properties.

In general, the ability to digest lignocellulose is no small feat, as lignocellulose is remarkably recalcitrant and resistant to degradation [49]. From a materials science perspective, the adaptive ability of saprophytic fungi to transform different kinds of organic matter through digestion and growth, while enabling the recirculation of nutrients that were previously locked within the substrate, is compelling [19, 41]. Furthermore, the nature of the substrate has been shown to play an essential role in fungal composition, morphology and growth pattern [33, 44, 45, 49, 51]. This aspect of tuneability and versatility is key for materials production, as not only can a variety of substrates be digested, but also the composition and morphology of the substrate influences the properties of the resulting materials and also of the mycelium itself [51].

2 Fusion

The growing network of mycelium acts as a natural glue that holds together partially digested substrate particles [52]. This phenomenon can be explained in part by the quorum sensing mechanisms employed by the cells, which regulate the production of adhesive compounds that form a sticky extracellular matrix surrounding the hyphae, helping them adhere to the substrate [53, 54, 55]. This allows the hyphae to penetrate or to otherwise engulf the surrounding substrate, binding the substrate within its 3D network structure [56, 57]. This growth pattern results in complex matrices consisting of a highly entangled and porous mycelium structure, interspersed by substrate [39, 56]. In nature, this feature is critical to maintaining soil structure and facilitating nutrient cycling [58], whereas in materials science the fusion of the substrate within a fungal network while it grows is an essential characteristic that enables the production of cohesive composite macro-objects from non-cohesive elements [59, 60]. The fusion of substrate within the growing network exists in a fine balance with digestion to enable the production of mycelium-bound composites from a variety of lignocellulosic materials, often sourced from waste streams. This balance dictates the cohesiveness of the final material and its composition, specifically the ratio of mycelium to substrate, both of which play an important role in mechanical performance.

3 Life

The living state of fungi is arguably the most significant characteristic of fungi for the production of materials [39, 61]. The fungal cell wall exhibits dynamism and structural adaptability, changing not only throughout maturation and in response to different substrates but also rapidly adapting to environmental fluctuations such as temperature and humidity changes [19, 62]. Similarly, fungal growth patterns vary based on nutritional availability and environmental conditions, with additional diversity observed across different phyla, species and strains [19, 26, 28]. The application of filamentous fungi as living biomaterial factories leverages the active and quick adaptation of fungi for the generation of materials with different properties and for different applications [56]. While different species can be targeted for specific materials outcomes, the genetic modification of fungi is also a powerful tool for engineering specific material properties [63, 64, 65]. Thus, environmental and/or genetic approaches can be applied to elicit changes in fungal phenotype, which in turn can elicit changes in material properties. From a materials perspective, it is exciting that we can shape our materials from living cells, while they are alive, taking advantage of their specific genetic script and ability to morph and adapt to different cues, or even that we are able to alter materials' properties by varying something as simple as growth time. Importantly, while this aspect of life is fascinating from a materials generation point of view, in most cases, except for self-healing or otherwise responsive applications based on living materials, once the material of interest is obtained, the life is terminated by heating and drying to yield a stable form.

These natural features of mycelium – *digestion*, *fusion* and *life* – make it attractive for the production of functional materials [39, 66, 67, 68]. As mentioned, mycelium can generate materials by partially breaking down the substrate, binding together particles that might not otherwise bind and embedding them into its network structure [5]. As such, the fungal fermentation of different substrates can yield a wide variety of 3D mycelium-bound composites, which are typically lightweight and highly porous, with the mycelium acting as both the matrix and binder, and the substrate as a reinforcing or otherwise functional filler [69]. Mycelium-bound composites have been proposed for many applications, including construction [70], packaging [5], fashion [4, 71], filtration [72], food [73] and more, including emerging applications such as electronic devices [74], robotics [75] and in off-planet activities [76]. So, how do we actually create materials using filamentous fungi?

Mycelium is grown in moist environments (from 20% to 99% humidity; although the lower limit is usually closer to 40%), where the amount of water present is the key distinguisher between solid-state and liquid fermentation [77, 78, 79]. Conveniently, both solid- and liquid-state fermentations can be accomplished under rather mild conditions of temperature, usually around room temperature, which is economically attractive and enables low-energy production processes [80].

Since its industrial application in 1963, liquid fermentation has been used to produce edible, medicinal and cosmetic fungal products, using various species such as *Ganoderma lucidum*, *Pleurotus ostreatus*, *Tremella fuciformis* and *Lentinus edodes* [19, 81]. Liquid fermentation uses a liquid culture medium, where the substrate is solubilized or dispersed in water [81]. This approach improves nutrient accessibility, but as oxygen is poorly soluble in water, aeration is required [19, 81, 82]. The homogeneity of the growth media in liquid fermentation allows better control over processing conditions and, consequently, over product quality [81].

Liquid fermentation generates two main product streams: a liquid fraction and a solid biomass fraction. While the compounds in the liquid fraction are frequently used in the production of value-added products, such as pharmaceuticals, supplements and enzymes, the solid biomass fraction is less frequently valorized [39, 83]. Although liquid fermentation is less commonly used to produce mycelium materials, it has been applied to prepare foods [84], films [85], leathers [71], papers [86] and yarns [87], and to extract compounds from fungal biomass, such as chitin [88].

Solid-state fermentation involves growing mycelium on a moist solid substrate, having a relative humidity between 20% and 70% [79]. The general process, depicted in Figure 1.2, includes substrate preparation, inoculation with living mycelium, incubation and drying [89]. First, if needed, the substrate is mechanically broken down into small and uniform particles, such as wood chips, recalling that a wide variety of substrates and substrate mixtures are possible, including supplementation to fine-tune pH and nutrient profile. The substrate is then homogenized and sterilized to ensure a uniform and hospitable feeding ground for the mycelium.

At this stage, the substrate is inoculated with living mycelium, usually in the form of spawn, as depicted in Figure 1.2. Mycelium spawn is somewhat analogous to plant seeds, as the spawn is the source from which the mycelium grows. Spawn consists of living mycelium, packaged together with lignocellulose nutrition, such as grains. These 'mycelium seeds' are then scattered throughout the solid media and left to develop into mycelium during incubation. Alternatively, liquid inoculum

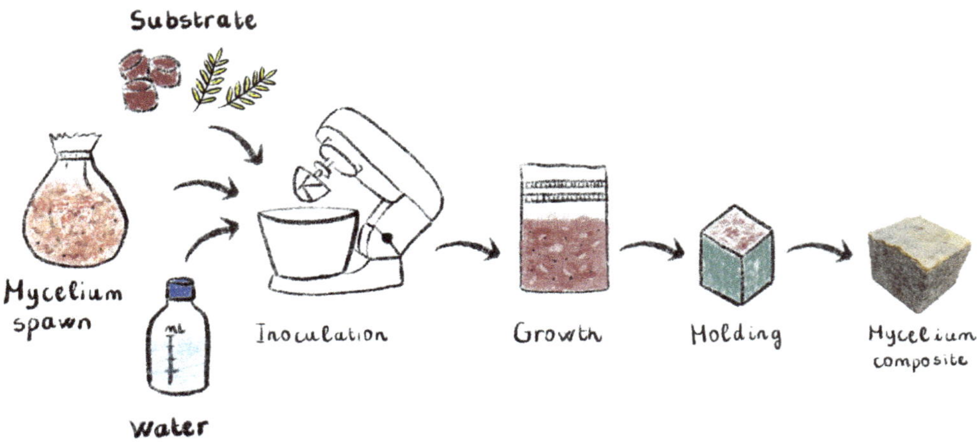

Figure 1.2 Schematic illustration of solid-state fermentation, including a photograph of a typical mycelium-bound composite. Image credit: Chiara Moretti.

(from liquid fermentation) can be used instead of spawn, arguably providing a more uniform inoculation.

The incubation phase itself is usually performed in two steps [89]. The first step is designed for mycelium activation and expansion within the substrate. This step is typically performed in breathable bags and ranges in duration from 7 to 14 days, depending on the mycelium species, substrate type and growth conditions [77]. Once growth in the bag is complete, the mycelium-colonized substrate is homogenized again and distributed into moulds for a second incubation step. During this step, the contents will adapt to the mould shape, as the mycelium continues to colonize the substrate. This key feature of *shapeability* is yet another important lever from a materials science standpoint, since the shape of the composite can be controlled through the selection of the growth container, lending to essentially limitless geometries. Finally, once the desired material is obtained the mycelium is usually terminated by a heating and drying treatment to stop further growth.

Two categories of materials can be produced in this way: the mycelium-bound composites described above and pure mycelium sheets. The first category uses the whole material and results in porous composites, such as the one depicted in Figure 1.2 [89, 90]. In the second category, the surface skin or tissue of mycelium that forms at the air interface is removed to give pure mycelium mats that resemble leather, paper or plastic after drying [90]. Pure mycelium materials can also be produced by liquid fermentation, requiring a more extensive dewatering process.

The approaches described here are seemingly straightforward but have certain requirements, such as a sterile substrate and environment to prevent contamination by other organisms [91]. Growth is generally conducted in the dark to promote mycelium development over fruiting body formation [19, 92], as light can trigger the growth of fruiting bodies in some species, especially once the mycelium has fully colonized the substrate and is ready to fruit [19, 28]. Furthermore, as mentioned previously, oxygen is needed considering that filamentous fungi prefer to operate under aerobic conditions [19, 28]. In solid fermentation, having a variety of particle sizes of the substrate as well as not packing the material too tightly into the mould are key to adequate oxygenation [57, 93]. Even from this basic description, it might be obvious that this process has its challenges, specifically related to uniformity – of substrate, air and inoculum dispersion – ultimately affecting the reproducibility of the materials themselves.

Still, mycelium materials from solid-state fermentation remain highly attractive for certain applications owing to the features outlined earlier and also their non-toxic (species-dependent), renewable and biodegradable nature [94], with a circular lifecycle that enables the return of nutrients to the soil at end of life by composting [90]. Another feature briefly mentioned earlier is its potential to valorize and upcycle waste, for example sourcing substrates from food and agricultural side and waste streams [51]. Mycelium naturally invades, degrades and consumes organic waste materials, such as coffee husks, sawdust, wheat bran, straw and bagasse [95]. This wide nutritional versatility is a powerful materials design strategy as it is feasible to directly target specific final properties through careful selection of the substrate [51]. To further refine material properties, functional additives, such as plasticizers or crosslinking agents, are commonly employed, as are finishing treatments such as pressing to densify [71, 96].

Equipped with an understanding of the basic biological features that enable the production of mycelium materials and how these materials are practically produced, we now highlight some applications that we find particularly exciting.

One of the first mainstream explorations of mycelium as a functional material was in 2006, by Eben Bayer and Gavin McIntyre, who founded Ecovative in 2007, a pioneering company in mycelium-based materials [97]. Their early research focused on developing mycelium for insulation and construction applications [98]. One key advantage of using mycelium-bound composites is their low density, making them significantly lighter than conventional materials such as cement [99]. In addition to their light weight, mycelium's porous and fibrous structure provides excellent acoustic and thermal insulation properties, comparable to those of synthetic foams [100]. In acoustic applications, mycelium's network disrupts sound wave propagation by internal friction [39]. By adjusting the density of the composite, acoustic performance can be optimized [101]. Mycelium-bound composite panels made from agricultural residues exhibit excellent sound absorption – the sound pressure level at 1,000 Hz, a common frequency of automotive road noise, was reduced by 70% when using a type of mycelium-bound composite [102]. Researchers have further enhanced acoustic properties by engineering perforations into mycelium panels [89]. By adjusting the number, size, depth and distribution of cavities, it was possible to fine-tune absorption across a broad frequency range [89]. Additionally, mycelium composites made with substrates such as straw and hemp have shown thermal insulating properties similar to commercial products, such as expanded polystyrene foams [103].

In addition to insulating construction materials, mycelium has been used in packaging applications, mainly for replacing stress-protecting packaging foams [104]. Mycelium's ability to take the shape of nearly any container makes it an ideal material for custom-fit packaging, providing optimal protection [5]. Ecovative demonstrated the use of mycelium in protective packages, showing its potential in the substitution of commercial plastic materials [97]. Moreover, compared with traditional packaging materials such as expanded polystyrene, mycelium materials can have superior mechanical and fire-retardant properties [104, 105].

Beyond packaging, mycelium is gaining interest as a leather-like material, owing to its interwoven structure and suede-like texture [4]. As mentioned, fungal mycelium can form a soft skin on the surface of the substrate, which can be peeled off, dried and processed into a textile material [106]. This pure mycelium mat can be tailored to achieve desired properties, such as water absorption, elasticity and texture [68]. Interestingly, the use of mycelial mats as textiles is not new and has been documented in the Indigenous peoples of North America [107]. More recently, mycelium textiles have been demonstrated in shoes, handbags, yoga mats, hats and apparel, by fashion brands such as Adidas and

Hermès [4]. After post-treatment, these textiles can be enhanced with unique features and customizable colors [108], with mechanical properties comparable to animal leather, for example tear abrasion [109].

Mycelium naturally functions as a filter in soil and water, breaking down or absorbing toxic substances to help maintain a healthy surrounding environment [19, 110]. This bioremediation capability has inspired the development of mycelium-based filtration membranes [111]. For instance, mycelial pellets have shown the ability to absorb dyes and heavy metals from wastewater streams [111]. This ability is largely thanks to the production of extracellular polymeric substances, which contain functional groups that can bind pollutants, bacteria and nanoparticles, allowing for their immobilization and removal [111]. The creation of membranes for efficient and targeted separation processes from mycelium biomass has also been studied [112]. A successful membrane surface for oil and water separation was achieved by engineering mycelium growth [92]. In addition to its use in water treatment, the porous structure of mycelium makes it suitable for air filtration applications, improving indoor air quality [113].

Fungi are also emerging as promising candidates for meat alternatives owing to their versatility, nutritional value, sustainability and texture. Certain species of filamentous fungi even contain higher protein contents than commercialized edible mushrooms [114]. The company Ecovative has recently leveraged mycelium into ready to eat meatless bacon alternatives [115]. By controlling the fermentation process, fungi can develop a meat-like texture and appearance, features that might sway consumer acceptance [116]. A more entrenched example, first launched in 1985, is Quorn, which uses liquid fermentation to grow *Fusarium venenatum*, harvesting the protein-rich biomass and using it in a variety of vegan and vegetarian products [84].

Building on the idea of using living fungi, mycelium has recently been proposed as an interface for biohybrid robots, to exploit its dynamic adaptation to different stimuli and environments [61, 75]. Mycelium generates electrical signals in response to various external stimuli, such as light, temperature and humidity [75]. These signals were measured and used to guide and control robotic systems [75]. Another application arising from the living nature of the mycelium is living fungal batteries [74], which rely on sugars and microbes to produce power. Fungal cells are mixed into a printing ink (based on cellulose) together with simple sugars which act as a nutrient source, and using 3D printing it has been possible to design arbitrary shapes and to deposit the amount of material needed for a specific amount of energy that one wants to store in the batteries [74]. These cells can produce electricity to power a temperature sensor for several days and could be used in agriculture or in environmental research [74]. Furthermore, these batteries are reported to be completely non-toxic and biodegradable [74]. They need feeding instead of charging, and once their work is done, they digest themselves from the inside [74]. The batteries can be stored in a dried, dormant state and activated by the addition of water and nutrients [74].

Mycelium is also emerging as a promising material for sustaining missions in space [117]. NASA is actively exploring its use for extraterrestrial habitats through the Myco-architecture off planet project [117]. The concept involves using mycelium bricks to directly grow buildings on the Moon and Mars [76]. To do so, astronauts could travel with compact lightweight materials containing dormant mycelium [76]. Upon arrival, by adding water and organic waste generated during the mission, the mycelium can be grown into functional structures [118]. At the end of their lifecycle, these myco-materials could be repurposed as fertilizer for mission farming [76].

Mycelium-bound composites are very promising for many different applications, but there are still some challenges to overcome. One of the main issues is the lack of

Figure 1.3 Imagining a DIY fermentation set up in your kitchen. Image credit: Inyoung Yonah Lee.

standardized and comparable methods for producing mycelium-based materials [119]. Species, strains, substrates and protocols vary significantly across studies [120, 121]. However, a lack of standardization is to be expected in any newly emerging field and is perhaps even beneficial at this exploratory stage. Another critical point is that the production of uniform and reproducible composites at scale is challenging since conventionally used feedstocks are highly heterogeneous and can vary seasonally [121]. Consumer acceptance is also a potential obstacle, as are concerns over long-term durability, especially for textiles, packaging and construction uses [122]. While promising, mycelium remains under-researched and under-utilized in the field of materials science. Addressing these challenges and exploring new use cases will be key for integrating mycelium composites into our daily lives.

Adventurous readers might now be feeling buoyed to start producing their own biodegradable mycelium-bound composites, and why not? In ancient times, fermentation was do-it-yourself (DIY) and low tech; after all, ancient peoples were not equipped with fancy laboratory equipment. Nowadays, there are so many resources available for the curious home grower, and even a basic kitchen set-up can be spruced up to accomplish the task. Figure 1.3 illustrates a kitchen featuring all the various tools that can be used for a proper home inoculation.

Happy fermenting!

Acknowledgements

The authors gratefully acknowledge the Chair in Sustainable Materials at EPFL, co-funded by BASF, Logitech, Nestlé and SIG, and funding from the Swiss National Science Foundation (Project number: 200021_214912) and the ETH Domain Joint Initiative Proteins for a Sustainable Future.

Note

Statement of authorship: All authors contributed equally to this chapter.

References

1. Sender R, Fuchs S, Milo R. Revised estimates for the number of human and bacteria cells in the body. PLoS Biol. 2016;14(8):e1002533.
2. Van Court R, Wiseman M, Meyer K, Ballhorn D, Amses K, Slot J, et al. Diversity, biology, and history of psilocybin-containing fungi: suggestions for research and technological development. Fungal Biol. 2022;126(4): 308–19.
3. Dupont J, Dequin S, Giraud T, Le Tacon F, Marsit S, Ropars J, et al. Fungi as a source of food. Microbiol Spectr. 2017;5(3):10–1128.
4. Amobonye A, Lalung J, Awasthi MK, Pillai S. Fungal mycelium as leather alternative: A sustainable biogenic material for the fashion industry. Sustain Mater Technol. 2023;38:e00724.
5. Abhijith R, Ashok A, Rejeesh CR. Sustainable packaging applications from mycelium to substitute polystyrene: a review. Mater Today Proc. 2018;5(1):2139–45.
6. Santomartino R, Averesch NJH, Bhuiyan M, Cockell CS, Colangelo J, Gumulya Y, et al. Toward sustainable space exploration: a roadmap for harnessing the power of microorganisms. Nat Commun. 2023;14(1):1391.
7. Chai Y, Bai M, Chen A, Peng L, Shao J, Luo S, et al. Valorization of waste biomass through fungal technology: Advances, challenges, and prospects. Ind Crops Prod. 2022;188: 115608.
8. Ray RC, Paramithiotis S, Thekkangil A, Nethravathy V, Rai AK, Martin JGP. Food Fermentation and Its Relevance in the Human History. In: Martin JGP, De Dea Lindner J, Melo Pereira GV de, Ray RC, editors. Trending Topics on Fermented Foods. Springer Nature Switzerland; 2024. pp. 1–57.
9. McGovern PE, Zhang J, Tang J, Zhang Z, Hall GR, Moreau RA, et al. Fermented beverages of pre- and proto-historic China. Proc Natl Acad Sci. 2004;101(51):17593–8.
10. Stanbury PF, Whitaker A, Hall SJ. Chapter 1 – An introduction to fermentation processes. In: Stanbury PF, Whitaker A, Hall SJ, editors. Principles of Fermentation Technology (Third Edition). Butterworth-Heinemann; 2017. pp. 1–20.
11. Wang P, Jia SL, Liu GL, Chi Z, Chi ZM. Aureobasidium spp. and their applications in biotechnology. Process Biochem. 2022;116:72–83.
12. Callister Jr WD, Rethwisch DG. Materials Science and Engineering: An Introduction. John Wiley & Sons; 2020.
13. Ashby MF. Chapter 1 – Introduction: material dependence. In: Ashby MF, editor. Materials and the Environment (Third Edition). Butterworth-Heinemann; 2021. pp. 1–19.
14. Heckman DS, Geiser DM, Eidell BR, Stauffer RL, Kardos NL, Hedges SB. Molecular evidence for the early colonization of land by fungi and plants. Science. 2001;293(5532):1129–33.
15. Loron CC, François C, Rainbird RH, Turner EC, Borensztajn S, Javaux EJ. Early fungi from the Proterozoic era in Arctic Canada. Nature. 2019;570(7760):232–5.
16. Phukhamsakda C, Nilsson RH, Bhunjun CS, De Farias ARG, Sun YR, Wijesinghe SN, et al. The numbers of fungi: Contributions from traditional taxonomic studies and challenges of metabarcoding. Fungal Divers. 2022;114(1): 327–86.
17. Hawksworth DL. The magnitude of fungal diversity: the 1.5 million species estimate revisited. Mycol Res. 2001;105(12):1422–32.
18. Wu B, Hussain M, Zhang W, Stadler M, Liu X, Xiang M. Current insights into fungal species diversity and perspective on naming the environmental DNA sequences of fungi. Mycology. 2019;10(3):127–40.
19. Moore D, Robson GD, Trinci AP. *21st century guidebook to fungi*. Cambridge University Press; 2020.

20. Wijayawardene NN, Hyde KD, Mikhailov KV, Péter G, Aptroot A, Pires-Zottarelli CL, et al. Classes and phyla of the kingdom Fungi. Fungal Divers. 2024;1–165.
21. Kun RS, Gomes ACS, Hildén KS, Cerezo SS, Mäkelä MR, de Vries RP. Developments and opportunities in fungal strain engineering for the production of novel enzymes and enzyme cocktails for plant biomass degradation. Biotechnol Adv. 2019;37(6):107361.
22. Naranjo-Ortiz MA, Gabaldón T. Fungal evolution: diversity, taxonomy and phylogeny of the Fungi. Biol Rev. 2019;94(6):2101–37.
23. McGinnis MR, Tyring SK. Introduction to Mycology. In: Baron S, editor. Medical Microbiology. 4th ed. Galveston (TX): University of Texas Medical Branch at Galveston; 1996.
24. Souza AC, Taborda CP. Epidemiology of dimorphic fungi. In: Zaragoza O, Casadevall A, editors. Encyclopedia of Mycology. Elsevier; 2021. pp. 613–23.
25. Steinberg G, Peñalva MA, Riquelme M, Wösten HA, Harris SD. Cell biology of hyphal growth. Microbiol Spectr. 2017;5(2):10–1128.
26. Kavanagh K. Fungi: biology and applications. John Wiley & Sons; 2017.
27. Borkovich K, Ebbole DJ. Cellular and molecular biology of filamentous fungi. American Society for Microbiology Press; 2010.
28. Watkinson SC, Boddy L, Money N. The fungi. Academic Press; 2015.
29. Samson RA. Cellular constitution, water and nutritional needs, and secondary metabolites. In: Viegas C, Viegas AC, Pinheiro C, Brandão J, Samson RA, editors. Environmental Mycology in Public Health. Elsevier; 2016. pp. 5–15.
30. Letti LAJ, Vítola FMD, de Melo Pereira GV, Karp SG, Medeiros ABP, da Costa ESF, et al. Solid-state fermentation for the production of mushrooms. In: Current developments in biotechnology and bioengineering. Elsevier; 2018. pp. 285–318.
31. Rajarathnam S, Shashirekha M. Mushrooms and Truffles: Classification and Morphology. In: *Encyclopedia of Food Sciences and Nutrition*. Elsevier; 2003. pp. 4040–8.
32. Jong T, Stack CM, Moffitt MC, Morton CO. An Introduction to the Influence of Nutritional Factors on the Pathogenesis of Opportunist Fungal Pathogens in Humans. *Pathogens*. 2025;14(4):335.
33. Moretti S, Goddard ML, Puca A, Lalevée J, Di Marco S, Mugnai L, et al. First description of non-enzymatic radical-generating mechanisms adopted by *Fomitiporia mediterranea*: an unexplored pathway of the white rot agent of the esca complex of diseases. J Fungi. 2023;9(4):498.
34. Penttilä M, Saloheimo M. Lignocellulose breakdown and utilization by fungi. In: Oliver RP, Schweizer M, editors. Molecular Fungal Biology. Cambridge University Press; 1999. p. 272–93.
35. Chevalier L, Pinar M, Le Borgne R, Durieu C, Peñalva MA, Boudaoud A, et al. Cell wall dynamics stabilize tip growth in a filamentous fungus. PLoS Biol. 2023;21(1):e3001981.
36. Klein D, Paschke M. Filamentous fungi: the indeterminate lifestyle and microbial ecology. Microb Ecol. 2004;47:224–35.
37. Gutiérrez–Medina B, Vázquez-Villa A. Visualizing three-dimensional fungal growth using light sheet fluorescence microscopy. Fungal Genet Biol. 2021;150:103549.
38. Wille ECG, Bento CRDC. Filamentous Fungi Growth as Metaphor for Mobile Communication Networks Routing. Adv Electr Comp Eng. 2021;21(2):59–66.
39. Manan S, Ullah MW, Ul-Islam M, Atta OM, Yang G. Synthesis and applications of fungal mycelium-based advanced functional materials. J Bioresour Bioprod. 2021;6(1):1–10.
40. Brand A, Gow NA. Mechanisms of hypha orientation of fungi. Curr Opin Microbiol. 2009;12(4):350–7.
41. Harris SD. Hyphal Growth. eLS; 2012.
42. Ledoux C, Chapeland-Leclerc F, Ruprich-Robert G, Bobée C, Lalanne C, Herbert É, et al. Prediction and experimental evidence of different growth phases of the Podospora anserina hyphal network. Sci Rep. 2023;13(1):8501.
43. Várnai A, Mäkelä MR, Djajadi DT, Rahikainen J, Hatakka A, Viikari L. Carbohydrate-binding modules of fungal cellulases: occurrence in nature, function, and relevance in industrial biomass conversion. Adv Appl Microbiol. 2014;88:103–65.
44. Niego AGT, Rapior S, Thongklang N, Raspé O, Hyde KD, Mortimer P. Reviewing the contributions of macrofungi to forest ecosystem processes and services. Fungal Biol Rev. 2023;44: 100294.
45. Viegas C, Pinheiro AC, Sabino R, Viegas S, Brandão J, Veríssimo C. Environmental mycology in public health: fungi and mycotoxins risk assessment and management. Academic Press; 2015.
46. Vigneshwaran N, Satyamurthy P. Nanocellulose production using cellulose degrading fungi. In: Prasad R, editor. Advances and applications through fungal nanobiotechnology. Springer; 2016. pp. 321–31.
47. Dos Santos TL, Tavares OCH, de Abreu Lopes S, Elias SS, Berbara RLL, García AC. Environmental implications of the organic matter structure for white-rot fungus *Pleurotus eryngii* growth in a tropical climate. Fungal Biol. 2021;125(11):845–59.
48. Goodell B, Qian Y, Jellison J. Fungal decay of wood: soft rot – brown rot – white rot. In Development of Commercial Wood Preservatives. ACS Symposium Series, Vol. 982. American Chemical Society; 2008. pp. 9–31.

49. Segers B, Nimmegeers P, Spiller M, Tofani G, Grojzdek EJ, Dace E, et al. Lignocellulosic biomass valorisation: A review of feedstocks, processes and potential value chains, and their implications for the decision-making process. RSC Sustain. 2024;2(12):3730–49.
50. Eslyn W, Kirk T, Effland M. Changes in the chemical composition of wood caused by six soft-rot fungi. Phytopathology. 1975;65(4): 473–6.
51. Hernando AV, Sun W, Abitbol T. 'You Are What You Eat': How Fungal Adaptation Can Be Leveraged toward Myco-Material Properties. Glob Chall. 2024;8(3):2300140.
52. Sun W, Tajvidi M, Howell C, Hunt CG. Insight into mycelium-lignocellulosic bio-composites: Essential factors and properties. Compos Part Appl Sci Manuf. 2022;161:107125.
53. Wang D, Zeng N, Li C, Li Z, Zhang N, Li B. Fungal biofilm formation and its regulatory mechanism. Heliyon. 2024;10(12).
54. Harding MW, Marques LL, Howard RJ, Olson ME. Can filamentous fungi form biofilms? Trends Microbiol. 2009;17(11):475–80.
55. Wongsuk T, Pumeesat P, Luplertlop N. Fungal quorum sensing molecules: role in fungal morphogenesis and pathogenicity. J Basic Microbiol. 2016;56(5):440–7.
56. Sierra LAB, Mendes-Pereira T, García GJY, Werkhaizer CQ, de Rezende JB, Rodrigues TAB, et al. Current situation and future perspectives for the use of fungi in the biomaterial industry and proposal for a new classification of fungal-derived materials. PeerJ Mater Sci. 2023;5:e31.
57. Houette T, Maurer C, Niewiarowski R, Gruber P. Growth and mechanical characterization of mycelium-based composites towards future bioremediation and food production in the material manufacturing cycle. Biomimetics. 2022;7(3):103.
58. Frąc M, Hannula SE, Bełka M, Jędryczka M. Fungal biodiversity and their role in soil health. Front Microbiol. 2018;9:707.
59. Shao G, Zhang L, Xu D, Jin Y, Wu F, Yang N, et al. Green and sustainable bioboards: Biomanufacturing of mycelium-based composite boards with tunable properties. Chem Eng J. 2025;503:158382.
60. Wang Y, Hausner G, Rout PR, Yuan Q. Investigation of fungal mycelium-bound bio-foams from agricultural wastes as sustainable and eco-conscious packaging innovations. J Clean Prod. 2025;501:145206.
61. Gantenbein S, Colucci E, Käch J, Trachsel E, Coulter FB, Rühs PA, et al. Three-dimensional printing of mycelium hydrogels into living complex materials. Nat Mater. 2023;22(1):128–34.
62. Zhao L, Schaefer D, Xu H, Modi SJ, LaCourse WR, Marten MR. Elastic properties of the cell wall of *Aspergillus nidulans* studied with atomic force microscopy. Biotechnol Prog. 2005;21(1): 292–9.
63. Wang Q, Hu Z, Li Z, Liu T, Bian G. Exploring the application and prospects of synthetic biology in engineered living materials. Adv Mater. 2023;2305828.
64. Zheng L, Zhang M, Zhao W. Enhanced mycelium biomass and polysaccharide production in genetically modified *Pleurotus ostreatus* using agricultural wastes. Int J Biol Macromol. 2024;278:134318.
65. Schaak D. Bio-manufacturing process. US11920126B2, 2024.
66. Jo C, Zhang J, Tam JM, Church GM, Khalil AS, Segrè D, et al. Unlocking the magic in mycelium: Using synthetic biology to optimize filamentous fungi for biomanufacturing and sustainability. Mater Today Bio. 2023;19:100560.
67. Gandia A, van den Brandhof JG, Appels FV, Jones MP. Flexible fungal materials: shaping the future. Trends Biotechnol. 2021;39(12):1321–31.
68. Haneef M, Ceseracciu L, Canale C, Bayer IS, Heredia-Guerrero JA, Athanassiou A. Advanced materials from fungal mycelium: fabrication and tuning of physical properties. Sci Rep. 2017;7(1):41292.
69. Muiruri JK, Chuan Yeo JC, Zhu Q, Ye E, Loh XJ, Li Z. Sustainable Mycelium-Bound Biocomposites: Design Strategies, Materials Properties, and Emerging Applications. ACS Sustain Chem Eng. 2023;11(18):6801–21.
70. Jones M, Mautner A, Luenco S, Bismarck A, John S. Engineered mycelium composite construction materials from fungal biorefineries: A critical review. Mater Des. 2020;187:108397.
71. d'Errico A, Schröpfer M, Mondschein A, Safeer AA, Baldus M, Wösten HA. Cross-linking impacts the physical properties of mycelium leather alternatives by targeting hydroxyl groups of polysaccharides and amino groups of proteins. Heliyon. 2024;10(16).
72. Zheng Y, Lian Y, Bao H, Guo H, Hu Y, Zhao J, et al. Janus microbial membrane constructed by multiple biological self-assembly for photothermal conversion. Chem Eng J. 2023;472: 145003.
73. Holt RR, Munafo Jr JP, Salmen J, Keen CL, Mistry BS, Whiteley JM, et al. Mycelium: A nutrient-dense food to help address world hunger, promote health, and support a regenerative food system. J Agric Food Chem. 2023;72(5):2697–707.
74. Reyes C, Fivaz E, Sajó Z, Schneider A, Siqueira G, Ribera J, et al. 3D Printed Cellulose-Based Fungal Battery. ACS Sustain Chem Eng. 2024;12(43):16001–11.
75. Mishra AK, Kim J, Baghdadi H, Johnson BR, Hodge KT, Shepherd RF. Sensorimotor control of robots mediated by electrophysiological measurements of fungal mycelia. Sci Robot. 2024;9(93):eadk8019.

76. Brandić Lipińska M, Maurer C, Cadogan D, Head J, Dade-Robertson M, Paulino-Lima IG, et al. Biological growth as an alternative approach to on and off-Earth construction. Front Built Environ. 2022;8:965145.
77. Barrios-González J, Tarragó-Castellanos MR. Solid-state fermentation: special physiology of fungi. In: Mérillon JM, Ramawat K, editors. Fungal Metabolites. Springer; 2017. pp. 319–47.
78. Mitchell DA, Luz LF de L, Krieger N, Berovič M. 2.25 – Bioreactors for Solid-State Fermentation. In: Moo-Young M, editor. Comprehensive Biotechnology (Second Edition). Academic Press; 2011. pp. 347–60.
79. Krishna C. Solid-state fermentation systems – an overview. Crit Rev Biotechnol. 2005;25(1–2): 1–30.
80. Livne A, Wösten HA, Pearlmutter D, Gal E. Fungal mycelium bio-composite acts as a CO2-sink building material with low embodied energy. ACS Sustain Chem Eng. 2022;10(37): 12099–106.
81. Bakratsas G, Polydera A, Katapodis P, Stamatis H. Recent trends in submerged cultivation of mushrooms and their application as a source of nutraceuticals and food additives. Future Foods. 2021;4:100086.
82. Hamza A, Shankar MP, Chowdary US, Ghanekar S, Sahoo S, Krishnaiah CV, et al. Submerged production of mycelium biomass and bioactive compounds from *P. ostreatus* in a controlled fermentation medium. Food Humanity. 2024;2:100302.
83. Merillon JM, Ramawat KG. Co-evolution of secondary metabolites. Springer; 2020.
84. Whittaker JA, Johnson RI, Finnigan TJ, Avery SV, Dyer PS. The biotechnology of quorn mycoprotein: past, present and future challenges. Gd Chall Fungal Biotechnol. 2020;59–79.
85. Attias N, Reid M, Mijowska SC, Dobryden I, Isaksson M, Pokroy B, et al. Biofabrication of nanocellulose–mycelium hybrid materials. Adv Sustain Syst. 2021;5(2):2000196.
86. Mousavi SN, Ramamoorthy SK, Hakkarainen M, Zamani A. Production of Mycelium-Based Papers from Carrot Pomace and Their Potential Applications for Dye Removal. J Polym Environ. 2024;32(9):4716–32.
87. Svensson SE, Ferreira JA, Hakkarainen M, Adolfsson KH, Zamani A. Fungal textiles: Wet spinning of fungal microfibers to produce monofilament yarns. Sustain Mater Technol. 2021;28:e00256.
88. Ifuku S, Nomura R, Morimoto M, Saimoto H. Preparation of Chitin Nanofibers from Mushrooms. Materials. 2011;4(8):1417–25.
89. Sun W, Strässle Zúñiga SH, Philippe V, Rinaldi L, Abitbol T. Mycelium-Bound composites from agro-industrial waste for broadband acoustic absorption. Mater Des. 2025;250:113591.
90. Rathinamoorthy R, Sharmila Bharathi T, Snehaa M, Swetha C. Mycelium as sustainable textile material – review on recent research and future prospective. Int J Cloth Sci Technol. 2023;35(3):454–76.
91. Vaghari H, Anarjan N, Najian Y, Jafarizadeh-Malmiri H. Sterilization Process. In: Berenjian A, editor. Essentials in Fermentation Technology. Springer International Publishing; 2019. pp. 85–103.
92. Yusef HM, Allam ME. The effect of light on growth and sporulation of certain fungi. Mycopathol Mycol Appl. 1967;33(2):81–9.
93. Rahardjo YS, Sie S, Weber FJ, Tramper J, Rinzema A. Effect of low oxygen concentrations on growth and α-amylase production of *Aspergillus oryzae* in model solid-state fermentation systems. Biomol Eng. 2005;21(6):163–72.
94. Arifin YH, Yusuf Y. Mycelium fibers as new resource for environmental sustainability. Procedia Eng. 2013;53:504–8.
95. Alemu D, Tafesse M, Mondal AK. Mycelium-Based Composite: The Future Sustainable Biomaterial. Int J Biomater. 2022;2022:1–12.
96. Appels FVW, Camere S, Montalti M, Karana E, Jansen KMB, Dijksterhuis J, et al. Fabrication factors influencing mechanical, moisture- and water-related properties of mycelium-based composites. Mater Des. 2019 Jan;161:64–71.
97. Kim Y, Ruedy DR. Mushroom packages: an ecovative approach in packaging industry. In: Dhiman S, Marques J, editors. Sustainable Development and Environmental Stewardship: Global Initiatives Towards Engaged Sustainability. Springer; 2023. pp. 199–223.
98. Zeller P, Zocher D. Ecovative's breakthrough biomaterials. Fungi Mag. 2012;5(1):51–6.
99. Alaneme KK, Anaele JU, Oke TM, Kareem SA, Adediran M, Ajibuwa OA, et al. Mycelium based composites: A review of their bio-fabrication procedures, material properties and potential for green building and construction applications. Alex Eng J. 2023;83:234–50.
100. Zhang X, Hu J, Fan X, Yu X. Naturally grown mycelium-composite as sustainable building insulation materials. J Clean Prod. 2022;342:130784.
101. Mamtaz H, Fouladi MH, Al-Atabi M, Narayana Namasivayam S. Acoustic absorption of natural fiber composites. J Eng. 2016; 2016(1):5836107.
102. Pelletier M, Holt G, Wanjura JD, Bayer E, McIntyre G. An evaluation study of mycelium based acoustic absorbers grown on agricultural by-product substrates. Ind Crops Prod. 2013;51:480–5.
103. Gezer ED, Kuştaş S. Acoustic and thermal properties of mycelium-based insulation materials produced from desilicated wheat straw – Part B. BioResources. 2024;19(1):1348–64.

104. Sivaprasad S, Byju SK, Prajith C, Shaju J, Rejeesh C. Development of a novel mycelium biocomposite material to substitute for polystyrene in packaging applications. Mater Today Proc. 2021;47:5038–44.

105. Joshi K, Meher MK, Poluri KM. Fabrication and Characterization of Bioblocks from Agricultural Waste Using Fungal Mycelium for Renewable and Sustainable Applications. ACS Appl Bio Mater. 2020;3(4):1884–92.

106. Sahu S, Bhardwaj P, Singh G, Bhalla A, Arya SK. A pioneering review on *Ganoderma lucidum*–derived leather: taking a step towards a cruelty-free leather manufacturing. Environ Sci Pollut Res. 2025;32(10):5730–46.

107. Blanchette RA, Haynes DT, Held BW, Niemann J, Wales N. Fungal mycelial mats used as textile by indigenous people of North America. Mycologia. 2021;113(2):261–7.

108. Crawford A, Miller SR, Branco S, Fletcher J, Stefanov D. Growing mycelium leather: a paste substrate approach with post-treatments. Res Dir Biotechnol Des. 2024;2:e6.

109. Jones M, Gandia A, John S, Bismarck A. Leather-like material biofabrication using fungi. Nat Sustain. 2021;4(1):9–16.

110. Harms H, Schlosser D, Wick LY. Untapped potential: exploiting fungi in bioremediation of hazardous chemicals. Nat Rev Microbiol. 2011;9(3):177–92.

111. Wang L, Yu T, Ma F, Vitus T, Bai S, Yang J. Novel self-immobilized biomass mixture based on mycelium pellets for wastewater treatment: A review. Water Environ Res. 2019;91(2): 93–100.

112. Cavalcante J, Szekely G. Surface engineering of a superamphiphilic, self-growing fibrous Janus membrane prepared from mycelium. J Mater Chem A. 2023;11(45):24598–607.

113. Lee T, Choi J. Mycelium-composite panels for atmospheric particulate matter adsorption. Results Mater. 2021;11:100208.

114. Barzee TJ, Cao L, Pan Z, Zhang R. Fungi for future foods. J Future Foods. 2021;1(1):25–37.

115. Huling R. Applications of fungi for alternative protein. In: Fungi and Fungal Products in Human Welfare and Biotechnology. Springer; 2023. pp. 365–95.

116. Wang X, Pei Y, Wu J, Zhong X, Liu H, Xue Y. Innovative mycelium-based food: Advancing One Health through nutritional insights and environmental sustainability. Compr Rev Food Sci Food Saf. 2025;24(2):e70166.

117. Rothschild LJ, Maurer C, Lipinska MB, Senesky D, Paulino-Lima I, Snyder J, et al. Mycotecture Off Planet: Fungi as A Building Material on The Moon and Mars. [accessed 15 Apr 2025]. Available at: https://ntrs.nasa.gov/citations/20210026908

118. Cortesão M, Schütze T, Marx R, Moeller R, Meyer V. Fungal Biotechnology in Space: Why and How? In: Nevalainen H, editor. Grand Challenges in Fungal Biotechnology. Springer International Publishing; 2020. pp. 501–35.

119. Bitting S, Derme T, Lee J, Van Mele T, Dillenburger B, Block P. Challenges and opportunities in scaling up architectural applications of mycelium-based materials with digital fabrication. Biomimetics. 2022;7(2):44.

120. Vandelook S, Elsacker E, Van Wylick A, De Laet L, Peeters E. Current state and future prospects of pure mycelium materials. Fungal Biol Biotechnol. 2021;8(1):20.

121. Madusanka C, Udayanga D, Nilmini R, Rajapaksha S, Hewawasam C, Manamgoda D, et al. A review of recent advances in fungal mycelium based composites. Discov Mater. 2024;4(1):13.

122. Camilleri E, Narayan S, Lingam D, Blundell R. Mycelium-based composites: An updated comprehensive overview. Biotechnol Adv. 2025;79:108517.

123. Wan C, Li Y, Fungal Pretreatment of Lignocellulosic Biomass, Biotechnology Advances. 2012;30(6):1447–57.

CHAPTER 2

Mycelium as a Building Component

THIBAUT HOUETTE and PETRA GRUBER

1 The place of material manufacturing for architecture in the current cycle of fungi

1.1 Current usage of fungi in human society

Fungi provide many services beneficial to humans, from nutrition to medicine and bioremediation. Paul Stamets describes a circular model in which fungi are utilized to generate compost, decontaminate polluted sites (i.e., mycoremediation), produce soil to grow plants (i.e., fertilizer), create organic materials (i.e., biomass), to grow mushrooms (i.e., for food and medicine), which in turn regenerate compost and so on [1] (Figure 2.1). Fungi secrete enzymes to break down complex substances before ingesting the chemical elements. As a result, fungi serve to both break down and take up harmful chemicals, while plants can only take them up. With the growing field of mycelium-based materials, their manufacturing can be included in this circular model. During fungal growth, our

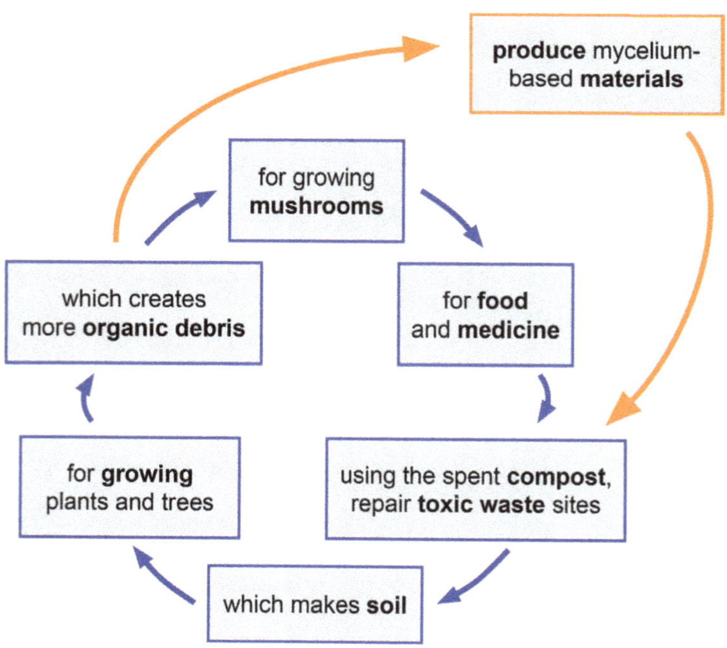

Figure 2.1 Cycle of fungi (adapted from Stamets, 2005, p. 103 – in blue).

Thibaut Houette and Petra Gruber, 'Mycelium as a Building Component' in: *Mycology for Architecture: Methods and Applications of Mycelium Materials in the Built Environment*. Pelagic Publishing (2026). © Thibaut Houette and Petra Gruber. DOI: 10.53061/XPOT4821 current focus is on producing fruiting bodies for food and medicine. Therefore, the production of materials, which utilize the mycelium and not the fruiting bodies, could be combined with the current model in a unified process [2].

1.2 Adding manufacturing building materials

As we continue to push Earth beyond our planetary boundaries[1] (seven out of nine in 2025), the construction field continues to negatively impact the environment from resource consumption and expansion over natural ecosystems to waste production [3, 4, 5]. The entire life cycle of materials used by society needs to be analysed to reduce our impact on the environment. Materials produced with renewable resources and through the growth of living cells, such as Engineered Living Materials, are therefore of increasing interest for all fields, including building construction [6]. Such materials allow the manufacturers to also benefit from growing qualities of living systems, which cannot be mimicked by our current technology [7].

The wide array of qualities of fungal mycelium and large range of applications make mycelium-based materials one of the most successful large-scale living materials [6, 8]. By searching through its environment for food, a fungal mycelium forms a branched network, which is used as a 3D living binder in material production. In its simplest form, the biocomposite material contains a fungus and an organic substrate, making it fully biodegradable. A variety of organic substrates can be used, including agricultural waste and invasive organisms to benefit the local environment. Despite the infrastructure required to keep the growth conditions constant in a clean environment, the overall material lifecycle requires few resources and has an overall low impact on the environment if grown from waste products compared to equivalent traditional materials. Lastly, many variables can tune the material properties, thus being adaptable to a wide range of building applications, as will be described in the next sections of this chapter.

2 Interesting material properties for architectural applications

2.1 Material cycle and morphing ability

The production of mycelium-based materials is compatible with the current usages of fungi. The mycoremediation can be included at two steps of the process: during the growth of the materials (i.e., on a contaminated substrate) or at the end of the material's life if the mycelium is still alive (i.e., discarded in a contaminated site). To reduce stress on natural over-exploited resources, local organic waste is used for mycelial growth and, at the end of their life, mycelium-based materials serve to produce new batches of materials or compost. To go beyond, the use of organic matter from invasive species combines the manufacture of materials while cleaning the environment and preserving the native biodiversity.

The amorphous inoculated mixture is grown inside a clean environment to avoid the introduction and growth of micro organisms competing with the spawn introduced and to give it a shape. Simple shapes can be produced (e.g., bricks and panels) and, with

1 These seven crossed boundaries are: climate change (2009), biosphere integrity (2009), biogeochemical flows (2009 for nitrogen and 2015 for phosphorus), land-system change (2015), novel entities (2023), freshwater change (2023), and ocean acidification (2025). Years in parentheses display when each boundary assessed was evaluated as crossed. As of 2025, the remaining two assessed but not crossed boundaries are: stratospheric ozone depletion and atmospheric aerosol loading.

Figure 2.2 Hanging model from Growing As Building GrAB project exhibition, Angewandte Innovation Lab AIL at University of Applied Arts Vienna, Austria. Image credit: Photo by Bruno Stubenrauch, 2015, Copyright: GrAB.

the rise of 3D printing, more organic morphologies have also been explored [7, 9, 10]. Yet removing the material from the scaffold and material shrinkage in the drying phase need to be anticipated for the latter. Curved scaffolds can also be used, but the drying process will result in uneven shrinking, adding strain to the material and potentially generating cracks on the side under tension [7]. Flexible scaffolds can adapt to the size variations throughout the process and even fuse with the grown mixture (Figure 2.2). Such systems could even follow the funicular method to only apply compression to the final structure, as in Gaudí's hanging models used for the Sagrada Familia [11]. The thickness of the material, along with its density, will directly impact oxygen access to the inner parts, and is therefore limited unless artificially aerated [12].

2.2 Wide range of properties

Mycelium-based materials are biocomposites and benefit from the properties of both the aggregate substrate and the matrix, the mycelium. The binding ability of the mycelium network enables the material to withstand compression, tension and bending loads in ranges similar to traditional building materials. Many composites impact the environment negatively as their components are not biodegradable and cannot be separated to be recycled. On the other hand, materials composed of water, mycelium and organic matter are entirely biodegradable and can have other purposes at the end of their life as a building material.

Mycelium-based composites (MBCs) are highly heterogeneous in terms of particle distribution, growth and resulting properties. Based on their mixing and placement in moulds, the inner distribution of materials is uneven, as seen in the large error bars across mechanical studies [2]. A gradient of mycelium growth is also observed throughout material thickness owing to lower oxygen access in the inner parts. Gas exchange limits the thickness of such materials, but researchers have experimented with ways to pump oxygen through the internal parts. Heterogeneity is often seen as a disadvantage in material design, as it comes with weaker areas reducing the material quality and more difficult prediction and design of structures. However, better understanding of mycelium growth and leveraging this heterogeneity could lead to the design of adaptive composites or hierarchical systems and improve current properties or lead to emerging ones.

Uncompacted MBCs are lightweight and highly porous, making them easy to handle and efficient for insulative purposes. Compacting them increases material density and mechanical strength for structural applications. The surface is covered by pure mycelium as it grows over the substrate, resulting in a soft and delicate skin, similar to mushrooms (Figure 2.3). Fungal hyphae are hydrophobic, making the surface of this myceliated material repel water, increasing its durability. However, as the surface cover is not entirely sealed, the composite starts to deteriorate relatively fast in humid conditions [7]. Keeping the mycelium alive in an outdoor environment slightly increases outdoor durability. Because of the symbiotic relationships between fungi and plants, the deterioration of the myceliated façade panels grown with mycorrhizal fungi could potentially serve as a base for more durable green wall systems. In this case, the fungal and plant species are

Figure 2.3 Close-up of the fungal hydrophobic skin. Image credit: Petra Gruber, 2019.

selected according to the local environment and biodiversity. As the durability of MBCs is lower than required in conventional façade systems, coatings can be added to enhance waterproofing. However, depending on the chemical composition of the coating, the final product may not remain biodegradable.

Finally, the long-term sustainability of MBCs produced on a large scale is still under study, but the following advantages and drawbacks are identified. Compared with traditional building materials, they provide a high multifunctional potential owing to their tuneable properties; contribute to the larger cycle of fungal usage in society; upcycle organic waste; allow cleaning of the environment; are biodegradable or reusable for producing new generations of materials; and require low amounts of energy. Their main drawbacks result from the need to maintain constant growth conditions in terms of humidity, temperature and sterility; and generate standardized properties from heterogeneous materials with unexpected growth results, leading to the need to produce duplicates.

In conclusion, a similar circular supply chain can lead to a wide array of materials produced. The same manufacturing equipment and setting serves to generate a variety of material types, by adjusting several variables.

2.3 Control of material properties through adjusting manufacturing variables

MBCs describe a wide array of materials produced with mycelium growing over a substrate. Throughout the manufacturing process, a range of variables can be adjusted to impact the final material properties. However, the manufacturing process and tests evaluating material properties are not yet fully understood or standardized. Therefore, the effect of the process on the material properties is difficult to exhaustively comprehend and predict.

2.3.1 Recipe

Diverse mycelium species show divergent characteristics owing to their anatomy and physical structure. In basidiomycetes, the main taxonomic division used for MBCs, three categories have been described: monomitic (i.e., generative), dimitic (i.e., generative and skeletal) and trimitic (i.e., generative, binding and skeletal) [12]. Trimitic species are supposed to produce higher mechanical strength than monomitic ones, as observed with materials grown from *Trametes versicolor* (trimitic) resulting in higher compressive, tensile and bending strength than those from *Pleurotus ostreatus* [13, 14]. Furthermore, hyphal growth speed also varies between fungal species. A faster growing mycelium reduces the manufacturing time and resulting resources required to maintain the growth conditions. Based on projects published, the three most common mycelium strains used for making materials are, in order, *Ganoderma lucidum*, *Pleurotus ostreatus* and *Trametes versicolor* [15, 16]. Widely understood thanks to their utilization in medicine and food production, these strains were rapidly adopted in the manufacturing materials field. As a result, producing materials, food and medicine in a circular model with such species is increasingly favourable.

The substrate's composition and particle size actively contribute to the material properties of the composite [2, 12, 16]. Organic substrate is provided as a food source for the fungi, and nutritional additives, such as wheat and rice bran, generic food, tea and tofu waste and olive products, can be included to increase concentration of sugar, nitrogen, fat and minerals in order to speed up mycelial growth [17]. Adjusting the percentage of the various ingredients allows the manufacturers to balance the various functions of interest. With particles larger than a couple of centimetres, the substrate is often chipped and even sieved to increase material density, in most cases leading to increased Young's modulus. However, a substrate too dense may reduce internal mycelial growth because of reduced oxygen access. For instance, materials grown from a substrate with micro-particles increases density and resulting Young's and elastic moduli, compared to those without micro-particles. The use of fibres as substrate would be interesting in terms of the structural potential, but poses a challenge to achieve even dispersal in the mixing process. However, controlling such arrangement is interesting as adding fibres oriented in the direction of loading led to an increase in Young's modulus under compression, while adding some perpendicular to it decreases Young's modulus and ultimate strength [18]. In some cases, substrate particle size can have a larger effect on compressive Young's modulus than density [19], but this is not true across all studies. For structural purposes it is easily possible to integrate larger structures such as nets, fabric or grids into the material production, as the mycelium grows around or through them.

2.3.2 Growth process

Multiple phases of growth can serve to generate more homogeneous material properties while saving growth time in the final scaffolds, as seen in the Ecovative process [20]. As the mycelium grows over the surface of the substrate, it creates a protective layer against the growth of competing species. Therefore, moving the inoculated substrate during the growth process will break this protective layer and may introduce contamination in the growth environment. Such a process should therefore be conducted in a controlled setting.

2.3.3 Post-growth treatments

A diversity of treatments is applied to the grown material during or after the drying phase to adjust its final properties. For instance, the same myceliated material can be used as insulation, owing to its high porosity, low thermal conductivity and low weight, or compacted for a structural application. Compaction is employed to increase density and resulting mechanical strength, despite breaking some of the mycelium network. Hot pressing increases homogeneity, strength and stiffness compared with cold pressing, resulting in similar density and elastic moduli as cork and specific types of wood [14]. Depending on the application, the material is then either dried or baked. Drying keeps the mycelium alive in a dormant phase, while baking kills it. In terms of process, baking requires a higher temperature but for a shorter period of time, increasing the speed of de-wetting. Both treatments generate a significant shrinkage of around 5% to 15%, depending on the material shape and measured direction, resulting from the loss of water throughout the composite matrix [2]. Baking was seen to result in slightly increased Young's and elastic modulus compared with drying [2]. However, dormant panels (i.e., dried) were more durable in an outdoor setting, probably because they stayed alive and were able to continue their growth once implemented [7]. Owing to the reduced internal growth, cutting should be avoided as it will expose substrate to humidity and other organisms without the protective and cohesive mycelium surface. Mycelium-based sandwich composites can also be produced with bioresins for applications targeting bending resistance [21].

2.3.4 Application environments

The relative humidity of the application environment will also affect the behaviour of the material, taking in and letting out the moisture. For an exhibition in an indoor art gallery

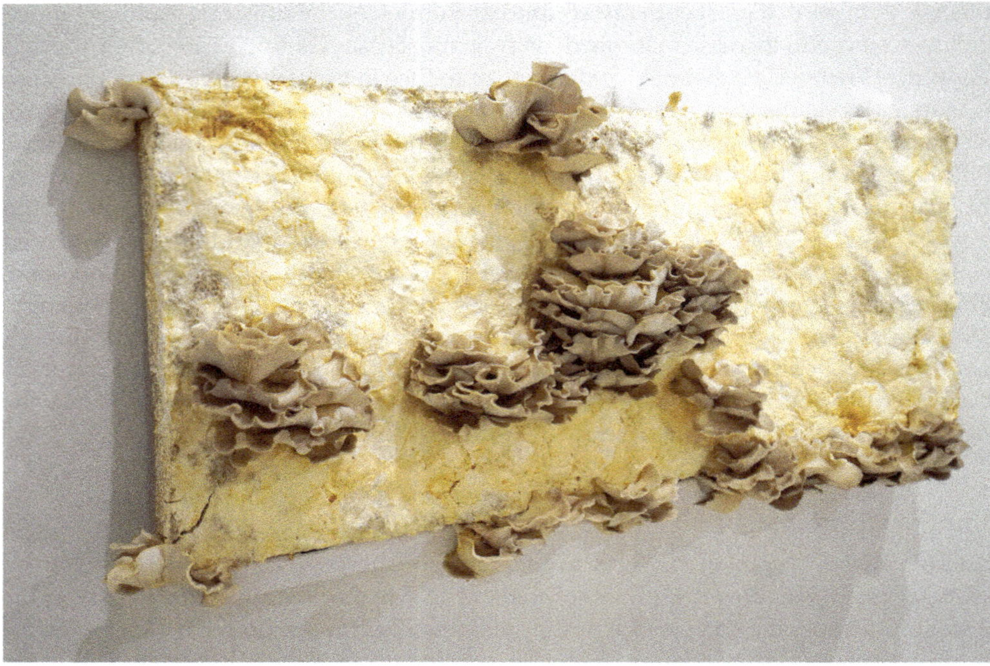

Figure 2.4 Large, architectural-scale panel, alive with mushrooms. Mycoterial Panel, Gruber/Houette, Myers School of Arts Faculty Exhibition, Akron, USA, September 2019. Image credit: Petra Gruber, 2019.

at the University of Akron, a living myceliated panel grown from *Pleurotus ostreatus*, measuring 130 cm in width, 50 cm in height and 8 cm in depth, shown in Figure 2.4, shrank by around 7.8% in height over 25 days after installation. The material was not dried, so as to display a living material which was producing edible fruiting bodies throughout the exhibition.

In conclusion, many properties of MBCs are desirable for building applications, and the ability to tailor them throughout the manufacturing process makes these materials promising for a diversity of building applications.

3 Replacing traditional building materials

3.1 Design prototypes, demonstrators and art

Art exhibitions and pavilions have played a key role in demonstrating the potential of MBCs. Phil Ross introduced mycotectures, vaults composed of myceliated blocks, at the 2009 Kunsthalle Düsseldorf's Eat Art exhibition. Simple brick-shaped MBCs were then combined to build the first large scale installation, the Hy-fi tower, designed by David Benjamin for the MoMA's PS1 space in 2014. More complex mycelial designs were produced through 3D printed and tensile scaffolds in the Growing as Building project by Barbara Imhof and Petra Gruber in 2015. Combined with parametric modelling, MBCs only loaded in compression were assembled by Dirk Hebel and the Block Research Group to form the Mycotree for the 2017 Seoul Biennale of Architecture and Urbanism. These projects have also been used to change the negative public perception of fungi, often seen as damaging or dirty [22]. In collaboration with Claudia Naményi and redhouse studio, the authors produced a myceliated panel from *Pleurotus ostreatus* and exposed it still alive at the 2019 Myers School of Arts Faculty exhibition. Throughout the exhibition, the panel was evolving; emerging fruiting bodies were collected daily for consumption and to reduce health risks linked to spore inhalation (Figure 2.4). In conclusion, artworks emphasize the potential of MBCs but also explore the artistic side of this growing living material, including the aesthetics of its impressive growth patterns.

3.2 Insulation

MBCs' high porosity and low weight are heavily explored as a sound sustainable alternative to traditional insulative materials, which often display a significant negative environmental impact [23]. Unlike amorphous biosourced insulative materials, mycelium generates a coherent material, which remains biodegradable. The mycelium is killed to avoid uncontrolled growth within building layers which are often not accessible.

3.3 Acoustic

In addition to being highly porous, MBCs possess soft heavily textured surfaces owing to the growth of the complex mycelium network. As a result, MBCs excel at absorbing sounds, making them very promising as acoustic panels [24, 25].

3.4 Firewall

Mycelium exhibits flame-retardant properties, making it promising for firewall applications. Its fire resistance outperforms comparable synthetic thermoplastic polymers (PMMA and PLA), while being more economical and sustainable [26].

3.5 Façade

Despite its hydrophobic state, the porous composite expands and shrinks under dry–wet cycles, leading to cracks. This focuses its use on indoor applications. Yet temporary structures have explored the potential of MBCs exposed to weathering: Mycelium Shell Pavilion by Beetles 3.3 – Yassin Areddia Designs, built in 2017, and LIWAS, a Living Wall System by Petra Gruber, Thibaut Houette and redhouse studio in 2018 (Figure 2.5). Under these conditions, research suggests that drying (rather than baking) improves durability [7]. One hypothesis is that being alive allows adaptability over environmental fluctuations. But the advantages of alive material have not been exhaustively studied yet. Owing to symbiotic relationships between fungi and plant roots in the rhizosphere, the potential of MBCs as a base for a green wall system was explored in the LIWAS project. Local seeds were placed on the myceliated panels. However, throughout the observation time, the only plant which was seen growing on the panels was another grass species, deposited naturally. Living façades from MBCs show promise but require further exploration.

Compaction at the end of mycelial growth increases material density, resulting in higher Young's modulus. MycoHAB, a collaboration between redhouse studio, Standard Bank Group and MIT, built MycoHouse 1.0, the first self-supporting mycelium house structure in Namibia, in 2024. Oyster fungi were grown over a substrate made from invasive plant species compacted to form building blocks. Despite utilizing fungal species not recognized for their mechanical performance, MycoHAB produced food, along with the building structure itself and provided insulation for the house – all while sequestering carbon and helping cleanse the environment of invasive species.

Figure 2.5 LIWAS Living Wall System installation by Gruber/Houette/Rupp at the University of Akron Biology Fieldstation, during long-term exposure to the elements. Image credit: Thibaut Houette, November 2018.

3.6 Space

MYCO-NIAC, a collaboration between NASA and redhouse studio, explores MBCs' potential for space applications. The team found promising radiation absorbance by testing samples outside the International Space Station. In addition, their growth ability has a significant advantage over traditional materials. A small mycelium strain could be launched on a rocket (i.e., where space is highly limited) and grown on site with resources

Figure 2.6 Close-up of mycelium fibres bridging material gaps. Image credit: Thibaut Houette, 2019.

found or produced at the building location. A circular system combining astronauts' waste, algae and mycelium is being explored to upcycle waste, generate oxygen and building materials.

3.7 Going beyond: living materials

Most building applications focus on implementing dead MBCs for their static characteristics linked to the absence of growth and lack of potential health hazard of spore release. However, alive materials can benefit from the qualities of growth: adaptation and response, resilience, dynamism and movement, differentiation and continuous functionality [7]. Self-healing materials, such as self-healing concrete through microbially induced calcite precipitation, are becoming of increasing interest, and the ability of MBCs to self-heal could be evaluated (Figure 2.6). Food production can also happen throughout the material utilization as seen in the panel exhibited at Myers School of Arts or in the LIWAS project. In the latter, panels were installed in an outdoor setting in Ohio, USA. Fruiting bodies were growing on the MBCs over six months after installation, and after the snowy winter. Furthermore, keeping the fungus dormant allows flexibility. For instance, mycelial growth could be started in a controlled environment. After a first growth period, it could be installed on site and continue to grow by adjusting to environmental conditions; it could merge with the rest of the building, performing its function while benefiting from the properties of life. In conclusion, MBCs represent a promising avenue to explore the future of living architecture.

4 Future steps to implement MBCs in buildings

Before widely implementing MBCs in buildings, several hurdles remain in terms of better understanding the material, optimizing the production process and opening the realm of architectural applications [7]. The heterogeneous properties of such grown composite materials reduce the predictability of their behaviour under various stresses. Since cutting MBCs is detrimental, the mould size must account for the non-uniform and hard to predict shrinking during drying/baking to generate desired final material size. Health hazards for humans and other species owing to the growth, installation and potential spore release from large quantities of MBCs remain to be evaluated. Scaling up production without utilizing nutritional additives is challenging because of the need for a clean environment. The production requires new infrastructure, because there are no equivalent products that allow reuse, in comparison with yeast or bacteria growth and fermentation processes. Fungi have been grown purposefully mostly for their fruiting bodies, but not for the mycelium specifically, often seen as a waste product. Despite the impact of conventional building materials, there is limited pressure to shift from cheap environmentally damaging fossil-based materials and invest in innovative cleaner alternatives. Fungi are negatively connoted in society and regarded as a biological system to avoid in buildings. Building regulations are a major hurdle to the integration of such innovative, uncommon, living, heterogeneous materials, especially for applications with structural requirements.

While almost all studies explore the use of fungal mycelium, some species of bacteria also produce mycelium. While trying to find a local mycelium species to grow for an outdoor application, the LIWAS team grew a diversity of mycelium species found in a local forest. The fastest growing one was further identified as a bacterium through DNA testing. Therefore, the exploration of MBCs not only stops at fungal mycelium but is also open to explore the potential of bacterial mycelium and their respective characteristics.

References

1. Stamets P. *Mycelium Running: How Mushrooms Can Help Save the World*. First. Berkeley: Ten Speed Press; 2005.
2. Houette T, Maurer C, Niewiarowski R, Gruber P. Growth and Mechanical Characterization of Mycelium-Based Composites towards Future Bioremediation and Food Production in the Material Manufacturing Cycle. Vol. 7, Biomimetics. 2022.
3. Sakschewski B, Caesar L, Andersen L, Bechthold M, Bergfeld L, Beusen A, Billing M, Bodirsky B L, Botsyun S, Dennis D, Donges J F, Dou X, Eriksson A, Fetzer I, Gerten D, Häyhä T, Hebden S, Heckmann T, Heilemann A, Huiskamp W N, Jahnke A, Kaiser J, Kitzmann N, Krönke J, Kühnel D, Laureanti N C, Li C, Liu Z, Loriani S, Ludescher J, Mathesius S, Norström A, Otto F, Paolucci A, Pokhotelov D, Rafiezadeh Shahi K, Raju E, Rostami M, Schaphoff S, Schmidt C, Steinert N J, Stenzel F, Virkki V, Wendt-Potthoff K, Wunderling N, Rockström J. Planetary Health Check 2025: A Scientific Assessment of the State of the Planet, Potsdam: Potsdam Institute for Climate Impact Research (PIK); 2025. Available from: https://doi.org/10.48485/pik.2025.017
4. United Nations Environment Programme. 2021 Global Status Report for Buildings and Construction: Towards a Zero-emission, Efficient and Resilient Buildings and Construction Sector. Nairobi, Kenya; 2021.
5. United States Environmental Protection Agency. Advancing Sustainable Materials Management: 2018 Fact Sheet. Washington, DC; 2020.
6. Nguyen PQ, Courchesne NMD, Duraj-Thatte A, Praveschotinunt P, Joshi NS. Engineered Living Materials: Prospects and Challenges for Using Biological Systems to Direct the Assembly of Smart Materials. Advanced Materials [internet]. 2018 ;30(19):1704847. Available from: http://doi.wiley.com/10.1002/adma.201704847
7. Houette T, Foresi B, Maurer C, Gruber P. Growing Myceliated Facades – Manufacturing and exposing experimental panels in a facade setting. In: Facade Tectonics World Congress 2020 [internet]. 2020 [accessed on 2020 Sep 8]. Available from: https://www.researchgate.net/publication/341271351_Growing_Myceliated_Facades_-_Manufacturing_and_exposing_experimental_panels_in_a_facade_setting
8. Haneef M, Ceseracciu L, Canale C, Bayer IS, Heredia-Guerrero JA, Athanassiou A. Advanced Materials From Fungal Mycelium: Fabrication and Tuning of Physical Properties. Sci Rep [internet]. 2017 ;7(1):41292. Available from: http://www.nature.com/articles/srep41292

9. Javadian A, Le Ferrand H, Hebel DE, Saeidi N, Javadian A, Le Ferrand H, et al. Application of Mycelium-Bound Composite Materials in Construction Industry: A Short Review. SOJ Materials Science & Engineering [internet]. 2020 ;7(2):1–9. Available from: https://doi.org/10.3929/ethz-b-000543782
10. Imhof B, Gruber P. Built to grow : blending architecture and biology. Birkhäuser; 2015. 175 p.
11. Meirelles CRM, Beraldo PN, Nazareth SBM. Geometric Shape, Structure and Material in Antoni Gaudí's Work: The Colònia Güell Crypt and the Templo Expiatorio de la Sagrada Familia. Journal of Civil Engineering and Architecture. 2019;13:608–21.
12. Jones M, Mautner A, Luenco S, Bismarck A, John S. Engineered mycelium composite construction materials from fungal biorefineries: A critical review. Vol. 187, Materials and Design. Elsevier Ltd; 2020. p. 108397.
13. Lelivelt RJJ. The mechanical possibilities of mycelium materials [internet]. Eindhoven University of Technology; 2015 [accessed on 2020 Sep 27]. Available from: https://research.tue.nl/en/studentTheses/the-mechanical-possibilities-of-mycelium-materials
14. Appels FVW, Camere S, Montalti M, Karana E, Jansen KMB, Dijksterhuis J, et al. Fabrication factors influencing mechanical, moisture- and water-related properties of mycelium-based composites. Mater Des. 2019 Jan 5;161:64–71. https://doi.org/10.1016/j.matdes.2018.11.027
15. Yang L, Park D, Qin Z. Material Function of Mycelium-Based Bio-Composite: A Revi ew. Front Matter [internet]. 2021 ;8:737377. https://doi.org/10.3389/fmats.2021.737377
16. Attias N, Danai O, Abitbol T, Tarazi E, Ezov N, Pereman I, et al. Mycelium bio-composites in industrial design and architecture: Comparative review and experimental analysis. J Clean Prod. 2020; https://doi.org/10.1016/j.jclepro.2019.119037
17. Soh E, Saeidi N, Javadian A, Hebel DE, Le Ferrand H (2021) Effect of common foods as supplements for the mycelium growth of *Ganoderma lucidum* and *Pleurotus ostreatus* on solid substrates. PLoS ONE 16(11): e0260170. https://doi.org/10.1371/journal.pone.0260170
18. Rigobello A, Ayres P. Compressive behaviour of anisotropic mycelium-based composites. Sci Rep [internet]. 2022 ;12(1):1–13. Available from: https://www.nature.com/articles/s41598-022-10930-5
19. Elsacker E, Vandelook S, Brancart J, Peeters E, De Laet L. Mechanical, physical and chemical characterisation of mycelium-based composites with different types of lignocellulosic substrates. PLoS One. 2019;14(7).
20. Winiski J, Van Hook S, Lucht M, McIntyre G. Process for Solid-state Cultivation of Mycelium on a Lignocellulose Substrate [internet]. US9914906B2, 2016 [accessed on 2020 Apr 27]. Available from: https://patents.google.com/patent/US9914906B2/en
21. Jiang L, Walczyk D, McIntyre G, Bucinell RB. A New Approach to Manufacturing Biocomposite Sandwich Structures: Mycelium-Based Cores. In: Volume 1: Processing [internet]. ASME; 2016. p. V001T02A025. Available from: http://proceedings.asmedigitalcollection.asme.org/proceeding.aspx?doi=10.1115/MSEC2016-8864
22. Karana E, Blauwhoff D, Hultink EJ, Camere S. When the Material Grows: A Case Study on Designing (with) Mycelium-based Materials. International Journal of Design [internet]. 2018 [accessed on 2020 Dec 11];12(2):119–36. Available from: http://www.ijdesign.org/index.php/IJDesign/article/view/2918
23. Zhang X, Hu J, Fan X, Yu X. Naturally grown mycelium-composite as sustainable building insulation materials. J Clean Prod. 2022 Mar 15;342:130784.
24. Pelletier MG, Holt GA, Wanjura JD, Bayer E, McIntyre G. An evaluation study of mycelium based acoustic absorbers grown on agricultural by-product substrates. Ind Crops Prod [internet]. 2013 ;51:480–5. Available from: https://www.sciencedirect.com/science/article/pii/S0926669013005050
25. Walter N, Gürsoy B. A Study on the Sound Absorption Properties of Mycelium-Based Composites Cultivated on Waste Paper-Based Substrates. Biomimetics 2022, Vol 7, Page 100 [internet]. 2022 ;7(3):100. Available from: https://www.mdpi.com/2313-7673/7/3/100/htm
26. Jones M, Bhat T, Kandare E, Thomas A, Joseph P, Dekiwadia C, et al. Thermal Degradation and Fire Properties of Fungal Mycelium and Mycelium – Biomass Composite Materials. Scientific Reports 2018 8:1 [internet]. 2018 ;8(1): 1–10. Available from: https://www.nature.com/articles/s41598-018-36032-9

CHAPTER 3

Mycelium as Building Material: From Bricks to Stressed Skin Structures

JAN WURM

1 Introduction: mycelium materials for sustainable and affordable living solutions

Current construction techniques come at a high cost both for current and future generations. Following the disruption of international supply chains during the pandemic, Associated Builders and Contractors reports that building material costs have increased by about 38% since 2020 [1], with substantial risk for further increases owing to the implementation of disruptive trade tariffs during the second Trump presidency. Rising material costs contribute to increased project costs that, in combination with the limited availability of affordable housing stock, contribute to the current housing crises in Europe that are particularly impacting low-income earners, key workers and young people. In regard to the environmental costs, construction consumes almost half of all extracted resources on our planet, making for about a third of today's waste generation. Construction materials alone contribute about 12% of total greenhouse gas emissions. At the same time, following the report of SYSTEMIQ on efficient and balanced space use in 2023, around 16% of dwellings in urban and pan-urban areas are not occupied and 35% are under-occupied [2]. Initiatives such as Empty Spaces to Homes project, led by Habitat for Humanity International, underline that the key to providing affordable sustainable places to live and work is to employ affordable and renewable bio-based materials to transform vacant and under-utilized spaces [3].

With the need for the adaptive reuse of empty spaces, the most inner layer of the building, including all visible and not structural elements such as suspended ceilings, wall finishes and partition walls, is becoming a key focus point for adaptive design interventions. This layer is responsible for approximately 15% of the total embodied carbon emissions of a contemporary building. Owing to the high percentage of composites being used, architectural fit-out materials often become waste at the end of their relatively short lifecycle of five to seven years [4].

To address the need for affordable, performing, sustainable materials, the construction industry is seeking for bio-based alternatives for interior design to reduce the carbon and resource footprint alongside procurement costs. In this context rapidly renewable and regionally produced materials such as mycelium-based composites (MBCs) can offer innovative solutions. Recent research has demonstrated that the technical performance of MBCs, for example for acoustic and thermal insulation as well as for acoustic absorption, is comparable with conventional materials at lower environmental costs [5]. This opens

Jan Wurm, 'Mycelium as Building Material: From Bricks to Stressed Skin Structures' in: *Mycology for Architecture: Methods and Applications of Mycelium Materials in the Built Environment*. Pelagic Publishing (2026). © Jan Wurm.
DOI: 10.53061/TORJ9102

new horizons for architects to design engaging living habitats within a competitive, carbon neutral and circular economy in Europe.

2 Building with mycelium: substrates, components and structures

Since the pioneering work of artist Phil Ross, the use of mycelium composites for building construction – which he coined as 'mycotecture' – is associated with the discrete building element of bricks. His seminal work 'Mycotectural Alpha' from 2009 (Figure 3.1) makes reference to the archetype of purely compression loaded masonry construction [8]. Manifesting a parabolic vault, the mycelium bricks demonstrate a surprising compressive strength for a grown and porous material reaching up to a tenth of a fired clay brick [5]. At the same time, the soft and porous (and somewhat alienating) texture of the outgrowing fruiting bodies of *Ganoderma lucidum* questions our conceptions of the relationship we as humans have with the built and living environment.

Phil Ross has underlined the versatility of moulded mycelium components for construction: 'Mycelium is like plastic and can be grown to express a number of different qualities within a single object' [7]. Agricultural side streams of a wide range of organic lignocellulosic materials such as hemp, straw and flax are inoculated with white

Figure 3.1 'Mycotectural Alpha' by Phil Ross, exhibited at Kunsthalle Düsseldorf as part of the Eat Art exhibit in 2009. Image credit: Phil Ross.

rot-basidiomycete fungi such as *Trametes versicolor*, *Pleurotus ostreatus* and *Ganoderma lucidum*, and are positioned in a mould, where in moist and warm conditions the mycelial network of the fungi binds together the fibrous substrates, creating a solid block-shaped composite material. The principle of fabricating mycelium components in moulds governs their use in construction today, from organically shaped acoustic absorber panels to segmented spatial branching structures made out of self-supporting mycelium components [6].

The stiffness and strength of MBCs can be varied by selecting and combining fungal strains, growth conditions and substrate types, but generally the porous and low-density nature of MBCs cultivated on loose substrates displays limited mechanical performance [9]. The short length of loose fibres and their random orientation in relation to the governing direction of applied loads limit the structural capacity of the composite. The low tensile and bending strength and the high costs for fabricating moulding and for additional substructure hinder larger scale applications of moulded mycelium components.

Despite the proven positive effect that oriented fibres and textiles integrated in the substrate can have on in-plane strength [10], the use of surface substrates such as textiles and mats are a highly under-investigated area, to date mostly used as lost-formwork for the positioning of loose substrates [11]. Increasing the in-plane strength and structural capacity of MBCs by utilizing surface substrates would set the basis for the design and construction of form-active structures, such as curved and folded shell structures. In the field of product design, designer Jonas Edvard has already demonstrated the feasibility of such an approach, using a technique for cultivating MBCs on substrate mats that he coined 'Mycelium textiles' [12]. In 2019, the author started a collaboration with Jonas Edvard to explore the translation of this technique to the scale of architectural fit-out elements, resulting in the design and installation of Myx Sail, an acoustic sail 3 m wide and 4 m long installed at the Danish Design Museum in 2021 (Figure 3.2). On-site testing demonstrated that the combination of surface substrates such as hemp mats and reinforcing burlap on the top and loose substrates on the bottom enhances the mechanical, acoustic and aesthetic qualities of MBCs [13]. The suspended catenary of the sail showcases the in-plane strength of this novel type of MBC and its potential for form-active mycelium structures. Based on their thin and slender nature, the author suggests the term 'mycelial stressed skin structures' for in-plane loaded MBC components cultivated on surface substrates.

3 Construction principles of stressed skin mycelium structures

Since 2022, the research team of the BioDigital Construction Lab at the Research Group Architectural Engineering at the Department of Architecture of KU Leuven explores the use of surface substrates, such as mats or textiles, instead of moulded loose substrates, and the related structural, performative and aesthetic potential of stressed skin mycelium structures for indoor environments.

The following three key design and construction principles are framing the ongoing research to synthesize the selection of substrates with the design and fabrication of mycelium components.

3.1 Selection of surface substrates

Surface substrates such as woven and non-woven fabrics made from hemp and flax or other natural lignocellulosic fibres provide a good feedstock for white rot-basidiomycete

Figure 3.2 Myx Sail by Jonas Edvard in collaboration with Arup, exhibited at the Danish Design Museum Copenhagen, 2022–4. Image credit: Jan Wurm.

fungi, with *Pleurotus ostreatus* and *Ganoderma lucidum* being the most common species used for material production [6]. Owing to their fibre volume, homogeneity and water retention capacity, non-woven, non-directional mats are well suited for cultivating MBC components (see Figure 3.3). A wide range of industrially produced mat substrates are on the market; these are mostly used as thermal or acoustic insulation fleeces and are usually made of short hemp, flax or coconut fibres. Hemp mats are particularly well suited as surface substrates, owing to their availability, affordability and technical characteristics.

As they are made from agricultural side streams, mats present an affordable half-finished product, with the market price for a 10 mm mat being around 10 € /m^2 in the spring of 2025. When industrial hemp plants are harvested for fibres, the remaining material including shorter fibres, inner hurds or core can be processed into mats. The staple fibres are cleaned to remove impurities and then processed to mats in a wet- or dry-laid fabrication process. In a wet-laid process the short fibres are mixed in a water tank and homogeneously positioned on a belt, then compressed and sometimes combined with other natural fibres, resins or binders. In the needle-punch process, hemp fibres of a

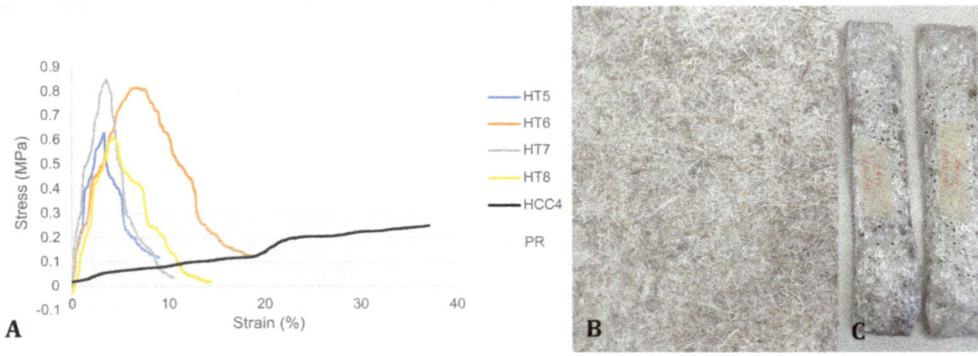

Figure 3.3 Tensile test with MBC samples cultivated with *Pleurotus ostreatus* on hemp mats at 75% moisture level. A: Stress-strain curves. B: close-up of the hemp mat surface substrate before cultivation. C: test specimen during cultivation. Image credit: BioDigital ConstructionLab, KU Leuven.

staple length of up to 70 mm are fed as tufts via mixing and laying belts, which transfers an initially still voluminous formation to a needling machine. The needling machine binds the fibres with a stitch density of up to 1,000 stitches/cm² to form a homogeneous non-woven needled fleece or needle felt. The uniform placement of the staple fibres and their mechanical entanglement lead to a tensile strength of 0.1 MPa [14].

Indicative tests carried out suggest that mycelium cultivation with *G. lucidum* increases the tensile strength of these mats of up to a factor of ten, demonstrating their structural potential for stressed skin structures (see Figure 3.3). At strains of around 5%, the samples of cultivated hemp mats (HT5–HT8) display a tensile strength of up to three times higher than those of conventional MBC (PR) and up to ten times higher than the non-cultivated control sample (HCC4).

The wet- or dry-laid process limits the production width of non-woven mats to a maximum of 1.20 m, which also limits the size of stressed skin components, requiring a modular approach for larger scale structures.

3.2 Prismatic and polyhedral folding and bending to generate form-active surface components

The relative homogeneity, elasticity and tensile integrity of surface substrates allow their geometric manipulation using physical form-finding techniques and the mould-free fabrication of MBCs. Suspending the surface substrate with defined support conditions under its own weight and applying additional loads simulating relevant load scenarios in service creates funicular geometries for specific load cases. When utilizing the plastic deformation of the material and/or cutting the substrates to size, double curved geometries with even higher geometric stiffness can be generated.

Maintaining these support and load conditions during the growth stage of the MBC allows the need for complex shaped moulds to be omitted. The forming of the mycelium matrix during the biofabrication and subsequent drying of the component stabilizes the force-moulded shape. Once stabilized and removed from the suspension mould, the component can perform as a compression loaded shell structure in service. As a basic example, the catenary of a surface substrate suspended under its own dead-load represents the form of a parabolic vault in compression in a gravity loaded system. This approach to form-finding of shell structures, exemplified in the works of pioneers such as Antoni Gaudí and Frei Otto leverages nature as a fundamental source of inspiration, using material constraints, loading and support parameters as a catalyst for designing material-efficient stressed skin structures [15].

In a real world scenario, even an indoor structure has to withstand different load cases, with the deviation between load case-specific catenaries leading to a degree of bending stress in the component that was moulded under a single load pattern. To increase the bending stiffness and robustness of the component, the component can be composed not of a single, but of two or more surface substrates. An additional substrate layer laminated on the full surface can double the structural depth, but even more efficiently can increase the out-of-plane stiffness if it creates folds perpendicular to the surface, acting as integrated stiffening ribs. A further substrate layer can encapsulate the ribbed layer on both sides, forming effectively a sandwich build-up. Both the sectional and the overall geometry become defining characteristics for a principal structural typology of mycelial stressed skin components. The geometry of folded, single curved catenaries or synclastic or anticlastic double curved shells can be beneficial for room acoustics, as their absorbing surfaces can be oriented to multiple noise sources in the room.

This intuitive and empirical form-finding approach using the elasticity and plasticity of the surface substrates as a medium to develop both the overall and section stiffness by manipulating the shape of the surface substrates represents a valuable design tool for mycelial stressed skin structures at a stage when numerical design models and digital design tools are not yet established.

3.3 Strengthening connections of stressed skin components

For creating larger structural assemblies, the stressed skin mycelium components need to be connected to harness the in-plane strength of the module's surfaces. Forming curved or double curved shell or space frame structures creates tension and/or compression forces in-plane of the mycelium skins. With the overall geometry and structural composition avoiding bending forces, the connections need to transfer tension, compression or shear forces, as anticipated during the force-moulding of the component. The connections between stressed skin components are neuralgic weak points, as linear connections channel the forces flowing across the surface along the edge and a point connection concentrates the forces at a single node. Local strengthening of the surface substrate is required to keep forces in the surface uniform. Depending on the vector of forces, yarns, strings or ropes are integrated along the edges to increase tensile strength or slender sawn timber sections are used to improve compression strength. Inoculating both the surface substrates and the lignocellulosic materials selected for the strengthening allows the mycelium matrix to bond both materials together during the growth stages. This composite effect ensuring a homogeneous load transfer can be enhanced by maximizing the contact surface between the surface substrate and strengthening materials, for example by wrapping the surface substrate edges around the inserts. At the nodal points, these linear enforcements need to be interconnected, to provide a controlled geometric interface with adjacent modules.

4 The Myx Arch

4.1 Introduction

The Myx Arch demonstrates the principles described here for the design, construction and fabrication of mycelial stressed skin surface structures as a full-scale prototype.

Figure 3.4 Prototypes of mycelial stressed skin components designed and fabricated by Master's students in the academic year 2023/4 as part of the course 'Circular building(s) with bio-based materials' at KU Leuven. Image Credit: BioDigital ConstructionLab, KU Leuven.

The concept design was developed during the academic research seminar 'Circular building(s) with bio-based materials', part of the Master's programme at the Faculty of Architecture at KU Leuven between October 2023 and January 2024. A group of 20 international students explored the architectural potential of surface substrates by implementing the form-finding principles for the design of a temporary shelter that would activate vacant building spaces. To allow the synthesis of architectural design ideas with physical form-finding and the biofabrication process, the students had access to a wet lab at the Department of Biology. Through a series of experiments and scaled models, a pair of students developed a concept design for a self-supporting shelter that synthesized aspects of design, fabrication and assembly, resulting in a full-scale prototype of a single component (see Figure 3.4). From all the proposals, which showcased the wide range of possible modular stressed skin structures, the design by Silke Smeets and Carmen Smit was selected for realization. This full-scale prototype was showcased during the New European Bauhaus Festival in Brussels in April 2024, the 600-year anniversary of the KU Leuven in February 2025 and during the International Conference on Structures and Architecture (ICSA) in Antwerp in July 2025.

4.2 The pyramidal surface module

A pyramid composed of a square base and four equilateral triangles, which geometrically can be described more precisely as a semi-octahedron, represents the basic structural module of the arch. The surfaces of the pyramid were cultivated on 10 mm thick hemp fibre mats with a density of approximately 1,300 g/m^2, resulting in a remarkably lightweight hollow component. The standard width of the hemp substrate mat defined the dimensions of the module. The mat width of 1 m defined the size of the pyramid, as the entire developed surface of the pyramid fitted on it, resulting in an edge length of 40 cm and a height of around 28 cm. The surface mat was processed into the stressed skin module in a two stage process, as described in section 4.4. Before the first growth stage, a woven flax textile was placed on the outer surface of the square base and sewed on with flax yarn to improve the homogeneity and robustness of the biofabricated surface. Between the first and second growth stage, flax yarn was also used for stitching together the edges of the triangular faces overlapping by 20 mm with the diagonals of the system. In addition, to control dimensional stability of the square base and to increase its compressive strength, a prefabricated frame of local solid beech wood with a 20 x 30 mm section was positioned on the inside of the two-ply surface substrate of the base. Similar to a canvas, the fabrics were stretched over the edges of the frame to tension and stabilize the surface. At the tip of the pyramid, the flax yarns stitching together the four edges were connected to the laser cut timber node. The flanges of the timber insert were laminated between the overlapping edges of the triangular mats to facilitate an even load transfer from the edges to the node. Following the drying and stabilization of the module, the nodes were interconnected by the continuous 8 mm bottom hemp cord that stabilizes the arch (Figure 3.5).

Figure 3.5 Detail view of one of the 11 mycelial stressed skin modules, showing the nodal connector at the tip of the pyramid connecting the pre-stressed hemp rope as the continuous bottom cord with the strengthened diagonal edges, transferring tensile forces to the top cord formed by the square base plates of the pyramidal modules. Image credit: Jan Wurm.

The materiality of the module showcases a palette of regional bio-based materials. The hemp for the production of the mats and rope was from the north of France, while the linen textile provided by the company Flaxco was made from flax fibres harvested in West Flanders. All timber components used either for the reinforcement, the connecting dowels between timber elements or the supporting substructure were made from local beech grown in the Sonian forest around Brussels and were processed by Sonian Wood Coop.

4.3 The arch structure

The structure of the arch is based on the principal geometry of a regular spaceframe [16]. The circular arch composed of eleven pyramidal modules has a height of 1.50 m and a span of 3 m. The fold lines between the triangular planes of the pyramid act as diagonals, while the square base plate forms the top cord. The additional continuous tensile member connecting the tips of the pyramids complements the structural system as the bottom cord. Overall stability is achieved by post-tensioning this 8 mm hemp tow, introducing regular radial force into the system, leading to controlled compression forces in the top cord, compensating for asymmetrical loads.

So that the prototype is more easily experienced by visitors when presented as an exhibit, the arch is lifted about 75 cm from the ground by timber-framed pedestals at either end. To close-loop any horizontal forces at the support points of the arch, the consoles at either end are interconnected by a pair of wooden cross-ties. The consoles and struts are made of solid sections of locally harvested solid beech wood. Lifting the bases of the arch allows the cross-ties to be used as a support for a modular table display, inviting visitors to engage (Figure 3.6).

Figure 3.6 The Myx Arch exhibited at the 600-year anniversary on Heverlee campus of KU Leuven in February 2025. Image credit: KU Leuven.

4.4 The two-stage fabrication process of the mycelium components

The pre-configured fabrics composed of hemp mats and flax fabrics are first sterilized in an industrial autoclave at approximately 120°C. The fabrics are then inoculated with grain spawn of *G. lucidum*. In the first growth stage the fabrics are packed in breathable polyethylene bags at about 24°C and a moisture content of about 70% and positioned in an incubator for seven days, during which the mycelium propagates across the substrate mats. The pre-cultivated mats are then removed from the growbags and the pyramidal modules are assembled in the quasi-sterile environment of a laminar flow bench. The pre-assembled timber frames and node connectors are placed and the overlapping diagonal edges are stitched together. Each pyramid is then suspended inside an incubator supported at the square base with the tip facing down. By adding additional weight to the tip, the gravity pulls the textiles in a symmetric shape aligning with the centre of gravity, effectively force-moulding the surface substrates to the regular shape of a semi-octahedra without the need of a surface mould. After a second growth stage of three to four days, the element is removed from the incubator and dried at about 30°C for three days to halt the growth process. Owing to the stretch of the surface substrate during the force-moulding, the height of the stabilized pyramid increased to between 40 and 50 cm in comparison with the nominal height of 28 cm.

5 Outlook

Growing from the roots of mycotecture that explores the tectonic qualities of mycelium composites, the use of surface substrates instead of loose substrates unlocks new architectural and structural expressions, increasing the formal freedom and flexibility of mycelium materials without the need for costly surface moulds. With the mycelium surfaces providing key spatial functions such as acoustic and thermal comfort, stressed skin mycelium structures could provide a set of innovative technologies to transform empty spaces into sustainable, affordable and engaging places to live.

Substantial research and innovation activities are needed to bring mycelial stressed skin components to market. The current work of the author focuses on novel liquid inoculation techniques that would allow to employ digital printing techniques and robotic fabrication to control the growth of mycelium across the surface substrates, and subsequently to configure their physical properties in response to project needs [17]. Validated mechanical material models for surface substrates in dry, wet and stabilized stages need to be developed, for example to calculate and control the stretch during the force-moulding which would inform digital design tools to complement the physical form-finding. Based on the basic understanding of cultivation parameters and material properties, future innovation activities need to focus on the development of semi-automated manufacturing processes for mycelial stressed skin components. To create a market, the supply side needs to be strengthened, and key actors and stakeholders from the cultivation of the fibres to the processing of the surface substrates and the biofabrication of the mycelial stressed skin need to form regional value chains to match the demand for the transformation of empty and under-utilized spaces into affordable, sustainable and engaging housing solutions.

Translating the qualities of mycelium surfaces into the architectural habitat would go some way towards solving the fact that the built and living environment is a root cause for the current environmental crisis, by blurring not only geometric and organic forms but also human and microbial agencies.

References

1. Townley, M. (28 Feb 2025) Overcoming construction inflation: How to beat rising construction costs. Buildertrend. Accessed at: https://buildertrend.com/blog/tips-for-inflation-construction-costs/ (31 Mar 2025)
2. Okatz, J., Ragot, C., Georgarakis, E., Herrmann, S. (Jan 2023) Efficient and balanced space use: shaping vibrant neighbourhoods and boosting climate progress in Europe. SYSTEMIQ. Accessed at: https://www.systemiq.earth/wp-content/uploads/2023/01/SYSTEMIQ-urban-spaces.pdf (31 Mar 2025)
3. Habitat for Humanity (Aug 2024) Strategies for Affordable Housing and Thriving Communities – Habitat for Humanity's Manifesto on Priorities of the European Union for 2024–2029. Habitat for Humanity. Accessed at https://www.habitat.org/media/25506/view (31 Mar 2025)
4. Carroll, C., Souza, Y. A., Salter, E., Hunziker, R., Giovanetti, L. De, Contucci, V. (Oct 2022) Net-zero buildings – Where do we stand. ARUP. Accessed at https://www.arup.com/globalassets/downloads/insights/net-zero-buildings-where-do-we-stand.pdf (31 Mar 2025)
5. Alaneme, K. K., Anaele, J. U., Oke, T. M., Kareem, S. A., Adediran, M., Ajibuwa, O. A., Anabaranze, Y. O. (15 Nov 2023) Mycelium based composites: A review of their biofabrication procedures, material properties and potential for green building and construction applications; Alexandria Engineering Journal; Volume 83: pp. 234–250. https://doi.org/10.1016/j.aej.2023.10.012
6. Jin, Y., De, G., Wilson, N., Qin, Z., Dong, B. (Jan 2025) Towards Carbon-Neutral Built Environment: A Critical Review of Mycelium-Based Composites. Energy and Built Environment. https://doi.org/10.1016/j.enbenv.2025.01.004
7. Ross, P. (2017) Fungal Mycelium Bio-Materials. Cultivated Building Materials: Industrialized Natural Resources for Architecture and Construction.
8. Maurer, Ch. (12 Jun 2024) Building with Mycotecture: Deep Dive into Fungal Solutions for Sustainable Architecture. Mycostories. Accessed at: https://www.mycostories.com/post/building-with-mycotecture-deep-dive-into-fungal-solutions-for-sustainable-architecture (31 Mar 2025)
9. Elsacker, E.V. (20 Apr 2021) Mycelium matters – an interdisciplinary exploration of the fabrication and properties of mycelium-based materials; PhD Thesis published at VUB Press. ISBN 978 94 6117 168 9
10. Rigobello, A., Colmo, C., Ayres, P. (25 Apr 2022) Effect of Composition Strategies on Mycelium-Based Composites Flexural Behaviour. Biomimetics 2022, 7(2), 53. https://doi.org/10.3390/biomimetics7020053
11. Kaiser, R., Bridgens, B., Elsacker, E.V., Scott, J. (14 Jul 2023) BioKnit: development of mycelium paste for use with permanent textile formwork. Frontiers in Bioengineering and Biotechnology 11: 1229693. https://doi.org/10.3389/fbioe.2023.1229693
12. Jonas Edvard Studio (2018) MYX – Mycelium Textile Project. Accessed at https://www.jonasedvard.com/myx (31 Mar 2025)
13. Dwan, A., Edvard Nielsen, J., Wurm, J. (4 Apr 2024) Room acoustics of Mycelium Textiles: the Myx Sail at the Danish Design Museum. Research Directions: Biotechnology Design. 2024;2:e7. https://doi.org/10.1017/btd.2024.2
14. Hanffaser Uckermark (2023) Technisches Datenblatt Hanfvlies. Hanffaser Uckermark. Accessed at: https://www.hanffaser.de/Downloads/Techn_DB_H_Vlies.pdf (31 Mar 2025)
15. Otto, F., Rasch, B. (19 Jan 1996) Finding Form: Towards an Architecture of the Minimal. Edition Axel Menges. ISBN 9783930698660
16. Emde, H. (1978) Geometrie der Knoten-Stab-Tragwerke. Strukturforschungszentrum Würzburg
17. Biront, A., Sillen, M., Van Dijck, P., Wurm, J. (8 Jan 2025) Growth Propagation of Liquid Spawn on Non-Woven Hemp Mats to Inform Digital Biofabrication of Mycelium-Based Composites. Biomimetics 2025 10(1), 33. https://doi.org/10.3390/biomimetics10010033

CHAPTER 4

Mycelium-Based Composites as/in Building Skins: Potential, Constraints and Way Forward

KUMAR BISWAJIT DEBNATH

1 The search for ecological skins

Building envelopes serve as the crucial interface between constructed spaces and their surrounding environments, playing a significant role in governing thermal performance, regulating daylight access, ensuring adequate ventilation and providing structural protection [1]. Under increasingly extreme and volatile climate change conditions, [2] depletion of natural resources [3] and with the global push towards net-zero [4], all in an escalating global carbon emissions context, architects and engineers face mounting challenges in designing and constructing building skins that not only excel in functional efficiency and climate resilience but also uphold high standards of material responsibility and sustainability in terms of embodied carbon.

Among the innovative approaches and alternative materials gaining attention in this context are mycelium-based composites (MBCs) [5]. MBCs are derived from the root networks of fungi, exhibiting remarkable potential as a biogenic alternative to conventional synthetic building materials, which are often carbon-intensive. MBCs can be cultivated using agricultural and industrial biomass waste as their substrate [6], which not only reduces waste but also contributes to a lower embodied energy profile. Furthermore, natural biodegradability at the end of their lifecycle positions them as pivotal representatives of a circular material economy, which seeks to minimize environmental impact through responsible resource use.

Despite the promising attributes of MBCs, significant challenges and uncertainties persist regarding their application in the construction industry. Questions remain about their structural integrity, particularly under varying environmental conditions, as well as their susceptibility to moisture, which can affect durability. Additionally, integrating these materials into long-lasting building systems poses design and engineering challenges that must be addressed.

This chapter provides an overview of the potential applications of MBCs in façade and building skin designs. It will begin by delving into the unique biological properties of mycelium that make it suitable as a building material. Following that, we will review contemporary research and relevant case studies that highlight innovative applications of MBCs. The chapter will also explore the constraints associated with their fabrication and application, reframing these limitations as opportunities for creativity and innovation

Kumar Biswajit Debnath, 'Mycelium-Based Composites as/in Building Skins: Potential, Constraints and Way Forward' in: *Mycology for Architecture: Methods and Applications of Mycelium Materials in the Built Environment*. Pelagic Publishing (2026). DOI: 10.53061/FQST2725 within the realm of climate-responsive architecture. Ultimately, the goal is to foster a deeper understanding of how MBCs can advance sustainable architectural practices while meeting the complex demands of modern building design.

2 Properties for building skin applications

Mycelium is the vegetative (root-like) part of fungi, consisting of a network of fine white filaments called hyphae [7]. When introduced into a substrate – typically agricultural waste such as straw, husks or sawdust – mycelium colonizes and binds the substrates together. Once the desired form is achieved, growth is halted by drying or heat treatment, resulting in a lightweight solid composite [5]. The following selected properties of MBCs highlight their potential building applications:

- Lightweight and low density: Typically, MBCs are lighter than many conventional materials, such as concrete or wood. Foam-like MBCs reported densities between 50–200 kg/m^3 [8] and 100–220 kg/m^3 [9], and on-pressed MBCs typically showed densities ranging from 0.06 to 0.22 g/cm^3 [10].
- Thermal insulation: MBCs have low thermal conductivity, often ranging from 0.03 to 0.07 $W/m{\cdot}K$, comparable to expanded polystyrene [11].
- Biodegradability: At the end of their lifecycle, MBCs can decompose naturally without toxic by-products, making them a prime candidate for circular construction models.
- Fire resistance: MBCs exhibit slow combustion and emit low smoke, contributing to passive fire protection in buildings [12].
- Acoustic performance: Their porous structure offers sound-absorbing properties ideal for envelope and partition systems.

3 Opportunities for climate-responsive design

3.1 Thermal comfort and passive design

The thermal insulation properties of MBCs make them suitable for architectural building skin applications [9]. In regions with significant diurnal temperature fluctuations or where cooling and heating demands are increasing owing to extreme weather events (here in the context of Vadodara, India, and Sydney, Australia), MBCs can help alleviate (0.4–3.2% reduction compared with base case) the load on mechanical heating and cooling systems (see Figure 4.1) [13]. By effectively regulating indoor temperatures through passive strategies, these materials enhance occupant comfort without relying heavily on electrical or gas-driven heating, ventilation and air conditioning systems. Studies have shown that an external perforated building screen made of MBCs (Bio-jaali) can reduce the active cooling energy demand by ~50% and lower indoor temperature by up to ~15°C under controlled hot-humid conditions [14]. The naturally porous microstructure of MBCs minimizes thermal bridging – an issue where heat is transferred through buildings via conductive materials – resulting in energy-efficient façade solutions that can maintain comfortable internal environments across varying climatic conditions. Studies have shown, through early-stage and small-scale experiments, that elephant skin-inspired biomimetic MBC panels may offer climate-responsive passive cooling solutions [15].

3.2 Carbon footprint

The production process of MBCs involves minimal industrial processing, making them a more sustainable alternative to traditional building materials. Unlike conventional

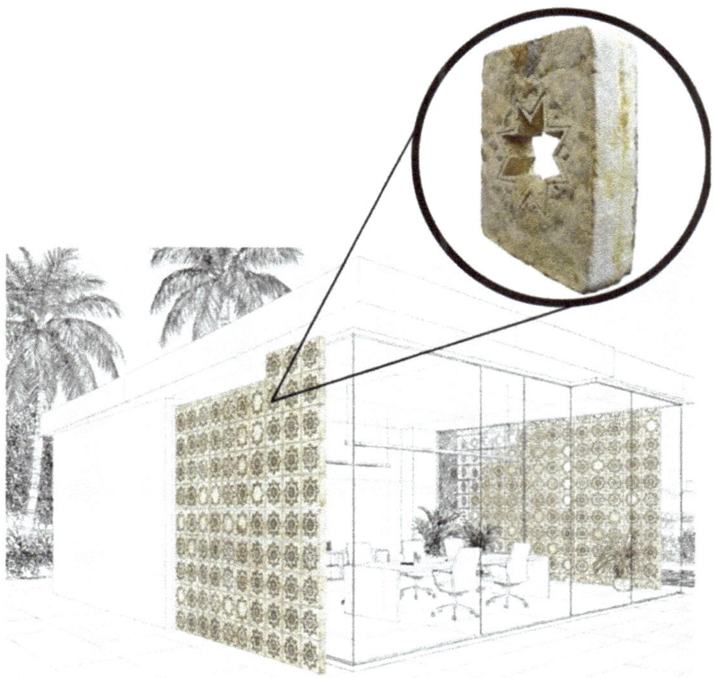

Figure 4.1 Example of a building screen made with MBC, with the star-shaped perforation reinventing traditional Islamic Architecture screen designs with circular and low-carbon bio-based material.

cement-based materials, MBCs require significantly lower energy inputs for production and release far fewer greenhouse gases during their lifecycle. An intriguing aspect of MBCs is the biological growth of mycelium, a process known as biofabrication, which is a low-carbon-intensive process [5]. The incubation stage accounts for the largest share (73%) of the overall energy demand [16]. MBCs have a significantly lower environmental impact than traditional fossil fuel-based materials, such as concrete, sand-lime and facing bricks, potentially reducing greenhouse gas emissions by up to 2.5 times compared with concrete and by three to six times compared with sand-lime and facing bricks [17]. Another study found that the embodied carbon of an MBC was –39.5 kg CO_2eq/m^3, indicating that it acts as a CO_2 sink [16]. MBCs as insulation material boast the lowest emissions per cubic metre of material, yielding the minimal cumulative emission value during production compared with EPS (231.2 kg CO_2eq/m^3), XPS (271.8 kg CO_2eq/m^3), polyurethane (560.5 kg CO_2eq/m^3), and phenolic foam (1136 kg CO_2eq/m^3) [11]. As a result, MBCs contribute to reducing the overall carbon footprint associated with the use of construction materials. In comprehensive lifecycle assessments, MBCs consistently outperform traditional insulating and panel materials, demonstrating a significantly lower global warming potential and reinforcing their status as an environmentally responsible choice.

3.3 Material circularity and regenerative design

One of the standout features of MBCs is their ability to promote material circularity. At the end of their lifecycle, these composites can be composted, returning nutrients to the soil and promoting ecological health. This approach effectively reverses the linear 'take–make–dispose' model that underscores most conventional construction materials, which often contribute to landfill waste. In stark contrast, building skins crafted from MBC not only minimize construction waste but also embody principles of regenerative design. This fosters an architectural practice that nurtures ecosystems, contributing positively to the environment rather than depleting it, and promotes a sustainable approach to resource use.

3.4 Biophilic and aesthetic potential

MBCs offer a unique aesthetic that distinguishes them from synthetic cladding systems. The organic textures and earthy hues of mycelium create a visually appealing façade that resonates with biophilic design principles. This design philosophy emphasizes our innate connection to nature, fostering emotional and psychological well-being through visual and tactile interactions with natural materials. MBCs can be produced in various forms, from monolithic panels that create striking visual statements to modular tiles that allow for creative flexibility in design. This variety introduces a material softness and warmth that humanizes building envelopes, creating spaces that are not only functional but also inviting and harmonious with the surrounding environment.

4 Constraints and limitations

The use of MBCs in indoor (insulation or wall panel) and external envelope systems holds significant promise; however, several critical limitations hinder their deployment, underscoring the need for innovative design solutions.

- **Moisture sensitivity:** MBCs are inherently hygroscopic, meaning they can absorb moisture in humid environments or upon exposure to liquid water [5]. This moisture absorption can lead to compromised mechanical strength and promote the growth of microorganisms. To mitigate these issues, protective coatings, hydrophobic treatments or hybrid skin systems may be necessary. However, these solutions can potentially diminish biodegradability or increase the material's embodied energy.
- **Mechanical strength and load-bearing capacity:** Common MBCs have lower compressive and tensile strengths than conventional materials, such as concrete, timber or brick. The compressive strength can range from 0.001 to 1.029 MPa for mycelium composite foams without hybridization, and 0.71 to 4.4 MPa for MBCs without hybridization [18]. As a result, they are generally unsuitable for use as structural skin elements unless reinforced. Nevertheless, MBCs can effectively serve as insulation layers or cladding panels in non-load-bearing roles.
- **UV and environmental degradation:** Untreated MBCs are vulnerable to degradation from exposure to ultraviolet (UV) radiation and varying weather conditions [19]. There is a lack of long-term studies focusing on UV resistance or environmental degradation under real-world weather conditions. Current protective strategies often include coatings that introduce synthetic content, posing challenges to the material's sustainability and circularity.
- **Standardization and building code integration:** One of the significant hurdles for MBCs is the absence of standardized performance metrics within most global construction codes. Their novel nature and heterogeneity create regulatory uncertainties that restrict large-scale adoption. Efforts to establish testing protocols, certification systems and standardized design practices are still underway.
- **Fabrication and scale:** While the fabrication of MBC at a small scale is becoming increasingly accessible, scaling up to building-sized components presents challenges.

Issues such as the complexity of moulds, drying times, contamination risks and energy requirements for sterilization can hinder progress. Advances in digital design and robotic fabrication may provide solutions, but they currently require interdisciplinary collaboration and adequate infrastructure.

5 Design and fabrication strategies for integrating MBC in building skins

Given the previously mentioned advantages, opportunities and constraints, MBCs could be integrated into building skins to enhance climate resilience and design innovation.

- **Composite and hybrid envelope systems:** MBCs may be used not in isolation but as part of a composite system, such as:
  - **Bio-SIP panels (structural insulated panels):** MBC insulation core sandwiched between weather-resistant outer layers (e.g., timber or bioplastics).
  - **Ventilated façades:** MBC panels used as inner skins within double-skin façades, protected from direct exposure while contributing insulation and acoustics.
  - **Interior cladding:** Using MBCs on the interior side of the building envelope to preserve performance while reducing exposure.
- **Modular and replaceable panels:** The relatively short lifespan of untreated MBC panels should be viewed as an opportunity rather than a drawback. These panels can be intentionally designed for periodic replacement and composting, enhancing their sustainability. A modular system allows for selective upgrades while ensuring the integrity of the larger envelope system remains intact. Moreover, this shorter lifespan presents a viable solution for sustainable skin applications in temporary structures, such as pavilions or installations for short-term events.
- **Surface treatments and bio-coatings:** To enhance environmental and weather resistance, surface treatments can be applied – ideally, bio-based ones. Research into chitosan coatings, natural waxes and fungal-derived resins is ongoing [20, 21]. These can extend longevity while preserving biodegradability and indoor air quality.
- **Parametric and biodigital design and fabrication:** Digital tools can optimize MBC form and performance. Parametric modelling allows designers to manipulate porosity, density and geometry for thermal or acoustic performance. Biodigital fabrication (combining robotic and biofabrication) or 3D-printed moulds can produce intricate forms that leverage MBCs' low weight and high workability.
- **Adaptive and responsive systems:** Future development may bring about biologically active skin that interacts with its environment. MBCs, when left partially alive in symbiosis with other organisms, could theoretically serve as sensing, regulating or even healing systems. Though speculative, this aligns with research in living architecture and biodigital hybrids.

6 Climate resilience and ecological integration

- **Redefining resilience:** Conventional resilience in building materials often implies permanence and resistance. However, in the context of the Anthropocene, resilience may mean adaptability, biodegradability and low impact in the event of failure. MBCs' impermanence becomes an asset: they decay cleanly, signal failure early and can be locally replaced.

- **Localization and decentralized production:** MBCs can be grown locally, using regional biomass waste streams. This decentralizes production, reduces transport emissions and supports climate-resilient economies. The architecture of the future might be grown on-site, tied intimately to local ecologies and communities.
- **Ecological literacy in design:** Using MBCs in building skins necessitates an ecological approach, considering lifespan, decomposition, maintenance and regeneration. It asks architects and builders to think like ecologists and choreograph buildings not as static objects but as evolving ecosystems.

7 Towards living skins

MBCs offer a promising opportunity in the pursuit of climate-responsive and ethically sourced architecture. While they may not be a universal solution, their inherent qualities – fragility, impermanence and variability – should be seen as catalysts for innovation in building skin typologies. These materials encourage us to explore designs that can breathe, decay and regenerate.

By challenging conventional approaches, MBCs raise thought-provoking questions: What if building skins were deliberately designed to be temporary? What if maintenance practices included composting? What if buildings were designed to reflect seasonal changes? Embracing the limitations of MBCs can inspire creative solutions and lead us toward architecture that aligns with biological and planetary systems.

To fully harness the potential of this material, future development must encourage interdisciplinary collaboration among architects, biologists, engineers and material scientists. Standardization can adapt, digital tools can improve fabrication processes and clients can be educated about new performance paradigms. The path forward for buildings of the future is encouraging; biology will not only inspire design but also play a vital role in the construction process.

References

1. S. B. Sadineni, S. Madala and R. F. Boehm, 'Passive building energy savings: A review of building envelope components,' *Renewable and sustainable energy reviews*, vol. 15, no. 8, pp. 3617–3631, 2011.
2. K. B. Debnath, D. Jenkins, S. Patidar, A. D. Peacock and B. Bridgens, 'Climate change, extreme heat, and South Asian megacities: Impact of heat stress on inhabitants and their productivity,' *ASME Journal of Engineering for Sustainable Buildings and Cities*, vol. 4, no. 4, p. 041006, 2023.
3. C. Lu and K. Wang, 'Natural resource conservation outpaces and climate change: Roles of reforestation, mineral extraction, and natural resources depletion,' *Resources Policy*, vol. 86, p. 104159, 2023.
4. S. Fankhauser, S. M. Smith, M. Allen, K. Axelsson, T. Hale, C. Hepburn, J. M. Kendall, R. Khosla, J. Lezaun, E. Mitchell-Larson, M. Obersteiner, L. Rajamani, R. Rickaby, N. Seddon and T. Wetzer, 'The meaning of net zero and how to get it right,' *Nature Climate Change*, vol. 12, no. 1, pp. 15–21, 2022.
5. M. Jones, A. Mautner, S. Luenco, A. Bismarck and S. John, 'Engineered mycelium composite construction materials from fungal biorefineries: A critical review,' *Materials & Design*, vol. 187, p. 108397, 2020.
6. K. B. Debnath, C. Ton-That, N. Biloria and A. Shahil, 'Exploring the potential of Australian agricultural and industrial wastes from the source to bio-fabricating viable mycelium-based composites,' in *4th International Conference of Sustainable Building Materials (ICSBM 2025)*, 10–13 Aug, Eindhoven, The Netherlands, 2025.
7. D. Raslan, E. Elsacker, K. B. Debnath and M. Dade-Robertson, 'Improving the Physical Properties and Mechanical Performance of Mycelium-Based Composites with Bio-Welding,' *SSRN 4996618*, 2024.
8. P. P. Dias, L. B. Jayasinghe and D. Waldmann, 'Investigation of Mycelium-Miscanthus composites as building insulation material,' *Results in Materials*, vol. 10, p. 100189, 2021.

9. X. Zhang, J. Hu, X. Fan and X. Yu, 'Naturally grown mycelium-composite as sustainable building insulation materials,' *Journal of Cleaner Production*, vol. 342, p. 130784, 2022.
10. F. V. Appels, S. Camere, M. Montalti, E. Karana, K. M. Jansen, J. Dijksterhuis, P. Krijgsheld and H. A. Wösten, 'Fabrication factors influencing mechanical, moisture-and water-related properties of mycelium-based composites,' *Materials & Design*, vol. 161, pp. 64–71, 2019.
11. S. Al-Qahtani, M. Koç and R. J. Isaifan, 'Mycelium-based thermal insulation for domestic cooling footprint reduction: a review,' *Sustainability*, vol. 15, no. 17, p. 13217, 2023.
12. M. Jones, T. Bhat, E. Kandare, A. Thomas, P. Joseph, C. Dekiwadia, R. Yuen, S. John, J. Ma and C.-H. Wang, 'Thermal degradation and fire properties of fungal mycelium and mycelium-biomass composite materials,' *Scientific reports*, vol. 8, no. 1, p. 17583, 2018.
13. K. B. Debnath, L. Thomas, N. Biloria and C. Burton, 'Bio-retrofit: Potential of Mycelium-based Composites to Future-proof India and Australia's Residential Buildings against Extreme Climate Events post-2040,' in *Building Simulation 2025*, Brisbane, Australia, Aug 24–27, 2025, 2025.
14. K. B. Debnath, N. Pynirtzi, J. Scott, C. Davie and B. Bridgens, 'Bio-jaali: Reimagining vernacular passive cooling screens with mycelium-based composites,' Energy and Buildings, vol. 357, pp. 117104, 2026.
15. E. Soh, N. Loh, J. H. Teoh, A. Jain and H. Le Ferrand, 'Biodegradable mycelium tiles with elephant skin inspired texture for thermal regulation of buildings,' *Energy and Buildings*, vol. 328, p. 115187, 2025.
16. A. Livne, H. A. Wösten, D. Pearlmutter and E. Gal, 'Fungal mycelium bio-composite acts as a CO2-sink building material with low embodied energy,' *ACS Sustainable Chemistry & Engineering*, vol. 10, no. 37, pp. 12099–12106, 2022.
17. S. Akromah, N. Chandarana and S. J. Eichhorn, 'Mycelium composites for sustainable development in developing countries: The case for Africa,' *Advanced Sustainable Systems*, vol. 8, no. 1, p. 2300305, 2024.
18. S. Womer, T. Huynh and S. John, 'Hybridizations and reinforcements in mycelium composites: a review,' *Bioresource Technology Reports*, vol. 22, p. 101456, 2023.
19. T. Houette, B. Foresi, C. Maurer and P. Gruber, 'Growing Myceliated Facades-Manufacturing and exposing experimental panels in a facade setting,' in *Proceedings of the Facade Tectonics World Congress*, 2020.
20. M. Jones, A. Gandia, S. John and A. Bismarck, 'Leather-like material biofabrication using fungi,' *Nature Sustainability*, vol. 4, no. 1, pp. 9–16, 2021.
21. K. K. Alaneme, J. U. Anaele, T. M. Oke, S. A. Kareem, M. Adediran, O. A. Ajibuwa and Y. O. Anabaranze, 'Mycelium based composites: A review of their bio-fabrication procedures, material properties and potential for green building and construction applications,' *Alexandria Engineering Journal*, vol. 83, pp. 234–250, 2023.

CHAPTER 5

Mycelium in Space Applications

MONIKA BRANDIĆ LIPIŃSKA

1 Introduction: biotechnologies in space applications

Designing habitats for space exploration presents a unique challenge: there are no traditional construction rules, tested methods or building codes. The extreme conditions of space demand novel approaches to sustainability, resilience and adaptability, effectively making it a pioneering endeavour. This absence of precedent also provides a rare opportunity to explore unconventional materials and construction strategies, potentially revolutionizing how we approach habitat design both in space and on Earth.

Among all extreme environments, space is inherently hostile to life. It is characterized by hypogravity or microgravity, vacuum or low atmospheric pressure, extreme temperature fluctuations and high radiation exposure owing to the absence of a protective magnetic field. The vast distance from Earth further constrains resupply, making long-term missions reliant on self-sufficient, resource-efficient systems. As plans for sustained lunar and Martian missions progress, developing habitats that ensure not only survival but also astronaut well-being becomes increasingly critical.

Space habitats are complex, integrated systems composed of interdependent subsystems, including structural elements, life support, power management, logistics and communication networks. The successful operation of these habitats depends on the seamless interplay of these components, making their design a highly multidisciplinary challenge. The harsh space environment and constraints of long-duration missions demand innovative approaches to both habitat construction and resource utilization.

One promising approach in space architecture is integrating biotechnologies, particularly those leveraging microorganisms, to address resource-limited challenges. These biological processes can support long-term human presence in space, offering solutions for in-situ resource utilization (ISRU), habitat fabrication, waste recycling and environmental control [1]. By creating resilient, adaptable systems similar to cellular metabolism, these living systems could perform vital functions such as waste recycling, oxygen production and detoxification. Biomaterials, such as mycelium, can be grown in situ, reducing reliance on Earth-based resources. Their ability to self-replicate, self-heal and adapt to environmental conditions makes them ideal for space applications, enabling multifunctional habitat design that supports both structural integrity and life-support functions such as air purification and radiation shielding [2, 3].

Fungal mycelium, in particular, exhibits exceptional versatility. It is able to form lightweight yet strong composites that can be tailored for specific mission requirements and – uniquely – can function as an active biological agent integrated into habitat

Monika Brandić Lipińska, 'Mycelium in Space Applications' in: *Mycology for Architecture: Methods and Applications of Mycelium Materials in the Built Environment*. Pelagic Publishing (2026). © Monika Brandić Lipińska. DOI: 10.53061/AJQM9832 systems. This dual role as both material and living system opens new possibilities for creating self-adapting, regenerative environments in space, highlighting its promise as a next-generation biomaterial for extraterrestrial applications.

2 Rationale for using mycelium in space applications

The lifecycle of mycelium unfolds through four principal stages, each fulfilling distinct ecological functions. When applied to space habitats, this progression provides a conceptual framework for linking biological processes with architectural performance and life-support needs. The continuous rhythm of growth, decomposition and renewal can be harnessed to sustain equilibrium within closed systems. In this context, mycelium is not limited to serving as a static construction material but can be envisioned as a versatile, adaptive agent – capable of shifting between structural, metabolic, informational and sensory roles within an integrated bioregenerative environment.

Cultivating mycelium directly on the Moon or Mars could, in principle, facilitate the in-situ development of both structural elements and interior furnishings. Mycelium's inherent properties – such as self-healing, resilience and potential integration into life-support systems – make it an attractive candidate. However, the application of living organisms as construction agents also introduces significant uncertainties. The following subsections highlight the opportunities and limitations associated with mycelium in space architecture.

2.1 In-situ growth and self-healing

A key advantage of mycelium-based construction lies in its ability to grow in situ, reducing the need to transport large volumes of building materials from Earth. Only lightweight starter cultures would need to be launched, after which mycelium could expand on site, consuming habitat-derived resources. This approach aligns with the ISRU principles, which are critical for long-duration missions. Moreover, mycelium's capacity to sporulate and regenerate allows for self-repair of micro-cracks and defects. Bio-welding techniques could seal joints and interfaces, potentially enabling autonomous maintenance of habitat structures while reducing reliance on external energy-intensive repair methods.

However, mycelial growth is highly dependent on controlled environmental conditions – humidity, temperature, oxygen supply and substrate availability – that may be challenging to achieve outside laboratory settings. While self-healing has been demonstrated in controlled conditions, its effectiveness in real-scale habitat structures remains limited, especially for larger cracks or under fluctuating environmental stressors. Thus, while promising, in-situ growth is not yet a robust or predictable solution.

2.2 Safety and durability

Certain mycelium species, particularly melanin-rich fungi, show exceptional resilience in extreme environments, offering protection against harmful radiation. These fungi, such as *Cryptococcus neoformans*, have demonstrated the ability to withstand simulated Martian conditions, providing radiation shielding more effective than charcoal [4]. Melanin plays a key role in this resilience by converting electromagnetic energy into chemical processes that support life [5, 6]. The ability of melanized fungi to thrive under ionizing radiation highlights their potential to support microbial survival beyond Earth, extending the boundaries of terrestrial life [5].

Beyond radiation shielding, mycelium-based materials can also enhance habitat durability by forming rigid structural barriers that provide thermal insulation and impact resistance against micrometeorites. These biological materials can create protective layers that regulate temperature fluctuations and mitigate structural damage, which is critical for maintaining long-term habitat integrity in space environments. The combination of radiation shielding, thermal protection and impact resistance positions mycelium as a multifunctional material capable of significantly improving the safety and resilience of extraterrestrial habitats.

Despite these properties, uncertainties remain regarding the long-term durability of mycelium-based composites in space. Mechanical strength and impact resistance may be insufficient for larger-scale structures, particularly when compared with engineered alternatives such as sintered regolith. Continuous exposure to radiation, vacuum conditions or extreme thermal cycling could degrade structural integrity over time. Furthermore, the balance between maintaining living composites (for self-repair) and inactivating them (for stability) introduces trade-offs that complicate long-term safety.

2.3 Integration with life-support systems

Mycelium-based materials also offer numerous advantages for life-support systems in space habitats, thanks to their biological functions, including waste degradation, resource recycling and potential contributions to heat and electricity generation [7, 8]. Mycelium can form symbiotic relationships with algae and cyanobacteria, such as *Spirulina*, enabling nutrient exchange and oxygen production, which are crucial for sustaining closed-loop systems (9–12). Additionally, it can serve as decomposers, breaking down organic waste, including human and plant waste, converting it into fertile soil or structural materials. This contributes to a circular bioeconomy within space habitats.

As part of food production, mycelium can produce protein-rich mushrooms, offering a renewable food source for astronauts, with the ability to grow efficiently in stacked, vertical configurations. Its pharmaceutical properties also provide potential health benefits, such as producing antibiotics and bioactive compounds such as beta-glucans [13]. Furthermore, NASA is investigating fungi-based methods to convert carbon-rich asteroid materials into soil, facilitating soil-based cultivation for larger space settlements [14].

Mycelium can also potentially be integrated into fungal electronics and biosensors. Living mycelium networks exhibit self-regulation, adaptation and autonomous growth, making them suitable for responsive, interactive environments in space habitats [15, 16]. Moreover, engineered mycelium-bacteria systems, such as *Bacillus subtilis*, can function as biosensors to monitor oxygen levels, structural integrity and environmental factors such as temperature, humidity, chemical concentrations and light [8, 17].

Integrating mycelium into life-support systems would, in most cases, require maintaining it in a living state. This demands tight environmental regulation – controlling factors such as humidity, temperature, oxygen and nutrients – which can be very resource intensive. Moreover, decomposition processes may generate undesirable by-products, complicating waste management within closed systems. Concepts such as fungal electronics and biosensors remain largely experimental, with unresolved questions about durability, reliability and suitability for mission-critical applications. Biosafety is another concern: preventing the emergence of pathogenic strains or uncontrolled proliferation is essential in a confined area. Thus, while mycelium demonstrates these functions in terrestrial ecosystems, translating them into controlled, artificial systems for space applications requires substantial research and technological development before they can become functional, predictable and reliable components of long-duration missions.

2.4 Sensory enrichment

Utilizing mycelium in space architecture offers a unique opportunity to address the psychological challenges associated with prolonged space missions. By incorporating mycelial components into astronauts' living spaces, it is possible to introduce a range of tactile experiences that enhance psychological well-being [18]. The organic and responsive nature of mycelium allows for the creation of environments that finely tune tactile stimuli, thereby contributing to the overall comfort of mission participants [19]. Moreover, mycelium can integrate various multisensory aspects of natural environments, such as colour variation, form diversity, light modulation and textural variation, into the built environment [20]. This capability is particularly important for long-duration missions, where the effects of reduced gravity can diminish haptic sensations, making tactile feedback even more essential for maintaining psychological comfort.

3 Mycelium adaptation to the habitat requirements

As a living organism, mycelium has specific requirements for growth, development and performance. For it to function as a structural or metabolic component of extraterrestrial habitats, these biological needs must be reconciled with the environmental realities of space.

Table 5.1 illustrates how the biological needs of mycelium align – or conflict – with Martian conditions, taken here as an example, and highlights the adaptations required to enable growth and functionality in situ.

Table 5.1 How the biological needs of mycelium align – or conflict – with Martian conditions.

Mycelium Requirements	Martian Environmental Conditions	Required Adaptations
Stable pressure (~1 atm) for gas exchange	~0.6 kPa (less than 1% of Earth's atmosphere)	Pressurized growth chambers; research into minimum viable pressure for fungal growth
Optimal temperature range (20–30°C; growth slows below 10°C)	–125°C to +20°C, avg. –60°C	Thermal insulation and active heating systems
High relative humidity (>70%)	Extremely low (~0.03% RH)	Humidity-controlled growth systems with recycled water
Aerobic metabolism (~21% O_2) and tolerance to moderate CO_2	Atmosphere 95% CO_2, 0.13% O_2	Oxygen enrichment and CO_2 control, potentially with algae/cyanobacteria cohabitation
Organic substrate (cellulose, lignin, agricultural waste)	Regolith abundant but toxic (perchlorates, no organics)	Substrate supplementation via greenhouse waste; detoxification or bioremediation of regolith
Sensitivity to UV and ionizing radiation	High radiation (~250 mSv/year)	Shielded cultivation chambers; use of melanin-rich, radiation-resistant fungi.
Growth in 1 g gravity	Mars' gravity is 0.38 g	Further research on morphology, material properties under reduced gravity
Requires controlled sterility during colonization	Mars is largely sterile, but contamination risks exist	Closed bioreactors; strict sterilization; planetary protection protocols

Adapting mycelium for space habitats, however, involves not only creating controlled growth environments but also tailoring the fungus itself – via species selection, genetic engineering or synthetic biology – to fulfil specific functional roles within a habitat system. This adaptation challenge operates across multiple scales:

Habitat scale: ensuring overall safety, resilience and functionality through radiation shielding, thermal regulation and structural stability.
Human scale: addressing crew health and comfort through protection against extreme temperatures and radiation, and providing nutritional or sensory benefits.
Material scale: translating these needs into material properties such as thermal stability, UV resistance, impact durability and biocompatibility.
Molecular scale: engineering traits such as high melanin content or enzymatic activity to enhance resistance, radiation absorption or waste recycling.

Advances in genetic engineering and synthetic biology hold the potential to expand the capabilities of mycelium, enabling the creation of resilient, adaptive building materials tailored for extraterrestrial environments [8]. We can imagine that such approaches could generate living materials with traits specifically designed for space habitats – for example, enhanced tolerance to radiation, perchlorates or desiccation, as well as engineered metabolic pathways for recycling or nutrient production. However, translating these possibilities into practical applications remains highly speculative, requiring careful consideration of biosafety, planetary protection protocols and predictable performance.

Cultivating mycelium on Mars or other planetary bodies will require specialized growth chambers that precisely regulate temperature, humidity, atmospheric composition and pressure, while providing substrates derived from habitat biomass or engineered regolith composites. Strict containment measures will also be essential to prevent forward contamination in compliance with planetary protection standards.

Mycelium has demonstrated growth in microgravity aboard the International Space Station (ISS), though with altered hyphal orientation and morphology. The effects of reduced gravity (0.38 g on Mars, 0.16 g on the Moon) on large-scale composite fabrication, structural stability and material performance remain poorly understood. Similarly, while some fungi exhibit bioremediation potential against perchlorates, these functions are still at a laboratory stage and require significant research before they can be applied for ISRU-based construction.

Adapting mycelium for Martian or lunar environments will therefore require extensive engineering of both the organism and its growth environment. Achieving predictable, scalable and safe performance will depend on advances in controlled-environment bioreactors, bioengineering and integration with closed-loop life-support systems.

4 Myco-architecture off-planet project

The Myco-Architecture Off-Planet project serves as a case study in the utilization of mycelium for space habitat design and development. Focused on creating self-sufficient habitats, the project explores how mycelium can be used as a core material in construction for extraterrestrial environments. The project aims to leverage mycelium's unique properties to reduce reliance on Earth-based resources and enhance the long-term viability of habitats on the Moon and Mars [2, 21].

The Myco-Architecture Off-Planet project developed two distinct approaches for habitat construction: [1] biomass-based approach and [2] a regolith-based approach. Both approaches utilize in-situ resources, enabling the construction of habitats with minimal material transport from Earth. By harnessing mycelium's potential, the project offers complementary strategies that ensure sustainability and provide a foundation for future space habitat development.

4.1 Lunar habitat: biomass-based approach

The inflatable, biomass-based construction method developed by NASA's team, led by Dr Lynn Rothschild and redhouse studio, employs a deployable inflatable system as the habitat's core structural framework. This method serves several critical purposes: it defines the habitat's geometry, supports mycelium growth by facilitating nutrient circulation and provides protection during the initial development phase.

The design features enclosures similar to biological cells, with an interconnected circulatory system that nurtures mycelium growth. These enclosures serve multiple functions: thermal protection, containment of inflation gases to maintain structural integrity and a vascular network that supplies water and oxygen for algae and mycelium. The inflatables, initially transported in a compact, stowed form, expand upon deployment to form the habitat's base structure, acting as living shells that grow and adapt to the environment.

As mycelium is seeded into the structure, it gradually replaces the air and water within the cavities, solidifying into a durable, biomass-based shell. This biological process not only aids in construction but also enhances sustainability by producing oxygen, generating heat and even providing electricity. Key features such as plumbing, stovetops and flooring can be integrated as the mycelium grows in situ, creating a functional and durable habitat comparable to traditional Earth-built structures.

The development process begins with the deployment of the inflated habitat, where water is distributed through an integrated vascular system to pre-seeded algae. These algae produce oxygen and sugars when heated, which the growing mycelium consumes. As the mycelium biomass increases, the substrate is dehydrated to encourage further growth. Oxygen is then reintroduced, allowing the fungi to metabolize the algal biomass and convert it into chitin, forming the habitat's rigid structure.

To enhance durability and radiation protection, the chitinous structure is compressed by pneumatic membranes, offering additional support and improving radiation shielding. A nutrient-rich hydrogel-coated scaffold supports the mycelium's three-dimensional growth, gradually releasing nutrients to sustain its expansion throughout the habitat's lifecycle [25].

4.2 Martian habitat: Regolith-based approach

Given Mars' greater distance from Earth than the Moon, resource constraints are more pronounced. The creation of mycelium-based structures, which typically involves growing mycelium on an organic substrate, is a resource-intensive process. On Mars, this becomes particularly challenging owing to the scarcity of resources and harsh environmental conditions.

To address these challenges, the myco-architecture approach for creating Martian habitats has been developed with an emphasis on in-situ resource use. By leveraging regolith as the primary structural material and using mycelium primarily as a binding agent, this approach reduces the need for extensive biomass production. Mycelium stabilizes and reinforces the regolith, minimizing reliance on external resources and optimizing the habitat construction process [2, 22, 23].

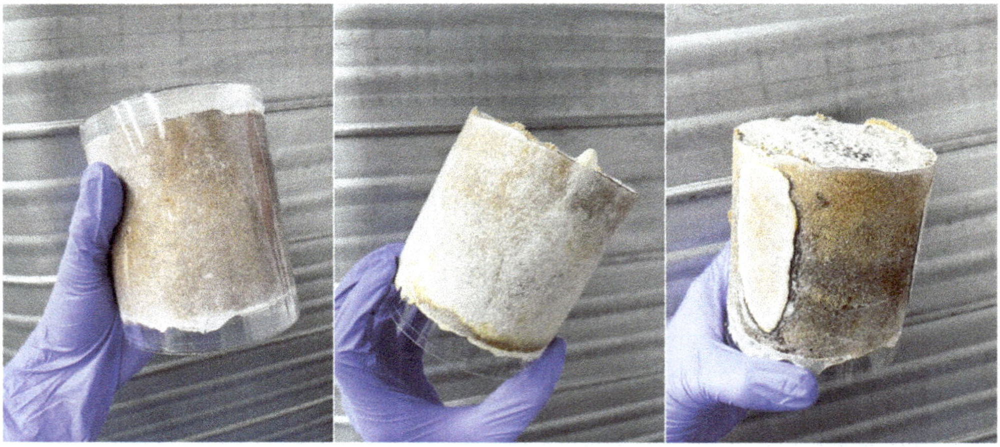

Figure 5.1 Mycelium–regolith composites, at different stages of the growth process. Image credit: Monika Brandić Lipińska.

Mycelium growth requires specific environmental conditions, such as temperature, humidity and oxygen levels. At certain thicknesses, insufficient oxygen can inhibit growth, making large-scale production more challenging. Given that this approach involves cultivating mycelium on regolith – a non-organic material that further complicates the environment for mycelial growth – the focus has shifted towards developing mycelium–regolith composites (Figure 5.1) rather than pursuing large-scale, monolithic applications.

The fabrication process for mycelium–regolith composites is tailored to the resource-limited conditions on Mars. The majority of structural components will be made from regolith, a readily available and abundant material. While an organic substrate is necessary for mycelium growth, it can be sourced locally from materials such as greenhouse waste, human waste or biomass, including algae and cyanobacteria. The process prioritizes minimizing biomass use, focusing primarily on inorganic materials to reduce dependence on organic inputs. The presence of carbon dioxide in the Martian atmosphere and water on Mars further supports the viability of in-situ resource utilization for habitat construction [24]. Initially, mycelium starter cultures would be transported from Earth to initiate growth, after which the mycelium would grow in situ within an enclosed environment. These components would then be assembled into a structure.

To optimize energy efficiency and minimize external control in this resource-limited environment, the approach leverages self-assembly principles, enabling autonomous assembly and reducing reliance on external interventions. The components are randomly assembled and, following jamming principles, form an aggregate structure (see Figure 5.2). However, this process depends on gravity to allow the components to lock and jam together. In reduced gravity conditions, such as on Mars, this principle is impaired, so while it aids in initial structure formation, further stabilization may be required. A key aspect of this approach is maintaining the viability of the mycelium–regolith components to facilitate bio-welding once assembled into an aggregate structure. The bio-welding process naturally fuses the components into a cohesive whole, reinforcing and stabilizing the structure without external intervention. Mycelium, in this approach, enables the transition from loose regolith to structural components and, ultimately, into a unified habitat structure.

Figure 5.2 Mycelium–regolith composites assembled into an aggregate structure and bio-welded. Image credit: Monika Brandić Lipińska.

5 Discussion: toward self-sustaining, bio-integrated environments

Creating habitats that function like biological systems requires not only new construction methodologies but also careful consideration of material choices. Mycelium presents a compelling opportunity in this context owing to its natural abilities, such as self-assembly, material transformation and environmental interaction, which could complement or even partially replace mechanical systems. In principle, mycelium could support habitats that evolve alongside astronauts and their needs, much like natural ecosystems. However, translating this vision into reality remains highly challenging and speculative.

Research into mycelium-based solutions in space architecture is still in its early stages. While mycelium demonstrates promising growth and adaptive capabilities under controlled and extreme conditions, the long-term behaviour of mycelium in space habitats, including under reduced gravity, radiation or closed-loop life-support systems, remains largely unknown. The ultimate goal of creating self-growing, symbiotic habitation systems, where mycelium interacts with other biological components and coexists with humans in a functional, regenerative environment, will require extensive experimental validation and careful engineering.

Concrete steps toward implementation are limited but emerging. Experiments aboard the ISS have already demonstrated mycelial growth in microgravity, providing initial insights into altered morphology and hyphal orientation. Future missions could focus on testing larger-scale mycelial composites, integrating them with in-situ resources such as regolith and assessing their structural and regenerative properties in situ.

Beyond technical feasibility, several critical barriers must be addressed. Regulatory and planetary protection concerns require strict containment protocols to prevent forward contamination, limiting experimental flexibility, while the high cost of space experiments restricts the scale and duration of testing. Additionally, the unpredictability of living materials poses challenges for reliability and safety in mission-critical habitats. Addressing these issues will require interdisciplinary collaboration between synthetic biology, materials science, aerospace engineering and space policy.

Mycelium research also has strong cross-domain potential. Insights gained from space-focused investigations, such as optimizing growth in low-resource environments or enhancing resilience under extreme conditions, can inform Earth-based construction, particularly for regenerative, sustainable architecture. Conversely, advances in terrestrial mycelium applications may accelerate space-related experimentation. Leveraging this dual-purpose approach can strengthen research outcomes while improving cost-effectiveness and innovation potential.

In conclusion, mycelium offers exciting possibilities for advancing bio-based, adaptive habitats. However, its implementation must be approached with caution, emphasizing rigorous experimentation, engineering controls and careful integration with existing habitat systems. Future progress will depend not only on technological advances but also on navigating regulatory frameworks, securing funding and systematically testing biological performance under realistic space conditions. By doing so, we can move toward habitats that are not only resilient and responsive but also scientifically validated and operationally safe – paving the way for the next generation of self-sustaining, bio-integrated environments both on Earth and in space.

References

1. Santomartino R, Averesch NJH, Bhuiyan M, Cockell CS, Colangelo J, Gumulya Y, et al. Toward sustainable space exploration: a roadmap for harnessing the power of microorganisms. Nat Commun. 2023;14(1). doi:10.1038/s41467-023-37070-2.
2. Brandić Lipińska M, Maurer C, Cadogan D, Head J, Dade-Robertson M, Paulino-Lima IG, et al. Biological growth as an alternative approach to on and off-Earth construction. Front Built Environ. 2022;8:1–17. doi:10.3389/fbuil.2022.965145.
3. Hall L, Rothschild L. Mycotecture Off Planet [internet]. NASA; 2021 [accessed 2024 Sep 19]. Available from: https://www.nasa.gov/general/mycotecture-off-planet/
4. Dadachova E, Bryan RA, Huang X, Moadel T, Schweitzer A, Aisen P, et al. Ionizing radiation changes the electronic properties of melanin and enhances the growth of melanized fungi. PLoS ONE. 2007.
5. Casadevall A, Cordero RJB, Bryan R, Nosanchuk J, Dadachova E. Melanin, radiation, and energy transduction in fungi. The Fungal Kingdom. 2017:509–514. doi:10.1128/9781555819583.ch23.
6. Cordero RJ, Casadevall A. Functions of fungal melanin beyond virulence. Fungal Biol Rev. 2017;31(2):99–112. doi:10.1016/j.fbr.2016.12.003.
7. Cohen MM, Flynn MT, Matossian RL. Water Walls Architecture: Massively Redundant and Highly Reliable Life Support for Long-Duration Exploration Missions. Glob Space Explor Conf. 2012:1–14.
8. Waters SM, Robles-Martínez JA, Nicholson WL. Exposure of Bacillus subtilis to low pressure (5 kilopascals) induces several global regulons, including those involved in the SigB-mediated general stress response. Appl Environ Microbiol. 2014. doi:10.1128/AEM.00885-14.
9. Das AAK, Bovill J, Ayesh M, Stoyanov SD, Paunov VN. Fabrication of living soft matter by symbiotic growth of unicellular microorganisms. J Mater Chem B. 2016. doi:10.1039/c5tb02489g.
10. Du ZY, Zienkiewicz K, van de Pol N, Ostrom NE, Benning C, Bonito GM. Algal–fungal symbiosis leads to photosynthetic mycelium. eLife. 2019;8:1–22. doi:10.7554/eLife.47815.
11. Podolich O, Zaets I, Kukharenko O, Orlovska I, Reva O, Khirunenko L, et al. Kombucha multimicrobial community under simulated spaceflight and Martian conditions. Astrobiology. 2017;17(5):459–469. doi:10.1089/ast.2016.1480.
12. Sanders WB. Lichens: the interface between mycology and plant morphology. BioScience. 2001;51(12):1025–1035. doi:10.1641/0006-3568(2001)051[1025:LTI-BMA]2.0.CO;2.

13. Patel S, Goyal A. Recent developments in mushrooms as anti-cancer therapeutics: a review. 3 Biotech. 2012;2(1):1–15. doi:10.1007/s13205-011-0036-2.

14. Hall L. Myco-architecture off planet: growing surface structures at destination [internet]. NASA; 2018 [cited 2024 Sep 19]. Available from: https://www.nasa.gov/general/myco-architecture-off-planet-growing-surface-structures-at-destination/

15. Adamatzky A. Towards fungal computer. Interface Focus. 2018;8(6). doi:10.1098/rsfs.2018.0029.

16. Adamatzky A, Ayres P, Beasley AE, Chiolerio A, Dehshibi MM, Gandia A, et al. Fungal electronics. BioSystems. 2022;212:104588. doi:10.1016/j.biosystems.2021.104588.

17. Beasley AE, Powell AL, Adamatzky A. Fungal photosensors. arXiv. 2020 Mar.

18. Tang TC, An B, Huang Y, Vasikaran S, Wang Y, Jiang X, et al. Materials design by synthetic biology. Nat Rev Mater. 2021;6(4):332–350. doi:10.1038/s41578-020-00265-w.

19. Arias D, Otto C. Defining the scope of sensory deprivation for long duration space missions. 2013.

20. Franco LS, Shanahan DF, Fuller RA. A review of the benefits of nature experiences: more than meets the eye. Int J Environ Res Public Health. 2017;14(8):1–27. doi:10.3390/ijerph14080864.

21. Tavares F. Could future homes on the Moon and Mars be made of fungi? [internet]. NASA; 2020 [cited 2024 Sep 19]. Available from: https://www.nasa.gov/centers-and-facilities/ames/could-future-homes-on-the-moon-and-mars-be-made-of-fungi/

22. Brandić Lipińska M, Dade-Robertson M, Zhang M, Rothschild LJ. DROP THE BASE: Biological, ISRU-Based Aleatory Construction System for Martian Habitat [internet]. Calgary: 2023 Int Conf Environ Syst; 2023 [cited 2024 Sep 19]. Available from: https://hdl.handle.net/2346/94678

23. Brandić Lipińska M, Dade-Robertson M, Theodoridou M, Rothschild LJ. Alien technology for alien worlds: design for biological construction of living habitation on Mars. Cambridge Open Engage. 2025. doi:10.33774/coe-2025-l4641.

24. Liu H, Yao Z, Fu Y, Feng J. Review of research into bioregenerative life support system(s) which can support humans living in space. Life Sci Space Res. 2021. doi:10.1016/j.lssr.2021.09.003.

25. Axpe E, Chan D, Offeddu GS, Chang Y, Merida D, Hernandez HL, Appel EA. A Multiscale Model for Solute Diffusion in Hydrogels. Macromolecules. 2019. doi:10.1021/acs.macromol.8b01757.

CHAPTER 6

For Mycoboscus: Sustainability, New Species and a Desirable Future with Mycelium Materials

ADRIEN RIGOBELLO

1 Introduction

Imagine being at the point in time when the development of leather work or woodworking is initiated, with all the accompanying cultural capacity and a timely sustainable potential. Mycelium-based composites (MBCs) represent just that: a nascent and exciting craft that might just outlive extractive industrial capitalism and may surprise us more than once in its creative uses. MBCs are an emerging technique that we have just begun to explore in a wider socio-cultural context; that is to say, under conditions that allow the affordable democratization of its knowledge.

So far, the terms 'MBCs' or 'mycelium materials' have been used in most cases to describe the results of this practice. I argue that expanding semantics and the epistemological methods allow us to valorize the coexistence of practices that can describe, open and enrich this new territory. Besides engineering goal-oriented language may live a plethora of verbs and nouns, that might also favor sensible craft, embodied knowledge and expansive, transformative and inclusive possibilities.

The meaning of 'sustainability' is shaped by local definitions, influencing how this technique is ecologically situated. This, in turn, reflects a particular ontology of nature and defines the ethical conditions of more-than-human relationships in the surrounding environment.

The actual sustainability of MBCs is indeed challengeable and not self-evident despite its appearance. First, I will present an overview of current production practices and of the main pathways of wood rot fungi spore production, their ecological significance and risks. I then propose a review of the existing literature on MBCs' lifecycle assessments and discuss them critically to understand what the actual environmental impacts encompass. Based on this information, it appears clear that despite their sustainable image, precautions must be taken when designing mycelium materials and communicating about them.

Building on recommendations from this critical analysis, I introduce in a third part a perspective for a craft informed by principles of absolute sustainability. Here, I refer to this craft as that of 'mycoboscus', which leverages a local-to-local cultivation and distribution model. This invites us to use local species and presents specific challenges and opportunities. I discuss methods for identifying and working with local species and expand the repertoire of fungi that may be considered for collaboration in mycoboscus

Adrien Rigobello, 'For Mycoboscus: Sustainability, New Species and a Desirable Future with Mycelium Materials' in: *Mycology for Architecture: Methods and Applications of Mycelium Materials in the Built Environment*. Pelagic Publishing (2026). © Adrien Rigobello. DOI: 10.53061/UNDM4876 craft work. Finally, I briefly discuss how such mycoboscus craft-oriented work eventually exceeds pragmatism with its potential for an expansive eco-social contribution.

2 Spores

MBCs utilize the capacity of wood-rot fungi to degrade lignin and other cellulosic components of plants to produce composite materials resembling particle board products. Because they do not use glues containing formaldehyde—a recognized carcinogen which can be present in particle board products, for example [1]—and are generally exclusively composed of plant and fungal material, these materials are deemed sustainable and compostable [2]. Figure 6.1 illustrates a typical production flow.

This field of material engineering has been developing in the past twenty years, since 2006, making it very recent. It has been focusing on the development of functional properties generally relating to thermal insulation, acoustic absorption, packaging, structural panels, coffins and surfboards. These applications target existing design use cases and functionally position MBCs as alternative materials to polystyrene (expanded or extruded), cardboard and laminated wood. In most applications, such as single-use packaging for example, yet not exclusively, MBCs' end of life is presented as an evident case of compostability and thus 'sustainability.' As has been discussed by other researchers, this might present ecological risks [3], which I try to comprehensively summarize here.

MBCs are generally produced by using fungi of the Basidiomycota phylum. These fungi mostly produce fruiting bodies, in which meiosis occurs leading to sexual spore production (Figure 6.2). Basidiospores produced this way contain a single haploid nucleus—a single set of chromosomes. Once they germinate, they develop into monokaryotic mycelia, which may fuse between compatible sexual groups via plasmogamy, leading to a subsequent growth in the form of a dikaryotic mycelium where each cell will hold two nuclei, one of each group. Dikaryotic mycelia can be identified under a microscope by the presence of clamp connections. They can form fruiting bodies, where karyogamy, the fusion of the two nuclei into a diploid nucleus—nucleus containing two sets of chromosomes—will take place within each basidium in the fruiting body.

The presence of sexual spores and fruiting body formation capacity can represent a health risk as a respiratory hazard and an ecological risk with the potential spread of the species in the environment [3]. Monokaryotic basidiomycete mycelia generally do not produce fruiting bodies, and the isolation of such strains can therefore help mycelium material cultivators to alleviate health risks.

While their main mode of reproduction is sexual, basidiomycetes may also produce asexual spores within both mono- and dikaryotic mycelia, depending on species. The two forms of mycelium-produced asexual spores are oidia and chlamydospores. Oidia form as hyphae break up into cells and develop into spores. Oidia are thin-walled and cannot survive in unfavorable environmental conditions for long. Chlamydospores are particularly resistant to unfavorable environmental conditions. They act as a 'back-up plan' for the fungus, and result from a thickening of the cell wall of a hyphal compartment or can be formed on a specific chlamydoconidium apparatus.

Most *Ganoderma* species have been reported to exhibit the capacity to produce chlamydospores [4, 5] showing a higher resistance to elevated temperatures ($\geq 40°C$) [6]. These are among the most sought-after species for MBC production, along with *Trametes versicolor* strains which have also been shown to have this capacity [7], although rarely. Certain *Pleurotus* species have also been reported to do so, such as *P. ostreatus*,

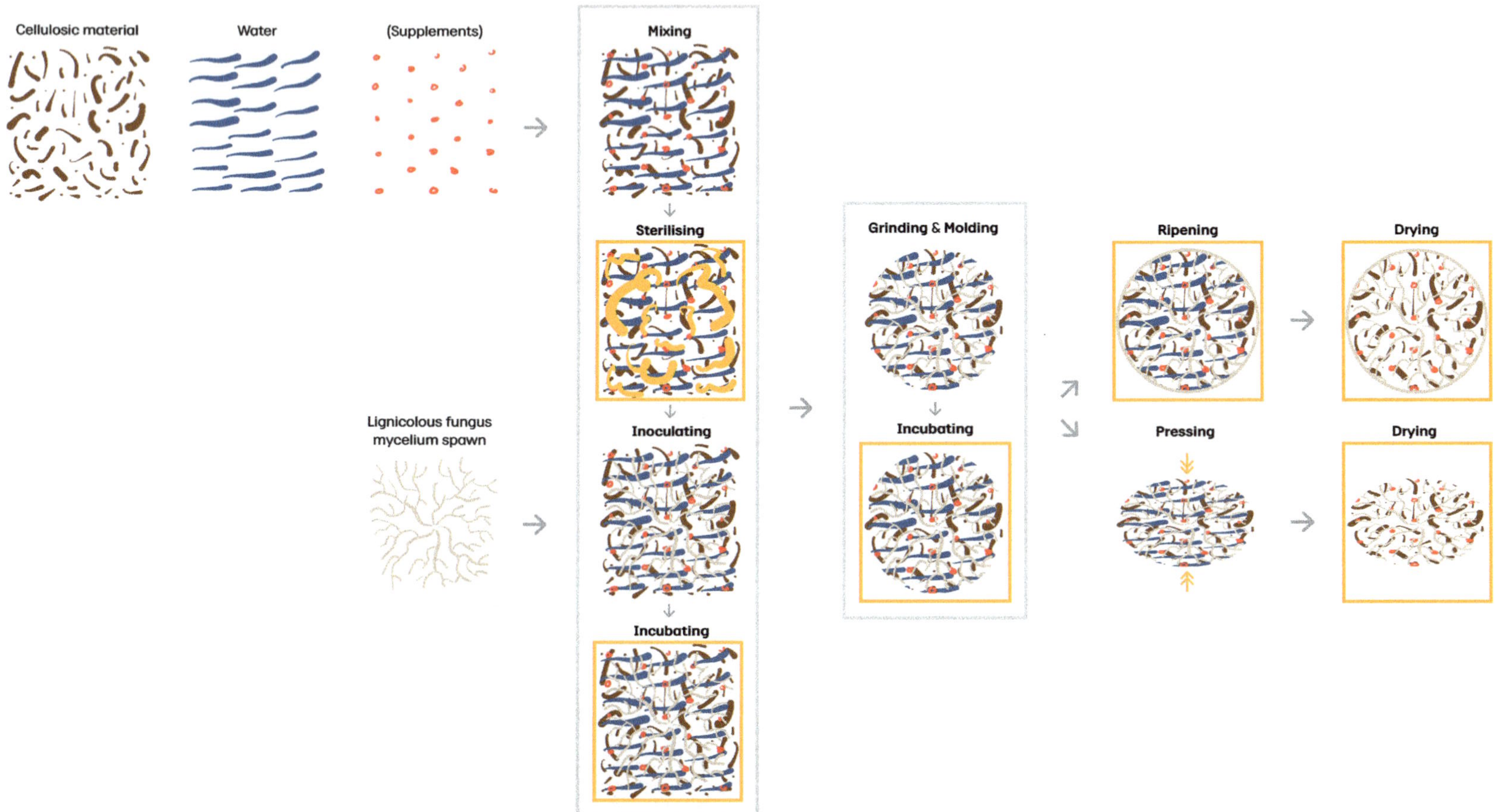

Figure 6.1 Typical production flow of MBC. Illustration credit: Author.

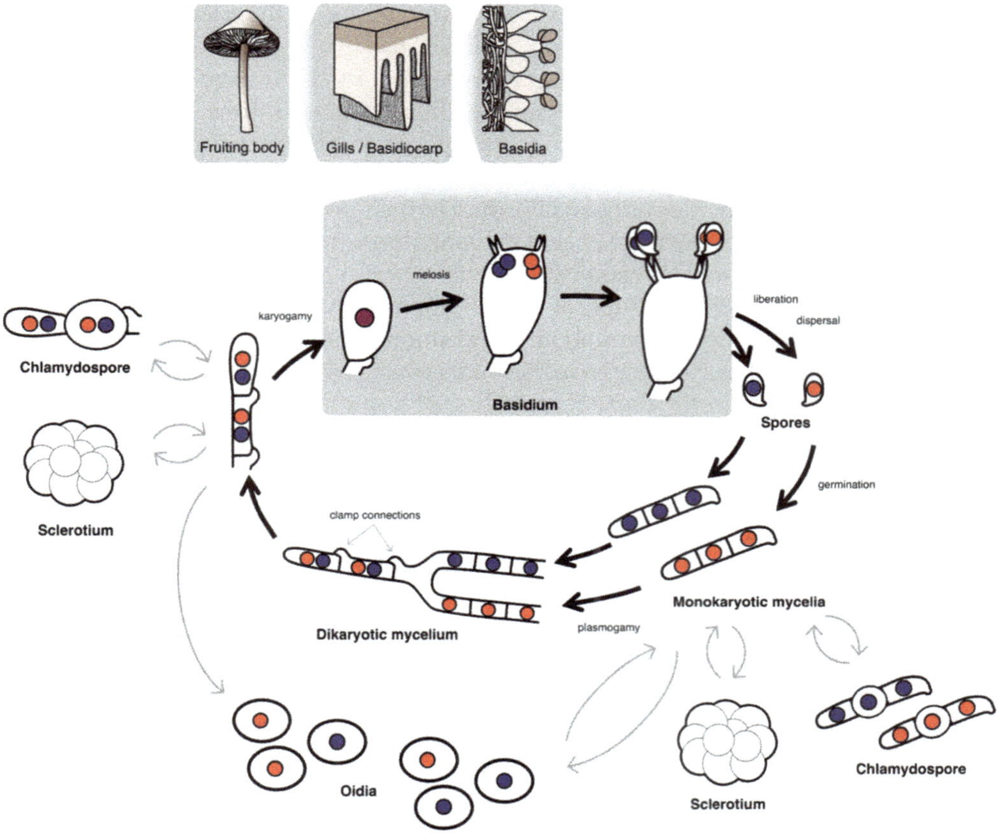

Figure 6.2 Basidiomycota lifecycle, emphasizing specialized reproductive structures such as chlamydospore and oidia. Illustration credit: author, adapted from Schematic of a typical basidiocarp (Wikimedia Commons, Devibort, 2006), Lifecycle of basidiomycete (Wikimedia Commons, Silicon Prophet and M. Piepenburg, 2020) and Lifecycle of C. cinereus (Kües, 2000).

P. pulmonarius and *P. dryinus* [8, 9, 10], which is correlated to *P. ostreatus* having the expected genetic toolkit for this [11, 12].

This is a significant aspect for the practice. While products such as single-use packaging made of MBCs are advertised for their compostability, users are incentivized to spread the remaining products, although they most likely contain active chlamydospores. While white rot fungal species can be saprotrophs—feeding off dead matter—some of the most ecologically competitive ones can also act as pathogens and infect living plants, such as *Ganoderma* spp. [13, 14, 15] among other species [3]. Introducing spores of this species, for instance, can affect local ecologies of decomposers [16] and negatively impact seedling density [17]. A similar concern has been recently raised by the United Kingdom's Royal Horticultural Society, which, seeing the boom of home grown kits for edible mushrooms, warns of the potential ecological effects of spreading competitive strains with little genetic variation [18].

To appreciate the importance of genetic diversity, we can imagine that picking two wood rot fungal species is like picking a cow and a horse; they feed off the same yet are very different. Now picking two strains, such as chanterelles from different locations in a forest, is rather like picking two humans; they feed off the same yet are very different. Beyond short-term ecosystem disruption and potential ripple effects on microbial, insect and plant wildlife and crops, genetic diversity is recognized by the scientific community as worthy of important ecological conservation efforts [19].

In emerging industrial practices, represented by Ecovative Design, which currently has a near market monopoly in the United States and the European Union via its subsidiaries for MBC-based products, one strain of North America-originated plant pathogen *G. sessile* has been reported to be in use, to the best of my knowledge. Furthermore, the worldwide commercialization of their popular product, the 'Grow-It-Yourself' kit, relies on the presence of active chlamydospores. The company sells a dried, lightweight, pre-inoculated hemp substrate with spores that are reactivated by the user by moisturization and the addition of flour (minerals). As the horticultural scientific community raises serious concerns about distribution of the genetic patrimony of a few strains via sexual and asexual reproduction, this emerging industrial practice presents an even higher risk, with the asexual spreading of a single pathogenic strain in a way that disregards geography and ecology. The global market for polystyrene represents about 16 million tons per year; should MBC solutions start reaching those volumes and planetary scale, the ecological effects could be devastating.

The inactivation conditions for the chlamydospores of *Ganoderma* spp., *Trametes* spp. and *Pleurotus* spp., the three most popular families in MBC practice, are currently mostly unreported for the species in use. As the practice develops, future research must investigate these inactivation conditions for chlamydospores to avoid spreading potentially pathogenic and competitive strains selected for their industrial competitiveness [20]. The screening of species for use in MBC production should then involve risk assessment.

Restricting the chlamydospore expression capacity of fungi might be possible, although this is tied to large pools of over two thousand differentially expressed genes that were identified under chlamydospore-producing conditions [21, 22]. Modifying the growth-medium profile has nonetheless been shown to influence chlamydospore synthesis in the ascomycete *Trichoderma harzianum* Tr-92 [21], and genetic modification is also being studied to suppress chlamydospores' synthesis capacity [23, 24].

Fungi of the phylum Ascomycota can also be encountered in mycelium material practices [25], although rarely. The wood rot capabilities of taxonomic orders indicate a potential research interest for this phylum. Most ascomycetes produce asexual spores, mitospores. A number of ascomycete fungi are only known in their asexual forms [26]. The risks of spreading a limited genetic patrimony are slightly different here as these are not chlamydospores, although some ascomycetes can produce these spore structures.

Spore production is therefore an important and defining aspect of MBC practice. It is recommendable that this aspect is addressed rigorously by any practitioner aiming to advance the field or by a commercial venture.

3 Lifecycle assessments

Have we found the panacea to the environmental crisis partly created by synthetic polymers? As we have seen, the ecological risks presented by the spore presence are considerable, and the actual 'sustainability' of MBCs also presents grey areas.

Four lifecycle assessments (LCAs) have been conducted for MBCs recently and offer a rich perspective over the characteristics of MBC production processes. These assessments integrate both the process and choice of input materials. Despite having varying scopes of study and hypotheses of work, they may help to inform our practices.

In these studies, I note some discrepancies in scope, as some researchers include the pre-culture (petri dish strain isolation) and grain spawn preparation phases in their scope of study. All have evaluated the substrate inoculation, incubation, and drying phases. Here, I choose not to comment on the design of the unit of study and its effect on LCAs and note that the choice of the species impacts the functional characteristics of the resulting product, the choice of the substrates, of incubation durations, and therefore of the final LCA. Country-specific electric mixes were used for the calculations (Germany, the Netherlands and Italy).

For the production process, the electrical energy mobilized for the sterilization of substrates and drying of the units of study are the most significantly impacting categories—along with grain and substrate acquisition. The use of electricity weighs heavily, reportedly over 80% of the CO_2 emissions equivalent for producing a unit [27]. This is assigned to various phases such as sterilization, incubation and drying. The drying phase can represent about 50% to 60% of the total electrical consumption, and the incubation phase about 15% to 40% [27, 28, 29]. It is not just in CO_2 eq., also called global warming potential, that the impact of electrical energy can be accounted for: this energetic use can represent about 60% of the contribution to acidification potential and over 70% of the photochemical ozone creation potential of a unit [27].

The choice of substrate also affects the material environmental impact. Using hemp by-products can weigh about 95% of the eutrophication potential of a unit and about 85% in water scarcity potential [27]. Using rapeseed straw instead of hemp can lead to an increase of 33% in acidification potential and over 20% increase in eutrophication potential, which relates to a different use of fertilizers in the cultivation of these crops. Yet using rapeseed straw can lead to a decrease of 17% in land-use potential [27]. If we were to use poplar wood, compared with hemp, the unit would result in a 50% reduction in eutrophication potential, as no fertilizers would be needed, but land use would increase by over 30% [27]. The choice of the substrate is therefore important to assess indexes of sustainability, and choosing is a matter of compromise rather than there being a potential best solution.

Scaling to industrial measures affects some of these indexes. In such cases, an industrially produced unit made of hemp may see a potential halving of its acidification potential, smog potential and water scarcity potential, and a reduction by almost a factor three in climate change potential. Such a unit would see its eutrophication and land-use potentials scale linearly [27].

An alternative method of assessment has been proposed, in which biogenic CO_2 resulting from photosynthesis of the substrate (i.e., carbon fixation) and metabolic CO_2 released by the enzymatic oxidation of cellulose is accounted for. When calculating a LCA in this way, MBCs can be CO_2 eq. negative [30]. This method is still very sensitive to the incubation time and energy use: above two weeks of incubation, the unit results in positive CO_2 eq. and is not carbon negative anymore.

When the input substrate is considered as a waste instead of a by-product, it would impact the account of CO_2 eq. in the calculation and make it negative because it is fixated carbon. If, for example, hemp by-products or rapeseed straw are used as substrates—for which fertilizers are used in cultivation—considering them as waste would lead to the nullification and concealment of the impact on many LCA metrics [27, 30]. Conversely, when considering input materials as by-products, only with an electric mix fully powered by renewable sources would a MBC be environmentally competitive with kenaf, cork, cotton stalk or straw solutions for insulation [28]. The LCAs of MBCs are clearly highly impacted by the electric energy that is consumed to produce them and the choice of substrates. The absolute sustainability of MBCs in the future will depend on the development of manufacturing methods that reduce energy consumption and opportunistically use substrate as waste rather than by-products. As demonstrated by the eutrophication, acidification and land-use potentials resulting from the selection of substrates, establishing a link with a by-product supply needs to be carefully considered. We can also note that, while there is a need to reduce the energy consumed for drying [28], we can foresee that the necessary inactivation of chlamydospores (under conditions that have yet to be investigated) will require minimal energy use at the drying stage.

4 Mycoboscus craft and species selection

MBCs can be produced in conditions of absolute sustainability by using locally available waste and making efforts to reduce the use of electrical energy during production. While chlamydospore inactivation conditions are yet to be elucidated for each species, it seems that the drying phase might need to be extended (as well as its environmental impact). An effective mitigation strategy rests in the use of locally foraged species to craft a local distribution of products. This would indeed limit the impact of the accidental spread of the fungus in the wild, as it would supposedly have already been present.

The affordability of the craft also invites everyone to try and learn about these fungal partners. This suggests craft practices that may be influenced by local geographical and social conditions and can lead to exciting new design landscapes, or 'design terroirs,' where design objects would inherit aesthetic qualities akin to those of cheese, wine or other specialty foods [31].

In view of the development of a craft issuing from these values of openness, affordability and full sustainability, I use the term 'mycoboscus' in place of MBC or mycelium material to designate a production issued from the eco-socially situated art of working with wood-rot fungi.

Devising a local network of input material sources and actors for the design of a mycoboscus involves the identification of a local fungal strain and its nutritional preferences. Generally, white rot fungi are preferred, and they are over-represented in forests worldwide (Figure 6.3).

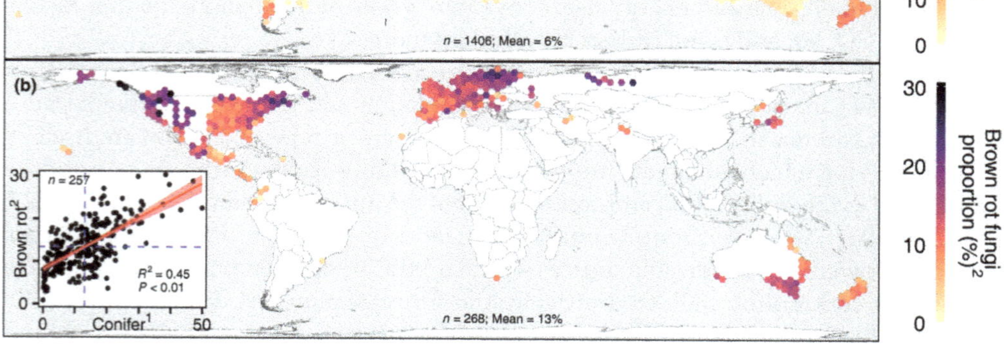

Figure 6.3 White-rot fungi are over-represented in forests worldwide and can be generally predicted in all forests. White-rot fungi and deciduous trees are generally a great match. Here we see the global distribution comparison between two fungal decay modes (brown rot (BR) and white rot (WR)) and tree type (deciduous and conifer). Reproduced [32].

Table 6.1 List of species referenced in this chapter and characteristics. A: ascomycetes, B: basidiomycetes.

Species	Order	Life-history strategy	Decay mode	Distribution
Auricularia delicata	Auriculariales (B)	Saprotroph	White rot	Pantropical, temperate southern hemisphere
Dichomitus squalens	Polyporales (B)	Saprotroph	White rot	Northern hemisphere
Fomitiporia mediterranea	Hymenochaetales (B)	Saprotroph, Plant pathogen	White rot	Mediterranea, Iran
Ganoderma spp.	Polyporales (B)	Saprotrophs, Plant pathogens, Parasites	White rot	Global
Ganoderma lucidum	Polyporales (B)	Saprotroph, Parasite	White rot	Global, temperate and subtropical regions
Ganoderma sessile	Polyporales (B)	Saprotroph, Plant pathogen	White rot	Eastern North America
Hypoxylon spp.	Xylariales (A)	Saprotrophs	White rot	Global, temperate and tropical regions
Irpex lacteus	Polyporales (B)	Saprotroph	White rot	Temperate regions
Phanerochaete chrysosporium	Polyporales (B)	Saprotroph	White rot	Restricted, temperate woodland
Phlebia tremellosa	Meruliaceae (B)	Saprotroph	White rot	Global
Pleurotus spp.	Agaricales (B)	Saprotrophs	White rot	Global except Antarctica
Pleurotus dryinus	Agaricales (B)	Saprotroph	White rot	Northern hemisphere
Pleurotus ostreatus	Agaricales (B)	Saprotroph	White rot	Global except Antarctica
Pleurotus pulmonarius	Agaricales (B)	Saprotroph	White rot	Temperate and subtropical forests
Punctularia strigosozonata	Corticiales (B)	Saprotroph	White rot	Global
Schizophillum commune	Agaricales (B)	Saprotroph	Uncertain wood decay	Global
Stereum hirsutum	Russulales (B)	Saprotroph, Plant pathogen	White rot	North America
Trametes spp.	Polyporales (B)	Saprotrophs, Plant pathogens, Parasites	White rot	Global
Trametes versicolor	Polyporales (B)	Saprotroph, Parasite	White rot	Global
Trichoderma asperellum	Hypocreales (A)	Endophyte	White rot	Global
Trichoderma harzianum	Hypocreales (A)	Saprotroph, Parasite	Litter-degrading, Mycoparasitism	Global
Xylaria spp.	Xylariales (A)	Saprotrophs	White rot	Global, terrestrial and marine environments

About 80 fungal species have been reported in mycelium materials studies. Getting information on their use through literature is a great entry point to local cultivation. Many other species have yet to be evaluated for their use in mycoboscus cultivation. One can initially identify them from their genome, by comparing their degradation capacities to the substrates at hand [33]. Higher gene reads for certain enzymes allowing lignin degradation, class II peroxidases, are for example identified from sequencing in model species *Phanerochaete chrysosporium* and *Auricularia delicata, Fomitiporia mediterranea, Dichomitus squalens* and *Punctularia strigosozonata* [34]. These species are good candidates to consolidate or expand the repertoire. Moreover, *P. chrysosporium* has been previously reported to have a radial extension 2.6 to 9.2 times higher than that of a *G. lucidum* strain [35]. *Phlebia tremellosa* or *D. squalens*—a species preferring softwood—could also be of interest to practitioners [36, 37]. It can be noted that *Schizophillum commune* is considered a white rot species although it has a limited lignin decay capacity, with no gene reads for important lignin degrading enzymes [34].

Despite the enthusiasm that an amateur 'mycoboscist' might show for a good foraging session from which they will craft unique designs, they would probably want to prevent any health and environmental risks here too. Some species, such as *Irpex lacteus* and *S. commune*, have been reported to cause pulmonary abscess in immuno-suppressed persons and display potential opportunistic human pathogenicity [38]. Saprotrophs can be plant pathogens, such as white rot *Stereum hirsutum* that is found as a pathogen in peach trees. *Xylaria* spp. and *Hypoxylon* spp. are wood-decaying ascomycetes, and as for *Trichoderma asperellum* that was already reported in a MBC study, their conidia formation capability may represent a health hazard. Fungi may also produce volatile compounds that repel or attract animals and insects depending on their cultivation conditions, and the widespread democratization of their use could impact insect biodiversity and infestation risks [39, 40, 41].

Developing such craft practice already involves an attentive presence, alternating between ecosystem immersion and extended phenomenological observation. One might experience diverse fungal behaviors; that is, as a result of environmental changes, genetics or epigenetics. Widespread species reproducing sexually have a higher genetic variability. This may impact the reproducibility of MBC or mycoboscus studies and call for local protocol adaptations. To illustrate, it was reported that the degradation of birch samples by five strains of species *P. chrysosporium* resulted in widely ranging lignin loss of 23–73%, glucose loss of 10–50% and xylose loss of 28–58% for an overall weight reduction of 17–46% [37].

Between species, we can even note different fungal 'character traits'. For example, looking at the difference in gene reads in *G. lucidum* G0119, *P. ostreatus* PC15 and *T. versicolor* strains [33], different decay behaviors may express phenotypically, and different fungal characters appear. *G. lucidum* G0119 stands out as it reflects a greater versatility in strategies for decomposing (amorphous) cellulosic compounds, appearing as a gourmet fungus, taking its time to feed deeply off its substrate. *P. ostreatus* PC15 seems more equipped to degrade more recalcitrant lignin, and its crystalline cellulose degradation strategies are sensibly more numerous than those of the two other strains. This one craves for quick food. *T. versicolor* is distinct for its relatively high glycosyltransferases reads, pointing to a great diversity of functions involved in cell wall synthesis among other activities, which can influence its decay activity considering a potential influence on membrane porosity or mycelium morphologies [42]. Glycosyltransferases cover various functions such as cell wall synthesis (e.g., chitin synthases, β-1,3-glucan synthases or β-mannosyltransferases), which affects tip extension, adhesion, interaction with the medium, secretion or secondary metabolites tailoring such as pigments. *T. versicolor* may therefore incidentally remind us of a peacock. Looking at the fruiting bodies of these three fungi will reflect these three character traits, where the gourmet has slower grown and fleshy, lacquered fruits; the quick food craving one has very lightweight fruits and a quickly sporulating life-history strategy; and the peacock one has a magnificent adornment.

In addition to their impact on reproducibility, aesthetics and performance, genetic variation and mutation present an environmental risk, as a species can see a change from endophytic to pathogenic lifestyle with a marginal genetic mutation [43].

Beyond life-history strategies of fungi, their strain-specific genomes and life history can influence the specific decaying capabilities for a same cultivation protocol and substrate. The genetic selection and *in vitro* gene regulation by media modifying over generations of sexual reproduction may lead to species gene expression modification and induce evolution by phenotypical selection.

One ends up being confronted by an issue of two-sided complexity in mycoboscus practice: on the one hand, working with local strains implies having to source and isolate a fungus which might not be showing 'normal,' expected behavior (e.g., varying growth rates, diets); on the other hand, a substrate composed of opportunistically found materials (e.g., wastes, invasive species) reflects a varied and complex chemical, structural and mechanical cultivation environment with thermodynamic and fluid mechanics effects [33]. As efforts are made to understand, model and describe these phenomena, ranging from genomic to macro-scales [42], their combined complexity might also be approached by the development of a resilient local capacity located between an individual person's experience and their community knowledge [31]. Beyond the laboratory, the craftsperson may develop a sensory experience of a locally foraged fungus and try out various combinations of found materials, either following known principles [33] or more by chance. Seasonality might also dramatically influence one's experience in the absence of a controlled environment, or even rain events.

While the laboratory is convenient for such experiments, the very craft experience is that of an attunement not just to a given fungus but to the landscape. While the economics of this inefficient process might seem unresolved and potentially ludicrous, woodwork, textile weaving, fermentation or ceramic crafts' persistence reminds us that there is something invaluable and unformalizable about this sort of activity.

5 Discussion and conclusion

MBCs are a promising class of materials with very particular characteristics owing to their organic nature. One such characteristic is the presence of spores, and some research efforts focus on their functional embedding, by leveraging chlamydospores to optimize logistics in the case of the company Ecovative or narrating their presence as a potential avenue towards biologically 'self-healing' materials [44], for example. It is a remarkable property of such new materials'—and perhaps inspiration for material recycling? Chlamydospores' presence, associated with the bioreceptive organic nature of MBCs, has also been leveraged in the study of the functionalization of genetically modified bacteria of a species that can be found opportunistically growing on a MBC [45].

Here, I argue that a goal-oriented material engineering focus may obscure critical health and environmental risks. I propose a co-existing practice, focused on craft activities and following a local-to-local model of production and distribution. This practice of mycoboscus craft recognizes fungi as living beings rather than reservoirs of functionalities awaiting objectivization in scientism, recognizes the centrality of attentive craft as an ethical aesthetics and operates an epistemic turn striving towards fair environmental ethics.

Working with locally foraged species and local waste and reducing electric energy use: these simple recommendations may radically change the potential of mycelium materials. An amended process flow with these modifications is presented in Figure 6.4. There we see a proto-vision for practices sensitive to climate and geography, where

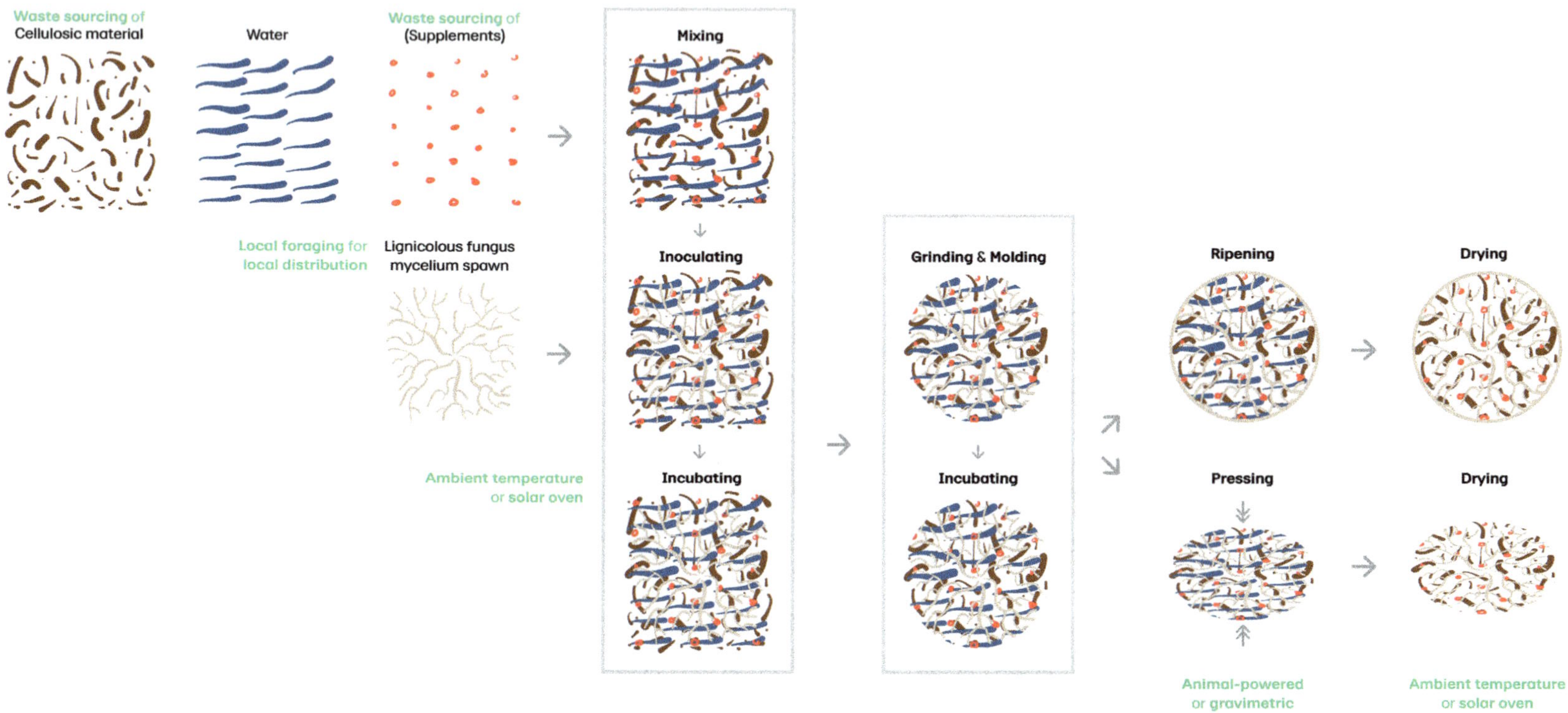

Figure 6.4 Sustainable mycoboscus process flow. Illustration credit: Author.

strategically primitive machines are used, and unique expressions of sensibly receptive dwelling are brought about.

Forasmuch as it might be beneficial to replace the environmental catastrophe that is polystyrene packaging by MBCs, this might turn into a catastrophe in its own right if the very model of centralized industrial production, romanticization and commodification continues to be seen as our future. The mycoboscus craft aims to contribute to a possible alternative path. In that spirit, I invite the multiplicity of other situated, militant practices to responsibly explore the possibilities for completely sustainable eco-social dwelling.

At least 2.01 billion tons of municipal solid waste is produced every year globally. I estimate that 45% of this could be used as substrate for mycobosci, and more when combined with inorganic materials. This is a chance to invent a relational economy of shared locality; as it is so affordable, the opportunity will not be going away any time soon.

Acknowledgements

Adrien Rigobello has received funding from the European Union's Horizon 2020 research and innovation program FET OPEN 'Challenging current thinking' under grant No. 858132.

References

1. Swenberg, James A., et al. 'Formaldehyde carcinogenicity research: 30 years and counting for mode of action, epidemiology, and cancer risk assessment.' *Toxicologic Pathology*, vol. 41, no.2, 2012, pp. 181–9.
2. Van Wylick, Aurélie, et al. 'Mycelium composites and their biodegradability: An exploration on the disintegration of mycelium-based materials in soil.' *Construction Technologies and Architecture*, vol. 1, 6 Jan. 2022, pp. 652–9. https://doi.org/10.4028/www.scientific.net/cta.1.652.
3. Van den Brandhof, Jeroen G., and Wösten Han A. B.. 'Risk assessment of fungal materials.' *Fungal Biology and Biotechnology*, vol. 9, no. 1, 24 Feb. 2022. https://doi.org/10.1186/s40694-022-00134-x.
4. Moradali, M. F., et al. 'The genus *Ganoderma* (Basidiomycota) in Iran.' *Mycotaxon*, vol. 99, Jan. 2007, pp. 251–69. www.cabdirect.org/abstracts/20073166001.html.
5. Stalpers, Joost A. 'The genus *Ptychogaster*.' *Karstenia*, vol. 40, 2000, pp. 167–80.
6. Loyd, Andrew L., et al. 'Cultural characterization and chlamydospore function of the *Ganodermataceae* present in the Eastern United States.' *Mycologia*, vol. 111, no. 1, Jan. 2019, pp. 1–12. https://doi.org/10.1080/00275514.2018.1543509.
7. Freitag, Michael. 'Measuring extracellular enzymes in pure and mixed cultures of *Trametes versicolor* (L.:Fr.) Pilat and *Trichoderma harzianum* Rifai.' Master's thesis, 1989. Corvallis: Oregon State University.
8. Kamzolkina O. V. et al. '[Micromorphological features of *Pleurotus pulmonarius* (Fr.) Quel. and *P. ostreatus* (Jacq.) P. Kumm. strains in pure and binary culture with yeasts].' *Tsitologiia*, vol. 48, no. 2, 2006, pp. 153–60. PMID: 16737183.
9. Chong, S. F., Zakaria, Z., Low, J. Z. 'Investigation of chlamydospores and mycelium liquid spawn storage period of grey oyster mushroom (*Pleurotus pulmonarius*) and effect on growth rate.' *International Journal of Advances in Science Engineering and Technology*, vol. 7, no. 2, 1 May 2019, pp. 50–2.
10. Maftoun, P. et al. 'The edible mushroom *Pleurotus* spp.: I. biodiversity and nutritional values.' *International Journal of Biotechnology for Wellness Industries*, vol. 4, 2015, pp. 67–83.
11. Hou L. et al. 'Transcriptome Analysis Revealed That Cell Wall Regulatory Pathways Are Involved in the Tolerance of *Pleurotus ostreatus* Mycelia to Different Heat Stresses.' *Journal of Fungi*, 2025, vol. 11, no. 4, p. 266. https://doi.org/10.3390/jof11040266
12. Wright R. et al. 'The genome sequence of the oyster mushroom, *Pleurotus ostreatus* ((Jacq.) P. Kummer, 1871).' *Wellcome Open Res*, vol. 8, 26 Jun 2023, p. 277. https://doi.org/10.12688/wellcomeopenres
13. Kües, Ursula, et al. 'Genome Analysis of Medicinal *Ganoderma* spp. with Plant-pathogenic and Saprotrophic Life-styles.' *Phytochemistry*, vol. 114, Feb. 2015, pp. 18–37. https://doi.org/10.1016/j.phytochem.2014.11.019.

14. Paterson, R. R. M. '*Ganoderma* Disease of Oil palm – A White Rot Perspective Necessary for Integrated Control.' *Crop Protection*, vol. 26, no. 9, Jan. 2007, pp. 1369–76. https://doi.org/10.1016/j.cropro.2006.11.009.
15. Hushiarian, Roozbeh, et al. 'Detection and control of *Ganoderma boninense*: strategies and perspectives.' SpringerPlus, vol. 2, no. 1, Oct. 2013. https://doi.org/10.1186/2193-1801-2-555.
16. Fukasawa, Yu, and Kimiyo Matsukura. 'Decay Stages of Wood and Associated Fungal Communities Characterise Diversity–decomposition Relationships.' *Scientific Reports*, vol. 11, no. 1, Apr. 2021. https://doi.org/10.1038/s41598-021-88580-2.
17. Fukasawa, Yu. 'Effects of Wood Decomposer Fungi on Tree Seedling Establishment on Coarse Woody Debris.' *Forest Ecology and Management*, vol. 266, Dec. 2011, pp. 232–8. https://doi.org/10.1016/j.foreco.2011.11.027.
18. Horton, Helena. 'Mushroom-growing Boom Could Cause Biodiversity Crisis, Warn UK Experts.' The Guardian, 31 May 2024. www.theguardian.com/science/article/2024/may/31/mushroom-growing-boom-could-cause-biodiversity-crisis-warn-uk-experts.
19. DeWoody, J. Andrew, et al. 'The Long-standing Significance of Genetic Diversity in Conservation.' *Molecular Ecology*, vol. 30, no. 17, Jul 2021, pp. 4147–54. https://doi.org/10.1111/mec.16051.
20. Balaeş, Tiberius, et al. 'Mycelium-Composite Materials – A Promising Alternative to Plastics?' *Journal of Fungi*, vol. 9, no. 2, Feb. 2023, p. 210. https://doi.org/10.3390/jof9020210.
21. Yuan, Min, et al. 'Whole RNA-sequencing and gene expression analysis of *Trichoderma harzianum* Tr-92 under chlamydospore-producing condition.' *Genes & Genomics*, vol. 41, no. 6, Apr. 2019, pp. 689–99. https://doi.org/10.1007/s13258-019-00812-y.
22. Liu, Lei, et al. 'Whole transcriptome analysis provides insights into the molecular mechanisms of chlamydospore-like cell formation in *Phanerochaete chrysosporium*.' *Frontiers in Microbiology*, vol. 11, Dec. 2020, https://doi.org/10.3389/fmicb.2020.527389.
23. Sun, Zhan-Bin, et al. 'Identification of genes related to chlamydospore formation in *Clonostachys rosea* 67–1.' *MicrobiologyOpen*, vol. 8, no. 1, Apr. 2018. https://doi.org/10.1002/mbo3.624.
24. Böttcher, Bettina, et al. '*Candida* species rewired hyphae developmental programs for chlamydospore formation.' *Frontiers in Microbiology*, vol. 7, Oct. 2016. https://doi.org/10.3389/fmicb.2016.01697.
25. Tacer-Caba, Zeynep, et al. 'Comparison of novel fungal mycelia strains and sustainable growth substrates to produce humidity-resistant biocomposites.' *Materials & Design*, vol. 192, Apr. 2020, p. 108728. https://doi.org/10.1016/j.matdes.2020.108728.
26. Deacon, J. W. *Fungal Biology*. Hoboken: John Wiley and Sons, 2005.
27. Stelzer, Lisa, et al. 'Life cycle assessment of fungal-based composite bricks.' *Sustainability*, vol. 13, no. 21, Oct. 2021, p. 11573. https://doi.org/10.3390/su132111573.
28. Carcassi, Olga Beatrice, et al. 'Carbon footprint assessment of a novel bio-based composite for building insulation.' *Sustainability*, vol. 14, no. 3, Jan. 2022, p. 1384. https://doi.org/10.3390/su14031384.
29. Volk, Rebekka, et al. 'Life cycle assessment of mycelium-based composite materials.' *Resources Conservation and Recycling*, vol. 205, Apr. 2024, p. 107579. https://doi.org/10.1016/j.resconrec.2024.107579.
30. Livne, Achiya, et al. 'Fungal mycelium biocomposite acts as a CO_2-sink building material with low embodied energy.' *ACS Sustainable Chemistry & Engineering*, vol. 10, no. 37, Sep 2022, pp. 12099–106. https://doi.org/10.1021/acssuschemeng.2c01314.
31. Rigobello, Adrien, and Joshua D. Evans. 'Design Terroir: an eco-social, relational, bioregional approach to design.' *Proceedings of DRS*, Jun 2024. https://doi.org/10.21606/drs.2024.269.
32. Simpson, Hunter J., et al. 'Global field collection data confirm an affinity of brown rot fungi for coniferous habitats and substrates.' *New Phytologist*, vol. 242, no. 6, Apr. 2024, pp. 2775–86. https://doi.org/10.1111/nph.19723.
33. Rigobello, Adrien, and Phil Ayres. 'Design strategies for mycelium-based composites.' *Fungi and Fungal Products in Human Welfare and Biotechnology*, 2023, pp. 605–35. https://doi.org/10.1007/978-981-19-8853-0_20.
34. Perez, Rolando, et al. 'Enabling community-based metrology for wood-degrading fungi.' *Fungal Biology and Biotechnology*, vol. 7, no. 1, Mar. 2020. https://doi.org/10.1186/s40694-020-00092-2.
35. Daly, Paul, et al. '*Dichomitus squalens* partially tailors its molecular responses to the composition of solid wood.' *Environmental Microbiology*, vol. 20, no. 11, Sep 2018, pp. 4141–56. https://doi.org/10.1111/1462-2920.14416.
36. Blanchette, Robert A., et al. 'Selection of white-rot fungi for biopulping.' *Biomass*, vol. 15, no. 2, Jan. 1988, pp. 93–101. https://doi.org/10.1016/0144-4565(88)90099-6.

37. Floudas, Dimitrios, et al. 'The Paleozoic origin of enzymatic lignin decomposition reconstructed from 31 fungal genomes.' *Science*, vol. 336, no. 6089, Jun 2012, pp. 1715–19. https://doi.org/10.1126/science.1221748.

38. Buzina, Walter, et al. 'The polypore mushroom *Irpex lacteus*, a new causative agent of fungal infections.' *Journal of Clinical Microbiology*, vol. 43, no. 4, Apr. 2005, pp. 2009–11. https://doi.org/10.1128/jcm.43.4.2009-2011.2005.

39. Bajwa, Dilpreet S., et al. 'Enhancement of termite (*Reticulitermes flavipes* L.) resistance in mycelium reinforced biofiber-composites.' *Industrial Crops and Products*, vol. 107, Jul 2017, pp. 420–6. https://doi.org/10.1016/j.indcrop.2017.06.032.

40. Korpi, Anne, et al. 'Microbial volatile organic compounds.' *Critical Reviews in Toxicology*, vol. 39, no. 2, Feb. 2009, pp. 139–93. https://doi.org/10.1080/10408440802291497.

41. Kabbaj, W., et al. 'Comparison of volatile compound production in fruit body and in mycelium of *Pleurotus ostreatus* identified by submerged and solid-state cultures.' *Applied Biochemistry and Biotechnology*, vol. 102–103, no. 1–6, Jan. 2002, pp. 463–70. https://doi.org/10.1385/abab:102-3:1-6:463.

42. Carlström, Vilhelm, et al. 'Modelling of morphogenesis to support the design of fungal-based engineered living materials.' *Research Directions Biotechnology Design*, vol. 2, Jan. 2024, https://doi.org/10.1017/btd.2024.10.

43. Saikkonen, K. 'Forest structure and fungal endophytes.' *Fungal Biology Reviews*, vol. 21, no. 2–3, May 2007, pp. 67–74. https://doi.org/10.1016/j.fbr.2007.05.001.

44. Elsacker, Elise, et al. 'Fungal engineered living materials: the viability of pure mycelium materials with self-healing functionalities.' *Advanced Functional Materials*, vol. 33, no. 29, Apr. 2023, https://doi.org/10.1002/adfm.202301875.

45. McBee, Ross M., et al. 'Engineering living and regenerative fungal–bacterial biocomposite structures.' *Nature Materials*, vol. 21, no. 4, Dec. 2021, pp. 471–8. https://doi.org/10.1038/s41563-021-01123-y.

CHAPTER 7

Myco-Lab Spaces: From Education to Industry

ENA NAITO, GABRIELA FARIAS, MANUEL ARIAS BARRANTES and BARBARA POLLINI

1 Introduction

1.1 Designing with mycelium

With the rising environmental crisis, the interest in biotechnologies is increasing. Mycelium, among other living organisms, is gaining a lot of attention as an emerging material in research, education and industry as a sustainable alternative to petroleum-based materials [7, 8] and animal leather [9]. Particular characteristics of mycelium, such as fast growth [7], light weight [10] heat resistance [11], biodegradability [12] and insulating properties [13], make this organism relevant for the development of sustainable solutions for greener architecture and circular design products [14, 15].

Currently, experimentation with mycelium material production has been conducted in a variety of settings, from scientific laboratories (e.g., biology laboratories) and industrial laboratories to design studio environments (e.g., design workshops or design studios) and community laboratories (e.g., biohacking or DIY biolabs), or a combination of these infrastructures [16, 17, 18, 19]. The number of opportunities to work with mycelium has grown significantly in the past years [16]; nevertheless, further development and knowledge are still needed as the field is still in its infancy.

In this chapter, we provide infrastructural guidance for individuals interested in incorporating mycelium into their design and/or teaching practices. In particular, the chapter draws on Aalto University's infrastructures and experiences in establishing a basic laboratory for working and experimenting with mycelium using accessible tools as well as specialized equipment. Our goal with this chapter is to support and inspire other communities interested in mycelium-making, offering insights into the processes, challenges and opportunities encountered through our own work and experiences. Furthermore, we aim to foreground the importance of collaboration between designers, researchers and industry partners.

1.2 Context: Aalto University's biodesign infrastructures

Currently, Aalto University has a network of infrastructures to support interdisciplinary research across art, design and science, especially for biodesign, material design and bioart activities. Students and researchers can benefit from such a diversity of places and

Ena Naito, Gabriela Farias, Manuel Arias Barrantes and Barbara Pollini, 'Myco-Lab Spaces: From Education to Industry' in: *Mycology for Architecture: Methods and Applications of Mycelium Materials in the Built Environment*. Pelagic Publishing (2026).
DOI: 10.53061/UDJR5356

settings, which can support multiple approaches in biodesign research and education (Figure 7.1).

CHEMARTS – a long-term collaboration between the School of Chemical Engineering (CHEM) and the School of Arts, Design and Architecture (ARTS) at Aalto University – started in 2011. A dedicated space, the CHEMARTS Lab, was established in 2017. The partnership aims to explore bio-based materials with an interdisciplinary approach for sustainable innovation [20]. The main goal is to inspire students and researchers to collaborate on bio-based materials and develop new, sustainable design concepts for their future applications.

Biofilia – the Base for Biological Arts – is a bioart unit and the first lab space created at Aalto University under ARTS, launched in 2012. This is a biosafety level one (BSL-1) lab dedicated to biology exploration in art and design. This space provides a distinctive platform for interdisciplinary research and education to foster cultural discussions and innovations on topics related to the manipulation of life and biological processes, considering both practical and theoretical aspects, as well as philosophical and ethical dimensions.

The latest established infrastructure is the BioMakerStudio, opened in 2024, which combines biodesign with elements of maker-space and design studio approaches. As a low-threshold biofabrication infrastructure, students, researchers and communities are able to use hands-on prototyping methods to work with growing materials and living systems. Importantly, the BioMakerStudio provides space for longitudinal and exploratory project-based work in biodesign.

Figure 7.1 Biodesign infrastructures at Aalto University: CHEMARTS Lab, Biofilia, BioMakerStudio. Image credits: Mikko Raskinen, Ena Naito.

These three workspaces nourish different approaches in biodesign that intersect and complement each other. For instance, in terms of mycelium design, it is possible to inoculate with a laminar flow hood in Biofilia, to incubate and test new fabrication techniques at the BioMakerStudio and to focus more on systematic biomaterials experiments in CHEMARTS. The university environment for material research is made even more lively by the proximity to VTT, one of Europe's leading research institutions (state-owned and operating under the guidance of the Finnish Ministry of Economic Affairs and Employment), with the aim of advancing the utilization and commercialization of research and new technologies.

2 Basics of a myco-lab

2.1 Mycelium composites in a myco-lab

Regardless of whether it is a DIY lab in a repurposed space, a professional lab or something in between, there are fundamental infrastructural requirements to support mycelium explorations. These conditions can be achieved through different strategies that align with the (physical, social and financial) resources and goals of each space [17].

In this section, we focus on the setting of the BioMakerStudio at Aalto University as a bottom-up access space to invite low-threshold mycelium experiments. We describe the equipment and spaces necessary for the creation of mycelium-based composites (MBCs), describing in parallel how to organize and manage the workspace for this activity.

In entry-level myco-labs, it is often more practical to begin with mycelial spawns for MBCs rather than liquid cultures. This approach is more cost-effective and provides an accessible introduction to aseptic techniques for substrate preparation. Moreover, when mycelium is removed from its natural environment and grown in a lab environment, it requires a level of hygiene and security standards to avoid mycelium competing with microbial contaminants [21].

Although there are various techniques and methods, the process of growing MBCs generally follows a set of fundamental steps. The process of cultivation takes place in three stages: inoculation, growth and ceasing [22]. Thanks to open-source equipment, low-tech methods have been developed to make growing MBCs available outside professional lab environments [23]. Even if one does not have access to professional lab space, one can gather tools from home, local hardware stores and grocery stores to build a DIY lab [24].

To start working with fungi, the mycelium can be obtained in different ways:

- Commercial mycelium strain on colonized substrates such as grow kits (e.g., Ecovative's Grow.bio kit).[1]
- Commercial mycelium spawn or liquid cultures (e.g., Helsieni is a resource in Finland; popular suppliers in Europe include Glucksplize,[2] and Mycelia).[3]

---

1 Some companies provide pre-inoculated substrate kits for DIY experiments with mycelium. One example is Ecovative's Grow.bio kits. https://grow.bio/products/grow-it-yourself-material

2 https://gluckspilze.com/home.

3 Mycelium spawns are obtained from a mushroom culture placed on grain; once it has completely grown on these grains, it can be further used to 'seed' mushroom compost or other solid substrates. Helsieni starter kit is an example of mycelium spawn. https://holvi.com/shop/Helsieni/product/279be523ab164f2b975fe0251b9ce55a/

Spawn also exists in liquid mediums, namely a suspension of living mycelial hyphae floating in an enriched medium. This suspension can be used for further multiplication, both in solid and liquid states. Several companies that sell mycelium strains offer this alternative, for example Mycelia. https://mycelia.be

- Isolation from live mushrooms. For instance, it is possible to sample commercial and wild species by taking a small piece of mushroom and transferring it to a petri dish and following the process outlined for cultivation [25].[4]

Previous research has identified certain species that perform better in MBCs [26, 27]. These include:

- *Ganoderma lucidum* – commonly known as Reishi – grows quickly, is more resistant to contamination and offers interesting coloration and texture.
- *Pleurotus ostreatus* – commonly known as Oyster mushroom – common fungal species in the food sector, but known also for the higher compressive strength [28]. This substrate is often available from local mushroom farmers when they can no longer be used for fruiting. However, since it is available when it is already older, there is a chance it is less reliable for making strong composites.
- *Trametes versicolor* – commonly known as Turkey tail – is a strong candidate for composites owing to its dense, fibrous growth and because of its wide accessibility in North America, Europe and Asia [25, 29].

When planning a myco-lab space, it is essential to consider the workflows of prototyping with mycelium (see Figure 2). Designated zones aligning with the sequence of workflows not only support the effective cultivation of mycelium but also create a more intuitive and user-friendly environment for new learners. In addition, a well-organized layout enhances safety. It is recommended to make any necessary signage early on in establishing the lab space to identify differentiation between shared space, 'messy' space and what needs to be kept as clean as possible between use times. Moreover, large workbenches create versatile spaces to accommodate a variety of uses, including workshops, classes and individual projects [17].

2.2 Fundamental workflow steps

In this section, we present the requirements for establishing a low-threshold myco-lab based on the workflows and the infrastructure of the BioMakerStudio, proposing a lab design according to necessary fundamental steps:[5] I. Preparation Zone, II. Grow Zone, III. Dry Zone (Figure 7.2).

2.2.1 Preparation zone (inoculation): preparing mycelium culture and substrate

This zone is for creating (semi-)aseptic conditions, crucial for preparing the grain spawns and substrates, inoculating with the mycelium strain and transferring them into a pre-designed mould. Maintaining sterility prevents unwanted contamination by competing microorganisms that may disrupt the cultivation process.

Maintaining an aseptic environment is necessary during two specific steps: (1) when sterilizing media (Petri dishes, grain spawn, substrate) using an autoclave and (2) inoculating the media with mycelium, using a laminar air flow hood (LAF hood). For

4 See Open Access sources, such as https://lamycosphere.com/en-pb/blogs/the-future-is-fungi/cloning-mushrooms-advance-guide-for-mycological-growers?srsltid=AfmBOoq6JBfFOB31OrjeNDC6ycbomYCAFsCbX3baC7qKOhAEBDjo1Lw8.

5 Adapted from Open Access BioFabLab Concept by CoCoon Consortium, see https://www.cocoon.bio/_files/ugd/f434f2_9e7201c32ae44f6488130d6ff0aaac2d.pdf.

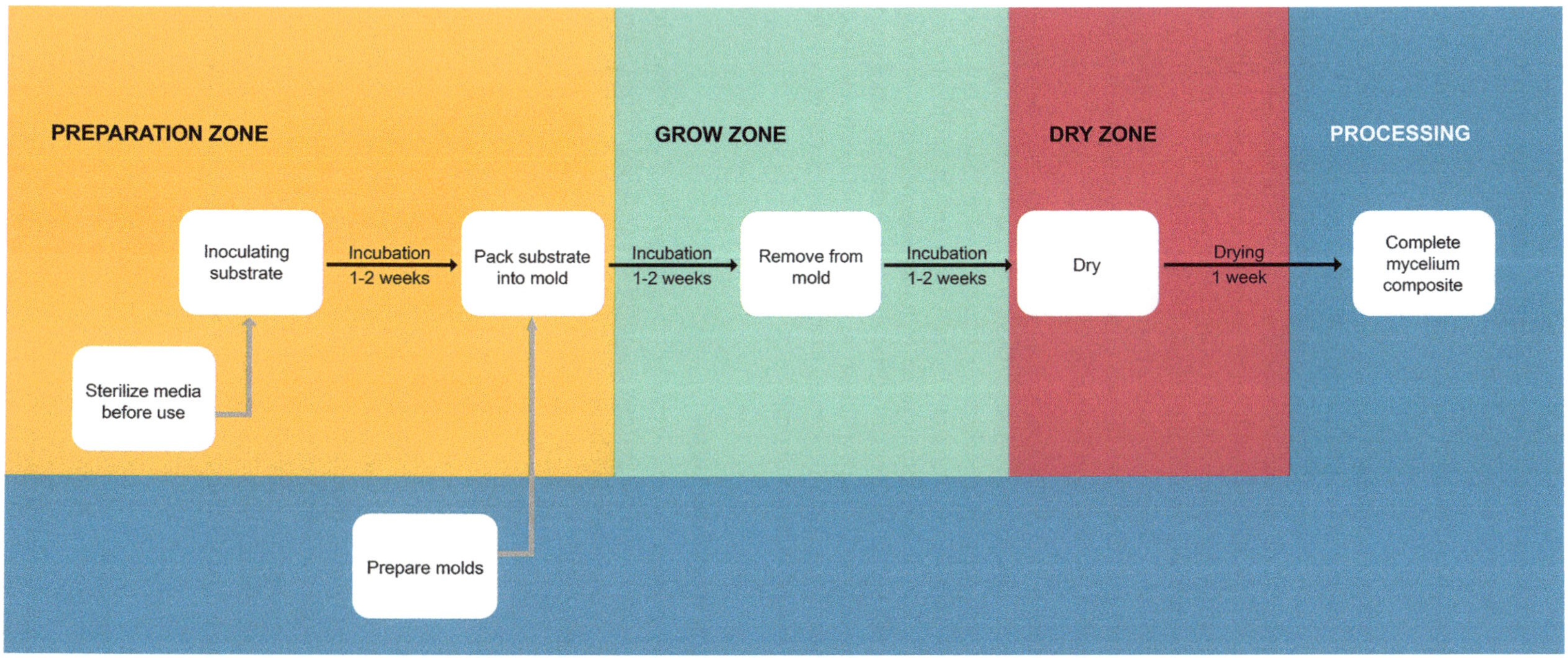

Figure 7.2 The zones align with fundamental steps of making mycelium composites.

Table 7.1 Items for professional versus low-tech lab

ZONE	Step in process	Professional standard	Low-tech alternative
Preparation Zone	Sterilizing media	Autoclave	Pressure cooker
	Inoculating media	LAF	SAB
	Sterilizing worksurfaces	Certified ethanol	Isopropyl alcohol
Growing Zone	Incubating mycelium	Incubator	Home incubation tent or mini-greenhouse
Drying Zone	Drying mycelium	Low-temperature oven	At home oven or mini-greenhouse for drying
Storage	Storing mycelial strains	Freezer (capacity for -70°C to -80°C) for best preservation	Home freezer, mini-fridge, insulated box for temporary preservation

starting a professional lab, these two steps are essential and the corresponding tools – especially an autoclave and the LAF hood – are critical investments. However, in DIY lab set-ups, viable alternatives exist (see Table 7.1).

Professional labs typically use 70% ethanol to eliminate surface bacteria [30], but certified ethanol is often limited to professional settings. DIY labs can substitute with 70% isopropyl alcohol. This is crucial for clearing working surfaces before and after working, as well as any pre-designed moulds and containers.

For sterilizing media in professional lab environments, autoclaves are the standard. They efficiently reach temperatures above 100°C, which kill competing organisms in substrates while preserving nutrients. To effectively treat substrates for mycelium, an autoclave must reach 121°C for 15 minutes at the core [31]. If a professional autoclave is unavailable, a pressure cooker can serve as a practical alternative for sterilizing media. Pressure cookers are far more accessible and can also treat substrates, requiring 15 psi/1 bar to achieve the same results as an autoclave [31].

Once the substrate has cooled after sterilization, it is mixed with mycelium grain spawn in a sterile environment. The grain spawn contains small amounts of living mycelium, which will colonize the substrate over time.

To inoculate media, a LAF hood creates a sterile workspace with continuous flow of air through high-efficiency particulate air (HEPA) filters, which prevents airborne contaminants from entering the culture. However, a LAF hood can be expensive and requires installation directly into the built infrastructure. One DIY option is to build still air boxes (SAB) – an enclosed workspace that relies on undisturbed air to minimize contamination risks [30] (see Figure 7.3). A SAB is typically made from a clear storage container with holes cut in the side for access to work in the enclosed space. In addition to the SAB, disposable gloves, spray bottles, paper towels, isopropyl alcohol and masks are recommended to create the clean environment required for working with mycelium in this stage of cultivation.

2.2.2 Grow zone (growth): letting mycelium incubate under optimal conditions

Mycelium requires certain environmental conditions to grow. Although this varies across the strains, most species thrive at around 25–30°C, humidity 60–65% [32]. The period of growth spans from a week to several months [32], depending on varying factors such as the species, the substrates, the preparation methods and the environmental conditions.

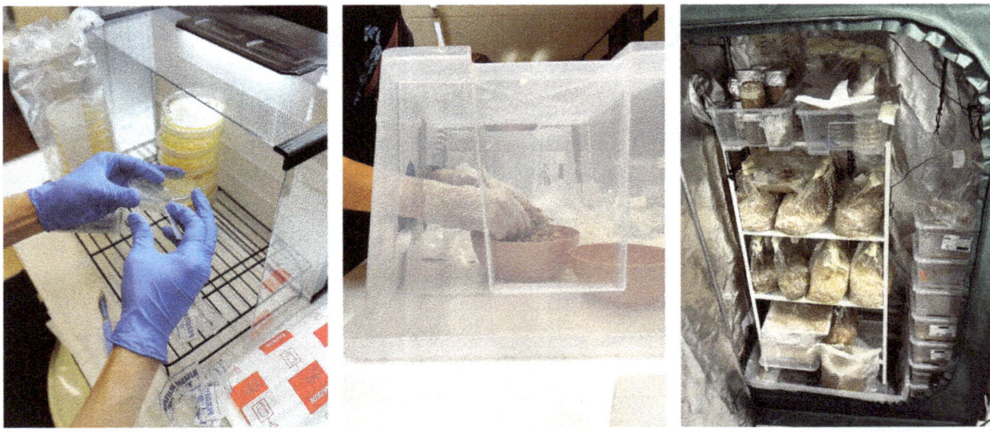

Figure 7.3. (L–R) Small LAF hood, SAB, Mycelium grow tent. Image credits: Ena Naito and Gabriela Farias.

Digital incubators with programmable variables (temperature, humidity, CO_2 levels, light conditions, etc.) allow maintaining more precise and controllable growing environments [17]. Similar conditions can also be created in grow tents or mini-greenhouses [33] (see Figure 7.3).

The time required to cultivate a substrate mature enough for composite making can vary widely. It depends on several factors, including the proportion of spawn used during inoculation – using 3% versus 10% spawn results in significantly different numbers of inoculation points for the strain to colonize [34]. Temperature is another key factor; for example, *Ganoderma lucidum* thrives at 25–28°C, while *Pleurotus ostreatus* prefers slightly cooler conditions [35]. Additionally, keeping the grow tent dark encourages the mycelium to grow rather than fruit.

2.2.3 Dry zone (ceasing): stopping the growth of mycelium

Once the substrate has been fully colonized by mycelium, it can be taken out of the mould and dried. Studies have shown that drying mycelium at lower temperatures causes less shrinkage and warping, whereas higher temperatures can cause more, as the rate of moisture evaporation also increases [36]. A dehydrator (or an oven) set between 60°C and 80°C continuously for 12–48 hours can be used for drying the mycelium composite, with 60°C resulting in better final results [36].

2.3 Further considerations: storing, processing and disposal

In addition to these workflow zones, myco-labs require sufficient and suitable storage spaces. To store mycelium by slowing down its growth, a fridge can be used for liquid cultures and pre-grown cultures on petri dishes [37]. For longer-term storage, mycelial strains are best preserved in a -70°C to -80°C freezer [38, 39]. The use of glycerol can aid in long-term storage acting as a cryoprotectant and inhibiting the development of ice crystals that can harm fungal cells during the freezing and thawing process, thus preserving their viability [40, 41].

Additional tools for pre- and post-processing of mycelium composites can also be provided. These may include 3D bio (or clay) printers (e.g., 22, 42), a laser cutter (e.g., 43, 44), electronic workstations (e.g., 43) and mould-making tools and supplies (e.g., 45).

Once physical samples are obtained, mechanical testing instruments are useful for evaluating, comparing and optimizing them, considering, for example, tensile strength, elasticity and ductility. Additionally, different types of microscopes can cater to a wide range of uses. For instance, basic stereo microscopes are useful for observing surface textures of mycelium samples, whereas scanning electron microscopes allow for microstructural analysis [46].

Another crucial aspect to consider in the production and post-production phases is the safe disposal of biowaste. If the strain of mycelium used is not native to the specific environment, introducing non-native mycorrhizal fungi can become invasive rather than beneficial [47]. For non-local strains, it is recommended that the mycelium is autoclaved before disposal. Consequently, more research is needed to understand how large-scale recycling of mycelium bioproducts might impact the soils they are intended to enrich [47, 48].

3 From education to collaboration with industry

3.1 Working in a scientific lab

Biodesign education plays a fundamental role in shaping future designers. These hands-on biofabrication spaces allow students to learn about the basics of growing mycelium-based materials, collaborate with peers and become familiar with commonly used tools and equipment. Beyond educational settings, institutions such as VTT have seen the advantage of collaborating with students from Aalto University and the importance of opening career paths for designers in the field.

The role of VTT as a government-owned and not-for-profit institution is to promote the wide-ranging utilization and commercialization of research in several areas, including biotechnology. VTT works together with industry partners or other external stakeholders to commercialize innovations. This institution also develops its own technologies for commercialization by other companies or provides researchers with in-house incubation programs to spin-out these technologies into actual start-ups. Some of VTT's most successful spin-outs are Solar Foods, Spinnova, Infinited Fiber and Onego Bio.

Collaboration at VTT takes place in the form of projects and different research activities. For those working towards the start-up route, the activities differ as they might be more focused on investment and the creation of a business. The projects are usually funded via customer contracts, academic grants and European Union programmes (e.g., Horizon Europe, European Defence Fund, etc.). All team members usually have specific roles, and other specialists are commonly involved depending on findings and project needs. The role of design in these projects is still relatively small, but owing to previous collaboration with designers in past years, some design methods and processes such as workshops for new ideas or potential applications, material hands-on experimentation in the lab and prototyping have become more common for material-related projects.

In terms of infrastructure and equipment for mycelium research, designers usually have access to the laboratories to conduct experiments and have an active role in meetings and development decisions. However, specialized tools to study fungal growth at cellular and molecular level, as well as mechanical testing devices to assess material strength, flexibility and durability, are usually handled by scientists or internal collaborators who are specialized in those areas.

3.2 Advantages of design in scientific environments

There are various advantages of inviting designers to collaborate with scientists in the context of mycelium-based materials [49]. The collaboration described in this section is based on the process known as submerged liquid fermentation,[6] a method commonly used for making food and beverages but becoming popular for material making as it allows the transition from lab scale to industrial settings [50]. Based on previous experiences from the collaboration between scientists and designers at VTT, it has been learned that inviting designers to collaborate with scientists can help to advance material research and development especially in: (1) appearance, (2) prototyping, (3) testing scalability and commercialization and (4) assessing sustainability.

1. Appearance: The material's feasibility, usability and desirability can be explored alongside technical developments [51]. For example, for the making of mycelium non-woven fabrics at VTT, design knowhow has helped to extend the list of bio-based additives that could result in interesting material properties, explore bio-based dyes, apply surface finishings and incorporate design tools to laser cut and engrave on the material (Figure 7.4).
2. Prototyping: In scientific environments, prototypes make ideas more concrete and easier to discuss, as they make scientific research tangible and place it into a specific context. Prototyping also facilitates rapid testing and iteration which help identify potential bottlenecks at early stages of any material development [16]. For example, the creation of prototypes at VTT helps to explore potential uses for mycelium-based leather-like materials, identify material development needs and showcase the work at different conferences, museums and design events worldwide (Figure 7.5).
3. Testing scalability and commercialization: Design can help to test scalability and explore different paths for commercialization. For example, involving designers in different phases of material research and development allows them to understand better equipment and processes used for industrial production [50]. As limited bio-based material supply can challenge prototyping and mass production, designers can help to unlock new ideas and become familiar with limitations when prototyping and thinking of potential applications. VTT's research on the continuous production of non-woven mycelium fabrics highlights the importance of this type of collaboration (Figure 7.6).

Figure 7.4 Mycelium-based leather-like material samples. Image credits: Manuel Arias Barrantes

6 Submerged liquid fermentation is the use of bioreactors for synthesizing mycelium in a nutrient-rich liquid medium to and thus obtain as a final result mycelium pulp for further processing.

Figure 7.5 Prototype created for Dutch Design Week 2024 using mycelium-based leather-like material samples. Image credit: Manuel Arias Barrantes

Figure 7.6 Infrastructure and equipment for the production of mycelium-based leather-like materials at VTT Technical Research Centre of Finland. Image credits: Manuel Arias Barrantes.

It is important to note that under this type of scientific setting, creativity and design operate within the boundaries of technical and commercial feasibility.

4. Assessing sustainability: Designers can help to assess sustainability. For example, designers can work with scientists to establish circular economy models within a product's lifecycle alongside experts. This involves assessing how materials can be manufactured, reused and recycled (e.g., lifecycle assessments). This collaboration can result in the creation of unique business models for the future.

4 Conclusion

Each setting presented in this chapter empowers the designer in distinct ways to reveal the diverse potentials of working with mycelium. Myco-labs can materialize in different forms – ranging from traditional laboratory settings to classroom environments and even outdoor spaces 'in the wild'. The designs of physical spaces will inevitably shape the kinds of projects, practices and people that are supported. For instance, low-threshold spaces, such as the BioMakerStudio, enable entry-level access to biodesign work (i.e., to understand basic tools and how mycelium composites are grown), support open-ended inquiry and allow freedom for design exploration. In contrast, industry settings such as VTT prioritize precise and systematic experimentation, target innovations and optimize the efficiency of material production. In industry, designers contribute to exploring a material's feasibility, usability and desirability alongside technical developments. Providing both scientific and design disciplines with the opportunity to influence material development from an early stage can drive effective innovation and/or commercialization efforts and better user acceptance. As academic, industrial and community interest in mycelium continues to grow, so does the need for infrastructure to support its research, development and access. This makes it crucial to expand the number of such spaces and diversify the spaces that support mycelium exploration.

Acknowledgements

Special thanks go to all the professors, researchers and technicians who support the unique Aalto infrastructure and make biodesign education and research possible. We thank professors Pirjo Kääriäinen and Tapani Vuorinen for pioneering interdisciplinary material design research with the project CHEMARTS; Biofilia's laboratory manager Larisa Chernyaeva for her passion and invaluable guidance in the lab; the *CoCoon – Co-Creating Greener Futures* (EU grant no. 101087204) and team members for supporting the continued development of the BioMakerStudio; The Centre of Excellence in Life-Inspired Hybrid Materials (project number 346106), for building a community for both designers and scientists to interact and share knowledge, the Finnish Cultural Foundation and the Center for Young Synbio Scientists for supporting doctoral research and VTT Technical Research Centre of Finland and Senior Scientist Géza Szilvay and Professor Merja Penttilä for their appreciation and support to designers working in scientific environments for the industry.

We finally acknowledge all the students, researchers and communities who work in and pass by these spaces, enriching and contributing to biodesign research and knowledge at Aalto University.

References

1. Camilleri E, Narayan S, Lingam D, Blundell R. Mycelium-based composites: An updated comprehensive overview. Biotechnology Advances. 2025 Mar 1;79:108517.
2. Albuhamad AMA. Introduction to Spectrophotometry [internet]. Available from: https://www.researchgate.net/publication/377237891_Introduction_to_Spectrophotometry
3. Agilent. Agilent. n.d. Liquid Chromatography/Mass Spectrometry Fundamentals. Available from: https://www.agilent.com/en/product/liquid-chromatography-mass-spectrometry-lc-ms/lcms-fundamentals
4. Brandt E, Binder T. Experimental design research: International association of societies of design research 2007: emerging trends in design. In Hong Kong; 2007.
5. Cross N, editor. Designerly Ways of Knowing. In: Designerly Ways of Knowing [internet]. London: Springer; 2006. pp. 1–13. Available from: https://doi.org/10.1007/1-84628-301-9_1
6. Jönsson L, Mäkelä M, Hansen F, Ahde-Deal P. Investigating design research landscapes through exhibitions. In: NODEM 2013 / International Conference on Design and Digital Heritage Stockholm, Sweden, 1-4122013 [internet]. Interactive Institute; 2013. pp. 113–20. Available from: https://research.aalto.fi/en/publications/investigating-design-research-landscapes-through-exhibitions

7. Angelova GV, Brazkova MS, Krastanov AI. Renewable mycelium based composite – sustainable approach for lignocellulose waste recovery and alternative to synthetic materials – a review. Zeitschrift für Naturforschung C. 2021 Nov 1;76(11–12):431–42.
8. Mojumdar A, Behera HT, Ray L. Mushroom Mycelia-Based Material: An Environmental Friendly Alternative to Synthetic Packaging. In: Vaishnav A, Choudhary DK, editors. Microbial Polymers: Applications and Ecological Perspectives [internet]. Singapore: Springer; 2021. pp. 131–41. Available from: https://doi.org/10.1007/978-981-16-0045-6_6
9. Amobonye A, Lalung J, Awasthi MK, Pillai S. Fungal mycelium as leather alternative: A sustainable biogenic material for the fashion industry. Sustainable Materials and Technologies. 2023 Dec 1;38:e00724.
10. Zhang M, Zhang Z, Zhang R, Peng Y, Wang M, Cao J. Lightweight, thermal insulation, hydrophobic mycelium composites with hierarchical porous structure: Design, manufacture and applications. Composites Part B: Engineering. 2023 Nov 1;266:111003.
11. Jones M, Mautner A, Luenco S, Bismarck A, John S. Engineered mycelium composite construction materials from fungal biorefineries: A critical review. Materials & Design. 2020 Feb 1;187:108397.
12. Wylick AV, Elsacker E, Yap LL, Peeters E, Laet L de. Mycelium Composites and their Biodegradability: An Exploration on the Disintegration of Mycelium-Based Materials in Soil. Construction Technologies and Architecture. 2022;1:652–9.
13. Zhang X, Hu J, Fan X, Yu X. Naturally grown mycelium-composite as sustainable building insulation materials. Journal of Cleaner Production. 2022 Mar 15;342:130784.
14. Attias N, Danai O, Abitbol T, Tarazi E, Ezov N, Pereman I, et al. Mycelium bio-composites in industrial design and architecture: Comparative review and experimental analysis. Journal of Cleaner Production. 2020 Feb 10;246:119037.
15. Elsacker E, Vandelook S, Brancart J, Peeters E, Laet LD. Mechanical, physical and chemical characterisation of mycelium-based composites with different types of lignocellulosic substrates. PLOS ONE. 2019 Jul 22;14(7):e0213954.
16. Ihls J, Pollini B. From observation to operation: the role of lab spaces in biodesign practice. Biotechnology Design. 2025 Jan;4:e2.
17. Naito E, Botero A. Growing Conditions for Growing: Infrastructures for Designing with Living Organisms in Fabrication Settings. In: Nordic Conference on Human-Computer Interaction [internet]. Uppsala Sweden: ACM; 2024. pp. 1–14. Available from: https://dl.acm.org/doi/10.1145/3679318.3685389
18. Pollini B. From biodesigners to designers in lab: testing the nuances of an emerging profession through autoethnography. Research Directions: Biotechnology Design. 2024 Jan;2:e20.
19. Naito E. Biodesigning with Living Organisms in Fablabs? Towards a Research Agenda. In: Mateus A, Leonor S, Paoliello C, editors. What's Around Design? Strategic and Speculative Biodesign For Sustainable Future. Springer Nature; 2026.
20. Kaarianen P, Tervinen L, Vuorinen T, Riutta N, Suorlahti E. The chemarts cookbook. Espoo, Finland: Aalto School of Art and Design; 2020. 192 p. (Aalto University publication series C).
21. Alaneme KK, Anaele JU, Oke TM, Kareem SA, Adediran M, Ajibuwa OA, et al. Mycelium based composites: A review of their bio-fabrication procedures, material properties and potential for green building and construction applications. Alexandria Engineering Journal. 2023 Nov 15;83:234–50.
22. Mohseni A, Vieira FR, Pecchia JA, Gürsoy B. Three-Dimensional Printing of Living Mycelium-Based Composites: Material Compositions, Workflows, and Ways to Mitigate Contamination. Biomimetics. 2023 Jun;8(2):257.
23. Parisi S, Ayala Garcia C, Rognoli V. Designing Materials Experiences through Passing of Time. Material-Driven Design Method applied to Mycelium-based Composites. 2016.
24. Meyer M. Domesticating and democratizing science: A geography of do-it-yourself biology. Journal of Material Culture. 2013 Jun;18(2):117–34.
25. Stamets P. Mycelium Running: How Mushrooms Can Help Save the World. Clarkson Potter/Ten Speed; 2005. 362 p.
26. Le Ferrand H. Critical review of mycelium-bound product development to identify barriers to entry and paths to overcome them. Journal of Cleaner Production. 2024 Apr 15;450:141859.
27. Sydor M, Cofta G, Doczekalska B, Bonenberg A. Fungi in Mycelium-Based Composites: Usage and Recommendations. Materials (Basel). 2022 Sep 9;15(18):6283.
28. Kohphaisansombat C, Jongpipitaporn Y, Laoratanakul P, Tantipaibulvut S, Euanorasetr J, Rungjindamai N, et al. Fabrication of mycelium (oyster mushroom)-based composites derived from spent coffee grounds with pineapple fibre reinforcement. Mycology. 2024 Oct 1;15(4):665–82.
29. Ajibola OO, Nolasco-Hipolito C, Carvajal-Zarrabal O, Salleh SF, Adeyinka GC, Adefegha SA, et al. Turkey tail mushroom (*Trametes versicolor*): an edible macrofungi with immense medicinal properties. Current Opinion in Food Science. 2024 Aug;58:101191.

30. Crawford A. Designer's Guide to Lab Practice. Taylor & Francis; 2023. 161 p.
31. Oei P. Mushroom Cultivation [internet]. 4th ed. ECO Consult Foundation; 2016. 520 p. Available from: https://mycotrop.com/nl/-magic-mushrooms/911-mushroom-cultivation-4th-edition-peter-oei.html
32. Alemu D, Tafesse M, Mondal AK. Mycelium-Based Composite: The Future Sustainable Biomaterial. International Journal of Biomaterials. 2022;2022(1):8401528.
33. Yang L, Xu R, Joardar A, Amponsah M, Sharifi N, Dong B, et al. Design and build a green tent environment for growing and charactering mycelium growth in lab. Lab Chip. 2023 Sep 13;23(18):4044–51.
34. Idowu OO, Kadiri M, Otunla CA. Influence of inoculation method and spawn level on biological efficiency of *Pleurotus ostreatus*. Journal of Applied Sciences and Environmental Management. 2016 Nov 2;20(3):542.
35. Schröder LMU, Wals AEJ, van Koppen CSA (Kris). Analysing the state of student participation in two Eco-Schools using Engeström's Second Generation Activity Systems Model. Environmental Education Research. 2020 Aug 2;26(8):1088–111.
36. Farrahnoor A, Hishamuddini ADN, Yusoff H, Hyie KM, Maideen NC, Fohimi NAM, et al. Influence of Drying Temperature in the Oven on Physical, Morphology and Mechanical Properties of Mycelium Composite. Pertanika Journal of Science & Technology [internet]. 2025 Jan;33(1). Available from: https://www.researchgate.net/publication/388403989_Influence_of_Drying_Temperature_in_the_Oven_on_Physical_Morphology_and_Mechanical_Properties_of_Mycelium_Composite
37. Ng W, Barati B, Karana E. Healing with Fungi: Unique Aesthetic Expressions for Mycelium-Based Materials Through Patch and Mend. In: Bruyns G, Wei H, editors. With Design: Reinventing Design Modes. Singapore: Springer Nature; 2022. pp. 3253–67.
38. Pasarell L, McGinnis MR. Viability of fungal cultures maintained at -70 degrees C. Journal of Clinical Microbiology. 1992 Apr;30(4):1000–4.
39. Zimberoff L. Technically Food: Inside Silicon Valley's Mission to Change What We Eat. Abrams; 2021. 179 p.
40. Kitamoto Y, Suzuki A, Shimada S, Yamanaka K. A new method for the preservation of fungus stock cultures by deep-freezing. Mycoscience. 2002;43(2):143–9.
41. Pasrija R, Kumari D. Handling, Culturing, Preservation and Revival of Fungal Pathogens. In: Pasrija R, Kumari D, editors. Protocols for Studying Pathogenic Fungi [internet]. Singapore: Springer Nature; 2025. pp. 27–52. Available from: https://doi.org/10.1007/978-981-96-7047-5_2
42. Luo D, Yang J, Peek N. 3D-Printed Mycelium Biocomposites: Method for 3D Printing and Growing Fungi-Based Composites. 3D Printing and Additive Manufacturing [internet]. 2025 Jan 23; Available from: https://www.liebertpub.com/doi/full/10.1089/3dp.2023.0342
43. Vasquez ESL, Vega K. From plastic to biomaterials: prototyping DIY electronics with mycelium. In: Adjunct Proceedings of the 2019 ACM International Joint Conference on Pervasive and Ubiquitous Computing and Proceedings of the 2019 ACM International Symposium on Wearable Computers [internet]. New York, NY, USA: Association for Computing Machinery; 2019. pp. 308–11. (UbiComp/ISWC '19 Adjunct). Available from: https://dl.acm.org/doi/10.1145/3341162.3343808
44. Vasquez ESL, Vega K. Myco-accessories: sustainable wearables with biodegradable materials. In: Proceedings of the 23rd International Symposium on Wearable Computers [internet]. London United Kingdom: ACM; 2019. pp. 306–11. Available from: https://dl.acm.org/doi/10.1145/3341163.3346938
45. Farias G. Giving sense to mycelium biomaterials through community-driven biodesign [internet] [Master's Thesis]. [Helsinki]: Aalto University; 2024. Available from: https://aaltodoc.aalto.fi/handle/123456789/132441
46. Crawford A, Miller SR, Branco S, Fletcher J, Stefanov D. Growing mycelium leather: a paste substrate approach with post-treatments. Research Directions: Biotechnology Design. 2024 Jan;2:e6.
47. Dickie IA, Bufford JL, Cobb RC, Desprez-Loustau ML, Grelet G, Hulme PE, et al. The emerging science of linked plant–fungal invasions. New Phytologist. 2017;215(4):1314–32.
48. Gan JK, Soh E, Saeidi N, Javadian A, Hebel DE, Le Ferrand H. Temporal characterization of biocycles of mycelium-bound composites made from bamboo and *Pleurotus ostreatus* for indoor usage. Sci Rep. 2022 Nov 12;12(1):19362.
49. Arias Barrantes M. Lab-grown futures: Design exploration for the development of fungi as a leather-like material [internet] [Master's Thesis]. [Helsinki]: Aalto University; 2019. Available from: https://aaltodoc.aalto.fi/handle/123456789/38464
50. Szilvay G, Arias Barrantes M. VTT. 2021. Mycelium leather: sustainable alternative for leather. Available from: https://www.vttresearch.com/en/news-and-ideas/alternative-leather-and-synthetic-leather-vtt-succeeded-demonstrating-continuous
51. Karana E, Barati B, Rognoli V, Zeeuw van der Laan A. Material Driven Design (MDD): A Method to Design for Material Experiences. International Journal of Design [internet]. 2015;9(2). Available from: http://www.ijdesign.org/index.php/IJDesign/article/view/1965

CHAPTER 8

Challenges of Facilitating Workshops with Mycelium-Based Materials for Architecture and Design Students

JESSICA DIAS

1 Introduction

Mycelium-based materials are gaining significant attention in architecture and design owing to their sustainability, biodegradability and potential as an alternative to conventional materials. As interest in mycelium grows, institutions, maker spaces and laboratories must find ways to educate students and professionals on how to collaborate with these living systems safely, creatively and mindfully.

Unlike static materials such as wood, metal and plastic, mycelium is a dynamic, living organism that requires specific environmental conditions to grow and thrive. Working with it from an early cultivation stage demands specialized tools and equipment, resources that many design institutions are not yet equipped to provide. This presents unique pedagogical challenges, as it requires both scientific knowledge and tools and a space for creative experimentation.

However, existing educational settings often fall short of meeting these needs. Design and art studios foster creativity but typically lack the sterility and structure necessary for advanced cultivation. Traditional laboratories offer controlled environments yet restrict the organic, exploratory nature of the design process. The ideal space lies somewhere in between, a hybrid environment that balances scientific precision with creative flexibility.

Equally important in an educational context is framing our relationship with other living organisms. Beyond practical tools, encouraging discussions around ethics, coexistence and the responsibilities of working with biological materials can provide a strong foundation for meaningful engagement with mycelium for design.

As an architecture student at the Institute of Advanced Architecture of Catalunya, I explored ways to refine the aesthetic qualities of mycelium composites for my final thesis project (Figures 8.1 and 8.2). This experience introduced me to the challenges of working with living materials, particularly with limited resources and no formal biological training. Through years of self-teaching, experimentation and transdisciplinary collaboration, I have advanced my practice and now host workshops and seminars for those eager to explore mycelium-based materials.

Jessica Dias, 'Challenges of Facilitating Workshops with Mycelium-Based Materials for Architecture and Design Students' in: *Mycology for Architecture: Methods and Applications of Mycelium Materials in the Built Environment*. Pelagic Publishing (2026). © Jessica Dias. DOI: 10.53061/JNYV2056

Figure 8.1 Render of 3D printed scaffold, inspired by mushroom gills and mycelium branching. Project: Hyperarticulated mycomorph – Jessica Dias 2017, thesis advisor: Marcos Cruz, Institute of Advanced Architecture of Catalonia.

Figure 8.2 3D-printed cross section of scaffold inoculated with *P. ostreatus*. Image credit: Jessica Dias.

This chapter reflects on my journey as both a practitioner and an educator, examining the challenges of workspace sterility, the ethics of interspecies design and the constraints of short-term educational formats, alongside the unpredictable nature of biological processes. I discuss the barriers that highlight the need for innovative, self-organizing learning experiences that embrace the complexities of working with living systems, while also sharing the solutions I have discovered along the way.

2 Introducing a philosophical and ethical framework for working with living materials

Working with mycelium is rooted in a planet-centric design philosophy, where it is crucial to prioritize the well-being of not only people but also the environment and the organisms we collaborate with.

Introducing this concept to undergraduate and Master's students early on by encouraging them to reflect on key questions can help establish a foundation for a mutually beneficial relationship with mycelium. Unlike bioplastics with agar or gelatine bases, mycelium is a living organism that is constantly growing. Its behaviour unfolds over time, meaning its aesthetic and functional properties are shaped by its environment. While some biomaterials, such as agar or gelatine-based bioplastics, offer instant results, mycelium requires patience.

It is important to recognize that while we use this living system to develop more sustainable materials, we are still, to some extent, controlling it for our own benefit. Acknowledging this dynamic can serve as a reminder that these organisms are alive, with their own needs and natural behaviours. Encouraging students to ask questions – such as what it means to be part of a living mycelial network and about what its habitats and needs are – can deepen their understanding of these complex relationships.

Providing a nurturing space of care and coexistence for these organisms is essential. Mycelium senses its surroundings and responds to the energy of its collaborator. Learning to support and care for these organisms fosters a deeper connection with and an empathy towards the natural world.

As designers and makers, we must also consider whether our work supports biodiversity, enriches or harms the natural environment and remains truly sustainable beyond the final outcome. Even when a project aims to provide a sustainable solution, the process itself may involve significant environmental costs, such as single-use plastics and toxic waste. How can we minimize these impacts within our practice?

There are no definitive answers to these questions. However, when introducing living systems to new practitioners, these considerations help emphasize that we are working with something alive, dynamic and with its own agenda. We are merely guiding these organisms in a way that supports a more sustainable future – they are the ones doing the real work.

In an educational setting, such as a workshop or seminar that introduces these concepts to students, it can be useful to provide further reading to help deepen their understanding of these complex relationships and establish a strong theoretical foundation for future practice. Such readings can include *Mycelium Running* by Paul Stamets, a mycology bible with sound research and data, and *Let's Become Fungal!* by Yasmine Ostendorf-Rodríguez for deeper understanding of Indigenous use of fungi.

3 A brief overview of mycelium-based practices in an educational context

3.1 The emergence of mycelium in design education

The use of mycelium and mushroom-based materials spans deep into Indigenous communities worldwide that have long harnessed fungi for various purposes – from textiles and fire starters to medicinal applications. However, the modern application of mycelium as a sustainable material emerged around 2007 when pioneers such as Ecovative and MycoWorks began developing mycelium composites for packaging and textiles.

Today, these materials are being explored globally by artists, architects, fashion designers and makers. While some engage with mycelium in collaborative research settings, others take a more hands-on approach, learning through trial and error. University students, in particular, are fascinated by the potential of this material, for example as a replacement for synthetic plastics, toxic construction materials and leather-like textiles, yet many enter these courses with little knowledge of biological practices. As an emerging field, resources remain limited as there are few formal textbooks or structured training programmes dedicated specifically to mycelium as a design material.

While academic studies and scientific papers provide a strong foundation and in-depth knowledge, the lack of practical, hands-on training makes it difficult for students to grasp the nuances of working with living organisms. Although various online courses are available, they often focus on mushroom cultivation for consumption rather than exploring mycelium as a material, leaving a gap in design-based applications.

New Master's and design programmes focused on sustainable materials are emerging globally, often incorporating mycelium through short seminars or one-day workshops. While some institutions, such as Central Saint Martins, London (MA Biodesign), and ELISAVA, Barcelona, are developing spaces tailored to working with living organisms and biomaterials, many still lack the necessary facilities to support mycelium cultivation. With multiple students sharing the same workspace, the risk of cross-contamination increases. This highlights the need for a globally accessible foundation of knowledge and tools to standardize mycelium-based design education. Institutions can begin adapting their existing infrastructure at relatively low cost to accommodate these materials more effectively.

3.2 Adapting existing infrastructure for mycelium cultivation

Many students and makers assume that working with mycelium requires laboratory-grade equipment. However, simple tools, such as a Bunsen burner, scalpel or DIY glove box, can be sufficient for beginners. Sterility remains a primary challenge, as mycelium is highly susceptible to contamination. While professional laboratories use laminar flow hoods and high-efficiency particulate air (HEPA) filtered environments, students can minimize contamination risks through aseptic techniques such as sterilizing tools with alcohol and flame, working quickly to reduce exposure and handling cultures in controlled spaces.

For an existing space to meet the basic requirements for working with mycelium, there should be a non-porous workspace, ideally stainless steel or plastic, an enclosed space with minimal airflow and minimal amount of porous surfaces. This may not always be easy to attain, in which case assembling a glove box or DIY flow hood (inexpensive) can be an easy and interactive way to engage students and teach them the importance

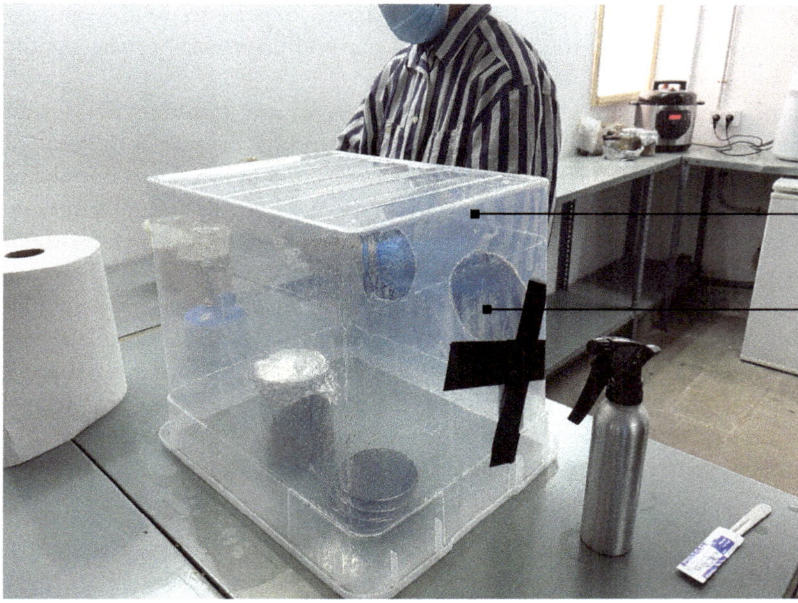

Figure 8.3 DIY still air box / glove box, Image credit: Jessica Dias.

of sterile protocols (Figure 8.3). Other essential tools and equipment would be a fridge for storing materials, a pressure cooker or autoclave, for sterilizing media and tools, and some basic laboratory glassware such as beakers and bottles. If a laboratory incubator is unattainable, then a grow tent with temperature controls such as heat mats, is a great starting point to ensure stable growth of mycelium.

4 Educating with various mycelium-based materials

4.1 Composites

A good entry point for working with mycelium materials is mycelium composites, which typically consist of a cellulose-rich substrate combined with a strong mycelium strain such as *Ganoderma* or *Trametes versicolor*. Students can explore how mycelium responds to different substrates and how composite properties vary depending on the materials used. Encouraging students to experiment with local waste streams, such as spent coffee grounds, dehydrated food waste or sawdust, to connect industrial waste to material generation, can be a great way to promote the development of a circular economy – which is often considered a crucial aspect of sustainable design. Mycelium-based composites are also adaptable: they can be extruded, moulded, Computer Numerical Control (CNC) machined and scaled up, making them a great material for architecture and design students to realize their projects, especially with access to digital fabrication tools.

Short workshops and seminars help provide essential knowledge that can be expanded through independent experimentation. While they cannot cover the vast potential of mycelium-based materials in just a few hours, they offer an introduction that sparks further exploration. These sessions typically utilize pre-grown mycelium substrates, such as hemp combined with *Ganoderma lucidum*, which can be sourced from suppliers such as Grown.bio or prepared in advance by facilitators. Using pre-grown materials allows for a more interactive and fast-paced learning experience while ensuring successful growth even in non-sterile conditions, making mycelium workshops accessible beyond laboratory settings.

Simplified solutions enable students to work with mycelium materials for design without requiring extensive tools and equipment. Pre-grown kits, originally developed by Ecovative to allow makers and myco-enthusiasts to experiment with the material and develop new applications, are now offered by various companies with different substrate options. They are easy to use and require only basic aseptic techniques, making them ideal for introductory workshops.

Alternatively, grain spawn for species such as *Ganoderma lucidum*, *Trametes versicolor* and *Pleurotus* strains can be purchased online. Skipping the initial cultivation phase and transferring grains directly to a substrate allows students to engage more immediately with the organism, experiment with different substrates and accelerate the design process while still gaining hands-on experience. Substrate pasteurization can be done using simple methods, such as soaking in boiled water for two to four hours, further lowering the barrier to entry.

4.2 Mycelium sheets or leathers

Working with mycelium sheets or leathers, however, presents additional challenges outside a laboratory context. The risk of contamination is higher, requiring better sterility and adequate incubation spaces. A more accessible approach is growing mycelium on a natural fabric sheet placed over a pre-myceliated fine substrate. This reduces contamination risks since the mycelium is already healthy and well developed. Additionally, reinforcing mycelium sheets with natural fibres creates a stronger final product. This can be a great way to introduce the concept of mycelium leathers to students in a relatively quick and practical way. For leather-like materials, purchasing a heat press can be an advantage to help reinforce the material, creating more uniform and rigid sheets.

In most cases, a one-day introductory workshop wouldn't usually begin with mycelium leathers as it requires a general understanding of mycelium and laboratory practices. It could, however, be adapted to a two- or three-day course, where basic principles are introduced on day one and participants have already handled the material, thus applying this knowledge to the more advanced practice of working with mycelium.

5 The challenges of facilitating workshops and seminars

5.1 Time constraints of traditional course structures

A major limitation in educational settings is the incompatibility of mycelium's growth cycle with traditional course structures. Unlike conventional materials, mycelium requires weeks to fully develop, posing challenges for short-term workshops and semester-based courses. Since design programmes prioritize material applications over cultivation principles, it is more practical for students to start with pre-grown materials while learning about growth processes in parallel. Facilitators can use hybrid models, allowing students to apply mycelium materials to design explorations while simultaneously cultivating their own strains.

Despite its slow growth, mycelium's ability to self-assemble into intricate networks offers a unique creative process. Observing the formation of hyphae strands, the thread-like structures that make up fungal networks, can be one of the most rewarding aspects of working with this material. However, mycelium does not always behave predictably; it has its own agenda, growing in ways that may not align with a designer's initial intentions. For example, two identical experiments, using the same strain, materials and conditions, can still yield different results. One sample might grow more 'fluffy' or exhibit subtle variations in texture and colour, while the other might not grow at all. This unpredictability often stems from the difficulty of perfectly controlling every variable. Even slight differences can lead the organism to mutate or become vulnerable to contaminants, ultimately hindering healthy growth. This makes working with mycelium a form of collaboration between human and organism. Experienced practitioners learn to manipulate environmental factors such as humidity, airflow and substrate composition to guide mycelial growth, producing novel forms and textures.

One of the main challenges in short workshops is maintaining engagement when results are not immediate. To keep the dynamic energy of mycelium present in an introductory workshop, facilitators can provide a range of samples that showcase different growth stages. Demonstrating mycelium development on an agar plate, in grain jars and within substrate bags helps students visualize its progression. Live demonstrations of propagation techniques, such as transferring cultures from one stage to another, can further enrich the learning experience. However, owing to time constraints, individual hands-on participation in these steps may require smaller class sizes or specialized facilities.

5.2 Student safety: biohazards and tools

Although mycelium itself is safe to handle, working with fungal cultures in non-sterile environments introduces potential biohazards. Airborne spores from contaminated cultures can pose respiratory risks, especially for individuals with mould sensitivities. Additionally, the tools required for cultivating mycelium, such as scalpels and Bunsen burners, pressure cookers and heat presses, introduce physical safety concerns.

To mitigate these risks, facilitators should implement safety protocols, including the use of personal protective equipment such as gloves, masks and safety glasses. Proper ventilation or air filtration can further reduce the risk of airborne contamination; an affordable home-scale air purification system can be a nice addition to a DIY laboratory set-up. Additionally, students should perform an induction on all laboratory tools and equipment to ensure safe use.

6 Fungal matters: an introductory guidebook to mycelium-based materials

Recognizing the need for structured educational resources, Fungal Matters (@fungal.matters, https://www.youtube.com/@FungalMatters) was born, an online course developed with Catherine Euale. After years of conducting short workshops, we realized that many students and professionals struggled to gain a deeper understanding of mycelium materials owing to the limitations of one- or two-day sessions. This led us to consolidate our knowledge and experience into an accessible online format. We are also in the process of publishing a book to further democratize learning about mycelium-based materials, ensuring that designers, architects and makers worldwide can integrate these materials into their practices.

Fungal Matters for Bio-based Design, forthcoming from Routledge, will address the challenges of working with mycelium materials outside a laboratory context, discuss the ethics of interspecies design and provide step-by-step protocols of working with a range of different mycelium-based materials and post-processing techniques. We hope it will be a great foundation for beginners who want to learn the practice and more about the wonders of fungi.

7 Conclusion

The integration of mycelium-based materials into architecture and design education presents both exciting opportunities and significant challenges. As a living, dynamic material, mycelium requires a different approach to teaching and experimentation than conventional materials, demanding patience, adaptability and an understanding of biological processes. While academic research provides a strong theoretical foundation, practical, hands-on engagement remains essential for developing a deeper appreciation of the material's potential and constraints.

Workshops, seminars and emerging Master's programmes are beginning to address this need, yet many institutions still lack the appropriate infrastructure and resources to support mycelium cultivation. Simplified solutions, such as pre-grown kits, online courses and hybrid learning models, have helped lower the barrier to entry, but more standardized educational frameworks are needed to ensure that students receive comprehensive training. Additionally, ethical considerations and sustainability must be embedded within these learning experiences, encouraging practitioners to reflect on their role in shaping the material's future while respecting the living organisms they work with.

Despite the challenges, the growing enthusiasm for mycelium-based design signals a shift towards more ecologically responsible materials and fabrication methods. As facilitators, educators and practitioners, it is our responsibility to create learning environments that embrace both scientific precision and creative exploration. By continuing to develop accessible educational resources, fostering global knowledge exchange and rethinking traditional workshop models, we can empower a new generation of designers to collaborate meaningfully with mycelium and other living systems.

The work is far from complete, but each step towards integrating mycelium into mainstream design education brings us closer to a future where biological materials are not just alternatives but essential components of sustainable architecture and innovation. Through continuous learning, collaboration and ethical engagement, we can cultivate a deeper connection with the fungal kingdom and unlock the full potential of mycelium-based materials.

Acknowledgements

IAAC and Fabricademy for their support in exploring the world of fungal materials, Catherine Euale for her enthusiasm and drive to realize our project 'Fungal Matters' and everyone who has helped me along my journey, especially Osmose Studios and Marcelo Aguirre.

Further reading

1. *Earth Emotions: New Words for a New World* by Glenn Albrecht, Cornell University Press, 2019.
2. *Entangled Life* by Merlin Sheldrake, Random House, 2020.
3. *Fungipedia, A Brief Compendium of Mushroom Lore* by Lawrence Millman, Princeton University Press, 2019.

4. *Let's Become Fungal! Mycelial Teachings and the Arts* by Yasmine Ostendorf-Rodríguez, Valiz, 2023.
5. *Matters of Care: Speculative Ethics in More than Human Worlds* by Maria Puig de la Bellacasa, University of Minnesota Press, 2017.
6. *Microbes and Other Shamanic Beings* by César E. Giraldo Herrera, Palgrave Macmillan, 2018.
7. *More Than Human Rights: An Ecology of Law, Thought and Narrative for Earthly Flourishing* by César Rodríguez-Garavito, NYU Press / Dejusticia, 2022.
8. *Mushrooms for Dyes, Paper, Pigments & Myco-Stix™* by Miriam Rice, Mushrooms for Color, 2007.
9. *Mushroom People* by Broccoli Magazine, 2022.
10. *Mycelium Running* by Paul Stamets, Ten Speed Press, 2005.
11. *Radical Mycology: A Treatise on Seeing and Working with Fungi* by Peter McCoy, Chthaeus Press, 2016.
12. *Spell of the Sensuous* by David Abram, Pantheon / Vintage, 1996.
13. *Staying with the Trouble* by Donna Haraway, Duke University Press, 2016.
14. *Symbiotic Planet* by Lynn Margulis, Basic Books, 1998.
15. *The Fungal Pharmacy: The Complete Guide to Medicinal Mushrooms and Lichens of North America* by Robert Rogers, North Atlantic Books, 2011.
16. *The Little Book of Mushrooms* by Alex Dorr, Adams Media, 2023.
17. *The Mushroom at the End of the World: On the Possibility of Life in Capitalist Ruins* by Anna Lowenhaupt Tsing, Princeton University Press, 2015.

CHAPTER 9

Design With the Living: Pedagogical Methods

ELISA STROBEL DO NASCIMENTO

'Does mycelium want to be a brick?'

This provocation by Dade-Robertson in *Living Construction* [1] is a nice way to open the discussion about teaching and learning design with living organisms. When we design with entities such as fungi or plants – there is a constant negotiation between everyone`s interests in the process [2] and new 'designerly sensibilities' [3] must be developed to mediate this interaction.[1] A lot could happen 'when species meet' [4],[2] and, we might add, when they meet through design. Teaching is all about facilitating and mediating the opportunities for this meeting.

This chapter begins by introducing some of the challenges in teaching and learning design with the living,[3] followed by a quick overview of some current teaching practices in formal biodesign education. Finally, some strategies for reaching further are presented, planned for when relevant infrastructure, such as a lab or a studio space, is still not in place.

1 Challenges in teaching and learning design with the living

At first glance, it is possible to observe that, in education, design with the living is being addressed under two main perspectives: the more-than-human design approaches and the living materials/biodesign approaches.

In the first case, the other non-human living entity is considered a user/client/stakeholder/actor in the design process or project. More-than-human design approaches are often found under the terms multispecies, interspecies, transspecies or cross-species

1 Serena Camere and Elvin Karana [3] make reference to Nigel Cross's 2006 *Designerly Ways of Knowing* – where he advocates for a uniqueness in design research, particular to design nature and its ability. Camere and Karana [3] argue that a new set of 'designerly sensibilities' would be needed to work in the intersection of design and biology, such as a 'scientific mind'.

2 In reference to Donna Haraway's *When Species Meet* [4], it is important to seek an understanding that the alterity of other non-human living beings, although often the embodiment of our biodesigned objects (often addressed as 'living materials') must be respected and cared for. Haraway tells us how we may produce meanings and worlds with other non-human species.

3 This term alludes to the symposia co-organized by the Design Museum, the Design & Living Systems Lab (Central Saint Martin's UAL), The Bio ID Lab (Bartlett School of Architecture, UCL) and the British Council in 2020 (https://designmuseum.org/whats-on/talks-courses-and-workshops/design-with-the-living-2020) and 2021 (https://designmuseum.org/whats-on/talks-courses-and-workshops/design-with-the-living-2021). It intends to comprise biodesign, but also other design situations where human and non-human living beings are involved.

Elisa Strobel do Nascimento, 'Design With the Living: Pedagogical Methods' in: *Mycology for Architecture: Methods and Applications of Mycelium Materials in the Built Environment*. Pelagic Publishing (2026). DOI: 10.53061/DUEI9593 design [5]. Examples of such designs would be the pigeon house capsule developed by Matali Crasset and presented by Donna Haraway [6], which would attract pigeons, providing them shelter while deterring them from unwanted areas; the videogame for human–cat interaction by Michelle Westerlaken and Stefano Galeni [7]; or Rasa Weber`s artificial reefs [8].

In the second case, the other living entity is either addressed as a material, used to obtain the material of the final product or used as part of the product/service system itself. These approaches might be found in the literature under the keywords biodesign, design with growing/active/living materials, living objects, growing design and biofabrication, among other terms. Examples are Full Grown's furniture, which is directly formed while trees grow [9]; Mush mycelium composite products available to the general public [10]; or bacterial cellulose leathers, such as the ones developed by Polybion [11]. Here, design with the living is often understood as a collaboration with other non-human living entities, implying that other organisms are actual co-creators of the design, along with us, the humans. For instance, Carol Collet reflected on the idea of human–non-human collaboration in biodesign, exemplified by a mycelium rubber she grew which spontaneously developed flower patterns [12]. However, this understanding of a collaboration might be debated, as the non-human living entity usually dies at the end of the biodesign process.

The differences between designing with the living and what designers are used to, and the complexity another living being brings to the equation, are exactly where the challenges of teaching and learning such a practice begin. For this chapter, the focus will be on the second approach, which will be referred to as biodesign from now on – even if the term is also polysemic within the design discipline.

To begin with, unpredictability is widely reported in biodesign – as now the 'material' has agency [3]. Zolotovsky, for instance, found that her bacterial cultures for cellulose often delivered different outcomes even if growing conditions were supposed to be the same – she attributed these results to spontaneous mutations during growth [13]. This could be positive, though, and Myers suggests designers could learn to empathize with the other living entity and negotiate form with it, surrendering 'a small amount of control' to them [14].

How do we teach such a profoundly personal, tacit relational skill? Camere and Karana refer to these new skills and competencies for biodesign as 'new designerly sensibilities' [3]. Similarly, Metcalfe has previously written about developing new 'sensitivities' towards non-humans within multispecies 'designerly' practices [15]. The new designerly sensibilities lay in the intersection with biology, and involve an 'informed eye' and a 'scientific mind' to understand what the other organisms are 'saying' [3].

Another skill students need to develop for biodesigning is collaboration with other disciplines (such as biology itself). Challenges involve diplomatic reaching out to scientists, maintaining effective communication throughout the project, addressing issues related to crediting, the need for a shared vocabulary and managing 'conflicting working styles and standards' [14, 16].

Living materials have different sensory properties that arise from their biological activity. When teaching biodesign, it must be considered that biodesigned products are still not very popular in everyone`s day-to-day life. This means it is highly probable that most students will not have a repertoire for this kind of aesthetics yet, and their process of getting acquainted with and experimenting with the textures, smells, colors and forms in biodesign must be mediated. Biodesign is so unique that Myers even foresees the emergence of a 'legible formal language' for it [14].

Finally, scalability and feasibility should also be addressed in teaching biodesign, once most biodesign research is at technology readiness levels one to three,[4] as Dade-Robertson observed [1], often at the proof-of-concept level. It is nice for young designers to develop a perspective for a product/concept to be 'brought to life' in the current production, circulation and consumption context.

As briefly summarized here, facilitating and mediating the student`s encounter with other non-human living entities through design is a task that entails many challenges. Methods for the biodesign practice are still being consolidated, and new things are expected to be learned every day. In the next session, a quick overview of some current teaching practices in formal design education for biodesign will be outlined.

2 How is it being taught?

In 2023, we conducted a systematic analysis within the 'Design with the Living: Learning to Work Together' project at the Design Department of the Federal University of Paraná in Brazil, aiming to better understand how biodesign was being taught in formal education at the time [17]. This mapping provided some insight by depicting biodesign teaching and learning strategies, some of which are repeated by different universities.

Through specific keywords in the Google search engine (in this case in English – other results could be found in Spanish and so forth), we found a masterclass for professionals (TU Delft), eight courses within undergraduate curricula (The University of Western Australia, University of Cincinnati, two at University of California Davis, University of Sydney, University of Economics of Izmir, University of Pennsylvania), one major (University of Sydney), four Master's (Arizona State University, University of the Arts London, University College London, The University of Western Australia) and two PhD programmes (University of Western Australia, Arizona State University).[5]

By consulting the information available for each of those courses/programmes, some common strategies may be perceived among them. Laboratory practice is a very strong one, laying the ground and basis for some of the courses/programmes. At Arizona State University, for example, students even rotate between laboratories to define a research interest and find an advisor. The University of the Arts London offers a biosafety level one (BSL-1) laboratory. Course leader Nancy Diniz explained that the Grow Lab is a central reference for the MA Biodesign programme, which is crucial in training students in a new language of visualization and representation which is not apprehensible by the naked eye [18]. Diniz argues that these are seen as a 'new way of drawing', a new design language in development in biodesign practices [18].

Many courses/programmes detail an interdisciplinary experience related to the biology repertoire developed through laboratory practice. The University of the Arts London, for example, includes a biologist in the teaching team and also offers an 'international network and knowledge exchange with industry partners'.

Furthermore, ethical implications are a main topic of the courses/programmes. Ethical discussions in biodesign usually address genetic modifications, intellectual property, contamination issues through a possible spreading of harmful microorganisms

4 This refers to a NASA proposition for maturity levels for new technologies. Dade-Robertson discusses biodesigned solutions within this framework – Level 1 is 'Basic principles observed and reported'; Level 2 is 'Technology/concept and/or application formulated; and 3 is 'Critical function, proof of concept established'(1).

5 This study was limited by the results retrieved in this specific search engine given the keywords used. It did not retrieve PhD programmes structured around the candidate's proposals. We recognize this limitation.

and the treatment/handling of the other non-human living entities that are being worked with.

Besides laboratory practice, project/studio structures have also been adopted. Learning by doing and solving problems that the students themselves often propose seems to be a very prolific strategy. TU Delft and the University of California, Davis, focus on market-driven/application-driven solutions. In addition to that, the result of a prototype is highly desirable. Five courses/programmes mention prototyping, which usually involves students growing the concept themselves. In addition, the University of Sydney and the University of the Arts London explicitly address communication skills for the project's results.

One last strategy to highlight is participation in the Biodesign Challenge (BC), an international educational competition for high schools and universities. Four of the universities offered activities oriented toward this competition. The University of California, Davis, has specific dedicated courses for that.

It is important to highlight that biodesign is also being addressed within the 'traditional' courses in design curricula. This presence is traceable through published papers, such as Elvin Karana's research group practice in TU Delft, in the context of teaching and learning materials.

Biodesign education does not happen only in formal education but also in other constellations, such as in open labs and by independent researchers. It is interesting to observe, however, how biodesign is being institutionalized through the universities, indicating it is being prioritized in the agenda.

3 How to reach further

Many educational institutions still lack access to a lab or a proper studio space, and teaching and learning strategies for biodesign must be adapted so they may reach further. I will briefly share our experience at the Design Department of the Federal University of Paraná in Brazil [19], outlining the elements of a framework for facilitating teaching and learning biodesign in such a limited context. The framework aims to introduce design students to their first experience in this practice. In our context, we worked together with third-year product design students.

All the pedagogical strategies previously mapped in formal education were somehow adapted to this framework – except for the participation in the BC. Two alternative context-spaces were foreseen without a studio space and a laboratory: the classroom and the students' homes. Six main elements laid the framework`s foundations: [1] concepts; [2] repertoire; [3] project methodology; [4] practice; [5] management; and [6] reflections. Concepts is the element that addresses biodesign terminology, where terms such as 'epigenetics' are explained. Repertoire provides biodesign examples and case studies, so students can picture possibilities. Project methodology outlines the main project phases of biodesign case studies, drawing a parallel to design methodologies students are already familiar with, such as the Double Diamond from the Design Council. In practice a hands-on biodesign project is developed. Management and reflections are addressed mainly through a structured project journal, so students can track their progress, mindfully relate to the other living being they are designing with and reflect on the process.

Some supporting materials assist the framework's application, such as an application script, a pre-structured project journal (for the students to keep), material samples for tests, a low-tech grow-it-yourself kit (with fungi), a project evaluation rubric (for written feedback for the students) and the framework's evaluation rubric (to assess and improve it). These materials are now available online [19].

The focus was on hands-on activities that relate to a studio/project structure. Students were challenged to develop a product design with mycelium in six weeks. The framework tried to include lab work by adapting it to be developed in the students' kitchens. This was brought to fruition through existing low-tech material design and do-it-yourself practices. Fungi were chosen as friendly and relatively forgiving organisms for students to work with for the first time, and they can be grown at home using a pressure cooker for substrate sterilization, for instance. This strategy made prototyping and tinkering experiments possible.

To create a more interdisciplinary environment, a bioprocess engineering and biotechnology intern, Caroline Aimi Ueda Fukuda, and the CEO of a mycelium startup company, Neomatter, Luiz Eduardo Piá de Andrade, were invited to participate in the project. They participated in content expositions about the species being worked with and how to grow mycelium at home. They also gave feedback on the students' projects throughout the course.

Students were asked for their products to be market-driven/application-driven solutions. Incorporating some marketing tools in the framework, such as a perception positioning map (Figure 9.1), helped in this challenge. These tools contributed to their thinking about the products in a production–circulation–consumption context. To improve their communication skills, students presented their projects twice during the six weeks, and developed a product flyer to potentially market and release the product.

Regarding the framework`s elements, some concepts were presented at the beginning. These concepts are aimed at helping the students navigate important keywords, making it

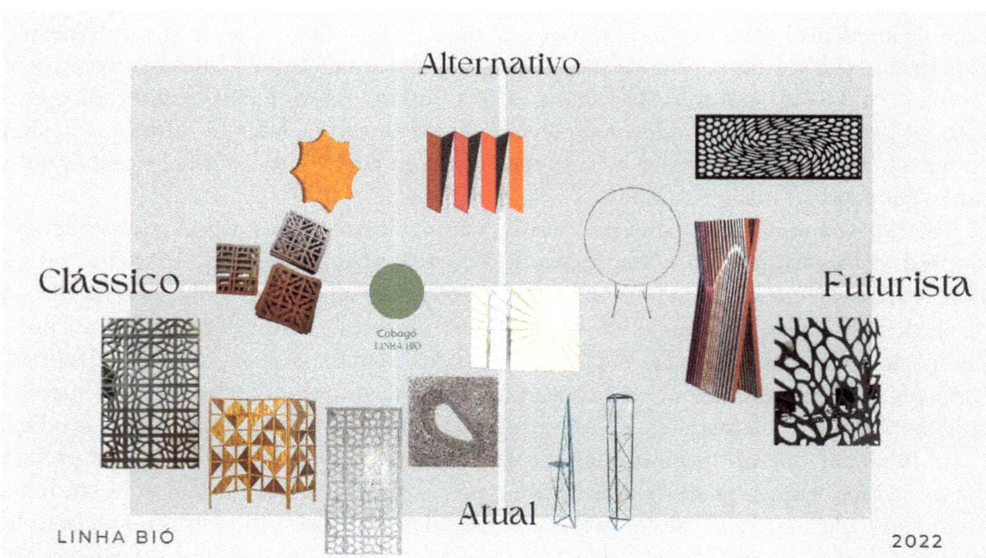

Figure 9.1 Example of a perception position map developed by the students for the project Cobogó Bió, showcased later in this chapter. Existing products are distributed along two axes according to perceptual qualities, such as Classic, Alternative, Futuristic and Contemporary, as is the case in this example. The green dot shows where the students aimed their biodesigned product. Image by Caio César Franco Pael Zanolla, Carolina Muller Friedrich, Victória Perkoski da Silva and Yasmine Rudek Rodrigues.

easier for them to retrieve relevant papers and resources. Additionally, references such as Camere and Karana's MA2E4 toolkit were used to become acquainted with the aesthetic and sensory characteristics of mycelium.[6]

Moreover, a repertoire of examples and a project methodology grounded in existing case studies was introduced, including those derived from interviews with professional biodesigners. Design phases were identified together with the interviewees and were illustrated in 'translation' to design methods already known by the students, such as the Double Diamond. Figure 9.2 illustrates the Mush Eco case study – a company in Brazil producing commercial mycelium products in partnership with Furf Design Studio.

Along with the practice element, project management was encouraged and mediated – students had to self-organize in teams and were aided by the structured project journal, with defined deadlines for project presentations and receiving feedback.

Finally, reflections aimed to help students assume a critical thinking posture in their practice. Reflective provocations, including ethical dilemmas and empathy exercises, were proposed for discussion in the groups.

Many of the products students designed in this experience would fall into the 'demonstrator' category [3], being mostly archetypal objects, such as chairs, lamps and jewellery – not fully exploring the material`s potential. However, this is not necessarily negative, especially for the students` first experience with biodesigning. It might be seen as a way to link something still unknown (working with a living material) with something one is familiar with (a widely known and archetypal object).

Other results were less archetypal, such as a beehive and a 'cobogó', which is a traditional Brazilian hollow construction element; this is illustrated in Figure 9.3. Cobogó Bió was the proposal developed by Caio César Franco Pael Zanolla, Carolina Muller Friedrich, Victória Perkoski da Silva and Yasmine Rudek Rodrigues. In the sequence, Figure 9.4, shows the production process for the prototype.

At the end, students wrote their feedback and impressions about the framework and the experience they had. Dade-Robertson notes that 'Instinctively we feel that there is something different about life' [1], and this was reflected in the students` feedback. In retrospect, they worried that mycelium would not grow and had trouble estimating the time. In the uncertainties they faced, they learned the tacit and deeply personal relational skills that may only be developed 'when species meet'.

4 Final considerations

We still have a long way to go in designing with the living education. Adapted biodesign teaching and learning methods are key for diffusing this important practice and expanding the research agenda. In a country like Brazil, for example, many opportunities for design to contribute to biodiversity remain unexplored – hence, the lack of a lab or a more structured setting shouldn't stop biodesign from being presented to students.

6 This is a toolkit developed within the context of the Materials Experience Lab (https://materialsexperiencelab.com/index.php/2022/06/05/ma2e4-toolkit-experiential-characterization-of-materials/#:~:text=The%20Ma2E4%20Toolkit%20aims%20to,levels%20interrelate%20in%20the%20experience) – aimed at supporting the understanding of materials according to sensorial, interpretive (meanings), affective (emotions) and performative (actions, performances) levels.

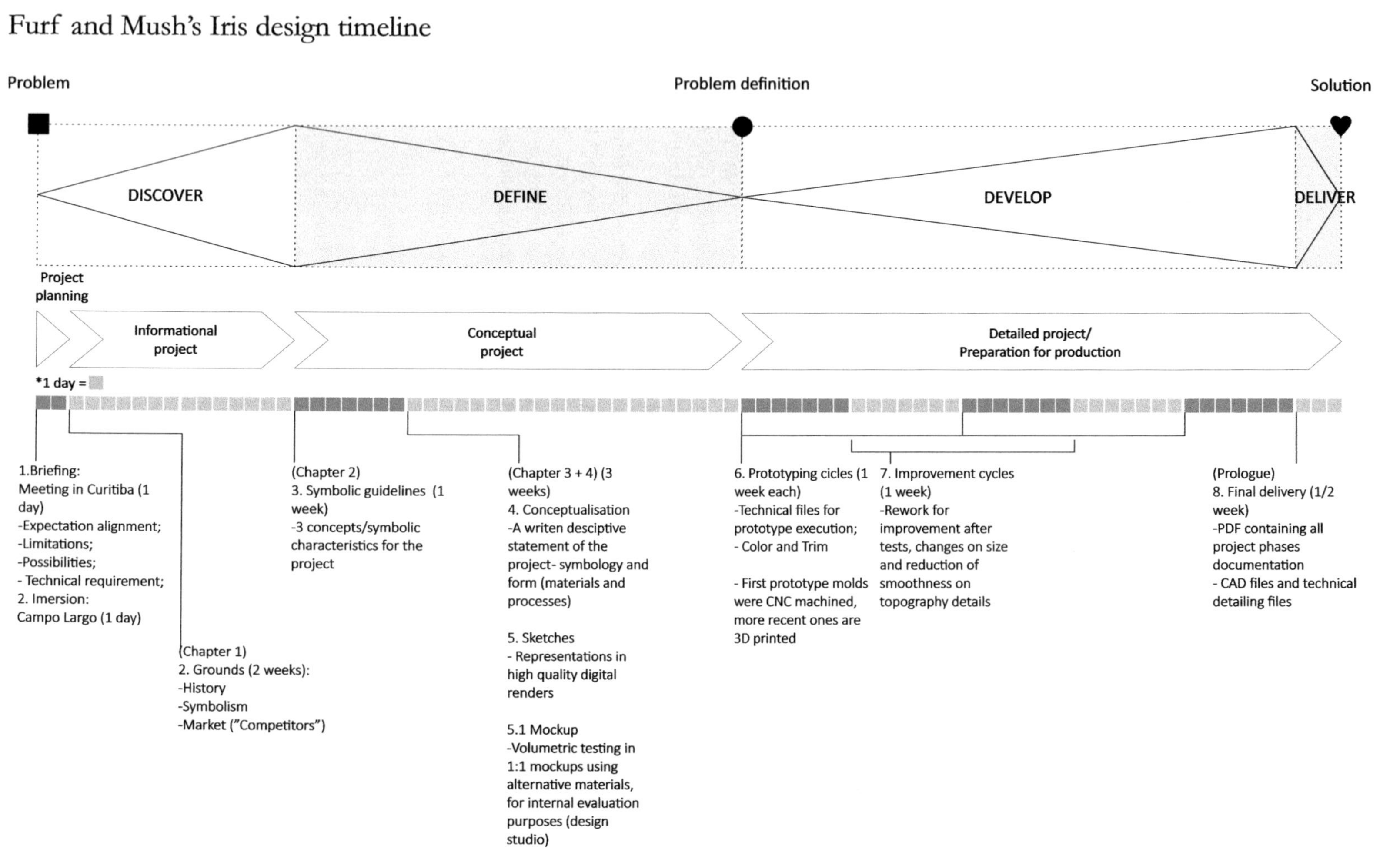

Figure 9.2 Mush and Furf process for designing with mycelium. Image credit: Elisa Strobel do Nascimento.

Figure 9.3 Left: Rendering. Right: Cobogó Bió prototype. Image by Caio César Franco Pael Zanolla, Carolina Muller Friedrich, Victória Perkoski da Silva and Yasmine Rudek Rodrigues.

Figure 9.4 Cobogó Bió prototyping process. Image credits: Caio César Franco Pael Zanolla, Carolina Muller Friedrich, Victória Perkoski da Silva and Yasmine Rudek Rodrigues.

We can often be unaware of our material world. Sometimes, we are not conscious of the matter that makes our daily lives, even if it can directly influence the way we move. We don't wash a crystal glass the same way we do a metal cup. Ultimately, there seems to be a risk of this invisibilization being transposed to non-human living entities when we treat them as materials.

A more integrated approach for more-than-human perspectives to biodesign would be a preferable scenario. New designerly sensibilities that encompass a radical relationality and efforts of care for other non-human living entities could be very helpful in this matter, as could incorporating design for systems [20, 21]. I understand this as a natural consequence of a more reflective biodesign practice, in which species can meet through design.

Acknowledgements

I would like to thank Prof. Dr Adriano Heemann and Prof. Aguinaldo dos Santos, PhD, for the insights and support. The research was financed in part by the Coordenação de Aperfeiçoamento de Pessoal de Nível Superior – Brasil (CAPES) – Finance Code 001. I also acknowledge the support of the Cluster of Excellence 'Matters of Activity. Image Space Material' funded by the Deutsche Forschungsgemeinschaft (DFG, German Research Foundation) under Germany's Excellence Strategy – EXC 2025 – 390648296.

References

1. Dade-Robertson M. Living Construction. 1st ed. New York: Routledge; 2021.
2. Collet C. Designing our future bio-materiality. AI and Society. 2020, 36(4): 1–12. doi: 10.1007/s00146-020-01013-y
3. Camere S, Karana E. Fabricating materials from living organisms: An emerging design practice. J. Clean. Prod. 2018,186: 570–584. doi: 10.1016/j.jclepro.2018.03.081
4. Haraway D. When Species Meet. Minneapolis: University of Minnesota Press; 2008.
5. Strobel do Nascimento, E. Design with the living – analysis and a preliminary proposal for the trio multispecies design, interspecies design, and transpecies design. Estudos em Design. 2024 32(2): 130–147. doi: https://doi.org/10.35522/eed.v32i2.1980
6. Haraway D. Staying with the Trouble: Making Kin in the Chthulucene. Durham: Duke University Press; 2016.
7. Westerlaken M.; Gualeni S. Felino: the philosophical practice of making an interspecies video game. In: Philosophy of Computer Games Conference Proceedings; 2014; Istanbul. Istanbul: The Game Philosophy Network; 2014. pp. 1–12.
8. Weber R. Kiki Reef Prototype in Stareso. 2023 [cited 2025 Jun 01]. Available from: https://www.rasaweber.com/kiki.
9. Munro G, Munro A. Full Grown. 2025 [cited 2025 Jun 1]. Available from: https://fullgrown.co.uk/.
10. Mush Shop. 2025 [cited 2025 Jun 1]. Available from: https://loja.mush.eco/.
11. Polybion. 2025 [cited 2025 Jun 1]. Available from: https://www.polybion.bio/.
12. Collet C. 'Grow-made' textiles. In: Karana E, Giaccardi E, Nimkulrat N, Niedderer K, Camere S, editors. Alive. Active. Adaptive. International Conference on Experiential Knowledge and Emerging Materials; 2017 Jun; Delft. Delft: TU Delft Open; 2017. pp. 24–37.
13. Zolotovsky K. BioConstructs – Methods for Bio-Inspired and Bio-Fabricated Design [Master Thesis]. Cambridge (MA): Massachusetts Institute of Technology; 2012.
14. Myers W. Biodesign. Nature, science, creativity. Revised and expanded edition. High Holborn: Thames & Hudson; 2018.
15. Metcalfe DJ. Multispecies Design [Ph.D. Dissertation]. London: University of Arts London and Falmouth University; 2015.
16. Cho T. Getting Started. In: Myers W. Biodesign. Nature, science, creativity. High Holborn: Thames & Hudson; 2018. pp. 266–268.

17. Strobel do Nascimento E., Heemann A. How to teach interspecies design? Formal education in biodesign. e-Revista Logo. 2023, 12: 1–20. doi10.26771/e-revista.logo/2023.01
18. Diniz N. Living Systems Lab | A symposium by Central Saint Martins – UAL [video file]. 2020 Nov [cited 2025 Mar 07]. Available from: https://www.youtube.com/watch?v=9vqWfE8cFwQ&list=LL&index=437
19. Strobel Do Nascimento E. Design with the Living: Learning to Work Together [Ph.D. Dissertation]. Curitiba (PR): Universidade Federal do Paraná; 2023.
20. Escobar A, Osterweil M, Sharma K. Relationality. An Emergent Politics of Life Beyond the Human. 1st ed. London: Bloomsbury Visual Arts; 2024.
21. Pasold A. Advanced Growing Materials. In: Ferraro V., Pasold A editors. Emerging Materials & Technologies. New Approaches in Design Teaching Methods of Four Exemplified Areas. Milano: FrancoAngeli; 2020. pp. 126–152.

CHAPTER 10

Methodological Framework for Designing with Fungi

DILAN OZKAN and MARTYN DADE-ROBERTSON

1 Rethinking mycelium's material properties

When introducing mycelium as a material to designers, architects or students, they often ask '*What are the properties of mycelium?*' expecting to find specific tensile or compressive strength values, as they would in a catalogue. My response is usually another question: '*How did you grow it?*' Although its properties can be experienced much like those of catalogue materials, they are formed by growth and can be altered by adjusting environmental conditions and cultivation methods. Unlike inert materials with unchanging, inherent properties, mycelium materials are processual. This means that their properties are not given in advance but emerge through cultivation, shaped by biological behaviour, environmental conditions and fabrication strategies.

By adjusting how it grows, through substrate choices, environmental conditions and additive strategies, its physical form and function can be intentionally shaped. Depending on these variables, mycelium can become leather-like, foam-like or rigid, with properties such as porosity, density, hydrophobicity, strength or thermal resistance [1, 2, 3]. This versatility makes it suitable for applications in construction, packaging and textiles. However, because these properties come from biological plasticity rather than fixed states, working with mycelium requires a fundamental shift: from static, outcome-oriented material thinking to process-based methodologies that integrate both scientific control and designerly interpretation. This raises a critical question: How do we work with a material whose behaviour is contingent, variable and co-produced through cultivation? It demands a new dynamic, in which the designer collaborates with biological processes rather than simply acting upon them.

This process-based approach resonates with material-led design methods, which treat materials not as passive but as active participants that interact, change and shape outcomes. A material-led approach involves identifying a material's capacities, resistances and effects across different scales, with or without human intervention and then innovating around these qualities so that form, function and meaning emerge through a dialogue with the material [4].

With mycelium, however, the challenge is different: the task is not to design the material itself but to design its growth. The focus, therefore, shifts from discovering fixed traits to exploring the processes that bring those traits into being. Methodologies should enable this exploration, enabling designers to work with growth as a generator of diverse and adaptable properties.

Dilan Ozkan and Martyn Dade-Robertson, 'Methodological Framework for Designing with Fungi' in: *Mycology for Architecture: Methods and Applications of Mycelium Materials in the Built Environment*. Pelagic Publishing (2026).
 DOI: 10.53061/CIGK8060

To navigate this complexity and answer how to innovate through a process that drives material properties, rather than through a material with inherent ones, this chapter focuses on methodologies. To establish desirable qualities that are not simply found but actively co-created with the organism, it proposes a mixed methodology that integrates two primary frameworks:

- Probing the material process and exploring ideas without a preconfigured objective, using a research through design (RtD) methodology.
- Combining open-ended material exploration with more rigorous scientific methods to systematically map and understand cause-and-effect relationships.

These methodologies are not merely tools for achieving specific outcomes; together they form the basis of a new design theory that acknowledges biological agency and generates material knowledge through iterative co-creation.

2 Designing variability

Understanding why mycelium's properties are variable, rather than fixed, requires engaging with a specific biological principle: plasticity. In philosophy, plasticity is described as 'the capacity of the living being to receive form and to give form' [5]. In biological terms, it refers to the ability of organisms or cells to alter their phenotype in response to environmental conditions [6]. Crucially, these changes occur within the same genes, without mutations or long-term genetic shifts (as seen in Figure 10.1).

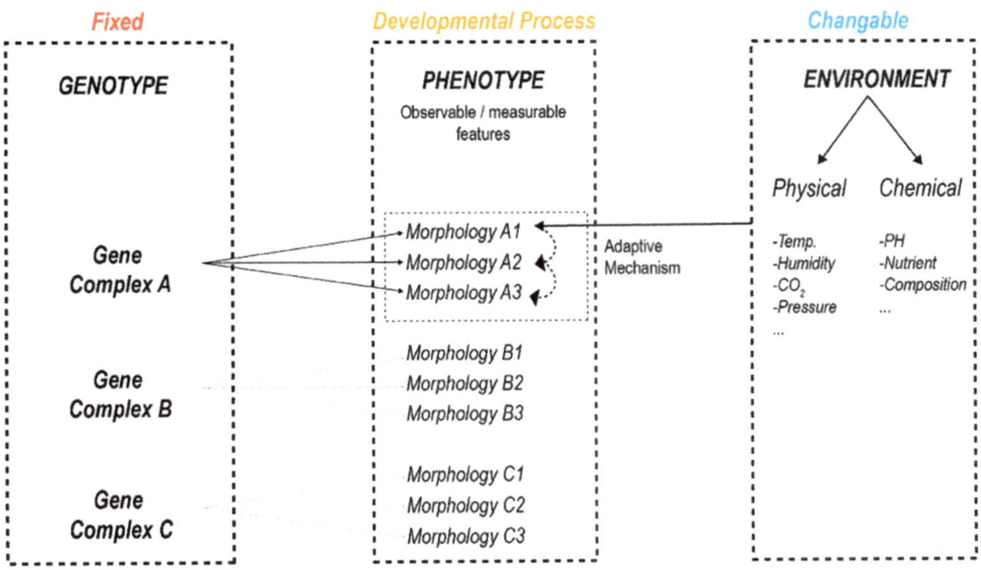

Figure 10.1 A diagram showing the relation between genes, morphology and environmental factors. Achieving a particular morphology in biology involves a developmental process. Individual cells undergo this process, and some transition into multicellular entities through the activation of a single genotype's capacity and its associated developmental pathways [7]. The developmental process continues until a point when significant changes in the capacity for further development become limited.

This phenomenon, known as developmental plasticity, enables genetically identical organisms to express a range of morphological outcomes, depending on the conditions they experience during growth [8].

For material designers and researchers, this has profound implications. Morphological and functional variability is not a by-product of cultivation but a design parameter that can be intentionally leveraged. By systematically adjusting growth conditions such as humidity, temperature, CO_2 levels, substrate composition and additives, designers can influence the way mycelium develops structurally and functionally [9]. This mode of working does not rely on direct mechanical shaping or static moulding, but rather on guiding developmental processes, making the act of design inseparable from the act of cultivation. As Dade-Robertson notes, such methods use 'nature's own agencies', using biological responsiveness rather than imposed constraints to shape material outcomes [10]. Here, the designer becomes a co-cultivator, who sets the stage for biological growth and opens possibilities for indirect design strategies, where properties are not imposed but emerge through the interaction between biological agency and carefully controlled environmental parameters.

For example, *Ganoderma lucidum* cultivated in liquid culture under dynamic (shaking, ~five days) and static (~ten days) conditions can yield soft, leather-like films. Altering temperature or agitation speed can even induce self-healing behaviours [11]. When grown on solid substrates such as sawdust, hemp or straw, the same species can produce lightweight, insulating foams or dense, load-bearing composites [12]. These radically different outcomes underscore the shift in perspective. The key design question is not 'What properties does mycelium have?' but 'What properties are needed, and how might growth be tuned to achieve them?' Consequently, a new methodological need emerges, one in which biological agency is treated as a partner rather than only a resource.

3 Mixed methods methodology

As mycelium's material properties are shaped by environmental factors, biological agency and the designer's interventions, the choice of methodology becomes central to the design process. Unlike conventional materials research, where standard testing protocols are well established, mycelium design requires accommodating uncertainty, responding to variability and leveraging emergence as a driver for discovery.

In this context, two dominant methodologies emerge: RtD and scientific methodology. Each offers strengths but also has limitations when applied in isolation. Therefore, mixed methods methodology, also known as integrated methodology, becomes essential. It enables taking a holistic perspective on research, combining various methodologies and methods, different data and research strategies [13]. Rather than treating scientific and designerly methods as opposites, it helps to frame them as complementary: rigorous experimentation establishes reliable baselines, while iterative, practice-based exploration captures emergent phenomena. Together they recognize biological agency as an active participant in design.

This integrated methodology reframes the knowledge production, allowing quantitative and qualitative methods to operate together. It incorporates literature review, research through design, prototypes, biomaterial probes, material-led design and visualization techniques to study living materials, test ideas and generate data for evaluation.

3.1 Research through design methods

RtD is an approach within the broader field of design research.[1] It focuses on using the design process and requires the active participation of an intentional observer in the design [15]. The process begins by addressing and analysing the cause of a problem, moving on to setting goals to suggest solutions, and ending with the potential improvements of the problematic situation. In short, analysis, projection and know-how generation compose the RtD scheme [15]. RtD employs various methods such as probes, prototypes and visualizations to generate new knowledge and produce tangible outcomes.

3.1.1 Prototypes

In order to test and evaluate the use of mycelium materials, experimentation and hands-on engagement from the designer are often required. Design actions such as prototyping, modelling and experimenting are central to this process as they help assess outcomes and guide subsequent iterations. Prototypes are used to gain experience, discuss future scenarios and test theories to develop a meaningful understanding [16]. They function as 'landmarks for reference', informing discussions, shaping hypotheses and guiding further development [16].

3.1.2 Probes

Design probes, in contrast, are used to investigate the matter and to inspire ideas. The biomaterial probes directly engage with biological systems and generate new questions about designing with biological systems [17]. They act as proxies, offering insights into how mycelium materials can participate in shaping outcomes and whether their developmental processes can be employed as part of the fabrication method. Rather than serving as an end goal, probes function as a means to test design suggestions and simulate the designer's thinking as a form of feedback [17]. Instead of simply illustrating data, these probes provide a context to question how to design with living material engagement. In this way, they serve as tools to gain insights, investigate imagined scenarios and deepen understanding of both material behaviours and design opportunities.

Although both a probe and a prototype are used in the early stages of the design process, the difference between them is that prototypes get more refined over time. While prototypes are physical manifestations of ideas/investigation, probes remain a conceptually investigative tool and are materials designed to provoke or elicit responses [18]. Therefore, prototypes are used to test products, while probes are used to test ideas.

3.1.3 Visualization tools

Visualization methods, such as photography, microscopic imaging, 3D scanning and digital modelling, play a critical role in both recording and interpreting material behaviour.

---

1 Glanville introduces the idea that design research is the scientific attitude of designing by perceiving a situation as a contextual challenge that requires understanding and a solution. He further suggests that it should be conducted through a systematic analysis of well-founded pieces of evidence [14]. Consequently, rather than simply using the existing theories, this approach tries to create tacit knowledge by designing. In this case, tacit knowledge is not created by theory alone; instead, design research unites theory and practice. While using theory, an observer is positioned outside the system to describe it, and in applied practice, the observer is incorporated actively into the system [14].

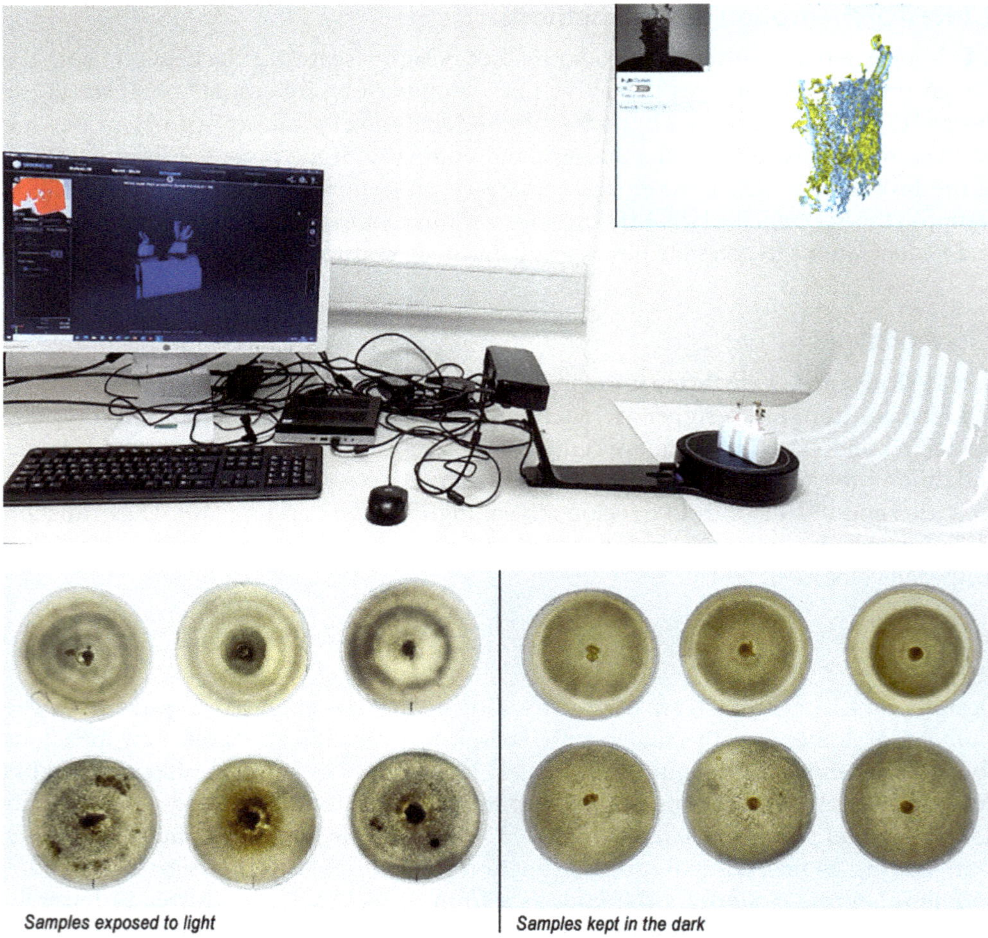

Figure 10.2 This figure presents two visualization methods – 3D scanning and photography – used to capture fungal growth. Above: 3D scanning with the EinScan-SE desktop scanner. Below: Photographs documenting pattern variations in mycelium growth. Image credit: Dilan Ozkan.

In addition to simply recording open-ended exploration of the material, they help to build a strategic understanding of outcomes and support methodical exploration of the relationship between form and context [19]. As the process shifts from simply recording results to representing them, graphic design begins to shape how results are perceived and communicated. Bolton describes such visual thinking methods as a blended research approach that fosters a 'designerly way of thinking' to address complex and ill-defined problems [20]. In this way, visualization methods function not only as records but also as active design tools, mapping relations between parameters and outcomes, informing decision-making and revealing patterns in living material behaviour that offer insights to support other designers working with biological systems.

3.2 Scientific methods

Scientific methods offer rigour and a hypothesis-driven approach to working with living materials. They involve the design of controlled experiments in which variables are deliberately manipulated and outcomes are measured through repeatable, quantifiable methods. Controlled experiments, used in scientific research, usually examine the impact of a planned modification on a certain quantifiable performance indicator [21]. In mycelium design, this means adjusting parameters such as substrate, humidity or CO_2 levels to determine their measurable effects on material properties.

However, the concept of 'experiment' differs significantly between the fields of science and design [22]. In science, an experiment traditionally serves as a tool to produce data in order to test a hypothesis; in contrast, within design practice, an experiment commonly refers to a process that has no limits or end and has a goal of formulating new ideas [23]. This distinction becomes especially relevant when working with non-standardized, responsive materials such as mycelium.

3.3 Integration of methods

Mixed methods research integrates qualitative and quantitative viewpoints to provide comprehensive understanding while also validating findings through the convergence of different research approaches [24]. This integration creates feedback loops in which data, insights and interpretations flow dynamically between scientific rigour and designerly exploration.

It reflects the realities of working with mycelium, where plasticity often leads to unexpected outcomes. For example, an irregular surface pattern observed in a controlled experiment might be reframed as a material probe, provoking new questions that guide the next cycle of testing. Similarly, patterns that emerge during design probe experiments can inspire targeted studies, demonstrating how scientific and design methodologies continually inform and refine one another.

This hybrid methodology also reframes the designer–organism relationship: mycelium is treated not as a passive material but as an active participant in form and function. By recognizing biological agency, variability becomes a generative design parameter rather than an error to control. The outcome is a process that is both rigorous and adaptive, capable of building cumulative knowledge while remaining responsive to the dynamics of living growth.

4 Conclusion

Mycelium's plasticity, its capacity to translate growth conditions into diverse material properties, redefines it as a processual material. Unlike conventional materials with fixed, catalogue-like values, mycelium demands approaches that balance scientific rigour with designerly exploration. This chapter has argued for a hybrid methodology that negotiates between control and emergence, quantification and intuition, positioning designers and scientists as co-creators with biological systems.

Such a methodology reframes design as an iterative dialogue with growth rather than a linear path to predetermined outcomes. Initial explorations may begin with open-ended probes, evolve into targeted optimization and loop back as unexpected behaviours provoke new questions. In this cycle, experiments generate not only data but also hypotheses, interpretations and design opportunities. What emerges is neither fully engineered nor fully wild, but a form of collaboration that recognizes biological agency while extending design practice.

By embracing this hybrid, iterative approach, designers move beyond simply applying mycelium 'as provided' and instead unlock what it is additionally capable of becoming. Over time, these cycles can transform situated experiments into broader principles for working with living materials, embedding knowledge not just in analytical results but in artefacts, methods and ways of thinking.

References

1. Madusanka C, Udayanga D, Nilmini R, et al. A review of recent advances in fungal mycelium-based composites. Discov Mater. 2024;4:13. https://doi.org/10.1007/s43939-024-00084-8
2. Elsacker E, Vandelook S, Brancart J, Peeters E, De Laet L. Mechanical, physical and chemical characterisation of mycelium-based composites with different types of lignocellulosic substrates. PLoS One. 2019;14(7):e0213954. https://doi.org/10.1371/journal.pone.0213954
3. Teeraphantuvat T, Jatuwong K, Jinanukul P, Thamjaree W, Lumyong S, Aiduang W. Improving the physical and mechanical properties of mycelium-based green composites using paper waste. Polymers. 2024;16(2):262. https://doi.org/10.3390/polym16020262
4. Correa I, Holbert N. Material-led research: a posthumanist methodology for maker education practitioners. Digital Culture & Education. 2023;14(5):191–213.
5. Žukauskaitė A. Organism-oriented ontology. Edinburgh: Edinburgh University Press; 2023.
6. Skipper M, Weiss U, Gray N. Plasticity. Nature. 2010;465(7299):703. https://doi.org/10.1038/465703a
7. Lafuente E, Beldade P. Genomics of developmental plasticity in animals. Front Genet. 2019;10:720. https://doi.org/10.3389/fgene.2019.00720
8. Moczek AP, Sultan S, Foster S, Ledón-Rettig C, Dworkin I, Nijhout HF, Abouheif E, Pfennig DW. The role of developmental plasticity in evolutionary innovation. Proc Biol Sci. 2011;278(1719):2705–13. https://doi.org/10.1098/rspb.2011.0971
9. Sultan SE. Phenotypic plasticity in plants: a case study in ecological development. Evol Dev. 2003;5(1):25–33. https://doi.org/10.1046/j.1525-142X.2003.03005.x
10. Dade-Robertson M. Living construction. 1st ed. London: Routledge; 2020. https://doi.org/10.4324/9780429431807
11. Elsacker E, Zhang M, Dade-Robertson M. Fungal engineered living materials: the viability of pure mycelium materials with self-healing functionalities. Adv Funct Mater. 2023;33(34):2301875. https://doi.org/10.1002/adfm.202301875
12. Vasatko H, Gosch L, Jauk J, Stavric M. Basic research of material properties of mycelium-based composites. Biomimetics. 2022;7(2):51. https://doi.org/10.3390/biomimetics7020051
13. Plowright PD. Making architecture through being human: a handbook of design ideas. 1st ed. London: Routledge; 2019. https://doi.org/10.4324/9780429261718
14. Glanville R. The sometimes uncomfortable marriages of design and research. In: Rodgers PA, Yee J, editors. The Routledge companion to design research. London: Routledge; 2018. pp. 9–22. https://doi.org/10.4324/9781315758466-3
15. Jonas W. A cybernetic model of design research. In: Rodgers PA, Yee J, editors. The Routledge companion to design research. London: Routledge; 2014. pp. 23–37. https://doi.org/10.4324/9781315758466-4
16. Stappers PJ, Sleeswijk Visser F, Keller I. The role of prototypes and frameworks for structuring explorations by research through design. In: Rodgers PA, Yee J, editors. The Routledge companion to design research. London: Routledge; 2014. pp. 163–174. https://doi.org/10.4324/9781315758466-16
17. Ramirez-Figueroa C. Biomaterial probe – design engagements with living systems [dissertation]. Newcastle upon Tyne (UK): Newcastle University; 2017.
18. Sanders EBN, Stappers PJ. Probes, toolkits and prototypes: three approaches to making in codesigning. CoDesign. 2014;10(1):5–14. https://doi.org/10.1080/15710882.2014.888183
19. Harland R. Seeking to build graphic theory from graphic design research. In: Rodgers PA, Yee J, editors. The Routledge companion to design research. London: Routledge; 2014. pp. 87–97. https://doi.org/10.4324/9781315758466-9
20. Bolton S. The visual thinking method. In: Rodgers PA, Yee J, editors. The Routledge companion to design research. London: Routledge; 2014. pp. 129–130. https://doi.org/10.4324/9781315758466-26
21. Anderson AJB. Controlled experiments. In: Interpreting data. Boca Raton (FL): CRC Press; 2019. pp. 183–190. https://doi.org/10.1201/9780203750315-16
22. Ginsberg AD. Design as the machines come to life. In: Ginsberg AD, Calvert J, Schyfter P, Elfick A, Endy D, editors. Synthetic aesthetics: investigating synthetic biology's designs on nature. Cambridge (MA): MIT Press; 2014. pp. 22–70.
23. Gauch HG. Introduction. In: Scientific method in practice. Cambridge: Cambridge University Press; 2003. pp. 1–30. https://doi.org/10.7312/zhao12754-003
24. Johnson RB, Onwuegbuzie AJ, Turner LA. Toward a definition of mixed methods research. J Mixed Methods Res. 2007;1(2):112–33. https://doi.org/10.1177/1558689806298224

CHAPTER 11

Designing with Mycelium: Observations from Practice

ABHINAV CHAUDHARY and SAVANNAH WILLITS

Time and time again, mycelium composites have come to surprise us.

In the beginning we were taken aback by mycelium's tenacity for growth, and then again by its ability to fuse with itself and with other natural materials, such as wood and fabric. By the time we felt well acquainted with the material it surprised us again with its strength and realized potential as an architectural material. It is likely that mycelium, and fungi at large, will continue to confound and astound us as we establish stronger symbiotic relationships with the natural world (Figure 11.1).

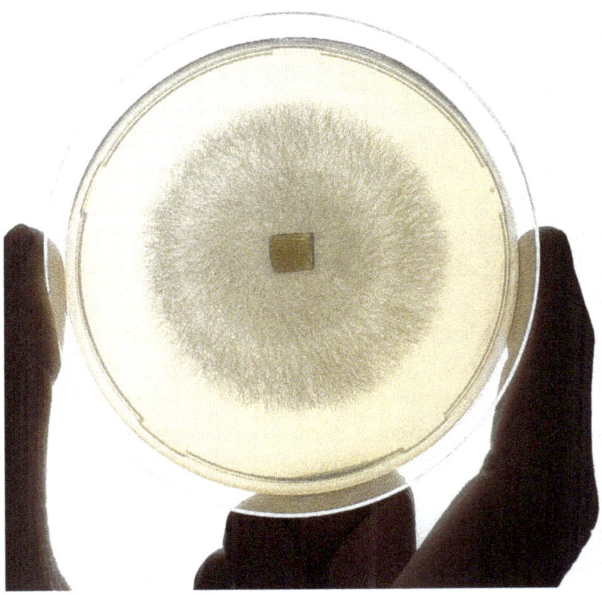

Figure 11.1 Mycelium growing in a culture medium in a Petri dish. Image credit: PLP Architecture.

Research for this chapter was conducted by PLP Labs: Abhinav Chaudhary, Savannah Willits, Deepthi Ravi, Jen Andersson, Harjit 'Ram' Sembi, Ron Bakker, Michael Polisano and Darshil Shah.

Abhinav Chaudhary and Savannah Willits, 'Designing with Mycelium: Observations from Practice' in: *Mycology for Architecture: Methods and Applications of Mycelium Materials in the Built Environment*. Pelagic Publishing (2026).
 DOI: 10.53061/MPSV7530

1 Background

PLP Lab's material research into mycelium is underpinned by the eco-philosophy of Glenn Albrecht, the author of *Earth Emotions: New Words for a New World*. In his work, he proposes the Symbiocene, a future epoch that takes place after the Anthropocene where 'human culture, habitat and technology will seamlessly re-integrate back into lifecycles and processes' [1]. During this epoch, Albrecht suggests architecture will be replaced by sumbiotecture, which places living together and collaboration at the heart of design rather than control. We are slowly beginning to see this transition with the rise of more-than-human design and biodesign, as the focus shifts towards constructing environments where all species can thrive. Albrecht considers symbiosis to be neither a lofty nor a radical ideal for it is far more mad to be enthusiastic, or complacent, perpetrators of our own ecocide [1].

At the same time, our materials research is contextualized by and operates within a global, corporate architecture firm. The ability to scale and the cost and viability of mycelium continue to be looming question marks over our work. Above all, accreditation remains an equally necessary and challenging step towards the adoption of mycelium in the building industry. Until this occurs, commercial mycelium composites are out of reach of our practice and of global clients who enlist us to build the future.

As the self-imposed deadlines agreed upon over multiple COPs loom closer, the built environment is still responsible for approximately 40% of annual carbon emissions with no indication of slowing down [2]. Therefore, regenerative materials such as mycelium composites are important 'carbon savers' especially for long-term asset managers who are surveying their stock, seeking compliance with the Paris Agreement and want to ensure the longevity of their asset. Within this context, we are acutely aware just how imperative it is to re-integrate the built environment within the natural environment as well as the policy and commercial challenges that inhibit us from doing so.

Ultimately, as architects and designers, we are beholden to people, the planet and our collective future. Beyond achieving Building Research Establishment Environmental Assessment Methodology (BREEAM), Leadership in Energy and Environmental Design (LEED) and WELL Building Standard and commitments to UK Architects Declare and the RIBA 2030 Climate Challenge, PLP Architecture conducts research via its research and development arm, PLP Labs, to work towards a more equitable and ecologically sane future (Figure 11.2).

PLP Labs serves the wider studio's architectural and urban work, as well as external clients with bespoke research; by operating outside the confines of traditional architecture. The group explores emerging social, environmental and technological ideas by conducting interdisciplinary research. Composed of a revolving group of in-house innovators and technical experts, we collaborate with leading experts from around the world.

Developing bio-based design, specifically with mycelium-based composites, is one of the many ongoing projects and is a cornerstone of the firm's exploration of speculative future materials. The multidisciplinary team that leads the project includes architects, product designers, model makers and sustainability specialists. PLP Labs dedicates one full-time team member to lead and coordinate such initiatives. While the initial, more speculative part of the research is carried out outside project work hours, once a clear brief is defined, more focused research is carried out within project parameters such as projected timelines and resource allocation.

This work plays a strategic role beyond its immediate outcomes: it frames the studio's broader thinking around sustainable innovation, informs thought leadership, supports

Figure 11.2 Modular Symbiocene demonstrator block. Image credit: PLP Architecture.

architecture tenders and serves as a compelling conversation point with clients and collaborators. As with much of the PLP Labs research, its value lies in setting a forward-looking agenda that influences projects across the firm and adds to broader industry conversations.

Research at PLP Labs is self-initiated and often conducted in partnership with academic collaborators, but it is fundamentally oriented towards commercial practice. Unlike academic research, which typically aims to expand theoretical knowledge, the projects seek to tackle relevant, pressing challenges in industry and create applicable, tangible results. The outputs of the work, such as material samples, augmented software to design in 3D space or public toolkits, are designed to inform practice. We share our findings with clients and broader industry audiences through reports, films, presentations and/or public events, supplementing traditional modes of academic discourse.

Although embedding research practitioners in the architecture and urban design firm is unique, conducting research is not. Architecture inherently involves research from understanding materials and systems to exploring context and culture, yet much of this knowledge is produced in isolation and remains under-utilized. By making research a formalized and visible part of the practice, PLP Labs captures and shares insights that would otherwise be lost. Positioning the firm at the forefront of design innovation gives a competitive edge and demonstrates added value to clients. Ultimately, research becomes a driver of future readiness across the studio.

2 Early experimentation

2.1 Mycelium composites

We developed a first-hand understanding of mycelium composites through hands-on experimentation, which was crucial for grasping the material's capabilities and potential. Our initial approach was small-scale; we began by growing mycelium cubes in casts. These early experiments helped us understand the growth process, its ability to fuse together and the challenges of cutting it apart.

Gradually, we introduced organic materials into the growing process, incorporating elements such as wooden dowels and muslin cloth. This allowed us to test and observe tolerances between aggregating components, control over surface finishes and webbing properties of mycelium biocomposites.

Our research took an intriguing turn when we discovered that mycelium could be grown within 3D-printed wood formwork rather than traditional casts, enabling the creation of intricate, complex shapes. Earlier tests had suggested that mycelium could grow over holes, bridge gaps and even 'climb' organic materials such as wood. To explore this further, we designed an experiment using a 3D-printed wood filament structure with a perforated surface. Our goal was to determine the maximum gap size mycelium could bridge and whether it would grow vertically on the perforated framework.

Over three weeks, the mycelium successfully bonded with the wood filament and did indeed 'climb' onto the perforated shape, which acted as a scaffold. Mycelium was able to grow over and cover perforations between 5 mm and 7 mm in diameter, confirming its adaptability to intricate geometries (Figure 11.3).

Figure 11.3 Mycelium-based composite with 3D printed wood formwork. Image credit: PLP Architecture.

2.2 Lost formwork

This discovery led to the development of the lost formwork technique, in which 3D-printed wood formwork serves as both a substrate and a cast for mycelium growth. This method has several advantages, including reduced shrinkage important for modular design, higher rate of design precision and reduced waste of the plastic cast.

As the mycelium conforms to the printed formwork, it becomes possible to produce complex angles and overhangs that would otherwise be difficult to achieve. This technique was instrumental in creating the Rooted Vase and the subtle ripples on the mycelium block showcased at Clerkenwell Design Week (Figure 11.4).

Despite its advantages, integrating mycelium with 3D-printed wood filament comes with trade-offs. The bonding and drying process takes longer – whereas a standard mycelium sample fully develops in ten days to two weeks, this technique requires three to four weeks. The drying must also be done at a lower temperature otherwise the outermost mycelium layer will start to crack. However, when these conditions are met the process seamlessly combines the beauty and texture of biophilia through printing technology.

2.3 Testing: compression and three-point bending

Alongside our design experimentation, we conducted compression and three-point bending tests on three types of samples:

- loosely packed substrate [LP];
- densely packed substrate [DP];
- densely packed substrate within 3D-printed wood formwork [3DP].

Figure 11.4 Mycelium-based composite vase, PLP Labs. Image Credit: PLP Architecture.

The 3DP formwork did not significantly affect the compression strength; in fact, it was slightly weaker than both the LP and DP samples. This suggests the compressive strength is reduced as the mycelial skin is formed outside the timber formwork. It may also be the case that owing to the slightly flexible nature of the timber printed formwork, the substrate packing has some gaps and cavities left over. However, the 3DP formwork made a substantial difference in tensile strength, as the formwork provided additional flexibility and support.

This implies the sacrificial shell of the 3DP formwork provided some tensile reinforcement, likely because of the flexible nature of the PLA/wood filament used in the 3D printing process.

For the compression tests, five samples were created for each type, each measuring 100 × 100 × 100 mm. Each sample was subjected to a load of up to 10 kN (approximately 1000 kg of force) over a period of five minutes, until failure.

At peak compression loading, the LP and DP samples displayed behaviour similar to foam-like materials, bulging in the middle prior to failure. This behaviour suggests that the strength of the samples is influenced by their density and internal structure. The 3DP samples acted slightly differently and had more of a smooth, uniform surface. At peak load, these samples fractured along the seams of the 3D-printed formwork, indicating that the compression was contained within the formwork. After testing, we observed that the samples slightly sprang back after compression.

The three-point bending test followed a similar approach, with five samples of each type, each measuring 200 × 30 × 30 mm. The test applied a load at the centre of each sample while the two side rollers held the sample in place.

Owing to the nature of the bending test and the smaller dimensions of the samples, the loads applied were much lower than in the compression tests. Still, during the testing, the samples deflected quickly under low loads. We observed subtle bending, along with a minor surface crack at the centre bottom of the samples. The DP samples exhibited much higher load capacities than the LP samples, and the 3DP outperformed the other two sample types by a large margin.

2.4 Lifecycle assessment

In addition to design and physical testing, we explored the environmental impact of 3D-printed wood and mycelium by conducting a lifecycle assessment (LCA) to compare the material's carbon footprint.

We calculated the embodied carbon of our mycelium composite material to be 5.8 $kgCO_2/m^2$, this is substantially less carbon than clay bricks (206 $kgCO_2/m^2$) and concrete (94.8 $kgCO_2/m^2$). This is based on a 10 × 10 × 10 cm, 153 g sample. The LCA methodology includes transportation emissions for both the substrate and filament used in 3D printing. Owing to the low density of these materials, transport emissions are relatively insignificant compared with other factors. Manufacturing emissions (A3 stage) come from preparing the formwork and sterilizing/dehydrating the finished product. This stage has the highest carbon emissions owing to the energy required for extended processing times.

As per the above methodology, a mycelium block produces almost zero carbon emissions per unit and could even be carbon negative. Carbon capture during the plant growth phase (A1 stage) helps offset emissions from later stages of production. Additionally, if the mycelium block is recycled or returned to the soil, it could have a net negative carbon impact. In comparison, a mycelium block offers a clear carbon advantage over a clay brick or solid concrete block, based on their standard cubic metre sizes.

However, the comparison between a mycelium block to a clay brick or a concrete block is not as straightforward at this time for two key reasons. First, it is difficult to compare the emissions of materials that are widely used, such as clay and concrete, to more niche materials such as mycelium. Second, the carbon emissions of clay or concrete, as reported in Environmental Product Declarations, come from industrial-scale production, while mycelium is still in the early stages of commercial development.

Hypothetically, industrial-scale production of mycelium blocks would require more electricity, but it could also benefit from renewable energy sources and vertical farming practices. Just as the industrial production of clay bricks and concrete blocks has been optimized over the past century through automation and prefabrication, we assume that similar efficiencies will be achieved for biomaterials with time and development.

3 Design showcases

In 2023, we showcased our innovative mycelium 3D-printed wood technique at major design festivals in London, including Clerkenwell Design Week, London Design Festival and London Design Biennale. At Clerkenwell Design Week, we debuted as what Glenn Albrecht described as a 'Symbiocene Demonstrator'. Our installation featured a modular system of 84 mycelium biocomposite blocks, which had been grown in our office over the course of three months (Figure 11.5).

The unique shapes of these blocks allowed for a variety of configurations, such as partitions, seating, planters and tables, all displayed at the event. The installation was located at the heart of Clerkenwell Design Week, just off Clerkenwell Road in St John's Square, and was attended by a record-breaking 37,725 visitors. During these showcases, we shared our journey of experimentation and collaboration with the public, introducing many to the concept of the Symbiocene.

The installation consisted of three layers, starting with a foam tile layer and MDF board beneath to even out the cobblestone ground and stabilize the structure. A cork pattern was then laid over the MDF board for aesthetic purposes, complementing the

Figure 11.5 PLP Labs team members at Clerkenwell Design Week 2023. Image credit: PLP Architecture.

modular mycelium biocomposite blocks. The largest structure on site was a partition made of 38 mycelium biocomposite blocks. The partition followed a repeating, interlocking hexagon pattern, with the short sides of the blocks connected to form the hexagons. At the 2023 London Design Festival and London Design Biennale, we presented a smaller version of the partition, featuring the same hexagon pattern. This unique block shape allowed for even more configurations beyond the furniture showcased at the events.

4 Material assessment

Mycelium biocomposites present a promising opportunity for integrating biomaterials into architectural design and construction. While there are significant challenges to widespread commercial adoption, the potential for further development is considerable. Mycelium biocomposites have also garnered widespread interest within architectural practice and academic research, as evidenced by the increasing number of studies and design experiments.

Despite the initial excitement, consumer trust and industry validation remain major obstacles. Concerns persist regarding the material's fire rating, structural performance, aesthetic appeal and issues such as allergic reactions and odour. External durability, fire resistance and waterproofing are some of the key hurdles to overcome before mycelium products can be used globally.

Therefore, it makes sense to explore the use of mycelium alongside more familiar natural materials such as timber and clay. Further research into bio-based coatings and genetic engineering may offer solutions to some of these challenges over time. However, a widespread shift away from fossil fuel-based materials will only occur if there is strong demand from the ground up.

Mycelium remains an intriguing opportunity, especially for carbon-neutral, large-scale commercial production. The process requires minimal energy, and mycelium biocomposites can be mass-produced using locally available agricultural waste and regional mushroom species. Growing conditions are relatively simple to replicate with the right expertise. Additionally, mycelium can be fabricated into various shapes and sizes, being form agnostic – an exciting prospect for designers.

Mycelium products are already available on the market, such as Mogu's Foresta system, and have proven to be a viable building material, as demonstrated by MycoHab developed by Red House Studio Architecture. Countless other examples exist, from lampshades, textiles and packaging to panelling and furniture by firms located around the globe. To this effect, every time we try to draw a line under it and move on, there is something else exciting happening in the realm of fungi and mycelium.

5 Conclusion

Our exploration of mycelium biocomposites within architectural design highlights both the immense potential and the pressing challenges of integrating biomaterials within the Architecture Engineering Construction (AEC) industry. Our role as a research design collaborative within an architectural practice provides us a unique perspective in developing materials and the opportunity to work towards practice adoption.

Through hands-on experimentation, material testing and real-world installations, we have demonstrated the viability of mycelium in creating adaptable, low-carbon and aesthetically compelling structures. The lost formwork technique, in particular, opens new possibilities for design precision and modularity, aligning with the principles of sustainable and regenerative design. The lifecycle assessment estimation of mycelium's carbon footprint to be 5.8 $kgCO_2/m^2$ is promising for the AEC industry's options for carbon conscious materials.

Yet despite its promise, mycelium remains on the fringes of commercial viability. Questions of durability, fire resistance and large-scale manufacturing must be addressed before it can be widely adopted. Industry accreditation will be critical in bridging this gap, as will continued research into hybrid applications with timber, clay and bio-based coatings.

Looking ahead, as an international practice we have the responsibility and means to introduce accredited regenerative materials into mainstream projects. This is critical and timely as the built environment grapples with its culpability in the climate crisis. As a research group positioned outside academia, we aim to continue pushing the boundaries of material research to engage international clients and inform major commercial projects. Connecting regenerative and speculative future materials to real world practice is important and pressing work now more than ever.

The industry integration of mycelium will take time, but mycelium offers a tangible path towards a more ecologically conscious built environment. Collaboration between the AEC industry, material scientists, ecologists and biologists, industry leaders, accreditation bodies and policymakers will be essential in overcoming the barriers to adoption. By fostering cross-disciplinary partnerships and advocating for regulatory frameworks that support and celebrate the qualities of biomaterials it may be possible to accelerate their transition from experimental to practical applications. In the meantime, we look forward to continuing learning from, collaborating with and being surprised by living materials.

References

1. Albrecht GA. *Earth Emotions: New Words for a New World*. Cornell University Press; 2019.
2. World Green Building Council (2022). *Embodied Carbon*. [online] World Green Building Council. Available at: https://worldgbc.org/climate-action/embodied-carb

CHAPTER 12

Mycelium Materials and CNC Biofabrication: Process and Perception

PHILLIP GOUGH and YUXIN GONG

1 Introduction

The pressing need for a transition to a sustainable, circular economy can hardly be overstated. Research in sustainable design includes not only design and fabrication with sustainable materials, but also processes for creating novel materials themselves. This introduces technical and ethical challenges for designers, as the livingness of the material becomes an attribute for exploration [1]. User perception of novel biomaterials extends these questions, as unfamiliarity of novel materials will impact the user experience. Though these challenges persist, the benefit of some living organisms, such as fungi, to transform low-value waste into useful materials presents an exciting opportunity. Composite mycelium materials (CMMs) are typically formed by filling moulds with inoculated substrate. Individual moulded pieces can be fused together through the natural process of anastomosis (where two fungal networks join into one); however, this fabrication process is limited owing to the inherent nature of fabrication through moulding.

Digital fabrication through automated additive (e.g., 3D printing) or subtractive (e.g., cutting or milling) procedures using Computer Numerical Control (CNC) has allowed designers to translate an intangible digital design to physical objects [2]. Several studies have demonstrated the use of various biomaterials in additive manufacturing, presenting a compelling value proposition in view of the entire lifecycle, by providing sustainable material options [3]. A systematic review conducted in 2023 on the use of biomaterials for achieving UN Sustainable Development Goal 9 identified digital fabrication,[1] including 3D printing, as an opportunity to maximize desirable qualities (such as biodegradability and mechanical properties) of biomaterials [4]. Analysis of the mechanical properties of the biomaterials often includes optimization of the fabrication process, such as compensation for material shrinkage when cured [3]. In a circular economy, localized waste materials vary or species of organisms employed to produce the biomaterial may be unavailable. This may have impacts on the design opportunities that are relevant across different contexts.

1 Sustainability Development Goal 9 relates to 'Innovation, Industry and Infrastructure' and is focused on objectives relevant to the private sector. These include sustainability, resilience, reliability, inclusive development of infrastructure and industrialization through scientific and technological innovation and policy development.

Phillip Gough and Yuxin Gong, 'Mycelium Materials and CNC Biofabrication: Process and Perception' in: *Mycology for Architecture: Methods and Applications of Mycelium Materials in the Built Environment*. Pelagic Publishing (2026).
 DOI: 10.53061/SJOX6884

The primary contribution of this chapter is results from an initial study that combines a novel combination of CNC fabrication with a fungi-based biomaterial. As researchers at a school of Architecture, Design and Planning, we have access to three key components that we frequently use to create CMMs. First, sawdust from CNC fabrication in our school (primarily Jelutong, *Dyera costulata*, a tree listed as a 'least-concern' species [5]); second, waste coffee grounds from local cafés and; third, *Ganoderma steyaertanum*, also known as Australian Reishi, a white-rot fungus that is endemic to our region. Like many Reishi species, this particular mushroom has a bioaffordance [6] of binding together organic material as it grows and producing a thick outer layer of mycelium, which is typically white and brown. It also produces an edible fruiting body, which is valued for medicinal purposes, but by controlling the growth environment, the production of the fruiting body can be slowed or prevented, leaving the colonized substrate material not completely consumed. In addition to the description of CNC biofabrication, we also contribute a study of the initial responses to the material from participants, using the Material to Experience in four levels (Ma2E4) toolkit [7], a holistic approach to characterization of materials beyond their technical qualities through structured user workshops.

So, is living CMM a candidate for digital fabrication? Our hypothesis was that a thoroughly colonized Reishi CMM would be able to regrow an outer layer of mycelium over a 3D design after CNC milling. This creates opportunities for biofabrication, through the ability to prepare stock biomaterial that can be formed on demand, eliminating the need for single-use moulds, and, with advanced six-axis robotic fabrication, forming shapes that are not able to be created by moulding. These could also include interlocking forms that can be combined into single, complex designs. Past research has shown that CMM can be used in digital fabrication, such as laser cutting [8], but this damages the mycelium outer layer, which presents an infection risk from wild mould on the now exposed sawdust or coffee grounds. We posit that regrowth may be successful, as working with CMM often involves breaking up the substrate after a period of bulk growth, which may be when using smaller moulds than the sample or before adding inclusions, such as waste coffee grounds if starting from commercially-available substrate (e.g., see [9]). CMM can be grown in moulds after the substrate is broken down by mechanical means, such as in a food processor, if the substrate is fully colonized and the processor is sanitized with 70% isopropyl alcohol. Partially breaking down of the CMM through CNC milling is feasible but presents risks owing to infection of ubiquitous mould spores. Treating the mycelium block like the stock material frequently used in CNC machining, we wished also to understand the types of designs that would work well using CMM CNC fabrication.

2 Methods

2.1 CMM sample preparation

We created a CMM from locally available and culturally relevant waste. As shown in Table 12.1, we targeted the waste from CNC machining in our school to demonstrate a circular approach. Additionally, we wished to understand how the material deforms as it regrows and dries, to determine what types of designs are best suited to mycelial growth. This deformation is due to two factors, mycelium growing along the outside machined surface and shrinkage owing to drying, as CMM is about two-thirds water before drying.

Table 12.1 CMM preparation and fabrication process. Ratios for each material are shown at the top and the process is shown underneath.

Ingredient	Role	Mass (g)
Sawdust (Jelutong)	Substrate	500
Spent Coffee Grounds	Nutrient Additive	300
Water	Hydration	1,600
Ganoderma steyaertanum on grain spawn	Mycelium species	75
Stage	**Process**	**Time/Temperature**
Sterilization	Substrate is sieved in a colander to remove large particles.	121°C (15 psi) for 35 minutes.
	Water and coffee grounds mixed in a large growbag and sterilized.	
Bulk growth	After the growbag is allowed to cool, the mycelium species is added and mixed thoroughly.	30°C for 3 weeks (approx.)
	The bag is sealed and pressed into a rectangular cottle board mould.	
	Mixture and cottle boards are placed in an incubator.	
Machining	The block of mycelium is CNC milled according to the design. Excess sawdust is manually removed with a brush.	n/a
Regrowth	The block is returned to the incubator and placed in a sealed container or wrapped in aluminium foil to reduce moisture loss and reduce infection risk.	1–2 weeks.
Drying	After a layer of mycelium is grown over the outside, it is left open in the incubator to dehydrate.	1 week – or until approximately a third of total mass.

Four forms were designed, machined and grown. We used an Axiom Precision CNC with a 6 mm cutting tool for all processes, which was programmed with a 4 mm maximum plunge depth using Autodesk Fusion. To minimize the time that the substrate was exposed to air, and ubiquitous mould spores, we did not perform finishing cuts, relying on rough cutting profiles to create the shapes. The patterns were designed with this consideration (e.g., pocket cuts could not be smaller than 6 mm, and there would be a 3 mm radius on all internal pockets corners). The CNC fabrication in our school workshop secures stock material to the work bed using adhesive tape. As this would not work on the wet mycelium, the blocks were secured to a piece of MDF with screws, which was clamped to the CNC bed. The designs therefore have rectangular zones in each corner to provide sufficient clearance to safely mill the material. A total of four blocks were machined in two experiments. The first experiment allowed us to benchmark the performance of our CMM mixture using two blocks. The second was intended to employ features that grew successfully in a new design.

2.2 Fabrication evaluation

In the first experiment, two blocks were grown to test five basic geometric elements – line, ramp, angle, curve and step. Each element included multiple features with different widths and depths to compare performance and identify the optimal geometric elements and design dimensions for future pattern designs, shown in Figure 12.1.

- Line: 10 features with two depths and three wall thicknesses (Design T1 section A–A).
- Ramp: 9 features with three depths and five widths (Design T1 section B–B).
- Angle: 10 features with two depths × five angles (Design T1 sections C–C and D–D).
- Step: 12 features with three step widths × four step heights (Design T2 section B–B and C–C).
- Curve: 8 features with a curved profile × three depths (Design T2 Section D–D and E–E)

The second evaluation utilized ramped and curved cut features to create polyhedral and wave designs on the surface of two CMM blocks. In the polyhedron design a pattern is made up of ramps (Figure 12.2), with a depth of 24 mm. The central large square has a side length of 150 mm and is composed of 12 symmetrical faces. The side grooves have dimensions of 40 × 30 mm. The wave pattern is composed of curved shapes (again see Figure 12.2), with a height of 28 mm. Each concave/convex section has dimensions of 50 × 30 mm.

A summative evaluation of the CNC machining process includes two approaches. First, a quantitative evaluation records changes in dimension at four stages: post-processing, mycelium regrowth, compression and final drying to benchmark the material's change through the machining and drying process, compared with the ideal 3D model that was created. The measurements were taken with digital vernier callipers after machining and after drying. A second, qualitative evaluation focuses on surface quality, shape outline and aesthetics. Since the realization of the physical material is not perfect as it is grown, the quantitative data alone cannot fully reflect the design's performance. We identified qualities to evaluate on a five-point scale, ranging from -2 (ineffective) to 2 (effective) as a heuristic approach to evaluating the appropriateness of the design feature as rendered by the material.

2.3 User evaluation

After the researcher's characterization, we conducted a small user study, to further explore a more comprehensive understanding of the material. The Ma2E4 toolkit by Camere and Karana [7] is used to guide the user testing, which evaluates the experiential characteristics of against four types of experience with the material: performative, sensorial, affective and interpretive. Our informants completed a workbook as they experienced the material, with an expanded list of prompts to the material included in the original Ma2E4 workbook [7] to include the biological nature of the material we provided. The questions and responses from participants are included in the Appendix to this chapter.

3 Results

3.1 Experiment 1

Regrowth after CNC machining was successful, and the CMM 'sawdust' (material removed from the CMM growth) was able to be pressed into small moulds to grow again. The thoroughly colonized material showed resistance to wild mould infection even though substantial parts of the external mycelium layer were removed by the

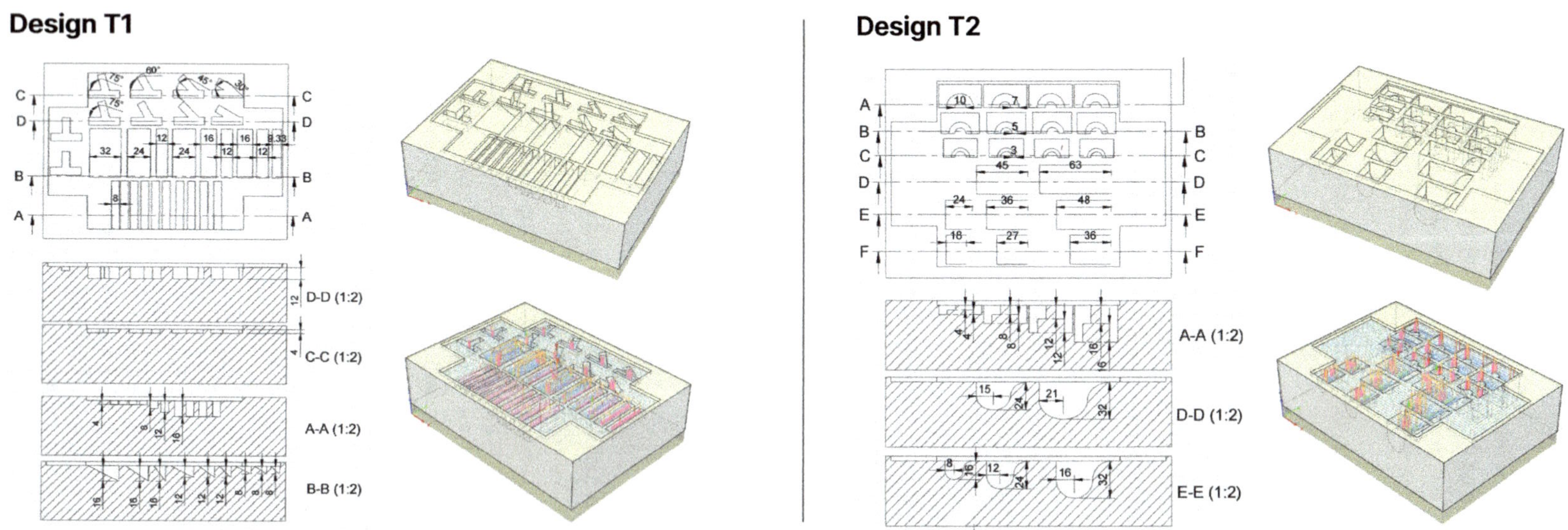

Figure 12.1 Experiment 1 designs. Left: features of Design T1 to compare wall thickness (section A–A), ramps (B–B) and angled cuts (sections C–C and D–D), Right: Design T2 showing stepped (Section A–A) and curved (sections D–D and E–E) features. Dimensions shown are in mm. The design and toolpath are shown to the right of each design.

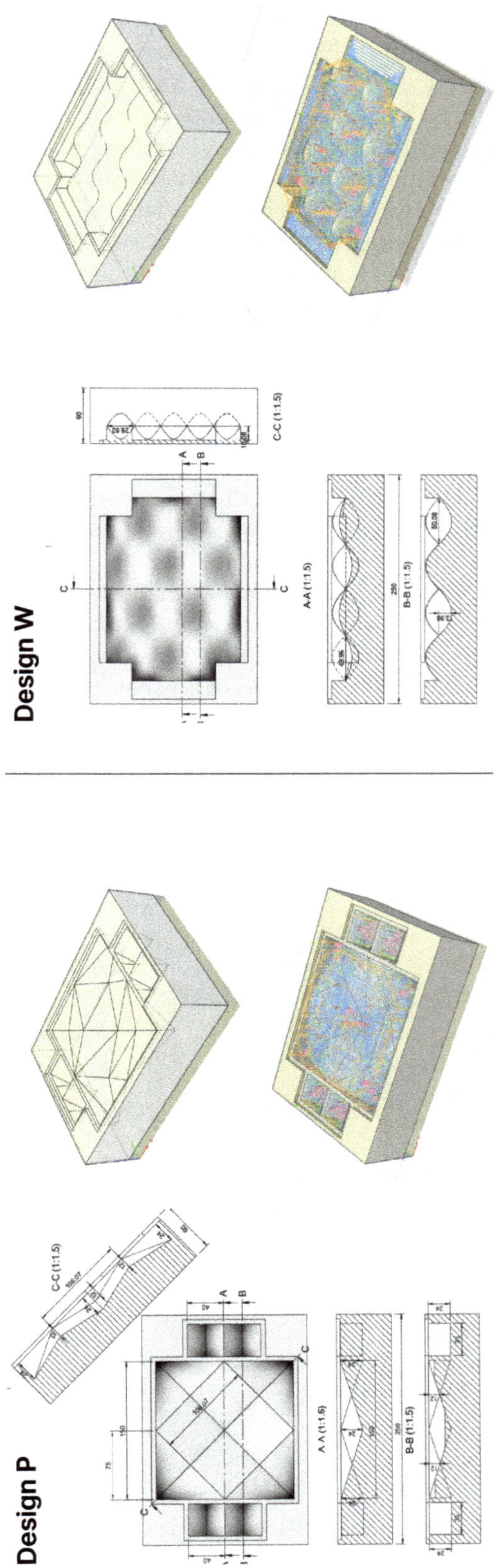

Figure 12.2 Experiment 2 designs. Left: Design P showing a polyhedron design, Right: Design W showing a wave pattern design.

process, and the wet, live CMM was screwed to a piece of MDF for the process while being cut with a metal bit. The material itself shows behaves as an elastic material, as has been shown in previous literature [9], especially as it was cut before drying, and the sawdust mixture we are using is fine in comparison with hemp-based materials used in some commercial kits [8]. This led to deformation of the design after growing, as the material deformed as the cutting bit pressed against the material. After machining, the blocks were wrapped in aluminium foil to reduce water loss and returned to the incubator. Figure 12.3 shows the progress of growth after CNC machining and cleaning. The first stage of the CNC machining was a clearing process that created a level surface from which the designs were cut. Sample T1 shows this clearing cut did not create a perfectly level surface, as it is difficult to create a flat surface inside the cottle boards. After this process, T2 was flipped, and the design was carved from the underside of the material.

The samples were kept in the incubator until machining, so the fungus was active and showed visible regrowth after 24 hours. After five days of growth, the mycelium network had bound together across the small gaps in the designs. This was then pressed down with a plastic spatula to flatten the mycelium against the surface of the substrate before the CMM was dried by being left uncovered in the incubator. Figure 12.4 shows the three measured features deformed after being dried.

A qualitative review of individual features determined that small features, such as narrow walls, were not reproduced effectively by the material and were difficult to identify even when we knew where the features were. Features such as ramped and curved surfaces were rendered by the regrowth of the material in an acceptable manner and were developed into the 3D models that were carved for the second experiment.

3.2 Experiment 2 and user studies

The application of these features to the second pair of designs demonstrates an application of ramped and curved surface features. Based on the changes in dimension

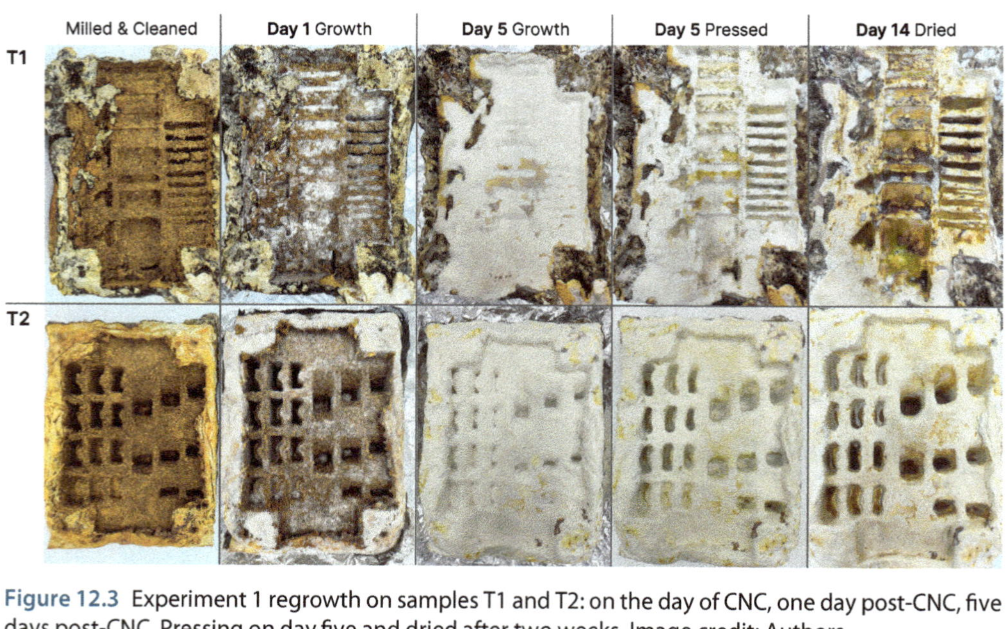

Figure 12.3 Experiment 1 regrowth on samples T1 and T2: on the day of CNC, one day post-CNC, five days post-CNC, Pressing on day five and dried after two weeks. Image credit: Authors.

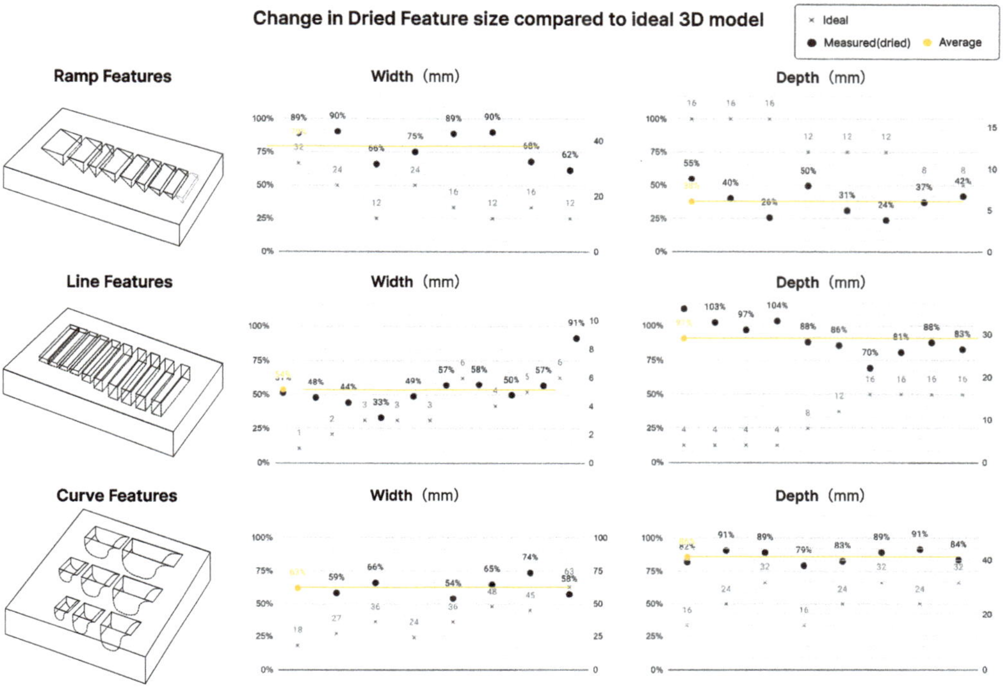

Figure 12.4 Features for all experiments isolated: ramp, line and curve features compared with the ideal 3D model by width and depth as measured after drying.

Table 12.2 Qualitative rating used to determine features to use in second experiment.

	Dry dimension compared with Ideal dimensions			Rating				
Feature	**Width**	**Depth**	**Observations**	**-2**	**-1**	**0**	**1**	**2**
Experiment 1								
Line	54%	91%	Distorted shape, unstable structure	○	○	□	○	○
Ramp	79%	38%	Clear shape, even surface	○	○	○	○	□
Angle	n/a	n/a	Vague shape, difficult to recognize	□	○	○	○	○
Curve	63%	86%	Clear shape, even surface	○	○	○	□	○
Step	n/a	n/a	Vague shape, difficult to recognize	□	○	○	○	○
Experiment 2								
Polyhedron pattern	95%	103%	Concave sections clear even surface, convex sections difficult to machine	○	○	○	□	○
Wave Pattern	96%	114%	Concave sections have a clear shape and even surface, convex sections difficult to machine effectively.	○	○	○	○	□

from the initial experiment, we were able to transform the design in software before producing the CNC toolpaths. This created a more accurate rendering of the ideal form in the material, as shown in Table 12.2, where position and horizontal dimensions of elements were within 5% of the 3D model, though depth of features was less consistent across the two designs.

Figure 12.5 Experiment 2 results. A: CNC machining the blocks. B: Machined samples before cleaning. C: Machined samples after brushing the surface to remove excess. D: After two days' growth in the incubator. E: After three days' growth in the incubator and pressing the surface. F: After drying in the incubator. Image credit: Authors.

Figure 12.6 Experience map synthesizing the user responses collected during the Ma2E4 workshops.

The workshop using the Ma2E4 toolkit revealed insights into four material experiences with CMMs. Eleven participants were recruited for these one-hour workshops and were given time to explore interactions with Design P and Design W before responding to the survey. The responses are synthesized into a map in Figure 12.6, and responses to each experience are included in the Appendix. Most notably, the responses from participants described their affective responses as intense but pleasant, selecting terms such as surprise and curiosity. This points to a positive response, which participants agreed that, at a sensorial level, the material was naturally eco-friendly, and at an interpretive level, also appeared hand-crafted.

4 Discussion and conclusion

The design considerations for using CMM in digital fabrication are similar to forming CMMs by casting the material into a mould. The key considerations are the selection and sterilization of the substrate materials, and selection and cultivation of fungal species. The CNC process introduced some additional considerations, such as deformation at the time of machining (owing to the elasticity of the material) and during drying. The challenge is that in any context, the mixture of substrate materials will impact the benchmark, so further research is needed to create digital tools to help designers compensate for unique material properties, such as material profiles or plug-ins that can be used inside CAD software, which are optimized for general CMM performance and tunable by the user. While previous attempts to CNC machine dried mycelium were not satisfactory [8], our substrate, mostly based on fine sawdust, was able to perform well while it was still alive. We were surprised that after less than one week, mycelium growth on the milled surface was so prolific that it needed manual post-processing for the designed forms to be visible. While this was the case, deformation due to drying meant that detailed features were not effectively produced in our approach. However, this approach could be developed further, as CMMs have been shown in previous research to create highly detailed forms, also using Reishi mushrooms [9].

The positive tactile experience users reported with the material shows that there are opportunities to create interactive experiences that leverage the interpretive and effective responses that users had to the CMM samples. The response at an interpretive level of the material being 'hand-crafted' presents an interesting dichotomy between the designer, using digital designs, the fungi, an active participant in the representation of the design, and the user, who experiences the product. This expression of autonomy by the fungi leverages circular and natural processes in the creation of user experiences, but requires a reconsideration of the position of the designer.

CMMs present an interesting candidate for digital fabrication using CNC machining. Our initial study demonstrates that the selected Reishi species can bind waste timber sawdust from the CNC process to create a block of material that can be CNC machined. We were able to compensate for deformation in drying by benchmarking the material mix and create designs that prompted surprise and curiosity from users through the material experience. Further resources such as digital tools and models to simulate the performance of the material will create further opportunities for a circular approach to fabrication using CMMs.

5 Limitations and future work

This project describes an initial exploration of digital fabrication and perception of CMMs using CNC machining. It utilizes waste materials that are relevant to our local economic and cultural context, which is consumed by *Ganoderma steyaertanum* fungi. We used a three-axis CNC to demonstrate that the process can be used to reuse substrate that is based on waste from CNC machining. In the future, we intend to expand this by utilizing six-axis machining to create complex forms that are difficult to replicate without digital fabrication technologies.

Acknowledgements

The user studies conducted for this workshop were approved by the University of Sydney, Human Research Ethics Committee approval number 2023/223.

References

1. Kim R, Risseeuw C, Groutars EG, Karana E. Surfacing Livingness in Microbial Displays: A Design Taxonomy for HCI. In: Proceedings of the 2023 CHI Conference on Human Factors in Computing Systems Hamburg Germany: ACM; 2023. pp. 1–21. Available from: https://dl.acm.org/doi/10.1145/3544548.3581417
2. Caneparo L. Digital Fabrication in Architecture, Engineering and Construction. Dordrecht: Springer Netherlands; 2014. Available from: https://link.springer.com/10.1007/978-94-007-7137-6
3. Dritsas S, Halim SEP, Vijay Y, Sanandiya NG, Fernandez JG. Digital fabrication with natural composites: Design and development towards sustainable manufacturing. Constr Robot. 2018 Dec;2(1–4):41–51.
4. Gera R, Chadha P, Banerjee SP, Sharma M, Pandey AK, Kampani S, et al. A narrative review on use of biomaterials in achieving SDG 9: Build resilient infrastructure, promote sustainable industrialization and foster innovation. Tummala SK, Kosaraju S, Bobba PB, Singh SK, editors. E3S Web Conf. 2023;391:01180.
5. Ariffin AZA, International Union for Conservation of Nature (IUCN). IUCN Red List of Threatened Species. 2023 [accessed 2025 Jan 8]. IUCN Red List of Threatened Species: Dyera costulata. Available from: https://www.iucnredlist.org/species/33212/67736172
6. Gough P, Yoo S, Tomitsch M, Ahmadpour N. Applying Bioaffordances through an Inquiry-Based Model: A Literature Review of Interactive Biodesign. Int J Human–Computer Interact. 2021 Apr 5;37(17):1–15.
7. Camere S, Karana E. Experiential Characterization of Materials: toward a toolkit. In Storni, C., Leahy, K., McMahon, M., Lloyd, P. and Bohemia, E. (eds.), Design as a catalyst for change - DRS International Conference 2018, 25–28 Jun, Limerick, Ireland. 2018. pp. 1687–1705 Available from: https://doi.org/10.21606/drs.2018.508
8. Vasquez ESL, Vega K. From plastic to biomaterials: prototyping DIY electronics with mycelium. In: Adjunct proceedings of the 2019 ACM international joint conference on pervasive and ubiquitous computing and proceedings of the 2019 ACM international symposium on wearable computers. New York, NY, USA: Association for Computing Machinery; 2019. pp. 308–11. (UbiComp/ISWC '19 adjunct). Available from: https://doi.org/10.1145/3341162.3343808
9. Gough P, Perera PB, Kertesz MA, Withana A. Design, mould, grow!: a fabrication pipeline for growing 3D designs using myco-materials. In: Proceedings of the 2023 CHI conference on human factors in computing systems. New York, NY, USA: Association for Computing Machinery; 2023. (Chi '23). Available from: https://doi.org/10.1145/3544548.3580958

Appendix: Workshop materials

The figures included here include user responses to the materials collected during the workshop.

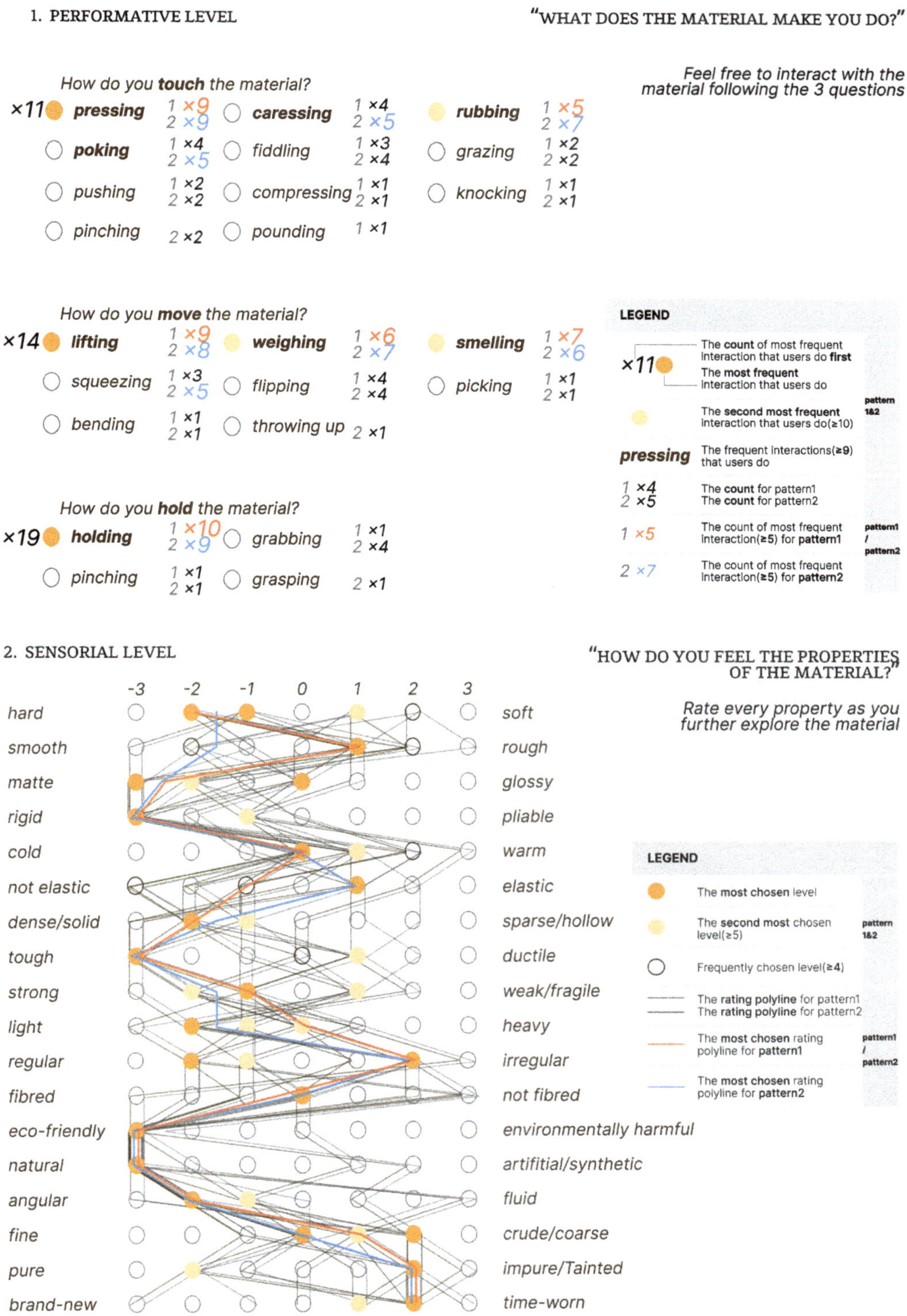

3. AFFECTIVE LEVEL

"WHAT EMOTIONS DOES THE MATERIAL ELICIT?"

List of Emotions

- ×9 **surprise**
- ×8 **curiosity**
- ×5 **comfort**
- ×4 **safety**
- ×4 **relaxation**
- ×3 amusement
- ×3 fascination
- ×3 respect
- ×3 reliability
- ×3 gentleness
- ×2 attraction
- ×6 **confusion**
- ×2 doubt
- ×1 tension/nervousness
- ×1 anxiety
- ×1 disappointment
- ×1 worry
- ×1 distrust
- ×1 fear

LEGEND

- The **most chosen** one
- The **second most** chosen one
- safety — The **primary emotion**
- **safety** — Frequently chosen emotion

pattern 1&2

Put all your feelings into the space of diagram and choose primary one(s)

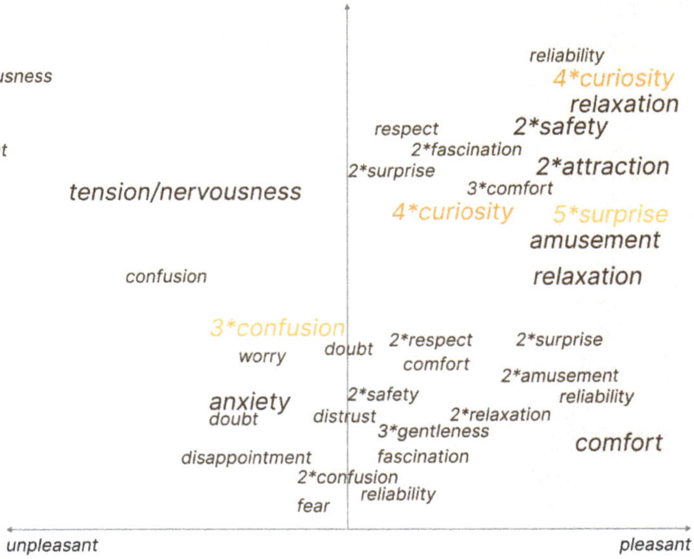

4. INTERPRETIVE LEVEL

"WHAT DO YOU ASSOCIATE WITH THE MATERIAL? HOW WOULD YOU DESCRIBE??"

Set of Meanings

Further imagine the scene in your mind for every association

×7 **natural** like... | ×6 **hand-crafted** | ×5 **cozy** | ×3 **toy-like** | ×3 **wild**

×3 **food-like** | ×3 **nostalgia** | ×2 meaningful | ×2 elegant like... | ×2 sober | ×2 feminine

×2 futuristic like... | ×1 calm | ×1 strange | ×1 unnatural | ×1 messy | ×1 simple

CHAPTER 13

Mycelial Skin in Biodigital Fabrication: Hybrid Pathways for Upcycling Fired Clay

BAHAREH BARATI and FENGYUAN WANG

1 Introduction

Soil is not merely a substrate but a structurally dynamic and biologically active system, shaped in part by the behaviour of fungi [1]. Fungi and soil are in continual interplay: the filamentous structure of mycelium both adapts to and reorganizes the spatial heterogeneity of the soil environment. Fungi promote soil aggregation through adhesion and enmeshment, and alter its porosity and hydrophobicity through biochemical activity. These ecological dynamics suggest a model of material reciprocity, where growth and structure co-emerge.

Clay, as a dense, mineral-rich form of soil, presents a particularly compelling case for exploring these dynamics within material design. Long valued for its plasticity and thermal properties, clay has historically been central to human practices of shaping and making. More recently, designers and researchers have begun to reimagine clay not simply as an inert medium, but as a biologically active host. For example, Back to Dirt by Studio Alea [2] uses raw clay as an organic mould for fungal growth, while Studio bioMATTERS [3] combines terracotta, industrial waste and mycelium to develop bioglazed surfaces on 3D-printed vessels. These works signal a broader shift toward sustainable and regenerative material ecologies [4, 5].

Our research extends this trajectory by focusing on a material that has been largely excluded from regenerative cycles: fired ceramic waste. Unlike air-dried clay, which can be rehydrated and reshaped, fired clay undergoes irreversible transformations, losing plasticity, increasing in density and becoming significantly less porous [6].

This project proposes a novel route for upcycling fired ceramic waste by leveraging the surface-binding and structural properties of 'mycelial skin': a thin, biologically active fungal layer that forms at the interface between substrate and environment [7]. Unlike bulk mycelium composites or engineered pure mycelium materials [8], mycelial skin functions as a performative interface, simultaneously contributing to mechanical reinforcement, surface cohesion and aesthetic expression.

To activate this potential, we draw on the emerging domain of biodigital fabrication, which brings biological growth and digital manufacturing into constructive dialogue. Since the 1990s, additive manufacturing of ceramics has evolved across industrial domains, from healthcare to construction [9, 10]. In particular, extrusion-based 3D printing of clay has enabled the production of intricate geometries using low-energy, low-waste

Bahareh Barati and Fengyuan Wang, 'Mycelial Skin in Biodigital Fabrication: Hybrid Pathways for Upcycling Fired Clay' in: *Mycology for Architecture: Methods and Applications of Mycelium Materials in the Built Environment*. Pelagic Publishing (2026). DOI: 10.53061/NDPI3349 techniques [11]. However, despite growing interest in this area, the formulation of printable clay pastes, especially those incorporating recycled or bio-integrated materials, remains under-documented in design research [10]. Moreover, the bioreceptivity of such pastes for fungal colonization is still largely unexplored [12].

In response, this study combines parametric clay printing, recycled ceramic additives and fungal cultivation to explore how *Ganoderma lucidum* can colonize and reinforce 3D-printed clay bodies containing up to 35% fired ceramic waste. In the sections that follow, we detail our biodigital fabrication process, evaluate the resulting samples through experiential and structural characterization, and reflect on the implications for hybrid material systems.

2 Optimizing clay paste for mycelial growth

This phase of the study focused on developing a clay-based paste suitable for both 3D printing and fungal colonization. The goal was to balance the structural requirements of additive manufacturing with the biological conditions needed for supporting mycelial skin formation.

Locally sourced fired ceramic waste, primarily composed of fired China clay and Ball Clay, was fragmented, sieved and milled to obtain fine particles suitable for paste formulation. This material was blended with virgin clay to form the base matrix. The initial formulations, adapted from Weng et al. [13], combined 25–50% recycled content with virgin bentonite. A mixture of organic additives, including glucose, flour, chitosan, yeast and calcium carbonate, was added to support fungal growth. The dry ingredients were blended and gradually hydrated with dechlorinated water to achieve a printable consistency.

Early tests, shown in Figure 13.1, involved shaping small clay bodies, sterilizing them by autoclaving and inoculating with *Ganoderma lucidum*. Among the first formulations, the paste with 25% recycled clay supported more extensive mycelial skin growth than the 50% variant. To reduce moisture loss and sustain colonization, glycerol was introduced as a hygroscopic agent. A formulation with 30% recycled clay and 7% glycerol produced denser and more cohesive skin (Figure 13.1 middle), though it required 16 days for full colonization. Further trials comparing 14% glycerol with a control revealed a trade-off: higher glycerol improved surface richness but delayed colonization (up to 36 days). These results indicate that glycerol concentration critically influences colonization dynamics, affecting both growth rate and aesthetic outcomes by modulating water availability and fungal metabolism.

Figure 13.1 Snippets from the early trials. Left: coiled clay forms used for iterative formulation testing. Middle: mycelial skin formed on 30% recycled clay with 7% glycerol. Right: overview of fully colonized hand-formed samples. Image credit: Bahareh Barati, Fengyuan Wang.

3 3D printing with 35% fired clay paste

Building on the optimized hand-formulated clay paste, which supported thicker mycelial skin formation, we proceeded to adapt the material for 3D printing applications. We used two 3D printers: [1] Z-Morph using the paste extruder head and [2] WASP clay printer. An overview of the clay formulation, hydration level and printer parameters are presented in Table 13.1.

Initial tests with the Z-Morph used the default setup for soft bio-pastes, driven by a stepper motor (NEMA 17) with a holding torque of ~0.4–0.5 Nm. However, our paste was too viscous for extrusion under these settings. We subsequently increased the hydration incrementally, and determined that at approximately 47% water content, the paste achieved a balance between consistent extrusion flow and post-print structural stability.

We printed two hollow cubes (90 × 90 × 90 mm) and manually carved rhombic perforations into the printed bodies, measuring 4–5 mm and 9 mm respectively, to visually assess the bridging behaviour of the mycelial skin. The decision to carve the perforations post-printing (rather than incorporating them directly into the digital 3D models) was driven by motor limitation, which required a larger nozzle diameter (6 mm) and the paste's high hydration level. These factors reduce precision and make it difficult to reliably produce clean perforations directly during printing. Both samples were fully colonized by mycelial skin within nine days, although base layer expansion was observed during incubation.

To achieve higher print fidelity and explore how mycelial skin might bridge perforated geometries (as observed in earlier hand-coiled trials), we used the WASP clay printer, which is designed for extruding thicker pastes using compressed air (Figure 13.3). We used the same hydration level as the hand-formed clay samples, approximately 43% of the dried weight. Because our formulation had higher hydration and lower viscosity than conventional ceramic mixes, we deviated from the standard settings: the flow rate was reduced to 10% and internal pressure was set to 0.4 MPa.

A key challenge during this phase was air entrapment, likely caused by the 1% yeast content in the final formulation. While yeast-driven gas release is well documented in baking, its behaviour in ceramic extrusion contexts remains unexplored. To mitigate bubble formation, the paste was allowed to rest at room temperature for two to three hours, facilitating gas escape. The mixture was then gently stirred in a single direction to further minimize trapped air before loading into the syringe.

Table 13.1 An overview of the optimized clay formulation and printer parameters.

Recycled Clay Formulation (extrudable and bioreceptive to fungi)	**Printer parameters** (parameters correspond to the slicing software)	
Recycled Clay: 35%	*Z-Morph*	*WASP*
Bentonite: 49%	[parameters set in Cura slicer]	[parameters set in Simplify 3D]
China Clay: 5%	**Nozzle diameter**: 6 mm	**Nozzle diameter**: 3 mm
Glucose: 2%	**Line width**: 5 mm	**Extrusion multiplier**: 0.2
Calcium Carbonate: 1%	**Layer height**: 2 mm	**Layer height**: 1 mm
Chitosan: 1%	**Initial layer height**: 1.2 mm	**Print speed**: 20 mm/s
Flour: 6%	**Print speed**: 6 mm/s	**Infill**: 0%
Yeast: 1%		
Glycerine: 3% of the total powder weight		
Hydration: Dechlorinated water weight/ paste weight x 100.	(47% hydration)	(43% hydration)

Ingredients		Ratio
Recycled Clay		35
Bentonite		49
China Clay		5
Glucose		2
Calcium Carbonate		1
Chitosan		1
Flour		6
Yeast		1
Glycerin (percentage of total powder weight)		3
Hydration	Handcraft	43
	ZMorph	47
	WASP	43

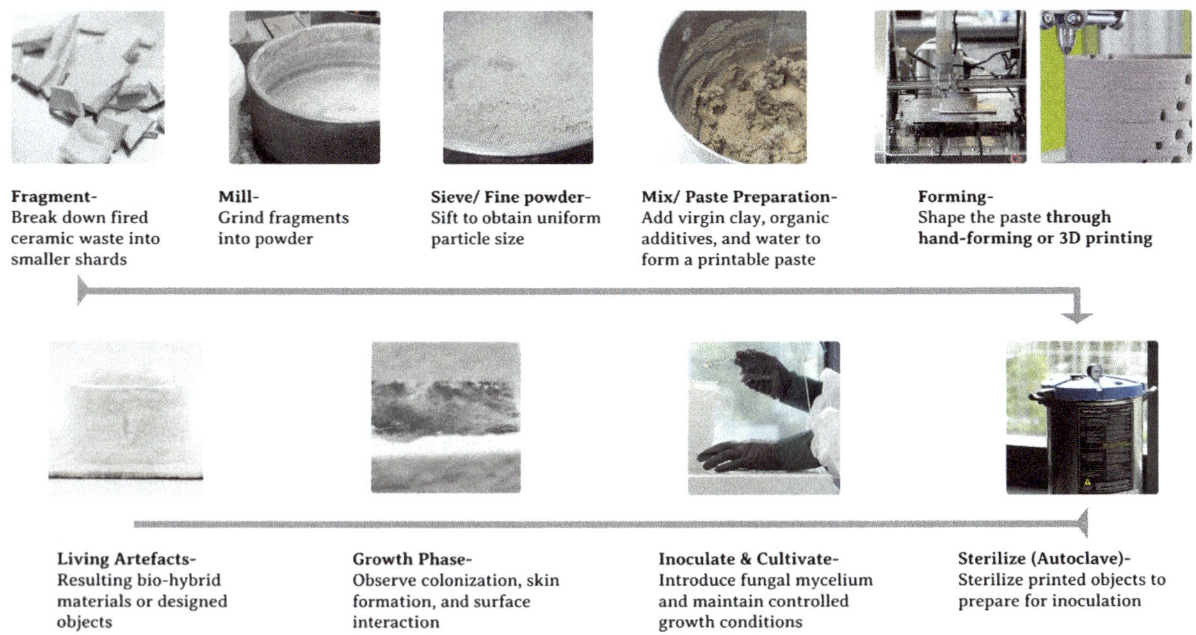

Figure 13.2 Operational workflow of the biodigital fabrication process. Left: overview of the clay formulation. Right: the experimental sequence, from material preparation and paste formulation to shaping (handcraft or 3D printing), sterilization, inoculation, incubation and the final development of bio-hybrid materials. Image credit: Bahareh Barati, Fengyuan Wang.

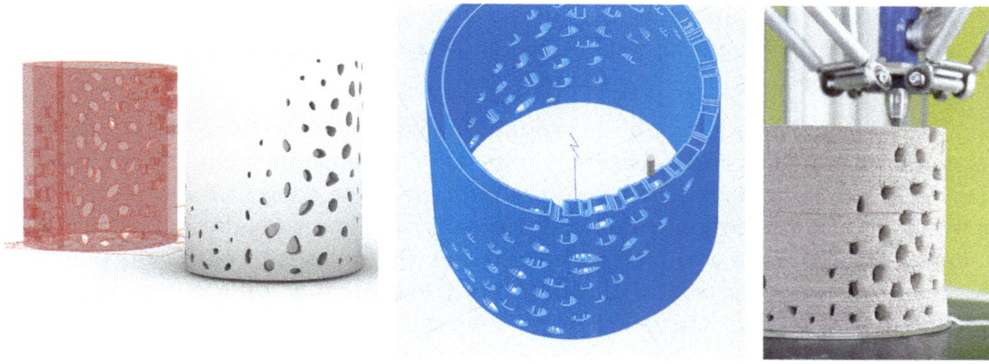

Figure 13.3 3D printing recycled clay paste with WASP clay printer. Left: digital 3D model created in Rhino, Grasshopper. Middle: printing toolpath preview generated in Simplify3D slicer. Right: the same digital model is being printed with WASP clay printer. Image credit: Bahareh Barati, Fengyuan Wang.

The digital 3D models were developed using the Grasshopper plug-in (Figure 13.3 left), enabling parametric control over geometric variables such as perforation size and distribution (ranging from 2.5 mm to 6.5 mm). Perforations smaller than 3.9 mm often became clogged during extrusion, resulting in unintended filling of the voids, which promoted further refinement of the geometry and slicing parameters. To ensure structural stability, wall thickness was set at 10 mm and the parameters were adjusted to generate double wall perimeters, with two extrusion lines forming each layer, providing reinforcement around the perforations (Figure 13.3 middle). Continuous extrusion across perforations led to material accumulation at restart points, which was manually corrected during printing. To reduce this issue, both flow rate and print speed were limited to 80% of their base settings.

4 Mycelial skin thickness, bridging capability and appearance

Our experiments resulted in the formation of mycelial skin across the recycled clay surfaces, which significantly transformed the visual and tactile qualities of the printed and hand-shaped samples. The experiential dimensions of materials, how they look, feel, behave and elicit responses from users, have been widely explored in design literature [14]. Material-driven approaches argue that such experiences are not incidental but central to how designers discover and shape new affordances, enabling novel functionalities or interactions [15].

In our case, these emergent qualities were not solely the result of clay hand-shaping, 3D printing or fungal growth in isolation, but rather the outcome of ongoing negotiations among these processes [16]. In the following, we characterize three experiential qualities observed in the hybrid samples:

1. Thickness variation of the mycelial skin, influenced primarily by the particle size, recycled clay content and gylcerine concentration in the formulation. These variations then in turn affected the tactile response.
2. The ability of the mycelium to span perforated geometries was related to the size and distribution of the perforations. Translucency of the mycelial skin over the perforations varied with thickness and bridging success, influencing the visual perception of the hybrid surface.

3. Under controlled and otherwise identical growth conditions, surface appearance and pigmentation patterns were primarily determined by growth time and glycerine concentration.

The skin thickness and adhesion to the clay body were tested qualitatively with scratching using nails and tweezers (Figure 13.4). Parameters including the particle size distribution, content of the fired clay and glycerine concentration, were found to influence water retention and overall viscosity, which in turn affected colonization patterns. Notably, even when recipes and growth conditions were nominally identical, the resulting samples exhibited subtle yet consequential differences in growth behaviour, such as the extent to which mycelium skin could span openings of equivalent dimensions.

A clear progression was observed across formulations. Samples incorporating 50% recycled clay with 60 mesh failed to form a coherent layer of skin, leaving the clay texture perceptible, or produced only an extremely thin and fragile layer. With 25% recycled clay (60 mesh) a soft but continuous skin was produced, representing an improvement in cohesion. In contrast, the recipe containing 30% recycled clay (120 mesh) with 7% glycerine generated a distinctly thicker, leather-like skin with stronger adhesion (Figure 13.4D).

We observed that the mycelial skin was capable of bridging perforations and surface indentations – a phenomenon we refer to as bridging behaviour. This behaviour first emerged unexpectedly during our early trials and was characterized by the mycelium

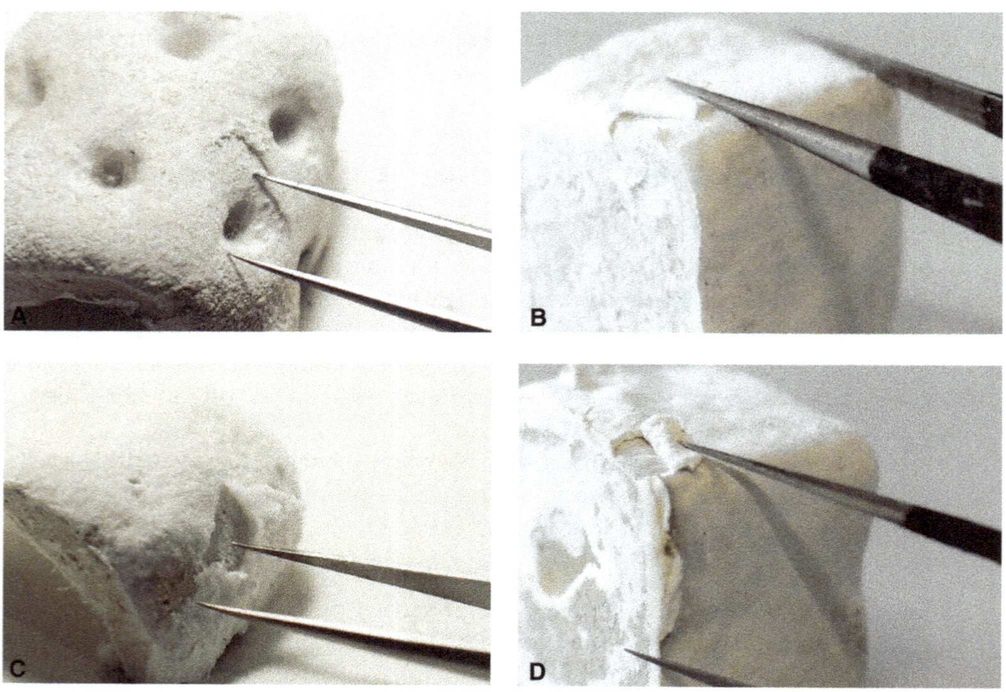

Figure 13.4 Testing the thickness and adhesion of mycelial skin with tweezers. Samples illustrate varying degrees of skin formation. Sample A and C: weakly attached and barely coherent layers on 50% recycled clay substrates. Sample B: thicker layer on 25% recycled clay surface. D: leather-like skins with stronger adhesion on 30% recycled clay with 7% glycerol. Image credit: Bahareh Barati, Fengyuan Wang.

Figure 13.5 Bridging behaviour of the mycelial skin. Left: 3D-printed sample with hand-carved rhombic perforations. Middle: Sample after eight days of incubation, showing complete bridging of the holes by mycelial skin. Right: Translucency of the dried mycelial skin revealed when held against backlight. Image credit: Bahareh Barati, Fengyuan Wang.

forming an extremely thin and transparent structure, or a continuous surface layer across small voids (Figure 13.5). Upon discovering this bioaffordance, we deliberately designed 3D-printed geometries to explore the boundary conditions of this bridging effect, including hole size, spacing and distribution. In some cases, the resulting mycelial layer exhibited partial translucency, particularly when samples were held against a backlight (Figure 13.5 right). The ability of bridging was found to depend on both the formulation and the size of the openings. While the recipe influenced the intrinsic capacity of the mycelium to bridge gaps, the perforations' size determined whether complete coverage could be achieved. Once bridging was established, growth duration and conditions further influenced the thickness of the layer, which in turn influenced its translucency.

The mycelial skins formed on our samples exhibited a range of surface textures and visual characteristics, shaped by both biological growth and material composition. In several cases, the mycelial skin appeared fluffy and velvet-like during active growth (Figure 13.6 middle), softening the ceramic body's tactile quality. Some samples developed irregular reddish-brown pigmentation (Figure 13.6 left), which, according to Lujan et al., may indicate fungal stress or shifts in metabolic activity [17]. While the exact cause of this pigmentation remains uncertain, it contributed to a distinctly organic and expressive aesthetic, suggesting potential avenues for intentional patterning or visual differentiation in future work.

Textural variation was also influenced by the interaction between mycelial growth and the geometry of the printed forms. In certain samples, the mycelium fully covered the rimmed extrusion pattern of the 3D-printed walls, producing a smooth, continuous surface (Figures 13.6 left and right). In others, the underlying print layers remained subtly visible beneath the fungal layer, adding a layered visual depth. This variation was conditioned by inherent surface texture of the clay substrates, which either facilitated or disrupted uniform coverage.

Following the air-drying process, these textures underwent further transformation. The once-fluffy skin observed under humid conditions became smoother and more compact, developing a leather-like appearance while still retaining some of the foam-like, springy quality characteristic of thicker mycelial growth. The material exhibits a soft, compliant and partially elastic texture after drying. Notably, wrinkles and microfolds often appeared around the perforations (Figure 13.6 right), likely caused by differential shrinkage between the pure mycelial skin spanning the holes and the denser hybrid clay–mycelium body.

Figure 13.6 Appearance and texture of the biodigital fabricated samples. Left: Brown pigmentation and partial translucency of the mycelial skin bridging perforations, visible against backlight. Middle: Velvety mycelium texture on a 3D-printed sample (WASP printer) under humid, pre-drying conditions; the mycelial skin subtly reveals the underlying rimmed pattern of the printed structure. Right: Fully colonized and air-dried sample printed with the WASP clay printer, showing a leather-like surface with slight wrinkling around the perforations. Image credit: Bahareh Barati, Fengyuan Wang.

5 The effects of mycelial skin on structural performance

To evaluate the structural impact of mycelial skin on the recycled clay composites, we performed uniaxial compression tests (Figure 13.7). This method was selected for its relevance to potential load-bearing and structurally supportive applications in bio-integrated design and fabrication.

Four cube-shaped samples were prepared, varying in composition and biological treatment:

- Sample A: 35% recycled clay, no mycelial skin.
- Sample B: 25% recycled clay, no mycelial skin.
- Sample C: 25% recycled clay, with mycelial skin.
- Sample D: 30% recycled clay, with thicker mycelial skin.

All samples were air-dried under controlled conditions to reduce variability caused by moisture content. Compression tests were conducted using an Instron hydraulic universal testing machine, applying a steadily increasing axial load until failure [18].

The results revealed a pronounced improvement in compressive strength for the biologically colonized samples (Figure 13.7). Sample C, which differed from Sample B only by the presence of a mycelial skin, showed a 166.8% increase in compressive strength, confirming the reinforcement effect of fungal colonization. Sample D demonstrated the highest peak load, reaching 0.62 kN, slightly outperforming Sample C at 0.49 kN. In contrast, the uncolonized samples (A and B) plateaued below 0.31 kN.

This performance increase can be attributed to the mycelial skin acting as a cohesive, load-distributing layer. In Samples C and D, the mycelium fully encased the clay bodies, forming a biologically grown shell that enhanced both surface integrity and internal cohesion. This hybrid fungal–ceramic interface likely limited microcracking and edge failure under load, contributing to the improved structural resilience of the composite.

Additional observations support this conclusion: one uncolonized, 3D-printed model collapsed during air-drying, while its mycelium-colonized counterpart maintained structural integrity throughout drying and testing, likely because of the binding and bridging function of the fungal network.

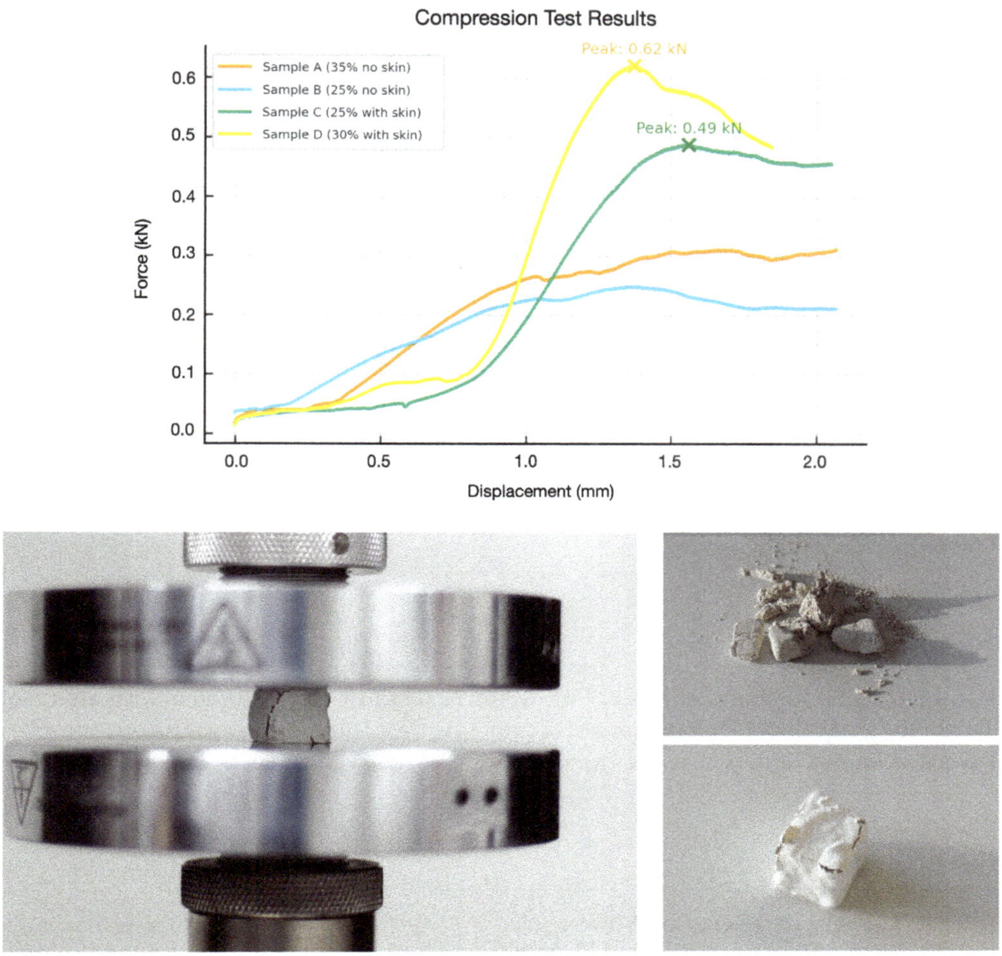

Figure 13.7 Compression test setup and results for recycled clay composites with and without mycelial skin. Left: Sample D positioned in the Instron hydraulic universal testing machine. Right, both images: Post-test condition of Sample B (uncolonized) and Sample D (with thick mycelial skin). Top: Graph of compression test results, demonstrating a twofold increase in compressive strength for samples encased in mycelial skin compared with their uncolonized counterparts. Image credit: Bahareh Barati, Fengyuan Wang.

6 Limitations and future work

This chapter explored the formation and potential of mycelial skin as a distinct bioaffordance, emerging from the reciprocal interaction between fungal growth and digitally fabricated clay substrates [16]. Our work extends prior research on digital clay formulations that incorporate fired clay [13] and bio-based dough [19] for shrinkage optimization and rheological feasibility, demonstrating how mycelial skins can enhance not only the aesthetic and tactile qualities of hybridized clay objects but also their mechanical performance. The observed improvements in compressive strength suggest that biomediated surface reinforcement offers a promising alternative for sustainable, regenerative material systems in architecture and design.

We acknowledge several limitations and open variables in the current study. Inoculation was performed using solid mycelium plugs, but alternative methods (e.g., slurry inoculation or liquid cultures) might yield different colonization dynamics and growth rates. While we explored multiple clay paste formulations, our approach was experimental rather than systematic. Our results suggest that glycerol concentration plays a critical role not only in water retention and skin thickness but also in pigmentation dynamics over time. Although we observed colour changes and surface evolution, we did not conduct a controlled study of aging processes (e.g., transitions from young white skins to mature brown layers [9]), which may have significant implications for both aesthetic expression and structural performance.

Finally, this study focused primarily on material and geometrical parameters of the substrate, and did not investigate environmental factors such as humidity, CO_2 concentration or light exposure, which have been shown to affect fungal morphology and metabolism [9]. A more comprehensive assessment of mycelial skin formation will require the integration of biological, material and environmental controls, and could open new design possibilities through the orchestration of living matter within computational fabrication systems.

Acknowledgements

We would like to thank the Building Materials group at Eindhoven University of Technology, especially Dr Yan Luo, for conducting the compression tests on our samples. We also acknowledge ArtCeramics for providing the fired clay materials used in this study.

References

1. Ritz K, Young IM. Interactions between soil structure and fungi. Mycologist. 1 May 2004;18(2):52–9.
2. Back to dirt | FAIRE 2021 | Projects [internet]. [cited 2025 Sep 7]. Available from: https://www.faireparis.com/en/projets/faire-2021/back-to-dirt-2157.html
3. Melendez F. bioMATTERS combines mycelium, clay and industrial waste to 3D print vessels and bowls [internet]. designboom | architecture & design magazine. 2024 [cited 2025 Jul 14]. Available from: https://www.designboom.com/design/biomatters-mycelium-clay-waste-materials-3d-printed-vessels-bowls-02-08-2024/
4. Oxman N. Towards a material ecology. In: Proceedings of the 32nd Annual Conference of the Association for Computer Aided Design in Architecture (ACADIA); 2012 Oct; San Francisco.
5. Nicenboim I, Karana E, McQuillan H, Devendorf L, Kakehi Y, Bell F, et al. Regenerative Material Ecologies in HCI. In: Proceedings of the Extended Abstracts of the CHI Conference on Human Factors in Computing Systems [internet]. New York, NY, USA: Association for Computing Machinery; 2025 [cited 2025 Apr 28]. pp. 1–5. (CHI EA '25). Available from: https://dl.acm.org/doi/10.1145/3706599.3716303
6. Kadir AA. An overview of wastes recycling in fired clay bricks. International Journal of Integrated Engineering [internet]. 2012 Oct 1 [cited 2025 Apr 27];4(2). Available from: https://penerbit.uthm.edu.my/ojs/index.php/ijie/article/view/224
7. Danninger D, Pruckner R, Holzinger L, Koeppe R, Kaltenbrunner M. MycelioTronics: Fungal mycelium skin for sustainable electronics. Science Advances. 2022 Nov 11;8(45):eadd7118.
8. Elsacker E, Zhang M, Dade-Robertson M. Fungal Engineered Living Materials: The Viability of Pure Mycelium Materials with Self-Healing Functionalities. Advanced Functional Materials. 2023;33(29):2301875
9. Wan Y, Wu T, Huang G. State-of-the-art research progress and challenge of the printing techniques, potential applications for advanced ceramic materials 3D printing. Materials Today Communications. 2024 Aug 1;40:110001.
10. Wolf A, Rosendahl PL, Knaack U. Additive manufacturing of clay and ceramic building components. Automation in Construction. 2022 Jan 1;133:103956.
11. Chen Z, Li Z, Li J, Liu C, Lao C, Fu Y, et al. 3D printing of ceramics: A review. Journal of the European Ceramic Society. 2019 Apr 1;39(4):661–87.

12. Guillitte O. Bioreceptivity: a new concept for building ecology studies. Science of The Total Environment. 1 May 1995;167(1):215–20
13. Weng T, Chen C, Li J, et al. Feasibility of Using Recycled Waste Ceramics in Ceramic 3D Printing Slurry [Internet]. [cited 2025 Sep 10]. Available from: https://www.ntcri.gov.tw/home/zh-tw/informationlist_4/7422116
14. Karana E, Pedgley O, Rognoli V, editors. Copyright. In: Materials Experience [internet]. Boston: Butterworth-Heinemann; 2014 [cited 2025 Sep 8]. p. iv. Available from: https://www.sciencedirect.com/science/article/pii/B9780080993591120016
15. Barati B, Karana E. Affordances as Materials Potential: What Design Can Do for Materials Development. International Journal of Design. 2019;13(3):105–23.
16. Mahler L, Barati B. Growing tessellations: Negotiation and wayfinding in interdisciplinary practice. Int J Design. Forthcoming.
17. Lujan P, Sanogo S, Puppala N, Randall J. Factors affecting mycelium pigmentation and pathogenicity of *Sclerotinia sclerotiorum* on Valencia peanut. Can J Plant Sci. 2016 Jun 96(3):461–73.
18. Callister WD Jr, Rethwisch DG. Materials Science and Engineering: An Introduction. John Wiley & Sons; 2020. 994 p.
19. Bell F, McClure E, Friedman-Gerlicz C, Ta R, Buechley L. Designing 3D Printed Ceramics from a Clay with Tunable Shrinkage. In 2024 [cited 2025 Sep 4]. Available from: https://dl.designresearchsociety.org/drs-conference-papers/drs2024/researchpapers/27

CHAPTER 14

Additive Manufacturing Strategies for Enabling Large-Scale Mycelium-Based Composites

EDA ÖZDEMIR, PHILIPP EVERSMANN, ANDREA ROSSI and NADJA NOLTE

1 Introduction

Mycelium-based composites (MBCs) have received significant attention in recent decades due to their environmental benefits and potential as a sustainable building material. Yet the widespread and feasible adoption of MBCs in architecture still faces several key challenges. One of the primary obstacles is structural stability. While MBCs are effective in small-scale applications, such as acoustic panels [1], they often lack the mechanical strength required for larger architectural components, such as construction panels or walls [2]. Therefore, relatively small modules are predominantly used in mycelium-based construction, as in acoustic components, floor tiles [3] and building insulation panels [4]. Furthermore, small modules mitigate the inherent mechanical limitations of mycelium, which are highly susceptible to moisture and have a relatively low tensile strength compared with conventional construction materials [5]. The drying phase of components can also result in significant shrinking owing to the high humidity levels needed during production. This can lead to dimensional inaccuracies that can accumulate in assemblies of multiple components.

Large-scale mycelium construction, on the other hand, presents a range of compelling advantages over small-scale uses, particularly in the context of sustainable and regenerative architecture. One of the most significant benefits is the material's extremely lightweight nature, which makes it well suited for prefabrication. Large mycelium components can be grown off-site, then easily transported and installed, reducing on-site labour and transportation emissions. Unlike smaller components, which require numerous joints and complex assembly, large panels or wall elements can be installed with minimal joining. This is especially beneficial in acoustic and insulating applications, where the presence of joints often compromises performance. In terms of production, large-scale mycelium elements can be grown directly into reusable moulds, eliminating the need for excessive formwork and generating little waste compared with traditional construction materials. This not only improves material efficiency but also reduces costs and environmental impact. Moreover, as architects and designers explore the structural and aesthetic potential of mycelium at scale, it enables the development of a new architectural language, characterized by organic forms, natural textures and a distinctly biophilic aesthetic.

Eda Özdemir, Philipp Eversmann, Andrea Rossi and Nadja Nolte, 'Additive Manufacturing Strategies for Enabling Large-Scale Mycelium-Based Composites' in: *Mycology for Architecture: Methods and Applications of Mycelium Materials in the Built Environment*. Pelagic Publishing (2026). DOI: 10.53061/EWWO5490

Large-scale applications, however, also present substantial challenges. For upscaling and prefabrication of building elements, manual fabrication processes need to be automated for both cost efficiency and consistent material quality. Uniform growth throughout the material must be assured and structural instability owing to the material's limited load-bearing capacity should be overcome. Due to these reasons, large monolithic components require reinforcement to ensure their stability. For improving the structural capacities of mycelium, reinforcement strategies such as the integration of a selection of fibrous substrates and additional natural fibres [6], woven and rigid reinforcements and densification of mycelium are explored [7].

One of the challenges in scaling MBCs is the need for formwork during the growth phase, which is required by both small- and large-scale elements. The sustainability and practicality of these moulds remain critical issues since they are often made from non-biodegradable plastics and generate waste, particularly when they are only used once for unique or complex geometries [8]. The biological nature of mycelium introduces additional constraints on mould design, differentiating it from conventional cast materials. Since mycelium growth depends on access to oxygen, moulds must incorporate ventilation features such as breathing holes or porous surfaces [5]. Ensuring uniform oxygen exposure becomes particularly challenging for large, complex geometries, where airflow variations can lead to uneven growth, material weakness or contamination [2]. Unlike rigid moulds used in industrial-scale construction, mycelium moulds must also allow for controlled humidity and temperature while maintaining the material's biological activity [9].

These limitations have hindered the widespread adoption of MBCs in architecture, requiring innovative fabrication approaches to unlock their full potential. To address these challenges, digital fabrication methods, particularly various forms of additive manufacturing, have been explored to improve scalability and structural integrity while maintaining the sustainability benefits of the material.

We explored three additive manufacturing (AM) processes to advance the integration of mycelium in construction. First, we developed lightweight wood lattice structures, designed to reinforce and guide mycelium growth and enhance the overall structural stability of the composite material. These lattices can be fabricated using automated robotic processes, enabling the production of elements up to 3 m in length (Section 2). Second, we implemented integrated formwork technologies, including the 3D printing of thin shell moulds that support mycelium growth while improving dimensional accuracy and surface quality (Section 3). Lastly, we experimented with hybrid formworks combining veneer lattices and natural fabrics. These stay-in-place systems enable the creation of complex, 3D-curved geometries while remaining fully biodegradable (Section 4). Together, these methods demonstrate the potential of AM to streamline the production of MBCs, expand design possibilities and support sustainable construction practices.

2 Reinforcement: robotically fabricated wood-veneer lattices

When fabricating large-scale MBCs, reinforcement becomes essential because of the material's inherent limitations in handling shear and bending forces. While MBCs offer sustainability and biodegradability, their mechanical properties alone are insufficient for applications that require both strength and durability. Additionally, interior architectural elements often need to meet varying structural criteria, further necessitating a reinforcement strategy.

To address these challenges, a novel composite material that integrates a mycelium matrix with a wood veneer reinforcement lattice was developed. Similar to the role of steel reinforcement in concrete, these veneer lattices provide increased strength and structural scaffolding for MBCs. The reinforcement lattices are produced using a custom AM method, which deposits continuous veneer strips in either 2D or 3D layouts. Once the lattices are fabricated, mycelium-inoculated substrate is introduced. During growth, mycelium binds the substrate to the reinforcement lattice, forming a biocomposite material with enhanced mechanical properties. The positioning and orientation of the veneers can be adjusted to optimize the composite's structural performance.

To manufacture the veneer reinforcement lattices, a robotic fabrication method was developed, featuring a bespoke end-effector capable of continuously extruding and welding veneer strips. The end-effector consists of a veneer extrusion mechanism, a stepper motor for precise control, a cutting blade actuated by a pneumatic cylinder and a guiding system that accommodates veneer strips of varying widths [10, 11]. This tool was mounted on an ABB IRB4600 industrial robot and controlled through an Arduino board, with programming executed via the Grasshopper plug-in Robot Components [12], and the communication between the tool and the robot controller was established through the Funken serial protocol toolkit [13].

The fabrication process follows a series of sub-steps: a single veneer strip is extruded, one at a time, and the material is cut whenever a change in extrusion direction is required. This approach enables the creation of complex tool paths, multidirectional reinforcement patterns and controlled anisotropy in the final composite.

To create a uniform reinforcement lattice, first, an orthogonal 2D grid is created, then the diagonal 3D waves are added, creating cross-bracings in each cell. Lastly, the top orthogonal 2D grid is connected to the rest, and the fabrication of the reinforcement lattices is completed. Such lattices can be further optimized by incorporating information about stress concentrations in various areas and adjusting the veneer density accordingly (Figure 14.1).

A key challenge in this process was developing a reliable method for binding the veneer strips together, as most bio-based adhesives have long open times, making them

Figure 14.1 Optimized veneer lattice. Image credit: EDEK, University of Kassel.

Figure 14.2 Ultrasonic welding process. Image credit: EDEK, University of Kassel.

unsuitable for AM applications. To overcome this, an ultrasonic welding technique was used for the purpose of joining thin veneer strips. This process entails the utilization of a welding horn and generator that produce high-frequency vibrations while pressing the veneer strips together. These vibrations generate thermal energy, which melts the lignin within the veneer, fusing the strips without additional adhesives [14]. This method ensures a strong and efficient bond while maintaining the biodegradability of the composite (Figure 14.2).

A proof-of-concept partition wall prototype was fabricated to assess the effectiveness of the proposed strategies. Designed to meet the structural and acoustic requirements of an office partition, the wall incorporated a customized internal reinforcement lattice and a tailored surface design. The final prototype measured 2.4 m by 0.8 m with a thickness of 12 cm (Figure 14.3). Computer numerical control milled multiplex panels coated with a release agent were used for the moulding of the bespoke design. The result successfully demonstrated that integrating this reinforced lattice system with MBCs enables the fabrication of large-scale architectural elements that achieve both sustainability and improved structural performance.

3 Moulding and accuracy: 3D printed shells for interlocking assemblies

3D printing has emerged as a promising technique to fabricate temporary formworks for casting mouldable materials, which optimizes material usage and reduces waste [15, 16]. This technology makes the production of extremely thin formworks possible, providing a method to shape fluid and aggregate materials using minimal resources. As in recent years several bio-based and/or compostable materials for AM have been proposed [17, 18]; this paves the way for a low-impact avenue to increase the geometric possibilities of cast elements.

Figure 14.3 Veneer-reinforced mycelium partition wall prototype. Image credit: EDEK, University of Kassel.

When looking at possible assembly systems for MBCs, the lower durability and the increased need for maintenance of these materials, owing to their short lifespans compared with their synthetic competitors, suggests the use of dry-joining principles. This way, the structure can be disassembled easily to repair and replace components when needed, without creating additional waste or damaging the rest of the structure. However, dry-joining assemblies rely mostly on kinematic interlocking and friction at the interfaces to transfer loads, and therefore require high precision to function correctly. For this reason, MBCs present an inherent challenge in their application in dry-joined systems due to the significant and non-uniform shrinkage during and after their production.

To address this, a novel biofabrication method integrating AM with mycelium cultivation was developed. This approach enabled the creation of larger structural assemblies that can withstand multidirectional loads. The process involved the use of fused deposition modelling to print an outer shell using polylactic acid (PLA) bioplastic infused with wood particles. This printed shell served as a lost formwork, supporting the cultivation of MBC within. The wood particles in the PLA facilitated mycelial growth by allowing penetration through the printed shell, establishing a strong bond between the two materials. While the PLA shell ensured the dimensional accuracy required for the kinematic interlocking of blocks, mycelium provided the necessary compressive strength (Figure 14.4A).

To demonstrate the proposed method, a block-based assembly with geometrically interlocking joints was designed. Therefore, the concept of topologically interlocking assemblies (TIA) was adopted, a sub-class of dry-joining systems. The geometry and the arrangement of each module is designed in such a way that each module supports and is supported by its neighbours, without needing any extra binders [19]. The assembly of the wall is achieved exclusively through friction between the interfaces, providing a fully reversible structure. Based on this principle, the system was successfully employed to

Figure 14.4 A: Single module, B: Assembled TIA wall. Image credit: EDEK, University of Kassel.

fabricate a full-scale partition wall prototype, measuring 2.40 m in height and composed of 112 interlocking blocks. These blocks were produced using small-scale consumer 3D printers and were assembled entirely through mechanical interlocking, held in place by an external timber compression frame (Figure 14.4B).

Given that PLA bioplastics require industrial composting for degradation, ongoing research is focused on replacing them with fully compostable alternatives, such as wood-based or cellulose-based pastes, to further enhance the environmental sustainability of this approach.

4 Moulding and reinforcement: stay-in-place formworks made of wood-veneer lattices and natural textiles

Current standard methods for producing MBCs primarily rely on thermoformed plastic moulds with breathing holes to support growth. While effective for simple geometries, these methods are limited in their ability to produce structurally sound, non-planar MBC components without extensive formwork resources. Unlike conventional mouldable materials, working with mycelium presents unique challenges owing to its biological nature. Oxygen exposure, for example, is a critical factor in ensuring consistent growth [20].

Building upon previous research in lattice reinforcement [10] and fabric integration [21, 22, 23], an innovative approach that combines wood veneer reinforcement lattices with a woven natural jute textile was explored. This hybrid system acts as a stay-in-place formwork, eliminating the need for disposable moulds while enhancing the structural integrity of MBC components. The combined lattice and textile not only provide a support structure but also maintain geometric accuracy throughout the mycelium growth phase, enabling the fabrication of complex, curved architectural elements.

While the veneer lattice follows the same robotic fabrication principles as the previously described reinforcement grids, the introduction of textile integration required slight modifications to the production process. The fabrication was carried out as follows: two identical 2D grids were robotically fabricated and stacked, with a jute textile encapsulated between them through ultrasonic welding (Figure 14.5A). Diagonal waves were then integrated to provide structural depth. The lattice was subsequently inverted, and the assembly was completed by joining the 2D grids on the opposite side. To enable welding, holes were made in the textile at the centre of each cell, allowing the ultrasonic welding horn to pass through (Figure 14.5B). A handheld ultrasonic welding gun was then employed to align and secure the nodes and wave peaks, ensuring correct alignment of the veneer strips within the lattice. Varying the distance between veneers on the top and bottom 2D grids forced the lattice to curve during fabrication, enabling the production of non-planar components. While further studies are needed for the precise calibration of this forming process and to allow the direct translation of a curved digital design into a curved reinforcement lattice and formwork, this initial study already demonstrates the feasibility of such an approach [24].

Following fabrication, the prototype was cultivated for seven days under controlled conditions. The filled formwork was placed in a growing tent and loosely wrapped with stretch film to preserve the high moisture content on the surface. No specific measure was taken to avoid the risk of contamination, as producing large scale elements would realistically be performed in non-sterile conditions. The hybrid formwork successfully preserved its geometry throughout the growth phase, demonstrating its effectiveness in shaping MBCs. Furthermore, it significantly reduced the deformation and shrinkage

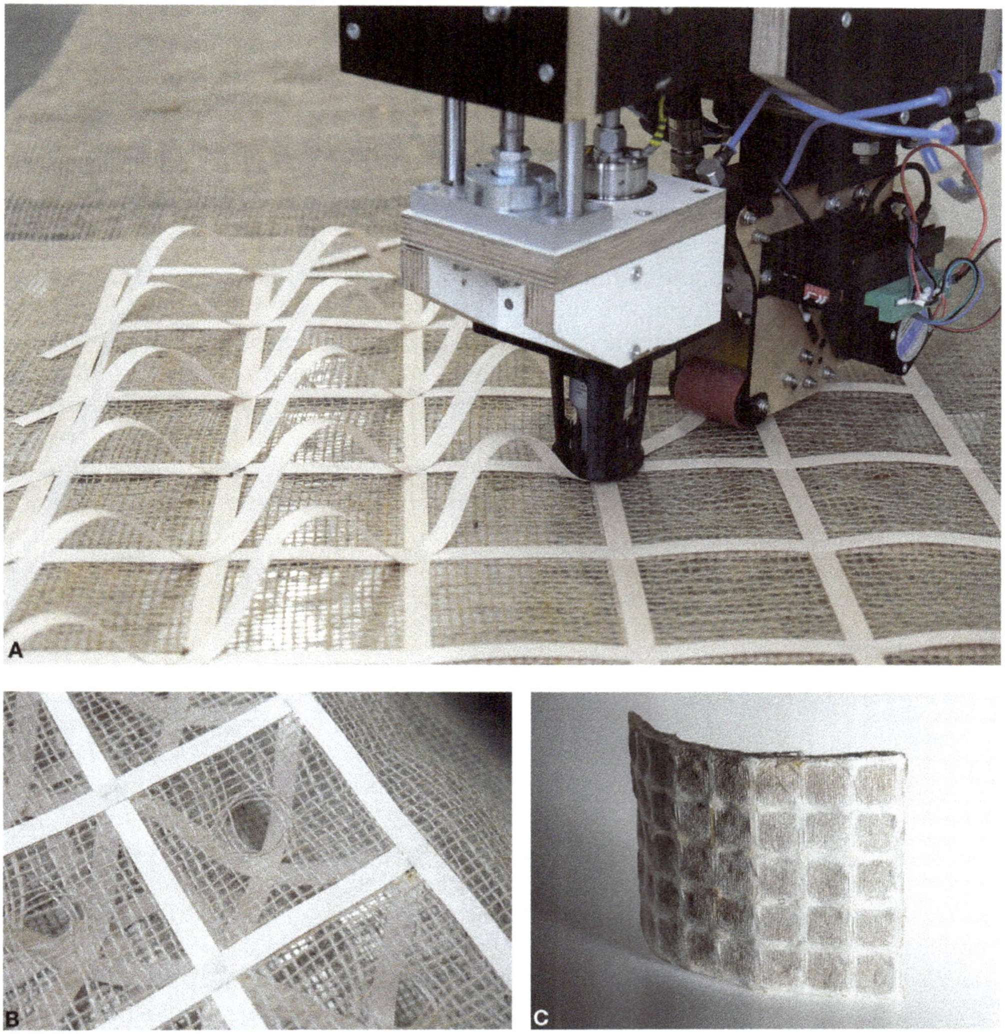

Figure 14.5 A: 2D grid welding with fabric in between. B: Welding hole on the fabric. C: Growing curved components. Image credit: EDEK, University of Kassel.

usually happening during the drying phase. By integrating internal reinforcement with stay-in-place formwork, the system provided both structural support and stability. The veneer lattice contributed to load-bearing capacity, while the textile acted as a scaffold, maintaining the integrity of the form during mycelium growth. These results confirm the potential of this approach for producing complex MBC components with enhanced structural reliability (Figure 14.5C).

5 Discussion

While MBCs offer a relevant alternative to a variety of harmful construction materials, they also draw attention to a series of significant challenges related to both the unique properties they possess and the production processes they require. The presented studies attempted to tackle some of these issues by merging conventional biofabrication processes with advanced digital fabrication techniques. This integration not only broadens the potential applications of MBCs but also enhances their structural performance and enables material-efficient shaping methods to achieve non-standard geometries.

Unlike conventional materials, which often rely on energy-intensive processing and generate significant waste, mycelium composites can be grown in controlled environments with minimal energy inputs. However, one of the key challenges in scaling up their use is the need for efficient formwork strategies. Traditional moulds, often made from synthetic materials, can be costly, wasteful and restrictive in terms of design flexibility. By leveraging computational design and digital fabrication, including AM and robotic fabrication, novel formwork solutions can be developed that minimize waste and facilitate the production of complex geometries without relying on single-use moulds.

Integrating digital fabrication with biofabrication presents a promising direction for sustainable construction, enabling precise control over material deposition, reinforcement integration and geometric customization. The continued development of digitally fabricated MBCs has the potential to transform the construction industry, offering a renewable, low-carbon alternative to conventional materials. By leveraging computational design, AM and robotic fabrication, strategies to address long-standing challenges related to scalability, structural performance and design flexibility can be developed, and new possibilities for large-scale applications can be unlocked.

Acknowledgements

The 'HOME' project was funded by Forschungsinitiative Zukunft Bau des Bundesinstitutes für Bau-, Stadt- und Raumforschung (BBSR), grant number 10.08.18.7-21.48. The follow-up project, 'HOME II – Robotergestützte Herstellung von Myzel-Verbundwerkstoffen', was also funded by BBSR with the grant number 10.08.18.7-23.09, and the project 'Robotic BioConstruction – Towards Large-Scale Circular Fabrication Processes with Biomaterials in Architecture' was funded by the Volkswagen Stiftung, with the Momentum grant number 9C269. The list of project contributors is as follows: Eda Özdemir, Andrea Rossi, Ole Weyhe, Nadja Nolte, Nazanin Saeidi, Alireza Javadian, Shibo Ren, Albert Dwan, Saar Driessen, Jessica Watts, Claudia di Noi, Richa Saini, Zoë Kaufmann, Ayleen Hümmerrich, Angelique Keil, Dirk E. Hebel, Jan Wurm and Philipp Eversmann.

References

1. ARUP. Foresta. 2025. Available from: https://www.arup.com/projects/foresta/
2. Islam MR, Tudryn G, Bucinell R, Schadler L, Picu RC. Morphology and mechanics of fungal mycelium. Sci Rep. 2017;7(1):13070. https://doi.org/10.1038/s41598-017-13295-2
3. MOGU. Official website. 2025. Available from: https://mogu.bio/
4. MycoTile. Official website. 2025. Available from: https://mycotile.eco/
5. Girometta C, Picco AM, Baiguera RM, Dondi D, Babbini S, Cartabia M, et al. Physico-Mechanical and Thermodynamic Properties of Mycelium-Based Biocomposites: A Review. Sustainability. 2019;11(1):281. https://doi.org/10.3390/su11010281
6. Voutetaki ME, Mpalaskas AC. Natural Fiber-Reinforced Mycelium Composite for Innovative and Sustainable Construction Materials. Fibers. 2024;12(7):57. https://doi.org/10.3390/fib12070057
7. Womer S, Huynh T, John S. Hybridizations and reinforcements in mycelium composites: A review. Bioresour Technol Rep. 2023;22:101456. https://doi.org/10.1016/j.biteb.2023.101456
8. Elsacker E, Vandelook S, Van Wylick A, Ruytinx J, De Laet L, Peeters E. A comprehensive framework for the production of mycelium-based lignocellulosic composites. Sci Total Environ. 2020;725:138431. https://doi.org/10.1016/j.scitotenv.2020.138431

9. Haneef M, Ceseracciu L, Canale C, Bayer IS, Heredia-Guerrero JA, Athanassiou A. Advanced Materials From Fungal Mycelium: Fabrication and Tuning of Physical Properties. Sci Rep. 2017;7(1):41292. https://doi.org/10.1038/srep41292

10. Özdemir E, Saeidi N, Javadian A, Rossi A, Nolte N, Ren S, et al. Wood-Veneer-Reinforced Mycelium Composites for Sustainable Building Components. Biomimetics. 2022;7(2):39. https://doi.org/10.3390/biomimetics7020039

11. Rossi A, Javadian A, Acosta I, Özdemir E, Nolte N, Saeidi N, et al. HOME: Wood-Mycelium Composites for CO2-Neutral, Circular Interior Construction and Fittings. IOP Conf Ser Earth Environ Sci. 2022;1078(1):012068. https://doi.org/10.1088/1755-1315/1078/1/012068

12. Deetman A, Rumpf G, Wannemacher B, Dawod M, Akbar Z, Rossi A. Robot Components: Intuitive Robot Programming for ABB Robots inside of Rhinoceros Grasshopper. Zenodo. 2024. https://doi.org/10.5281/ZENODO.5773814

13. Stefas A, Rossi A, Tessmann O. Funken: Serial Protocol Toolkit for Interactive Prototyping. eCAADe Conference. 2018. https://doi.org/10.52842/conf.ecaade.2018.2.177

14. Troughton MJ, editor. Chapter 2 – Ultrasonic Welding. In: Handbook of Plastics Joining. 2nd ed. Boston: William Andrew Publishing; 2009. pp. 15–35. https://doi.org/10.1016/B978-0-8155-1581-4.50004-4

15. Burger J, Lloret-Fritschi E, Scotto F, Demoulin T, Gebhard L, Mata-Falcón J, et al. Eggshell: Ultra-Thin Three-Dimensional Printed Formwork for Concrete Structures. 3D Print Addit Manuf. 2020;7(2):48–59. https://doi.org/10.1089/3dp.2019.0197

16. Leschok M, Dillenburger B. Dissolvable 3DP Formwork. In: ACADIA 2019: Ubiquity and Autonomy. Austin (TX); 2019. pp. 188–97. https://doi.org/10.52842/conf.acadia.2019.188

17. Dielemans G, Lachmayer L, Recker T, Atanasova L, Hechtl CM, Matthäus C, et al. Mobile Additive Manufacturing: A Case Study of Clay Formwork for Bespoke in Situ Concrete Construction. In: Buswell R, Blanco A, Cavalaro S, Kinnell P, editors. Third RILEM International Conference on Concrete and Digital Fabrication. Cham: Springer; 2022. pp. 15–21. https://doi.org/10.1007/978-3-031-06116-5_3

18. Cutajar S, Dielemans G, Krakovska E, Dorresteijn E, Mai I, Lowke D, et al. Upscaling earth formworks: 3D printing strategies for material optimised reinforced concrete structures. Constr Robot. 2024;8(1):3. https://doi.org/10.1007/s41693-024-00120-2

19. Tessmann O, Rossi A. Geometry as Interface: Parametric and Combinatorial Topological Interlocking Assemblies. J Appl Mech. 2019;86(11):111002. https://doi.org/10.1115/1.4044606

20. Ghazvinian A, Khalilbeigi Khameneh A, Mottaghi E, Gursoy B. A Computational Framework for the Design and Fabrication of Spatial Structures with Mycelium-based Composites. 2021. [Preprint]

21. Jiang L, Walczyk D, McIntyre G, Bucinell R, Tudryn G. Manufacturing of biocomposite sandwich structures using mycelium-bound cores and preforms. J Manuf Process. 2017;28:50–9. https://doi.org/10.1016/j.jmapro.2017.04.029

22. Kaiser R, Bridgens B, Elsacker E, Scott J. BioKnit: development of mycelium paste for use with permanent textile formwork. Front Bioeng Biotechnol. 2023;11:1229693. https://doi.org/10.3389/fbioe.2023.1229693

23. Agraviador A, Scott J, Kaiser R, Elsacker E, Hoenerloh A, Topcu A, et al. BioKnit: Computation and material investigation in the design of biohybrid textiles towards architectural integration. 2022. [Conference paper/Preprint]

24. Özdemir E, Rossi A, Eversmann P. MycoCurva: stay-in-place fabric formworks for curved veneer-reinforced mycelium building components. Archit Struct Constr. 2025;5(1):16. https://doi.org/10.1007/s44150-025-00134-6

CHAPTER 15

Textile Templating for Myco-Architecture

ROMY KAISER-STYLE and JANE SCOTT

1 What is textile templating?

This chapter outlines the core principles, potential applications and future influence of textile templating in relation to the biofabrication of mycelium-based composites (MBCs) in myco-architecture. This chapter illustrates how the textile templating approach effectively addresses key challenges in myco-architecture. Through analysis of a series of prototypes and their features, the chapter emphasizes how this method advances mycofabrication, unlocking its full potential for architectural and biofabricated applications.

2 Definition of textile templating

Textile templating is a fabrication strategy that solves several design and fabrication challenges related to the structural and functional characteristics when building and designing with living materials, such as mycelium, which are often encountered in myco-architecture. Rooted in the principles of templating [1], this work introduces textile systems as templates for bioassembly, as a tool to address challenges within mycofabrication, such as the biofabrication of large-scale structures and complex geometries [2].

Templates are proposed as a fabrication support mechanism to bridge scales between living organisms and architectural structures [1]. They involve a digital, pre-designed format that serves as a guide for shaping emerging, living materials. Digital fabrication tools are key to precisely embedding information in templates by accurately distributing and allocating materials hierarchically within a template. These template designs are used to support the organization of living materials through bioassembly, equipping hybrid living materials (HLMs) [7] with features such as functional gradients, structural complexity and environmental interactions. The intersection of material bottom-up and top-down design enables the template to serve as a blueprint for material processes, guiding the generation of patterns, order, arrangements or shapes, and defining a structure's functionality, material qualities and metabolism.

Textile templating draws on this concept; however, it integrates textile principles into the template design, utilizing textile materials and harnessing their underlying, embedded textile logic and principles.

Romy Kaiser-Style and Jane Scott, 'Textile Templating for Myco-Architecture' in: *Mycology for Architecture: Methods and Applications of Mycelium Materials in the Built Environment*. Pelagic Publishing (2026). © Romy Kaiser-Style and Jane Scott. DOI: 10.53061/FTFS8318

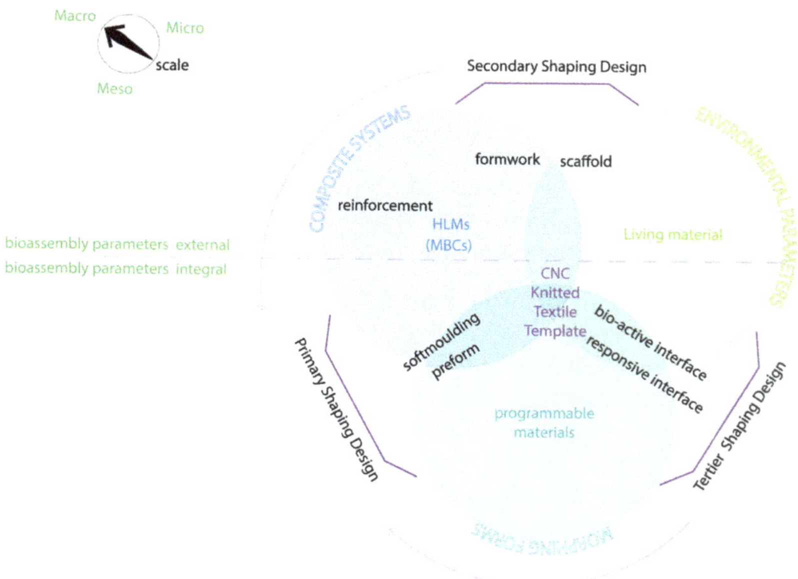

Figure 15.1 Positioning of textile templating within fabrication approaches. Image credit: Romy Kaiser-Style.

Textile materials and their properties are valued for the beneficial functionality in composite applications across various scales, ranging from biocompatibility for scaffolds in tissue engineering [3] up to large-scale reinforcement of concrete formworks from shell to tubular applications [4]. Recent advances in fabrication – especially digitally fabricated textile forms and systems based on embedded textile logic – are recognized for their ability to provide seamless, highly tunable and programmable material interfaces. These go beyond material or smart data interfaces, incorporating environmentally responsive and temporal features by embedding pre-coded information within a computer numerical control (CNC) knitted textile structure [5].

The integration of these underlying textile principles transforms templating into textile templating, offering the intersection of several material-related fabrication approaches. These approaches, spanning (soft-)moulding and scaffolding, also include the integration of behavioural design features such as programmable, morphing or living attributes into fabrication processes, as depicted in Figure 15.1. Behavioural design is important for bioassembly and biofabrication because of the agency of living materials and their ability to design behaviours in response to environmental stimuli. However, current design and fabrication systems lack controllable and reliable behavioural integration in the design of living materials [6].

3 Mycofabrication features of textile templating

HLMs are predicted to be a promising approach for overcoming biological scaling issues and advancing the integration of living and bioassembled materials into current material applications, including scale-up for architectural purposes [7, 8, 9]. In this context, templates, including textile templates, can be classified as such hybrid material approaches, allowing for the integration of design components and features through engineering of the template design.

The proposed fabrication requirements of textile templates comprise:

- material carrier design,
- multi-functionality,
- support of biological performance.

These three requirements need to be considered during the material design process [7, 8, 9]. They can be achieved through the application of several overlapping design parameters, broadly summarized by the following sub-themes.

Material carriers that allow containment (in the form of moulds, soft moulds) or adhesion (scaffolds) and contribute also to the fabrication potential through reinforcing and positioning of additional material systems. Since scaling HLMs requires significant life-preserving features, carrier materials can also protect living systems from harsh conditions during scaling.

Material functionality is not only considered as the material performance but also addressed through specifically tuned material mixes and additives, and further includes the complexity and smartness of material systems, with the integration of new functionalities such as multifunctionality and 4D capabilities as responsive or adaptive features.

Biological growth requirements are crucial to ensure viability and preservation of biological functions as much as possible during scaling and beyond, in conjunction with the previous points.

Addressing these aspects within current mycofabrication is still challenging. Conventional casting techniques don't provide the complexity of simultaneously integrating several features. This is evident in the constant challenge to overcome scaling issues related to structural and living features and beyond small-scale lab fabrications, for the production of inert MBC blocks or bricks [10, 11].

Based on material-based and material-logic capabilities, 3D knitted textile pre-forms provide a textile template capable of designing biofabrication features through the systematic embedding of information across textile scales.

Figure 15.2 provides an overview of the potential design parameters that can be embedded in a textile template, considering the scales from fibre to structure of the textile system for the 3D knitted template design. It displays the range of available design and material features of the textile templating approach, addressing key fabrication and material requirements through the design of a 3D knitted textile as a hybrid support system. The parameters are colour-coded according to the biochemical/biological (green), physical (blue) and mechanical (lilac) aspects that influence the mycofabrication process. These parameters affect the behaviour of the living material, and facilitate further development and 'activation' of these design features during biofabrication and bioassembly processes.

4 How textile templating can address the current challenges of myco-architecture

A commonly identified bottleneck within architectural applications of current MBCs is the low load-bearing capacity. To address this, research is required that incorporates growth strategies for material strengthening, reinforcements, substrate compositions and integration of specific growth parameters, as well as structural fabrication approaches that contribute through geometrical advances [11]. Current mycofabrication

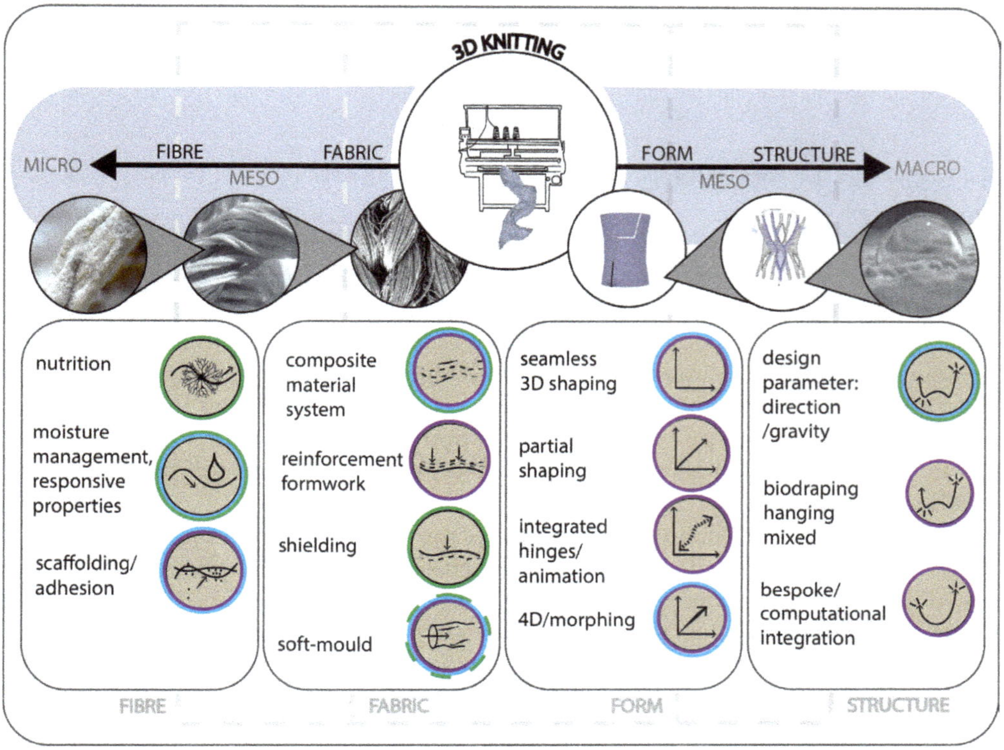

Figure 15.2 Overview of biofabrication parameters according to the textile scale, which can be embedded within the textile template design. Image credit: Romy Kaiser-Style

techniques have not yet fully reflected and integrated these fabrication requirements for growth and assembly strategies in myco-architecture. This necessitates adaptations in fabrication techniques, material designs and geometric possibilities, which are inherently interconnected and influence each other's success at the architectural scale [12].

CNC knitted textile systems bring together separate design aspects within a single design system, including the tuned textile pre-form design and tailored adjacent workflows [2, 13]. This integration enables workflows that contribute to structural and geometrical optimization, allowing for the biofabrication of complex, large-scale mycelium-based prototypes. The systematic integration of various fabrication requirements not only enhances material properties, such as mechanical strength and the capability to form complex shapes, but also enables the integration of several coherent shaping steps, which in turn increases the complexity of the prototype across multiple levels. Specifically, it increases system complexity by allowing multi-step fabrication within one workflow. In Figure 15.1, this multi-step system is illustrated as primary shaping design (related to the direct shaping and design of the material carrier as pre-form itself), secondary shaping design (related to behavioural responses of the material carrier and the hybrid material system, allowing for the incorporation of behaviour design) and tertiary shaping design (any further changes based on environmental responses of any component of the HLM system).

Currently, examples of the primary shaping design approach are used only in the static version of mycofabrication of component-based CNC knitted moulds [14]. In contrast, secondary shaping design approaches still lack a fabrication process that offers the necessary control for both design and structural performance in large-scale applications [15], primarily because they neglect the systematic integration of several simultaneous complexity levels.

5 Designing complexity in textile templated myco-architecture

The Living Textiles Research Group at Newcastle University, UK, has developed a series of prototypes demonstrating textile-templated MBC structures at an architectural scale. Each prototype presents different variations and features within fabrication processes, while achieving versatile material properties, aesthetics and integration into the broader ecosystem. The inclusion of tailored multi-material systems promotes living attributes, enabling robust growth processes outside laboratory conditions and facilitating local, on-site fabrication. This achieves complex material behaviours in real-life environmental and large-scale settings.

In the following sections, these examples are presented and contrasted, illustrating the achievable prototype features and their approaches to integration of stages of complexity based on unifying carrier system design, behavioural design and environmental interactions to expand integration of performance and functional features. The examples introduced here present the relevance and significance of the textile templating strategy and demonstrate trends for feasibility and further approaches for myco-architecture.

5.1 Primary complexity: integrating multi-material carrier systems

To meet growth requirements and unlock the scaling potential of MBCs for architectural structures, designing the material carrier system for enabling multi-material systems is crucial. Textile templating offers a flexible textile carrier system designed to support, hold and protect, while incorporating a tailored material mixture that includes substrate, nutrition and the living organism. This is shown by the development of a specific mycelium-based paste called mycocrete, which is applied to a knitted formwork, transforming the geometric, structural and scaling possibilities. The specific, highly viscous formulation provides not only fabrication requirements and high structural material strength but also ideal mycelium growth conditions with high moisture and nutritional content. This nurtures and shields the mycelium, enabling on-site biofabrication [2].

Mycocrete has enabled two separate fabrication strategies as part of a multi-material system: first, plastering as demonstrated in the example of the Living Room Prototype by Scott et al. [16] (Figure 15.3); and secondly, injection, applied using a cartridge gun in the BioKnit Prototypes (Figure 15.4). While both fabrication strategies lead to excellent MBC growth, they demonstrate different potential and features in the resulting material performance.

Plastering approaches create a coherent material surface; utilizing a canopy of textile formwork in a shell-like manner. For the Living Room Prototype (Figure 15.3), mycocrete was plastered directly onto the textile formwork, allowing for the growth of a robust, self-supporting structure 4 m in diameter. Here, the primary shape of the 3D tensioned textile formwork (with integrated shaping of doubly curved hyperbolic forms) is biosolidified through mycelium growth. To accommodate the plastering application, a highly tactile textile surface created from hand-knitted wool tops was chosen for the application of the mycocrete composition. The coverage of the MBC layer is single-sided, bonding with the high tactility of the thick wool formwork, creating a composite with high

Figure 15.3 The Living Room Prototype (a), the adjusted material ingredients to waste-streams: raw wool for hand-knitted formwork (b), raw material paper sludge (c) and adapted mixed mycocrete paste (d), the fabrication process of plastering (e) and a close-up of a final material sample (f). Collage created by the authors with image courtesy of ©Hub for Biotechnology in the Built Environment, Living Textiles Research Group.

insulation and sound-absorbing properties. The Living Room Prototype demonstrated that plastering offers a highly adaptable process: mycocrete can be applied effectively by hand, and the composition of the material ingredients (both within the textile formwork and the mycocrete formulation) can be suited to local resources and waste streams (Figure 15.3). For example, the Living Room Prototype integrated waste resources including waste wool, sourced from local farmers and paper sludge, a by-product of paper production [17], into the material system design. Expert mycelium knowledge and a tacit understanding of both materials and processes is required to successfully adapt the mycocrete formulation. This has been particularly important to achieve the correct consistency while changing the formulation to incorporate new waste streams. Alongside this, digital design tools support the accommodation of design and fabrication parameters for textile pre-form manufacturing, form finding and structural development to realize large-scale grown architectures.

This example highlights the necessity and versatility of mycocrete as a multi-material carrier system with focus on the material composition. It offers radical material adjustments towards holistic, circular systems. The Living Room Prototype incorporates a sustainable perspective through an overarching textile ecosystem approach by considering the complexity of holistic material systems [18].

5.2 Secondary complexity: integrating behavioural material activation

At the secondary level, textile templates allow for complex form design, influencing not only the behaviour of the mycelium growth but also the behaviour of the textile formwork system in relation to its chosen environment. Here, the use of CNC knitting to fabricate tubular pre-forms enables mycocrete to be injected into the textile formwork, creating highly complex geometrical biofabricated forms. This is demonstrated in the BioKnit Prototype, a 1.8 x 2.0 m self-supporting, freestanding slender vault [13] as depicted in Figure 15.4c. The CNC approach produces seamless, conformal features that incorporate precise geometry, branching shapes, gaps, lattice and a grid structure (Figure 15.4). The versatility of a CNC knit pre-form, used in combination with the mycocrete injection process, allows for customizable designs, significantly improving structural properties of MBCs. These self-supporting structures demonstrate the transformation of MBCs towards functioning myco-architectures.

To achieve this, physics simulation integrated within computational workflows [17] facilitates design control of the behaviour of the textile formwork in response to fundamental forces such as gravity (see BioKnit Prototype and BioKnit Arch, Figure 15.4c and i). This is translated into catenary logic by hanging the prototype during growth [13] and represents a secondary shaping stage. This approach incorporates environmental material responses, specifically the soft moulding of the textile formwork during the biofabrication process. It therefore expands the complexity of the material fabrication and outcome from a primary into a secondary shaping stage. This effect can enhance the structural performance through geometrical refinement for large-scale designs, counteracting the low load-bearing capacity of traditional MBCs. This approach demonstrates technological maturity by successfully transitioning into prototypes designed for commission in public buildings. The BioKnit Arch Prototype (Figure 15.4i), designed to fit precisely on the balustrade of the Design Museum London with minimal tolerances, exemplifies the next stage of fabrication control: precise dimensions and bespoke design requirements [17].

The additional design complexity through secondary shaping integrations serves as a tangible tool for versatile design features. This is also visible in the case of the Mycoknit CompoSITe Chair (Figure 15.4d). Here, rather than using hanging as the secondary shaping stage, the chair form is generated by draping the mycocrete-filled pre-form during

Figure 15.4 Mixed mycocrete paste (a) and injection process into textile formwork (b); BioKnit (c), Mycoknit CompoSITe chair (d) and BioKnit Arch Prototype (i) with fabrication process (g and h). Close-up of design detail of MycoKnit CompoSITe chair (e) and mini-dome shape based on textile pre-form with integrated shaping and hinges mechanism (f). Collage created by the authors with image courtesy of ©Hub for Biotechnology in the Built Environment, Living Textiles Research Group.

growth. The use of draping as the secondary shaping stage creates different aesthetic and functional features, including new surface textures and patterns [19]. Furthermore, the textile design capabilities within the tubular pre-form of the textile template can embed actuatable physical and mechanical form features. This proposed unlocking of additional adaptive and dynamic fabrication features includes techniques such as partial shaping and hydraulic morphing of knitted pre-forms, as well as the integration of flexible parts or hinge design during or post-biosolidification (Figure 15.4f) [19]. While these features currently remain mainly of a mechanical nature, they are promising for integrating a wider range of strategies to increase the complexity of forms over the lifespan of the object or myco-architecture.

5.3 Tertiary complexity: towards living integrations across material lifespans

Beyond design facilitated through mechanical-leaning actuations, textile templates are recognized for activating, controlling and modifying biological (mycelium) growth. In this context, the bespoke material composition, such as in mycocrete, as used in the Living Room or BioKnit prototypes, is crucial. The shielding aspect provided by the material carrier and textile system (particularly in a tubular formwork) ensures that essential growth requirements are met at scale in semi-controlled growth environments, outside isolated laboratory spaces.

These examples demonstrate how the increased complexity required to support living systems can be achieved by merging multi-materials, behavioural and environmental response mechanisms and growth parameters. The integration of the biological, the digital and the physical within the design of a textile pre-form will begin to realize the full potential of the textile template.

Textile templating uses the inherent hierarchy within textile materials as a system to integrate the necessary mechanical and physical carrier solutions within the textile pre-form. For example, yarn selection can directly influence growth parameters. Yarns with high moisture retention properties or with high nutritional value for mycelium support and enhance growth. The growth of hyphae towards a water source can be directed and promoted using the moisture uptake of a fabric. In addition, yarns can act as physical scaffolding to encourage growth directions and mycelial bonding. Alternatively, varied growth patterns can be produced by mycelium in response to different textile fibre substrates (as shown in Figure 15.5) [20].

While the examples shown here illustrate these principles at a small scale, it is recognized that design and fabrication parameters across all scales influence the growth success at a large-scale. Tertiary complexity can be further employed to contribute towards the regulation of growing and living stages of myco-architecture, to manage and maintain key parameters including nutrition, moisture, temperature and airflow distribution. This will significantly improve growth success and viability stages for myco-architecture and combat the difficulties presented by hard-to-control multiple parameters in large-scale environments. By managing these factors more locally and precisely within the textile template, the template is equipped to support the needs of the living organisms through active management and maintenance. This presents a future outlook for textile templates to activate environmental, physical and biological viability parameters for mycelium growth throughout its lifespan. This highlights that the strategic pre-design of textile templates has the potential to play a crucial role within the mycofabrication process, owing to their capability to act as a proactive growth template and allow for controlled responses during bioassembly.

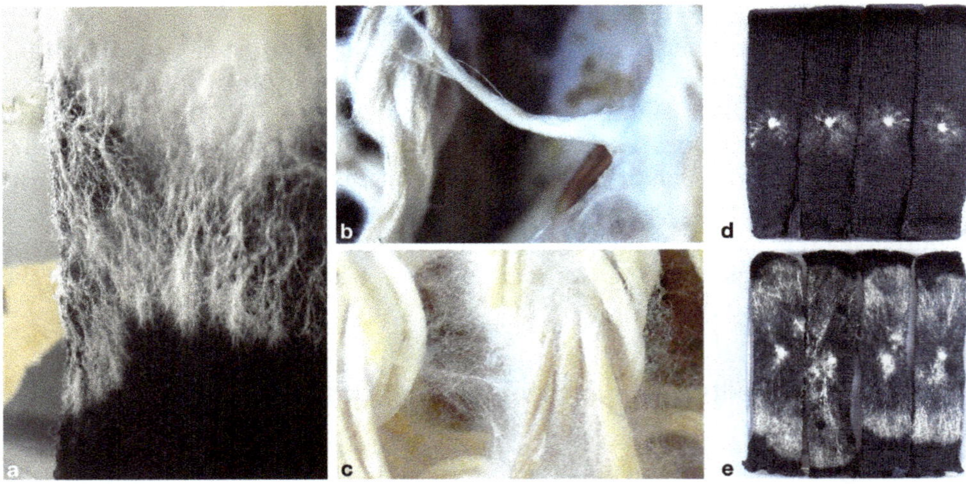

Figure 15.5 Mycelial hyphal growth responding to environmental factors provided and guided through purely textile templating systems. Hyphae crawling towards the water source along woollen textile (a). Hyphae bridging along paper yarn (b), colonization between loops of paper yarn (c); seeded textile strips wool (d) and linen (e), showing different growth developments after identical growth conditions based on the nutritional value of the textile substrate. Collage created with pictures © Romy Kaiser-Style.

6 The future potential of textile templating for myco-architecture

In summary, textile templating presents a unique way to organize complex interrelations essential for scaling growth requirements. It uses material hierarchies to generate intersecting multi-material carrier solutions, behavioural material and environmental living material responses. The textile templating opportunities introduced here conceptualize a living, temporal dimension within mycelium-based architectures and provide a viable design space that enables the design control of MBCs to evolve over their lifespan. While large-scale prototypes as introduced in this chapter already demonstrate a vast spectrum of expressions, aesthetical and functional choices at scale, trends towards further complex features paint a future picture of further conformal integration within the textile template. This represents a tangible opportunity for adaptive and living material developments. Textile templating will benefit from further digital fabrication and simulation developments and achieve even further complexity as it evolves towards living textile systems. The exemplified applications show the trend towards a living approach crossing the boundaries of sustainable material development towards a sustainable system approach, facilitating renewable and even regenerative features. This inclusive approach facilitated through textile templates shows a high potential to equip the developing field of myco-architecture with a tangible, sustainable transformation of current architectural approaches.

References

1. Oxman, N. (2015). Templating design for biology and biology for design. Architectural Design, 85(5), 100–107. https://doi.org/10.1002/ad.1961
2. Kaiser, R., Bridgens, B., Elsacker, E., & Scott, J. (2023). BioKnit: development of mycelium paste for use with permanent textile formwork. Frontiers in Bioengineering and Biotechnology, 11. https://doi.org/10.3389/fbioe.2023.1229693
3. Jiao, Y., Li, C., Liu, L., Wang, F., Liu, X., Mao, J., & Wang, L. (2020). Construction and application of textile-based tissue engineering scaffolds: a review. Biomaterials Science, 8(13), 3574–3600. https://doi.org/10.1039/D0BM00157K
4. Veenendaal, D., West, M., & Block, P. (2011). History and overview of fabric formwork: Using fabrics for concrete casting. In Structural Concrete, 12(3), 164–177. Wiley-Blackwell. https://doi.org/10.1002/suco.201100014
5. Scott, J. (2023). Digital Fabrication and Fabric Innovation. In T. Stonorov (ed.), Fabric [ated]: Fabric Innovation and Material Responsibility in Architecture (pp. 28–39). Taylor & Francis.
6. Camere, S., & Karana, E. (2018). Fabricating materials from living organisms: An emerging design practice. Journal of Cleaner Production, 186, 570–584. https://doi.org/10.1016/j.jclepro.2018.03.081
7. Srubar, W. V. (2021). Engineered Living Materials: Taxonomies and Emerging Trends. In Trends in Biotechnology, 39(6). https://doi.org/10.1016/j.tibtech.2020.10.009
8. An, B., Wang, Y., Huang, Y., Wang, X., Liu, Y., Xun, D., Church, G. M., Dai, Z., Yi, X., Tang, T. C., & Zhong, C. (2023). Engineered Living Materials for Sustainability. In Chemical Reviews, 123(5), 2349–2419. American Chemical Society. https://doi.org/10.1021/acs.chemrev.2c00512
9. Lantada, A. D., Korvink, J. G., & Islam, M. (2022). Taxonomy for engineered living materials. In Cell Reports Physical Science, 3(4). Cell Press. https://doi.org/10.1016/j.xcrp.2022.100807
10. Jones, M., Mautner, A., Luenco, S., Bismarck, A., & John, S. (2020). Engineered mycelium composite construction materials from fungal biorefineries: A critical review. In Materials and Design, 187. Elsevier Ltd. https://doi.org/10.1016/j.matdes.2019.108397
11. Bitting, S., Derme, T., Lee, J., Van Mele, T., Dillenburger, B., & Block, P. (2022). Challenges and Opportunities in Scaling up Architectural Applications of Mycelium-Based Materials with Digital Fabrication. In Biomimetics 7.2. MDPI. https://doi.org/10.3390/biomimetics7020044
12. Ghazvinian, A. (2021). A Sustainable Alternative to Architectural Materials: Mycelium-based Bio-Composites. https://api.semanticscholar.org/CorpusID:235129148
13. Agraviador, A., Scott, J., Kaiser, R., Elsacker, E., Hoenerloh, A., Topcu, A., & Bridgens, B. (2022). BioKnit: Computation and material investigation in the design of biohybrid textiles towards architectural integration.
14. Yogiaman, C., Pambudi, C. P., Jayashankar, D. K., Chia, P., Quek, Y., & Tracy, K. (2020). Knitted bio-material assembly. Proceedings of the 40th Annual Conference of the Association for Computer Aided Design in Architecture: Distributed Proximities, ACADIA 2020, 1.
15. Caracaracollective. (2017). Grown Structures. https://www.caracaracollective.com/grown-structures (Retrieved 09.06.2023).
16. Scott, J., Bridgens, B, Ozkan, D., Kaiser, R., Agraviador, A. (2024). The Living Room: New Expressions of Biohybrid Textile Architecture. In Fabricate 2024 (pp. 32–39). Eds: Ayres, P., Thomsen, M.R, Sheil,B. & Skavara, M. UCL Press. https://doi.org/10.2307/jj.11374766.8
17. Scott, J., Agraviador, A., Bridgens, B., Kaiser, R., Perry, O., & Ozkan, D. (2024). BioKnit Arch: Precision Knit Mycelium BioFabrication. ACADIA 2024: Designing Change - Proceedings Volume 1 for the 2024 Association for Computer Aided Design in Architecture Conference; Vol. 1). Association for Computer Aided Design in Architecture, 1, 419–429.
18. Scott, J. (2022). Design Ecosystems Fellowship Report 2022, The Living Textiles Ecosystem.
19. Kaiser, R., Dade-Robertson, M., Elsacker, E., & Scott, J. (2024). Knit-Mycelium Hybrids: Textile Design Strategies for Biofabrication Systems Based on Mycocrete Injection in 3D Knitted Tubular Formworks. Journal of Textile Design Research and Practice, 12(1–2), 1–26. https://doi.org/10.1080/20511787.2024.2389613
20. Kaiser, R., Bridgens, B., Elsacker, E., Scott, J. (2025). Textile Templating: Knit Design Strategies for Mycofabrication. Selection for Special Issue ICSA25 – Accepted for: Springer Nature – Architecture, Structures and Construction (Under review – In progress)

Glossary

A1, A3 carbon emissions: The embodied carbon emissions associated with a building's lifecycle. This includes the carbon emissions from raw material extraction (A1) and manufacturing processes (A3).

Anastomosis: Connections between passageways or paths, such root networks of separate mycelia joining together to create a single, larger network.

Ascomycota phylum: Biological division of fungi characterized by the presence of specialized spore-bearing structures called ascus.

Basidiomycota phylum: One of the fifteen phyla of the fungal kingdom with about 30,000 known species, it is distinct for its dikaryotic mycelium showing clamp connections and basidium spore-bearing structure.

Bioaffordance: Properties of a living thing relevant to people's interaction with a design that can be created by leveraging said properties.

Biological fabrication (or biofabrication): In the design field, the process of growing biomaterials or living materials by combining design and engineering principles with the unique properties to produce complex materials and artefacts from organisms, cells, molecules, extracellular matrices and biomaterials [1, 2, 3].

Biomaterials: Natural materials designed to interact with biological systems, often used in medicine or sustainable design.

Bioplastics: Biodegradable or compostable, plastic-like material made from renewable biological sources such as corn starch, algae or gelatine.

Bioreactor: A controlled system or vessel designed to cultivate living organisms, such as fungi or algae, for research or practical applications.

Bioreceptivity: The potential of the material to be colonized by living organisms.

Bioremediation: The use of organisms to neutralize or remove contaminants from an environment, such as using fungi to detoxify Martian regolith (e.g., perchlorates).

Bio-welding: A process where living organisms, such as mycelium, are used to fuse or bind components, creating a cohesive structure without external adhesives.

Carbon conscious materials: Refers to materials that absorb, retain or produce little to no carbon during production, including waste or by-products that are repurposed.

Chromatography: Liquid chromatography/mass spectrometry (LC/MS) is an analytical technique that combines the separation power of liquid chromatography with the direct mass measurement of a mass spectrometer as the detector. Liquid chromatography can separate a wide range of compounds, while the mass detector provides valuable information about molecular weight, structure, identity, quantity and purity [4].

CO_2 sink: A system or material that absorbs and/or stores more carbon dioxide than it releases.

Circular model: Describes a system that strives for maximum reuse and keeping resources as long as possible in a cyclic process by turning wastes into raw materials.

Closed-loop life- support system: A habitat system designed to recycle resources such as air, water and nutrients, minimizing reliance on resupply from Earth by integrating biological, chemical and physical processes.

Designerly exploration: Creative, iterative experiments and open-ended inquiries, using the methods of design practice through materials, technologies and/or expressions.

Design with the living: In reference to the symposia hosted by the Design Museum, in 2020 and 2021. The term is used here to encompass biodesign as well as other design practices that involve interactions between human and non-human living beings.

Eutrophication: Unbalanced nutrient accumulation in an environment, typically nitrogen, characterized by excessive growth of plants and algae asphyxiating the ecosystem.

End-effector: Consists of a veneer extrusion mechanism, a stepper motor for precise control, a cutting blade actuated by a pneumatic cylinder and a guiding system that accommodates veneer strips of varying widths.

Endophyte: Typically, a microbe living within a host plant for part of its lifecycle without causing apparent disease.

Engineered living materials: Engineered materials made up of living cells that constitute or assemble the material itself and regulate its functional performance after production [5].

Fermentation: In the context of mycelium cultivation, metabolic processes where fungi enzymatically break down organic substrates for growth and nutrient absorption.

Form-finding: Physical process of determining the shape that a structure will naturally take under specific conditions, such as forces, constraints or environmental factors.

Grain spawn: Sterilized grains (e.g., rye or millet) that have been myceliated and are used to inoculate bulk substrates such as hemp or straw.

Heterotroph: Describing an organism that obtains nutrients by consuming organic compounds from other organisms, rather than producing its own food.

Hydrophobic: Describes the property of a material surface that repels water or is not readily wetted by it (static water-contact angle greater than 90°).

Hygroscopic agent: A material or substance that attracts, absorbs and retains moisture from its surroundings.

Hyphae: Multicellular thread-like structures that form the vegetative body of filamentous fungi.

In-situ resource utilization (ISRU): The use of local extraterrestrial resources (e.g., lunar regolith, Martian soil, local water) to support habitat construction, life support or other mission-critical activities, reducing reliance on Earth-supplied materials.

Lignocellulosic materials: Organic materials composed primarily of lignin, cellulose and hemicellulose such as wood, straw, agricultural residues and certain grasses; they serve as substrates for MBC.

More-than-human relationships: Ensemble of performative socio-technical interactions between identifiable beings and objects that are constitutive of a specific ecosystem; the expression acknowledges an intent to challenge anthropocentrism and user-centeredness in favour of ecocentrism.

Myceliated: A substrate infused with and covered by mycelium.

Mycoremediation: The process of using fungi to break down or eliminate toxins from the environment [6].

Regolith: Loose, unconsolidated rock and dust covering planetary surfaces (e.g., Moon or Mars).

Rheological feasibility: The suitability of a material's flow properties, such as viscosity and yield stress, for extrusion-based printing processes.

Pre-grown materials: Substrates such as hemp or straw that have been inoculated and myceliated before being broken up and shaped into products.

Processual systems: The systems in which material form and properties emerge result from process qualities.

Scalability: Manufacturing of monolithic, building element scale components out of mycelium composites.

Saprophytic: Describing organisms that obtain nutrients through extracellular digestion of dead or decomposing organic matter.

Spectrophotometry: A quantitative measurement based on the reflection or transmission properties of a material. It is a method used to measure how much a chemical substance absorbs light by measuring the intensity of light as a beam of light passes through sample solution [7].

Submerged liquid fermentation: The use of bioreactors for synthesizing mycelium in a nutrient-rich liquid medium to obtain as a final result mycelium pulp for further processing [8].

Substrates: An organic fibrous material such as agricultural waste or straw providing the nutrient and supporting structure for mycelium to grow, forming a mycelium-based composite.

Symbiocene: A proposed era that envisions a harmonious and mutually beneficial coexistence between humans and the natural world, emphasizing interconnectedness and sustainability.

Ultrasonic welding: The utilization of a welding horn and generator that produce high-frequency vibrations while pressing the veneer strips together.

Technology readiness level: Refers to a NASA proposition for maturity levels for new technologies [9].

References

1. Zolotovsky K. Biological fabrication: A new paradigm for design and manufacturing of the built environment [Doctoral dissertation]. Massachusetts Institute of Technology; 2017.
2. Camere S, Karana E. Fabricating materials from living organisms: An emerging design practice. Trends Biotechnol. 2018;36(2):107–10.
3. Mironov V, Trusk T, Kasyanov V, Little S, Swaja R, Markwald R. Biofabrication: a 21st century manufacturing paradigm. Biofabrication. 2009;1(2):022001.
4. Agilent. Agilent. n.d. Liquid Chromatography/Mass Spectrometry Fundamentals. Available from: https://www.agilent.com/en/product/liquid-chromatography-mass-spectrometry-lc-ms/lcms-fundamentals
5. Srubar WV. Engineered living materials: Taxonomies and emerging trends. Trends Biotechnol. 2021;39(6):574–87.
6. Stamets P. Mycelium Running: How Mushrooms Can Help Save the World. First. Berkeley: Ten Speed Press; 2005.
7. Albuhamad AMA. Introduction to Spectrophotometry [internet]. Available from: https://www.researchgate.net/publication/377237891_Introduction_to_Spectrophotometry
8. Szilvay G, Arias Barrantes M. VTT. 2021. Mycelium leather: sustainable alternative for leather. Available from: https://www.vttresearch.com/en/news-and-ideas/alternative-leather-and-synthetic-leather-vtt-succeeded-demonstrating-continuous
9. NASA. Technology readiness levels [Internet]. NASA; 2023 [accessed 02 Mar 2026]. Available from: https://www.nasa.gov/directorates/somd/space-communications-navigation-program/technology-readiness-levels/

Index

www.ingramcontent.com/pod-product-compliance
Ingram Content Group UK Ltd.
Pitfield, Milton Keynes, MK11 3LW, UK
UKHW051837260726
473621UK00005B/10

9 781784 276515